Monitoring Forest Carbon Sequestration with Remote Sensing

Monitoring Forest Carbon Sequestration with Remote Sensing

Editors

Huaqiang Du
Wenyi Fan
Mingshi Li
Weiliang Fan
Fangjie Mao

MDPI • Basel • Beijing • Wuhan • Barcelona • Belgrade • Manchester • Tokyo • Cluj • Tianjin

MDPI

Editors
Huaqiang Du
Zhejiang A&F University
Hangzhou, China

Wenyi Fan
Northeast Forestry University
Harbin, China

Mingshi Li
Nanjing Forestry University
Nanjing, China

Weiliang Fan
Zhejiang A&F University
Hangzhou, China

Fangjie Mao
Zhejiang A&F University
Hangzhou, China

Editorial Office
MDPI
St. Alban-Anlage 66
4052 Basel, Switzerland

This is a reprint of articles from the Special Issue published online in the open access journal *Remote Sensing* (ISSN 2072-4292) (available at: https://www.mdpi.com/journal/remotesensing/special_issues/monitoring_forest_carbon_sequestration_remote_sensing).

For citation purposes, cite each article independently as indicated on the article page online and as indicated below:

LastName, A.A.; LastName, B.B.; LastName, C.C. Article Title. *Journal Name* **Year**, *Volume Number*, Page Range.

ISBN 978-3-0365-7208-6 (Hbk)
ISBN 978-3-0365-7209-3 (PDF)

Contents

remote sensing

Article

Retrieval of Boreal Forest Heights Using an Improved Random Volume over Ground (RVoG) Model Based on Repeat-Pass Spaceborne Polarimetric SAR Interferometry: The Case Study of Saihanba, China

Yu Mao [1], Opelele Omeno Michel [1,2], Ying Yu [1], Wenyi Fan [1,*], Ao Sui [1], Zhihui Liu [1] and Guoming Wu [1]

1 Key Laboratory of Sustainable Forest Ecosystem Management-Ministry of Education, School of Forestry, Northeast Forestry University, Harbin 150040, China; maoyu@nefu.edu.cn (Y.M.); michel.opelele@unikin.ac.cd (O.O.M.); yuying@nefu.edu.cn (Y.Y.); suiao_2019@nefu.edu.cn (A.S.); liuzh@nefu.edu.cn (Z.L.); wgm@nefu.edu.cn (G.W.)

2 Department of Natural Resources Management, Faculty of Agricultural Sciences, University of Kinshasa, 117 Kinshasa XI, Mont Amba District, Kinshasa 01031, Democratic Republic of the Congo

* Correspondence: fanwy@nefu.edu.cn; Tel.: +86-139-4605-5384

Abstract: Spaceborne polarimetric synthetic aperture radar interferometry (PolInSAR) has the potential to deal with large-scale forest height inversion. However, the inversion is influenced by strong temporal decorrelation interference resulting from a large temporal baseline. Additionally, the forest canopy induces phase errors, while the smaller vertical wavenumber (k_z) enhances the sensitivity of the inversion to temporal decorrelation, which limits the efficiency in forest height inversion. This research is based on the random volume over ground (RVoG) model and follows the assumptions of the three-stage inversion method, to quantify the impact of repeat-pass spaceborne PolInSAR temporal decorrelation on the relative error of retrieval height, and develop a semi-empirical improved inversion model, using ground data to eliminate the interference of coherence and phase error caused by temporal decorrelation. Forest height inversion for temperate forest in northern China was conducted using repeat-pass spaceborne L-band ALOS2 PALSAR data, and was further verified using ground measurement data. The correction of temporal decorrelation using the improved model provided robust inversion for mixed conifer-broad forest height retrieval as it addressed the over-sensitivity to temporal decorrelation resulting from the inappropriate k_z value. The method performed height inversion using interferometric data with temporal baselines ranging from 14 to 70 days and vertical wavenumbers ranging from 0.015 to 0.021 rad/m. The R^2 and RMSE reached 0.8126 and 2.3125 m, respectively.

Keywords: forest height; synthetic aperture radar (SAR); interferometry; random volume over ground (RVoG) model; three-stage inversion method

Citation: Mao, Y.; Michel, O.O.; Yu, Y.; Fan, W.; Sui, A.; Liu, Z.; Wu, G. Retrieval of Boreal Forest Heights Using an Improved Random Volume over Ground (RVoG) Model Based on Repeat-Pass Spaceborne Polarimetric SAR Interferometry: The Case Study of Saihanba, China. *Remote Sens.* **2021**, *13*, 4306. https://doi.org/10.3390/rs13214306

Academic Editor: Klaus Scipal

Received: 14 September 2021
Accepted: 23 October 2021
Published: 26 October 2021

1. Introduction

Forest ecosystems are the main components of terrestrial ecosystems [1]. Estimating the distribution and change of biomass and carbon storage in forest ecosystems can help to understand the relationship between carbon sources and carbon sinks, and the changing trends in terrestrial ecosystems [2–4]. Forest height is an essential parameter for representing the vertical structure of the forest. It provides a significant reference value in estimating forest carbon storage and plays a key role in evaluating forest stand quality and climate impact [5–7]. Remote sensing is an essential forest monitoring method that has allowed the development of various forest height retrieval technologies. Microwave remote sensing has attracted much attention due to its intense penetration into the atmosphere and forest canopy, and independence from weather conditions [8–12]. Meanwhile, spaceborne synthetic aperture radar (SAR) is widely used to observe forest heights of various forest types due to its advantages of the all-weather and all-season observation [4]. The use of spaceborne SAR data are of great significance in the study of forest parameters of northern temperate forests, and could play a role in achieving the upcoming global forest biomass and carbon cycle detection mission [13,14].

The SAR interferometry is used to calculate the vertical height of the ground using the phase difference between two sensors. This method is particularly important in terrain survey and evaluating terrain deformation [15–17]. For a complex distributed scattering unit such as a forest, it is often difficult to accurately separate the canopy phase and the ground phase using interferometric information alone (even when a high-frequency imaging system, such as C and X band, is used) [4,11]. PolInSAR has strong penetrability in the low frequency band (such as L-band and P-band), and can effectively retrieve the distribution of forest height through the canopy [18,19]. The difference in sensitivity of polarization to different components of the forest combined with the interferometric technique [20] allows inversion of forest height by distinguishing canopy scattering centers from ground scattering centers through the difference in ground scattering contribution ratio.

However, the position of scattering center is not entirely located on top of the canopy and the ground, making it difficult to extract forest height information [11]. A variety of interferometric coherence models such as the interferometric water cloud model (IWCM) [21–23], third-order Fourier-Legendre (FL) polynomial inversion mode [24], random volume over ground model (RVoG) [25–27], and the two-level method (TLM) [28] have been used to extract forest parameters. At present, the RVoG is the widely accepted model due to its simplicity and high accuracy [27]. Additionally, it is a relatively robust inversion model [29,30] whose limitations and potential errors in terms of temporal decorrelation and terrain interference can be addressed. As a result, several improved models and methods have been developed [9,31–33] to improve its applicability [34]. Meanwhile, the most commonly used method is the three-stage inversion method [35]. This method significantly improves the efficiency of parameter inversion through geometric analysis and strengthens the control of errors in the inversion process. The three-stage inversion method has been successfully applied in parameter inversion of different wavebands and different forest types [36]. Recent studies have shown that there are still more improved models for the RVoG model, which are necessary to make it more adaptable to various inversion conditions and increase the inversion accuracy. After comparing a variety of forest height retrieval methods, Chen et al. found that there was a significant increase in the retrieval accuracy when the S-RVoG model was used to retrieve forest heights by ALOS PALSAR data after introducing the normalized vegetation index [37]. Shi et al. improved the RVoG+VTD model using dual baselines by P-band E-SAR data, which consequently improved forest height inversion accuracy [38]. Xing et al. added a temporal-decorrelated adaptive estimation process based on the expectation maximum (EM) algorithm to the RVoG+VTD model and converted the Euclidean distance to a generalized distance to extract the magnitude features more efficiently. The accuracy of forest height estimated by this method was significantly improved compared with the original model and was closer to the Lidar data [39].

Nevertheless, the RVoG model considers the effect of forest height and volume decorrelation, but other decorrelation caused by any other factors introduces errors in the inversion process [9,40–42]. Therefore, it is still affected by several limiting factors, with the temporal decorrelation being the most significant factor [43–46]. To correct for temporal decorrelation, several improved models based on RVoG have been proposed. The most widely used models include RVoG+VTD model [35] and RMoG model [32,43]. Although these models have shown some effectiveness in the inversion process, they still have some limitations. The RVoG+VTD model fixes the extinction coefficient parameter in the model and introduces a parameter representing temporal decorrelation to correct for the effect of temporal decorrelation during the inversion process. However, this means that the role of the extinction coefficient in the inversion is neglected, and the choice of the initial value of this parameter will have an impact on the model accuracy. The RMoG model considers the interaction between temporal decorrelation and volume decorrelation, and inverts both simultaneously in the model. The model is more advanced and effective, but the model introduces more positional parameters, which increases the difficulty of the model and the requirement of observation quantity. In addition, these models were validated only on airborne data, but did not take into account the case of repeat-pass spaceborne data. The repeat-pass spaceborne SAR data has a larger temporal baseline than airborne SAR, and the impact of decorrelation is more difficult to ignore [23,47]. When using these models to invert forest heights on repeat-pass spaceborne PolInSAR, forest heights in the study area are seriously overestimated. Based on the characteristics of repeat-pass spaceborne InSAR, Lei et al. proposed a nonlinear iterative model based on the RMoG model to invert the forest height [41,48]. Although the model yields high accuracy results in repeat-pass spaceborne SAR, it uses coherence for inversion and neglects the effect of temporal decorrelation on the interferometric phase. However, due to the range of effective vertical wavenumbers, the interferometric phase of the repeat-pass spaceborne SAR is more significantly affected by temporal decorrelation. The effective vertical wavenumber (k_z) plays an essential role in linking the forest height to interferometry [40]. However, the range of vertical wavenumber is faced with limitations when inverting the forest height through the RVoG model [49]. For instance, k_z with an excessively large or excessively small value will increase the interference of decorrelation and cause considerable deviations in the inversion results [50]. The k_z value of the repeat-pass spaceborne PolInSAR data are often lower than the scope of inversion, making it difficult to invert the forest height using the RVoG model.

The present research is based on the RVoG model and the three-stage inversion method. It aimed at analyzing the effects of coherence and phase errors of temporal decorrelation on repeat-pass spaceborne PolInSAR inversion performance using theoretical and experimental approaches. This study further recommends a method improved by the RVoG model and suitable for repeat-pass spaceborne PolInSAR data. The research theoretically analyzed the inversion accuracy of the improved model, and used the world's only repeat-pass spaceborne L-band ALOS-2 PolInSAR data to test and verify the temperate forests of Hebei Province, China. Finally, a semi-empirical improved model based on RVoG model was established by using ground data as prior data to correct for the interference caused by temporal decorrelation to the forest height inversion obtained from repeat-pass spaceborne PolInSAR data.

The structure of this paper is as follows: Section 2 introduces the study area and the acquisition process of ground measurement data. The basic conditions and preprocessing of the ALOS2 PALSAR data set have been introduced as well. In Section 3, we present the theoretical background of the RVoG model and the three-stage inversion method. The effect of temporal decorrelation on forest height inversion and the corresponding theoretical model have also been explained in this section. An improved model for temporal decoherence is proposed in Section 4. The paper presents theoretical background and the inversion process of the model, and performs a theoretical error analysis of the model. Section 5 uses the Saihanba ground measurement data and ALOS2 PALSAR data to retrieve the forest height and evaluate the results. Some shortcomings of the model and its robustness under different temporal baselines are discussed in Section 6, and recommendations made for the selection of future PolInSAR spatial baselines. Finally, the conclusion is presented in Section 7.

2. Research Materials and Theoretical Models

2.1. Study Area and Sample Site Data Collection

The study was conducted in Saihanba Forest, a typical temperate forest in northern China. The area is located in the transition zone from Yanshan Mountain to Inner Mongolia in Hebei Province (117°E, 42°N), and is the largest plantation forest in the world. The forest sites in the area are rugged and mountainous, with an average elevation of 1500 to 2067 m, making ground surveys difficult (Figure 1). The complex climatic conditions and the vulnerability of conventional imagery to weather conditions in the region make it necessary to conduct microwave observation studies in the region [51].

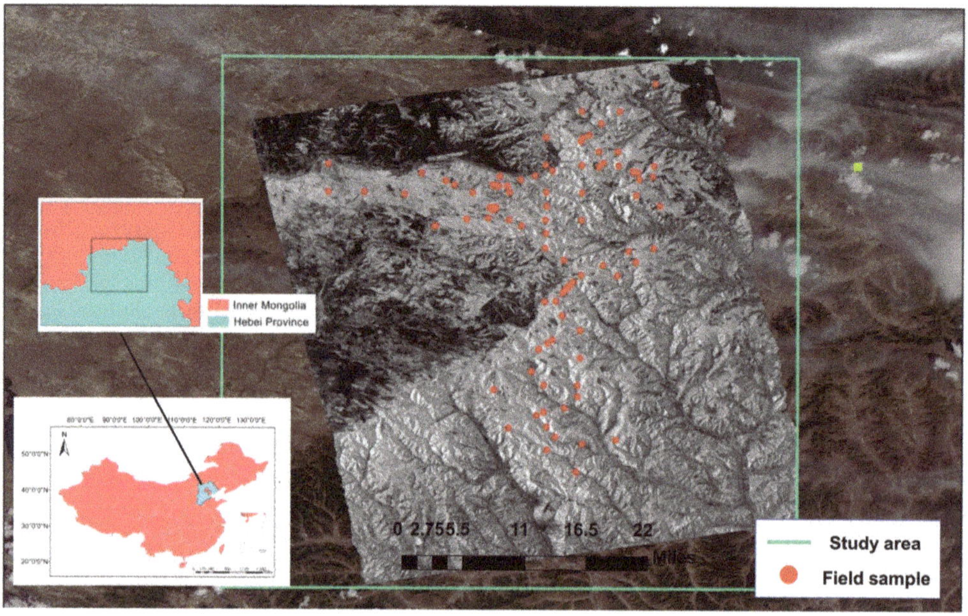

Figure 1. Map of the study area. The area is located in the Saihanba Forest in Hebei Province, China. The red points represent the distribution of the measured samples. The image in the study area is an interferometric image under HV polarization of PolInSAR data.

The measured data from the plots was collected in the same study area as the PolInSAR data and randomly sampled within the study area. There are various types of forests in this area, and the main are temperate mixed conifer-broad forests. *Larix principis-rupprechtii* Mayr. and *Betula platyphylla* Suk. are the dominant species in the area. In addition, they include *Picea asperata* Mast., *Pinus sylvestris* var. *mongolica* Litv., and some broad-leaved species (Figure 2). During the field measurements, we conducted a careful field inspection before selecting the sample sites in order to make the samples better represent the overall condition of the forest area. Among the final collected samples, 28 plots were measured at fixed points on a kilometer grid, and the remaining 69 plots were randomly sampled, hoping to represent the real situation of the forest as much as possible. The proportion of stand types in the random sampling samples was ensured to be similar to the proportion of overall stand types in the forest area as far as possible, and the forest types in the sample sites are shown in Table 1. Furthermore, the forest management in Saihanba area is in good condition, with obvious differences in stand age and density. Due to tending and thinning, most of the forests gradually decrease in density as the forest age increases. In order to restore the true forest condition, at least 30% of the samples were guaranteed to be young (high-density stands) or old growth (low-density stands) during the sampling process.

Figure 2. Forest conditions in the study area and the sample plot survey. (**a**) The diamond-shaped sample with an area of 0.06 ha. Moreover, (**b**,**c**) show the determination of sample plots and measuring heights of trees using ultrasonic height gauge. Furthermore, (**d**–**h**) show forest conditions in the study area, where: (**d**) is the pure forest of *Larix principis-rupprechtii* Mayr.; (**e**) is the pure forest of *Pinus sylvestris* var. *mongolica* Litv.; (**f**) is the mixed forest of *Larix principis-rupprechtii* Mayr. and *Betula platyphylla* Suk.; (**g**) is the pure forest of *Picea asperata* Mast.; and (**h**) is the broadleaf mixed forest.

Table 1. Forest types and the collected random sample sizes.

Forest Types	Sample Size	Forest Types	Sample Size
Pure forest of *Larix principis-rupprechtii* Mayr.	36	Pure forest of Pinus tabuliformis var. mukdensis	2
Pure forest of *Betula platyphylla* Suk.	12	Mixed coniferous forest	4
Pure forest of *Picea asperata* Mast.	4	Mixed broad-leaved forest	4
Pure forest of *Pinus sylvestris* var. *mongolica* Litv.	2	Coniferous and broad-leaved mixed forest	5

When collecting samples, we avoided the forest edge and large empty windows, and chose a relatively central position in the small forest class to ensure that the samples represent the real situation of the surrounding small forests as much as possible. After determining the center point, a distance of 17.32 m was measured along the four positive directions to determine the location of the four corner points, enclosing a diamond-shaped sample with an area of 0.06 ha (Figure 2a). The trees in the sample plot were inspected for each log, and the height of each log was measured with a Vertex IV ultrasonic height gauge. In this study, the forest height is defined as the average tree height of the sample plot. This assumption takes into account the value of average tree height in forest surveys on the one hand, and on the other hand provides support for the subsequent estimation of forest biomass. According to the field measurement results, the forest height was between 0 and 30 m, and the average height was 17.98 m. The Kolmogorov–Smirnov test was performed on the forest height sample data to test for normal distribution. The results proved that the sample data obeyed the normal distribution and has good representativeness (Figure 3).

2.2. PolInSAR Data

The PolInSAR datasets from the study area were in five scenes of ALOS2 PALSAR repeat-pass fully polarized synthetic aperture radar data (Figure 4). The 1.1-level L-band SLC data were developed by Japan Aerospace Exploration Agency (JAXA). This data were used in this study because it is among the few commercial L-band segment satellite-based fully polarized SAR data that is publicly available worldwide. Additionally, the L-band is less sensitive to forest vertical heterogeneity and allows accurate inversion of forest height from empirical structure functions [52]. Therefore, it is important to study independent inversion of forest height from this data before the opening of the new low-band satellite-based SAR system.

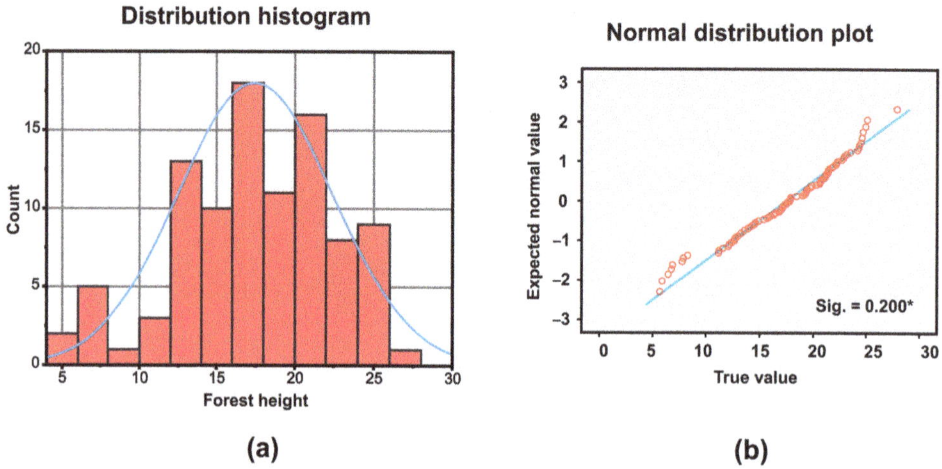

Figure 3. (**a**) Histogram of the distribution of the sample forest heights. The blue curve is the normal distribution curve. (**b**) Scatterplot of samples, normal distribution. The red points are the true distribution of the samples, and the blue line represents the standard normal distribution. The significance value Sig. was verified at the 0.05 level (*). And the samples showed a significant normal distribution after they were tested for normal distribution using the k–s method.

The datasets were collected over the study area from July to September 2020, representing the growing season of the forests in northern China. This was done within the same period the field observation data of the sample site was collected. The average zenith angle was 27.8°, the range pixel spacing was 5.66 m, the azimuthal pixel spacing was 2.86 m, the general observational area was 4944.62 km^2, and the average height of the sensor from the Earth's surface was 636.56 km. The spatial baseline length between different data ranged from 80.2 to 170.4 m. k_z ranged from 0.012 to 0.021 rad/m, and the temporal baseline between two adjacent data were 14 days. The images acquired on different dates were combined as the interferometry primary and secondary data. Four groups of PolInSAR interferometric pairs with different temporal baselines and vertical wavenumbers were set up (Table 2). The inversion of each pair of interferometric data were done independently to compare and verify the results. The inversion process was repeated in each pair of interferometric data.

Each set of interferometric pairs was pre-processed to remove decorrelation geometrically using GAMMA software [53,54]. Meanwhile, ionosphere-induced phase drift and path delays are the main sources of error in ALOS2 repeat-pass spaceborne PolInSAR when performing interferometry [55,56]. In this study, the ionospheric effect was eliminated using the distance splitting spectroscopy method (Figure 5). Terrain correction was performed using 30 m resolution SRTM DEM data [57,58].

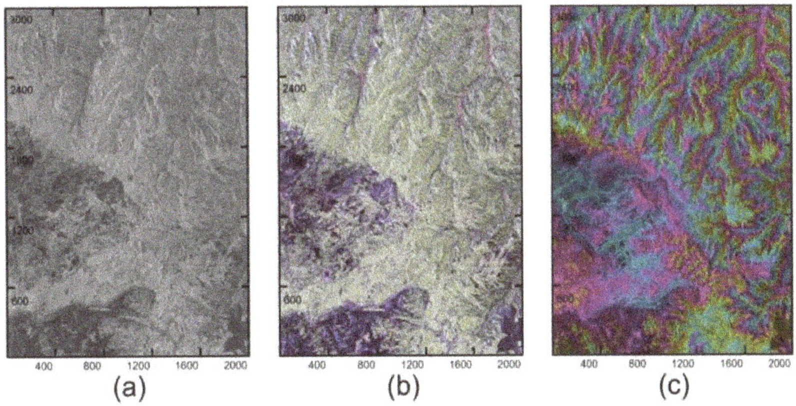

Figure 4. (**a**) Grayscale image of the original L-band image under HV polarization; (**b**) PolInSAR image Pauli-based false color image; (**c**) interferometric DEM of the on-board data.

Table 2. PolInSAR interferometric datasets.

Data Sets	Date of Image 1	Date of Image 2	Average Vertical Wavenumber	Temporal Baseline/Day
0711-0725	11 July 2020	25 July 2020	0.015	14
0905-0919	5 September 2020	19 September 2020	0.018	14
0808-0919	8 August 2020	19 September 2020	0.018	42
0711-0919	11 July 2020	19 September 2020	0.021	70

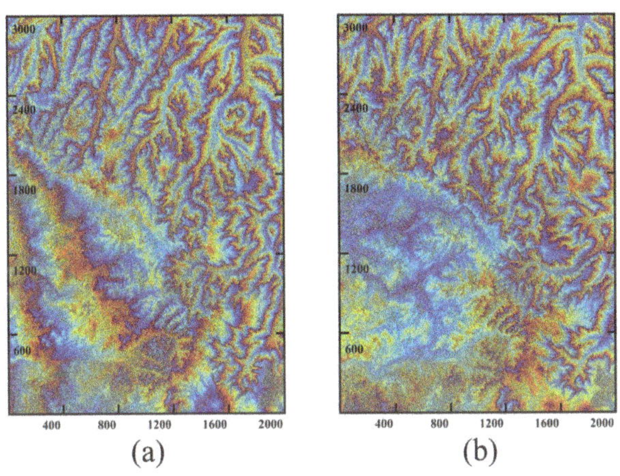

Figure 5. Interferograms before (**a**) and after (**b**) reducing the phase drift and path delay caused by the ionosphere.

3. Theoretical Analysis of Forest Height Inversion

3.1. RVoG Model and Three-Stage Inversion Method

PolInSAR data combines both interferometry and polarization properties [59]. The complex interferometric coherence $\gamma_{Obs}(\vec{\omega})$ is obtained by combining the matrices $s_1(\vec{\omega})$

and $s_2(\vec{\omega})$ of the primary and secondary images at a particular polarization $\vec{\omega}$ and can be expressed as follows:

$$\gamma_{Obs}(\vec{\omega}) = \frac{\langle s_1(\vec{\omega})s_2^*(\vec{\omega})\rangle}{\sqrt{\langle s_1(\vec{\omega})s_1^*(\vec{\omega})\rangle\langle s_2(\vec{\omega})s_2^*(\vec{\omega})\rangle}} \tag{1}$$

where * represents the conjugate of the SAR image and $\langle\rangle$ represents the expected value [28,29]. The magnitude of the complex coherence $\gamma_{Obs}(\vec{\omega})$ ($|\gamma_{Obs}|$) represents the coherence between two images (i.e., degree of similarity between two images), and its value ranges from 0 to 1.

Previous studies have shown that the complex coherence obtained by Equation (1) is still affected by several decorrelations even after eliminating system induced decorrelation [2,54]. The observed interferometric coherence can be modeled as a combination various contribution [54,55] and illustrated as follows:

$$\gamma_{Obs} = \gamma_{SNR}\gamma_{Tmp}\gamma_{vol} \tag{2}$$

Here, γ_{SNR} represents the decorrelation effect from thermal noise; γ_{Tmp} is temporal decorrelation; and γ_{vol} is volumetric decorrelation, which is widely used to forest height inversion. γ_{SNR} can be eliminated during image preprocessing. However, due to the constraints of a variety of factors, γ_{Tmp} can introduce bias between γ_{Obs} and γ_{vol}; it not only affects the phase of γ_{Obs}, but also further reduces the overall coherence. For northern forests with lower forest heights, γ_{Tmp} may sometimes mask the influence of γ_{Obs}, especially for repeat-pass spaceborne PolInSAR data with a temporal baseline of several days [28].

γ_{vol} is included into the model to allow calculation of the forest height. The RVoG model combines the forest height with the scattering properties by treating the scattering as a volume scattering and ground scattering contributions through the assumption of the forest as a random homogeneous scatterer [28]. The model expresses the forest volume scattering complex coherence as follows:

$$\gamma_{vol} = e^{i\varphi_0}\frac{\gamma_v + m(\vec{\omega})}{1+m(\vec{\omega})} \tag{3}$$

where $e^{i\varphi_0}$ is the ground scattering contribution, m is the effective ground-to-volume amplitude ratio, and γ_v is the volume scattering complex coherence without the ground contribution. γ_v can be expressed by the mean extinction coefficient σ and the forest height h_v as follows:

$$\gamma_v = \frac{2\sigma}{\cos\theta\left(e^{\frac{2\sigma h_v}{\cos\theta}}-1\right)}\int_0^{h_v} e^{-ik_z z}e^{-\frac{(2\sigma z)}{\cos\theta}}dz \tag{4}$$

where k_z is the vertical wavenumber.

$$k_z = \alpha\frac{2\pi\Delta\theta}{\lambda\sin\theta_0} \approx \alpha\frac{2\pi B_\perp}{\lambda R\sin\theta_0} \tag{5}$$

θ_0 is the radar incidence angle, $\Delta\theta$ is the incidence angle difference between the two images induced by the spatial baseline, λ is the wavelength, $B_\perp$ is perpendicular component of the spatial baseline, R is the slant range, and α is an integer constant that is equal to 2 for monostatic acquisition and 1 for bistatic acquisition.

Since the ground scattering ratio is a parameter affected by polarization, Equation (3) can be observed as a straight line in the complex plane (Figure 6). The three-stage inversion method is applicable to PolInSAR developed from this geometric property [38]. The first stage of the inversion method uses different ground contributions contained in the different PolInSAR data polarizations, which fall at different positions in the complex plane (blue shade in Figure 6), and the coherence line of Equation (3) can be fitted. In the second stage, the two intersection points of the coherence line and the unit circle of the complex plane are determined as candidate points for the ground coherence point. The actual ground coherence point is determined through screening, and its influence is eliminated. In the third stage, a 2D lookup table of extinction coefficient and forest height (red points in Figure 6) can be established according to Equation (4). Generally, it is assumed that the complex coherence point under HV polarization is mainly volume-only scattering in InSAR, and the complex coherence in PolInSAR coherence optimization represents volume scattering, which allows forest height inversion, and determination of extinction coefficient value in the lookup table.

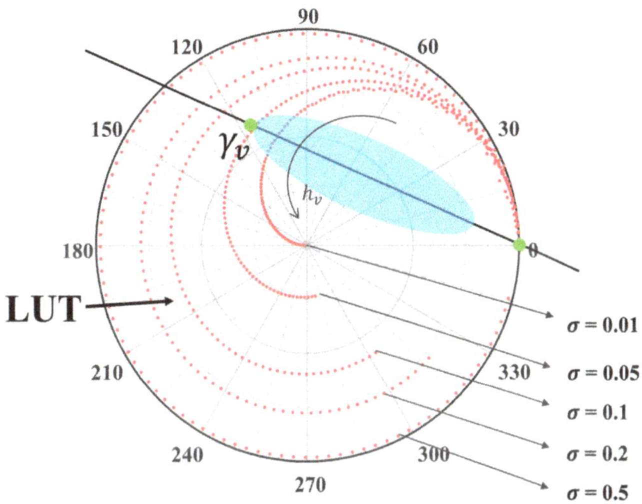

Figure 6. Schematic diagram of the three-stage inversion method. The black straight line is the fitted coherence line, the blue shaded area is the coherence region composed of different polarized coherence points, the green points indicate volume-only scattering and ground coherence points, and the red points indicate the 2D lookup table calculated by the RVoG model, in which the average extinction coefficient values of the points in each curve are the same, but gradually moves away from the phase origin as the forest height increases. The different curves represent different average extinction coefficients, and the closer they are to the circumference, the larger the value of extinction coefficient. Moreover, θ_0 in the figure is 45°, h_v ranges from 0 to 50 m, and the interval between two adjacent points is 0.5 m.

3.2. Temporal Decorrelation

The temporal decorrelation of interferometric image often causes obvious errors to the inversion results, regardless of whether RVoG model or other interferometric inversion models are used. Temporal decorrelation reduces the coherence and causes a phase shift of the interferometric data so that the forest height inversion model is greatly affected. The error sources of temporal decorrelation have a complex structure [42] and are influenced by combination of factors that are difficult to quantify. As a result, temporal decorrelation is difficult to remove when pre-processing image data.

In general, the degree of temporal decorrelation is described by the temporal baseline, which is the time interval between primary and secondary image observations. Data with larger temporal baselines undoubtedly face greater temporal decorrelation. The temporal baseline of repeat-pass spaceborne SAR data for the same observation area tends to be more significant compared to airborne PolInSAR data. This study's temporal baseline of ALOS2 repeat-pass spaceborne SAR varies from a few days to tens of days. Therefore, there is a larger temporal decorrelation contribution in the repeat-pass spaceborne SAR interferometric data. In addition, the sensitivity of different images to the same temporal decorrelation of the interference varies and PolInSAR data with smaller k_z has a more pronounced response to temporal decorrelation. For the InSAR, k_z is defined by the angular difference between the primary and secondary sensors (Equation (7)) and is a factor that indicates the sensitivity of the interferometric phase to changes in terrain (height) [49]. For the PolInSAR inversion forest height model, the smaller k_z makes the forest height more sensitive to changes in the coherence and phase of the interferometric data during inversion [49]. Therefore, when using interferometric images with small k_z to retrieve forest height, there may be huge errors in the inversion results even if there is a weak temporal decorrelation factor [50]. The situation is reflected in the absence of intersection between the observed complex coherence in the unit circle and the LUT when using the three-stage inversion method for height inversion [60]. Meanwhile, the k_z of repeat-pass spaceborne PolInSAR data tend to be lower, making the inversion results (which already contain a large temporal decorrelation factor) less accurate. Therefore, it is necessary to adopt an effective correction for temporal decorrelation inversion model when retrieving forest height from repeat-pass spaceborne PolInSAR data.

There are several improved models for temporal decorrelation including the RVoG+VTD model [60], the RMoG model [35,43], and the semi-empirical iterative model for dielectric constant and random motion modeling using Gauss-Newton iterative optimization model [42,48].

The RVoG+VTD model demonstrates that, temporal decorrelation shifts the volume-only coherence points in the complex plane unit circle during the inversion using the three-stage method. As such, there is no intersection between the height-extinction LUT and the volume-only coherence point. Therefore, correction terms for the shifted volume-only coherence points are achieved by fixing the extinction coefficient and using the SINC function. Meanwhile, RMoG model is different from the RVoG+VTD model since the effect of temporal decorrelation on volumetric decorrelation is not considered to be a multiplicative relationship. This model quantifies the cause of temporal decorrelation as a stochastic motion parameter that varies with the forest canopy. This was the first method to attempt to model the direct extraction of forest height from the mixed effect of temporal decorrelation and volumetric decorrelation. Additionally, a ten-dimensional parametric model was developed from the observations under different polarization channels. Another study decomposed temporal decorrelation into the temporal effects of dielectric constant variation and random motion, used coherence to build an empirical model, and extracted the parameters using the Gauss-Newton iterative method [42,48]. The forest height model can use the repeat-pass spaceborne InSAR data in L-band with large temporal baseline.

4. Improved Inversion Model

In this study, a new inversion method has been proposed to enable forest height inversion by empirical iteration. Corrections were performed to degrade coherence and phase shift caused by error sources such as temporal decorrelation. In addition, the inversion accuracy of the improved model was simulated, and its geometric process was analyzed theoretically.

4.1. Theoretical Background

The interferometric phase obtained from the interferometry can be used to calculate the topographic height of the ground surface.

In repeat-pass interferometry, the same sensor makes two recordings of the same ground target at a certain time interval to form an interferometric data pair (Figure 7). The first recording is referred to as the primary image (marked as s_1), and the second recording is referred to as the secondary image (marked as s_2).

$$\begin{aligned} s_1 &= a_1 e^{i\varphi_1} \\ s_2 &= a_2 e^{i\varphi_2} \end{aligned} \quad (6)$$

The primary and secondary images are conjugated and multiplied to obtain the interferometric phase values.

$$\phi = \arctan(s_1 s_2{}^*) = \varphi_1 - \varphi_2 \tag{7}$$

When phase ambiguity is not considered, the interferometric phase can be expressed as follows:

$$\phi = -\tfrac{4\pi}{\lambda}\Delta R \tag{8}$$

where ΔR is the difference between s_1 and s_2 radar wave propagation distances.

The interfering phase ϕ consists of the following five main components:

- Flat Earth phase ϕ_{flat} due to reference ellipsoid.
- Topographic phase ϕ_{topo} due to terrain undulation.
- The deformation phase caused by the deformation of the ground surface during the two imaging sessions.
- The phase difference caused by atmospheric disturbances.
- The phase difference due to noise.

In this study, the ground surface was considered undeformed. The atmospheric and noise disturbances were ignored.

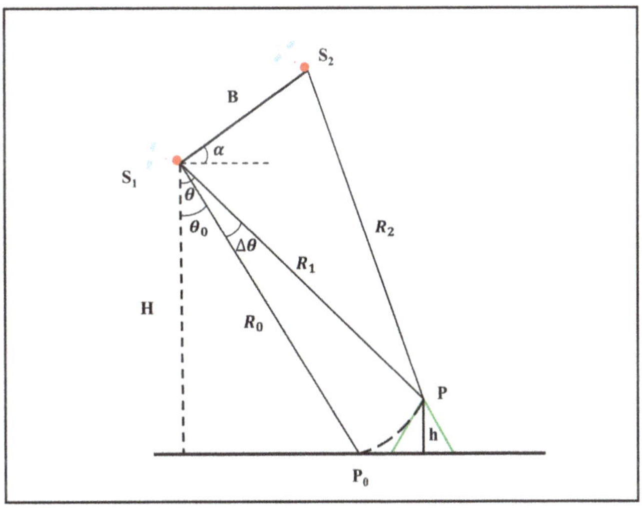

Figure 7. Schematic illustration of Interferometric geometry.

After eliminating the above interference factors, only the flat earth phase and the topographic phase remained in the interferometric phase. The geometric principle of interferometry is explained using a previous example where the topographic height h was measured at the surface target point P in Figure 7 [16]. Point P in the figure is the target point of interferometry. P_0 is the point on the reference ellipsoid, and it allows equal distance from s_1 to P and P_0.

When the sensor is far enough from the ground target, the component $B_{\|}$ of the spatial baseline (between s_1 and s_2) that is parallel to the line R_1 (between s_1 and P) can be approximately equal to ΔR ($B_{\|} \approx \Delta R$). The topographic phase at point P is expressed as follows:

$$\phi_{topo} = -\tfrac{4\pi}{\lambda}B_{\|} = -\tfrac{4\pi}{\lambda}B\sin(\theta - \alpha) \tag{9}$$

B is the spatial baseline between s_1 and s_2, θ is the angle between the line connecting s_1 to the ground target point P and the vertical direction, and α is the angle between the baseline B and the horizontal direction. The flat earth phase at point P_0 is as follows:

$$\phi_{flat} = -\frac{4\pi}{\lambda} B_{\parallel}^0 = -\frac{4\pi}{\lambda} B \sin(\theta_0 - \alpha) \tag{10}$$

$B_{\parallel}^0$ is the component of the spatial baseline B that is parallel to the line R_0 between s_1 and P_0, θ_0 is the angle between lines s_1 and P_0 and the vertical direction, and ϕ_{flat} is the flat earth phase corresponding to the point P. Meanwhile, the phase difference $\Delta\phi$ between P and P_0 can be expressed as follows:

$$\Delta\phi = \phi_{topo} - \phi_{flat} = -\frac{4\pi}{\lambda} B(\sin(\theta - \alpha) - \sin(\theta_0 - \alpha)) \tag{11}$$

The angle $\Delta\theta$ between θ_0 and θ is very small, due to the long distance between the sensor and the ground target. Thus, Equation (11) can be simplified as follows:

$$\Delta\phi = -\frac{4\pi}{\lambda} B \cos(\theta_0 - \alpha)\Delta\theta = -\frac{4\pi}{\lambda} B_{\perp}^0 \Delta\theta \tag{12}$$

$B_{\perp}^0$ is the component of the spatial baseline B, which is perpendicular to R_0. In the geometric relationship illustrated in Figure 7, the value of the height of point P is computed as shown below:

$$\begin{gathered} h = H - R_1 \cos\theta = R_1 \Delta\theta \sin\theta - \Delta R_1 \cos\theta \\ \Delta\theta = \frac{h + \Delta R_1 \cos\theta}{R_1 \sin\theta} \end{gathered} \tag{13}$$

h is the height of the ground target point P from the horizontal plane, H is the height of s_1 from the horizontal plane, R_1 is denoted as the radar wave propagation distance of s_1, and ΔR_1 is the difference between R_1 and R_0. Therefore, the topographic phase and height can be expressed as shown in the equation below:

$$\Delta\phi = -\frac{4\pi B_{\perp}^0}{\lambda R_1 \sin\theta} \cdot (h + \Delta R_1 \cos\theta) = -k_z \cdot h - \frac{4\pi B_{\perp}^0}{\lambda R_1 \tan\theta} \cdot \Delta R_1 \tag{14}$$

The first term on the right side of the equation is the terrain phase given the terrain height, while the second term is the flatland phase considering the zero change in elevation. After removing the flat earth phase, we obtain a linear relationship between height and terrain phase, and it is linked by k_z.

The above equation showed terrain height measurement by interferometric phase. However, there is a significant difference in measurements when using PolInSAR to measure forest height. Meanwhile, the interferometric phase between the top point of the canopy and the underlying surface point should be included in the calculation of forest height. However, the observed phase at the top of the canopy had two-phase contributions in addition to the five components mentioned in Equation (8). The first contribution is the shift of the phase center caused by the penetration of low-frequency SAR into the forest canopy [53]. The second contribution is the phase shift caused by random movement of the canopy during the imaging of the primary and secondary images [43]. Both of these contributions interfere with the forest height measurement and should be eliminated during the inversion process.

The geometry of the forest height measurement by interferometric phase is illustrated in Figure 8. P_1 is the phase center at the top of the canopy. However, the phase shift caused by the random motion of the canopy causes the phase center of the canopy to shift to P_2 when being observed. The shift in phase center that is caused by low-frequency SAR penetration makes it possible for the observed phase to lie anywhere between P_2 and P_2'. P_0 is the corresponding point on the reference ellipsoid. The interferometric phase and the height from the horizontal plane at P_2 are then described as follows:

$$\Delta\phi_1 = -\frac{4\pi}{\lambda} B_{\perp}^0 \Delta\theta_1 \tag{15}$$

$$h_1 = H - R_1' \cos\theta \tag{16}$$

The relationship between interferometric phase and height at point P_2 is as follows:

$$\Delta\phi_1 = -\frac{4\pi B_\perp^0}{\lambda R_1' \sin\theta} \cdot (h_1 + \Delta R_1' \cos\theta) = -k_z \cdot (h + \Delta h) - \frac{4\pi B_\perp^0}{\lambda R_1' \tan\theta} \cdot \Delta R_1' \tag{17}$$

where $\Delta R_1'$ is the difference between R_1' and R_0'. Δh is the height difference between P_1 and P_2, and the offset of the measured height of the canopy. The second term on the right-hand side of the equation is the flat-earth phase of P_2. The above two interference factors influence the height deviation, so it is necessary to quantify the height error of the two effects. Previous studies have described random motion as Gaussian function that varies with height and uniformly when in the vertical direction [35,43]. There was neither deformation on the surface nor phase difference caused by random motion on the surface. When only the offsets produced by the random motion on the canopy are taken into account, the height offset can be considered as a linear function of the canopy height and can be expressed as follows:

$$\Delta h = \delta_r(h) = \frac{\delta_r}{h_r} h \tag{18}$$

where δ_r denotes the standard deviation of the motion at a certain reference height h_r, and ε_0 represents the variation of this offset from the height.

$$\Delta h = \varepsilon_0 \cdot h \tag{19}$$

Previous studies have focused on the modeling of random motion and understanding the relationship between interferometric coherence and random motion. However, the effect phase shift errors on the height inversion results may be more pronounced. Therefore, this study attempts to correct this error and improve accuracy in the height inversion results.

Meanwhile, the phase bias caused by low-frequency SAR penetration makes the observed phase lie between the top of the canopy and a half of the height [53]. Therefore, the height deviation obtained by coupling the two factors can be simplified as a linear function that changes with the vertical target height:

$$\Delta h = \varepsilon_0 \cdot h - d \tag{20}$$

where d is the distance between the scattering center and the underlying surface after the canopy phase shift $h/2 \le d \le h$. The height error was brought into the canopy phase without the flat-earth phase in order to obtain the relationship between the observed canopy phase and the true canopy height. The equation is as follows:

$$\phi_{vol} = k_z \cdot ((1 + \varepsilon_0)h - d) = \varepsilon k_z h - k_z d = \varepsilon k_z h + \varphi_e \tag{21}$$

where ε is the correction term for temporal decorrelation due to random motion, and φ_e is the corrected phase for the phase center shift. Noteworthy, $\varepsilon \ge 1, -\pi \le \varphi_e \le \pi$.

Temporal decorrelation not only causes phase shift but also leads to a reduction of factors affecting interferometric coherence, such as dielectric constant with temporal baseline. This also has a more pronounced effect on the PolInSAR inversion of forest height and, therefore, measures are needed to reduce this interference. In order to address the apparent temporal decorrelation of coherence amplitude, phase interference, and canopy phase center shift suffered during the repeat-pass spaceborne PolInSAR inversion, this study proposes a new inversion method to achieve the inversion of forest height by empirical iteration. First, to address the interference of temporal decorrelation on the interferometric coherence and correct the overall coherence, this study introduces a correction term $|\gamma_e|$ based on the RVoG model (Equation (4)):

$$\hat{\gamma} = |\gamma_e|\cdot\frac{2\sigma}{\cos\theta\left(e^{\frac{2\sigma h_v}{\cos\theta}}-1\right)}\int_0^{h_v} e^{-ik_z z}e^{-\frac{(2\sigma z)}{\cos\theta}}dz \tag{22}$$

With respect to the offset phase caused by the random motion and microwave penetration factors in the temporal coherence, the offset phase value of the height modeling in Equation (21) is introduced into Equation (22) as follows:

$$\begin{aligned}\hat{\gamma} &= |\gamma_e|\cdot\frac{2\sigma}{\cos\theta\left(e^{\frac{2\sigma h_v}{\cos\theta}}-1\right)}\int_0^{h_v} e^{-i(\varepsilon k_z z+\varphi_e)}e^{-\frac{(2\sigma z)}{\cos\theta}}dz \\ &= \gamma_e\cdot\frac{2\sigma}{\cos\theta\left(e^{\frac{2\sigma h_v}{\cos\theta}}-1\right)}\int_0^{h_v} e^{-i(\varepsilon\cdot k_z)z}e^{-\frac{(2\sigma z)}{\cos\theta}}dz\end{aligned} \tag{23}$$

where $\gamma_e = |\gamma_e|\cdot e^{i\varphi_e}$.

In previous studies, the physical model construction method [35,42] decomposed the temporal decorrelation into factors such as dielectric constant and random motion, and then modeled together with volume decorrelation to extract forest height. These models are undoubtedly advanced and effective, but often require complex iterative processes and have many model parameters, which increases the uncertainty of the inversion. This study attempts to use an empirical model to achieve fast and efficient inversion, and achieve results that are similar to previous temporal–decoherent models. Therefore, the model is built based on the three-stage inversion method, and the geometric properties of the model are used to improve efficiency of the inversion. The error factors encountered during the inversion are integrated into Equation (23) as a correction term ε on the phase, and a complex correction term γ_e. They are both brought into the iterative process to ensure that complete error sources are considered in the model.

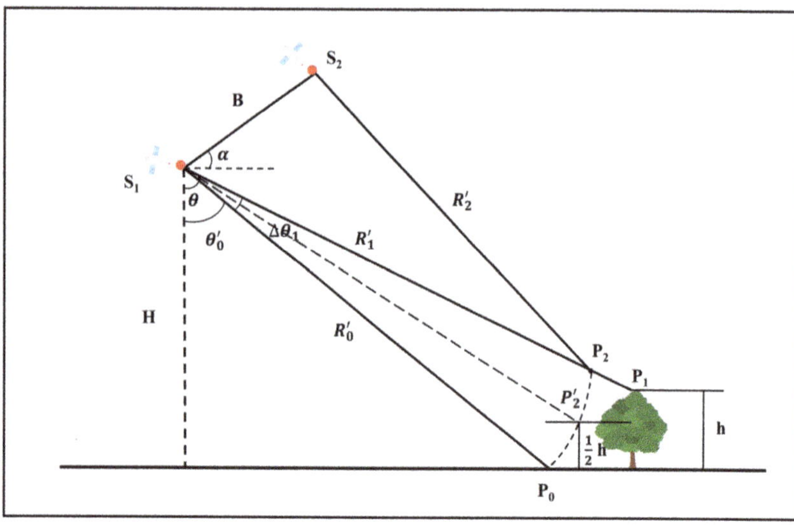

Figure 8. Schematic diagram showing the phase shift of the observed target. P_2 is the component of the observed phase on the line of sight affected by random motion, P_2' is the height reduction at the center of the P_2 phase caused by microwave penetration, and P_1 is the true phase point.

The model is performed under the assumption that P_2 is in line with s_1P_1, but this does not affect the validity of the model. This study focuses on the height difference in the vertical direction between P_2 and P_1, and does not require obtaining of the specific true phase point. Therefore, even if the two are not on the same line, the inversion results in another point with the same height on the line. Moreover, since this is an empirical model, the real data are used as prior information. Therefore, forest height can be calculated based on this assumption.

4.2. Iterative Process of the Improved Model

The iterative process of the improved model follows the flow of the three-stage inversion method. First, the polarized interferometric information needs to be extracted from the pre-processed PolInSAR data. Each set of polarized interferometric data pairs contains the original complex data under full polarization, and the Pauli basis scattering vector for each set of images is as follows:

$$\begin{aligned} k_1 &= \tfrac{1}{\sqrt{2}}\left(s_{HH_1} + s_{VV_1}\ s_{HH_1} - s_{VV_1}\ 2s_{HV_1}\right)^T \\ k_2 &= \tfrac{1}{\sqrt{2}}\left(s_{HH_2} + s_{VV_2}\ s_{HH_2} - s_{VV_2}\ 2s_{HV_2}\right)^T \end{aligned} \tag{24}$$

where s represents the scattering matrix elements collected twice at different polarizations, H represents horizontal polarization, and V represents vertical polarization. $(\cdot)^T$ represents the transpose of the matrix. With the Pauli basis vector, we can obtain the T coherence matrix and the Ω_{12} matrix as follows.

T coherence matrix:

$$T = \tfrac{1}{2}(T_{11} + T_{22}) = \tfrac{1}{2}\left(\langle k_1 k_1{}^H\rangle + \langle k_2 k_2{}^H\rangle\right) \tag{25}$$

The Ω_{12} matrix:

$$\Omega_{12} = \langle k_1 k_2{}^H\rangle \tag{26}$$

where $\langle\cdot\rangle$ denotes the mathematical expectation, and $(\cdot)^H$ is the conjugate transpose.

With the T matrix and Ω_{12} matrix, we can calculate the coherence and the phase of the primary and secondary images with different polarization and coherence optimization. This study employs the common and easily extracted complex coherence values, including three basic polarization types HH, HV, VV, four linear combinations of different polarizations (HH+VV, HH-VV, HHVV, HV+VH), three circular polarizations (LL, LR, RR), three Opt coherence optimizations (Opt1, Opt2, Opt3), and two PD coherence optimizations (PD High, PD Low). Due to the specific polarization correlation of the ground scattering contribution, the 15 complex coherences have different ground scattering ratio coherence.

The improved model extracted the volume-only scattering complex coherence method in general agreement with the three-stage inversion method. During the inversion, 15 types of complex coherences were projected into the unit circle of the complex plane. The least-squares method was used to fit the phase trunk using the different scattering ratios of the earth's surface under different polarizations. The two intersections of the fitted coherence line and the unit circle of the complex plane allowed determination of the candidate points for ground coherence. The external DEM phase in this image was compared with the candidate points and filter to identify the true ground coherence points and remove their influence. After the removal of terrain phase, the coherent region of each pixel was obtained, and the volume-only complex coherent points in each pixel were screened by comparing the distance between the optimized complex coherent points and the ground coherent points.

Once the volume-only scattering complex coherence is scattered, a 2D look-up table of forest height-extinction coefficients is a requirement in the flow of the three-stage inversion method. The forest height and extinction coefficient values are obtained according to the intersection of the LUT and the volume-only complex coherence.

In this study, the process of improving the model is illustrated in Figure 9. After introducing ε and γ_e into the RVoG model (through Equation (23)) as unknown parameters, a new set of LUTs was generated when the amplitude $|\gamma_e|$ and the phase φ_e of ε and γ_e had different values. A triple iteration of the LUT was then performed after setting the range for the parameters. The intersection of each set of LUT with the volume-only decorrelation performed the forest height inversion under the set of parameters. Meanwhile, iteration added unknown parameters to the model. Therefore, the improved model needs the control condition of iteration, which is the true height measured on the ground. In this study, 25% of the ground-measured forest heights was randomly selected as the priori data for the iterations. The best parameter values in the iteration range and the corresponding inversion results were obtained by calculating and comparing the RMSE of each set of inversion heights with the ground data.

The improved model performs height inversion for all regions in the image, which was consistent with the original inversion method. The iterative parameters ε and γ_e in the model are calculated at the image level, and only one parameter result is iterated for the same group of interferometric pairs. Therefore, the real ground measurement data used in the inversion need at least one true value that is a true reflection of the forest height in the image. This is particularly important in controlling the inversion error in order to understand forest height inversion from the image and create a balance in the model.

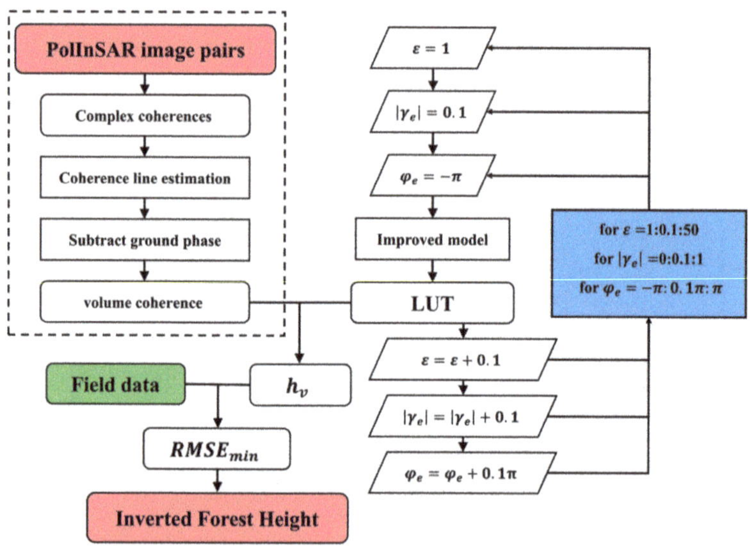

Figure 9. The inversion flow chart.

4.3. Theoretical Analysis of the Improved Model

In this study, the improved model is an empirical model. Besides the control of ground data, selecting the initial range of empirical parameters also has an important impact on the results. To discuss the feasibility of the improved inversion model, and establish whether the model satisfies the inversion conditions, this study theoretically tested the improved model using simulated data.

Previous reports have shown that the magnitude of the vertical wavenumber k_z determines the sensitivity of the inversion results to temporal decorrelation interference [49,50]. The simulated data in the present study was therefore created to invert the study areas with different mean tree heights (h_{True}) using PolInSAR with different k_z. When a certain degree of temporal decorrelation interference is present, the use of improved model to compute the relative height error ($|h - h_{True}|/h_{True} \times 100\%$) can reduce the relative error to less than 15%.

The distribution of error in the inversion results is shown in Figure 10. As the control variable, the interference size of γ_{Tmp} to the simulated data are fixed at $0.5 \times e^{0.3i}$. The average extinction coefficient σ and the observation Angle θ are also fixed. For each given h_{True}, the corresponding volume-only complex coherence γ_v for different k_z can be calculated using Equation (4). When γ_v is multiplied with the set γ_{Tmp}, the relative errors of the improved model inversion results under different k_z can be simulated in the actual inversion process. Based on Figure 10, it can be noted that when no improvement is made to the model (when ε is 1), the inversion error gradually increases with the decrease of k_z. When h_{True} = 30 m, k_z value larger than 0.05 is needed to determine forest height within 15% of the inversion error. Moreover, when h_{True} is smaller, the inversion can only be accurate with larger k_z. As ε increases, there exists one or more intervals of ε value for the compensated three-stage improvement method to accurately invert forest height within 15%, regardless of the size of k_z taken from the interferometric data. Besides, when ε increases and reaches the next interval suitable for inversion, the range of this interval will be larger than that of the previous interval. This moderates the inversion error change, and makes the model more adaptable to the forest height change in the observed area. However, when the ε value selected for inversion is too large, the inversion error will be very unstable. The error of the inversion result changes rapidly and loses regularity when the change of k_z and ε are not apparent. Inversion in this range will undoubtedly reduce the accuracy of inversion. Therefore, it should also try to avoid selecting too large ε to avoid fluctuations in the accuracy during the inversion.

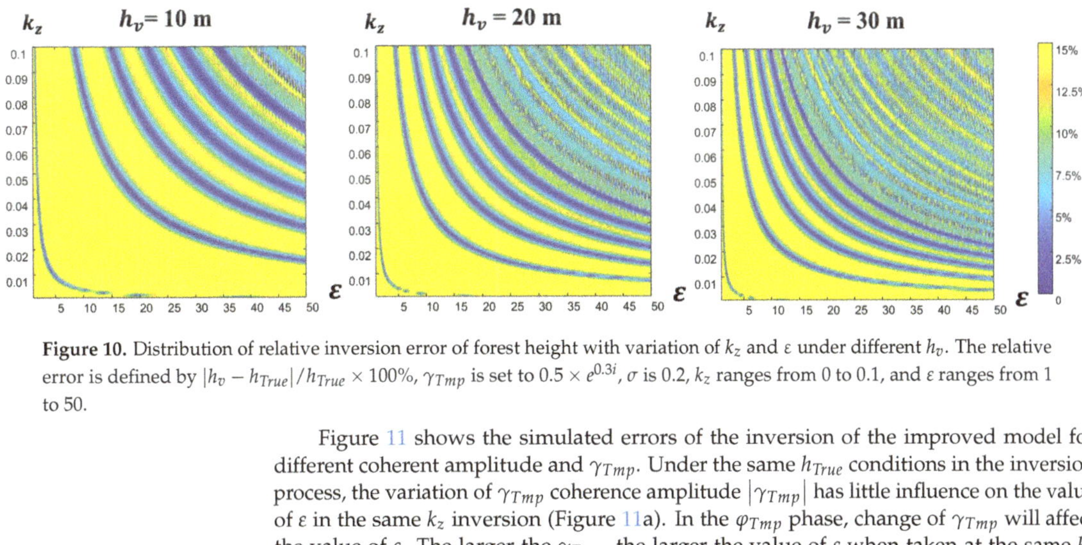

Figure 10. Distribution of relative inversion error of forest height with variation of k_z and ε under different h_v. The relative error is defined by $|h_v - h_{True}|/h_{True} \times 100\%$, γ_{Tmp} is set to $0.5 \times e^{0.3i}$, σ is 0.2, k_z ranges from 0 to 0.1, and ε ranges from 1 to 50.

Figure 11 shows the simulated errors of the inversion of the improved model for different coherent amplitude and γ_{Tmp}. Under the same h_{True} conditions in the inversion process, the variation of γ_{Tmp} coherence amplitude $|\gamma_{Tmp}|$ has little influence on the value of ε in the same k_z inversion (Figure 11a). In the φ_{Tmp} phase, change of γ_{Tmp} will affect the value of ε. The larger the γ_{Tmp}, the larger the value of ε when taken at the same k_z (Figure 11b). Therefore, it can be proved that the influence of decorrelation on the inversion accuracy is mainly due to the phase change caused by γ_{Tmp}. Based on our results, small k_z was found to be more sensitive to phase change and highly likely to cause significant errors, which was consistent with previous research results [49,50]. In addition, the decrease in coherence amplitude caused by temporal decorrelation had an effect on the inversion results, and should not be ignored.

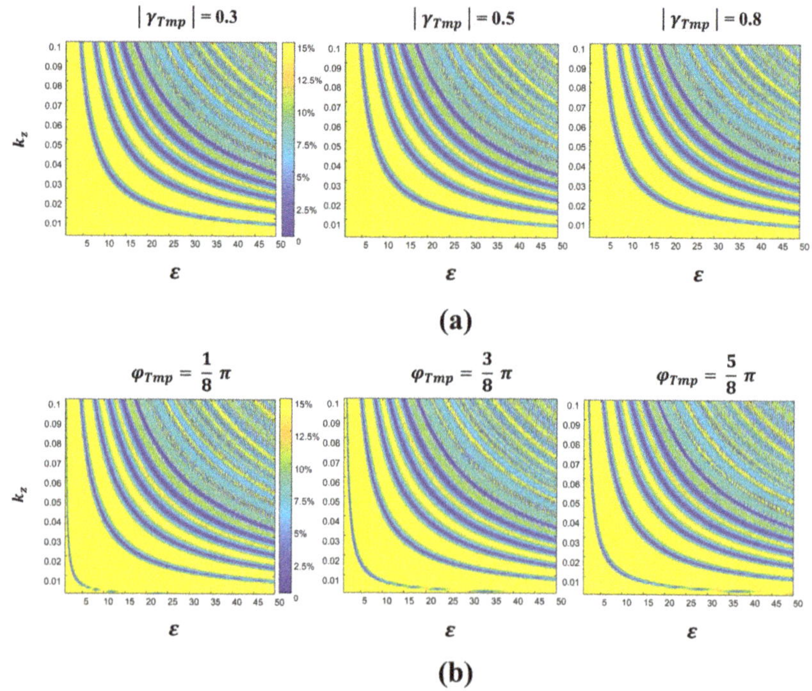

Figure 11. Distribution of relative forest height inversion errors with k_z and ε for different temporal decorrelation effects. h_{True} is set to 20 m, σ is 0.2, k_z ranges from 0 to 0.1, and ε ranges from 1 to 50. (**a**) Plots of forest height errors for different γ_{Tmp} coherence amplitudes, with phase set to 0. (**b**) Plots of forest height errors for different γ_{Tmp} phases, with amplitude set to 0.5.

The most significant difference between the improved model in this study and previous models is that it is an improvement on the three-stage inversion method, which is based on the geometric properties of the model. Therefore, there is need to analyze the geometric significance of the parameters expressed in the inversion process in order to understand each parameter's intrinsic logic in improving the accuracy of the inversion.

The correction term ε on the phase was corrected for the phase shift with height, and it ranged between 1 and 50 as shown in Figure 10. When the phase is corrected by using ε, the sensitivity of h_v to complex coherence changes in the complex plane unit circle is reduced, the phase value at the maximum height is increased, and the coherence amplitude is decreased (Figure 12). In this case, it confirmed that there is an intersection point between the height-extinction coefficient curve and the volume-only coherence point. This method achieved accurate inversion results. Meanwhile, the magnitude of the complex correction term $|\gamma_e|$ is the correction for the reduction of coherence due to decorrelation ($|\gamma_e| < 1$), while the phase φ_e is the correction for the change of scattering center ($-\pi < \varphi_e < \pi$). In the complex plane unit circle, $|\gamma_e|$ moves the curve in the lookup table closer to the center of the circle (Figure 13a), and φ_e causes changes in the starting phase of the lookup table curve (Figure 13b). Therefore, the constant change of parameters during the iteration process generates new inversion results.

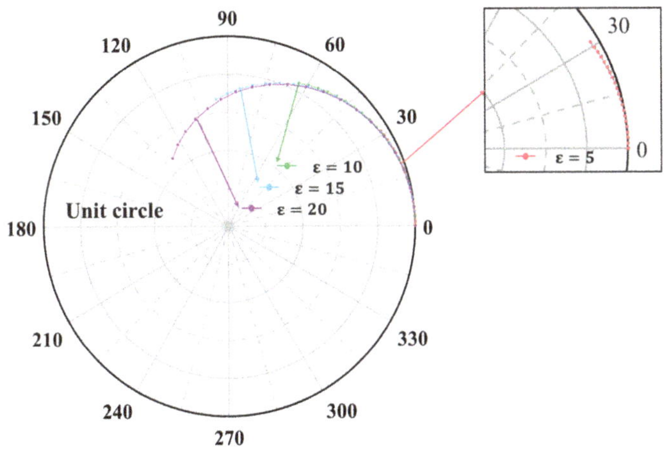

Figure 12. Schematic diagram of inversion with ε under the improved model. In this figure, σ is 0.2, h_v ranges from 0 to 30 m, the interval between two adjacent points of the same curve is 1 m, and k_z is set to 0.01. ε differs between different curves, and the interval between two adjacent points gradually increases as ε increases. When h_v reaches 30 m, the phase gradually moves away from the origin. During the inversion process, an increase in ε will expand the phase when h_v reaches the upper limit of the phase of LUT. Besides, the sensitivity of h_v gradually reduces changes in γ_v, making it more suitable for forest height inversion.

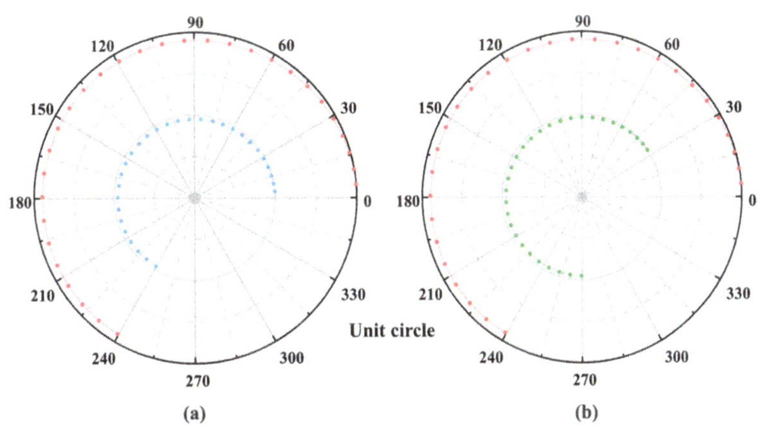

Figure 13. The complex correction term γ_e changes the position of the LUT in the complex plane unit circle. (**a**) In order for $|\gamma_e|$ to reduce the coherence amplitude of the LUT points, the curve composed of red points is the LUT calculated by the original RVoG model, and the curve composed of blue points is the LUT after $|\gamma_e|$ correction, and its more suitable for the inversion of low coherence cases. (**b**) Shows the LUT after φ_e is introduced on the basis of (**a**), and the curve composed of green points is the LUT after φ_e correction, which is more suitable for inversion of data with obvious phase center shift. σ is 0.2, k_z is 0.01, ε is 15, and the forest height range is 0–30 m with 1 m interval between two points.

5. Results

A total of 25% of the ground measured height was randomly selected as the prior data to control the iterative process. The inversion process follows the inversion procedure in Section 4.2. After quantifying the inversion error (Figures 10 and 11), the initial range limit of ε was set to 1–50, the initial range limit of $|\gamma_e|$ was set to 0–1, and the phase φ_e was set to

$-\pi\sim\pi$. Figure 14 shows scatter plots of height results of the iterations using the improved model, with prior true data. Table 3 shows the iterative results and the inversion accuracy.

It can be noted that the inversion results of the improved model were all within 15% RMSE compared with the true values, and there was a good fit. The iteration parameter ε ranged from 20 to 40, indicating that the random motion of the canopy produces a more significant interference with the phase error of the repeat-pass spaceborne PolInSAR data. Moreover, the decorrelation correction term of the parameter γ_e became larger as the temporal baseline increased. This indicates that the improved model has a better correction for the interference of temporal decorrelation. However, the magnitude of ε does not increase with the temporal baseline across the data but decreases with increasing k_z. This also shows that the smaller the k_z, the greater the sensitivity of the data to temporal decorrelation, leading to a decrease in inversion accuracy.

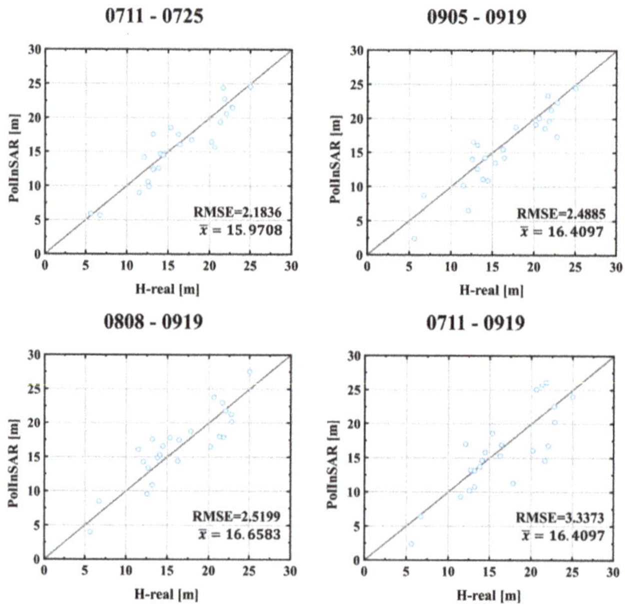

Figure 14. Scatter plots of the height inversions based on the improved model and prior true height. RMSE and $\overline{x}$ are the root mean square error and the mean value of the inversion results, respectively.

Table 3. Iteration parameters and inversion accuracy.

Data Sets	Parameters		Inversion Accuracy		
	ε	γ_e	RMSE	R^2	RSD
0711-0725	31.1	$0.60 \times e^{i\cdot 0.1\pi}$	2.1836	0.8355	32.66%
0905-0919	24.0	$0.75 \times e^{i\cdot 0.2\pi}$	2.4885	0.8154	30.98%
0808-0919	26.5	$0.49 \times e^{-i\cdot 0.3\pi}$	2.5199	0.7712	30.47%
0711-0919	19.9	$0.48 \times e^{-i\cdot 0.6\pi}$	3.3373	0.6941	30.99%

The inversion accuracy was found to be suitable for height inversion. However, all the relative standard deviations (RSDs) of the models were around 30% (Table 3), indicating that the inverse performance of the models still exhibits some volatility and randomness on a few samples. This phenomenon may be because the effects arising from temporal decorrelation were not completely eliminated by semi-empirical iterations. The calibration of temporal coherence could improve the deviation of coherence, but cannot solve the increase of phase line fitting variance caused by low γ_{vol} coherence [49]. In addition, since the conditions of forest height and coherence between different elements in the same image are not the same, the size of the correction term is subject to various constraints. As such, the use of the same correction term may inevitably lead to inconsistent parameter effects of different pixel values. Nevertheless, the errors in the inversion results are still within reasonable limits and do not affect the forest height distribution or hinder further studies.

To test the robustness of the improved model, the iterative model was validated using an additional 75% of the ground truth data. Performance of the model was further evaluated by introducing the nonlinear least squares inverse model based on repeat-pass spaceborne InSAR [42] for comparison. In this model, the coupling coherence of volume decorrelation and temporal decorrelation is taken as the image observation, and the random motion of forest and the change of dielectric constant are taken into account under a certain temporal baseline. The model can be expressed as follows:

$$\begin{aligned}|\gamma_{v+t}| &= S_{scene}\cdot exp\left(-\frac{1}{2}\left(\frac{4\pi\delta_r\alpha}{\lambda h_r}\right)^2 {h_v}^2\right)\\ &\approx S_{scene}\cdot sinc\left(\frac{h_v}{C_{scene}}\right),\ h_v < \pi\cdot C_{scene}\end{aligned} \quad (27)$$

where S_{scene} is a non-negative real value less than or equal to 1 and C_{scene} represents the random motion level of the volume scatterers. Consistent with the original study, HV polarization was used in this study as the polarization to represent the volume scatterer. The same modeling data as the improved model in this study were used for training, and the model parameters were obtained by nonlinear least squares iteration. The scatter plots of the height inversions using models and the validation data are shown in Figure 15. The validation accuracy of the four data sets is also shown in Table 4.

Table 4. The validation accuracy of the four data sets.

Data Sets	Validation Accuracy of the Improved Model			Validation Accuracy of the Nonlinear Least Squares Model		
	RMSE	R^2	RSD	RMSE	R^2	RSD
0711-0725	2.7305	0.7401	29.06%	3.2597	0.6342	29.94%
0905-0919	2.3125	0.8126	30.58%	3.3024	0.6782	32.51%
0808-0919	3.1490	0.6871	32.33%	3.8472	0.6007	33.32%
0711-0919	4.1016	0.5978	34.51%	4.1194	0.5522	34.69%

Based on Figure 15, it can be noted that the inversion accuracy of the improved model is relatively close for different interferometric pairs. Therefore, the model can still perform forest height inversion within 15% accuracy even when the temporal baseline is different from k_z. This indicates that the inversion of the improved model is more robust and can be applied to repeat-pass spaceborne PolInSAR data with larger temporal baselines and smaller k_z. In contrast, the nonlinear least squares model had lower inversion accuracy and a larger randomness due to the lack of correction phase. The accuracy of the model decreased more obviously when the data with large temporal baseline was inversed. Moreover, the lack of polarization information may be another reason for the decrease in accuracy.

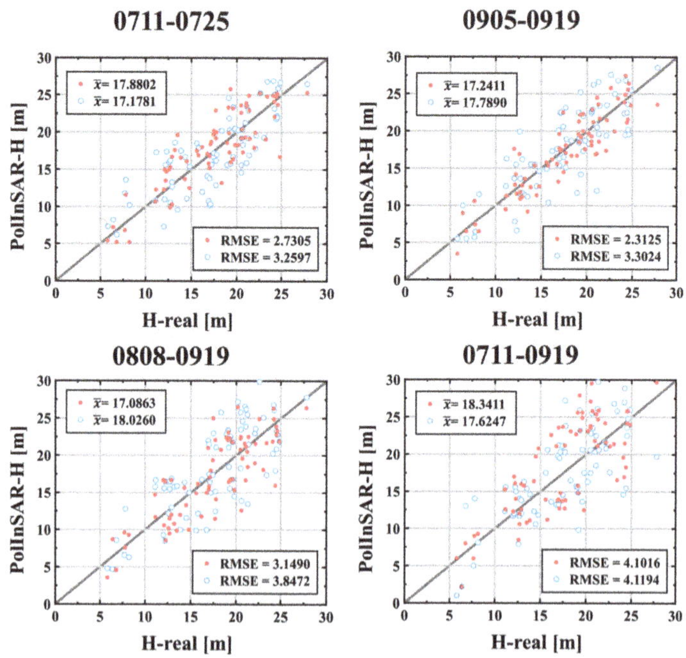

Figure 15. Scatter plots of the height inversions based on the improved model and the validation data. The red points are the inversion heights of the improved models proposed in this study. The blue points are the nonlinear least squares model inversion heights. RMSE and $\bar{x}$ are the root mean square error and the mean value of the inversion results, respectively.

6. Discussion

By improving the model inversion of forest height, the significant errors caused by repeat-pass spaceborne PolInSAR temporal decorrelation are reduced, increasing the scope of application of this data in forest height inversion. Based on the results of the improved model in Section 4 under different interferometric data, the following conclusions can be drawn:

1. The correction of temporal decorrelation can improve the robustness and accuracy of the inversion and meet the needs of remote sensing for forest height inversion.
2. A more accurate forest height inversion of common SAR data can be performed using the improved model, but there may still be a small degree of error in the inversion results.
3. Data with large temporal baselines should be carefully selected when using models for height inversion.

6.1. Inversion Performance of the Model

Temporal decorrelation causes abrupt changes in the interferometric phase on the one hand and decreases in coherence on the other. In Figure 16, the coherence of the larger temporal baseline is lower than that of the smaller temporal baseline, which also proves that for repeat-pass spaceborne PolInSAR data, temporal decorrelation is the main source of inversion errors [61]. Besides, the coherence of the 0711-0919 data are slightly lower than that of the 0808-0919 dataset, so it can be shown that the increase of the temporal baseline will not cause more loss of decorrelation when the loss of coherence reaches a certain level. Combined with the empirical parameters obtained by inversion of the improved model (Table 3), it can be seen that the continuous increase of the temporal baseline causes more changes in the phase of the observed complex coherence, compared to decreases in the coherence. The changes in phase are also affected by the sensitivity of k_z to temporal decorrelation in the inversion process, and correction for this error can be better achieved using an improved model. In addition, the magnitude of coherence varies at the same temporal baseline, which can be caused by various factors, such as wind-induced motion, weather-induced changes in dielectric constant, etc. [9,26,43,62]. The coherence is not uniformly distributed in the same image, resulting from the joint action of temporal decorrelation and volume decorrelation. The validated improved inversion model summarizes the temporal decorrelation in the same data set as a complex parameter, which can also control the error of inversion results within a relatively small range.

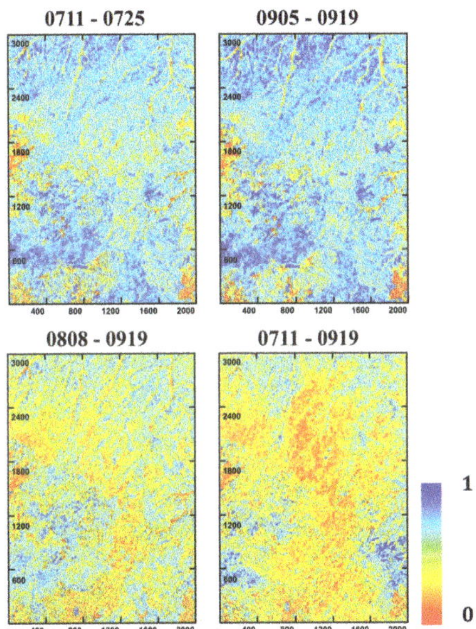

Figure 16. Coherence images for different interferometric data; 0711-0725 and 0905-0919 data sets have higher coherence, 0808-0919 data set has lower coherence, and 0711-0919 data set has the lowest coherence. The data with larger temporal baselines have lower coherence, and temporal decorrelation decreases the coherence. The spatial distribution of coherence is not uniform and receives a combination of volume decorrelation and temporal decorrelation.

Figure 17 shows the forest height inversion results of the improved model with different interference datasets. The inversion results show that the improved inversion method is robust for the forest height inversion under different k_z and temporal baselines. Although there was some biasness between the data sets, this error may have been influenced by abrupt changes in the ground phase in addition to the reasons mentioned in Section 5. Moreover, the inversion effect does not decrease significantly for the data sets 0808-0919 and 0711-0919 with larger temporal baselines. The temporal baseline of the 0711-0919 data set was found to be larger, and it showed a better inversion effect than 0808-0919. This could be because the difference in the coherence between the two sets of data were little. Meanwhile, the data are less sensitive to temporal decorrelation interference due to the large original k_z of 0711-0919 data, resulting in the preference in the overall inversion result. Therefore, for repeat-pass spaceborne PolInSAR, the improved model can be relatively robust to forest height inversion.

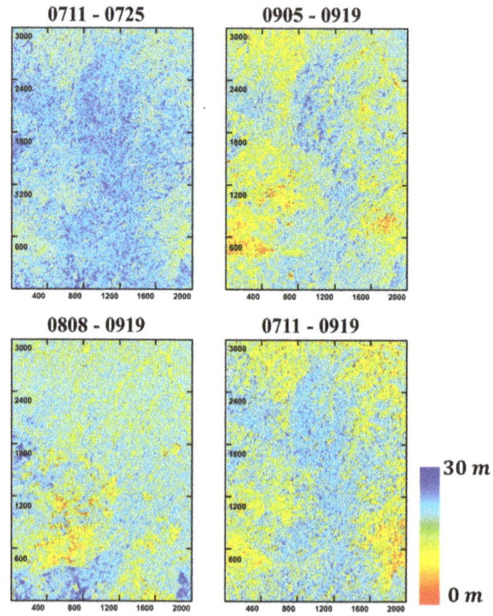

Figure 17. Inversion of forest height using different data.

6.2. Error Analysis of Inversion Results

Errors in the inversion results and the effect of residual temporal decorrelation on the model are the common factors that affect the accuracy of the inversion results. However, other factors outside the model may also cause the problem of accuracy degradation. This study was based on the assumption of the traditional three-stage method [38]. Besides the temporal decorrelation source, the inversion results are still faced with two aspects of errors: residual ground contribution in the volume scattering complex coherence and the estimated shift of the real ground phase [63]. The inversion process usually assumes that the ground scattering contribution of volume-only scattering complex coherence is 0. However, when there is a strong interaction at the dihedral angle of the earth, the polarization channels with low ground contribution will still be affected by the ground scattering center, increasing the variance of the inversion results. Thus, the accuracy of the fitting is reduced [38].

Previous studies have attempted to optimize this phenomenon and can be categorized into three groups based on their principles: separation of the contribution of volume scattering based on polarization decomposition [8,64], optimization by coherence to minimize the effect of ground scattering contributions [39,45,65], and methods that introduce other data sources, such as different baselines [66], and different frequency bands [67,68]. At present, these methods can only reduce the error of ground contribution to a certain extent. Based on the repeat-pass spaceborne PolInSAR data with low coherence, the improved model introduces partial coherence optimization to help in the model calculation. This method is very valuable in research. However, removal of ground contribution to the inversion effect deserves further in-depth study.

Even after terrain correction, the influence of ground phase estimation is still inevitable [11]. The ground phase images were estimated from different interferometric data during the three-stage inversion (Figure 18). The ground phase of all interferometric pairs, as distinguished from the interferometric DEM (Figure 4c), was affected by abrupt changes resulting to incorrect ground point selection during the inversion. The phase error was particularly prominent in 0711-0725, where the k_z of this data set was smaller than the other groups. Thus, the spatial baseline of this data set was shorter than the other groups during interferometry. In addition, the abrupt changes of terrain phase were more obvious in rugged terrain, which was consistent with previous findings [69]. In contrast, the abrupt ground phase error for 0711-0919 was not significant, indicating that the estimation error of the ground phase was not significantly affected by the temporal baseline when compared to spatial baseline. This also supported previous conclusions about the same [32,61].

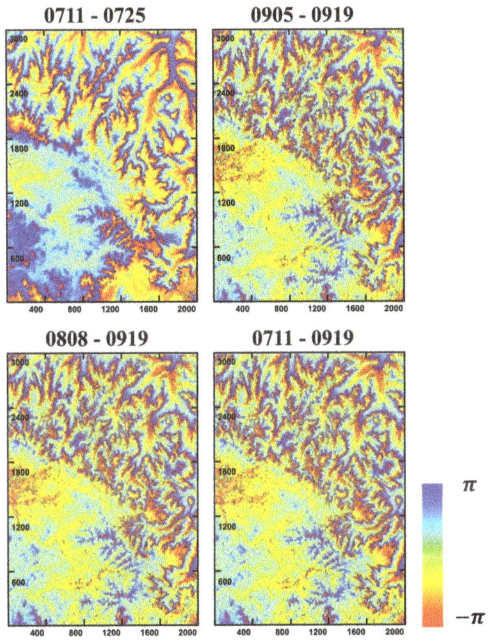

Figure 18. Phase estimation of three-stage inversion method for different interferometric pairs. There are relatively obvious phase point anomalies in the estimates of 0711-0725, which are more evident in the rugged terrain type (upper right corner). Thus, it can be demonstrated that the interferometric data with shorter spatial baselines are more prone to errors in the ground phase estimates.

The forest average height serves as the standard forestry table measurement parameter and plays an important role in forest inventory and subsequent biomass estimation [70]. However, it has also been suggested that forest dominance height is closer to the effective height of the remote sensing signal [63], and the difference tends to be more pronounced for forests with lower heights. In addition, differences in forest species and density may also have an impact on the inversion results. The reasons are mainly the differences arising from the canopy cover and the ratio of canopy to bare ground. In the general boreal temperate forests of China, the distribution of forest species and density is complex. They were not strictly classified and studied separately in this study. This will undoubtedly have some influence on the inversion accuracy, which is worthy of our further study. It is worth noting that although the sample size does not bias the fit results, it may still cause a reduction in correlation [52].

The model proposed in this study is only for the fuzzy inversion of forest height caused by temporal decorrelation. It does not focus on the inversion errors caused by other factors that may exist. As such, the model is only suitable for data whose main error source is temporal decorrelation. In addition, this study is a regional study, and the model has only been verified in the height inversion of forests in northern China. Therefore, the inversion of the model in other regional forests requires further studies.

6.3. Suitable Range of Spatial Baseline for Forest Height Inversion

In this study, height inversion of repeat-pass spaceborne PolInSAR data revealed that the degree of influence of temporal decorrelation on the inversion accuracy depends not only on the length of the temporal baseline, but also on the sensitivity of the data to temporal decorrelation, such as the magnitude of k_z. Given the upcoming large-scale forest biomass observation mission, the multifaceted parameters of PolInSAR data should be carefully considered. The spaceborne SAR sensor has a relatively fixed flight trajectory when compared to the airborne SAR, and it is relatively difficult to change its space parameters. Therefore, this study aims at providing more insights about the spatial parameter suitable for the forest height of the spaceborne SAR inversion. This will help future studies in establishing a suitable spaceborne inversion system.

Given the sensitivity of the ground target height to the interferometric phase, the phase shift caused by temporal decorrelation causes the most interference to the forest height calculation. Therefore, the correction term ε in the model (Equation (23)) has the most significant impact on the inversion accuracy. In Equation (23), the function of the ε parameter is to correct the phase deviation. However, from another perspective, the inversion accuracy can be measured by converting ε to the size of k_z. As such, the inversion accuracy can be improved by using the data suitable for k_z.

In this study, we used L-band ALOS2 repeat-pass spaceborne PolInSAR data to retrieve the height of temperate forest in the Saihanba region of northern China, and obtained the ε parameters under different k_z and temporal baselines. When the k_z value of the spaceborne data reached $\varepsilon \cdot k_z$ of the corresponding height, the data were considered suitable for inversion of the forest height using the polarized interferometric model. The suitable k_z for different data pairs can be obtained from Tables 2 and 3, as shown in Table 5.

Table 5. Suitable k_z for different data pairs.

Data Sets	Temporal Baselines (Day)	Original k_z	Suitable k_z	Suitable Vertical Baseline
0711-0725	14	0.015	0.465	2391.7 m
0905-0919	14	0.018	0.432	2723.7 m
0808-0919	42	0.018	0.477	3007.4 m
0711-0919	70	0.021	0.418	2635.4 m

The analysis of Table 5 shows that a larger vertical baseline is more suitable for repeat-pass spaceborne PolInSAR inversion of forest height, and a similar conclusion was reached in a previous study upon analysis of airborne data [11,50]. This is because when the k_z is larger, the baseline decorrelation caused by the spatial baseline between the primary and secondary images is greater. The larger k_z results obtained in the study indicate that the disturbance caused by baseline decorrelation can be hardly considered when compared to temporal decorrelation. Conversely, smaller k_z may cause abrupt changes in the phase estimation of the underlying surface, causing a decrease in the inversion accuracy. In addition, the suitable k_z of different temporal baselines slightly differed for the same area data, indicating that the sensitivity of k_z to temporal decorrelation is more important than that of temporal baseline. Therefore, more attention should be paid to the use of suitable k_z when using repeat-pass spaceborne SAR data interferometry to measure forest height.

However, according to previous studies, the choice of k_z was negatively correlated to the forest height in the observation area [11]. The k_z value could be appropriately lowered when the forest height of the study area was higher. According to the average forest height in this study, the suitable k_z exceeded the fuzzy interval of 2π. Since the relationship between k_z of the data and the ground height has to be within the 2π fuzzy interval, there may not be a suitable value of k_z for inversion if the inversion model is not improved. Therefore, the correction of the ε parameter for the phase is essential when the temporal decorrelation of the data has a significant effect.

7. Conclusions

The performance study of forest height through repeat-pass spaceborne PolInSAR inversion will effectively improve large-scale forest height estimation efficiency. To address the limitations of the current inversion of forest height with repeat-pass L-band spaceborne PolInSAR data, a theoretical analysis of the effects of phase deviation and temporal decorrelation on the inversion performance and an improved inversion method are proposed based on the RVoG model. In this respect, this study makes three main conclusions: (1) the correction of temporal decorrelation can improve the robustness and accuracy of the forest height inversion. (2) A more accurate forest height inversion of common SAR data can be performed using the improved model, but there may still be a small degree of error in the inversion results. (3) Data with large temporal baselines should be carefully selected when using models for height inversion.

The repeat-pass spaceborne sensors have a more extended temporal baseline than the airborne sensors, which exposes the repeat-pass spaceborne PolInSAR to more excellent decorrelation effects when inverting forest heights in the study area. Besides, the vertical wavenumber of the repeat-pass spaceborne sensor is lower than the interval suitable for inversion, making the inversion height more sensitive to decorrelation. When affected by these two factors, the spaceborne PolInSAR data often loses the ability to invert the forest height using the traditional three-stage inversion method.

To avoid the interference of both factors as much as possible, we propose a semi-empirical improvement model that controls the iterations by ground true data, based on the three-stage inversion method. Through theoretical analysis, it was found that in general cases, there is a correction term to make the error between the inversion results and the true value to converge to less than 15%. Moreover, there is more than one correction term for any reasonable range of specified forest height and k_z can accurately invert the forest height under the influence of temporal decorrelation.

This method uses ALOS2 repeat-pass spaceborne L-band PolInSAR data with large temporal baseline to perform accurate forest height inversion in the mixed conifer-broad forest of the Saihanba of northern China. For different interferometric pairs, the RMSEs of inversion results were less than 15%. For the interferometric data of the larger temporal baseline, the inversion results were slightly lower than other interferometric pairs because of the low coherence caused by decorrelation. Nevertheless, this study is based on the assumptions of the traditional three-stage inversion method. Therefore, there is need for further research to improve the accuracy of this method using previous improved models, such as ground contribution and terrain impact.

The present study achieves adequate accuracy in forest height inversion by repeat-pass spaceborne PolInSAR data through the improved model. However, accuracy degradation may occur during inversion of heterogeneous forests because the results are affected by height variation. Further research is therefore needed to establish whether general patterns can be summarized to avoid the above problems by examining data from different forest height intervals.

Author Contributions: Conceptualization, W.F. and Y.Y.; investigation, A.S., Z.L., G.W.; experimentation, data processing, validation, and manuscript writing, Y.M.; editing, O.O.M. All authors have read and agreed to the published version of the manuscript.

Funding: This research was funded by the National Natural Science Foundation of China (contract no. 31971654) and the Civil Aerospace Technology Advance Research Project (contract no. D040114).

Data Availability Statement: Data was obtained from the Japan Aerospace Exploration Agency and are available at https://www.eorc.jaxa.jp/ALOS-2/en/about/palsar2.htm (accessed on 12 September 2021).

Conflicts of Interest: The funders had no role in the design of the study; in the collection, analyses, or interpretation of data; in the writing of the manuscript, or in the decision to publish the results.

References

1. Purves, D.; Pacala, S. Forests in Flux. *Science* **2013**, *342*, 776. [CrossRef]
2. Hall, F.G.; Bergen, K.; Blair, J.B.; Dubayah, R.; Houghton, R.; Hurtt, G.; Kellndorfer, J.; Lefsky, M.; Ranson, J.; Saatchi, S.; et al. Characterizing 3D vegetation structure from space: Mission requirements. *Remote Sens. Environ.* **2011**, *115*, 2753–2775. [CrossRef]
3. Houghton, R.A.; Hall, F.; Goetz, S. Importance of biomass in the global carbon cycle. *J. Geophys. Res. Space Phys.* **2009**, *114*, 302. [CrossRef]
4. Lei, Y.; Treuhaft, R.; Gonçalves, F. Automated estimation of forest height and underlying topography over a Brazilian tropical forest with single-baseline single-polarization TanDEM-X SAR interferometry. *Remote Sens. Environ.* **2021**, *252*, 112132. [CrossRef]
5. Gu, C.; Clevers, J.G.; Liu, X.; Tian, X.; Li, Z.; Li, Z. Predicting forest height using the GOST, Landsat 7 ETM+, and airborne LiDAR for sloping terrains in the Greater Khingan Mountains of China. *ISPRS J. Photogramm. Remote Sens.* **2018**, *137*, 97–111. [CrossRef]
6. Huang, H.; Liu, C.; Wang, X.; Biging, G.S.; Chen, Y.; Yang, J.; Gong, P. Mapping vegetation heights in China using slope correction ICESat data, SRTM, MODIS-derived and climate data. *ISPRS J. Photogramm. Remote Sens.* **2017**, *129*, 189–199. [CrossRef]
7. Schlund, M.; Baron, D.; Magdon, P.; Erasmi, S. Canopy penetration depth estimation with TanDEM-X and its compensation in temperate forests. *ISPRS J. Photogramm. Remote Sens.* **2019**, *147*, 232–241. [CrossRef]
8. Aghabalaei, A.; Ebadi, H.; Maghsoudi, Y. Forest height estimation based on the RVoG inversion model and the PolInSAR decomposition technique. *Int. J. Remote Sens.* **2019**, *41*, 2684–2703. [CrossRef]
9. Hajnsek, I.; Kugler, F.; Lee, S.-K.; Papathanassiou, K.P. Tropical-Forest-Parameter Estimation by Means of Pol-InSAR: The INDREX-II Campaign. *IEEE Trans. Geosci. Remote Sens.* **2009**, *47*, 481–493. [CrossRef]
10. Koch, B. Status and future of laser scanning, synthetic aperture radar and hyperspectral remote sensing data for forest biomass assessment. *ISPRS J. Photogramm. Remote Sens.* **2010**, *65*, 581–590. [CrossRef]
11. Liao, Z.; He, B.; van Dijk, A.I.; Bai, X.; Quan, X. The impacts of spatial baseline on forest canopy height model and digital terrain model retrieval using P-band PolInSAR data. *Remote Sens. Environ.* **2018**, *210*, 403–421. [CrossRef]
12. Wang, M.; Sun, R.; Xiao, Z. Estimation of Forest Canopy Height and Aboveground Biomass from Spaceborne LiDAR and Landsat Imageries in Maryland. *Remote Sens.* **2018**, *10*, 344. [CrossRef]
13. Huang, H.; Liu, C.; Wang, X.; Zhou, X.; Gong, P. Integration of multi-resource remotely sensed data and allometric models for forest aboveground biomass estimation in China. *Remote Sens. Environ.* **2019**, *221*, 225–234. [CrossRef]
14. Liao, Z.; He, B.; Quan, X.; van Dijk, A.I.; Qiu, S.; Yin, C. Biomass estimation in dense tropical forest using multiple information from single-baseline P-band PolInSAR data. *Remote Sens. Environ.* **2019**, *221*, 489–507. [CrossRef]
15. Ferretti, A.; Prati, C.; Rocca, F. Nonlinear subsidence rate estimation using permanent scatterers in differential SAR interferometry. *IEEE Trans. Geosci. Remote Sens.* **2000**, *38*, 2202–2212. [CrossRef]
16. Graham, L.C. Synthetic interferometer radar for topographic mapping. *Proc. IEEE* **1974**, *62*, 763–768. [CrossRef]
17. Huang, L.; Hajnsek, I. Polarimetric Behavior for the Derivation of Sea Ice Topographic Height From TanDEM-X Interferometric SAR Data. *IEEE J. Sel. Top. Appl. Earth Obs. Remote Sens.* **2021**, *14*, 1095–1110. [CrossRef]
18. Schlund, M.; von Poncet, F.; Hoekman, D.H.; Kuntz, S.; Schmullius, C. Importance of Bistatic SAR Features from TanDEM-X for Forest Mapping and Monitoring. *Remote Sens. Environ.* **2014**, *151*, 16–26. [CrossRef]
19. Zhang, L.; Duan, B.; Zou, B. Research on Inversion Models for Forest Height Estimation Using Polarimetric SAR Interferometry. *Int. Arch. Photogramm. Remote Sens. Spatial Inf. Sci.* **2017**, *42*, 659–663. [CrossRef]
20. Garestier, F.; le Toan, T. Forest Modeling for Height Inversion Using Single-Baseline InSAR/Pol-InSAR Data. *IEEE Trans. Geosci. Remote Sens.* **2010**, *48*, 1528–1539. [CrossRef]
21. Askne, J.; Santoro, M.; Smith, G.; Fransson, J. Multitemporal repeat-pass sar interferometry of boreal forests. *IEEE Trans. Geosci. Remote Sens.* **2003**, *41*, 1540–1550. [CrossRef]
22. Askne, J.I.; Soja, M.J.; Ulander, L.M. Biomass estimation in a boreal forest from TanDEM-X data, lidar DTM, and the interferometric water cloud model. *Remote Sens. Environ.* **2017**, *196*, 265–278. [CrossRef]
23. Hagberg, J.; Ulander, L.M.H.; Askne, J. Repeat-pass SAR interferometry over forested terrain. *IEEE Trans. Geosci. Remote Sens.* **1995**, *33*, 331–340. [CrossRef]
24. Zhang, B.; Fu, H.; Zhu, J.; Peng, X.; Xie, Q.; Lin, D.; Liu, Z. A Multibaseline PolInSAR Forest Height Inversion Model Based on Fourier–Legendre Polynomials. *IEEE Geosci. Remote Sens. Lett.* **2021**, *18*, 687–691. [CrossRef]
25. Cloude, S.; Papathanassiou, K. Polarimetric SAR interferometry. *IEEE Trans. Geosci. Remote Sens.* **1998**, *36*, 1551–1565. [CrossRef]
26. Treuhaft, R.N.; Madsen, S.N.; Moghaddam, M.; van Zyl, J.J. Vegetation characteristics and underlying topography from interferometric radar. *Radio Sci.* **1996**, *31*, 1449–1485. [CrossRef]
27. Treuhaft, R.N.; Siqueira, P.R. Vertical structure of vegetated land surfaces from interferometric and polarimetric radar. *Radio Sci.* **2000**, *35*, 141–177. [CrossRef]
28. Soja, M.J.; Persson, H.; Ulander, L.M.H. Estimation of Forest Height and Canopy Density From a Single InSAR Correlation Coefficient. *IEEE Geosci. Remote Sens. Lett.* **2014**, *12*, 646–650. [CrossRef]
29. Fu, H.; Wang, C.; Zhu, J.; Xie, Q.; Zhang, B. Estimation of Pine Forest Height and Underlying DEM Using Multi-Baseline P-Band PolInSAR Data. *Remote Sens.* **2016**, *8*, 820. [CrossRef]
30. Xie, Q.; Zhu, J.; Wang, C.; Fu, H.; Lopez-Sanchez, J.M.; Ballester-Berman, J.D. A Modified Dual-Baseline PolInSAR Method for Forest Height Estimation. *Remote Sens.* **2017**, *9*, 819. [CrossRef]

31. Cloude, S.R. Robust parameter estimation using dual baseline polarimetric SAR interferometry. In Proceedings of the IEEE International Geoscience and Remote Sensing Symposium, Toronto, ON, Canada, 24–28 June 2002; IEEE: Piscataway, NJ, USA, 2003; Volume 2, pp. 838–840.
32. Lavalle, M.; Hensley, S. Extraction of Structural and Dynamic Properties of Forests From Polarimetric-Interferometric SAR Data Affected by Temporal Decorrelation. *IEEE Trans. Geosci. Remote Sens.* **2015**, *53*, 4752–4767. [CrossRef]
33. Lu, H.; Suo, Z.; Guo, R.; Bao, Z. S-RVoG model for forest parameters inversion over underlying topography. *Electron. Lett.* **2013**, *49*, 618–620. [CrossRef]
34. Ghasemi, N.; Tolpekin, V.; Stein, A. A modified model for estimating tree height from PolInSAR with compensation for temporal decorrelation. *Int. J. Appl. Earth Obs. Geoinf.* **2018**, *73*, 313–322. [CrossRef]
35. Cloude, S.R.; Williams, M.L. A coherent EM scattering model for dual baseline POLInSAR. In Proceedings of the 2003 IEEE International Geoscience and Remote Sensing Symposium, Toulouse, France, 21–25 July 2003; IEEE: Piscataway, NJ, USA, 2004; Volume 3, pp. 1423–1425.
36. Managhebi, T.; Maghsoudi, Y.; Zoej, M.J.V. A Volume Optimization Method to Improve the Three-Stage Inversion Algorithm for Forest Height Estimation Using PolInSAR Data. *IEEE Geosci. Remote Sens. Lett.* **2018**, *15*, 1214–1218. [CrossRef]
37. Chen, W.; Zheng, Q.; Xiang, H.; Chen, X.; Sakai, T. Forest Canopy Height Estimation Using Polarimetric Interferometric Synthetic Aperture Radar (PolInSAR) Technology Based on Full-Polarized ALOS/PALSAR Data. *Remote Sens.* **2021**, *13*, 174. [CrossRef]
38. Shi, Y.; He, B.; Liao, Z. An improved dual-baseline PolInSAR method for forest height inversion. *Int. J. Appl. Earth Obs. Geoinf.* **2021**, *103*, 102483. [CrossRef]
39. Xing, C.; Zhang, T.; Wang, H.; Zeng, L.; Yin, J.; Yang, J. A Novel Four-Stage Method for Vegetation Height Estimation with Repeat-Pass PolInSAR Data via Temporal Decorrelation Adaptive Estimation and Distance Transformation. *Remote Sens.* **2021**, *13*, 213. [CrossRef]
40. Ahmed, R.; Siqueira, P.; Hensley, S.; Chapman, B.; Bergen, K. A survey of temporal decorrelation from spaceborne L-Band repeat-pass InSAR. *Remote Sens. Environ.* **2011**, *115*, 2887–2896. [CrossRef]
41. Lee, S.-K.; Kugler, F.; Papathanassiou, K.; Hajnsek, I. Quantification and compensation of temporal decorrelation effects in polarimetric SAR interferometry. In Proceedings of the 2012 IEEE International Geoscience and Remote Sensing Symposium, Munich, Germany, 22–27 July 2012; IEEE: Piscataway, NJ, USA, 2012; pp. 3106–3109.
42. Lei, Y.; Siqueira, P. Estimation of Forest Height Using Spaceborne Repeat-Pass L-Band InSAR Correlation Magnitude over the US State of Maine. *Remote Sens.* **2014**, *6*, 10252–10285. [CrossRef]
43. Lavalle, M.; Simard, M.; Hensley, S. A Temporal Decorrelation Model for Polarimetric Radar Interferometers. *IEEE Trans. Geosci. Remote Sens.* **2011**, *50*, 2880–2888. [CrossRef]
44. Neumann, M.; Ferro-Famil, L.; Reigber, A. Estimation of Forest Structure, Ground, and Canopy Layer Characteristics from Multibaseline Polarimetric Interferometric SAR Data. *IEEE Trans. Geosci. Remote Sens.* **2009**, *48*, 1086–1104. [CrossRef]
45. Papathanassiou, K.; Cloude, S. The effect of temporal decorrelation on the inversion of forest parameters from Pol-InSAR data. In Proceedings of the IGARSS 2003, 2003 IEEE International Geoscience and Remote Sensing Symposium, Toulouse, France, 21–25 July 2003; IEEE: Piscataway, NJ, USA, 2004; Volume 3, pp. 1429–1431.
46. Simard, M.; Denbina, M. An Assessment of Temporal Decorrelation Compensation Methods for Forest Canopy Height Estimation Using Airborne L-Band Same-Day Repeat-Pass Polarimetric SAR Interferometry. *IEEE J. Sel. Top. Appl. Earth Obs. Remote Sens.* **2018**, *11*, 95–111. [CrossRef]
47. Lee, S.-K.; Fatoyinbo, T.; Osmanoğlu, B.; Sun, G. Polarimetric SAR interferometry evaluation in mangroves. In Proceedings of the 2014 IEEE Geoscience and Remote Sensing Symposium, Quebec City, QC, Canada, 13–18 July 2014; IEEE: Piscataway, NJ, USA, 2014; pp. 4584–4587.
48. Lei, Y.; Siqueira, P.; Torbick, N.; Ducey, M.; Chowdhury, D.; Salas, W. Generation of Large-Scale Moderate-Resolution Forest Height Mosaic With Spaceborne Repeat-Pass SAR Interferometry and Lidar. *IEEE Trans. Geosci. Remote Sens.* **2018**, *57*, 770–787. [CrossRef]
49. Kugler, F.; Lee, S.-K.; Hajnsek, I.; Papathanassiou, K.P. Forest Height Estimation by Means of Pol-InSAR Data Inversion: The Role of the Vertical Wavenumber. *IEEE Trans. Geosci. Remote Sens.* **2015**, *53*, 5294–5311. [CrossRef]
50. Du, K.; Lin, H.; Wang, G.; Long, J.; Li, J.; Liu, Z. The Impact of Vertical Wavenumber on Forest Height Inversion by PolInSAR. In Proceedings of the 2018 Fifth International Workshop on Earth Observation and Remote Sensing Applications (EORSA), Xi'an, China, 18–20 June 2018; IEEE: Piscataway, NJ, USA, 2018; pp. 1–5.
51. Jiang, L.; Wang, J.; Cui, H.; Wang, G.; Zhao, T.; Zhao, S.; Chai, L.; Liu, X.; Yang, J. In situ soil moisture and temperature network in genhe watershed and saihanba area in China. *Data Brief* **2020**, *31*, 105693. [CrossRef]
52. Hosseini, S.; Garestier, F. Pol-InSAR sensitivity to hemi-boreal forest structure at L- and P-bands. *Int. J. Appl. Earth Obs. Geoinf.* **2021**, *94*, 102213. [CrossRef]
53. Gatelli, F.; Guamieri, A.M.; Parizzi, F.; Pasquali, P.; Prati, C.; Rocca, F. The wavenumber shift in SAR interferometry. *IEEE Trans. Geosci. Remote Sens.* **1994**, *32*, 855–865. [CrossRef]
54. Zebker, H.; Villasenor, J. Decorrelation in interferometric radar echoes. *IEEE Trans. Geosci. Remote Sens.* **1992**, *30*, 950–959. [CrossRef]
55. Gomba, G.; Parizzi, A.; De Zan, F.; Eineder, M.; Bamler, R. *Toward Operational Compensation of Ionospheric Effects in SAR Interferograms: The Split-Spectrum Method*; IEEE: Piscataway, NJ, USA, 2016; Volume 54, pp. 1446–1461.

56. Liang, C.; Fielding, E.J. Measuring Azimuth Deformation with L-Band ALOS-2 ScanSAR Interferometry. *IEEE Trans. Geosci. Remote Sens.* **2017**, *55*, 2725–2738. [CrossRef]
57. Shimada, M.; Isoguchi, O.; Tadono, T.; Isono, K. PALSAR Radiometric and Geometric Calibration. *IEEE Trans. Geosci. Remote Sens.* **2009**, *47*, 3915–3932. [CrossRef]
58. Woods, J.W.; Biemond, J. Comments on "A Model for Radar Images and Its Application to Adaptive Digital Filtering of Multiplicative Noise". *IEEE Trans. Pattern Anal. Mach. Intell.* **1984**, *PAMI-6*, 658–659. [CrossRef]
59. Rosen, P.A.; Hensley, S.; Joughin, I.R.; Li, F.K.; Madsen, S.N.; Rodriguez, E.; Goldstein, R.M. Synthetic aperture radar interferometry. *Proc. IEEE* **2000**, *88*, 333–380. [CrossRef]
60. Cloude, S.; Papathanassiou, K. Three-stage inversion process for polarimetric SAR interferometry. *IEE Proc.-Radar Sonar Navig.* **2003**, *150*, 125–134. [CrossRef]
61. Lei, Y.; Siqueira, P. An Automatic Mosaicking Algorithm for the Generation of a Large-Scale Forest Height Map Using Spaceborne Repeat-Pass InSAR Correlation Magnitude. *Remote Sens.* **2015**, *7*, 5639–5659. [CrossRef]
62. Thiel, C.; Schmullius, C. Investigating ALOS PALSAR Interferometric Coherence in Central Siberia at Unfrozen and Frozen Conditions: Implications for Forest Growing Stock Volume Estimation. *Can. J. Remote Sens.* **2013**, *39*, 232–250. [CrossRef]
63. Mette, T.; Kugler, F.; Papathanassiou, K.; Hajnsek, I. Forest and the Random Volume over Ground—Nature and Effect of 3 Possible Error Types. In Proceedings of the European Conference on Synthetic Aperture Radar (EUSAR), Hamburg, Germany, 6–9 June 2016; pp. 1–4.
64. Ballester-Berman, J.D.; Lopez-Sanchez, J.M. Applying the Freeman–Durden Decomposition Concept to Polarimetric SAR Interferometry. *IEEE Trans. Geosci. Remote Sens.* **2010**, *48*, 466–479. [CrossRef]
65. Xie, Q.; Zhu, J.; Wang, C.; Fu, H. Boreal forest height inversion using E-SAR PolInSAR data based coherence optimization methods and three-stage algorithm. In Proceedings of the 2014 Third International Workshop on Earth Observation and Remote Sensing Applications (EORSA), Changsha, China, 11–14 June 2014; IEEE: Piscataway, NJ, USA, 2014; pp. 145–150. [CrossRef]
66. Babu, A.; Kumar, S. TREE CANOPY HEIGHT ESTIMATION USING MULTI BASELINE RVOG INVERSION TECHNIQUE. *ISPRS Int. Arch. Photogramm. Remote Sens. Spat. Inf. Sci.* **2018**, *XLII-5*, 605–611. [CrossRef]
67. Khati, U.; Singh, G.; Kumar, S. Potential of Space-Borne PolInSAR for Forest Canopy Height Estimation over India—A Case Study Using Fully Polarimetric L-, C- and X-Band SAR Data. *IEEE J. Sel. Top. Appl. Earth Obs. Remote Sens.* **2018**, *11*, 2406–2416. [CrossRef]
68. Li, X.; Guo, H.; Li, Z.; Wang, L. Inversion of vegetation height using SIR-C dual frequency polarimetric sar interferometry data. In Proceedings of the IEEE International Geoscience and Remote Sensing Symposium, Anchorage, AK, USA, 20–24 September 2004; IEEE: Piscataway, NJ, USA, 2004; Volume 5, pp. 3132–3135. [CrossRef]
69. Le Toan, T.; Quegan, S.; Davidson, M.; Balzter, H.; Paillou, P.; Papathanassiou, K.; Plummer, S.; Rocca, F.; Saatchi, S.; Shugart, H.; et al. The BIOMASS mission: Mapping global forest biomass to better understand the terrestrial carbon cycle. *Remote Sens. Environ.* **2011**, *115*, 2850–2860. [CrossRef]
70. Mette, T.; Papathanassiou, K.; Hajnsek, I.; Zimmermann, R. Forest biomass estimation using polarimetric SAR interferometry. In Proceedings of the IEEE International Geoscience and Remote Sensing Symposium, Toronto, ON, Canada, 24–28 June 2002; IEEE: Piscataway, NJ, USA, 2002; Volume 2, pp. 817–819. [CrossRef]

remote sensing

Article

Spatiotemporal Evolution of the Carbon Fluxes from Bamboo Forests and their Response to Climate Change Based on a BEPS Model in China

Fangfang Kang [1,2,3], Xuejian Li [1,2,3], Huaqiang Du [1,2,3,*], Fangjie Mao [1,2,3], Guomo Zhou [1,2,3], Yanxin Xu [1,2,3], Zihao Huang [1,2,3], Jiayi Ji [1,2,3] and Jingyi Wang [1,2,3]

1 State Key Laboratory of Subtropical Silviculture, Zhejiang A & F University, Hangzhou 311300, China; 2019103032007@stu.zafu.edu.cn (F.K.); 2017303661004@stu.zafu.edu.cn (X.L.); maofj@zafu.edu.cn (F.M.); zhougm@zafu.edu.cn (G.Z.); xuyanxin@stu.zafu.edu.cn (Y.X.); 2018103241008@stu.zafu.edu.cn (Z.H.); 2019103032006@stu.zafu.edu.cn (J.J.); 2019103032012@stu.zafu.edu.cn (J.W.)
2 Key Laboratory of Carbon Cycling in Forest Ecosystems and Carbon Sequestration of Zhejiang Province, Zhejiang A & F University, Hangzhou 311300, China
3 School of Environmental and Resources Science, Zhejiang A & F University, Hangzhou 311300, China
* Correspondence: duhuaqiang@zafu.edu.cn

Abstract: Carbon flux is the main basis for judging the carbon source/sink of forest ecosystems. Bamboo forests have gained much attention because of their high carbon sequestration capacity. In this study, we used a boreal ecosystem productivity simulator (BEPS) model to simulate the gross primary productivity (GPP) and net primary productivity (NPP) of bamboo forests in China during 2001–2018, and then explored the spatiotemporal evolution of the carbon fluxes and their response to climatic factors. The results showed that: (1) The simulated and observed GPP values exhibited a good correlation with the determination coefficient (R^2), root mean square error (RMSE), and absolute bias (aBIAS) of 0.58, 1.43 g C m^{-2} day^{-1}, and 1.21 g C m^{-2} day^{-1}, respectively. (2) During 2001–2018, GPP and NPP showed fluctuating increasing trends with growth rates of 5.20 g C m^{-2} yr^{-1} and 3.88 g C m^{-2} yr^{-1}, respectively. The spatial distribution characteristics of GPP and NPP were stronger in the south and east than in the north and west. Additionally, the trend slope results showed that GPP and NPP mainly increased, and approximately 30% of the area showed a significant increasing trend. (3) Our study showed that more than half of the area exhibited the fact that the influence of the average annual precipitation had positive effects on GPP and NPP, while the average annual minimum and maximum temperatures had negative effects on GPP and NPP. On a monthly scale, our study also demonstrated that the influence of precipitation on GPP and NPP was higher than that of the influence of temperature on them.

Keywords: bamboo forest; BEPS model; gross primary productivity; net primary productivity; spatiotemporal evolution; climate change

Citation: Kang, F.; Li, X.; Du, H.; Mao, F.; Zhou, G.; Xu, Y.; Huang, Z.; Ji, J.; Wang, J. Spatiotemporal Evolution of the Carbon Fluxes from Bamboo Forests and their Response to Climate Change Based on a BEPS Model in China. *Remote Sens.* **2022**, *14*, 366. https://doi.org/10.3390/rs14020366

Academic Editor: Hubert Hasenauer

Received: 24 November 2021
Accepted: 11 January 2022
Published: 13 January 2022

1. Introduction

Dynamic change in the carbon cycles of terrestrial ecosystems is a core component of climate change and regional sustainable development [1]; it plays an important role in the global carbon balance. Because of the impacts of various environmental and biological factors (such as climate change, vegetation distribution, and land-use change), the carbon cycles of terrestrial ecosystems show significant spatial heterogeneity [2]. Forest ecosystems are an important component of terrestrial ecosystems and play an important role in improving and maintaining the ecological environment, in addition to regulating the global carbon balance [3,4]. Therefore, it is essential to quantify carbon fluxes in forest ecosystems and explore their response to environmental factors in the carbon cycles of terrestrial ecosystems. Carbon flux [5] is the basis for determining the carbon source/sink of an ecosystem and plays an important role in the global carbon cycle and carbon balance. Ecosystem productivity is an important indicator for quantitatively describing the carbon sequestration capacity of an ecosystem, which mainly includes gross primary production (GPP) and net primary production (NPP) [6]. GPP refers to the amount of organic carbon fixed by photosynthesis per unit time and unit area of green plants [7,8]. It reflects the carbon sequestration ability of vegetation, and is the largest carbon flux in terrestrial ecosystems [9]. NPP is the organic matter or energy remaining for vegetation growth after deducting the organic matter consumed by vegetation autotrophic respiration (RA) on the basis of GPP [10]. It can directly reflect the production capacity and ecological environment quality of surface vegetation in the natural environment, and is an important indicator for evaluating the carbon sink of ecosystems, in addition to regulating ecological processes [11,12].

The methods to obtain carbon fluxes of forest ecosystems mainly include sample site inventory, eddy covariance technology, and model simulation. Sample site inventory can estimate carbon fluxes more accurately, but the estimation of the carbon fluxes of forest ecosystems requires long-term field measurements, which consume substantial amounts of time and labor [13]; therefore, it limits the estimation of carbon fluxes in forest ecosystems. Eddy covariance technology has the advantages of being a long-term, continuous, and non-destructive method [14], and has been widely used to estimate the carbon fluxes of forest ecosystems [15]. However, the number of flux observation sites is limited and the area of effective observation is very small. There are uncertainties in expanding it from the site to the regional scale, and it is often hindered by topography and climate conditions; therefore, eddy covariance technology has limitations in studying carbon fluxes in forest ecosystems at the regional scale.

Model simulation is an important method of evaluating carbon fluxes in forest ecosystems [16,17]. Remote sensing for earth observation technology has the characteristics of real-time, dynamic, and large-area synchronous monitoring, in addition to rich information [13]. It readily records the dynamic changes in environmental conditions, vegetation distribution patterns and activities, and land use in the form of electromagnetic information. This provides the necessary parameters of vegetation (such as NDVI and LAI) and environmental variables for the carbon flux model, and becomes a powerful method to study the distribution, seasonal change, and interannual change in carbon fluxes [18]. Therefore, the application of remote sensing data in the model estimation helps achieve cross-scale simulation of the carbon cycle process and reflect the spatial distribution and dynamic changes in the carbon budget at the regional and global scales. It increases the reliability and operability of vegetation carbon flux estimation and has become an important research topic [18–22]. Ecological process models simulate the effects of biological vegetation processes such as canopy photosynthesis, absorption, transpiration, and changes in soil moisture content on carbon fluxes, and have become an important method for carbon flux simulation. Common ecological process models include the Biome-BGC model [23], the BEPS model [24], and the InTEC model [25]. In recent years, scholars have studied the carbon fluxes of forest ecosystems in different regions using different ecological process models combined with remote sensing data. For example, Du et al. [26] used an improved Biome-BGC model with remote sensing data to simulate the above-ground carbon storage of bamboo forests in Zhejiang Province from 2003 to 2014, and analyzed its spatiotemporal patterns and influencing factors. Zhang et al. [27] used remote sensing data and the BEPS model to study the spatiotemporal distribution characteristics of GPP and NPP in terrestrial ecosystems in East Asia. Zheng et al. [28] used the InTEC model to simulate the NEP of the forests in Zhejiang Province during 1985–2015, and analyzed the response of climatic factors such as temperature, precipitation, relative humidity, and radiation.

The BEPS model is an ecological process model based on the FOREST-BGC model [29]. It integrates multi-source data as model inputs and is a good choice for simulating terrestrial ecosystem productivity with higher accuracy on larger spatial scales. The BEPS model successfully solved the problem of spatiotemporal scale conversion by using remote sensing data. Additionally, it solved the overestimation problem of the FOREST-BGC model by introducing the clumping index and advanced canopy transmission model [30]. Compared with other ecological process models, it may have the most potential to adequately address the spatiotemporal dynamics of carbon fluxes because of its strong theoretical basis and practical applicability [31]. Previously, it was used to simulate the productivity of the boreal forest ecosystem in Canada [32]. Presently, the model has been frequently modified and improved, and has been widely used to simulate the carbon fluxes of different regional terrestrial ecosystems at various spatiotemporal scales [33–38].

Bamboo belongs to a family of perennial graminaceous plants. There are approximately 150 genera and 1225 species of bamboo forests in the world, and the total area of bamboo forests worldwide accounts for more than 30 million ha [39], making it "the second largest forest in the world". China is located in the center of bamboo distribution in the world. It has the richest bamboo resources in the world in terms of the number of species (more than 500 varieties of 39 species) and area [40]. According to the ninth National Forest Resources Inventory (2014–2018), China's bamboo forest area is 6,411,600 ha [41], accounting for approximately 20% of the world's bamboo forest area. Compared with the eighth National Forest Resources Inventory (2009–2013), their area of bamboo forest has increased by more than 400,000 ha. It is known as the "Bamboo Kingdom" [40,42]. Bamboo forests have a great carbon sequestration capacity and differ from other forests in mitigating climate change, and their impact on global climate change has become an important concern [43,44]. Several scholars have explored carbon cycling in bamboo forests, and have synthesized information concerning primary production [45,46], carbon stocks [47,48], and biomass [49,50].

Although relevant studies on bamboo forests' carbon cycles have been conducted, the characteristics of bamboo forests' carbon dynamics and their response to changing environmental conditions are still poorly understood [47,51,52]. Previous studies mainly focused on estimating carbon fluxes at the scale of sites, regions, and provinces, while relatively few studies have done so at the national scale. In addition, some studies lack simulations of physiological and ecological processes, leading to massive errors in the estimated results [53]. Therefore, the study of carbon fluxes from bamboo forests in China is essential for the study of the carbon cycles of forest ecosystems under the global climate background. The objectives of this study include (1) driving the BEPS model to simulate the carbon fluxes of bamboo forests in China from 2001 to 2018; (2) exploring the spatiotemporal evolution of bamboo forests' carbon fluxes in China and the driving influence of climate change on carbon fluxes of bamboo forests in China.

2. Materials and Methods

2.1. Study Area

China has a vast territory and diverse climate types (Figure 1). The country has a north–south temperature gradient and an east–west precipitation gradient driven by the summer monsoon [54]. Bamboo forests are a unique and important forest type in subtropical regions of China, and are widely distributed across Zhejiang, Fujian, Jiangxi, Hunan, Sichuan, Anhui, Hubei, Guangdong, Guangxi, and other provinces.

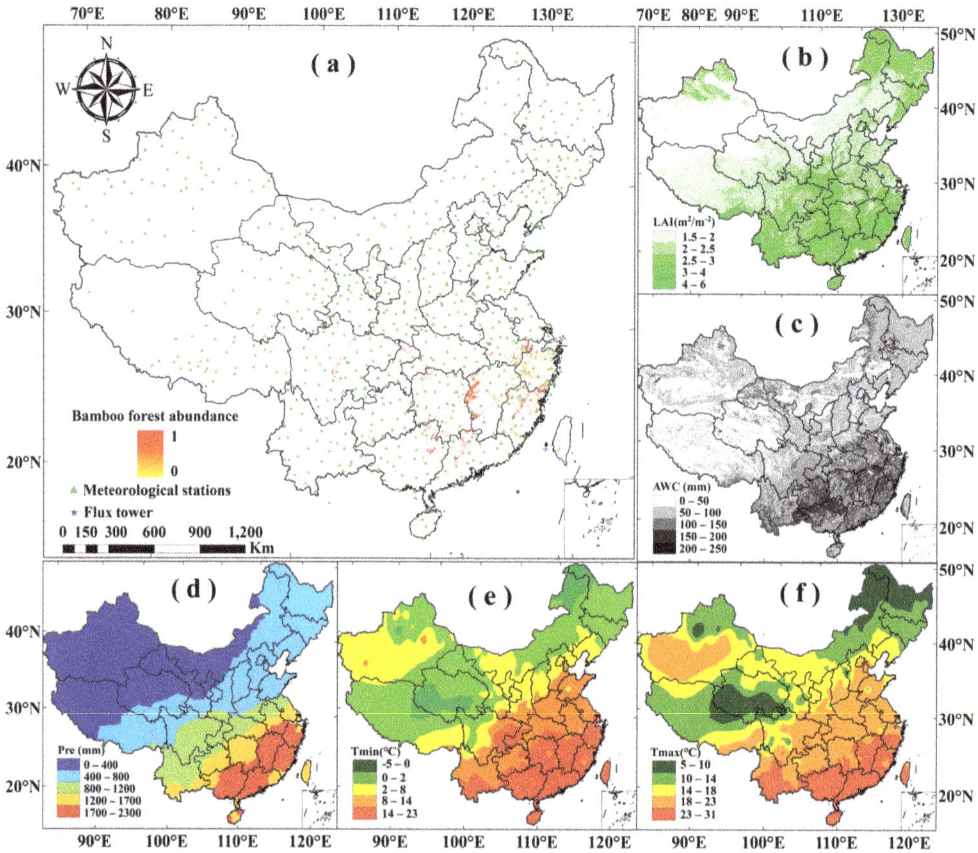

Figure 1. Study area and spatial distribution of data: (**a**) distribution of bamboo forests and meteorological stations, (**b**) leaf area index (LAI), (**c**) available soil water-holding capacity (AWC), (**d**) precipitation (Pre), (**e**) minimum temperature (T_{min}), and (**f**) maximum temperature (T_{max}).

2.2. Flux Measurement Sites

The flux observation sites are located in Zhejiang Province, which are the Anji Moso bamboo flux measurement site (30.46° N, 119.66° E) and the Lin'an Lei bamboo flux measurement site (30.30° N, 119.58° E) (Figure 1a). The height of the observation tower at Anji was 40 m, and the vegetation type around the flux tower for 1 km × 1 km was dominated by 1–4-year-old Moso bamboo forests. The height of the observation tower at Lin'an was 20 m, and the vegetation types around the flux tower were mainly 2–3-year-old Lei bamboo forests. The carbon flux data were continuously measured by an eddy covariance system of flux measurement sites. The system consists of an open-path infrared CO_2/H_2O gas analyzer (Li-7500, LiCor Biosciences Inc., Lincoln, NE, USA) and a three-dimensional sonic anemometer (CAST3, Campbell Scientific Inc., Logan, UT, USA). According to the principle of the eddy covariance system, 30-min carbon flux data were calculated online and stored.

2.3. Data Acquisition and Processing

The required BEPS model input data included bamboo forest information in China, MODIS leaf area index (MODIS LAI) (Table 1), the available soil water-holding capacity

(AWC), daily meteorological data, and the biological parameters of bamboo forests. All data were reprojected to the WGS84 coordinate system with a spatial resolution of 1 km.

Table 1. MODIS data and descriptions.

MODIS	Abbreviation	Time	Spatial Resolution	Time Resolution	To Use
MOD13A2	NDVI	2018	1000 m	16 days	Extract the bamboo forest
MOD09A1	REF	2018	500 m	8 days	Extract the bamboo forest
MOD15A2	LAI	2001–2018	1000 m	8 days	Model input

2.3.1. MODIS Data and Preprocessing

MODIS is a new generation of optical and infrared remote sensing instruments that "integrate image and spectrum" in the current world. It is widely used in the carbon cycles of terrestrial ecosystems because of its high time and spectral resolutions. This study uses MODIS normalized difference vegetation index (MODIS NDVI) (MOD13A2), MODIS land surface reflectance (MODIS REF) (MOD09A1), and MODIS LAI (MOD15A2) from NASA (https://ladsweb.modaps.eosdis.nasa.gov, accessed on 13 May 2020) to extract information on bamboo forests in China and simulate the carbon fluxes of bamboo forests in China. The MODIS data are shown in Table 1.

The MODIS Reprojection Tool (MRT) was used to preprocess MODIS data, such as mosaicking, format conversion, reprojection, and resampling. MOD09A1 was reprojected to the WGS84 coordinate system, and the spatial resolution was resampled to 1 km using the nearest neighborhood method. After resampling, these data were clipped to the boundaries of China.

2.3.2. Bamboo Forest Distribution Data of China

The distribution information of Chinese bamboo forests in 2003, 2008, 2014, and 2018 was extracted. The information on Chinese bamboo forests from 2003, 2008, and 2014 has been extracted in our previous study [42]. On this basis, we extracted information on Chinese bamboo forests from 2018. The flow chart of bamboo extraction is shown in Figure 2.

The main process is as follows: First, a total of 23 multi-temporal MODIS NDVI images are available. In order to further improve the MODIS NDVI data quality, these 23 images were composited into 12 multi-temporal images by selecting a maximum of two corresponding pixels of two neighboring MODIS NDVI images as the value of a new pixel ($NDVI_{max12}$) [55]. Then, a minimum noise fraction (MNF) transform [56] was employed to convert the $NDVI_{max12}$ data to obtain the principal component variables of NDVI $_{max12}$ data (NDVI $_{max12}$ MNF), and the first six bands with a cumulative contribution rate greater than 90% ($NDVI_{max12}$ $MNF_{1\text{-}6}$) were retained for classification. Second, according to the image texture and spectral information features, the five types of samples (forest, farmland, water, bare land, and residential land) were selected by visual interpretation [57], and then the study area was classified by the maximum likelihood classification (MLC). On this basis, the forest information in China was extracted by masking. Third, using the forest information in China to extract the normalized difference vegetation of forests ($NDVI_{forest\_12}$) and the land surface reflectance of forests (REF_{forest_7}), MNF was then performed on them to obtain the principal component variables of $NDVI_{forest\_12}$ data ($NDVI_{forest\_12}$ MNF) and REF_{forest_7} data (REF_{forest_7} MNF). We retained the bands with a cumulative contribution rate greater than 85%, that is, the first nine bands of $NDVI_{forest\_12}$ MNF data ($NDVI_{forest\_12}$ $MNF_{1\text{-}9}$) and the first five bands of REF_{forest_7} MNF data (REF_{forest_7} $MNF_{1\text{-}5}$). On this basis, according to the training samples of bamboo forests, broad-leaved forests, and coniferous forests, the corresponding attribute values were extracted as the characteristic variables to construct a decision tree model, and the information on Chinese bamboo forests was extracted by using the constructed decision tree model (Figure 3). In this study, 85 bamboo forest survey samples from Zhejiang Province in 2019 and 440 bamboo forest samples from China selected from Landsat 8 images in 2018 were used as bamboo forest verification samples for point-by-point verification. Finally, the least-squares mixed-pixel decomposition method [58] was used to obtain the abundance information of bamboo forests in China. The results were presented in Figure 1a.

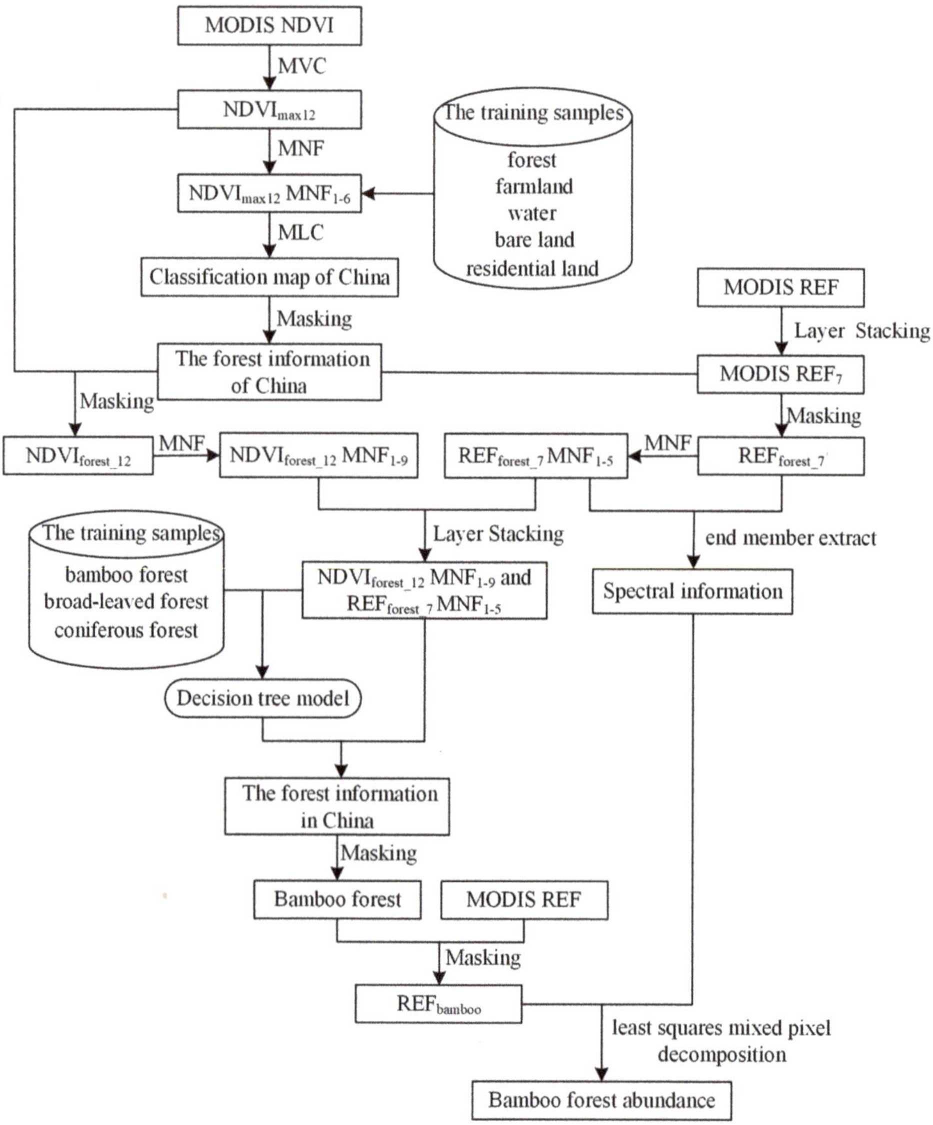

Figure 2. Flow chart of bamboo extraction.

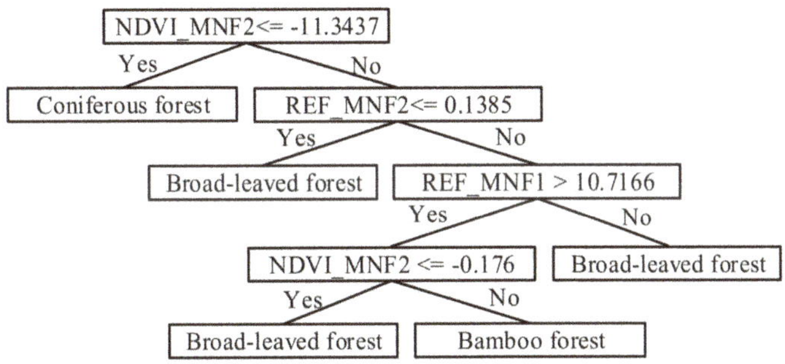

Figure 3. The optimal decision tree.

It was verified that the accuracy of bamboo forest extraction was 76.54–81.56%, and that the extracted area was close to the inventory area of forest resources (Table 2), which laid the foundation for the simulation of GPP and NPP for bamboo forests in China. Only the information of bamboo forests in 2003, 2008, 2014, and 2018 was extracted; therefore, the bamboo forest information of the unclassified year was replaced by a similar year from which the bamboo forest information was extracted.

Table 2. Extraction accuracy evaluation and the comparison of estimated and inventory bamboo forest area of China.

Year	Classification Accuracy Evaluation				Bamboo Forest Area (10^4 ha)	
	Bamboo Forest Samples	Correctly	Incorrectly	User's Accuracy (%)	Estimate	Inventory
2003	387 [42]	309	78	79.84	486.56	495.32 [42]
2008	414 [42]	328	86	79.23	545.14	548.73 [42]
2014	536 [42]	435	101	81.16	639.22	610.65 [42]
2018	525	402	123	76.54	669.83	656.08 [41,42]

Note: the results of the ninth National Forest Resources Survey do not have data from Taiwan, so the bamboo forest area of Taiwan is based on the results of the eighth National Forest Resources Survey.

2.3.3. MODIS LAI Data

Leaf area index (LAI) is an important input parameter for simulating the carbon cycles of forest ecosystems, and is closely related to the photosynthesis, steaming, water utilization, and productivity formation of vegetation [59]. Remote sensing technology is an important method for obtaining a large-scale LAI. However, MODIS LAI data are susceptible to the influences of factors such as the atmosphere, which leads to an irregular reduction in data. To reduce data noise and improve data quality, the locally adjusted cubic-spline capping (LACC) [60] algorithm was used to smooth the clipped MODIS LAI data. Then, the smoothed MODIS LAI data were assimilated by the particle filter (PF) algorithm [61]. The assimilated MODIS LAI data were shown in Figure 1b.

2.3.4. Soil Data

The soil texture data map was provided by the Chinese Academy of Sciences (http://www.soil.csdb.cn, accessed on 11 December 2020). AWC is an important factor in terms of plant growth, affecting stomatal conductance and photosynthesis [62]. In this study, based on the empirical relationship, an AWC map with a 1 km resolution was obtained from a soil data thematic map. The spatial distribution of AWC was shown in Figure 1c.

2.3.5. Meteorological Data

Meteorological data from 2001 to 2018 were obtained from the National Meteorological Information Center of the China Meteorological Administration (http://data.cma.cn, accessed on 7 August 2020), and mainly included minimum temperature (T_{min}), maximum temperature (T_{max}), precipitation (Pre) solar radiation, and relative humidity. These meteorological factors are the main environmental factors in the carbon–water cycles [63]. The inverse distance weighting method was used to interpolate the data of 824 meteorological sites observed in the study area (Figure 1a) into spatial data with a 1 km resolution to obtain the grid cells of daily scale meteorological data of the study area. They are shown in Figure 1d–f. Among them, the temperature was corrected by the digital elevation method, and it was assumed that the temperature decreased by 6.5 °C for each per-kilometer increase in altitude. Solar radiation was simulated based on the measurements of sunshine duration at each site, following the methods of Ju et al. [64]. Monthly and annual meteorological data based on interpolated daily scale meteorological data were obtained.

2.3.6. Biological Parameters

The major biological parameters of bamboo forests used in the BEPS model are shown in Table 3. The clumping index (Ω) and specific leaf area (S_{area}) came from the measured data of the flux observation station. The maximum carboxylation rate at 25 °C (V_m) and the Q_{10} for leaves, stems, and roots were calculated based on an iteration method. The initial value of the four parameters was established according to Chen et al. [65], and the iteration range for each parameter was set as ± 100%. The iteration step was defined as 1 for V_m and 0.1 for the other three parameters. The average carbon storage of leaves, stems, and roots was calculated using the methods of Zhou and Jiang [66]. Bamboo forests are a special type of forest. The photosynthesis capacity of bamboo forests is similar to C_3 trees [67]. Therefore, for constant parameter values, we referred to Feng et al. [62] to simulate the carbon cycle of bamboo forests.

Table 3. Major biological parameters used as inputs into the BEPS model for simulating the CO_2 fluxes of bamboo forests.

Symbol	Unit	Description	Value	Reference
Ω	-	Clumping index	0.5	Measurement
S_{area}		Specific leaf area	27	Measurement
$V_{m,25}$	umol $m^{-2}s^{-1}$	Maximum carboxylation rate at 25 °C	50	Iteration
$Q_{10,leaf}$	-	Q_{10} for leaf	1.4	Iteration
$Q_{10,stem}$	-	Q_{10} for stem	1.3	Iteration
$Q_{10,root}$	-	Q_{10} for root	1.2	Iteration
M_{leaf}	kg C m^{-2}	Average carbon storage of leaf	0.15	[66]
M_{stem}	kg C m^{-2}	Average carbon storage of stem	1.76	[66]
M_{root}	kg C m^{-2}	Average carbon storage of root	1.15	[66]

2.4. BEPS Model Simulation and Evaluation

2.4.1. BEPS Model Description

The BEPS model is mainly composed of four parts: energy transmission, carbon cycle, water cycle, and physiological regulation sub-models [68]. It combines ecology, plant physiology, meteorology, and other disciplines to simulate the relationship between the photosynthesis, respiration, carbon distribution, water balance, and energy balance of vegetation [63], which demonstrates the combination of remote sensing data and ecological process models. The main feature of this model is that the instantaneous Farquhar photosynthetic model at the leaf scale is converted into the daily total photosynthetic model through the integration of stomatal conductance to realize the time scale expansion. Then, according to the principle of light transmission in the canopy, the vegetation canopy leaves were divided into shaded and sunlit leaves to simulate the radiation budget of the corresponding leaves. This helps achieve the expansion from the leaf scale to the canopy space scale. Detailed descriptions of the BEPS model can be found in Liu et al. [69] and Chen et al. [65]. The main simulation process of the model is as follows:

(1) The LAI_{sunlit} and LAI_{shade} are calculated as follows:

where LAI_{canopy} is the total LAI of the canopy; LAI_{sunlit} and LAI_{shade} are the canopy LAIs of sunlit and shaded leaves, respectively; θ is the daily mean solar zenith angle; and Ω is the clumping index:

$$LAI_{sunlit} = 2\cos\theta[1 - \exp(-0.5\Omega LAI/\cos\theta)] \quad (1)$$

$$LAI_{shade} = LAI_{canopy} - LAI_{sunlit} \quad (2)$$

(2) The photosynthesis rate is calculated as follows:

$$A = \min(W_c, W_j) - R_d \quad (3)$$

$$W_c = V_m \frac{C_i - \Gamma}{C_i + K_c(1 + O_2/K_o)} \quad (4)$$

$$W_j = J \frac{C_i - \Gamma}{4(C_i + 2\Gamma)} \quad (5)$$

$$R_d = 0.015V_m \quad (6)$$

where A is the net photosynthesis rate; W_c and W_j are the Rubisco-limited and RuBP-limited gross photosynthesis rates, respectively; R_d is the daytime leaf dark respiration; V_m is the maximum carboxylation rate at 25 °C; C_i and O_2 are the intercellular CO_2 and oxygen concentrations in the atmosphere, respectively; Γ is the CO_2 compensation point, without dark respiration; K_c and K_o are the Michaelis–Menten constants for CO_2 and O_2, respectively; and J is the electron transmission rate.

(3) The total canopy photosynthesis rate is evaluated as follows:

$$A_{canopy} = A_{sunlit}LAI_{sunlit} + A_{shade}LAI_{shade} \quad (7)$$

where A_{canopy} is the total photosynthesis rate of the canopy; A_{sunlit} and A_{shade} are the photosynthesis rates of sunlit and shaded leaves, respectively; and LAI_{sunlit} and LAI_{shade} are the LAIs of sunlit and shaded leaves, respectively.

(4) The GPP and NPP values are determined as follows:

$$GPP = A_{canopy} \times L_{day} \times F_{GPP} \quad (8)$$

$$NPP = GPP - R_a \quad (9)$$

$$R_a = R_m + R_g = R_{m,i} + R_{g,i} \quad (10)$$

where GPP is gross primary productivity; NPP is net primary productivity; R_a is the autotrophic respiration of the vegetation; L_{day} is the length of the day; F_{GPP} is a scale factor for converting photosynthesis into GPP; R_m and R_g are the maintenance breathing rate and growth respiration rates, respectively; i is the different parts of vegetation (i = 1, 2, and 3 for leaves, stems, and roots, respectively); and $R_{m,i}$ and $R_{g,i}$ are the maintenance and growth respiration rates of different parts, respectively.

2.4.2. Evaluation of Simulation Results

In this study, the results of the BEPS model simulation were evaluated with precision using the determination coefficient (R^2), root mean square error (RMSE), and absolute bias (aBIAS). The formulas for the calculation are as follows [13]:

$$R^2 = 1 - \frac{\sum_{i=1}^{n}(m_i - o_i)^2}{\sum_{i=1}^{n}(o_i - \overline{o_i})^2} \quad (11)$$

$$RMSE = \sqrt{\frac{1}{n}\sum_{i=1}^{n}(m_i - o_i)^2} \quad (12)$$

$$\text{aBIAS} = \frac{1}{n}\sum_{i=1}^{n}|m_i - o_i| \tag{13}$$

where m_i is the simulated value; o_i is the observed value; and $\overline{o_i}$ is the average value of the observed value. Generally, the larger the R^2 value, the smaller the RMSE and aBIAS values, the higher the accuracy, and vice versa.

2.5. *Spatiotemporal Evolution Analysis of Carbon Fluxes*

2.5.1. Variation Coefficient of Carbon Fluxes

The variation coefficient (CV) is the ratio of the standard deviation to the average, which reflects the stability of a set of data. The higher the value of the CV, the more unstable the data, that is, the greater the fluctuation, and vice versa. To analyze the spatial fluctuations in carbon fluxes of bamboo forests during 2001–2018, the CVs of the GPP and NPP of each pixel were calculated as follows [70]:

$$\text{CV} = \sqrt{\frac{\frac{1}{n-1}\sum_{i=1}^{n}\left(P_i - \overline{P}\right)^2}{\overline{P}}} \tag{14}$$

In Equation (14), CV is the variation coefficient; n = 18, and is the number of monitoring years; P_i is the value of each pixel of the GPP or NPP image in the i-th year (where i = 1, 2, ..., n); and $\overline{P}$ is the average value of each pixel of GPP or NPP. According to the calculation results, by performing the Jenks natural breaks classifications in ArcGIS software [71] the results of the CV were divided into five levels: low fluctuation ($CV <= 0.1246$), lower fluctuation ($0.1246 < CV \leq 0.2342$), medium fluctuation ($0.2342 < CV \leq 0.4132$), higher fluctuation ($0.4132 < CV \leq 0.7364$), and high fluctuation ($CV > 0.7364$).

2.5.2. Trend Slope of Carbon Fluxes

To quantitatively study the trends of carbon fluxes of bamboo forests in China from 2001 to 2018 a linear regression analysis was used to calculate the trends of GPP and NPP of each pixel, as follows [72,73]:

$$\text{slope} = \frac{n \times \sum_{i=1}^{n}(i \times P_i) - \sum_{i=1}^{n} i \times \sum_{i=1}^{n} P_i}{n \times \sum_{i=1}^{n} i^2 - \left(\sum_{i=1}^{n} i\right)^2} \tag{15}$$

In Equation (15), slope is the trend slope; n = 18, and is the number of monitoring years; and P_i is the GPP or NPP of bamboo forests in the i-th year, (i = 1, 2, ..., n). The value of the trend slope indicates the rate of increase or decrease. When slope > 0, the GPP and NPP increase, and when slope < 0, the GPP and NPP decrease.

To analyze whether the variation trend of the GPP and NPP was significant, the F-test was used to test the significance of the variation trend of GPP and NPP. The variation trend was divided into five levels: significantly reduced (slope < 0, $p < 0.01$), reduced (slope < 0, $0.01 < p < 0.05$), basically stable ($p > 0.05$), increased (slope > 0, $0.01 < p < 0.05$), and significantly increased (slope > 0, $p < 0.01$).

2.6. *Analysis of Spatiotemporal Responses of Carbon Fluxes to Climate Change*

2.6.1. Partial Correlation Analysis of Carbon Fluxes to Climate Change

A correlation analysis reveals the closeness of the relationship between the study variables. Partial correlation analysis refers to the calculation of the correlation between two variables without considering the influence of other variables [74]. Partial correlation analysis can better reflect the impact of a single climate factor on carbon fluxes. Therefore, this study uses a pixel-based partial correlation analysis to calculate the partial correlation coefficients (PPCs) of GPP and NPP with climatic factors, and analyze the response between carbon fluxes and climatic factors. To determine the PCCs, we first calculated the correlation coefficient using the following formula [75]:

$$R_{xy} = \frac{\sum_{i=1}^{n}[(x_i - \overline{x})(y_i - \overline{y})]}{\sqrt{\sum_{i=1}^{n}(x_i - \overline{x})^2 \sum_{i=1}^{n}(y_i - \overline{y})^2}} \quad (16)$$

In Equation (16), R_{xy} is the correlation coefficient between variables x and y; n is the number of study years; xi and yi represent the values of variables x and y in the i-th year, respectively, (i = 1, 2, ..., n); and $\overline{x}$ and $\overline{y}$ represent the mean value of variables x and y, respectively. The range of R_{xy} is [−1,1]; when $R_{xy} > 0$ the two variables are positively correlated, and when $R_{xy} < 0$ the two variables are negatively correlated. The larger the absolute value of R_{xy} the higher the correlation, and vice versa. Generally, 0.3 and 0.6 are the distinction points of the absolute value of the correlation coefficient, as the weak correlation ($0 < |R_{xy}| \leq 0.3$), low correlation ($0.3 < |R_{xy}| \leq 0.6$), and significant correlation ($0.6 < |R_{xy}| \leq 1$).

Based on the evaluated correlation coefficient, the PPC was calculated as follows [72]:

$$R_{ab,cd} = \frac{R_{ab,d} - R_{ac,d} \times R_{bc,d}}{\sqrt{(1 - R_{ac,d}^2) \times (1 - R_{bc,d}^2)}} \quad (17)$$

In Equation (17), $R_{ab,cd}$ represents the PCC between variables a and b when variables c and d are fixed; $R_{ab,d}$, $R_{ac,d}$, and $R_{bc,d}$ represent the PCC between variables a and b, variables a and c, and variables b and c, respectively, when the variable d is fixed. The higher the PCC, the greater the influence of the variable on GPP and NPP. A *t*-test was used to test the significance of the PPC.

2.6.2. Path Analysis of Climate Change to Carbon Fluxes

In order to analyze the direct and indirect effects of climate factors (temperature and precipitation) on carbon fluxes of bamboo forests, path analysis [76] was used to calculate the direct and indirect path coefficients of temperature and precipitation on the carbon fluxes. The formulas for the calculation are as follows:

$$P_{i \to y} = \frac{b_i S_i}{S_y} \quad (18)$$

$$P_{j \to i \to y} = r_{ij} P_{i \to y} \quad (19)$$

where $P_{i \to y}$ is the direct path coefficient, b_i is the regression coefficient, S_i is the standard deviation of variable i, S_y is the standard deviation of variable y, $P_{j \to i \to y}$ is the indirect path coefficient of variable j acting on variable y through variable i, and r_{ij} is the correlation coefficient between variable i and variable j.

3. Results

3.1. BEPS Model Validation

In this study, the observed carbon flux data from the Anji site and the Lin'an site during 2011–2014 were used to validate the BEPS model. The daily scale carbon flux data were obtained by accumulating the observed 30-min carbon flux data. The evaluation results are shown in Figure 4, where the R^2, RMSE, and aBIAS were 0.58, 1.43 g C m^{-2} day^{-1}, and 1.21 g C m^{-2} day^{-1}, respectively. There was a good correlation between the simulated and observed values of GPP. Therefore, the BEPS model could be considered to be suitable to simulate the productivity of bamboo forests in China.

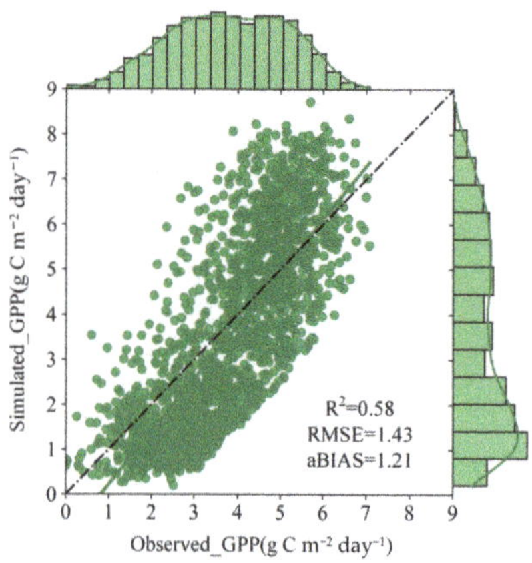

Figure 4. Comparison of the simulated and observed values of GPP.

3.2. Spatiotemporal Evolution of Carbon Fluxes from Bamboo Forests in China

3.2.1. Temporal Evolution Trend

The variation trends of the monthly and annual average GPP and NPP during 2001–2018 are shown in Figure 5. The GPP and NPP exhibited similar temporal variation characteristics. At the monthly scale, the average values of the GPP and NPP of bamboo forests showed unimodal changes. At the annual scale, the average values of GPP and NPP were 904.02 g C m^{-2} yr^{-1} and 716.88 g C m^{-2} yr^{-1}, respectively, and the ranges in variation were 764.42–994.61 g C m^{-2} yr^{-1} and 600.03–788.25 g C m^{-2} yr^{-1}, respectively. The annual average values of GPP and NPP were the lowest in 2003 and the highest in 2007. During the statistical period, the overall variation trends of GPP and NPP were similar, showing an increasing trend, and the increasing trend was not significant ($p > 0.05$); the growth rates were 5.20 g C m^{-2} yr^{-1} and 3.88 g C m^{-2} yr^{-1}, respectively.

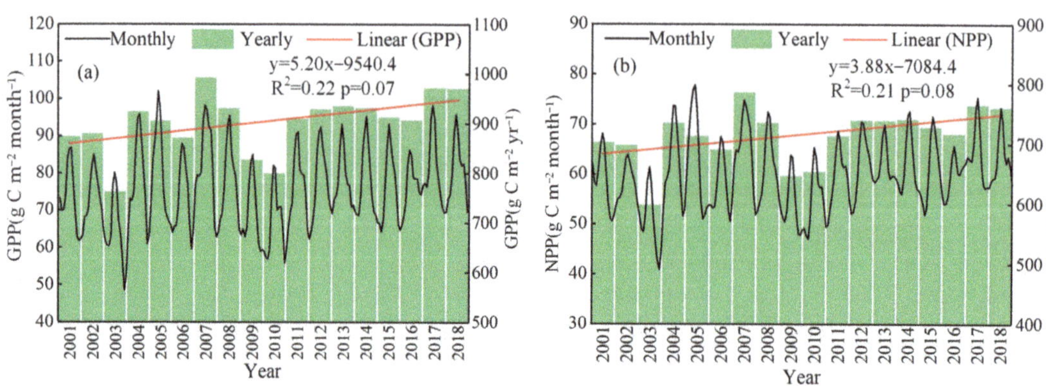

Figure 5. Monthly and annual variation trends of bamboo forests' (**a**) GPP and (**b**) NPP in China from 2001 to 2018.

3.2.2. Spatial Distribution Characteristics

The spatial distribution of the mean GPP and NPP values of bamboo forests in China is shown in Figure 6. From Figure 6 we can see that the mean GPP and NPP values had strong spatial heterogeneity. On the whole, the GPP and NPP present a distribution characteristic of being more in the south and east, and less in the north and west. In addition, the spatial distributions of GPP and NPP were compared and it was found that the high GPP and NPP values of bamboo forests were mainly concentrated in northwestern Zhejiang, central Fujian, western Jiangxi, and so on, and the proportion of high-value distribution was gradually increasing. The low GPP and NPP values were mainly distributed in Guizhou, Shanxi, Yunnan, and other regions where the distribution of bamboo forests is relatively scattered.

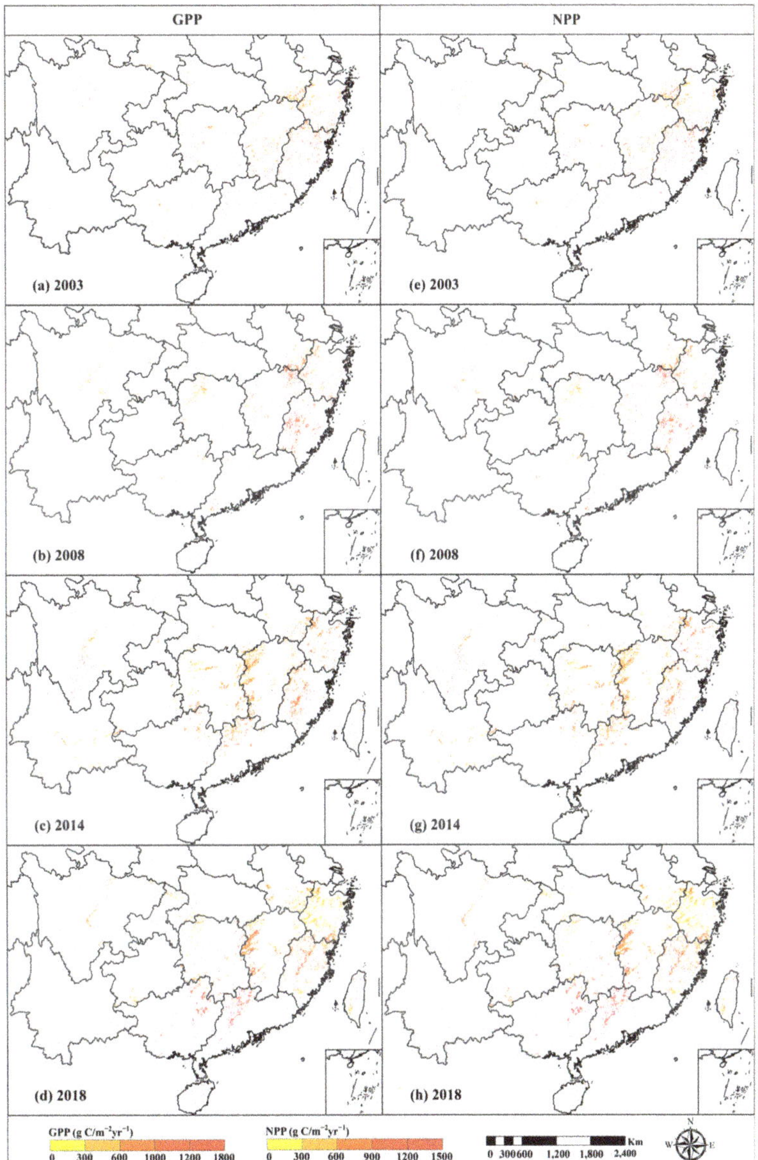

Figure 6. Spatial distribution of (**a–d**) GPP and (**e–h**) NPP of bamboo forests in China during different periods.

3.2.3. Analysis of the Fluctuation in Carbon Fluxes

To analyze the fluctuations in carbon fluxes from bamboo forests in China from 2001 to 2018, we calculated the CVs of GPP and NPP and prepared a spatial distribution diagram of the fluctuations based on the CV classification results, as shown in Figure 7. The zones with low fluctuations in GPP and NPP accounted for the largest proportion, at 43.71% and 42.37%, respectively. These were followed by zones with lower fluctuation, at 36.37% and 37.77%, respectively. The areas of higher fluctuation and high fluctuation were considerably small and scattered, among which the areas of higher fluctuation accounted for 3.78% and 3.81%, respectively, and those of high fluctuation accounted for 1.36% for both. In addition, by comparing Figure 7a,b, it was found that the spatial fluctuations in GPP and NPP exhibited evident consistency, where GPP and NPP had low fluctuations, and vice versa.

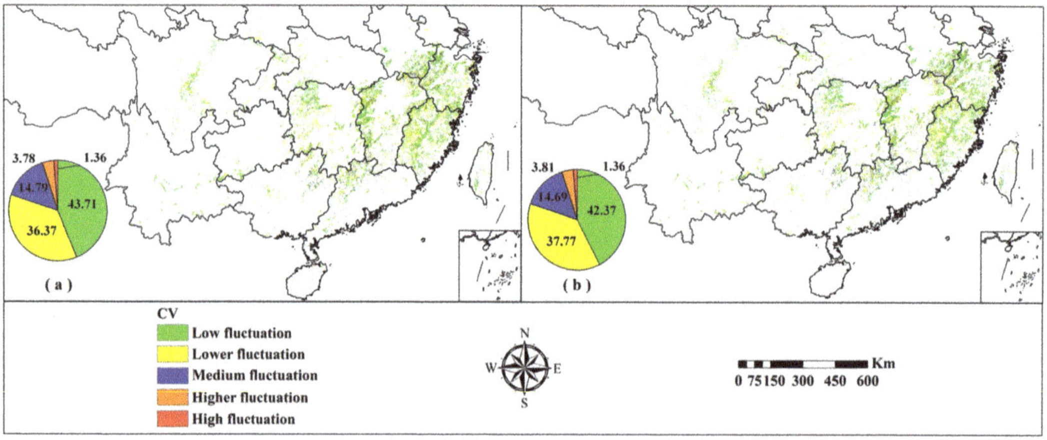

Figure 7. Spatial distribution of the variation coefficients (CVs) of the (**a**) GPP and (**b**) NPP of bamboo forests in China from 2001 to 2018.

3.2.4. Analysis of the Trend Slope of Carbon Fluxes

The spatial distribution of the trend slope and the significance of GPP and NPP from 2001 to 2018 are shown in Figure 8.

Figure 8a,c show the spatial distribution of the trend slope and the significance of GPP, respectively. From Figure 8a,c, it can be seen that GPP exhibits an increasing trend ($slope_{gpp} > 0$) at 57.58% and a significant increasing trend ($slope_{gpp} > 0$, $p < 0.01$) at 30.32%, mainly distributed in northwestern Zhejiang, western Jiangxi, central Fujian, southwest Anhui, and central Sichuan. GPP exhibits a decreasing trend ($slope_{gpp} < 0$) at 42.42% and a significant decreasing trend ($slope_{gpp} < 0$, $p < 0.01$) at 20.53%, mainly distributed in southwestern Zhejiang, eastern Jiangxi, eastern Anhui, and western Guangdong.

Figure 8b,d show the spatial distribution of the trend slope and the significance of NPP, respectively. As shown in Figure 8b,d, NPP shows an increasing trend ($slope_{npp} > 0$) at 57.56% and a significant increasing trend ($slope_{npp} > 0$, $p < 0.01$) at 30.32%. NPP shows a decreasing trend ($slope_{npp} < 0$) at 42.44% and a significant decreasing trend ($slope_{npp} < 0$, $p < 0.01$) at 20.54%. By comparing Figure 8c,d, it can be seen that the regions with significantly increased and decreased NPP are consistent with the regions that had significantly increased and decreased GPP.

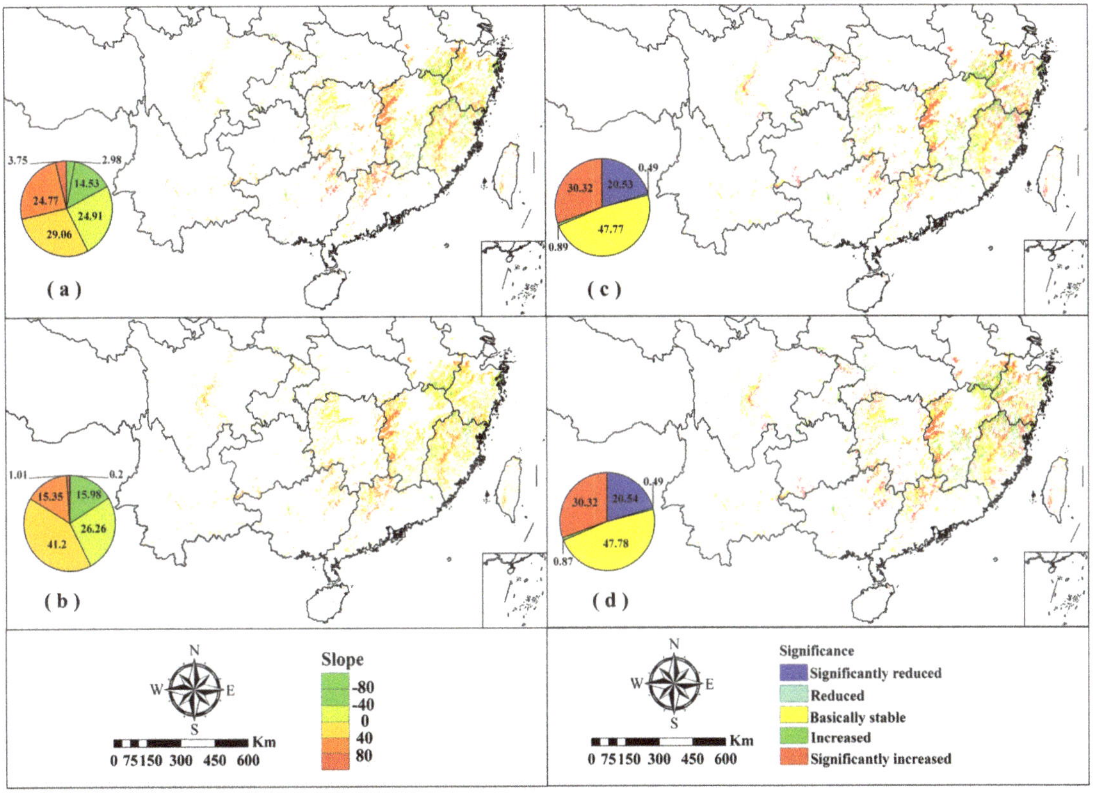

Figure 8. Spatial distribution of trend changes in (**a**) GPP and (**b**) NPP, significant changes in the (**c**) GPP and (**d**) NPP of bamboo forests in China from 2001 to 2018.

In summary, the spatial distribution of the trend slope of GPP and NPP was similar, the spatial distribution range of the increasing trend was larger than the spatial distribution range of the decreasing trend, and the areas of approximately 30% showed a significant increasing trend, indicating that the carbon fluxes of bamboo forests in China had been gradually increasing over the past 20 years.

3.3. Analysis of Climate Drivers of Carbon Fluxes of Spatiotemporal Evolution

3.3.1. Partial Correlation between Carbon Fluxes and Climate Factors

Climatic factors are important environmental factors that affect the growth of bamboo forests. To quantitatively analyze the influence of climatic factors on carbon fluxes of bamboo forests, the PPC of GPP and NPP with the Pre, T_{min}, and T_{max} of bamboo forests in China from 2001 to 2018 were calculated. The results are presented in Figure 9.

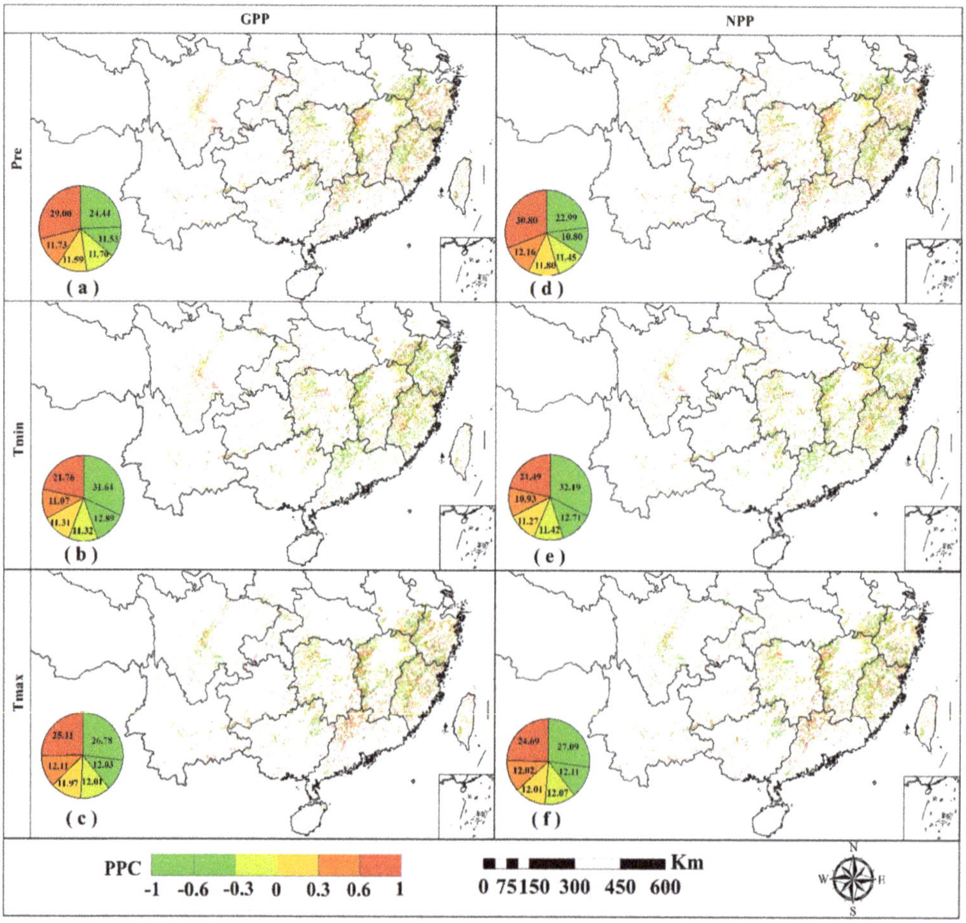

Figure 9. Spatial distribution of partial correlation coefficient (PPC) values of GPP with (**a**) precipitation (Pre), (**b**) minimum temperature (T_{min}), and (**c**) maximum temperature (T_{max}); NPP with (**d**) precipitation (Pre), (**e**) minimum temperature (T_{min}), and (**f**) maximum temperature (T_{max}) of bamboo forests in China from 2001 to 2018.

Figure 9a,d show the spatial distribution of the PCC of GPP and NPP with Pre, respectively. The proportions of the study area with positive correlations of GPP and NPP with Pre were 52.32% and 54.76%, respectively, mainly distributed in central Zhejiang, northwestern Jiangxi, Chongqing, and Sichuan. The proportions with negative correlations were 47.68% and 45.24%, respectively, mainly distributed in southeast Anhui, northwestern Zhejiang, and eastern Guangxi. Overall, Pre was mainly positively correlated with GPP and NPP, that is, the amount of precipitation considerably promoted the growth of bamboo forests. The proportions with significant ($p < 0.05$) correlations of the PCC of GPP and NPP with Pre were only 5.74% and 5.67%, respectively.

Figure 9b,e show the spatial distribution of the PPC of GPP and NPP with T_{min}, respectively. The areas where the GPP and NPP were positively correlated with T_{min} were 44.15% and 43.68%, respectively, and were mainly distributed in northwestern Zhejiang, central Hunan, and Hubei. Meanwhile, in 55.85% and 56.32% of the areas the GPP and NPP, respectively, showed a negative correlation with T_{min}, and were mainly distributed in central Zhejiang, Guangdong, and Guangxi. By comparing Figure 9a,d, in addition to Figure 9b,e, it was observed that the PPC of GPP and NPP with Pre and T_{min} had opposite spatial distribution patterns. Where GPP and NPP were positively correlated with Pre the correlation was negative with T_{min}, and vice versa. The proportions with significant ($p < 0.05$) correlations of the PCC of GPP and NPP with T_{min} were only 6.13% and 6.10%, respectively.

Figure 9c,f show the spatial distribution of the PCC of GPP and NPP, respectively, with T_{max}. GPP and NPP were positively correlated with T_{max}, accounting for 49.18% and 48.72%, respectively, mainly distributed in Guangdong, Guizhou, and western Jiangxi, and they negatively correlated with T_{max}, accounting for 50.82% and 51.28%, respectively, mainly distributed in Guangxi, Anhui, Yunnan, and western Hunan. Overall, there was a nonsignificant negative correlation of GPP and NPP with T_{max}, which indicated that high temperature somewhat affected the growth of bamboo forests. The proportions with significant ($p < 0.05$) correlations of the PCC of GPP and NPP with T_{max} were only 5.13% and 5.20%, respectively.

In summary, a certain correlation existed for the GPP and NPP of bamboo forests in China with precipitation and temperature, and, overall, they were positively correlated with Pre, negatively correlated with T_{min}, and had an insignificant negative correlation with T_{max}. In addition, there were evident spatial differences in the correlation of GPP and NPP with climatic factors, and the PPC with Pre and T_{min} exhibited complementary characteristics.

3.3.2. The Impact of Climate Factors on Carbon Fluxes on a Monthly Scale

The variations in GPP and NPP with temperature and precipitation on a monthly scale are shown in Figure 10. The values of GPP and NPP exhibited different characteristics owing to the influence of hydrothermal conditions. From February to July, with the temperature and precipitation gradually increasing, bamboo forests entered the growing season; therefore, the values of GPP and NPP showed a rapid increase trend. After August, the decrease in GPP and NPP was caused by the gradual decrease in temperature and precipitation, in addition to the fall of bamboo leaves. In December, January, and February the temperature and precipitation are lower, and the values of GPP and NPP were also decreased to the smallest values of the year. In summary, the values of GPP and NPP are closely related to temperature and precipitation, and good hydrothermal conditions are conducive to the growth of bamboo forests.

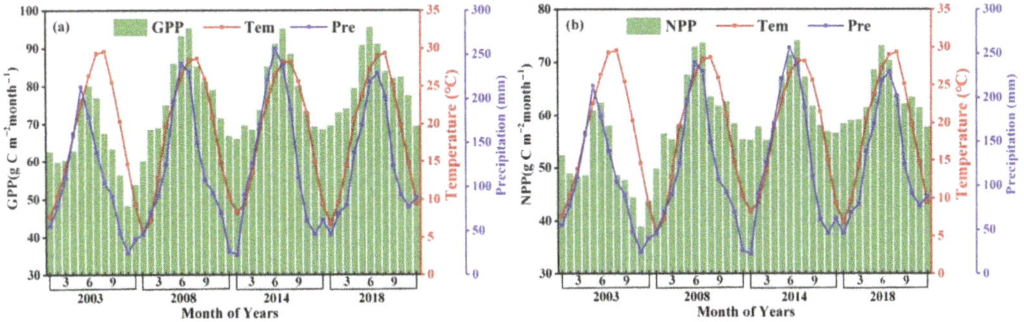

Figure 10. Variation trends of bamboo forests' (**a**) GPP and (**b**) NPP with temperature and precipitation on a monthly scale.

To further analyze the impact of temperature and precipitation on carbon fluxes of bamboo forests, we conducted a path analysis of the impact of temperature and precipitation on GPP and NPP on a monthly scale. The results are shown in Table 4. It can be seen that temperature and precipitation have a significant correlation with GPP and NPP. According to the correlation coefficient and partial correlation coefficient, the influence of precipitation on GPP and NPP is higher than that of the influence of temperature on them.

In addition, according to the direct path coefficient and indirect path coefficient, the direct influence of precipitation on GPP and NPP is higher than that of the direct influence of temperature on GPP and NPP.

Table 4. Path analysis of temperature and precipitation to GPP and NPP.

	Climate Factors	Correlation Coefficient	Direct Path Coefficient	Indirect Path Coefficient		Partial Correlation Coefficient
				→ Temperature	→ Precipitation	
GPP	Temperature	0.649 **	0.387 **	-	0.26	0.399
	Precipitation	0.659 **	0.415 **	0.24	-	0.423
NPP	Temperature	0.562 **	0.301 **	-	0.15	0.293
	Precipitation	0.602 **	0.412 **	0.11	-	0.386

Note: **, $p < 0.01$.

4. Discussion

The simulated value of the carbon fluxes of bamboo forests had a good correlation with the observed value of the flux observation station (Figure 4), and the R^2 was 0.58. Other than that, in order to further prove the reliability of this study we compared the simulated NPP with a related study (see Table 5). It can be seen from Table 5 that our simulated mean value of NPP was slightly lower than that of related studies. Due to the fact that structures, mechanisms, and input parameters varied for different models, there are variances in the simulation results of different models. Additionally, there may be differences due to different study areas and periods. Of course, this study also has some shortcomings, the following aspects of which can be analyzed. Firstly, the simulated results of the BEPS model largely depend on the quality of the input data; deficiencies in the input data will affect the accuracy of the simulation results. The resolution of the data in this study is low, so there may be limitations in simulating the carbon flux of bamboo forests in China. Secondly, in this study, the bamboo forest abundance data were used to drive the BEPS model, which solved the influence of mixed pixels on the carbon flux simulation to some extent. However, the phenomenon of "different objects with same spectrums" in remote sensing images will affect the result of bamboo forest extraction. Thirdly, we only used the observed data of two carbon flux observation stations to verify the simulated results of bamboo forest carbon fluxes in China. Therefore, there are limitations on the spatial scale. Finally, the carbon fluxes of bamboo forests were not only affected by climate factors but also by human activities and geographic factors (such as slope, aspect, and elevation). This study only considered the impact of climate factors, so there may still be a certain gap between the simulated results and the real situation.

Table 5. Comparison of the simulated NPP results in this study with the simulated results of other studies.

Site	Model	Mean NPP (g C m^{-2} y^{-1})	Reference
China	BEPS	716.88	This study
Tianmu Mountain, Zhejiang	CASA	740	[77]
Anji, Zhejiang	Triplex-Flux	835.58	[45]
Fujian	BEPS	788.6	[78]

Due to the rapid growth of bamboo forests and their high ecological, economic, and social value, some areas promoted the reclamation of wasteland and the plantation of bamboo forests [79], which increased the total area of bamboo forests in China. Therefore, the GPP and NPP of bamboo forests also increased. As shown in Figure 5, certain fluctuations occurred in the annual average GPP and NPP values of bamboo forests, which might be related to climate change. For example, in 2003 there was less precipitation and large-scale drought occurred in the summer (Figure 10), which was not conducive to the growth of bamboo forests, leading to low GPP and NPP values in that year. In 2009 and 2010 the values of GPP and NPP were low, which may be related to the natural large-scale low-temperature, snow, and ice disaster in South China in 2008 [80].

This study found that the spatial distribution range of carbon fluxes of bamboo forests was increasing larger than that which was decreasing. The areas with increases were mainly distributed in northwestern Zhejiang, western Jiangxi, central Fujian, and other regions. The reason for the increase in carbon fluxes may be that under the combined influence of favorable factors (such as a warm climate, abundant precipitation, and sufficient radiation) the growth ability of bamboo forests is relatively strong. In addition, these regions have significantly developed bamboo industries and advanced bamboo forest management techniques; therefore, bamboo forests in these regions have increased rapidly with higher productivity. The areas with decreased GPP and NPP were mainly distributed in southwest and central Zhejiang, eastern Jiangxi, northeastern Fujian, and eastern Guangdong. On the one hand, bamboo forests might be reduced due to urban expansion in some areas. On the other hand, because bamboo forests mostly have a scattered distribution, when they are distributed across a small area the difficulty of bamboo forest extraction would be increased, affecting the simulation results of carbon fluxes from bamboo forests.

Climate change has an important impact on vegetation growth. An evident coupling relationship was observed between vegetation and climatic factors [81]. At present, many scholars have analyzed the effects of climatic factors on the carbon fluxes of different vegetation from different spatiotemporal scales, and have found that there is a correlation between carbon fluxes and climatic factors [82,83]. Bamboo forests have a warm and humid climate and are very sensitive to hydrothermal changes. Related scholars have conducted studies on the impact of climatic factors on the carbon fluxes of bamboo forests. For example, Li et al. [4] analyzed the relationship between the carbon fluxes and climatic factors (temperature and precipitation) of bamboo forests in Zhejiang Province from 2011 to 2015, and found that lower precipitation and higher temperatures may have a negative impact on the carbon fluxes from bamboo forests. Chen et al. [84] used eddy correlation technology to continuously observe the carbon fluxes of bamboo forests in Anji, and found that high temperature and drought caused a significant decrease in the carbon fluxes of bamboo forests. These results are consistent with the results of this study on the driving influence of climatic factors and the carbon fluxes of bamboo forests.

5. Conclusions

This study utilized remote sensing data to drive the BEPS model to simulate the carbon fluxes from bamboo forests in China during 2001–2018, and analyzed the spatiotemporal evolution pattern of carbon fluxes and the response of climatic factors to these changes. Our study showed that the simulated values had a good correlation with the observed values, and the R^2, RMSE, and aBIAS were 0.58, 1.43 g C m^{-2} day^{-1}, and 1.21 g C m^{-2} day^{-1}, respectively. It provided a feasible way for the study of bamboo forest carbon cycles on a large spatial scale. In addition, our study also suggested that climate change was a driver that affected the spatiotemporal dynamic evolution of carbon fluxes in bamboo forests, and its driving effect exhibited evident spatial variations. This provided a theoretical basis of bamboo forests to cope with climate change.

However, this study still has some limitations. For example, (1) the low resolution of the data limited the simulation of bamboo forest carbon flux; (2) fewer flux observation sites may lead to certain deficiencies in verifying the model simulation results; and (3) we only considered the impact of climatic factors (temperature and precipitation) on the carbon fluxes of bamboo forests. In the future, these limits can be further improved to better simulate the carbon fluxes of bamboo forests in China.

Author Contributions: Conceptualization, H.D.; Data Curation, F.K., X.L. and F.M.; Formal Analysis, F.K., H.D., X.L., F.M., Y.X. and Z.H.; Funding Acquisition, G.Z., H.D. and F.M.; Investigation, Y.X., Z.H., J.J., J.W. and F.K.; Methodology, F.K. and X.L.; Supervision, G.Z.; Project Administration, H.D.

and G.Z.; Validation, X.L. and F.K.; Writing—Original Draft Preparation, F.K.; Writing—Review and Editing, H.D. All authors have read and agreed to the published version of the manuscript.

Funding: The authors gratefully acknowledge the support of the National Natural Science Foundation of China (No. 32171785, U1809208, 31901310), the State Key Laboratory of Subtropical Silviculture (No. ZY20180201), Zhejiang Provincial Collaborative Innovation Center for Bamboo Resources and High-Efficiency Utilization (No. S2017011), and the Key Research and Development Program of Zhejiang Province (2021C02005).

Institutional Review Board Statement: Not applicable.

Informed Consent Statement: Not applicable.

Data Availability Statement: MODIS data comes from NASA (https://ladsweb.modaps.eosdis.nasa.gov), accessed on 11 December 2020.

Acknowledgments: The authors gratefully acknowledge the support of various foundations. The authors are grateful to the editor and anonymous reviewers, whose comments have contributed to improving the quality of this study.

Conflicts of Interest: The authors declare that they have no competing interests.

References

1. Li, G.D.; Zhang, J.H.; Chen, C.; Tian, H.F.; Zhao, L.P. Research progress on carbon storage and flux in different terrestrial ecosystem in China under global climate change. *Ecol. Environ. Sci.* **2013**, *22*, 873–878.
2. Yu, G.R.; Zhu, X.J.; Fu, Y.L.; He, H.L.; Wang, Q.F.; Wen, X.F.; Li, X.R.; Zhang, L.M.; Zhang, L.; Su, W.; et al. Spatial patterns and climate drivers of carbon fluxes in terrestrial ecosystems of China. *Glob. Chang. Biol.* **2013**, *19*, 798–810. [CrossRef] [PubMed]
3. Payne, N.J.; Cameron, D.A.; Leblanc, J.-D.; Morrison, I.K. Carbon storage and net primary productivity in canadian boreal mixedwood stands. *J. For. Res.* **2019**, *30*, 1667–1678. [CrossRef]
4. Li, X.; Du, H.; Mao, F.; Zhou, G.; Han, N.; Xu, X.; Liu, Y.; Zhu, D.; Zheng, J.; Dong, L.; et al. Assimilating spatiotemporal MODIS LAI data with a particle filter algorithm for improving carbon cycle simulations for bamboo forest ecosystems. *Sci. Total Environ.* **2019**, *694*, 133803. [CrossRef] [PubMed]
5. Litton, C.M.; Raich, J.W.; Ryan, M.G. Carbon allocation in forest ecosystems. *Glob. Chang. Biol.* **2007**, *13*, 2089–2109. [CrossRef]
6. Tian, Z.; Zan, M.; Wang, J. Studies on temporal and spatial variations of ecosystem productivity in poyang lake basin based on modis data. *Ecol. Environ. Sci.* **2018**, *27*, 1933–1942.
7. Fang, J.Y.; Ke, J.H.; Tang, Z.Y.; Chen, A.P. Implications and estimations of four terrestrial productivity papameters. *Acta Phytoecol. Sin.* **2001**, *25*, 414–419.
8. Chen, S.; Zhang, Y.; Wua, Q.; Liua, S.; Songb, C.; Xiao, J.; Bandh, L.E.; Vose, J.M. Vegetation structural change and co2 fertilization more than offset gross primary production decline caused by reduced solar radiation in China. *Agric. For. Meteorol.* **2021**, *296*, 108207. [CrossRef]
9. Musavi, T.; Migliavacca, M.; Reichstein, M.; Kattge, J.; Wirth, C.; Black, T.A.; Janssens, I.; Knohl, A.; Loustau, D.; Roupsard, O. Stand age and species richness dampen interannual variation of ecosystem-level photosynthetic capacity. *Nat. Ecol. Evol.* **2017**, *1*, 48. [CrossRef]
10. Chen, G.Z.; Li, X.H.; Jiao, L.F.; Wang, J.Q.; Gu, K.K. Spatial-temporal variation of vegetation net primary productivity in huainan coal mine area from 2000 to 2012. *Ecol. Environ. Sci.* **2017**, *26*, 196–203.
11. Chi, D.; Wang, H.; Li, X.; Liu, H.; Li, X. Assessing the effects of grazing on variations of vegetation npp in the xilingol grassland, china, using a grazing pressure index. *Ecol. Indic.* **2018**, *88*, 372–383. [CrossRef]
12. Mao, F.J.; Du, H.Q.; Li, X.J.; Ge, H.L.; Zhou, G.M. Spatiotemporal dynamics of bamboo forest net primary productivity with climate variations in southeast China. *Ecol. Indic.* **2020**, *116*, 106505. [CrossRef]
13. Liu, T.; Mao, F.; Li, X.; Xing, L.; Dong, L.; Zheng, J.; Zhang, M.; Du, H. Spatiotemporal dynamic simulation on aboveground carbon storage of bamboo forest and its influence factors in Zhejiang province, China. *Chin. J. Appl. Ecol.* **2019**, *30*, 1743–1753.
14. Hinko-Najera, N.; Isaac, P.; Livesley, S.J.; Beringer, J.; Isaac, P.; Arndt, S.K. Net ecosystem carbon exchange of a dry temperate eucalypt forest. *Biogeosci. Discuss.* **2017**, *2016*, 3781–3800. [CrossRef]
15. Ji, X.; Lu, J.; Yang, J.; Jiang, J.; Wang, D.; He, X.; Fang, W. Carbon flux variation characteristics and its influencing factors in coniferous and broad-leaved mixed forest in fengyang mountain. *J. Northeast For. Univ.* **2019**, *47*, 51–57.
16. Xie, X.Y.; Li, A.N.; Jin, H.A. The simulation models of the forest carbon cycle on a large scale: A review. *Acta Ecol. Sin.* **2018**, *38*, 41–54.
17. Yuan, Q.; Wu, S.; Zhao, D.; Dai, E.; Chen, L.; Zhang, L. Modeling net primary productivity of the terrestrial ecosystem in China from 1961 to 2005. *J. Geogr. Sci.* **2014**, *24*, 3–17. [CrossRef]
18. Zhao, G.S.; Wang, J.B.; Fan, W.Y.; Ying, T.Y. Vegetation net primary productivity in northeast China in 2000–2008: Simulation and seasonal change. *Chin. J. Appl. Ecol.* **2011**, *22*, 621.

19. Cao, M.; Yu, G.R.; Liu, J.Y.; LI, K.R. Multi-scale observation and cross-scale mechanistic modeling on terrestrial ecosystem carbon cycle. *Sci. China Ser. D Earth Sci.* **2005**, *48*, 17–32.
20. Cui, X.; Feng, Q.S.; Liang, T.G. Research progress on remote sensing based net primary productivity of terrestrial vegetation. *Pratacult. Sci.* **2007**, *10*, 36–42.
21. Koju, U.A.; Zhang, J.; Maharjan, S.; Bai, Y.; Zhang, S.; Yao, F. Analysis of spatiotemporal dynamics of forest net primary productivity of nepal during 2000–2015. *Int. J. Remote Sens.* **2020**, *41*, 4336–4364. [CrossRef]
22. Zhang, M.; Zeng, Y.N. Net primary production estimation by using fusion remote sensing data with highspatial and temporal resolution. *J. Remote Sens.* **2018**, *22*, 143–152.
23. Mao, F.; Li, P.; Zhou, G.; Du, H.; Xu, X.; Shi, Y.; Mo, L.; Zhou, Y.; Tu, G. Development of the biome-bgc model for the simulation of managed moso bamboo forest ecosystems. *J. Environ. Manag.* **2016**, *172*, 29–39. [CrossRef]
24. Zhang, F.; Chen, J.M.; Chen, J.; Gough, C.M.; Martin, T.A.; Dragoni, D. Evaluating spatial and temporal patterns of modis gpp over the conterminous U.S. Against flux measurements and a process model. *Remote Sens. Environ.* **2012**, *124*, 717–729. [CrossRef]
25. Wang, B.; Li, M.; Fan, W.; Yu, Y.; Jia, W. Impacts of change in atmospheric co2 concentration on larix gmelinii forest growth in northeast China from 1950 to 2010. *Forests* **2019**, *10*, 454. [CrossRef]
26. Du, H.Q.; Mao, F.J.; Zhou, G.M.; Li, X.; Xu, X.J.; Ge, H.L.; Cui, L.; Liu, Y.L.; Zhu, D.; Li, Y.G. Estimating and analyzing the spatiotemporal pattern of aboveground carbon in bamboo forest by combining remote sensing data and improved biome-bgc model. *IEEE J. Sel. Top. Appl. Earth Obs. Remote Sens.* **2018**, *11*, 2282–2295. [CrossRef]
27. Zhang, F.M.; Wei-Min, J.U.; Chen, J.M.; Wang, S.Q.; Gui-Rui, Y.U.; Han, S.J. Characteristics of terrestrial ecosystem primary productivity in east Asia based on remote sensing and process-based model. *Chin. J. Appl. Ecol.* **2012**, *23*, 307.
28. Zheng, J.; Mao, F.; Du, H.; Li, X.; Zhou, G.; Dong, L.; Zhang, M.; Han, N.; Liu, T.; Xing, L. Spatiotemporal simulation of net ecosystem productivity and its response to climate change in subtropical forests. *Forests* **2019**, *10*, 708. [CrossRef]
29. Running, S.W.; Coughlan, J.C. A general model of forest ecosystem processes for regional applications i. Hydrologic balance, canopy gas exchange and primary production processes. *Ecol. Model.* **1988**, *42*, 125–154. [CrossRef]
30. Bunkei, M.; Yang, C.; Chen, J.; Wang, Q.; Satoshi, K.; Masayuki, T. Accurate estimation of net primary productivity of terrestrial ecosystem at a regional scale. *Acta Geogr. Sin.* **2004**, *59*, 80–87.
31. Mo, X.; Chen, J.M.; Ju, W.; Black, T.A. Optimization of ecosystem model parameters through assimilating eddy covariance flux data with an ensemble kalman filter. *Ecol. Model.* **2008**, *217*, 157–173. [CrossRef]
32. Liu, J.; Chen, J.M.; Chen, J.C. Net primary productivity mapped for canada at 1-km resolution. *Glob. Ecol. Biogeogr.* **2002**, *11*, 115–129. [CrossRef]
33. Kang, Z.; Zhang, S.; Bai, Y.; Henchiri, M.; Zhang, J. Spatio-temporal changes of grassland net primary productivity (npp) in inner Mongolia and its response to drought. *Acta Agrestia Sin.* **2021**, *29*, 156–165.
34. Li, X.J.; Mao, F.J.; Du, H.Q.; Zhou, G.M.; Xu, X.J.; Li, P.H.; Liu, Y.L.; Cui, L. Simulating of carbon fluxes in bamboo forest ecosystem using beps model based on the lai assimilated with dual ensemble kalman filter. *Chin. J. Appl. Ecol.* **2016**, *27*, 3797–3806.
35. Liu, Y.; Zhou, Y.; Ju, W.; Wang, S.; Wu, X.; He, M.; Zhu, G. Impacts of droughts on carbon sequestration by China's terrestrial ecosystems from 2000 to 2011. *Biogeosciences* **2014**, *11*, 2583–2599. [CrossRef]
36. Matsushita, B.; Tamura, M. Integrating remotely sensed data with an ecosystem model to estimate net primary productivity in east Asia. *Remote Sens. Environ.* **2002**, *81*, 58–66. [CrossRef]
37. Wang, S.; Lei, Z.; Chen, J.; Ju, W.; Feng, X.; Wu, W. Relationships between net primary productivity and stand age for several forest types and their influence on China's carbon balance. *J. Environ. Manag.* **2011**, *92*, 1651–1662. [CrossRef]
38. Zhou, Y.; Zhu, Q.; Chen, J.M.; Wang, Y.Q.; Liu, J.; Sun, R.; Tang, S. Observation and simulation of net primary productivity in Qilian mountain, western China. *J. Environ. Manag.* **2007**, *85*, 574–584. [CrossRef]
39. Du, H.Q.; Mao, F.J.; Li, X.J.; Zhou, G.M.; Zhou, Y.F. Mapping global bamboo forest distribution using multisource remote sensing data. *IEEE J. Sel. Top. Appl. Earth Obs. Remote Sens.* **2018**, *11*, 1458–1471. [CrossRef]
40. Zhou, B.; Fu, M.; Xie, J.; Yang, X.; Li, Z. Ecological functions of bamboo forest: Research and application. *J. For. Res.* **2005**, *16*, 143–147.
41. Li, Y.M.; Feng, P.F. Bamboo resources in china based on the ninth national forest inventory data. *World Bamboo Ratt.* **2019**, *17*, 45–48.
42. Cui, L.; Du, H.Q.; Zhou, G.M.; Li, X.J.; Mao, F.J.; Xu, X.J.; Fan, W.L.; Li, Y.G.; Zhu, D.; Liu, T.Y.; et al. Combination of decision tree and mixed pixel decomposition for extracting bamboo forest information in China. *J. Remote Sens.* **2019**, *23*, 166–176.
43. Li, Y.; Li, Y.; Chang, S.X.; Yang, Y.; Fu, S.; Jiang, P.; Luo, Y.; Yang, M.; Chen, Z.; Hu, S. Biochar reduces soil heterotrophic respiration in a subtropical plantation through increasing soil organic carbon recalcitrancy and decreasing carbon-degrading microbial activity. *Soil Biol. Biochem.* **2018**, *122*, 173–185. [CrossRef]
44. Zhou, G.M.; Meng, C.F.; Jiang, P.K.; Xu, Q.F. Review of carbon fixation in bamboo forests in China. *Bot. Rev.* **2011**, *77*, 262.
45. Chen, S.; Jiang, H.; Cai, Z.; Zhou, X.; Peng, C. The response of the net primary production of moso bamboo forest to the on and off-year management: A case study in Anji county, Zhejiang, China. *For. Ecol. Manag.* **2018**, *409*, 1–7. [CrossRef]
46. Zhang, M.; Chen, S.; Jiang, H.; Peng, C.; Zhang, J.; Zhou, G. The impact of intensive management on net ecosystem productivity and net primary productivity of a lei bamboo forest. *Ecol. Model.* **2020**, *435*, 109248. [CrossRef]
47. Li, P.H.; Zhou, G.M.; Du, H.Q.; Lu, D.S.; Mo, L.F.; Xu, X.J.; Shi, Y.J.; Zhou, Y.F. Current and potential carbon stocks in moso bamboo forests in China. *J. Environ. Manag.* **2015**, *156*, 89–96. [CrossRef] [PubMed]

48. Du, H.Q.; Sun, X.Y.; Han, N.; Mao, F.J. Rs estimation of inventory parameters and carbon storage of moso bamboo forest based on synergistic use of object-based image analysis and decision tree. *Chin. J. Appl. Ecol.* **2017**, *28*, 11.
49. Li, X.; Du, H.; Mao, F.; Zhou, G.; Chen, L.; Xing, L.; Fan, W.; Xu, X.; Liu, Y.; Cui, L. Estimating bamboo forest aboveground biomass using enkf-assimilated modis lai spatiotemporal data and machine learning algorithms—Science direct. *Agric. For. Meteorol.* **2018**, *256–257*, 445–457. [CrossRef]
50. Du, H.Q.; Cui, R.; Zhou, G.M.; Shi, Y.J.; Xu, X.J.; Fan, W.L.; Lü, Y.L. The responses of moso bamboo (*Phyllostachys heterocycla* var. Pubescens) forest aboveground biomass to landsat tm spectral reflectance and ndvi. *Acta Ecol. Sin.* **2010**, *30*, 257–263. [CrossRef]
51. Xu, X.; Zhou, G.; Liu, S.; Du, H.; Mo, L.; Shi, Y.; Jiang, H.; Zhou, Y.; Liu, E. Implications of ice storm damages on the water and carbon cycle of bamboo forests in southeastern China. *Agric. For. Meteorol.* **2013**, *177*, 35–45. [CrossRef]
52. Xiao, F.M.; Fan, S.H.; Wang, S.L.; Guan, H.Y.; YU, X.J.; Shen, Z.Q. Estimation of the carbon balance in moso bamboo and chinese fir plantation ecosystem. *Sci. Silvae Sin.* **2010**, *46*, 59–65.
53. Wang, X.C.; Wang, C.K. Fundamental concepts and field measurement methods of carbon cycling in forest ecosystems: A review. *Acta Ecol. Sin.* **2015**, *35*, 4241–4256.
54. Hou, H.-Y. Vegetation of china with reference to its geographical distribution. *Ann. Mo. Bot. Gard.* **1983**, *70*, 509–549. [CrossRef]
55. Lu, D.; Tian, H.; Zhou, G.; Ge, H. Regional mapping of human settlements in southeastern China with multisensor remotely sensed data. *Remote Sens. Environ.* **2015**, *112*, 3668–3679. [CrossRef]
56. Shang, Z.; Zhou, G.; Du, H.; Xu, X.; Shi, Y.; Lü, Y.; Zhou, Y.; Gu, C. Moso bamboo forest extraction and aboveground carbon storage estimation based on multi-source remotely sensed images. *Int. J. Remote Sens.* **2013**, *34*, 5351–5368. [CrossRef]
57. Zhang, D.M.; Yuan, Q.; Wang, J.T. Discuss a method of assistant remotely sensed imagery visual interpretation on grassland—remotely sensed image and thematic map semitransparent overlap. *Remote Sens. Techonol. Appl.* **2006**, *21*, 560–564.
58. Lu, D.; Moran, E.; Batistella, M. Linear mixture model applied to amazonian vegetation classification. *Remote Sens. Environ.* **2003**, *87*, 456–469. [CrossRef]
59. Zhang, T.; Sun, R.; Zhang, R.; Zhang, L. Simulation of water and carbon fluxes in harvard forest area based on data assimilation method. *Chin. J. Appl. Ecol.* **2013**, *24*, 2746–2754.
60. Chen, J.M.; Feng, D.; Chen, M. Locally adjusted cubic-spline capping for reconstructing seasonal trajectories of a satellite-derived surface parameter. *IEEE Trans. Geosci. Remote Sens.* **2006**, *44*, 2230–2238. [CrossRef]
61. Li, X.J.; Mao, F.J.; Du, H.Q.; Zhou, G.M.; Xu, X.J.; Han, N.; Sun, S.B.; Gao, G.L.; Chen, L. Assimilating leaf area index of three typical types of subtropical forest in China from modis time series data based on the integrated ensemble kalman filter and prosail model. *ISPRS J. Photogramm. Remote Sens.* **2017**, *126*, 68–78. [CrossRef]
62. Feng, X.; Liu, G.; Chen, J.M.; Chen, M.; Liu, J.; Ju, W.M.; Sun, R.; Zhou, W. Net primary productivity of China's terrestrial ecosystems from a process model driven by remote sensing. *J. Environ. Manag.* **2007**, *85*, 563–573. [CrossRef]
63. Xue, M.; Chen, Y.Z.; Yan, M.; Li, Z.Y.; Wang, X.Q.; Xu, H.S.; Zhang, Z.P.; Tian, X. Simulation and spatio-temporal variation analysis of net primary productivity in northeast China. *J. Fuzhou Univ. Nat. Sci. Ed.* **2018**, *46*, 821–830.
64. Ju, W.; Chen, J.M.; Black, T.A.; Barr, A.G.; Liu, J.; Chen, B. Modelling multi-year coupled carbon and water fluxes in a boreal aspen forest. *Agric. For. Meteorol.* **2006**, *140*, 136–151. [CrossRef]
65. Chen, J.M.; Liu, J.; Cihlar, J.; Goulden, M.L. Daily canopy photosynthesis model through temporal and spatial scaling for remote sensing applications. *Ecol. Model.* **1999**, *124*, 99–119. [CrossRef]
66. Zhou, G.; Jiang, P. Density, storage and spatial distribution of carbon in phyllostachy pubescens forest. *Sci. Silvae Sin.* **2004**, *40*, 20–24.
67. Huang, Q.M.; Yang, D.D.; Gao, A.S. A study on photosynthesis of bamboo. *Sci. Silvae Sin.* **1989**, *25*, 4.
68. Li, D.Q.; Ju, W.M.; Zheng, G.; Liu, Y.B.; Zan, M.; Zhang, C.H.; Huang, J.L. Comparison of estimated forest biomass increment rate based on a process-based ecological model and forest inventory data. *Ecol. Environ. Sci.* **2013**, *22*, 1647–1657.
69. Liu, J.; Chen, J.M.; Cihlar, J.; Park, W.M. A process-based boreal ecosystem productivity simulator using remote sensing inputs. *Remote Sens. Environ.* **1997**, *62*, 158–175. [CrossRef]
70. Li, J.; Zhang, J.; Liu, C.L.; Yang, X.C. Spatiotemporal variation of vegetation coverage in recent 16 years in the border region of china, laos, and myanmar based on modis-ndvi. *Sci. Silvae Sin.* **2019**, *55*, 10.
71. Li, N.Q.; Xu, G.Y. Grid analysis of land use based on natural breaks (jenks) classification. *Bull. Surv. Mapp.* **2020**, 106–110. [CrossRef]
72. Liu, H.; Li, X.; Mao, F.; Zhang, M.; Zhu, D.e.; He, S.; Huang, Z.; Du, H. Spatiotemporal evolution of fractional vegetation cover and its response to climate change based on modis data in the subtropical region of China. *Remote Sens.* **2021**, *13*, 913. [CrossRef]
73. Wu, S.S.; Yao, Z.J.; Jiang, L.G.; Wang, R.; Liu, Z.F. The spatial-temporal variations and hydrological effects of vegetation npp based on modis in the source region of the Yangtze river. *J. Nat. Resour.* **2016**, *31*, 39–51.
74. Chen, Y.H.; Li, X.B.; Shi, P.J. Variation in ndvi driven by climate factors across China, 1983–1992. *Acta Phytoecol. Sin.* **2001**, *25*, 716–720.
75. Zhang, X.L.; Xiao, W.H.; Wang, Y.C. Temporal-spatial variations of npp and its climatic driving mechanism in the three gorges reservoir area based on modified casa model. *Acta Ecol. Sin.* **2021**, *41*, 3488–3498.
76. Jiang, X.D.; Li, Y.X. Path analysis on the meteorological factors impacting soil respiration rate of wheat field. *J. Anhui Agric. Sci.* **2009**, *9*, 74–769.

77. Chen, S.L.; Jiang, H.; Jin, J.X.; Wang, Y. Changes in net primary production in the Tianmu mountain nature reserve, China, from 1984 to 2014. *Int. J. Remote Sens.* **2017**, *38*, 211–234. [CrossRef]
78. Jiang, H.; Wang, X.X.; Sun, W.J. Simulation by remote sensing and temporal-spatial analysis of forest ecosystem net primary productivity in Fujian province, China. *J. Geo-Inf. Sci.* **2010**, *12*, 580–586. [CrossRef]
79. Bai, S.B.; Zhou, G.M.; Wang, Y.X.; Yu, S.Q.; Li, Y.H.; Fang, F.Y. Stand structure change of phyllostachys pubescens forest expansion in Tianmushan national nature reserve. *J. West China For. Sci.* **2012**, *41*, 77–82.
80. Liu, G.; Sun, R.; Xiao, Z.Q.; Cui, T.X. Analysis of spatial and temporal variation of net primary productivity and climate controls in China from 2001 to 2014. *Acta Ecol. Sin.* **2017**, *37*, 4936–4945.
81. Xu, Y.Q.; Xiao, F.J.; Yu, L. Review of spatio-temporal distribution of net primary productity in forest ecosystem and its responses to climate change in China. *Acta Ecol. Sin.* **2020**, *40*, 4710–4723.
82. Zhang, L.; Xiao, J.; Li, J.; Wang, K.; Lei, L.; Guo, H. The 2010 spring drought reduced primary productivity in southwestern China. *Environ. Res. Lett.* **2012**, *7*, 045706. [CrossRef]
83. Zhou, W.; Mu, F.Y.; Gang, C.C.; Guan, D.J.; He, J.F.; Li, J.L. Spatio-temporal dynamics of grassland net primary productivity and their relationship with climatic factors from 1982 to 2010 in China. *Acta Ecol. Sin.* **2017**, *37*, 4335–4345.
84. Chen, X.F.; Jiang, H.; Niu, X.D.; Zhang, J.M.; Fang, C.Y. Effect of seasonal high temperature and drought on carbon flux of bamboo forest ecosystem in subtropical region. *Chin. J. Appl. Ecol.* **2016**, *27*, 335.

remote sensing

MDPI

Article

The Potential of Fully Polarized ALOS-2 Data for Estimating Forest Above-Ground Biomass

Zhihui Liu [1], Opelele Omeno Michel [1,2], Guoming Wu [1], Yu Mao [1], Yifan Hu [1] and Wenyi Fan [1,*]

[1] Key Laboratory of Sustainable Forest Ecosystem Management—Ministry of Education, School of Forestry, Northeast Forestry University, Harbin 150040, China; liuzh@nefu.edu.cn (Z.L.); michel.opelele@unikin.ac.cd (O.O.M.); wgm@nefu.edu.cn (G.W.); maoyu@nefu.edu.cn (Y.M.); yifan_hu2021@nefu.edu.cn (Y.H.)

[2] Department of Natural Resources Management, Faculty of Agricultural Sciences, University of Kinshasa, 117 Kinshasa XI, Mont-Amba District, Kinshasa 01031, Congo

* Correspondence: fanwy@nefu.edu.cn; Tel.: +86-139-4605-0384

Abstract: SAR data have a longer wavelength and stronger penetrating power compared with traditional optical remote sensing. Therefore, SAR data are more suitable for the estimation of the above-ground biomass (AGB) of forests. This study was aimed at evaluating the sensitivity of L-band full polarization data to AGB. L-band data were improved to estimate the saturation point produced by AGB, and were found to be suitable for estimating a wide range of AGB. This study extracted backscattering coefficients, polarization decomposition variables, and terrain factors. New parameters were constructed from these variables, and their performance in predicting AGB was evaluated. Significant variables found with AGB were added to the multivariate linear model. A statistical analysis showed the presence of multicollinearity between the variables. Therefore, ridge regression, random forest method (RF), and principal component analysis (PCA) were introduced to solve the problem of collinearity. In all the three methods, the saturation of the ridge regression model was low, reaching it at 150 t/ha. Better accuracy was obtained with the RF model. No obvious saturation incident was detected in the model established using the principal component analysis. This could be attributed to the low biomass levels observed in our study area. This model provided accurate results (adjusted $r^2 = 0.90$ rmse = 14.24 t/ha), indicating that L-band data have the potential to estimate AGB. Additionally, suitable variables and models were selected in this study, with the principal component analysis being more helpful in combining various SAR parameters. The achievement of these accurate results could be attributed to the synergy among variables.

Keywords: backscatter coefficients; polarization decomposition; collinearity; ridge regression; RF; PCA

Citation: Liu, Z.; Michel, O.O.; Wu, G.; Mao, Y.; Hu, Y.; Fan, W. The Potential of Fully Polarized ALOS-2 Data for Estimating Forest Above-Ground Biomass. *Remote Sens.* **2022**, *14*, 669. https://doi.org/10.3390/rs14030669

Academic Editor: Klaus Scipal

Received: 21 December 2021
Accepted: 26 January 2022
Published: 30 January 2022

1. Introduction

Carbon sequestration capacity is an important manifestation of forest functions. Forest above-ground biomass (AGB) is a consequential evaluation index of carbon sequestration capacity. Therefore, it is necessary to estimate AGB to understand the carbon sequestration capacity in a particular area [1]. Previous studies have shown that the use of the backscatter coefficient of airborne L-band SAR data could not significantly improve the ability to estimate AGB [2]. In large survey areas, the AGB root mean square error (RMSE) estimated from HH polarization has been found to be about 30% [3,4]. Similarly, the use of L-band data to predict the Indian tropical forest had a higher accuracy with RMSE = 16.06 t/ha [5]. Previous studies have used the random forest (RF) method to estimate the AGB with RMSE = 18.9 t/ha. Additionally, a regression model was built to estimate the boreal forest AGB with RMSE = 37.3 t/ha [6,7]. These studies obtained tree height by laminar analysis of SAR data and then calculated AGB with RMSE = 36.3 t/ha [8]. However, different AGB estimation methods have been found to provide discordant results even when the same SAR data are analyzed. Meanwhile, variability of forests has been found to be among the factors affecting the accuracy of AGB estimation. Several studies have estimated the forest biomass of tree trunks using radar backscatter coefficients. The sensitivity of the radar backscatter coefficients to AGB depends on the wavelength; the longer the wavelength, the higher the sensitivity [9,10]. A National Aeronautics and Space Administration (NASA) study in Landers Pine Forest showed that the dynamic range of the radar backscatter coefficient was greater in the P-band, followed by the L-band, which had the highest HV polarization and VH polarization sensitivity [11,12]. HH polarization has been found to be suitable for sparse areas, while HV polarization is suitable for dense areas [13]. The use of radar data to estimate the AGB of planted forests tends to have more accurate estimations than in other forest types [14,15]. However, the applicability of this method in natural forests is still uncertain. Previous studies have estimated a saturation point when using backscatter coefficients to estimate forest biomass, and reasonable results have been achieved when the biomass is less than 150 t/ha [16]. Lower saturation points have been recorded in AGB of complex tropical forests and different forest types, while pure forests and swampy areas have shown higher saturation points [17,18]. This suggests that the difference in estimated AGB saturation point is affected by area, forest density, and tree species composition. The ratio combination of different polarization channels increased the saturation point when estimating AGB, leading to more accurate results [19,20]. Meanwhile, the ability to estimate AGB using radar backscatter coefficients has been found to be limited [21]. In addition, this method of estimating AGB has a lower saturation point, which limits its application.

Other reasons that affect the accuracy of AGB estimates include model variability and different parameters. Some previous studies have not discussed the synergistic effects when estimating AGB using backscattering coefficients or decomposition parameters. Improving the saturation point of the estimated AGB has also been difficult. However, some polarization decomposition methods have proved suitable for estimating the AGB of forests. Three simple scattering mechanisms have been used to describe SAR observation results. These mechanisms achieved acceptable accuracy, which proved that the decomposition method is suitable for estimating vegetation biomass [22,23]. The azimuth offset compensation of SAR data before polarization decomposition partially improves the accuracy of AGB estimation [24]. In addition, the VanZyl three-component decomposition and Yamaguchi three-component decomposition obtain more accurate results [25,26]. The polarization decomposition method estimated a higher saturation point for AGB than the backscatter coefficient. Most researchers have used linear and nonlinear regression models to predict AGB [27–31]. Although these studies have optimized the model, there has been limited focus on the parameters. Meanwhile, the accuracy obtained by using different decomposition methods to estimate AGB varies greatly [32–34]. This shows that different polarization decomposition methods are suitable for different types of ground features. However, combining the polarization decomposition parameters and the water cloud model to predict AGB achieves better results [35]. Additionally, the use of multiple polarization decomposition parameters to establish a multivariate model could slightly improve accuracy [36]. The RF method has been found to obtain accurate results in estimating AGB [37–39]. However, its applicability to small sample sizes remains uncertain. Therefore, the choice of the model is an important factor affecting accuracy.

Previous studies did not select the most suitable variables for forest AGB estimation. In addition, there is still lack of in-depth studies on the relationships between variables, making it difficult to reasonably utilize SAR data. As such, it is difficult to improve the saturation point of the estimated AGB.

At present, AGB can be estimated using long-band SAR data, although the saturation points and estimation accuracy can still be improved. The present study not only estimated the AGB based on backscatter coefficient and polarization decomposition parameters, but also combined the two to establish the potential of long-wavelength full-polarization data to estimate forest biomass. Unlike previous studies, this study used variables from SAR data to construct parameters that were more sensitive to AGB. All variables that were significantly related to AGB were combined. A model that was more suitable for the estimation of AGB was also selected.

Specifically, this study was aimed at:

(1) Using the original channel backscatter coefficients to establish a univariate model to estimate AGB. The ratio of backscatter coefficients was calculated and a univariate model established. The impact of topographical factors on AGB was also analyzed.
(2) Selecting the most suitable polarization decomposition method and polarization decomposition parameters. Polarization decomposition parameters were used to construct a stronger estimation ability for the new parameters, and a model was established with AGB.
(3) Comparing the ability of ridge regression, RF and the PCA method to resolve a high-dimensional variable set. The focus was on establishing a model, and predicting the AGB at the regional scale by using all the relevant parameters.

2. Study Area and Data

2.1. SAR Data

The SAR data used in this study were the ALOS-2 PALSAR full-polarization observation data obtained on 8 August 2020. The selected image covered northern China (Figure 1). The image is from a 1.1-level L-band radar developed by the Japan Aerospace Exploration Agency (JAXA). The average zenith angle was 27.8°, the radar center frequency was 1.27×10^3 MHz, the range resolution was 5.66 m, and the azimuth resolution was 2.86 m. The pixel size was16.19 m^2. The overall observation area was 4494.62 km^2, and the average height of the sensor from the Earth's surface was 634.24 km^2.

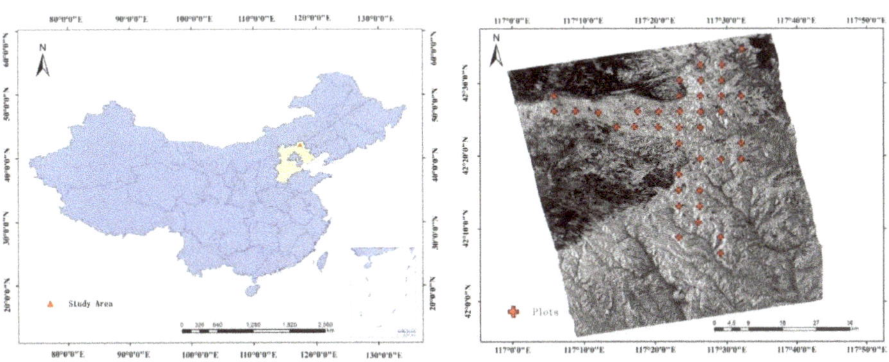

Figure 1. SAR data illustration of sample sites.

2.2. Field Data

The study area was a typical temperate forest in northern China. The forest is located in Hebei Province, North China (117E, 42N). This area is located in the transition zone from Yanshan Mountain to Inner Mongolia. Except for the mountain, the rest of the area consists of plains and cities. Altitude ranges from 1171 m to 1960 m asl. This area is characterized by a mixed forest of coniferous and broad-leaved trees, with North China larch (*Larix principis-rupperchtii* Mayr) and white birch (*Betula platyphylla* Suk) being the main species. A total of 38 fixed plots were used in the image. The field data used in this study were obtained through field surveys in 2020. In order to avoid interference, the measurements were carried out at a distance of more than 30 m from non-forest areas. Field surveys included measuring tree species composition, and measuring diameter at breast height (DBH) at a distance of 1.3 m from the ground. All trees with a DBH of less than 2.5 cm were eliminated. Tree height was measured using Vertex IV and Transponder T3. The coordinates of the center point of the plot were determined using the Unistreng RTK-G10.

We ensured that the center point coordinate error of the sample plot was within 10 cm. The individual allometry equation of local tree species was used to calculate the AGB of the forest for each sample [40,41]. Based on the measured results, the minimum biomass above the forest was 4.24 t/ha and the maximum was 185.08 t/ha. The plots were separated at equal intervals, and each plot had an area of 0.06 ha. The shape of each set of field data was a rhombus, with a diagonal length of 17.3 m. The area of the plot was 149.645 m^2. The AGB level in this area was found to be more suitable for this study. Meanwhile, microwave remote sensing observation methods were more suitable for forest biomass estimation, considering the complex geological and climatic conditions in the area. The actual biomass is shown in Table 1.

Table 1. Statistical data of the plots.

Number	AGB (t/ha)	Number	AGB (t/ha)	Number	AGB (t/ha)	Number	AGB (t/ha)
01	129.252	11	80.171	21	172.128	31	54.836
02	142.776	12	174.936	22	166.592	32	167.011
03	139.653	13	147.477	23	98.417	33	165.710
04	58.508	14	153.008	24	104.223	34	144.930
05	166.223	15	185.083	25	151.198	35	109.724
06	4.248	16	177.771	26	164.641	36	95.931
07	113.727	17	163.469	27	180.735	37	24.281
08	79.486	18	157.807	28	102.843	38	33.292
09	29.471	19	65.888	29	150.238		
10	132.427	20	90.952	30	114.632		

3. Methods

The processing steps for field inventory and ALOS-2 PALSAR-2 data are shown in Figure 2.

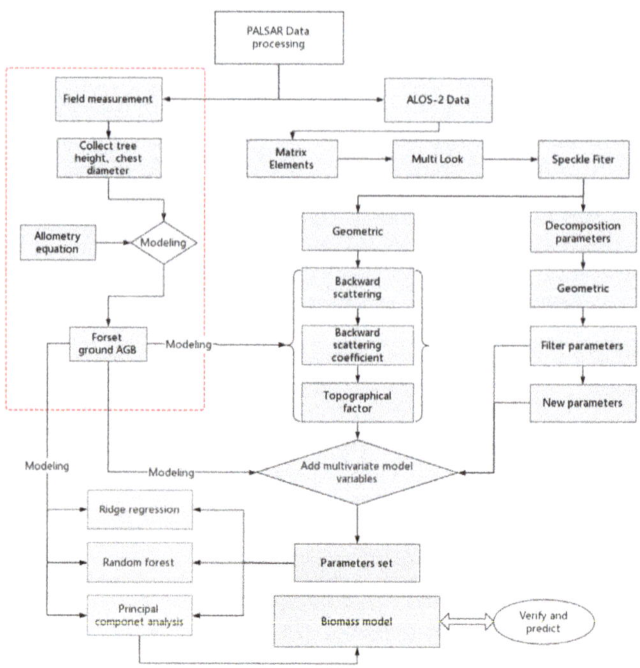

Figure 2. Flowchart.

3.1. SAR Data Processing

Calibration: Conversion of the amplitude data record in the original image into a backscatter coefficient was not affected by changes in surface parameters. In order to expand the dynamic range of the scattering coefficient, the amplitude data record was expressed in decibels, as follows:

$$\sigma^0_{i,j} = 10 \cdot \lg\left(DN^2_{i,j}\right) + CF_1. \quad (1)$$

where $\sigma^0_{i,j}$ is backscatter coefficients, $DN_{i,j}$ is the gray value of the pixel, and CF_1 is the calibration factor [42].

The four backscatter coefficients (σ_{HH}, σ_{HV}, σ_{VH}, and σ_{VV}) were obtained by radiation calibration.

Filter denoising: Given that SAR is a coherent system, speckle noise becomes an inherent feature that interferes with image readings. The present experiment used a refined Lee filter [43]. At the same time, multi-look processing also had a noise suppressing effect. Multi-look processing improved the effectiveness of feature information extraction by averaging the pixels of the SAR image azimuth and distance. A 4 × 9 multi-look process was performed on the original image to ensure that the pixels closed to the square and matched the area in the plot. The above process was run in the Gamma software [44].

Decomposition parameter acquisition: Since the research object was a distributed target, it was found suitable for incoherent decomposition. Three polarization decomposition methods suitable for forests were selected [45,46]. These methods included Yamaguchi three-component decomposition, eigenvalue-based H/A/alpha decomposition, and eigenvector-based H/A/alpha decomposition [47–53]. The Yamaguchi three-component decomposition method decomposes the echo signal into three scattering mechanisms, and volume scattering in the layered random medium provides good results [51]. H/A/alpha decomposition contains information about the dominance relationship between scattering mechanisms. Among them, the scattering entropy (H) not only represents the specific gravity of different scattering mechanisms in the whole scattering process, but also describes the randomness of the scattering process. The degree of heterogeneity in different directions (A) characterizes the degree of influence of the other two scattering mechanisms, which do not dominate the result when H increases. Scattering angle (α) describes the degree of freedom inside the target [54–56]. The polarization decomposition parameters were obtained using PolSARpro 6.0.2 [57].

Geocoding: Since SAR is a side-view system, it causes nonlinear distortion in areas with large terrain undulations. Therefore, SAR images cannot transform into a reference coordinate system by polynomial correction or affine transformation. The present study combined the imaging characteristics of the sensor and the ground morphology. It exploited external DEM data (SRTM V2 30 m resolution) and used a strict-range Doppler to geocode SAR image data. This process was run using Gamma software.

Thirty-five original polarization decomposition parameters were obtained through three polarization decompositions shown in Table 2 [49,50].

3.2. Backscatter Coefficient and Its Combination

The correlations between backscatter coefficients (σ_{HH}, σ_{HV}, σ_{VH}, σ_{VV}) and AGB were analyzed. The radar satisfied the reciprocity of a single station, thus the cross-polarization channels were averaged (σ_X replaces $\sigma_{(HV+VH)/2}$). Each variable was used to establish univariate linear models with AGB. The model was built using Matlab-2014b [58]. Meanwhile, different combinations of backscatter coefficients had different sensitivities to AGB [19]. We combined the backscatter coefficients to find the parameters with more significant correlations. A total of 26 different combinations were created using backscatter coefficients, and correlation analysis was performed for the 26 combinations. Significant variables were selected to establish univariate linear models with AGB.

Table 2. Decomposition parameters.

Method	Parameter	
Yamaguchi three-component decomposition	Odd scattering component of Yamaguchi 3 decomposition ($Yamaguchi_{Odd}$) Even scattering component of Yamaguchi 3 decomposition ($Yamaguchi_{Dbl}$) Scattering component of Yamaguchi 3 decomposition volume ($Yamaguchi_{Vol}$)	
H/A/alpha eigenvalue set decomposition	Eigenvalue	anisotropy, ansiotropy_lueneburg, anisotropy 12 asymetry, derd, derd_norm, entropysh, entropy 1 entropy 2, entropy 3, entropy 4, entropy 5, I1, I2, I3, p1, p2, p3, prdestal, polarisation_fraction rvi, serd, serd_norm
H/A/alpha eigenvector set decomposition	Eigenvector	alpha, alpha1, alpha 2, alpha 3 beta, beta 1, beta 2, beta 3 delta, delta 1, delta 2, delta 3 gamma, gamma1, gamma 2, gamma 3

3.3. Terrain Factors

Topography is an important factor that affects AGB [59–61]. This study obtained slope, aspect, and elevation using DEM, which were combined in Arcmap 10.7 as shown in Figure 3 [62]. The correlation between the extracted terrain factors and AGB was also analyzed.

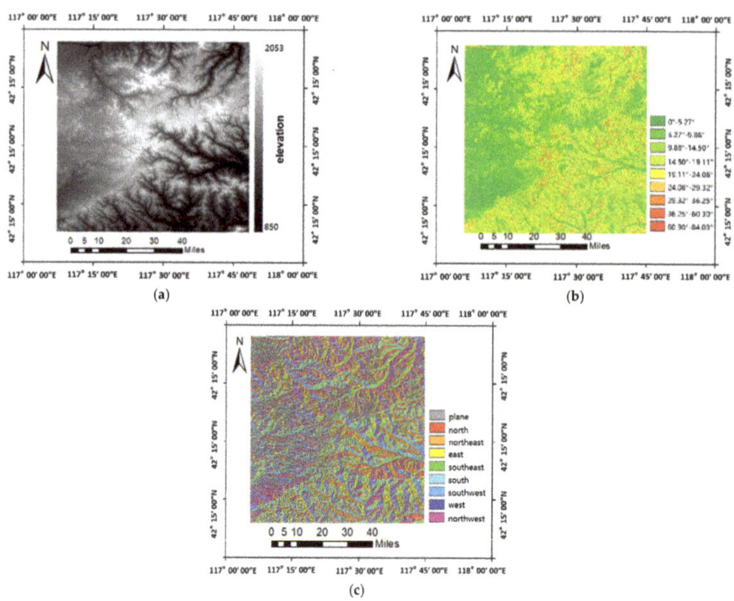

Figure 3. (**a**) Elevation; (**b**) slope; (**c**) aspect.

3.4. Constructing New Parameters

Previous studies have shown that the proportional combination of volume scattering, secondary scattering, and surface scattering has a certain sensitivity to forest canopy structure [63]. They derived a relationship between the growing stock volume (GSV) and polarimetric decomposition powers. In addition, these studies have found that the volumetric scattering power and GSV in different samples were all positively correlated, while the surface scattering and GSV were all negatively correlated. Therefore, it was concluded that the GSV and ratio of the three scattering powers have a certain sensitivity [64]. In the present study, there was a close relationship between AGB and GSV. The products of the different scattering mechanisms of the Yamaguchi three-component decomposition and the other two scattering mechanisms were found to possess a ratio relationship. This study established new parameters with reference to the above-mentioned study. Each new parameter with AGB was used to build a linear model. The construction of the new parameters was as follows:

- Ground scattering—scattering parameter ratio;

$$R_1 = \frac{\text{Yamaguchi}_{\text{Odd}}}{\text{Yamaguchi}_{\text{Dbl}} \times \text{Yamaguchi}_{\text{Vol}}}$$

- Even-scattering molecular parameters;

$$R_2 = \frac{\text{Yamaguchi}_{\text{Dbl}}}{\text{Yamaguchi}_{\text{Odd}} \times \text{Yamaguchi}_{\text{Vol}}}$$

- Volume-scattering molecular parameters.

$$R_3 = \frac{\text{Yamaguchi}_{\text{Vol}}}{\text{Yamaguchi}_{\text{Dbl}} \times \text{Yamaguchi}_{\text{Odd}}}$$

3.5. Multivariate Linear Model

The parameters obtained through the three polarization decomposition methods were not all applicable to the study of the forest. Therefore, we analyzed the correlation between decomposition variables and AGB to obtain significant correlation variables. A multivariate linear model with all the significant correlation variables was set up to predict AGB. The model variance inflation factor (VIF) test showed that the variables had significant multicollinearity.

3.6. Ridge Regression

Due to the complexity of the radar signal, there was a degree of information overlap between the variables, resulting in collinearity. This study attempted to use the ridge regression model to solve the collinearity problem. Ridge regression is a regularization method for the regression analysis of ill-posed problems.

3.7. Random Forest

The RF method is a classifier that consists of multiple decision trees. It belongs to the Bagging ensemble learning algorithm. This method was used to collect multiple sub-datasets from the original dataset and train multiple different decision trees. The prediction results of multiple decision trees were then averaged to obtain the final result. This method was not affected by collinearity between variables.

3.8. Principal Component Analysis

Principal component analysis (PCA) is suitable for populations of high-dimensional variables with a certain correlation between samples. There was collinearity in the above parameter set. Nevertheless, PCA is more suitable for removing collinearity [65]. The principal components express the additive combination with the variance of each sample. The linear combination is added when the current component is not enough to represent the information of the original parameter set. The principal component was calculated as follows:

$$F_P = a_{1i} \times Z_{x1} + a_{2i} \times Z_{x2} + \cdots + a_{pi} \times Z_{xp} \quad (2)$$

$$A = (a_{ij})p \times m = (a_1, a_2, \cdots a_m), R_{ai} = \gamma_i a_i. \quad (3)$$

The eigenvector corresponding to the covariance matrix is $a_{1i}, a_{2i} \cdots a_{pi} (i = 1, \cdots m)$. Z_{x1}, $ZX_2 Z_{x2}, \cdots$, Z_{xp} are the standard variables, R is the correlation coefficient matrix, and γ_i, a_i are the corresponding eigenvalue and eigenvectors.

We used IBM-SPSS 23.0 to perform principal component analysis on the dataset [66]. The two principal components were used to build a multivariate model to estimate AGB.

3.9. Verification and Prediction

Due to the small number of samples in this study, the cross-leave-one-out method was used for verification [67]. In total, 37 samples were used to model, and one sample was used for verification, resulting in 38 models. Through this method, the predicted AGB of SAR variables was obtained. Some of the evaluation indicators were used to describe the difference between the true AGB and the predicted AGB. The indicators selected in this study were goodness of fit (R^2), root mean square error (RMSE), mean absolute error (MAE), mean absolute percentage error (MARE), mean error (ME), and mean relative error (MRE) [68]. The best model was selected based on these indicators. The predicted biomass map was then obtained. The accuracy of the final model was referred to as the accuracy of the biomass map.

4. Results

4.1. Backscatter Coefficient and Its Combination

The correlation analysis between the backscatter coefficient and AGB (Table 3) showed that the horizontal cross-polarization in this region was more sensitive to AGB [9,11]. A univariate linear model was established between the backscatter coefficient and AGB (Figure 4). The accuracy of the model was low, and there were large deviations in estimating low-level AGB. The backscatter coefficient estimation of AGB was found to produce a saturation point, which led to greater limitations in the estimation results. The most accurate linear models were the σ_X and AGB. The formula of this model was AGB $= 280.394 + 12.591 \times \sigma_X$.

Table 3. AGB backscatter coefficients correlation analysis.

Correlation Coefficient	Backscatter Coefficients		
	σ_{HH}	σ_X	σ_{VV}
Person coefficient	0.497 **	0.680 **	0.425 **

** Statistical significance: Statistical significance represents a significant correlation between the variables.

The backscatter coefficients were combined to determine whether they had the potential to improve AGB estimation (Table 4).

In this study, the combination of poorly correlated backscatter coefficients was not significant. The correlation of the three significantly correlated and newly combined variables ($\sigma_{HH\times VV}$, $\sigma_{HH\times X}$ and $\sigma_{HH\times X\times VV}$) was better than the σ_{HH} and σ_{VV} polarization channels. A univariate model with AGB was established for the three new variables (Figure 5). The most accurate were the $\sigma_{HH\times X\times VV}$ and AGB linear models. The formula of this model was AGB $= 152.822 + 0.039 \times \sigma_{HH\times X\times VV}$. However, it was found that the accuracy of these models was the same as that of the backscatter coefficient model. The saturation point of this model was about 125 t/ha.

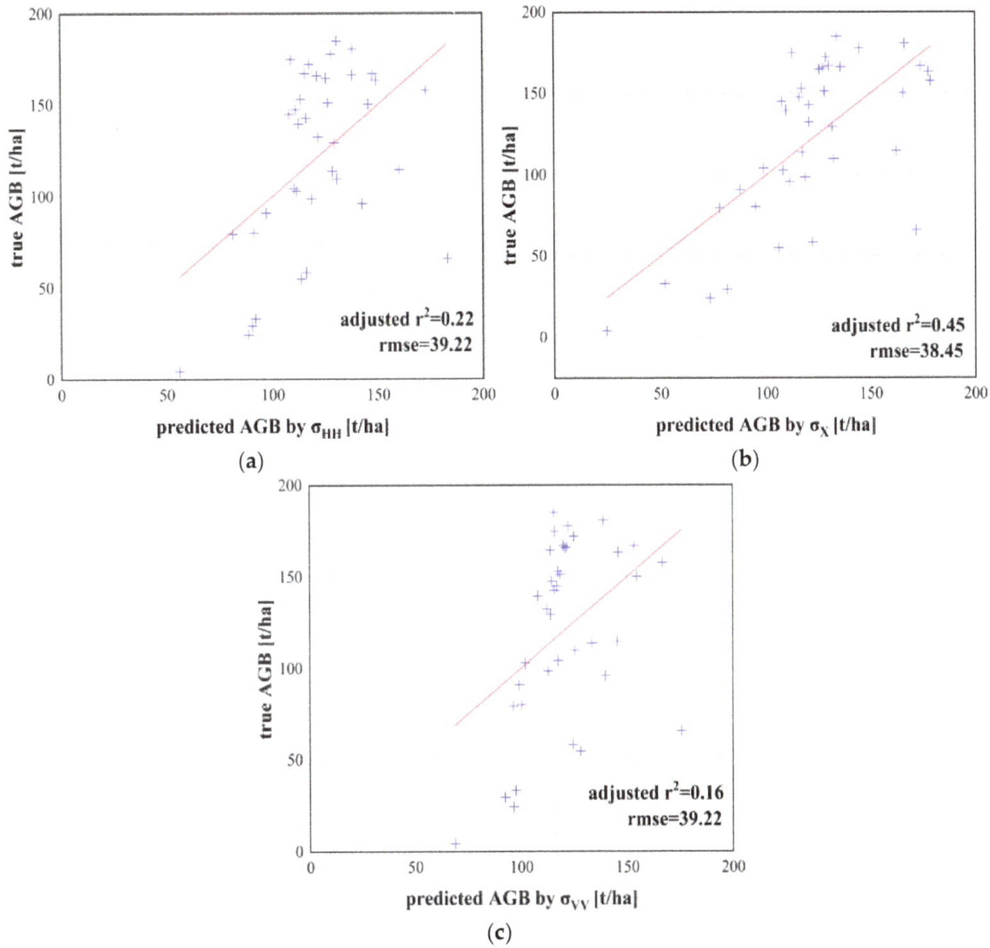

Figure 4. (**a**) Models established by σ_{HH} and AGB; (**b**) models established by σ_X and AGB; (**c**) models established by σ_{VV} and AGB.

Table 4. AGB backscatter coefficients correlation analysis.

Parameter	Pearson Coefficient	Parameter	Pearson Coefficient	Parameter	Pearson Coefficient
$\sigma_{HH/VV}$	−0.154	$\sigma_{(VV+X)/HH}$	−0.060	σ_{VV+X}	0.102
$\sigma_{HH/X}$	−0.200	$\sigma_{(VV+X)/X}$	−0.093	σ_{HH-VV}	−0.217
$\sigma_{VV/X}$	−0.093	$\sigma_{(VV+X)/VV}$	0.144	σ_{HH-X}	0.243
$\sigma_{(HH+VV)/HH}$	0.043	$\sigma_{(HH+VV+X)/HH}$	−0.060	σ_{X-VV}	−0.263
$\sigma_{(HH+VV)/VV}$	−0.154	$\sigma_{(HH+VV+X)/X}$	0.143	$\sigma_{HH-VV-X}$	−0.232
$\sigma_{(HH+VV)/X}$	−0.148	$\sigma_{(HH+VV+X)/VV}$	−0.148	$\sigma_{HH\times VV}$	−0.637 **
$\sigma_{(HH+X)X}$	−0.200	σ_{HH+VV}	0.285	$\sigma_{X\times HH}$	−0.631 **
$\sigma_{(HH+X)/HH}$	−0.066	σ_{HH+X}	0.117	$\sigma_{HH\times X\times VV}$	0.666 **
$\sigma_{(HH+X)/VV}$	0.143	$\sigma_{HH+VV+X}$	0.244		

** Statistical significance: Statistical significance represents a significant correlation between the variables.

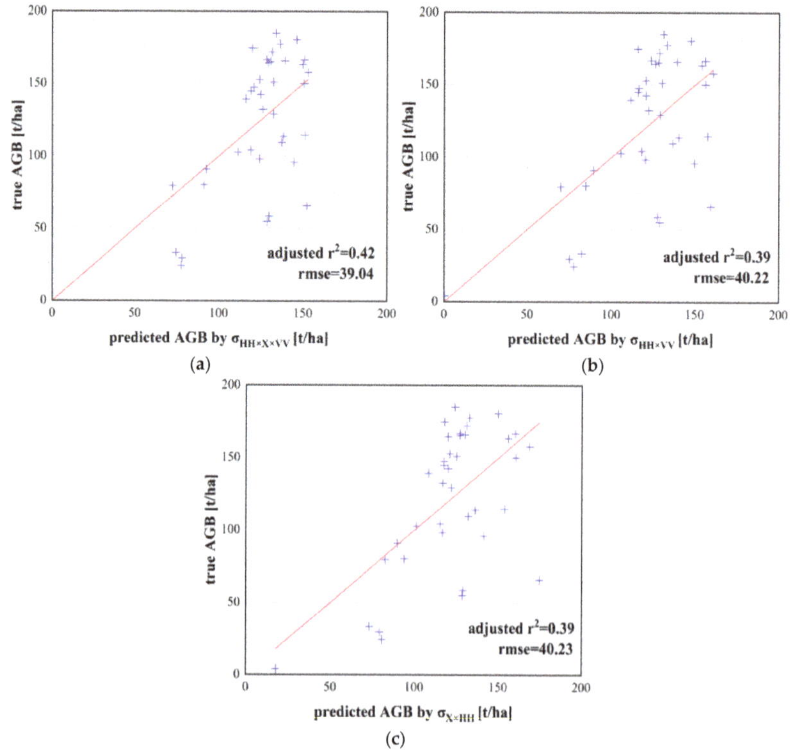

Figure 5. (**a**) Models established by $\sigma_{HH \times X \times VV}$ and AGB; (**b**) models established by $\sigma_{HH \times VV}$ and AGB; (**c**) models established by $\sigma_{X \times HH}$ and AGB.

4.2. Influence of Topographical Factors

Correlation analysis revealed that slope was the most important factor affecting AGB in this study (Table 5). During the field investigation, the slope of the study area was found to change greatly. However, the effect of aspect on AGB was not obvious, possibly due to the small number of samples.

Table 5. Topographical factors—AGB correlation analysis.

Parameters	Topographical Factors		
	Slope	**Elevation**	**Aspect**
Pearson coefficient	0.417 **	0.162	0.223

** Statistical significance: Statistical significance represents a significant correlation between the variables.

4.3. New Parameters and AGB Estimation

The correlation between the three newly constructed parameters and AGB was determined, and it is illustrated in Table 6. The univariate model with R_1 as the independent variable produced better AGB estimation results. The formula of this model was $AGB = 181.427 - 3.822 \times R_1$. This model achieved the highest accuracy among all univariate models. The new parameters R_2 and R_3 predicted that the AGB results were poor (Figure 6). The saturation point of the model was relatively high (140 t/ha) when compared to the backscattering coefficient model.

Table 6. Parameter correlation analysis.

Correlation Coefficient	New Parameter		
	R_1	R_2	R_3
Pearson coefficient	−0.756 **	−0.322	0.190

** Statistical significance: Statistical significance represents a significant correlation between the variables.

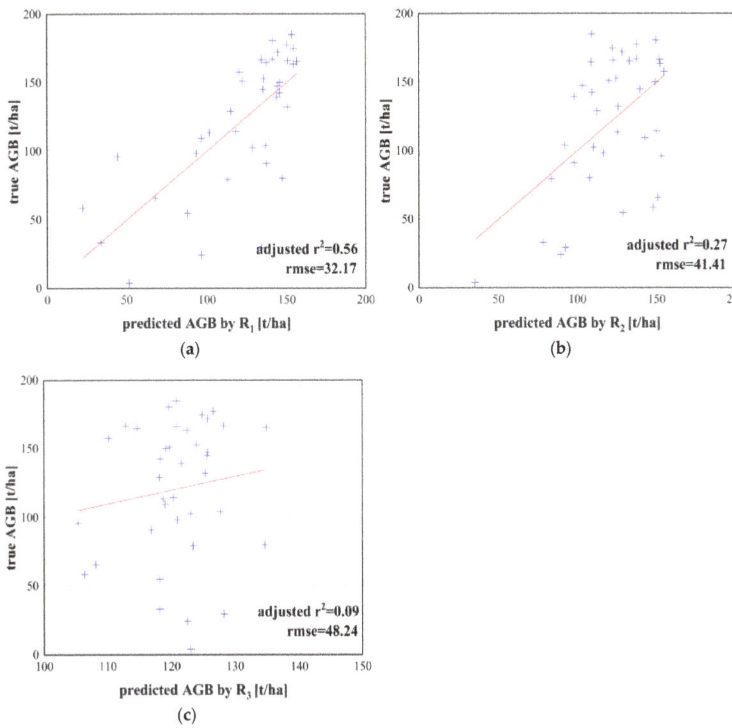

Figure 6. (**a**) Models established by R_1 and AGB; (**b**) models established by R_2 and AGB; (**c**) models established by R_3 and AGB.

4.4. Multivariate Linear Model

A correlation analysis of the polarization decomposition variables and AGB was performed, and it is summarized in Table 7 (only relevant significant variables are displayed). The variable R_1 had a stronger correlation with AGB compared to the Yamaguchi scattering mechanism. This showed that there was still a relationship between the mechanisms of polarization decomposition.

Table 7. AGB decomposition parameter correlation analysis.

Correlation Coefficient	Decomposition Parameter				
	Entropysh	Entropy 1	Entropy 2	Entropy 3	Gamma 3
Pearson coefficient	0.672 **	0.596 **	0.617 **	0.696 **	0.439 **
	I2	I3	$\text{Yamaguchi}_{\text{Vol}}$	$\text{Yamaguchi}_{\text{Dbl}}$	
Pearson coefficient	0.667 **	0.635 **	0.623 **	0.697 **	

** Statistical significance: Statistical significance represents a significant correlation between the variables.

All the variables that were significantly correlated with AGB were used to build a multivariate linear model (Figure 7). These variables included three original backscatter coefficients, three combined backscatter coefficients, slope, nine polarization decomposition variables, and R_1. However, the VIF test proved that there was multicollinearity between them (Table 8). The joint hypothesis F value of the model was 17.536, and there was no significant saturation point (sig) ≤ 0.001. This model provided poor residual test results (Figure 7). Meanwhile, existing studies have showed that AGB cannot be estimated with a simple multivariate model [69,70]. Although this model provided reasonable results, it could not predict AGB in large areas. The formula of this model was $AGB = -412.481 - 1.992 \times \sigma_{HH\times VV} - 612.789 \times \text{Yamaguchi}_{vol} + 279.684 \times \text{Yamaguchi}_{Dbl} - 819.198 \times \text{Entropy } 2 + 1359.506 \times I3 + 1693.609 \times I2 + 1051.944 \times \text{Entropy } 3 + 123.159 \times \text{Entropy } 1 - 196.097 \times \text{Entropysh} - 1.167 \times R_1 + 0.062 \times \text{Gamma } 3 - 3.714 \times \sigma_{HH\times X} - 49.921 \times \sigma_{HH} - 5.658 \times \sigma_X - 412.481 \times \sigma_{VV} - 0.154 \times \sigma_{HH\times X\times VV}$.

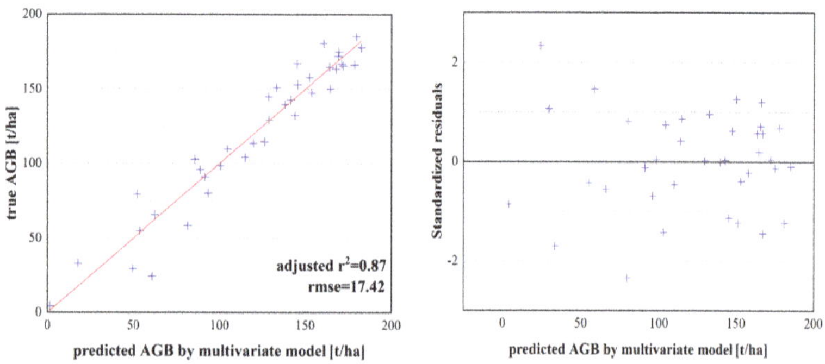

(a) (b)

Figure 7. (**a**) Models established by all the variables that were significantly correlated; (**b**) residual analysis graph.

Table 8. Collinearity analysis.

Variable	Dimension	Sig	Vif
σ_{HH}	1	0.607	84.422
σ_X	2	0.251	689.347
σ_{VV}	3	0.925	150.481
$\sigma_{HH\times VV}$	4	0.150	346.537
$\sigma_{HH\times X}$	5	0.448	2571.564
$\sigma_{HH\times X\times VV}$	6	0.126	444.793
entropysh	7	0.656	1798.225
entropy 1	8	0.648	218.010
entropy 2_2	9	0.070	335.184
entropy 3_3	10	0.015	153.786
gamma 3	11	0.648	1.585
I2	12	0.535	295.838
I3	13	0.935	792.565
R_1	14	0.999	17.052
Yamaguchi_{Vol}	15	0.898	1125.782
Yamaguchi_{Dbl}	16	0.531	59.219

4.5. Ridge Regression Model

Ridge regression was used to estimate AGB and solve the collinearity problem. Variables in the ridge regression model were consistent with the multivariate model. The variables were standardized before ridge regression. However, the ridge regression model was found to have a poor fitting effect (Figure 8). This method solved the collinearity between the variables. However, the normality of the residuals of the model was poor. Based on these results, the model estimated that the AGB saturation point was low (~145 t/ha). We determined the ridge parameter (K) = 0.141 based on the variance expansion factor method. The formula of this model was $AGB = 83.396 - 0.161 \times \sigma_{HH \times VV} - 12.726 \times Yamaguchi_{vol} + 405.828 \times Yamaguchi_{Dbl} + 2.125 \times Entropy\ 2 - 13.587 \times I3 + 78.432 \times I2 + 50.428 \times Entropy\ 3 - 1.316 \times Entropy\ 1 + 0.805 \times Entropysh - 1.737 \times R_1 + 0.077 \times Gamma\ 3 - 0.065 \times \sigma_{HH \times X} - 1.446 \times \sigma_{HH} + 1.366 \times \sigma_X - 2.56 \times \sigma_{VV} + 0.009 \times \sigma_{HH \times X \times VV}$.

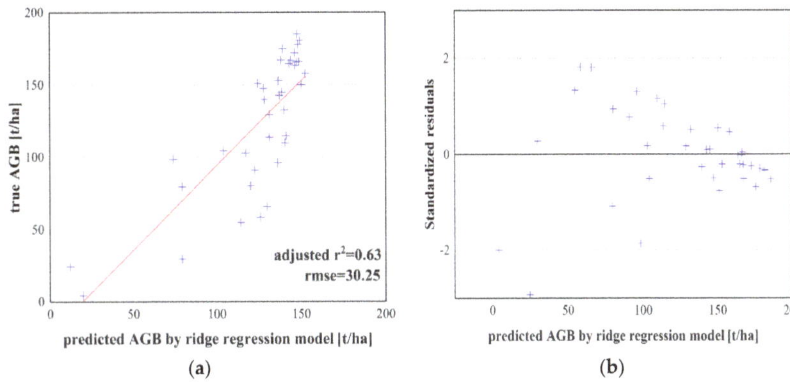

Figure 8. (**a**) Ridge regression model; (**b**) residual analysis graph.

4.6. Random Forest

There were 200 decision trees in this dataset. In order to reduce the result volatility caused by bootstrap sampling, all the models were trained 50 times and the average was obtained. We obtained the predicted AGB and computed the residuals (Figure 9). The saturation point of this model was about 155 t/ha.

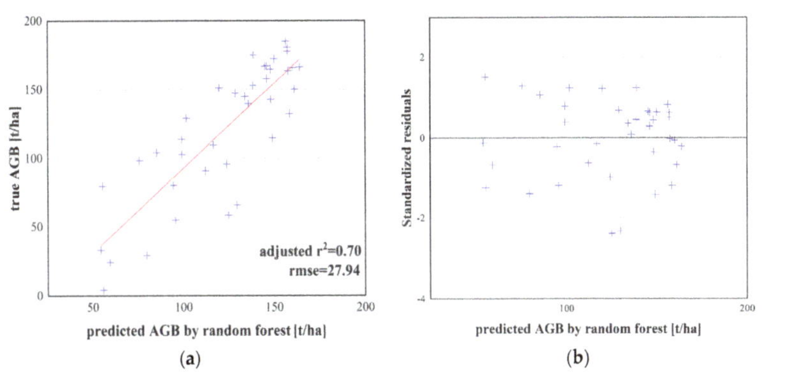

Figure 9. (**a**) Random forest model; (**b**) residual analysis graph.

4.7. Principal Component Analysis

Although the above results showed that multiple variables had the ability to improve AGB estimation, obtaining a stable model was still a challenge. However, the problem of collinearity can be solved through PCA. This method uses the same variable set as ridge regression. PCA processing was performed on the new parameter set. In addition, the Kaiser–Meyer–Olkin and Bartlett's tests were performed (Table 9), and the suitability of Kaiser–Meyer–Olkin sampling was between 0 and 1. A larger value indicated that it was convenient for PCA. Bartlett's spherical significance test showed that the selection of the parameter population was suitable for PCA. We calculated the principal components by the principal component coefficients (Table 10). Two principal component variables were extracted by default. The cumulative variance described 80.606% of the original parameters (Table 11). Additionally, the AGB was estimated using a multivariate model (Figure 10). The formula of this model was AGB $= 120.895 + 36.028 \times$ Factor1 $- 31.266 \times$ Factor2. The F value of the multiple regressions model was 164.421, sig ≤ 0.001 (Table 12), and it passed the significance test, indicating a significant improvement in accuracy when compared with other multivariate models. PCA solved the problem of collinearity and the residuals of the model were normal. The residuals were evenly distributed, and thus the variance was considered to be homogeneous (Figure 9). All the results obtained through this model were acceptable.

Table 9. Kaiser–Meyer–Olkin and Bartlett's test.

Kaiser–Meyer–Olkin Sampling Suitability		**0.746**
Bartlett's Test	Approximated chi-square	1753.496
	Degree of freedom	136
	Significance	0.000

Table 10. Principal component coefficient.

Variable	**Principal Component Coefficient**	
	Factor 1	**Factor 2**
σ_{HH}	0.904	−0.310
σ_{X}	0.962	−0.305
σ_{VV}	0.900	−0.331
$\sigma_{HH \times VV}$	−0.915	0.237
$\sigma_{HH \times X}$	−0.933	0.209
$\sigma_{HH \times X \times VV}$	0.877	−0.185
entropysh	0.982	0.053
entropy 1	0.972	−0.056
entropy 2_2	0.966	0.082
entropy 3_3	0.953	−0.002
gamma 3	0.227	0.458
I2	0.924	0.240
I3	0.905	0.239
R_1	−0.349	−0.795
Yamaguchi_{Vol}	0.900	0.229
Yamaguchi_{Dbl}	0.847	0.287

Table 11. Illustration of the total variance.

Component	**Eigenvalue**			**Cumulative**		
	Aggregate	**Variance (%)**	**Total (%)**	**Aggregate**	**Variance (%)**	**Total (%)**
Factor 1	12.161	71.538	71.538	12.161	71.538	71.538
Factor 2	1.541	9.068	80.606	1.541	9.068	80.606

Table 12. Regression coefficients of the model.

Parameter	**Unstandardized Coefficient**	**Student's Test Value**	**Sig**	**Vif**
Constant	112.635	39.083	0.000	
Factor 1	34.427	10.982	0.000	1.000
Factor 2	−29.648	−9.458	0.000	1.000

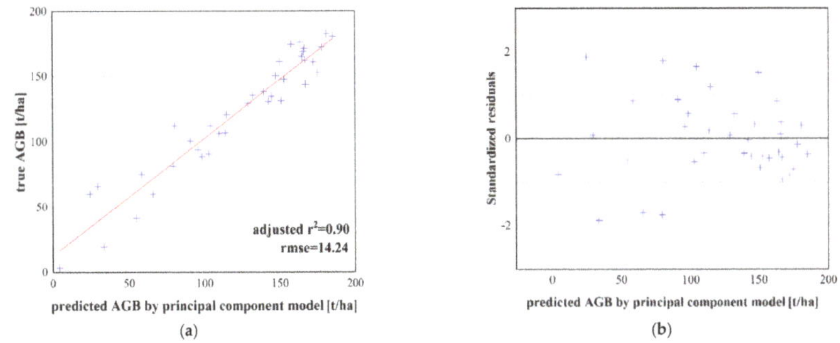

Figure 10. (**a**) Models established by principal component analysis; (**b**) residual analysis graph.

4.8. Model Selection and Prediction of the Study Area AGB

We evaluated the relationship between the predicted AGB of each model and the real AGB (Table 13). This study concluded that the principal component model was the best model when compared with the evaluation indexes of the models. As such, the principal component model was used to predict the AGB in the study area (Figure 11). All the variable matrices required by the principal components were entered in Matlab2014b. All the matrices were additively combined according to the principal component coefficients to obtain the principal components. Finally, large-area AGB prediction was based on the multivariate model formula of the principal component. The accuracy of the biomass map was 88%.

Table 13. Model evaluation index.

Type	Model	Evaluation Index					
		Adjusted r^2	RMSE (t/ha)	ME (t/ha)	MAE (t/ha)	MARE (%)	MRE (%)
Unary model	σ_X and AGB model	0.45	38.45	1.17	37.22	45.51	17.71
	$\sigma_{HH\times X\times VV}$ and AGB model	0.42	39.04	3.15	32.94	44.30	21.25
	R_1 and AGB model	0.56	32.17	1.82	25.69	39.17	8.77
	Multivariate model	0.87	17.42	−0.12	14.82	27.35	15.25
	Ridge regression model	0.63	30.25	−0.04	23.18	36.22	17.32
	Principal component model	0.90	14.24	−0.023	10.96	18.92	5.03
	Random forest model	0.70	27.94	−1.99	23.04	23.21	17.44

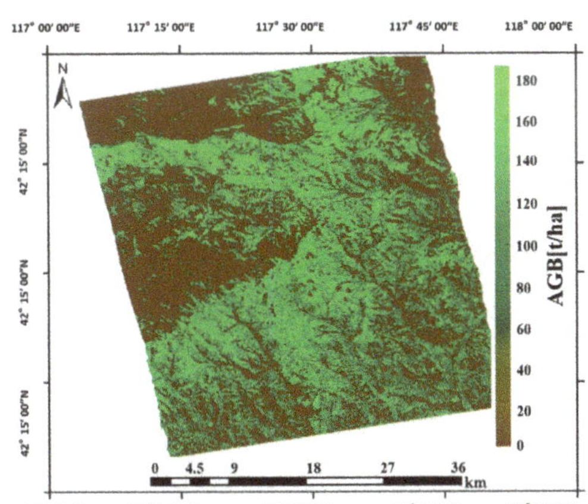

Figure 11. AGB prediction results.

5. Discussion

The use of large-scale remote sensing analyses to accurately estimate AGB is of great significance to global carbon-neutrality research. This study used a combination of backscatter coefficients, terrain parameters, and polarization decomposition parameters to estimate AGB. The adjusted r^2 increased from 0.45 to 0.90 through different processes. The accuracy of the processing results increased with the progress in various steps. The RMSE was 14.24 t/ha, as shown in step 4.6. Meanwhile, the ground information carried by the backscatter coefficient was limited. Therefore, AGB could not be accurately estimated. It was found that the combination of backscatter coefficients was more effective than HH and VV polarizations. Given that the topography of the study area was relatively complex, the slope was considered as an important factor affecting AGB. This study selected the polarization decomposition parameters that were suitable for forest AGB estimation. The univariate and the multivariate models were then compared. The results showed that the multivariate model estimates a high saturation point of the AGB. The saturation point of the backscattering coefficient was about 120 t/ha, and no obvious saturation point was estimated by the multivariate model. The saturation point of the variable R_1 was about 160 t/ha, which was higher than the backscattering coefficient model. This showed that the polarization decomposition parameters carried more ground information. The principal component analysis was found to be more suitable for the collinearity variable sets. As such, we used two principal components to build a multivariate model for estimating AGB without collinearity. No saturation point was found in this model, suggesting that the saturation point had been effectively improved. Finally, the principal component multivariate model was used to predict the AGB in our study area.

Previous reports have shown that the accuracy of the non-parametric model and the linear model is consistent in estimating AGB [71]. The accuracy of the multivariate model was proved to be higher than that of the univariate model [2]. Our findings were consistent with these previous findings. In the backscatter coefficient, cross-polarization had the strongest correlation with AGB [9,11]. The backscatter coefficient previously estimated the saturation point of AGB to be about 100 t/ha [17,18]. However, the saturation point of the backscattering coefficients in this study was higher than that previously reported. Many factors have been found to affect the saturation point of AGB [21]. For instance, variation in environmental conditions in a given study area was found to play a key role in causing variation in the saturation point [19]. Therefore, the different environments in our study area could have affected the saturation point. Reports have shown that the ratio of polarization backscattering coefficients has a high correlation with AGB [11]. However, the present study could not verify this finding. There is a possibility that no suitable combination of backscattering coefficients was found. Among several polarization decomposition methods, the most relevant parameter was $\text{Yamaguchi}_{\text{Dbl}}$. $\text{Yamaguchi}_{\text{Dbl}}$ represented the secondary scattering between forest trunks, and 90% of the forest AGB was tree trunks. These results are consistent with previously reported L-band characteristics [72]. Meanwhile, the Yamaguchi three-component decomposition corresponded to the physical model. In some aspects, the performance was better than the characteristic decomposition parameters. The several scattering mechanisms of the Yamaguchi three-component decomposition make it impossible to correctly distinguish land units. However, H/A/alpha decomposition provides a different decomposition method. The H/A/alpha polarization decomposition theorem is based on the coherence matrix analysis of eigenvalues and eigenvectors. The decomposed parameters describe the main relationship between the scattering mechanisms [45]. This suggests a lack of conflict between the two polarization decomposition methods. Previous studies showed a certain connection between the scattering mechanisms [64]. This finding is supported by the variable R_1 reported in the present study. In the present study, the RF method was found to be less effective and unsuitable for small sample studies. This was consistent with previous reports [39]. The additive model was found to be suitable for estimating AGB, as previously reported [27–29].

The L-band was used to obtain higher accuracy. This study showed that SAR data have the potential to estimate AGB, and are not limited by the saturation point of the estimated AGB. The unary model proved to be unsuitable for accurately estimating AGB. The multivariate model was proved to have a higher saturation point. However, the results estimated by principal component analysis were the closest to the real AGB. This study showed that AGB could be accurately estimated by one SAR image. However, the present research did not achieve such results; the obtained results were more suitable for the prediction of AGB in large areas.

Future studies could, however, use more penetrating P-band data, and select study areas with high biomass levels, such as tropical rain forests. Such studies may therefore enhance the ability to estimate AGB based on model selection. The present study employed an estimation technique that obtains better accuracy at the biomass level. It proved that estimating AGB using a combination of long-wave SAR data parameters and non-remote sensing factors can address actual needs. Although previous studies investigated polarization decomposition methods, the present study obtained better results due to the choice of images and reasonable processing methods used.

6. Conclusions

This study investigated the effectiveness of using an L-band image to estimate AGB. The obtained results could be widely applied to estimate AGB. The problem of complex radar signals that generate high-dimensional parameter sets was also addressed, further emphasizing the wide applicability of the method. The key findings of this study were as follows:

(1) The use of the backscatter coefficient to estimate AGB was more limited. The multivariate model provided better estimation capabilities than the univariate model. However, there was collinearity among the variables.
(2) The backscatter coefficient estimated that the AGB saturation point was low. The variable R_1 improved the estimation of the saturation point.
(3) The Earth-scattering ratio was more suitable for estimating AGB. This indicated that there was a degree of information complementarity between the variables. The combined backscatter coefficient was weak at estimating AGB.
(4) The model established by combining the backscatter coefficients, terrain factors, and polarization decomposition parameters achieved high accuracy. The principal component analysis method was suitable for analyzing SAR data to estimate AGB. The final model effectively improved the saturation point of AGB.

It is noteworthy that the study did not require a large amount of SAR data to accurately estimate AGB. L-band PALSAR data can be used in most areas of the world, making this research widely applicable to the estimation of AGB in forest-covered areas. However, the level of AGB in this study was not among the highest recorded in the world. Therefore, the applicability of this method in areas with high biomass levels is still uncertain.

Author Contributions: W.F. and Z.L. developed the research, conceived of, and designed the experiments. Z.L., Y.M., Y.H. and G.W. collected data for analysis. Z.L. and O.O.M. led the writing of the manuscript. W.F., O.O.M. and Z.L. contributed to the editing, revision, and discussion of the paper. All authors contributed critically to the drafts and gave final approval for publication. All authors have read and agreed to the published version of the manuscript.

Funding: This work was supported by the National Natural Science Foundation of China (Contract No. 31971654) and the Civil Aerospace Technology pre research project (Contract No. D040114).

Acknowledgments: We thank JAXA and the Kyoto & Carbon Initiative for providing ALOS-2 PALSAR-2 data. We also thank the Northeast Forestry University (NEFU) for logistic support during the field surveys.

Conflicts of Interest: The authors declare no conflict of interest. The founding sponsors had no role in the design of the study; in the collection, analyses, or interpretation of data; in the writing of the manuscript; or in the decision to publish the results.

References

1. Metcalf, C.J.E.; Graham, A.L.; Huijben, S.; Barclay, V.C.; Long, G.H.; Grenfell, B.T.; Read, A.F.; Bjørnstad, O.N. Partitioning Regulatory Mechanisms of Within-Host Malaria Dynamics Using the Effective Propagation Number. *Science* **2011**, *333*, 984–988. [CrossRef] [PubMed]
2. Tanase, M.A.; Panciera, R.; Lowell, K. Airborne multi-temporal L-band polarimetric SAR data for biomass estimation in semi-arid forests. *Remote Sens. Environ.* **2014**, *145*, 93–104. [CrossRef]
3. Blomberg, E.; Ferro-Famil, L.; Soja, M.J. Forest Biomass Retrieval From L-Band SAR Using Tomographic Ground Backscatter Removal. *IEEE Geosci. Remote Sens. Lett.* **2018**, *15*, 1030–1034. [CrossRef]
4. Sinha, S.; Jeganathan, C.; Sharma, L.K. Nathawat, M.S.; Das, A.K.; Mohan, S. Developing synergy regression models with space-borne ALOS PALSAR and Landsat TM sensors for retrieving tropical forest biomass. *J. Earth Syst. Sci.* **2016**, *125*, 725–735. [CrossRef]
5. Pulliainen, J.T.; Kurvonen, L.; Hallikainen, M.T. Multitemporal behavior of L- and C-band SAR observations of boreal forests. *IEEE Trans. Geosci. Remote Sens.* **1999**, *37*, 927–937. [CrossRef]
6. Baghdadi, N.; le Maire, G.; Bailly, J.-S.; Ose, K.; Nouvellon, Y.; Zribi, M.; Lemos, C.; Hakamada, R. Evaluation of ALOS/PALSAR L-Band Data for the Estimation of Eucalyptus Plantations Aboveground Biomass in Brazil. *IEEE J. Sel. Top. Appl. Earth Obs. Remote Sens.* **2015**, *8*, 3802–3811. [CrossRef]
7. Atwood, D.K.; Andersen, H.E.; Matthiss, B. Impact of Topographic Correction on Estimation of Aboveground Boreal Biomass Using Multi-temporal, L-Band Backscatter. *IEEE J. Sel. Top. Appl. Earth Obs. Remote Sens.* **2014**, *7*, 3262–3273. [CrossRef]
8. Caicoya, A.T.; Pardini, M.; Hajnsek, I.; Papathanassiou, K. Forest Above-Ground Biomass Estimation from Vertical Reflectivity Profiles at L-Band. *IEEE Geosci. Remote Sens. Lett.* **2015**, *12*, 2379–2383. [CrossRef]
9. Le Toan, T.; Beaudoin, A.; Riom, J.; Guyon, D. Relating forest biomass to SAR data. *IEEE Trans. Geosci. Remote Sens.* **1992**, *30*, 403–411. [CrossRef]
10. Rignot, E.; Way, J.; Williams, C.; Viereck, L. Radar estimates of aboveground biomass in boreal forests of interior Alaska. *IEEE Trans. Geosci. Remote Sens.* **1994**, *32*, 1117–1124. [CrossRef]
11. Ranson, K.; Sun, G. Mapping biomass of a northern forest using multifrequency SAR data. *IEEE Trans. Geosci. Remote Sens.* **1994**, *32*, 388–396. [CrossRef]
12. Eini-Zinab, S.; Maghsoudi, Y.; Sayedain, S.A. Assessing the performance of indicators resulting from three-component Freeman–Durden polarimetric SAR interferometry decomposition at P-and L-band in estimating tropical forest aboveground biomass. *Int. J. Remote Sens.* **2019**, *41*, 433–454. [CrossRef]
13. Golshani, P.; Maghsoudi, Y.; Sohrabi, H. Relating ALOS-2 PALSAR-2 Parameters to Biomass and Structure of Temperate Broadleaf Hyrcanian Forests. *J. Indian Soc. Remote Sens.* **2019**, *47*, 749–761. [CrossRef]
14. Asari, N.; Suratman, M.N.; Jaafar, J. Modelling and mapping of above ground biomass (AGB) of oil palm plantations in Malaysia using remotely-sensed data. *Int. J. Remote Sens.* **2017**, *38*, 4741–4764. [CrossRef]
15. Trisasongko, B.H.; Paull, D. A review of remote sensing applications in tropical forestry with a particular emphasis in the plantation sector. *Geocarto Int.* **2018**, *35*, 317–339. [CrossRef]
16. Ranson, K.J.; Sun, G.; Weishampel, J.F.; Knox, R.G. Forest biomass from combined ecosystem and radar backscatter modeling. *Remote Sens. Environ.* **1997**, *59*, 118–133. [CrossRef]
17. Kasischke, E.S.; Melack, J.M.; Dobson, M.C. The use of imaging radars for ecological applications—A review. *Remote Sens. Environ.* **1997**, *59*, 141–156. [CrossRef]
18. Yu, Y.; Saatchi, S. Sensitivity of L-Band SAR Backscatter to Aboveground Biomass of Global Forests. *Remote Sens.* **2016**, *8*, 522. [CrossRef]
19. Sarker, M.L.R.; Nichol, J.; Ahmad, B.; Busu, I.; Rahman, A.A. Potential of texture measurements of two-date dual polarization PALSAR data for the improvement of forest biomass estimation. *ISPRS J. Photogramm. Remote Sens.* **2012**, *69*, 146–166. [CrossRef]
20. Lal, P.; Kumar, A.; Saikia, P.; Das, A.; Patnaik, C.; Kumar, G.; Pandey, A.C.; Srivastava, P.; Dwivedi, C.S.; Khan, M.L. Effect of vegetation structure on above ground biomass in tropical deciduous forests of Central India. *Geocarto Int.* **2021**, *1*, 1–17. [CrossRef]
21. Baig, S.; Qazi, W.A.; Akhtar, A.M.; Waqar, M.M.; Ammar, A.; Gilani, H.; Mehmood, S.A. Above Ground Biomass Estimation of *Dalbergia sissoo* Forest Plantation from Dual-Polarized ALOS-2 PALSAR Data. *Can. J. Remote Sens.* **2017**, *43*, 297–308. [CrossRef]
22. Freeman, A.; Durden, S.L. A three-component scattering model for polarimetric SAR data. *IEEE Trans. Geosci. Remote Sens.* **1998**, *36*, 963–973. [CrossRef]
23. Zhang, Z.; Wang, Y.; Sun, G.; Ni, W.; Huang, W.; Zhang, L. Biomass Retrieval Based on Polarimetric Target Decomposition. In Proceedings of the IEEE International Geoscience and Remote Sensing Symposium, Vancouver, BC, Canada, 24–29 July 2011; pp. 1942–1945.
24. Bharadwaj, P.S.; Kumar, S.; Kushwaha, S.; Bijker, W. Polarimetric scattering model for estimation of above ground biomass of multilayer vegetation using ALOS-PALSAR quad-pol data. *Phys. Chem. Earth Parts A/B/C* **2015**, *83-84*, 187–195. [CrossRef]

25. Van Zyl, J.J.; Arii, M.; Kim, Y. Model-Based Decomposition of Polarimetric SAR Covariance Matrices Constrained for Nonnegative Eigenvalues. *IEEE Trans. Geosci. Remote Sens.* **2011**, *49*, 3452–3459. [CrossRef]
26. Verma, A.; Haldar, D. SAR polarimetric analysis for major land covers including pre-monsoon crops. *Geocarto Int.* **2019**, *36*, 2224–2240. [CrossRef]
27. Townsend, P.A. Estimating forest structure in wetlands using multitemporal SAR. *Remote Sens. Environ.* **2002**, *79*, 288–304. [CrossRef]
28. Santos, J.R.; Freitas, C.C.; Araujo, L.S.; Dutra, L.V.; Mura, J.C.; Gama, F.F.; Soler, L.S.; Sant'Anna, S.J. Airborne P-band SAR applied to the aboveground biomass studies in the Brazilian tropical rainforest. *Remote Sens. Environ.* **2003**, *87*, 482–493. [CrossRef]
29. Mougin, E.; Proisy, C.; Marty, G.; Fromard, F.; Puig, H.; Betoulle, J.; Rudant, J. Multifrequency and multipolarization radar backscattering from mangrove forests. *IEEE Trans. Geosci. Remote Sens.* **1999**, *37*, 94–102. [CrossRef]
30. Hyde, P.; Nelson, R.; Kimes, D.; Levine, E. Exploring LiDAR–RaDAR synergy—predicting aboveground biomass in a southwestern ponderosa pine forest using LiDAR, SAR and InSAR. *Remote Sens. Environ.* **2007**, *106*, 28–38. [CrossRef]
31. Wolpert, D.H.; Macready, W.G. An Efficient Method to Estimate Bagging's Generalization Error. *Mach. Learn.* **1999**, *35*, 41–55. [CrossRef]
32. Fu, W.; Guo, H.; Li, X. Relating Forest Biomass to the Polarization Phase Difference of the Double-Bounce Scattering Component. *IEEE Geosci. Remote Sens. Lett.* **2020**, *18*, 2048–2051. [CrossRef]
33. Sharifi, A.; Amini, J. Forest biomass estimation using synthetic aperture radar polarimetric features. *J. Appl. Remote Sens.* **2015**, *9*, 097695. [CrossRef]
34. Waqar, M.M.; Sukmawati, R.; Ji, Y.Q.; Sumantyo, J.T.S.; Segah, H.; Prasetyo, L.B. Retrieval of Tropical Peatland Forest Biomass from Polarimetric Features in Central Kalimantan, Indonesia. *Prog. Electromagn. Res. C* **2020**, *98*, 109–125. [CrossRef]
35. Huang, X.; Ziniti, B.; Torbick, N.; Ducey, M.J. Assessment of Forest above Ground Biomass Estimation Using Multi-Temporal C-band Sentinel-1 and Polarimetric L-band PALSAR-2 Data. *Remote Sens.* **2018**, *10*, 1424. [CrossRef]
36. Tanase, M.A.; Panciera, R.; Lowell, K.; Hacker, J.; Walker, J.P. Estimation of forest biomass from L-band polarimetric decomposition components. In Proceedings of the 2013 IEEE International Geoscience and Remote Sensing Symposium—IGARSS, Melbourne, VIC, Australia, 21–26 July 2013; pp. 949–952.
37. Gleason, C.J.; Im, J. Forest biomass estimation from airborne LiDAR data using machine learning approaches. *Remote Sens. Environ.* **2012**, *125*, 80–91. [CrossRef]
38. Mutanga, O.; Adam, E.; Cho, M.A. High density biomass estimation for wetland vegetation using WorldView-2 imagery and random forest regression algorithm. *Int. J. Appl. Earth Obs. Geoinf.* **2012**, *18*, 399–406. [CrossRef]
39. Fassnacht, F.; Hartig, F.; Latifi, H.; Berger, C.; Hernández, J.; Corvalán, P.; Koch, B. Importance of sample size, data type and prediction method for remote sensing-based estimations of aboveground forest biomass. *Remote Sens. Environ.* **2014**, *154*, 102–114. [CrossRef]
40. Sun, Z.; Liu, L.; Peng, S. Age-related modulation of the nitrogen resorption efficiency response to growth requirements and soil nitrogen availability in a temperate pine plantation. *Ecosystems* **2016**, *19*, 689–709. [CrossRef]
41. Wang, C. Biomass allometric equations for 10 co-occurring tree species in Chinese temperate forests. *For. Ecol. Manag.* **2006**, *222*, 9–16. [CrossRef]
42. Shimada, M.; Isoguchi, O.; Tadono, T. PALSAR radiometric and geometric calibration. *IEEE Trans. Geosci. Remote Sens.* **2009**, *47*, 3915–3932. [CrossRef]
43. Lee, J.-S.; Grunes, M.; De Grandi, G. Polarimetric SAR speckle filtering and its implication for classification. *IEEE Trans. Geosci. Remote Sens.* **1999**, *37*, 2363–2373. [CrossRef]
44. Gatelli, F.; Guamieri, A.M.; Parizzi, F. The wavenumber shift in SAR interferometry. *IEEE Trans. Geosci. Remote Sens.* **1994**, *32*, 855–865. [CrossRef]
45. Varghese, A.O.; Suryavanshi, A.; Joshi, A.K. Analysis of different polarimetric target decomposition methods in forest density classification using C band SAR data. *Int. J. Remote Sens.* **2016**, *37*, 694–709. [CrossRef]
46. Van Zyl, J.J. Application of Cloude's Target Decomposition Theorem to Polarimetric Imaging Radar Data. In Proceedings of the SPIE Volume 1748, Radar Polarimetry, San Diego, CA, USA, 22 July 1992; SPIE: Bellingham, WA, USA, 1993. [CrossRef]
47. Freeman, A. Fitting a Two-Component Scattering Model to Polarimetric SAR Data from Forests. *IEEE Trans. Geosci. Remote Sens.* **2007**, *45*, 2583–2592. [CrossRef]
48. Yamaguchi, Y.; Moriyama, T.; Ishido, M.; Yamada, H. Four-component scattering model for polarimetric SAR image decomposition. *IEEE Trans. Geosci. Remote Sens.* **2005**, *43*, 1699–1706. [CrossRef]
49. Yamaguchi, Y.; Yajima, Y.; Yamada, H. A Four-Component Decomposition of POLSAR Images Based on the Coherency Matrix. *IEEE Geosci. Remote Sens. Lett.* **2006**, *3*, 292–296. [CrossRef]
50. Yamaguchi, Y.; Sato, A.; Boerner, W.-M.; Sato, R.; Yamada, H. Four-Component Scattering Power Decomposition with Rotation of Coherency Matrix. *IEEE Trans. Geosci. Remote Sens.* **2011**, *49*, 2251–2258. [CrossRef]
51. Cui, Y.; Yamaguchi, Y.; Yang, J.; Park, S.-E.; Kobayashi, H.; Singh, G. Three-Component Power Decomposition for Polarimetric SAR Data Based on Adaptive Volume Scatter Modeling. *Remote Sens.* **2012**, *4*, 1559–1572. [CrossRef]
52. Cloude, S.R.; Pottier, E. Concept of polarization entropy in optical scattering. *Opt. Eng.* **1995**, *34*, 1599–1610. [CrossRef]
53. Touzi, R. Target Scattering Decomposition in Terms of Roll-Invariant Target Parameters. *IEEE Trans. Geosci. Remote Sens.* **2007**, *45*, 73–84. [CrossRef]

54. Hyde, P.; Dubayah, R.; Walker, W.; Blair, J.B.; Hofton, M.; Hunsaker, C. Mapping forest structure for wildlife habitat analysis using multi-sensor (LiDAR, SAR/InSAR, ETM+, Quickbird) synergy. *Remote Sens. Environ.* **2006**, *102*, 63–73. [CrossRef]
55. Zheng, D.; Rademacher, J.; Chen, J.; Crow, T.; Bresee, M.; Le Moine, J.; Ryu, S.-R. Estimating aboveground biomass using Landsat 7 ETM+ data across a managed landscape in northern Wisconsin, USA. *Remote Sens. Environ.* **2004**, *93*, 402–411. [CrossRef]
56. Cloude, R.S.; Pottier, E. An entropy based classifification scheme for land applications of polarimetric SARs. *IEEE Trans. Geosci. Remote Sens.* **1997**, *35*, 68–78. [CrossRef]
57. Anconitano, G.; Lavalle, M.; Arabini, E. Sensitivity to soil moisture by applying a model-based polarimetric decomposition to a time-series of airborne radar L-band data over an agricultural area. *Microw. Remote Sens. Data Process. Appl.* **2021**, *11861*, 1186105. [CrossRef]
58. Simons-Legaard, E.; Legaard, K.; Weiskittel, A. Predicting aboveground biomass with LANDIS-II: A global and temporal analysis of parameter sensitivity. *Ecol. Model.* **2015**, *313*, 325–332. [CrossRef]
59. Mette, T.; Papathanassiou, K.; Hajnsek, I. Biomass estimation from polarimetric SAR interferometry over heterogeneous forest terrain. In Proceedings of the IEEE International Geoscience and Remote Sensing Symposium, Anchorage, AK, USA, 20–24 September 2004.
60. Sun, G.; Ranson, K.; Kharuk, V. Radiometric slope correction for forest biomass estimation from SAR data in the Western Sayani Mountains, Siberia. *Remote Sens. Environ.* **2002**, *79*, 279–287. [CrossRef]
61. Ojoyi, M.; Mutanga, O.; Odindi, J.; Abdel-Rahman, E.M. Application of topo-edaphic factors and remotely sensed vegetation indices to enhance biomass estimation in a heterogeneous landscape in the Eastern Arc Mountains of Tanzania. *Geocarto Int.* **2015**, *31*, 1–21. [CrossRef]
62. Aykut, T. Determination of groundwater potential zones using Geographical Information Systems (GIS) and Analytic Hierarchy Process (AHP) between Edirne-Kalkansogut (northwestern Turkey). *Groundw. Sustain. Dev.* **2021**, *12*, 100545. [CrossRef]
63. Lee, J.-S. Grunes, M.; Pottier, E.; Ferro-Famil, L. Unsupervised terrain classification preserving polarimetric scattering characteristics. *IEEE Trans. Geosci. Remote Sens.* **2004**, *42*, 722–731. [CrossRef]
64. Chowdhury, T.A.; Thiel, C.; Schmullius, C.; Stelmaszczuk-Górska, M. Polarimetric Parameters for Growing Stock Volume Estimation Using ALOS PALSAR L-Band Data over Siberian Forests. *Remote Sens.* **2013**, *5*, 5725–5756. [CrossRef]
65. Martínez, A.M.; Kak, A.C. PCA versus LDA. *IEEE Trans. Pattern Anal. Mach. Intell.* **2001**, *23*, 228–233. [CrossRef]
66. Castaño-Díaz, M.; Barrio-Anta, M.; Afif-Khouri, E.; Cámara-Obregón, A. Willow Short Rotation Coppice Trial in a Former Mining Area in Northern Spain: Effects of Clone, Fertilization and Planting Density on Yield after Five Years. *Forests* **2018**, *9*, 154. [CrossRef]
67. Wong, T.T. Performance evaluation of classification algorithms by k-fold and leave-one-out cross validation. *Pattern Recognit.* **2015**, *48*, 2839–2846. [CrossRef]
68. Quint, T.C.; Dech, J. Allometric models for predicting the aboveground biomass of Canada yew (*Taxus canadensis* Marsh.) from visual and digital cover estimates. *Can. J. For. Res.* **2010**, *40*, 2003–2014. [CrossRef]
69. Solberg, S.; Astrup, R.; Gobakken, T.; Næsset, E.; Weydahl, D.J. Estimating spruce and pine biomass with interferometric X-band SAR. *Remote Sens. Environ.* **2010**, *114*, 2353–2360. [CrossRef]
70. Wold, S.; Esbensen, K.; Geladi, P. Principal component analysis. *Chemom. Intell. Lab. Syst.* **1987**, *2*, 37–52. [CrossRef]
71. Neumann, M.; Saatchi, S.S.; Ulander, L.M.H.; Fransson, J.E.S. Assessing Performance of L- and P-Band Polarimetric Interferometric SAR Data in Estimating Boreal Forest Above-Ground Biomass. *IEEE Trans. Geosci. Remote Sens.* **2012**, *50*, 714–726. [CrossRef]
72. Kobayashi, S.; Omura, Y.; Sanga-Ngoie, K.; Widyorini, R.; Kawai, S.; Supriadi, B.; Yamaguchi, Y. Characteristics of Decomposition Powers of L-Band Multi-Polarimetric SAR in Assessing Tree Growth of Industrial Plantation Forests in the Tropics. *Remote Sens.* **2012**, *4*, 3058–3077. [CrossRef]

remote sensing

Article

Combining Sample Plot Stratification and Machine Learning Algorithms to Improve Forest Aboveground Carbon Density Estimation in Northeast China Using Airborne LiDAR Data

Mingjie Chen [1], Xincai Qiu [2], Weisheng Zeng [3] and Daoli Peng [1,*]

[1] State Forestry and Grassland Administration Key Laboratory of Forest Resources & Environmental Management, College of Forestry, Beijing Forestry University, Beijing 100083, China; mjchen@bjfu.edu.cn
[2] Intelligent Forestry Key Laboratory of Haikou City, College of Forestry, Hainan University, Haikou 570228, China; qiuxc9109@hainanu.edu.cn
[3] Academy of Inventory and Planning, National Forestry and Grassland Administration, Beijing 100714, China; zengweisheng@afip.com.cn
* Correspondence: dlpeng@bjfu.edu.cn

Abstract: Timely, accurate estimates of forest aboveground carbon density (AGC) are essential for understanding the global carbon cycle and providing crucial reference information for climate-change-related policies. To date, airborne LiDAR has been considered as the most precise remote-sensing-based technology for forest AGC estimation, but it suffers great challenges from various uncertainty sources. Stratified estimation has the potential to reduce the uncertainty and improve the forest AGC estimation. However, the impact of stratification and how to effectively combine stratification and modeling algorithms have not been fully investigated in forest AGC estimation. In this study, we performed a comparative analysis of different stratification approaches (non-stratification, forest type stratification (FTS) and dominant species stratification (DSS)) and different modeling algorithms (stepwise regression, random forest (RF), Cubist, extreme gradient boosting (XGBoost) and categorical boosting (CatBoost)) to identify the optimal stratification approach and modeling algorithm for forest AGC estimation, using airborne LiDAR data. The analysis of variance (ANOVA) was used to quantify and determine the factors that had a significant effect on the estimation accuracy. The results revealed the superiority of stratified estimation models over the unstratified ones, with higher estimation accuracy achieved by the DSS models. Moreover, this improvement was more significant in coniferous species than broadleaf species. The ML algorithms outperformed stepwise regression and the CatBoost models based on DSS provided the highest estimation accuracy (R^2 = 0.8232, RMSE = 5.2421, RRMSE = 20.5680, MAE = 4.0169 and Bias = 0.4493). The ANOVA of the prediction error indicated that the stratification method was a more important factor than the regression algorithm in forest AGC estimation. This study demonstrated the positive effect of stratification and how the combination of DSS and the CatBoost algorithm can effectively improve the estimation accuracy of forest AGC. Integrating this strategy with national forest inventory could help improve the monitoring of forest carbon stock over large areas.

Keywords: aboveground carbon density; LiDAR; stratified estimation; machine learning algorithm; Northeast China

Citation: Chen, M.; Qiu, X.; Zeng, W.; Peng, D. Combining Sample Plot Stratification and Machine Learning Algorithms to Improve Forest Aboveground Carbon Density Estimation in Northeast China Using Airborne LiDAR Data. *Remote Sens.* **2022**, *14*, 1477. https://doi.org/10.3390/rs14061477

Academic Editors: Huaqiang Du, Wenyi Fan, Mingshi Li, Weiliang Fan and Fangjie Mao

Received: 14 February 2022
Accepted: 16 March 2022
Published: 18 March 2022

1. Introduction

Covering about 30% of the earth land area, forest ecosystems are a huge global carbon reservoir with carbon stocks of about 861 ± 66 Pg C [1]. Over 80% of vegetation above-ground carbon in terrestrial ecosystems and more than 70% of global soil organic carbon are stored in forest ecosystems [2–4]. As carbon is naturally exchanged between forests and the atmosphere through photosynthesis, respiration, decomposition and combustion, forest ecosystems play a key role in the global carbon cycle [5–7]. To better understand and regulate the mechanisms of the global carbon cycle, we require accurate estimation and monitoring of forest aboveground carbon density (AGC). Forest AGC is an important indicator of the fundamental characteristics of forest ecosystems and the basis for evaluating the structural function and carbon sink capacity of forests [8,9]. Moreover, the current need to mitigate the impact of climate change on the global ecosystems raises the importance of quantifying forest carbon exchange and carbon stock from local to global scales [10,11]. Quantitative and accurate estimation of forest AGC is also required by many international climate change adaptation and mitigation policies (e.g., the United Nations Framework Convention on Climate Change (UNFCCC), the Kyoto Protocol, the Reducing Emissions from Deforestation and Forest Degradation (REDD+) program and the carbon neutrality policy) [12–15].

Traditionally, forest AGC is obtained through ground surveys, which are generally recognized to be the most accurate method [16,17]. However, the field-measured method is usually labor-intensive and time-consuming, and it is difficult to carry out at large scales or in remote areas [12]. The advent of remote-sensing technology, particularly light detection and ranging (LiDAR), has overcome these limitations to some extent. LiDAR technology is considered to be the most accurate remote-sensing-based estimation tool for forest aboveground biomass (AGB) and carbon stock [18]. As an active remote-sensing technology, LiDAR has the greatest advantage over other sensors in the ability to accurately capture the vertical structure information of forest vegetation, which plays an important role in forest AGC estimation. Due to its high spectral saturation point, LiDAR can also overcome the data saturation problem in optical and radar data. Metrics from LiDAR data (e.g., height and density) are highly correlated with forest AGB and AGC, and have been reported to provide good estimation results in several studies across various geographical areas [19–23]. To date, the most common approach to estimate forest AGC based on LiDAR data is achieved by establishing statistical regression models between LiDAR metrics and ground survey data. These regression models can be divided into two main categories: parametric and non-parametric algorithms. The parametric algorithms that have been widely used include multiple linear regression, stepwise regression, partial least squares regression, etc. Parametric algorithms have a clear model structure and strong interpretability of model parameters, but need to obey strict statistical assumptions and are hardly generalized. The non-parametric machine-learning algorithms, such as artificial neural networks (ANNs), support vector machines (SVMs), K-nearest neighbors (K-NNs), random forest (RF) and Cubist have attracted great interests in recent years [24–26]. Compared with parametric algorithms, non-parametric algorithms determine the model structure in a data-driven manner and are insensitive to noisy data. Due to the flexibility of non-parametric algorithms, they may be more suitable for modeling complex non-linear forest carbon-stock estimates [18]. Recently, two novel decision-tree-based ensemble algorithms, extreme gradient boosting (XGBoost) and categorical boosting (Catboost), have excelled in several machine-learning competitions and attracted much attention. Although XGBoost and CatBoost have outperformed other machine-learning algorithms in various fields [27–30], these two algorithms have rarely been used in forest AGC estimation, and the performance remains to be examined.

Stratified estimation is suggested to be an effective approach to reduce variance and improve the accuracy of estimates without increasing the sample size [31]. The main purpose of stratification is to group heterogeneous components within populations into strata so that the within-stratum variance will be significantly smaller than the overall variance, resulting in a better estimate result. This method has been proven to be useful in forest AGB estimation, and the stratification methods range from forest type and topography to site quality [32–34]. Among these methods, stratification based on forest type has been frequently used and has shown positive effects, as forest AGB and AGC vary with different stand structures and species composition [35]. However, other studies have reported only slight improvements when using forest-type stratification [36–38]. The mixed results raise the need for further research on the effects of stratification in forest AGC estimates and provide guidance for appropriate stratification methods. Moreover, limited by the number of sample plots, few studies have explored the effect of finer stratification (e.g., dominant species stratification) on forest AGC estimates.

The northeast forest region, together with the southern forest region and the southwest forest region, are known as the three major forest regions in China. As the largest natural forest area in China, the northeast forest region is the largest carbon pool, with its forest aboveground carbon stock reaching more than 1/4 of the country's total [39,40]. Despite that several studies have developed remote-sensing-based forest AGB estimation models at the regional level, forest-type level or species level in the northeast forest region [41,42], species-level carbon-stock estimation models based on LiDAR data in the northeast forest region of China have not been reported. Finer descriptions of forest ecosystems and structures, such as specific-species characteristics, are needed to meet the new challenges posed by forest management and monitoring [43]. Species-level information is of great value to support refined forest management and sustainable development. Moreover, the species-level approach makes full use of existing forest inventory information and avoids the additional costs of ground surveys.

Here, in order to fill the above gaps, we used airborne LiDAR data, stepwise regression and four machine learning algorithms (RF, Cubist, XGBoost and CatBoost) to develop forest AGC estimation models based on forest type and dominant species stratification in the northeastern forest regions of China. We hypothesized that the accuracy of forest AGC estimation can be substantially increased by combining finer stratification (dominant species stratification) and non-parametric machine learning algorithms. To examine this assumption, the performance of estimation models was compared (a) between stratification and non-stratification; (b) between forest type stratification (FTS) and dominant species stratification (DSS); (c) within the strata; and (d) between multiple stepwise regression and four machine learning algorithms, RF, Cubist, XGBoost and CatBoost. The objectives of this study were threefold: (1) to examine the effect of stratification on forest AGC estimation and explore appropriate stratification methods; (2) to evaluate the application of machine-learning algorithms in forest AGC estimation, especially the performance of the two novel decision tree-based ensemble algorithms, XGBoost and CatBoost; and (3) to establish species-level forest AGC estimation models in the northeastern forest regions of China.

2. Materials and Methods

2.1. Study Area

We conducted this study in the forest regions of Northeast China, across three provinces, Heilongjiang, Jilin and Inner Mongolia. The study area covered 12 areas in six forest regions (Figure 1), including the Daxinganling in Inner Mongolia, the Daxinganling in Heilongjiang, the Yichun, the Songhua River, the Mudanjiang and the Changbai Mountain (longitude 119°36′—134°05′E, latitude 41°37′—53°33′N). The climate in most of the region is temperate monsoon, with a cold monsoon climate in the north. The average annual precipitation ranges from 400 to 1000 mm, and the average annual temperature varies between −2 and 2.6 °C. The northeast forest region is surrounded by mountains to the east, north and west, with an average altitude distribution of 500–2500 m. The northeast forest region is one of the richest forest areas in China, with a total forest area of about 680,000 km^2 and a total forest volume stock of about 3.2 billion m^3, accounting for 37% of the country's total forest area and 30% of the country's total forest volume stock, respectively (Pan et al., 2011). The Northeast Forest Region contains three zonal vegetation types: cold-temperate coniferous forests, temperate mixed-coniferous forests and warm-temperate deciduous broadleaf forests. The main coniferous species include Larch (*Larix gmelinii*), Camphor Pine (*Pinus sylvestris var. mongolica*), Red Pine (*Pinus koraiensis*) and Spruce (*Picea asperata*); the main broadleaf species are Poplar (*Populus davidiana*), White Birch (*Betula platyphylla*), Oak (*Quercus mongolica*) and Elm (*Ulmus pumila*).

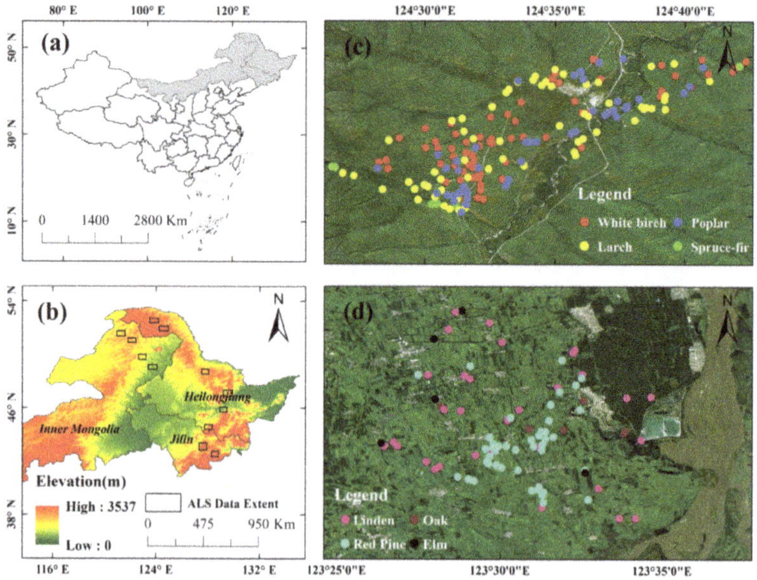

Figure 1. (**a**) Location of the Heilongjiang, Jilin and Inner Mongolia three provinces in China. (**b**) Location of the study area with 12 ALS data areas highlighted in black. (**c**,**d**) Field plots' distribution in two ALS data areas.

2.2. Data Source and Preprocessing

2.2.1. Field Measurements Data and Forest AGC Calculation

The field survey was conducted from September 2019 to November 2019. The dominant species, origin, age group and depression of each stand and the diameter at breast height (DBH), tree height, age and canopy cover of the individual tree that DBH $\geq$ 6 cm within each plot were measured and recorded by using traditional measuring instruments in the forest inventory. Based on the latest national forest resources inventory results, the distribution of dominant tree species, traffic conditions and other factors in the northeast region, 12 areas covering the target species were typically selected as aerial flight areas for obtaining LiDAR data. A total of 1600 sample plots were randomly collected in these areas, covering five typical forest areas, namely Da Hinggan Ling, Xiao Hinggan Ling, Wanda Mountain, Zhangguangcai Mountain and Changbai Mountain. The sample plots were circular, with a radius of 13.82 m and an area of about 600 m^2. The quality of the sample plot survey was checked to ensure that the error in DBH measurement was less than 3% and the error in tree height measurement was less than 5%. To ensure the geographic match between the field data and the LiDAR data, clear markers were set up at the center of each sample plot, and Real-Time Kinematic (RTK) technology was used to accurately locate the center of the sample plot, ensuring that the horizontal and vertical coordinates of the sample plot were positioned to within 1 m.

The individual tree data obtained from the field survey were statistically summarized, and the outliers were removed according to the criterion of triple standard deviation; and the data of dead trees were removed to calculate the mean area at breast height, mean diameter at breast height, mean tree height and stand density of the sample plots. After screening, a total of 1587 plots were selected. The AGB was calculated by applying species-specific allometric equations, and then the aboveground carbon stock was received by multiplying by the species-specific mean carbon conversion factor. The allometric equations and carbon-conversion factors for each tree species are shown in Table 1, with reference to [44,45]. Finally, the individual tree aboveground carbon stock in each plot was summed up and converted into hectares to obtain the forest AGC at plot level.

Table 1. Allometric equations and mean carbon conversion factors used in this study.

Tree Species	Allometric Equation	Mean Carbon Conversion Factors
Picea asperata	$AGB = 0.08070 \times D^{2.25957} \times H^{0.25663}$	0.4804
Abies fabri	$AGB = 0.06945 \times D^{2.05753} \times H^{0.50839}$	0.4805
Larix gmelinii	$AGB = 0.06848 \times D^{2.01549} \times H^{0.59146}$	0.4742 (Natural forest) 0.4674 (Plantation)
Pinus koraiensis	$AGB = 0.027847 \times D^{1.810004} \times H^{0.905002}$	0.4809
Populus davidiana	$AGB = 0.02884 \times D^{2.8785}$	0.4956 (Natural forest) 0.4761 (Plantation)
Ulmus pumila	$AGB = 0.0607 \times D^{2.4316} + 0.0678 \times D^{1.9623} + 0.0148 \times D^{1.9816}$	0.4648
Betula platyphylla	$AGB = 0.06807 \times D^{2.10850} \times H^{0.52019}$	0.4656
Quercus mongolica	$AGB = 0.06149 \times D^{2.14380} \times H^{0.58390}$	0.4802
Tilia tuan	$AGB = 0.01275 \times D^{2.0188} \times H^{1.0094} + 0.00182 \times D^{1.9492} \times H^{0.9746} + 0.00024 \times D^{1.9814} \times H^{0.9907}$	0.4677

2.2.2. Design of Sample Plot Stratification

The stratification of sample plots was based on the species information recorded in the field data. In DSS, the criterion for stratification was that one or several tree species account for more than 70% of the entire sample plot in volume stock. The sample plots were therefore stratified (a) to coniferous forests and broadleaf forests (b) to three dominant coniferous tree species, namely Spruce–Fir, Larch and Red Pine, and five dominant broadleaf tree species, namely of Poplar, Elm, Linden, Oak and White Birch. According to Reference [38], strata with smaller populations may return higher prediction errors, which, in turn, can affect the total prediction error. Therefore, to minimize the impact of strata size on the estimation results, we deliberately kept the sample sizes of the eight dominant species strata on a comparable level (approximately 200 sample plots per strata). The detailed information and summary statistics for the forest AGC of each stratification are provided in Tables 2 and 3. The distribution of the dominant species in the study area is shown in Figure 2. This map was generated from Sentinel-2A images and RF classifier.

Table 2. Overview and distribution of forest AGC of the forest-type-based stratification sample plots.

Forest Type	Number Of Plot			Forest AGC (Mg/ha)		
	Total	Training Plot	Validation Plot	Range	Mean	Standard Deviation
Coniferous forests	591	473	118	1.40–82.30	26.23	13.09
Broadleaf forests	996	795	201	0.52–79.83	26.19	11.98
All forests (non-stratification)	1587	1267	320	0.52–82.30	26.20	12.35

Table 3. Overview and distribution of forest AGC of the dominant-species-based stratification sample plots.

Dominant Species	Tree Species Composition	Number of Plot			Forest AGC (Mg/Ha)		
		Total	Training Plot	Validation Plot	Range	Mean	Standard Deviation
Picea asperata and *Abies fabri*	*Picea asperata* dominant forests or *Abies fabri* dominant forests with a small mixture of *Larix gmelinii*	197	158	39	2.29–82.30	30.73	15.35
Larix gmelinii	Pure or *Larix gmelinii* dominant forests with a small mixture of *Betula platyphylla and Populus davidiana*	197	158	39	1.40–56.13	25.33	12.11
Pinus koraiensis	Pure or *Pinus koraiensis* dominant forests with a small mixture of *Larix gmelinii*	197	158	39	1.44–49.13	22.64	9.96
Populus davidiana	Pure or *Populus davidiana* dominant forests with a small mixture of *Larix gmelinii*	209	167	42	0.52–79.83	34.36	17.44
Ulmus pumila	*Ulmus pumila* dominant forests with a small mixture of *Populus davidiana*	199	159	40	5.81–48.09	23.12	7.62
Betula platyphylla	Pure or *Betula platyphylla* dominant forests with a small mixture of *Larix gmelinii*	203	162	41	1.82–52.63	22.17	9.74
Quercus mongolica	*Quercus mongolica dominant forests* with a small mixture of *Pinus tabuliformis*	196	157	39	2.27–65.42	25.86	12.07
Tilia tuan	*Tilia tuan dominant forests* with a small mixture of *Larix gmelinii*	200	160	40	5.74–42.26	21.71	7.71

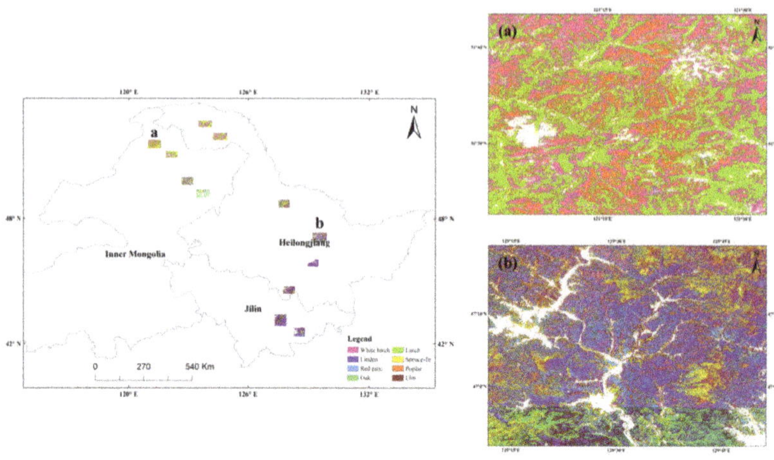

Figure 2. Dominant species map of the study area: (**a**,**b**) show the spatial distribution of dominant species in two areas at a larger scale.

2.2.3. Airborne Laser Scanning Data

In order to minimize the impact of forest condition change and errors caused by the time mismatch between field measurements and LiDAR data, airborne LiDAR data were acquired in twelve regions within the six forest regions in September and October 2019, with a total aerial area of 1043 km^2. Using RIEGL VUX-1UAV airborne laser scanner (RIEGL, Horn, Austria) mounted on a medium rotorcraft UAV platform (Siwei Spatial Data, Beijing, China), with a maximum pulse emission frequency of 550 kHz, a beam divergence angle of 0.5 mrad, a spot diameter of 50 mm, an average point density of about 4 points/m^2, an average ground point distance of about 1 m, a measurement accuracy of 10 mm, a flight height of about 100 m and a flight speed of 70–110 km/h.

The raw ALS data were processed in the TerraScan modules running on the Microstation platform (TerraSolid, Ltd., Helsinki, Finland) and the LiDAR 360 software (GreenValley, Beijing, China). The main preprocessing procedures include (a) route leveling; (b) point cloud denoising; (c) point cloud filtering—an improved TIN (triangulated irregular network) densification filtering algorithm [46] was used to classify the raw point clouds into the ground or non-ground points; (d) DEM generation, interpolation of the classified ground points using the TIN algorithm [47] to generate DEM; (e) point cloud data normalization—the absolute elevation of each point was subtracted from the DEM, and the height of the point cloud was normalized to remove the elevation effects of the terrain; (f) point cloud data clipping—the point cloud data corresponding to the sample plot was clipped out according to the coordinates of the sample center and the radius information to facilitate the extraction of LiDAR variables; and (g) LiDAR metrics extraction—the 32 LiDAR metrics were extracted from the normalized point clouds within each sample plot with a threshold of 2 m to exclude shrubs and grasses.

2.3. Methods

In this study, we integrated sample plot stratification and ML algorithms to establish forest AGC estimation models based on airborne LiDAR data in the forest regions of Northeast China. Figure 3 showed the framework of the methods for this study. Field measurement data and ALS data were first under preprocessing to obtain plot-level forest AGC and normalized LiDAR data within plots. To explore the effect of stratification in forest AGC estimation, the initial sample plots were stratified into three groups: non-stratification, FTS and DSS (Section 2.2.2). Thirty height-related metrics and 2 canopy-related metrics were extracted from normalized LiDAR data, and Pearson correlation analysis and Boruta algorithms were used to perform variables selection (Section 2.3.1). Then forest AGC estimation models were built based on Stepwise regression and four ML algorithms (RF, Cubist, XGBoost and CatBoost), and independent validation sample plots were used to evaluate the established models (Sections 2.3.2 and 2.3.5). The analysis of variance (ANOVA) was applied to identify the important factors in forest AGC modeling (Section 2.3.4). Finally, based on the model validation and ANOVA results, the optimal stratification approach and algorithms, and the important factors in forest AGC estimation were derived.

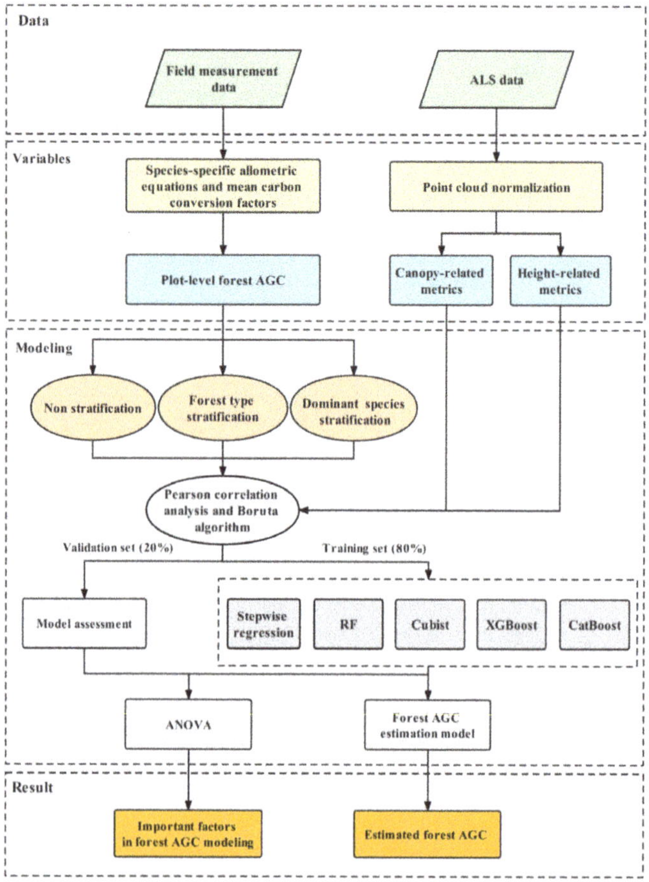

Figure 3. Flowchart of the methods for forest AGC estimation by combining sample plots stratification and ML algorithms using ALS data.

2.3.1. Model Variables Extraction and Selection

Height-related variables and canopy-related variables derived from LiDAR data are suggested to be useful at plot-level estimation and show a high correlation with forest AGB and AGC [48,49]. The height metrics directly describe the vertical height and geometry character of the trees, the density metrics reflect the return density of the trees, the canopy metrics depict canopy structure and the intensity metrics refer to the energy backscattered from the feature to the LiDAR sensor [50,51]. In this study, we extracted 30 height-related and 2 canopy-related variables based on normalized point cloud data with a threshold of 2 m. The detailed information and description of LiDAR metrics are shown in Table 4.

Table 4. Summary of the metrics extracted from ALS data used in this study.

LiDAR Metrics	Description
CC	Canopy cover
Canopy_relief_ratio	Canopy relief ratio
H_1, H_5, H_10, H_20, H_30, ... H_80, H_90, H_95, H_99	Height percentiles. Vertical distribution of point cloud height: 1%, 5%, 10%, 20%, 30%, ... , 80%, 90%, 95%, 99% quantile
H_max	Maximum height
H_min	Minimum height
H_mean	Mean height
H_median	Median of height
H_madmedian	Median of median absolute deviation of height
H_sqrt_mean_sq	Generalized means for the 2nd power of height
H_curt_mean_cube	Generalized means for the 3rd power of height
H_AIH_IQ	Interquartile distance of cumulative height
H_IQ	Interquartile distance of height
H_skewness	Skewness of height
H_kurtosis	Kurtosis of height
H_aad	Average absolute deviation of height
H_cv	Coefficient of variation of height
H_stddev	Standard deviation of height
H_variance	Variance of height

Although forest AGC is influenced by various factors, not all variables are useful in forest AGC modeling, due to the information redundancy issue. Identifying optimal variables is challenging but the key to establishing a forest AGC estimation model. In this study, Pearson correlation analysis and the Boruta algorithm were used to perform variable selection. The Pearson correlation analysis was first used to select the LiDAR metrics that most correlated with forest AGC. Then, the Boruta algorithm was used to further identify the optimal variables. The core idea of the Boruta algorithm is to construct a shadow feature by randomly mixing the original object feature values to determine whether the importance result of any given feature is significant or not, and then to classify all feature objects in a random forest classification using an extended aggregate with random samples. The maximum Z score among shadow attributes (MZSA) was found and then a two-sided test was performed for each feature object with unassigned importance. Features significantly below the MZSA were considered "unimportant" and features significantly above the MZSA were considered "important". This process was repeated until all attributes were assigned importance values, resulting in the optimal set of feature variables [52]. All of these procedures were performed in R 4.1.0 using the Boruta packages [52].

2.3.2. Modeling Algorithms

Stepwise regression and four machine-learning algorithms, namely RF, Cubist, XGBoost and CatBoost, were used for forest AGC modeling in this study. Stepwise regression is a parametric algorithm to screen variables and establish the optimal regression equation. In the modeling process of stepwise regression, the predictive variables are input into the regression equation one by one according to the given statistical standard. At each step of the analysis, the predictive variables with the highest correlation with the dependent variables first enter the regression equation, and then the variables are introduced into the model one by one, and the F-test is carried out to judge whether the variable can be selected. Stepwise regression has been widely applied in forest AGB and AGC estimation, as it can remove the variables causing multicollinearity [53–55].

RF is an improved machine learning integration algorithm based on decision trees; it was first proposed by Breiman et al. in 2001. Its advantages over traditional decision tree algorithms are that it is insensitive to noisy data, can deal with discrete or continuous data sets and can handle huge datasets [21]. The basic principle of RF is that multiple decision trees are integrated into a single but powerful model, using the "bagging" idea [56], and the Bootstrap resampling technique is used to generate a new training sample set from N original training samples by repeatedly selecting a random k (k < N) sample set. In the whole sampling process, some samples may be taken more than once, while some of the training data will not be sampled. This part of the training data is called "out-of-bag" (OOB) data; the OOB data are not involved in the model-fitting process, but are used to examine the generalization of the model. As randomness can effectively reduce model variance, the RF algorithm can achieve good generalization and low variance resistance without additional "pruning" of the decision trees [57].

Cubist is a rule-based decision-tree model extending from the earlier M5 model, based on which a regression tree is constructed, and generating a linear regression model at the end nodes of the tree, with predictions based on linear regression results at the end nodes rather than on discrete values. The final model of Cubist is a set of multivariate models associated with a set of rules associated with it, where each rule corresponds to a multivariate linear expression. Cubist also uses a boosting-like scheme known as committees, which uses the results of the training set to adjust and create subsequent trees, and then averages the predictions of all committees to generate the final predictions [26]. In addition, the predictions generated by the model rules can be adjusted by using the neighborhoods defined by the parameter neighbors in the training-set data, as this enables Cubist to predict outside of the sample coverage [58].

XGBoost is an ensemble learning algorithm based on the Gradient Boosting Decision Tree (GBDT) framework proposed by Chen and Guestrin in 2016 [59] that has won numerous awards in Kaggle machine-learning competitions and has received widespread attention in recent years. The algorithm is based on the idea of "Boosting" to generate a number of decision trees in turn, combining all the predictions of a set of weak learners to develop a strong learner through an additive training strategy. In contrast to the general GBDT algorithm, the XGBoost algorithm performs a second-order Taylor expansion on the objective function, using the second-order derivatives to accelerate the convergence of the model during training. At the same time, a regularization term is added to the objective function to control the complexity of the tree in order to obtain a simpler model and avoid overfitting [60]. Thus, XGBoost is a flexible and highly scalable tree-structured boosting algorithm with the advantages of being able to handle sparse data, greatly increase the speed of the algorithm, and reduce computational memory in training on very large scale datasets.

CatBoost is a novel gradient boosting decision-tree algorithm developed by Dorogush et al. [61] that belongs to the same boosting family as XGBoost, both being an improved implementation in the framework of the GBDT algorithm. CatBoost uses oblivious trees as base predictors, with fewer parameters and high accuracy, which can also handle categorical features well. In addition, CatBoost has solved the statistical problems of Gradient Bias and Prediction shift that all existing gradient boosting algorithms face by proposing a new and improved gradient boosting algorithm, order boosting, to reduce the occurrence of overfitting and thus improve the algorithm's accuracy and generalization. The basic idea is, firstly, the CatBoost model correlates the category features to account for the different bases of category features, including calculating the frequency of category occurrences and considering different combinations of category features to construct the regression tree. Secondly, to solve the prediction drift problem caused by gradient bias, random permutations are generated in the training dataset, and gradients are obtained based on it. For training distinct models, different permutations are used; thus, overfitting will not happen. Compared with existing GBDT algorithms, the advantages of CatBoost are the following: (a) using an innovative algorithm that automatically treats categorical features as numerical features, (b) combining category features and making full use of the connections between features greatly enriches the feature dimension and (c) the use of a fully symmetric tree model reduces overfitting and improves the accuracy and generalization of the algorithm [57,62].

Forest AGC estimation is largely dependent on the relationship between tree height and AGC due to the allometric relationships of the tree. Complex forest structure can affect the relationship between forest AGC and tree height and thus interfere with the estimation results. In theory, separate modeling of forest AGC for different dominant species can mitigate this interference and improve the algorithm's estimation performance. Therefore, in this study, we assume that the finer the stratification and the simpler the forest structure, the better the algorithm's estimation performance will be. Moreover, the estimation performance of different algorithms may be various due to the differences in forest structure among species. To verify these hypothesizes, three different scenarios were designed: (1) forest AGC models were established based on five algorithms without stratification, resulting in a total of 5 models; (2) forest AGC models were established based on FTS with five algorithms, resulting in a total of 10 models; and (3) forest AGC models were established based on DSS with five algorithms, resulting in a total of 40 models.

2.3.3. Hyperparameter Optimization in Machine Learning Algorithm

Four machine learning algorithms, RF, Cubist, XGBoost and CatBoost, were used in this study. In a machine-learning algorithm, the predicted results and model performance are largely determined by the hyperparameters of the model. A set of hyperparameters should be tuned for each algorithm to obtain the best model performance. The hyperparameters of different machine learning vary greatly, and it is difficult to adjust the parameters manually. Therefore, grid search technology was used to perform hyperparameter tuning automatically. Hyperparameter tuning was performed on the RF, Cubist, XGBoost and CatBoost algorithms based on the lowest RMSE of the model obtained by repeating the 10-fold cross-validation method five times on the training dataset, respectively, to ensure the robustness in the modeling process. All of these procedures were performed in R 4.1.3, using the Caret packages. Details about various hyperparameters and their corresponding grid values are presented in Table 5.

Table 5. Hyperparameter tuning ranges for four machine learning algorithms.

Algorithm	Hyperparameter	Description	Value Ranges
RF	mtry	the number of predictor variables randomly sampled at each split	(1–*n*) *n* refers to the number of predictor variables
	ntree	the number of trees	(100–1000) at intervals of 50
Cubist	committees	the number of trees	(1–100) at intervals of 1
	neighbors	controls the rule-based model predictions	(0–9) at intervals of 1

Table 5. *Cont.*

Algorithm	Hyperparameter	Description	Value Ranges
XGBoost	max_depth	the depth of the tree	(1–10) at intervals of 1
	eta	the learning rate	(0.01–0.5) at intervals of 0.01
	gamma	minimum loss reduction of the tree	(0–1) at intervals of 0.1
	colsample_bytree	the number of predictor variables supplied to a tree	(0–1) at intervals of 0.1
	min_child_weight	minimum number of instances	(1–10) at intervals of 1
	subsample	the number of observations supplied to a tree	(0–1) at intervals of 0.1
CatBoost	depth	the depth of the tree	(1–10) at intervals of 1
	learning_rate	the learning rate	(0.01–0.5) at intervals of 0.01
	l2_leaf_reg	coefficient at the L2 regularization term of the cost function	(0–5) at intervals of 0.1
	rsm	the percentage of features to use at each split selection	(0–1) at intervals of 0.1

2.3.4. Statistical Analysis

The two-way analysis of variance (ANOVA) was used to quantify the effect of each factor on the estimation error and to identify the key factors in forest AGC estimation. These factors include the stratification method (non-stratification, FTS and DSS), the regression algorithm (stepwise regression, RF, Cubist, XGBoost and CatBoost) and their interactions. To better show how each factor explains the total variance, we calculated the eta-squared (η^2), the proportion of the sum of squares of each factor to the total sum of squares. The ANOVA was performed in R 4.1.0.

2.3.5. Model Validation

To compare the estimation performance of stepwise regression and four machine-learning algorithms in this study, coefficient of determination (R^2, Equation (1)), root mean square error (RMSE, Equation (2)), relative root mean square error (RRMSE, Equation (3)), mean absolute error (MAE, Equation (4)) and Bias (Equation (5)) were employed. The hold-out method was used for calculating the model performance metrics, and the field measurement data of each stratification were randomly split into a training set (80% of the total) and a validation set (the remaining 20%). The training set was used to train and establish the model, while the validation set was not involved in the model establishing process but acted as an independent sample to evaluate the model performance. After hyperparameter optimization, the best models were built based on the training set, and the model performance metrics were calculated based on the validation set. The higher R^2, lower RMSE, RRMSE, MAE and Bias values imply a higher prediction accuracy and better estimation results:

$$R^2 = 1 - \frac{\sum_{i=1}^{n}(y_i - \hat{y}_i)^2}{\sum_{i=1}^{n}(y_i - \overline{y})^2} \quad (1)$$

$$\text{RMSE} = \sqrt{\frac{\sum_{i=1}^{n}(y_i - \hat{y}_i)^2}{n}} \quad (2)$$

$$\text{RRMSE} = \frac{\text{RMSE}}{\overline{y}} \times 100 \quad (3)$$

$$\text{MAE} = \frac{\sum_{i=1}^{n}|y_i - \hat{y}_i|}{n} \quad (4)$$

$$\text{Bias} = \frac{\sum_{i=1}^{n}(\hat{y}_i - y_i)}{n} \quad (5)$$

where n is the number of sample plots, $\hat{y}_i$ is the predicted forest AGC, y_i is the field measurement forest AGC and $\overline{y}$ is the mean of field measurement forest AGC.

3. Results

3.1. Comparative Analysis of Forest AGC Estimation Results

3.1.1. Forest AGC Estimation Results Based on FTS

To evaluate the effect of FTS and non-parametric machine-learning algorithms in establishing the forest AGC estimation models, 15 forest AGC estimation models were developed by using stepwise regression and four machine-learning algorithms (RF, Cubist, XGBoost and CatBoost) based on non-stratification and two stratified datasets (coniferous forests and broadleaf forests), respectively. The model performance and evaluation results for the stratified and the unstratified models are shown in Table 6.

Table 6. Performance of forest AGC estimation model based on non-stratification and FTS in the validation datasets.

Forest Type	Model	R^2	RMSE (Mg/ha)	RRMSE (%)	MAE (Mg/ha)	Bias (Mg/ha)
All forests (non-stratification)	Stepwise	0.3948	9.7867	39.0596	7.3902	0.8163
	RF	0.4213	9.5699	38.1947	7.1368	0.8704
	Cubist	0.4119	9.6471	38.5028	7.0665	−0.6283
	XGBoost	0.4392	9.4209	37.5998	7.0208	0.0435
	CatBoost	0.4411	9.4052	37.5374	7.0520	0.8851
Coniferous forests	Stepwise	0.3911	9.4519	38.2231	7.0240	0.4808
	RF	0.5853	7.8005	31.5447	5.9307	0.2851
	Cubist	0.5304	8.3004	33.5663	6.5400	−0.0962
	XGBoost	0.6017	7.6441	30.9124	5.7157	0.1689
	CatBoost	0.6073	7.5907	30.6961	5.7559	−0.1662
Broadleaf forests	Stepwise	0.3577	9.9602	41.2753	7.7378	2.0755
	RF	0.4249	9.4252	39.0582	7.0348	1.5388
	Cubist	0.3818	9.7718	40.4946	7.2849	0.6979
	XGBoost	0.4585	9.1452	37.8982	6.8907	1.7294
	CatBoost	0.4745	9.0093	37.3350	6.8652	1.6480

According to the results illustrated in Table 6, the FTS models improved the performance and predicted accuracy when applying machine-learning algorithms, as evidenced by an increase in R^2 and a decrease in RMSE, RRMSE and MAE, while the reversed results were achieved in stepwise regression models. Compared to the unstratified models, a significant improvement was observed in the coniferous-forest-stratified models, while only a slight improvement in the broadleaf-forest-stratified models, indicating that FTS provided a more positive effect in coniferous forests than broadleaf forests. Overall, four machine learning algorithms outperformed stepwise regression, regardless of the datasets used. The CatBoost models achieved the best estimation performance in all the three datasets, with the highest R^2 (0.4411 in all forests, 0.6073 in coniferous forests and 0.4745 in broadleaf forests), lowest RMSE (9.4052 in all forests, 7.5907 in coniferous forests and 9.0093 in broadleaf forests), RRMSE (37.5374 in all forests, 30.6961 in coniferous forests and 37.3350 in broadleaf forests) and MAE (6.8652 in broadleaf forests), followed by XGBoost, RF, Cubist and stepwise regression. The Bias of the CatBoost models in the three datasets were 0.8851, −0.1662 and 1.6480 Mg/ha, respectively, suggesting a general overestimation of forest AGC in unstratified and broadleaf forest models, as well as a general underestimation of forest AGC in coniferous forest models.

The improvement provided by the FTS models can be further evidenced in the scatter plots between the field-measurement forest AGC and model estimated values (Figure 4).

Figure 4 shows the correlation between the estimated forest AGC and the reference data based on FTS is better compared to the non-stratification ones except the models using stepwise regression. Moreover, a significant underestimation is observed when the forest AGC is larger than 40 Mg/ha in all the 15 models, while a significant overestimation is observed when the forest AGC is lower than 10 Mg/ha in unstratified and broadleaf forests models. However, the extent of overestimation and underestimation is reduced when using FTS.

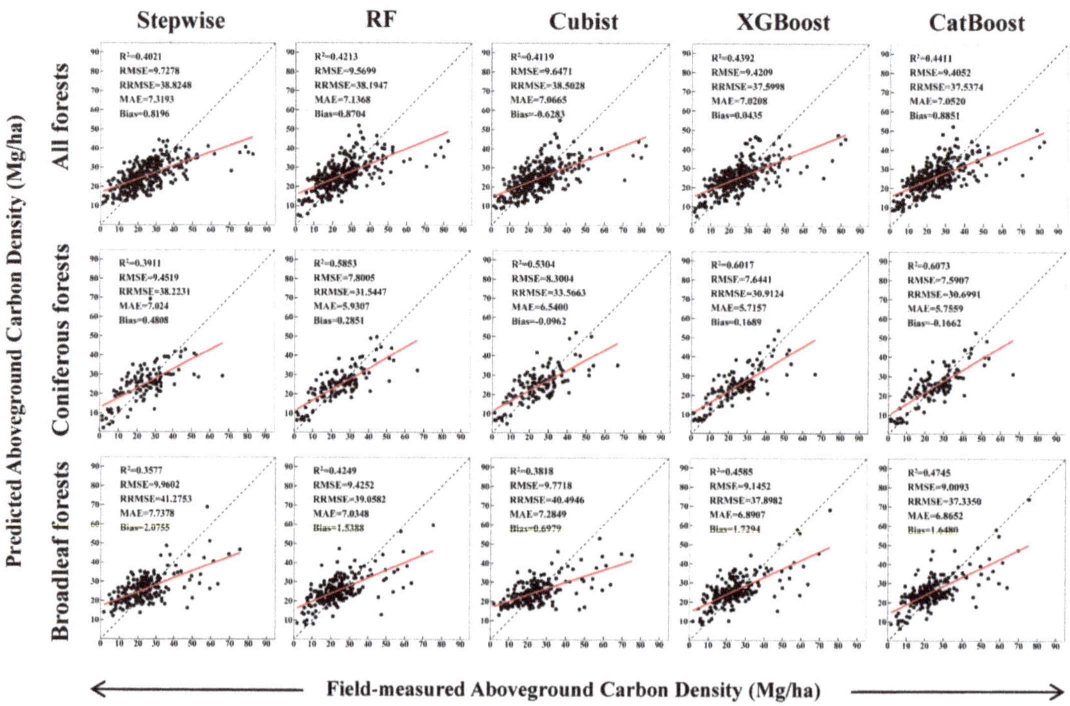

Figure 4. Scatter plots of the field-measured (x-axis) and predicted forest AGC (y-axis) using stepwise regression and four different ML models based on FTS in the validation datasets.

3.1.2. Aboveground Carbon Density Estimation Results Based on DSS

To examine the influence of DSS and ML algorithms in the forest AGC estimation, we compared and analyzed the model validation results of the forest AGC models established by using stepwise regression and four machine-learning algorithms (RF, Cubist, XGBoost and CatBoost) based on eight DSS datasets (Spruce-Fir, Larch, Red Pine, Poplar, White Birch, Oak, Linden and Elm), respectively, resulting in a total of 40 models. The results of model performances are summarized in Table 7 and Figure 5.

Table 7. Performance of forest AGC estimation models based on DSS in the validation datasets.

Dominant Species	Model	R^2	RMSE (Mg/ha)	RRMSE (%)	MAE (Mg/ha)	Bias (Mg/ha)
Spruce–Fir	Stepwise	0.7371	6.8977	23.4290	5.3067	−0.1559
	RF	0.7547	6.6623	22.6294	4.9116	0.1994
	Cubist	0.7493	6.7361	22.8801	5.2763	0.4992
	XGBoost	0.7936	6.1119	20.7600	4.5688	−0.3968
	CatBoost	0.8175	5.7463	19.5181	4.2701	1.0252
Larch	Stepwise	0.6931	6.5371	28.4119	4.9649	1.7802
	RF	0.6273	7.2045	31.3124	5.8318	1.9752
	Cubist	0.6854	6.6184	28.7652	5.2080	0.5859
	XGBoost	0.7047	6.4125	27.8701	4.8272	1.1372
	CatBoost	0.7304	6.1274	26.6309	4.7103	1.1988
Red Pine	Stepwise	0.7864	4.8843	21.8278	3.6780	−1.0045
	RF	0.8351	4.2915	19.1786	3.2918	−0.7201
	Cubist	0.8014	4.7098	21.0482	3.8554	−1.0005
	XGBoost	0.8509	4.0810	18.2380	3.3971	−0.1736
	CatBoost	0.8699	3.8113	17.0328	3.2853	0.1476
Poplar	Stepwise	0.6751	8.9241	23.6450	6.8659	−0.9275
	RF	0.7607	7.6595	20.2943	6.0103	−0.0022
	Cubist	0.7486	7.8506	20.8007	6.1131	0.5429
	XGBoost	0.7778	7.3812	19.5569	5.8989	0.1414
	CatBoost	0.8054	6.9076	18.3022	5.2377	−0.0178
White Birch	Stepwise	0.7211	5.3155	24.7447	4.1372	0.2416
	RF	0.7407	5.0642	23.5747	3.7654	0.2466
	Cubist	0.7662	4.8671	22.6570	3.5408	−0.2407
	XGBoost	0.7636	4.8943	22.7840	3.5005	0.0718
	CatBoost	0.7852	4.6653	21.7180	3.6770	−0.1229
Oak	Stepwise	0.6362	6.6328	27.7826	4.8668	0.9758
	RF	0.7468	5.5342	23.1808	4.0921	0.1669
	Cubist	0.7386	5.6229	23.5524	3.9071	0.1812
	XGBoost	0.7652	5.3294	22.3229	4.0862	−0.5591
	CatBoost	0.7903	5.0355	21.0920	3.8465	0.3638
Linden	Stepwise	0.3224	6.5837	30.2533	5.0754	0.7719
	RF	0.5294	5.4869	25.2136	4.1952	0.4577
	Cubist	0.4821	5.7557	26.4485	4.2222	0.3208
	XGBoost	0.5450	5.3949	24.7906	4.1490	0.2983
	CatBoost	0.6327	4.8474	22.2750	3.8665	0.5140
Elm	Stepwise	0.5362	4.8298	20.4512	3.9670	0.9204
	RF	0.5959	4.5080	19.0887	3.7378	1.2237
	Cubist	0.5448	4.7845	20.2596	3.9858	1.1691
	XGBoost	0.6308	4.3089	18.2456	3.5939	0.9103
	CatBoost	0.6906	3.9446	16.7032	3.1906	0.5471

Figure 5 illustrates the estimation accuracy of forest AGC varies with different dominant species. In terms of algorithm performance, estimation models based on DSS show similar trends to those based on FTS; that is, the four machine-learning algorithms outperform the stepwise regression, with the CatBoost models achieving the highest estimation accuracy, followed by XGBoost, RF, Cubist and stepwise regression. The detailed information of the model evaluation results can be found in Table 7. Table 7 shows the 40 models for eight different dominant species with R^2 varying from 0.3224 to 0.8699, RMSE varying from 3.8113 to 8.9241, RRMSE varying from 16.7032 to 31.3124, MAE varying from 3.1906 to 6.8659 and Bias varying from −1.0045 to 1.9752. Relatively high estimation accuracy was achieved in all eight dominant species, with the CatBoost model based on DSS for Red Pine achieving the best estimation accuracy (R^2 = 0.8699, RMSE = 3.8133, RRMSE = 17.0328, MAE = 3.2853 and Bias = 0.1476). In terms of Bias, no single algorithm is optimal in all dominant species, with the highest mean Bias (1.2755) being observed in the Larch models, indicating a more significant overestimation of forest AGC in the Larch, regardless of the algorithm used. Overall, the models established based on DSS achieved much higher estimation accuracy compared to the unstratified models (Table 6), and this improvement is more significant in the Spruce–Fir, Larch, Red Pine, Poplar, White Birch and Oak models. The estimated forest AGC of eight dominant species models based on the CatBoost algorithm was shown in Figure 6. The mean estimated forest AGC ranged from 21.36 to 37.72 Mg/ha in eight dominant species, with the estimated forest AGC of Poplar and Spruce–Fir being significantly higher than the other dominant species, and the estimated forest AGC of the remaining dominant species were at a comparable level.

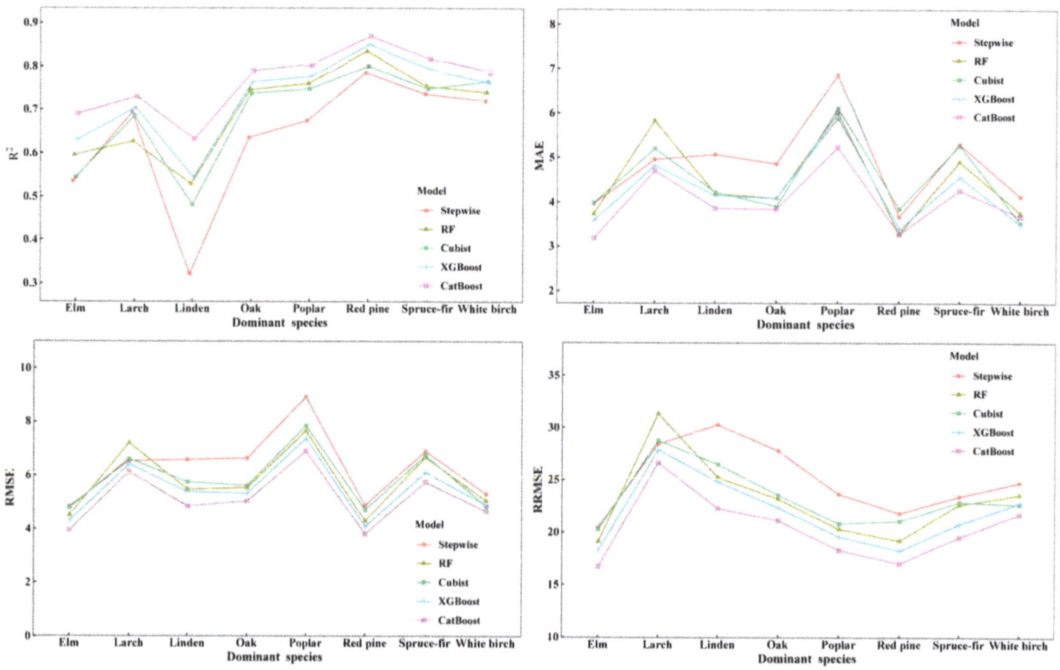

Figure 5. Model estimation accuracy evaluation results based on the validation datasets using stepwise regression and four ML algorithms in eight different dominant species.

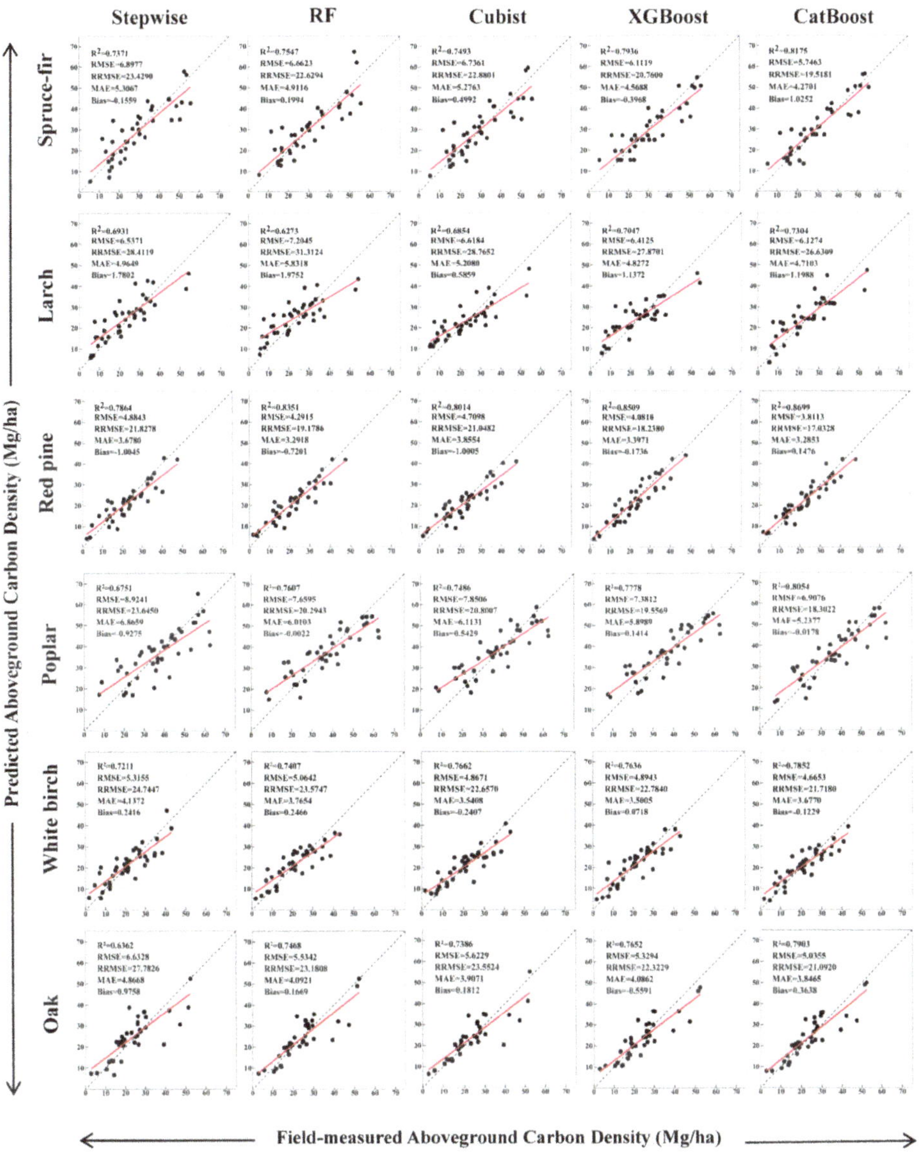

Figure 6. *Cont.*

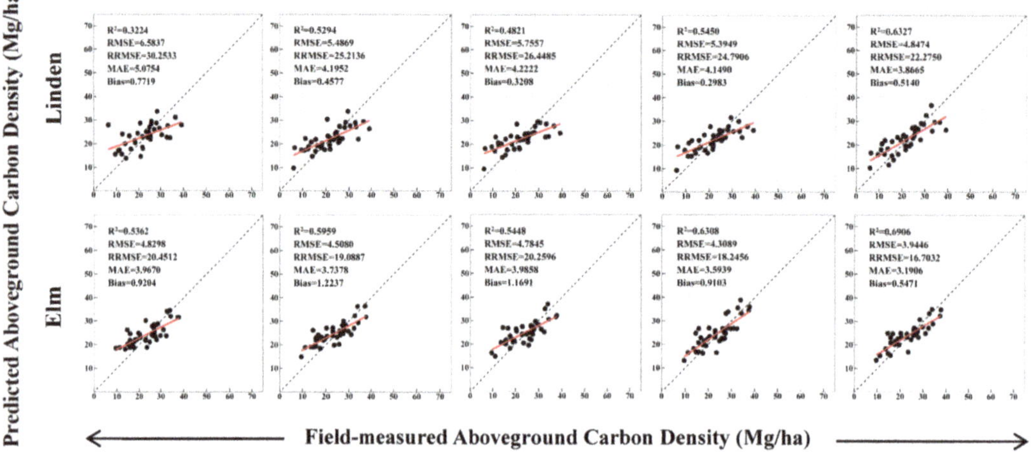

Figure 6. Model estimation accuracy evaluation results based on the validation datasets using stepwise regression and four ML algorithms in eight different dominant species.

The scatter plots between the field-measurement forest AGC and estimated values for the eight dominant species are provided in Figure 7. Figure 7 shows that the linear relationships between the estimated and measured values of forest AGC are relatively better in Spruce–Fir, Larch, Red Pine, Poplar, White Birch and Oak models, while relatively poor linear relationships are observed in Linden and Elm. A significant underestimation is observed when the forest AGC is larger than 40 Mg/ha in Larch, Poplar and Oak models, while a significant overestimation is observed when the forest AGC is lower than 20 Mg/ha in Larch, Poplar, Linden and Elm models. Compared to the unstratified models (Figure 4), the linear relationships between the estimated and measured values of forest AGC and the extent of overestimation and underestimation are greatly improved in all 40 models established based on DSS. The forest AGC estimation models based on DSS have achieved much better estimation performance than unstratified ones.

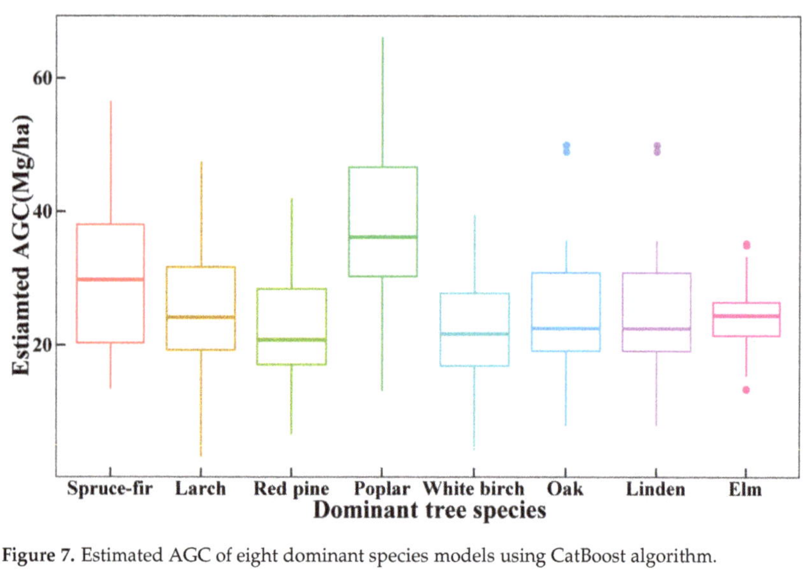

Figure 7. Estimated AGC of eight dominant species models using CatBoost algorithm.

3.1.3. Comparative Analysis of Forest AGC Estimation Results Based on FTS and DSS

To further explore the optimal stratification method in forest AGC estimation, the overall estimation accuracy of models based on non-stratification, FTS and DSS were summarized in Table 8. Generally, the CatBoost models based on DSS have achieved the best estimation accuracy (R^2 = 0.8232, RMSE = 5.2421, RRMSE = 20.5680, MAE = 4.0169 and Bias = 0.4493), while the stepwise regression models based on FTS provided the worst estimation accuracy (R^2 = 0.3700, RMSE = 9.7752, RRMSE = 40.1415, MAE = 7.4738 and Bias = 1.4856). The comparative results illustrate that the estimation accuracy of models based on DSS is significantly higher than that of models based on FTS, regardless of the algorithm used, with R^2 increased from 0.3700~0.5223 to 0.7309~0.8232, RMSE reduced from 8.5121~9.7752 to 5.2421~6.4663, RRMSE reduced from 34.9546~40.1415 to 20.5680~25.3713, MAE reduced from 6.4549~7.4738 to 4.0169~4.8700 and Bias reduced from 0.4042~1.4856 to 0.1803~0.4493. As the CatBoost models based on DSS have provided the highest estimation accuracy, they were chosen for mapping the spatial distribution of the estimated forest AGC in the study area (Figure 8). Moreover, compared to the non-stratification models, a significant improvement was observed in DSS models, while only a slight improvement was observed in FTS models.

Table 8. Summary of the overall estimation accuracy of non-stratification, FTS and DSS models on the validation datasets.

Stratification Method	Model	R^2	RMSE (Mg/ha)	RRMSE (%)	MAE (Mg/ha)	Bias (Mg/ha)
Non-stratification	Stepwise regression	0.3948	9.7867	39.0596	7.3902	0.8163
	RF	0.4213	9.5699	38.1947	7.1368	0.8704
	Cubist	0.4119	9.6471	38.5028	7.0665	−0.6283
	XGBoost	0.4392	9.4209	37.5998	7.0208	0.0435
	CatBoost	0.4411	9.4052	37.5374	7.0520	0.8851
FTS	Stepwise regression	0.3700	9.7752	40.1415	7.4738	1.4856
	RF	0.4826	8.8590	36.3788	6.6264	1.0751
	Cubist	0.4353	9.2548	38.0042	7.0094	0.4042
	XGBoost	0.5101	8.6205	35.3995	6.4561	1.1522
	CatBoost	0.5223	8.5121	34.9546	6.4549	0.9769
DSS	Stepwise regression	0.7309	6.4663	25.3713	4.8700	0.3162
	RF	0.7737	5.9307	23.2698	4.5070	0.4091
	Cubist	0.7705	5.9719	23.4313	4.5200	0.2599
	XGBoost	0.7984	5.5975	21.9624	4.2611	0.1803
	CatBoost	0.8232	5.2421	20.5680	4.0169	0.4493

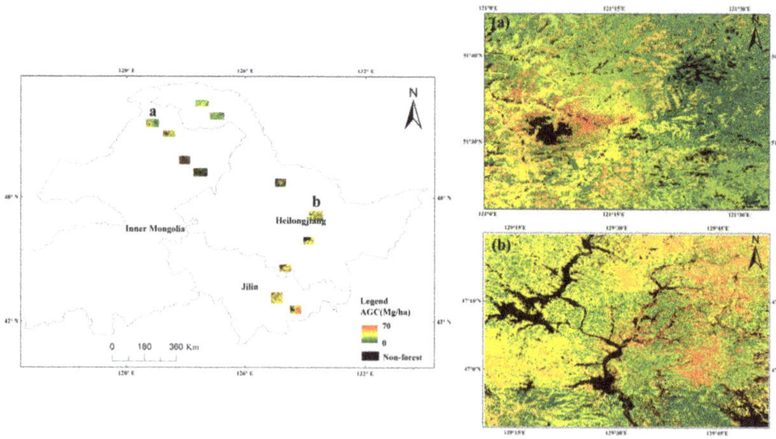

Figure 8. Forest AGC estimation map in the study area retrieved by the CatBoost models based on DSS. (**a**,**b**) Spatial distribution of estimated forest AGC in two areas at a larger scale.

Further comparison of scatter plots of field-measured forest AGC and estimated values between FTS models (Figure 4) and DSS models (Figure 6) illustrates that the linear relationships between the estimated and measured values of forest AGC and the extent of overestimation and underestimation are greatly improved in all 40 models established based on DSS. In summary, the forest AGC estimation model established by each dominant species has a higher predictive ability and applied potential than the models constructed by each forest type.

3.2. *Variable Importance Analysis*

The variable importance for forest AGC estimation models was evaluated by the PredictionValuesChange method based on CatBoost in the DSS models. The relative importance of the 10 highest ranked variables was shown in Figure 9, revealing that the important variables vary in different dominant species models. The height percentile metrics have achieved the highest relative importance in most of the DSS models, accounting for more than 80% in the Larch model, more than 70% in the Spruce–Fir model, more than 40% in the Oak model, more than 30% in the Red Pine, Poplar, White Birch and Linden model and more than 25% in the Elm model. Canopy-related metrics are also useful in the forest AGC estimation, with the canopy relief ratio metric being the most important variable in the White Birch models and the fifth and sixth important variable in Linden and Poplar model. In general, the height-related metrics and canopy-related metrics play an important role in forest AGC estimation, with height-related metrics being more important. The variables importance analysis results demonstrate that the important variables for the models vary with dominant species, illustrating the necessity to identify optimal model variables for forest AGC estimation models in different dominant species.

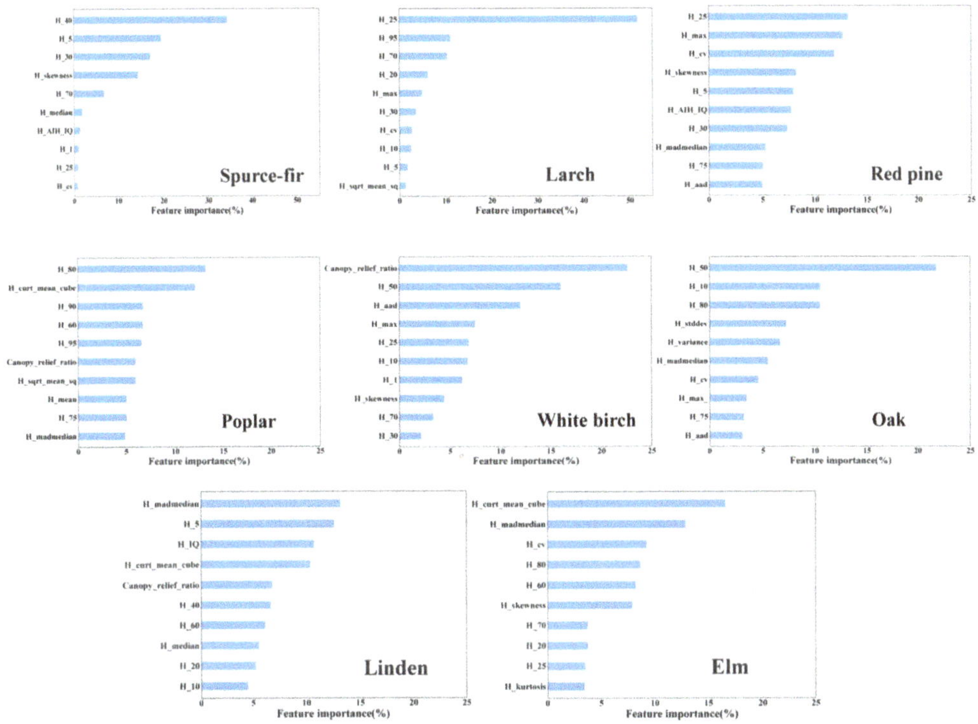

Figure 9. Relative importance of the 10 highest ranked variables of CatBoost models based on DSS.

4. Discussion

4.1. Variables Selection in Forest AGC Estimation

Identifying suitable variables is a prerequisite and key to building a high-precision forest AGC estimation model. The commonly used variables derived from LiDAR data in forest AGB and AGC estimation can be divided into four categories: height, density, intensity and canopy metrics [63–65]. In this study, the initial LiDAR dataset contained 30 height-related and two canopy-related variables, without considering density and intensity variables. It is based on the prior knowledge that density and intensity metrics are often influenced by many other factors, including transmitted power, range, angle of incidence, atmospheric transmittance, environmental parameters and the structural characteristics of the target itself [66], resulting in the density and intensity values obtained for the same feature on different flight routes varying significantly and making it difficult to reflect the true character of the feature. Moreover, several studies have proposed that LiDAR intensity values must be calibrated before they can be applied to forest AGB and AGC estimation [67,68], but to date, no standard approach for LiDAR intensity correction has been established. Then Pearson correction analysis and the Boruta algorithm were used to further provide auxiliary information on variable selection for each dataset. Feature selection based on expert knowledge allows for the selection of the most useful variables in AGC estimation from an empirical perspective, while correlation analysis and automated feature selection algorithms provide the best set of variables from a statistical perspective. It was also suggested in Reference [11] that the inclusion of expert knowledge in variables selection would make the model more ecologically meaningful and generalized than those using only automatic feature-selection algorithms, such as stepwise regression, RFE and Boruta.

The variable importance analysis results showed that the height percentile metrics are the most important in most cases, which is consistent with several previous studies [20,58], revealing the high corrective between height percentile metrics and forest AGC. In addition, the variable importance results also demonstrated that the important variables are different in different dominant species. Thus, there are possibly two ways to improve the large-scale forest AGC estimation: One is to select optimal variables for a specific study area. In this case, considering the experience and previous effort on AGC estimation in variables selection of the specific study region may be more effective before modeling. The other is to examine the potential generic indicators that are independent of geographical and environmental conditions, e.g., the TCH metric derived from ALS data [69] and LiDAR biomass index (LBI) obtained from TLS data [70]. However, the extent to which these indicators are effective remains to be tested, and more studies should be carried out on the model transferability to provide accurate forest AGC estimation on a large scale.

4.2. The Role of Stratification in Forest AGC Estimation

Our study indicated that both FTS and DSS could improve the estimation accuracy of forest AGC compared to non-stratification estimation, which confirmed the effectiveness of stratification in forest AGC estimation and was consistent with previous studies [71,72]. The essence of stratified estimation is to aggregate observations of target variables into more homogeneous strata or levels than the whole. Forest AGC varies greatly across different forest types and dominant species, as forest AGC is related to a variety of factors, such as forest structure, species composition, stand characteristics and site factors. The heterogeneity between different forests makes the relationship between forest AGC and tree height becoming particularly complex and limit the estimation accuracy of LiDAR data. Stratifying the sample plots into forest types or dominant species can reduce forest heterogeneity arisen from the interference of other factors in AGC estimates, thus improving the correlation between forest AGC and LiDAR metrics. Moreover, allometric models and carbon conversion factors are developed at the tree species level, and thus the AGC estimation models should be established on individual forest type or dominant species to reduce the uncertainty [16].

A two-way ANOVA was used to explore the important factors in forest AGC estimation. The ANOVA results (Table 9) showed that the stratification method had the most significant effect on the estimation error, explaining 53% of the total variance in R^2, 66% of the RMSE, 77% of the RRMSE and 64% of the MAE. The regression algorithm and its corresponding interactions had a marginal impact on estimation accuracy, explaining less than 10% of the total variance in R^2, RMSE, RRMSE and MAE, respectively. The ANOVA results proved that a stratification of the sample plots is of greater importance than the modeling algorithm, which was inconsistent with Reference [38]. The discrepancy may be contributed to the differences in stratification method, sample size and the study area; thus, more studies should be conducted to further examine the generalizability of our results.

Table 9. ANOVA of the R^2, RMSE, RRMSE and MAE respective to the stratification method, regression method and their interaction.

Factor	Df	R^2 SumSq	η^2	RMSE SumSq	η^2	RRMSE SumSq	η^2	MAE SumSq	η^2
Stratification	2	0.65	0.53	123.45	0.66	2171.4	0.77	63.89	0.64
Regression method	4	0.10	0.08	8.39	0.05	131.1	0.05	4.51	0.05
Stratification: regression method	8	0.01	0.01	0.68	0.00	11.5	0.00	0.50	0.01
Residuals	40	0.47		53.52		511.3		30.57	

4.3. FTS versus DSS

The comparative results between FTS and unstratified estimates show that significant improvement was obtained in AGC estimation models based on coniferous forest, while only marginal improvement was obtained in AGC estimation models based on broadleaf forest. One possible explanation for this is to be found in the substantial differences in tree crowns and distribution of branches and leaves across different broadleaf tree species in this study, which heavily affects the penetration of the laser pulses and thus influences the relationships between LiDAR metrics and forest AGC [73]. It is also mentioned by Reference [21] that AGB modeling based on coniferous forest provided poorer estimation performance due to the difference in crown size, shape and the relationship between the AGB and canopy height of the Masson pine and Chinese fir. Therefore, stratifying the sample plots into the coniferous forest and broadleaf forest may not be sufficient to reduce the heterogeneity within strata and provide better estimation performance. In addition, a higher estimation accuracy was obtained in the coniferous forest than in the broadleaf forest, as is consistent with several previous studies [74,75]. The difference may be attributed to the fact that broadleaf tree species tend to have more biomass in the branches and weaken the relationships between height and forest AGC [76].

Further comparison of the estimation performance between the FTS and DSS models illustrated that a substantially higher R^2, RMSE, RRMSE, MAE and Bias were observed in DSS models, and this is in line with previous studies [21,77]. The results demonstrate that DSS is a more recommended approach for stratification estimation. The improvement provided by DSS can be attributed to the fact that the relationships between tree height and forest AGC are the same in individual tree species, as they share similar canopy structures and AGB distribution. Stratifying sample plots into dominant species can provide highly homogenous strata and minimize the within-strata variance, leading to a better forest AGC estimation. However, there are also several studies reporting that only minor improvements in estimation performance were obtained when the same data were used to construct individual forest type or species strata for estimation [38,78,79]. The difference in results may be attributed to inconsistent sample sizes across different studies and small sample sizes within strata in most studies. Higher uncertainty and prediction errors may be produced with fewer within-strata sample sizes, and these, in turn, affect the total prediction error. For example, the Douglas fir and maple had the highest RMSD value for 261% and 315%, which accounted for the smallest number of overall sample plots (7.0% and 5.7%) [80]; the subtropical *Picea abies* forest (SPAF) had the highest RMSE (82.7 ± 28.2 Mg/ha) and bias (−36.8 ± 19.5 Mg/ha) with the smallest number of reference data (16) [81]. It is also mentioned by Reference [82] that estimates of standard errors can be biased in the case of small sample sizes within strata. In this study, the within-strata sample plot sizes of each dominant species were kept at around 200, which is a comparable and relatively large level, making the estimation results more robust and representative.

4.4. Machine-Learning Algorithms for Forest AGC Estimation

Modeling algorithms have been suggested to be an important factor for the accurate estimation of forest AGB and AGC [83]. However, to date, no single algorithm has been optimal in all cases. Therefore, identifying a proper algorithm has been a critical step to constructing AGC estimation models. In this study, the estimation performances of one parametric approach (Stepwise regression) and four non-parametric machine learning algorithms (RF, Cubist, XGBoost and CatBoost) were compared. The results showed that four machine-learning algorithms outperform stepwise regression in most cases, thus confirming previous findings that non-parametric machine-learning algorithms were suggested to be more suitable for forest AGB and AGC estimation than the parametric algorithm [22,24,25]. We attribute the better performance of ML algorithms to the fact that the relationships between forest AGC and the LiDAR metrics are likely nonlinear and complex, especially in those forests with complex stand structures and tree species composition, and this makes it difficult to model these relationships through parametric algorithms with a fixed model structure. However, overestimation of forest AGC at low AGC values and underestimation of forest AGC at high values are still common in ML algorithms. Moreover, the hyperparameter tuning methods and tuning ranges vary with study area and input data, which greatly limit the model transferability of ML algorithms. Moreover, we found that, when forest AGC estimation models were established based on DSS, a significant improvement was observed in stepwise regression models, implying that the relationships between forest AGC and the LiDAR metrics are expected have a more linear association at the species level.

Among the four ML algorithms, two novel boosting-based ensemble algorithms, XGBoost and CatBoost, have provided better forest AGC estimation accuracy than others, and the CatBoost algorithm outperformed other algorithms in all datasets. Before this study, XGBoost and CatBoost algorithms have not been used for forest AGC estimation, but there have been several studies on forest AGB estimation. Pham et al. [84] combined a genetic algorithm (GA) and XGBoost to achieve the best estimation of mangrove AGB than other four ML algorithms (RF, SVM, GBRT and CatBoost); Zhang et al. [85] compared and evaluated the performance of eight ML algorithms (MARS, RF, SVM, GBRT, ANN, SGB, ERT and CatBoost) in forest AGB estimation, and the results showed that CatBoost provided the best performance with an R^2 of 0.72, an RMSE of 45.63 Mg/ha, a bias of 0.06 Mg/ha, and a relative RMSE of 25%. Luo et al. [86] examined the different combinations of three feature selection methods and three ML algorithms (RF, XGBoost and CatBoost) in forest AGB estimation and found that combining RFE and CatBoost obtained the highest estimation accuracy. The compared results in this study were consistent with these previous studies and further demonstrated the superiority and application potential of XGBoost and CatBoost in forest AGC estimation. Compared with XGBoost, CatBoost has achieved better estimation with fewer hyperparameters, higher model efficiency and slighter overestimation and underestimation problems, making CatBoost a more recommended algorithm in forest AGC estimation. However, more studies should be carried out to further examine the application potential of CatBoost across various forest types within different geographical environments.

4.5. Species-Level Forest AGC Estimation

In this study, we established eight species-level forest AGC estimation models by using CatBoost algorithms and achieved satisfactory estimation accuracy. Our species-level estimation accuracy (R^2 = 0.63~0.87) was significantly higher than that of Fu et al. (R^2 = 0.14~0.56) [42] and Zhang et al. (R^2 = 0.01~0.47) [87], which linked field measurement plots and MODIS data to map species-level biomass in Northeast China. High estimation accuracy has been achieved in Spruce–Fir, Larch, Red Pine, Poplar, White Birch and Oak, while relatively low-estimation accuracies were achieved for Linden and Elm. The discrepancy may be explained by allometric equations and mean carbon conversion factors used for Linden and Elm. The sample plots of Linden and Elm spanned six flight regions and Heilongjiang and Jilin two provinces, with a difference of more than 10 degrees in latitude between north and south. However, the allometric equations and mean carbon conversion factors used for Linden and Elm in this study were not established for a specific region but for the whole Northeast China region. The differences in hydrothermal conditions caused by the latitude could have a significant effect on the growth of Linden and Elm, and these difference, in turn, increase the uncertainty and errors of allometric equations and mean carbon-conversion factors. Moreover, the relatively low-point cloud density of the LiDAR data used in this study (4 points/m^2) may not be enough to fully capture the structure information, leading to the poorly structured Linden and Elm models. To our knowledge, species-level forest AGC estimation models in northeast forest regions of China based on LiDAR data have not yet been reported in studies. Species-level AGC estimation models can provide important basic information for large-scale forest resource monitoring, but they pose new challenges in terms of sample size and accurate forest classification products. The lack of spectral information from LiDAR sensors makes it difficult to achieve accurate dominant species maps based on LiDAR data. Therefore, using LiDAR as a sampling tool and fusing LiDAR with other sensors (e.g., hyperspectral and optical) to acquire dominant species area and build forest AGC models could be a potential solution [88,89].

4.6. Uncertainty Analysis and Limitations

Identifying and understanding the uncertainty of the remote sensing-based forest AGC estimation models is necessary for improving forest AGC estimation accuracy and establishing standard estimation designs and procedures [90]. In addition to the errors and uncertainties introduced by the variable selection methods, model algorithms themselves, there are a number of external factors that can contribute to uncertainty in this study. (1) The first factor is the allometric equations and mean carbon conversion factors used for estimating plot-level forest AGB and forest AGC. The errors in allometric equations have been regarded as a common and primary source of uncertainty in forest AGB and AGC estimations [91–93]. The sample plots in this study were located in three provinces, Heilongjiang, Jilin and Inner Mongolia, while the species-specific allometric equations and mean carbon conversion factors used were developed for the entire Northeast China region. The allometric equations depend on the assumptions of the allometric relationships between diameter at breast height (DBH) and tree height (H), and these allometric relationships may vary with environment and stand structure, resulting in different forest AGB estimations and great uncertainty. The uncertainty propagates and accumulates with the error in the carbon conversion factors, influencing the final estimation accuracy of forest AGC. (2) The second factor is the errors from small trees shrubs and herbs. In this study, the trees smaller than 6 cm in DBH, as well as shrubs and herbs, were not recorded in the ground survey, which could be captured by the LiDAR data. The cumulative AGC of these small trees, shrubs and herbs may become a non-negligible part of the total and thus introduce errors into the forest AGC estimates. (3) The third factor is the effect of point density. The point density used in this study was 4 points/m^2, which is low-density point cloud data. Previous studies have demonstrated that the ability of LiDAR to estimate vegetation height decreases with lower point density [94,95]. The relative low point density in this study has limited the detection of the vegetation canopy and the number of points that penetrate to the ground, which may affect the DEM generation and the canopy and height-based forest AGC estimation. (4) The fourth factor is the edge effect and geolocation error. The effect of edge effect may be attributed to the fact that the field measurement is based on the position of the stem while the LiDAR data capture the tree crown and height information within the whole specific region. Therefore, some trees detected by LiDAR data may not be recorded by the ground survey, thus contributing to the uncertainty in the forest AGC estimation. The field sample plots are usually located by consumer-grade GPS whose positional accuracy largely depends on the open conditions of the environment, leading to location error from 1 to 10 m in the complex environment of forest [96]. The mismatch of geographic location between LiDAR data and sample plots data may provide great uncertainty and error in forest AGC estimation. (5) The fifth factor is the error from field measurement. In this study, tree metrics, such as DBH and tree height, were measured manually, using traditional tools. It is usually difficult to locate the treetop in forests with high canopy closure and complex structures. Therefore, the quality and accuracy of these metrics are largely determined by the quality and skill level of the surveyors, which may introduce errors and uncertainty into the results. The advent of advanced technologies, such as ground-based LiDAR and backpack LiDAR, promises to act as a new alternative to reduce uncertainty and improve the accuracy of ground survey.

Some sources of uncertainty, such as the edge effects and geographical location errors, are difficult to quantify empirically and statistically, as it is impossible to find an ideal sample free of the effects of edge effects and geographical location errors. The advent of simulation studies promises to be a powerful tool to solve the present limitations and better quantify and understand uncertainties in forest AGB and AGC estimations. For example, Knapp et al. [97] quantified the effect of border effects by using the bottom-up simulation method, and the simulation results showed that the edge effects decreased with increasing plot sizes, with the edge effects being most significant at the 10 m scale and having no influence at the 100 m scale. There are also several studies using similar simulation methods to successfully qualify the uncertainty introduced by the geolocation error [98], allometric equations [92] and forest structure [99]. Future studies should consider multiple uncertainties simultaneously and quantify the weight of each component to better understand the uncertainty in the entire process of forest AGC estimation.

5. Conclusions

In this study, we retrieved the potential of integrating sample plots stratification and non-parametric machine-learning algorithms for forest AGC modeling in the forest regions of Northeast China. Four major conclusions can be drawn:

(1) The ANOVA result showed that the stratification method had a more important effect on forest AGC estimation than the regression algorithm. Both FTS and DSS were effective in improving the estimation accuracy of forest AGC compared to non-stratified models, demonstrating the positive role of stratification in forest AGC estimation. Compared to the non-stratified models, the estimation accuracy of forest AGC was significantly improved in coniferous species, while marginal improvement was observed in the broadleaf species.
(2) Compared with FTS, models based on DSS achieved greater improvements, indicating that DSS is a better stratification estimation method for forest AGC.
(3) Regardless of the stratification method used, of the five algorithms, the four non-parametric ML algorithms outperformed parametric stepwise regression, with the CatBoost algorithm obtaining the best estimation performance, followed by XGBoost, RF, Cubist and stepwise regression.
(4) The most important LiDAR metrics for forest AGC estimation were the height percentiles and the canopy relief ratio.
(5) The CatBoost models based on DSS achieved the highest estimation accuracy, with R^2 = 0.8232, RMSE = 5.2421, RRMSE = 20.5680, MAE = 4.0169 and Bias = 0.4493. The estimation values of the best forest AGC estimation model for the eight dominant species ranged from 21.36 to 37.72 Mg/ha, with the Poplar having the highest forest AGC and the White Birch having the lowest.

The main contribution of this study is the successful combination of DSS and the CatBoost algorithm to improve the estimation performance of forest AGC and to obtain the first high-precision species-level forest AGC estimation models based on the CatBoost algorithm in the forest regions of Northeast China. Integrating this strategy with the national forest inventory or accurate remote-sensing-based wall-to-wall dominant species classification products is expected to provide a new solution to reduce the uncertainty and improve the estimation accuracy of large-scale forest carbon stock.

Author Contributions: Methodology, data curation, formal analysis, writing—original draft preparation and review and editing, M.C.; software and data curation, X.Q.; investigation, W.Z.; conceptualization, project administration and writing—review and editing, D.P. All authors have read and agreed to the published version of the manuscript.

Funding: This research was funded by the National Key R&D Program of China (2016YFD0600205) and the China National Land Survey and Planning Institute Bidding Project (GXTC-A-19070081).

Acknowledgments: This work was supported by the National Key R&D Program of China (2016YFD0600205) and the China National Land Survey and Planning Institute Bidding Project (GXTC-A-19070081). We would like to thank Xiaoyao Li for his advice on language and manuscript. We would also like to thank the editors and the anonymous reviewers for their valuable comments.

Conflicts of Interest: The authors declare no conflict of interest. The funders had no role in the design of the study; in the collection, analyses or interpretation of data; in the writing of the manuscript; or in the decision to publish the results.

References

1. Pan, Y.; Birdsey, R.A.; Fang, J.; Houghton, R.; Kauppi, P.E.; Kurz, W.A.; Phillips, O.L.; Shvidenko, A.; Lewis, S.L.; Canadell, J.G.; et al. A Large and Persistent Carbon Sink in the World's Forests. *Science* **2011**, *333*, 988–993. [CrossRef] [PubMed]

2. Six, J.; Callewaert, P.; Lenders, S.; De Gryze, S.; Morris, S.J.; Gregorich, E.G.; Paul, E.A.; Paustian, K. Measuring and Understanding Carbon Storage in Afforested Soils by Physical Fractionation. *Soil Sci. Soc. Am. J.* **2007**, *66*, 1981–1987. [CrossRef]
3. Lin, B.; Ge, J. Valued Forest Carbon Sinks: How Much Emissions Abatement Costs Could Be Reduced in China. *J. Clean. Prod.* **2019**, *224*, 455–464. [CrossRef]
4. Santini, N.S.; Adame, M.F.; Nolan, R.H.; Miquelajauregui, Y.; Pinero, D.; Mastretta-Yanes, A.; Cuervo-Robayo, A.P.; Eamus, D. Storage of Organic Carbon in the Soils of Mexican Temperate Forests. *For. Ecol. Manag.* **2019**, *446*, 115–125. [CrossRef]
5. García, M.; Riaño, D.; Chuvieco, E.; Danson, F.M. Estimating Biomass Carbon Stocks for a Mediterranean Forest in Central Spain Using LiDAR Height and Intensity Data. *Remote Sens. Environ.* **2010**, *114*, 816–830. [CrossRef]
6. Kuuluvainen, T.; Gauthier, S. Young and Old Forest in the Boreal: Critical Stages of Ecosystem Dynamics and Management under Global Change. *For. Ecosyst.* **2018**, *5*, 26. [CrossRef]
7. Zhao, M.; Yang, J.; Zhao, N.; Liu, Y.; Wang, Y.; Wilson, J.P.; Yue, T. Estimation of China's Forest Stand Biomass Carbon Sequestration Based on the Continuous Biomass Expansion Factor Model and Seven Forest Inventories from 1977 to 2013. *For. Ecol. Manag.* **2019**, *448*, 528–534. [CrossRef]
8. Fang, J.; Guo, Z.; Hu, H.; Kato, T.; Muraoka, H.; Son, Y. Forest Biomass Carbon Sinks in East Asia, with Special Reference to the Relative Contributions of Forest Expansion and Forest Growth. *Glob. Chang. Biol.* **2014**, *20*, 2019–2030. [CrossRef] [PubMed]
9. Mitchard, E.T.A. The Tropical Forest Carbon Cycle and Climate Change. *Nature* **2018**, *559*, 527–534. [CrossRef] [PubMed]
10. Le Toan, T.; Quegan, S.; Davidson, M.W.J.; Balzter, H.; Paillou, P.; Papathanassiou, K.; Plummer, S.; Rocca, F.; Saatchi, S.; Shugart, H.; et al. The BIOMASS Mission: Mapping Global Forest Biomass to Better Understand the Terrestrial Carbon Cycle. *Remote Sens. Environ.* **2011**, *115*, 2850–2860. [CrossRef]
11. Lu, D.; Chen, Q.; Wang, G.; Liu, L.; Li, G.; Moran, E. A Survey of Remote Sensing-Based Aboveground Biomass Estimation Methods in Forest Ecosystems. *Int. J. Digit. Earth* **2016**, *9*, 63–105. [CrossRef]
12. Lin, C.; Thomson, G.; Popescu, S.C. An IPCC-Compliant Technique for Forest Carbon Stock Assessment Using Airborne LiDAR-Derived Tree Metrics and Competition Index. *Remote Sens.* **2016**, *8*, 528. [CrossRef]
13. Huang, S.; Ramirez, C.; Kennedy, K.; Mallory, J. A New Approach to Extrapolate Forest Attributes from Field Inventory with Satellite and Auxiliary Data Sets. *For. Sci.* **2017**, *63*, 232–240. [CrossRef]
14. Li, Z.; Zan, Q.; Yang, Q.; Zhu, D.; Chen, Y.; Yu, S. Remote Estimation of Mangrove Aboveground Carbon Stock at the Species Level Using a Low-Cost Unmanned Aerial Vehicle System. *Remote Sens.* **2019**, *11*, 1018. [CrossRef]
15. Xie, B.; Cao, C.; Xu, M.; Bashir, B.; Singh, R.P.; Huang, Z.; Lin, X. Regional Forest Volume Estimation by Expanding LiDAR Samples Using Multi-Sensor Satellite Data. *Remote Sens.* **2020**, *12*, 360. [CrossRef]
16. Lu, D. The Potential and Challenge of Remote Sensing-based Biomass Estimation. *Int. J. Remote Sens.* **2006**, *27*, 1297–1328. [CrossRef]
17. Chave, J.; Réjou-Méchain, M.; Búrquez, A.; Chidumayo, E.; Colgan, M.S.; Delitti, W.B.C.; Duque, A.; Eid, T.; Fearnside, P.M.; Goodman, R.C.; et al. Improved Allometric Models to Estimate the Aboveground Biomass of Tropical Trees. *Glob. Chang. Biol.* **2014**, *20*, 3177–3190. [CrossRef]
18. Zolkos, S.G.; Goetz, S.J.; Dubayah, R. A Meta-Analysis of Terrestrial Aboveground Biomass Estimation Using Lidar Remote Sensing. *Remote Sens. Environ.* **2013**, *128*, 289–298. [CrossRef]
19. Brovkina, O.; Novotny, J.; Cienciala, E.; Zemek, F.; Russ, R. Mapping Forest Aboveground Biomass Using Airborne Hyperspectral and LiDAR Data in the Mountainous Conditions of Central Europe. *Ecol. Eng.* **2017**, *100*, 219–230. [CrossRef]
20. Cao, L.; Pan, J.; Li, R.; Li, J.; Li, Z. Integrating Airborne LiDAR and Optical Data to Estimate Forest Aboveground Biomass in Arid and Semi-Arid Regions of China. *Remote Sens.* **2018**, *10*, 532. [CrossRef]
21. Jiang, X.; Li, G.; Lu, D.; Chen, E.; Wei, X. Stratification-Based Forest Aboveground Biomass Estimation in a Subtropical Region Using Airborne Lidar Data. *Remote Sens.* **2020**, *12*, 1101. [CrossRef]
22. Poorazimy, M.; Shataee, S.; McRoberts, R.E.; Mohammadi, J. Integrating Airborne Laser Scanning Data, Space-Borne Radar Data and Digital Aerial Imagery to Estimate Aboveground Carbon Stock in Hyrcanian Forests, Iran. *Remote Sens. Environ.* **2020**, *240*, 111669. [CrossRef]
23. Chan, E.P.Y.; Fung, T.; Wong, F.K.K. Estimating Above-Ground Biomass of Subtropical Forest Using Airborne LiDAR in Hong Kong. *Sci. Rep.* **2021**, *11*, 1751. [CrossRef]
24. Gleason, C.J.; Im, J. Forest Biomass Estimation from Airborne LiDAR Data Using Machine Learning Approaches. *Remote Sens. Environ.* **2012**, *125*, 80–91. [CrossRef]
25. Li, M.; Im, J.; Quackenbush, L.J.; Liu, T. Forest Biomass and Carbon Stock Quantification Using Airborne LiDAR Data: A Case Study Over Huntington Wildlife Forest in the Adirondack Park. *IEEE J. Sel. Top. Appl. Earth Obs. Remote Sens.* **2014**, *7*, 3143–3156. [CrossRef]
26. John, R.; Chen, J.; Giannico, V.; Park, H.; Xiao, J.; Shirkey, G.; Ouyang, Z.; Shao, C.; Lafortezza, R.; Qi, J. Grassland Canopy Cover and Aboveground Biomass in Mongolia and Inner Mongolia: Spatiotemporal Estimates and Controlling Factors. *Remote Sens. Environ.* **2018**, *213*, 34–48. [CrossRef]
27. Huang, W.; Dolan, K.; Swatantran, A.; Johnson, K.; Tang, H.; O'Neil-Dunne, J.; Dubayah, R.; Hurtt, G. High-Resolution Mapping of Aboveground Biomass for Forest Carbon Monitoring System in the Tri-State Region of Maryland, Pennsylvania and Delaware, USA. *Environ. Res. Lett.* **2019**, *14*, 095002. [CrossRef]

28. Dos Reis, A.A.; Werner, J.P.S.; Silva, B.C.; Figueiredo, G.K.D.A.; Antunes, J.F.G.; Esquerdo, J.C.D.M.; Coutinho, A.C.; Lamparelli, R.A.C.; Rocha, J.V.; Magalhães, P.S.G. Monitoring Pasture Aboveground Biomass and Canopy Height in an Integrated Crop–Livestock System Using Textural Information from PlanetScope Imagery. *Remote Sens.* **2020**, *12*, 2534. [CrossRef]
29. Sun, H.; He, J.; Chen, Y.; Zhao, B. Space-Time Sea Surface PCO2 Estimation in the North Atlantic Based on CatBoost. *Remote Sens.* **2021**, *13*, 2805. [CrossRef]
30. Ahirwal, J.; Nath, A.; Brahma, B.; Deb, S.; Sahoo, U.K.; Nath, A.J. Patterns and Driving Factors of Biomass Carbon and Soil Organic Carbon Stock in the Indian Himalayan Region. *Sci. Total Environ.* **2021**, *770*, 145292. [CrossRef]
31. McRoberts, R.E.; Gobakken, T.; Næsset, E. Post-Stratified Estimation of Forest Area and Growing Stock Volume Using Lidar-Based Stratifications. *Remote Sens. Environ.* **2012**, *125*, 157–166. [CrossRef]
32. Zhao, P.; Lu, D.; Wang, G.; Liu, L.; Li, D.; Zhu, J.; Yu, S. Forest Aboveground Biomass Estimation in Zhejiang Province Using the Integration of Landsat TM and ALOS PALSAR Data. *Int. J. Appl. Earth Obs. Geoinf.* **2016**, *53*, 1–15. [CrossRef]
33. Shao, G.; Shao, G.; Gallion, J.; Saunders, M.R.; Frankenberger, J.R.; Fei, S. Improving Lidar-Based Aboveground Biomass Estimation of Temperate Hardwood Forests with Varying Site Productivity. *Remote Sens. Environ.* **2018**, *204*, 872–882. [CrossRef]
34. Silveira, E.M.O.; Espírito Santo, F.D.; Wulder, M.A.; Acerbi Júnior, F.W.; Carvalho, M.C.; Mello, C.R.; Mello, J.M.; Shimabukuro, Y.E.; Terra, M.C.N.S.; Carvalho, L.M.T.; et al. Pre-Stratified Modelling plus Residuals Kriging Reduces the Uncertainty of Aboveground Biomass Estimation and Spatial Distribution in Heterogeneous Savannas and Forest Environments. *For. Ecol. Manag.* **2019**, *445*, 96–109. [CrossRef]
35. Gao, Y.; Lu, D.; Li, G.; Wang, G.; Chen, Q.; Liu, L.; Li, D. Comparative Analysis of Modeling Algorithms for Forest Aboveground Biomass Estimation in a Subtropical Region. *Remote Sens.* **2018**, *10*, 627. [CrossRef]
36. Tonolli, S.; Dalponte, M.; Neteler, M.; Rodeghiero, M.; Vescovo, L.; Gianelle, D. Fusion of Airborne LiDAR and Satellite Multispectral Data for the Estimation of Timber Volume in the Southern Alps. *Remote Sens. Environ.* **2011**, *115*, 2486–2498. [CrossRef]
37. Kulawardhana, R.W.; Popescu, S.C.; Feagin, R.A. Fusion of Lidar and Multispectral Data to Quantify Salt Marsh Carbon Stocks. *Remote Sens. Environ.* **2014**, *154*, 345–357. [CrossRef]
38. Latifi, H.; Fassnacht, F.E.; Hartig, F.; Berger, C.; Hernández, J.; Corvalán, P.; Koch, B. Stratified Aboveground Forest Biomass Estimation by Remote Sensing Data. *Int. J. Appl. Earth Obs. Geoinf.* **2015**, *38*, 229–241. [CrossRef]
39. Fang, J.; Chen, A.; Peng, C.; Zhao, S.; Ci, L. Changes in Forest Biomass Carbon Storage in China Between 1949 and 1998. *Science* **2001**, *292*, 2320–2322. [CrossRef]
40. Tian, Y.; Huang, H.; Zhou, G.; Zhang, Q.; Tao, J.; Zhang, Y.; Lin, J. Aboveground Mangrove Biomass Estimation in Beibu Gulf Using Machine Learning and UAV Remote Sensing. *Sci. Total Environ.* **2021**, *781*, 146816. [CrossRef]
41. Zhang, Y.; Liang, S.; Sun, G. Forest Biomass Mapping of Northeastern China Using GLAS and MODIS Data. *IEEE J. Sel. Top. Appl. Earth Obs. Remote Sens.* **2014**, *7*, 140–152. [CrossRef]
42. Fu, Y.; He, H.S.; Hawbaker, T.J.; Henne, P.D.; Zhu, Z.; Larsen, D.R. Evaluating *k*-Nearest Neighbor (*k*NN) Imputation Models for Species-Level Aboveground Forest Biomass Mapping in Northeast China. *Remote Sens.* **2019**, *11*, 2005. [CrossRef]
43. Vaglio Laurin, G.; Puletti, N.; Hawthorne, W.; Liesenberg, V.; Corona, P.; Papale, D.; Chen, Q.; Valentini, R. Discrimination of Tropical Forest Types, Dominant Species, and Mapping of Functional Guilds by Hyperspectral and Simulated Multispectral Sentinel-2 Data. *Remote Sens. Environ.* **2016**, *176*, 163–176. [CrossRef]
44. Jia, W. *Forest Biomass and Carbon Stock of Each Stand Type in the Northeast Forest Region*; Heilongjiang Science and Technology Press: Harbin, China, 2015.
45. *LY/T 2654-2016*; Tree Biomass Models and Related Parameters to Carbon. National Forestry and Grassland Administration of China: Beijing, China, 2016.
46. Zhao, X.; Guo, Q.; Su, Y.; Xue, B. Improved Progressive TIN Densification Filtering Algorithm for Airborne LiDAR Data in Forested Areas. *ISPRS J. Photogramm. Remote Sens.* **2016**, *117*, 79–91. [CrossRef]
47. Axelsson, P. DEM Generation from Laser Scanner Data Using Adaptive TIN Models. *Int. Arch. Photogramm. Remote Sens.* **2000**, *33*, 110–117.
48. Knapp, N.; Fischer, R.; Huth, A. Linking Lidar and Forest Modeling to Assess Biomass Estimation across Scales and Disturbance States. *Remote Sens. Environ.* **2018**, *205*, 199–209. [CrossRef]
49. de Oliveira, C.P.; Caraciolo Ferreira, R.L.; Aleixo da Silva, J.A.; de Lima, R.B.; Silva, E.A.; da Silva, A.F.; Silva de Lucena, J.D.; Tavares dos Santos, N.A.; Correa Lopes, I.J.; de Lima Pessoa, M.M.; et al. Modeling and Spatialization of Biomass and Carbon Stock Using LiDAR Metrics in Tropical Dry Forest, Brazil. *Forests* **2021**, *12*, 473. [CrossRef]
50. Luo, S.; Wang, C.; Xi, X.; Pan, F.; Qian, M.; Peng, D.; Nie, S.; Qin, H.; Lin, Y. Retrieving Aboveground Biomass of Wetland *Phragmites australis* (Common Reed) Using a Combination of Airborne Discrete-Return LiDAR and Hyperspectral Data. *Int. J. Appl. Earth Obs. Geoinf.* **2017**, *58*, 107–117. [CrossRef]
51. Wang, D.; Wan, B.; Liu, J.; Su, Y.; Guo, Q.; Qiu, P.; Wu, X. Estimating Aboveground Biomass of the Mangrove Forests on Northeast Hainan Island in China Using an Upscaling Method from Field Plots, UAV-LiDAR Data and Sentinel-2 Imagery. *Int. J. Appl. Earth Obs. Geoinf.* **2020**, *85*, 101986. [CrossRef]
52. Kursa, M.B.; Rudnicki, W.R. Feature Selection with the Boruta Package. *J. Stat. Softw.* **2010**, *36*, 1–13. [CrossRef]
53. Sun, G.; Ranson, K.J.; Guo, Z.; Zhang, Z.; Montesano, P.; Kimes, D. Forest Biomass Mapping from Lidar and Radar Synergies. *Remote Sens. Environ.* **2011**, *115*, 2906–2916. [CrossRef]

54. Kronseder, K.; Ballhorn, U.; Boehm, V.; Siegert, F. Above Ground Biomass Estimation across Forest Types at Different Degradation Levels in Central Kalimantan Using LiDAR Data. *Int. J. Appl. Earth Obs. Geoinf.* **2012**, *18*, 37–48. [CrossRef]
55. Ku, N.-W.; Popescu, S.C. A Comparison of Multiple Methods for Mapping Local-Scale Mesquite Tree Aboveground Biomass with Remotely Sensed Data. *Biomass Bioenergy* **2019**, *122*, 270–279. [CrossRef]
56. Breiman, L. Random Forests. *Mach. Learn.* **2001**, *45*, 5–32. [CrossRef]
57. Huang, G.; Wu, L.; Ma, X.; Zhang, W.; Fan, J.; Yu, X.; Zeng, W.; Zhou, H. Evaluation of CatBoost Method for Prediction of Reference Evapotranspiration in Humid Regions. *J. Hydrol.* **2019**, *574*, 1029–1041. [CrossRef]
58. de Almeida, C.T.; Galvão, L.S.; Aragão, L.E.D.O.C.E.; Ometto, J.P.H.B.; Jacon, A.D.; Pereira, F.R.D.S.; Sato, L.Y.; Lopes, A.P.; Graça, P.M.L.D.A.; Silva, C.V.D.J.; et al. Combining LiDAR and Hyperspectral Data for Aboveground Biomass Modeling in the Brazilian Amazon Using Different Regression Algorithms. *Remote Sens. Environ.* **2019**, *232*, 111323. [CrossRef]
59. Chen, T.; Guestrin, C. Xgboost: A scalable tree boosting system. In Proceedings of the 22nd ACM SIGKDD International Conference on Knowledge Discovery and Data Mining, San Francisco, CA, USA, 13–17 August 2016; pp. 785–794.
60. Li, Y.; Li, M.; Li, C.; Liu, Z. Forest Aboveground Biomass Estimation Using Landsat 8 and Sentinel-1A Data with Machine Learning Algorithms. *Sci. Rep.* **2020**, *10*, 9952. [CrossRef]
61. Prokhorenkova, L.; Gusev, G.; Vorobev, A.; Dorogush, A.V.; Gulin, A. CatBoost: Unbiased Boosting with Categorical Features. *arXiv* **2019**, arXiv:1706.09516.
62. Hancock, J.T.; Khoshgoftaar, T.M. CatBoost for Big Data: An Interdisciplinary Review. *J. Big Data* **2020**, *7*, 94. [CrossRef]
63. de Souza Pereira, F.R.; Kampel, M.; Gomes Soares, M.L.; Duque Estrada, G.C.; Bentz, C.; Vincent, G. Reducing Uncertainty in Mapping of Mangrove Aboveground Biomass Using Airborne Discrete Return Lidar Data. *Remote Sens.* **2018**, *10*, 637. [CrossRef]
64. Liu, K.; Shen, X.; Cao, L.; Wang, G.; Cao, F. Estimating Forest Structural Attributes Using UAV-LiDAR Data in Ginkgo Plantations. *ISPRS J. Photogramm. Remote Sens.* **2018**, *146*, 465–482. [CrossRef]
65. Shi, Y.; Wang, T.; Skidmore, A.K.; Heurich, M. Important LiDAR Metrics for Discriminating Forest Tree Species in Central Europe. *ISPRS J. Photogramm. Remote Sens.* **2018**, *137*, 163–174. [CrossRef]
66. Kashani, A.G.; Olsen, M.J.; Parrish, C.E.; Wilson, N. A Review of LIDAR Radiometric Processing: From Ad Hoc Intensity Correction to Rigorous Radiometric Calibration. *Sensors* **2015**, *15*, 28099–28128. [CrossRef] [PubMed]
67. Hoefle, B.; Pfeifer, N. Correction of Laser Scanning Intensity Data: Data and Model-Driven Approaches. *ISPRS J. Photogramm. Remote Sens.* **2007**, *62*, 415–433. [CrossRef]
68. Javier Mesas-Carrascosa, F.; Luisa Castillejo-Gonzalez, I.; Sanchez de la Orden, M.; Garcia-Ferrer Porras, A. Combining LiDAR Intensity with Aerial Camera Data to Discriminate Agricultural Land Uses. *Comput. Electron. Agric.* **2012**, *84*, 36–46. [CrossRef]
69. Bouvier, M.; Durrieu, S.; Fournier, R.A.; Renaud, J.-P. Generalizing Predictive Models of Forest Inventory Attributes Using an Area-Based Approach with Airborne LiDAR Data. *Remote Sens. Environ.* **2015**, *156*, 322–334. [CrossRef]
70. Wang, Q.; Pang, Y.; Chen, D.; Liang, X.; Lu, J. Lidar Biomass Index: A Novel Solution for Tree-Level Biomass Estimation Using 3D Crown Information. *For. Ecol. Manag.* **2021**, *499*, 119542. [CrossRef]
71. Zhao, P.; Lu, D.; Wang, G.; Wu, C.; Huang, Y.; Yu, S. Examining Spectral Reflectance Saturation in Landsat Imagery and Corresponding Solutions to Improve Forest Aboveground Biomass Estimation. *Remote Sens.* **2016**, *8*, 469. [CrossRef]
72. Liu, Y.; Gong, W.; Xing, Y.; Hu, X.; Gong, J. Estimation of the Forest Stand Mean Height and Aboveground Biomass in Northeast China Using SAR Sentinel-1B, Multispectral Sentinel-2A, and DEM Imagery. *ISPRS J. Photogramm. Remote Sens.* **2019**, *151*, 277–289. [CrossRef]
73. Heurich, M.; Thoma, F. Estimation of Forestry Stand Parameters Using Laser Scanning Data in Temperate, Structurally Rich Natural European Beech (*Fagus sylvatica*) and Norway Spruce (*Picea abies*) Forests. *Forestry* **2008**, *81*, 645–661. [CrossRef]
74. Nelson, R.; Short, A.; Valenti, M. Measuring Biomass and Carbon in Delaware Using an Airborne Profiling LIDAR. *Scand. J. For. Res.* **2004**, *19*, 500–511. [CrossRef]
75. Clark, D.B.; Kellner, J.R. Tropical Forest Biomass Estimation and the Fallacy of Misplaced Concreteness. *J. Veg. Sci.* **2012**, *23*, 1191–1196. [CrossRef]
76. Nelson, R.F.; Hyde, P.; Johnson, P.; Emessiene, B.; Imhoff, M.L.; Campbell, R.; Edwards, W. Investigating RaDAR–LiDAR Synergy in a North Carolina Pine Forest. *Remote Sens. Environ.* **2007**, *110*, 98–108. [CrossRef]
77. Sarrazin, M.J.D.; van Aardt, J.A.N.; Asner, G.P.; McGlinchy, J.; Messinger, D.W.; Wu, J. Fusing Small-Footprint Waveform LiDAR and Hyperspectral Data for Canopy-Level Species Classification and Herbaceous Biomass Modeling in Savanna Ecosystems. *Can. J. Remote Sens.* **2011**, *37*, 653–665. [CrossRef]
78. Labrecque, S.; Fournier, R.A.; Luther, J.E.; Piercey, D. A Comparison of Four Methods to Map Biomass from Landsat-TM and Inventory Data in Western Newfoundland. *For. Ecol. Manag.* **2006**, *226*, 129–144. [CrossRef]
79. Tipton, J.; Opsomer, J.; Moisen, G. Properties of Endogenous Post-Stratified Estimation Using Remote Sensing Data. *Remote Sens. Environ.* **2013**, *139*, 130–137. [CrossRef]
80. Breidenbach, J.; Nothdurft, A.; Kändler, G. Comparison of Nearest Neighbour Approaches for Small Area Estimation of Tree Species-Specific Forest Inventory Attributes in Central Europe Using Airborne Laser Scanner Data. *Eur. J. For. Res.* **2010**, *129*, 833–846. [CrossRef]
81. Zhang, R.; Zhou, X.; Ouyang, Z.; Avitabile, V.; Qi, J.; Chen, J.; Giannico, V. Estimating Aboveground Biomass in Subtropical Forests of China by Integrating Multisource Remote Sensing and Ground Data. *Remote Sens. Environ.* **2019**, *232*, 111341. [CrossRef]

82. Westfall, J.A.; Patterson, P.L.; Coulston, J.W. Post-Stratified Estimation: Within-Strata and Total Sample Size Recommendations. *Can. J. For. Res.* **2011**, *41*, 1130–1139. [CrossRef]
83. Feng, Y.; Lu, D.; Chen, Q.; Keller, M.; Moran, E.; dos-Santos, M.N.; Bolfe, E.L.; Batistella, M. Examining Effective Use of Data Sources and Modeling Algorithms for Improving Biomass Estimation in a Moist Tropical Forest of the Brazilian Amazon. *Int. J. Digit. Earth* **2017**, *10*, 996–1016. [CrossRef]
84. Pham, T.D.; Yokoya, N.; Xia, J.; Ha, N.T.; Le, N.N.; Nguyen, T.T.T.; Dao, T.H.; Vu, T.T.P.; Pham, T.D.; Takeuchi, W. Comparison of Machine Learning Methods for Estimating Mangrove Above-Ground Biomass Using Multiple Source Remote Sensing Data in the Red River Delta Biosphere Reserve, Vietnam. *Remote Sens.* **2020**, *12*, 1334. [CrossRef]
85. Zhang, Y.; Ma, J.; Liang, S.; Li, X.; Li, M. An Evaluation of Eight Machine Learning Regression Algorithms for Forest Aboveground Biomass Estimation from Multiple Satellite Data Products. *Remote Sens.* **2020**, *12*, 4015. [CrossRef]
86. Luo, M.; Wang, Y.; Xie, Y.; Zhou, L.; Qiao, J.; Qiu, S.; Sun, Y. Combination of Feature Selection and CatBoost for Prediction: The First Application to the Estimation of Aboveground Biomass. *Forests* **2021**, *12*, 216. [CrossRef]
87. Zhang, Q.; He, H.S.; Liang, Y.; Hawbaker, T.J.; Henne, P.D.; Liu, J.; Huang, S.; Wu, Z.; Huang, C. Integrating Forest Inventory Data and MODIS Data to Map Species-Level Biomass in Chinese Boreal Forests. *Can. J. For. Res.* **2018**, *48*, 461–479. [CrossRef]
88. Wulder, M.A.; White, J.C.; Nelson, R.F.; Næsset, E.; Ørka, H.O.; Coops, N.C.; Hilker, T.; Bater, C.W.; Gobakken, T. Lidar Sampling for Large-Area Forest Characterization: A Review. *Remote Sens. Environ.* **2012**, *121*, 196–209. [CrossRef]
89. Campbell, M.J.; Dennison, P.E.; Kerr, K.L.; Brewer, S.C.; Anderegg, W.R.L. Scaled Biomass Estimation in Woodland Ecosystems: Testing the Individual and Combined Capacities of Satellite Multispectral and Lidar Data. *Remote Sens. Environ.* **2021**, *262*, 112511. [CrossRef]
90. Chen, Q.; Laurin, G.V.; Valentini, R. Uncertainty of Remotely Sensed Aboveground Biomass over an African Tropical Forest: Propagating Errors from Trees to Plots to Pixels. *Remote Sens. Environ.* **2015**, *160*, 134–143. [CrossRef]
91. Chave, J.; Condit, R.; Aguilar, S.; Hernandez, A.; Lao, S.; Perez, R. Error Propagation and Scaling for Tropical Forest Biomass Estimates. *Philos. Trans. R. Soc. B Biol. Sci.* **2004**, *359*, 409–420. [CrossRef] [PubMed]
92. Rammig, A.; Heinke, J.; Hofhansl, F.; Verbeeck, H.; Baker, T.R.; Christoffersen, B.; Ciais, P.; De Deurwaerder, H.; Fleischer, K.; Galbraith, D.; et al. A Generic Pixel-to-Point Comparison for Simulated Large-Scale Ecosystem Properties and Ground-Based Observations: An Example from the Amazon Region. *Geosci. Model Dev.* **2018**, *11*, 5203–5215. [CrossRef]
93. Xu, Q.; Man, A.; Fredrickson, M.; Hou, Z.; Pitkanen, J.; Wing, B.; Ramirez, C.; Li, B.; Greenberg, J.A. Quantification of Uncertainty in Aboveground Biomass Estimates Derived from Small-Footprint Airborne LiDAR. *Remote Sens. Environ.* **2018**, *216*, 514–528. [CrossRef]
94. Disney, M.I.; Kalogirou, V.; Lewis, P.; Prieto-Blanco, A.; Hancock, S.; Pfeifer, M. Simulating the Impact of Discrete-Return Lidar System and Survey Characteristics over Young Conifer and Broadleaf Forests. *Remote Sens. Environ.* **2010**, *114*, 1546–1560. [CrossRef]
95. Garcia, M.; Saatchi, S.; Ferraz, A.; Silva, C.A.; Ustin, S.; Koltunov, A.; Balzter, H. Impact of Data Model and Point Density on Aboveground Forest Biomass Estimation from Airborne LiDAR. *Carbon Balance Manag.* **2017**, *12*, 4. [CrossRef] [PubMed]
96. Hernández-Stefanoni, J.L.; Reyes-Palomeque, G.; Castillo-Santiago, M.Á.; George-Chacón, S.P.; Huechacona-Ruiz, A.H.; Tun-Dzul, F.; Rondon-Rivera, D.; Dupuy, J.M. Effects of Sample Plot Size and GPS Location Errors on Aboveground Biomass Estimates from LiDAR in Tropical Dry Forests. *Remote Sens.* **2018**, *10*, 1586. [CrossRef]
97. Knapp, N.; Huth, A.; Fischer, R. Tree Crowns Cause Border Effects in Area-Based Biomass Estimations from Remote Sensing. *Remote Sens.* **2021**, *13*, 1592. [CrossRef]
98. Frazer, G.W.; Magnussen, S.; Wulder, M.A.; Niemann, K.O. Simulated Impact of Sample Plot Size and Co-Registration Error on the Accuracy and Uncertainty of LiDAR-Derived Estimates of Forest Stand Biomass. *Remote Sens. Environ.* **2011**, *115*, 636–649. [CrossRef]
99. Roedig, E.; Knapp, N.; Fischer, R.; Bohn, F.J.; Dubayah, R.; Tang, H.; Huth, A. From Small-Scale Forest Structure to Amazon-Wide Carbon Estimates. *Nat. Commun.* **2019**, *10*, 5088. [CrossRef] [PubMed]

remote sensing

MDPI

Article

Stand Canopy Closure Estimation in Planted Forests Using a Geometric-Optical Model Based on Remote Sensing

Xiguang Yang [1,2], Ping He [1,2], Ying Yu [1,2,*] and Wenyi Fan [1,2]

[1] School of Forestry, Northeast Forestry University, Harbin 150040, China; yangxiguang@nefu.edu.cn (X.Y.); 2416603957@nefu.edu.cn (P.H.); fanwy@nefu.edu.cn (W.F.)

[2] Key Laboratory of Sustainable Forest Ecosystem Management—Ministry of Education, Northeast Forestry University, Harbin 150040, China

* Correspondence: yuying@nefu.edu.cn

Abstract: Canopy closure, which is the ratio of the vertical projection area of the crowns to the area of forest land, can indicate the growth and tending situation of a forest and is of great significance for forest management planning. In this study, a geometric-optical model (GOST model) was used to simulate the canopy gap fraction of a forest. Then, a canopy closure estimation method using the gap fraction was discussed. In this study, three typical planted forest farms (the Mengjiagang (MJG), Gaofeng (GF), and Wangyedian (WYD) forest farms) containing the most commonly planted tree species in the north and south regions of China were selected, and field measurements were executed. The results show that the gap fraction ($P_{vg\text{-}c}$) had a higher correlation with the average projected area of the tree crowns, and the relationship was an exponential function, with R^2 and RMSE values of 0.5619 and 0.0723, respectively. Finally, the applicability and accuracy of this method were evaluated using line transects, and a fisheye camera measured the canopy closure. The accuracy of the canopy closure estimated by the $P_{vg\text{-}c}$ was 86.69%. This research can provide a reference for canopy closure estimation using a geometric-optical model.

Keywords: canopy closure; the GOST model; fisheye camera photos; transects; LAI

Citation: Yang, X.; He, P.; Yu, Y.; Fan, W. Stand Canopy Closure Estimation in Planted Forests Using a Geometric-Optical Model Based on Remote Sensing. *Remote Sens.* **2022**, *14*, 1983. https://doi.org/10.3390/rs14091983

Academic Editor: Lars T. Waser

Received: 16 March 2022
Accepted: 19 April 2022
Published: 20 April 2022

1. Introduction

Globally, planted forests are an important type of forest. According to a number of studies, planted forest areas have continued to increase due to industrial demands for wood shifting from natural forests to planted forests [1,2]. Planted forests reduce harvesting from natural forests by 26%, and they have significant ecological benefits [3]. In order to improve the management level of China's planted forests, it is essential to plan the afforestation process and to scientifically arrange the forest management strategy. At the same time, it is also important to monitor and evaluate the resources in planted forests precisely and meticulously [4].

Canopy closure, which is the ratio of the vertical projection area of tree crowns to the ground area [5–8], plays a very important role in the state of the forest ecosystem as well as in environmental evaluation, and it is widely used in forestry evaluations [9]. Canopy closure is an important index that reflects the spatial structure of an ecosystem and the tree stand density, and it is an important investigation indicator for planted forests [10,11]. Accurate canopy closure measurements and estimations provide an important reference that can be used to evaluate the quality of a plantation.

Canopy closure can be measured using a wide variety of ground-based techniques. These ground-based methods mainly include ocular estimates, hemispherical photography, transects, sample points, the line intercept method, canopy projections, the visual observation method, and canopy instrument analysis, among others [12]. Among these methods, transects and sample points have a higher measurement accuracy than other ground-based methods [13,14]. Hemispherical photography is also a commonly used method that is used to estimate the canopy closure [13]. The development of digital cameras and their increasing availability has made hemispherical photography more widely available for forest inventory evaluations [15]. However, the cost and required resources for hemispherical photography still preclude many forest managers from using it as a monitoring tool [11]. In addition, canopy closure is underestimated by using the instruments, such as fisheye cameras, due to gaps in the canopy [13]. Macfarlane et al. concluded that future investigations of fisheye camera methods should concentrate on obtaining an accurate gap fraction distribution method that can separate the effects of foliage angle distribution from those of foliage clumping [16]. The transect method is considered to be the most reliable method for investigating canopy closure, and it can directly verify the canopy closure estimated by remote sensing images [17]. However, even though ground-based methods always have better accuracy, human and material resource inputs need to be considered. Additionally, it is difficult to obtain the canopy closure distribution at a regional or larger scale [18].

As an efficient and low-cost data resource, remote sensing is regarded as one of the most effective ways to estimate canopy closure in regional or large areas [19–22]. Aerial platforms are used to obtain photos of the forest region during the early stages, and originally, these aerial photos were obtained to estimate forest information such as the canopy closure, LAI, and forest volume [23–25]. With the development of laser ranging technology, light detection and ranging (LiDAR) data have been used to estimate forest structural parameters, such as tree height as well as tree crown and single tree properties [26,27]. Placing LiDAR sensors on the aerial platforms is a good option for tree canopy closure estimation [28]. With the emergence of various types of LiDAR sensors on the market, more and more research on canopy closure estimation using LIDAR data has been published [29–31]. Additionally, the high-precision canopy height model (CHM) is extracted using LIDAR data and image segmentation, and canopy width extraction technology is used to estimate canopy closure [32,33]. Bode et al. proposed a canopy closure estimation approach using a light penetration index (LPI) based on airborne LiDAR data at watershed scales [34]. However, huge economic costs must be considered, especially when aerial platforms are being used to estimate the canopy closure in large areas [35]. Additionally, the costs make it very difficult to determine the forest inventory at a high frequency and in a large-scale region [36]. Compared to aerial platforms, satellite remote sensing platforms have advantages, such as good real-time, large-scale, low-cost, accurate, and multiple types of sensors. Satellite remote sensing images have been widely used to determine the forest inventory at both local and large scales [37–39].

To estimate canopy closure in a large area, statistical or physical models are used to estimate the forest canopy closure. Statistical models represent economical and efficient canopy closure estimation methods that establish the relationship between remote sensing variables and measured field data [40–42].

Among the various statistical models, multiple linear regression models are usually used to establish the relationship between the canopy closure and the remote sensing variables obtained from remote sensing images, such as spectral information, the vegetation index, and texture information [43–45]. However, the method does not fully consider the influence of the spatial resolution of remote sensing images on the accuracy of canopy closure estimations [46,47]. Additionally, the effect of mixed pixels on the accuracy of canopy closure estimations is ignored in this method. The pixel binary model considers mixed pixels, it cannot separate arbors and shrubs, and it is also affected by the background soil type, meaning that it is difficult for this method to provide accurate stand canopy closure information [48–51].

Physical models are other types of canopy closure estimation models that are based on the energy transmission procedures of remote sensing. Typical physical models include the geometric-optical model and the radiative transfer model [52]. The Li–Strahler geometric-optical model is suitable for canopy closure estimations in flat study areas [53,54]. Additionally, the core of this method is its ability to accurately extract proportions of four scene components—sunlit foliage, sunlit ground, shaded foliage, and shaded ground—in a pixel [52,55]. It is suitable for simulating canopy reflectance by considering forest distribution and structure. The PROSAIL model is one of the most famous radiative transfer models used for canopy reflection simulation and combines the PROSPECT leaf model and SAIL canopy model (arbitrary oblique leaf-scattering) [56–58]. The PROSAIL model is also a commonly used radiation transmission model for canopy closure estimation [59,60]. The geometric-optical model and the radiation transmission model are more complex and require more parameters, but parameter sensitivity analysis can reduce the complexity of these models. Additionally, canopy closure estimation models that are established with fewer parameters are less affected by the region and measured dataset, resulting in them having better robustness [61].

In summary, there is a lot of uncertainty and unknowns when estimating canopy closure using remote sensing techniques. In this study, we considered the canopy from two perspectives—an opaque canopy body or with gaps in the tree crown. Then, the Geometric-Optical Model for Sloping Terrains (GOST) model, which is a geometric-optical model that considers terrain effects on the input parameters, was used to simulate the canopy gaps from these two perspectives. Finally, the stand-scale canopy closure was estimated in planted forests. The estimated results were tested using the measured canopy closure results obtained from transects and fisheye camera photos to answer the following questions: (1) For remote sensing data with high spatial resolution, which parameter is appropriate to estimate the stand-scale canopy closure with the highest accuracy? (2) For remote sensing data with low or medium spatial resolution, how can a geometric-optical model be used to establish a better robustness canopy closure model while considering the mixed pixel problem? (3) How can we evaluate the feasibility and accuracy of the canopy closure method presented in this study in a planted forest?

2. Materials and Methods

2.1. Study Area

In China, the main forest types include evergreen broadleaf planted forests, deciduous broad-leaved planted forests, evergreen coniferous planted forests, and deciduous needleleaf planted forests. To enhance the accuracy of the proposed method and to achieve better robustness, research regions that included most of the forest types were selected from around China. The tree species that were planted in these forest farms were also the most commonly planted species in China. The Wangyedian Forest Farm, Gaofeng Forest Farm, and Mengjiagang Forest Farm were chosen for this study because they represent the typical plantation types in the north and south regions of China (Figure 1).

The Mengjiagang Forest Farm (MJG) is located in Heilongjiang Province, which is located in the northeast part of China. It is located at 130°32′–130°52′E, 46°20′–46°30′N. The relative height of the mountain on which it is located is 168 to 575 m, and the slope is 10–20°. The total area of the forest farm is 1.7×10^4 km^2, and the plantation accounts for 76.7% of the total area and mainly includes larch (*Larix gmelinii* (Rupr.) Kuzen.), mongolica (*Pinus sylvestris* var. *mongolica* Litv.), and Korean pine (*Pinus koraiensis* Sieb. et Zucc.), with these tree species accounting for about 80% of the plantation population.

The Gaofeng Forest Farm (GF) is located in Guangxi Province, which is located in the south of China. It is located at 108°08′–108°53′E, 22°49′–23°15′N. The forest area is 4.7×10^4 km^2, and the forest coverage is 83.7%. The relative height of the mountain on which the forest area is located is 150 to 400 m, and the slope is 20–30°. The forest farm is mainly composed of a secondary plantation. The main tree species are Masson pine (*Pinus massoniana Lamb*), fir (*Cunninghamia lanceolata* (Lamb.) Hook.), and fast-growing eucalyptus (*Eucalyptusrobusta Smith*).

The Wangyedian Forest Farm (WYD) is located in the Inner Mongolia Autonomous Region in the north of China. It is located at 118°09′–118°30′E, 41°35′–41°50′N. The forest area is about 2.5×10^4 km^2, and the forest coverage rate is over 80%. The elevation is between 800 m to 1890 m. The relative height of the mountain is around 200 to 400 m and has a slope of 15–35°. The area of the forest farm is composed of 47% planted forest and

53% natural forest, and the main tree species include oil pine (*Pinus tabuliformis* Carrière) and larch (*Larix gmelinii* (Rupr.) Kuzen.). Oil pine is present in a proportion of 51%, and the proportion of larch is 47%. The natural forests in this area are mainly composed of white birch (*Betula platyphylla*), black birch (*Betula dahurica*), mountain apricot (*Armeniaca sibirica*), aspen (*Populus davidiana*), elm (*Ulmus pumila*), hazelnut (*Corylus heterophylla*), and other tree species.

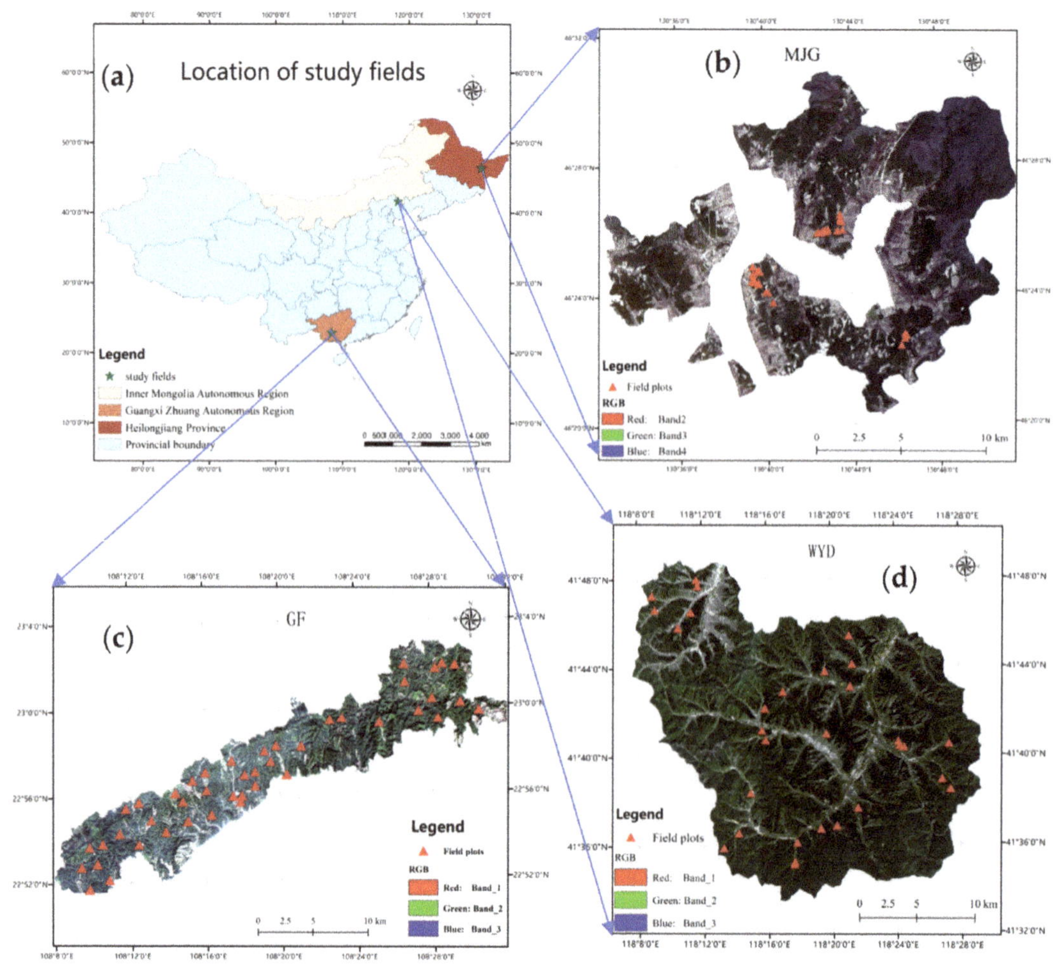

Figure 1. (**a**) The location of the study area and the field plots: (**b**) Mengjiagang Forest Farm (MJG); (**c**) Gaofeng Forest Farm (GF); (**d**) Wangyedian Forest Farm (WYD).

2.2. Field Data

A total of 102 plots were set up in the WYD, GF, and MJG. When setting up the plots, it was necessary to ensure that the plots were fully representative of the stand, that they were not scattered across different forest types, and that they were evenly distributed at different levels of the slope and aspect in each forest farm. A total of 30 plots were set up in the WYD with plot areas of 25 × 25 m, of which 18 plots were oil pine forests and 12 were larch forests. Additionally, there were 43 plots in the GF with plot areas of 20 × 20 m, with eucalyptus forests comprising most of the plots. There were 29 plots in the MJG with plot areas of 30 × 20 m. All of the plots were larch forests. The data were normalized to a minimum size of 20 × 20 m to ensure the consistency of the data.

The measured forest parameters included diameter at breast height, tree height, stem height, crown width, and tree species. All of the trees with a diameter at breast height above 5 cm in the plot were measured and recorded in one plot. Then, the GPS coordinates of each plot were recorded. The statistical information of the three research areas can be found in Table 1.

Table 1. Statistical information regarding the measurements in the three study areas.

Study Area	Plot Type	Number of Plots	Average Tree Height (m)	Average Crown Width (m)	Plant Number Density (Plants/hm^2)	Average Canopy Closure
WYD	Oil Pine	18	11.5	3.34	1552	0.64
	Larch	12	14.4	2.80	1568	0.6
MJG	Larch	29	13.3	3.40	934	0.6
GF	Eucalyptus	43	13.6	2.00	1825	0.57

The canopy closure, leaf area index (LAI), and clumping index of each plot were measured at the same time. The canopy closure was measured using transects and fisheye camera photos. Two transect lines were laid along the diagonals of the plot, and the vertical projection lengths of the crowns along each transect were recorded. The average ratio of the total projection lengths of the crowns along the two diagonals to the diagonal length was the canopy closure.

The LAI was measured using fisheye camera photos. All of the photos were taken with a Nikon Coolpix 4500. The photos were taken at the four corners and in the center of each plot. The locations in which the images were collected can be found in Figure 2a. The images were analyzed using digital hemispherical photography (DHP). The edges of the photos with large amounts of distortion were removed and processed via binarization. Then, the canopy pixel value was recorded as 1, the gap was recorded as 0, and the ratio of the number of pixels with a value of 1 to the total number of pixels was the canopy closure. Finally, the average canopy closure of the five images was the canopy closure of the plot obtained using the fisheye camera method. At the same time, the effective LAI was also able to be automatically calculated using DHP. Details of this method can be found in [62].

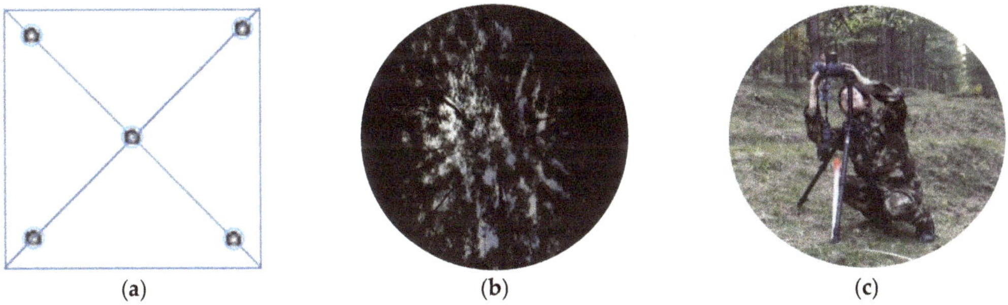

Figure 2. (**a**) The location of the fisheye camera photographs in one plot; (**b**) a sample of a fisheye camera photo; (**c**) a scene from the field measurements.

The clumping index of each plot was measured by using a TRAC instrument. Two line transects that were 20 m in length were measured. Then, the TRAC-based PPFD gradient values along the transects perpendicular to the incident directions of the solar beams were collected. In the end, the clumping index of each plot was calculated using TRACWin software. Details can be found in [63].

2.3. Methods

2.3.1. The GOST Model and Canopy Gap Fraction Simulation

The GOST model is a geometric-optical model for sloping terrains developed based on the four-scale model [64]. The four-scale model is one of the most popular geometric-optical models and can be used to simulate the bidirectional reflectance distribution function of forest canopies on flat surfaces [65]. By considering the structure of the canopy at four scales, viz., tree groups, tree crowns, branches, and shoots, the bidirectional reflectance characteristics of the forest canopies can be simulated. The four-scale model defines how the canopy reflectance in one pixel is a linear combination of the signals from four components: the sunlit and shaded foliage and the sunlit and shaded backgrounds. The total canopy reflectance is as follows:

$$R = R_T \times P_T + R_G \times P_G + R_{ZT} \times Z_T + R_{ZG} \times Z_G \quad (1)$$

where R_T is the reflectivity of the sunlit foliage; R_G is the reflectivity of the sunlit background; R_{ZT} is the reflectivity of the shaded foliage, R_{ZG} is the reflectivity of the shaded background; P_T, P_G, Z_T, and Z_G are the sensor-viewing probabilities of the four scene components, respectively (Figure 3).

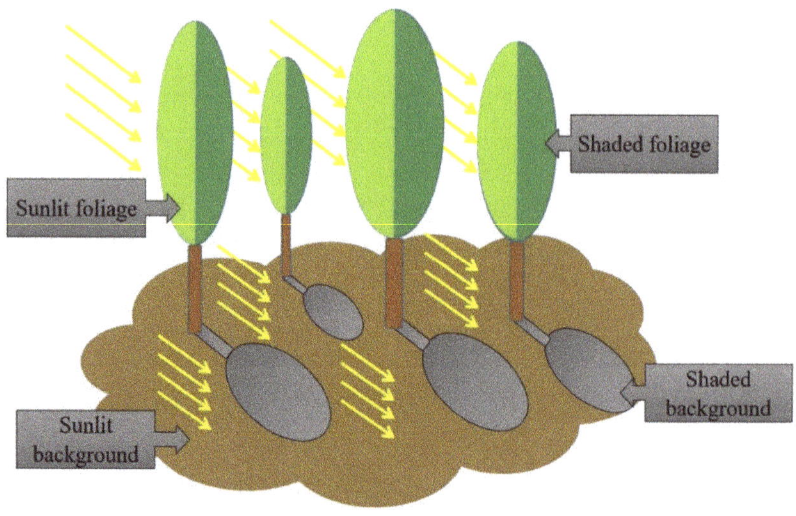

Figure 3. Schematic illustration of sunlit foliage and background and shaded foliage and background.

In the four-scale model, the probability of seeing the ground represents the ground that can be seen between tree crowns. It is a function of the projected tree crown area and the spatial distribution characteristics of trees in a quadrat. It can be calculated as follows, based on the method in [65]:

$$P_{vg\text{-}c} = \sum_{i=0}^{k} P_N(i)\left[1 - \frac{V_g}{A}\right]^i \quad (2)$$

where $P_{vg\text{-}c}$ is the probability of seeing the ground, i is the number of trees in a region with the area A, and V_g is the ground surface not seen by the viewer because of one tree. $P_N(i)$ is the probability of having i trees in A, and it is determined by the Neyman distribution. In this equation, the tree crowns are assumed to be opaque, and the gaps within the tree crowns are not considered.

If the gaps in the crowns and the overlaps in the crowns are considered, then Equation (2) can be written as follows:

$$P_{vg} = \sum_{i=1}^{k} P_{tj}(V_g) P_{gap}^{j}(\theta_v) + P_{t0} \tag{3}$$

where

$$P_{gap}^{j}(\theta_v) = \prod\nolimits_{1}^{j} P_{gap}(\theta_v) \tag{4}$$

This is the gap probability inside the trees.

$$P_{gap}(\theta_v) = e^{-G(\theta_v)L_0\Omega_E/\gamma_E}. \tag{5}$$

The gap probability $P_{gap}(\theta_v)$. is calculated based on the method developed by Li and Strahler [53], but the foliage clumping effect is considered in the four-scale model. $G(\theta_v)$. is the function of the foliage angle distribution $G(\theta_v) = 0.5$ [65] in this research. L_0 is the LAI. Ω_E . is the clumping index of the shoots within the tree crowns. γ_E is the ratio of the needle to shoot area.

In Equation (2), $P_{tj}(V_g)$ is the probability of having j trees intercepting the view line. Additionally, it can be calculated using a negative binomial function.

$$P_{tj}(V_g) = \sum_{i=j}^{k} P_N(i) \left[\frac{(i+j-1)!}{(i-1)!j!}\right] \left[1 - \frac{V_g}{A}\right]^i \left[\frac{V_g}{A}\right]^j \tag{6}$$

In the case of $j = 0$, P_{t0} is equal to Equation (2).

The canopy gap fraction determines the contribution of the under surface to the reflectance measured above the canopy. Additionally, the canopy gap fraction can be calculated using the tree crown projection made on the ground surface. It is based on the method described above for calculating the shadow area. After calculating the shadow area projected by a single tree onto the ground, the sunlit crown proportion seen by the viewer can be computed using the total surface area of the tree visible to the viewer projected to a plane perpendicular to the view line. It is not difficult to determine that the canopy gap fraction is highly related to canopy closure according to the method in the four-scale model.

In addition, topography is an important factor that has a serious effect on the bidirectional reflectance distribution function of forest canopies. Therefore, in order to make the four-scale model more suitable to simulate the bidirectional reflectance distribution function of the forest canopies on slopes, the GOST model, which is a geometric-optical model that considers the oblique topographical factors, was used in this study [64].

The GOST model was used to simulate the canopy reflectance for two different gap fractions. In one case, the gap fraction ($P_{vg\text{-}c}$) was calculated by Formula (2) during the assumption of the opaque canopy bodies. In the other case, the gap fraction (P_{vg}) was calculated using Formula (3), and gaps and overlaps were observed in the crowns. The output of the GOST model was the canopy reflectance and probability of seeing the four scene components under different view angles (Equation (1)). The four components were the sunlit foliage, the sunlit ground, the shaded foliage, and the shaded ground, and the mixed pixel decomposition problem was able to be solved as well. A database of the reflectance under various gap fractions was established, allowing the relationships between the canopy gap fraction and the stand canopy closure to be discussed. When the canopy was an opaque body, the estimated canopy closure was the true stand canopy closure, and the canopy closure measured by the traditional transects was used to verify the estimated results. When the gaps in the crowns were considered, the estimated canopy closure was compared to those measured results obtained via fisheye camera photos. This study discusses the feasibility of estimating the canopy closure using the GOST model and the spatial resolution effects of the remote sensing images.

2.3.2. Canopy Closure Estimation Based on the GOST Model

The GOST model was selected and used to simulate the canopy reflectance for a complex canopy structure. When the canopy reflectance was able to be estimated accurately, it was found to be closely related to the gap fraction function. This was because the canopy gap fraction was used to describe the light passing through the canopy. As such, the canopy gap fraction can describe canopy closure. A higher canopy gap fraction means less canopy closure, and vice versa. Additionally, it is also the reason why the canopy gap fraction can be used to estimate the canopy closure.

Two gap fractions were included in the GOST model. In one case, the canopy was assumed to be an opaque canopy body. Additionally, the gaps within the tree crown were not considered. Since canopy closure does not consider the gaps in the canopy, if the viewing direction was vertical, then the gap fraction of the viewing canopy gap fraction is the percentage of light passing through the canopy and projected to the ground surface. Additionally, the stand canopy closure should be $1 - P_{vg\text{-}c}$.

$$CC_1 = 1 - P_{vg\text{-}c} \tag{7}$$

where CC_1 is the canopy closure of the opaque canopy bodies and $P_{vg\text{-}c}$ is the gap fraction of the opaque canopy bodies.

In the other case, the gap fraction was assumed to be a tree crown with gaps. Generally, it was assumed that there were gaps in the canopy for the canopy closure values that were obtained by the fisheye camera images or estimated from the remote sensing data; therefore, the estimated canopy closure should be $1 - P_{vg}$. As such, the canopy closure could be calculated as follows:

$$CC_2 = 1 - P_{vg} \tag{8}$$

where CC_2 and P_{vg} were the canopy closure and gap fraction when there were tree crown gaps, respectively.

The input parameters of the GOST model were determined to be $\{x_1, x_2, \cdots x_n\}$, and then the $P_{vg\text{-}c}$ and P_{vg} could be simulated under various input parameters using the GOST model, and the database for the $P_{vg\text{-}c}$ and P_{vg} and the input parameters was established. The relationships among the $P_{vg\text{-}c}$, P_{vg}, and the inputs were able to be established using a statistical method.

$$P_{vg\text{-}c} = F(x_1, x_2, \cdots x_n) \tag{9}$$

$$P_{vg} = G(x_1, x_2, \cdots x_n) \tag{10}$$

where $F(x_1, x_2, \cdots x_n)$ and $G(x_1, x_2, \cdots x_n)$ were the functions of the $P_{vg\text{-}c}$ and P_{vg} related inputs, respectively. Once the relationship between the gap fraction and the inputs was established, then the gap fraction could be estimated, and the canopy closure was able to be inverted based on Equations (7) and (8).

2.3.3. Sensitivity Analysis of the GOST Model Parameters

Geometric-optical models have a large number of input variables, but we only need to know the key input variables. It is crucial to carry out a sensitivity analysis on the parameters for simulation in a complex model, especially for a model with multiple input parameters. The inputs of the GOST model included plot parameters (size of plot, number of trees in plot, LAI, slope, aspect, solar zenith angle, solar azimuth angle, view zenith angle, view azimuth angle), tree structural parameters (radius of the crowns, stem height, crown height, half apex angle, clumping index for shoots in the crown), and spectral parameters (leaf reflectivity, leaf transmittance, and ground reflectance). Therefore, the sensitivity analysis was an important process in this study. The function (Equation (11)) of the sensitivity analysis was used to determine the sensitivity of the inputs affecting the gap fraction of the canopy. It was very helpful to improve the accuracy and robustness of the model for the canopy closure estimation [66,67].

$$F' = \frac{\sum_{j=1}^{n}\left(P_0^j - P_{pert}^j\right)^2}{P_0^j} \tag{11}$$

where P_0^j is the $P_{vg\text{-}c}$, and P_{vg} is calculated based on the measured parameters (reference). P_{pert}^j is the $P_{vg\text{-}c}$ and P_{vg} when the parameter was disturbed, and F' is the sensitivity of the input parameters [68].

The sensitivity of the tree structure parameters, leaf reflectivity, leaf transmittance, and ground reflectivity was analyzed in this study. Additionally, the topography effects were corrected in the GOST model, so there was no need for a slope sensitivity analysis.

According to the results of the sensitivity analysis, the input parameter dataset, which was sensitive to canopy closure, was determined as $\{x_1, x_2, \cdots x_n\}$. The insensitive inputs were set as the average values of the measurements in the plot. Then, the $P_{vg\text{-}c}$ and P_{vg} were able to be simulated under various input parameters using the GOST model, and the database between the $P_{vg\text{-}c}$ and P_{vg} and the input parameters was established. The relationship between the $P_{vg\text{-}c}$, P_{vg}, and the inputs was able to be established, and the canopy closure could then be inverted based on Equations (7) and (8).

2.3.4. Validation

In this study, the R^2 and RMSE were selected to evaluate the precision of the canopy closure estimation model. These were calculated using Equations (12) and (13).

$$R^2 = 1 - \frac{\sum_{i=1}^{n}(y_i - \hat{y}_i)^2}{\sum_{i=1}^{n}(y_i - \overline{y})^2} \tag{12}$$

$$\text{RMSE} = \sqrt{\frac{\sum_{i=1}^{n}(y_i - \hat{y}_i)^2}{n}} \tag{13}$$

where y_i was the measured value of the sample i, $\hat{y}_i$ was the predicted value of the sample i, $\overline{y}$ was the average of all the samples, and n was the number of samples.

3. Results

3.1. Results of Sensitivity Analysis of the Parameters in the GOST Model

Since the canopy closure was the ratio of the vertical projection area of the crowns to the area of forest land and was not affected by the slope, the incident angle of the view direction was equal to 0 degrees. Without considering the influence of the slope, the sensitivity of the slope was 0. The sensitivity of the radius of the crown, stem height, crown height, half apex angle, clumping index for shoots in the crown, shape of crown, LAI, tree number, leaf reflectivity, leaf transmittance, ground reflectivity, solar azimuth angle, and view azimuth angle were calculated, and the results are shown in Table 2.

The radius of the crown and the number of plants had a greater influence on the $P_{vg\text{-}c}$ (the gap fraction between the crowns), and the other parameters had no influence. This was because the canopy was considered to be an opaque canopy. Additionally, the canopy closure was only affected by the radius of the crown and individual trees, and it was not related to the leaf area index and clumping index. Therefore, when the radius of the crown and the number of trees were determined, the vertical projection area of the canopy on the forest could be estimated. However, when the gaps in the crown were considered, the LAI had the greatest effect on the P_{vg} (total gap fraction of the canopy). The clumping index had the next largest effect, followed by the radius of the crown and the number of trees.

Table 2. Sensitivity analysis of the parameters in the GOST model.

Parameter	F' (P_{vg})	F' ($P_{vg\text{-}c}$)
Radius of Crown (m)	0.118	0.684
Stem Height (m)	0	0
Crown Height (m)	0	0
Half Apex Angle (rad)	0	0
Clumping Index	0.283	0
Shape of Crown	0	0
LAI (m/m)	0.571	0
Number of Trees (num./ha)	0.022	0.501
Leaf Reflectivity	0	0
Leaf Transmittance	0	0
Ground Reflectivity	0	0
Solar Azimuth Angle (°)	0	0
View Azimuth Angle (°)	0	0

3.2. Estimation of the Canopy Gap Fraction Based on the GOST Model

The results of the parameter sensitivity analysis show that the number of trees (n), radius of the crown (d), and LAI (l) had a greater impact on $P_{vg\text{-}c}$ and P_{vg} simulation. While n, d, and l were set as the measured values, other insensitive parameters were set as the average value of all of the sample plots. Then, the $P_{vg\text{-}c}$ and P_{vg} of all of the plots were simulated by the GOST model. The max of the simulated $P_{vg\text{-}c}$ was 0.6671, the min value of the $P_{vg\text{-}c}$ was 0.2051, and the mean value was 0.4589. Additionally, the max, min, and mean values of the P_{vg} were 0.7439, 0.3011, and 0.5641, respectively. Additionally, the value of $P_{vg\text{-}c}$ was obviously smaller than the simulated P_{vg} value.

When the $P_{vg\text{-}c}$ and P_{vg} of all of the plots were simulated, the relationship between the $P_{vg\text{-}c}$ and P_{vg} and the field-measured data was analyzed. In our statistical analysis, we found that the linear relationships among the $P_{vg\text{-}c}$ and P_{vg} and the single parameters of the number of trees and the radius of the crown were quite insignificant, respectively. In contrast, this combination of $n \times d^2$ variables had a better linear relationship with the canopy gap fraction. Additionally, the $n \times d^2$ variable was a variable because it represented the projected area of the forest canopy and was physically significant for $P_{vg\text{-}c}$ and P_{vg} estimation compared to the single variables d or n.

As such, the relationships among the $P_{vg\text{-}c}$, P_{vg}, and $n \times d^2$ were established (Table 3). The established models passed the significance test.

Table 3. Statistical models of P_{vg} and $P_{vg\text{-}c}$ estimation using $n \times d^2$.

Dependent Variable	Independent Variable	Model	R^2	RMSE
$P_{vg\text{-}c}$	$n \times d^2$	$y = 0.623e^{-0.002x}$	0.5619	0.0723
P_{vg}	$n \times d^2$	$y = 0.6732e^{-0.001x}$	0.3138	0.0813

The results show that the linear relationship between $n \times d^2$ and the $P_{vg\text{-}c}$ was significant, with the coefficient of determination being 0.5619 (Figure 4). Additionally, the reason for this could be that the $P_{vg\text{-}c}$ only considered the gaps between the canopies and ignored the gaps in the crown. The variable $n \times d^2$ represented the projected area of all of the canopies in the plot, taking the canopy as an opaque body without considering the gaps in the crown. Therefore, $n \times d^2$, which represents the projected area of the forest canopy, can describe the canopy closure well. Therefore, the linear relationship between $n \times d^2$ and the $P_{vg\text{-}c}$ was more significant, and the coefficient of determination was larger than the one in Table 3.

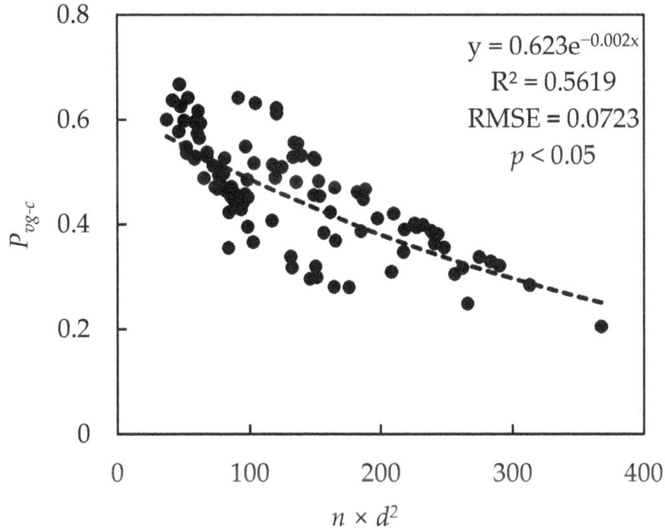

Figure 4. Scattering plot between $P_{vg\text{-}c}$ and $n \times d^2$.

The relationship between $n \times d^2$ and the P_{vg} can be seen in Figure 5. The coefficient of determination and the RMSE of the canopy closure estimation model were 0.3138 and 0.0813, respectively. Compared to the model established with the $P_{vg\text{-}c}$, these results do not seem to be as good as the results above. This is because the gap between the canopy was represented by the P_{vg} in the GOST model. Additionally, $n \times d^2$ represented the projections of the canopy without considering the gaps in the canopy. Therefore, the linear relationship between these two variables was weakened. At the same time, the results that were estimated using this model underestimate the canopy due to the gaps in the canopy represented by the P_{vg}.

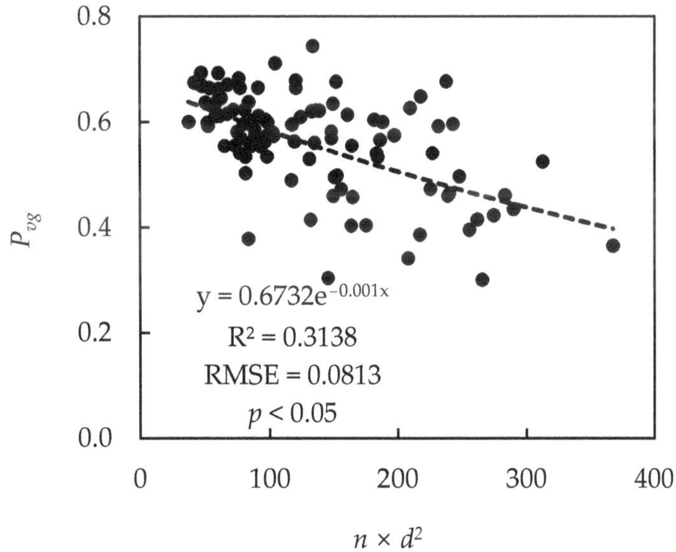

Figure 5. Scattering plot between P_{vg} and $n \times d^2$.

3.3. Estimation of the Canopy Gap Fraction Based on the LAI

The canopy gap fraction can be estimated based on the number of trees and the radius of the crown. However, it was difficult to estimate the number of trees and the radius of the crown from most of the middle-resolution remote sensing data, and it turned out to be impossible. To solve this shortage, the LAI was used to solve this problem. Based on the equation to calculate the canopy gap fraction (Equation (5)), the LAI had a significant relationship with the canopy gap fraction. Additionally, the LAI was a key parameter for solving this problem. The LAI is a commonly used forest structural parameter, and there are many LAI estimation methods that are highly accurate [69,70]. This means that the LAI can be obtained from remote sensing images easily and that it can be used to estimate the canopy closure for middle-resolution remote sensing images. The relationships among the $P_{vg\text{-}c}$, P_{vg}, and LAI (l) were established and are shown in Table 4. These models passed the significance test.

Table 4. Statistical models of P_{vg} and $P_{vg\text{-}c}$ estimation using the LAI.

Dependent Variable	Independent Variable	Model	R^2	RMSE
$P_{vg\text{-}c}$	l	$y = 0.5762e^{-0.071x}$	0.2597	0.0901
P_{vg}	l	$y = 0.739e^{-0.08x}$	0.5467	0.0654

The results show that the linear relationships among the $P_{vg\text{-}c}$, P_{vg}, and LAI were significant, with the coefficients of determination being 0.2597 and 0.5467, respectively (Table 4). An exponential function between the $P_{vg\text{-}c}$ and LAI was observed, with the coefficient of determination being 0.2597, shown in Figure 6. Although this model passed the significance test, the accuracy of the simulation was not as good as expected. Additionally, these fitting results are also not as good as the results that were obtained using the variable $n \times d^2$.

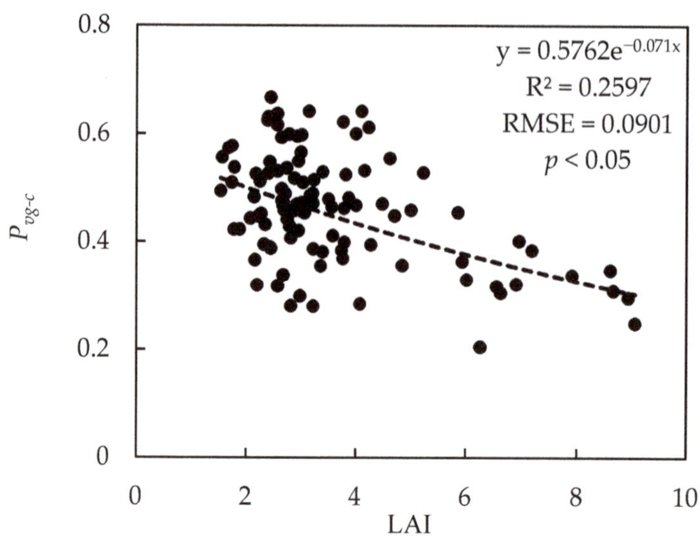

Figure 6. Scattering plot between $P_{vg\text{-}c}$ and LAI.

The linear relationship between the P_{vg} and LAI was significant, with the coefficients of determination being 0.5467 and RMSE of 0.0654, respectively (Table 4). An exponential function between the P_{vg} and LAI is also visible in Figure 7. These results are consistent with Equation (5), but the relationship between the gap fraction and the LAI was affected by other parameters of Equation (5). The reason for this is that the real exponential function was not as significant as Equation (5). In addition, compared to the results of the $P_{vg\text{-}c}$, the P_{vg} had a better linear relationship with the LAI. This could be related to the LAI, which was defined as half of the total leaf area of all of the canopies in the plot per unit of ground area [71]. As such, the gaps in and between the canopies were considered. Therefore, the linear relationship between the P_{vg} and the LAI was better than that between the $P_{vg\text{-}c}$ and the LAI. At the same time, compared to the variable $n \times d^2$, the LAI was more suitable for P_{vg} estimation, achieving R^2 and RMSE values of 0.5467 and 0.0654, respectively.

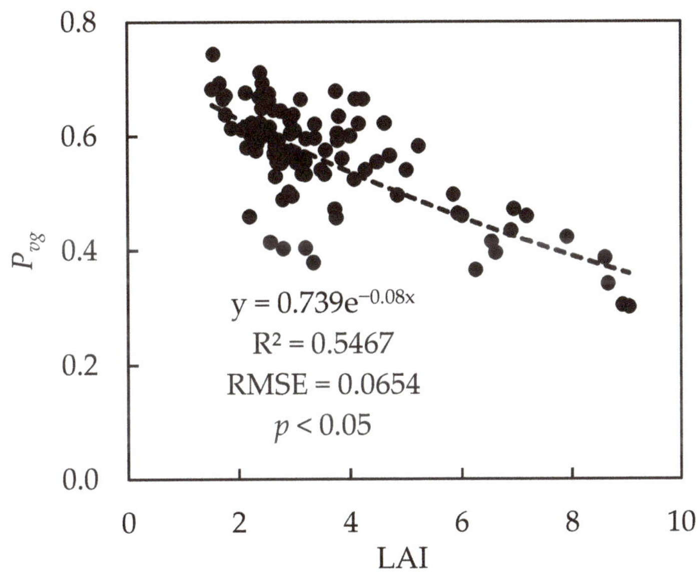

Figure 7. Scattering plot between P_{vg} and LAI.

3.4. Verification of Estimation of Canopy Closure Based on $P_{vg\text{-}c}$

Based on the definition of canopy closure, only the gaps between the canopies were considered, so the value of $1 - P_{vg\text{-}c}$ should be equal to the canopy closure. A linear regression model was established between the canopy closure and $1 - P_{vg\text{-}c}$. The R^2 and RMSE values were 0.5216 and 0.0832, respectively (Figure 8), and the precision of the canopy closure estimation was 86.69%. At the same time, the established model passed the significance test. It can be seen that the canopy closure measured by the line transects had a significantly good relationship with $1 - P_{vg\text{-}c}$. Among the traditional methods for measuring canopy closure, the line transect method was more accurate. Therefore, it was very effective and feasible to estimate canopy closure using the GOST model.

3.5. Verification of the Estimation of the Canopy Closure Based on P_{vg}

The canopy closure calculated based on the fisheye camera photos included the gaps in the canopy. Therefore, calculating the canopy closure using the fisheye camera photos can verify the canopy closure estimated based on the P_{vg}. The R^2 and RMSE values of the linear regression model were 0.1418 and 0.3295, respectively (Figure 9). The average accuracy of the canopy closure estimation was 73.2%. The model also passed the significance test. The results show that the relationship does not appear to be good. This could be because the fisheye camera images were taken at a 180 degree angle, resulting in the deformation increasing as the view angle became larger, while the GOST model calculated the P_{vg} when the view angle was 90 degrees [72]. The reason for this was that the projected area of the crown between these two styles was different, so the linear relationship was not as significant.

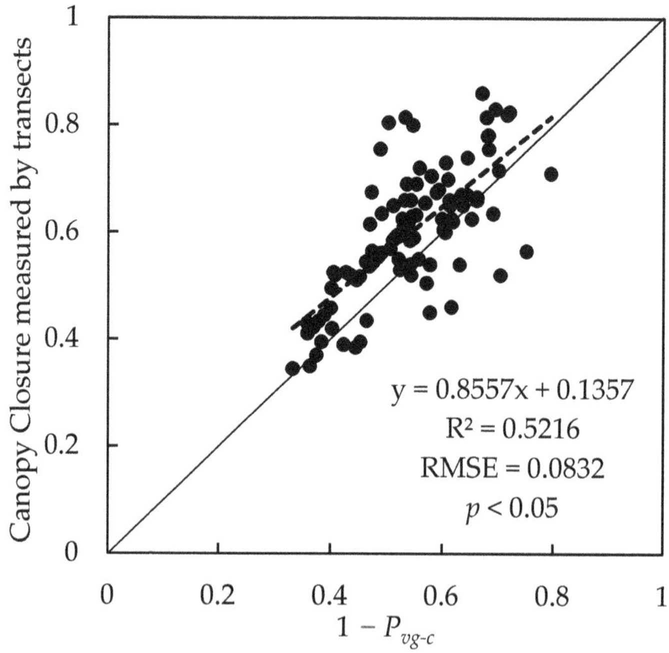

Figure 8. Scattering plot between the measured canopy closure and $1 - P_{vg\text{-}c}$ (the black solid line is $y = x$).

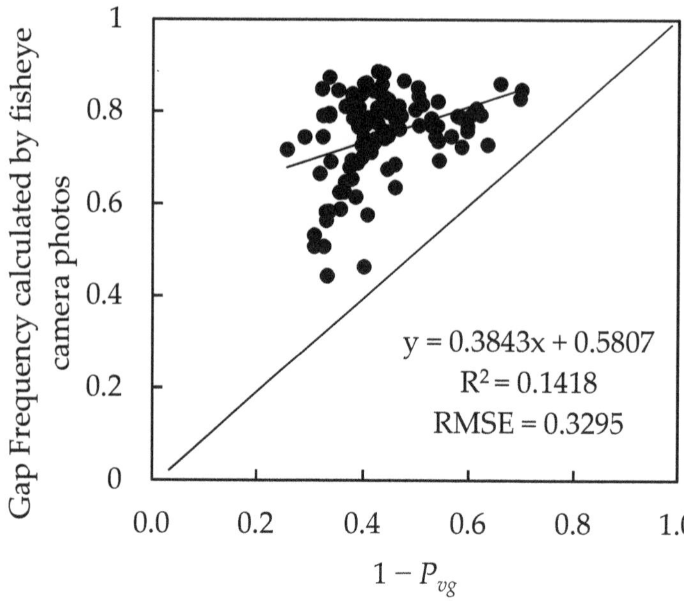

Figure 9. Scatter plot between canopy closure measured by fisheye camera images and $1 - P_{vg}$.

4. Discussion

Canopy closure is a multipurpose ecological indicator that is used to assess the light conditions and forest floor microclimate as well as to distinguish types of plant and animal habitats [13]. Canopy closure also determines light interception and is crucial for understanding forest carbon fixation and responses and feedback to climate change [73]. Unfortunately, the in situ measurement of canopy closure is a time-consuming and laborious process, whilst advances in non-destructive, indirect techniques have been made [15]. Indirect optical methods derive canopy closure by measuring the canopy gap fraction or via transmittance [74]. These in situ measurements are able to capture the canopy at the plot (and site) scale, but these methods cannot adequately characterize spatiotemporal dynamics [32].

More and more remote sensing applications involve the estimation of canopy closure. The least squares method is one of the most commonly used methods for canopy closure estimation. However, the linear relationship is data dependent. This means that accuracy will be affected by the data quality and that applications will be limited by the area of the research region [75]. Other studies have used the random forest and Cubist models to estimate the canopy closure. However, the precision of these models is not accurate enough because a single variable is used for regression analysis [10]. In contrast, non-parametric models have always had better estimation results, and the accuracy of the modeling and estimation would be limited by the number of the samples, as small samples weaken the predictive ability of the model [76,77]. To make up for these shortages, we established a canopy closure estimation method that uses the gap fraction based on the geometric-optical GOST model, and the results were discussed.

A canopy closure estimation method that uses the gap fraction was represented in this study. The canopy closure $P_{vg\text{-}c}$ was highly correlated with the average projected area of the canopy. The $P_{vg\text{-}c}$ was estimated based on the number of plants in the plot and the average radius of the crowns, and the accuracy of the estimated canopy closure was 86.69%, while the accuracy of the canopy closure estimated by the P_{vg} was 7%. In the residual analysis of the canopy closure model, the $P_{vg\text{-}c}$, P_{vg}, and $n \times d^2$ (Figures 10 and 11) show that the residuals of the two models were distributed within a reasonable range and relatively evenly. The results show that the accuracy of the $P_{vg\text{-}c}$ estimation of the model was higher than that of the residual plots for $P_{vg\text{-}c}$ and $n \times d^2$. The residuals were distributed between ± 0.2. However, the P_{vg} was not estimated as well as the $P_{vg\text{-}c}$ was (Figure 11). The residuals of the P_{vg} were distributed between -0.2 and 0.3. It can be seen that estimating the canopy closure $P_{vg\text{-}c}$ between the canopies using the number of plants and the average radius of the crowns produced better results.

For most of the remote sensing images with a high spatial resolution, the number of plants and the radius of the crown were able to be extracted easily by means of image segmentation technology, and the parameters of these two variables can be considered an approach to canopy closure estimation using remote sensing images with a high spatial resolution. The relationship between the $P_{vg\text{-}c}$ and $n \times d^2$ was the best optimal method, obtaining high-precision canopy closure estimation results. The line transect method was more accurate compared to other methods. As such, the scattering plot for the estimated canopy closure determined and measured by the gap fraction can be found in Figure 12. The canopy closure that was estimated by the $P_{vg\text{-}c}$ was consistent with the measured canopy closure. However, the canopy closure that was estimated by the P_{vg} deviated from the line $y = x$. Additionally, the canopy closure was underestimated.

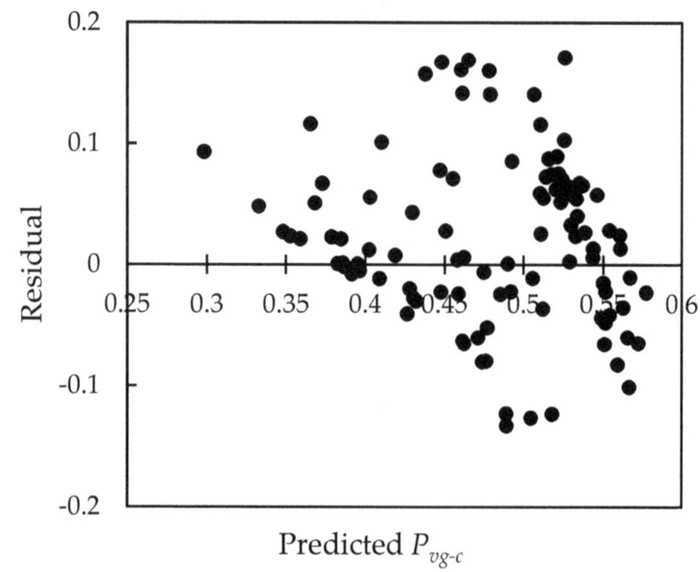

Figure 10. The residual plot of the estimated $P_{vg\text{-}c}$ and $n \times d^2$.

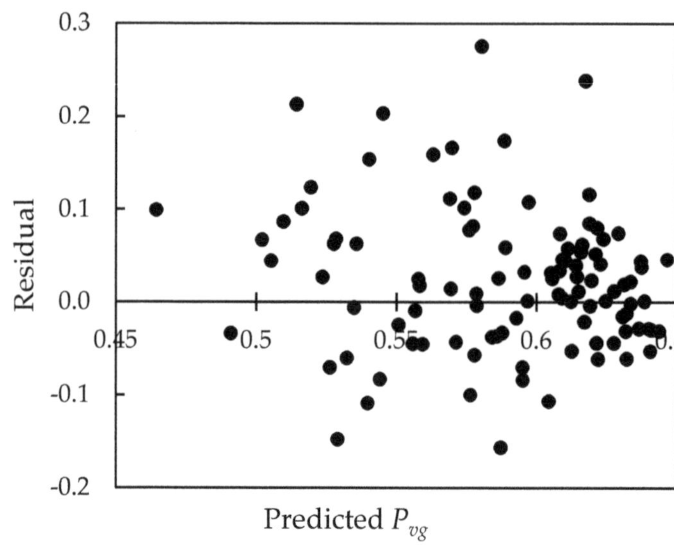

Figure 11. The residual plot of the estimated P_{vg} and $n \times d^2$.

Compared to traditional statistical models, the model in the present study was not affected by the study area or the dataset, and there was no need to establish models in different study areas. The canopy closure estimation method simulated the probability of being able to see the ground by considering the mechanism of the light transmission process in the canopy and the gaps or opaque canopy in the viewing direction, which can also be considered individually using the geometric-optical model (the GOST model), and the estimated results also have high accuracy. In other words, this means that this estimated method was more robust compared to statistical methods. If remote sensing data with medium or low spatial resolution are used, then the parameters used for canopy estimation cannot be inverted accurately because of the mixed pixel effect [44]. Additionally, most of the parameters, such as the number of the trees and the radius of the crown could be obtained, so this method was difficult to implement. Instead, some medium-level variables, such as the LAI, should be used for canopy closure estimation even though the accuracy varies from study to study [78–82]. In this study, we evaluated the canopy closure estimation efficiency when using the LAI. The values of the residuals were more discrete from the residual $P_{vg\text{-}c}$ and LAI plots (Figure 13). Additionally, the figure shows that the accuracy of the $P_{vg\text{-}c}$ estimation was better when the canopy closure was greater than 0.4 and when the value of residuals was lower than 0.1. When the canopy closure was greater than 0.4, the accuracy of the $P_{vg\text{-}c}$ prediction was obviously lower.

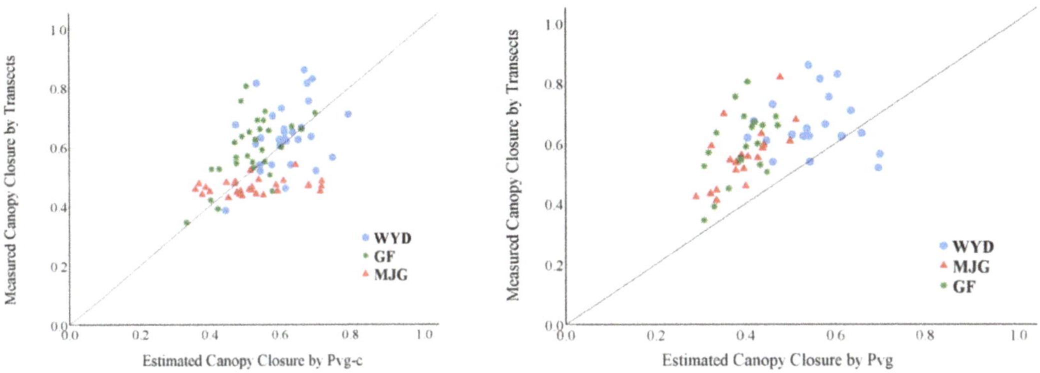

(a) (b)

Figure 12. (**a**) The scattering plot of the canopy closure measured and estimated by $P_{vg\text{-}c}$; (**b**) the scattering plot of the canopy closure measured and estimated by P_{vg}.

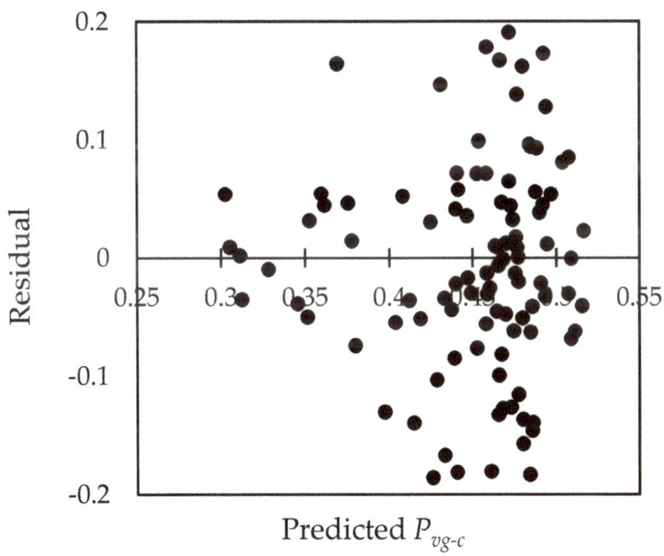

Figure 13. The residual plot of the estimated $P_{vg\text{-}c}$ and LAI.

The situation with the P_{vg} was similar to that of the $P_{vg\text{-}c}$. The residuals of the P_{vg} became significantly discrete from the residual P_{vg} and LAI plots when the canopy closure was larger than or less than 0.45 (Figure 14). The most possible reason for this is that more gaps in the crown could be detected when using remote sensors with a low or medium spatial resolution, and the efficiency of the mixed pixels enhanced the signals of the forest floor or those of other features. Inversely, the larger amount of canopy closure meant smaller gaps in the crown, and the detected probability of the signal outside of the canopy was lower compared to larger gaps, weakening the mixed pixel effect. As such, it was necessary to consider the efficiency of the mixed pixels in the images when the LAI was used to estimate the canopy closure. In comparison, a geometric-optical model was used to decompose the mixed pixels into four components, allowing the forest components in the pixels to be more accurately distinguished. That was the advantage of the method using a physical model. At the same time, the physical canopy closure estimation procedure using a geometric-optical model was clearer than the traditional statistical model or the pixel binary model, and the robustness and accuracy of the model were self-evident. In addition, the results also indicate that the LAI can be used to estimate the P_{vg} with good accuracy. Once high-quality LAI data were obtained, the canopy closure could also be estimated, especially for remote sensing images with a medium or low spatial resolution. This method provided a way to estimate the canopy closure using the LAI. Additionally, this relationship was derived from geometric-optical models and was not data dependent, which is the case using statistical theory. Additionally, this method can effectively avoid the shortcomings of canopy estimation models that are based on statistical methods.

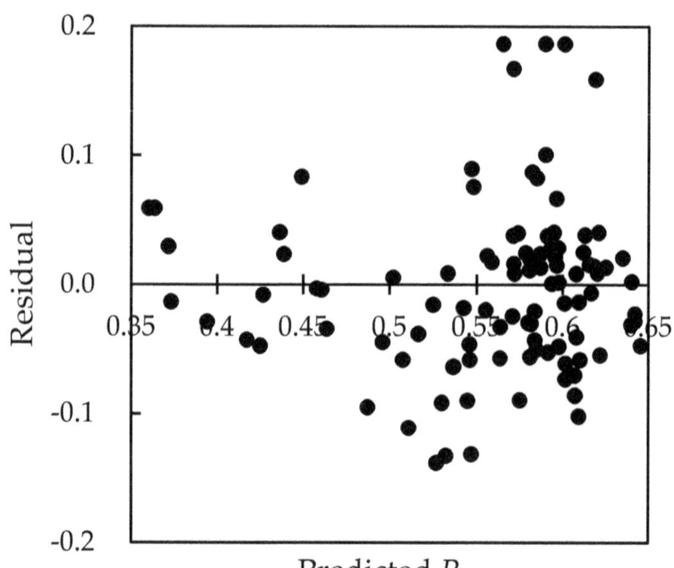

Figure 14. The residual plot of the estimated P_{vg} and LAI.

5. Conclusions

Canopy closure is an important forest inventory parameter, and it plays an important role in forestry production and management and forest health evaluation. In this study, the GOST model was used to simulate the characteristics of forest canopy gap fractions by considering an opaque canopy ($P_{vg\text{-}c}$) and gaps in the tree crown (P_{vg}). Additionally, exponential models for estimating the canopy closure of a plantation based on the gap fraction were established. The results show the following:

(1) It was feasible to estimate canopy closure based on the GOST model, and the feasible method was proved with sample data measured from three different regions in China.
(2) Compared to the LAI, $n \times d^2$ had a better relationship with the gap fractions simulated using the GOST model. Therefore, when remote sensing images or LiDAR data of the study area with high spatial resolution were available, the crown recognition method could be used to obtain the number of plants and the average radius of the crowns in the plot, so the gap fraction $P_{vg\text{-}c}$ and the forest canopy closure could be accurately estimated and predicted in the research area.
(3) When the number of plants and the average radii of the crowns in the plot could not be extracted using remote sensing images, especially when only medium- or low-spatial resolution remote sensing data were available, the LAI, a medium parameter, could be used to estimate the canopy closure with an acceptable level of accuracy. This also provided a new a canopy closure estimation approach using medium- or low-spatial resolution remote sensing data. This study can provide a reference for canopy closure estimation using geometric-optical models.

Author Contributions: Y.Y. conceptualized and designed the experiments; P.H. performed the experiments and analyzed the data; X.Y. and P.H. wrote the paper; X.Y., Y.Y. and W.F. reviewed and edited the paper. All authors have read and agreed to the published version of the manuscript.

Funding: This research was funded by the National Natural Science Foundation of China, grant numbers: 31870621 and 31971580; the Fundamental Research Funds for the Central Universities of China, grant numbers: 2572021BA08, 2572019BA10, and 2572019CP12; the China Postdoctoral Science Foundation, grant number: 2019M661239; National Forestry and Grassland Data Center-Heilongjiang platform, grant number: 2005DKA32200-OH.

Conflicts of Interest: The authors declare no conflict of interest.

References

1. Payn, T.; Carnus, J.-M.; Freer-Smith, P.; Orazio, C.; Nabuurs, G.-J. Third International Congress on Planted Forests: Planted Forests on the Globe-Renewable Resources for the Future. *N. Z. J. For. Sci.* **2014**, *44*, S1. [CrossRef]
2. Payn, T.; Carnus, J.-M.; Freer-Smith, P.; Kimberley, M.; Kollert, W.; Liu, S.; Orazio, C.; Rodriguez, L.; Silva, L.N.; Wingfield, M.J. Changes in planted forests and future global implications. *For. Ecol. Manag.* **2015**, *352*, 57–67. [CrossRef]
3. Buongiorno, J.; Zhu, S. Assessing the impact of planted forests on the global forest economy. *N. Z. J. For. Sci.* **2014**, *44*, S2. [CrossRef]
4. Brockerhoff, E.G.; Jactel, H.; Parrotta, J.A.; Ferraz, S.F.B. Role of eucalypt and other planted forests in biodiversity conservation and the provision of biodiversity-related ecosystem services. *For. Ecol. Manag.* **2013**, *301*, 43–50. [CrossRef]
5. Gschwantner, T.; Schadauer, K.; Vidal, C.; Lanz, A.; Tomppo, E.; di Cosmo, L.; Robert, N.; Duursma, D.E.; Lawrence, M. Common tree definitions for national forest inventories in Europe. *Silva Fenn.* **2009**, *43*, 303–321. [CrossRef]
6. Jennings, S.B.; Brown, N.D.; Sheil, D. Assessing forest canopies and understorey illumination: Canopy closure, canopy cover and other measures. *Forestry* **1999**, *72*, 59–74. [CrossRef]
7. IPCC. *Good Practice Guidance for Land Use, Land-Use Change and Forestry*; Institute for Global Environmental Strategies (IGES): Hayama, Japan, 2003.
8. Xu, B.; Gong, P.; Pu, R. Crown closure estimation of oak savannah in a dry season with Landsat TM imagery: Comparison of various indices through correlation analysis. *Int. J. Remote Sens.* **2003**, *24*, 1811–1822. [CrossRef]
9. Hua, Y.; Zhao, X. Multi-Model Estimation of Forest Canopy Closure by Using Red Edge Bands Based on Sentinel-2 Images. *Forests* **2021**, *12*, 1768. [CrossRef]
10. Chen, G.; Lou, T.; Jing, W.; Wang, Z. Sparkpr: An Efficient Parallel Inversion of Forest Canopy Closure. *IEEE Access* **2019**, *7*, 135949–135956. [CrossRef]
11. Smith, A.M.; Ramsay, P.M. A comparison of ground-based methods for estimating canopy closure for use in phenology research. *Agric. For. Meteorol.* **2018**, *252*, 18–26. [CrossRef]
12. Fiala, A.C.S.; Garman, S.L.; Gray, A.N. Comparison of five canopy cover estimation techniques in the western Oregon Cascades. *For. Ecol. Manag.* **2006**, *232*, 188–197. [CrossRef]
13. Korhonen, L.T.; Korhonen, K.; Rautiainen, M.; Stenberg, P. Estimation of forest canopy cover: A comparison of field measurement techniques. *Silva Fenn.* **2006**, *40*, 577–588. [CrossRef]
14. Paletto, A.; Tosi, V. Forest canopy cover and canopy closure: Comparison of assessment techniques. *Eur. J. For. Res.* **2009**, *128*, 265–272. [CrossRef]

15. Brown, L.A.; Ogutu, B.O.; Dash, J. Tracking forest biophysical properties with automated digital repeat photography: A fisheye perspective using digital hemispherical photography from below the canopy. *Agric. For. Meteorol.* **2020**, *287*, 107944. [CrossRef]
16. Macfarlane, C.; Hoffman, M.; Eamus, D.; Kerp, N.; Higginson, S.; McMurtrie, R.; Adams, M. Estimation of leaf area index in eucalypt forest using digital photography. *Agric. For. Meteorol.* **2007**, *143*, 176–188. [CrossRef]
17. Vales, D.J.; Bunnell, F.L. Comparison of methods for estimating forest overstory cover. I. Observer effects. *Can. J. For. Res.* **1988**, *18*, 606–609. [CrossRef]
18. Li, J.; Mao, X. Comparison of Canopy Closure Estimation of Plantations Using Parametric, Semi-Parametric, and Non-Parametric Models Based on GF-1 Remote Sensing Images. *Forests* **2020**, *11*, 597. [CrossRef]
19. Chopping, M.; Moisen, G.G.; Su, L.; Laliberte, A.; Rango, A.; Martonchik, J.V.; Peters, D.P.C. Large area mapping of southwestern forest crown cover, canopy height, and biomass using the NASA Multiangle Imaging Spectro-Radiometer. *Remote Sens. Environ.* **2008**, *112*, 2051–2063. [CrossRef]
20. Smith, A.M.S.; Falkowski, M.J.; Hudak, A.T.; Evans, J.S.; Robinson, A.P.; Steele, C.M. A cross-comparison of field, spectral, and lidar estimates of forest canopy cover. *Can. J. Remote Sens.* **2009**, *35*, 447–459. [CrossRef]
21. Hill, M.J.; Román, M.O.; Schaaf, C.B.; Hutley, L.; Brannstrom, C.; Etter, A.; Hanan, N.P. Characterizing vegetation cover in global savannas with an annual foliage clumping index derived from the MODIS BRDF product. *Remote Sens. Environ.* **2011**, *115*, 2008–2024. [CrossRef]
22. Chopping, M.; North, M.; Chen, J.; Schaaf, C.B.; Blair, J.B.; Martonchik, J.V.; Bull, M.A. Forest Canopy Cover and Height from MISR in Topographically Complex Southwestern US Landscapes Assessed with High Quality Reference Data. *IEEE J. Sel. Top. Appl. Earth Obs. Remote Sens.* **2012**, *5*, 44–58. [CrossRef]
23. Tuominen, S.; Pekkarinen, A. Local radiometric correction of digital aerial photographs for multi source forest inventory. *Remote Sens. Environ.* **2004**, *89*, 72–82. [CrossRef]
24. Lisein, J.; Pierrot-Deseilligny, M.; Bonnet, S.; Lejeune, P. A Photogrammetric Workflow for the Creation of a Forest Canopy Height Model from Small Unmanned Aerial System Imagery. *Forests* **2013**, *4*, 922–944. [CrossRef]
25. Navarro, J.A.; Tomé, J.L.; Marino, E.; Guillén-Climent, M.L.; Fernández-Landa, A. Assessing the transferability of airborne laser scanning and digital aerial photogrammetry derived growing stock volume models. *Int. J. Appl. Earth Obs. Geoinf.* **2020**, *91*, 102135. [CrossRef]
26. Edson, C.; Wing, M.G. Airborne Light Detection and Ranging (LiDAR) for Individual Tree Stem Location, Height, and Biomass Measurements. *Remote Sens.* **2011**, *3*, 2494–2528. [CrossRef]
27. Kato, A.; Moskal, L.M.; Schiess, P.; Swanson, M.E.; Calhoun, D.; Stuetzle, W. Capturing tree crown formation through implicit surface reconstruction using airborne lidar data. *Remote Sens. Environ.* **2016**, *113*, 1148–1162. [CrossRef]
28. Moeser, D.; Roubinek, J.; Schleppi, P.; Morsdorf, F.; Jonas, T. Canopy closure, LAI and radiation transfer from airborne LiDAR synthetic images. *Agric. For. Meteorol.* **2014**, *197*, 158–168. [CrossRef]
29. Korhonen, L.; Korpela, I.; Heiskanen, J.; Maltamo, M. Airborne discrete-return LIDAR data in the estimation of vertical canopy cover, angular canopy closure and leaf area index. *Remote Sens. Environ.* **2011**, *115*, 1065–1080. [CrossRef]
30. Parent, J.R.; Volin, J.C. Assessing the potential for leaf-off LiDAR data to model canopy closure in temperate deciduous forests. *ISPRS J. Photogramm. Remote Sens.* **2014**, *95*, 134–145. [CrossRef]
31. Wallace, L.; Lucieer, A.; Malenovskỳ, Z.; Turner, D.; Vopěnka, P. Assessment of forest structure using two UAV techniques: A comparison of airborne laser scanning and structure from motion (SfM) point clouds. *Forests* **2016**, *7*, 62. [CrossRef]
32. Zhu, X.; Skidmore, A.K.; Wang, T.; Liu, J.; Darvishzadeh, R.; Shi, Y.; Premierd, J.; Heurich, M. Improving leaf area index (LAI) estimation by correcting for clumping and woody effects using terrestrial laser scanning. *Agric. For. Meteorol.* **2018**, *263*, 276–286. [CrossRef]
33. Granholm, A.-H.; Lindgren, N.; Olofsson, K.; Nyström, M.; Allard, A.; Olsson, H. Estimating vertical canopy cover using dense image-based point cloud data in four vegetation types in southern Sweden. *Int. J. Remote Sens.* **2017**, *38*, 1820–1838. [CrossRef]
34. Bode, C.A.; Limm, M.P.; Power, M.E.; Finlay, J.C. Subcanopy Solar Radiation model: Predicting solar radiation across a heavily vegetated landscape using LiDAR and GIS solar radiation models. *Remote Sens. Environ.* **2014**, *154*, 387–397. [CrossRef]
35. Griffin, A.M.R. Using LiDAR and Normalized Difference Vegetation Index to Remotely Determine LAI and Percent Canopy Cover at Varying Scales. Ph.D. Thesis, Texas A&M University, College Station, TX, USA, 2010.
36. Halperin, J.; LeMay, V.; Coops, N.; Verchot, L.; Marshall, P.; Lochhead, K. Canopy cover estimation in miombo woodlands of Zambia: Comparison of Landsat 8 OLI versus RapidEye imagery using parametric, nonparametric, and semiparametric methods. *Remote Sens. Environ.* **2016**, *179*, 170–182. [CrossRef]
37. Muukkonen, P.; Heiskanen, J.P. Estimating biomass for boreal forests using ASTER satellite data combined with standwise forest inventory data. *Remote Sens. Environ.* **2005**, *99*, 434–447. [CrossRef]
38. Puliti, S.; Breidenbach, J.; Schumacher, J.; Hauglin, M.; Klingenberg, T.F.; Astrup, R. Above-ground biomass change estimation using national forest inventory data with Sentinel-2 and Landsat. *Remote Sens. Environ.* **2021**, *265*, 112644. [CrossRef]
39. Hemmerling, J.; Pflugmacher, D.; Hostert, P. Mapping temperate forest tree species using dense Sentinel-2 time series. *Remote Sens. Environ.* **2021**, *267*, 112743. [CrossRef]
40. Homolová, L.; Malenovský, Z.; Clevers, J.G.P.W.; García-Santos, G.; Schaepman, M.E. Review of optical-based remote sensing for plant trait mapping. *Ecol. Complex.* **2013**, *15*, 1–16. [CrossRef]

41. Yuan, Y.; Wang, X.; Yin, F.; Zhan, J. Examination of the Quantitative Relationship between Vegetation Canopy Height and LAI. *Adv. Meteorol.* **2013**, *2013*, 1–6. [CrossRef]
42. Ozdemir, I. Linear transformation to minimize the effects of variability in understory to estimate percent tree canopy cover using RapidEye data. *GIScience Remote Sens.* **2014**, *51*, 288–300. [CrossRef]
43. Wolter, P.T.; Townsend, P.A.; Sturtevant, B.R. Estimation of forest structural parameters using 5 and 10 meter SPOT-5 satellite data. *Remote Sens. Environ.* **2009**, *113*, 2019–2036. [CrossRef]
44. Kahriman, A.; Gunlu, A.; Karahalil, U. Estimation of Crown Closure and Tree Density Using Landsat TM Satellite Images in Mixed Forest Stands. *J. Indian Soc. Remote Sens.* **2014**, *42*, 559–567. [CrossRef]
45. Carreiras, J.M.B.; Pereira, J.M.C.; Pereira, J.S. Estimation of tree canopy cover in evergreen oak woodlands using remote sensing. *For. Ecol. Manag.* **2006**, *223*, 45–53. [CrossRef]
46. Montesano, P.M.; Neigh, C.S.R.; Sexton, J.; Feng, M.; Channan, S.; Ranson, K.J.; Townshend, J.R. Calibration and Validation of Landsat Tree Cover in the Taiga–Tundra Ecotone. *Remote Sens.* **2016**, *8*, 551. [CrossRef]
47. Hadi; Korhonen, L.; Hovi, A.; Rönnholm, P.; Rautiainen, M. The accuracy of large-area forest canopy cover estimation using Landsat in boreal region. *Int. J. Appl. Earth Obs. Geoinf.* **2016**, *53*, 118–127. [CrossRef]
48. Tong, S.; Zhang, J.; Ha, S.; Lai, Q.; Ma, Q. Dynamics of Fractional Vegetation Coverage and Its Relationship with Climate and Human Activities in Inner Mongolia, China. *Remote Sens.* **2016**, *8*, 776. [CrossRef]
49. Xiao, Q.; Tao, J.; Xiao, Y.; Qian, F. Monitoring vegetation cover in Chongqing between 2001 and 2010 using remote sensing data. *Environ. Monit. Assess.* **2017**, *189*, 493. [CrossRef]
50. Ding, Y.; Zheng, X.; Zhao, K.; Xin, X.; Liu, H. Quantifying the Impact of NDVIsoil Determination Methods and NDVIsoil Variability on the Estimation of Fractional Vegetation Cover in Northeast China. *Remote Sens.* **2016**, *8*, 29. [CrossRef]
51. Yang, S.-W.; Dong, B.; Liu, L.; Sun, L.; Sheng, S.-W.; Wang, Q.; Peng, W.; Wang, X.; Zhang, Z.; Zhao, J. Research on Vegetation Coverage Change in Sheng Jin Lake Wetland of Anhui Province. *Wetlands* **2015**, *35*, 677–682. [CrossRef]
52. Zeng, Y.; Schaepman, M.E.; Wu, B.; Clevers, J.G.P.W.; Bregt, A.K. Scaling-based forest structural change detection using an inverted geometric-optical model in the Three Gorges region of China. *Remote Sens. Environ.* **2008**, *112*, 4261–4271. [CrossRef]
53. Li, X.; Strahler, A.H. Geometric-optical bidirectional reflectance modeling of the discrete crown vegetation canopy: Effect of crown shape and mutual shadowing. *IEEE Trans. Geosci. Remote Sens.* **1992**, *30*, 276–292. [CrossRef]
54. Zeng, Y.; Schaepman, M.E.; Wu, B.; Clevers, J.G.P.W.; Bregt, A.K. Quantitative forest canopy structure assessment using an inverted geometric-optical model and up-scaling. *Int. J. Remote Sens.* **2009**, *30*, 1385–1406. [CrossRef]
55. Wang, C.; Du, H.; Xu, X.; Han, N.; Zhou, G.; Sun, S.; Gao, G. Multi-scale crown closure retrieval for moso bamboo forest using multi-source remotely sensed imagery based on geometric-optical and Erf-BP neural network models. *Int. J. Remote Sens.* **2015**, *36*, 5384–5402. [CrossRef]
56. Jacquemoud, S.; Baret, F. PROSPECT: A model of leaf optical properties spectra. *Remote Sens. Environ.* **1990**, *34*, 75–91. [CrossRef]
57. Verhoef, W. Light scattering by leaf layers with application to canopy reflectance modeling: The SAIL model. *Remote Sens. Environ.* **1984**, *16*, 125–141. [CrossRef]
58. Jiao, Q.; Sun, Q.; Zhang, B.; Huang, W.; Ye, H.; Zhang, Z.; Zhang, X.; Qian, B. A Random Forest Algorithm for Retrieving Canopy Chlorophyll Content of Wheat and Soybean Trained with PROSAIL Simulations Using Adjusted Average Leaf Angle. *Remote Sens.* **2022**, *14*, 98. [CrossRef]
59. Ding, Y.; Zhang, H.; Li, Z.; Xin, X.; Zheng, X.; Zhao, K. Comparison of fractional vegetation cover estimations using dimidiate pixel models and look-up table inversions of the PROSAIL model from Landsat 8 OLI data. *J. Appl. Remote Sens.* **2016**, *10*, 36022. [CrossRef]
60. Ding, Y.; Zhang, H.; Zhao, K.; Zheng, X. Investigating the accuracy of vegetation index-based models for estimating the fractional vegetation cover and the effects of varying soil backgrounds using in situ measurements and the PROSAIL model. *Int. J. Remote Sens.* **2017**, *38*, 4206–4223. [CrossRef]
61. Gu, C.-Y.; Du, H.-Q.; Zhou, G.-M.; Han, N.; Xu, X.-J.; Zhao, X.; Sun, X.-Y. Retrieval of leaf area index of moso bamboo forest with Landsat Thematic Mapper image based on PROSAIL canopy radiative transfer model. *J. Appl. Ecol.* **2013**, *24*, 2248–2256.
62. Liu, Z.; Jin, G.; Zhou, M. Evaluation and correction of optically derived leaf area index in different temperate forests. *Iforest-Biogeosci. For.* **2016**, *9*, 55–62. [CrossRef]
63. Ma, L.; Zheng, G.; Wang, X.; Li, S.; Lin, Y.; Ju, W. Retrieving forest canopy clumping index using terrestrial laser scanning data. *Remote Sens. Environ.* **2018**, *210*, 452–472. [CrossRef]
64. Fan, W.; Chen, J.M.; Ju, W.; Zhu, G. GOST: A Geometric-Optical Model for Sloping Terrains. *IEEE Trans. Geosci. Remote Sens.* **2014**, *52*, 5469–5482. [CrossRef]
65. Chen, J.; Leblanc, S. A four-scale bidirectional reflectance model based on canopy architecture. *IEEE Trans. Geosci. Remote Sens.* **1997**, *35*, 1316–1337. [CrossRef]
66. Verstraete, M.M.; Pinty, B.; Myneni, R.B. Potential and limitations of information extraction on the terrestrial biosphere from satellite remote sensing. *Remote Sens. Environ.* **1996**, *58*, 201–214. [CrossRef]
67. Yang, X.G.; Fan, W.Y.; Yu, Y. Estimation of Forest Canopy Chlorophyll Content Based on PROSPECT and SAIL Models. *Spectrosc. Spectr. Anal.* **2010**, *30*, 3022–3026. [CrossRef]
68. Gu, C.; Du, H.; Mao, F.; Han, N.; Zhou, G.; Xu, X.; Sun, S.; Gao, G. Global sensitivity analysis of PROSAIL model parameters when simulating Moso bamboo forest canopy reflectance. *Int. J. Remote Sens.* **2016**, *37*, 5270–5286. [CrossRef]

69. Li, C.; Song, J.; Wang, J. Modifying Geometric-Optical Bidirectional Reflectance Model for Direct Inversion of Forest Canopy Leaf Area Index. *Remote Sens.* **2015**, *7*, 11083–11104. [CrossRef]
70. Xu, J.; Quackenbush, L.J.; Volk, T.A.; Im, J. Forest and Crop Leaf Area Index Estimation Using Remote Sensing: Research Trends and Future Directions. *Remote Sens.* **2020**, *12*, 2934. [CrossRef]
71. Fieber, K.D.; Davenport, I.J.; Tanase, M.A.; Ferryman, J.M.; Gurney, R.J.; Walker, J.P.; Hacker, J.M. Effective LAI and CHP of a Single Tree from Small-Footprint Full-Waveform LiDAR. *IEEE Geosci. Remote Sens. Lett.* **2014**, *11*, 1634–1638. [CrossRef]
72. Song, G.-Z.M.; Chao, K.-J.; Doley, D.; Yates, D. Sky-canopy border length, exposure and thresholding influence accuracy of hemispherical photography for complex plant canopies. *Bot. Stud.* **2018**, *59*, 19–59. [CrossRef]
73. Richardson, A.D.; Keenan, T.F.; Migliavacca, M.; Ryu, Y.; Sonnentag, O.; Toomey, M. Climate change, phenology, and phenological control of vegetation feedbacks to the climate system. *Agric. For. Meteorol.* **2013**, *169*, 156–173. [CrossRef]
74. Fang, H.; Li, W.; Wei, S.; Jiang, C. Seasonal variation of leaf area index (LAI) over paddy rice fields in NE China: Intercomparison of destructive sampling, LAI-2200, digital hemispherical photography (DHP), and AccuPAR methods. *Agric. For. Meteorol.* **2014**, *198*, 126–141. [CrossRef]
75. Verrelst, J.; Camps-Valls, G.; Muñoz-Marí, J.; Rivera, J.P.; Veroustraete, F.; Clevers, J.G.P.W.; Moreno, J. Optical remote sensing and the retrieval of terrestrial vegetation bio-geophysical properties—A review. *ISPRS J. Photogramm. Remote Sens.* **2015**, *108*, 273–290. [CrossRef]
76. Melin, M.; Korhonen, L.; Kukkonen, M.; Packalen, P. Assessing the performance of aerial image point cloud and spectral metrics in predicting boreal forest canopy cover. *ISPRS J. Photogramm. Remote Sens.* **2017**, *129*, 77–85. [CrossRef]
77. Temesgen, H.; Ver Hoef, J.M. Evaluation of the spatial linear model, random forest and gradient nearest-neighbour methods for imputing potential productivity and biomass of the Pacific Northwest forests. *For. Int. J. For. Res.* **2014**, *88*, 131–142. [CrossRef]
78. Propastin, P.; Erasmi, S. A physically based approach to model LAI from MODIS 250m data in a tropical region. *Int. J. Appl. Earth Obs. Geoinf.* **2010**, *12*, 47–59. [CrossRef]
79. Korhonen, L.; Hadi; Packalen, P.; Rautiainen, M. Comparison of Sentinel-2 and Landsat 8 in the estimation of boreal forest canopy cover and leaf area index. *Remote Sens. Environ.* **2017**, *195*, 259–274. [CrossRef]
80. Zhang, D.; Liu, J.; Ni, W.; Sun, G.; Zhang, Z.; Liu, Q.; Wang, Q. Estimation of Forest Leaf Area Index Using Height and Canopy Cover Information Extracted from Unmanned Aerial Vehicle Stereo Imagery. *IEEE J. Sel. Top. Appl. Earth Obs. Remote Sens.* **2019**, *12*, 471–481. [CrossRef]
81. Ling, C.; Liu, H.; Ji, P.; Hu, H.; Wang, X.; Hou, R. Estimation of Vegetation Coverage Based on NDVI Index of UAV Visible Image-Using the Shelterbelt Research Area as An Example. *For. Eng.* **2021**, *37*, 57–66.
82. Wang, K.; Peng, X.; Zhang, Y.; Luo, Z.; Jiang, D. A Hyperspectral Classification Method for Agroforestry Vegetation Based on Improved U-Net. *For. Eng.* **2022**, *38*, 58–66.

remote sensing

MDPI

Article

A Modified Two-Steps Three-Stage Inversion Algorithm for Forest Height Inversion Using Single-Baseline L-Band PolInSAR Data

Jianshuang Zhang [1,2], Yangjian Zhang [1], Wenyi Fan [2,*], Liyuan He [3], Ying Yu [2] and Xuegang Mao [2]

1 Key Laboratory of Ecosystem Network Observation and Modeling, Institute of Geographic Sciences and Natural Resources Research, Chinese Academy of Sciences, Beijing 100101, China; zhangjs@igsnrr.ac.cn (J.Z.); zhangyj@igsnrr.ac.cn (Y.Z.)

2 Key Laboratory of Sustainable Forest Ecosystem Management-Ministry of Education, School of Forestry, Northeast Forestry University, Harbin 150040, China; yuying@nefu.edu.cn (Y.Y.); maoxuegang@nefu.edu.cn (X.M.)

3 Department of Biology, San Diego State University, San Diego, CA 92182, USA; lhe2@sdsu.edu

* Correspondence: fanwy@nefu.edu.cn

Abstract: Forest height inversion with Polarimetric SAR Interferometry (PolInSAR) has become a research hotspot in the field of radar remote sensing. In this paper, we systematically studied a modified two-step, three-stage inversion simulating the L-band (L = 23 cm) full-polarization interferometric SAR data with an average forest height of 18 m using ESA PolSARpro-SIM software. We applied this method to E-SAR L-band single-baseline full PolInSAR data in 2003. In the first step, we modified the three-stage inversion algorithm based on phase diversity (PD)/maximum coherence difference (MCD) coherence optimization methods, corresponding to PD, MCD, respectively. In the second step, we introduced the coherence amplitude inversion term and modified the fixed weight to the variable of ε times the ground scattering ratio, which improved the accuracy of forest height inversion. The mean of forest height inversion by the HV method was the lowest (15.83 m) and the RMSE was the largest (4.80 m). The PD method was superior to the HV method with RMSE (4.60 m). The MCD method was slightly better than using the PD method with the smallest RMSE (4.43 m). After adding the coherence amplitude term, the RMSE was improved by 0.15 m, 0.14 m, and 0.08 m, respectively. The smallest RMSE was obtained by MCD, followed by the PD and HV methods. Although the robustness of the forest height inversion algorithm was reduced, the underestimation was improved and the RMSE was reduced. Due to the complexity of the real SAR E-SAR L-band single-baseline full PolInSAR data and the small sample sizes, the three-stage inversion methods based on coherent optimization were lower than the three-stage in-version method. After introducing the coherent magnitude term, the overestimation of the forest height was significantly weakened in HVWeight, PDweight, and MCDWeight, and PDWeight was optimal. The modified two-step, three-stage inversion algorithm had significant effects in alleviating forest height underestimation and overestimation, improving the accuracy of forest height inversion, and laying a foundation for the upcoming L-band SAR satellite generation, new SAR and LIDAR systems combined with RPAs (remotely piloted aircrafts)/UAVs (unmanned aerial vehicles) for small areas mapping initiatives, and promoting the depth and breadth of the SAR applications of the new SAR system.

Keywords: forest height inversion; three-stage algorithm; coherence optimization; complex coherence amplitude inversion

Citation: Zhang, J.; Zhang, Y.; Fan, W.; He, L.; Yu, Y.; Mao, X. A Modified Two-Steps Three-Stage Inversion Algorithm for Forest Height Inversion Using Single-Baseline L-Band PolInSAR Data. *Remote Sens.* **2022**, *14*, 1986. https://doi.org/10.3390/rs14091986

Academic Editor: Huaqiang Du

Received: 6 March 2022
Accepted: 18 April 2022
Published: 21 April 2022

1. Introduction

Forest height is an important biophysical parameter [1], and its spatial distribution is of great significance for forest resource management, forest biomass estimation, and regional and global carbon cycle research [2–4]. The measurement methods of forest height include ground survey, photogrammetry, Lidar, and so on. Since the measurement of large-scale forest height is difficult, it has not been well quantified [5]. Although forest ground survey data can also obtain forest heights, its application in large-scale research is limited due to the large amount of human and material resources, and the lack of observations in remote regions. Remote sensing is the most effective method for real-time forest inversion on a regional and global scale. Although LIDAR can achieve high accuracy, it is limited by the atmosphere, mist, and clouds, especially in areas with cloudy, rainy, foggy, and snowy weather. Moreover, the acquisition cost of LIDAR is too high to obtain information on the vertical structure of the forest, and there is no information on the horizontal structure of the forest. Optical remote sensing is not only influenced by clouds, fog, and snow, but also cannot penetrate the forest to the ground. Therefore, it is of great scientific significance to estimate forest height based on synthetic aperture radar (SAR), which combines the properties of both the real-time forest inversion at regional and global scales provided by remote sensing and the forest penetration to the ground by microwaves [6].

Among many forest height inversion methods widely used, Polarimetric SAR Interferometry (PolInSAR) combines the comprehensive advantages of interferometry and polarization information. PolInSAR is not only sensitive to the vegetation spatial distribution with SAR Interferometry (InSAR), but also has the property of Polarimetric SAR (PolSAR), which is sensitive to the shape and direction of the vegetation [7].

Cloude and Paphthanassious first proposed PolInSAR technology [8], and afterwards, formally proposed the concept and extended the random volume over ground (RVoG) [9] to the full PolInSAR, thus laying the foundation of forest height inversion with PolInSAR [10]. The applicability of the RVoG model for forest height inversion has been studied for cases of different forest density [11]. Due to the high computational complexity of the six-dimensional nonlinear parameter method [12], the PolInSAR forest height inversion was simplified to a three-stage method based on the RVOG model [13]. By reason of its simplicity and generality, the three-stage inversion algorithm has low calculation cost and been widely used. The three-stage inversion algorithm was validated in various activities at different frequencies and configurations [14–17]. Some literature has improved the three-stage algorithm by considering the effect of terrain or extinction coefficient successively, which ultimately improved the accuracy of forest height inversion [18,19].

However, the three-stage inversion algorithm accuracy is affected by the accuracy of the estimated topographic phase [20] and pure volume scattering complex coherence [21]. One of the main reasons is the irrational selection of polarized channels, resulting in the ineffective separation of ground and volume scattering phase centers. Therefore, several coherence optimization methods based on optimization theory [10,22–26] have emerged to solve this problem. Xie Q. et al. combine the three-stage inversion algorithm with the five effective separation phase centers generated by singular value decomposition (SVD) [10] and phase diversity (PD) [26] coherence optimization methods for forest height inversion, which improves the forest height accuracy [27]. Lavalle compared the max of phase with the max of magnitude over the coherence boundary and studied the dependence of L-band PolInSAR complex coherence on forest height inversion [28]. Fu W.X. et al. extend the three-stage algorithm and SVD [10] to the dual-polarized PolInSAR and, furthermore, propose a search method based on the RVoG model for solving pure volume scattering complex coherence on the fuzzy line segment, which has similar inverted forest height to full polarization [6]. Based on the RVoG model, Lin D.F. et al. have proposed a new method based on TSVD decomposition to directly estimate "pure" volume scattering complex coherence. Compared with the three-stage method, the RMSE of the inverted forest height is improved by 48.6% [29].

Aiming to reduce the underestimation of forest height, Cloude proposed a hybrid model combining interference coherence amplitude with the phase difference method [30–32]. Additionally, due to the underestimation of forest height by the estimating signal parameters by the use of the rotation invariance techniques (ESPRIT) method, the total least square ESPRIT (TLS-ESPRIT) method is combined with the coherence amplitude inversion method to improve the accuracy of forest height inversion [33,34]. Using a novel hybrid inversion algorithm based on covariance matrix decomposition, a forest vegetation parameter inversion experiment [35] was carried out on the L-band airborne data of the simulation and SIR-C/X-SAR system. Chen E.X. et al. verify and compare several available forest height inversion methods, such as the SINC, three stage SINC inversion (TSS), and phase and coherence inversion (PCI) methods, using E-SAR repeated orbits data and corresponding ground heights, besides analyzing the effect of SVD on the SINC method for forest height inversion [36]. Some scholars have estimated the forest height by combing the three stage inversion algorithm with coherence optimization [37,38]. Aghabalaei et al. demonstrated the capability for forest height estimation with the single-baseline L-band compact PolInSAR (C-PolInSAR) in the Remningstorp, southern Sweden [39]. They developed a novel four-stage algorithm with volumetric temporal decorrelation to improve the forest height estimation accuracy using the repeat-pass PolInSAR data of Gabon Lope Park acquired in the AfriSAR campaign of the German Aerospace Center (DLR) [40]. In recent years, some researchers have investigated the potential of forest height mapping with spotlight-mode data with TanDEM-X (TDM) combined with in situ measurements [41,42]. They aim to demonstrate the potential of space-borne PolInSAR datasets at the L-, C-, and X-band frequencies (ALOS-2/PALSAR-2, TerraSAR-X, and RadarSAT-2) for forest height estimation, with the RMSE of 5.4 m, 12.8 m, and 7.6 m, respectively [43]. Chen et al. have developed a new extended Fourier–Legendre series approach for combing Global Ecosystem Dynamics Investigation (GEDI) LiDAR waveforms with TanDEM-X data to improve forest height estimation [44]. Some papers use Tomographic SAR (TomoSAR) technology for forest 3D structure mapping [5,45,46].

In summary, although forest height inversion by single-baseline PolInSAR has been widely used at home and abroad with good precision, it still presents the phenomenon of underestimating and overestimating the forest height, to which there is no good solution yet. In this paper, without considering the influence of residual motion, baseline and coregistration error, topography, temporal, and signal-to-noise ratio (SNR) decorrelation sources, we used ESA PolSARproSIM software to simulate single-baseline PolInSAR data. Furthermore, we applied this method to E-SAR L-band single-baseline full PolInSAR data in 2003. The purpose of this paper was to alleviate alleviated the problem of the overestimation or underestimation of the three-stage inversion algorithm with the introduction of an automatic weight adjustment method, which was an improvement on the original three-stage method. First, for the problem of incomplete coherence separation in the three-stage inversion method, we introduced the PD/MCD method in coherence optimization. Then, for forest height underestimation and overestimation problems, we introduced the complex coherence amplitude inversion method and the ground scattering proportional variable to automatically select weight. We compared the performance of these methods of the forest height, laying the foundation for selecting the algorithm based on single-baseline PolInSAR forest height inversion and providing exploration to develop a better inversion method.

2. Study Data

Due to the difficulty in getting airborne SAR images in forested areas and data simulated by ESA PolSARproSIM software based on Maxwell's wave propagation equation and scattering model without considering the influence of residual motion, baseline and coregistration error, topography, and temporal and signal-to-noise ratio (SNR) decorrelation sources [47], we thus used PolSARproSIM software to simulate the L-band (L = 23 cm) for single-baseline full-polarized PolInSAR data. The parameters were set as follows: the slopes of range and azimuth were both 0, regardless of topography; the radar platform height was 3000 m; the horizontal baseline was 10 m; the vertical baseline was 1 m; and the incidence angle was 45°. The center frequency was 1.3 GHz; the azimuth resolution was 1.5 m; the slant resolution was 1.0607 m; the forest type was coniferous; and the average forest height was 18 m. Since the product of the vertical wavenumber and the forest height is less than 2π, there is no ambiguous problem of forest height inversion [48]. Figure 1 presents the overall scenario of the study area, with the forest height of 18 m in height in the middle area and non-forest in the remaining areas. The tree height standard deviation would be under programmer control, and typically set to 5% of the mean value. The process of tree location map generation began by determining the number of trees in the scene, then initializing them in a regular pattern, and realizing their heights and global crown radii. The trees were subsequently "shuffled" around in random, collision-avoiding walks using Monte Carlo techniques, to reach a more realistic distribution of tree positions. The tree height would be drawn from a normal distribution. There were 137 trees in a stand of radius with 30 m, as shown in Figure 1. Although we could not get a clear value of RMSE from this information, we knew it had a low RMSE.

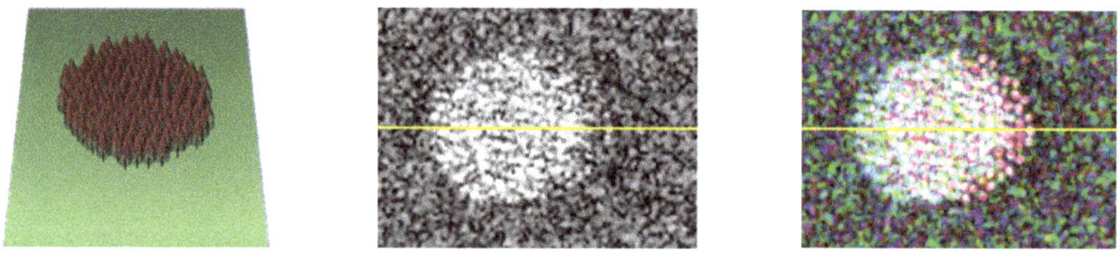

(a) (b) (c)

Figure 1. Image of the simulate forest scene image. (**a**) Forest scene; (**b**) Power image of HV; (**c**) Pauli image.

3. Methods

3.1. RVoG Model

The RVoG model [9] considered vegetation as a common process of vegetation volume scatter decay and surface scatter. The vegetation layer was assumed to be an isotropic medium, so we used a single extinction coefficient to represent overall attenuation. The vegetation's structural function decayed exponentially in the vertical direction.

The RVoG model was established by Treuhaft [9], and it was the theoretical basis of the forest height inversion algorithm. Regardless of other decoherence factors, only considering volume scattering coherence, we obtained the following explicit equation for the complex coherence [13]:

$$
\begin{aligned}
\gamma(\omega) &= \exp(j\varphi_0)\tfrac{\gamma_v+m(w)}{1+m(w)} = \exp(j\varphi_0)\left(\gamma_v + \tfrac{m(w)}{1+m(w)}(1-\gamma_v)\right) \\
&= \exp(j\varphi_0)\left(\gamma_v + L_{w_s}(1-\gamma_v)\right) \quad 0 \le L_{w_s} \le 1
\end{aligned}
\tag{1}
$$

where w is a three-component unitary complex vector defining the choice of polarization, φ_0 is the ground phase center, $m(w)$ is the ground-to-volume scattering ratio, which was only a function of polarization, and γ_v is pure volume complex coherence. Formula (2) is thus as follows

$$
\gamma_v = \frac{\int_0^{h_v} e^{(2\sigma z)/\cos\theta} e^{jk_z z} dz}{\int_0^{h_v} e^{(2\sigma z)/\cos\theta} dz} = \frac{p}{p_1}\frac{e^{p_1 h_v}-1}{e^{p h_v}-1} \qquad \begin{cases} p = \frac{2\sigma}{\cos\theta} \\ p_1 = p + ik_Z \\ k_Z = \frac{4\pi\Delta\theta}{\lambda\sin\theta} \approx \frac{4\pi B_n}{\lambda H\tan\theta} \end{cases}
\tag{2}
$$

where σ is the mean wave extinction in the medium, z is the scatter position, h_v is forest height, θ was the mean angle of incidence, $\Delta\theta$ is the apparent angular separation of the baseline from the scattering point, H is senor altitude, k_z is vertical wavenumber, and B_n is the vertical baseline.

L_{w_s} is the ground ratio, which had a relation with $m(w)$:

$$L_{w_s} = \frac{m(w)}{1+m(w)} = \frac{\frac{ground}{volume}}{1+\frac{ground}{volume}} = \frac{ground}{ground+volume}$$

G/G + V (ground/(ground + volume)) was also used to indicate the ground scattering ratio. Equation (1) represents a straight line in a complex plane passing through a point γ_v with a slope of $1-\gamma_v$.

When taking different extreme values of $m(w)$, $\gamma_{w_v}\gamma_{w_s}$ were reached as follows:

$$\begin{cases} \gamma_{w_s} = e^{j\varphi_0}\frac{\gamma_v+m(w)}{1+m(w)} \\ \gamma_{w_v} = e^{j\varphi_0}\gamma_v \end{cases} \tag{3}$$

where γ_{w_v} is the complex coherence corresponding to a pure volume scattering mechanism for the top forest canopy, and γ_{w_s} is the complex coherence corresponding to the surface scattering mechanisms near the ground surface under the forest canopy.

According to Formula (3), the formula for calculating the ground scattering ratio is as shown in (4):

$$L_{w_s} = \frac{-B-\sqrt{B^2-4AC}}{2A} \tag{4}$$

in which $A = |\gamma_{w_v}|^2 - 1$, $B = 2\mathrm{Re}((\gamma_{w_s}-\gamma_{w_v})\cdot\gamma^*_{w_v})$, and $C = |\gamma_{w_s}-\gamma_{w_v}|^2$.

Figure 2 shows pure volume complex coherence changes with forest height and extinction in a complex plane according to the RVoG model. Figure 2 was used as a look-up table for the three-stage inversion algorithm described later, laying the basis for forest height estimation. When knowing pure volume scattering complex coherence, we can use this lookup table to inverse the forest height and extinction coefficient. Meanwhile, the estimation of pure volume scattering complex coherence was very important, so we introduced coherence optimization to make its estimation more accurate.

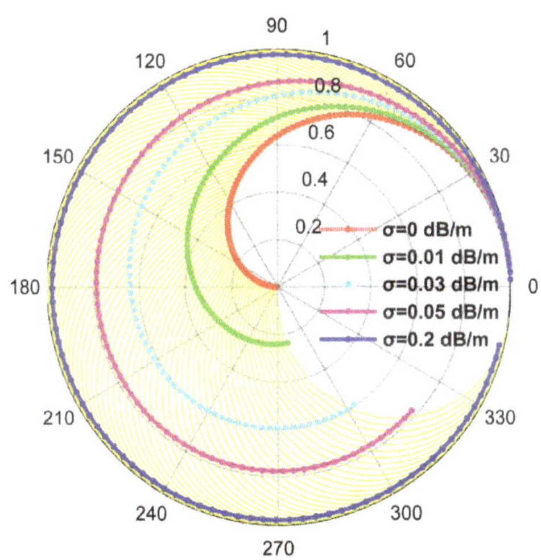

Figure 2. Volume complex coherence changes with forest height and extinction at k_z = 0.1154. Note: Calculate k_z = 0.1154 and *ha* = 54.4470 m from the setting parameters. Different color curves corresponded to different extinction coefficients. With larger extinction or strong ground scattering, the pure volume complex coherence amplitude was prone to saturation (such as the blue curve), but the phase was not saturated. When the extinction coefficient was in the range of 0~2 dB/m (such as yellow curves), the pure volume complex coherence changed with forest height h_v (0~*ha* m, step is

0.5 m). When the forest height was the same (any yellow curve) and pure volume complex coherence scattering amplitude was large, this was caused by surface scattering or the strong attenuation of vegetation. Complex coherence fell with increasing vegetation height, as a consequence of volume decorrelation.

3.2. Methods

In this paper, without considering baseline and coregistration error, topography, or temporal and signal-to-noise ratio (SNR) decorrelation sources, we introduced PD/MCD coherence optimization to solve for incomplete coherence separation in the three-stage inversion [13]. Additionally, we introduced complex coherence amplitude inversion [36] to solve the underestimation or overestimation problem and introduced the ground scatter ratio to automatically select the weight to improve the three-stage method. We compared and analyzed these methods combined with the topographic phase affecting forest height. A technical roadmap is shown in Figure 3.

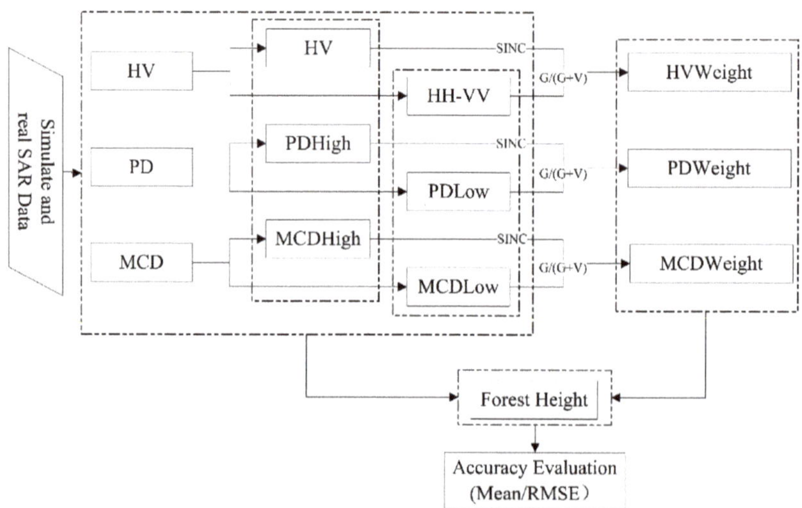

Figure 3. Forest Height Inversion Technology Roadmap.

The symbol meanings are shown in Table 1. In addition, SINC: introduced the corresponding complex coherence amplitude inversion; G/(G + V): ground scatter ratio of the corresponding parameter estimation, namely L_{w_s}. Weight: Based on three-stage inversion algorithm, it was further improved by introducing the corresponding complex coherence amplitude inversion and modified constant weight to the ε times of the ground scatter ratio. To correct the pure volume complex coherence, we projected it onto a coherence line to invert forest height.

Table 1. Parameters corresponding to various methods.

Method	Use Complex Coherence	Pure Volume Complex Coherence	Point Closest to the Topographic Phase
HV/HVWeight	A, B	HV	HH-VV
PD/PDWeight	A, C	PDHigh	PDLow
MCD/MCDWeight	A, D	MCDHigh	MCDLow

Note: A: HV + VH/HH + VV/VV/HH/LL/LR/RR/Opt1/Opt2/Opt3; B: HH − VV/HV; C: PDHigh/PDLow; D: MCDHigh/MCDLow; Where HV/VV/HH: linear polarization; HH + VV/HH − VV/HV + VH: Pauli base polarization; LL/LR/RR: circular polarization; Opt1/Opt2/Opt3: SVD; PDHigh/PDLow: PD; MCDHigh/MCDLow: MCD.

The process of three-stage inversion [13] was as follows:

Stage 1: Least squares line fit: Least squares were used to linearly fit the real and imaginary components of the polarization interference complex coherence and found the best fit straight line within the complex unit circle.

Stage 2: Vegetation bias removal: The true topographic phase was estimated with the coherence ranking order algorithm and removed it.

Stage 3: Height and extinction estimation: According to Formula (2), a look-up table (LUT) was established, where pure volume coherence changed with forest height and extinction coefficient (Figure 2). According to the topographic phase estimated from stage 2, it was easy to determine the coherence $\hat{\gamma}_v$ farthest from it in observation data. By comparing $\hat{\gamma}_v$ with LUT, forest height and extinction can be obtained without additional iterative optimization algorithms.

The three-stage inversion method was based on the assumption that the ground-to-volume scatter ratio in one of the polarization channels is zero. To simplify the problem, it was generally assumed that the ground-to-volume scatter ratio in the HV channel was zero, i.e., the estimated pure volume coherence was obtained from the estimated HV channel coherence. In addition, since the estimated HV and HH-VV channel coherence did not reach the maximum separation, there were some errors in forest height inversion, so PD and MCD coherence optimization was introduced to invert forest height. Several studies had explained specific steps of PD and MCD coherence optimization [25,26]. However, even with coherence optimization, the volume phase center may lie anywhere between half-way and the top height layer. Hence, the true forest height will still be underestimated [31]. The retrieved forest height may be overestimated or underestimated depending on the selected vertical baseline. Therefore, at least a coherence amplitude correction term can be employed to partially compensate for this underestimate or overestimate problem [36]. Based on a hybrid method proposed by Cloude (such as Formula (5)) [31,32], this study compensated for the "compression" phenomenon of forest top height not considered in the three-stage method based on coherence optimization, and modified constant weight ε to a variable weight $\varepsilon \cdot L_{w_s}$, as shown in Formula (6)

$$h_v = \frac{\arg(\gamma_{w_v}) - \varphi_0}{k_z} + \varepsilon \cdot \frac{2\sin^{-1}(|\gamma_{w_v}|)}{k_z} \quad (5)$$

$$h_v = \mathrm{h}_{ThreeStage} + \varepsilon \cdot L_{w_s} \cdot \frac{2\sin c^{-1}(|\gamma_{w_v}|)}{k_z} \quad (6)$$

where $\varepsilon \cdot L_{w_s}$ is a weight.

The first term represents the forest height inversed by the three-stage method with PolSARpro, which affects the accuracy of the method according to the linear equations of different scattering fittings. The second term is the coherence amplitude correction term with PolSARpro, which was solved for the first term underestimation or overestimated problem, but only considering the pure volume scatter complex coherence. The second part affects the accuracy of the method according to the ε times of the ground scattering ratio of the SINC inversion method, where the ground scattering ratio L_{w_s} is the performance of different scattering from the same vegetation/canopy or similar scattering from different types of vegetation, and the accuracy of the method is affected by volume scattering in the SINC inversion method. Therefore, we used the ground scatter ratio to modify it, and because the highest value of L_{w_s} was approximately 0.8, we added an adjustment factor ε to modify it again with MATLAB. The general rule is that the RMSE between the retrieved forest heights of these two parts and the true stand average height is the smallest to invert the forest heights. Therefore, $\varepsilon \cdot L_{w_s}$ made the full expression as robust as possible to changes in the structure function. However, in practical applications, forest height inversion by single PolInSAR data required ideal baselines. The vertical baseline depended on the platform, target geometry, forest height, and forest vertical structure, and the length of the vertical baseline determined the sensitivity of the interferometric phase difference to different forest heights. The retrieved forest height may be overestimated or underestimated depending on the selected vertical baseline. The value of ε ($-\alpha \sim \alpha$, $\alpha = 1/L_{w_s}$) was obtained according to the minimum RMSE of forest height. Negative values of ε corresponded to overestimation, and positive values corresponded to underestimation. The flow chart is shown in Figure 4 with MATLAB software.

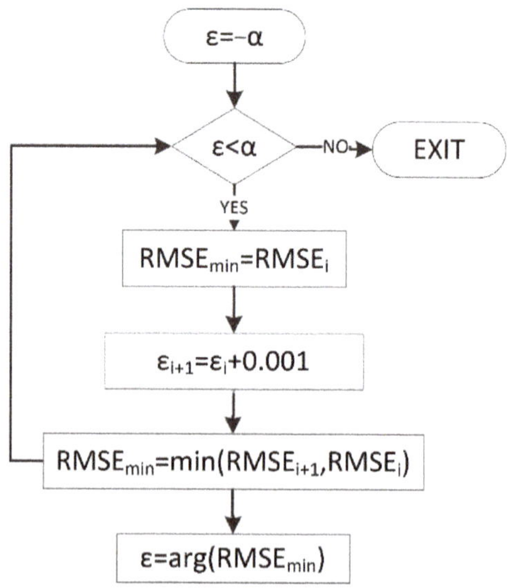

Figure 4. Flow chart for automatically selecting ε.

In this paper, we improved the three-stage inversion method based on the PD/MCD coherence optimization method. The PD and MCD methods were compared with the three-stage inversion algorithm (HV method) to invert forest height. We further inverted the forest height according to Formula (6), and they were HVweight, PDweight, and MCDweight. To modify the pure volume scattering complex coherence, we projected it to the coherence line to invert the forest height. When using real data, we need to consider the impact of these errors and we cannot systematically evaluate this algorithm. The simulation data could be used to systematically evaluate the algorithm, laying the foundation for the future application of the airborne L-band SAR data to invert the forest height. However, if the decorrelation source was not considered, the result of the three-stage inversion algorithm for forest height inversion may be high, and therefore the Weight method was applicable with negative values of ε.

4. Experimental Results

4.1. Topographic Phase

Even if there was a slow terrain change, the coherence phase would change rapidly, so forest height inversion must take the topographic phase into account [20]. Figure 5 shows the ground scatter ratio. Figure 6 shows an image of the estimated topographic phase. Figure 7 present a profile at azimuth = 47 (yellow line in the Figure 1b) and the statistical histogram of the estimated topographic phase.

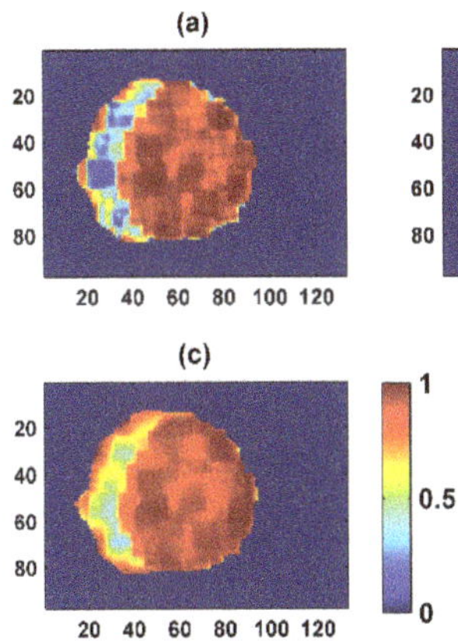

Figure 5. Image of the estimated L_{w_s} according to Formula (4), namely, the ground scatter ratio. (**a**) HV/HVWeight: L_{w_s} calculated from complex of HH-VV and HV channel; (**b**) PD/PDWeight: L_{w_s} calculated from a complex of Phase Diversity; (**c**) MCD/MCDWeight: L_{w_s} calculated from complex of Maximum Coherence Difference.

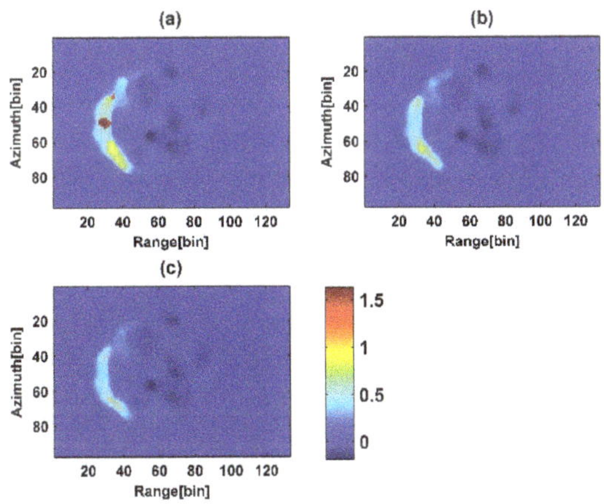

Figure 6. Image of estimated topographic phase. (**a**) HV/HVWeight method; (**b**) PD/PDWeight method; (**c**) MCD/MCDWeight method.

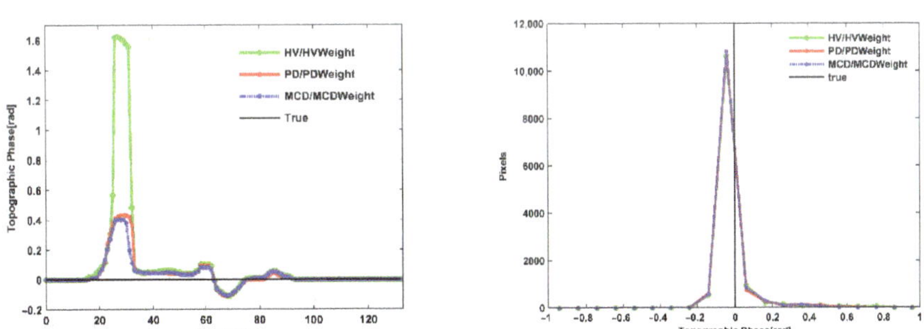

(**a**) (**b**)

Figure 7. Profile and statistical histogram of the estimated topographic phase. (**a**) Profile of the estimated topographic phase; (**b**) Statistical histogram of the estimated topographic phase.

It can be seen from Figure 5 that the overall ground scatter ratio trend was basically the same: small in the near direction and large in the far direction. After coherence optimization, the estimated ground scatter ratio improved. After PD and MCD coherence optimization, the ground scatter ratio was much better than the HV/HVWeight method, and the MCD/MCDWeight method was better than the PD/PDWeight method.

It can be seen from Figures 5 and 6 that in the region with a small ground scattering ratio in Figure 5, the estimated topographic phase was obviously high, and the error was large. In Figure 6, the estimated topographic phases all had negative values, the HV/HVWeight method values were significantly higher, and the difference in topographic phase estimated by the PD/PDWeight and MCD/MCDWeight methods was small. The profile and the statistical histogram of the topographic phase in Figure 7 further illustrated it. As seen from the profile of Figure 7a, the overall trend of the four methods was basically the same, and the proximity to the topographic phase from small to large was MCD/MCDWeight, PD/PDWeight, and HV/HVWeight methods. The statistical histogram in Figure 7b shows that there was no significant difference between the three methods.

Since the true topographic phase was 0, it was easy to cause positive and negative cancellation, so we used the arithmetic mean of absolute values for quantitative analysis. Table 2 lists the mean of the absolute values and the RMSE of the estimated topographic phase. It can be seen that the HV/HVWeight method had the worst estimation and the largest error. The PD/PDWeight method was greatly improved compared with the HV/HVWeight method but was inferior to the MCD/MCDWeight method. The estimation of the MCD/MCDWeight method was optimal, and the error was minimal.

Table 2. Estimation results of the topographic phase.

Method	Topographic Phase (Rad)	
	ABSMEAN	RMSE
HV/HVWeight	0.041	0.140
PD/PDWeight	0.032	0.095
MCD/MCDWeight	0.029	0.081

4.2. Forest Height

Figure 8 present an image of the inversed forest height, (A) the improved three-stage inversion algorithm based on coherence optimization, and (B) the forest height inversion after introducing the coherence amplitude. Figure 9 shows a profile at azimuth = 47 (yellow line in the Figure 1b) and the statistical histogram of the inversed forest height. Table 3 shows the image of inversed forest height inversion.

Table 3. Results of forest height inversion.

Method	Forest Height (m)	
	MEAN	RMSE
HV/HVWeight	15.83/16.29	4.80/4.65
PD/PDWeight	16.16/16.73	4.60/4.46
MCD/MCDWeight	16.19/16.71	4.43/4.35

In the area where the ground scatter ratio was small, the inversed forest height is low, and even there, the forest height cannot be inversed from Figures 5 and 8. The HV method had the worst forest height inversion. After PD and MCD coherence optimization, the forest height inversion results were much better than the HV method, and the MCD method was better than the PD method. The decreasing order of missing values in different methods was as follows: HV, PD, MCD methods. The corresponding method in Figure 8B had the same trend as in Figure 8A.

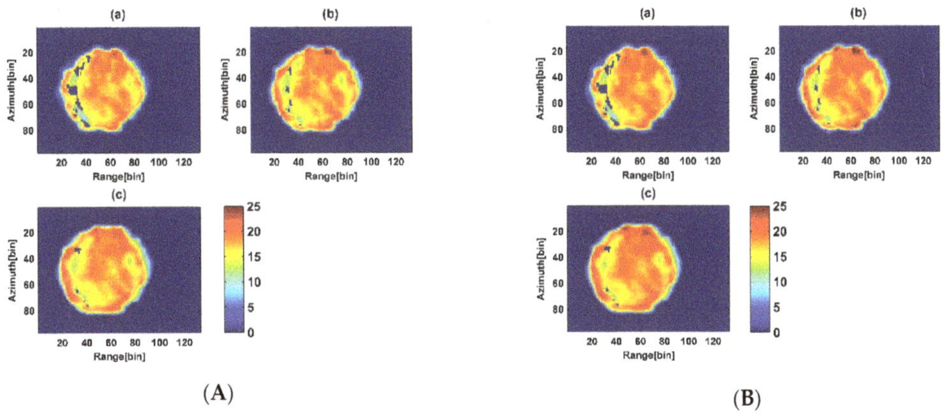

Figure 8. Improved three-stage method for forest height inversion. (**A**) Three-stage method for forest height inversion based on coherence optimization (**a**) HV (**b**) PD (**c**) MCD (**B**) After introduction coherence amplitude for Forest height inversion (**a**) HVWeight (**b**) PDWeight (**c**) MCDWeight.

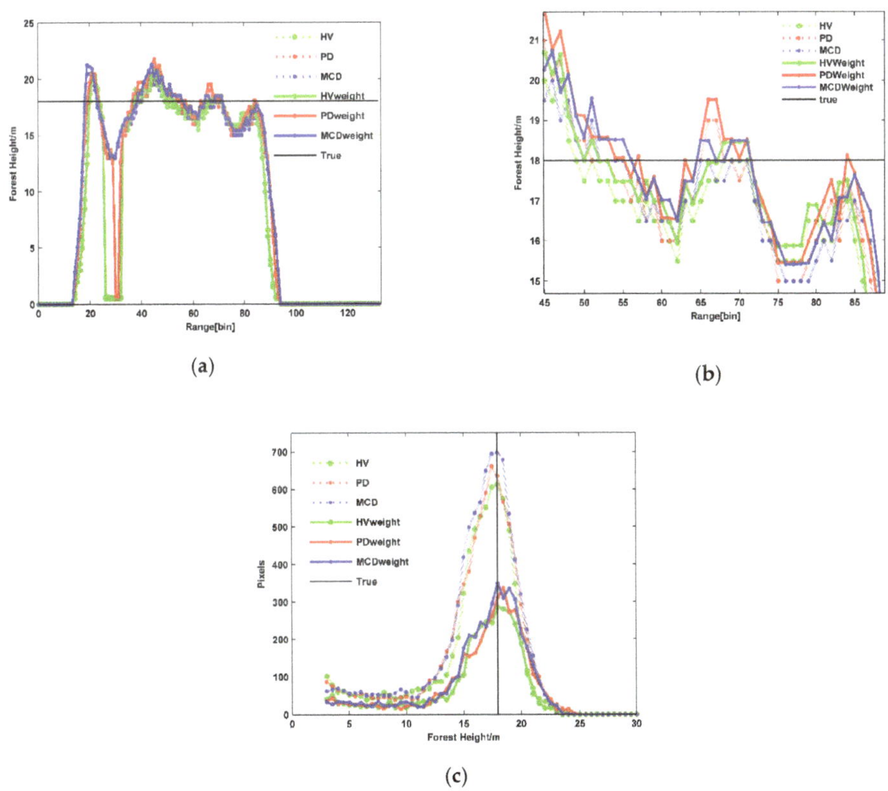

Figure 9. Profile and statistical histogram of the inversed forest height. (**a**) Profile of the inversed forest height; (**b**) Enlargement image of Figure (**a**) in the range at 45~90; (**c**) Statistical histogram of the inversed forest height.

The mean of forest height inversion by the HV method was the lowest (15.83 m) and the RMSE was the largest (4.80 m). The PD method maximized the separation of complex coherence phases representing canopy scattering and ground scattering, which making its mean (16.19 m) and its RMSE (4.60 m) superior to the HV method. The MCD method maximized the distance between canopy scattering and ground scattering complex coherence in the complex plane, its mean (16.19 m) was slightly better than using the PD method, and the RMSE was the smallest (4.43 m).

Figure 9b shows a profile of the inversed forest height enlargement image from Figure 9a in the range near 45~90. After adding ε times the ground scatter ratio as the weight, the forest height was equivalent to a proportionality increase. After adding the coherence amplitude term from Table 3, the mean of forest heights inversed by the HV, PD, and MCD methods were increased by 0.46 m, 0.57 m, and 0.52 m, respectively. The RMSE was improved by 0.15 m, 0.14 m, and 0.08 m, respectively, and the improvement of the HV method was obviously best, followed by the PD method. After adding the coherence amplitude inversion method from Figure 9c and Table 3, although the robustness of the forest height inversion algorithm was reduced, the underestimation was improved and the RMSE was reduced. Among these methods, the mean forest height inversion by the MCD method was the closest to the true value, followed by the PD and HV methods. The smallest RMSE was obtained by MCD, followed by the PD and HV methods. The MCD method made the coherence distance between the complex coherence maximum in the complex plane and obtained the accurate pure volume scatter coherence, such that the mean forest height inversion was the highest and its RMSE was the lowest. The PD method maximized the phase of complex coherence and obtained the accurate pure volume scatter coherence so that the mean and the RMSE of forest height inversion were lower than the MCD method. The mean of forest height inversion by the HV method was the largest difference from the true value, and its RMSE was the largest.

5. Discussion

The estimated topographic phase may be caused by the following aspects: (1) The ground scatter ratio: topographic phase was good and the error was large in the area where the ground scattering ratio was small. Combining Figures 5 and 6, the estimated topographic phase error in Figure 6 was relatively large where the ground scattering ratio of Figure 5 was small. The estimated topographic phase error was significantly reduced where the ground scattering ratio of Figure 5 was large. (2) The difference of the complex coherence combination and sample number used to fit the coherence line: Among these methods, the coherence line was fitted according to the least squares method, in addition to jointly using a group of complex coherences, and the HV/HVWeight method also used B group complex coherences, the PD method with C group complex coherences. The MCD/MCDWeight method in addition used D group complex coherences. The sample number of HV/HVWeight/PD/PDWeight/MCD/MCDWeight methods was the same, and only the B/C/D group complex coherence was different. C/D was the complex coherence obtained after PD and MCD coherence optimization, respectively. PD coherence optimization maximized the phase center of complex coherence in the complex plane. MCD coherence optimization made the distance of complex coherence maximum in the complex plane. Figure 10 shows the complex coherence and the fitted coherence line used by these methods in a complex plane. The estimated topographic phase in Figure 10a from small to large was: MCD/MCDWeight, PD/PDWeight, and HV/HVWeight. Squares of different colors represented different complex coherences in Figure 10. Straight lines of different colors were coherence lines fitted by different methods according to corresponding complex coherences. (a) When the azimuth was 70 and the range was 40 in the image, various coherence lines were fit by different methods. The intersection of the coherence line fit by the MCD/MCDWeight method and the complex unit circle (the point close to the x real axis), i.e., the topographic phase, was closest to the real topographic phase, followed by PD/PDWeight and HV/HVWeight. (b) When the azimuth was 51 and the range was 44 in the image, because the distance between the complex coherence representing ground and volume scattering was enough large, and the distance in the normal coherence line direction was small, coherence lines fitted by these methods were the same, and the intersections with the unit circle were also the same, i.e., the topographic phases were the same. It could be seen that when the sample number of complex coherence was the same, the coherence line fit by the PD/PDWeight method was more accurate than the HV/HVWeight method. The PD/PDWeight method was lower than the MCD/MCDWeight method because it was considered complex coherence amplitude information. In Figure 10b, the estimated topographic phase is the same, and it can be seen that the greater the distance between volume and ground complex coherence in the complex plane, and the smaller the distance in the coherence line normal direction, the higher the accuracy of the topographic phase. The topographic phase was not affected by complex coherence when the direction of the complex coherence line and its normal reached a certain value, even without coherence optimization, the same topographic phase could be obtained. (3) The different selection of points closest to the ground phase: the HV/HVWeight/PD/PDWeight/MCD/MCDWeight methods select complex coherence closest to the ground phase, respectively. MCDLow was optimal, followed by PDLow and HH-VV.

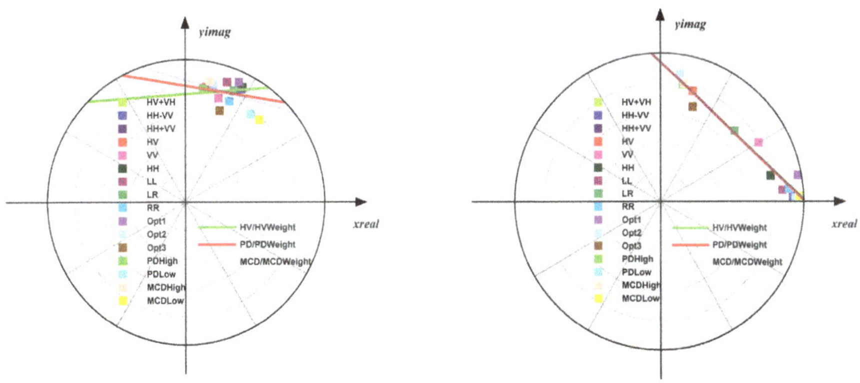

(**a**) (**b**)

Figure 10. Comparison of complex coherence and fitted coherence straight lines in a complex plane. (**a**) azimuth = 70 range = 40; (**b**) azimuth = 51 range = 44.

Without using coherence optimization, the topographic phase estimated by HV/HVWeight method was the worst. PD coherence optimization maximized the difference between coherence phase centers in the complex plane. Compared with the HV/HVWeight method, the topographic phase estimation accuracy was greatly improved. The MCD coherence optimization maximized the complex coherence distance in the complex plane, comprehensively considering phase and amplitude information. The obtained complex coherence MCDLow representing ground scattering was better than PD coherence optimization, and closer to the real ground phase. The topographic phase estimated by the MCD/MCDWeight method was optimal and closest to the true topographic phase. When applied to real SAR data with lots of field data, we can acquire the same conclusion as the simulation data. However, when the sample sizes were small, there may be some differences from this simulation conclusion. Therefore, in order to obtain higher accuracy, it was necessary to carry out a variety of optimization methods in single-baseline L-band PolInSAR technology with real SAR data.

With analyzing the effect of the estimated topographic phase on the forest height, we compared three-stage inversion methods based on coherence optimization. The HV method selected HV channel complex coherence as the canopy scattering complex coherence, and its phase center may be located at any position between half-way and the top forest height [31,32]. The estimated topographic phase was the highest, and the forest height inversion had the largest error. Due to the observed propagation error of the topographic phase, and the selection of channel as forest canopy scattering complex coherence, with the lowest forest height mean and largest error, the HV method estimated topographic phase accuracy performed worse. The PD method maximized the complex coherence phase center of canopy and ground scattering, which made it more accurate than the coherence line fitted using the HV and HH-VV channel complex coherence used by the HV method. Therefore, the mean of topographic phase absolute value was better than the HV method, and pure volume scatter complex coherence (PDHigh) was more accurate than the HV complex coherence. The mean forest height was increased by 0.33 m compared with the HV method, whereas the RMSE increased by 0.2 m. The MCD method made the distance between complex coherence of canopy and ground scattering maximum in the complex plane. The topographic phase and forest height were very similar to the PD method. However, overall, the MCD method, while comprehensively considering the coherence amplitude and phase, had a better forest height inversion than the PD method, as well as the smallest RMSE. The result estimated by the MCD method obtained the optimal topographic phase and more accurate pure volume coherence. Therefore, forest inversion based on the lookup table was also more accurate, and the mean of the inversed forest height was closest to the true value. Compared with the HV method, the forest height inversion underestimation was improved after coherence optimization, and the RMSE was reduced [27,36]. The PD method was based on phase information, which improved forest height inversion to some extent. The MCD method was based on coherence amplitude and phase, which was better than the PD method.

Figure 11 shows the relationship between the weighted forest height statistical indicator (MEAN/RMSE/SD) and ε. Compared with the three-stage inversion algorithm based on coherence optimization, after introducing the coherence amplitude term and modifying fixed weights to weight variables, the mean of forest height increased while the RMSE decreased [36]. As the ε increases, the average bias of the forest height increased first and then decreased, but the final reduced value was greater than the original, and the minimum value was near 0.1~0.2. The RMSE of forest height decreased first and then increased and finally decreased with the ε increased, with a minimum value near 0.1. After introducing the coherence amplitude inversion term, the standard deviation of forest height increased, i.e., the algorithm robustness was no better than the three-stage inversion method. We selected the ε corresponding to the RMSE minimum value with comprehensive consideration.

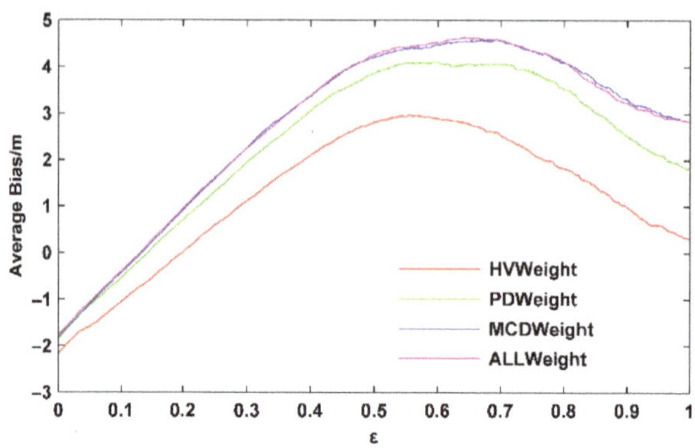

(a)

Figure 11. *Cont.*

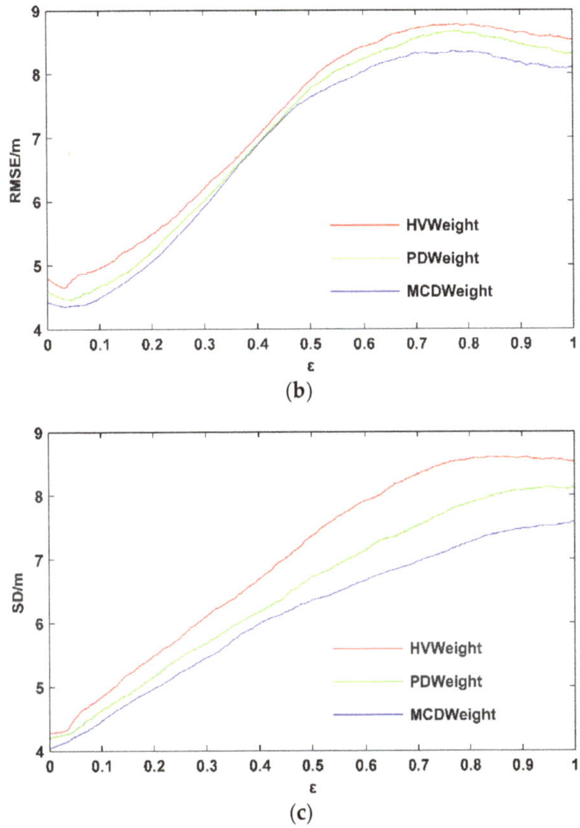

Figure 11. Forest height statistical indicator changes with the ε. (**a**) Average bias; (**b**) RMSE; (**c**) Standard deviation.

Compared with the HV method, the underestimation of forest height inversion was improved after coherence optimization, and the RMSE was reduced [27,36]. The PD method was based on phase information, which improved forest height inversion to some extent. The MCD method was based on coherence amplitude and phase, which was better than the PD method. After introducing the coherence amplitude term and modifying fixed weights to weight variables, the corresponding method (weight) was the same as the overall trend of the three-stage inversion algorithm based on coherence optimization.

Considering time and signal-to-noise ratio decorrelation, this study could invest forest height from the airborne L-band single-baseline full polarization SAR data in the flat terrain. When applied to real data, the inverted forest height was combined with the average height of the forest stand to minimize its RMSE to automatically select an appropriate ε. The smaller the RMSE, the higher the accuracy of the corresponding method. At the same time, a y = x straight line equation could be fitted between the inverted forest height and the measured forest stand height. The closer the fitted straight line equation was to y = x, the more accurate the corresponding inversion method was.

The accuracy of the algorithm was related to the system parameters of the aircraft and the structural parameters of the forest (forest density, height), etc. Since only some parameters could be fixed to analyze the relative error of the RVoG model inversion of forest height, an uncertainty could not be given for quantifying all involved parameters. Due to the most influential factor being the vertical wave number k_z of the forest, the forest height relative error changed with the forest height and vertical wavenumber k_z when the time decorrelation was 0.8, and extinction coefficient was fixed at 0, 0.1, and 0.5 dB/m, respectively. Given the values of h_v, σ and k_z, the vegetation pure volume coherence γ_v (h_v, σ, k_z) was calculated according to Formula (2) for L-band data. Given a fixed time decorrelation intensity γ_t, $\gamma_o = \gamma_v\ (h_v, \sigma, k_z)\ \gamma_t$ was then calculated. For each generated γ_o sample, the forest height h_{vi} was inverted by the three-stage method within the 2π ambiguity elevation, and the relative error of forest height was analyzed. The relative error of forest height was $\mid h_{vi} - h_v \mid / h_v \times 100\%$. Figure 12 shows the forest height relative error changes with the forest height and vertical wavenumber k_z under three extinction coefficient levels (0.0, 0.1, and 0.5 dB/m) at γ_t = 0.8, respectively.

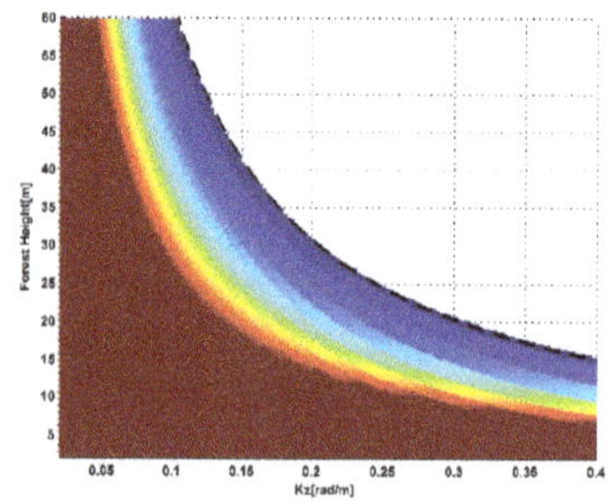

(**a**)

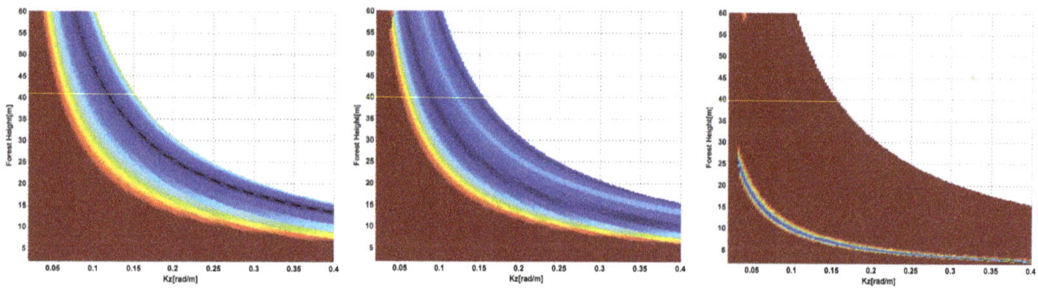

(**b**)

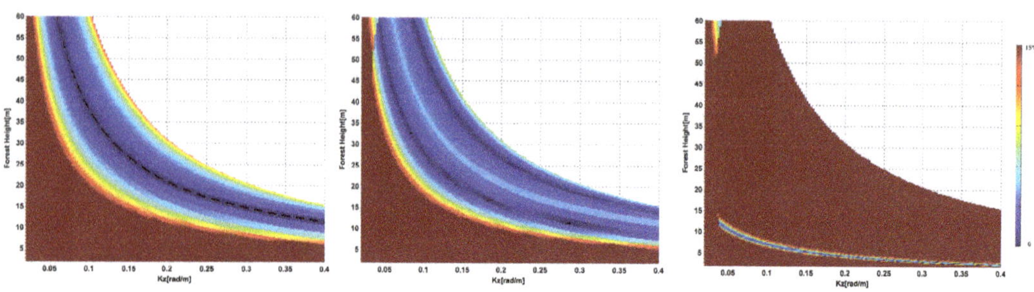

(**c**)

Figure 12. When the time decorrelation was 0.8, and extinction coefficient was fixed at 0, 0.1, 0.5 dB/m, respectively, the forest height relative error changed with the forest height and vertical wavenumber k_z. (**a**) σ = 0 dB/m; (**b**) σ = 0.1 dB/m left: $\varepsilon \cdot L_{w_s} = 0$ middle: $\varepsilon \cdot L_{w_s} = 0.04$ right: $\varepsilon \cdot L_{w_s} = 0.5$; (**c**) σ = 0.5 dB/m left: $\varepsilon \cdot L_{w_s} = 0$ middle: $\varepsilon \cdot L_{w_s} = 0.04$ right: $\varepsilon \cdot L_{w_s} = 0.5$.

It was shown that for a given vertical wavenumber k_z, the inversion performance of forest height was the best only in a certain height range. The forest height relative error was larger in places with low forest height, and it decreased with increasing forest height (Figure 12).

The simulation results showed that in order to achieve 10% accuracy in the forest height range from 8 m to 60 m, various baselines (vertical wavenumber k_z) had to be required, and the number of baselines required depended on the extinction coefficient. Among all extinction coefficient levels, the forest height ranged from 8 m to 60 m, and the three baselines were sufficient to make the forest height relative error better than 10% (on the left of Figure 12). For example, if the forest height ranged from 8 m to 40 m, the three baselines were sufficient to make the forest height relative error better than 10% at $\sigma = 0.1$ dB/m (on the left of Figure 12b), and only the two baselines were sufficient at $\sigma = 0.5$ dB/m (on the left of Figure 12c). After introducing the overestimation and underestimation terms (on the middle and right of Figure 12b,c), the accuracy of forest height inversion could be improved by selecting the appropriate weight coefficient $\varepsilon \cdot L_{w_s}$, and even higher forest height inversion accuracy could be obtained by using one baseline (on the middle and right of Figure 12b,c). An uncertainty model that quantifies all involved parameters will be the subject of future work.

ALOS-2 with an enhanced PALSAR instrument launched in 2014, where ALOS left in 2011, and will build L-band SAR data for monitoring the global environment. However, ALOS-2 has a strong temporal decoherence effect, leading the coherence in the forest to be too low to make the forest height estimation with POLInSAR impossible. The upcoming TanDEM-L with spaceborne monostatic and bistatic SAR imagery solved the problem of time decoherence very well. We therefore expected that our results would be valuable for a wide range of future research topics, including all future airborne and spaceborne SAR with the upcoming low frequencies forest missions, ALOS-4, NISAR (NASA-ISRO Synthetic Aperture Radar), and Tandem-L (all L-band), as well as BIOMASS. An unprecedented combination of sensors will be seen in the next few years, e.g., BIOMASS links to the Global Ecosystem Dynamics Investigation (GEDI) and NISAR missions will be particularly important for measuring forest structure parameter, such as forest height and biomass. The in-situ data for GEDI, BIOMASS, and NISAR collaborated by the ESA-NASA, will further help to achieve more forest height inversion performance. Meanwhile, LIDAR data with a relatively fine scale and accurate map of forest height and biomass represents an important complement to in situ, airborne data. In situ data, when combined with LIDAR and GEDI data, will allow forest height inversion on canopy structure and even biomass with POLInSAR to be estimated. This study is expected to mitigate the overestimation and underestimation problem of forest height inversion for the upcoming L-band SAR satellite generation, new SAR and LIDAR systems combined with RPAs (Remotely Piloted Aircrafts)/UAVs (Unmanned Aerial Vehicles) for small areas mapping initiatives, and to promote the depth and breadth of SAR applications of the new SAR system.

6. Real SAR Data

The SAR data in the Traunstein were the E-SAR L-band single-baseline full PolInSAR data obtained by the German Aerospace Center (DLR) in 2003. The study area was a plantation forest, and the terrain was relatively flat with only some small slopes. The altitude of the aircraft was about 3000 m, the space baseline was about 5 m, the time baseline was 20 min, and the central incidence angle was 45°. The range resolution was 1.5 m, and the azimuth resolution was 3 m with four looks. The data were precisely registered, and flat phase and effective wavenumbers were provided.

The computational effort was 2~3 times that required of the original three-stage inversion method. The time of the SINC function inversion method was similar to that of the three-stage method, in which the time for the automatic weight selection method with one forest stand was about 45.70 s. Finally, the three weight methods with one forest stand with Matlab software for calculation took about 3.04 s.

The measured data and ground-truth data concerning the stand height of the sample plot are shown in Figure 13. The forest height images obtained by the three-stage inversion methods are shown in Figure 13A, and the three-stage methods introducing the coherent magnitude term are shown in Figure 13B. The ground-truth data used in the study mainly included the boundaries of eight stands and the average height of dominant trees, which were obtained by the Munich Forest Harvest Scientific Committee through field surveys, as shown in Figure 13(Ad,Bd). Table 4 shows the RMSE of various methods for the corresponding eight forest stands with real data. The result showed that coherent optimization methods may also not achieve the best accuracy. After the modified two-step, three-stage inversion algorithm is carried out, the RMSE can always be minimized, and the number of the minimum RMSE obtained by the PD coherent optimization method is greater. In order to obtain a better RMSE, it was necessary to use coherent optimization methods. L_{w_s} was the response to forest stand structure, because the scattering mechanisms were different for different forest stands and different forest heights. The approach performed well in the case of different plant densities and different plant height variability with simulation forest relative error and real forest stands. The fitting equations between the forest height estimated by six methods and the stand height of the plot are shown in Table 5. Unlike the simulated data, the three-stage inversion method overestimated the forest height, which may be caused by the vertical baseline of the data. Simulation data showed that the three-stage method would underestimate forest height, but in practical applications, forest height could produce overestimation and underestimation owing to the length of the vertical baseline determined by the sensitivity of the interferometric phase difference to different forest heights. Forest height inversion by PolInSAR data requires ideal baselines. Due to the complexity of the real SAR data and the small sample sizes, the three-stage inversion method based on coherent optimization was lower than the three-stage inversion method. After introducing the coherent magnitude term, the overestimation of the forest height was significantly weakened in HVWeight, PDweight, and MCDWeight, and PDWeight was optimal. Compared with the original three-stage method, the inversion accuracy of simulated data increased by up to 9.38%, and 59.85% with real data at most.

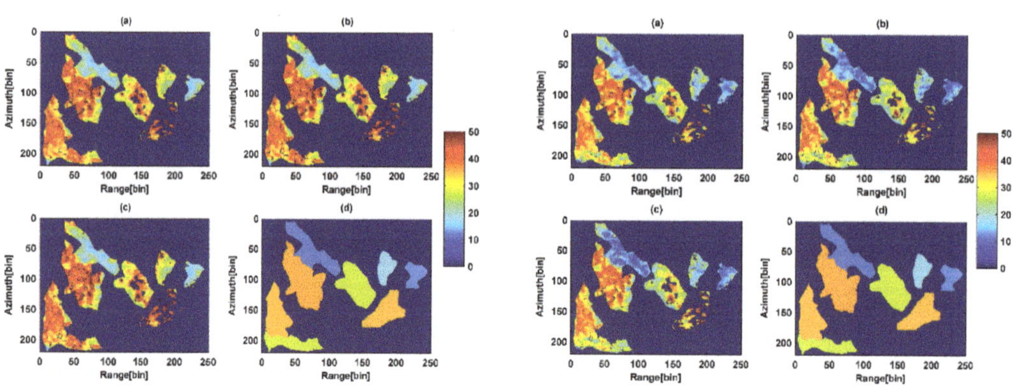

(A) (B)

Figure 13. Improved three-stage method for forest height inversion of ESA PolInSAR data. (**A**) Three-stage method for forest height inversion based on coherence optimization (**a**) HV (**b**) PD (**c**) MCD (**d**) The average stand forest height of field measurement (**B**) Introduction of coherence amplitude for Forest height inversion (**a**) HVWeight (**b**) PDWeight (**c**) MCDWeight (**d**) The average stand forest height of field measurement.

Table 4. The RMSE of various methods for the corresponding 8 forest stands with real data.

Forest Stand	RMSE (m)					
	HV	HVWeight	PD	PDWeight	MCD	MCDWeight
1	7.04	6.37	7.17	6.47	7.51	6.76
2	11.67	5.12	11.73	5.07	11.73	4.71
3	7.75	4.08	7.72	3.10	7.70	3.57
4	11.12	5.58	11.12	4.26	11.19	4.43
5	8.19	5.99	8.03	4.94	8.03	5.55
6	8.75	5.72	8.94	5.02	9.10	4.90
7	5.73	5.19	6.05	5.34	6.29	5.38
8	7.41	4.50	8.44	4.12	8.48	4.67

Table 5. The fitting equation between the retrieved forest height and the field measurement stand forest height.

Method	Retrieved Forest Height and Field Measurement Stand Forest Height	
	Equation	RMSE (m)
HV	$0.76292x + 12.54232$, $R^2 = 0.9557$	1.69865
PD	$0.78461x + 12.17832$, $R^2 = 0.9523$	1.81638
MCD	$0.7950x + 12.03076$, $R^2 = 0.9549$	1.78694
HVWeight	$0.99053x + 1.15906$, $R^2 = 0.9892$	1.07157
PDWeight	$1.00693x - 0.05627$, $R^2 = 0.9990$	0.33435
MCDWeight	$1.09135x - 2.54197$, $R^2 = 0.9792$	1.64526

7. Conclusions

Compared with the forest height inversion accuracy with the simulation and real SAR data, it was necessary to use coherent optimization methods to obtain a better RMSE for the forest height inversion using single-baseline L-band PolInSAR data. Using ε times the ground scattering ratio as the weight alleviates the underestimation and overestimation phenomena of the forest height estimation and reduces the RMSE to some extent, but the robustness of the forest height inversion is reduced due to the introduction of the coherence amplitude term.

This study can invest forest height from the airborne L-band single-baseline full polarization SAR data in the flat terrain. Since this study only simulates coniferous forests with a forest height of 18 m and a forest density of 500, and applies and validates these methods with small real data, other scholars can apply these methods with more airborne L-band SAR data to better explain the applicability and limitations of these methods.

Due to the inherent characteristics of SAR images, shadows, overlays, and top-to-bottom overlaps may occur with large terrain fluctuations. Single-baseline PolInSAR cannot solve these problems temporarily. Therefore, this study mainly considers areas with flat terrain, not taking the terrain into account. This study does not consider the effect of slope on forest height inversion, and mainly focuses on solving the problem of the overestimation and underestimation of forest height inversion by the three-stage method through a modified two-step, three-stage inversion algorithm. When the slope of the terrain is not very high, the R-RVoG model can be used to invert the forest height. In the case of a higher slope, the multi-baselines TomoSAR method can be used to invert the forest height more accurately, which is what we want to do in the near future.

Author Contributions: J.Z. conceived the idea, designed and performed the experiments, produced the results, and drafted the manuscript. W.F. contributed to discuss with the idea and results, and revision of the article. Y.Z., L.H., Y.Y. and X.M. contributed to the discussion of the results and revision of the article. All authors have read and agreed to the published version of the manuscript.

Funding: This research was funded by the National Natural Science Foundation of China (contract no. 31971654, 41725003), the Civil Aerospace Technology Advance Research Project (contract no. D040114), and the National Natural Science Foundation of China (contract no. 42101067).

Institutional Review Board Statement: Not applicable.

Informed Consent Statement: Not applicable.

Conflicts of Interest: The authors declare no conflict of interest.

References

1. Woodhouse, I.H. Predicting backscatter-biomass and height-biomass trends using a macroecology model. *IEEE Trans. Geosci. Remote Sens.* **2006**, *44*, 871–877. [CrossRef]
2. Houghton, R.A.; Hall, F.; Goetz, S.J. Importance of biomass in the global carbon cycle. *J. Geophys. Res.* **2015**, *114*, 1–13. [CrossRef]
3. Wright, S.J. Tropical forests in a changing environment. *Trends Ecol. Evol.* **2005**, *20*, 553–560. [CrossRef] [PubMed]
4. Pan, Y.; Birdsey, R.A.; Fang, J.; Houghton, R.; Kauppi, P.E.; Kurz, W.A.; Phillips, O.L.; Shvidenko, A.; Lewis, S.L.; Canadell, J.G.; et al. A large and persistent carbon sink in the world's forests. *Science* **2011**, *333*, 988–993. [CrossRef]
5. Minh, D.H.T.; Le Toan, T.; Rocca, F.; Tebaldini, S.; Villard, L.; Rejou-Mechain, M.; Phillips, O.L.; Feldpausch, T.R.; Dubois-Fernandez, P.; Scipal, K.; et al. SAR tomography for the retrieval of forest biomass and height: Cross-validation at two tropical forest sites in French Guiana. *Remote Sens. Environ.* **2016**, *175*, 138–147. [CrossRef]
6. Fu, W.X.; Guo, H.D.; Li, X.W.; Tian, B.S.; Sun, Z.C. Extended Three-Stage Polarimetric SAR Interferometry Algorithm by Dual-Polarization Data. *IEEE Trans. Geosci. Remote Sens.* **2016**, *54*, 2792–2802.
7. Sagues, L.; Lopez-Sanchez, J.M.; Fortuny, J.; Fabregas, X. Polarimetric radar interferometry for improved mine detection and surface clutter rejection. *IEEE Trans. Geosci. Remote Sens.* **2001**, *39*, 1271–1278. [CrossRef]
8. Cloude, S.; Papathanassiou, K. Polarimetric optimisation in radar interferometry. *Electron. Lett.* **1997**, *33*, 1176–1178. [CrossRef]
9. Treuhaft, R.N.; Madsen, S.N.; Moghaddam, M.; Zyl, J.J. Vegetation characteristics and underlying topography from interferometric radar. *Radio Sci.* **1996**, *31*, 1449–1485. [CrossRef]
10. Cloude, S.R.; Papathanassiou, K.P. Polarimetric SAR interferometry. *IEEE Trans. Geosci. Remote Sens.* **1998**, *36*, 1551–1565. [CrossRef]
11. Wang, C.C.; Wang, L.; Fu, H.Q.; Xie, Q.H.; Zhu, J.J. The Impact of Forest Density on Forest Height Inversion Modeling from Polarimetric InSAR Data. *Remote Sens.* **2016**, *8*, 291. [CrossRef]
12. Papathanassiou, K.P.; Cloude, S.R. Single-baseline polarimetric SAR interferometry. *IEEE Trans. Geosci. Remote Sens.* **2001**, *39*, 2352–2363. [CrossRef]
13. Cloude, S.R.; Papathanassiou, K.P. Three-stage inversion process for polarimetric SAR interferometry. *IEE Proc-Radar Sonar Navig.* **2003**, *150*, 125–134. [CrossRef]
14. Ballester-Berman, J.D.; Lopez-Sanchez, J.M.; Fortuny-Guasch, J. Retrieval of biophysical parameters of agricultural crops using polarimetric SAR interferometry. *IEEE Trans. Geosci. Remote Sens.* **2005**, *43*, 683–694. [CrossRef]
15. Praks, J.; Kugler, F.; Papathanassiou, K.P.; Hajnsek, I.; Hallikainen, M. Height estimation of boreal forest: Interferometric model-based inversion at L- and X-band versus HUTSCAT profiling scatterometer. *IEEE Geosci. Remote Sens. Lett.* **2007**, *4*, 466–470. [CrossRef]
16. Garestier, F.; Dubois-Fernandez, P.C.; Champion, I. Forest Height Inversion Using High-Resolution P-Band Pol-InSAR Data. *IEEE Trans. Geosci. Remote Sens.* **2008**, *46*, 3544–3559. [CrossRef]
17. Lee, S.K.; Kugler, F.; Papathanassiou, K.P.; Hajnsek, I. Quantifying Temporal Decorrelation over Boreal Forest at L- and P-band. In Proceedings of the European Conference on Synthetic Aperture Radar, Friedrichshafen, Germany, 2–5 June 2008; pp. 1–4.
18. Zhang, Q.; Liu, T.D.; Ding, Z.G.; Zeng, T.; Long, T. A Modified Three-Stage Inversion Algorithm Based on R-RVoG Model for Pol-InSAR Data. *Remote Sens.* **2016**, *8*, 861. [CrossRef]
19. Managhebi, T.; Maghsoudi, Y.; Zoej, M.J.V. An Improved Three-Stage Inversion Algorithm in Forest Height Estimation Using Single-Baseline Polarimetric SAR Interferometry Data. *IEEE Geosci. Remote Sens. Lett.* **2018**, *15*, 887–891. [CrossRef]
20. Bai, L.; Hong, W.; Cao, F. Estimation error of topographic phase based on RVoG model using POLinSAR data. In Proceedings of the Progress in Electromagnetics Research Symposium (PIERS ONLINE), Nantes, France, 13–17 July 1998; pp. 306–310.
21. Hashjin, S.S.; Khazai, S.; Sadeghi, A. A Method to Select Coherence Window Size for forest height estimation using PolInSAR Data. In Proceedings of the ISPRS, Tehran, Iran, 5–8 October 2013; pp. 505–508.
22. Neumann, M.; Ferro-Famil, L.; Reigber, A. Multibaseline polarimetric SAR interferometry coherence optimization. *IEEE Geosci. Remote Sens. Lett.* **2008**, *5*, 93–97. [CrossRef]
23. Colin, E.; Titin-Schnaider, C.; Tabbara, W. An interferometric coherence optimization method in radar polarimetry for high-resolution imagery. *IEEE Trans. Geosci. Remote Sens.* **2006**, *44*, 167–175. [CrossRef]
24. Gomezdans, J.L.; Quegan, S. Constraining Coherence Optimisation in Polarimetric Interferometry of Layered Targets. In Proceedings of the POLinSAR 2005 Workshop ESRIN, Frascati, Italy, 17–21 January 2005; pp. 1–6.

25. Flynn, T.; Tabb, M.; Carande, R. Coherence region shape extraction for vegetation parameter estimation in polarimetric SAR interferometry. In Proceedings of the IEEE International Geoscience and Remote Sensing Symposium, Toronto, ON, Canada, 24–28 June 2002; pp. 2596–2598.
26. Tabb, M.; Orrey, J.; Flynn, T.; Carande, R. Phase diversity: A decomposition for vegetation parameter estimation using polarimetric SAR interferometry. In Proceedings of the Fourth European Synthetic Aperture RADAR Conference, Cologne, Germany, 4–6 June 2002; pp. 721–724.
27. Xie, Q.; Zhu, J.; Wang, C.; Fu, H. Boreal forest height inversion using E-SAR PolInSAR data based coherence optimization methods and three-stage algorithm. In Proceedings of the International Workshop on Earth Observation and Remote Sensing Applications, Changsha, China, 11–14 June 2014; pp. 145–150.
28. Lavalle, M.; Solimini, D.; Pottier, E.; Desnos, Y.L. Comparison of models of POLINSAR coherence for forest height retrieval using POLINSAR simulated data. *Matrix* **2008**, *54*, 2792–2802.
29. Lin, D.F.; Zhu, J.J.; Fu, H.Q.; Xie, Q.H.; Zhang, B. A TSVD-Based Method for Forest Height Inversion from Single-Baseline PolInSAR Data. *Appl. Sci.* **2017**, *7*, 435. [CrossRef]
30. Cloude, S.R. Polarization coherence tomography. *Radio Sci.* **2006**, *41*, 1–27. [CrossRef]
31. Cloude, S.R. POL-InSAR training course. In Proceedings of the Advance Training Course on Land Remote Sensing, Scotland, UK, January 2008; pp. 1–44.
32. Cloude, S.R. *Polarisation Applications in Remote Sensing*; Oxford University Press Inc.: New York, NY, USA, 2010.
33. Minh, N.P.; Wang, C.; Zou, B.; Nguyen, Q.T.; Le, V.N. Forest Height Extraction from PolInSAR Image Using a Hybrid Method. *Int. J. Signal Process. Image Process. Pattern Recognit.* **2014**, *7*, 257–274. [CrossRef]
34. Minh, N.P.; Zou, B.; Lu, D. Accuracy improvement method of forest height estimation for PolInSARImage. In Proceedings of the International Conference on Audio, Language and Image Processing, Shanghai, China, 16–18 July 2012; pp. 594–598.
35. Minh, N.P.; Zou, B. A Novel Algorithm for Forest Height Estimation from PolInSAR Image. *Int. J. Signal Process. Image Process. Pattern Recognit.* **2013**, *6*, 15–32.
36. Chen, E.X. Comparison of Methods to Derive Forest Height from Polarimetric SAR Interferometry. In Proceedings of the Proc Dragon, Beijing, China, 21–25 April 2008; pp. 1–9.
37. Managhebi, T.Y.; Maghsoudi; Zoej, M.J.V. A volume optimization method to improve the three-stage inversion Algorithm for Forest Height Estimation Using PolInSAR Data. *IEEE Geosci. Remote Sens. Lett.* **2018**, *15*, 1214–1218. [CrossRef]
38. Huu, C.T.; Minh, N.P.; Xuan, M.T.; Le, V.N.; Dang, C.H.; Nghia, P.M. An improved volume coherence optimization method for forest height estimation using PolInSAR images. In Proceedings of the 2019 3rd International Conference on Recent Advances in Signal Processing, Telecommunications & Computing (SigTelCom), Hanoi, Vietnam, 21–22 March 2019.
39. Aghabalaei, A.; Ebadi, H.; Maghsoudi, Y. Forest height estimation by means of compact PolInSAR data. *Remote Sens. Appl. Soc. Environ.* **2021**, *23*, 100552. [CrossRef]
40. Xing, C.; Zhang, T.; Wang, H.M.; Zeng, L.; Yin, J.; Yang, Y. A novel four-stage method for vegetation height estimation with repeat-pass PolInSAR data via temporal decorrelation adaptive estimation and distance transformation. *Remote Sens.* **2021**, *13*, 213. [CrossRef]
41. Soja, M.J.; Karlson, M.; Bayala, J.; Bazié, H.R.; Sanou, J.; Tankoano, B.; Eriksson, L.E.B.; Reese, H.; Ostwald, M.; Ulander, L.M.H. Mapping tree height in Burkina Faso Parklands with TanDEM-X. *Remote Sens.* **2021**, *13*, 2747. [CrossRef]
42. Lei, Y.; Treuhaft, R.; Gonçalves, F. Automated estimation of forest height and underlying topography over a Brazilian tropical forest with single-baseline single-polarization TanDEM-X SAR interferometry. *Remote Sens. Environ.* **2021**, *252*, 112132. [CrossRef]
43. Khati, U.; Singh, G.; Kumar, S. Potential of space-borne PolInSAR for forest canopy height estimation over India—A case study using fully polarimetric L-, C-, and X-band SAR data. *IEEE J. Sel. Top. Appl. Earth Obs. Remote Sens.* **2018**, *11*, 2406–2416. [CrossRef]
44. Chen, H.; Cloude, S.R.; White, J.C. Using GEDI Waveforms for improved TanDEM-X forest height mapping: A combined SINC + Legendre approach. *Remote Sens.* **2021**, *13*, 2882. [CrossRef]
45. Lavalle, M.; Simard, M.; Solimini, D.; Pottier, E. Height-dependent temporal decorrelation for POLINSAR and TOMOSAR forestry applications. in Synthetic Aperture Radar (EUSAR). In Proceedings of the 2010 8th European Conference on VDE, Aachen, Germany, 7–10 June 2010.
46. Pardini, M.; Cazcarra-Bes, V.; Papathanassiou, K.P. TomoSAR mapping of 3D forest structure: Contributions of L-band configurations. *Remote Sens.* **2021**, *13*, 2255. [CrossRef]
47. Lucas, R.M.; Lee, A.C.; Williams, M.L. Enhanced Simulation of Radar Backscatter from Forests Using LiDAR and Optical Data. *IEEE Trans. Geosci. Remote Sens.* **2006**, *44*, 2736–2754. [CrossRef]
48. Shi, L. Vegetation Height und Underlying Ground Altitude Estimation Based on Multi-Baseline POLINSAR Images. Ph.D. Thesis, Wuhan University, Wuhan, China, 2013.

remote sensing

Article

Estimating Aboveground Biomass of Two Different Forest Types in Myanmar from Sentinel-2 Data with Machine Learning and Geostatistical Algorithms

Phyo Wai [1,2], Huiyi Su [3] and Mingshi Li [1,3,*]

[1] Co-Innovation Center for Sustainable Forestry in Southern China, Nanjing Forestry University, Nanjing 210037, China; forestergis2014@gmail.com
[2] Forest Department, Ministry of Natural Resources and Environmental Conservation, Naypyitaw 15015, Myanmar
[3] College of Forestry, Nanjing Forestry University, Nanjing 210037, China; suhuiyi@njfu.edu.cn
* Correspondence: nfulms@njfu.edu.cn

Abstract: The accurate estimation of spatially explicit forest aboveground biomass (AGB) provides an essential basis for sustainable forest management and carbon sequestration accounting, especially in Myanmar, where there is a lack of data for forest conservation due to operational limitations. This study mapped the forest AGB using Sentinel-2 (S-2) images and Shuttle Radar Topographic Mission (SRTM) based on random forest (RF), stochastic gradient boosting (SGB) and Kriging algorithms in two forest reserves (Namhton and Yinmar) in Myanmar, and compared their performance against AGB measured by the traditional methods. Specifically, a suite of forest sample plots were deployed in the two forest reserves, and forest attributes were measured to calculate the plot-level AGB based on allometric equations. The spectral bands, vegetation indices (VIs) and textures derived from processed S-2 data and topographic parameters from SRTM were utilized to statistically link with field-based AGB by implementing random forest (RF) and stochastic gradient boosting (SGB) algorithms. Followed by an evaluation of the algorithmic performances, RF-based Kriging (RFK) models were employed to determine the spatial distribution of AGB as an improvement of accuracy against RF models. The study's results showed that textural measures produced from wavelet analysis (WA) and vegetation indices (VIs) from Sentinel-2 were the strongest predictors for evergreen forest reserve (Namhton) AGB prediction and spectral bands and vegetation indices (VIs) showed the highest sensitivity to the deciduous forest reserve (Yinmar) AGB prediction. The fitted models were RF-based ordinary Kriging (RFOK) for Namhton forest reserve and RF-based co-Kriging (RFCK) for Yinmar forest reserve because their respective R^2, whilst the RMSE values were validated as 0.47 and 24.91 AGB t/ha and 0.52 and 34.72 AGB t/ha, respectively. The proposed random forest Kriging framework provides robust AGB maps, which are essential to estimate the carbon sequestration potential in the context of REDD+. From this particular study, we suggest that the protection/disturbance status of forests affects AGB values directly in the study area; thus, community-participated or engaged forest utilization and conservation initiatives are recommended to promote sustainable forest management.

Keywords: SRTM; random forest; stochastic gradient boosting; random forest Kriging; wavelet analysis

Citation: Wai, P.; Su, H.; Li, M. Estimating Aboveground Biomass of Two Different Forest Types in Myanmar from Sentinel-2 Data with Machine Learning and Geostatistical Algorithms. *Remote Sens.* **2022**, *14*, 2146. https://doi.org/10.3390/rs14092146

Academic Editors: Klaus Scipal and Guangsheng Chen

Received: 3 March 2022
Accepted: 26 April 2022
Published: 29 April 2022

1. Introduction

Global climate change, which has become a major environmental problem for all nations, mainly results from anthropogenic fossil fuel combustion and land-use changes [1]. As the biggest carbon pool of terrestrial ecosystems, forests play a major role in the global carbon sequestration process, which contributes to climate change mitigation [2]. Overall, forests store approximately 45% of terrestrial carbon globally [1], most of which is stored in trees in the major form of aboveground biomass (AGB) that accounts for 44% of the total biomass through the process of photosynthesis [3]. Understanding the spatial and temporal dynamics of AGB is the most critical step in quantifying carbon stocks and fluxes from forests [4]. Hence, it is necessary to develop robust methods to estimate AGB for calculating carbon stocks, which is an essential indicator of the 'reducing emission from deforestation and forest degradation' (REDD)+ initiative as well as sustainable forest management [5,6].

AGB estimation with direct field surveys or that uses of species-specific allometric equations based on destructive sampling is accurate and widely used. However, these methods are difficult, expensive, time-consuming and not viable for wide regions due to limited samples and incomplete spatial coverage [7]. Remote sensing images have become efficient data sources for AGB predictions in diverse-scale landscapes by providing different spatial, spectral and temporal resolutions of images. Recent studies have used various remote sensing data types to map the biomasses of different forests [8–16]. Currently, the prevailing high-resolution light detection and ranging (LiDAR) systems can capture the three-dimensional information of vertical forest structures and are well suited for forest biomass estimation since they reduce the spectral saturation problem [17,18] against those optical remote sensing images. However, LiDAR data have an operational limitation for large area estimation due to their increased imaging cost, data processing cost and spatial limitation [19]. Additionally, LiDAR systems do not provide infrared signals which undermine their capability to analyze the vegetation status. On the other hand, low-resolution satellite images (e.g., AVHRR, MODIS and SPOT Vegetation) have an advantage for AGB prediction in large areas since only one scene covers a wide area of interest. However, their accuracy is the lowest compared to moderate or high spatial data due to their plot-pixel matching differences [20,21]. Therefore, the medium resolution satellite images (e.g., Sentinel and Landsat) were increasingly applied for forest AGB estimation at different spatial scales for their free accessibility and high suitability to landscape scale analysis [22,23].

The European Space Agency (ESA) launched the Sentinel-2A (S-2) multispectral satellite in 2015 following the SPOT and Landsat missions to monitor terrestrial surface [24]. The S-2 has a wide swath at 290 km with 13 multispectral bands including four bands at 10 m, six bands at 20 m, and three bands at 60 m, respectively. Therefore, it provides data on land surface reflectance for many different wavelengths, such as Landsat 8. Some bands of S-2 have better resolution over Landsat 8 (10 m overrides 30 m of Landsat 8). Additionally, the level-1 TOA radiance or reflectance product from the S-2 satellite (S-2 L1C) can be improved to a level-2 product as the BOA reflectance (S-2 L2A) using free atmospheric correction tools in the Sentinel Application Platform (SNAP) software [25]. Particularly, for the presence of longer wavelength red-edge bands in S-2 data, it is extremely useful in vegetation monitoring [26]. Although many studies have explored the application of S-2 optical satellite imagery in biomass mapping, there is still a present saturation problem since it lacks the capability to vertically penetrate dense forests [27]. For example, an optimal combination of reflectance, VIs and textural variables of S-2 has been employed in existing AGB mapping in an attempt to reduce this saturation effect [28].

S-2 derived VIs can minimize atmospheric saturation to some extent and better distinguish vegetation characteristics (e.g., moisture content) and forest AGB, especially in the relatively simply structured forests. Two studies by Adamu et al. [29] and Nuthammachot et al. [30] proved that VIs from S-2 were best correlated with forest AGB. However, the sensitivity of VIs to AGB varies among forest types and structures [31]. In a canopy complex forest where spectral features cannot identify the vegetation structure (e.g., canopy depth), texture features may improve the prediction accuracy of AGB thanks to their sensitivity to the horizontal arrangement patterns of canopies and their shadows. For example, Li et al. explored the performances of S-2 textures in AGB estimation for the mature broadleaved forest with complex canopy layers [32]. Pandit et al. found that if proper processing techniques were used, texture features could mitigate the saturation problem of S-2 spectral data to some extent in the mature forest AGB prediction [33]. Different texture processing techniques including the principal component analysis (PCA), gray level co-occurrence matrix (GLCM) and wavelet analysis (WA) were applied in existing scientific works. PCA has been proven to be a method of reducing the dimensionality and increasing the interpretability of satellite datasets [34] and can hold over 80% of all representative original information [22]; thus, these generated principal components may have a stronger relationship with biomass than individual original spectral bands, which are somehow independent of different biophysical conditions. Recent studies have claimed that textural measures derived from the GLCM and WA of satellite images could improve forest cover differentiation and forest AGB estimation [35,36]. Additionally, topographic features such as elevation, slope, and aspect are significantly related with the growth and distribution pattern of forests, and thus are of great significance in AGB estimations. Recent research findings have proven that variables derived from SRTM data at 30 m resolution are helpful for estimating forest biomass and have great influence over the spatial distribution of AGB [37]. Elevation is especially well correlated with forest AGB since it provides information about the forest distribution and site attributes [38]. As proven by numerous studies for different forest types with varying terrains, the topography may depend on the reflectivity of the specific forest site and affect the AGB estimation [39].

Empirical regression techniques were widely used in the early studies of remote sensing-based AGB estimation, considering the normality of the modeling datasets [40,41]. Therefore, simple linear or multiple linear regression models for practical AGB estimation are limited because of the complex non-linear relationships between forest AGB and remote sensing variables. Non-parametric models, also called machine-learning algorithms, do not require a strictly linear assumption between the response and covariates due to the independence of their data distribution. Machine learning models such as artificial neural network (ANN), support vector machine (SVM), random forest (RF), and stochastic gradient boosting (SGB) are popular non-parametric methods for identifying complex relationships between the predictors and forest AGB [42,43]. Among these, RF and SGB are efficient machine learning methods proposed by Breiman [44] and Friedman [45] and have been successfully used in forest AGB estimations. Although these models estimating AGB well, the major drawback of these models lies in their ignorance of the spatial autocorrelation of sample plots [35]. Kriging interpolation provides the linear unbiased prediction of variables based on the variogram model and is best applied to minimize the spatial variation error between samples in AGB estimation [46].

Although these existing studies have had varying degrees of success in estimating forest AGB in different forest ecosystems with varying structural complexities, they regard AGB as an independent spatial biophysical variable when creating AGB prediction models, whereas as one of the more important variables in biogeochemical cycles, forest AGB not only has its own randomness in distribution, but also has structurized characteristics in space. Thus, the modeling techniques of the classic statistics applied in the vast majority of existing works that do not consider the auto-correlation information of forest AGB cannot adequately capture spatial variations in AGB, which necessitates an improvement upon existing modeling means by introducing geostatistical analytical methods, such as Kriging interpolation.

In recent decades, radical demographic, economic and social changes in Myanmar have placed considerable pressure on its forest resources [47]. According to the Food and Agricultural Organization (FAO), in 2015, the forest area in Myanmar was 42.92% of the total country, which had decreased from 45.04% in 2010 [48,49]. To cover this loss, the forestry sector in Myanmar has been implementing the Myanmar reforestation and rehabilitation plan (MRRP) under the REDD+ scheme starting from 2017 to 2026 in the areas of forest degradation. As a REDD+ scheme, Myanmar predicts its forest reference emission level (FREL) based on preliminary information from the reference year 2005 to 2017, which needs to be periodically updated by integrating carbon improvement from reforestation programs based on new knowledge, methods and trends in the future [50]. Hence, the reliable method and data sources for mapping AGB by forest types are essential for Myanmar's future FREL calculation of REDD+ since local AGB maps are also the basis for the extension of estimates to larger areas using remote sensing approaches. To date, however, no systematic research has been conducted to predict the spatial distribution of AGB by forest types, especially in inaccessible areas of northern and eastern Myanmar because of operational limitations and the consequent lack of data, technology and appropriately qualified individuals [47]. In this context, the optimal integration of remote sensing data and modeling algorithms may fill this gap.

The overall goal of this study was to evaluate the performance improvement of overlaying geostatistical interpolation onto machine learning modeling based on S-2 and SRTM in mapping the AGB of two forest reserves in Myanmar. Additionally, the robust AGB maps generated from this work were also expected to support the strategic development of carbon sequestration-aimed forestry management efforts in Myanmar.

2. Materials and Methods

2.1. Study Area

Two forest reserves in Myanmar, namely Namhton (NH) and Yinmar (YM), were selected as case studies (Figure 1). They are located in the northern and central-eastern parts of the country and have been formally protected by the Forest Department and by the 1992 Forest Law since 1995 and 2003, respectively.

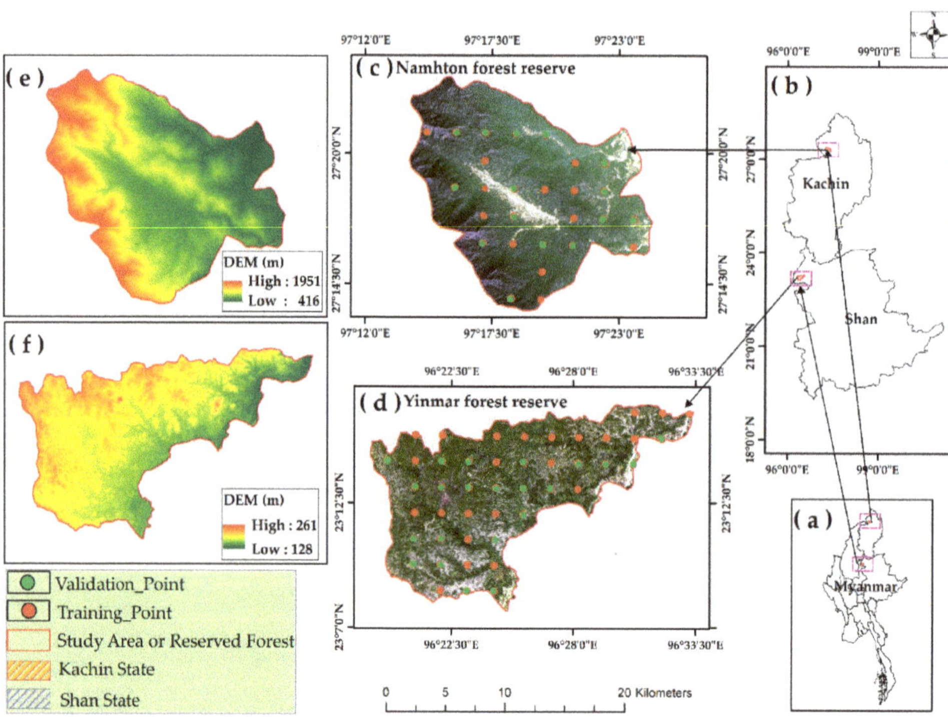

Figure 1. Location of the study area: (**a**) country boundary; (**b**) location of the two forest reserves; (**c**) the Sentinel-2 true color image (collected in 26 January 2017) attached with field sample plots in NH; (**d**) the Sentinel-2 true color image (collected in 5 February 2017) attached with field sample plots in YM; (**e**) the DEM of NH; and (**f**) the DEM of YM.

The NH forest reserve area is approximately 19,418 hectares and the dense evergreen forest type dominates this region, geographically spanning from 97°13′00″E, 27°23′30″N to 97°24′30″E, 27°13′00″N. It is situated in the Putao Township, Myitkyina District, of northern Kachin State. The terrain in the region is mountainous, with an altitude ranging from 416 m to 1951 m. The annual average temperature of the area is 22.13 °C, with its average annual precipitation ranging from 1218 mm to 2800 mm; snowfall used to be heavy in the northern part, at higher altitudes of the mountainous regions. Evergreen tree species such as *Quercus glauca, Macaranga denticulata, Michelia champaca, Shorea assamica, and Ficus cuspidata*, are typical tree species of NH. A village called "Namhton Ku" is located in the center of this reserved forest as an encroachment, where forest-related field data collection was not performed.

The YM is a deciduous forest reserve-dominated region with an area of 20,163 hectares which is located between latitude 23°7′30″N–23°17′00″N and longitude 96°18′30″E–96°33′30″E, in Moemaik Township, Kyaukme District of Shan State. The mean annual temperature of this region is approximately 27 °C with an average annual rainfall ranging from 1000 mm to 1500 mm. The terrain is relatively flat, ranging from 128 m to 261 m according to the analysis of the SRTM DEM data. The vegetation in the YM region is predominantly Dipterocapaceae species such as *Dipterocarpus alatus, Hopea odorata, Dipterocarpus tuberculatus*, and *Shorea obtuse*. The vegetation in this area is mainly deciduous, losing its leaves during the dry season. Additionally, bush fires, which frequently occur during the dry season, further reduce the available foliage. Biomass or carbon stocks during the dry season months can therefore dramatically differ from those during the rainy season for the same area.

2.2. Data Collection and Processing Methods

2.2.1. Sentinel-2 Images Pre-Processing and Indices Extraction

The study site is located in two eco-regions (northern and central eastern) of Myanmar. Frequent rains and cloud contamination exist in the study sites, which highly restrict the availability of images collected in the peak season of vegetation growth (i.e., June–September). Therefore, two S-2 L1C MSI satellite images, with tile numbers of T47RLL and T46QHL acquired on 26 January 2017 and 5 February 2017, respectively, were downloaded from the European Space Agency. Available online: https://www.scihub.copernicus.eu (accessed 21 March 2022). These images are composed of 100 km^2 tiles with UTM/WGS84 projection. The descriptive information of the images is summarized in Table 1. The atmospheric correction of the two S-2 L1C scenes was performed with the Sen2Cor plugin in SNAP software to reduce the atmospheric, adjacency, and slope effects [51]. In the process, TOA reflectance images were converted into surface reflectance images with aerosol-free and noise reduction. Then, all 20 m spectral bands were resampled to 10 m using the nearest neighbor strategy. Bands 1, 9, and 10 were not suitable for AGB estimation and excluded from the analysis [52]. The images and spectral response curves for a test vegetation pixel before and after atmospheric correction are shown in Figure 2.

Table 1. Descriptive information of the images used in the analysis.

Image/Product	Tile Number and Acquisition Date	Cloud %	Bands Used for Modeling	Spatial Resolution (m)	Central Wavelength (nm)
S-2 L1C Product	T47RLL on 26 January 2017	3.63	B2 (blue)	10	490
			B3 (green)	10	560
			B4 (red)	10	665
	T46QHL on 5 February 2017	0.18	B5 (red edge)	20	705
			B6 (red edge)	20	740
			B7 (red edge)	20	783
			B8 (NIR)	10	842
			B8A (red edge)	20	865
			B11 (SWIR1)	20	1610
			B12 (SWIR2)	20	2190

The individual application of spectral values in a predictive model could not give a reliable estimation compared to when using combined VIs. In addition to the spectral bands, VIs were calculated based on the original reflectance bands in the raster calculator tool. The plot-level vegetation index mean values were extracted using the zonal statistics tool of ArcGIS, and using the plot size (0.08 ha) to match the AGB calculations. In this study, the normalized difference vegetation index (NDVI) [53], red-edge normalized difference vegetation index (RENDVI) [54], weighted difference vegetation index (WDVI) [55], enhanced vegetation index (EVI) [56], red-edge enhanced vegetation index (REEVI) [57], soil-adjusted vegetation index (SAVI) [58], green-normalized vegetation index (GNDVI) [59], normalized difference water index (NDWI) [60], simple ratio (SR) [61], normalized difference vegetation index with bands 4 and 5 (NDI45) [62] and meris terrestrial chlorophyll index (MTCI) [63] were calculated. The detailed formulas for VIs calculation are described in Table 2.

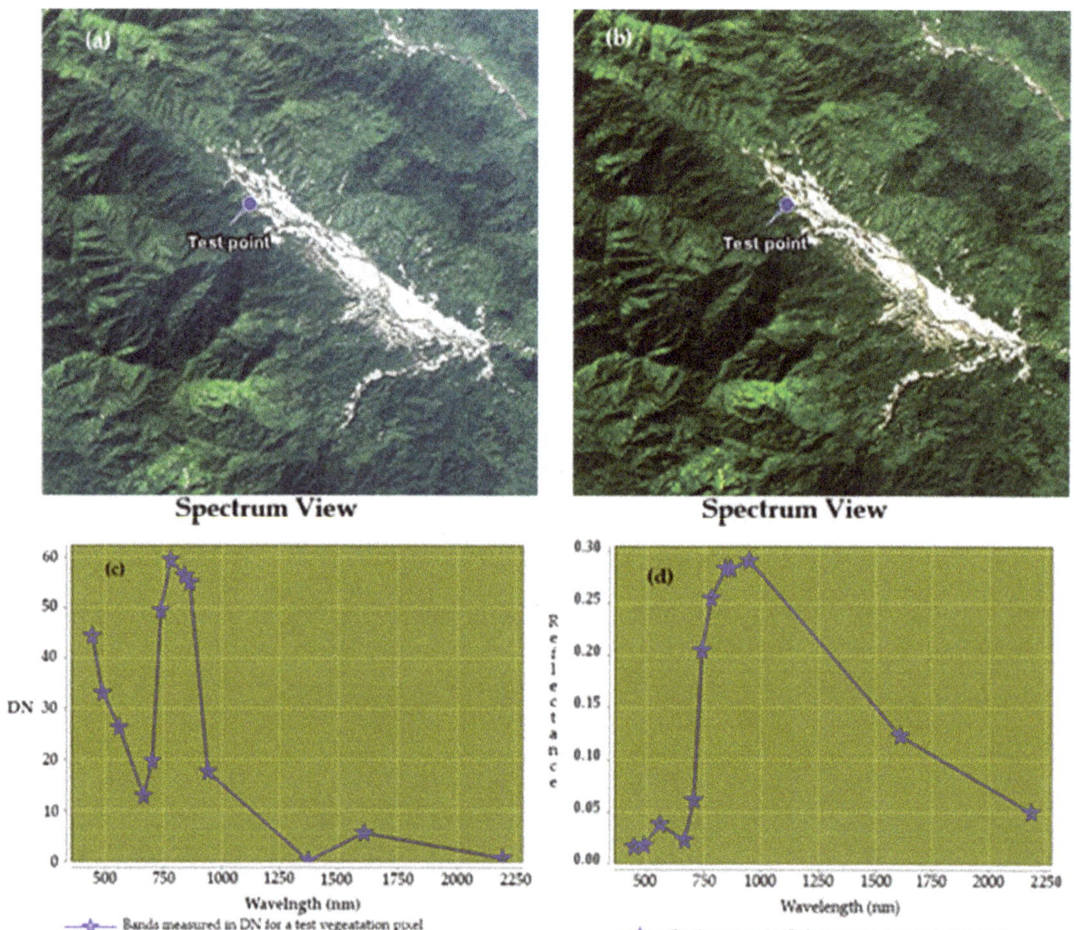

Figure 2. The original Sentinel-2 L1C image (**a**) and atmospherically corrected image L2A (**b**), and the original (**c**) and corrected (**d**) spectral curves for a test vegetation pixel.

Table 2. List of Sentinel-2-derived variables and topographic factors used in AGB modeling.

Satellite Data		Bands and Indices	Formula
Sentinel-2 Level-2A 10 m-resolution	Multispectral bands	Band 2	BLUE
		Band 3	GREEN
		Band 4	RED
		Band 5	RE1
		Band 6	RE2
		Band 7	RE3
		Band 8	NIR
		Band 8A	RE4
		Band 11	SWIR1
		Band 12	SWIR2
	Vegetation indices (Broad bands)	NDVI	NIR − RED/NIR + RED
		SAVI	1.5 × (NIR − RED)/(NIR + RED + L)
		EVI	2.5 × (NIR − RED/NIR + 2.4RED + 1)
		GNDVI	(NIR − GREEN)/(NIR + GREEN)
		WDVI	(NIR − 0.5 × RED)
		SR	(NIR/RED)
		NDWI	NIR − SWIR2/NIR + SWIR2
		NDI45	(RE1 − RED)/(RE1 + RED)
		MTCI	(RE2 − RE1)/(RE1 − RED)
	Vegetation indices (Narrow red-edge bands)	RENDVI	NIR − RE1/NIR + RE1
		REEVI	2.5 × (NIR − RE1/NIR + 2.4RE1 + 1)
Resampled SRTM DEM (10 m)	Elevation	Ele	-
	Slope	Slope	-
	Aspect	Asp	-

2.2.2. SRTM Data Pre-Processing and Variables Extraction

SRTM topographic data were downloaded from the USGS EROS Data Center. Available online: https://www.earthexplorer.usgs.gov/ (accessed 23 March 2022). These elevation data offer worldwide coverage of void-filled data at a resolution of one arc-second (approximately 30 m) and a high-resolution global dataset. These topographic data were first reprojected into UTM/WGS84 since the projection system of these data are GCS/WGS84. To match with S-2 spectral bands, they were also resampled to 10 m spatial resolution using the nearest neighbor method in the ArcGIS package. Then, from this resampled dataset, two forest reserves boundaries were clipped and the elevation, slope and aspect were similarly extracted using the zonal statistics tool in ArcGIS (Table 2).

2.2.3. Texture Features Extraction

Principal component analysis (PCA) can be used to remove correlated or redundant information in the satellite images and simultaneously reduce their dimensionality [34]. The first three principal components were produced as the potential image variable for modeling AGB. The first principal component (PC1) was used for texture extraction as it contained over 80% of the original spectral information. When extracting textural features, the gray level co-occurrence matrix (GLCM) method and wavelet decomposition method were applied. Among these, the GLCM textures including the mean, variance, homogeneity, contrast, dissimilarity, entropy, second moment, and correlation were extracted with different window sizes ($3 \times 3, 5 \times 5, 7 \times 7$) from the PC1 image in the ENVI5.3 package (Table 3). In this study, the GLCM-based textures derived from a 7×7 window size were selected as predictive variables for AGB estimation after the correlations of different window sizes' textures with the measured AGB were tested.

Table 3. Gray level co-occurrence matrix-based textural measures extracted in the current work.

Data	GLCM Texture	Formula	Reference
PC1 image from 10 m resolution bands of S-2 L2A	mean	$\sum_{i,j=0}^{N-1} iP_{i,j}$	Robert [64]
	variance	$\sum_{i,j=0}^{N-1} iP_{i,j}(1-\mu i)$	
	homogeneity	$\sum_{i,j=0}^{N-1} iP_{i,j}/(1+(i-j)2)$	
	contract	$\sum_{i,j=0}^{N-1} iP_{i,j}(i-j)$	
	dissimilarity	$\sum_{i,j=0}^{N-1} iP_{i,j}\lvert i-j\rvert$	
	entropy	$\sum_{i,j=0}^{N-1} iP_{i,j}InP_{i,j}$	
	second moment	$\sum_{i,j=0}^{N-1} iP_{i,j^2}$	
	correlation	$\sum_{i,j=0}^{N-1}(i(\sum_{i,j=0}^{N-1} ijP_{i,j^2}-\mu i\mu i)/\sigma i^2\sigma i^2)$	

Additionally, the wavelet analysis was also considered to be an effective means of extracting textures. Wavelet transformation is a multi-resolution analysis tool for image signal processing, which has two distinct abilities: subtle variation in spectral features in the original data can be detected at different scales and the useful information can be represented by fewer wavelet features by compressing data [51]. The wavelet analysis produces four basic components including the approximation image, horizontal detail, vertical detail and diagonal detail images of which the latter three are usually regarded as helpful textural measures. In this study, the Coiflect discrete wavelet function was chosen after repeated tests with different mother wavelets (the Haar wavelet, Daubechies (dbN) and Symlets (symN)) in the Matlab package because it had the highest correlation with AGB. Thus, based on the first principal component (PC1), a three-level decomposition strategy was implemented through programming in the Matlab environment to generate 9 detailed images as independent textural variables for AGB modeling. Finally, two types of textures derived from GLCM-based and wavelet analysis were included in the AGB modeling in this study. The Coiflet wavelet-based decomposition procedure is summarized in Figure 3.

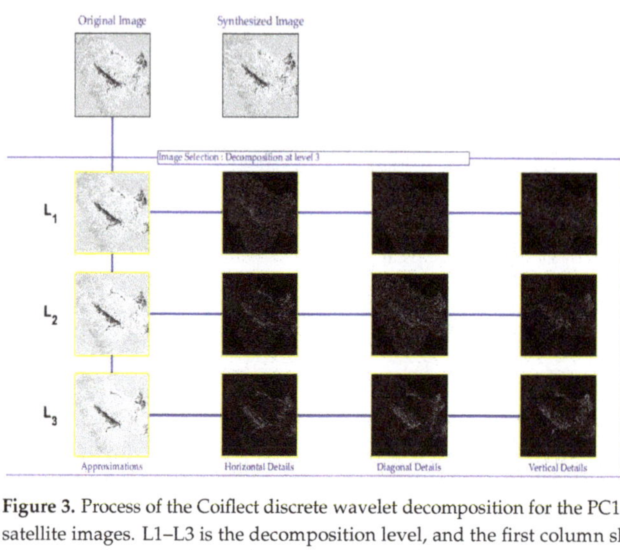

Figure 3. Process of the Coiflect discrete wavelet decomposition for the PC1 of the original Sentinel-2 satellite images. L1–L3 is the decomposition level, and the first column shows the approximation images; the second column shows the images of horizontal details; the third column shows the images of diagonal details and the fourth column shows the images of vertical details.

2.2.4. Field Forest Inventory Data

Myanmar's Forest Department conducted the national forest inventory (NFI) data collection in February and March of 2017 with the financial support of the Finnish government and according to the guidelines of the FAO technical team. These secondary forest inventoried data were used to estimate and infer the plot-level AGB of the study area. The sampling design of Myanmar's NFI is shown in Figure 4. The systematic sampling method was constructed according to the eco-regions of Myanmar in order to cover all forest types. However, due to inaccessibility, no sample plots were collected for some areas in the NM (Figure 1c). Each sampling cluster comprises four 0.08-hectare circular subplots (Elbow, East, North and Northeast) in which the distance between adjacent subplots is 50 m. Within these circular subplots, all trees above 10 cm diameter at breast height (DBH) were measured to record data, including the DBH, tree height (H), and crown width. The diameter tape and Leica laser finder were used to measure the DBH and H of trees. Other forest parameters such as shrub cover, sapling cover, bamboo coverage, humus depth, litter coverage, and tree bark thickness were also collected in all sample plots. Ultimately, data collection was performed in 88 subplots in NH (evergreen forest) and 170 subplots in YM (deciduous forest).

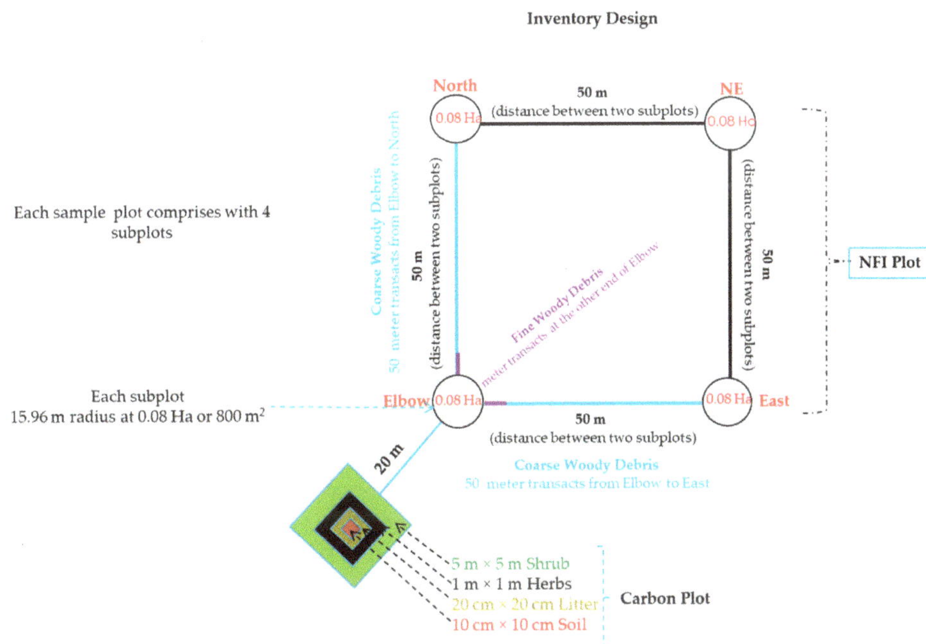

Figure 4. Field forest inventory sampling design in Myanmar. **Source:** Planning and Statistics Division, Forest Department of Myanmar.

2.2.5. Allometric Equation and Calculated AGB

Since the species-specific allometric equations are not available for the study region, we had to use the unpublished national-level coarse allometric equations for evergreen forest and broad-leaved forest to calculate the plot-level AGB. The AGB formulas were as follows:

$$\text{For NH Evergreen, } AGB = \rho_1 \times \exp\left(-1.499 + 2.148\ln(DBH) + 0.207(\ln(DBH))^2 - 0.0281(\ln(DBH))^3\right) \quad (1)$$

$$\text{For YM Deciduous, } AGB = \rho_2 \times \exp\left(-0.667 + 1.78\ln(DBH) + 0.207(\ln(DBH))^2 - 0.0281(\ln(DBH))^3\right) \quad (2)$$

where, AGB is in kg per tree, DBH is in cm, ρ_1 and ρ_2 are the basic wood density parameters for evergreen forest and deciduous broad-leaved forest, respectively.

Using the national level biomass expansion factors Equations (1) and (2), AGB was computed for each tree and summed up by each plot to obtain plot-level AGB. Finally, the plot-level field AGB values were converted from kg/plot into ton/ha. Therefore, the AGB unit in this study was t/ha. Table 4 shows the descriptive statistics of the field-observed AGB.

Table 4. Descriptive statistics of the field-measured AGB (t/ha) for the two forest reserves.

Forest Type	Number of Sample Plots	AGB (t/ha) Value Range	Median	Mean	Std. Deviation	Number of Sample Plot Used in Modeling	
						Training	Validation
Evergreen	88	0.57–151.64	38.40	49.00	39.206	71	17
Deciduous	170	2.74–215.24	100.02	98.00	51.73	140	30

2.3. *Aboveground Biomass Detection Methods*

2.3.1. Prediction Model Establishment

In this study, Random Forest (RF) and stochastic gradient boosting (SGB) models were first performed for AGB prediction. Then, based on the better one (RF or SGB), the predicted residuals (the difference between the observed AGB and the model-predicted AGB) were further analyzed and compared using ordinary Kriging (OK) and the co-Kriging (CK) to separate the structured components hidden in the residuals, followed by adding the better structured components onto the better model predictions to obtain the final AGB predictions. The detailed modeling approaches are summarized as follows:

Random Forest and Stochastic Gradient Boosting Models

Parametric and non-parametric models have been utilized either alone or combined with environmental variables for remote sensing-based AGB mapping. Nevertheless, choosing the suitable variables set and modeling algorithm is critical for the improving the accuracy of prediction model.

The RF model is a bagging algorithm which enhances accuracy and reduces overfitting and bias [65]. SGB is a boosting ensemble method with low sensitivity to outliers, with the ability to deal with unbalanced training datasets [44]. Both models are non-parametric modeling approaches which have a performance superior to those of other machine learning techniques such as the K-nearest neighbor (KNN), support vector machine (SVM), and the multivariate adaptive regression splines (MARS) [66,67], which is increasingly being applied to satellite-based biomass mapping [9].

In view of these advantages, in the current study, the RF and SGB models were performed as the first attempt to predict the AGB of the two forest reserves (NH and YM). The RF model was implemented in the "randomForest" package [68] within R Studio. This package supports the chart that illustrates the GI-index and OOB error rate to determine the most important modeling variables. From this comprehensive chart, preference variables can be selected for a prediction model to reduce the complexity and load of computation. In the RF regression analysis, the variables' importance ranking was determined by out-of-bag (OOB) error and node-purity percent (IncNodePurity). The first variable importance analysis was calculated by randomly permuting each predictor variable and computing the associated reduction in predictive performance using the out-of-bag (OOB) error. The second most important variable was estimated by determining the decrease in node impurities attributable to each predictor variable. Larger InNodePurity and %InMSE indicate higher model accuracy in terms of ranking variable importance. Parameters such as the number of trees (ntree), the number of variables used to split the tree at each node (mtry), and node size are adjustable for the RF model. For RF-prediction models in this study, after multiple tests, the ntree, mtry and nodesize were specified to be 600, 4, and 5 for the evergreen forest (NH) and 1400, 2, and 5 for the deciduous forest (YM), respectively.

The SGB algorithm is based on the combination of the regression tree and boosted algorithms to predict the response variable. This algorithm also reduces the chance of overfitting by introducing an element of stochasticity due to its flexibility and high predictive performance. Unlike RF, a tree is constructed from a different random sub-sample of the dataset in the SGB model, producing an incremental improvement in the model. In this study, all steps of the SGB analysis were implemented in the "gbm" package in the R-studio [69]. The "vip" function in the "gbm" package was also used for the selection of important variables for the SGB model. The SGB model adjustment includes the distribution, interaction depth, bagging fraction, shrinkage rate, and training fraction. The out-of-bag method was used for determining the optimal number of boosting iterations. The maximum iteration tree was stopped at 420 in the evergreen forest AGB (NH) SGB-prediction model and 600 at the deciduous forest AGB (YM) SGB-prediction model, as more iteration tree numbers could no longer improve models' accuracy. The interaction depth, also known as the maximum number of possible interactions, was set to 3 and 4 nodes for the NH and YM SGB-models, respectively. The bagging fraction controlling the fraction of the training data, the shrinkage rate controlling the learning speed of the algorithm, and the training fraction randomly selected for calculating each tree were set to 10, 0.03, and 10 for both SGB models (NH and YM).

Once the RF and SGB models were created for the two forest reserves, their predictive performances in AGB estimation were compared to determine the better model; then, based on this model, all the residuals resulting from the model were further analyzed by implementing OK and CK autocorrelation algorithms to separate the structurized component or trend item hidden in the residuals.

Random Forest-Based Kriging Model

In this study, RF performed better than SGB in both the NH and YM AGB prediction models in terms of accuracy evaluations; thus, to improve the accuracy of RF models or finding the spatial correlation of AGB samples, RF-based OK and CK analyses (RFOK and RFCK) were also performed as subsequent steps. Since the RF model does not consider spatial autocorrelation among the AGB sample plots, AGB is actually a typical item with relatively high spatial autocorrelation; thus, a combination of RF and Kriging (RFK) was potentially an effective and more reliable means of determining the spatial distribution of AGB in this study. Specifically, a regression-Kriging technique was used to extract the structured components of the residuals obtained from the RF regression [70]. As the procedure of Kriging interpolation, the modeling semivariogram is important to determine the accuracy and reliability of the estimates. Kriging includes ordinary Kriging (OK) and co-Kriging (CK) in which OK is a suitable interpolation method for the uneven distribution of terrain and climatic variation events, while CK is the best method for improving the accuracy of target prediction [22]. OK is a linear estimation method suitable for inherently stationary random fields which satisfies the isotropic hypothesis [71] and fully considers spatial parametric non-stationarity as well as the effects of environmental variables derived from the benefits of RF. It is a widely used geostatistical technique that generates an optimal unbiased estimated surface employing a semivariogram based on regionalized variables. The interpolation formula of OK is as follows:

$$Z_{OK}{}^{*}(x_0) = \sum_{i=1}^{n} \lambda_i Z(x_i) \tag{3}$$

where $Z_{OK}{}^{*}(x_0)$ is the residual value of the AGB to be estimated at location x_0, n is the number of sample points used for interpolation, $Z(x_i)$ is the AGB residual of site i, and λ_i is the weighting coefficient at point i.

CK is an improvement over the OK method and deals with multivariate problems [22]. Since the study areas include two reserved forests with different terrains, AGB is definitely affected by elevation. Thus, the elevation factor was used as a co-variable for interpolation. This was expressed as follows:

$$Z_{2,CK}^*(x_0) = \sum_{i=1}^{N_1} \lambda_{1i} Z_1(x_{1i}) + \sum_{j=1}^{N_2} \lambda_{2j} Z_2(x_{2j}) \quad (4)$$

where $Z_{2,CK}^*(x_0)$ is the residual values of AGB to be estimated, $Z_1(x_{1i})$ is the AGB residual of the site i, λ_{1i} is the weighting coefficient of site i, $Z_2(x_{2j})$ is the elevation of site j, and λ_{2j} is the weighting coefficient of site j.

Variograms are an effective tool for analyzing the spatial variation and structure of target predictors for the reliable estimation of AGB. The performance of the semivariogram was assessed by the coefficient of determination (R^2) and the root mean square error (RMSE). The larger the R^2, the smaller the RMSE and nugget effect, and the better the fitting performance was.

In RF-Kriging modeling, the estimated residual value of each sample point was calculated by subtracting the RF-derived predicted AGB value from the field-observed AGB value. It can be calculated as follows:

$$Z(x_i) = C(x_i) - \hat{C}_{RF}(x_i) \quad (5)$$

Final AGB prediction by RFOK or RFCK method was acquired by the following equation:

$$\hat{C}_{RFOK/RFCK}(x_i) = \hat{C}_{RF}(x_i) + \hat{Z}_K(x_i) \quad (6)$$

where $\hat{C}_{RFOK/RFCK}(x_i)$ is the predicted AGB at site *i* using RFOK or RFCK, where $\hat{Z}_K(x_i)$ is the AGB residual value of site i, $C(x_i)$ is the observed AGB of site i, $\hat{C}_{RF}(x_i)$ is the RF-based predicted AGB at site i.

Finally, the maximum livelihood classifier in ENVI Classic 5.3 was applied to classify the forest and non-forest areas of the study area. The resulting classified forest area was used as a mask to obtain the forest AGB maps of the study area.

2.4. Accuracy Assessment

The training and validation sets were determined as 80 and 20% of the sampled data using stratified sampling for all statistical analyses. Model performances were assessed based on the coefficient of determination-R^2, the mean absolute error (MAE), the root mean square error (RMSE), the RMSE%, bias, and bias%, based on the validation set (20% of samples). In addition, the relative improvement (RI) index for assessing RFOK and RFCK over RF models was also performed (7):

$$RI = \frac{RMSE_{RF} - RMSE_{RFOK/RFCK}}{RMSE_{RF}} \quad (7)$$

where, $RMSE_{RF}$ is the root mean square error from RF predicted model, $\widehat{RMSE}_{RFOK/RFCK}$ is the root mean square error from RFOK and RFCK models, respectively.

3. Results

3.1. Variable Importance and Selections

In the RF variable importance analysis, the %IncMSE by OOB error and %InNodePurity by Gini index are indicators of the importance ranking. The topmost important variables picked up by the evergreen NH and deciduous YM RF models are shown in Figure 5a,b. For the generalizing the model and reducing the computation load, only the top 10 variables were selected as the final RF model inputs. S-2 derived the reflectance, Vis and textures from the wavelet decomposition, and the topographic variables were included in the list. In the RF model, topographic variables and textures from WA were major contributors to the evergreen NH forest AGB estimation, while VIs showed better sensitivity to deciduous AGB.

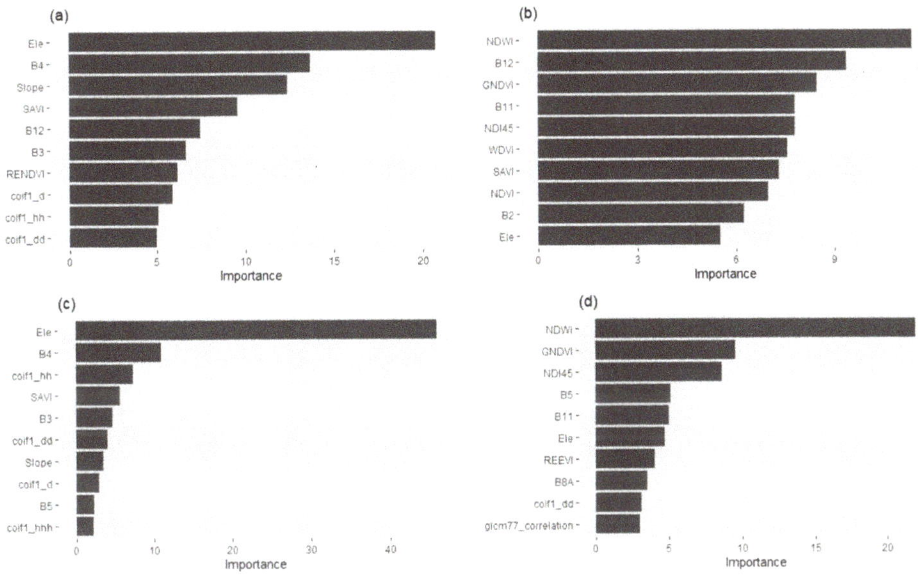

Figure 5. The importance ranking of the top 10 variables identified by (**a**) evergreen RF model; (**b**) deciduous RF model; (**c**) evergreen SGB model; and (**d**) deciduous SGB model.

In the variables analysis for the SGB model, the lowest RMSE could be found for an interaction depth of 3, a shrinkage ratio of 0.03, and 420 iterations in the evergreen forest reserve. For the deciduous forest reserve, an interaction depth of 4, shrinkage ratio of 0.03, and 600 trees were the best parameters for model fitting. Figure 5c,d show the most important variables of SGB models for evergreen NH and deciduous YM AGB estimation. In SGB models, topographic variables and textures from WA were identified as major contributors to the NH evergreen AGB estimation, while VIs had better sensitivity for the deciduous AGB prediction. According to the variable importance analyses in the RF and SGB models, the variables did not differ a great deal in terms of their sensitivity to AGB in same forest types.

3.2. Validation Metrics for RF and SGB Models

Among the sample plots, 80% were used for training the models (RF and SGB). For the RF evergreen model training, the optimum model accuracy was obtained from the following parameters, ntree = 600, mtry = 4 with nodesize = 5, considering the important predictors. The adjusted parameters values of ntree = 1400, mtry = 2 and nodesize = 5 were selected for the RF deciduous model. The RMSEs values of 11.17, 17.36 t/ha and corresponding RMSE%s 23.06 and 17.45 were obtained while the R^2 values were 0.95 and 0.97 for evergreen and deciduous training models. It was also observed that the bias and bias% of the evergreen training models were −0.59 and −1.20%, while the deciduous model obtained −0.011 and −0.11%, respectively. The smallest RMSE value was observed in the evergreen forest since the AGB values of the training plots in this type were smaller than those of the deciduous forest (Table 4).

Depending on the shrinkage ratio, the interaction depth, and the number of regression trees, the two SGB model performances were evaluated. The SGB evergreen model resulted in an R^2 value of 0.98 and an RMSE of 4.62 t/ha with the corresponding RMSE%, bias and bias% of 9.45, −1.24 and −2.54, respectively. In the trained model of SGB for the deciduous forest, the R^2, RMSE, RMSE%, bias and bias% were 0.97, 8.83, 8.88, −3.02 and −3.04.

In addition to model training, 20% of the sample plots were used for model validation. Figure 6 and Table 5 show the validation metrics for RF and SGB models. The R^2 values of the NH evergreen RF model and NH SGB model were 0.47 and 0.35, respectively, while RMSEs values of those models were 25.45 and 32.02 t/ha, respectively. Meanwhile, the R^2 and RMSE values of the YM deciduous RF model and YM SGB model were estimated to be 0.38, 0.35 and 40.23, 41.85 t/ha. In the scatter plots of Figure 6, the points along the fitted lines showing the correlation of predicted and observed AGB are scattered in both NH and YM, which showed large AGB values which were underestimated and small AGB values which were overestimated. Nevertheless, by considering the validation models' metrics, the RF models had higher R^2, lower RMSE, and bias values than SGB models in the biomass estimation at each growth stage, indicating that the RF models can provide more accurate biomass estimations than SGB; thus, the residuals from the RF model were further analyzed to attempt to extract the structured components from the residuals to possibly improve the ultimate prediction accuracy of AGB.

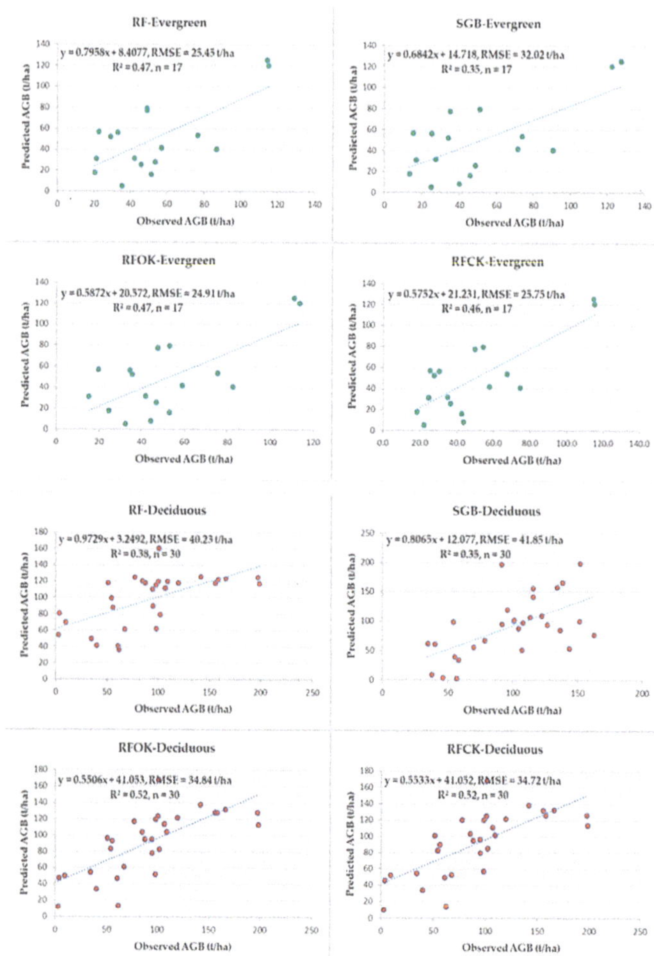

Figure 6. The scatter plots for validation metrics of RF, SGB, RFOK, and RFCK models.

Table 5. Validation metrics of RF, SGB, RFOK, and RFCK models based on 20% of the sampled data.

Forest Type	Model	R^2	RMSE (t/ha)	RMSE%	MAE (t/ha)	Bias	Bias%	RI
NH Evergreen	RF	0.47	25.45	51.44	22.45	0.15	0.29	-
NH Evergreen	SGB	0.35	32.02	64.72	27.03	2.48	5.03	-
NH Evergreen	RFOK	0.47	24.91	50.34	22.19	3.35	6.76	0.021
NH Evergreen	RFCK	0.46	25.75	52.04	57.67	−49.27	−99.57	−0.011
YM Deciduous	RF	0.38	40.23	44.09	33.07	6.18	6.77	-
YM Deciduous	SGB	0.35	41.85	45.88	33.00	6.91	7.58	-
YM Deciduous	RFOK	0.52	34.84	38.19	27.51	0.30	0.33	0.134
YM Deciduous	RFCK	0.52	34.72	38.06	27.47	0.06	0.06	0.137

3.3. Semivariogram Analysis Results of RF-Derived Residuals

Residuals of the RF-predicted AGB were derived by subtracting the RF-predicted AGB from the field-measured AGB. Table 6 shows the statistics of the residuals. The mean residual values for the evergreen and deciduous AGB were −0.91 and −1.60 t/ha, respectively. In addition, the residual value range of the deciduous forest (82.43 t/ha) was much higher than that of the evergreen forest (34.52 t/ha). The standard deviation of the evergreen residuals (15.38 t/ha) was smaller than that of the deciduous (23.12 t/ha). The residual values were shown in different colors and sizes based on their distribution in Figure 7. As shown in the histogram distribution in Figure 7, the residual values of the evergreen forest were not close to normal distribution while deciduous residuals were approximately normally distributed. Nevertheless, the procedure of the semivariogram analyses for both forest reserves was performed for testing the accuracy of the models.

Table 6. Residual's statistics derived from the RF model prediction.

Forest Type	Residual Mean (t/ha)	Std. Deviation (t/ha)	Value Range (t/ha)	Skewness	Kurtosis
NH Evergreen	−0.91	15.38	−45.90–34.52	−0.31	3.57
YM Deciduous	−1.60	23.12	−75.97–82.43	−0.03	4.50

After the normality of the residuals was verified in SPSS, these residual values could be used for semivariogram analysis based on the ordinary Kriging (OK) and the co-Kriging (CK) in the semivariance analysis. The model with the largest R^2 and smallest RMSE was determined as the optimal analytical function of the semivariogram. As a result, Gaussian function was picked up (Table 7). Furthermore, elevation was used as a co-variable in the CK model for the better estimation of AGB. Table 7 and Figure 8 show the modeled semivariogram and semivariance models using OK and CK analyses. Evergreen forest models were poorly fitted; the R^2 values were much lower than those of deciduous models (0.19-OK and 0.15-CK versus 0.58-OK and 0.62-CK). The nugget value in the OK model of the evergreen AGB residuals was slightly smaller than that of the corresponding CK model, indicating stronger spatial homogeneity. In the CK model of deciduous residuals, the nugget value was also smaller than that of the OK model. Moreover, the ratio of nugget and sill (nugget/sill) determined the variation of spatial autocorrelation between the AGB sample plots. From the OK models, the smaller nugget/sill ratio (0.95) exhibited stronger spatial homogeneity than the CK model with nugget/sill (0.99) for evergreen residuals. In the deciduous residuals, CK gave a smaller nugget/sill (0.74), indicating a stronger spatial correlation by considering elevation as a co-variable. This meant that deciduous forest AGB varied closely in space with the terrain variations. The improvement was not obvious in the evergreen OK and CK models compared to the evergreen RF model since the residuals were not normally distributed because of the existence of a spatial gap between the evergreen sample plots that restrict the principle of spatial correlation (Figure 7a). Thus, the OK model could not serve to fit the spatial patterns of the residuals in the evergreen forest AGB, while the CK model performed better in terms of fitting the spatial patterns of the residuals in the deciduous forest AGB in the study area when considering their R^2 and RMSE values. Nevertheless, these residuals were used to interpolate their structured components in the evergreen and deciduous forests, respectively.

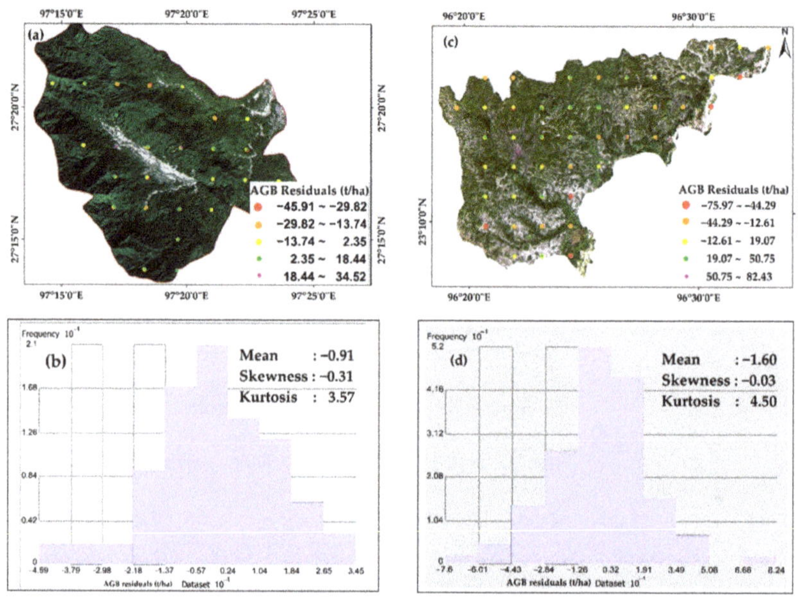

Figure 7. Spatial distributions of the residuals and corresponding frequency histograms; (**a**,**c**) are RF-based AGB-residuals for NH evergreen and YM deciduous, respectively; (**b**,**d**) are frequency histograms of AGB-residuals for NH evergreen and YM deciduous, respectively.

Table 7. Parameter estimations for the semivariogram analysis based on Gaussian function.

Model Parameter	Theoretical Model	Nugget	Sill	Nugget/Sill	Range (m)	R^2	RMSE (t/ha)
NH Evergreen OK	Gaussian	138.53	145.09	0.95	99.57	0.19	12.42
NH Evergreen CK	Gaussian	249.63	249.75	0.99	9611	0.15	18.69
YM Deciduous OK	Gaussian	245.93	324.58	0.76	10123	0.58	24.38
YM Deciduous CK	Gaussian	239.97	326.26	0.74	9890	0.62	22.65

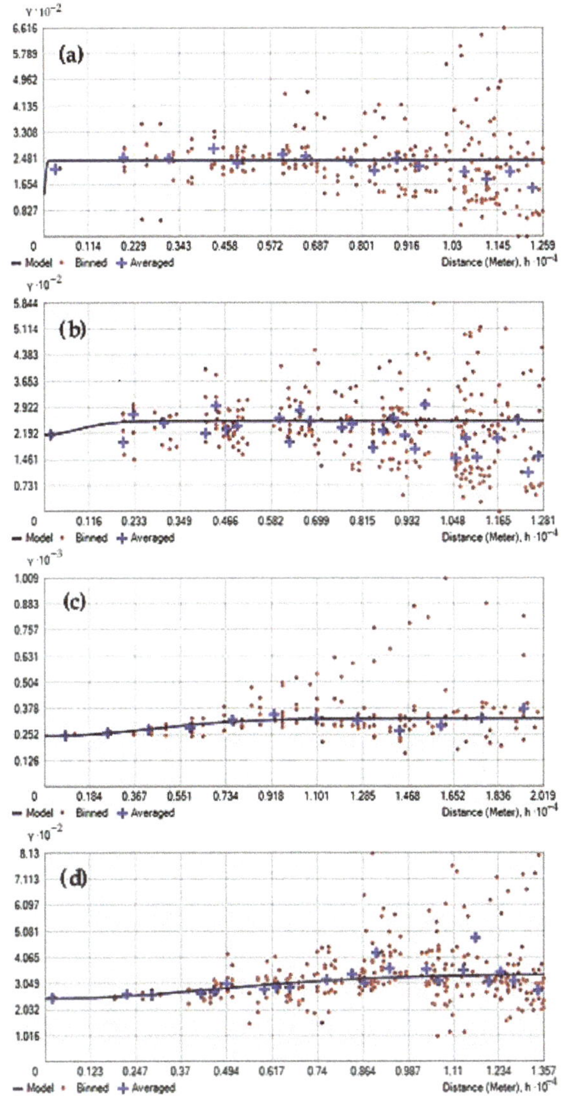

Figure 8. Empirical semivariograms and covariance models for the two forest reserves' RF-derived residuals; (**a**,**c**) are the semivariogram models for NH evergreen and YM deciduous using OK analysis; (**b**,**d**) are the semivariogram models for NH evergreen and YM deciduous using CK analysis with a covariable of elevation. The vertical axis is the $\frac{1}{2}$ variance (γ) and covariance (C) of the two positions as the distance increases.

3.4. Forest AGB Mapping Results Based on RFOK and RFCK Models

The predicted AGB of the two forest reserves were obtained from the RFOK for the evergreen and RFCK for the deciduous forests, and the validation performances with 20% sample data based on Equation (6) were correspondingly derived. Table 5 shows the validation accuracy improvements of RFOK and RFCK in relation to the initial RF model.

Although the R^2 value (0.47) of the RFOK model for evergreen did not increase compared to the original RF model, its RMSE, RMSE% and MAE all decreased with different magnitudes, and its RI value was 0.02, indicating a slight improvement in AGB prediction. However, the RFOK had the worst predictive performance in the evergreen forest. For the deciduous forest, the RFCK model outperformed other two RFOK and RF models, particularly in relation to the RF, as the RFCK's R square value increased from 0.38 to 0.52 and its RMSE decreased from 40.23 t/ha to 34.72 t/ha, with an RI value of 13.7%. However, compared to RFOK, RFCK only took on a very tiny improvement in the prediction accuracy.

In addition to the model evaluation, the generalization ability of the model was also considered. The AGB value range of the predicted map could reflect the model's robustness to some extent. The range of AGB values predicted for the evergreen using the RF model was 88.75–129 t/ha. The AGB prediction value range from RFOK was 94.3–139.83 t/ha and the RFCK had a value range of 94.06–139.62 t/ha, respectively. The AGB prediction value range in the deciduous was 40–176 t/ha, 30.41–187.84 t/ha, and 32.88–185.65 t/ha for RF, RFOK, and RFCK. These variations in the value range clearly indicated an improved generalization ability of RFOK and RFCK, with higher robustness. The largest evergreen AGB values were found in the northern and western boundaries of NH while the AGB in the central and eastern parts were sparsely distributed. In this area, the low AGB values were found close to a village and flat area. Deciduous AGB was covered with large values in some parts of northern YM, and they were evenly distributed in the reserved area, except in the southeastern boundary which is close to a village. The AGB maps derived from all performed models are shown in Figures 9 and 10.

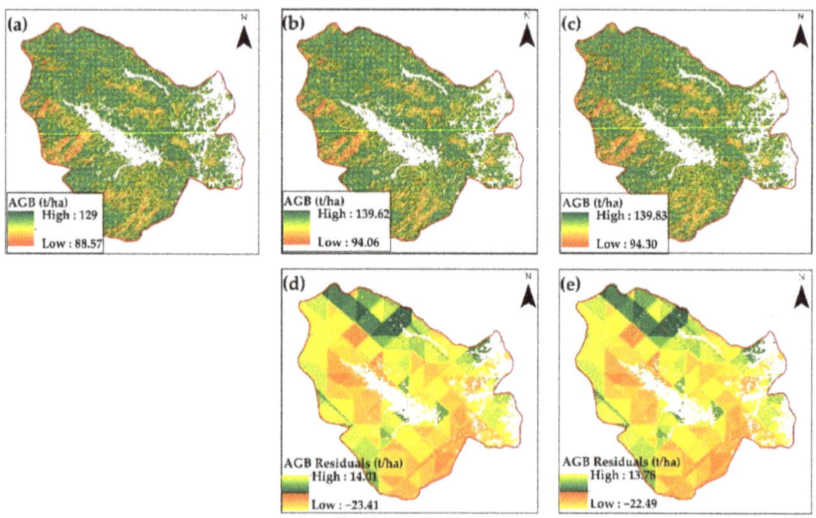

Figure 9. The estimated AGB of evergreen forests in the NM from (**a**) RF, (**b**) RFOK, and (**c**) RFCK models, and the calculated residuals for (**d**) RFOK and (**e**) RFCK models.

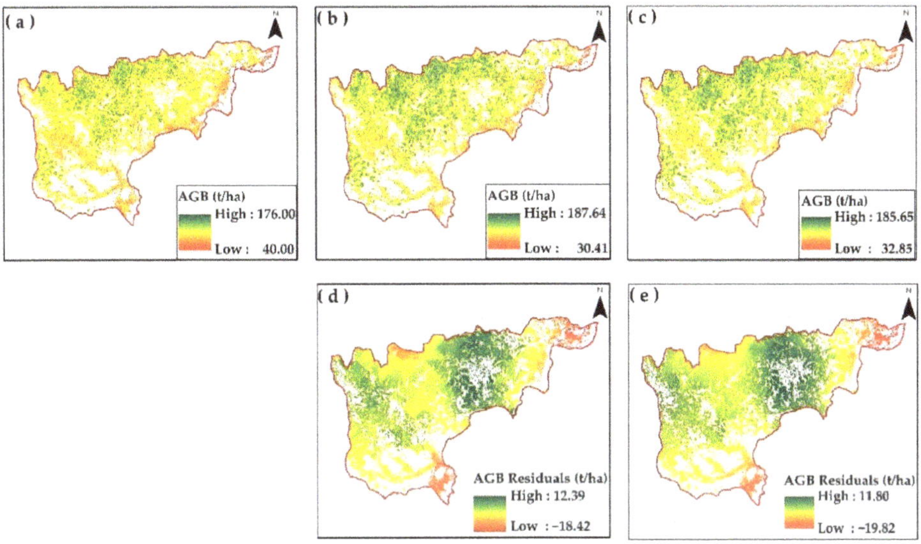

Figure 10. The estimated AGB of deciduous forests in the YM from (**a**) RF, (**b**) RFOK, and (**c**) RFCK models, and the calculated residuals for (**d**) RFOK and (**e**) RFCK models.

4. Discussions

To the best of our knowledge, this is the first study evaluating the capability of S-2 satellite data for AGB mapping in evergreen and deciduous forest reserves in Myanmar. AGB mapping in these forest reserves is very fundamental for carbon strategies, especially for Myanmar REDD+ mechanisms which need to report the carbon improvement of forests from reforestation programs for future FREL submission. Moreover, the prediction methods proposed in this study are noticeably relevant for future forest management practices in Myanmar, which previously exhibited a lack of robust AGB estimation methods for forest types. We proved that THE combined use of S-2, its derivatives, and topographic parameters, in tandem with proper modeling techniques, could improve AGB estimation.

4.1. Sensitivity of Sentinel-2 Derivatives to AGB

The correct selection of variables contributing to a model is critical for AGB estimation. In most AGB mapping studies with S-2 derivatives, four novel wavebands in the red-edge region and near-infrared region (NIR) (B5, B6, B7 and B8A) showed good performance [16,72]. These reflectance regions offer unprecedented spectral signatures which are highly sensitive to the biophysical and biochemical responses of vegetation that are critical for measuring vegetation characteristics such as biomass [72]. However, in this study, the classical and short-wave infrared bands (B2, B3, B4, B11, and B12) outperformed these aforementioned spectral bands. The excellent performance of these bands is that carbon and nitrogen-containing metabolites reach their reflectance peak in wavebands between 440 nm and 570 nm due to the nature of the forests in the study area (B2 and B3) and concurred with the previous study [73]. Forest canopy in the evergreen forest can uptake maximum chlorophyll absorption due to non-deciduous phenomena. B3 and B4 in the S-2 have strong sensitivity to chlorophyll in evergreen vegetation, while B2 can effectively distinguish vegetation and soil background in the deciduous forest where the reflectivity of soil is apparent because of the simple canopy structure. Even though vegetation can reflect the maximum energy at NIR despite the fact that it is unable to provide any information on the soil under the vegetation, SWIR bands in S-2 can distinguish the vegetation and soil to some extent. A recent finding by Chen et al. verified that SWIR spectral bands (B11) could efficiently detect the moisture content of vegetation [37]. Moreover, Dang et al. proved that the broad bands (B11, B12) of S-2 were the best response variables of AGB prediction with an R^2 of 0.81 and an RMSE of 36.67 Mg t/ha [6]. Hence, the high sensitivity of SWIR bands to biomass observed in this study seems plausible, which is in agreement with the existing studies.

The VIs used in this study (NDVI, GNDVI, SAVI, SR, RENDVI, NDI45, and NDWI) were primary contributors to the AGB estimation of forests since they have the ability to maximize the sensitivity of vegetation characteristics and minimize soil background reflectance and atmospheric effects. Balidoy et al. proved that the SR and NDVI of S-2 data were the most effective biomass predictors, providing the highest accuracy (R^2 = 0.89; RMSE = 5.69 Mg/ha) [74]. Ghosh et al. found that the effectiveness of NDVI, GNDVI, SAVI indices of S-2 for dense tropical AGB mapping had an R^2 value of 0.6 and an RMSE value of 79.45 t/ha for the teak forest [35]. The findings of Pandit et al. were entirely consistent with those of this study. The set of 24 variables including NDVI, GNDVI, RENDVI, SAVI, and SR produced overall plausible and strongly explained variable values, with R^2 = 0.81 and RMSE = 1.07 kg/m [72]. Although VIs produced from traditional broad bands can reduce the saturation problem in simple canopy forest, they are less sensitive to complex forest stands with high biomass values [75]. In this regard, the red-edge bands derived indices are highly sensitive to such kind of dense vegetation structures and relatively less prone to spectral saturation. For example, the standard NDVI from B4 and B8 is less effective than RENDVI from red-edge bands (B5, B8) in AGB estimation [76] and hence red-edge-derived indices can be effectively applied in dense vegetation cover (e.g., RENDVI is sensitive to the NH evergreen forest AGB in the current work). SWIR bands are related with nitrogen, lignin, and cellulose, capable of retrieving canopy structural attributes and biomass. Canopy water content index from SWIR bands (e.g., NDWI in this work) is highly sensitive to deciduous forest AGB but not to evergreen AGB because canopy structure in deciduous forest is relatively simple, which is in agreement with the previous findings of Ewald [77]. They pointed out that in the very dense canopy plantation, SWIR indices could not effectively improve AGB estimation compared to other indices. This study claims that red-edge indices are suitable for complex canopy AGB retrievals while SWIR indices are useful for simple canopy forest AGB estimation.

The textural variables are strongest candidates of evergreen forest AGB, especially the textures (coif1-d, coif1-dd, and coif1-hh) extracted from Coiflect wavelet analysis of the PC1 image. They could obviously improve the AGB estimation as the horizontal structures of the evergreen forest can be effectively characterized by them. Previous research had shown that texture measures have the potential to improve AGB estimation, especially for complex vegetation structures where canopies' reflectance values tend to be saturated but the horizontal structures represented by textural indices still have differences. If proper processing techniques are used, textural measures could improve the prediction accuracy of AGB models. According to Eckert et al., textures were much better to capture the various forest canopy structures of the forest strata than the spectral reflectance or band ratios, due to their sensitivity to the spatial aspects of the canopy shadow [78]. Su et al. proved the excellent performance of textures from the PC1 image for AGB prediction of sub-tropical forest where saturation problem occurred [22]. Moreover, Cutler et al. argued that the textures extracted from GLCM method and the WA of satellite images yielded better results in the AGB estimation and forest type classification [79]. Our results concur with aforementioned studies. However, in this study, GLCM-based textures were not highly correlated with AGB. The reason might be that the window size (7×7) could not reduce the border effects of pixels to attain original spectral values and thus, texture window size determination should be considered in accordance with the types of satellite data in future studies. We conclude that the wavelet decomposition analysis of satellite images might improve evergreen forest AGB estimation because it produces more suitable textures to effectively depict evergreen forest horizontal structures to reduce the saturation problem of spectral signals.

4.2. Sensitivity of Topographic Variables to AGB

Topographic features (elevation and slope) were also strong predictors of AGB in this study. The vital role of topography is also an important factor influencing AGB because water and sunlight storage change in function of topography [80]. Chen et al. proved that elevation was the strongest factor for complex AGB estimation in China [37]. Slope inclination was also a strong factor affecting AGB in the evergreen forest, since the terrain in this area is mountainous with different sunlight and water detention levels by its varying geomorphological characteristics while deciduous forest is not affected by slope by growing its trees over sandy soil flat surface. This finding is consistent with the proof of Hamere et al., which claimed that AGB carbon, BGB carbon, and the total carbon density trend showed a decrease as the slope increased due to the little vegetation cover in very steep slope areas [81]. In addition, in the accessible flat area of nearby settlements and stream banks, the anthropogenic effects might decrease AGB values. Therefore, topographic factors ultimately affect the AGB observed in this study.

4.3. Comparison between Models

This study evaluated the two modeling techniques (RF and SGB) for AGB mapping in two forest reserves and indicated a saturation problem of S-2 d satellite data, thus causing the presence of bias in the AGB prediction models. The NH evergreen RF model estimated the smaller AGB value of 129 t/ha than the observed AGB value of 151.64 t/ha. Similarly, the estimation of the YM deciduous forest AGB value (176 t/ha) was smaller than the field-observed AGB value (215.24 t/ha). The scatterplots in the validation metrics of this study indicated the limitation of the classical wavelength bands in the S-2 MSI sensor when dealing with saturation in high biomass stands. From the important variables ranking, the candidates of the classical wavelength region and topography such as B3, B4, Ele, Slope, and one vegetation index SAVI in the evergreen forest, and B11, B12, NDWI, and GNDVI in the deciduous forest were high ranked while no red-edge reflectance was correlated with AGB. This ranking affects the performance of models since the improvement in red-edge bands features was relatively larger than that in classical bands and topographic variables. This assumption was proven by previous studies of Forkuor et al. [76] and Nuthammachot et al. [30]. Additionally, Chen et al. suggested that the broadleaved forests with AGB values above 160 t/ha could be underestimated because of the saturation problem in S-2 satellite data [82]. The observed AGB value in the YM deciduous forest was 215.24 t/ha and thus should agree with the finding of Chen. An almost similar ranking in variables was observed in the two SGB models but the two models could not improve the estimation when considering their performances in prediction. Thus, these SGB models occurred and similar saturation problem was found in the RF models.

In order to optimize the estimation, the Kriging interpolation analysis of the residuals from RF models (RFK) was further employed since the RF showed better performance than SGB in this study. Our study claimed that the ordinary Kriging of RF's residuals (RFOK) performed better than other tested models in NH evergreen, while co-Kriging of RF's residuals (RFCK) with covariance elevation (Ele) provides a better result than other models in YM deciduous AGB prediction. The limited contribution of the accuracy of the RFOK model for the NH evergreen forest reserve was due to the poor spatial autocorrelation between AGB samples which occurred due to the spatial gaps between the sample plots. In the YM deciduous forest, RFCK achieved good prediction results, however, the spatial correlations of the current AGB were also weaker than previous studies of Su et al. and Chen et al. on forest AGB mapping based on the integration of multi-sensor and Advanced Land Observing Satellite (ALOS) data [22,37]. It was denoted that AGB residuals from the integration of the MSI and SRTM data model in this study obtained a lower spatial correlation than that built by the integration of MSI and ALOS indices. This also proves that AGB estimation from the combined MSI and SRTM data was only suitable for the simple structure forest stand (e.g., deciduous forest in this study).

Overall, we conclude that the accuracies of AGB prediction could be improved to a certain extent by Kriging methods by reducing the spatial heterogeneity between AGB samples. In the future, the accuracy of evergreen forest AGB estimation in this study might be improved with LiDAR data since it can penetrate the forest canopy to a certain depth, so that its variables are suitable for extracting vertical vegetation structures with their sensitivity to the biomass of vegetation and the roughness of land cover surfaces. The determinant of the spatial setting and a sufficient number of sample plots should be considered in future AGB studies to maintain the Kriging accuracy. Additionally, the study area is located in the regions where frequent rains and clouds exists, highly restricting the availability of images collected in the vegetation growth peak season (e.g., June–September). Therefore, we had to use cloud-free images acquired in January or February, which falls outside the ideal time window for characterizing vegetation attributes. This limitation may affect the accuracy of forest AGB prediction model.

In addition, an allometric equation for in situ AGB calculation may be another factor undermining accuracy. This study used a national-level coarse allometric equation which was based on an existing inventory dataset and pantropical equation (Chave et al., 2005, Chen et al., 2013, and IPCC, 2003) for AGB calculation since there have been no species-specific equations developed for this study area. In the near future, developing species-specific allometric equations through limited destructive sampling should prioritize the carbon accounting and climate change response studies in Myanmar because these create more accurate in situ plot-level AGB measurements, laying a solid foundation for the remote sensing-based regional estimation of AGB. In general, AGB estimated from the RF model could yield acceptable results of validated R^2 = 0.47, RMSE = 25.45 t/ha for evergreen and R^2 = 0.38, RMSE= 40.23 t/ha for deciduous from S-2 derivatives, topographic variables and ancillary information.

4.4. Effects of Forest Management on AGB in the Study Sites

Population growth has led to a high demand for forest products, unsustainable forest management practices, and high deforestation rates, thus causing forest cover loss. The extent of the forest cover loss depends on the forest protection status with different rules applying to public and reserved areas [83]. Forest protection typically reduces the conversion of natural land cover types to alternative uses and often results in positive outcomes (including reduced deforestation rates and the maintenance of forest cover) compared to unprotected sites. In Myanmar, intact forests are gradually decreasing to only 38% of the country's forests due to the rapid political and economic changes. The study site comprises two protected forests under Myanmar forest law. However, the expansion of the human population and the need for more agricultural lands tend to encroach into these areas, especially in the evergreen forest presently under study. Encroachment in protected forests for agricultural lands, food, and fuels is directly correlated with loss in AGB values. On the other hand, deciduous forests' AGB values might be following a decreasing trend because the demands for commercial timber species are increasing and AGB sources are gradually decreasing. In this context, the spatial agreement of AGB was observed in the estimated AGB maps derived from the RFOK evergreen and RFCK deciduous models. For example, the small AGB values were estimated in the forest area closest to the villages and cultivated lands while large AGB values were distributed in the high-elevation forest of the NH evergreen forest reserve. A similar finding was observed in the YM deciduous forest. To sustain the forest AGB in these areas, community-based forest management is suggested to reduce these pressures as this would meet the needs of forestry products for forest dwellers.

4.5. Attainment for SDG and REDD+

The UN SDGs set out the commitment of the international community to rid the world of poverty and hunger and achieve sustainable development in its three dimensions—the economic, social and environmental facets. In addition to using standardized national official data sources as the basis for monitoring and reporting on these goals and targets, geospatial information and global/regional datasets have been identified as viable replacement and complementary data sources for achieving these SDGs [84]. As the indicator of SDG 15, the accurate estimation and monitoring of aboveground biomass stocks need to be achieved. In this regard, the results of this study are applicable and useful for the attainment of SDG 15, especially for Myanmar where freely available optical data are preferred to map biomass stocks, and will greatly assist in the deriving, monitoring, and reporting of carbon stock changes in a timely and accurate manner.

Myanmar has been implementing the REDD+ project to achieve SDG goals and targets through sustainable forest management practices since 2013. The REDD+ objective is to find an accurate method of biomass estimation that is also cost-effective. Based on the results obtained in this study, S-2-derived derivatives (spectra, VIs, textures) and the topographic features of SRTM (elevation, slope) have potential in the forest biomass estimation of two forest reserves. In addition, the methods used in this study are viable and compatible software has been developed (e.g., SNAP), in such a way that REDD+ can apply it at a larger scale, including the national and regional levels. The outcomes of this study can surely assist the evaluation of carbon stock changes via reforestation programs that will be included in the upcoming FREL calculation under the REDD+ agenda of Myanmar.

5. Conclusions

This study investigates the performance of S-2 MSI derivatives and SRTM DEM topographic data with field ancillary information based on two machine learning models (RF and SGB) in mapping the forest AGB of two forest reserves (namely the NH evergreen and YM deciduous forests) in Myanmar. In addition, the RF-based Kriging (RFK) was employed for improving the prediction accuracy to find a spatial correction between the AGB samples. Based on these findings, it is concluded that:

(1) S-2-derived reflectance, VIs, and textures are effective in predicting the AGB of the two forests if the proper processing techniques are applied;
(2) The RFOK model in the evergreen forest and RFCK model in the deciduous forest provided a more realistic spatial distribution of AGB by considering the spatial correlation than the RF and SGB models with R^2 = 0.47, RMSE = 24.91 t/ha and R^2 = 0.52, RMSE = 34.72 t/ha due to their spatial correlation between AGB sample plots;
(3) The extraction of textures from wavelet analysis (WA) is suggested to improve estimation for the forests with a complex structure and saturation problems;
(4) In future studies, the accuracy may be improved by combining both the active and passive remotely sensed data to characterize complex forest structures to better estimate the forest AGB and understand their spatial distributions.

Author Contributions: Conceptualization, M.L.; methodology, P.W. and H.S.; software, P.W. and H.S.; validation, P.W. and H.S.; formal analysis, P.W.; investigation, P.W.; resources, M.L. and P.W.; data curation, P.W. and H.S.; writing—original draft preparation, P.W.; writing—review and editing, M.L.; visualization, P.W.; supervision, M.L.; project administration, M.L.; funding acquisition, M.L. All authors have read and agreed to the published version of the manuscript.

Funding: This research was jointly funded by the Natural Science Foundation of China, grant number 31971577, the Priority Academic Program Development of Jiangsu Higher Education Institutions (PAPD).

Data Availability Statement: Not applicable.

Acknowledgments: Special thanks need to go to the European Space Agency for the provision of the remotely sensed images used in the work and the Forest Department of Myanmar for providing the forest inventory data.

Conflicts of Interest: The authors declare no conflict of interest.

References

1. Bonan, G.B. Forests and Climate Change: Forcings, Feedbacks, and the Climate Benefits of Forests. *Science* **2008**, *320*, 1444–1449. [CrossRef] [PubMed]
2. Wolosin, M.; Harris, N. *Tropical Forests and Climate Change: The Latest Science*; World Resources Institute: Washington, DC, USA, 2018.
3. Rodríguez-Veiga, P.; Wheeler, J.; Louis, V.; Tansey, K.; Balzter, H. Quantifying Forest Biomass Carbon Stocks from Space. *Curr. For. Rep.* **2017**, *3*, 1–18. [CrossRef]
4. Gibbs, H.K.; Brown, S.; Niles, J.O.; Foley, J.A. Monitoring and Estimating Tropical Forest Carbon Stocks: Making REDD a Reality. *Environ. Res. Lett.* **2007**, *2*, 045023. [CrossRef]
5. Basuki, T.M.; van Laake, P.E.; Skidmore, A.K.; Hussin, Y.A. Allometric Equations for Estimating the Above-Ground Biomass in Tropical Lowland Dipterocarp Forests. *For. Ecol. Manag.* **2009**, *257*, 1684–1694. [CrossRef]
6. Dang, A.T.N.; Nandy, S.; Srinet, R.; Luong, N.V.; Ghosh, S.; Senthil Kumar, A. Forest Aboveground Biomass Estimation Using Machine Learning Regression Algorithm in Yok Don National Park, Vietnam. *Ecol. Inform.* **2019**, *50*, 24–32. [CrossRef]
7. Lu, D.; Chen, Q.; Wang, G.; Liu, L.; Li, G.; Moran, E. A Survey of Remote Sensing-Based Aboveground Biomass Estimation Methods in Forest Ecosystems. *Int. J. Digit. Earth* **2016**, *9*, 63–105. [CrossRef]
8. Li, Y.; Li, M.; Li, C.; Liu, Z. Forest Aboveground Biomass Estimation Using Landsat 8 and Sentinel-1A Data with Machine Learning Algorithms. *Sci. Rep.* **2020**, *10*, 9952. [CrossRef] [PubMed]
9. Shen, W.; Li, M.; Huang, C.; Tao, X.; Wei, A. Annual Forest Aboveground Biomass Changes Mapped Using ICESat/GLAS Measurements, Historical Inventory Data, and Time-Series Optical and Radar Imagery for Guangdong Province, China. *Agric. For. Meteorol.* **2018**, *259*, 23–38. [CrossRef]
10. Mon, M.S.; Myint, A.A. Estimating above Ground Biomass of Tropical Mixed Deciduous Forests Using Landsat ETM+ Imagery for Two Reserved Forests in Bago Yoma Region, Myanmar. In *Proceedings of the Multi-Scale Forest Biomass Assessment and Monitoring in the Hindu Kush Himalayan Region: A Geospatial Perspective*; Murthy, M.S.R., Wesselman, S., Gilani, H., Eds.; International Centre for Integrated Mountain Development: Patan, Nepal, 2015; pp. 165–177.
11. Madugundu, R.; Nizalapur, V.; Jha, C.S. Estimation of LAI and Above-Ground Biomass in Deciduous Forests: Western Ghats of Karnataka, India. *Int. J. Appl. Earth Obs. Geoinf.* **2008**, *10*, 211–219. [CrossRef]
12. Naveenkumar, J.; Arunkumar, K.S.; Sundarapandian, S. Biomass and Carbon Stocks of a Tropical Dry Forest of the Javadi Hills, Eastern Ghats, India. *Carbon Manag.* **2017**, *8*, 351–361. [CrossRef]
13. Gasparri, N.I.; Parmuchi, M.G.; Bono, J.; Karszenbaum, H.; Montenegro, C.L. Assessing Multi-Temporal Landsat 7 ETM+ Images for Estimating above-Ground Biomass in Subtropical Dry Forests of Argentina. *J. Arid Environ.* **2010**, *74*, 1262–1270. [CrossRef]
14. Li, M.; Tan, Y.; Pan, J.; Peng, S. Modeling Forest Aboveground Biomass by Combining Spectrum, Textures and Topographic Features. *Front. For. China* **2008**, *3*, 10–15. [CrossRef]
15. Shen, W.; Li, M.; Huang, C.; Wei, A. Quantifying Live Aboveground Biomass and Forest Disturbance of Mountainous Natural and Plantation Forests in Northern Guangdong, China, Based on Multi-Temporal Landsat, PALSAR and Field Plot Data. *Remote Sens.* **2016**, *8*, 595. [CrossRef]
16. Castillo, J.A.A.; Apan, A.A.; Maraseni, T.N.; Salmo, S.G. Estimation and Mapping of Above-Ground Biomass of Mangrove Forests and Their Replacement Land Uses in the Philippines Using Sentinel Imagery. *ISPRS J. Photogramm. Remote Sens.* **2017**, *134*, 70–85. [CrossRef]
17. Xue, B. Lidar and Machine Learning Estimation of Hardwood Forest Biomass in Mountainous and Bottomland Environments. Master's Thesis, University of Arkansas, Fayetteville, AR, USA, 2015.
18. Pham, T.D.; Yokoya, N.; Bui, D.T.; Yoshino, K.; Friess, D.A. Remote Sensing Approaches for Monitoring Mangrove Species, Structure, and Biomass: Opportunities and Challenges. *Remote Sens.* **2019**, *11*, 230. [CrossRef]
19. Laurin, G.V.; Chen, Q.; Lindsell, J.A.; Coomes, D.A.; Del Frate, F.; Guerriero, L.; Pirotti, F.; Valentini, R. Above Ground Biomass Estimation in an African Tropical Forest with Lidar and Hyperspectral Data. *ISPRS J. Photogramm. Remote Sens.* **2014**, *89*, 49–58. [CrossRef]
20. Yuan, X.; Li, L.; Tian, X.; Luo, G.; Chen, X. Estimation of Above-Ground Biomass Using MODIS Satellite Imagery of Multiple Land-Cover Types in China. *Remote Sens. Lett.* **2016**, *7*, 1141–1149. [CrossRef]
21. Blackard, J.A.; Finco, M.V.; Helmer, E.H.; Holden, G.R.; Hoppus, M.L.; Jacobs, D.M. Mapping US Forest Biomass Using Nationwide Forest Inventory Data and Moderate Resolution Information. *Remote Sens. Environ.* **2008**, *112*, 1658–1677. [CrossRef]
22. Su, H.; Shen, W.; Wang, J.; Ali, A.; Li, M. Machine Learning and Geostatistical Approaches for Estimating Aboveground Biomass in Chinese Subtropical Forests. *For. Ecosyst.* **2020**, *7*, 64. [CrossRef]
23. López-Serrano, P.M.; Cárdenas Domínguez, J.L.; Corral-Rivas, J.J.; Jiménez, E.; López-Sánchez, C.A.; Vega-Nieva, D.J. Modeling of Aboveground Biomass with Landsat 8 OLI and Machine Learning in Temperate Forests. *Forests* **2019**, *11*, 11. [CrossRef]
24. Addabbo, P.; Focareta, M.; Marcuccio, S.; Votto, C.; Ullo, S.L. Contribution of Sentinel-2 Data for Applications in Vegetation Monitoring. *Acta IMEKO* **2016**, *5*, 44–54. [CrossRef]
25. Gascon, F.; Ramoino, F.; Deanos, Y. Sentinel-2 Data Exploitation with ESA's Sentinel-2 Toolbox. *EGU Gen. Assem.* **2017**, *19*, 19548.
26. Imran, A.B.; Khan, K.; Ali, N.; Ahmad, N.; Ali, A.; Shah, K. Narrow Band Based and Broadband Derived Vegetation Indices Using Sentinel-2 Imagery to Estimate Vegetation Biomass. *Glob. J. Environ. Sci. Manag.* **2020**, *6*, 97–108. [CrossRef]

27. Li, H.; Kato, T.; Hayashi, M.; Wu, L. Estimation of Forest Aboveground Biomass of Two Major Conifers in Ibaraki Prefecture, Japan, from PALSAR-2 and Sentinel-2 Data. *Remote Sens.* **2022**, *14*, 468. [CrossRef]
28. Safari, A.; Sohrabi, H. Integration of Synthetic Aperture Radar and Multispectral Data for Aboveground Biomass Retrieval in Zagros Oak Forests, Iran: An Attempt on Sentinel Imagery. *Int. J. Remote Sens.* **2020**, *41*, 8069–8095. [CrossRef]
29. Adamu, B.; Ibrahim, S.; Rasul, A.; Whanda, S.J.; Headboy, P.; Muhammed, I.; Maiha, I.A. Evaluating the Accuracy of Spectral Indices from Sentinel-2 Data for Estimating Forest Biomass in Urban Areas of the Tropical Savanna. *Remote Sens. Appl. Soc. Environ.* **2021**, *22*, 100484. [CrossRef]
30. Nuthammachot, N.; Askar, A.; Stratoulias, D.; Wicaksono, P. Combined Use of Sentinel-1 and Sentinel-2 Data for Improving above-Ground Biomass Estimation. *Geocarto Int.* **2022**, *37*, 366–376. [CrossRef]
31. Taddese, H.; Asrat, Z.; Burud, I.; Gobakken, T.; Ørka, H.O.; Dick, Ø.B.; Næsset, E. Use of Remotely Sensed Data to Enhance Estimation of Aboveground Biomass for the Dry Afromontane Forest in South-Central Ethiopia. *Remote Sens.* **2020**, *12*, 3335. [CrossRef]
32. Li, L.; Zhou, X.; Chen, L.L.; Chen, L.L.; Zhang, Y.; Liu, Y. Estimating Urban Vegetation Biomass from Sentinel-2A Image Data. *Forests* **2020**, *11*, 125. [CrossRef]
33. Pandit, S.; Tsuyuki, S.; Dube, T. Exploring the Inclusion of Sentinel-2 MSI Texture Metrics in above-Ground Biomass Estimation in the Community Forest of Nepal. *Geocarto Int.* **2019**, *35*, 1832–1849. [CrossRef]
34. Jolliffe, I.T.; Cadima, J.; Cadima, J. Principal Component Analysis: A Review and Recent Developments. *Philos. Trans. R. Soc. A Math. Phys. Eng. Sci.* **2016**, *374*, 20150202. [CrossRef] [PubMed]
35. Ghosh, S.M.; Behera, M.D. Aboveground Biomass Estimation Using Multi-Sensor Data Synergy and Machine Learning Algorithms in a Dense Tropical Forest. *Appl. Geogr.* **2018**, *96*, 29–40. [CrossRef]
36. Ouma, Y.O.; Tetuko, J.; Tateishi, R. Analysis of Co-occurrence and Discrete Wavelet Transform Textures for Differentiation of Forest and Non-forest Vegetation in Very-high-resolution Optical-sensor Imagery. *Int. J. Remote Sens.* **2008**, *29*, 3417–3456. [CrossRef]
37. Chen, L.; Wang, Y.; Ren, C.; Zhang, B.; Wang, Z. Optimal Combination of Predictors and Algorithms for Forest Above-Ground Biomass Mapping from Sentinel and SRTM Data. *Remote Sens.* **2019**, *11*, 414. [CrossRef]
38. Wang, Y.; Zhang, X.; Guo, Z. Estimation of Tree Height and Aboveground Biomass of Coniferous Forests in North China Using Stereo ZY-3, Multispectral Sentinel-2, and DEM Data. *Ecol. Indic.* **2021**, *126*, 107645. [CrossRef]
39. Simard, M.; Zhang, K.; Rivera-Monroy, V.H.; Ross, M.S.; Ruiz, P.L.; Castañeda-Moya, E.; Twilley, R.R.; Rodriguez, E. Mapping Height and Biomass of Mangrove Forests in Everglades National Park with SRTM Elevation Data. *Photogramm. Eng. Remote Sens.* **2006**, *72*, 299–311. [CrossRef]
40. Ahmad, A.; Gilani, H.; Ahmad, S.R. Forest Aboveground Biomass Estimation and Mapping through High-Resolution Optical Satellite Imagery—A Literature Review. *Forests* **2021**, *12*, 914. [CrossRef]
41. Zhang, J.; Lu, C.; Xu, H.; Wang, G. Estimating Aboveground Biomass of Pinus Densata-Dominated Forests Using Landsat Time Series and Permanent Sample Plot Data. *J. For. Res.* **2019**, *30*, 1689–1706. [CrossRef]
42. Ye, Q.; Yu, S.; Liu, J.; Zhao, Q.; Zhao, Z. Aboveground Biomass Estimation of Black Locust Planted Forests with Aspect Variable Using Machine Learning Regression Algorithms. *Ecol. Indic.* **2021**, *129*, 107948. [CrossRef]
43. Torres, C.; Almeida, D.; Soares, L.; Eduardo, L.; Cruz, D.O.; Pierre, J.; Balbaud, H.; Daniele, A.; Rocha, F.; Pereira, D.S.; et al. Combining LiDAR and Hyperspectral Data for Aboveground Biomass Modeling in the Brazilian Amazon Using Di Ff Erent Regression Algorithms. *Remote Sens. Environ.* **2019**, *232*, 111323. [CrossRef]
44. Breiman, L. Random Forests. *Mach. Learn.* **2001**, *45*, 5–32. [CrossRef]
45. Friedman, J.H. Stochastic Gradient Boosting. *Comput. Stat. Data Anal.* **2002**, *38*, 367–378. [CrossRef]
46. Chen, L.; Wang, Y.; Ren, C.; Zhang, B.; Wang, Z. Assessment of Multi-Wavelength SAR and Multispectral Instrument Data for Forest Aboveground Biomass Mapping Using Random Forest Kriging. *For. Ecol. Manag.* **2019**, *447*, 12–25. [CrossRef]
47. Wang, C.; Myint, S.W. Environmental Concerns of Deforestation in Myanmar 2001–2010. *Remote Sens.* **2016**, *8*, 728. [CrossRef]
48. Ministry of Natural Resources and Environmental Conservation; Forest Department. *Forestry in Myanmar*; Winn, U.O., Ed.; Ministry of Natural Resources and Environmental Conservation: Naypyitaw, Myanmar, 2020; Volume 53, ISBN 9788578110796.
49. FAO; Forest Department, Myanmar. *Country Report: Forest Resource Assessment 2015, Myanmar*; FAO: Rome, Italy, 2014.
50. Forest Department, Myanmar, UN-REDD Program. *Forest Reference Level (FRL) of Myanmar*; Ministry of Natural Resources and Environmental Conservation: Naypyitaw, Myanmar, 2018.
51. Banskota, A.; Wynne, R.H.; Kayastha, N. Improving Within-Genus Tree Species Discrimination Using the Discrete Wavelet Transform Applied to Airborne Hyperspectral Data. *Int. J. Remote Sens.* **2011**, *32*, 3551–3563. [CrossRef]
52. Maung, W.S. Assessing the Natural Recovery of Mangroves after Human Disturbance Using Neural Network Classification and Sentinel-2 Imagery in Wunbaik Mangrove Forest, Myanmar. *Remote Sens.* **2021**, *13*, 52. [CrossRef]
53. Rouse, J.W., Jr.; Haas, R.H.; Schell, J.A.; Deering, D.W. Monitoring Vegetation Systems in the Great Plains with Erts. In *Third ERTS Symposium*; NASA SP-351; NASA: Washington, DC, USA, 1974; Volume 351.
54. Karlson, M.; Ostwald, M.; Reese, H.; Sanou, J.; Tankoano, B.; Mattsson, E. Mapping Tree Canopy Cover and Aboveground Biomass in Sudano-Sahelian Woodlands Using Landsat 8 and Random Forest. *Remote Sens.* **2015**, *7*, 10017–10041. [CrossRef]
55. Richardson, A.J.; Wiegand, C.L. Distinguishing Vegetation from Soil Background Information. *Photogramm. Eng. Remote Sens.* **1977**, *43*, 1541–1552.

56. Sims, D.A.; Gamon, J.A. Relationships between Leaf Pigment Content and Spectral Reflectance across a Wide Range of Species, Leaf Structures and Developmental Stages. *Remote Sens. Environ.* **2002**, *4257*, 337–354. [CrossRef]
57. Jiang, Z.; Huete, A.R.; Didan, K.; Miura, T. Development of a Two-Band Enhanced Vegetation Index without a Blue Band. *Remote Sens. Environ.* **2008**, *112*, 3833–3845. [CrossRef]
58. Alam, M.J.; Rahman, K.M.; Asna, S.M.; Muazzam, N.; Ahmed, I.; Chowdhury, M.Z. Comparative Studies on IFAT, ELISA & DAT for Serodiagnosis of Visceral Leishmaniasis in Bangladesh. *Bangladesh Med. Res. Counc. Bull.* **1996**, *22*, 27–32. [PubMed]
59. Gitelson, A.A.; Merzlyak, M.N. Signature Analysis of Leaf Reflectance Spectra: Algorithm Development for Remote Sensing of Chlorophyll. *J. Plant Physiol.* **1996**, *148*, 494–500. [CrossRef]
60. Gao, B.C. NDWI—A Normalized Difference Water Index for Remote Sensing of Vegetation Liquid Water from Space. *Remote Sens. Environ.* **1996**, *58*, 257–266. [CrossRef]
61. Birth, G.S.; Mcvey, G.R. Measuring the Color of Growing Turf with a Reflectance Spectrophotometer Is Grown Primarily. *Agron. J.* **1968**, *60*, 2–5. [CrossRef]
62. Delegido, J.; Verrelst, J.; Alonso, L.; Moreno, J. Evaluation of Sentinel-2 Red-Edge Bands for Empirical Estimation of Green LAI and Chlorophyll Content. *Sensors* **2011**, *11*, 7063–7081. [CrossRef] [PubMed]
63. Taylor, P.; Dash, J.; Curran, P.J. International Journal of Remote The MERIS Terrestrial Chlorophyll Index. *Int. J. Remote Sens.* **2014**, *25*, 5403–5413. [CrossRef]
64. Haralick, R.M.; Shanmugam, K.; Dinstein, I. Textural Features for Image Classification. *IEEE Trans. Syst. Man Cybern.* **1973**, *SMC-3*, 610–621. [CrossRef]
65. Liu, Y.; Gong, W.; Hu, X.; Gong, J. Forest Type Identification with Random Forest Using Sentinel-1A, Sentinel-2A, Multi-Temporal Landsat-8 and DEM Data. *Remote Sens.* **2018**, *10*, 946. [CrossRef]
66. Wu, C.; Shen, H.; Shen, A.; Deng, J.; Gan, M.; Zhu, J.; Xu, H.; Wang, K. Comparison of Machine-Learning Methods for above-Ground Biomass Estimation Based on Landsat Imagery. *J. Appl. Remote Sens.* **2016**, *10*, 035010. [CrossRef]
67. Zhang, Y.; Ma, J.; Liang, S.; Li, X.; Li, M. An Evaluation of Eight Machine Learning Regression Algorithms for Forest Aboveground Biomass Estimation from Multiple Satellite Data Products. *Remote Sens.* **2020**, *12*, 4015. [CrossRef]
68. Liaw, A.; Wiener, M. Classification and Regression by RandomForest. *R News* **2002**, *2*, 18–22.
69. Elith, J.; Leathwick, J.R.; Hastie, T. A Working Guide to Boosted Regression Trees. *J. Anim. Ecol.* **2008**, *77*, 802–813. [CrossRef] [PubMed]
70. Hengl, T.; Heuvelink, G.B.M.; Rossiter, D.G. About Regression-Kriging: From Equations to Case Studies. *Comput. Geosci.* **2007**, *33*, 1301–1315. [CrossRef]
71. Cressie, N. The Origins of Kriging 1. *Math. Geol.* **1990**, *22*, 239–252. [CrossRef]
72. Pandit, S.; Tsuyuki, S.; Dube, T. Estimating Above-Ground Biomass in Sub-Tropical Buffer Zone Community Forests, Nepal, Using Sentinel 2 Data. *Remote Sens.* **2018**, *10*, 601. [CrossRef]
73. Ferwerda, J.G.; Skidmore, A.K.; Mutanga, O. Nitrogen Detection with Hyperspectral Normalized Ratio Indices across Multiple Plant Species. *Int. J. Remote Sens.* **2005**, *26*, 4083–4095. [CrossRef]
74. Baloloy, A.B.; Blanco, A.C.; Candido, C.G.; Argamosa, R.J.L.; Dumalag, J.B.L.C.; Dimapilis, L.L.C.; Paringit, E.C. Estimation of mangrove forest aboveground biomass using multispectral bands, vegetation indices and biophysical variables derived from optical satellite imageries: Rapideye, planetscope and sentinel-2. *ISPRS Ann. Photogramm. Remote Sens. Spat. Inf. Sci.* **2018**, *4*, 29–36. [CrossRef]
75. Mngadi, M.; Odindi, J.; Mutanga, O. The Utility of Sentinel-2 Spectral Data in Quantifying above-Ground Carbon Stock in an Urban Reforested Landscape. *Remote Sens.* **2021**, *13*, 4281. [CrossRef]
76. Forkuor, G.; Zoungrana, J.B.B.; Dimobe, K.; Ouattara, B.; Vadrevu, K.P.; Tondoh, J.E. Above-Ground Biomass Mapping in West African Dryland Forest Using Sentinel-1 and 2 Datasets—A Case Study. *Remote Sens. Environ.* **2020**, *236*, 111496. [CrossRef]
77. Ewald, F.; Poblete-Olivares, J.; Rivero, L.; Lopatin, J.; Galleguillos, M. Using Sentinel-2 and Canopy Height Models to Derive a Landscape-Level Biomass Map Covering Multiple Vegetation Types. *Int. J. Appl. Earth Obs. Geoinf.* **2021**, *94*, 102236. [CrossRef]
78. Eckert, S. Improved Forest Biomass and Carbon Estimations Using Texture Measures from WorldView-2 Satellite Data. *Remote Sens.* **2012**, *4*, 810–829. [CrossRef]
79. Cutler, M.E.J.; Boyd, D.S.; Foody, G.M.; Vetrivel, A. Estimating Tropical Forest Biomass with a Combination of SAR Image Texture and Landsat TM Data: An Assessment of Predictions between Regions. *ISPRS J. Photogramm. Remote Sens.* **2012**, *70*, 66–77. [CrossRef]
80. Su, Y.; Guo, Q.; Xue, B.; Hu, T.; Alvarez, O.; Tao, S.; Fang, J. Spatial Distribution of Forest Aboveground Biomass in China: Estimation through Combination of Spaceborne Lidar, Optical Imagery, and Forest Inventory Data. *Remote Sens. Environ.* **2016**, *173*, 187–199. [CrossRef]
81. Yohannes, H.; Soromessa, T. Carbon Stock Analysis along Slope and Slope Aspect Gradient in Gedo Forest: Implications for Climate Change Mitigation. *J. Earth Sci. Clim. Chang.* **2015**, *6*, 6–11. [CrossRef]
82. Chen, L.; Ren, C.; Zhang, B.; Wang, Z.; Xi, Y. Estimation of Forest Above-Ground Biomass by Geographically Weighted Regression and Machine Learning with Sentinel Imagery. *Forests* **2018**, *9*, 582. [CrossRef]

83. Tun, K.; Stefano, J.; Volkova, L. Forest Management Influences Aboveground Carbon and Tree Species Diversity in Myanmar's Mixed Deciduous Forests. *Forests* **2016**, *7*, 217. [CrossRef]
84. Petersen, K.; Varela, J.B. INDC ANALYSIS: An Overview of the Forestry Sector. In *INDC ANALYSIS: An Overview of the Forestry Sector*; World Wide Fund for Nature: Gland, Switzerland, 2015; pp. 1–9.

remote sensing

MDPI

Article

Dynamic Simulation of Land Use/Cover Change and Assessment of Forest Ecosystem Carbon Storage under Climate Change Scenarios in Guangdong Province, China

Lei Tian [1,2], Yu Tao [3], Wenxue Fu [2], Tao Li [1], Fang Ren [1] and Mingyang Li [1,*]

[1] College of Forestry, Nanjing Forestry University, Nanjing 210037, China; tianlei@njfu.edu.cn (L.T.); litao3014@njfu.edu.cn (T.L.); renfang@njfu.edu.cn (F.R.)

[2] Key Laboratory of Digital Earth Science, Aerospace Information Research Institute, Chinese Academy of Sciences, Beijing 100094, China; fuwx@aircas.ac.cn

[3] Anhui Province Key Laboratory of Physical Geographical Environment, Chuzhou 239000, China; 2014210363@mail.chzu.edu.cn

* Correspondence: lmy196727@njfu.edu.cn

Abstract: Exploring the spatial distribution of land use/cover change (LUCC) and ecosystem carbon storage under future climate change scenarios can provide the scientific basis for optimizing land resource redistribution and formulating policies for sustainable socioeconomic development. We proposed a framework that integrates the patch-generating land use simulation (PLUS) model and integrated valuation of ecosystem services and tradeoffs (InVEST) model to assess the spatiotemporal dynamic changes in LUCC and ecosystem carbon storage in Guangdong based on shared socioeconomic pathways and representative concentration pathways (SSP-RCP) scenarios provided by the Coupled Model Intercomparison Project 6 (CMIP6). The future simulation results showed that the distribution patterns of LUCC were similar under SSP126 and SSP245 scenarios, but the artificial surface expanded more rapidly, and the increase in forest land slowed down under the SPP245 scenario. Conversely, under the SSP585 scenario, the sharply expanded artificial surface resulted in a continuous decrease in forest land. Under the three scenarios, population, elevation, temperature, and distance to water were the highest contributing driving factors for the growth of cultivated land, forest land, grassland, and artificial surface, respectively. By 2060, the carbon storage in terrestrial ecosystems increased from 240.89 Tg in 2020 to 247.16 Tg and 243.54 Tg under SSP126 and SSP245 scenarios, respectively, of which forest ecosystem carbon storage increased by 17.65 Tg and 15.34 Tg, respectively; while it decreased to 226.54 Tg under the SSP585 scenario, and the decreased carbon storage due to forest destruction accounted for 81.05% of the total decreased carbon storage. Overall, an important recommendation from this study is that ecosystem carbon storage can be increased by controlling population and economic growth, and balancing urban expansion and ecological conservation, as well as increasing forest land area.

Keywords: carbon storage; climate change; land use/cover change; scenario simulation; PLUS model; InVEST model

Citation: Tian, L.; Tao, Y.; Fu, W.; Li, T.; Ren, F.; Li, M. Dynamic Simulation of Land Use/Cover Change and Assessment of Forest Ecosystem Carbon Storage under Climate Change Scenarios in Guangdong Province, China. *Remote Sens.* **2022**, *14*, 2330. https://doi.org/10.3390/rs14102330

Academic Editors: Huaqiang Du, Wenyi Fan, Mingshi Li, Weiliang Fan and Fangjie Mao

Received: 13 April 2022
Accepted: 9 May 2022
Published: 11 May 2022

1. Introduction

Global climate change, caused by emissions of greenhouse gases (GHG) such as carbon dioxide (CO_2) [1,2], has greatly affected ecosystems processes and patterns [3,4], with unpredictable implications on global ecology, human survival, and economic development, and has become one of the major challenges facing all of humanity [5–7]. With the accelerated pace of industrialization, the economic development driving force is gradually shifting from agriculture to industry and services, and urbanization levels are increasing, resulting in dramatic changes in land use/cover change (LUCC), which not only has a significant impact on terrestrial ecosystems functions, but also directly affects the carbon storage of terrestrial ecosystems [8–10]. Changes in carbon storage in terrestrial ecosystems have far-reaching implications for the global ecosystem's carbon cycle, concentration of CO_2 in the atmosphere, and global climate change [11]. At present, one of the most eco-friendly and efficient energy conservation ways to mitigate climate change and the greenhouse effect is to increase carbon storage in terrestrial ecosystems, as it would reduce the amount of CO_2 in the atmosphere and contribute a significant role in mitigating global warming [12,13]. Forests, as the main body of terrestrial ecosystems, contain the highest carbon storage of terrestrial ecosystems, which not only regulate the global carbon balance, improve and maintain the regional ecological environment [14,15], but also the change of forest ecosystem carbon storage largely affects the change of carbon storage in terrestrial ecosystems [16]. However, forest degradation and deforestation caused by human activities and climate change pose a significant challenge to sustainable development. In 2015, "Transforming Our World: The 2030 Agenda for Sustainable Development" proposed 17 sustainable development goals (SDGs) that aim to address the three dimensions of development—social, economic, and environmental—in an integrated manner. Among them, SDG 15: life on land, aims to protect and restore terrestrial ecosystems such as forest, wetland, dryland and mountain ecosystems, and to promote sustainable management of forests and halt deforestation, which contributes to increasing carbon storage in forest ecosystems and mitigating climate change [17].

As the world's largest developing country, China has experienced unprecedented urbanization and significant landscape change over the past several decades [18], with the urbanization rate increasing dramatically from 17.92% to 59.58% [19]. Rapid economic development and intensive land exploitation have resulted in a steady decrease in carbon storage of terrestrial ecosystems, which also further exacerbates climate warming [20,21], especially in reform and opening-up frontier provinces like Guangdong. According to the China forest resources report, the per capita forest coverage in Guangdong province is only 0.15 ha per person. Moreover, the area of arboreal forest in Guangdong province in 2018 was 7,809,800 ha, with a large proportion of young forests, which reached 51.29%. With the proposed goal of carbon neutrality in China, improving the carbon storage and carbon sequestration capacity of terrestrial ecosystems has become a topical issue for research from various disciplines. Indeed, as early in 1999, China has launched the Grain for Green Program (GCP) and aims to increase the forest cover and mitigate soil erosion by converting cultivated land to forest land [22,23]. Over the past two decades, China's GCP has contributed more than 4% of the global net increase in green area, with forests contributing 42% of the green area [24]. Therefore, accurate assessment of future changes in LUCC and terrestrial ecosystem carbon storage, especially forest ecosystems, is essential for optimizing regional ecosystems' service functions and formulating policies for sustainable socioeconomic development [25,26].

Previous studies have shown that LUCC, which affects the carbon storage of ecosystems, is influenced by a combination of climate change and socioeconomic development [27,28]. The latest Coupled Model Intercomparison Project 6 (CMIP6) has shown that by coupling shared socioeconomic pathways (SSP) and representative concentration pathways (RCP), it can provide multiple future global climate change scenarios for researchers [29,30], which could be used to predict future LUCC, changes in carbon storage, and dynamic distribution of ecosystems services, etc. For example, one study used a scenario-based land use change assessment framework to simulate the land use demand and spatial distribution of land use in China [31]. Wang et al. [28] have integrated the system dynamics (SD) model, patch-generating land use simulation (PLUS) model, and integrated valuation of ecosystem service and tradeoffs (InVEST) model into a framework to simulate the dynamic distribution of LUCC and carbon storage at the urban level. Another study predicted global soil erosion rates and assessed future global soil regulating services for the period of 2015–2070 under three SSP-RCP scenarios [32]. Furthermore, Li et al. [33] have simulated the spatial and temporal distribution of land use in Central Asia under the SSP-RCP scenarios based on future land use demand, and comprehensively evaluated the level of ecosystems services in the region. However, most of these studies have focused on the dynamics of LUCC at the global, national, or city scale, and these methods may not be effective for the effects of environmental variables on LUCC at regional scales, including population, economic, and climate variables. Thus, there is still a significant uncertainty in the assessment of future LUCC and ecosystem carbon storage changes at the regional scale.

Current land use simulation models—such as the CA (cellular automata)–Markov model [34], ANN-CA model [35], CLUS-S model [36,37], and FLUS model [33,38,39]—were widely applied to simulate the spatial distribution of LUCC. However, these models have certain limitations, i.e., not permitting the simulation of multiple land use types, particularly natural land use types, in a dynamic spatiotemporal manner, neither can effectively identify the factors affecting LUCC, which limits the application of LUCC simulations under future climate change scenarios. The recently developed PLUS model retains the advantages of adaptive inertial competition and roulette wheel competition mechanisms of the CA model, and can combine future predicted variables to calculate the development potential of each land use type by random forest (RF) algorithms, so that it can more accurately simulate changes of land use distribution [40]. Furthermore, the InVEST model was widely used to investigate the impact of dynamic distribution of LUCC on carbon storage in terrestrial ecosystems (including forest ecosystems) due to its simple input parameters, high generality and stability, and high confidence [28,34,39]. However, previous studies assumed that the forest carbon density does not change with time and is a constant [33,34,41], which is obviously not consistent with objective facts [42], and affects the accuracy of model predictions of forest ecosystem carbon storage. Therefore, it is essential to obtain accurate estimations of the values of future forest carbon density and use them as input parameters of the InVEST mode, as this could improve the accuracy of forest ecosystem carbon storage estimation. Indeed, the combination of the PLUS model and the InVEST model could more accurately estimate the changes of terrestrial ecosystem carbon storage caused by LUCC.

In this study, we used an integrated simulation framework of the PLUS model and InVEST model to simulate the spatiotemporal distribution patterns of LUCC in the study area based on future population, economy, climate variables, and land use demand under three SSP-RCP (SSP126, SSP245, and SSP585) scenarios, and quantitatively assessed the distribution changes of carbon storage. In particular, we aimed to: (1) simulate the spatial distribution of LUCC in Guangdong province during the period of 2020–2060 based on the PLUS model; (2) analyze the impact of each driving factor on LUCC distribution; and (3) assess the spatiotemporal distribution patterns of ecosystem carbon storage in the study area under different climate change scenarios. Overall, the results of this work provide a new insight that could provide policy makers with recommendations for future land resource reallocation and socioeconomic development policies in the study area, and to provide data to support increasing forest carbon sequestration and meeting carbon neutrality goals.

2. Materials and Methods

2.1. Study Area

The study area was Guangdong province, which is located in the southeast coastal areas in China, ranging from 20°13′N–25°31′N and 109°39′E–117°19′E, with a total area of 179,725 km^2 (Figure 1). The elevation of Guangdong province is high in the north and low in the south, and the elevation decreases gradually from the mountains in northern Guangdong to the coastal areas in the south, showing a geomorphic feature with mountains in the north [43], hills in the middle, and mainly plains in the south. Over the past four decades, the forest area of Guangdong province has increased from 59,840 km^2 in 1980 to 105,241 km^2 in 2020, with an annual growth rate of 1.90%, and the forest coverage rate was 58.66% [44,45]. In addition, according to the China forest resources report (2014–2018), the national forest coverage rate is 22.96% and the forest area is 2.2 million km^2. Guangdong province ranks eighth in terms of forest coverage, with Fujian province and Jiangxi province ranking the top two [46]. In 2018, the area that can be afforested in Guangdong province was 12,042 km^2. If all the afforestable areas in Guangdong province are afforested artificially, the maximum forest coverage in Guangdong province could reach 65.26%, which could increase nearly 6% from 2020. In contrast, by 2020, the cultivated land area of Guangdong province was 25,941 km^2, which decreased by 15,320 km^2 compared with 1980 [44]. The soil types in Guangdong province include limestone soils, purplish soils, fluvo-aquic soils, humid-thermo ferrditic, lateritic red earths, red earths, and yellow earths, etc. [47]. As China's largest economic province, Guangdong province has a resident population of 126 million in 2020 and regional gross domestic product (GDP) reached 11.07 trillion RMB, up 2.3% from the same period last year [44]. In general, carbon emissions strengthen as GDP rises, the huge population and GDP may represent huge per capita carbon emissions [48].

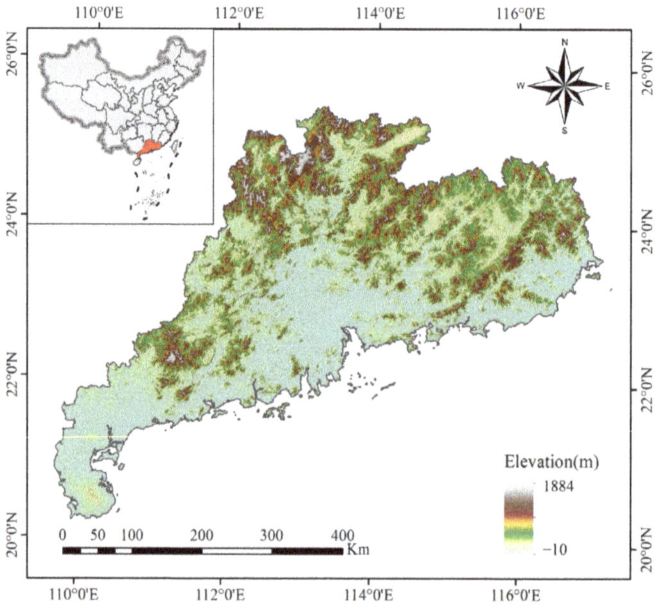

Figure 1. Location of Guangdong province together with the DEM.

Influenced by the southeast and southwest monsoon, the climate of Guangdong province from north to south is central subtropical, southern subtropical, and tropical climates, respectively [42]. The annual average temperature of Guangdong province is 22.3 °C. The average temperature is approximately 16 °C to 19 °C in January and 28 °C to 29 °C in July. The average annual precipitation in Guangdong ranges from 1300–2500 mm, with a provincial average of 1777 mm. The spatial distribution of rainfall basically also shows a tendency toward low precipitation in the north and high precipitation in the south. Adequate water and heat conditions have contributed to a wide variety of vegetation and vegetation communities in Guangdong province, including northern tropical seasonal rainforest, subtropical monsoon evergreen broadleaf forest, typical evergreen broadleaf forest in middle subtropics, coastal tropical mangroves, shrublands and grasslands, etc. [42].

To meet the goal of peaking carbon emissions and carbon neutrality, Guangdong province has designated the development goals and targets of the 14th Five-Year Plan: to build a model area for the convergence of rules, a concentration area for upscale elements, a source of scientific and technological industrial innovation, a linkage area for internal and external circulation, and a support area for security development, and to take the lead in exploring the effective paths conducive to the formation of a new development pattern. In indeed, steady increase in carbon storage in terrestrial ecosystems is one of the effective ways to reach the goal of carbon neutrality [28].

2.2. Data Acquisition and Preprocessing

The data for this work include LUCC data, socioeconomic data, and meteorological data. The data sources for the spatial data used in this study are shown in Table 1. Specifically, the 2000, 2010, and 2020 LUCC data were obtained from the GLOBELAND30 dataset (30 m spatial resolution) produced by the National Geomatics Center of China (http://www.globallandcover.com, accessed on 27 December 2021). We obtained GDP, population density, and soil types data (all with 1-km spatial resolution) from the Data Center for Resources and Environmental Sciences of the Chinese Academy of Sciences (https://www.resdc.cn, accessed on 28 December 2021). A digital elevation model (DEM) data (at 30 m spatial resolution) was obtained from the ASTER GDEM 30 M dataset of the Geospatial Data Cloud (http://www.gscloud.cn, accessed on 27 December 2021). The slope data was obtained by processing the DEM data using ArcGIS 10.7 software.

Table 1. Spatial driving factors of the land use change in this study.

Category	Data	Year [1]	Original Resolution	Data Resource
Land use/cover data	Land use/cover data	2000, 2010, 2020	30 m	GLOBELAND30 dataset
Socioeconomic driver	Population GDP	2015 2015	1000 m 1000 m	https://www.resdc.cn, accessed on 28 December 2021
	Distance to governments Distance to train stations	2020	30 m	https://lbs.amap.com, accessed on 27 December 2021
	Distance to highways Distance to primary roads Distance to secondary roads Distance to tertiary roads Distance to trunk roads	2020	30 m	OpenStreetMap (https://www.openstreetmap.org, accessed on 27 December 2021)
	Distance to settlements	2018	30 m	https://www.webmap.cn, accessed on 1 March 2022
Climatic and environmental driver	Distance to water	2020	30 m	Land use/cover in 2020
	DEM Slope	2009	30 m	ASTER GDEM 30 M dataset
	Soil types	1995	30 m	https://www.resdc.cn, accessed on 28 December 2021
	Average annual temperature Average annual precipitation	2000–2020	1000 m	http://www.geodata.cn, accessed on 27 December 2021

[1] The driving factors collected were allowed to be inconsistent with the time period of the land use data [49], but the time period was as close as possible to the time period of the LUCC data.

In addition, current roads vector data were obtained from the OpenStreetMap (https://www.openstreetmap.org, accessed on 27 December 2021). The location data of all levels of governments and train stations were obtained from the lbs.amp.com (https://lbs.amap.com, accessed on 27 December 2021). The settlement data were obtained from the National Catalogue Service for Geographic Information (https://www.webmap.cn, accessed on 1 March 2022). Temperature and precipitation data (both 1 km spatial resolution) were obtained from the National Earth System Science Data Center (http://www.geodata.cn, accessed on 27 December 2021). After a series of data preprocessing in ArcGIS 10.7 software—including projection transformation, Euclidean distance, resampling, and clipping—all of the above data were converted to raster data with the same projected coordinate system and a spatial resolution of 30 m.

2.3. Methods

The research framework of this paper consists of two parts: the PLUS model for simulating LUCC data and the InVEST model for estimating ecosystem carbon storage (Figure 2). Specifically, we used the PLUS model to simulate the distribution of LUCC in Guangdong province from 2020 to 2060 based on population, GDP, temperature, and precipitation data under different SSP-RCP scenarios, as well as the InVEST model to assess the spatiotemporal variation of ecosystem carbon storage caused by LUCC.

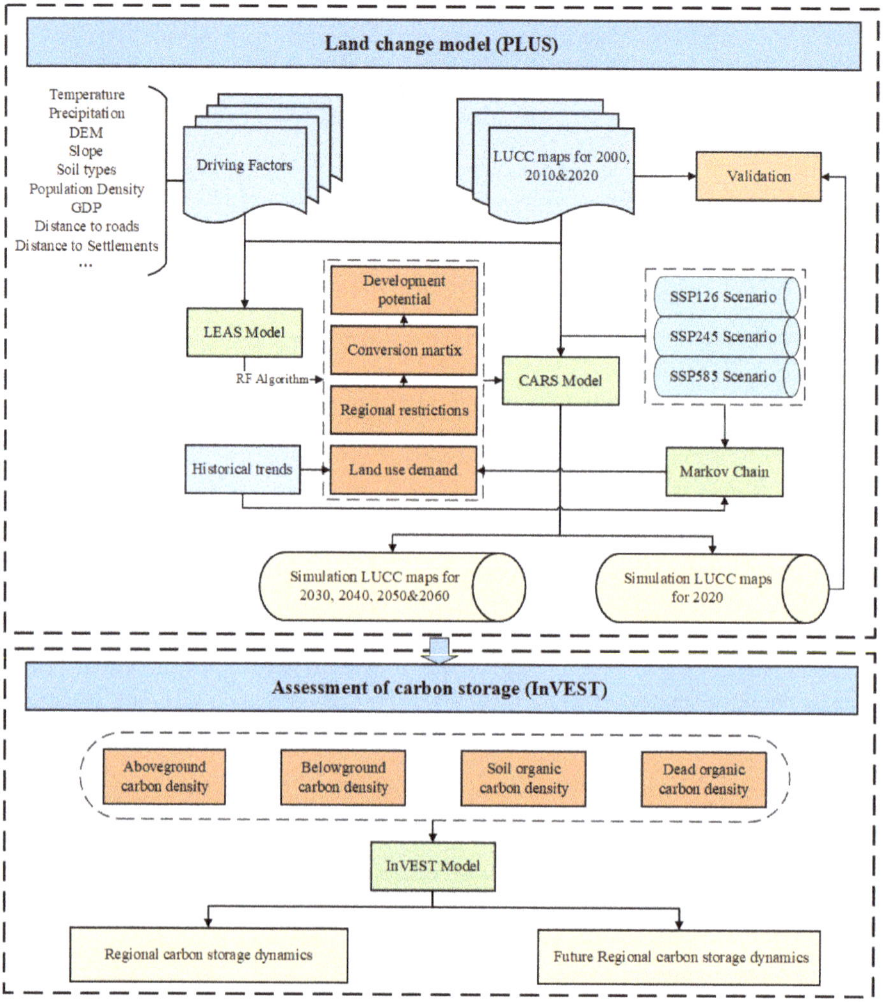

Figure 2. Research framework.

2.3.1. Future Climate Scenarios Based on the CMIP6

The Coupled Model Intercomparison Project (CMIP) has evolved over five phases into a major international multi-model climate research activity [50–52], which has not only introduced a new era in climate science research, but also facilitated national and international climate change assessments [29]. Compared to CMIP5, CMIP6 combines different SSP-RPC scenarios [53,54], which emphasizes the driving effect of different socioeconomic development patterns on climate change [30,33].

To consider a range of possible futures, we use simulations from three SSP-RCPs: SSP126 (integrated scenario of SSP1 and RCP2.6): sustainability—taking the green road, which presents sustainable socio-economic development with a low level of GHG emissions and emphasizes more inclusive development. Land use is strongly regulated, e.g., forest land is well preserved. SSP245 (integrated scenario of SSP2 and RCP4.5): middle of the road pathway, which represents the world follows a middle road of the socioeconomic and technological development, and with a medium level of GHG emissions. Land use change is incompletely regulated, i.e., forest land would still be potentially destroyed, although the probability is slowly decreasing over time. SSP585 (integrates scenario of SSP5 and RCP8.5): high-end forcing pathway, which is characterized by rapid resource-intensive development and material-intensive consumption patterns, as well as very high level of fossil fuel use and high GHG emissions [55,56].

In this work, we consider four driving factors that affect LUCC in future climate change scenarios, including population, GDP, temperature, and precipitation. The population [57] and GDP [58] data were obtained from the kilometer-scale grids data of the SSPs future climate change scenarios, respectively. Previous studies [59] have provided future temperature and precipitation data for SSP126, SSP245, and SSP585 scenarios based on the MRI-ESM2-0 model [60].

2.3.2. Simulation of LUCC under Different Scenarios Provided by CMIP6

The PLUS model is a simulation model of future land use/cover change integrated with a rule-mining framework based on a land expansion analysis strategy (LEAS) model and a CA based on multi-type random patch seeds (CARS) model [40]. At first, the LEAS model overlays the land use data from two periods, extracts the image elements with changed status from the later land use data, represents the change area of each land use type, and then uses the RF algorithm to explore the relationship between each land use type and multiple drivers to obtain the transition rules for each land use type, i.e., the development potential of each land use type. In the LEAS model, the number of regression trees refers to the number of trees generated by RF, sampling rate defaults to 0.01, indicating that 1% of the pixels will be used for model training, and mTry is the number of driving factors [40]. In this work, the number of regression trees, sampling rate, and mTry were determined to be 50, 0.01, and 16, respectively, after conducting several experiments.

Subsequently, for simulating the evolution of multiple land use types, the CARS model combines the traditional CA model with a patch generation and a descending threshold mechanism to perform future land use simulation based on the available LUCC data and the development potential of each land type. When the neighborhood effect of a single land use type is equal to zero, the mechanism generates 'seeds' to the development probability of each land use type. With the development potential restraints, PLUS will automatically generate simulated patches [40]. Previous studies have shown that the PLUS model can integrate the effects of various spatial factors with the dynamics of geographic units to simulate land use change in order to obtain higher accuracy and more realistic landscape patterns [28,61].

The demand for LUCC under different climate change scenarios (Figure 3) was estimated based on historical land use data (i.e., LUCC data for Guangdong province in 2000, 2010, and 2020) [33] and the Markov chains method [62,63], and used it as the future land use demand input parameter for the PLUS model. Historical data for 2020 were used to evaluate the accuracy of the land use demand. In addition, 16 types of factors affecting LUCC (Figure 4) as the predictor variables (including population density, GDP, distance to government, distance to settlements, distance to water, distance to train station, distance to highways, distance to other roads, DEM, slope, soil type, temperature, and precipitation) input into the RF model to determine the development potential of each land use type. We then obtained the simulation results of LUCC in 2020 by running the PLUS model based on 2000 and 2010 LUCC data and the above 16 driving factors, and compared it with the actual 2020 LUCC data (Figure 5) for assessing the accuracy of the model. The overall accuracy and Kappa coefficient were used to assess the simulation accuracy of the PLUS model. If the accuracy of the simulation results is sufficient, the driving factors and the land use demand of Guangdong province from 2020 to 2060 (at 10-year intervals) under different scenarios are input into the PLUS model to predict the spatiotemporal changes of future land use distribution based on the LUCC data in 2020.

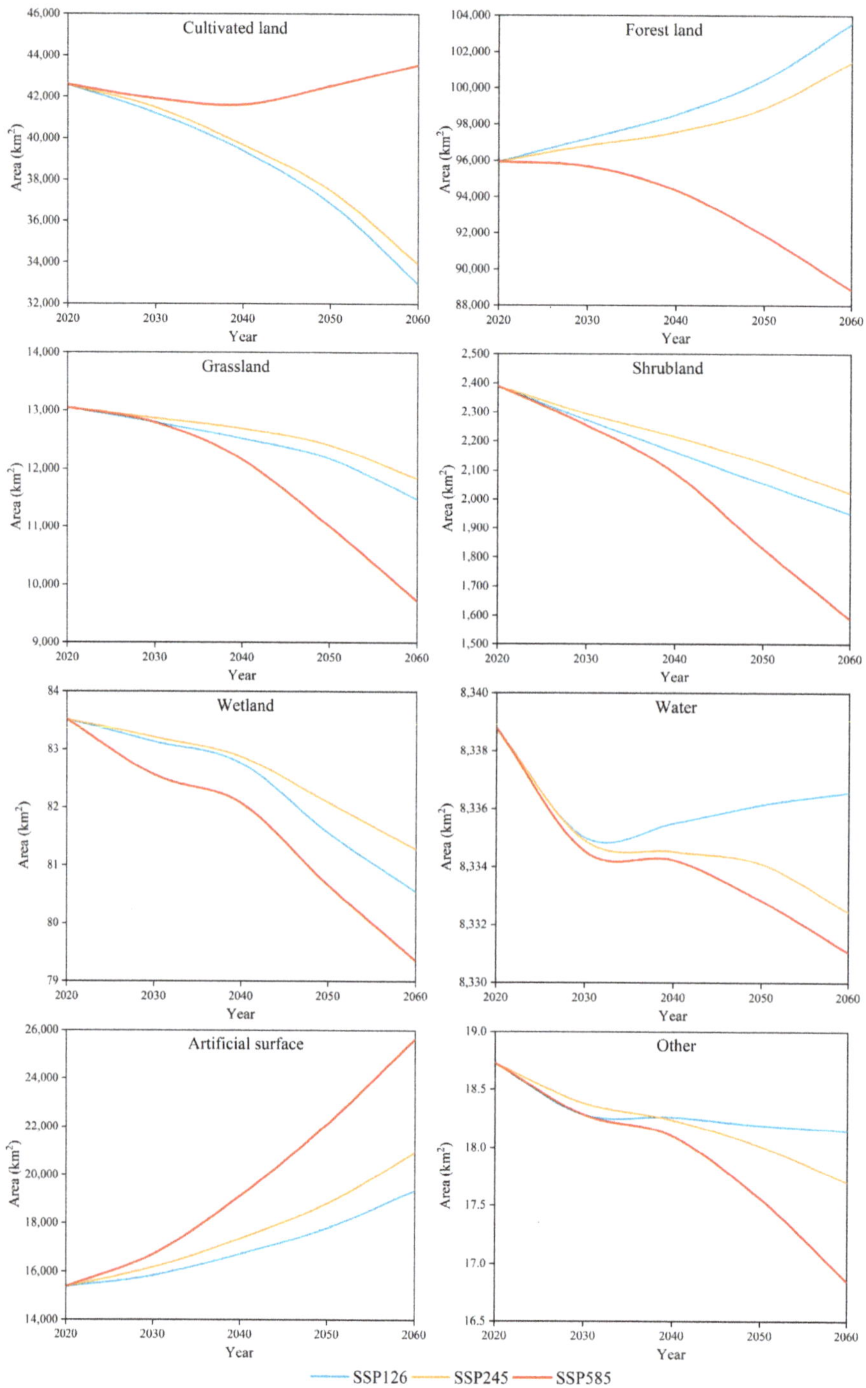

Figure 3. Demand prediction of each land use type under different scenarios.

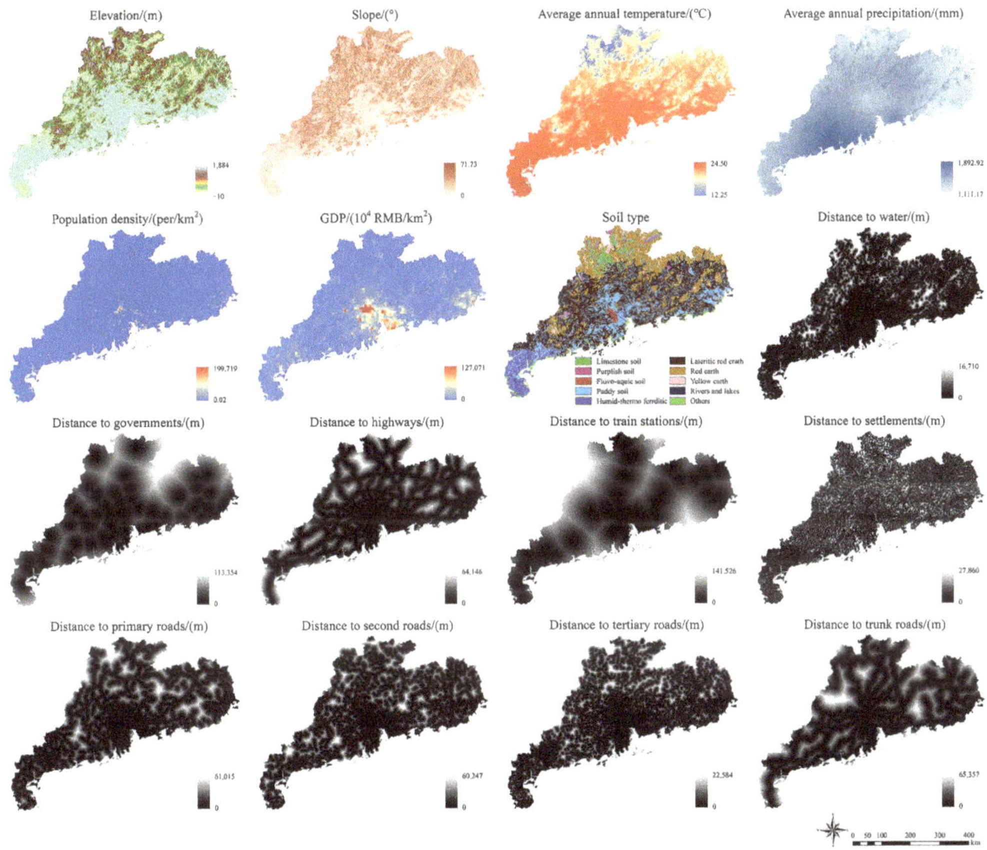

Figure 4. Sixteen types of driving factors affecting LUCC.

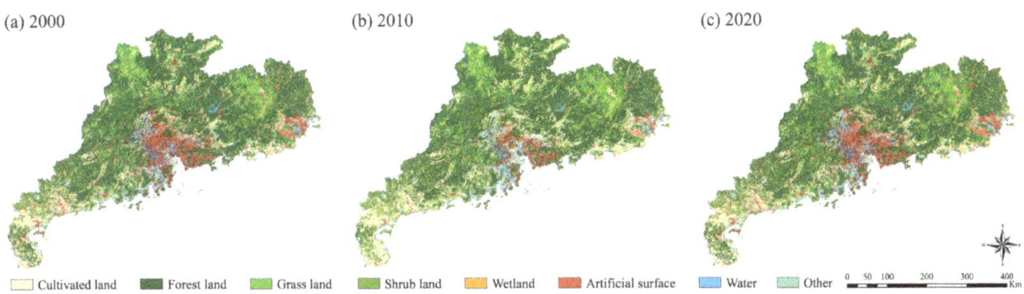

Figure 5. Historic LUCC data for 2000, 2010, and 2020.

2.3.3. Estimation of Carbon Storage Based on the InVEST Model

The Carbon Storage and Sequestration module of the InVEST model can spatially integrate land use change and terrestrial ecosystem carbon storage dynamics directly, making it possible to assess the impact of past to present land use change on terrestrial ecosystem carbon storage in the study area as well, as to simulate changes in terrestrial ecosystem carbon storage under future land use change scenarios [37,64]. Specifically, the InVEST model based on the average carbon density of four carbon pools (aboveground, belowground, soil, and dead organic matter) for each land use/cover type, and multiplied by their corresponding area to calculate ecosystem carbon storage [34]. The calculation formulas for carbon storage are

$$C_i = C_{above} + C_{below} + C_{soil} + C_{dead} \quad (1)$$

$$C_{total} = \sum_{i=1}^{n} C_i \times A_i, \ (i = 1, 2, \cdots, n) \quad (2)$$

where i represents the land use type, C_i represent the total carbon storage per unit area of each land cover type (kg/m^2), C_{above} is the aboveground carbon density, C_{below} is the belowground carbon density, C_{soil} is the soil organic carbon density, and C_{dead} is the dead organic carbon density. C_{total} is the total carbon storage of the ecosystems and A_i is the area of each land cover type. We obtained carbon density data for the four carbon pools of different land use types were obtained from previous studies (Table 2) [42,45,65], where C_{soil} refers to the soil organic carbon density at 1 m depth. Notably, the forest carbon density data are not constant, and we obtained a growth rate of 1.96% per decade for forest carbon density (including aboveground and belowground carbon density) in Guangdong province based on previous studies [42,45]. In addition, we assessed the economic value of sequestering a ton of carbon (1284.63 RMB, derive from social cost of CO_2 = 349.88 RMB), assuming the annual rate of change in the price of carbon to be zero and the market discount rate of 3% [41].

Table 2. Carbon densities of each land ues type (2020) used in InVEST model (kg/m^2).

Land Use Types	C_{above}	C_{below}	C_{soil}	C_{dead}	Sources
Cultivated land	1.45	0.10	7.95	0.10	[45,65]
Forest land	2.28	0.83	15.84	0.65	[42,45,65]
Grassland	0.11	0.52	6.28	0.19	[45,65]
Shrubland	0.31	0.20	8.14	0.70	[65]
Wetland	0	0	8.19	0	[65]
Water	0	0	0	0	/
Artificial surface [1]	0	0	0	0	/
Other	0.02	0	5.80	0	[65]

[1] Notably, the artificial surface mainly includes artificial infrastructure such as buildings, impervious surfaces and infrastructure, and cultivated land is not part of the artificial surface.

3. Results

3.1. Simulation of LUCC under Different Scenarios and Accuracy Assessment

The land use status in 2020 was simulated based on the LUCC data in 2000 and 2010 using the PLUS model, and the simulation results were compared with the actual LUCC data in 2020. The assessment results show that the overall accuracy of the PLUS model was 93.34%, and the Kappa coefficient was 0.89, which indicates that the PLUS model has a high simulation accuracy and could be reliably applied to predict future LUCC [33].

Subsequently, the spatial and temporal distribution of LUCC in 2030, 2040, 2050, and 2060 (Figure 6) under different climate change scenarios was simulated using the PLUS model based on the land use demand and LUCC data in 2010 and 2020, and the statistics for each type of land use are shown in Table 3. The results indicated that the distribution of LUCC data showed a significant difference under different climate change scenarios. Specifically, under the SSP126 scenario, cultivated land, grassland, and shrubland showed different degrees of decrease. In contrast, the artificial surface area was rapidly increasing, encroaching on the previous cultivated and grassland areas. The forest land was effectively preserved, the area increasing from 95,939.51 km^2 in 2020 to 103,583.88 km^2 in 2060, with a growth rate of 1.84% per decade. In addition, wetland areas have slowly decreased, while water and other land types remained essentially unchanged.

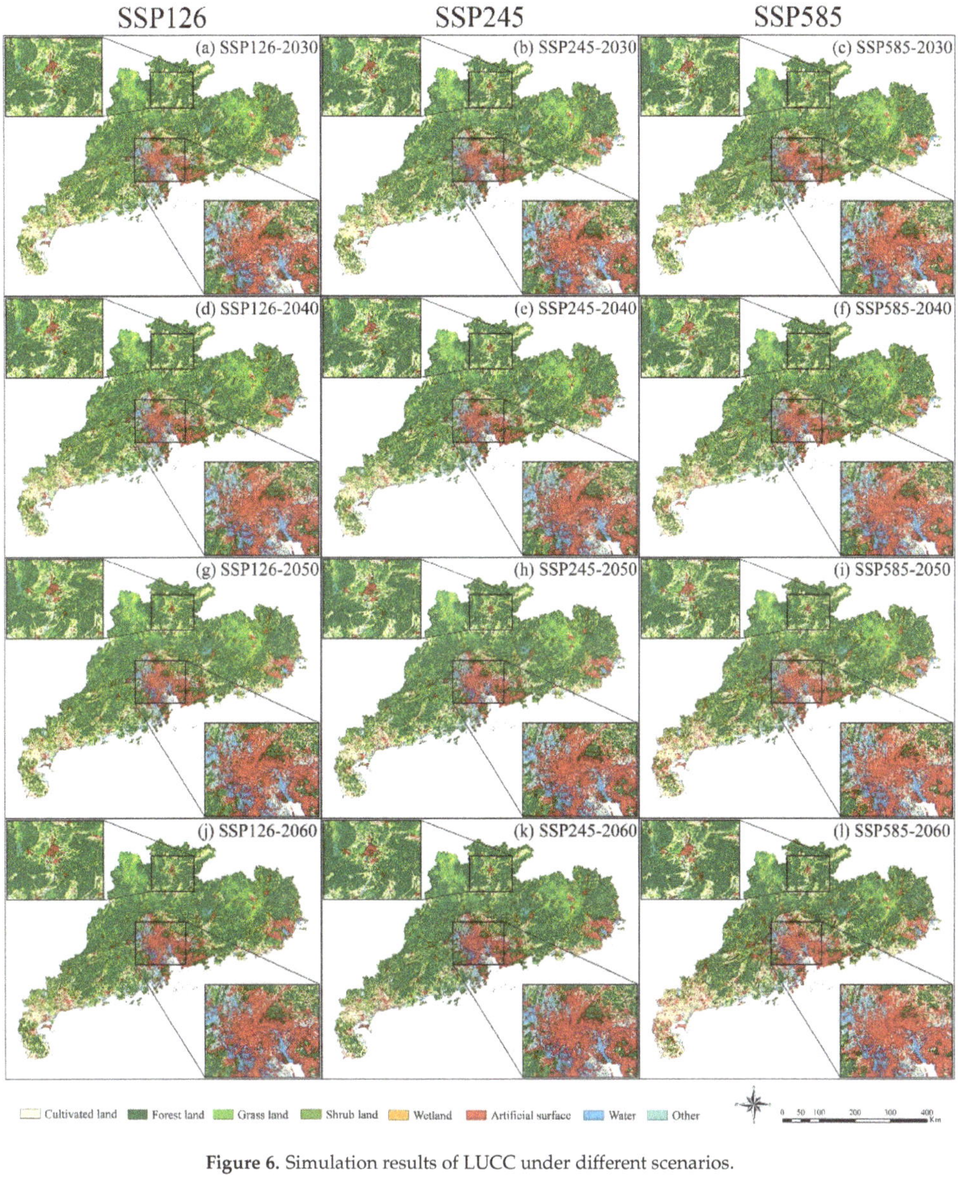

Figure 6. Simulation results of LUCC under different scenarios.

Table 3. Statistics of each land use type under different scenarios (km^2).

Land Use Types	SSP126				SSP245				SSP585			
	2030	2040	2050	2060	2030	2040	2050	2060	2030	2040	2050	2060
Cultivated land	41,209.63	39,389.27	36,854.29	32,964.14	41,478.84	39,679.46	37,193.65	33,125.88	41,907.50	41,609.79	42,518.38	43,550.52
Forest land	97,211.42	98,536.66	100,459.61	103,583.88	96,815.41	97,979.54	99,601.01	102,423.78	95,670.26	94,364.20	91,876.67	88,866.33
Grassland	12,796.31	12,515.64	12,174.49	11,480.47	12,867.80	12,692.54	12,404.31	11,835.97	12,792.12	12,155.74	11,004.67	9717.45
Shrubland	2272.03	2162.64	2054.88	1949.88	2294.53	2215.77	2126.88	2021.88	2254.03	2090.64	1829.88	1588.16
Wetland	83.14	82.76	81.56	79.75	83.21	82.87	82.09	81.28	82.56	82.07	80.66	79.36
Water	8336.35	8334.64	8334.64	8334.64	8336.56	8334.64	8334.64	8334.64	8334.63	8334.63	8334.63	8334.64
Artificial surface	15,851.87	16,739.16	17,801.37	19,368.14	15,884.30	16,775.99	18,018.44	19,937.72	16,719.66	19,123.86	22,116.58	25,625.74
Other	18.28	18.26	18.19	18.15	18.38	18.24	18.01	17.90	18.28	18.10	17.56	16.85

Under the SSP245 scenario, the expansion patterns of cultivated land, grassland, and shrubland were similar to the SSP126 scenario, but with a slower decreasing tendency than the SSP126 scenario. Forest land was also well preserved; however, its growth rate has also slowed to only 1.58% per decade. The slightly accelerated artificial surface expansion, and water bodies and other land types generally similar to the SSP126 scenario.

In contrast to the other scenarios, the area of cultivated land changes from a decreasing tendency in the period of 2020–2040 to an increasing tendency in the period of 2040–2060 under the SSP585 scenario. The area of grassland and shrubland showed a sharper decreasing tendency. As a result of the rapid expansion of the artificial surface, forest land was ineffectively preserved and its area shows a decreasing tendency with decrease of 8898.22 km^2 by 2060. Under the SSP585 scenario, the rapidest expansion of the artificial surface is observed in Guangdong province, where the artificial surface area expands nearly 1.7 times in 2060 compared to 2020.

3.2. Spatiotemporal Patterns of Carbon Storage

3.2.1. Spatiotemporal Variation of Carbon Storage in Terrestrial Ecosystems

Changes in terrestrial ecosystem carbon storage caused by LUCC under different scenarios from 2020 to 2060 in Guangdong province were assessed using the InVEST model (Figure 7). Significant differences in carbon storage under different scenarios (Table 4). Under SSP126 and SSP245 scenarios, carbon storage continuously increases positively and maintains a continuous tendency to increase. The carbon storage increases from 240.89 Tg in 2020 to 247.16 Tg (SSP126) and 245.33 Tg (SSP245) in 2060, with an increase of 6.27 Tg and 4.44 Tg, respectively. Compared to the SSP126 scenario, the increase in carbon storage is slightly lower under the SSP245 scenario. While the carbon storage shows a negative increase and continuously decreases under the SSP585 scenario, which decreases from 240.89 Tg in 2020 to 226.54 Tg in 2060, with a total decrease of 14.35 Tg.

As illustrated in Figure 7, under the SSP126 scenario, the area of carbon storage increase is mainly located in northern and western Guangdong, where the forest land area maintains growth. The area of carbon storage decrease is mainly the artificial surface expansion area, where cultivated land and grassland are destroyed. The carbon storage changes under the SSP245 scenario are similar to the SSP126 scenario, with a slightly smaller increase in carbon storage under the SSP245 scenario, which is caused by the smaller area of forest land growth that mainly influences carbon storage changes in terrestrial ecosystems under the SSP245 scenario. In contrast, the decreased area of carbon storage obviously increases and was significantly larger than the increased area of carbon storage under the SSP585 scenario, and the decreased area was mainly distributed in the area of artificial surface expansion and forest land reduction.

Moreover, the economic value of carbon sequestration in terrestrial ecosystems for the different scenarios is shown in Figure 8, with units of monetary value per grid cell (RMB). The positive values indicate that carbon is being sequestered, and negative values indicate that carbon is lost to the atmosphere. According to the economic view of the Kyoto Protocol, forest owners should realize revenue while reducing carbon emissions [41]. In this study, future and current carbon sequestration are treated equally, and the discount rate and the social value of sequestered carbon are assumed to be constant, which contributes to obtain the net present value (NPV) of sequestered carbon in any particular year. Under the SSP126 and SSP245 scenarios, the total economic value of carbon sequestration is 8.05 billion and 5.70 billion RMB in Guangdong province during the period of 2020–2060, respectively. Under the SSP585 scenario, the economic value loss due to carbon loss would be approximately 18.43 billion RMB in Guangdong province during the period of 2020–2060. This ecosystems service function expressed as a monetary value can be effective in raising awareness of the significance of ecosystems and biodiversity, and conveying it to policy makers [41].

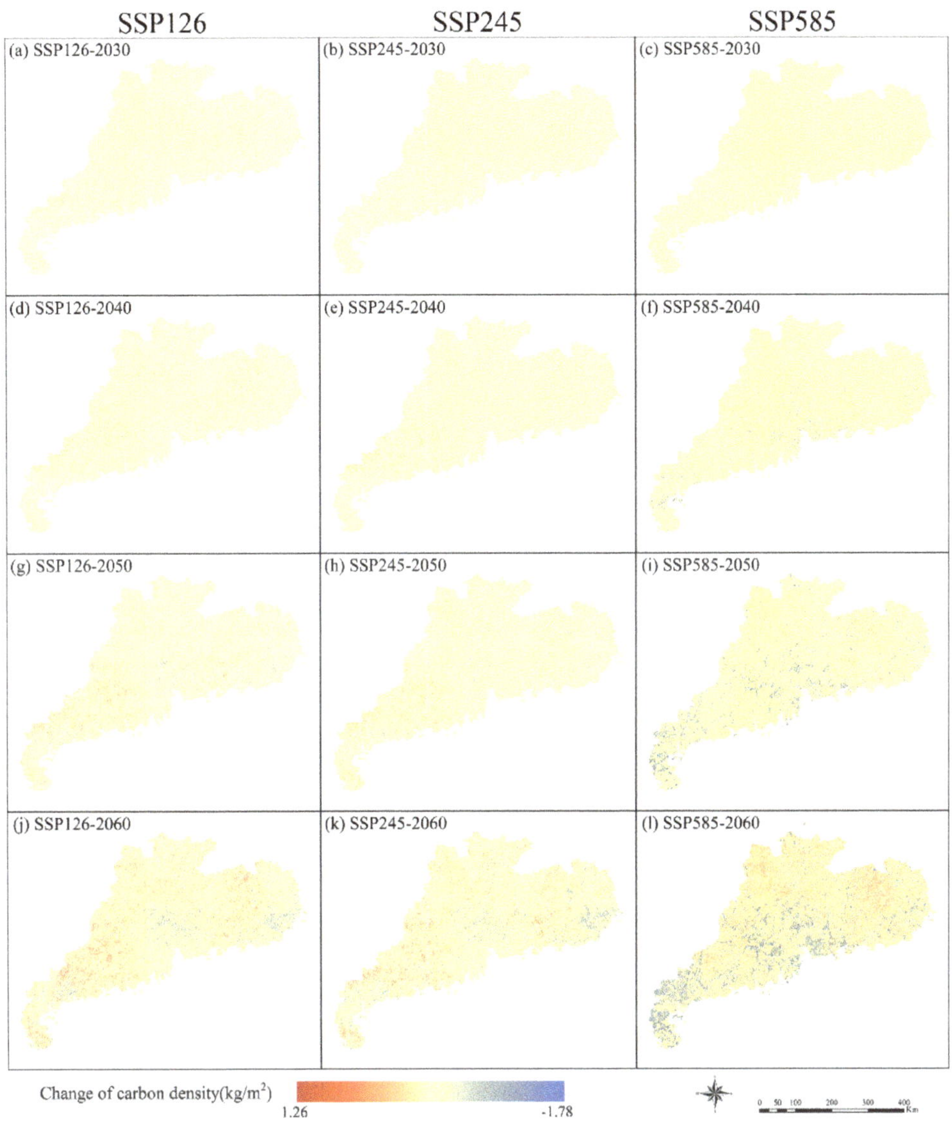

Figure 7. Distribution changes of carbon storage under each scenario compared to 2020.

Table 4. Carbon storage dynamic changes in terrestrial ecosystems under different scenarios during the period of 2020–2060.

Climate Scenarios	Total Carbon Storage (Tg)					Carbon Storage Change (Tg)				
	2020	2030	2040	2050	2060	2020–2030	2030–2040	2040–2050	2050–2060	2020–2060
SSP126	240.89	241.82	242.97	244.68	247.16	0.93	1.15	1.71	2.48	6.27
SSP245	240.89	241.37	242.32	243.54	245.33	0.48	0.95	1.22	1.79	4.44
SSP585	240.89	239.44	236.55	232.11	226.54	−1.45	−2.89	−4.44	−5.57	−14.35

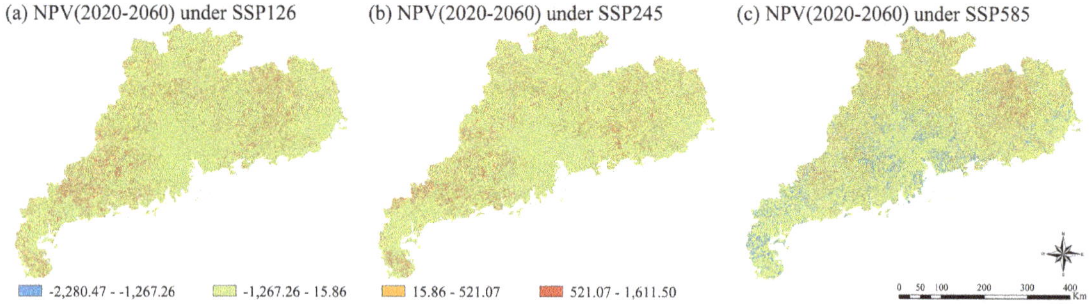

Figure 8. Net present value (unit: RMB) of Guangdong province in the period of 2020–2060 under different scenarios.

3.2.2. Spatiotemporal Variation of Carbon Storage in Forest Ecosystems

In this paper, forest ecosystem carbon storage accounts for approximately 78% of terrestrial ecosystem carbon storage. Thus, we individually assessed the changes in forest ecosystem carbon storage caused by LUCC (Table 5). The results showed that carbon storage in forest ecosystems had a similar change pattern to terrestrial ecosystems under the three scenarios, but it was more drastic than terrestrial ecosystems. By 2060, forest ecosystem carbon storage increases by 17.64 Tg and 15.34 Tg under SSP126 and SSP245 scenarios, respectively, with an annual increase of 0.44 Tg year^{-1} and 0.38 Tg year^{-1}, respectively. Under the SSP585 scenario, forest ecosystem carbon storage slightly increased and then rapidly decreased, with the total decrease of 11.64 Tg. In addition, forest ecosystem carbon storage accounts for up to 83.38% of carbon storage in terrestrial ecosystems by 2060 (SSP126 scenario). In the SSP585 scenario, the rapid expansion of the artificial surface encroached on previously forested land, grassland, wetlands, etc., which resulted in a total decrease in terrestrial ecosystem carbon storage of 14.35 Tg (Table 4), and the decreased carbon storage due to forest land destruction accounted for 81.05% of the total decreased carbon storage. Obviously, the changes in carbon storage in forest ecosystems largely determine changes in carbon storage in terrestrial ecosystems.

Table 5. Carbon storage dynamic changes in forest ecosystems under different scenarios during the period of 2020–2060.

Climate Scenarios	Total Carbon Storage (Tg)					Carbon Storage Change (Tg)				
	2020	2030	2040	2050	2060	2020–2030	2030–2040	2040–2050	2050–2060	2020–2060
SSP126	188.43	191.52	194.75	199.20	206.07	3.09	3.23	4.45	6.87	17.64
SSP245	188.43	190.74	193.65	197.50	203.77	2.31	2.91	3.84	6.27	15.34
SSP585	188.43	188.49	186.51	182.18	176.79	0.06	−1.98	−4.32	−5.39	−11.64

4. Discussion

4.1. Impact of Various Driving Factors on LUCC

In this study, we evaluated the dynamic distribution of LUCC in Guangdong province from 2020 to 2060 under three scenarios of SSP126, SSP245, and SSP585. The expansion of cultivated land, forest land, grassland, and artificial surface showed significant differences among the three scenarios. The importance ranking of the driving factors for growth of the four land use types in 2060 [40] is shown in Figure 9. The driving factors that ranked first in importance for cultivated land, forest land, grassland, and artificial surface were consistent under the three scenarios.

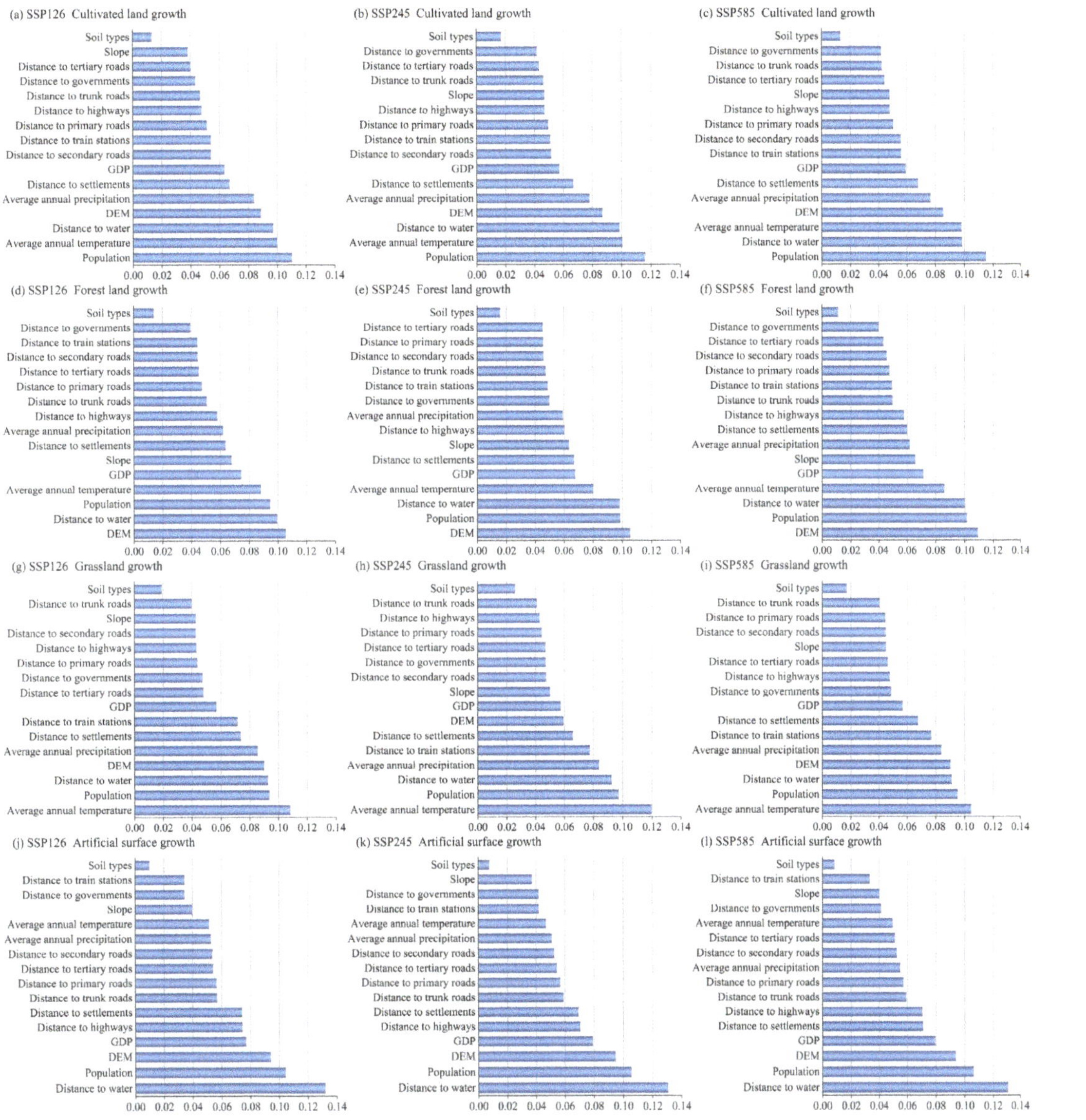

Figure 9. The importance of the contribution of each factor to the growth of four land use types.

For the cultivated land, we found that population density had the most influence on the growth of cultivated land. Population growth requires more food supply, and with a certain amount of food production, it requires additional land to supply food. Additionally, population dynamics and economic growth largely determine the future development of agricultural systems [66], including other basic socioeconomic conditions, such as technological changes in crops and livestock [67], investments in agricultural technology [68], and trade of agricultural goods [69]. Therefore, it is not difficult to understand that changes in cultivated land area are strongly influenced by population growth [70,71]. The main driving factors of forest land change are elevation, population, and distance to water. On the one hand, forest land in Guangdong province is mainly distributed in the higher altitude mountainous areas in northern and western Guangdong [42]; on the other hand, the impact of population density on forests is not negligible, the expansion of population not only needs forest land to provide more forestry products, but people need to enjoy the ecological service value attached to forest land [72]. In addition, water area is rarely converted to natural vegetation under natural factor conditions and forest land tends to expand to more ecologically healthy areas, which may explain the reason why distance to water bodies is one of the main driving factors of forest land growth [73].

The average annual temperature, population, and distance to water are the main factors influencing the growth of grassland. This indicates that grassland are more sensitive to temperature response [39], and areas strongly influenced by human activities also affect the growth of grassland [40]. The driving factors for artificial surface growth include distance to water, population density, and elevation. The water area hinders the urban expansion, which generally avoids or surrounds the water area by encroaching on cultivated land, grassland, or other land use types [28,61]. Increasing population density means that urban areas need to expand further to accommodate a greater number of people to survive. Indeed, urban expansion is generally influenced by elevation factors, as the difficulty and cost of urban construction was determined by topographical factors. In general, urban expansion avoids the large topographic undulations of mountainous areas [61]. As can be observed in Figure 6, the expanded artificial surface is mainly distributed in the areas with relatively low topographic fluctuations, which is consistent with the general pattern of urban development.

4.2. *Impact of LUCC on Carbon Storage*

This paper reveals the spatial distribution of carbon storage under different climate change scenarios during the period of 2020–2060 in Guangdong province, and the results showed that an obviously spatial heterogeneity in carbon storage changes (Figure 7). The changes in carbon storage are the result of a combination of climate change, population growth, economic development, and ecological interests. This comprehensive assessment helps us to improve our understanding of future changes in carbon storage, especially resulting from changes in LUCC.

4.2.1. Impact on Carbon Storage in Terrestrial Ecosystems

As expected, forest land, cultivated land, and shrubland accumulate more carbon storage than other land use types [39]. In our study, the highest carbon density was found in forest land, followed by cultivated land, shrubland, and grassland (Table 2). There are significant differences in the distribution of LUCC under different scenarios, which also result in the spatial heterogeneity of changes in carbon storage in terrestrial ecosystems. In general, the expansion of artificial surface and the decrease in forest area are the most significant reasons for the decrease in carbon storage in terrestrial ecosystems. The decrease in terrestrial ecosystem carbon storage due to the expansion of the artificial surface could be up to 186.45 Mg under the three scenarios. It seems profitable for urban expansion by providing more jobs and rapidly increasing GDP, but it will reduce regional ecosystem carbon storage in the long-term [74,75]. Therefore, balancing urban expansion and ecological conservation is an important measure to maintain sustainable development.

Rapid economic development and urbanization have seriously affected the quality of the regional ecosystems, resulting in the continuous degradation of forest land, grassland, and shrubland, further leading to a decline in terrestrial ecosystem carbon storage in the study area. This is consistent with previous findings that the accelerated economic development will lead to gradual ecological degradation, and further resulting in a continuous decline of carbon storage in terrestrial ecosystems [39]. Therefore, enhancing the quality of socio-economic development and promoting economic development from "high speed" to "high quality" could not only improve the value of regional ecosystems services, but also increase the carbon storage in the ecosystems [28]. In addition, rapid climate change and future socioeconomic and land use driving factor uncertainties may lead to very different

LUCC dynamic changes and consequences for changes in terrestrial ecosystem carbon storage based on LUCC [66]. Reducing the use of fossil fuels and increasing the use of clean energy for energy conversion, such as solar and wind energy resources, would mitigate the global warming effect, and prevent further degradation of forest land, grassland, and shrubland, hence maintaining the balance of carbon storage in terrestrial ecosystems.

4.2.2. Impact on Carbon Storage in Forest Ecosystems

Figure 10 shows the changes in forest ecosystem carbon storage under different future scenarios compared to 2020, with a gradual increase in forest ecosystem carbon storage under the SSP126 and SSP245 scenarios, while the forest ecosystem carbon storage increases by a minor amount in 2030 and then decreasing continuously under the SSP585 scenario. Specifically, under the SSP126 scenario, it is projected that 5897.75 km^2 of cultivated land will be converted to forest land by 2060, contributing 60.42 Mg of increased carbon storage, which is consistent with previous findings that the ecological engineering of Grain to Green could significantly increase the carbon sequestration in Chinese soil ecosystems through the conversion of cultivated land to forest land [76]. Additionally, 1522.73 km^2 of grassland and 426.26 km^2 of shrubland will be converted to forest land. Overall, the increase in carbon storage from conversion to forest land is expected to reach 84.36 Mg. Stable climatic conditions and lower socioeconomic development would encourage the expansion of forest land [77,78], and its propensity to expand towards more ecologically healthy areas [73]. Therefore, moderate urban expansion and lower GHG emissions are effective paths for increasing carbon storage in regional forest ecosystems [28].

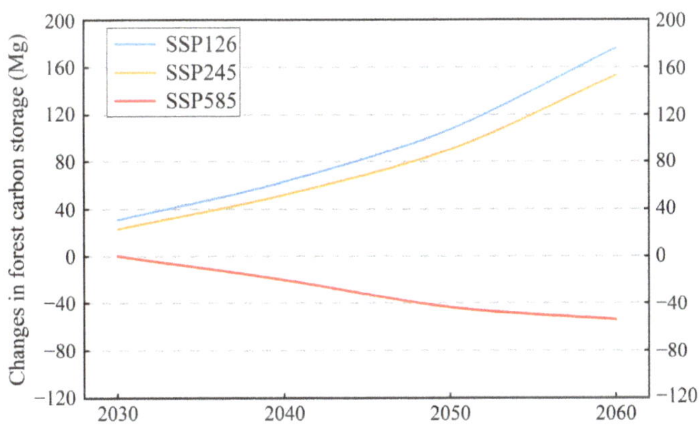

Figure 10. Changes in forest ecosystem carbon storage under different scenarios compared to 2020.

Furthermore, the pattern of forest ecosystem carbon storage change under the SSP245 scenario was roughly same with the SSP126 scenario, but its total carbon storage increase was lower than that of the SSP126 scenario. Under the SSP245 scenario, the increase in carbon storage was attributed to the conversion of cultivated land and grassland to forest land. Notably, under the SSP585 scenario, the rapidly expanding artificial surface and the continuously decreasing forest land resulted in 116.32 Mg of forest ecosystem carbon storage decrease by 2060, which is also one of the reasons for the decrease in forest ecosystem carbon storage in Central Asia [33]. Interestingly, the forest ecosystem carbon storage in 2030 has a minor increase under the scenario of decreasing forest land area, which is likely caused by the increase in carbon storage due to the increase in forest carbon intensity in 2030 offsetting the decrease in forest ecosystem carbon storage caused by the decreased area of forest land. Moderate GDP and lower population growth have maintained slight changes in LUCC and contributed to the growth of forest ecosystem carbon storage [79], and increasing the area of forest land and grassland and slowing urban expansion are effective measures to counteract decreasing carbon storage [34]. In addition, it can be seen that forest ecosystems have the greatest influence on carbon storage in terrestrial ecosystems, and increasing the area of forest land by artificial afforestation and maintaining the health and vitality of forest ecosystems can increase the carbon sequestration capacity of forest ecosystems.

4.3. Suggestions for Future Development

In the context of increased future climate and socioeconomic uncertainties, ecological environments are becoming increasingly fragile and natural vegetation land use types such as forest land are continuously degraded [28], which has resulted in a decrease in carbon storage in the study area. Therefore, it is particularly important for policy makers to formulate and implement policies related to socio-economic development and land use planning in order to optimize the land use structure and increase carbon storage.

The results of this study indicate that rapid economic growth will lead to a continuous decrease in ecosystem carbon storage and degradation of the ecological environment. Therefore, slowing down the rate of economic growth and reasonably planning urban development could improve the value of ecosystem services in the study area. Reducing the use of fossil fuels and increasing the proportion of clean energy use will not only mitigate the effects of climate change, but also prevent further degradation of forest land and grassland. In addition, various stakeholders should pursue the acceleration of the construction of provincial key public welfare forests, ecological demonstration villages, and demonstration rural road forestry networks, and programs to nurture unestablished forest land, replanting and replenishing them to encourage them to become forest land as soon as possible. Furthermore, insisting on the implementation of GCP, and artificial afforestation of unused land and forestable land, and maintaining the health and vitality of forest ecosystems, could improve the carbon sequestration capacity of forest ecosystems.

4.4. Strengths and Uncertainties

This paper provides a new approach for the future LUCC spatial simulation and carbon storage assessment based on population, GDP, and climate variables (temperature and precipitation), and land use demand under the SSP-RCP scenarios, combined with PLUS and InVEST models (Figure 2). We used the GDP, population, temperature, and precipitation change data generated by the SSP-RCP scenarios and future land use demand as simulation parameters for PLUS model, which produced a reasonable spatial distribution of LUCC (Figure 6). Unfortunately, the PLUS model assumes fixed transition rules during the LUCC simulations for each land use type, and these rules may change in the coming decades [40]. Moreover, only three climate change scenarios (SSP126, SSP245, and SSP585) generated by the MRI-ESM2-0 model were used in this work, and the differences in climate projections generated by different general circulation models (GCMs), which is one of the challenges for our future work [80,81].

Moreover, although the InVEST model has been widely used for multi-scale carbon storage assessment; however, this pattern also has limitations. For example, the InVEST model has a limitation that it cannot effectively estimate water and unused land carbon storage [39]. Indeed, the carbon loss due to the interconversion of each land use type and the seasonal variation of LUCC was not taken into account in the calculation of regional carbon storage in the InVEST model, which is also one of the sources of uncertainty in this work [82]. Furthermore, we collected carbon density data for all the land use types in the study area as much as possible, and assumed decadal trends in forest carbon density based on previous studies [42,45] to minimize uncertainty in carbon storage assessment. However, the carbon density values and their corresponding land use type areas can only approximately estimate the carbon storage of a regional ecosystems [83,84], and we will devote more efforts to address this challenge in future work.

In this study, we revealed a range of possible future spatiotemporal distribution patterns of LUCC and dynamic changes of carbon storage in Guangdong province, although with certain limitations. The results of this work can provide supporting data for responding to future climate change and formulating policies for sustainable socioeconomic development, and meeting the goals of carbon peaking and carbon neutrality.

5. Conclusions

By integrating the PLUS and InVEST models, we simulated the spatiotemporal dynamic distribution of LUCC and ecosystem carbon storage in Guangdong in the future (2020–2060) under the SSP126, SSP245, and SSP585 scenarios. The results of the future land use simulation indicated that land use changes varied under different scenarios. Under the SSP126 scenario, cultivated land, grassland, and shrubland were decreasing in varied degrees, the artificial surface was slightly expanded, and forest land was effectively protected; The overall change patterns of LUCC under the SSP245 scenario were similar to the SSP126 scenario, but the artificial surface expanded more rapidly and the increase in forest land slowed down under the SPP245 scenario; and under the SSP585 scenario, forest land is not effectively preserved and the artificial surface area sharply expanding, which encroaches on the previous grassland and forest land areas.

Under the three scenarios, population, elevation, temperature, and distance to water were the highest contributing driving factors for the growth of cultivated land, forest land, grassland, and artificial surface, respectively. During the period of 2020–2060, terrestrial ecosystem carbon storage in Guangdong province was increased from 240.89 Tg in 2020 to 247.16 Tg and 243.54 Tg in 2060 under SSP126 and SSP245 scenarios, respectively; and decreased under the SSP585 scenario, with a total decrease of 14.35 Tg. Forest ecosystem carbon storage is the main source of carbon storage increase, which can effectively offset the decrease in ecosystem carbon storage due to artificial surface expansion and other vegetation land type area reduction. Overall, forest land is the most influential land use type for carbon storage in terrestrial ecosystems, and the carbon sequestration capacity of forest ecosystems can be increased by increasing the area of forest land through artificial afforestation. Moreover, the results not only can provide a new insight into the redistribution of land resources and economic development strategies at the regional scale, but also support data to meet China's carbon neutrality goals.

Author Contributions: Conceptualization, L.T. and M.L.; Methodology, L.T. and M.L.; Software, L.T. and Y.T.; Validation, L.T., Y.T. and T.L.; Formal analysis, Y.T., T.L. and F.R.; Writing—original draft preparation, L.T.; Writing—review and editing, L.T., W.F. and M.L.; Funding acquisition, L.T. and M.L. All authors have read and agreed to the published version of the manuscript.

Funding: This research was funded by Postgraduate Research & Practice Innovation Program of Jiangsu Province, and the National Natural Science Foundation of China, grant number 31770679.

Data Availability Statement: Not applicable.

Acknowledgments: We are thankful for all of the helpful comments provided by the reviewers. The authors would like to thank the National Earth System Science Data Center, National Science & Technology Infrastructure of China (http://www.geodata.cn, accessed on 27 December 2021), and the Data Center for Resources and Environmental Sciences of the Chinese Academy of Sciences (https://www.resdc.cn, accessed on 28 December 2021) for providing data support.

Conflicts of Interest: The authors declare no conflict of interest.

References

1. Solomon, S.; Plattner, G.-K.; Knutti, R.; Friedlingstein, P. Irreversible climate change due to carbon dioxide emissions. *Proc. Natl. Acad. Sci. USA* **2009**, *106*, 1704–1709. [CrossRef] [PubMed]
2. Abeydeera, L.H.U.W.; Mesthrige, J.W.; Samarasinghalage, T.I. Global Research on Carbon Emissions: A Scientometric Review. *Sustainability* **2019**, *11*, 3972. [CrossRef]
3. Li, S.C.; Zhang, Y.L.; Wang, Z.F.; Li, L.H. Mapping human influence intensity in the Tibetan Plateau for conservation of ecological service functions. *Ecosyst. Serv.* **2018**, *30*, 276–286. [CrossRef]
4. Rodríguez-Echeverry, J.; Echeverría, C.; Oyarzún, C.; Morales, L. Impact of land-use change on biodiversity and ecosystem services in the Chilean temperate forests. *Landsc. Ecol.* **2018**, *33*, 439–453. [CrossRef]

5. Fang, J.; Zhu, J.; Wang, S.; Yue, C.; Shen, H. Global warming, human-induced carbon emissions, and their uncertainties. *Sci. China Earth Sci.* **2011**, *54*, 1458–1468. [CrossRef]
6. Cramer, W.; Guiot, J.; Fader, M.; Garrabou, J.; Gattuso, J.-P.; Iglesias, A.; Lange, M.A.; Lionello, P.; Llasat, M.C.; Paz, S.; et al. Climate change and interconnected risks to sustainable development in the Mediterranean. *Nat. Clim. Chang.* **2018**, *8*, 972–980. [CrossRef]
7. Tian, L.; Fu, W.X.; Tao, Y.; Li, M.Y.; Wang, L. Dynamics of the alpine timberline and its response to climate change in the Hengduan mountains over the period 1985–2015. *Ecol. Indic.* **2022**, *135*, 108589. [CrossRef]
8. Feddema, J.J.; Oleson, K.W.; Bonan, G.B.; Mearns, L.O.; Buja, L.E.; Meehl, G.A.; Washington, W.M. The Importance of Land-Cover Change in Simulating Future Climates. *Science* **2005**, *310*, 1674–1678. [CrossRef]
9. Newbold, T.; Hudson, L.N.; Hill, S.L.L.; Contu, S.; Lysenko, I.; Senior, R.A.; Börger, L.; Bennett, D.J.; Choimes, A.; Collen, B.; et al. Global effects of land use on local terrestrial biodiversity. *Nature* **2015**, *520*, 45. [CrossRef]
10. Li, S.C.; Bing, Z.L.; Jin, G. Spatially Explicit Mapping of Soil Conservation Service in Monetary Units Due to Land Use/Cover Change for the Three Gorges Reservoir Area, China. *Remote Sens.* **2019**, *11*, 468. [CrossRef]
11. Houghton, R.A. Revised estimates of the annual net flux of carbon to the atmosphere from changes in land use and land management 1850–2000. *Tellus B Chem. Phys. Meteorol.* **2003**, *55*, 378–390. [CrossRef]
12. Molotoks, A.; Stehfest, E.; Doelman, J.; Albanito, F.; Fitton, N.; Dawson, T.P.; Smith, P. Global projections of future cropland expansion to 2050 and direct impacts on biodiversity and carbon storage. *Glob. Chang. Biol.* **2018**, *24*, 5895–5908. [CrossRef] [PubMed]
13. Dybala, K.E.; Steger, K.; Walsh, R.G.; Smart, D.R.; Gardali, T.; Seavy, N.E. Optimizing carbon storage and biodiversity co-benefits in reforested riparian zones. *J. Appl. Ecol.* **2019**, *56*, 343–353. [CrossRef]
14. Payne, N.J.; Cameron, D.A.; Leblanc, J.-D.; Morrison, I.K. Carbon storage and net primary productivity in Canadian boreal mixedwood stands. *J. For. Res.* **2019**, *30*, 1667–1678. [CrossRef]
15. Li, X.; Du, H.; Mao, F.; Zhou, G.; Han, N.; Xu, X.; Liu, Y.; Zhu, D.; Zheng, J.; Dong, L.; et al. Assimilating spatiotemporal MODIS LAI data with a particle filter algorithm for improving carbon cycle simulations for bamboo forest ecosystems. *Sci. Total Environ.* **2019**, *694*, 133803. [CrossRef]
16. Kang, F.; Li, X.; Du, H.; Mao, F.; Zhou, G.; Xu, Y.; Huang, Z.; Ji, J.; Wang, J. Spatiotemporal Evolution of the Carbon Fluxes from Bamboo Forests and their Response to Climate Change Based on a BEPS Model in China. *Remote Sens.* **2022**, *14*, 366. [CrossRef]
17. Nerini, F.F.; Sovacool, B.; Hughes, N.; Cozzi, L.; Cosgrave, E.; Howells, M.; Tavoni, M.; Tomei, J.; Zerriffi, H.; Milligan, B. Connecting climate action with other Sustainable Development Goals. *Nat. Sustain.* **2019**, *2*, 674–680. [CrossRef]
18. Li, X.C.; Zhou, Y.Y.; Zhu, Z.Y.; Liang, L.; Yu, B.L.; Cao, W.T. Mapping annual urban dynamics (1985–2015) using time series of Landsat data. *Remote Sens. Environ.* **2018**, *216*, 674–683. [CrossRef]
19. Zhu, S.Y.; Kong, X.S.; Jiang, P. Identification of the human-land relationship involved in the urbanization of rural settlements in Wuhan city circle, China. *J. Rural Stud.* **2020**, *77*, 75–83. [CrossRef]
20. Wang, Z.; Xu, L.H.; Shi, Y.J.; Ma, Q.W.; Wu, Y.W.; Lu, Z.Q.; Mao, L.W.; Pang, E.Q.; Zhang, Q. Impact of Land Use Change on Vegetation Carbon Storage During Rapid Urbanization: A Case Study of Hangzhou, China. *Chin. Geogr. Sci.* **2021**, *31*, 209–222. [CrossRef]
21. Zhu, G.F.; Qiu, D.D.; Zhang, Z.A.X.; Sang, L.Y.; Liu, Y.W.; Wang, L.; Zhao, K.L.; Ma, H.Y.; Xu, Y.X.; Wan, Q.Z. Land-use changes lead to a decrease in carbon storage in arid region, China. *Ecol. Indic.* **2021**, *127*, 107770. [CrossRef]
22. Xiao, J. Satellite evidence for significant biophysical consequences of the "Grain for Green" Program on the Loess Plateau in China. *J. Geophys. Res. Biogeosci.* **2014**, *119*, 2261–2275. [CrossRef]
23. Cao, S.; Chen, L.; Liu, Z. An Investigation of Chinese Attitudes toward the Environment: Case Study Using the Grain for Green Project. *Ambio* **2009**, *38*, 55–64. [CrossRef]
24. Chen, C.; Park, T.; Wang, X.; Piao, S.; Xu, B.; Chaturvedi, R.K.; Fuchs, R.; Brovkin, V.; Ciais, P.; Fensholt, R.; et al. China and India lead in greening of the world through land-use management. *Nat. Sustain.* **2019**, *2*, 122–129. [CrossRef] [PubMed]
25. Mallapaty, S. How China could be carbon neutral by mid-century. *Nature* **2020**, *586*, 482–483. [CrossRef] [PubMed]
26. Zhao, X.; Ma, X.; Chen, B.; Shang, Y.; Song, M. Challenges toward carbon neutrality in China: Strategies and countermeasures. *Resour. Conserv. Recycl.* **2021**, *176*, 105959. [CrossRef]
27. Lai, L.; Huang, X.; Yang, H.; Chuai, X.; Zhang, M.; Zhong, T.; Chen, Z.; Chen, Y.; Wang, X.; Thompson, J.R. Carbon emissions from land-use change and management in China between 1990 and 2010. *Sci. Adv.* **2016**, *2*, e1601063. [CrossRef]
28. Wang, Z.; Li, X.; Mao, Y.; Li, L.; Wang, X.; Lin, Q. Dynamic simulation of land use change and assessment of carbon storage based on climate change scenarios at the city level: A case study of Bortala, China. *Ecol. Indic.* **2022**, *134*, 108499. [CrossRef]
29. Eyring, V.; Bony, S.; Meehl, G.A.; Senior, C.A.; Stevens, B.; Stouffer, R.J.; Taylor, K.E. Overview of the Coupled Model Intercomparison Project Phase 6 (CMIP6) experimental design and organization. *Geosci. Model Dev.* **2016**, *9*, 1937–1958. [CrossRef]
30. Cook, B.I.; Mankin, J.S.; Marvel, K.; Williams, A.P.; Smerdon, J.E.; Anchukaitis, K.J. Twenty-First Century Drought Projections in the CMIP6 Forcing Scenarios. *Earth's Futur.* **2020**, *8*, e2019EF001461. [CrossRef]
31. Dong, N.; You, L.; Cai, W.; Li, G.; Lin, H. Land use projections in China under global socioeconomic and emission scenarios: Utilizing a scenario-based land-use change assessment framework. *Glob. Environ. Chang.* **2018**, *50*, 164–177. [CrossRef]

32. Borrelli, P.; Robinson, D.A.; Panagos, P.; Lugato, E.; Yang, J.E.; Alewell, C.; Wuepper, D.; Montanarella, L.; Ballabio, C. Land use and climate change impacts on global soil erosion by water (2015–2070). *Proc. Natl. Acad. Sci. USA* **2020**, *117*, 21994–22001. [CrossRef] [PubMed]
33. Li, J.; Chen, X.; Kurban, A.; Van de Voorde, T.; De Maeyer, P.; Zhang, C. Coupled SSPs-RCPs scenarios to project the future dynamic variations of water-soil-carbon-biodiversity services in Central Asia. *Ecol. Indic.* **2021**, *129*, 107936. [CrossRef]
34. Zhao, M.; He, Z.; Du, J.; Chen, L.; Lin, P.; Fang, S. Assessing the effects of ecological engineering on carbon storage by linking the CA-Markov and InVEST models. *Ecol. Indic.* **2019**, *98*, 29–38. [CrossRef]
35. Yang, X.; Chen, R.; Zheng, X. Simulating land use change by integrating ANN-CA model and landscape pattern indices. *Geomat. Nat. Hazards Risk* **2016**, *7*, 918–932. [CrossRef]
36. Luo, G.; Yin, C.; Chen, X.; Xu, W.; Lu, L. Combining system dynamic model and CLUE-S model to improve land use scenario analyses at regional scale: A case study of Sangong watershed in Xinjiang, China. *Ecol. Complex.* **2010**, *7*, 198–207. [CrossRef]
37. Jiang, W.; Deng, Y.; Tang, Z.; Lei, X.; Chen, Z. Modelling the potential impacts of urban ecosystem changes on carbon storage under different scenarios by linking the CLUE-S and the InVEST models. *Ecol. Model.* **2017**, *345*, 30–40. [CrossRef]
38. Liu, X.; Liang, X.; Li, X.; Xu, X.; Ou, J.; Chen, Y.; Li, S.; Wang, S.; Pei, F. A future land use simulation model (FLUS) for simulating multiple land use scenarios by coupling human and natural effects. *Landsc. Urban Plan.* **2017**, *168*, 94–116. [CrossRef]
39. Li, J.; Gong, J.; Guldmann, J.-M.; Li, S.; Zhu, J. Carbon Dynamics in the Northeastern Qinghai–Tibetan Plateau from 1990 to 2030 Using Landsat Land Use/Cover Change Data. *Remote Sens.* **2020**, *12*, 528. [CrossRef]
40. Liang, X.; Guan, Q.; Clarke, K.C.; Liu, S.; Wang, B.; Yao, Y. Understanding the drivers of sustainable land expansion using a patch-generating land use simulation (PLUS) model: A case study in Wuhan, China. *Comput. Environ. Urban Syst.* **2021**, *85*, 101569. [CrossRef]
41. Babbar, D.; Areendran, G.; Sahana, M.; Sarma, K.; Raj, K.; Sivadas, A. Assessment and prediction of carbon sequestration using Markov chain and InVEST model in Sariska Tiger Reserve, India. *J. Clean. Prod.* **2020**, *278*, 123333. [CrossRef]
42. Li, T.; Li, M.-Y.; Tian, L. Dynamics of Carbon Storage and Its Drivers in Guangdong Province from 1979 to 2012. *Forests* **2021**, *12*, 1482. [CrossRef]
43. Li, T.; Li, M.Y.; Qian, C.H. Combining crown density to estimate forest net primary productivity by using remote sensing data. *J. Nanjing For. Univ.* **2021**, *45*, 153–160. [CrossRef]
44. Yang, X.H.; Zhao, Y.C.; Zhu, S.W.; Yang, X.T.; Wang, L.Y.; Li, Z.Q.; Liu, Z.H.; Yang, S.L.; Xiong, D.G.; Wang, G.X.; et al. *Guangdong Statistical Yearbook*; Guangdong Yearbook Press: Guangdong, China, 2021.
45. Fang, J.Y.; Zhu, J.X.; Li, P.; Ji, C.J.; Zhu, J.L.; Jiang, L.; Chen, G.P.; Cai, Q.; Su, H.J.; Feng, Y.H.; et al. *Carbon Budgets of Forest Ecosystems in China*; Science Press: Beijing, China, 2021.
46. State Forestry and Grassland Administration. *China Forest Resources Report (2014–2018)*; China Forestry Press: Beijing, China, 2019.
47. Guangdong Soil Survey Office. *Soil in Guangdong Province*; Science Press: Beijing, China, 1993.
48. Tang, R.; Zhao, J.; Liu, Y.; Huang, X.; Zhang, Y.; Zhou, D.; Ding, A.; Nielsen, C.P.; Wang, H. Air quality and health co-benefits of China's carbon dioxide emissions peaking before 2030. *Nat. Commun.* **2022**, *13*, 1–9. [CrossRef] [PubMed]
49. Long, Y.; Han, H.; Lai, S.-K.; Mao, Q. Urban growth boundaries of the Beijing Metropolitan Area: Comparison of simulation and artwork. *Cities* **2013**, *31*, 337–348. [CrossRef]
50. Meehl, G.A.; Boer, G.J.; Covey, C.; Latif, M.; Stouffer, R.J. The Coupled Model Intercomparison Project (CMIP). *Bull. Am. Meteorol. Soc.* **2000**, *81*, 313–318. [CrossRef]
51. Meehl, G.A.; Covey, C.; Delworth, T.; Latif, M.; McAvaney, B.; Mitchell, J.F.B.; Stouffer, R.J.; Taylor, K.E. The WCRP CMIP3 Multimodel Dataset: A New Era in Climate Change Research. *Bull. Am. Meteorol. Soc.* **2007**, *88*, 1383–1394. [CrossRef]
52. Taylor, K.E.; Stouffer, R.J.; Meehl, G.A. An Overview of CMIP5 and the Experiment Design. *Bull. Am. Meteorol. Soc.* **2012**, *93*, 485–498. [CrossRef]
53. O'Neill, B.C.; Tebaldi, C.; van Vuuren, D.P.; Eyring, V.; Friedlingstein, P.; Hurtt, G.; Knutti, R.; Kriegler, E.; Lamarque, J.-F.; Lowe, J.; et al. The Scenario Model Intercomparison Project (ScenarioMIP) for CMIP6. *Geosci. Model Dev.* **2016**, *9*, 3461–3482. [CrossRef]
54. Van Vuuren, D.P.; Edmonds, J.; Kainuma, M.; Riahi, K.; Thomson, A.; Hibbard, K.; Hurtt, G.C.; Kram, T.; Krey, V.; Lamarque, J.-F.; et al. The representative concentration pathways: An overview. *Clim. Chang.* **2011**, *109*, 5–31. [CrossRef]
55. Yun, X.; Tang, Q.; Li, J.; Lu, H.; Zhang, L.; Chen, D. Can reservoir regulation mitigate future climate change induced hydrological extremes in the Lancang-Mekong River Basin? *Sci. Total Environ.* **2021**, *785*, 147322. [CrossRef]
56. Hurtt, G.C.; Chini, L.; Sahajpal, R.; Frolking, S.; Bodirsky, B.L.; Calvin, K.; Doelman, J.C.; Fisk, J.; Fujimori, S.; Klein Goldewijk, K.; et al. Harmonization of global land use change and management for the period 850–2100 (LUH2) for CMIP6. *Geosci. Model Dev.* **2020**, *13*, 5425–5464. [CrossRef]
57. Chen, Y.; Guo, F.; Wang, J.; Cai, W.; Wang, C.; Wang, K. Provincial and gridded population projection for China under shared socioeconomic pathways from 2010 to 2100. *Sci. Data* **2020**, *7*, 83. [CrossRef] [PubMed]
58. Murakami, D.; Yoshida, T.; Yamagata, Y. Gridded GDP Projections Compatible with the Five SSPs (Shared Socioeconomic Pathways). *Front. Built Environ.* **2021**, *7*, 760306. [CrossRef]
59. Peng, S.; Ding, Y.; Wen, Z.; Chen, Y.; Cao, Y.; Ren, J. Spatiotemporal change and trend analysis of potential evapotranspiration over the Loess Plateau of China during 2011–2100. *Agric. For. Meteorol.* **2017**, *233*, 183–194. [CrossRef]

60. Yukimoto, S.; Koshiro, T.; Kawai, H.; Oshima, N.; Yoshida, K.; Urakawa, S.; Tsujino, H.; Deushi, M.; Tanaka, T.; Hosaka, M.; et al. MRI MRI-ESM2.0 model output prepared for CMIP6 CMIP. MRI MRI-ESM2.0 model output prepared for CMIP6 CMIP. *Earth Syst. Grid Fed.* **2019**, *10*. [CrossRef]
61. Zhai, H.; Lv, C.; Liu, W.; Yang, C.; Fan, D.; Wang, Z.; Guan, Q. Understanding Spatio-Temporal Patterns of Land Use/Land Cover Change under Urbanization in Wuhan, China, 2000–2019. *Remote Sens.* **2021**, *13*, 3331. [CrossRef]
62. Munsi, M.; Areendran, G.; Joshi, P.K. Modeling spatio-temporal change patterns of forest cover: A case study from the Himalayan foothills (India). *Reg. Environ. Chang.* **2012**, *12*, 619–632. [CrossRef]
63. Nor, A.N.M.; Corstanje, R.; Harris, J.A.; Brewer, T. Impact of rapid urban expansion on green space structure. *Ecol. Indic.* **2017**, *81*, 274–284. [CrossRef]
64. Liu, S.Y.; Hu, N.K.; Zhang, J.; Lv, Z.C. Spatiotemporal change of carbon storage in the Loess Plateau of northern Shaanxi, based on the InVEST Model. *Sci. Cold Arid. Reg.* **2018**, *10*, 240–250.
65. Xu, L.; He, N.P.; Yu, G.R. A dataset of carbon density in Chinese terrestrial ecosystems (2010s). *China Sci. Data* **2014**, *4*, 90–96.
66. Popp, A.; Calvin, K.; Fujimori, S.; Havlik, P.; Humpenöder, F.; Stehfest, E.; Bodirsky, B.L.; Dietrich, J.P.; Doelmann, J.C.; Gusti, M.; et al. Land-use futures in the shared socio-economic pathways. *Glob. Environ. Chang.* **2017**, *42*, 331–345. [CrossRef]
67. Havlík, P.; Valin, H.; Mosnier, A.; Obersteiner, M.; Baker, J.S.; Herrero, M.; Rufino, M.C.; Schmid, E. Crop Productivity and the Global Livestock Sector: Implications for Land Use Change and Greenhouse Gas Emissions. *Am. J. Agric. Econ.* **2012**, *95*, 442–448. [CrossRef]
68. Robinson, S.; van Meijl, H.; Willenbockel, D.; Valin, H.; Fujimori, S.; Masui, T.; Sands, R.; Wise, M.; Calvin, K.; Havlik, P.; et al. Comparing supply-side specifications in models of global agriculture and the food system. *Agric. Econ.* **2013**, *45*, 21–35. [CrossRef]
69. Schmitz, C.; Biewald, A.; Lotze-Campen, H.; Popp, A.; Dietrich, J.P.; Bodirsky, B.L.; Krause, M.; Weindl, I. Trading more food: Implications for land use, greenhouse gas emissions, and the food system. *Glob. Environ. Chang.* **2012**, *22*, 189–209. [CrossRef]
70. Tan, M.; Li, X.; Xie, H.; Lu, C. Urban land expansion and arable land loss in China—A case study of Beijing–Tianjin–Hebei region. *Land Use Policy* **2005**, *22*, 187–196. [CrossRef]
71. Zhang, Y.; Xie, H. Interactive Relationship among Urban Expansion, Economic Development, and Population Growth since the Reform and Opening up in China: An Analysis Based on a Vector Error Correction Model. *Land* **2019**, *8*, 153. [CrossRef]
72. Wang, Z.; Miao, Y.; Xu, M.; Zhu, Z.; Qureshi, S.; Chang, Q. Revealing the differences of urban parks' services to human wellbeing based upon social media data. *Urban For. Urban Green.* **2021**, *63*, 127233. [CrossRef]
73. Weisberg, P.J.; Lingua, E.; Pillai, R.B. Spatial Patterns of Pinyon–Juniper Woodland Expansion in Central Nevada. *Rangel. Ecol. Manag.* **2007**, *60*, 115–124. [CrossRef]
74. Eigenbrod, F.; Bell, V.; Davies, H.N.; Heinemeyer, A.; Armsworth, P.; Gaston, K.J. The impact of projected increases in urbanization on ecosystem services. *Proc. R. Soc. B Boil. Sci.* **2011**, *278*, 3201–3208. [CrossRef]
75. Xie, W.; Huang, Q.; He, C.; Zhao, X. Projecting the impacts of urban expansion on simultaneous losses of ecosystem services: A case study in Beijing, China. *Ecol. Indic.* **2018**, *84*, 183–193. [CrossRef]
76. Zhang, K.; Dang, H.; Tan, S.; Cheng, X.; Zhang, Q. Change in soil organic carbon following the 'Grain-for-Green' programme in China. *Land Degrad. Dev.* **2010**, *21*, 13–23. [CrossRef]
77. Angelsen, A.; Kaimowitz, D. Rethinking the Causes of Deforestation: Lessons from Economic Models. *World Bank Res. Obs.* **1999**, *14*, 73–98. [CrossRef] [PubMed]
78. Noss, R.F. Beyond Kyoto: Forest Management in a Time of Rapid Climate Change. *Conserv. Biol.* **2001**, *15*, 578–590. [CrossRef]
79. Yang, H.; Huang, J.L.; Liu, D.F. Linking climate change and socioeconomic development to urban land use simulation: Analysis of their concurrent effects on carbon storage. *Appl. Geogr.* **2020**, *115*, 102135. [CrossRef]
80. Kim, Y.-H.; Min, S.-K.; Zhang, X.; Sillmann, J.; Sandstad, M. Evaluation of the CMIP6 multi-model ensemble for climate extreme indices. *Weather. Clim. Extrem.* **2020**, *29*, 100269. [CrossRef]
81. Yazdandoost, F.; Moradian, S.; Izadi, A.; Aghakouchak, A. Evaluation of CMIP6 precipitation simulations across different climatic zones: Uncertainty and model intercomparison. *Atmospheric Res.* **2020**, *250*, 105369. [CrossRef]
82. Sharp, R.; Douglass, J.; Wolny, S.; Arkema, K.; Bernhardt, J.; Bierbower, W.; Chaumont, N.; Denu, D.; Fisher, D.; Glowinski, K.; et al. *InVEST 3.10.2 User's Guide, The Natural Capital Project*; Stanford University: Stanford, CA, USA; University of Minnesota: Minneapolis, MN, USA; The Nature Conservancy, and World Wildlife Fund: Arlington, VI, USA, 2020.
83. Xie, X.L.; Sun, B.; Zhou, H.Z.; Li, Z.P.; Li, A.B. Organic carbon density and storage in soils of china and spatial analysis. *Acta Pedol. Sin.* **2004**, *41*, 35–43.
84. Li, Y.G.; Han, N.; Li, X.J.; Du, H.Q.; Mao, F.J.; Cui, L.; Liu, T.Y.; Xing, L.Q. Spatiotemporal Estimation of Bamboo Forest Aboveground Carbon Storage Based on Landsat Data in Zhejiang, China. *Remote Sens.* **2018**, *10*, 898. [CrossRef]

remote sensing

Article

Estimation and Spatio-Temporal Change Analysis of NPP in Subtropical Forests: A Case Study of Shaoguan, Guangdong, China

Tao Li, Mingyang Li *, Fang Ren and Lei Tian

College of Forestry, Nanjing Forestry University, Nanjing 210037, China; litao3014@njfu.edu.cn (T.L.); renfang@njfu.edu.cn (F.R.); tianlei@njfu.edu.cn (L.T.)
* Correspondence: lmy196727@njfu.edu.cn; Tel.: +86-025-8542-7327

Abstract: Exploring the spatial and temporal dynamic characteristics of regional forest net primary productivity (NPP) in the context of global climate change can not only provide a theoretical basis for terrestrial carbon cycle studies, but also provide data support for medium- and long-term sustainable management planning of regional forests. In this study, we took Shaoguan City, Guangdong Province, China as the study area, and used Landsat images and National Forest Continuous Inventory (NFCI) data in the corresponding years as the main data sources. Random forest (RF), multiple linear regression (MLR), and BP neural network were the three models applied to estimate forest NPP in the study area. Theil–Sen estimation, Mann–Kendall trend analysis and the standard deviation ellipse (SDE) were chosen to analyze the spatial and temporal dynamic characteristics of NPP, whereas structural equation modeling (SEM) was used to analyze the driving factors of NPP changes. The results show that the performance of the RF model is better than the MLR and BP neural network models. The NPP in the study area showed an increasing trend, as the NPP was 5.66 $t{\cdot}hm^{-2}{\cdot}a^{-1}$, 7.68 $t{\cdot}hm^{-2}{\cdot}a^{-1}$, 8.17 $t{\cdot}hm^{-2}{\cdot}a^{-1}$, 8.25 $t{\cdot}hm^{-2}{\cdot}a^{-1}$, and 10.52 $t{\cdot}hm^{-2}{\cdot}a^{-1}$ in 1997, 2002, 2007, 2012, and 2017, respectively. Spatial aggregation of NPP was increased in the period of 1997–2017, and the center shifted from the mid-west to the southwest. In addition, the forest stand factors had the greatest effect on NPP in the study area. The forest stand factors and environmental factors had a positive effect on NPP, and understory factors had a negative effect. Overall, although forest NPP has fluctuated due to the changes of forestry policies and human activities, forest NPP in Shaoguan has been increasing. In the future, the growth potential of NPP in Shaoguan City can be further increased by continuously expanding the area proportion of mixed forests and rationalizing the forest age group structure.

Keywords: net primary productivity; remote sensing inversion; dynamic change; driving factors; Shaoguan City

Citation: Li, T.; Li, M.; Ren, F.; Tian, L. Estimation and Spatio-Temporal Change Analysis of NPP in Subtropical Forests: A Case Study of Shaoguan, Guangdong, China. *Remote Sens.* **2022**, *14*, 2541. https://doi.org/10.3390/rs14112541

Academic Editors: Huaqiang Du, Wenyi Fan, Weiliang Fan, Fangjie Mao and Mingshi Li

Received: 28 April 2022
Accepted: 23 May 2022
Published: 26 May 2022

1. Introduction

With the intensification of global climate change, the global carbon cycle has become one of the core issues in global climate change research [1–3]. As the main body of the terrestrial ecosystem and the largest carbon reservoir in the terrestrial ecosystem, the forest ecosystem fixes about two-thirds of the carbon in the whole terrestrial system every year, and its role in regulating global carbon balance, mitigating the rising concentration of greenhouse gases such as CO_2 in the atmosphere, and maintaining global climate is irreplaceable [4–6]. The net primary productivity (NPP) of forests is the amount of organic matter accumulated per unit area and per unit time, which is expressed as the portion of organic carbon fixed by photosynthesis minus the portion consumed by plants themselves through respiration [7]. Estimation of NPP is the basis for the study of the functioning of matter and energy in ecosystems, reflects the production capacity of vegetation communities under natural environmental conditions, and also directly affects the carbon stocks of above-ground plant parts, below-ground root parts, soil carbon pools, and the carbon interception of the whole forest ecology. As a component of the surface cycle, NPP can directly reflect the production capacity of forests under natural environmental conditions, as it is one of the important indicators for evaluating the sustainable development of the forest ecosystem and is the main indicator for judging the carbon sink of ecosystems and regulating ecological processes [8]. In the context of intensifying global climate change and human activities, considering the important role played by NPP in the global carbon cycle, carbon sequestration, carbon storage, and global change, the research on the NPP estimation of forest vegetation is carried out to provide a scientific basis for the making of a long-term sustainable forest management plan at the regional scale.

At the spatial scale, the estimation of forest NPP is divided into sample plot observations, regional simulation, and global simulation. The localization observation uses the extrapolation method, which is the extrapolation of sample survey points to the whole region. Although the rationality is insufficient, the estimation uses spatially measured data; thus, it can still be used as a reference for NPP estimation [9]. At the regional or global scale, because forest ecosystems are the most complex types of structural levels and functional behaviors on Earth, the impacts of environmental changes and human activities on forests and the feedback effects of forests are long-term. However, the availability of productivity observation data points on forest areas worldwide is extremely limited; therefore, the model estimation of productivity becomes an important research method [10,11]. Models for estimating NPP can be classified into four categories: the statistical model, parametric model, process model, and ecology and remote sensing coupling model. Statistical models include the Miami model, the Thornthwaite memorial model, etc. A large number of empirical studies have shown that the Thornthwaite memorial model is sensitive only to changes in precipitation. The Miami model, though sensitive to both precipitation and temperature, is more sensitive to temperature [12]; in addition, this model does not take into account the influence of structural changes of vegetation itself, nor the influence of atmospheric conditions and site conditions, and only uses mathematical methods to derive the relationship between NPP and major climatic factors. Therefore, the estimation results obtained from the model are approximate. Parametric models mainly include the CASA model, GLO-PEM model, VPM model, etc., among which the CASA model has become a mature model for estimating NPP. Potter et al. [13] used the CASA model to estimate global biomass and productivity in 1980. Wen [14] used the CASA model to estimate NPP as a vegetation growth indicator and combined it with other climate factors to study the global vegetation's response to climate warming from 1982 to 2013, but there are uncertainties in the simulation results for vegetation NPP because the parametric models cannot explain the mechanisms of productivity changes in terms of physiological ecology. The process model takes into account the physiological characteristics of the vegetation and environmental factors to simulate the vegetation growth and development process, but the model is complex and requires more parameters, which are the two main defects of the process model. The ecology and remote sensing coupling model combines the advantages of the plant ecophysiological process model and remote sensing parameter model, which can make a global-scale NPP estimation, and the model parameters can be obtained using remote sensing technology. However, this kind of model is more complex and requires more parameters, and subjective factors have a greater influence on the investigator and parameter determination.

From remote sensing data, information on forest cover status and forest spectral characteristics can be obtained at a large regional scale, although this information is closely related to forest productivity. Therefore, optical sensors and active sensors have been widely used in forest net primary productivity estimation [15]. Since the spatial resolution of Landsat time series stacks (LTSS) image elements is 30 m $\times$ 30 m and the image element size is close to the area of the forest management unit, it makes up for the shortage of single-period images in monitoring long-term forest dynamics, and is therefore often used for regional forest productivity estimation. Remote sensing estimation combined with ground survey data has become one of the important tools in regional scale forest productivity estimation.

Forest productivity is related to age and vegetation growth status, whereas the remote sensing characteristic variables extracted by traditional methods do not contain such information. Forest canopy density is the ratio of the forest vegetation to projected area observed from the air, and is numerically the same value as crown density as an indicator of stand density. It is one of the important components of stand structure, and it can be used to reflect the distribution of light, water, and other environmental factors entering the stand through the forest canopy. The magnitude of forest NPP is closely related to canopy density, and the accuracy of the model may be improved by introducing the canopy density variable into the forest NPP inversion model [16].

NPP varies spatially and temporally, and there are also many driving factors that cause spatial and temporal changes in NPP. Therefore, it is important to study the spatial and temporal dynamics of NPP and the driving factors for long-term sustainable forest management. The Theil–Sen median slope estimation and Mann–Kendall trend analysis are more robust to errors and can determine the significance of trends. Nyikadzino [17] used these methods to observe seasonal changes in precipitation in the Limpopo River Basin and found that rainfall in this area showed a non-significant decrease trend. Alhaji [18] used this method to study the temperature in Gombe State, Federal Republic of Nigeria and found that due to the effects of climate change, extreme weather caused a significant increase in the maximum and average temperature in the area. The SDE method takes into account three levels of center offset, directional offset, and angular offset, and can visually reflect the spatial variation results. Peng [19] studied the spatial and temporal distribution of PM2.5 concentrations in China from 1999 to 2011 using the SDE. In previous studies, the Theil–Sen median slope estimation, Mann–Kendall trend analysis, and SDE are mainly applied to the study of climate spatial and temporal variability over long time periods. At present, there are no relevant reports on the analysis of spatio-temporal dynamics and long-term change of forest NPP based on the Theil–Sen median slope estimation, Mann–Kendall trend analysis, and SEM analysis of driving factors. There are many drivers of spatial and temporal changes in forest NPP, which can be divided into two main categories: biotic natural factors and socio-economic factors. Wang [20] studied the changes in forest NPP and multi-level driving mechanisms in the Changbai Mountains, Northeast China, and the results showed that precipitation and vegetation cover were the key drivers.

Subtropical forests hold a special position in the global ecosystem and play an important role in the global terrestrial ecosystem material cycle and terrestrial carbon pool. China's subtropical forest is a unique forest ecosystem, characterized by rich forest types, a wide range of tree species, and high forest productivity [21]. Guangdong Province is located in the southeastern part of the Asian continent, south of the tropical sea, and is strongly influenced by the monsoon climate, with many typhoons and heavy rains in summer. Under this climate condition, the forest service functions of water conservation and soil conservation are particularly important. Since the influence of the Quaternary Ice Age is small, the flora in the province has had a long history of development, producing a rich variety of forest plant species and forming a flora with many ancient plants and relict plants [22,23]. Shaoguan City is located in the northern part of Guangdong Province and is an important part of the southern collective forest area. Due to the differences in topographic conditions, forest resource status, and economic development level, there is spatial and temporal heterogeneity of forest NPP and its drivers. Since the reform and opening up in 1978, the urbanization and industrialization process in Shaoguan City has been accelerated. The forests have been disturbed and damaged by human activities for a long time, the remnant native forest in the city has been decreasing, and the habitats for many wildlife have been deteriorating.

Until now, there are no relevant reports on the analysis of spatio-temporal dynamics and driving factors of forest NPP over a long time based on the Theil–Sen median slope estimation, Mann–Kendall trend analysis, SDE, and SEM. The main objectives of this study are as follows: (1) using Landsat images and NFCI data as the main information sources to introduce the stand structure factor of forest canopy density, which is closely related to productivity, to estimate the NPP in Shaoguan City in 1997, 2002, 2007, 2012, and 2017, and to explore more accurate NPP estimation methods from the three RF, MLR, and BP neural network models; (2) to reveal the spatio-temporal change trend of forest NPP in a typical forest prefecture in a Chinese subtropical region; and (3) to identify driving factors of forest NPP dynamics to provide a scientific basis for making sustainable forest management plans.

2. Materials and Methods

2.1. Study Area

Shaoguan City (Figure 1) is located in the northern part of Guangdong Province (23°53′~25°31′N, 112°53′~114°45′E). The whole territory spans 186.30 km from east to west and 173.40 km from north to south. Shaoguan's topography is high in the north and low in the south, with the highest peak of Shikenggang (1902 m asl) in the north of Guangdong and the lowest point (35 m asl) in the south. Shaoguan belongs to the central subtropical humid monsoon climate zone and has a pleasant climate. The average annual temperature is 21 °C, with the temperature increasing from north to south in winter, and the temperature is almost the same in summer. Rainfall is abundant, with an average annual rainfall of 1700 mm. March–August is the rainy season, September–February is the dry season, and there is snow in the north in winter.

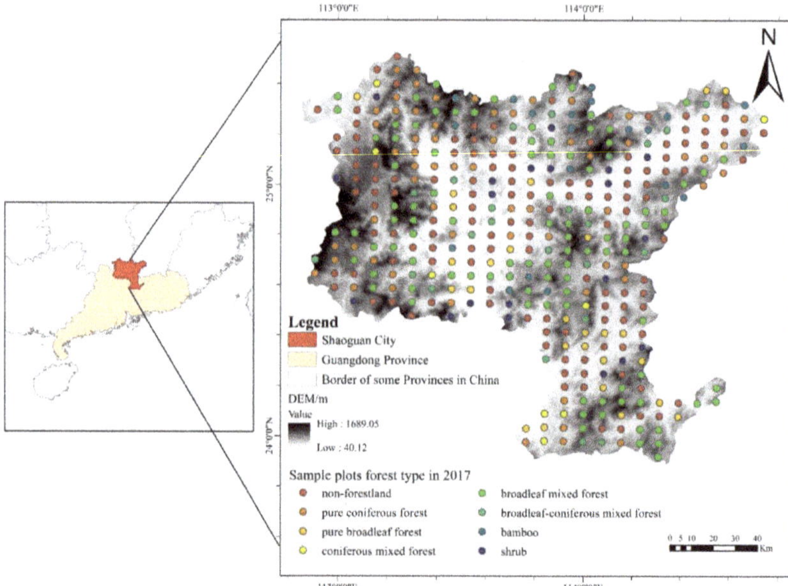

Figure 1. Location of Shaoguan City, Guangdong Province, together with the DEM.

Shaoguan is a national key forest area, being the important base of the timber forest, water source forest, and key moso bamboo in Guangdong Province, and is known as the biological gene pool of South China and the ecological shield of the Pearl River Delta. In 2021, the city had a forested area of 1,277,300 hm^2, with a forest coverage rate of 74.43%, forest greening rate of 74.90%, and stock volume of 96.52 million m^3, ranking first in Guangdong Province, which is known as "The World's Most Complete Oasis Preserved on the Same Latitude as the Tropic of Cancer". The forest in Shaoguan is dominated by broadleaf mixed forests and broadleaf pure forests. The broad-leaved pure forests mainly include oak (*Quercus*), eucalyptus (*Eucalyptus robusta*), camphor tree (*Cinnamomum camphora*), etc. These forests are followed by coniferous pure forests represented by horsetail pine (*Pinus massoniana*) and Chinese fir (*Cunninghamia lanceolata*), mixed coniferous forests, and bamboo forests represented by moso bamboo (*Phyllostachys heterocycle*) [24]. The specific information is shown in Table 1.

Table 1. Forest types in Shaoguan City.

Main Forest Types	Standard of Division	Typical Tree Species	Characteristic
pure coniferous forest	stand volume of single coniferous species ≥65%	*Cunninghamia lanceolata*	fast growth, high volume per unit area
		Pinus massoniana	wide distribution, main tree species for timber forest
pure broadleaf forest	stand volume of single broadleaf species ≥65%	*Eucalyptus robusta*	high proportion of young forests
		Acacia confusa	higher volume per unit area, mainly planted forests
		Cinnamomum camphora	grow faster, native hardwood species
broadleaf mixed forest	total stand volume of broadleaf species ≥65%		few natural broad-leaved mixed forests, the dominant tree species is not obvious
broadleaf-coniferous mixed forest	total stand volume of coniferous or broadleaf species accounting for 35–65%	*Pinus massoniana-Schima superba*	tree growth is higher than their respective pure forests
coniferous mixed forest	total stand volume of coniferous species ≥65%	*Cunninghamia lanceolata-Pinus massoniana*	less pests and diseases

Shaoguan City is rich in forest resources, and forest cover and forest stock volume are higher than the national average. Although Shaoguan City forest resources have had steady growth, due to frequent human activities, accelerating urbanization and industrialization, and irrational forestry policies, Shaoguan City forest resources still have certain problems, such as low forest quality and uneven forest age groups. Therefore, in this study, based on the Landsat series remote sensing images, we compare the estimation methods of forest NPP in the study area, explore the dynamic changes of forest NPP in long time series, and explore the driving factors affecting NPP.

2.2. Data Acquisition and Preprocessing

2.2.1. The Fixed Sample Data of National Forest Resources Continuous Inventory

The NFCI [25] takes provinces as the sampling population and adopts systematic sampling. According to the actual situation of each province, the sampling interval of each province is determined by the kilometer grid. Permanent sample plots are set up to conduct forest resource surveys. In Shaoguan City, there are 388 fixed sampling plots (25.82 × 25.82 m each) based on 6 km × 8 km spacing. The attributes of these plots include slope, slope direction, slope position, altitude, soil name, soil layer thickness, soil texture, humus thickness, average age, average diameter at breast height (DBH), average tree height, canopy density, tree species structure, live tree stock volume, and other investigation factors. The National Forestry and Grassland Administration is responsible for establishing the inventory plots and gathering data.

Before the estimation of NPP, the non-forestland sample plots with a stock volume of 0, such as water bodies and buildings, were removed from the NFCI. The NPP of forest consists of community growth (the sum of annual net growth of stems, branches, and roots of the tree layer and annual net growth of shrubs and the herbaceous layer) and annual withered volume. Based on the relationship between biomass and stock volume, and the function relationship between biomass, community growth, and annual withering, the biomass of different forest types was calculated based on the volume, the community growth and annual withering were calculated based on the biomass, and the NPP of the forest was finally obtained for each sample plot. The specific calculation formula is shown in Table 2.

Table 2. Relationship between forest volume, biomass, and NPP of typical tree species (excerpt) [26].

Forest Types	Relationship between Biomass and Volume	Relationship between Biomass and Community Growth	Relationship between Biomass and Annual Withering
coniferous and broadleaf mixed forest	B = V/(1.1731 + 0.0018 × V)	Y = B/(0.1038 × A + 0.0761 × B)	L = 3.46
deciduous broadleaf forest	B = V/(0.6539 + 0.0038 × V)	Y = B/(0.2393 × A + 0.0495 × B)	L = B/(18.2460 + 0.0366 × B)
broadleaf mixed forest	B = V/(0.5788 + 0.0020 × V)	Y = B/(0.3018 × A + 0.0331 × B)	L = B/(9.1028 + 0.0575 × B)
cypress forest	B = V/(1.0202 + 0.0022 × V)	Y = B/(0.1132 × A + 0.0745 × B)	L = B/(9.8381 + 0.1337 × B)
fir forest	B = V/(1.2917 + 0.0022 × V)	Y = B/(0.4598 × A + 0.0069 × B)	L = B/(10.1320 + 0.0874 × B)
Pinus massoniana forest	B = V/(1.4254 + 0.0004 × V)	Y = B/(0.4046 × A + 0.0098 × B)	L = B/(15.4510 + 0.0225 × B)
other warm pine forest	B = V/(1.3624 − 0.0003 × V)	Y = B/(0.2423 × A + 0.0581 × B)	L = B/(18.9050 + 0.0422 × B)
evergreen broadleaf forest	B = V/(0.7883 + 0.0026 × V)	Y = B/(0.2503 × A + 0.0226 × B)	L = B/(20.5070 + 0.0383 × B)
deciduous broadleaf forest	B = V/(0.6539 + 0.0038 × V)	Y = B/(0.2393 × A + 0.0495 × B)	L = B/(18.2460 + 0.0366 × B)

Note: B is biomass ($t \cdot hm^{-2}$), V is volume ($t \cdot hm^{-2}$), Y is the community growth($t \cdot hm^{-2}$), L is annual withering($t \cdot hm^{-2}$), A is average age (a).

2.2.2. Landsat Time Series Data

(1) Image pre-processing

Landsat images were pre-processed using ENVI 5.3 software. To eliminate the errors of the sensor, the images were first radiometrically calibrated [27], and then atmospheric correction [28] was performed using the FLAASH module of the ENVI software. The terrain in the study area is mainly mountainous and hilly, with large differences in elevation between the north and the south. The remote sensing images are influenced by the sensor orientation and the sun height and orientation, resulting in differences in brightness values due to the different illumination received by the shaded and sunny slopes [29]. The C-correction algorithm is used to correct the topography of the remote sensing images to eliminate the variation of radiance values caused by the topographic relief, so that the images can better reflect the spectral characteristics of the features [30]. Due to the failure of the Landsat-7 ETM+ on-board scan line corrector (SLC) after June 2003, data strips were lost in the images acquired after that date; thus, the 2007 and 2012 image data needed to be strip-repaired first and then pre-processed.

(2) Extraction of Feature Variables

According to previous studies [31], it is known that the productivity of forests is closely related to the conditions of forest stand, soil, topography, and climate. This study uses ENVI software to extract seven vegetation indices, including the normalized difference vegetation index (NDVI), ratio vegetation index (RVI), difference vegetation index (DVI), enhanced vegetation index (EVI), green vegetation index (GVI), perpendicular vegetation index (PVI) [32] and leaf area index (LAI) [33]. Tasseled cap transformation [34] was performed on the pre-processed Landsat remote sensing images, and the first three components of brightness, greenness, and wetness were chosen. Three window sizes of 3 × 3, 5 × 5, and 7 × 7 were selected when extracting texture features, and contrast, dissimilarity, mean, homogeneity, angular second moment, entropy, skewness, and correlation were calculated. Alternative independent variables of the NPP remote sensing estimation model are shown in Table 3.

Table 3. Alternative independent variables of NPP remote sensing estimation model.

Variable Type	Variable Name	Code	Variable Type	Variable Name	Code
single band	blue band	B2	stand structure	canopy density	FCD
	green band	B3	tasseled cap	brightness index	Bri
	red band	B4		greenness index	Gre
	near-infrared band	B5		wetness index	Wet
	shortwave infrared band 1	B6	topographic factor	slope	Slope
	shortwave infrared band 2	B7		elevation	DEM
texture feature	contrast	Bij_con	vegetation index	normalized difference vegetation index	NDVI
	dissimilarity	Bij_dis		ratio vegetation index	RVI
	mean	Bij_mea		difference vegetation index	DVI
	homogeneity	Bij_hom		enhanced vegetation index	EVI
	angular second moment	Bij_asm		green vegetation index	GVI
	entropy	Bij_ent		perpendicular vegetation index	PVI
	skewness	Bij_ske		leaf area index	LAI
	correlation	Bij_cor			

Note: texture feature code of Bij_xxx: i is the band of 2–7, j is 3 × 3, 5 × 5, or 7 × 7 texture window size.

2.3. Research Method

2.3.1. Calculation of Forest Canopy Density

According to JOSHI's [35] study, it is known that there are four methods for forest canopy density extraction: the artificial neural network (ANN), multiple linear regression technique (MLR), forest canopy density mapper (FCD), and maximum likelihood classification (MLC). Among the four methods, the average accuracy of forest canopy density obtained by the FCD model in three Southeast Asian countries was 92% [36]; thus, the FCD was chosen for this study. The FCD model is calculated using forest growth condition and is able to monitor temporal changes in forest canopy density. The FCD model is based on the Landsat remote sensing image extracted index and includes the advanced vegetation index (AVI), bare soil index (BI), and shadow index (SI). Compared with the NDVI, the AVI is more sensitive to the amount of vegetation, the BI increases with the increase of surface bareness, and the SI increases with the increase of forest density. Taking Landsat 8 OLI images as an example, the three indices and FCD were calculated as follows.

$$\text{AVI} = [(B_5+1)(65{,}536 - B_4)(B_5 - B_4)]^{\frac{1}{3}} \quad (1)$$

$$\text{BI} = \frac{(B_6+B_4)-(B_5+B_2)}{(B_6+B_4)+(B_5+B_2)} \times 100 + 100 \quad (2)$$

$$\text{SI} = [(65{,}536 - B_2)(65{,}536 - B_3)(65{,}536 - B_4)]^{\frac{1}{3}} \quad (3)$$

$$\text{FCD} = (VD \times SSI + 1)^{\frac{1}{2}} - 1 \quad (4)$$

where B_2–B_6 indicate the brightness values of blue, green, red, near-infrared, and shortwave infrared bands; AVI indicates the advanced vegetation index and AVI = 0 when $(B_5 - B_4)$ is less than 0; VD indicates the vegetation density value, which is synthesized from the vegetation index and shading index using a principal component analysis; and SSI indicates the scale shading index, which is calculated using the linear transformation function of the normalized shading index.

2.3.2. Remote Sensing Estimation Model

A random forest (RF) is a compositional supervised learning algorithm that uses a bootstrap method to extract multiple samples from the original samples, establishes a decision tree for each resampled sample, and then combines the decision trees to obtain the final prediction by voting [37]. A large number of theoretical and practical studies have proved that RF has a high prediction accuracy, good tolerance for outliers and noise, and can reduce the overfitting phenomenon because the samples of random forest-generated decision trees are randomly selected [38]. In this study, the RF model is executed by the randomForest function in the R language randomForest package. The number of decision trees (ntree) and the number of variables (mtry) to be extracted when splitting the decision trees need to be adjusted when using the model. The value of mtry is 1/3 of the number of independent variables when building the RF regression model, and the default value of mtry is 1 when the number of independent variables is less than 3 [39].

A multiple linear regression (MLR) has two or more independent variables. The multiple linear regression model uses the vegetation index, texture characteristics, original band, forest canopy density, and other factors as independent variables and the NPP of Shaoguan's fixed sample plots as dependent variables. The screening of independent variables is conducted using stepwise regression, of which the basic idea can be seen in Section 2.3.3 [40]. Stepwise regressions were performed using SPSS software, with stepwise criteria of F probabilities, and entry and deletion were set to be 0.05 and 0.1, respectively.

An artificial neural network (ANN) simulates neuronal activity with a mathematical model, and is an information processing system based on imitating the structure and function of neural networks in the brain. The basic idea of the back propagation (BP) neural network is that a learning process consists of forward propagation of the signal and backward propagation of the error. Forward propagation means that the input samples are processed in the input layer and then passed to the output layer after the hidden layer. If the actual output of the output layer does not match the expected output, it will be transferred to the back propagation of error, and the back propagation of error will back propagate the error of the output into the input layer in a certain form through the hidden layer, spread the error to all units in each layer, and then obtain the error signal of each unit in each layer, where the error will be used as the basis for correcting the cell weight. The process of adjusting the weights is the process of network learning and training until the network output error can be accepted and proceed to the preset learning times. This study used a three-layer structure of a BP neural network model, including the input layer, hidden layer, and output layer [41]. The BP neural network model in this study was constructed with the R language neuralnet function package, where the model is first built with the default number of nodes and hidden layers of the system, and the number of hidden layers is further increased according to the resultant error to improve the model accuracy.

2.3.3. Screening of Model Feature Variables

The RF defines two metrics to measure the importance of variables and that can be used to rank the variables: the first is %IncMSE, which is the percentage increase in prediction error per decision tree computed with out-of-bundle (OOB) data replacement; the second is IncNodePurity, which is the total reduction in node impurity when the decision tree nodes split, measured as the sum of squared residuals. Higher values of %IncMSE and IncNodePurity of the predictor variables indicate greater importance for model prediction [42]; thus, these two metrics were used to screen the variables added to the RF and BP neural network models, and this step was performed in the R software.

Stepwise regression is an important method for selecting the optimal explanatory variables for linear regression models and which mainly addresses the problem of how to select the explanatory variables when there are too many variables. Therefore, the explanatory variables selected for the regression model have a significant effect on the response variables [43]. The basic idea of stepwise regression is to introduce variables into the model one by one, perform an F-test after introducing each explanatory variable, and perform a t-test on the explanatory variables that have been selected one by one. When the explanatory variables initially introduced become no longer significant due to the introduction of later explanatory variables, they are removed to ensure that only significant variables are included in the regression equation before each new variable is introduced.

This is an iterative process that lasts until neither significant explanatory variables are selected into the regression equation nor insignificant explanatory variables are removed from the regression equation, to ensure that the final set of explanatory variables obtained is optimal [44]. However, in the process of stepwise regression, because there may be significant correlation between independent variables, it is easy to cause the problem of collinearity between model variables. In order to eliminate the influence of collinearity between variables on the model, the variance inflation factor (VIF) between variables is calculated to test whether there is collinearity between variables. According to the study, the VIF is usually larger than 7.5, which indicates that serious collinearity between the variables exists and it needs to be eliminated [45,46]. The final variable screening of the multiple linear regression model in this study was performed using stepwise regression, which was executed in SPSS software with the confidence level set at 95%, and the F-test probabilities of entry and deletion of predictor variables were set at 0.05 and 0.10, respectively.

The specific screening results are shown in Table 4.

Table 4. The selected predictor variables for LF, BP neural network, and MLR models.

Year	Predictor Variable of LF and BP	Predictor Variable of MLR
1997	DEM, B37_con, B27_con, B23_ske, B23_ent, B23_asm, B23_mea, B35_cor, Slope, B73_ent, B73_mea, B45_cor, B33_ent, B43_mea, B73_asm, B73_ske, B33_mea	DEM, B73_ske, B27_con
2002	B67_mea, FCD, B33_con, B47_con, B35_con, B25_hom, B63_dis, Slope, B55_mea, NDVI, B25_ske, B43_dis, B35_dis, B45_con, B27_ske, B77_con, DEM, B37_hom, B37_con, B23_hom, B57_mea, B27_con, B65_asm, B65_hom, B63_hom, B67_con	B37_con, B33_hom
2007	B63_dis, Wet, B25_con, B27_con, B45_con, B65_dis, B23_con, DEM, B2, B33_con, RVI, Bri, B4, B75_dis, FCD	RVI, B35_cor, B77_ent
2012	Wet, B2, DEM, B65_con, B47_ent, B73_ent, Slope, RVI, B73_mea, FCD, B27_con, B73_con, B4, B23_con, LAI, B45_mea, B3, B47_mea, B25_con, B53_hom	B25_con, B47_ske, B7, B55_dis, B45_ske
2017	B73_cor, B45_asm, B75_cor, B33_ske, B23_con, DEM, B53_hom, FCD, B2, B63_cor, B53_asm, RVI, B75_ent, B55_asm, NDVI, B45_con	B35_con, B75_cor, DEM, B27_con, B55_hom, B53_asm, Slope

Note: DEM is elevation, Slope is slope, FCD is canopy density, NDVI is difference normalized vegetation index, RVI is ratio vegetation index, LAI is leaf area index, Wet is wetness index, Bri is brightness index, B2 is blue light band, B3 is green light band, B4 is red light band. In Bij_xxx, i is band, where 2–7 are blue light band, green light band, infrared band, shortwave infrared band 1, and shortwave infrared band 2, respectively; j is texture window size, where 3, 5, and 7 denote 3×3, 5×5, and 7×7 windows, respectively; con is contrast, dis is dissimilarity, mea is mean, hom is homogeneity, asm is angular second moment, ent is entropy, ske is skewness, and cor is correlation.

2.3.4. Model Accuracy Evaluation

After the model is established, it is necessary to check the goodness of fit and applicability of the model, to analyze the advantages and disadvantages of the model, and finally, to choose the optimal model.

In this study, we used 10-fold cross-validation to verify the accuracy of the model [47]. The method divided the data into 10 parts, where 9 of them are used as training data and 1 as test data in turn, and the mean value of 10 times was used as an estimate of the accuracy of the model.

There are many indicators for evaluating estimation models, such as coefficient of determination (R^2), root mean square error (RMSE), mean absolute error (MAE), etc. [48]. These metrics are usually used to determine the strength of the model by performing a comparative analysis between predicted and measured values. In this study, R^2, RMSE, and MAE are used to evaluate the accuracy of the model. R^2 reflects the proportion of the total variation of the dependent variable that can be explained by the independent variable through the regression relationship, and its value interval is usually between (0, 1). MAE is the mean value of the absolute error, which can better reflect the actual situation of the prediction error, and is equal to 0 when the predicted value and the true value match exactly; the larger the error, the larger the numerical value.

In this study, the model prediction performance is mainly evaluated by calculating the R^2, RMSE, and MAE of the model for accuracy evaluation.

$$R^2 = 1 - \frac{\sum_{i=1}^{n}(y_i - \hat{y}_i)^2}{\sum_{i=1}^{n}(y_i - \overline{y})^2} \tag{5}$$

$$\text{RMSE} = \sqrt{\frac{\sum_{i=1}^{n}(y_i - \hat{y}_i)^2}{n-1}} \tag{6}$$

$$\text{MAE} = \frac{1}{n}\sum_{i=1}^{n}|\hat{y}_i - y_i| \tag{7}$$

In the formula, y_i is the actual observed value, $\hat{y}_i$ is the model prediction, $\overline{y}$ is the average of the actual values, and n is the sample size.

2.3.5. Theil–Sen Median Slope Estimation and Mann–Kendall Trend Analysis

The Theil–Sen median slope estimation (also known as Sen slope estimation) is a robust, nonparametric statistical trend calculation method which can reduce data outliers, and combined with the Mann–Kendall trend analysis (MK test) method is suitable for trend analysis of long time series data [49]. This method does not require the data to obey normal distribution, has a strong resistance to data errors with a more solid statistical theoretical basis for the test of significance level, and the results are more scientific and reliable.

The Sen slope estimate is calculated as

$$\beta = Median\left(\frac{NPP_j - NPP_i}{j-i}\right), \quad 1997 \le i \le j \le 2017 \tag{8}$$

In the formula: *Median ()* represents the median value; when $\beta > 0$, it indicates that the forest NPP shows an upward trend; when $\beta = 0$, it indicates that the forest NPP has no change; when $\beta < 0$, it indicates that the forest NPP shows a downward trend. The specific grading is shown in Table 5.

Table 5. Trend grading of MK test [50].

β	Z	Trend Grading
$\beta > 0$	$2.58 < Z$	extremely significant increase
	$1.96 < Z \le 2.58$	significant increase
	$1.65 < Z \le 1.96$	least-significant increase
	$Z \le 1.65$	non-significant increase
$\beta = 0$	Z	no change
$\beta < 0$	$Z \le 1.65$	non-significant decrease
	$1.65 < Z \le 1.96$	least-significant decrease
	$1.96 < Z \le 2.58$	significant decrease
	$2.58 > Z$	extremely significant decrease

The MK test is a non-parametric statistical test for trend of time series, used to judge the significance of the trend. The data of time series do not need to obey the normal distribution, independent of a few outliers and missing values. The MK test is more applicable to non-normally distributed data, and is usually used to explain the change in the trend of the time series of forest NPP. In the trend test, the original hypothesis H_0 indicates that no trend exists in data set x; the opposing hypothesis H_1 indicates that there is a monotonic trend in data set x.

Suppose $x_1, x_2, \ldots, x_n$ are time series variables and the constructed statistic is

$$S = \sum_{j=1}^{n-1}\sum_{i=j+1}^{n} sgn(x_i - x_j) \tag{9}$$

$$sgn(x_i - x_j) = \begin{cases} 1, & x_i - x_j > 0 \\ 0, & x_i - x_j = 0 \\ -1, & x_i - x_j < 0 \end{cases} \quad (10)$$

In the formula: x_i and x_j are the corresponding data (NPP) values of the *i*-th and *j*-th years, respectively, and $i > j$; *n* is the length of the data set. Then, there is

$$Z = \begin{cases} \frac{S-1}{\sqrt{Var(S)}}, & S > 0 \\ 0, & S = 0 \\ \frac{S+1}{\sqrt{Var(S)}}, & S < 0 \end{cases} \quad (11)$$

In the formula: *Z* is a normally distributed statistic; *Var*(*S*) is the variance. The original hypothesis is rejected if $|Z| \geq Z_{(\alpha/2)}$ at a given α significant level, i.e., there is a significant upward or downward trend in the time series data at the α significant level.

2.3.6. Standard Deviational Ellipse

The standard deviational ellipse [51–53] is an analysis method to characterize the spatial distribution characteristics, including the center coordinate, the rotation angle, and the standard deviation along the long axis (i.e., *y*-axis) and the short axis (i.e., *x*-axis). These elements, respectively, represent the relative position of the spatial distribution pattern of elements, the main trend direction of development, and the degree of dispersion in the main and secondary directions. In this study, ArcGIS was used to generate the standard deviation ellipse of NPP in the study area to identify the position of the center and the spatial movement trend of NPP from 1997 to 2017.

2.3.7. Structural Equation Model

The structural equation model (SEM) [9,54–58] is an advanced and robust multivariate statistical method that combines factor analysis and regression analysis, allowing hypothesis testing on a complex network of path relationships to analyze the relationship between measured variables and latent variables, as well as the relationship between each latent variable. The SEM is composed of a measurement model and structural model. The former is used to analyze the relationship between measurement variables and latent variables, and the latter is used to analyze the relationship between latent variables.

The SEM can study not only observable variables, but also the relationship of variables that cannot be observed directly [59–61]. The SEM can be evaluated from many aspects, such as model regression coefficient, load coefficient, and model fitting index. In this study, a chi-square degrees of freedom ratio (χ^2/df), comparative fit index (CFI), and root-mean-square error of approximation (RMSEA) were used to evaluate the model [62].

3. Results

3.1. NPP Estimation

The selected variables were brought into three models, including the RF, MLR, and BP neural network, to establish forest NPP remote sensing estimation models, which were validated using 10-fold cross-validation. The specific prediction accuracy evaluation is shown in Figure 2.

Comparing the prediction accuracy of the three estimation models, the R^2 of the RF (0.492–0.660) was higher than the MLR (0.307–0.532) and BP neural network (0.422–0.471) models in every year. RF sampling was performed twice. Firstly, the algorithm obtained a sampling set of training samples by random sampling with put-back. Then, a variable was randomly selected from all variables. Meanwhile, the best segmentation feature was selected as a node to build a classification and regression tree. The above reasons made the final model of RF have strong generalization and the highest prediction accuracy of NPP. Compared with traditional machine learning, the BP neural network requires more data to support. In this study, there were only 388 fixed sample plots in the study area, thus the prediction accuracy of NPP was low. The study area has large elevation differences and complex terrain; therefore, the linear correlation between the dependent variable and most of the characteristic variables was poor, resulting in the MLR having the lowest prediction accuracy of NPP. The optimal RF model was selected for spatial mapping of forest NPP in the study area. The mean NPP values in 1997, 2002, 2007, 2012, and 2017 were 5.66 t·hm^{-2}·a^{-1}, 7.68 t·hm^{-2}·a^{-1}, 8.17 t·hm^{-2}·a^{-1}, 8.25 t·hm^{-2}·a^{-1}, and 10.52 t·hm^{-2}·a^{-1}, respectively. The mean and standard deviation of the forest NPP was calculated for the five periods and the NPP was classified into five classes, including low, medium-low, medium, medium-high, and high, by using the mean value of NPP minus 1-fold standard deviation, plus 1-fold standard deviation, and plus 2-fold standard deviation. The NPP mapping of each year is shown in Figure 3.

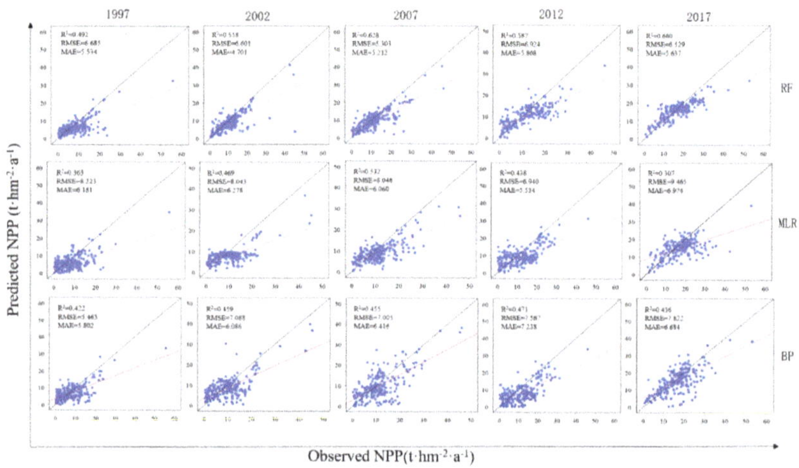

Figure 2. The performance of different inversion models in different years.

From the NPP grade classification mapping in 1997, 2002, 2007, 2012, and 2017, it can be seen that the distribution of high and low values of forest NPP was relatively consistent: the NPP of forests in the north, southwest, and west of the study area was higher, and the NPP of forests in the middle, northwest, and northeast was lower, which is relatively consistent with the altitude distribution of the study area. The northern, southwestern, and western parts of the study area have higher elevation, which are mostly mountainous, less densely populated, and less disturbed by human activities, and the forest vegetation can grow naturally, whereas the central, northwestern, and northeastern parts are mostly hilly areas and valley basins with lower elevation, which are densely populated and more urbanized, with low forest cover and more disturbance from human activities. It can be seen that the spatial distribution of forest NPP is highly consistent with the geomorphic characteristics and socio-economic conditions of the study area.

The forest NPP over 20 years, from 1997 to 2017, shows that the percent of low-grade NPP is gradually increasing, which is mainly due to the economic development and the expansion of the area of towns and cities. The area percentage of medium-low-grade NPP gradually decreased, as the forest with medium-low-grade NPP was farther away from the town compared to the low-grade NPP and was less affected by economic development and human activities. Meanwhile, with the implementation of forest land protection and management regulations, the forest of this grade was intensively managed and the forest NPP had been increasing, making the medium-low-grade NPP gradually develop to medium and medium-high grade. The area proportion of forest with medium-grade and high grade of NPP basically remained stable. The area proportion of medium-high-grade NPP is gradually increasing, as the forest of this grade is mainly located in the hills and river valley basin far away from the town. At the same time, with the enforcement of laws for forest resources and forest land protection, the phenomenon of indiscriminate logging was obviously reduced, making the forest NPP gradually increase. It should be noticed that the area proportion of low-grade NPP in the study area has increased significantly due to urbanization development.

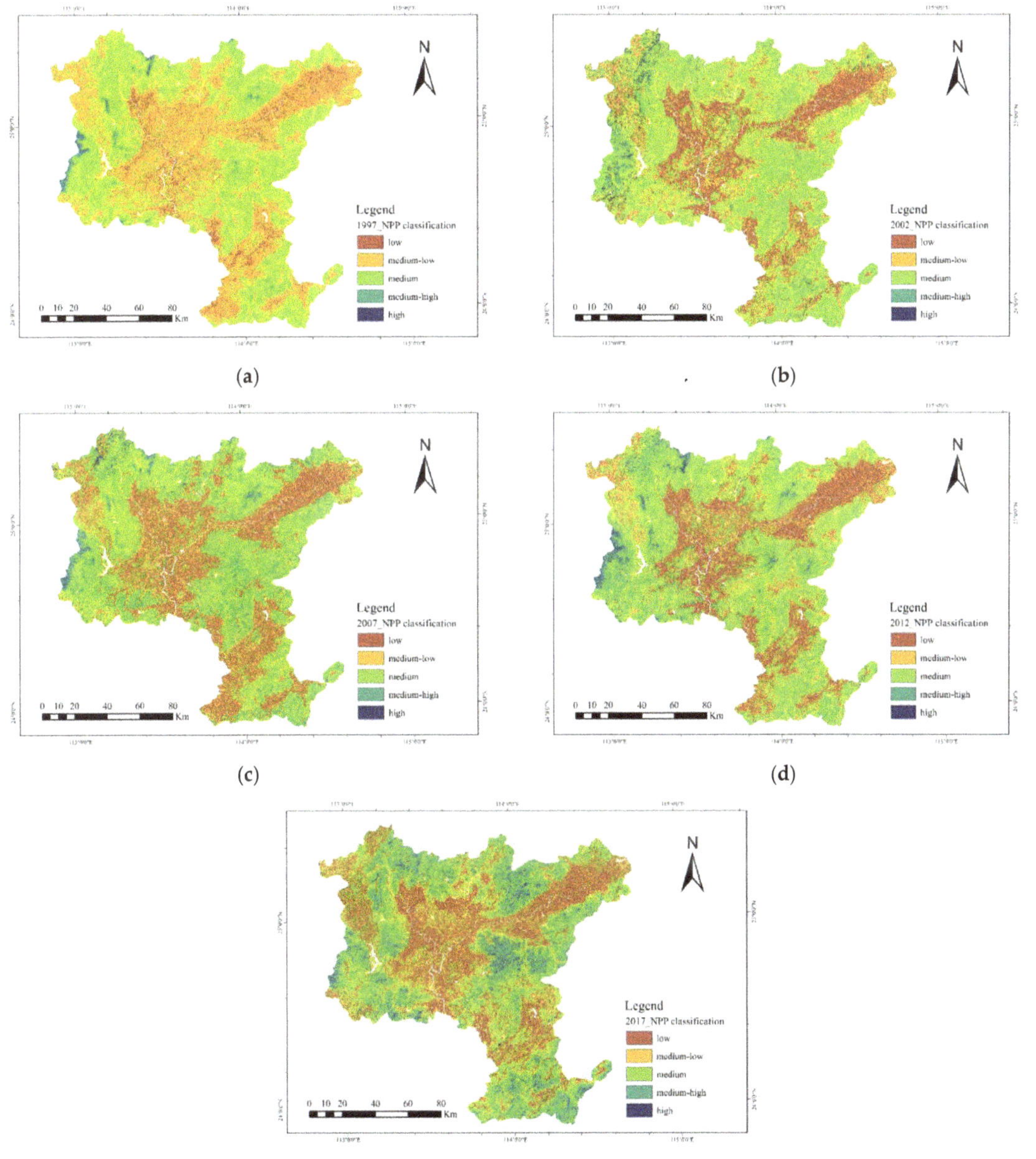

Figure 3. (**a**–**e**) are the spatial distribution maps of NPP classification in Shaoguan in 1997, 2002, 2007, 2012, and 2017, respectively.

3.2. NPP Spatial and Temporal Dynamics

3.2.1. Temporal Dynamics of NPP

Figure 4 shows the trend of the temporal dynamics of NPP in the study area from 1997 to 2017. Generally, the forest NPP in the study showed an upward trend from 1997 to 2017, with the increasing part being 69.86%, the non-significantly increasing part being 47.80%, the least-significant increase part being 7.37%, the significantly increasing part being 9.68%, and the extremely significantly increasing part being 5.02%. The areas of negative increase accounted for 30.12%, the areas of non-significant decrease accounted for 25.10%, the areas of least-significant decrease accounted for 1.59%, the areas of significant decrease accounted for 1.78%, and the areas of extremely significant decrease accounted for 1.67%. The areas with increasing NPP in the study area were mainly located in the mountainous and hilly areas in the west, southwest, and east-central parts of the study area; the areas with decreasing NPP were mainly located in the urban areas in the central, south-central, and northwest parts of the study area.

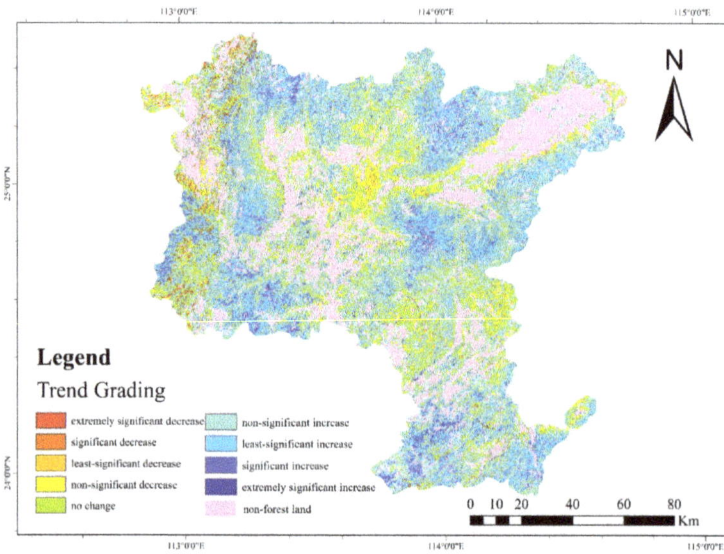

Figure 4. Temporal changes of NPP in Shaoguan from 1997 to 2017.

The NPP in the study area showed an upward trend during the 20 years from 1979 to 2017. In the areas close to the towns, there was a slight downward trend in NPP due to economic development and disturbance from human activities. In the areas far away from the towns, the active adjustment of stand species composition and stand structure made the overall NPP in the study area show an increasing trend.

3.2.2. Spatial Dynamics of NPP

Figure 5 shows the schematic diagram of the NPP standard deviation ellipse and the center point in the study area from 1997 to 2017. The short axis (i.e., x-axis) of the standard deviation ellipse of forest NPP from 1997 to 2002 became longer, the long axis (i.e., y-axis) became shorter, the flatness decreased, and the spatial aggregation increased, indicating that the forest NPP was gradually aggregated from the original uniform distribution and the distribution center shifted to the southeast. This is mainly due to the fact that around 1997, China entered a period of rapid economic development and economic overheating occurred, which led to an increase in forest logging in the low elevation areas of the study area. During the period of 1997–2002, economic rectification measures were gradually implemented, and the destruction of forest resources was slowly reduced. However, during this period, the forest vegetation in the low elevation areas was not immediately restored, and the forest vegetation productivity remained high in the high elevation areas, causing an enlargement in the productivity gap between forest with low elevation and forest with high elevation and an increase in aggregation. From 2002 to 2012, the short axis of the standard deviation ellipse of forest NPP in the study area gradually became shorter, the long axis also gradually became shorter, the flatness decreased, and the spatial aggregation became shorter, indicating that the distribution of forest NPP in the study area was gradually uniform and the distribution center of forest NPP still shifted to the southwest. This is due to the vigorous development of regional ecological construction that focused on forestry development. More afforestation and greening efforts were made in economically developed, low-altitude areas, resulting in a gradual reduction of the gap with high-altitude, economically backward areas. From 2012 to 2017, the short axis of the standard deviation ellipse of forest NPP in the study area became longer, the long axis became shorter, and the flatness decreased, indicating that the spatial aggregation of forest NPP in the study area increased and the center shifted to the southeast. During this period, several forest-related policies were made and forest protection measures were strengthened in areas with a higher altitude and more backward economy; thus, forests were less subjected to human interference and trees grew vigorously, whereas low altitude areas were susceptible to the conversion from high NPP forest land to built-up economic areas. Therefore, the phenomenon of increased spatial aggregation of forest NPP occurred. Generally, from 1997 to 2012, the spatial aggregation of forest NPP in the study area increased and the NPP distribution center shifted from the central-western region to the southwestern direction. From the above analysis, it can be seen that the spatial variation of forest NPP in the study area was more influenced by forestry policies and socio-economic development, such as logging, and the conversion of forest land to built-up economic areas also had some negative effects on it.

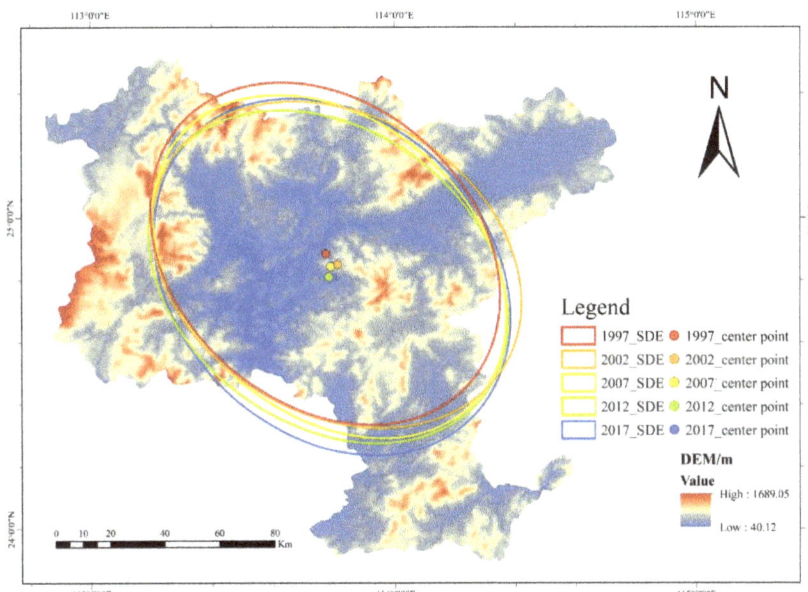

Figure 5. Spatial dynamic changes of NPP in Shaoguan from 1997 to 2017.

3.3. Driving Factors for NPP

Forest growth is influenced by environmental factors, such as topographic factors that can affect the distribution and allocation of light, precipitation, and forest soil, which affects forest productivity. Forest growth is also influenced by its own characteristics, such as understory vegetation, forest stand structure, etc. There is a certain relationship between these factors which affect forest stand growth and forest NPP through interaction. In this study, three types of drivers including environmental factors, understory factors, and forest stand factors in 2017 were chosen to analyze their interactions and effects on forest NPP. Five factors, including slope direction, slope position, slope gradient, elevation, and landform, reflected topographic factors; whereas ten factors, including soil type, soil texture, soil thickness, humus thickness, litter thickness, shrub height, shrub cover, herb height, herb cover, and vegetation cover, reflected understory factors; and nine factors, including dominant species, species structure, age group, average age, average diameter at breast height (DBH), average tree height, canopy density, naturalness, and stand volume, reflected stand factors.

The driving factors were inputted into the SED model, and only 21 driving factors were retained after repeated tests; finally, the optimal SED model was obtained (Figure 6). χ^2/df was 1.99, GFI was 0.954, and RMSEA was 0.064, indicating that the fit of the constructed SED model was basically ideal.

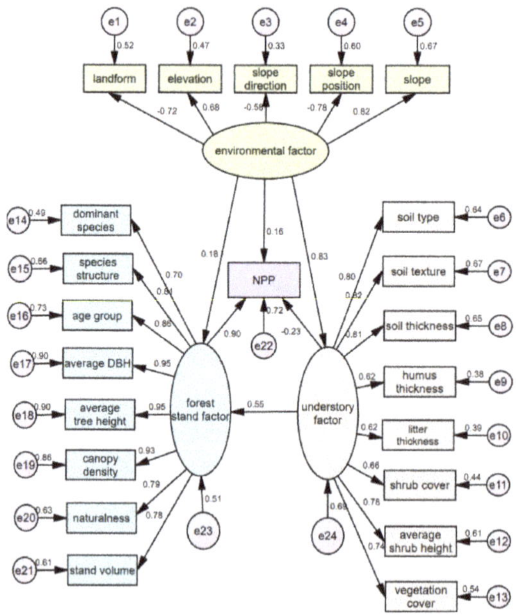

Figure 6. Schematic diagram of NPP structural equation model in Shaoguan in 2017.

As shown in Figure 6, the forest stand factor had the greatest influence on the forest NPP (the path coefficient of 0.90), followed by the understory factor which was negatively correlated with the forest NPP (the path coefficient of −0.23), whereas the environmental factor had the least influence (the path coefficient of 0.16), the environmental factor had a significant positive influence on both the forest stand factor and the understory factor with path coefficients of 0.18 and 0.83, respectively, and the understory factor had a significant positive influence on the forest stand factor with a path coefficient of 0.55. All the driving factors shown in Figure 6 reached significant levels. The standardized factor loading coefficients of environmental factors were slope (0.82), slope position (−0.78), landform (−0.72), elevation (0.68), and slope direction (−0.58), ranking in descending order. Among the forest stand factors, the highest standardized factor loading coefficients of 0.95 were for average DBH and average tree height, followed in descending order by canopy density (0.93), age group (0.86), species structure (0.81), naturalness (0.79), stand volume (0.78), and dominant species (0.70). The loading coefficients of standardized factors in the understory were, from largest to smallest, soil texture (0.82), soil thickness (0.81), soil type (0.80), average shrub height (0.78), vegetation cover (0.74), shrub cover (0.66), humus thickness (0.62), and litter thickness (0.62).

The average DBH and the average tree height in a forest stand directly determine the forest biomass and, thus, the forest NPP; therefore, there is a strong positive correlation between forest stand factor and NPP. The environmental factors including the slope, slope direction, slope position, and elevation directly affected the absorption of light, heat, and nutrients by the forest; thus, the environmental factors have a positive influence on the forest NPP. The understory factors showed a negative relationship with the forest NPP, because the understory plants competed with arbor trees for nutrients and living space, and the more vigorous the understory plants, the slower the arbor trees grew, which eventually reduced the growth rate of NPP.

4. Discussion

In this study, three models, including the RF, MLR, and BP neural network, were applied to estimate forest NPP in the study area. After the selection of feature variables and valuation of the performance of the three models, the optimal model with the highest prediction accuracy was used to predict forest NPP. Based on the NPP prediction results of the optimal model, a spatial and temporal dynamic analysis and driver analysis were conducted to evaluate the long-term effects of forestry policies, human economic activities, and urbanization processes on forest NPP, thus providing some scientific basis for long-term sustainable forest management planning at the regional scale.

Among the three estimation models, the performance of the RF model was better than the MLR and BP neural network models. Using the optimal model of RF, the NPP estimation results were obtained in the study area. The average NPP in Shaoguan increased from 5.66 $t{\cdot}hm^{-2}{\cdot}a^{-1}$ in 1997 to 10.52 $t{\cdot}hm^{-2}{\cdot}a^{-1}$ in 2017. The average increase of NPP in the study area was 0.24 $t{\cdot}hm^{-2}{\cdot}a^{-1}$. An increasing trend was seen in the tropical zone, where an average increase of about 0.15 ($\pm$0.71) $t{\cdot}hm^{-2}{\cdot}a^{-1}$ occurred. A change of more than 1.00 $t{\cdot}hm^{-2}{\cdot}a^{-1}$ was noted in the subtropical zone [63]. NPP by forest types were figured out using the sample plots data from 1997 to 2017 in the study area. The order of NPP from largest to smallest by forest types is: bamboo forest (21.90–28.77 $t{\cdot}hm^{-2}{\cdot}a^{-1}$), broad-leaved mixed forest (15.76–17.99 $t{\cdot}hm^{-2}{\cdot}a^{-1}$), broad-leaved forest (7.85–15.33 $t{\cdot}hm^{-2}{\cdot}a^{-1}$), mixed coniferous forest (11.82–15.07 $t{\cdot}hm^{-2}{\cdot}a^{-1}$), mixed coniferous forest (9.46-12.65 $t{\cdot}hm^{-2}{\cdot}a^{-1}$), coniferous forests (3.91–10.09 $t{\cdot}hm^{-2}{\cdot}a^{-1}$) and shrub forests (2.40–3.32 $t{\cdot}hm^{-2}{\cdot}a^{-1}$). Based on the NPP grading results, it can be seen that although the NPP of the study area increased during the 20 years from 1997 to 2017, the proportion of medium- and low-grade parts in the study area were higher. The Theil–Sen estimation and Mann–Kendall trend analysis showed that the area with increasing forest NPP in the Shaoguan area accounted for 69.86% and was mainly located in mountainous and hilly areas. The area with decreasing forest NPP accounted for 30.12% and was mainly in the built-up area of the town. The SDE showed that the spatial aggregation of NPP in the study area increased from 1997 to 2017, with the distribution center shifting to the southwest. The SEM showed that NPP was significantly positively correlated with forest stand factors and environmental factors, and negatively correlated with understory factors. The method summarized in this study, including the selection of feature variables, introduction of FCD, and spatio-temporal change analysis using Theil–Sen, Mann–Kendall, and SDE, can be applied to NPP estimation in other subtropical regions.

The average NPP in the same period was about 7.00 $t{\cdot}hm^{-2}{\cdot}a^{-1}$ [64] in southern China, 6.50 $\pm$ 3.00 $t{\cdot}hm^{-2}{\cdot}a^{-1}$ in Asia, and 5.89 $\pm$ 2.60 $t{\cdot}hm^{-2}{\cdot}a^{-1}$ in North America [65], which shows that the average NPP of the study area is higher than the above areas. The reason lies in the fact that the forests in the study area are dominated by young and middle-aged forests with faster growth rates. Besides, Shaoguan is a key forest area in China, with better forest land quality, higher forest management intensity, and more attention to forest protection. The mean value of NPP in South America was 9.21 $\pm$ 3.79 $t{\cdot}hm^{-2}{\cdot}a^{-1}$, the mean value of NPP in the tropics was 10.78 ± 3.40 $t{\cdot}hm^{-2}{\cdot}a^{-1}$ [65], and the mean value of NPP in the subtropics was about 12.76 $t{\cdot}hm^{-2}{\cdot}a^{-1}$ [66], which shows that the mean NPP in the study area is lower than in these regions and even lower than in the subtropics, mainly due to the high number of young and middle-aged forests and small storage volume per unit area; thus, the age group structure should be actively adjusted to increase the proportion of mature and over-mature forests, whereas the rotation period of trees should be extended. We should also choose long-lived native broadleaf species and high carbon-fixing efficient species for planting. At the same time, more mixed forests of conifers and broadleaf trees should be established, and stand thinning measures be taken in young and middle-aged forests, so that the trees get enough light and nutrients to improve photosynthetic efficiency. The growth of understory vegetation should be reasonably controlled, and a certain thickness of litter on forest soil should remain to maintain soil productivity. The spatial and temporal dynamics of NPP in the study area are greatly influenced by forestry policies and socio-economic conditions. For example, China implemented the return of some forest land to local ownership in 1981, cessation of cutting in 1985, and removal of agricultural pursuits on forest land in 2003 [67]. Therefore, we should continue to strengthen the implementation of forestry policies, such as returning farmland to forests, constructing beautiful countryside, and compensating for ecological benefits of public welfare forests, in order to continuously reduce the adverse effects of socio-economic development and urbanization on forest productivity.

The extraction and screening of model feature variables, the modeling method, and the introduction of FCD variables can be applied to the study of forest NPP in subtropical regions. In this study, we used a linear model and machine learning methods to estimate NPP. In the future, deep learning models such as convolutional neural networks (CNN) should be introduced to compare their performance with linear models and machine learning methods. In addition to forest canopy density, the forest stand structure factors, including tree height, forest age, and other stand structure factors, can be added to the feature variables of estimation models to research whether they can improve the prediction accuracy of NPP.

5. Conclusions

In this study, the prediction accuracy of forest NPP using the RF model was better than other machine learning models and linear models. The introduction of forest canopy density (FCD) improved the NPP modeling accuracy. The NPP in the study area has gradually increased, but the tree species composition and age group structure still remain unreasonable. The spatial variation of forest NPP in the study area is more influenced by forestry policies, social development, and human disturbance. The NPP in the study area is significantly influenced by stand factors, followed by understory factors and environmental factors.

Author Contributions: Conceptualization, T.L. and M.L.; methodology, T.L.; software, T.L.; validation, T.L., M.L. and F.R.; formal analysis, T.L.; writing—original draft preparation, T.L.; writing—review and editing, F.R., M.L. and L.T.; visualization, T.L.; supervision, M.L.; project administration, M.L.; funding acquisition, M.L. All authors have read and agreed to the published version of the manuscript.

Funding: This research was funded by the National Natural Science Foundation of China, grant number 31770679.

Conflicts of Interest: The authors declare no conflict of interest.

References

1. Stocker, T. *Climate Change 2013*; Cambridge University Press: Cambridge, UK, 2014.
2. Ravindranath, N.H.; Sathaye, J.A. *Climate Change and Developing Countries*; Kluwer Academic Publishers: Norwell, MA, USA, 2002; pp. 247–265. [CrossRef]
3. Tian, L.; Fu, W.; Tao, Y.; Li, M.Y.; Wang, L. Dynamics of the alpine timberline and its response to climate change in the Hengduan mountains over the period 1985–2015. *Ecol. Indic.* **2022**, *135*, 108589. [CrossRef]
4. Kramer, P.J. Carbon dioxide concentration, photosynthesis, and dry matter production. *Bio Sci.* **1981**, *31*, 29–33. [CrossRef]
5. Barkham, J.P.; Waring, R.H.; Schlesinger, W.H. Forest ecosystems: Concepts and management. *J. Ecol.* **1987**, *75*, 284. [CrossRef]
6. Dixon, R.K.; Brown, S.; Houghton, R.A.; Trexier, M.C.; Wisniewski, J. Carbon pools and flux of global forest ecosystems. *Science* **1994**, *263*, 185–190. [CrossRef] [PubMed]
7. Lieth, H.; Whittaker, R.H. *Primary Productivity of the Biosphere*; Ecological Studies; Springer: Berlin/Heidelberg, Germany, 1975. [CrossRef]
8. Wu, W.H.; Li, M.Y.; Bu, Z.H. Estimation of net primary productivity of vegetation in Jiangsu Province based on open datasets. *J. Northwest For. Univ.* **2010**, *25*, 146–151.
9. Li, T.; Li, M.Y.; Tian, L. Dynamics of carbon storage and its drivers in Guangdong Province from 1979 to 2012. *Forests* **2021**, *12*, 1482. [CrossRef]
10. Kicklighter, D.W.; Bondeau, A.; Schloss, A.L.; Kaduk, J.; Mcguire, A.D.; ThE. Participants OF. ThE. Potsdam NpP. Model Intercomparison. Comparing global models of terrestrial net primary productivity (NPP): Global pattern and differentiation by major biomes. *Glob. Chang. Biol.* **1999**, *5*, 16–24. [CrossRef]
11. Liu, J.; Chen, J.M.; Cihlar, J.; Chen, W. Net primary productivity distribution in the BOREAS region from a process model using satellite and surface data. *J. Geophys. Res. Atmos.* **1999**, *104*, 27735–27754. [CrossRef]
12. Wu, Z.F. Response of net primary productivity to climate warming in Northeast China. *Econ. Geogr.* **1997**, *4*, 49–55.
13. Potter, C.S. Terrestrial Biomass and the effects of deforestation on the global carbon cycle. *BioScience* **1999**, *49*, 769–778. [CrossRef]
14. Wen, Y.; Liu, X.; Du, G. Nonuniform time-lag effects of asymmetric warming on net primary productivity across global terrestrial biomes. *Earth Interact.* **2018**, *22*, 1–26. [CrossRef]
15. Xu, X.L.; Cao, M.K. Remote sensing estimation and application analysis of forest biomass. *Geo-Inf. Sci.* **2006**, *8*, 7. [CrossRef]
16. Li, T.; Li, M.Y.; Qian, C.H. Combining crown density to estimate forest net primary productivity by using remote sensing data. *J. Nanjing For. Univ.* **2021**, *45*, 153–160. [CrossRef]
17. Nyikadzino, B.; Chitakira, M.; Muchuru, S. Rainfall and runoff trend analysis in the Limpopo river basin using the Mann Kendall statistic. *Phys. Chem. Earth Parts A/B/C* **2020**, *117*, 102870. [CrossRef]
18. Alhaji, U.U.; Yusuf, A.S.; Edet, C.O. Trend analysis of temperature in gombe state using Mann Kendall trend Test. *J. Sci. Res. Rep.* **2018**, *20*, 1–9. [CrossRef]
19. Peng, J.; Chen, S.; Lü, H.; Liu, Y.; Wu, J. Spatiotemporal patterns of remotely sensed PM 2.5 concentration in China from 1999 to 2011. *Remote Sens. Environ.* **2016**, *174*, 109–121. [CrossRef]
20. Wang, C.; Jiang, Q.; Engel, B.; Mercado, J.A.V.; Zhang, Z. Analysis on net primary productivity change of forests and its multilevel driving mechanism—A case study in Changbai mountains in northeast China. *Technol. Forecast. Soc. Chang.* **2020**, *153*, 119939. [CrossRef]
21. Mi, X.; Feng, G.; Hu, Y.; Zhang, J.; Chen, L.; Corlett, R.T.; Hughes, A.C.; Pimm, S.; Schmid, B.; Shi, S.; et al. The global significance of biodiversity science in China: An overview. *Natl. Sci. Rev.* **2021**, *8*, nwab032. [CrossRef]
22. Lin, M.Z.; Ma, X.F.; Yang, M.Z.; Chen, Z.Y.; Jiang, X.D. Dynamic evaluation of forest eco-system services in Guangdong Province from 1987 to 2004. *Resour. Sci.* **2009**, *31*, 980–984.
23. Ma, X.F.; Lin, M.Z.; Qiao, M.H. The Evaluation of Forest Resources and development counter measures in Guangdong Province. *J. Taiyuan Norm. Univ.* **2007**, *2*, 123–127.
24. Shaoguan Municipal People's Government. Available online: https://www.sg.gov.cn/ (accessed on 20 April 2022).
25. *GB/T 38590-2020*; Technical Regulations for Continuous Forest Inventory. The National Forestry and Grassland Administration: Beijing, China, 2020.
26. Yu, C.; Wang, B.; Liu, H.; Yang, X.S.; Xiu, Z.Z. Dynamic change of net production and mean net primary productivity of China's forests. *For. Res.* **2014**, *27*, 542–550.
27. Zhang, J.B. *Principles and Applications of Remote Sensing*; Wuhan University Press: Wuhan, China, 2009.
28. Liang, S.; Fang, H.; Chen, M. Atmospheric correction of Landsat ETM+ land surface imagery. I. Methods. *IEEE Trans. Geosci. Remote Sens.* **2001**, *39*, 2490–2498. [CrossRef]
29. Unger Holtz, T.S. Introductory digital image processing: A remote sensing perspective, third edition. *Environ. Eng. Geosci.* **2007**, *13*, 89–90. [CrossRef]
30. Teillet, P.M.; Guindon, B.; Goodenough, D.G. On the slope-aspect correction of multispectral scanner data. *Can. J. Remote Sens.* **1982**, *8*, 84–106. [CrossRef]
31. Xiao, X.W. Study on Forest Biomass and Productivity in China. Ph.D. Thesis, Northeast Forestry University, Harbin, China, 2005.
32. Tian, Q.J.; Min, X.J. Progress of vegetation index research. *Adv. Earth Sci.* **1998**, *4*, 10–16.
33. Watson, D.J. Comparative physiological studies on the growth of field crops: I. variation in net assimilation rate and leaf area between species and varieties, and within and between years. *Ann. Bot.* **1947**, *11*, 41–76. [CrossRef]

34. Kauth, R.J.; Thomas, G.S. The Tasselled-cap—A graphic description of the spectral-temporal development of agricultural crops as seen by Landsat. In Proceedings of the Symposium on Machine Processing of Remotely Sensed Data, West Lafayette, IN, USA, 29 June–1 July 1976; pp. 41–51.
35. Joshi, C.; Leeuw, J.D.; Skidmore, A.K.; van Duren, I.C.; van Oosten, H. Remotely sensed estimation of forest canopy density: A comparison of the performance of four methods. *Int. J. Appl. Earth Obs. Geoinf.* **2006**, *8*, 84–95. [CrossRef]
36. Rikimaru, A. Landsat TM data processing guide for forest canopy density mapping and monitoring model. In Proceedings of the ITTO Workshop on Utilization of Remote Sensing in Site Assessment and Planning for Rehabilitation of Logged-Over Forests, Bangkok, Thailand, 30 July–1 August 1996; pp. 1–8.
37. Breiman, L. Random forests. *Mach. Learn.* **2001**, *45*, 5–32. [CrossRef]
38. Cutler, A.; Cutler, D.R.; Stevens, J.R. Random forests. In *Ensemble Machine Learning;* Springer: Berlin/Heidelberg, Germany, 2012; pp. 157–175. [CrossRef]
39. Pavlov, Y.L. Limit Distributions of the height of a random forest. *Theory Probab. Its Appl.* **1984**, *28*, 471–480. [CrossRef]
40. Kolehmainen, E.; Knuutinen, J. Multiple linear regression analysis of RF values of chlorinated catechols and guaiacols. *Chromatographia* **1981**, *14*, 626–628. [CrossRef]
41. Safitri, L.; Mardiyati, S.; Rahim, H. Forecasting the mortality rates of indonesian population by using neural network. *J. Phys. Conf. Ser.* **2018**, *974*, 012030. [CrossRef]
42. Breiman, L.; Cutler, A. *randomForest: Breiman and Cutler's Random Forests for Classification and Regression;* R Package Version 4.6-7; The R Foundation for Statistical Computing: Vienna, Austria, 2008.
43. Li, Y.C.; Li, C.; Li, M.Y.; Liu, Z.Z. Influence of variable selection and forest type on forest aboveground biomass estimation using machine learning algorithms. *Forests* **2019**, *10*, 1073. [CrossRef]
44. You, S.B.; Yan, Y. Stepwise regression analysis method and its application. *Stat. Decis.* **2017**, *14*, 31–35. [CrossRef]
45. Robert, M.O.B. A caution regarding rules of thumb for variance inflation factors. *Qual. Quant.* **2007**, *41*, 673–690. [CrossRef]
46. Bollinger, G.; Belsley, D.A.; Kuh, E.; Welsch, R.E. Regression diagnostics: Identifying influential data and sources of collinearity. *J. Mark. Res.* **1981**, *18*, 392. [CrossRef]
47. Singh, G.; Panda, R.K. Daily sediment yield modeling with Artificial Neural Network using 10-fold cross validation method: A small agricultural watershed, Kapgari, India. *Int. J. Earth Sci. Eng.* **2011**, *4*, 443–450.
48. Zheng, B.; Agresti, A. Summarizing the predictive power of a generalized linear model. *Stat. Med.* **2000**, *19*, 1771–1781. [CrossRef]
49. Sen, P.K. Estimates of the regression coefficient based on Kendall's Tau. *J. Am. Stat. Assoc.* **1968**, *63*, 1379. [CrossRef]
50. Wang, Y.W.; Zhao, Y.J.; Han, L.; Ao, Y. Spatiotemporal variation of vegetation net primary productivity and its driving factors from 2000 to 2015 in Qinling-Daba Mountains. *Chin. J. Appl. Ecol.* **2018**, *29*, 2373–2381. [CrossRef]
51. Zheng, D.F.; Hao, S.; Sun, C.Z.; Lv, L.T. Spatio-temporal pattern evolution of eco-efficiency and the forecast in mainland of China. *Geogr. Res.* **2018**, *37*, 1034–1046. [CrossRef]
52. Peng, J.; Wang, Y.L.; Zhang, Y.; Ye, M.T.; Wu, J.S. Research on the influence of land use classification on landscape metrics. *Acta Geogr. Sin.* **2006**, *61*, 157–168.
53. Cheng, Y.D.; Li, X.D.; Yang, J.Z. Research on NDVI variation characteristics and precipitation sensitivity of the Yuanjiang River Basin in Guizhou Province. *Acta Ecol. Sin.* **2020**, *40*, 1161–1174.
54. Roldán José, L. Review of composite-based Structural Equation Modeling: Analyzing latent and emergent variables. *Struct. Equ. Model. A Multidiscip. J.* **2021**, *28*, 823–825. [CrossRef]
55. Brandmaier, A.M. Permutation Distribution Clustering and Structural Equation Model Trees. Ph.D. Thesis, Saarland University, Saarbrücken, Germany, 2012.
56. Tempelaar, D.T.; Schim, V.D.L.S.; Gijselaers, W.H. A structural equation model analyzing the relationship of students' attitudes toward statistics, prior reasoning abilities and course performance. *Stat. Educ. Res. J.* **2007**, *6*, 78–102. [CrossRef]
57. Anderson, J.C.; Gerbing, W. Structural equation modeling in practice: A review and recommended two-step approach. *Psychol. Bull.* **1988**, *27*, 5–24. [CrossRef]
58. Muthén, B. A general structural equation model with dichotomous, ordered categorical, and continuous latent variable indicators. *Psychometrika* **1984**, *49*, 115–132. [CrossRef]
59. Hoyle, R.H. Introduction to the special section: Structural equation modeling in clinical research. *J. Consult. Clin. Psychol.* **1994**, *62*, 427–428. [CrossRef]
60. Carlsson, M.; Hamrin, E. Evaluation of the life satisfaction questionnaire (LSQ) using structural equation modelling (SEM). *Qual. Life Res.* **2002**, *11*, 415–425. [CrossRef]
61. Quintana, S.M.; Maxwell, S.E. Implications of recent developments in structural equation modeling for counseling psychology. *Couns. Psychol.* **1999**, *27*, 485–527. [CrossRef]
62. Wen, Z.L.; Kittai, H.; Marsh, H.W. Structural equation model: Cutoff criteria for goodness of fit indices and chi-square test. *Acta Psychol. Sin.* **2004**, *36*, 186–194.
63. Wang, Q.; Zhao, P.; Ren, H.; Kakubari, Y. Spatiotemporal dynamics of forest net primary production in China over the past two decades. *Glob. Planet. Chang.* **2008**, *61*, 267–274. [CrossRef]
64. Liu, G.; Sun, R.; Xiao, Z.Q.; Gui, T.X. Analysis of spatial and temporal variation of net primary productivity and climate controls in China from 2001 to 2014. *Acta Ecol. Sin.* **2017**, *37*, 4936–4945. [CrossRef]

65. Jiao, C.C.; Yu, G.R.; Zhan, X.Y.; Zhu, X.J.; Chen, Z. Spatial patterns of net primary productivity of global forest ecosystems and their regional characteristics. *Quat. Sci.* **2014**, *34*, 699–709. [CrossRef]
66. Li, H. Spatiotemporal Evolution of Fractional Vegetation Cover and Net Primary Productivity in the Subtropical Region and Climate Driving. Master's Thesis, Zhejiang A&F University, Hangzhou, China, 2021.
67. Yu, Y.T.; Wu, S.R.; Meng, G.; Zhang, X.F. Study on the Cooperation Networks Evolution of Forestry Policy-making Authorities in China. *For. Econ.* **2020**, *42*, 3–19.

remote sensing

MDPI

Article

Above-Ground Biomass Estimation of Plantation with Different Tree Species Using Airborne LiDAR and Hyperspectral Data

Linghan Gao [1,2,3,4], Guoqi Chai [1,2] and Xiaoli Zhang [1,2,*]

1 Precision Forestry Key Laboratory of Beijing, Beijing Forestry University, Beijing 100083, China; gaolh@nciae.edu.cn (L.G.); chaigq@bjfu.edu.cn (G.C.)
2 The Key Laboratory for Silviculture and Conservation of Ministry of Education, Beijing Forestry University, Beijing 100083, China
3 School of Remote Sensing and Information Engineering, North China Institute of Aerospace Engineering, Langfang 065000, China
4 Hebei Collaborative Innovation Center for Aerospace Remote Sensing Information Processing and Application, Langfang 065000, China
* Correspondence: zhangxl@bjfu.edu.cn

Abstract: Forest above-ground biomass (AGB) is an important index to evaluate forest carbon sequestration capacity, which is very important to maintain the stability of forest ecosystems. At present, the wide use of remote sensing technology makes it possible to estimate the large-scale forest AGB accurately and efficiently. Airborne hyperspectral remote sensing data can obtain rich spectral information and spatial structure information on the forest canopy with the characteristics of high spatial and hyperspectral resolution. Airborne LiDAR data can describe the three-dimensional structure characteristics of a forest and provide vertical structure information related to biomass. Based on the characteristics of the two data sources, this study takes Gaofeng forest farm in Nanning, Guangxi, as the study area, Chinese fir, pine tree, eucalyptus and other broadleaved trees as the research object, and constructs the AGB estimation models of different tree species by fusing airborne LiDAR and hyperspectral features. Firstly, spectral features, texture features, vegetation index, wavelet transform features and edge features are extracted from hyperspectral data. Canopy structure features, point cloud structure features, point cloud density features and terrain features are extracted from airborne LiDAR data. Secondly, the random forest (RF) method is used to screen the features of the two sets of data, and the features with the highest importance are selected. Finally, based on the optimal features of the two data sources, the forest AGB model is constructed using the multiple stepwise regression method. The results show that the texture features extracted by wavelet transform can be used for AGB modeling. The AGB of eucalyptus has high correlation with height features derived from airborne LiDAR, the AGB of other broadleaved trees mostly depends on the wavelet transform texture features from airborne hyperspectral data, while the AGB of Chinese fir and pine tree has high correlation with both height features and spectral features. Feature-fusion-based LiDAR and hyperspectral data can greatly improve the accuracy of the AGB models. The accuracy of the optimal AGB models of Chinese fir, pine tree, eucalyptus and other broadleaved trees is 0.78, 0.95, 0.72 and 0.89, respectively. In conclusion, more accurate estimation results can be obtained by combining active and passive remote sensing data for forest AGB estimation, which provides a solution for carbon storage assessment and forest ecosystem assessment.

Keywords: above-ground biomass (AGB); airborne LiDAR; airborne hyperspectral; wavelet transform; feature fusion

Citation: Gao, L.; Chai, G.; Zhang, X. Above-Ground Biomass Estimation of Plantation with Different Tree Species Using Airborne LiDAR and Hyperspectral Data. *Remote Sens.* **2022**, *14*, 2568. https://doi.org/10.3390/rs14112568

Academic Editors: Huaqiang Du, Wenyi Fan, Weiliang Fan, Fangjie Mao and Mingshi Li

Received: 25 April 2022
Accepted: 24 May 2022
Published: 27 May 2022

1. Introduction

Forests are important natural resources for maintaining ecological balance and stability. The changes in forest resources and reserves will directly affect the decisions of major national forestry planning [1]. At the same time, as a natural and renewable resource, the quantity and quality of forests will directly affect national economic construction and people's quality of life [2]. In 2007, a report from the IPCC (United Nations Intergovernmental Panel on Climate Change) pointed out that forestry has multiple benefits and has the dual functions of mitigating and adapting to climate change. This report is an economic and effective measure to increase carbon sequestration and reduce emissions in the next 30–50 years. The Paris Agreement also lists the forestry provisions separately, encourages countries to take actions to protect and enhance forest carbon pools and sinks after 2020, and continues to encourage developing countries to implement and support REDD+ (reducing deforestation, mitigating forest degradation and reducing greenhouse gas emissions). China also put forward the vision of "carbon peaking and carbon neutralization" in 2020, which shows that forest resources play an important role in global climate change and ecological balance. Therefore, it is necessary to monitor and assess the dynamic information of forest resources in time.

China is rich in forest resources and various forest types, which occupy an important position in the terrestrial ecosystem. At the same time, the forest community structure is complex and the biomass is high. More than 80% of the forest vegetation biomass is stored in the ecosystem. Therefore, the study of forest biomass estimation can better evaluate the problems of forest productivity and forest carbon cycle, and provide key data support for the study of global climate change and development trend. At this stage, plantation resources account for a large proportion of China's forest land resources, accounting for about 40% of the national forest land area, which is of great significance to the development of forest resources and the construction of the ecological environment in China. The plantation is mainly distributed in Southern China, and Eucalyptus, pine tree and Chinese fir are the main tree species. Their wide distribution range and high forest canopy density pose a great challenge to China's scientific forest management.

At present, remote sensing technology is developing rapidly and is widely used in forest inventory and large-scale real-time monitoring of forest resources [3–6]. It effectively solves the limitations of being time consuming and labor intensive of traditional manual inventory, can quickly and conveniently obtain a large number of forestry remote sensing basic data, and can realize large-scale and long-time monitoring of forest AGB [7–10]. Optical remote sensing is the most widely used and popular remote sensing data resource, which can provide spectral and texture features for forest AGB estimation [11,12], but optical remote sensing data also have many limitations. Multispectral optical images have the disadvantages of few spectral bands and narrow wavelength range, which has limitations in describing the physiological and ecological characteristics of forest vegetation [13]. A hyperspectral image adopts imaging spectral technology, which contains hundreds of bands in the imaging spectral domain, with a wide spectral range and a large number of bands, forming continuous spectral curve data, which can meet the needs of spectral information for forest AGB estimation [14–16]. However, hyperspectral images also have some problems, such as foreign objects with the same spectrum, different spectra of the same object and weak penetration to ground objects. Different tree species, shrubs and trees with different heights may have similar spectral information, which will affect the inversion accuracy of forest AGB [17,18]. As an active remote sensing technology, light detection and ranging (LiDAR) has incomparable advantages over traditional remote sensing and measurement methods in data acquisition [19]. The laser pulse can obtain the terrain information under the forest canopy through the forest cover, obtain the forest height information or stand density and other information closely related to biomass with high precision, and can be used for high-resolution three-dimensional reconstruction [20,21]. However, although LiDAR data have unique advantages that optical remote sensing data do not have, it is less precise than optical remote sensing data in canopy detection and spatial resolution. Therefore, the combination of these two data has the potential to accurately estimate forest parameters [22–25].

Baccini A et al. [26] took Africa as the study area, combined spaceborne laser GLAS data with MODIS data to retrieve the forest AGB in tropical Africa and generated the biomass distribution map. The results showed that there was a strong correlation between the height feature of GLAS data and AGB, with R^2 of 0.9. Boudreau J et al. [27] combined SRTM, ETM + and airborne LiDAR point cloud data to carry out forest AGB inversion in Quebec, Canada, and constructed an AGB model using the method of feature fusion, with R^2 of 0.65. Chen G et al. [28] used LiDAR data, QuickBird data ground survey data to retrieve the tree height, AGB and volume of some forest areas in Vancouver, Canada, using the support vector machine method. The research showed that the performance of SVR is better than that of the multiple regression method. Laurin et al. [29] combined hyperspectral data with LiDAR point cloud data, and estimated the forest AGB in tropical forest areas of Africa. The results showed that the R^2 of the model was 0.7, which improved the accuracy by 6% compared with using LiDAR data alone. Li et al. [30] studied the basic distribution of forest AGB in California, USA, by using LiDAR and multi-temporal MODIS data. The results showed that the accuracy was high and R^2 was 0.74. Luo et al. [22] took the forest area of Heihe River Basin in Heilongjiang Province as the study area, analyzed the forest AGB in this area, in combination with hyperspectral data and airborne LiDAR data, and concluded that the estimation accuracy using the two set of data was 0.893, and the accuracy using LiDAR data only was 0.872. In comparison, the combination of the two sets of data increases the estimation accuracy. Catherine T de A et al. [31] took the Amazon region of Brazil as the study area, combined hyperspectral data with LiDAR data to establish an AGB estimation model by screening indicative features from 333 features (45 from LiDAR and 288 from hyperspectral). The results showed that the model combining the two data sources can obtain more accurate forest biomass estimation value, and the R^2 of best model was 0.70. Wang et al. [32] used Sentinel-2 and airborne LiDAR data to inverse the AGB of mangrove forests in the northeast of Hainan Island. The results showed that the method based on a point line polygon framework proposed in the study can effectively estimate the AGB of this area, and verify the feasibility of this method in different mangrove types.

To sum up, estimating forest AGB based on active and passive remote sensing data can overcome the limitations of a single data source, give full play to the respective advantages of data and realize high-precision estimation of forest parameters. However, most forest AGB estimation research does not consider tree species or only a specific tree species, which makes the research for large-area forest AGB estimation with multiple tree species limited. At the same time, the extraction of feature parameters of optical remote sensing data is mostly the use of spectral reflectance features and gray-level co-occurrence matrix (GLCM) texture features, and there is no in-depth research on the spatial features of optical images. In view of the above problems, taking Guangxi Gaofeng forest farm as the research area, this study discusses the accuracy difference in estimating forest AGB using only single remote sensing data and combining active and passive remote sensing data. At the same time, the spatial and spectral features related to forest AGB are extracted from airborne hyperspectral imagery, and the different AGB modeling of four main dominant tree species in this area are discussed, in order to improve the estimation accuracy of forest AGB and provide a reference for forest carbon storage estimation and ecological assessment in China.

2. Materials and Methods

2.1. Materials

2.1.1. Study Area

The study area is located in Gaofeng forest farm in Nanning, Guangxi province. The geographical location is 108°19′30″~108°23′30″E and 22°56′~23°1′N. The forest farm is dominated by low hills, with an altitude of 70~875 m and a slope of 25~35°. The terrain has little fluctuation, low in the southeast and high in the northwest (Figure 1). The forest farm has a tropical monsoon climate with an annual average temperature of 12.5~28.2 °C, an average rainfall of 1304 mm, sufficient sunshine and an annual average sunshine time of 1550 h. The proportion of non-forest land and forest land is about 1:99, and the forest coverage rate of the whole forest farm is close to 90%. In addition to forest land, the forest farm also includes shrub, nursery, non-standing forest land after cutting, other sparse forest land and non-forested areas, with an area ratio of 1:1:56:29:13. The forest type has typical characteristics of forests in South China, with rich tree species, mainly planted forests. The tree species include *Eucalyptus Urophylla*, *Eucalyptus Grandis X Urophylla*, *Castanopsis Hystrix*, *Cunninghamia Lanceolata* (Chinese fir) and *Pinus Massoniana*. The proportion of Chinese fir, pine tree, eucalyptus and other broadleaved trees in the study area is 1:1:5:3; eucalyptus accounts for nearly half.

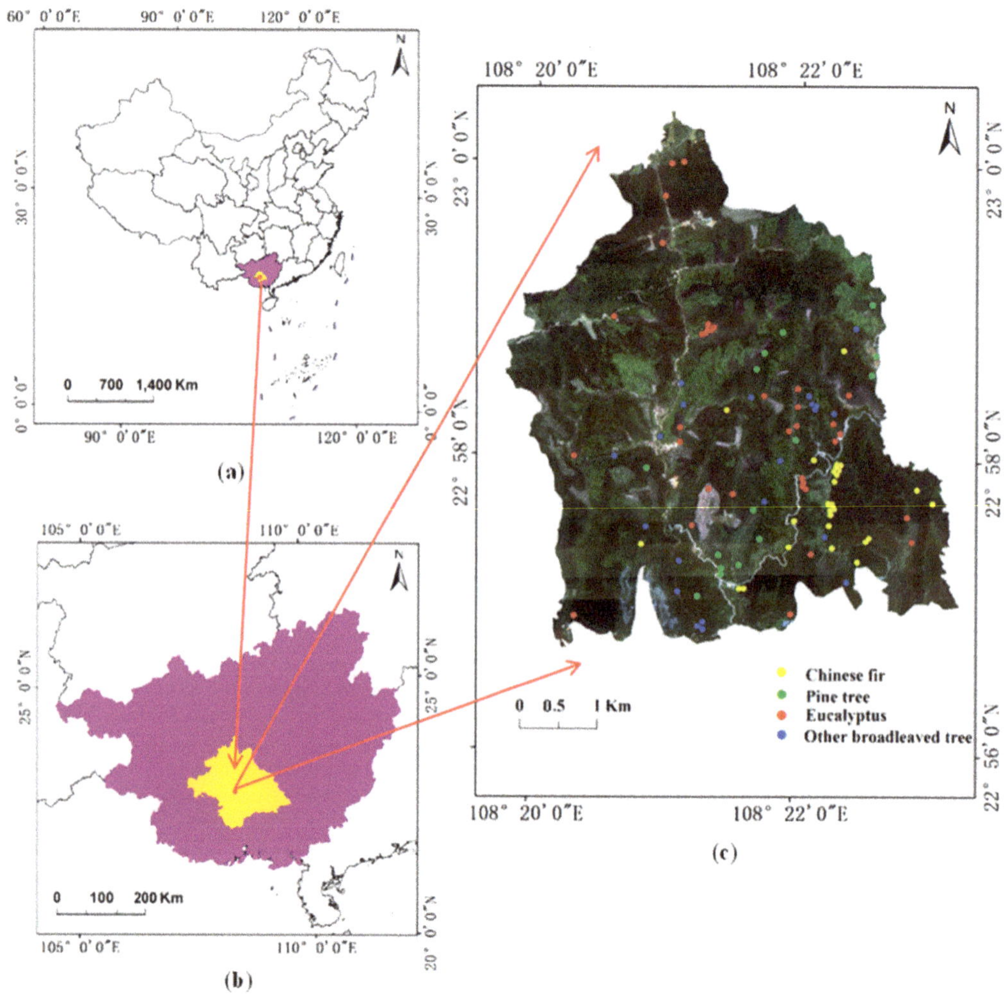

Figure 1. Location of the study area and the field survey plots ((**a**) is the location of Guangxi Province. (**b**) is the location of Nanning City. (**c**) is the distribution of each species sample plot, the base map is hyperspectral image of the study area).

2.1.2. Field Data

In January 2018, a field survey was conducted in Gaofeng forest farm, Nanning, Guangxi province, and the measured sample plot data were collected. According to the terrain and stand characteristics of the study area, sample plots of different sizes were set up. A total of 98 plots are arranged in the study area, including 27 Chinese fir plots, 15 pine tree plots, 35 eucalyptus plots and 21 other broadleaved tree plots. Other broadleaved trees mainly include *Dygoxyllum, Lllicium Linn, Magnolia Denudata, Magnoliaceae Glanca Blume* and *Erythrophleum Fordii Oliv*. The diameter at breast height (DBH) of each tree with DBH greater than 5 cm was measured using DBH ruler, the height of each tree was measured with laser altimeter, the coordinates of each tree were measured with total station, and the coordinates of the center and four corners of the sample plots were measured with RTK. The basic information of the sample plots is shown in Table 1.

Table 1. Sample plots information.

Tree Species	Forest Age (Year)	DBH (cm)	Tree Height (m)	Stem Density ($n \cdot ha^{-1}$)
Chinese fir	26 ± 5	26.9 ± 22.1	18.8 ± 13.4	2098 ± 1634
Pine Tree	14 ± 7	12.7 ± 6.6	6.2 ± 6.0	1211 ± 667
Eucalyptus	15 ± 13	17.8 ± 14.8	19.2 ± 17.9	1610 ± 1034
Other broadleaved tree	24 ± 16	27.2 ± 22.9	14.9 ± 10.7	1373 ± 806

Note: m ± n, m is the median of the tree parameters for each tree species, n is the maximum value that this parameter fluctuates up or down.

The AGB of each tree species is calculated using the allometric growth equation of AGB by the measured DBH and tree height. For tree species with more than 20 samples, the number of verification samples is set to a number greater than 10, and the rest are training samples. The tree species with less than 20 samples are modeled and verified by the leave-one method.

2.1.3. Remote Sensing Data

The remote sensing data were obtained by the institute of resource information, Chinese Academy of Forestry Sciences in February 2018 using the Yun-12 fixed wing UAV equipped with RIEGLLMS-Q680i laser scanning system (Horn, Austria) and AISA Eagle II sensor (Oulu, Finland) in sunny weather. The point density of airborne LiDAR point cloud data is 3.35 points/m^2, and the data format is .las. The hyperspectral data contains 125 bands, and the data format is .dat. The parameters of laser scanning system are shown in Gao Linghan et al. [33], and the detailed parameters of hyperspectral sensor are shown in Table 2.

Table 2. The main spatial parameters of hyperspectral system.

Parameters	Value
Spectral range (nm)	400~1000
Spectral resolution (nm)	3.3
Field angle (°)	37.7
Instantaneous field angle (mrad)	0.646
Focal length (mm)	18.1
Number of spatial pixels	1024
Spectral sampling interval (nm)	4.6
Quantized value (bits)	12
Number of bands	125

2.1.4. Data Preprocessing

The main preprocessing includes radiation calibration, atmosphere correction and terrain radiation correction [34]. Conversion formula was used to complete the radiometric calibration, so as to convert the DN value of the initial image into the radiance value. The fast atmospheric correction method was used to correct the hyperspectral image data, so as to eliminate the influence of atmosphere on the reflection of ground objects [35]. SCS + C correction model is used for terrain correction to eliminate the influence of surface roughness on ground reflectance or brightness [36].

The preprocessing of LiDAR point cloud data is mainly to realize elevation normalization [37]. Firstly, the threshold method is used to remove the noise points generated in the scanning process and retain the important point cloud data in the study area [38,39]. According to the measured forest height, crown width and terrain elevation, the forest threshold range is 0.3~50 m and the search radius is 3 m [40,41]. Secondly, the irregular triangulation filtering TIN algorithm was used to classify the point cloud data and distinguish the ground points and non-ground points [42]. Finally, the ground points were interpolated using TIN interpolation algorithm to generate digital elevation model (DEM) [43], and the non-ground points are interpolated using Kriging interpolation algorithm to generate digital surface model (DSM). The difference operation was performed between DSM and DEM to obtain the canopy height model (CHM). The point cloud data after elevation normalization are shown in Figure 2.

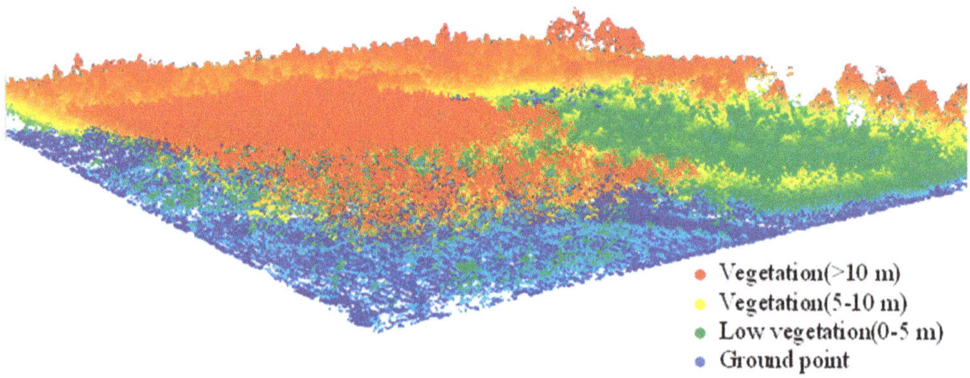

Figure 2. Elevation normalized point cloud of the study area.

2.2. Methods

2.2.1. Feature Variables Extraction from Hyperspectral Imagery

Hyperspectral imagery has higher spatial resolution and spectral resolution contains richer spectral information and spatial structure information and can obtain more features related to forest AGB. Firstly, the spectral reflectance features, first derivative and second derivative features of 125 bands of hyperspectral data were extracted, respectively (Table 3). Secondly, based on the previous literature [44–46], several typical vegetation indices characterizing vegetation coverage and biomass were extracted (Table 4), which included the indices related to atmospheric impedance and topographic characteristics and chlorophyll content and indices representing the characteristics of vegetation leaves. As such, 8 s-order texture features from band 19 (482 nm), band 34 (550 nm) and band 55 (645 nm) were extracted separately based on the GLCM method (Table 5) [47]. These three bands correspond to blue, green and red bands, respectively, with high definition, less interference and obvious ground feature information. In this way, a total of 24 object-based texture features was obtained. Finally, in order to extract more spatial information related to forest structure, wavelet transform [48] and mathematical morphology [49] were used to extract spatial texture features, transformed spectral features and edge features (Table 6).

Table 3. List of the spectral features for hyperspectral data.

Type	Name
Spectral reflectance	Band1, Band2. . .Band125
First derivative	X1st1, X1st2. . .X1st125
Second derivative	X2nd1, X2nd2. . .X2nd125

Table 4. List of the vegetation indices for hyperspectral data.

Type	Name	Formula
Broad-band greenness index	Normalized differential vegetation index (NDVI)	$NDVI = \frac{\rho_{800}-\rho_{680}}{\rho_{800}-\rho_{680}}$
	Enhanced vegetation index (EVI)	$EVI = 2.5\left(\frac{\rho_{800}-\rho_{680}}{\rho_{800}+6\rho_{680}-7.5\rho_{450}+1}\right)$
Narrow-band greenness index	Red edge normalized difference vegetation index ($NDVI_{705}$)	$NDVI_{705} = \frac{\rho_{750}-\rho_{705}}{\rho_{750}+\rho_{705}}$
	$NDVI_1$	$NDVI_1 = \frac{\rho_{750}-\rho_{740}}{\rho_{750}+\rho_{740}+0.0001}$
	$NDVI_2$	$NDVI_2 = \frac{\rho_{756}-\rho_{735}}{\rho_{756}+\rho_{735}}$
	Soil adjust vegetation index (SAVI)	$SAVI = \frac{1.5\times(\rho_{800}-\rho_{680})}{\rho_{800}+\rho_{680}+0.5}$
	SAVI2	$SAVI2 = \frac{1.5\times(\rho_{800}-\rho_{680})}{\rho_{800}+\rho_{680}+0.5}$
Light utilization index	Photochemical reflectance index (PRI)	$PRI = \frac{\rho_{531}-\rho_{570}}{\rho_{531}+\rho_{570}}$
Other indexes	Transformed chlorophyll absorption in reflectance index ($TCARI_{670.700}$)	$TCARI_{670.700} = 3[(\rho_{700}-\rho_{670}) - 0.2(\rho_{700}-\rho_{550}) \times (\rho_{700}/\rho_{670})]$
	Optimized soil-Adjusted vegetation index ($OSAVI_{670.800}$)	$OSAVI_{670.800} = (1+0.16) \times (\rho_{800}-\rho_{670})/(\rho_{800}+\rho_{670}+0.16)$
	Modified chlorophyll absorption in reflectance index (MCARI)	$MCARI = [(\rho_{700}-\rho_{670}) - 0.2(\rho_{700}-\rho_{550})](\rho_{700}/\rho_{670})$
	OSAVI	$OSAVI = (1+0.16) \times (\rho_{800}-\rho_{700})/(\rho_{800}+\rho_{700}+0.16)$
	SARVI	$SARVI = 1.5 \times (\rho_{800} - 2 \times \rho_{670} + \rho_{445})/(\rho_{800} + 2 \times \rho_{670} - \rho_{445} + 0.5)$

Table 5. List of the second-order texture indices calculated by GLCM for hyperspectral data.

Type	Name	Formula
Entropy	entropy1, entropy2...entropy125	$\sum_{ij=0}^{N-1} Pij \times (-lnPij)$
Second moment	second.moment1, second.moment2...second.moment125	$\sum_{ij=0}^{N-1} Pij2$
Variance	variance1, variance2...variance125	$\sum_{ij=0}^{N-1} Pij \times (1-mean)2$
Mean	mean1, mean2...mean125	$\sum_{ij=0}^{N-1} iPij$
Correlation	correlation1, correlation2...correlation125	$\sum_{ij=0}^{N-1} Pij \times [\frac{(i-mean)-(j-mean)}{\sqrt{variance_i \times variance_j}}]$
Homogeneity	homogeneity1, homogeneity2...homogeneity125	$\sum_{ij=0}^{N-1} i\frac{Pij}{1+(i-j)2}$
Contrast	contrast1, contrast2...contrast125	$\sum_{ij=0}^{N-1} iPij \times (i-j)2$
Dissimilarity	dissimilarity1, dissimilarity2...dissimilarity125	$\sum_{ij=0}^{N-1} iPij \times \lvert i-j \rvert$

Table 6. List of the spatial texture and transform features for hyperspectral data.

Types	Name	Describing
Spectral feature	(BT1, BT2...BT125)	Spectral features of two-dimensional wavelet transform
Texture feature	Horizontal texture (Hor1, Hor2...Hor125)	Horizontal texture of two-dimensional wavelet transform
	Vertical texture (Ver1, Ver2...Ver125)	Vertical texture of two-dimensional wavelet transform
	Approximate texture (App1, App2...App125)	Approximate texture of two-dimensional wavelet transform
	Diagonal texture (Dia1, Dia2...Dia125)	Diagonal texture of two-dimensional wavelet transform
Edge feature	(Edg1, Edg2...Edg125)	Edge texture of mathematical morphology analysis

2.2.2. Feature Variables Extraction from LiDAR

According to the data structure features of point cloud data and comprehensively considering the ecological and spatial structure indicators, the feature parameters of point cloud data were extracted from forest canopy information (including canopy density and leaf area index), point cloud structure information (including height percentile, height maximum and minimum), point cloud density information (including point cloud density parameters at different height levels of point cloud) and terrain information (including slope and aspect). For meaning and abbreviation of each feature variable of point cloud data see Gao Linghan et al. [33]. The important features of point cloud data include height percentile and cumulative height percentile. The height percentile refers to the height of X% points in a unit grid. The cumulative height percentile is the height sum of X% points in a unit grid.

2.2.3. Feature Variables of Three-Level Screening and Modeling

In this study, there are many hyperspectral feature parameters and few sample plots. Putting all features into the model will lead to data redundancy, supersaturation of modeling variables and low model accuracy. The following two-level screening scheme was designed for hyperspectral features: First, the extracted feature sets of 11 categories, such as spectral reflectance features, first derivative features, second derivative features, vegetation indices, GLCM texture features, spectral feature of wavelet transform, horizontal texture of wavelet transform, vertical texture of wavelet transform, diagonal texture of wavelet transform, approximate texture of wavelet transform and edge texture, are successively screened for each tree species according to RF method, and modeled separately by multiple stepwise regression method (MSR) [50]. Compare the model accuracy, eliminate the feature sets with model accuracy less than 0.5, and use the remaining feature sets for subsequent screening and modeling. Then, according to the first screening results, RF screening [51] is carried out again to select the corresponding top ranking features of each tree species. Finally, the optimal model of each tree species based on hyperspectral data is obtained by using the MSR method again. According to the number of training samples and the principle of moderate proportion, the proportion of training samples and independent variables is set as 4:1.

RF method was used to screen the best feature variables derived from airborne LiDAR point cloud data for each tree species. The screening results are shown in Gao Linghan et al. [33]. Then the optimal variables of each tree species screened by airborne hyperspectral and LiDAR data were fused, and the optimal variables of each tree species were screened again by RF method to realize the optimal feature fusion of the two data sources and obtain the final feature variable set. The AGB model of each tree species was established by MSR method to realize the AGB modeling of each tree species based on the feature fusion of multi-source data. The three-level screening and modeling process is shown in Figure 3.

The RF method is a popular feature-selection method, which can realize data reduction and optimization. The decline in target prediction accuracy after removing variables is indicated by %IncMSE, which is the growth of root mean square error rate. When the value is larger, the contribution of the variable is greater. Further, %IncMSE formula is shown in Gao Linghan et al. [33]. MSR method considers the variance contribution value of all variables when introducing variables and sorts them into a regression equation according to their importance. The final equation does not contain unnecessary independent variables. The coefficient of determination R^2, the root mean square error (*RMSE*) and mean absolute error (*MAE*) were used to compare the accuracy. The formula is as follows:

$$R^2 = 1 - \frac{mean(Xmodel, i - Xobs, i)^2}{mean(mean(Xobs, i) - Xobs, i)^2} \quad (1)$$

where: R^2 is the coefficient of determination, *Xobs*, *i* is the measured value, *Xmodel*, *i* is the estimated value, and *mean* is the average value.

$$RMSE = \left(mean(Xmodel, i - Xobs, i)^2\right)^{0.5} \quad (2)$$

$$MAE = \frac{1}{N}\sum_{i=1}^{N} |Xmodel, i - Xobs, i| \quad (3)$$

where: *RMSE* is the root mean square error, *MAE* is the mean absolute error, *N* is the number of samples.

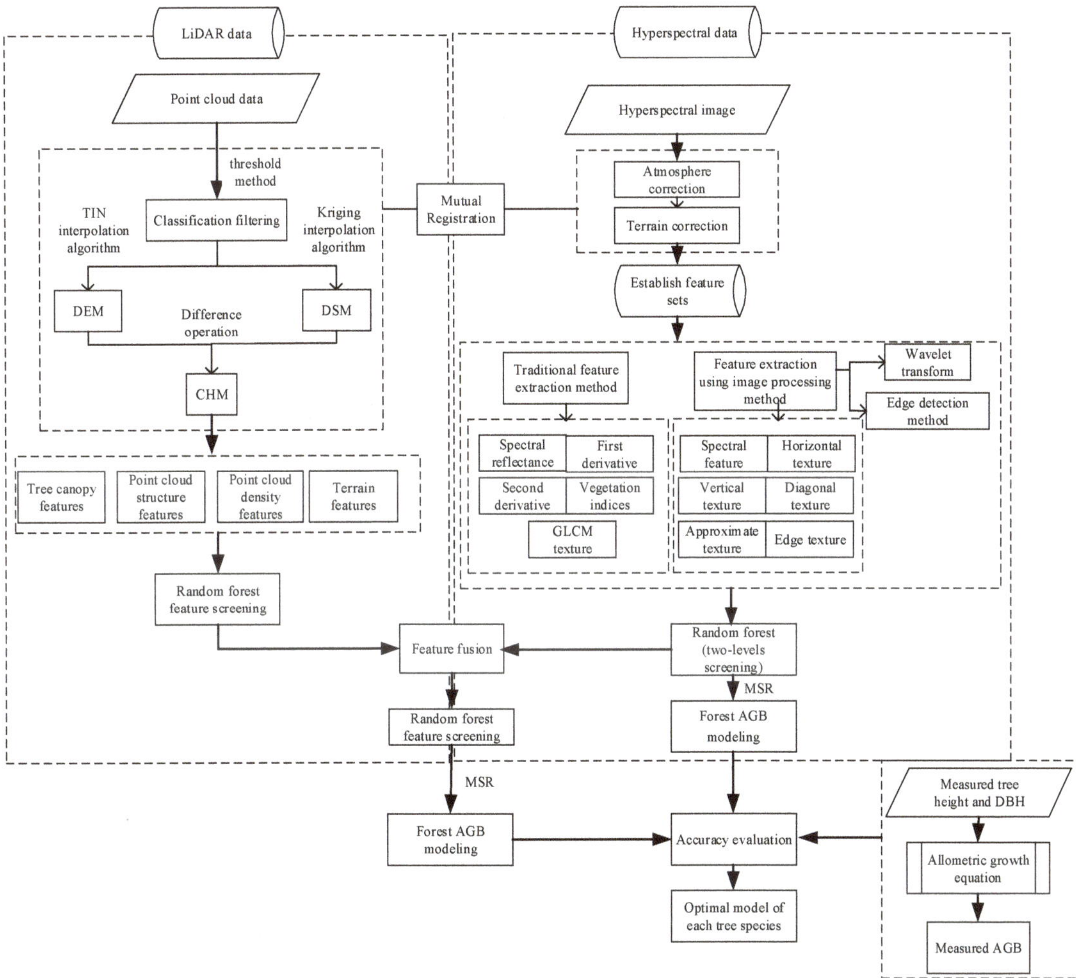

Figure 3. Three-level screening and modeling process.

3. Results

3.1. Hyperspectral Features Selection

The feature-screening results based on airborne LiDAR data are shown in Gao Linghan et al. [33]. The screening of hyperspectral features eliminates redundant feature parameters and obtains several feature parameters, with the highest correlation between 11 feature sets and AGB of each tree species using the RF method. Four, four, six and three feature parameters were selected for Chinese fir, pine tree, eucalyptus and other broadleaved trees, respectively. The screening results are shown in Figure 4.

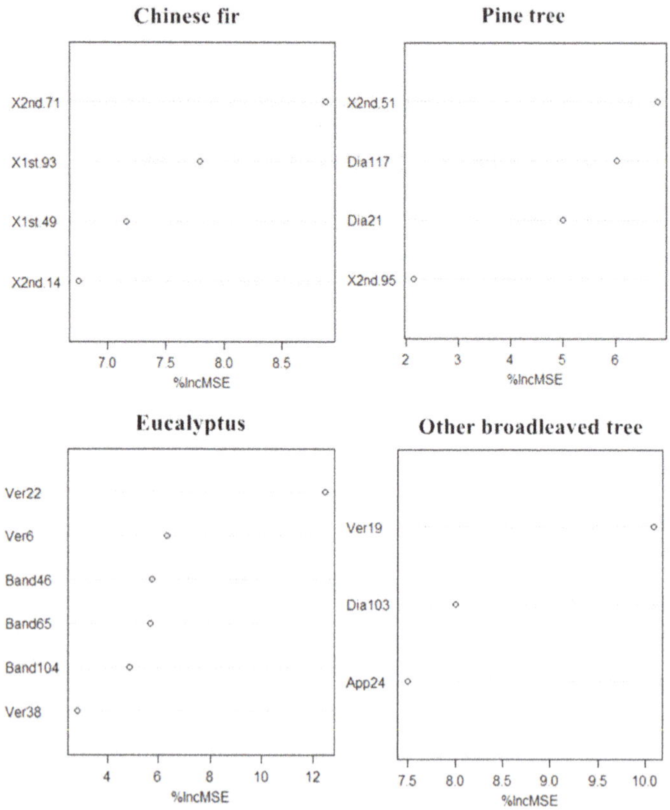

Figure 4. The feature importance ranking map after two-level screening of hyperspectral features.

It can be seen from Figure 4 that the AGB of Chinese fir has a strong correlation with the derivative features. The AGB of pine trees has a good correlation with the second derivative and the diagonal texture features of wavelet transform, and the texture features of GLCM are removed. The AGB of eucalyptus has a strong correlation with the spectral reflectance features and the vertical texture features of wavelet transform. The AGB of other broadleaved trees has a good correlation with the three texture features of wavelet transform. From the above results, it can be concluded that there is a good correlation between the texture features extracted by wavelet transform and forest AGB, which can be used as an important modeling variable for forest AGB estimation.

3.2. AGB Modeling Using Screened Hyperspectral Features

Based on the features of hyperspectral data screening, the AGB model was constructed by using the multiple stepwise regression method, and its accuracy is shown in Table 7.

Table 7. Accuracy of AGB model based on hyperspectral data.

Tree Species	Modeling after Two-Levels Screening	Training Accuracy R^2	Verification Accuracy R^2	RMSE (t/hm^2)	MAE (t/hm^2)
Chinese fir	Y = 94.98 + 46900.41 × 2nd-14 − 71056.38 × 1st-49 − 3306.91 × 1st-93	0.89	0.38	9.67	7.43
Pine tree	Y = 90.93 − 111797.85 × 2nd-51 − 19166.65 × 2nd-95 + 203276.65 × Dia117	0.84	0.79	20.02	14.37
Eucalyptus	Y = −54.84 + 25089.15 × Band46 − 14272.49 × Band65 + 508.42 × Band104 + 350052.48 × Ver6 + 1791491.57 × Ver22 − 378751.92 × Ver38	0.78	0.03	350.14	194.55
Other broadleaved tree	Y = 139.3 −3498000 × Ver19	0.89	0.13	128.47	94.53

It can be seen from Table 7 that the training accuracy of the four tree species is more than 0.75, but the verification accuracy is very different. The training accuracy of pine trees is the best, with accuracy of 0.84, and the verification accuracy is also about 0.8. The training accuracy of Chinese fir is the second. Although the training accuracy of eucalyptus and other broadleaved trees is high, the verification accuracy is very low, indicating that the effect of the model for these two tree species is not ideal, especially the training accuracy and verification accuracy of eucalyptus, as these are all very low. The reason for the deviation between training accuracy and verification accuracy may be that the selected modeling and verification sample plots are uneven, and many attempts can be made in subsequent research. The estimation accuracy of coniferous tree AGB based on hyperspectral data is good, while that of broadleaved tree AGB is low. From the selection of independent variables, it can be seen that coniferous tree AGB is mostly highly correlated with spectral features, broadleaved tree AGB is mostly highly correlated with texture features. Texture features are features after image transformation, and there will be some deviation in calculation, which also makes the estimation results of coniferous tree and broadleaved tree AGB different.

3.3. Feature Screening of LiDAR and Hyperspectral

Firstly, the optimal variables extracted from airborne LiDAR data are shown in Gao Linghan et al. [33]. These are basically the point cloud structure features. These features are fused with the optimal features extracted from hyperspectral data for subsequent feature screening. Then, the RF method was used for the three-level screening of features. The ranking results of the importance features of each tree species are shown in Figure 5.

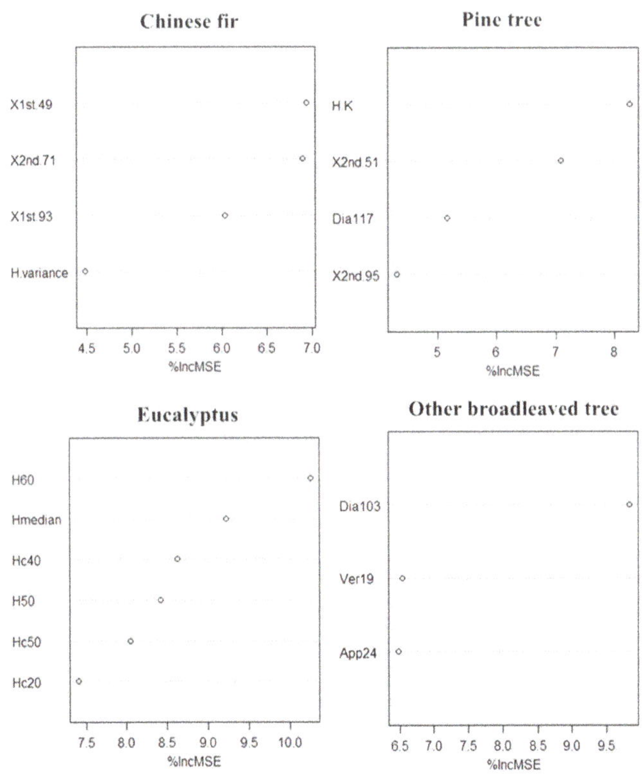

Figure 5. The feature importance ranking map after three-level screening of fused features.

As can be seen from Figure 5, the feature variables with high importance of Chinese fir and pine trees include features from two data sources, mainly height, spectral and texture features of wavelet transform. The important features of eucalyptus are only correlated with height metrics derived from LiDAR data, and the importance features of other broadleaved trees are texture features of wavelet transform based on hyperspectral data. In terms of tree species structure, eucalyptus belongs to tall trees, with straight and complete trunks and fewer branches. Branches are mostly concentrated in the tree canopy, and the biomass is concentrated in the trunk. Therefore, the biomass is mostly related to the height features. Other broadleaved trees mainly include *Castanopsis hystrix Miq.*, *Magnolia denudata Desr.*, *Illicium verum Hook.f.*, *Erythrophleum fordii Oliv.* and *Magnoliaceae glanca Blume*. There is little difference in the height of these tree species, but different tree species have different canopy structure, branch size and leaf shape. Therefore, the forest biomass is mostly related to some shape and texture features. Through the above screening, it can be seen that the AGB of different tree species has the strongest correlation with different features and different data sources.

3.4. AGB Modeling Using Features Fusion

According to the optimal features selected by the three-level screening strategy for each tree species, the AGB models of the four dominant tree species (group) were constructed. The AGB model of other broadleaved trees was constructed based on the features of hyperspectral data (Table 7). The model of eucalyptus was constructed based on the features from airborne LiDAR point cloud data, and the models of Chinese fir and pine tree were constructed based on the fused features of two data sources. The models and accuracy are shown in Table 8.

Table 8. Modeling accuracy of feature fusion.

Tree Species	Modeling after Three-Levels Screening	Training Accuracy R^2	Verification Accuracy R^2	RMSE (t/hm^2)	MAE (t/hm^2)
Chinese fir	Y = 96.25 − 5680.31 × 2nd-71 − 6762.93 × 1st-93 − 0.34 × H-variance	0.78	0.44	11.02	9.15
Pine tree	Y = 92.72 − 92027.59 × 2nd-51 − 9579.86 × 2nd-95 + 166851.96 × Dia117 − 5.62 × H-K	0.95	0.91	12.94	8.95
Eucalyptus	Y = −28.6 + 3.6 × H50 + 5.0 × Hc40	0.72	0.71	50.75	25.48
Other broadleaved tree	Y = 139.3 − 3498000 × Ver19	0.89	0.13	128.47	94.53

It can be seen from Table 8 that the training accuracy of Chinese fir and pine trees is 0.78 and 0.95, respectively, the verification accuracy is 0.44 and 0.91, respectively, and the RMSE is 11.02 and 12.94 t/hm^2, respectively. For the pine tree, compared with the modeling results from hyperspectral data, the training accuracy (0.84 based on hyperspectral data) and verification accuracy (0.79 based on hyperspectral data) of the fusion-based model has been greatly improved to 0.95 and 0.91, respectively. For Chinese fir, the training accuracy is slightly lower than that of hyperspectral feature-based modeling (R^2 is 0.89), but the verification accuracy (0.38 based on hyperspectral data) is improved to 0.44. The training accuracy of eucalyptus AGB model is 0.72 and the verification accuracy is 0.71. Compared with the modeling results based on hyperspectral data, the verification accuracy of eucalyptus is greatly improved and the training accuracy is reduced by 0.06. After three-level feature screening of other broadleaved trees, the optimal features obtained are the same as those extracted from hyperspectral data. Therefore, the final AGB model results are the same. In summary, AGB models of different tree species based on active and passive data greatly improved the accuracy of Chinese fir, pine tree and eucalyptus, and the AGB of other broadleaved trees has the highest correlation with hyperspectral features.

3.5. Forest Above-Ground Biomass Mapping of the Forest Farm

According to the class II survey data of forest resources in Guangxi Province, the distribution area of each tree species within the forest farm is extracted and the corresponding feature variables in each stand area are extracted. The AGB value of each tree species within the forest farm is estimated by using the optimal model of each tree species based on

LiDAR data, based on hyperspectral data and based on fused features. The AGB thematic maps of different tree species in the study area based on the optimal AGB model of different data sources are shown in Figure 6.

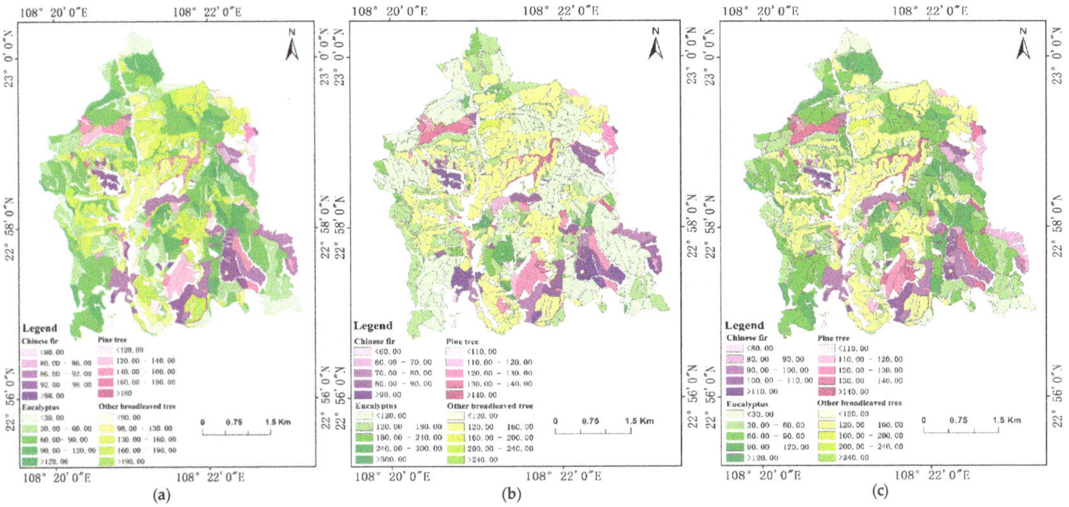

Figure 6. Distribution of forest AGB in the study area based on different data sources ((**a**) is the AGB map of each tree species based on airborne LiDAR data. (**b**) is the AGB map of each tree species based on airborne hyperspectral data. (**c**) is the AGB map of each tree species based on feature fusion.).

It can be seen from Figure 6 that the spatial distribution law of biomass of each tree species obtained by the three methods is the same. Chinese fir is mainly distributed in the south and central parts of the study area. Pine trees are distributed in a small range, mainly in the northwest and southeast of the study area. Eucalyptuses are mainly distributed in the east and west of the study area; there is a small distribution in the central and north region. Other broadleaved trees are evenly distributed in the central part and around the study area.

Comparing Figure 6b,c, the AGB of Chinese fir in b is mainly concentrated between approximately 70 and 100 t/hm^2, and the AGB of Chinese fir in c is mainly concentrated between approximately 75 and 120 t/hm^2. The AGB value of Chinese fir estimated based on airborne hyperspectral data is low, and the minimum AGB value of Chinese fir in b is 11.5 t/hm^2 and that in c is 77.5 t/hm^2. Compared with the measured AGB of Chinese fir, the smallest measured AGB of Chinese fir is 59.2 t/hm^2, and most of the Chinese firs in the forest farm are middle aged and mature forests. Therefore, the minimum value of Chinese fir AGB estimation after feature fusion is more accurate.

The AGB of pine trees in b is mainly concentrated between approximately 120 and 130 t/hm^2, and in c is mainly concentrated between approximately 120 and 140 t/hm^2. The maximum and minimum values in b are 160 t/hm^2 and 95 t/hm^2, respectively, and the maximum and minimum values in c are 170 t/hm^2 and 91 t/hm^2, respectively. It can be seen that the biomass estimation value of the area with high AGB is too small based on hyperspectral data.

The maximum AGB of eucalyptus in b is more than 300 t/hm^2, and that in c is more than 120 t/hm^2. By analyzing the AGB of eucalyptus in small class areas of b and c, respectively, it is found that the maximum AGB of eucalyptus in c is 150 t/hm^2, that in b is 700 t/hm^2. Compared with the measured AGB of eucalyptus, the measured maximum AGB of eucalyptus is 338.8 t/hm^2. It can be concluded that the AGB of eucalyptus in b has a serious oversaturation problem. It shows that the AGB of eucalyptus has little correlation with the features extracted based on hyperspectral data, such as spectral feature and texture feature. The height feature is the key feature to determine the AGB of eucalyptus.

The AGB models of other broadleaved trees in b and c are constructed based on airborne hyperspectral data, and the results are consistent. The AGB of other broadleaved trees in a is calculated based on the optimal model of LiDAR data. The AGB values of other broadleaved trees are mostly between approximately 130 and 190 t/hm^2, and the AGB values of other broadleaved trees in c are mostly between approximately 120 and 160 t/hm^2, indicating that the AGB values of other broadleaved trees calculated based on LiDAR data are generally greater than those calculated based on feature fusion.

In summary, feature fusion based on different data sources can avoid the problem of data value oversaturation. The estimation results of Chinese firs and pine trees based on feature fusion are better, the results of eucalyptuses based on LiDAR data are the best, and the estimation results of other broadleaved trees based on hyperspectral data are the best.

4. Discussion

4.1. Significance of Multi-Level Feature Screening

In this study, airborne LiDAR point cloud data and hyperspectral data were used to analyze the optimal feature variables of AGB modeling and the AGB estimation models of different tree species were established in a complex plantation in China. The three-level feature screening strategy was adopted in the feature screening of multi-source data. The airborne LiDAR features and hyperspectral features were screened, respectively, and then the fused features of the two data sources were screened. Finally, the selection of the optimal features was completed. At the same time, in the feature screening of hyperspectral data, two-level screening are also carried out. First, the feature screening was carried out based on different feature sets, and then, the final optimal hyperspectral features were screened based on the optimal feature sets. This hierarchical screening strategy can effectively avoid the problem of feature redundancy and effectively reduce irrelevant features in the case of few measured samples.

4.2. Selection of Optimal Feature Variables of Different Tree Species

The optimal feature variables of different tree species are related to the tree structures. Compared with most previous studies, the estimation accuracy of AGB is mostly related to vegetation index and point cloud height features [20,25]. In this study, the best features of Chinese firs and pine trees include spectral derivative features, point cloud height features and wavelet transform texture features. The best feature of eucalyptus is the height feature of point cloud, and the best feature of other broadleaved trees is the texture feature of wavelet transform. This shows that the optimal features of different tree species are different due to the specific vertical and canopy structure, and the texture features extracted by wavelet transform can be used for forest AGB modeling. In the subsequent forest AGB research, the corresponding remote sensing data can be selected according to different tree species to extract relevant feature variables.

4.3. Importance of Tree Species AGB Modeling

It is necessary to distinguish tree species to estimate the AGB models. Based on the optimal features of different data sources, using the MSR method to construct the AGB model by tree species can effectively avoid the problem that the previous AGB model is not targeted. More accurate mapping results can be obtained for forest AGB estimation and large-scale regional mapping with complex tree species composition and structural heterogeneity. At the same time, the canopy structure and tree shape of different tree species are different, and the carbon sequestration capacity is also different [21]. Distinguishing tree species to construct AGB models can improve the estimation accuracy of each tree species and also provide a more accurate reference basis for carbon reserve estimation.

4.4. Existing Problems and Future Research Directions

This study only studies the forest AGB model of Gaofeng forest farm in Nanning, Guangxi. There is no comparative analysis on whether the same tree species in other areas can use the model of this study. It can be extended in the follow-up study. At the same time, this study uses the method of feature fusion to combine the two data sources. Later, we can carry out further research on different data-source fusion methods.

5. Conclusions

This study explored the impact of a single remote sensing data source and active and passive remote sensing data fusion on the estimation accuracy of AGB of different tree species. In data feature extraction, according to the characteristics of different data sources, the feature set was constructed from tree canopy features, point cloud structure features, point cloud density features, terrain features, spectral reflectance, spectral derivative, GLCM texture, wavelet transform features and edge detection features. After three-level feature screening and modeling, the optimal models of AGB of different tree species were obtained. The results are as follows:

(a) Based on airborne hyperspectral data, the feature set was constructed by using multiple band combinations, wavelet transform and edge detection methods. Through two-level screening and modeling, it can be concluded that vegetation index and texture features based on GLCM have no obvious effect on improving the accuracy of the AGB model. Spectral features and texture features of wavelet transform play a decisive role in the construction of the AGB model. The AGB accuracy of the optimal models of the four tree species based on the optimal features of hyperspectral data was higher than 0.78, but the verification accuracy was very different. The verification accuracy of eucalyptus was only 0.03, which has the problem of over fitting. In conclusion, modeling using only hyperspectral data will have an impact on the estimation results of eucalyptus AGB. This is because for tall tree species, height features are also an important factor affecting the estimation accuracy of AGB.

(b) AGB models of different tree species were constructed based on multi-source feature fusion. From the results of feature screening, it can be concluded that the optimal features of Chinese firs and pine trees included the features of two data sources. Eucalyptus AGB had the best correlation with LiDAR point cloud data. The top features of other broadleaved trees were the features extracted from hyperspectral data. The training accuracy of the AGB model for each tree species was more than 0.72, and the verification accuracy was quite different. However, after feature fusion, the verification accuracy of Chinese firs and pine trees was improved. The results showed that AGB estimation and mapping in areas with complex tree species composition and high structural heterogeneity must be modeled by tree species. For coniferous trees, the AGB model constructed by combining airborne LiDAR height features and hyperspectral texture features had higher accuracy. The optimal features of the broadleaved tree AGB model will have different choices according to different tree species. For tall broadleaved trees, the AGB model based on airborne LiDAR height features had higher accuracy. Meanwhile, the AGB model for pure forests, such as Chinese firs, pine trees and eucalyptuses, can also be based on the above conclusions.

Author Contributions: L.G., Investigation, Methodology, Software, Validation, Visualization, Writing—original draft; G.C., Validation, Visualization; X.Z., Funding acquisition, Project administration, Resources. All authors have read and agreed to the published version of the manuscript.

Funding: This research was funded by the NSFC (32171779) and National Ministry of Science and Technology [grant number 2017YFD0600900].

Acknowledgments: We would like to thank the Gaofeng Forest Farm in Nanning City, Guangxi Province, for their aid during the field survey. We would also like to thank Erxue Chen and Lei Zhao from the Institute of Forest Resource Information Techniques CAF and Yueting Wang, Kaili Cao, Zhengqi Guo and Xuemei Zhou from Beijing Forestry University for their help in the field work.

Conflicts of Interest: The authors declare that they have no known competing financial interests or personal relationships that could have appeared to influence the work reported in this paper.

References

1. Fassnacht, F.; Hartig, F.; Latifi, H.; Berger, C.; Hernández, J.; Corvalán, P.; Koch, B. Importance of sample size, data type and prediction method for remote sensing-based estimations of aboveground forest biomass. *Remote Sens. Environ.* **2014**, *154*, 102–114. [CrossRef]
2. Castillo, J.A.A.; Apan, A.A.; Maraseni, T.N.; Salmo, S.G., III. Estimation and mapping of above-ground biomass of mangrove forests and their replacement land uses in the Philippines using Sentinel imagery. *ISPRS J. Photogramm. Remote Sens.* **2017**, *134*, 70–85. [CrossRef]
3. Ratle, F.; Camps-Valls, G.; Weston, J. Semisupervised neural networks for efficient hyperspectral image classification. *IEEE Trans. Geosci. Remote Sens.* **2010**, *48*, 2271–2282. [CrossRef]
4. García, M.; Riaño, D.; Chuvieco, E.; Danson, F.M. Estimating biomass carbon stocks for a Mediterranean forest in central Spain using LiDAR height and intensity data. *Remote Sens. Environ.* **2010**, *114*, 816–830. [CrossRef]
5. Propastin, P. Modifying geographically weighted regression for estimating aboveground biomass in tropical rainforests by multispectral remote sensing data. *Int. J. Appl. Earth Obs. Geoinf.* **2012**, *18*, 82–90. [CrossRef]
6. Kankare, V.; Vastaranta, M.; Holopainen, M.; Räty, M.; Yu, X.; Hyyppä, J.; Hyyppä, H.; Alho, P.; Viitala, R. Retrieval of forest aboveground biomass and stem volume with airborne scanning LiDAR. *Remote Sens.* **2013**, *5*, 2257–2274. [CrossRef]
7. Cutler, M.; Boyd, D.; Foody, G.; Vetrivel, A. Estimating tropical forest biomass with a combination of SAR image texture and Landsat TM data: An assessment of predictions between regions. *ISPRS J. Photogramm. Remote Sens.* **2012**, *70*, 66–77. [CrossRef]
8. Estornell, J.; Ruiz, L.; Velázquez-Martí, B.; Fernández-Sarría, A. Estimation of shrub biomass by airborne LiDAR data in small forest stands. *For. Ecol. Manag.* **2011**, *262*, 1697–1703. [CrossRef]
9. Güneralp, İ.; Filippi, A.M.; Randall, J. Estimation of floodplain aboveground biomass using multispectral remote sensing and nonparametric modeling. *Int. J. Appl. Earth Obs. Geoinf.* **2014**, *33*, 119–126. [CrossRef]
10. Wallis, C.I.; Homeier, J.; Peña, J.; Brandl, R.; Farwig, N.; Bendix, J. Modeling tropical montane forest biomass, productivity and canopy traits with multispectral remote sensing data. *Remote Sens. Environ.* **2019**, *225*, 77–92. [CrossRef]
11. Liu, Y.; Gong, W.; Xing, Y.; Hu, X.; Gong, J. Estimation of the forest stand mean height and aboveground biomass in Northeast China using SAR Sentinel-1B, multispectral Sentinel-2A, and DEM imagery. *ISPRS J. Photogramm. Remote Sens.* **2019**, *151*, 277–289. [CrossRef]
12. Wittke, S.; Yu, X.; Karjalainen, M.; Hyyppä, J.; Puttonen, E. Comparison of two-dimensional multitemporal Sentinel-2 data with three-dimensional remote sensing data sources for forest inventory parameter estimation over a boreal forest. *Int. J. Appl. Earth Obs. Geoinf.* **2019**, *76*, 167–178. [CrossRef]
13. Zhong, Y.; Zhang, L. An adaptive artificial immune network for supervised classification of multi-/hyperspectral remote sensing imagery. *IEEE Trans. Geosci. Remote Sens.* **2011**, *50*, 894–909. [CrossRef]
14. Brantley, S.T.; Zinnert, J.C.; Young, D.R. Application of hyperspectral vegetation indices to detect variations in high leaf area index temperate shrub thicket canopies. *Remote Sens. Environ.* **2011**, *115*, 514–523. [CrossRef]
15. Van der Meer, F.D.; Van der Werff, H.M.; Van Ruitenbeek, F.J.; Hecker, C.A.; Bakker, W.H.; Noomen, M.F.; Van Der Meijde, M.; Carranza, E.J.M.; De Smeth, J.B.; Woldai, T. Multi-and hyperspectral geologic remote sensing: A review. *Int. J. Appl. Earth Obs. Geoinf.* **2012**, *14*, 112–128. [CrossRef]
16. Halme, E.; Pellikka, P.; Mõttus, M. Utility of hyperspectral compared to multispectral remote sensing data in estimating forest biomass and structure variables in Finnish boreal forest. *Int. J. Appl. Earth Obs. Geoinf.* **2019**, *83*, 101942. [CrossRef]
17. Cooper, S.; Okujeni, A.; Pflugmacher, D.; van der Linden, S.; Hostert, P. Combining simulated hyperspectral EnMAP and Landsat time series for forest aboveground biomass mapping. *Int. J. Appl. Earth Obs. Geoinf.* **2021**, *98*, 102307. [CrossRef]
18. Koch, B. Status and future of laser scanning, synthetic aperture radar and hyperspectral remote sensing data for forest biomass assessment. *ISPRS J. Photogramm. Remote Sens.* **2010**, *65*, 581–590. [CrossRef]
19. Silva, C.A.; Klauberg, C.; Hudak, A.T.; Vierling, L.A.; Liesenberg, V.; Carvalho, S.P.E.; Rodriguez, L.C. A principal component approach for predicting the stem volume in Eucalyptus plantations in Brazil using airborne LiDAR data. *For. Int. J. For. Res.* **2016**, *89*, 422–433. [CrossRef]
20. Fassnacht, F.E.; Latifi, H.; Hartig, F. Using synthetic data to evaluate the benefits of large field plots for forest biomass estimation with LiDAR. *Remote Sens. Environ.* **2018**, *213*, 115–128. [CrossRef]
21. Cao, L.; Coops, N.C.; Sun, Y.; Ruan, H.; Wang, G.; Dai, J.; She, G. Estimating canopy structure and biomass in bamboo forests using airborne LiDAR data. *ISPRS J. Photogramm. Remote Sens.* **2019**, *148*, 114–129. [CrossRef]
22. Luo, S.; Wang, C.; Xi, X.; Pan, F.; Peng, D.; Zou, J.; Nie, S.; Qin, H. Fusion of airborne LiDAR data and hyperspectral imagery for aboveground and belowground forest biomass estimation. *Ecol. Indic.* **2017**, *73*, 378–387. [CrossRef]

23. Wu, Z.; Dye, D.; Vogel, J.; Middleton, B. Estimating forest and woodland aboveground biomass using active and passive remote sensing. *Photogramm. Eng. Remote Sens.* **2016**, *82*, 271–281. [CrossRef]
24. García, M.; Saatchi, S.; Ustin, S.; Balzter, H. Modelling forest canopy height by integrating airborne LiDAR samples with satellite Radar and multispectral imagery. *Int. J. Appl. Earth Obs. Geoinf.* **2018**, *66*, 159–173. [CrossRef]
25. Abutaleb, K.; Newete, S.W.; Mangwanya, S.; Adam, E.; Byrne, M.J. Mapping eucalypts trees using high resolution multispectral images: A study comparing WorldView 2 vs. SPOT 7. *Egypt. J. Remote Sens. Space Sci.* **2021**, *24*, 333–342. [CrossRef]
26. Baccini, A.; Laporte, N.; Goetz, S.; Sun, M.; Dong, H. A first map of tropical Africa's above-ground biomass derived from satellite imagery. *Environ. Res. Lett.* **2008**, *3*, 045011. [CrossRef]
27. Boudreau, J.; Nelson, R.F.; Margolis, H.A.; Beaudoin, A.; Guindon, L.; Kimes, D.S. Regional aboveground forest biomass using airborne and spaceborne LiDAR in Québec. *Remote Sens. Environ.* **2008**, *112*, 3876–3890. [CrossRef]
28. Chen, G.; Hay, G.J. A support vector regression approach to estimate forest biophysical parameters at the object level using airborne LiDAR transects and quickbird data. *Photogramm. Eng. Remote Sens.* **2011**, *77*, 733–741. [CrossRef]
29. Laurin, G.V.; Chen, Q.; Lindsell, J.A.; Coomes, D.A.; Del Frate, F.; Guerriero, L.; Pirotti, F.; Valentini, R. Aboveground biomass estimation in an African tropical forest with LiDAR and hyperspectral data. *ISPRS J. Photogramm. Remote Sens.* **2014**, *89*, 49–58. [CrossRef]
30. Li, L.; Guo, Q.; Tao, S.; Kelly, M.; Xu, G. LiDAR with multi-temporal MODIS provide a means to upscale predictions of forest biomass. *ISPRS J. Photogramm. Remote Sens.* **2015**, *102*, 198–208. [CrossRef]
31. De Almeida, C.T.; Galvao, L.S.; Ometto, J.P.H.B.; Jacon, A.D.; de Souza Pereira, F.R.; Sato, L.Y.; Lopes, A.P.; de Alencastro Graça, P.M.L.; de Jesus Silva, C.V.; Ferreira-Ferreira, J. Combining LiDAR and hyperspectral data for aboveground biomass modeling in the Brazilian Amazon using different regression algorithms. *Remote Sens. Environ.* **2019**, *232*, 111323. [CrossRef]
32. Wang, D.; Wan, B.; Liu, J.; Su, Y.; Guo, Q.; Qiu, P.; Wu, X. Estimating aboveground biomass of the mangrove forests on northeast Hainan Island in China using an upscaling method from field plots, UAV-LiDAR data and Sentinel-2 imagery. *Int. J. Appl. Earth Obs. Geoinf.* **2020**, *85*, 101986. [CrossRef]
33. Gao, L.; Zhang, X. Aboveground biomass estimation of plantation with complex forest stand structure using multiple features from airborne laser scanning point cloud data. *Forests* **2021**, *12*, 1713. [CrossRef]
34. Guo, B.; Gunn, S.R.; Damper, R.I.; Nelson, J.D. Customizing kernel functions for SVM-based hyperspectral image classification. *IEEE Trans. Image Process.* **2008**, *17*, 622–629. [CrossRef]
35. Fauvel, M.; Tarabalka, Y.; Benediktsson, J.A.; Chanussot, J.; Tilton, J.C. Advances in spectral-spatial classification of hyperspectral images. *Proc. IEEE* **2012**, *101*, 652–675. [CrossRef]
36. Soenen, S.A.; Peddle, D.R.; Coburn, C.A. SCS+ C: A modified sun-canopy-sensor topographic correction in forested terrain. *IEEE Trans. Geosci. Remote Sens.* **2005**, *43*, 2148–2159. [CrossRef]
37. Yang, X. Cover: Use of LIDAR elevation data to construct a high-resolution digital terrain model for an estuarine marsh area. *Int. J. Remote Sens.* **2005**, *26*, 5163–5166. [CrossRef]
38. Zhou, S.; Liu, X.; Wang, C.; Yang, B. Non-iterative denoising algorithm based on a dual threshold for a 3D point cloud. *Opt. Lasers Eng.* **2020**, *126*, 105921. [CrossRef]
39. Gorgens, E.B.; Valbuena, R.; Rodriguez, L.C.E. A method for optimizing height threshold when computing airborne laser scanning metrics. *Photogramm. Eng. Remote Sens.* **2017**, *83*, 343–350. [CrossRef]
40. Zhang, Y.; Lyu, X. A three-dimensional diffusion filtering model establishment and analysis for point cloud intensity noise. *J. Comput. Inf. Sci. Eng.* **2017**, *17*, 011010. [CrossRef]
41. Bayram, E.; Frossard, P.; Vural, E.; Alatan, A. Analysis of airborne LiDAR point clouds with spectral graph filtering. *IEEE Geosci. Remote Sens. Lett.* **2018**, *15*, 1284–1288. [CrossRef]
42. Liu, H.; Wu, C. Developing a scene-based triangulated irregular network (TIN) technique for individual tree crown reconstruction with LiDAR data. *Forests* **2019**, *11*, 28. [CrossRef]
43. Polat, N.; Uysal, M.; Toprak, A.S. An investigation of DEM generation process based on LiDAR data filtering, decimation, and interpolation methods for an urban area. *Measurement* **2015**, *75*, 50–56. [CrossRef]
44. Mutanga, O.; Skidmore, A.K. Narrow band vegetation indices overcome the saturation problem in biomass estimation. *Int. J. Remote Sens.* **2004**, *25*, 3999–4014. [CrossRef]
45. Behmann, J.; Steinrücken, J.; Plümer, L. Detection of early plant stress responses in hyperspectral images. *ISPRS J. Photogramm. Remote Sens.* **2014**, *93*, 98–111. [CrossRef]
46. Tong, X.; Duan, L.; Liu, T.; Singh, V.P. Combined use of in situ hyperspectral vegetation indices for estimating pasture biomass at peak productive period for harvest decision. *Precis. Agric.* **2019**, *20*, 477–495. [CrossRef]
47. Aasen, H.; Burkart, A.; Bolten, A.; Bareth, G. Generating 3D hyperspectral information with lightweight UAV snapshot cameras for vegetation monitoring: From camera calibration to quality assurance. *ISPRS J. Photogramm. Remote Sens.* **2015**, *108*, 245–259. [CrossRef]
48. Rai, H.M.; Chatterjee, K. Hybrid adaptive algorithm based on wavelet transform and independent component analysis for denoising of MRI images. *Measurement* **2019**, *144*, 72–82. [CrossRef]
49. Bi, B.; Zeng, L.; Shen, K.; Jiang, H. An effective edge extraction method using improved local binary pattern for blurry digital radiography images. *NDT E Int.* **2013**, *53*, 26–30. [CrossRef]

50. Chen, W.; Zhao, J.; Cao, C.; Tian, H. Shrub biomass estimation in semi-arid sandland ecosystem based on remote sensing technology. *Glob. Ecol. Conserv.* **2018**, *16*, e00479. [CrossRef]
51. Sun, Q. Research on the driving factors of energy carbon footprint in Liaoning province using random forest algorithm. *Appl. Ecol. Environ. Res.* **2019**, *17*, 8381–8394. [CrossRef]

remote sensing

MDPI

Article

Mapping Spatiotemporal Changes in Forest Type and Aboveground Biomass from Landsat Long-Term Time-Series Analysis—A Case Study from Yaoluoping National Nature Reserve, Anhui Province of Eastern China

Boxiang Yang [1,2], Yali Zhang [3], Xupeng Mao [2], Yingying Lv [4], Fang Shi [2] and Mingshi Li [1,2,*]

1 Co-Innovation Center for Sustainable Forestry in Southern China, Nanjing Forestry University, Nanjing 210037, China; ybx0329@njfu.edu.cn
2 College of Forestry, Nanjing Forestry University, Nanjing 210037, China; mxp@njfu.edu.cn (X.M.); shifang@njfu.edu.cn (F.S.)
3 School of Geographic Information and Tourism, Chuzhou University, Chuzhou 239000, China; zyl930624@163.com
4 Nanjing Institute of Environmental Sciences (NIES), Ministry of Environmental Protection (MEP), Nanjing 210042, China; lvyingying@nies.org
* Correspondence: nfulms@njfu.edu.cn

Citation: Yang, B.; Zhang, Y.; Mao, X.; Lv, Y.; Shi, F.; Li, M. Mapping Spatiotemporal Changes in Forest Type and Aboveground Biomass from Landsat Long-Term Time-Series Analysis—A Case Study from Yaoluoping National Nature Reserve, Anhui Province of Eastern China. *Remote Sens.* **2022**, *14*, 2786. https://doi.org/10.3390/rs14122786

Academic Editors: Klaus Scipal and Henning Buddenbaum

Received: 3 May 2022
Accepted: 8 June 2022
Published: 10 June 2022

Abstract: A natural reserve's forest is an important base for promoting natural education, scientific research, biodiversity conservation and carbon accounting. Dynamic monitoring of the forest type and forest aboveground biomass (AGB) in a nature reserve is an important foundation for assessing the forest succession stage and trend. Based on the Landsat images covering the National Nature Reserve of Yaoluoping in Anhui province spanning from 1987 to 2020, a total of 42 Landsat scenes, the forest cover product set was first developed by using the well-established vegetation change tracker (VCT) model. On this basis, a new vegetation index, NDVI_DR, which considers the phenological characteristics of different forest types, was proposed to distinguish coniferous forest from broad-leaved forest. Next, multiple modeling factors, including remote sensing spectral signatures, vegetation indices, textural measures derived from gray level co-occurrence matrix and wavelet analysis and topographic attributes, were compiled to model the AGB in 2011 by forest type separately by using the stochastic gradient boosting (SGB) algorithm. Then, using the 2011 Landsat image as the base, all the Landsat images in the other years involved in the modelling were relatively normalized by using the weighted invariant pixels (WIP) method, followed by an extrapolation of the 2011 AGB model to other years to create a time-series of AGB. The results showed that the overall accuracy of the VCT-based forest classification products was over 90%. The annual forest type classifications derived from NDVI_DR thresholding gained an overall accuracy above 92%, with a kappa coefficient above 0.8. The 2011 forest-type-dependent SGB-based AGB estimation model achieved an independent validation R^2 at 0.63 and an RMSE at 11.18 t/ha for broad-leaved forest, and 0.61 and 14.26 t/ha for coniferous forest. The mapped time-series of AGB showed a gradual increasing trend over the past three decades. The driving factors responsible for the observed forest cover and AGB changes were analyzed to provide references for reasonable protection and development. The proposed methodology is a reliable tool for evaluating the management status, which can be extended to other similar regions.

Keywords: Landsat time-series; VCT model; classifying forest types; stochastic gradient boosting; forest aboveground biomass

1. Introduction

Forest provides 80% of the global aboveground vegetation biomass [1]. Obtaining reliable long-term forest change information over wide regions in an efficient, low-cost and timely manner becomes one of the major missions of the modern remote sensing framework to satisfy the technical demands of sustainable forest management. Particularly, persistent and accurate estimation of forest AGB or carbon storage in Chinese ecosystems has been a technical and data basis for achieving the "carbon peak before 2030 and carbon neutrality before 2060" aim [2–4].

Due to poor timeliness and limited accessibility, traditional forest field inventory has difficulty meeting the requirements of temporal and spatial dynamic monitoring of forest resources in a wide region and a long time span [5]. With the rapid development of modern remote sensing technology and computerized image analysis algorithms, multi-temporal and multi-resolution remote sensing images provide an important data source for land cover change studies at the landscape, regional and global scales, of which forest change information extraction based on long-term time-series remote sensing observations has attracted more and more attention of scholars [6–10]. The early study of forest change by remote sensing is generally based on the pairwise comparison of images between two periods [11–13]. However, this method may miss those fast-change events, for example, forest loss caused by fire or harvesting and forest gain due to immediate regeneration following harvesting in southern China, due to the relatively long time interval of the two periods (e.g., 10 years). Therefore, to adequately characterize forest changes, we need to use dense time-series images, for example, yearly observations. The Landsat imagery has a medium spatial resolution (30 m) and long-term archive of data (around 50 years), which is suitable for creating a long-term dense time-series stack to sufficiently monitor forest change at the landscape scale. Thanks to its free availability and long historical archive, Landsat-based long-term forest change analysis has been widely implemented by using several reputational automated analytical algorithms, including Landtrendr [8], VCT [9] and Breaks For Additive Season and Trend Monitor (BFAST) [10]. Among them, VCT has attracted much attention in the long-term forest change analysis because of its advantages of automation, high efficiency, high accuracy, easy implementation and full utilization of time information.

Mapping the forest type from remote sensing is one of the important contents of forest resources investigation and monitoring, which can help assess the successional stages and trends of a particular ecosystem. Additionally, the existing studies have shown that the accuracy of separately modeling the AGB by different forest types is generally higher than that of modeling the AGB without differentiating forest types or combining all the forest types together [14–16]. However, in practice, spatially explicit data of forest type are not always available or not suited to support the AGB modeling. For example, although the existing land cover products, e.g., the American NLCD and China's land use datasets, generally contain forest types data (coniferous forest, broad-leaved forest and mixed forest classes), these products generally have a relatively long time cycle, every five years, to update; thus, they cannot adequately record forest type changes induced by frequent forest harvesting and regeneration events occurring within the time interval, let alone accurately capture the phenological differences of forest types [17,18]. Furthermore, these products generally have limitations on local-scale uses due to their inaccuracy at this scale [19]. At present, methods such as decision tree, support vector machine, random forest and neural network have been widely used in the field of forest type classification research to satisfy specific or personalized needs at the local or regional scale [20–23]. However, these methods generally require a large number of training samples to train the classification models, and the model's parameters must be elaborately tuned to gain a high accuracy of classifications, which result in these methods being less efficient and not easily implemented. Thus, developing accurate and efficient methods to map forest types to facilitate the accurate modelling of dense time-series AGB, e.g., annual AGB, is of high priority.

Forest AGB is an underlying indicator for evaluating forest carbon sequestration capacity and biodiversity carrying capacity. The traditional field method for AGB estimation is subject to temporal and spatial limitations [24]. Currently, remote sensing combined with statistical modeling technology acts as a promising alternative to provide macro, near-real-time and multi-scale forest AGB estimation products. The earlier empirical modeling methods include simple linear regression or multiple linear regression models. These methods require a strict normal distribution assumption regarding the used dataset, but remote-sensing-derived variables and other GIS layers often do not conform to this distribution, so their use is greatly limited [25,26]. With the rapid development of computer technology, artificial intelligence has attracted more and more attention of scholars [27]. Machine learning, as one of the important components of artificial intelligence, with the advantages of not requiring data distribution assumptions, nonlinear complex mapping and being insensitive to sample outliers [28], with high model estimation accuracy through self-learning [29,30], has been predominantly used in classification and modelling applications. For example, Lawrence et al. [31] compared the classification results of the SGB algorithm and decision classification trees based on three kinds of datasets, indicating that the SGB algorithm can improve the classification accuracy. Guneralp et al. [32] used Landsat 7 ETM+ and SPOT 5 data to compare the accuracy of SGB-modeled forest AGB, multivariate adaptive regression spline algorithm and Cubist algorithm and indicated that the SGB algorithm was more accurate after adding terrain data, such as elevation, slope and aspect. Dube et al. [33] constructed texture and spectral features based on Landsat images and found that the SGB algorithm had higher accuracy in AGB estimation than random forest. Therefore, coupling the SGB algorithm and Landsat data can generate more accurate AGB inversion results than other machine learning algorithms and classic statistical methods. However, it is noted that these modelling methods are just used in very limited time points because adequate field forest measurement sample plots collected in multiple years, such as the model training set, are not available; thus, dynamically mapping AGB in a dense-time-series manner remains extremely difficult or even impossible. Therefore, this challenge necessitates the development of an effective and reliable framework that integrates long-term Landsat observations, the SGB algorithm and in situ biomass measurements in one year to dynamically model AGB.

Yaoluoping National Nature Reserve is a typical representative base for biodiversity conservation in the Dabie Mountains, Anhui province of eastern China. It is known as the "gene bank of natural species". Understanding its changes in forest resources can help formulate more scientific and reasonable development and protection policies or actions. Therefore, the main objectives of this paper were to develop an efficient and reliable framework to generate multi-temporal AGB, and to investigate the driving factors responsible for forest temporal and spatial changes to recommend targeted policy suggestions for better management of the reserve. Specifically, the major contributions of the current work lie in: (1) developing an efficient and accurate image index, NDVI-DR, to map forest types in multiple years, and (2) devising a new framework that considers forest types, the SGB algorithm, Landsat time-series observations and one-year field biomass measurements to extrapolate the established 2011 biomass estimation model to other years in order to realize the dynamic generation of forest biomass products.

2. Materials and Methods

2.1. Study Area

The Yaoluoping National Nature Reserve, covering a total area of 123 km^2, is located in the northwest of Yuexi County, Anqing City, Anhui Province, with latitudes of 30°57′20″ to 31°06′10″N and longitudes of 116°02′20″ to 116°11′53″E (Figure 1). It has an average elevation of 800 m and belongs to the North subtropical monsoon zone. Because its location is between the Yangtze River and the Huaihe River, the reserve is affected by cyclones, with an abundant precipitation. According to the records of China Meteorological Administration, the annual mean temperature of the reserve is about 12.7°C and the annual rainfall is 1700 mm. Yaoluoping was approved by the State Council to establish a national nature reserve on 5 April 1994. In 1999, Yaoluoping National Nature Reserve was included into the "Chinese people and Biosphere" list, with a dominant orientation of "Forest ecology". More than 40 national key rare and endangered animal and plant species in the reserve had become important protection objects due to the inclusion of membership. The vegetation in the reserve is mainly deciduous broad-leaved forest and evergreen coniferous forest, and the dominant species are *Cunninghamia lanceolata (lamb.) Hook., Pinus massoniana Lamb., Pinus dabeshanensis Cheng et Law, Pinus taiwanensis Hayata, Quercus stewardii Rehd. Quercus variabilis Bl., Alnus trabeculosa Hand.-Mazz.* [34].

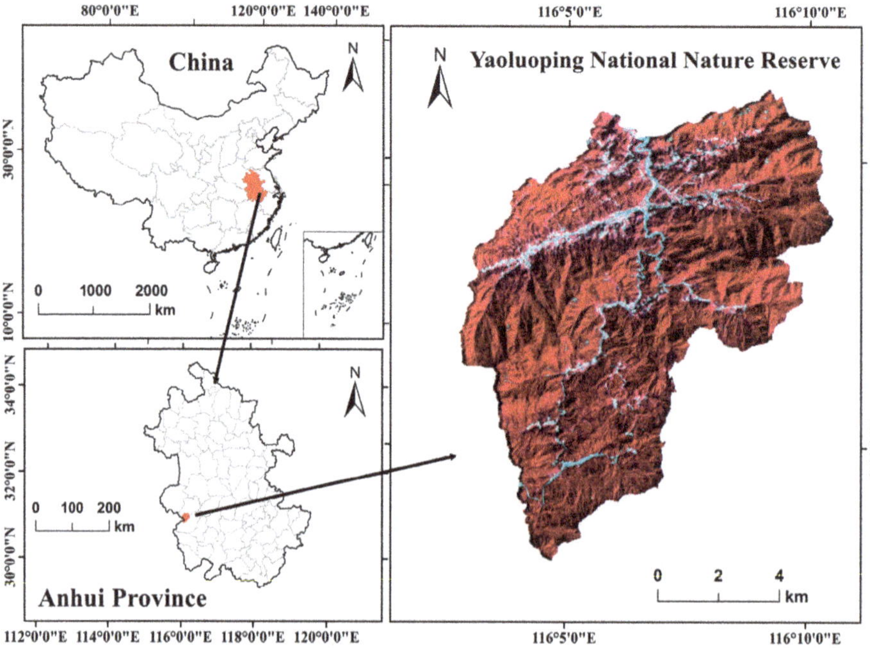

Figure 1. Geographical location map of the study area. The image on the right is the false color composite of the Landsat 8 OLI image acquired on 12 August 2020 covering the Yaoluoping National Nature Reserve.

2.2. Data and Preprocessing

The remote sensing data used in this study included 42 scenes of Landsat TM/OLI imagery from 1987 to 2020, with a WRS-2 path/row number 122/038. The detailed description of the images was summarized in Table 1. All images were downloaded from the USGS official portal (https://glovis.usgs.gov/, accessed on 20 January 2022). In order to explore the feasibility of using seasonal differences to distinguish coniferous forest from broad-leaved forest, images acquired in winter season were also downloaded for partial years of 1987, 1992, 1997, 2002, 2007, 2011, 2013, 2017 and 2020. Before placing the data order to USGS EROS, we directly requested the Landsat Level-2 products to lower the preprocessing complexity or workload. The Level-2 products are time-series observational data of sufficient length, consistency and continuity to record effects of climate change, and they are research-quality, applications-ready and generated for viable surface reflectance science data by USGS EROS data center. Specifically, Landsat 8 Operational Land Imager (OLI) surface reflectance products are generated using the Land Surface Reflectance Code (LaSRC) algorithm [35]. Landsat 5 TM surface reflectance products are generated using the Landsat Ecosystem Disturbance Adaptive Processing System (LEDAPS) algorithm [36].

Table 1. Description of the Landsat TM/OLI images used in the current work.

Acquisition Date	Satellite/Sensor	Cloud %	Acquisition Date	Satellite/Sensor	Cloud %
7 February 1987	Landsat 5 TM	0%	17 September 2004	Landsat 5 TM	0%
19 September 1987	Landsat 5 TM	0%	18 July 2005	Landsat 5 TM	35%
21 September 1988	Landsat 5 TM	9%	19 June 2006	Landsat 5 TM	1%
23 August 1989	Landsat 5 TM	33%	29 January 2007	Landsat 5 TM	0%
23 June 1990	Landsat 5 TM	40%	25 August 2007	Landsat 5 TM	23%
29 August 1991	Landsat 5 TM	1%	12 September 2008	Landsat 5 TM	49%
5 February 1992	Landsat 5 TM	9%	17 October 2009	Landsat 5 TM	0%
30 July 1992	Landsat 5 TM	1%	4 October 2010	Landsat 5 TM	5%
3 September 1993	Landsat 5 TM	8%	24 January 2011	Landsat 5 TM	2%
24 October 1994	Landsat 5 TM	0%	5 September 2011	Landsat 5 TM	69%
5 June 1995	Landsat 5 TM	1%	12 October 2013	Landsat 8 OLI	0.04%
25 July 1996	Landsat 5 TM	1%	25 January 2013	Landsat 8 OLI	2.1%
18 February 1997	Landsat 5 TM	0%	15 October 2014	Landsat 8 OLI	1.91%
29 August 1997	Landsat 5 TM	2%	2 October 2015	Landsat 8 OLI	0.01%
28 May 1998	Landsat 5 TM	23%	4 October 2016	Landsat 8 OLI	20.39%
2 July 1999	Landsat 5 TM	2%	25 February 2017	Landsat 8 OLI	11.84%
18 June 2000	Landsat 5 TM	0%	7 October 2017	Landsat 8 OLI	17.18%
23 July 2001	Landsat 5 TM	19%	8 September 2018	Landsat 8 OLI	0.77%
30 December 2001	Landsat 5 TM	2%	27 September 2019	Landsat 8 OLI	15.83%
28 September 2002	Landsat 5 TM	5%	18 February 2020	Landsat 8 OLI	2.12%
29 July 2003	Landsat 5 TM	16%	28 August 2020	Landsat 8 OLI	1.14%

Other auxiliary data used in this study included the 2012 vegetation distribution map of Yaoluoping National Nature Reserve, the 2011 forest resource inventory and planning data, the administrative boundary and functional zoning vector files, DEM data with a spatial resolution of 12.5 m and the Statistical Yearbooks of Anqing City in 2011 through 2019. Additionally, Google Earth maps were also collected to support the validation of classifications.

2.3. VCT-Based Forest Distribution Extraction in Yaoluoping Nature Reserve

VCT algorithm was developed by Huang et al. at the University of Maryland in 2010 [9], and it has been widely used and tested around the world in recent decades to characterize forest change patterns, with an average overall accuracy at about 85% [37–39]. Here, we directly applied this algorithm to create a time-series of forest cover maps for the reserve. Because the output of VCT disturbance year map contains 7 classes, to produce forest cover product, we needed to aggregate the 7 classes into forest and non-forest, two classes. Table 2 shows the detailed criteria for the aggregation.

Table 2. Definition and aggregation of the forest disturbance map.

Code	Class Description in VCT Model	Aggregated Class
0	Background	Abandoned
1	Persisting non-forest	Non-forest
2	Persisting forest	Forest
3	Persisting water	Non-forest
4	Probable forest with recent disturbance	Forest
5	Disturbed in this year	Non-forest
6	Post-disturbance non-forest	Non-forest

2.4. Development and Validation of NDVI_DR Model

To support subsequent AGB modelling by forest type, we must create a time-series of forest type distribution data for the reserve. The growth of vegetation is affected by various factors, such as temperature and precipitation, and presents different growth characteristics in different periods. Through analyzing the difference in vegetation spectral characteristics in different growth periods, different vegetation types can be effectively distinguished [40,41]. This work used the remote sensing images collected both in growing season (summer) and winter season of the same year to construct a new NDVI-based index to distinguish coniferous forest from broad-leaved forest. Equation (1) illustrates the specifics of the new index, NDVI-DR. Specifically, based on the 2011 forest resource inventory and planning data, we first randomly selected 200 coniferous forest stands and 200 broad-leaved forest stands, respectively, and, based on the gravity center of each stand, its 8 adjacent pixels (8 neighborhoods) in 8 directions of the central pixel were jointly considered. Thus, the average NDVI value of the 9 pixels (central pixel plus 8 adjacent pixels) was extracted as the modified NDVI value of the central pixel to minimize the potential locational errors between summer image and corresponding winter image. By using the NDVI_S (NDVI value in summer) and NDVI_W (NDVI value in winter) of the same location, a new image index, NDVI_DR, was calculated by following Equation (1). The change patterns of NDVI_S, NDVI_W and NDVI_DR for different central pixels were plotted in Figure 2. Figure 2 shows that the NDVI values of coniferous forest pixels and broad-leaved forest pixels in summer (NDVI_S) are higher than those in winter (NDVI_W). The mean NDVI of coniferous forest is 0.742 in summer and 0.617 in winter. The mean NDVI of broad-leaved forest is 0.571 in summer and 0.209 in winter. The NDVI values of coniferous forest in summer have little differences compared with those in winter, while the NDVI values of broad-leaved forest in summer have greater differences compared with those in winter. According to the above observations, a new NDVI_DR vegetation index was constructed following Equation (1). In this equation, the denominator is the difference between NDVI in summer and NDVI in winter, and the numerator is the NDVI value in winter.

$$\mathrm{NDVI\_DR} = (\mathrm{NDVI\_S} - \mathrm{NDVI\_W})/(\mathrm{NDVI\_W}) \quad (1)$$

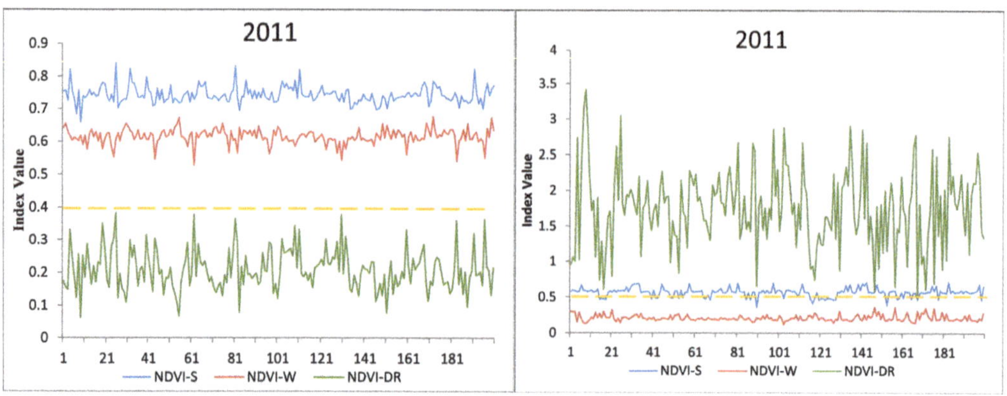

(a) (b)

Figure 2. Conceptual framework of NDVI_DR derived from Landsat TM imagery to distinguish coniferous forest from broad-leaved forest. (**a**): Coniferous forest. (**b**): Broad-leaved forest.

NDVI_DR can reflect the variation intensity of NDVI value in summer and winter. Figure 2 shows that, in coniferous forest, the value of NDVI_DR is generally less than or equal to 0.4, while, in broad-leaved forest, the value of NDVI_DR is greater than or equal to 0.5. Thus, the thresholds of 0.4 and 0.5 for NDVI_DR were determined as the final classification criteria for forest type identification.

To validate the effectiveness of NDVI_DR thresholding model, we used the 2013 Landsat 8 OLI images coupled with the 2012 vegetation distribution map and adopted fully the same process to extract the values of NDVI_DR for those 400 locations or pixels. After visually interpreting the 2012 vegetation distribution map, the forest types of those 400 pixels remaining unchanged were doubly confirmed. The extracted change patterns of those 400 locations based on the 2013 Landsat OLI image were summarized in Figure 3. Obviously, the proposed

NDVI_DR thresholding model remains stable; the thresholds of 0.4 and 0.5 are still effective in separating coniferous forest from broad-leaved forest. Thus, the following rules in Equation (2) were used to distinguish coniferous forest from broad-leaved forest by using Landsat observations in the current work. Equation (2) was written as :

$$\begin{cases} \text{NDVI\_DR} \leq 0.4 \rightarrow \text{Coniferous forest} \\ \text{NDVI\_DR} \geq 0.5 \rightarrow \text{broad} - \text{leaved forest} \end{cases} \quad (2)$$

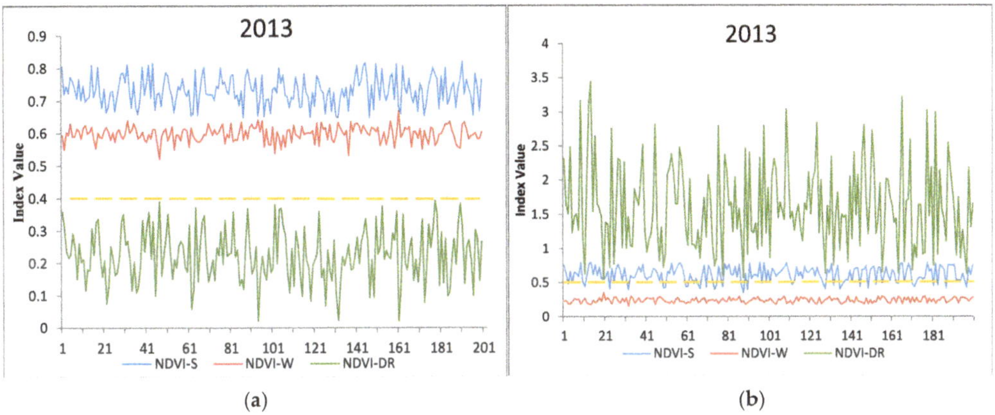

Figure 3. Validation framework of NDVI_DR derived from Landsat OLI imagery to distinguish coniferous forest from broad-leaved forest. (**a**): Coniferous forest. (**b**): Broad-leaved forest.

2.5. Forest AGB Modeling

2.5.1. Independent Variable

Based on the 2011 forest resources inventory and planning data, the per area AGB was derived as the independent variable for modelling. Specifically, according to the recorded attributes in the inventory data, including the dominant tree species, average diameter at breast height (DBH), average height, area and number of stems of each stand, the single tree-level AGB was calculated by using the allometric growth equations by tree species (Table 3), and then the total stand-level AGB was derived by summing up the AGB of each tree in the stand, followed by the calculation of per unit area AGB (t/ha) via dividing the total stand-level AGB by the stand area. To match the pixel size of Landsat image in the AGB modelling, it was necessary to convert the per unit area AGB into the pixel-level AGB, with the unit of t/900 m^2. After this conversion, all the field AGB measurements were used as the independent variable for model training (80%) and validation (20%) purposes to facilitate the modelling.

Table 3. Biomass allometric growth equations for the major tree species in Anhui Province.

Tree Species	Aboveground Biomass Formula
Cedarwood	$W_T = W_S + W_B + W_L = 0.00849(D^2 H)^{1.107230} + 0.00175(D^2 H)^{1.091916} + 0.00071D^{3.88664}$
Oak	$W_T = W_S + W_B + W_L = 0.00888(D^2 H)^{1.08} + 0.01(D^2 H)^{0.90} + 0.00378(D^2 H)^{0.94}$
Larch	$W_T = W_S + W_B + W_L = 0.099496(D^2 H)^{0.786530} + 0.098620(D^2 H)^{0.598367} + 0.294136(D^2 H)^{0.357506}$
Masson pine	$W_T = 0.01672(D^2 H)^{0.8559}$
Sclerophyllous broad-leaved forest	$W_T = 0.07112(D^2 H)^{0.910359078}$
Soft-leaved broad-leaved forest	$W_T = W_S + W_B + W_L = 0.012541(D^2 H)^{1.144} + 0.004786(D^2 H)^{1.006} + 0.047180(D^2 H)^{0.769}$

Note: W_T is the total AGB; W_S is the trunk biomass; W_B is the branch biomass; W_L is the leaf biomass; D is the DBH of trees; H is the height of trees. The AGB formula is derived from the main technical provisions of China's National Forest Inventory.

2.5.2. Development of Dependent Variables

In this work, four types of modelling features, including the original band transformations, vegetation indices, textural measures and terrain variables, were extracted as the potential dependent variables to support the establishment of AGB inversion model.

The Landsat multi-spectral imagery has abundant spectral information and different bands have different levels of correlation to AGB. In order to remove the redundant information among bands and to screen out those comprehensive features highly relating to AGB, this analysis adopted the KT transform to generate three new orthogonal features with explicit physical implications, named TCB (Brightness), TCG (Greenness), TCW (Wetness) [42]. Meanwhile, the TCD (Distance) [43] and TCA (Angle) [44] indices that reflect vegetation coverage and tree growth status [45] were also developed based on the three features (Table 4).

Vegetation index can enhance vegetation signature and accurately reflect vegetation growth and distribution [46]. The ratio vegetation index (RVI) [47] can eliminate the influence of terrain and shadow on vegetation analysis, and the normalized difference vegetation index (NDVI) can minimize the effects of atmosphere on vegetation and characterize vegetation density and vigor, and both indices show a good correlation with AGB [48,49]. Additionally, the $NDVI_C$ index, making use of its better resistance of short-wave infrared band to atmospheric condition changes, can unify different coverage types and reduce the influence of forest background signals [50,51]. These indices were derived by following the formula in Table 4.

Table 4. Vegetation indices used in the analysis and their calculation formulas.

Index	Formula
NDVI [48]	$NDVI = \frac{\rho_{nir} - \rho_{red}}{\rho_{nir} + \rho_{red}}$
$NDVI_C$ [50]	$NDVI_C = \frac{\rho_{nir} - \rho_{red}}{\rho_{nir} + \rho_{red}} \times \left(1 - \frac{\rho_{swir} - \rho_{swir_{min}}}{\rho_{swir_{max}} - \rho_{swir_{min}}}\right)$
RVI_{43} [47]	$RVI_{43} = \frac{\rho_{nir}}{\rho_{red}}$
RVI_{54} [47]	$RVI_{54} = \frac{\rho_{swir}}{\rho_{nir}}$
NDMI [52]	$NDMI = \frac{\rho_{nir} - \rho_{swir}}{\rho_{nir} + \rho_{swir}}$
mNDWI [52]	$mNDWI = \frac{\rho_{green} - \rho_{swir}}{\rho_{green} + \rho_{swir}}$
TCD [43]	$TCD = \sqrt{TCB^2 + TCG^2}$
TCA [44]	$TCA = \arctan\left(\frac{TCG}{TCB}\right)$

Note: ρ_{nir} is the spectral reflectance of near infrared band. ρ_{red} is the spectral reflectance of red band. ρ_{swir} is the spectral reflectance of short-wave infrared band. ρ_{green} is the spectral reflectance of green band.

The texture of remote sensing image refers to the recurring primitives or elements and their arrangement rules in the image, which is a unity of local variability and spatial correlation [53–55]. In this paper, 6 multispectral bands, $NDVI_C$ and RVI_{54} were used as the inputs for texture calculation by using the gray level co-occurrence matrix (GLCM) method [56]. Specifically, a 3×3 window size was selected, and the moving directions at 0°, 45°, 90° and 180° were considered, respectively, and the average of the four directions was taken as the final texture analysis result. To compensate for the potential limitations of GLCM method, the wavelet multi-scale decomposition was also implemented by programming in MATLAB environment to extract the high-frequency vertical, horizontal and diagonal details images as new textural features [57]. Firstly, the first principal component of the 2011 Landsat multi-spectral images was decomposed by using a biorthogonal wavelet base function in a three-layer recursive manner. Thus, another 9 images of details were obtained as new textural features to support the modeling. The GLCM-based textures and their calculation formula were summarized in Table 5 [58].

Table 5. Formulas for calculating texture index based on GLCM.

Texture Index	Formula
Mean, (ME)	$ME = \sum_{i,j=0}^{N-1} i \times P_{i,j}$
Variance, (VA)	$VA = \sum_{i,j=0}^{N-1} i \times P_{i,j}(i - ME)^2$
Homogeneity, (HO)	$HO = \sum_{i,j=0}^{N-1} i \times \frac{P_{i,j}}{1+(i-j)^2}$
Contrast, (CO)	$CO = \sum_{i,j=0}^{N-1} i \times P_{i,j}(i - j)^2$
Dissimilarity, (DI)	$DI = \sum_{i,j=0}^{N-1} i \times P_{i,j}\lvert i - j\rvert$
Entropy, (EN)	$EN = \sum_{i,j=0}^{N-1} i \times P_{i,j}(-\ln P_{i,j})$
Second Moment, (SM)	$SM = \sum_{i,j=0}^{N-1} i \times P_{i,j}{}^2$
Correlation (CR)	$CR = \sum_{i,j=0}^{N-1} i \times P_{i,j}\left[\frac{(i - ME)(j - ME)}{\sqrt{VA_i \bullet VA_j}}\right]$

Note: $P_{i,j} = \frac{V_{i,j}}{\sum_{i,j=0}^{N-1} V_{i,j}}$; $V_{i,j}$ represents the pixel brightness value at the position of row i and column j; N is the size or dimension of the moving window when the texture index is calculated.

Terrain features influence the distribution patterns of AGB to some extent [59]. Therefore, elevation, slope and aspect extracted from the 12.5 m resolution DEM were also considered as the modeling factors. These terrain factors were resampled to 30 m resolution by implementing the bi-linear interpolation resampling technique to match the Landsat pixel size.

2.5.3. Correlation Analysis

In this paper, 6 multispectral bands, 3 topographic factors, 6 vegetation indices, 5 KT transform and 73 texture features were selected as potential variables for biomass prediction. In order to ensure the accuracy of the model and reduce the workload, correlation analysis was needed to select the best combination of variables [60]. Assuming that there is no correlation or weak correlation between the selected variables and a linear relationship with the dependent variable, AGB, we could automatically select the variables by regression analysis, and this method was usually used in previous studies [61,62]. This analysis first used random forest importance ranking to select those important variables and then eliminated variables with high correlation between variables through Pearson correlation analysis [63]. The random forest package was run in R environment, and the characteristic factors with high correlation to biomass of broad-leaved forest and coniferous forest were, respectively, screened out as modeling variables.

2.5.4. SGB-Based AGB Modelling and Its Time Extrapolation

Stochastic gradient boosting (SGB) algorithm proposed by Friedman was adopted to estimate AGB in this work [64]. SGB takes into account the advantages of both boosting algorithm and bagging algorithm and has been widely used in regression and classification tasks [65]. It can avoid the problem of long calculation time due to large amount of data and can also improve the prediction accuracy, with better robustness to overfitting [33]. The "learning rate" parameter in the SGB algorithm determines the growth rate of modeling complexity. Generally, a smaller learning rate means that more regression trees will be generated and the contribution of each tree to the whole forest will be weaker and the modeling performance will be better [66]. The "depth of interaction" parameter determines the splitting number of each tree. This parameter value represents the number of nodes in each decision tree, and the maximum value is usually set to 10. Other important parameters include: tree complexity, shrinkage, distribution function, training ratio, etc. The modeling process was implemented by using the "caret" package in R environment. Through multiple comparisons, it was found that the optimal combination of parameters in SGB modeling was set as: interaction depth at 3, shrinkage at 0.01, ntrees at 500 and n.minobsinnode at 9. Thus, the 2011 AGB map of the reserve was produced from implementing the above-identified SGB parameters by forest types, followed by a spatial overlay analysis of both the estimated coniferous forest AGB and the predicted broad-leaved forest AGB.

At present, there are two main methods to observe long-term forest AGB changes using remote sensing images. The first is to build separate varying relationships between remote sensing images and corresponding forest AGB field sample measurements in different years. This method is relatively accurate, but it is subject to the constraints of realistic conditions, such as lack of historical field AGB measurement data. The second method assumes that there is a relatively stable relationship between forest AGB and remote sensing images over time. Through the relative radiometric normalization operations between the base image (e.g., the 2011 Landsat TM image of the current work) and the target images in different years, followed by the development of a fully same set of features as the base image based on the normalized target images, the relationship constructed for the base year is extended to other target years to retrieve varying AGB patterns in the same study area. The second method has been more widely used in estimating forest AGB thanks to its lower cost and stronger operability in a long-term monitoring period [67,68]. Here, the 2011 Landsat TM image was set as the base image, and Landsat TM/ETM+/OLI images acquired in 1987, 1992, 1997, 2002, 2007, 2017 and 2020, as the target images, were radiometrically normalized, respectively, band by band using the automated weighted invariant points (WIP) method [69]. Based on these normalized target images, a fully same set of features as the base year was developed. Then, the relationship created for the base year was extended to other target years to create a time-series of AGB maps for the reserve.

2.6. *Validation Method*

2.6.1. Forest Distribution Verification

According to the historical documentations of the reserve, forest was the largest land cover type in the reserve, occupying an area proportion of about 90%. Thus, to validate the accuracy of forest distribution products mapped from VCT, the stratified random sampling method was implemented. First, based on the 2011 forest resources inventory and planning data and the 2012 vegetation distribution map, 900 points were randomly generated in the forest area and 100 points were randomly generated in the non-forest area for the classifications in other years. Then, visually interpreting the corresponding year's Google Earth images was conducted to gain the ground truths to validate the classifications. For the year of 1997 or earlier, no Google Earth high resolution images were available, so the original Landsat images were directly visually interpreted. Finally, by comparing the classifications and the interpreted results, the overall accuracy and kappa coefficient were derived to evaluate the classification accuracy. The kappa coefficient calculation formula was as follows:

$$\text{kappa} = \frac{P_A - P_e}{1 - P_e} \tag{3}$$

where P_A refers to the relative observed agreement among raters, and P_e is the hypothetical probability of chance agreement.

2.6.2. Forest Type Distribution Verification

The broad-leaved forest dominated the reserve historically. Similarly, the stratified random sampling method was used to verify the forest type classification accuracy. Based on the 2011 forest resources inventory and planning data and the 2012 vegetation distribution map, 700 points were randomly generated in the broad-leaved forest area and 300 points were randomly generated in the coniferous forest area. Then, visually interpreting the corresponding year's Google Earth images was conducted to gain the ground truths to validate the classifications.

2.6.3. Forest AGB Modeling Verification

Twenty percent of the pixel-level AGB field measurements, as an independent dataset, were used to evaluate the SGB-based modelling of AGB by calculating the validation determination coefficient R^2 and root mean square error (RMSE). R^2 indicates the variance explanation degree of dependent variables to independent variable in the model. Generally speaking, larger R^2 and smaller RMSE mean higher prediction accuracy that the model has. The calculation formulas of R^2 and RMSE were shown in Equations (4) and (5):

$$R^2 = 1 - \frac{\sum_{i=1}^{n} (y_i - \hat{y}_i)^2}{\sum_{i=1}^{n} \left(y_i - \bar{y}\right)^2} \quad (4)$$

$$\text{RMSE} = \sqrt{\frac{\sum_{i=1}^{n} (y_i - \hat{y}_i)^2}{n}} \quad (5)$$

where $\hat{y}_i$ indicates the model-predicted AGB, y_i represents the measured AGB, $\bar{y}$ indicates the average value of the measured AGB and n means the sample size.

Based on the 2011 forest resources inventory data in 2011, 200 broad-leaved forest stands and 200 coniferous forest stands were randomly selected, of which 120 stands were used for modeling and 80 for independent validation. Additionally, to highlight the potential superior performance of separate modelling by forest types, these 400 stands were combined to construct a mixed AGB model without differentiating forest types, of which 240 stands were used for modeling and 160 for validation. Because the historical field records of AGB in the reserve were unavailable, we were unable to validate the AGB predictions produced from extrapolating the 2011 AGB models to other years.

3. Results

3.1. Forest Distribution Mapping and Validation

Table 6 shows the confusion matrix of the 2020 forest classifications. The results showed that an overall accuracy of 94.7%, accompanied by a kappa coefficient of 0.73, were observed in the 2020 forest classifications. Table 7 exhibits the verification results in other years. As a whole, the overall accuracy of the VCT-based forest classification products was over 90% and the kappa coefficient above 0.6, which meant that VCT produced reliable time-series forest distribution maps.

Table 6. Confusion matrix of forest extraction accuracy verification.

Predicter Results \ Actual Results	Forest	Non-Forest	Total
Forest	863	16	879
Non-Forest	37	84	121
Total	900	100	1000
OA = 0.947		Kappa = 0.731	

Table 7. Verification results of forest extraction accuracy.

Year	Overall Accuracy (%)	Kappa Coefficient
1987	93.2%	0.66
1992	93.1%	0.66
1997	92.8%	0.64
2002	90.5%	0.56
2007	93.9%	0.70
2011	93.1%	0.67
2017	90.2%	0.55
2020	94.7%	0.73

Figure 4 shows the spatio-temporal distribution patterns of the forest in the Yaoluoping National Nature Reserve during the period of 1987 to 2020. Obviously, the forest absolutely dominated the reserve in all the years and its areal proportion dropped from 92.92% in 1987 to 87.15% in 1997, then gradually fluctuated to 90.94% in 2020 (Figure 5). The lowest forest area was observed in 1997, and then forest area showed an upward fluctuating trend. The non-forest type only occupied a small area, principally distributed in the northern and central portions of the reserve (Figure 4), along with the valleys and the gentle-slope lands at relatively low altitudes in the study area.

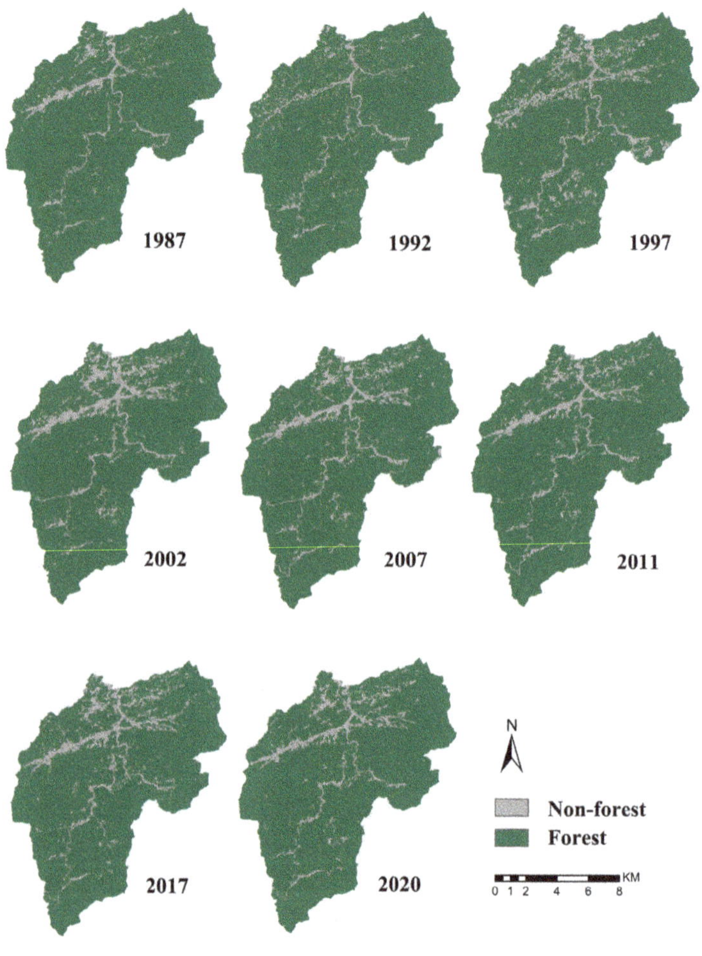

Figure 4. Forest distribution maps mapped from VCT model during the period 1987 to 2020.

3.2. Forest Type Distribution Mapping

Table 8 displays the validation accuracy statistics of the forest type classifications derived from the NDVI_DR thresholding. The overall accuracies at 94.5%, 94.0%, 94.2%, 92.5%, 93.5%, 94.0%, 92.0% and 94.8%, with corresponding kappa coefficient values at 0.87, 0.85, 0.86, 0.82, 0.84, 0.85, 0.81 and 0.87, were observed during the period of 1987 to 2020, respectively. Among them, the imaging times of the 2002 and 2017 summer images were on September 28 and October 7, respectively, which fell into late fall season and were not in the peak season of plant growth. This would lead to actual changes in the vegetation spectral reflectance, accordingly resulting in a relatively low accuracy of forest type classifications in these two years. However, overall, the DNVI_DR thresholding model provided a relatively high accuracy of forest type classifications in the reserve (Table 8).

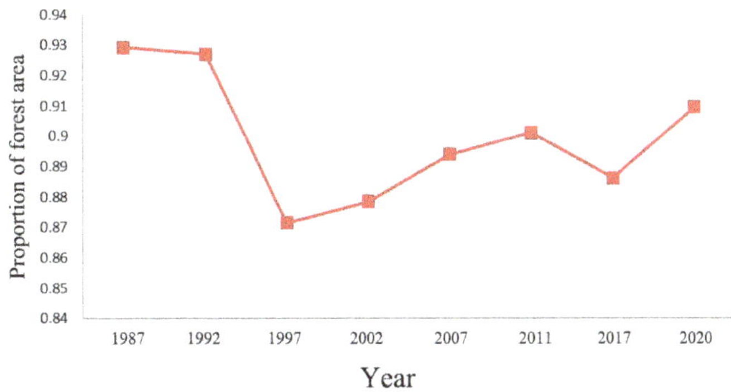

Figure 5. Changes in forest area proportion in Yaoluoping during the period 1987 to 2020.

Table 8. Verification results of forest type classifications derived from NDVI_DR thresholding.

Year	Overall Accuracy (%)	Kappa Coefficient
1987	94.5%	0.87
1992	94.0%	0.85
1997	94.2%	0.86
2002	92.5%	0.82
2007	93.5%	0.84
2011	94.0%	0.85
2017	92.0%	0.81
2020	94.8%	0.87

Figure 6 shows the spatio-temporal variations in the forest type distribution. Generally, broad-leaved forest dominated the reserve in space, with an apparent large-scale continuous distribution in all the years (Figure 6). Its areal share remained relatively stable (above 80% in all the years), with a small variation (Figure 7). Coniferous forest had a dispersing distribution over the reserve, mainly distributed in the northern and south-eastern portions of the reserve (Figure 6), and its areal share gradually decreased from 8.5% in 1987 to the lowest value at about 4.7% in 2011 (Figure 7), and then the proportion of coniferous forest area increased steadily. By 2020, the proportion slightly exceeded that in 1987.

3.3. Biomass Estimation Results

3.3.1. Variable Selection Results

Tables 9–11 show the combinations of variables determined by using random forest importance ranking coupled with Pearson correlation analysis, most suitable for broad-leaved forest, coniferous forest and the combination of both in AGB modeling. As shown in Table 9, the average texture information of SWIR and $NDVI_C$ had a high correlation with the broad-leaved forest AGB, while the absolute value of the correlation coefficient of other texture information was less than 0.3. As can be seen from Table 10, short-wave infrared, RVI_{54}, NDVI and $NDVI_C$ were highly correlated with coniferous forest AGB, and the correlation between texture features and coniferous forest AGB was very low; thus, they were unable to be selected. As can be seen from Table 11, when considering the combined AGB of the coniferous forest and broad-leaved forest together, the number of selected modeling variables was substantially less than that when distinguishing between the forest types, and the correlation coefficients were also apparently lower. Finally, according to the ranking of the correlation coefficient values, 10 modeling factors were selected for broad-leaved forest, coniferous forest and combined forest types together, respectively, to consider the operational practicability. The factors selected for broad-leaved forest were B7, B5, TCW, TCB, TCD and B5_mean, B7_mean, $NDVI_C$ _mean and $NDVI_C$; RVI_{54}, $NDVI_C$, B2, B5, B4, B7, TCD, NDVI, mNDWI and NDMI for coniferous forest; B2, B4, B5, B7, TCD, RVI_{54}, NDMI, mNDWI, NDVI and $NDVI_C$ for combined AGB modelling of both forest types.

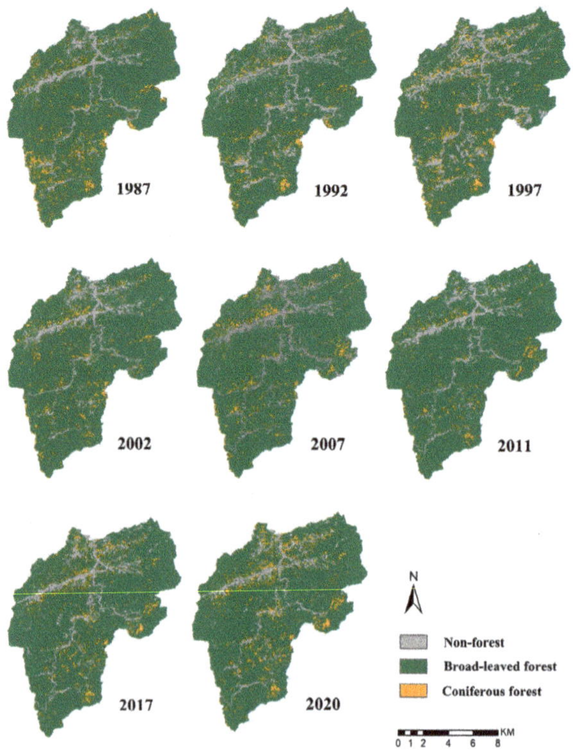

Figure 6. Changes in forest type distributions derived from NDVI_DR thresholding during the period 1987 to 2020.

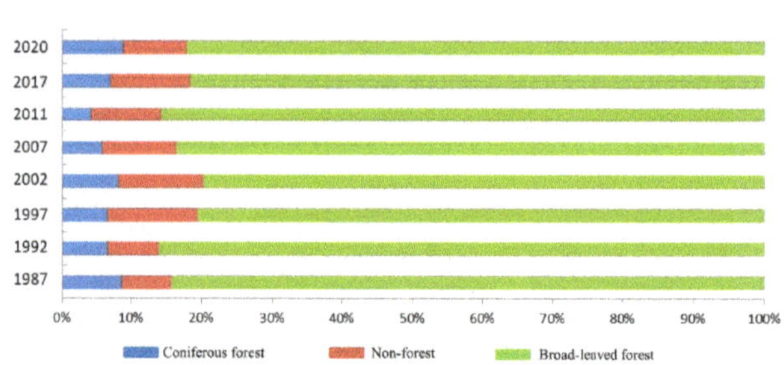

Figure 7. Changes in area proportion of forest types in the study area during the period 1987 to 2020.

Table 9. Characteristic variables with significant correlation to broad-leaved forest AGB.

Characteristic	Correlation Index	Characteristic	Correlation Index
B1	−0.152 *	RVI_{54}	0.294 **
B2	−0.314 **	NDMI	0.287 *
B3	−0.221 *	mNDWI	0.267 *
B4	−0.265 *	B5_mean	−0.473 **
B5	−0.469 **	B7_mean	−0.473 **
B7	−0.486 **	RVI_{54}_mean	0.211 *
TCW	−0.478 **	$NDVI_C$ _mean	0.452 **
TCB	−0.406 **	B2_ Correlation	0.166 *
TCG	−0.263 **	RVI_{54}_ Correlation	0.154 *
TCD	−0.341 **	DBH	0.326 **
NDVI	0.305 **	bio2.8	0.196 *
$NDVI_C$	0.466 **		

Note: B1, B2, B3, B4, B5, B7 mean the blue, green, red, NIR, SWIR 1, SWIR 2 band of Landsat 5 and Landsat 8 separately. * means significant at the level of 0.05 (the confidence level is 95%); ** means significant at the level of 0.01 (the confidence level is 99%).

Table 10. Characteristic variables with significant correlation to coniferous forest AGB.

Characteristic	Correlation Index	Characteristic	Correlation Index
B1	−0.138 *	RVI_{54}	0.324 **
B2	−0.301 **	NDMI	0.246 *
B3	−0.191 *	mNDWI	0.207 **
B4	−0.224 *	NDVI	0.312 **
B5	−0.435 **	$NDVI_C$	0.459 **
B7	−0.477 **	bio2.8	0.186 *
TCD	−0.264 **		

Note: * means significant at the level of 0.05; ** means significant at the level of 0.01.

3.3.2. Modeling Accuracy Evaluation

Figures 8–10 show the modeling and validation accuracy of broad-leaved forest, coniferous forest and combined forest types using SGB, respectively. The modeling and validation R^2 of broad-leaved forest AGB were at 0.68 and 0.63, and the corresponding RMSEs were at 7.53 and 11.19 t/hm^2, respectively. The modeling and validation R^2 of coniferous forest biomass were at 0.71 and 0.61, and the corresponding RMSEs were 4.46 and 14.27 t/hm^2, respectively. The modeling and validation R^2 of combined biomass were at 0.54 and 0.51, with the corresponding RMSEs at 18.91 and 20.47 t/hm^2, respectively. Obviously, modelling AGB by forest type was more accurate than modelling AGB without differentiating forest types.

Table 11. Characteristic variables with significant correlation to combined AGB of both forest types.

Characteristic	Correlation Index	Characteristic	Correlation Index
B2	−0.173 *	RVI_{54}	0.163 *
B4	−0.107 *	NDMI	0.142 *
B5	−0.227 *	mNDWI	0.101 *
B7	−0.248 **	NDVI	0.152 *
TCD	−0.134 *	$NDVI_C$	0.279 **

Note: * means significant at the level of 0.05; ** means significant at the level of 0.01.

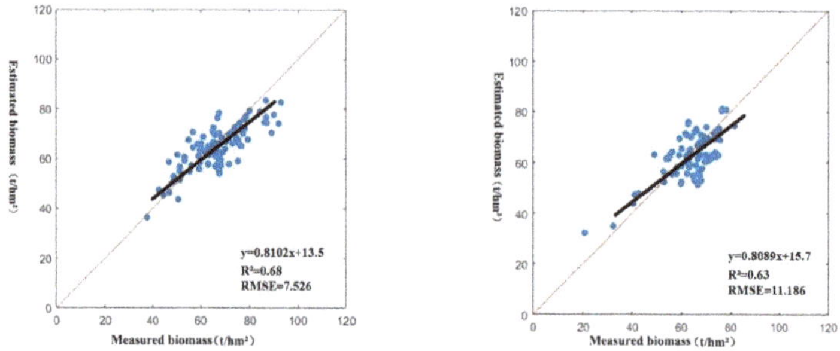

(a) (b)

Figure 8. Assessment of SGB-based broad-leaved forest biomass modelling and validation. (**a**): Modeling; (**b**): Validation.

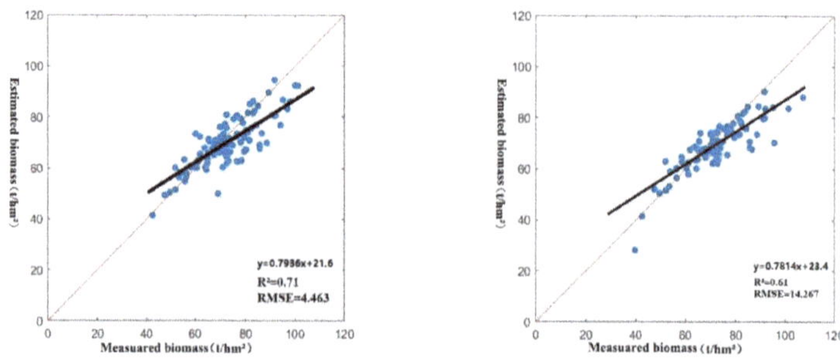

(a) (b)

Figure 9. Assessment of SGB-based coniferous forest biomass modelling and validation. (**a**): Modeling; (**b**): Validation.

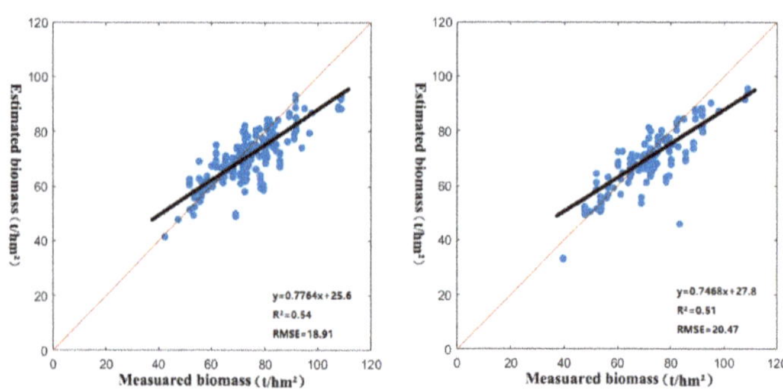

(a) (b)

Figure 10. Assessment of SGB-based combined biomass modelling and validation. (**a**): Modeling; (**b**): Validation.

3.3.3. AGB Extrapolation of SGB Model

In this work, the 2011 well-established and tested SGB-based AGB models for broad-leaved forest and coniferous forest were extrapolated to create a time-series of AGB based on the forest type maps created in Section 3.2 during the period of 1987 to 2020 (Figure 11). The total biomass and mean values of different forest types in each year were calculated, as shown in Table 12. In terms of temporal distribution, the forest AGB in the reserve showed a gradual increasing trend in the past 30 years. The average AGB values in the eight years were at 57.37, 61.56, 64.38, 70.14, 73.21, 75.52, 76.23 and 78.85 t/hm^2, respectively, which showed that the AGB of Yaoluoping National Nature Reserve was increasing continuously, and the growth rate was the fastest during the period of 1992 to 2002. As for different forest types, although the total AGB of coniferous forest was much lower than that of broad-leaved forest, the average biomass of coniferous forest was higher than that of broad-leaved forest (Table 12).

Table 12. Statistics of the mapped AGB in each year.

Year	Forest		Coniferous Forest		Broad-Leaved Forest	
	Mean (t/hm^2)	Summation (10,000 Tons)	Mean (t/hm^2)	Summation (10,000 Tons)	Mean (t/hm^2)	Summation (10,000 Tons)
1987	57.37	70.57	62.27	6.77	56.68	60.20
1992	61.56	75.72	64.84	5.41	60.12	65.20
1997	64.38	79.19	67.53	5.59	63.74	64.72
2002	70.14	86.27	72.03	7.37	69.45	69.76
2007	73.21	90.05	74.71	5.45	72.37	76.23
2011	75.52	92.89	76.43	4.08	74.72	80.84
2017	76.23	93.76	78.98	6.97	75.46	77.59
2020	78.85	96.99	80.45	8.88	78.85	81.64

The spatial distribution of the AGB in Figure 11 shows that the areas with low AGB values were mainly located near agricultural lands, buildings and water bodies in river valleys, and took on apparent linear features in the northern and central portions of the reserve, whereas high AGB values mainly occurred in the western and north-eastern areas of the reserve (Figure 11).

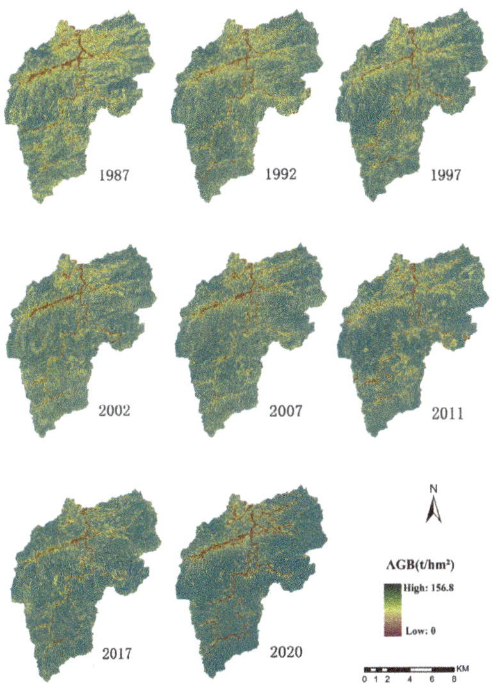

Figure 11. The time-series AGB distributions in Yaoluoping Nature Reserve from 1987 to 2020 mapped from extrapolating the 2011 SGB-based models.

4. Discussion

4.1. NDVI_DR Thresholding Model

Accurate identification of forest types is of great significance to forestry development planning and forestry policy formulation, helps to reliably evaluate the successive stage and trend of a specific forest ecosystem [3,4] and potentially improves the modelling accuracy of remote-sensing-based AGB estimation due to being able to separately model coniferous forest and broad-leaved forest [14–16]. In this paper, the NDVI_DR vegetation index that makes use of phenological or seasonal differences in the spectral signature of evergreen forest and deciduous forest was developed to classify forest types in the study area based on long-term time-series Landsat images, with an overall accuracy of above 92% and a kappa coefficient of about 0.85. The classification accuracy of the current NDVI_DR thresholding model is higher than other similar studies using Landsat for forest type classification. For example, Li et al. [70] used three machine learning approaches, including decision trees, random forest and support vector machines and Landsat images, to classify local forest communities at the Huntington Wildlife Forest (HWF). Among them, the SVM had the highest accuracy, with an overall accuracy of 88.2% and kappa coefficient of 0.793. Hill et al. [71] used two methods of low-pass spatial filtering to reduce the local spectral variation and image segmentation to implement supervised classifications of forest types in Peruvian Amazonia from Landsat TM data and gained an overall accuracy of about 90%. In their studies, they constructed various vegetation indices to express the spectral characteristics differences in different forest types, and they found that the spectral differences in different forest types characterized by diverse vegetation indices were not adequately separable. In contrast to these existing investigations, our NDVI_DR thresholding model does not require the development of diverse image features, e.g., spectral indices, image textures or contextual information; more importantly, there is no need to accurately tune the parameters of the classification models or algorithms. Conversely, our NDVI_DR model just calculates a derivative of a forest pixel's NDVI in the summer and corresponding NDVI in the winter to reflect the seasonal differences in the spectral signature of forest types, and it specifies stable thresholds to the derivative (less than 0.4 for coniferous forest and greater than or equal to 0.5 for broad-leaved forest, Figures 2 and 3) to classify forest types. Obviously, our model is efficient and easily implemented compared to the existing methods, and it substantially improves the identification accuracy of forest types by considering the seasonal differences of different forest types, which is in agreement with Zhu's [72] research results. Dong et al. [73] found that using seasonal time-series data has the potential to improve the accuracy for monitoring forest attributes, whereas the separability and stability of the NDVI_DR thresholding model have just been tested in the subtropical forest ecosystem by Landsat TM and OLI sensors, so its robustness in other ecoregions and other similar sensors, such as Sentinel-2 MSI, should be continuously verified in the near future to doubly confirm its generalization or popularization.

4.2. Forest AGB Modeling

At present, the main methods used in forest biomass prediction include random forest, support vector machine and a multiple linear regression model [20,21]. Nguyen et al. [74] developed a random-forest (RF)-based kNN model to produce annual maps of AGB from 1988 to 2017 for over 7.2 million ha of forests in Victoria, Australia. The modeling R^2 is between 0.37 and 0.59, and the RMSE is between 104.7 and 168.5 t/hm^2. Main-Knorn et al. [75] obtained the AGB from 1985 to 2010 through RF models based on Landsat time-series images and field data, showing an RMSE of 41.3 t/hm^2. Compared with the previous methods, the SGB model adopted in this paper has better robustness for outliers, inaccurate data, missing values and unbalanced datasets, and has relatively stable estimation results [64]. Dube [33] found that, when Landsat series images were selected as research data, the SGB algorithm had higher accuracy in forest biomass estimation than the random forest algorithm regardless of texture features, spectral features or if both of them were used. In addition, most existing forest AGB modelling and mapping studies did not distinguish between forest types but considered forests as a whole to retrieve the forest aboveground biomass in the entire region [76,77]. However, Fassnacht [78] actually found that there were fundamental differences in the NIR reflectance between coniferous forests and broad-leaved forests, and hardwood canopies could reflect 50% more in NIR than pine canopies due to different cellulose compactness or structures in their leaves. Thus, separate AGB modeling of different forest types may more adequately capture the respective variances in canopies' signatures of coniferous forest and broad-leaved forest. Actually, Figures 8–10 show the modeling R^2 of broad-leaved forest, coniferous forest and combined forest types at 0.68, 7.53 and 0.54, respectively, with the corresponding RMSEs at 7.53, 4.46 and 14.27 t/hm^2. This shows that AGB modeling by different forest types can achieve higher accuracy of AGB estimation than modelling without differentiating between forest types [14–16], which is consistent with Zheng's study [79].

However, due to the unavailability of historical records of field sample plots, we were unable to reliably validate the long-term time-series AGB maps generated by extrapolating the 2011 AGB prediction models into other years in the current analysis (Figure 11), although these AGB maps can provide basic data support for evaluating the ecosystem dynamic trends of the reserve and the effectiveness of management practices. Although we have assumed that there is a relatively stable relationship between forest AGB and remote sensing images in a specific area over time, in reality, natural disasters or anthropogenic disturbances may alter the forest spatial structure, species compositional structure and age structure over time and space, which may affect this stable relationship, thus affecting the accuracy of those extrapolated AGB estimations [80,81]. In the future, more efforts should be invested to ensure a sufficient validation of the historical AGB patterns generated from satellite image archives.

4.3. Driving Factors for Forest Area and AGB Changes

Our results showed that the forest areal proportion dropped from 92.92% in 1987 to 87.15% in 1997, and from 90.11% in 2011 to 88.6% in 2017 (Figure 5). It can be seen that coniferous forest is mainly distributed in the northern and south-eastern portions of the reserve, and the increased area of the coniferous forest in the reserve is significantly more than the reduced area in the past 30 years (Figure 6). This result is in agreement with the vegetation distribution of the Yaoluoping Nature Reserve studied by Xie et al. [82].

The coniferous forest decreased mainly in the southwest, northwest and southeast of the reserve between 1987 and 2002 (Figure 6). Major forest harvesting species, such as *Chinese fir* and *Pinus taiwanensis Hayata*, grew in the southwest and southeast of the reserve. The economic sources of the local residents in Yaoluoping Reserve mainly depend on timber and agricultural production. Thus, unrestricted timber harvesting of these coniferous species led to a significant reduction in area. Bahurudeen et al. [83] found that the coniferous forest had excellent materials and high economic value and was widely used in a variety of industries, which effectively explains the rapid decline in the coniferous forest in the reserve. In addition, in 2011, a large area of coniferous forest decreased due to natural disasters, such as landslides, in the southeastern part of the reserve. The increased area of coniferous forest was mainly distributed in the valley of the Baojia River basin, and the years were mainly from 1992 to 2007. In 2001, a zero-felling quota of commercial timber was implemented, and the felling of commercial timber was basically eradicated and the overall forest felling decreased dramatically. At the same time, a large number of tree species, such as *Chinese fir*, have been planted and renewed in the reserve. According to the records, the natural forest of Huangliyuan forest farm has been completely updated to the existing *Chinese fir* forest since 1993. All the events can reasonably explain the dynamic patterns of coniferous forest in the reserve during the period of 1987 to 2020.

The biomass inversion results showed that the forest AGB in Yaoluoping National Nature Reserve increased continuously from 1987 to 2020, but the coniferous forest biomass decreased correspondingly in those disturbed years (Table 12). This indicates that human factors are the main factors affecting the forest AGB. During the periods of rapid population density and GDP growth, forests, as the main source of income and livelihood, are greatly affected, and the forest AGB decreases accordingly. Rozendaal et al. [84] showed that human activities, especially logging disturbances, had a significant impact on forest biomass. In the future development process, the local government should properly consider the carrying capacity of the forest ecosystem to population density and establish an ecological compensation mechanism combined with its own characteristic forest industries so as to realize the goal of protecting natural resources and protecting the interests of the community. Liu et al. [85] took Yaoluoping National Nature Reserve as the research object and discussed how to effectively promote the development of the national nature reserve by using ecological compensation mechanisms based on the observed problems in the development process. For the scientific protection of the reserve, we can cultivate biological resources with economic benefits and carry out activities, such as ecotourism and popularization of science, to mitigate the impacts of human demands on forest resources. Xu et al. [86] quantified the biodiversity value of Yaoluoping Nature Reserve and confirmed the ecological and economic value of the reserve, which poses a high priority of sustainable natural resources management in the reserve. In addition, some deforested areas should be classified according to the degree of deforestation and site conditions, and the vegetation restoration should be organically combined with natural regeneration and artificial intervention to shorten the time of vegetation restoration to the forest community scientifically and effectively.

4.4. Limitations and Future Improvements

Although important results were obtained from this study, the following aspects still need to be further studied: (1) Although the constructed vegetation index NDVI_DR can better distinguish between coniferous forest and broad-leaved forest, its extraction effects for mixed forest and shrubs need further verification [72]. (2) More efforts should be made to ensure a reliable validation of historical AGB maps, such as collecting measurements from those permanent sample plots possibly deployed in the reserve. (3) Although the SGB algorithm can reduce the variance and bias and has high estimation accuracy, it is very sensitive to the change in the outliers in the training samples [64]. Thus, its tuned parameters need to be optimized in later research.

5. Conclusions

In this study, a VCT model was used to generate forest cover datasets in the study area, and NDVI-DR was developed to classify the forest types. On this basis, the SGB algorithm and extrapolation were applied to retrieve the AGB in the study area from 1987 to 2020. The findings from our study can provide potential insights for long-term forest remote sensing observations, forest type classification, and accurate AGB mapping. These findings also can inform similar management agencies of a carbon accounting data basis and provide informed actions on sustainable development with high ecological interests. Based on the findings, it is concluded that:

(1) The NDVI_DR thresholding provides an efficient and accurate classification method for distinguishing between coniferous forest and broad-leaved forest. The overall accuracy is above 92%, with a kappa coefficient above 0.8.

(2) The 2011 forest-type-dependent stochastic-gradient-boosting-based (SGB-based) AGB estimation model achieved an independent validation R square at 0.63 and an RMSE at 11.18 t/ha for broad-leaved forest, and 0.61 and 14.26 t/ha for coniferous forest. A time-series of AGB was generated by extrapolating the 2011 AGB models to other years, and the mapped AGB showed a gradual increasing trend over the past three decades.

(3) There is a significant correlation between human disturbance and AGB, especially irregular deforestation. Thus, we suggest that the local government should properly consider the carrying capacity of the forest ecosystem to population density and establish an ecological compensation mechanism combined with its own characteristic forest industries.

Author Contributions: Conceptualization, M.L.; methodology, B.Y., Y.Z., X.M. and M.L.; software, B.Y. and X.M.; validation, B.Y., Y.Z. and X.M.; formal analysis, B.Y., Y.Z. and Y.L.; investigation, B.Y., Y.Z., X.M. and Y.L.; resources, Y.L. and Y.Z.; data curation, B.Y., X.M., Y.Z. and F.S.; writing—original draft preparation, B.Y. and X.M.; writing—review and editing, B.Y., F.S. and X.M.; visualization, B.Y., X.M. and F.S.; supervision, Y.Z., Y.L., X.M., F.S. and M.L.; project administration, Y.Z. and M.L.; funding acquisition, M.L. All authors have read and agreed to the published version of the manuscript.

Funding: This research was jointly funded by the Natural Science Foundation of China, grant number 31971577, and the Priority Academic Program Development of Jiangsu Higher Education Institutions (PAPD).

Data Availability Statement: Not applicable.

Conflicts of Interest: The authors declare no conflict of interest.

References

1. Liu, N.; Wang, D.; Guo, Q. Exploring the Influence of Large Trees on Temperate Forest Spatial Structure from the Angle of Mingling. *For. Ecol. Manag.* **2021**, *492*, 119220. [CrossRef]
2. Ministry of Commerce, PRC. Circular of The State Council on the issuance of an Action plan to peak Carbon emissions by 2030. *Bull. State Counc. People's Repub. China* **2021**, *31*, 48–58. (In Chinese)
3. Diao, J.; Liu, J.; Zhu, Z.; Wei, X.; Li, M. Active Forest Management Accelerates Carbon Storage in Plantation Forests in Lishui, Southern China. *For. Ecosyst.* **2022**, *9*, 100004. [CrossRef]
4. Zhang, R.; Zhou, X.; Ouyang, Z.; Avitabile, V.; Qi, J.; Chen, J.; Giannico, V. Estimating Aboveground Biomass in Subtropical Forests of China by Integrating Multisource Remote Sensing and Ground Data. *Remote Sens. Environ.* **2019**, *232*, 111341. [CrossRef]
5. Cao, G. Application of Remote Sensing Technology in Forest Resources Investigation. *Remote Sens.* **2020**, *9*, 46. [CrossRef]
6. Meigs, G.W.; Kennedy, R.E.; Cohen, W.B. A Landsat Time Series Approach to Characterize Bark Beetle and Defoliator Impacts on Tree Mortality and Surface Fuels in Conifer Forests. *Remote Sens. Environ.* **2011**, *115*, 3707–3718. [CrossRef]
7. Röder, A.; Hill, J.; Duguy, B.; Alloza, J.A.; Vallejo, R. Using Long Time Series of Landsat Data to Monitor Fire Events and Post-Fire Dynamics and Identify Driving Factors. A Case Study in the Ayora Region (Eastern Spain). *Remote Sens. Environ.* **2008**, *112*, 259–273. [CrossRef]
8. Kennedy, R.E.; Yang, Z.; Cohen, W.B. Detecting Trends in Forest Disturbance and Recovery Using Yearly Landsat Time Series: 1. LandTrendr-Temporal Segmentation Algorithms. *Remote Sens. Environ.* **2010**, *114*, 2897–2910. [CrossRef]
9. Huang, C.; Goward, S.N.; Masek, J.G.; Thomas, N.; Zhu, Z.; Vogelmann, J.E. An Automated Approach for Reconstructing Recent Forest Disturbance History Using Dense Landsat Time Series Stacks. *Remote Sens. Environ.* **2010**, *114*, 183–198. [CrossRef]
10. Gao, Y.; Solórzano, J.V.; Quevedo, A.; Loya-Carrillo, J.O. How BFAST Trend and Seasonal Model Components Affect Disturbance Detection in Tropical Dry Forest and Temperate Forest. *Remote Sens.* **2021**, *13*, 2033. [CrossRef]
11. Singh, A. Review Article: Digital Change Detection Techniques Using Remotely-Sensed Data. *Int. J. Remote Sens.* **1989**, *10*, 989–1003. [CrossRef]
12. Mas, J.F. Monitoring Land-Cover Changes: A Comparison of Change Detection Techniques. *Int. J. Remote Sens.* **1999**, *20*, 139–152. [CrossRef]
13. Lu, D.; Mausel, P.; Brondízio, E.; Moran, E. Change Detection Techniques. *Int. J. Remote Sens.* **2004**, *25*, 2365–2401. [CrossRef]
14. Li, C.; Li, M.; Li, Y.; Qian, P. Estimating Aboveground Forest Carbon Density Using Landsat 8 and Field-Based Data: A Comparison of Modelling Approaches. *Int. J. Remote Sens.* **2020**, *41*, 4269–4292. [CrossRef]
15. Singh, M.; Malhi, Y.; Bhagwat, S. Biomass Estimation of Mixed Forest Landscape Using a Fourier Transform Texture-Based Approach on Very-High-Resolution Optical Satellite Imagery. *Int. J. Remote Sens.* **2014**, *35*, 3331–3349. [CrossRef]
16. Lu, D.; Chen, Q.; Wang, G.; Liu, L.; Li, G.; Moran, E. A Survey of Remote Sensing-Based Aboveground Biomass Estimation Methods in Forest Ecosystems. *Int. J. Digit. Earth* **2016**, *9*, 63–105. [CrossRef]
17. Li, C.; Wang, J.; Hu, L.; Yu, L.; Clinton, N.; Huang, H.; Yang, J.; Gong, P. A circa 2010 Thirty Meter Resolution Forest Map for China. *Remote Sens.* **2014**, *6*, 5325–5343. [CrossRef]
18. Xiao, X.; Boles, S.; Liu, J.; Zhuang, D.; Liu, M. Characterization of Forest Types in Northeastern China, Using Multi-Temporal SPOT-4 VEGETATION Sensor Data. *Remote Sens. Environ.* **2002**, *82*, 335–348. [CrossRef]
19. Loveland, T.R.; Reed, B.C.; Ohlen, D.O.; Brown, J.F.; Zhu, Z.; Yang, L.; Merchant, J.W. Development of a Global Land Cover Characteristics Database and IGBP DISCover from 1 Km AVHRR Data. *Int. J. Remote Sens.* **2000**, *21*, 1303–1330. [CrossRef]
20. Raczko, E.; Zagajewski, B. Comparison of Support Vector Machine, Random Forest and Neural Network Classifiers for Tree Species Classification on Airborne Hyperspectral APEX Images. *Eur. J. Remote Sens.* **2017**, *50*, 144–154. [CrossRef]
21. Noi, P.T.; Kappas, M. Comparison of Random Forest, k-Nearest Neighbor, and Support Vector Machine Classifiers for Land Cover Classification Using Sentinel-2 Imagery. *Sensors* **2017**, *18*, 18. [CrossRef]
22. Chunhui, Z.; Bing, G.; Lejun, Z.; Xiaoqing, W. Classification of Hyperspectral Imagery Based on Spectral Gradient, SVM and Spatial Random Forest. *Infrared Phys. Technol.* **2018**, *95*, 61–69. [CrossRef]

23. Tompoulidou, M.; Gitas, I.Z.; Polychronaki, A.; Mallinis, G. A GEOBIA Framework for the Implementation of National and International Forest Definitions Using Very High Spatial Resolution Optical Satellite Data. *Geocarto Int.* **2016**, *31*, 342–354. [CrossRef]
24. Pan, L.; Sun, Y.; Wang, Y.; Chen, L.; Cao, Y. Estimation of Aboveground Biomass in a Chinese Fir (*Cunninghamia lanceolata*) Forest Combining Data of Sentinel-1 and Sentinel-2. *J. Nanjing For. Univ. Nat. Sci. Ed.* **2021**, *44*, 149–156. [CrossRef]
25. Gao, Y.; Lu, D.; Li, G.; Wang, G.; Chen, Q.; Liu, L.; Li, D. Comparative analysis of modeling algorithms for forest aboveground biomass estimation in a subtropical region. *Remote Sens.* **2018**, *10*, 627. [CrossRef]
26. Ji, L.; Wylie, B.K.; Brown, D.R.N.; Peterson, B.; Alexander, H.D.; Mack, M.C.; Rover, J.; Waldrop, M.P.; McFarland, J.W.; Chen, X.; et al. Spatially Explicit Estimation of Aboveground Boreal Forest Biomass in the Yukon River Basin, Alaska. *Int. J. Remote Sens.* **2015**, *36*, 939–953. [CrossRef]
27. Liu, J.; Wang, X.; Wang, T. Automatic Identification of Tree Species Based on Deep Learning. *J. Nanjing For. Univ. Nat. Sci. Ed.* **2021**, *44*, 138–144. [CrossRef]
28. Han, H.; Wan, R.; Li, B. Estimating Forest Aboveground Biomass Using Gaofen-1 Images, Sentinel-1 Images, and Machine Learning Algorithms: A Case Study of the Dabie Mountain Region, China. *Remote Sens.* **2022**, *14*, 176. [CrossRef]
29. Frazier, R.J.; Coops, N.C.; Wulder, M.A.; Kennedy, R. Characterization of Aboveground Biomass in an Unmanaged Boreal Forest Using Landsat Temporal Segmentation Metrics. *ISPRS J. Photogramm. Remote Sens.* **2014**, *92*, 137–146. [CrossRef]
30. Lei, C.; Ju, C.; Cai, T.; Jing, X.; Wei, X.; Di, X. Estimating Canopy Closure Density and Above-Ground Tree Biomass Using Partial Least Square Methods in Chinese Boreal Forests. *J. For. Res.* **2012**, *23*, 191–196. [CrossRef]
31. Lawrence, R.; Bunn, A.; Powell, S.; Zambon, M. Classification of Remotely Sensed Imagery Using Stochastic Gradient Boosting as a Refinement of Classification Tree Analysis. *Remote Sens. Environ.* **2004**, *90*, 331–336. [CrossRef]
32. Güneralp, I.; Filippi, A.M.; Randall, J. Estimation of Floodplain Aboveground Biomass Using Multispectral Remote Sensing and Nonparametric Modeling. *Int. J. Appl. Earth Obs. Geoinf.* **2014**, *33*, 119–126. [CrossRef]
33. Dube, T.; Mutanga, O. Evaluating the Utility of the Medium-Spatial Resolution Landsat 8 Multispectral Sensor in Quantifying Aboveground Biomass in uMgeni Catchment, South Africa. *ISPRS J. Photogramm. Remote Sens.* **2015**, *101*, 36–46. [CrossRef]
34. Wang, Z.G.; Zhang, Z.X.; Wang, W.G.; Chu, J. Preliminary Analysis of Forest Community Structure of Yaoluoping National Nature Reserve in Yuexi County, Anhui Province, China. *Chin. J. Plant Ecol.* **2016**, *40*, 615–619. [CrossRef]
35. Skakun, S.; Vermote, E.F.; Roger, J.C.; Justice, C.O.; Masek, J.G. Validation of the LaSRC Cloud Detection Algorithm for Landsat 8 Images. *IEEE J. Sel. Top. Appl. Earth Obs. Remote Sens.* **2019**, *12*, 2439–2446. [CrossRef]
36. Masek, J.G.; Vermote, E.F.; Saleous, N.E.; Wolfe, R.; Hall, F.G.; Huemmrich, K.F.; Gao, F.; Kutler, J.; Lim, T.K. A Landsat Surface Reflectance Dataset for North America, 1990–2000. *IEEE Geosci. Remote Sens. Lett.* **2006**, *3*, 68–72. [CrossRef]
37. Shen, W.; Li, M.; Huang, C.; He, T.; Tao, X.; Wei, A. Local Land Surface Temperature Change Induced by Afforestation Based on Satellite Observations in Guangdong Plantation Forests in China. *Agric. For. Meteorol.* **2019**, *276*, 107641. [CrossRef]
38. Qiu, J.; Wang, H.; Shen, W.; Zhang, Y.; Su, H.; Li, M. Quantifying Forest Fire and Post-Fire Vegetation Recovery in the Daxing'anling Area of Northeastern China Using Landsat Time-Series Data and Machine Learning. *Remote Sens.* **2021**, *13*, 792. [CrossRef]
39. Tang, D.; Fan, H.; Yang, K.; Zhang, Y. Mapping Forest Disturbance across the China–Laos Border Using Annual Landsat Time Series. *Int. J. Remote Sens.* **2019**, *40*, 2895–2915. [CrossRef]
40. Gamon, J.A.; Peñuelas, J.; Field, C.B. A Narrow-Waveband Spectral Index That Tracks Diurnal Changes in Photosynthetic Efficiency. *Remote Sens. Environ.* **1992**, *41*, 35–44. [CrossRef]
41. Baret, F.; Guyot, G.; Major, D.J. TSAVI: A Vegetation Index which Minimizes Soil Brightness Effects on LAI and APAR Estimation. In Proceedings of the 12th Canadian Symposium on Remote Sensing Geoscience and Remote Sensing Symposium, Vancouver, BC, Canada, 10–14 July 1989; Volume 3, pp. 1355–1358. [CrossRef]
42. Liu, Q.; Liu, G.; Huang, C.; Xie, C. Comparison of Tasselled Cap Transformations Based on the Selective Bands of Landsat 8 OLI TOA Reflectance Images. *Int. J. Remote Sens.* **2015**, *36*, 417–441. [CrossRef]
43. Duane, M.V.; Cohen, W.B.; Campbell, J.L.; Hudiburg, T.; Turner, D.P.; Weyermann, D.L. Implications of Alternative Field-sampling Designs on Landsat-based Mapping of Stand Age and Carbon Stocks in Oregon Forests. *For. Sci.* **2010**, *56*, 405–416. [CrossRef]
44. Powell, S.L.; Cohen, W.B.; Healey, S.P.; Kennedy, R.E.; Moisen, G.G.; Pierce, K.B.; Ohmann, J.L. Quantification of Live Aboveground Forest Biomass Dynamics with Landsat Time-Series and Field Inventory Data: A Comparison of Empirical Modeling Approaches. *Remote Sens. Environ.* **2010**, *114*, 1053–1068. [CrossRef]
45. Ahmed, O.S.; Franklin, S.E.; Wulder, M.A.; White, J.C. Characterizing Stand-Level Forest Canopy Cover and Height Using Landsat Time Series, Samples of Airborne LiDAR, and the Random Forest Algorithm. *ISPRS J. Photogramm. Remote Sens.* **2015**, *101*, 89–101. [CrossRef]
46. Pflugmacher, D.; Cohen, W.B.; Kennedy, R.E.; Yang, Z. Using Landsat-Derived Disturbance and Recovery History and Lidar to Map Forest Biomass Dynamics. *Remote Sens. Environ.* **2014**, *151*, 124–137. [CrossRef]
47. Li, C.; Wang, J.; Wang, Q.; Wang, D.; Song, X.; Wang, Y.; Huang, W. Estimating Wheat Grain Protein Content Using Multi-Temporal Remote Sensing Data Based on Partial Least Squares Regression. *J. Integr. Agric.* **2012**, *11*, 1445–1452. [CrossRef]
48. Todd, S.W.; Hoffer, R.M.; Milchunas, D.G. Biomass Estimation on Grazed and Ungrazed Rangelands Using Spectral Indices. *Int. J. Remote Sens.* **1998**, *19*, 427–438. [CrossRef]
49. Mutanga, O.; Adam, E.; Cho, M.A. High Density Biomass Estimation for Wetland Vegetation Using Worldview-2 Imagery and Random Forest Regression Algorithm. *Int. J. Appl. Earth Obs. Geoinf.* **2012**, *18*, 399–406. [CrossRef]

50. Brown, L.; Chen, J.M.; Leblanc, S.G.; Cihlar, J. A shortwave infrared modification to the simple ratio for LAI retrieval in boreal forests: An image and model analysis. *Remote Sens. Environ.* **2000**, *71*, 16–25. [CrossRef]
51. Nemani, R.; Pierce, L.; Running, S.; Band, L. Forest Ecosystem Processes at the Watershed Scale: Sensitivity to Remotely-Sensed Leaf Area Index Estimates. *Int. J. Remote Sens.* **1993**, *14*, 2519–2534. [CrossRef]
52. Singh, K.V.; Setia, R.; Sahoo, S.; Prasad, A.; Pateriya, B. Evaluation of NDWI and MNDWI for Assessment of Waterlogging by Integrating Digital Elevation Model and Groundwater Level. *Geocarto Int.* **2015**, *30*, 650–661. [CrossRef]
53. Hawlick, R.M. Statistical and Structural Approaches to Texture. *Proc. IEEE* **1979**, *67*, 786–804. [CrossRef]
54. Tuominen, S.; Pekkarinen, A. Performance of Different Spectral and Textural Aerial Photograph Features in Multi-Source Forest Inventory. *Remote Sens. Environ.* **2005**, *94*, 256–268. [CrossRef]
55. Chica-Olmo, M.; Abarca-Hernandez, F. Computing Geostatistical Image Texture for Remotely Sensed Data Classification. *Comput. Geosci.* **2000**, *26*, 373–383. [CrossRef]
56. He, D.C.; Wang, L. Texture Unit, Texture Spectrum, and Texture Analysis. *IEEE Trans. Geosci. Remote Sens.* **1990**, *28*, 509–512. [CrossRef]
57. Garzelli, A. Possibilities and Limitations of the Use of Wavelets in Image Fusion. *IEEE Int. Geosci. Remote Sens. Symp.* **2002**, *1*, 66–68. [CrossRef]
58. Haralick, R.M.; Shanmugam, K.; Dinstein, I.H. Textural Features for Image Classification. *IEEE Trans. Syst. Man Cybern.* **1973**, *6*, 610–621. [CrossRef]
59. Barbosa, J.M.; Melendez-Pastor, I.; Navarro-Pedreño, J.; Bitencourt, M.D. Remotely Sensed Biomass over Steep Slopes: An Evaluation among Successional Stands of the Atlantic Forest, Brazil. *ISPRS J. Photogramm. Remote Sens.* **2014**, *88*, 91–100. [CrossRef]
60. Zhang, H.; Zhu, J.; Wang, C.; Lin, H.; Long, J.; Zhao, L.; Fu, H.; Liu, Z. Forest Growing Stock Volume Estimation in Subtropical Mountain Areas Using PALSAR-2 L-Band PolSAR Data. *Forests* **2019**, *10*, 276. [CrossRef]
61. McRoberts, R.E.; Chen, Q.; Walters, B.F. Multivariate Inference for Forest Inventories Using Auxiliary Airborne Laser Scanning Data. *For. Ecol. Manag.* **2017**, *401*, 295–303. [CrossRef]
62. You, R.; Zhu, N.; Deng, X.; Wang, J.; Liu, F. Variation in Wood Physical Properties and Effects of Climate for Different Geographic Sources of Chinese Fir in Subtropical Area of China. *Sci. Rep.* **2021**, *11*, 4664. [CrossRef] [PubMed]
63. Yao, X.; Sun, S.; Li, X.; Liu, R. Accuracy Evaluation of the CCI Remote Sensing Soil Moisture for Revealing Drought in Northeast China. *IOP Conf. Ser. Earth Environ. Sci.* **2018**, *185*, 012040. [CrossRef]
64. Friedman, J.H. Stochastic Gradient Boosting. *Comput. Stat. Data Anal.* **2002**, *38*, 367–378. [CrossRef]
65. Bauer, E.; Kohavi, R. An Empirical Comparison of Voting Classification Algorithms: Bagging, Boosting, and Variants. *Mach. Learn.* **1999**, *36*, 105–139. [CrossRef]
66. Breiman, L. Random Forests. *Mach. Learn.* **2001**, *45*, 5–32. [CrossRef]
67. Healey, S.P.; Yang, Z.; Cohen, W.B.; Pierce, D.J. Application of Two Regression-Based Methods to Estimate the Effects of Partial Harvest on Forest Structure Using Landsat Data. *Remote Sens. Environ.* **2006**, *101*, 115–126. [CrossRef]
68. Canty, M.J.; Nielsen, A.A.; Schmidt, M. Automatic Radiometric Normalization of Multitemporal Satellite Imagery. *Remote Sens. Environ.* **2004**, *91*, 441–451. [CrossRef]
69. Lin, C.H.; Lin, B.Y.; Lee, K.Y.; Chen, Y.C. Radiometric Normalization and Cloud Detection of Optical Satellite Images Using Invariant Pixels. *ISPRS J. Photogramm. Remote Sens.* **2015**, *106*, 107–117. [CrossRef]
70. Li, M.; Im, J.; Beier, C. Machine Learning Approaches for Forest Classification and Change Analysis Using Multi-temporal Landsat TM Images Over Huntington Wildlife Forest. *GIScience Remote Sens.* **2013**, *50*, 361–384. [CrossRef]
71. Hill, R.A. Image Segmentation for Humid Tropical Forest Classification in Landsat TM Data. *Int. J. Remote Sens.* **1999**, *20*, 1039–1044. [CrossRef]
72. Zhu, X.; Liu, D. Improving Forest Aboveground Biomass Estimation Using Seasonal Landsat NDVI Time-Series. *ISPRS J. Photogramm. Remote Sens.* **2015**, *102*, 222–231. [CrossRef]
73. Dong, J.; Kaufmann, R.K.; Myneni, R.B.; Tucker, C.J.; Kauppi, P.E.; Liski, J.; Buermann, W.; Alexeyev, V.; Hughes, M.K. Remote Sensing Estimates of Boreal and Temperate Forest Woody Biomass: Carbon Pools, Sources, and Sinks. Remote Sens. *Remote Sens. Environ.* **2003**, *84*, 393–410. [CrossRef]
74. Nguyen, T.H.; Jones, S.D.; Soto-Berelov, M.; Haywood, A.; Hislop, S. Monitoring Aboveground Forest Biomass Dynamics Over Three Decades Using Landsat Time-Series and Single-Date Inventory Data. *Int. J. Appl. Earth Obs. Geoinf.* **2020**, *84*, 101952. [CrossRef]
75. Main-Knorn, M.; Cohen, W.B.; Kennedy, R.E.; Grodzki, W.; Pflugmacher, D.; Griffiths, P.; Hostert, P. Monitoring Coniferous Forest Biomass Change Using a Landsat Trajectory-Based Approach. *Remote Sens. Environ.* **2013**, *139*, 277–290. [CrossRef]
76. Gómez, C.; White, J.C.; Wulder, M.A.; Alejandro, P. Historical Forest Biomass Dynamics Modelled with Landsat Spectral Trajectories. *ISPRS J. Photogramm. Remote Sens.* **2014**, *93*, 14–28. [CrossRef]
77. Karlson, M.; Ostwald, M.; Reese, H.; Sanou, J.; Tankoano, B.; Mattsson, E. Mapping Tree Canopy Cover and Aboveground Biomass in Sudano-Sahelian Woodlands Using Landsat 8 and Random Forest. *Remote Sens.* **2015**, *7*, 10017–10041. [CrossRef]
78. Fassnacht, K.S.; Gower, S.T.; MacKenzie, M.D.; Nordheim, E.V.; Lillesand, T.M. Estimating the Leaf Area Index of North Central Wisconsin Forests Using the Landsat Thematic Mapper. *Remote Sens. Environ.* **1997**, *61*, 229–245. [CrossRef]

79. Zheng, D.; Rademacher, J.; Chen, J.; Crow, T.; Bresee, M.; Moine, J.L.; Ryu, S.R. Estimating Aboveground Biomass Using Landsat 7 ETM+ Data Across a Managed Landscape in Northern Wisconsin, USA. *Remote Sens. Environ.* **2004**, *93*, 402–411. [CrossRef]
80. Agramont, A.R.E.; Maass, S.F.; Bernal, G.N.; Hernández, J.I.V.; Fredericksen, T.S. Effect of Human Disturbance on the Structure and Regeneration of Forests in the Nevado De Toluca National Park, Mexico. *J. For. Res.* **2012**, *23*, 39–44. [CrossRef]
81. Shi, F.; Liu, M.; Qiu, J.; Zhang, Y.; Su, H.; Mao, X.; Li, X.; Fan, J.; Chen, J.; Lv, Y.; et al. Assessing Land Cover and Ecological Quality Changes in the Forest-Steppe Ecotone of the Greater Khingan Mountains, Northeast China, from Landsat and MODIS Observations from 2000 to 2018. *Remote Sens.* **2022**, *14*, 725. [CrossRef]
82. Xie, Z.; Wu, G. The Vegetation Types and Distributions in Yaoluoping Natural Reserve of Anhui Province. *J. East China Norm. Univ. Nat. Sci.* **1995**, *3*, 93–102. (In Chinese)
83. Bahurudeen, A.; Moorthi, P.V.P. *Testing of Construction Materials*, 1st ed.; CRC Press: Boca Raton, FL, USA, 2021; pp. 95–128.
84. Rozendaal, D.M.; Chazdon, R.L. Demographic Drivers of Tree Biomass Change During Secondary Succession in Northeastern Costa Rica. *Ecol. Appl.* **2015**, *25*, 506–516. [CrossRef] [PubMed]
85. Liu, H.; Wang, K.; Wang, X.; Zhang, Y.; Wen, Y. Study on Ecological Compensation Mechanism of National Nature Reserve Based on the Case of Yaoluoping. *Environ. Sustain. Dev.* **2014**, *39*, 143–146. (In Chinese) [CrossRef]
86. Xu, H.; Qian, Y.; Zheng, L.; Peng, B.Z. Assessment of Indirect Use Values of Forest Biodiversity in Yaoluoping National Nature Reserve, Anhui Province. *Chin. Geogr. Sci.* **2003**, *13*, 277–283. [CrossRef]

remote sensing

Article

Spatial Scale Effect and Correction of Forest Aboveground Biomass Estimation Using Remote Sensing

Ying Yu [1,2], Yan Pan [1,2], Xiguang Yang [1,2,*] and Wenyi Fan [1,2]

1 School of Forestry, Northeast Forestry University, Harbin 150040, China; yuying@nefu.edu.cn (Y.Y.); 2020121174@nefu.edu.cn (Y.P.); fanwy@nefu.edu.cn (W.F.)
2 Key Laboratory of Sustainable Forest Ecosystem Management, Ministry of Education, Northeast Forestry University, Harbin 150040, China
* Correspondence: yangxiguang@nefu.edu.cn

Abstract: Forest biomass is critically important for forest dynamics in the carbon cycle. However, large-scale AGB mapping applications from remote sensing data still carry large uncertainty. In this study, an AGB estimation model was first established with three different remote sensing datasets of GF-2, Sentinel-2 and Landsat-8. Next, the optimal scale estimation result was considered as a reference AGB to obtain the relative true AGB distribution at different scales based on the law of conservation of mass, and the error of the scale effect of AGB estimation at various spatial resolutions was analyzed. Then, the information entropy of land use type was calculated to identify the heterogeneity of pixels. Finally, a scale conversion method for the entropy-weighted index was developed to correct the scale error of the estimated AGB results from coarse-resolution remote sensing images. The results showed that the random forest model had better prediction accuracy for GF-2 (4 m), Sentinel-2 (10 m) and Landsat-8 (30 m) AGB mapping. The determination coefficient between predicted and measured AGB was 0.5711, 0.4819 and 0.4321, respectively. Compared to uncorrected AGB, R^2 between scale-corrected results and relative true AGB increased from 0.6226 to 0.6725 for Sentinel-2, and increased from 0.5910 to 0.6704 for Landsat-8. The scale error was effectively corrected. This study can provide a reference for forest AGB estimation and scale error reduction for AGB production upscaling with consideration of the spatial heterogeneity of the forest surface.

Keywords: forest aboveground biomass (AGB); scale effect; random forest (RF); scale correction

Citation: Yu, Y.; Pan, Y.; Yang, X.; Fan, W. Spatial Scale Effect and Correction of Forest Aboveground Biomass Estimation Using Remote Sensing. *Remote Sens.* **2022**, *14*, 2828. https://doi.org/10.3390/rs14122828

Academic Editor: Klaus Scipal

Received: 28 April 2022
Accepted: 10 June 2022
Published: 13 June 2022

1. Introduction

Terrestrial ecosystems, covering approximately 30% of the Earth's land surface, play an important role in the global carbon cycle and climate changes [1]. Forests are a major contributor to the terrestrial carbon pool. Forests store approximately 45% of the carbon found in terrestrial ecosystems as living biomass and dead wood and litter [2,3]. At the same time, forests can sequester large amounts of carbon dioxide from the atmosphere and contribute approximately 50% of the global net primary production (NPP) and approximately 80% of terrestrial NPP [4–6]. Forests absorb atmospheric CO_2 through photosynthesis and remove nearly 3 billion tons of anthropogenic carbon every year [7]. As forests grow, around 30% of CO_2 emissions from fuel burning and net deforestation are absorbed [8]. Therefore, forest ecosystems can increase or decrease carbon sequestration by restoring or degrading vegetation [9]. If a forest is disturbed by fire, deforestation or other human factors, the carbon stored in the forest would be released back into the atmosphere; therefore, the accurate estimation of forest carbon stocks is essential for the study of the carbon exchange between terrestrial ecosystems and the atmosphere and its effects on ecosystem-level carbon cycling, feeding back to climate change [10,11].

Forest aboveground biomass, or aboveground biomass of trees (AGB), is defined as the mass of the living organic material, which includes the living stems, branches and leaves of vegetation, with units of mass per unit area [12]. The aboveground biomass constitutes the main portion of the carbon stock, and it is the most important and visible terrestrial ecosystem carbon pool. The AGB is intimately related to the emission of CO_2 caused by land use change and fire and the stored CO_2 in the atmosphere by vegetation growth. AGB is a key quantity estimating terrestrial carbon pools and is recognized as an Essential Climate Variable (ECV) by the Global Climate Observing System (GCOS) [13]. It is also used to monitor climate change by the United Nations Framework Convention on Climate Change (UNFCCC) and the Intergovernmental Panel on Climate Change (IPCC) [14]. In addition, vegetation biomass has also wider significance to human society in the form of food, materials, energy and other ecosystem services that can be found in forests and other ecosystems [15]. AGB is also important for monitoring the forest planting dynamics, evaluating forest management practices and assessing wood resources [15,16]. Accurately monitoring and reporting the forest aboveground biomass is essential to correctly budget carbon emissions and is beneficial for mitigating climate change through the reduction of greenhouse gas emissions [17].

Conventionally, the AGB estimation methods can be categorized into field measurement, ecological model simulation and approaches involving remote sensing [1]. Field measurement requires an intensive field inventory; the most important field forest-related parameters include the number of trees, tree density, the taxonomical information, tree height and the diameter at breast height (DBH) [18]. Field measurement is considered to provide the most accurate AGB estimation, combined with allometric equations [19]. The traditional approach of manual data collection is called field inventory in forestry [20]. It is an important method that is used in the monitoring and management of the forest. However, the conventional method of AGB inventory is destructive, labor-intensive, expensive, time-consuming and sometimes limited by poor accessibility [21]. It cannot provide the spatial and temporal distribution and explicit forest biomass information [1]. Hence, the field inventory method is practical only in relatively smaller areas [22].

Limited by the traditional methods of AGB estimation on a regional scale, remote sensing has been widely used to estimate AGB during the past few decades. On the one hand, remote sensing technology has the advantage of facilitating the collection of forest type characteristics and coverage information, which is greatly useful in field inventory [23]. On the other hand, remote sensing can provide the spectral characteristics of vegetation and offer the repeated historical observation required for change detection, and digital data can be easily stored and integrated with geographic information [22]. Remote sensing technology has been extensively used as an efficient and economical method for the large-scale estimation of forest biomass [24,25]. Various remote sensing data, such as optical, thermal, microwave, radar and LiDAR remote sensing data, have been used for AGB estimation [26,27]. The intermediate- and high-resolution remote sensing are usually used for biomass estimation at local or sub-regional scales, such as Landsat series, SPOT, WorldView-2 and GF series [28–30]. Moreover, MODIS and NOAA-AVHRR remote sensing data have been used to estimate AGB at a regional, national or global scale [31,32]. Optical remote sensing data are a common data source for biomass estimation, but data saturation in the optical regime is an important factor influencing the accuracy of biomass estimation in forests [33,34]. With the rapid development of light detection and ranging (LiDAR) technology, LiDAR has become a vital method for AGB estimation [35]. Compared to optical remote sensing, LiDAR can derive the canopy height information, which is strongly related to AGB at high levels, and it is considered the most promising technique for biomass estimation. However, LiDAR data are usually limited to use in small areas due to high costs [26].

In the last few decades, multiple-resource remote sensing data have been used to estimate AGB and are recognized as dominant data sources for AGB mapping [1]. However, current AGB estimations were retrieved by using different methods and remote sensing data with various spatial resolutions [36]. This would lead to significant disagreement when estimating AGB and reduce its value in global and national applications. There are many random and/or systematic errors that arise during AGB estimation, such as in field-based AGB estimation, matching field and remote sensing measurement, satellite-based measurement and global wall-to-wall extrapolation [37]. First, the uncertainty of the field-based AGB estimation is a matter of concern for remote sensing scholars. Chen et al. tracked the errors from the field-based uncertainty of AGB estimation based on airborne LiDAR remote sensing [38]. Longo et al. concluded that the field-based uncertainty contributed around 10% of the total pixel-level uncertainty of AGB prediction at a 0.16 ha resolution [39]. Second, the mismatch between field measurements and remote sensing images may lead to errors in AGB estimation. Moreover, it would reduce the AGB model error from 51% to 4% at 20 to 100 m resolution when using LiDAR remote sensing [40]. The spatial difference between field measurements and remote sensing data is a common problem that can increase the AGB estimation error in remote sensing studies [41]. Næsset et al. found that the field plot size would have an effect on the estimation of AGB and pointed out that the plot size should be considered when using remotely sensed data for AGB estimation [42]. Persson et al. analyzed the uncertainties of AGB estimation at the plot and stand level and concluded that the error of field-based AGB estimation cannot be ignored, such as measured errors, missed or double-counted trees, measurement errors, sample plot location positioning errors or erroneous registration of tree species [43]. Moreover, this uncertainty is difficult to predict in natural forests, but the error can be reduced with an increase in the plot size [44]. In addition, Shen pointed out that the distribution of survey sites may bring some uncertainty in AGB estimation [45]. Lastly, the issue of the scale mismatch between the calibration of the field measurement and remote sensing pixels is a significant challenge. When coarse-resolution remote sensing is used to map the local- or national-scale AGB, the numerous data of small field plots, such as national forest inventory datasets, will be used to establish the AGB estimation model. However, the area of the most of the field plots is less than 0.1 ha in size. Réjou-Méchain et al. analyzed the local spatial variability of AGB in plots with the size of 0.1 ha and 100 ha. Results show that the local spatial variability is large for standard plots, and the value of the local spatial variability is 46.3% for 0.1 ha subplots and 16.6% for 1 ha subplots, respectively [46]. Compared with large plots, small plots carry large errors due to AGB variability for a consistent pixel-to-plot size. Moreover, this will generate large sampling errors and produce significant bias when estimating an AGB map. Thus, field measurements that better match the remote sensing pixel resolution are a challenge, and a more reliable approach to minimizing this sampling error needs to be developed.

It is complex to assess and quantify of the AGB estimation errors by using remote sensing data. However, it is also important to identify the errors and assess the effect of the different spatial resolutions of remote sensing data on AGB estimation, as it contributes to the uncertainty of the RS-based estimation; it is also necessary to consider how to correct this error. In this study, the issue of the scale mismatch of AGB estimation for remote sensing of different spatial resolutions was discussed. Moreover, a novel method, which is based on the law of conservation of mass, was developed, and the errors of the scale effect of AGB estimation at various spatial resolutions were analyzed. To achieve this goal, an AGB estimation model was first established that referred to three different spatial resolutions of remote sensing data, and the accuracy of AGB estimation at three different scales was analyzed. Second, the optimal scale estimation result was considered as a reference AGB spatial distribution true value, and the mismatch error due to the spatial variability and non-linearity of the AGB estimation model was discussed at three different spatial resolutions of remote sensing. Third, the scale conversion method of the entropy-weighted index with the assumption of the average AGB constant in this region, which was used to correct the scale error of AGB estimation by using coarse-resolution remote sensing, was established and the weight coefficient was determined by analyzing the biases between estimated and true AGB values. Finally, the AGB map with coarse-resolution remote sensing was corrected. This research represents an approach to the scale error correction of AGB estimation by using coarse-resolution remote sensing, and it provides a reference for high-precision AGB mapping using coarse-resolution remote sensing at a regional, national or global scale.

2. Study Area and Data

2.1. Study Area

The study area is located at the Maoer Mountain Experimental Forest Farm (127°29′–127°44′E, 45°14′–45°29′N), Shangzhi City, Heilongjiang Province, Northeast China (Figure 1). The area of the Maoer Mountain Forest Farm is approximately 26.496 km^2. Maoer Mountain belongs to the offset of the Changbai Mountains and extends to the northwest offset of the Zhangguangcai Range. The study area is a low mountainous and hilly area. The terrain of the forest area gradually rises from south to north, with an average altitude of 300 m. The research region belongs to the mid-temperate continental monsoon climate zone. The annual average temperature is around 2.7 °C, and the annual precipitation is around 649 mm. The average temperature of the hottest month of July is 21.8 °C, and January is the coldest month, with average temperatures of −19.9 °C [47]. The average annual thermal amplitude is 41.7 °C. The average forest coverage rate is 95%, and the total forest volume is 3.5 million m^3. The main tree species are Korean pine (*Pinus koraiensis*) mixed with deciduous species including birch (*Betula* spp.), larch (*Larix* spp.), poplar (*Populus* spp.), sylvestris pine (*Pinus sylvestris*) and Mongolian oak (*Quercus* spp.) [47,48].

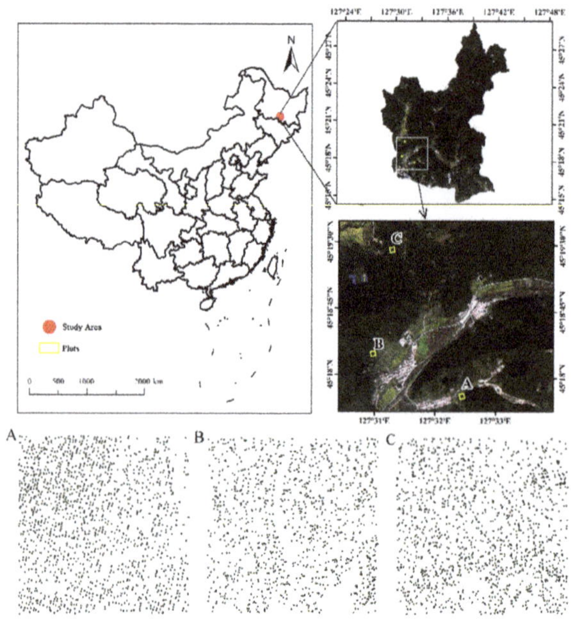

Figure 1. Study area and sample site location. (**A**) the distribution of single trees in the plot of the broadleaf forest; (**B**) the distribution of single trees in the plot of coniferous forest; (**C**) the distribution of single trees in the plot of the coniferous and broadleaf mixed forest.

2.2. Data

2.2.1. Remote Sensing Data and Pre-Processing

Three different spatial resolutions of remote sensing data, including GF-2, Sentinel-2 and Landsat-8 OLI, were used in this study.

Gaofen-2 (GF-2) is a civilian optical remote sensing satellite. The GF-2 satellite was launched by the China National Space Administration (CNSA) on 19 August 2014. GF-2 is the first satellite in China with a resolution below 1 m and captures high-resolution remote sensing images. It has been widely used in land use investigation, monitoring of the environmental atmosphere and water environment, urban planning, monitoring of disasters and resource surveys [49]. The GF-2 satellite platform is equipped with a panchromatic band with a 1 m spatial resolution and four multispectral band scanners with 4 m resolution, spatial including red (R), green (G), blue (B) and near-infrared (NIR). GF-2 can achieve a swath width of 45 km ground observation at one time and the revisiting time of GF-2 is 69 days. The remote sensing data were collected in August 2019.

The pre-processing of the GF-2 remote sensing images included the following: (1) radiation calibration for spectral channels was multiplied by gain and bias coefficients; (2) atmospheric correction was carried out by using the fast line-of-sight atmospheric analysis of the spectral hypercubes (FLAASH) model to obtain the surface reflectance; (3) geometric correction was performed. The GF-2 images were corrected based on a 1:10,000 topographic map with the method of polynomial and bilinear interpolation resampling. Then, the vegetation indices used in this study were calculated.

Sentinel-2A and 2B were designed by the European Space Agency (EAS) to meet the needs of the Copernicus program. The Sentinel-2A satellite was launched on 23 June 2015, followed by Sentinel-2B on 7 March 2017. The Sentinel-2A satellite has 13 bands covering the visible to shortwave infrared (SWIR) wavelength regions and it collects multispectral remote sensing data. The swath width of Sentinel-2A is 290 km and the revisiting time is 10 days. The spatial resolutions of Sentinel-2A data included four bands from visible and near-infrared (NIR) with a spatial resolution of 10 m, six bands from red-edge to shortwave infrared (SWIR) with a spatial resolution of 20 m and three atmospheric correction bands with a spatial resolution of 60 m, respectively [50]. Sentinel-2A Level-1C production with 10 m resolution was used in this study. The Sentinel-2A Level-1C data were obtained at the end of July 2019.

The pre-processing of the Sentinel-2A remote sensing images included the following: (1) resampling, in which all the bands were resampled to 10 m resolution; (2) atmospheric correction and terrain correction, which was carried out using the ESA SEN2COR processor to obtain the surface reflectance. Then, the vegetation indices used in this study were calculated.

The Landsat-8 Operational Land Imager (OLI) is an instrument in the Landsat series of satellite imagers. It was launched in February 2013. The Landsat-8 OLI continues the legacy of the Landsat series and adds two bands of the cirrus clouds and a coastal/aerosol (CA) band to detect water and aerosols in the blue region with a better resolution [51]. Landsat-8 OLI images consist of 11 spectra with a spatial resolution of 30 m. The images of Landsat-8 OLI data were acquired in July 2019.

The pre-processing of the Landsat-8 OLI remote sensing images included radiometric calibration, atmospheric correction, terrain correction and geometric correction. All the pre-processing of satellite data was conducted using ENVI 5.3 software (developed by Exelis Visual Information Solutions, Inc., Boulder, CO, USA).

2.2.2. Field Measurement

The ground data survey began in August 2019, and a total of 3 rectangular plots with a size of 100 m × 100 m were laid out (see Figure 1). The forest type of the sample plots included coniferous forests, broad-leaved forests and mixed forest types. The plot of the coniferous forest was mainly composed of Korean pine and larch, and the plot of the broad-leaved forest was mainly made up of birch and linden. Before the sample plot investigation, the GPS coordinates of the four corners and the central position of each sample plot were recorded by using a high-precision Differential Global Positioning System (DGPS; produced by Trimble Navigation Limited, Sunnyvale, CA, USA). Then, all trees were numbered and geolocated within each plot. The following forest parameters were then measured: diameter at breast height (DBH), tree height, under branch height, crown width, tree species. All of the trees with diameters at breast height greater than 5 cm in the sample plot were measured.

Stand aboveground biomass was calculated on the basis of established individual tree biomass models. A single-tree univariate additive biomass model established by Dong was used to estimate the single-tree aboveground biomass [52]. First, the biomass of the tree components of the tree stem, branch and leaf was calculated, and the total aboveground biomass of the single tree was the sum of the biomass of the tree components. It is important to note that the root biomass was not included in this study. Then, plot biomass could be derived from the sum of the all living trees' biomass in each sample plot. Finally, the stand aboveground biomass per ground area could be calculated.

For the comparison of the biomass remote sensing estimate at different scales, the large rectangular plot was divided into differently sized subplots. According to the spatial resolution of GF-2, Sentinel-2 and Landsat-8, we divided the 100 m × 100 m plot into several subplots with the size of 4 m ×4 m (GF-2), 10 m × 10 m (Sentinel-2) and 30 m × 30 m (Landsat-8), respectively. Some of the data were selected randomly to establish the AGB prediction model. The AGB statistical information of the selected subset with different sizes is shown in Table 1.

Table 1. The AGB statistical information of the field measurements at pixel scale (Unit: t/ha).

Index	GF-2 (n = 70)	Sentinel-2 (n = 70)	Landsat-8 (n = 55)
Mean	112.7623	98.4261	105.2296
Standard deviation	23.1844	31.0125	31.8978
Range	47.2768–181.5890	47.1336–174.2340	47.9736–193.5593

3. Methodology

3.1. Method of Aboveground Forest Biomass Estimation at Different Spatial Scales

3.1.1. Remote Sensing Variable Selection

The purpose of the AGB modeling was to construct the relationships between the variables extracted from remote sensing data and AGB. The first important step was selecting the variables for AGB estimation. To increase the number of candidates in the independent variable dataset, the spectral indices, vegetation indices, texture features, terrain factors and other parameters of the images of three different spatial resolutions from GF-2, Sentinel-2 and Landsat-8 were extracted as candidate characteristic variables [50,53]. There were a total 62 candidate remote sensing variables extracted from GF-2 satellite data, 57 candidate remote sensing variables extracted from Sentinel-2 satellite data and 63 candidate remote sensing variables extracted from Landsat-8 satellite data.

The significant relationships between the variables of the remote sensing data and AGB demonstrated the candidates for optical remote sensing data for AGB estimation and determined the accuracy of AGB estimation. Therefore, it was very important to screen variables from the remote sensing data carefully for AGB modeling [54]. For the first step, the Pearson correlation coefficient between candidate remote sensing variables and AGB field measurement was calculated, and those variables with a lower correlation coefficient ($R < 0.05$) were removed to improve the quality of the candidate remote sensing variables.

The variable importance in projection (VIP) score is often used to assess the importance of variables. In general, those variables with a greater VIP score are considered to be more important than those with smaller ones [55]. Therefore, we used the VIP score to evaluate the importance of the candidate remote sensing variables in the AGB modeling in the next step. The remaining variables screened in the first step were ranked according to the VIP, calculated with random forest, to screen the independent variables for a second time. Finally, three groups of remote sensing variables were successfully selected to prepare for AGB modeling at three different spatial resolutions of remote sensing. There were 8 remote sensing variables in each group, and the details can be found in Table 2.

Table 2. Selected variables for AGB modeling.

Sensor	Variable	Formular	Description
GF-2	ME2	$\sum_{i,j=0}^{N-1} iP_{ij}$	Mean of the four directional textural features of GF-2 band 2
	Var4	$\sum_{i,j=0}^{N-1} P_{ij}(i-ME)^2$	Sum variance of the gray-level co-occurrence matrix of GF-2 band 4
	Ho2	$\sum_{i,j=0}^{N-1} i\frac{P_{ij}}{1+(i-j)^2}$	Homogeneity of the gray-level co-occurrence matrix of GF-2 band 2
	B1	Blue, 450–520 nm	Reflectance of the GF-2 blue band
	B431 [56]	(B4 + B3)/B1	A vegetation index calculated by GF-2 band 1, 3 and 4
	B4	Near-infrared band, 770–890 nm	Reflectance of the GF-2 near-infrared band
	B13	B1 + B3 [56]	A vegetation index calculated by GF-2 band 1 and 3
	PX	Slope	Slope extracted from DEM data resampled to GF-2 spatial resolution
Sentinel-2	Var7	$\sum_{i,j=0}^{N-1} P_{ij}(i-ME)^2$	Sum variance of gray-level co-occurrence matrix of Sentinel-2 band 7
	Cor8	$\sum_{i,j=0}^{N-1} iP_{ij}\left[\frac{(i-ME)(j-ME)}{\sqrt{VA_iVA_j}}\right]$	The correlation texture between the grey levels and those neighboring pixels of Sentinel-2 band 8
	IRECI [57]	(B7 − B4)/(B5/B6)	Inverted red-edge chlorophyll index
	B3	Green, 560 nm	Reflectance of the Sentinel-2 green light band
	PX	Slope	Slope extracted from DEM data resampled to Sentinel-2 spatial resolution
	Wetness [58]	0.2578 × B2 + 0.2305 × B3 + 0.0883 × B4 + 0.1071 × B8 − 0.7611 × B11 − 0.5308 × B12 [59]	Tasseled Cap (KT) transformation wetness
	REIP [57]	700 + 40 × [(B4 + B7)/2 − B5]/(B6 − B5)	Red-edge infection point index
	Brightness [58]	0.351 × B2 + 0.3813 × B3 + 0.3437 × B4 + 0.7196 × B8 + 0.2396 × B11 + 0.1949 × B12 [59]	Tasseled Cap (KT) transformation brightness
Landsat-8	B4/Albedo [60]	B4/(0.246 × B2 + 0.146 × B3 + 0.191 × B4 + 0.304 × B5 + 0.105 × B6 + 0.008 × B7)	Band combination vegetation index
	PX	Slope	Slope extracted from DEM data resampled to Landsat-8 spatial resolution
	B4	Red, 640–670 nm	Reflectance of the Landsat-8 red light band
	ND563 [60]	(B5 + B6 − B3) × (B5 + B6 + B3)	Normalized difference vegetation index

Table 2. *Cont.*

Landsat-8	Cor5	$\sum_{i,j=0}^{N-1} iP_{ij}\left[\frac{(i-ME)(j-ME)}{\sqrt{VA_i VA_j}}\right]$	The correlation texture between the grey levels and those neighboring pixels of Landsat-8 band 5
	SM5	$\sum_{i,j=0}^{N-1} P_{i,j}$	Angular second moment of Landsat-8 band 5
	Var5	$\sum_{i,j=0}^{N-1} P_{ij}(i-ME)^2$	Sum variance of gray-level co-occurrence matrix of Landsat-8 band 5
	ME2	$\sum_{i,j=0}^{N-1} iP_{ij}$	Mean of the four directional textural features of Landsat-8 band 2

(Note: The variables of the $P(i, j)$ refer to the value at the position of (i,j) in a gray-level co-occurrence matrix, where i and j are the number of the rows and columns in the gray-level co-occurrence matrix. The variable of N is the number of rows or columns of the gray-level co-occurrence matrix. ME and VA are the mean and variance of the four directional textural features, respectively).

3.1.2. Method of AGB Modeling

In this study, a random forest (RF) algorithm was conducted to estimate the AGB of the research area. Then, multiple linear regression (MLR) was used to compare the accuracy of AGB estimation.

Random forest (RF) is a machine learning algorithm proposed by Leo Breiman (2001) [61]. Random forest was developed based on multiple regression trees; it shows that the relationship between an input relates to its dependent variable by using multiple regression trees [62]. The main advantage of RF is the ability to describe complex nonlinear relationships, such as in a complex ecological system. It is more effective than a linear regression model for multi-variable models. Thus, a random forest algorithm was selected to effectively predict forest AGB by using remote sensing data. Moreover, the number of regression trees was set to 1000 and the random state of the random forest algorithm was set to 10 in this study.

A traditional multiple linear regression (MLR) was applied as a baseline for AGB model accuracy comparison. A backward stepwise multiple linear regression was performed to establish the forest AGB retrieval model. The formula of the MLR is as follows:

$$y = \beta 0 + \beta 1 \times 1 + \beta 2 \times 2 + \ldots + \beta n \times n + \varepsilon \quad (1)$$

where y is the variable of the forest AGB; xi is a dataset of remote sensing variables; βi is a fitting parameter; ε is an error term.

Using random sampling, a total of 75% of the sample data were selected for the model establishment, while the remaining 25% of the sample data were employed for the accuracy evaluation.

3.2. Scale Error Calculation

Resampling of the field-measured data or remote sensing AGB production to a consistent spatial resolution with remote sensing data is a commonly used method of error evaluation. Thus, we used two upscaling paths to calculate the AGB from the coarse-resolution remote sensing data and compared the results of these two methods, after which the scale error could be determined (Figure 2).

Figure 2 is a schematic flowchart of the two upscaling methods of AGB estimation. The first method aimed to estimate AGB by using high-resolution remote sensing data and then aggregate the estimated AGB to coarse-resolution remote sensing (Path 1). We named this path "first inversion and then aggregation". The other method was to aggregate the characteristic variables of high-resolution remote sensing images, such as various vegetation indices, to coarse-resolution remote sensing and then estimate the AGB using the remote sensing inversion model of AGB. We named this path "first aggregation and then inversion".

The first path (Path 1) retrieved the AGB through high-resolution remote sensing and resampled the AGB results to coarse-resolution remote sensing by summation. First, we obtained the characteristic variable (V_i) from the high-resolution remote sensing image and established the AGB estimation model ($AGB_i = f_1(V_i)$); then, we inverted the AGB (AGB_i) from the high-resolution remote sensing image (Path 1 step one) and resampled the inverted AGB to coarse-resolution remote sensing by summation (Path 1, Step 2). The calculated AGB with the coarse resolution is the sum of the AGB estimated from high-resolution data. This is often referred to as a distributed algorithm. As a linear transformation, the statistical information of AGB at coarse resolution obtained by this upscaling method was consistent with AGB at the high resolution [63]. Thus, this AGB estimation can be considered as the relative true value of the biomass at the coarse-resolution pixel, defined as AGB_{exa}.

The second path (Path 2) aggregated the characteristic variable of the high-resolution remote sensing to coarse-resolution remote sensing (V_m) first (Path 2, Step 1). This meant that the high-resolution remote sensing should resample to the same spatial resolution with the coarse-resolution remote sensing. Then, an AGB estimation method ($AGB_{app} = f_2(V_m)$)

was established based on the resampled characteristic variable (V_m), and the AGB_{app} could be retrieved from the coarse-resolution remote sensing data (Path 2, Step 2). This is commonly referred to as the lumped algorithm. This process (Path 2, Step 1) can be understood as the imaging process of the coarse-spatial-resolution satellite sensor. The estimated AGB results contained the scale error due to the heterogeneity of the surface feature [64,65]. This can be considered as an estimation of the coarse-resolution pixel biomass, defined as AGB_{app}.

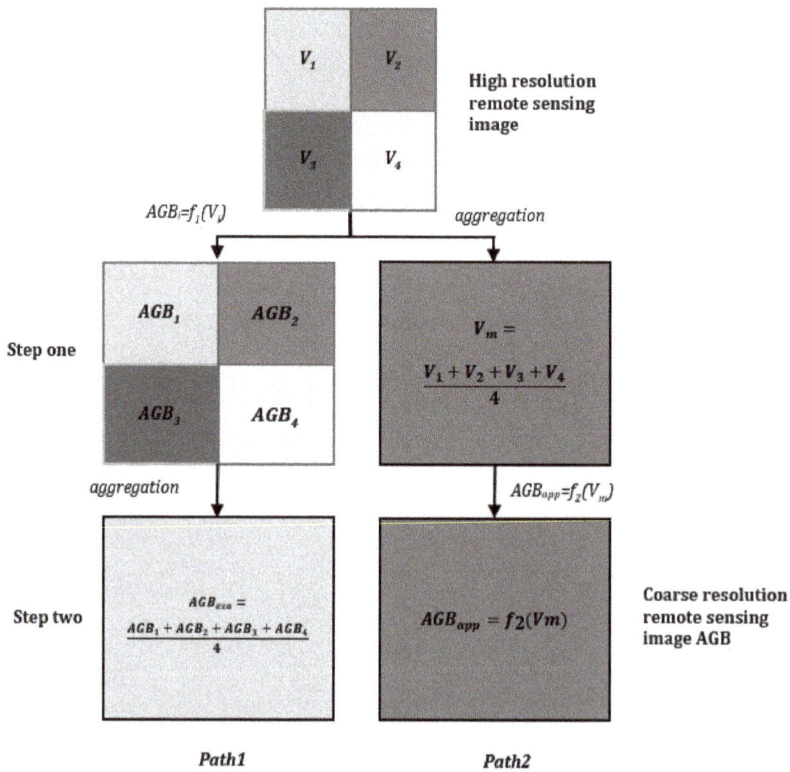

Figure 2. Schematic flowchart of the two upscaling methods of the AGB estimation. (f_i is the remote sensing inversion model of forest aboveground biomass; $V_{(i=1,2,3,4)}$ is the characteristic variable of the high-resolution remote sensing image; V_m is the average value of V_i, namely the pixel V value of the coarse-resolution image; AGB_{exa} is the biological value of the high-resolution image, namely the relative truth value; AGB_{app} is the biological value of the coarse-resolution image, namely the biological values with scale errors).

At present, it is recognized by the academic community that the spatial heterogeneity of surface features is the major reason for the scale effect [66,67]. In other words, it is assumed that multiple spatially heterogeneous, high-resolution pixels are contained in a single coarse-resolution pixel. Due to the heterogeneity of the surface feature, the estimated AGB will contain the error of the scale effect. Therefore, spatial heterogeneity serves as the main contributing factor to the scale error e. It can be calculated as follows:

$$e = AGB_{exa} - AGB_{app} \tag{2}$$

The authenticity test of the remote sensing products involved evaluating the accuracy of the AGB product. The field measurement or AGB estimation result obtained by using high-resolution remote sensing was usually used to verify the precision. Currently, the verification of the AGB remote sensing product at coarse resolution is usually performed according to the relationship between field measurements or generated high-spatial-resolution distribution map of AGB and coarse remote sensing estimated AGB result. To analyze the differences in AGB for various spatial resolutions of remote sensing, the forest biomass map using GF-2 was resampled to 10 m and 30 m resolutions using the distributed algorithm (Path 1). The biomass estimation error of the coarse resolution was calculated, and the scale effect from GF-2 to Sentinel-2 and Landsat-8 was analyzed.

3.3. Scale Error Measurement of Mixed Pixels

3.3.1. Determination of the True Mean Value

According to the law of conservation of mass, we assume that the total quality remains uniform and unchanged in any substance system (isolated system) isolated from the surroundings. A fixed study area or a remote sensing image of the research area also can be considered as an isolated material system. Any changes in the resolution within the region would not alter the total surface area of the research area or the mass of the total forest aboveground biomass. Thus, a hypothesis is proposed that the average of the AGB true value at any scale will remain constant. This true value of the AGB was defined as the mass of the forest aboveground biomass per unit surface area at a certain time and under specific spatial conditions.

To understand this, it can be assumed that the total amount of dry matter in a large region was 8 units of mass, and this value contained no measured or system error. We assumed that the surface area was also 8 units of area. The aboveground biomass (AGB) equated to 1 mass per area; the diagram is shown in Figure 3A.

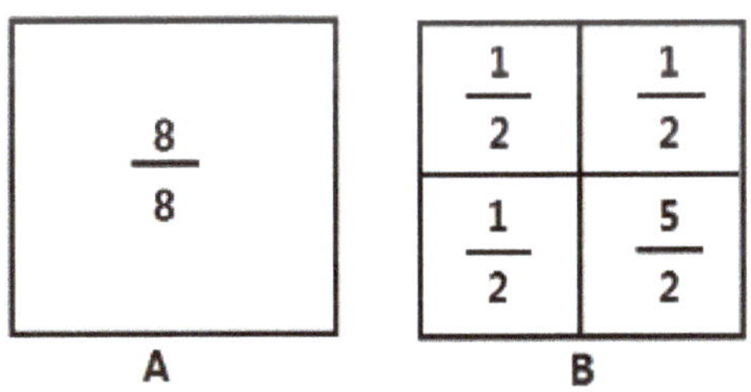

Figure 3. Schematic diagram of true mean AGB scale invariance. (**A**) the mass of the AGB per unit surface area of coarse resolution remote sensing data; (**B**) the mass of the AGB per unit surface area of high resolution remote sensing data.

Then, this large region was divided into 4 equal parts of 2 units of area in each part. The dry matter in each part was assumed as 1, 1, 1 and 5 units of mass due to the spatial heterogeneity. The AGB of each part was 1/2, 1/2, 1/2 and 5/2 mass per area (Figure 3B). However, the average AGB in this region was calculated as [(1/2) + (1/2) +(1/2) + (5/2)]/4 = 1 mass per area. Similarly, if the region was divided into more small units, the average AGB in this region was constant at 1 mass per area under the same condition.

According to the demonstration above, we assumed that the size of the total area was $N \times N$, the size of the remote sensing pixel was $n_j \times n_j$ at j-scale, and the AGB of the ith pixel could be expressed as follows:

$$AGB_{ni} = \frac{1}{n_j^2} \sum_{j=1}^{n_j^2} f(V_{j,i}) \tag{3}$$

where $f(V_{j,i})$ represented the biomass inversion model of the forest, and $V_{j,i}$ represented the modeled remote sensing variable of the ith pixel at j scale.

Average AGB in this region at the n-scale AGB_n could be expressed as

$$\begin{aligned} AGB_n &= \frac{1}{\left(\frac{N}{n_j}\right)^2} \sum_{i=1}^{\left(\frac{N}{n_j}\right)^2} AGB_{ni} \\ &= \frac{1}{\left(\frac{N}{n_j}\right)^2} \sum_{i=1}^{\left(\frac{N}{n_j^2}\right)^2} \left[\frac{1}{n_j^2} \sum_{j=1}^{n_j^2} f(V_{j,i}) \right] \\ &= \frac{1}{\left(\frac{N}{n_j}\right)^2} \frac{1}{n_j^2} \sum_{i=1}^{\left(\frac{N}{n_j}\right)^2} \left[\sum_{j=1}^{n^2} f(V_{j,i}) \right] \\ = AGB_m &= \frac{1}{\left(\frac{N}{m_j}\right)^2} \sum_{i_m=1}^{\left(\frac{N}{m_j}\right)^2} \left[\frac{1}{n_m^2} \sum_{m=1}^{n_m^2} f_m(V_{m,i_m}) \right] \end{aligned} \tag{4}$$

where AGB_m was the average AGB in this region at the m-scale. $f_m(V_{m,i_m})$ represented the biomass inversion model of the forest at m scale and V_{m,i_m} represented the modeled remote sensing variable of the i_mth pixel at m scale.

The various spatial resolutions between remote sensors composed the primary scale effect of the remote sensing image, but the relative true value of AGB at different spatial resolutions was constant, with the average AGB based on measured or other scales in this region. This has been demonstrated previously [68,69]. With the assumption of the average AGB being constant in this region, the high-resolution remote sensing data can be considered as a bridge to connect the field-measured data to the coarse-resolution remote sensing image. Then, the AGB_{exa} retrieved from the high-resolution image was aggregated to a coarse-resolution image pixel AGB_{nexa} (n represented pixel scale). The AGB_{nexa} can be considered as a relative true value of the AGB at the coarse spatial resolution scale. Moreover, compared with the AGB estimation for coarse remote sensing data, the error caused by the scale effect could be evaluated.

3.3.2. Method of Scale Error Correction

According to the scale error formula, the AGB corrected value (AGB^{ni}_{cor}) of each pixel at n-scale can be expressed as an estimated AGB (AGB^{ni}_{app}) of each pixel plus a scale error e^n_i at n-scale, which can be written as:

$$AGB^{ni}_{cor} = AGB^{ni}_{app} + e^n_i \tag{5}$$

e^n_i was the scale error of the ith pixel at n-scale, AGB^{ni}_{cor} was the AGB corrected value of the ith pixel when the pixel scale was n, AGB^{ni}_{app} was the estimated AGB of the ith pixel when the pixel scale was n.

If the estimated AGB ($AGB^{n}{}_{app}$) did not contain a scale error, it should be equal to the relative true AGB at n-scale ($AGB^{n}{}_{exa}$). When the scale error was included in each estimated AGB, the total average error could be considered as the scale effect on AGB estimation. The mean scale error can be calculated as follows:

$$\overline{e}^n = \overline{AGB}^n_{exa} - \overline{AGB}^n_{app} \tag{6}$$

$\overline{e}^n$ referred to the mean scale error of total pixels at n-scale in the research area. $\overline{AGB}^n_{exa}$ was the relative true mean value of the research area AGB, and $\overline{AGB}^n_{app}$ was the mean estimated AGB of the whole research area.

Deduced from the law of large numbers and the central limit theorem, the scale error arithmetic mean e^n_i of samples and its population arithmetic mean $\overline{e}^n$ had the same mathematical expectation and the scale error showed a normal distribution [47]. Then, e^n_i

could be seen as a fluctuation result of the scale error value $\bar{e}^n$ at the point of the pixel i. The range of the fluctuation v_i^n was determined by the difference in the population and sample mean and could be measured by the P_i weight. The fluctuations can be expressed as

$$v_i^n = P_i \times \bar{e}^n \tag{7}$$

Therefore, e_i^n was regarded as the result, which was impacted by the weight P_i of pixel i on the basis of the population value of scale error $\bar{e}^n$. Moreover, e_i^n can be rewritten as:

$$\begin{aligned} e_i^n &= \bar{e}^n + P_i \times \bar{e}^n \\ &= \bar{e}^n \times (1 + P_i) \end{aligned} \tag{8}$$

The coefficient of variation (CV), which was used to measure the relative variation of a random variable to its mean, has been widely used in remote sensing [70]. The coefficient of variation method can be used to evaluate the difference between objects by using the feature of remote sensing. Based on the heterogeneity of the land surface space, the information entropy index was selected to determine the index weight in this study [70]. The formula is as follows [71]:

$$P_i = \sum_{i=0}^{L} -W_i ln W_i \tag{9}$$

where P_i represented the weight of surface heterogeneity calculated by the information entropy from the high-resolution remote sensing data, and W_i was the probability of the occurrence of the i-th land use type. L was the number of land use types included in the high-resolution remote sensing data. According to the information entropy, the weighting index of coarse-resolution pixel-scale space variation was acquired and the scale error could be corrected.

3.3.3. Accuracy Evaluation

After completing the model establishment, four indices were applied for AGB model evaluation, including the determination coefficient (R^2), the root mean squared error (RMSE), the relative root mean squared error (rRMSE) and the mean absolute error (MAE). The equations were as follows [72]:

$$R^2 = 1 - \frac{\sum_{i=1}^{n} (y_i - \hat{y}_i)^2}{\sum_{i=1}^{n} (y_i - \bar{y})^2} \tag{10}$$

$$RMSE = \sqrt{\frac{\sum_{i=1}^{n} (y_i - \hat{y}_i)^2}{n}} \tag{11}$$

$$rRMSE = \sqrt{\frac{\frac{1}{n} \sum_{i=1}^{n} (y_i - \hat{y})^2}{\bar{y}}} \tag{12}$$

$$MAE = \frac{1}{n} \sum_{i=1}^{n} |y_i - \hat{y}_i| \tag{13}$$

where y_i represented the AGB measured value, $\hat{y}$ was the estimated AGB value, $\bar{y}$ referred to the mean value of AGB measured, and n was the number of the samples.

Moreover, another four indices were selected to evaluate the accuracy and efficiency of the upscaling-based method described in this study. These were the mean deviation error (MBE), the root mean squared error (RMSE), average absolute percentage error (MAPE) and determination coefficient (R^2) [73].

$$MBE = \frac{\sum_{i=1}^{n} (p_i - o_i)}{n} \tag{14}$$

$$RMSE = \sqrt{\frac{\sum_{i=1}^{n}(p_i - o_i)^2}{n}} \quad (15)$$

$$MAPE = \frac{100}{n} \times \sum_{i=1}^{n} \frac{|p_i - o_i|}{\overline{p}} \quad (16)$$

$$R^2 = 1 - \frac{\sum_{i=1}^{n}(p_i - o_i)^2}{\sum_{i=1}^{n}(p_i - \overline{o})^2} \quad (17)$$

In the above formula, p_i was the relative true value of AGB at a coarse resolution, $\overline{p}$ was the mean of the p_i, o_i was the estimated AGB after the scale effect correction for a coarse-resolution image, $\overline{o}$ was the mean of the o_i, i was the i-th pixel for a coarse-resolution image, n was the pixel number of a coarse-resolution image.

4. Results

4.1. Results of the AGB Modeling at Various Spatial Resolutions of Remote Sensing

AGB estimation modeling using remote sensing is an important method for large-scale biomass estimation and is a relevant field in remote sensing research [74,75]. Therefore, a large number of extracted vegetation indices, spectral indices and texture variables have been developed for AGB prediction [76–78]. To select the optimal variables, Pearson's correlation coefficient and the VIP score were calculated and eight remote sensing variables were selected; see Table 2. Then, the AGB estimation model was established by using the random forest algorithm and multiple linear regression (MLR) method, and Table 3 shows the performance of the six models using the two modeling methods (Table 3).

Table 3. The comparation of the AGB modeling accuracy with various remote sensing data.

Image	Resolution	Model	R^2	RMSE	MAE	rRMSE
GF-2	4 m	Multiple Linear	0.5110	19.4328	15.7262	0.1737
		Random Forest	**0.5943**	**17.5056**	**14.2677**	**0.1282**
Sentinel-2	10 m	Multiple Linear	0.5409	21.4195	17.7690	0.2442
		Random Forest	0.4971	20.2485	14.9274	0.2308
Landsat-8	30 m	Multiple Linear	0.3034	28.8477	23.7357	0.2778
		Random Forest	0.4235	28.6546	25.2203	0.2759

The modeling accuracy of the random forest model using GF-2 was calculated. The R^2, RMSE, MAE and rRMSE were 0.5943, 17.5056, 14.2677 and 0.1282, respectively. By contrast, the multiple linear regression model established between AGB and the GF-2 feature had lower accuracy (R^2 = 0.5110, RMSE = 19.4328, MAE = 15.7262, rRMSE = 0.1737). For Sentinel-2 images, the determinant coefficient of the random forest model was 0.4971, and it was smaller than that of the multiple linear regression model of 0.5409. However, the RMSE, MAE and rRMSE were larger than those of the multiple linear regression. For Landsat-8 images, the modeling accuracy was the worst compared with the other two satellite data AGB models, and the R^2, RMSE, MAE and rRMSE of the random forest model were 0.4235, 28.6546, 25.2203 and 0.2759, respectively. The multiple linear regression model showed worse accuracy, with R^2, RMSE, MAE and rRMSE of 0.3034, 28.8477, 23.7357 and 0.2778, respectively. Comparing the AGB modeling accuracy obtained with different modeling methods, it can be seen that the random forest model had better accuracy than the multiple linear regression model. Comparing the AGB modeling accuracy obtained with different remote sensing data, the performance of the random forest AGB model using GF-2 was the best, and it displayed a better estimation effect for the aboveground forest biomass in the research area.

To test the prediction efficiency of the model for independent samples, 25% of the sample data were used to preformed the AGB estimation. Then, the estimation accuracy was calculated, and the results can be found in Table 4.

Table 4. The comparison of the AGB prediction accuracy with various remote sensing data.

Image	Resolution	Model	R^2	RMSE	MAE	rRMSE
GF-2	4 m	Multiple Linear	0.4072	20.2781	16.9451	0.1759
		Random Forest	0.5711	16.9586	12.7153	0.1471
Sentinel-2	10 m	Multiple Linear	0.3344	23.6606	19.7257	0.2353
		Random Forest	0.4819	19.4657	14.3562	0.1936
Landsat-8	30 m	Multiple Linear	0.2892	32.8565	25.6158	0.3034
		Random Forest	0.4321	29.7677	28.0137	0.2749

The determinant coefficient between measured and estimated AGB using the random forest model with GF-2 was 0.5711, RMSE was 16.9586, MAE was 12.7153, and rRMSE was 0.1471. The R^2 when using the multi-linear regression model was 0.4072, the RMSE was 20.2781, MAE was 16.9451, and rRMSE was 0.1759. These values for Sentinel-2 using the random forest modeling method were 0.4819, 19.4657, 14.3562 and 0.1936, respectively. The multi-linear regression model results for Sentinel-2 were 0.3344, 23.6606, 19.7257 and 0.2353, respectively. The results of the random forest model for Landsat-8 were 0.4321, 29.7677, 28.0137 and 0.2749, respectively. The results of the multi-linear regression model for Landsat-8 were 0.2892, 32.8565, 25.6158 and 0.3034, respectively. It can be seen that the prediction accuracy of the random forest regression model was better than that of the multi-linear regression model.

A scatter plot of the measured and estimated AGB is shown in Figure 4. Compared with the AGB estimated by using the multi-linear regression model, AGB estimation by using the random forest model was distributed at nearly $y = x$, which indicates that the estimated AGB deviated less from the measured value, and it could perform with higher accuracy. The estimated AGB obtained using the multi-linear regression model showed a larger bias from the line of $y = x$. This meant that the model would produce a larger estimation error.

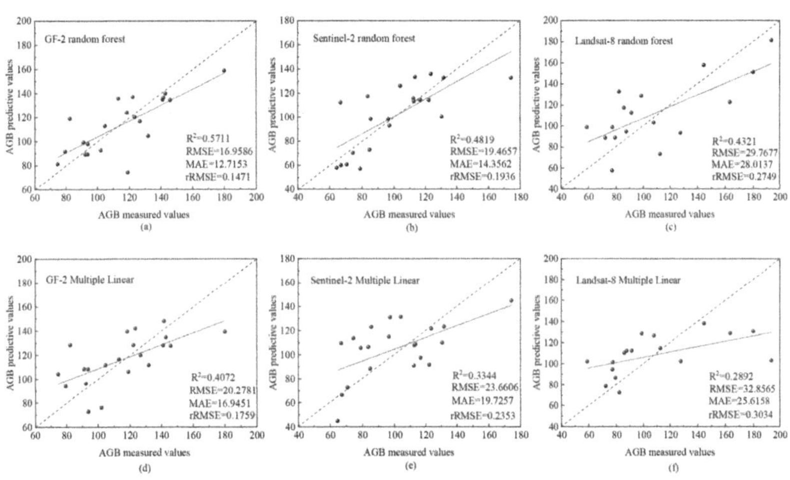

Figure 4. Scatter plot between measured and estimated AGB (black dotted line indicates $y = x$). (**a**) the scattering plot of measured and estimated AGB of GF-2 by using random forest model; (**b**) the scattering plot of measured and estimated AGB of Sentinel-2 by using random forest model; (**c**) the scattering plot of measured and estimated AGB of Landsat-8 by using random forest model; (**d**) the scattering plot of measured and estimated AGB of GF-2 by using multiple linear model; (**e**) the scattering plot of measured and estimated AGB of Sentinel-2 by using multiple linear model; (**f**) the scattering plot of measured and estimated AGB of Landsat-8 by using multiple linear model.

4.2. Retriveved AGB at Various Spatial Resolutions

First, the AGB prediction using the random forest model was performed for GF-2, Sentinel-2 and Landsat-8 images, and the AGB inversion results of GF-2 with 4 m spatial resolution ($AGB_{GF\text{-}2}$), Sentinel-2 with 10 m spatial resolution ($AGB_{Sentinel\text{-}2}$) and Landsat-8 with 30 m spatial resolution ($AGB_{Landsat\text{-}8}$) were obtained (Figure 5).

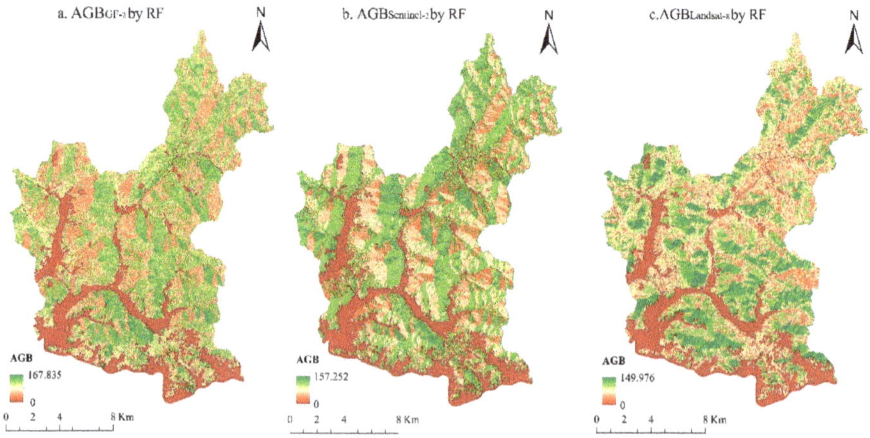

Figure 5. AGB estimation using the random forest model for various remote sensing data: (**a**) AGB estimation result using GF-2; (**b**) AGB estimation result using Sentinel-2; (**c**) AGB estimation result using Landsat-8. (Unit was t/ha).

The statistical information is summarized in Table 5. The mean of the AGB estimation of GF-2 ($AGB_{GF\text{-}2}$) was 101.30 t/ha. The standard deviation was 40.25 t/ha. The mean of the AGB estimation of Sentinel-2 ($AGB_{Sentinel\text{-}2}$) was 102.52 t/ha, with the standard deviation of 43.95 t/ha. The mean of the AGB estimation of Landsat-8 ($AGB_{Landsat\text{-}8}$) was 94.70 t/ha, with the standard deviation of 40.02 t/ha. The mean AGB estimation among GF-2 and Sentinel-2 had a similar value, but the standard deviation showed a significant difference. This meant that the AGB estimated with Sentinel-2 had a large deviation. Moreover, the mean of the AGB estimated by Landsat-8 had a significant difference compared with other two results. The differences among AGB estimation were mainly caused by the estimation error and scale effect.

Table 5. The statistical information of AGB estimation results using various remote sensing data.

Index	$AGB_{GF\text{-}2}$	$AGB_{Sentinel\text{-}2}$	$AGB_{Landsat\text{-}8}$
Mean	101.30	102.52	94.70
Standard deviation	40.25	43.95	40.02

Then, the relative true values of AGB at 10 m ($AGB_{exa\text{-}10}$) and 30 m ($AGB_{exa\text{-}30}$) spatial resolution were calculated based on the AGB estimation using GF-2, and the AGB distribution can be found in Figure 6. The statistical information is shown in Table 6. The results show that the relative true values of AGB among the various spatial resolutions were similar, without a significant difference, and this result was consistent with our assumption. However, the relative true value of the AGB estimation had a significant bias compared with the AGB estimated using remote sensing data. The mean of the AGB estimation of Landsat-8 ($AGB_{Landsat\text{-}8}$) was 94.70 t/ha, with the standard deviation of 40.02 t/ha. The mean of the relative true value of AGB ($AGB_{exa\text{-}30}$) at the same spatial resolution was 101.24 t/ha, with the standard deviation of 37.98 t/ha. This bias of the mean value of AGB estimation was obvious. This difference can be considered as the effect of the scale error on the AGB estimation. Compared with high-spatial-resolution GF-2 data, the surface spatial heterogeneity and mixed pixels would have a greater effect on one pixel of Landsat-8. Moreover, there may be more pixels with a single property in one pixel of GF-2. This scale effect led to errors in the estimation results for different spatial resolutions.

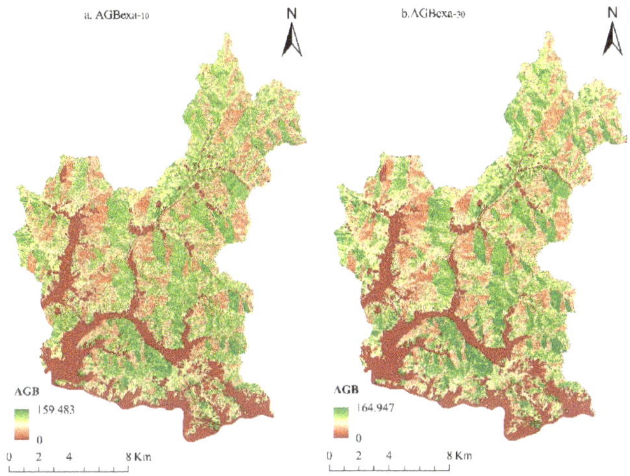

Figure 6. Relative values of the AGB extracted by using estimated AGB of GF-2 at 10 m and 30 m spatial resolution. (**a**) Relative value of the AGB of 10 m; (**b**) Relative value of the AGB of 30 m. (Unit was t/ha).

Table 6. The statistical information of relative true value of the AGB using various remote sensing data.

Index	$AGB_{GF\text{-}2}$	$AGB_{exa\text{-}10}$	$AGB_{exa\text{-}30}$
Mean	101.30	101.29	101.24
Standard deviation	40.25	39.31	37.98

4.3. Verification of the Scale Error Correction

To correct the scale error of the upscaling-based AGB estimation, a scale conversion method using the entropy-weighted index was developed based on the different land use types in one pixel of the 10 m and 30 m spatial resolution remote sensing. The $AGB_{cor\text{-}10}$ and $AGB_{cor\text{-}30}$ after scale effect correction were calculated. Comparing the relative true values of AGB at 10 m ($AGB_{exa\text{-}10}$) and 30 m ($AGB_{exa\text{-}30}$) calculated by GF-2 with the pre- and post-scale effect correction results of AGB estimation by the random forest model of Sentinel-2 and Landsat-8, the accuracy was calculated (Table 7).

Table 7. Accuracy of the scale error corrected AGB at various spatial resolutions.

Index	$AGB_{Sentinel\text{-}2}$	$AGB_{exa\text{-}10}$	$AGB_{Landsat\text{-}8}$	$AGB_{exa\text{-}30}$
MBE	11.4635	1.2378	6.0725	−6.0069
RMSE	16.3102	10.7745	9.0367	8.2139
MAPE	12.0822	7.4743	7.0241	6.3071
R^2	0.6226	0.6725	0.5910	0.6704

Comparing the relative true values of AGB with the pre- and post-correction results of AGB estimation, the MBE of the 10 m resolution corrected AGB decreased from 11.4635 to 1.2348 t/ha. The root mean squared error index of the corrected AGB of 10 m resolution had a significant improvement; the RMSE of the 10 m resolution corrected AGB decreased from 16.3102 to 10.7745 t/ha. MAPE decreased from 12.0822 to 7.4743 t/ha, and the R^2 was increased from 0.6226 to 0.6725. The scatter plots of the pre- and post-correction results are shown in Figure 7. After the scale error correction, the AGB showed a better linear relationship with the relative true value. This indicated that the scale conversion method using the entropy-weighted index had a good effect on scale error correction.

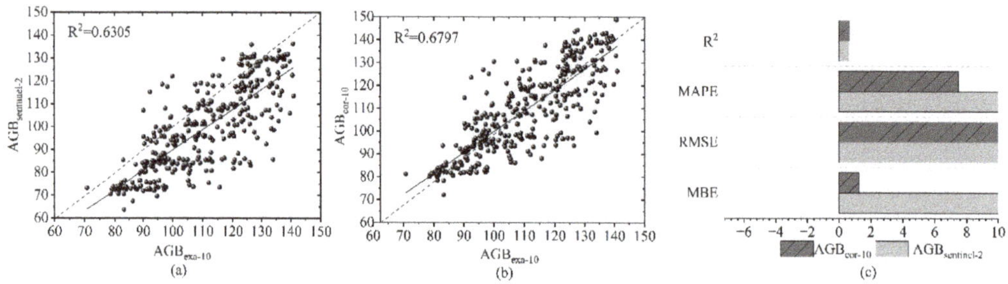

Figure 7. Scatter plot of the pre- and post-correction results and true AGB and the accuracy comparison of 10 m. (**a**) Scatter plot of the pre-correction results and true AGB at 10 m; (**b**) scatter plot of the post-correction results and true AGB at 10 m; (**c**) accuracy comparison.

Similar results were obtained for Landsat-8 AGB estimation. Comparing the relative true values of AGB with the pre- and post-correction results of Landsat-8 AGB estimation, the MBE of the 30 m resolution corrected AGB decreased from 6.0725 to −6.0069 t/ha. The RMSE decreased from 9.0367 to 8.2139 t/ha. MAPE decreased from 7.0241 to 6.3071 t/ha, and the R^2 was increased from 0.5910 to 0.6704. The scatter plots of the pre- and post-correction results of Landsat-8 AGB estimation are shown in Figure 8. There was a significant underestimation of the AGB from Landsat-8 data using the random forest model (Figure 8a). However, the scale error-corrected results were evenly distributed along the line of $y = x$, as seen in Figure 8b. This indicated that the underestimation was improved well by using the scale error correction method. The results also showed that this method can be used to correct the scale effect resulting from the heterogeneity of land use types caused by the various spatial resolutions.

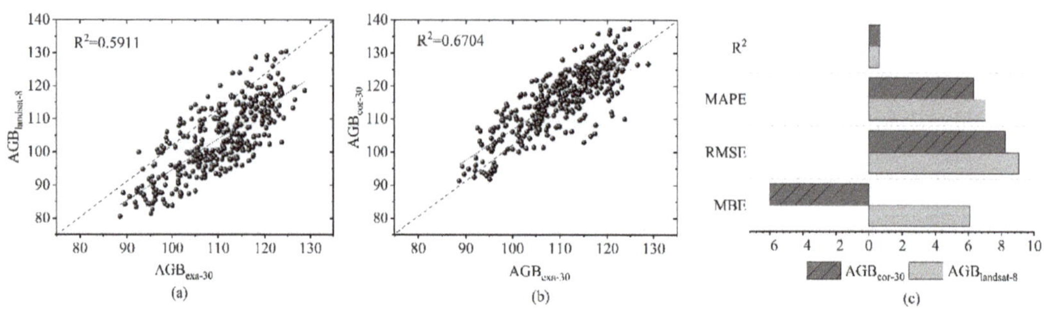

Figure 8. Scatter plot of the pre- and post-correction results and true AGB and the accuracy comparison of 30 m. (**a**) Scatter plot of the pre-correction results and true AGB at 30 m; (**b**) scatter plot of the post-correction results and true AGB at 30 m; (**c**) accuracy comparison.

Then, this scale error correction method was applied to the overall range of the study area, and the distribution of the AGB estimation based on remote sensing at the various coarse resolutions was obtained (Figure 9).

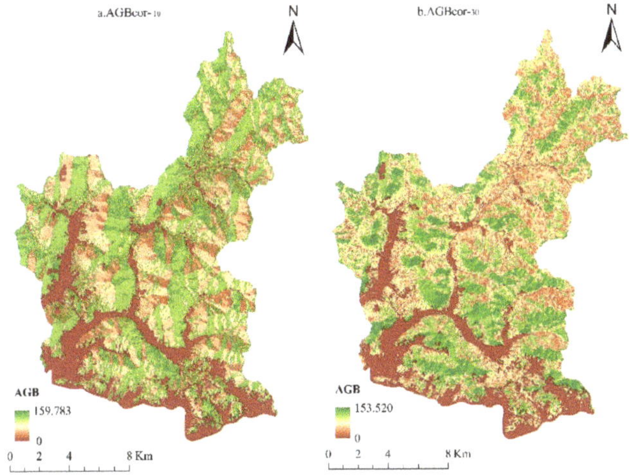

Figure 9. The scale error-corrected AGB distribution at 10 m and 30 m spatial resolution. (**a**) Corrected AGB at 10 m; (**b**) corrected AGB at 30 m. (Unit was t/ha).

5. Discussion

Forest biomass is critically important for forest dynamics in the carbon cycle [79]. However, it remains uncertain because large-scale AGB mapping applications from remote sensing data still carry large uncertainty [37,80]. In this study, a random forest model was devised to estimate the AGB at three different spatial scales (4 m, 10 m, 30 m). The determination coefficient between estimated and measured AGB for various remote sensing data using an independent dataset was 0.5711, 0.4819 and 0.4321, respectively. The same model and dataset were used, but the prediction accuracy of the AGB varied among different remote sensing data. The results generally demonstrated a tendency in which the accuracy of AGB estimation was decreased with the increase in the pixel size of the remote sensing data. In other words, there was a significant scale effect, which is the main problem associated with parameter estimation when using remote sensing. According to the results, this scale effect resulted in significant uncertainty in forest AGB estimation in this study [81].

Some scholars have focused on scale effect research and attempted to identify the reasons for the scale error. Chen found that this scale effect was caused by the surface heterogeneity. He noted that the nonlinearity of the retrieval algorithm and mixed pixels led to the scale effect of the inversion of land surface parameters [82]. Leeuwen et al. pointed out that spectral mixing would increase the error of the classification [83]. Therefore, we developed a scale error correction method using information entropy of the land use type and compared the corrected AGB results. The fitting R^2 of the AGB estimation after scale correction at a resolution of 10 m increased from 0.6226 to 0.6725, and the MBE, RMSE and MAPE were significantly decreased compared with the AGB results without correction. Compared with other similar research, the accuracy was increased [84,85]. The fitting R^2 of the AGB estimation at a resolution of 30 m before correction was 0.5910, and it increased to 0.6704 after correction. In contrast, Zhou concluded that the R^2 of AGB estimation only using Landsat-8 was 0.61 [86]. It was easily found that the correction of the scale effect can effectively improve the accuracy of AGB estimation, and our method presented good performance for scale error correction. Thus, it can be considered as an approach to correct the scale effect and improve the AGB estimation accuracy of coarse-resolution images.

To correct the scaling bias, scholars have developed many methods, such as statistical regression, the Taylor series expansion method, the wavelet fractal method, the fractal method and geostatistical theory [87]. The geostatistical method is commonly used for the upscaling of AGB over the feature space [88]. In this study, four geostatistical scale conversion methods, namely bilinear interpolation, the nearest neighbor method, cubic convolution and the Kriging interpolation method, were selected to upscale the AGB estimation from 4 m to 10 m and 30 m spatial resolution. Figure 10 shows the scatter plot of the relative true value and scale corrected using the geostatistical method for Sentinel-2 (Figure 10). The R^2 was 0.5981, 0.4354, 0.4445 and 0.6024. In contrast, the AGB results corrected by the Kriging interpolation method showed the best accuracy among these four methods. However, it was still inferior to our method, with R^2 of 0.6797. At the same time, the AGB estimation using the geostatistical scale was biased from the relative true value. The AGB value was overestimated when the AGB true value was small and vice versa. Moreover, the AGB bias was reduced using the method of this study.

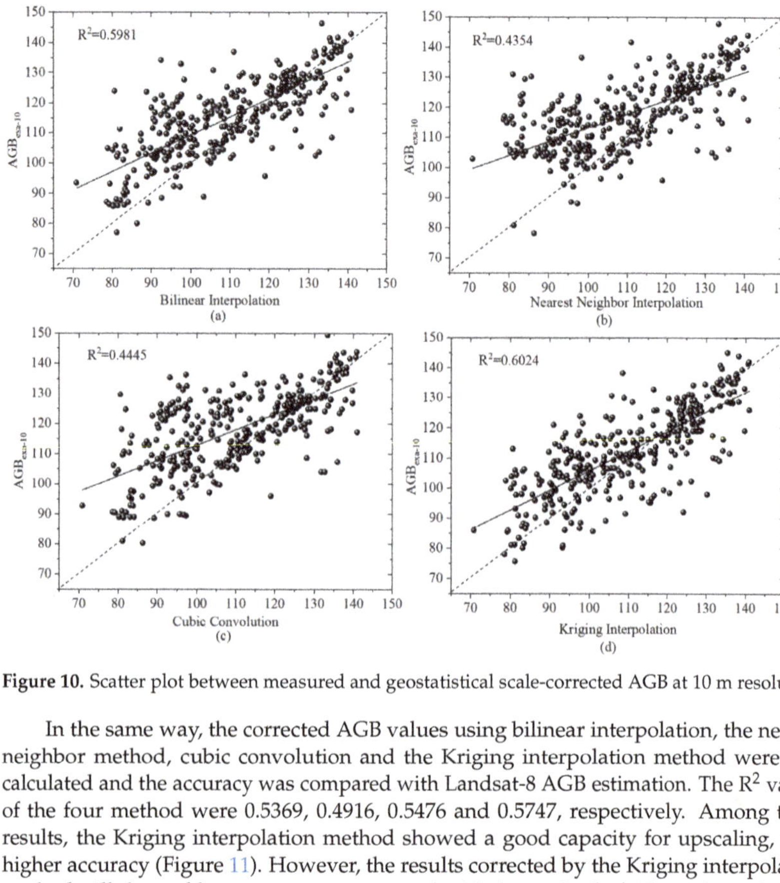

Figure 10. Scatter plot between measured and geostatistical scale-corrected AGB at 10 m resolution.

In the same way, the corrected AGB values using bilinear interpolation, the nearest neighbor method, cubic convolution and the Kriging interpolation method were also calculated and the accuracy was compared with Landsat-8 AGB estimation. The R^2 values of the four method were 0.5369, 0.4916, 0.5476 and 0.5747, respectively. Among these results, the Kriging interpolation method showed a good capacity for upscaling, with higher accuracy (Figure 11). However, the results corrected by the Kriging interpolation method still showed lower accuracy compared with the method of this study, with R^2 of 0.6704. The main reason was that a constant AGB value at different scales was selected, and it could be considered as a ruler, which was used to measure the scale error. After this, the scale conversion method with the entropy-weighted index was used to correct the scale error of the coarse-resolution image. Since the information entropy weight index considered the information entropy of the land use type, the heterogeneity of the surface feature could be fully considered. At the same time, this entropy weight index varied pixel by pixel and thus the correction index of the scale error for each pixel could be calculated, realizing the AGB scale error correction with various spatial resolutions.

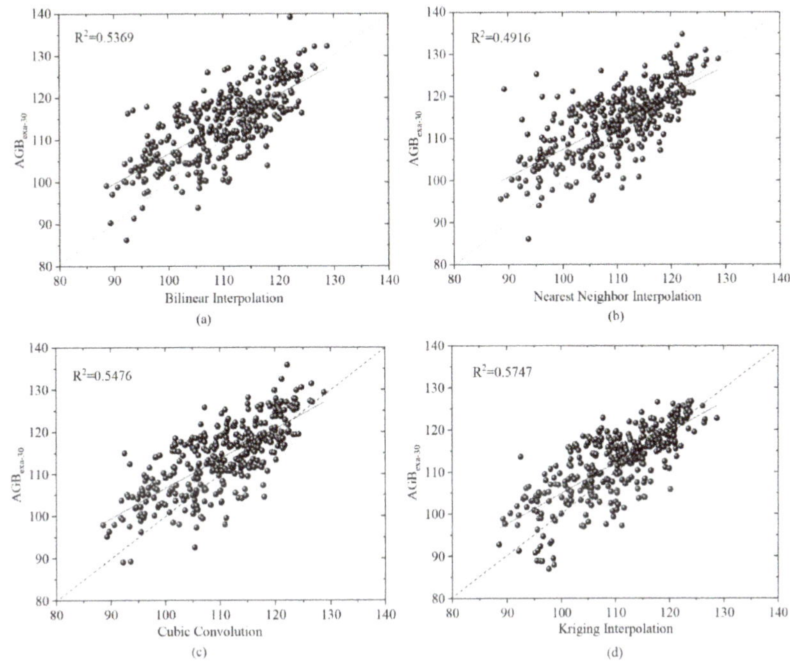

Figure 11. Accuracy evaluation among geostatistical scale-corrected AGB results at 30 m resolution.

It should be noted that some uncertainty may have existed in the current research. First, in the sample survey, there were many typical forest type survey sites selected, but the samples were still unable to cover the total research regions, so the number of the samples and the distribution of the survey sites many bring some uncertainty to the AGB estimation described in this paper. Moreover, the measurement error in the field investigation was not evaluated, and this will lead to the uncertainty of the stand aboveground biomass value calculated based on investigated data. Vegetation growth stages and seasonal differences should be considered for optical remote sensing data applications [89]. Shen et al. found that the vegetation index (VI) introduced large uncertainty in each season, and this affected the AGB estimation results [90,91]. In addition, the algorithm itself will have error transmission and introduce the uncertainty of the estimated AGB. All these issues need to be studied in further work.

6. Conclusions

In this study, a method of forest AGB modeling for three different types of remote sensing data was performed and the accuracy of AGB estimation was compared. Then, the error caused by the scale effect was analyzed and a method to correct this scale error was developed. Some valuable conclusions were as follows:

(1) The random forest model had better AGB estimation accuracy for three different spatial resolutions of remote sensing. This indicates that the nonlinear machine learning method would be promising candidate for AGB estimation.

(2) With the assumption of the law of conservation of mass, a scale error correction method using the information entropy of land use type was developed and successfully applied to the upscaling of AGB estimation for data of different resolution. Compared with other geostatistical interpolation methods, this method can obtain a high-accuracy AGB estimation and reduce the effect of the scale error on AGB estimation. The results indicated that this method can reduce the scale effect caused by the heterogeneity of the surface feature.

This research can provide a reference for AGB estimation and AGB upscaling methods at different spatial resolutions of remote sensing.

Author Contributions: Y.Y. conceived and designed the experiments; Y.P. performed the experiments and analyzed the data; X.Y. and Y.P. wrote the paper; X.Y., Y.Y. and W.F. reviewed and edited the paper. All authors have read and agreed to the published version of the manuscript.

Funding: This research was funded by the National Natural Science Foundation of China, grant numbers 31870621 and 31971580; the Fundamental Research Funds for the Central Universities of China, grant numbers 2572021BA08, 2572019BA10, and 2572019CP12; and the China Postdoctoral Science Foundation, grant number 2019M661239.

Conflicts of Interest: The authors declare no conflict of interest.

References

1. Zhang, Y.; Liang, S.; Yang, L. A review of regional and global gridded forest biomass datasets. *Remote Sens.* **2019**, *11*, 2744. [CrossRef]
2. Anderegg, W.R.; Schwalm, C.; Biondi, F.; Camarero, J.J.; Koch, G.; Litvak, M.; Ogle, K.; Shaw, J.D.; Shevliakova, E.; Williams, A. Pervasive drought legacies in forest ecosystems and their implications for carbon cycle models. *Science* **2015**, *349*, 528–532. [CrossRef] [PubMed]
3. Bonan, G.B. Forests and climate change: Forcings, feedbacks, and the climate benefits of forests. *Science* **2008**, *320*, 1444–1449. [CrossRef] [PubMed]
4. DeLucia, E.H.; Moore, D.J.; Norby, R.J. Contrasting responses of forest ecosystems to rising atmospheric CO2: Implications for the global C cycle. *Glob. Biogeochem. Cycles* **2005**, *19*, GB3006. [CrossRef]
5. Li, P.; Peng, C.; Wang, M.; Li, W.; Zhao, P.; Wang, K.; Yang, Y.; Zhu, Q. Quantification of the response of global terrestrial net primary production to multifactor global change. *Ecol. Indic.* **2017**, *76*, 245–255. [CrossRef]
6. Norby, R.J.; DeLucia, E.H.; Gielen, B.; Calfapietra, C.; Giardina, C.P.; King, J.S.; Ledford, J.; McCarthy, H.R.; Moore, D.J.; Ceulemans, R. Forest response to elevated CO_2 is conserved across a broad range of productivity. *Proc. Natl. Acad. Sci. USA* **2005**, *102*, 18052–18056. [CrossRef]
7. Canadell, J.G.; Raupach, M.R. Managing forests for climate change mitigation. *Science* **2008**, *320*, 1456–1457. [CrossRef]
8. Arora, V.K.; Melton, J.R. Reduction in global area burned and wildfire emissions since 1930s enhances carbon uptake by land. *Nat. Commun.* **2018**, *9*, 1326. [CrossRef]
9. Tang, X.; Zhao, X.; Bai, Y.; Tang, Z.; Wang, W.; Zhao, Y.; Wan, H.; Xie, Z.; Shi, X.; Wu, B. Carbon pools in China's terrestrial ecosystems: New estimates based on an intensive field survey. *Proc. Natl. Acad. Sci. USA* **2018**, *115*, 4021–4026. [CrossRef]
10. Carvalhais, N.; Forkel, M.; Khomik, M.; Bellarby, J.; Jung, M.; Migliavacca, M.; Saatchi, S.; Santoro, M.; Thurner, M.; Weber, U. Global covariation of carbon turnover times with climate in terrestrial ecosystems. *Nature* **2014**, *514*, 213–217. [CrossRef]
11. Hamilton, S.E.; Friess, D.A. Global carbon stocks and potential emissions due to mangrove deforestation from 2000 to 2012. *Nat. Clim. Chang.* **2018**, *8*, 240–244. [CrossRef]
12. Araza, A.; de Bruin, S.; Herold, M.; Quegan, S.; Labriere, N.; Rodriguez-Veiga, P.; Avitabile, V.; Santoro, M.; Mitchard, E.T.; Ryan, C.M. A comprehensive framework for assessing the accuracy and uncertainty of global above-ground biomass maps. *Remote Sens. Environ.* **2022**, *272*, 112917. [CrossRef]
13. O'Connor, B.; Bojinski, S.; Röösli, C.; Schaepman, M.E. Monitoring global changes in biodiversity and climate essential as ecological crisis intensifies. *Ecol. Inform.* **2020**, *55*, 101033. [CrossRef]
14. Rodríguez-Veiga, P.; Saatchi, S.; Tansey, K.; Balzter, H. Magnitude, spatial distribution and uncertainty of forest biomass stocks in Mexico. *Remote Sens. Environ.* **2016**, *183*, 265–281. [CrossRef]
15. Herold, M.; Carter, S.; Avitabile, V.; Espejo, A.B.; Jonckheere, I.; Lucas, R.; McRoberts, R.E.; Næsset, E.; Nightingale, J.; Petersen, R.; et al. The Role and Need for Space-Based Forest Biomass-Related Measurements in Environmental Management and Policy. *Surv. Geophys.* **2019**, *40*, 757–778. [CrossRef]
16. Hermans-Neumann, K.; Gerstner, K.; Geijzendorffer, I.R.; Herold, M.; Seppelt, R.; Wunder, S. Why do forest products become less available? A pan-tropical comparison of drivers of forest-resource degradation. *Environ. Res. Lett.* **2016**, *11*, 125010. [CrossRef]
17. Flade, L.; Hopkinson, C.; Chasmer, L. Allometric Equations for Shrub and Short-Stature Tree Aboveground Biomass within Boreal Ecosystems of Northwestern Canada. *Forests* **2020**, *11*, 1207. [CrossRef]
18. Piermattei, L.; Karel, W.; Wang, D.; Wieser, M.; Mokroš, M.; Surový, P.; Koreň, M.; Tomaštík, J.; Pfeifer, N.; Hollaus, M. Terrestrial Structure from Motion Photogrammetry for Deriving Forest Inventory Data. *Remote Sens.* **2019**, *11*, 950. [CrossRef]
19. Fang, J.; Guo, Z.; Hu, H.; Kato, T.; Muraoka, H.; Son, Y. Forest biomass carbon sinks in E ast A sia, with special reference to the relative contributions of forest expansion and forest growth. *Glob. Chang. Biol.* **2014**, *20*, 2019–2030. [CrossRef]
20. Liang, X.; Hyyppä, J.; Kaartinen, H.; Lehtomäki, M.; Pyörälä, J.; Pfeifer, N.; Holopainen, M.; Brolly, G.; Francesco, P.; Hackenberg, J.; et al. International benchmarking of terrestrial laser scanning approaches for forest inventories. *ISPRS J. Photogramm.* **2018**, *144*, 137–179. [CrossRef]

21. Mutanga, O.; Adam, E.; Cho, M.A. High density biomass estimation for wetland vegetation using WorldView-2 imagery and random forest regression algorithm. *Int. J. Appl. Earth Observ. Geoinf.* **2012**, *18*, 399–406. [CrossRef]
22. Tsitsi, B. Remote sensing of aboveground forest biomass: A review. *Trop. Ecol.* **2016**, *57*, 125–132.
23. Zhang, X.; Ni-Meister, W. Biophysical applications of satellite remote sensing. *Remote Sens. For. Biomass* **2014**, 63–98. [CrossRef]
24. Nian, V. The carbon neutrality of electricity generation from woody biomass and coal, a critical comparative evaluation. *Appl. Energy* **2016**, *179*, 1069–1080. [CrossRef]
25. Hayashi, M.; Saigusa, N.; Yamagata, Y.; Hirano, T. Regional forest biomass estimation using ICESat/GLAS spaceborne LiDAR over Borneo. *Carbon Manag.* **2015**, *6*, 19–33. [CrossRef]
26. Urbazaev, M.; Thiel, C.; Cremer, F.; Dubayah, R.; Migliavacca, M.; Reichstein, M.; Schmullius, C. Estimation of forest aboveground biomass and uncertainties by integration of field measurements, airborne LiDAR, and SAR and optical satellite data in Mexico. *Carbon Balance Manag.* **2018**, *13*, 5. [CrossRef]
27. Migolet, P.; Goïta, K.; Pambo, A.F.K.; Mambimba, A.N. Estimation of the total dry aboveground biomass in the tropical forests of Congo Basin using optical, LiDAR, and radar data. *GISci. Remote Sens.* **2022**, *59*, 431–460. [CrossRef]
28. Nguyen, T.H.; Jones, S.; Soto-Berelov, M.; Haywood, A.; Hislop, S. Landsat time-series for estimating forest aboveground biomass and its dynamics across space and time: A review. *Remote Sens.* **2019**, *12*, 98. [CrossRef]
29. Hlatshwayo, S.T.; Mutanga, O.; Lottering, R.T.; Kiala, Z.; Ismail, R. Mapping forest aboveground biomass in the reforested Buffelsdraai landfill site using texture combinations computed from SPOT-6 pan-sharpened imagery. *Int. J. Appl. Earth Observ. Geoinf.* **2019**, *74*, 65–77. [CrossRef]
30. Zhu, Y.; Liu, K.; Myint, S.W.; Du, Z.; Li, Y.; Cao, J.; Liu, L.; Wu, Z. Integration of GF2 optical, GF3 SAR, and UAV data for estimating aboveground biomass of China's largest artificially planted mangroves. *Remote Sens.* **2020**, *12*, 2039. [CrossRef]
31. Yuan, X.; Li, L.; Tian, X.; Luo, G.; Chen, X. Estimation of above-ground biomass using MODIS satellite imagery of multiple land-cover types in China. *Remote Sens. Lett.* **2016**, *7*, 1141–1149. [CrossRef]
32. Yu, R.; Yao, Y.; Wang, Q.; Wan, H.; Xie, Z.; Tang, W.; Zhang, Z.; Yang, J.; Shang, K.; Guo, X.; et al. Satellite-Derived Estimation of Grassland Aboveground Biomass in the Three-River Headwaters Region of China during 1982–2018. *Remote Sens.* **2021**, *13*, 2993. [CrossRef]
33. Lu, D.; Chen, Q.; Wang, G.; Liu, L.; Li, G.; Moran, E. A survey of remote sensing-based aboveground biomass estimation methods in forest ecosystems. *Int. J. Digit. Earth* **2016**, *9*, 63–105. [CrossRef]
34. Zhao, P.; Lu, D.; Wang, G.; Wu, C.; Huang, Y.; Yu, S. Examining spectral reflectance saturation in Landsat imagery and corresponding solutions to improve forest aboveground biomass estimation. *Remote Sens.* **2016**, *8*, 469. [CrossRef]
35. Hernández-Stefanoni, J.L.; Castillo-Santiago, M.Á.; Mas, J.F.; Wheeler, C.E.; Andres-Mauricio, J.; Tun-Dzul, F.; George-Chacón, S.P.; Reyes-Palomeque, G.; Castellanos-Basto, B.; Vaca, R. Improving aboveground biomass maps of tropical dry forests by integrating LiDAR, ALOS PALSAR, climate and field data. *Carbon Balance Manag.* **2020**, *15*, 1–17. [CrossRef]
36. Réjou-Méchain, M.; Tanguy, A.; Piponiot, C.; Chave, J.; Hérault, B. biomass: An R package for estimating above-ground biomass and its uncertainty in tropical forests. *Methods Ecol. Evolut.* **2017**, *8*, 1163–1167. [CrossRef]
37. Réjou-Méchain, M.; Barbier, N.; Couteron, P.; Ploton, P.; Vincent, G.; Herold, M.; Mermoz, S.; Saatchi, S.; Chave, J.; de Boissieu, F.; et al. Upscaling Forest Biomass from Field to Satellite Measurements: Sources of Errors and Ways to Reduce Them. *Surv. Geophy.* **2019**, *40*, 881–911. [CrossRef]
38. Chen, Q.; Laurin, G.V.; Valentini, R. Uncertainty of remotely sensed aboveground biomass over an African tropical forest: Propagating errors from trees to plots to pixels. *Remote Sens. Environ.* **2015**, *160*, 134–143. [CrossRef]
39. Longo, M.; Keller, M.; dos-Santos, M.N.; Leitold, V.; Pinage, E.R.; Baccini, A.; Saatchi, S.; Nogueira, E.M.; Batistella, M.; Morton, D.C. Aboveground biomass variability across intact and degraded forests in the Brazilian Amazon. *Glob. Biogeochem. Cycles* **2016**, *30*, 1639–1660. [CrossRef]
40. Mascaro, J.; Detto, M.; Asner, G.P.; Muller-Landau, H.C. Evaluating uncertainty in mapping forest carbon with airborne LiDAR. *Remote Sens. Environ.* **2011**, *115*, 3770–3774. [CrossRef]
41. Avitabile, V.; Camia, A. An assessment of forest biomass maps in Europe using harmonized national statistics and inventory plots. *For. Ecol. Manag.* **2018**, *409*, 489–498. [CrossRef] [PubMed]
42. Næsset, E.; Bollandsås, O.M.; Gobakken, T.; Solberg, S.; McRoberts, R.E. The effects of field plot size on model-assisted estimation of aboveground biomass change using multitemporal interferometric SAR and airborne laser scanning data. *Remote Sens. Environ.* **2015**, *168*, 252–264. [CrossRef]
43. Persson, H.J.; Ståhl, G. Characterizing Uncertainty in Forest Remote Sensing Studies. *Remote Sens.* **2020**, *12*, 505. [CrossRef]
44. Chambers, J.Q.; Negron-Juarez, R.I.; Marra, D.M.; Di Vittorio, A.; Tews, J.; Roberts, D.; Ribeiro, G.H.; Trumbore, S.E.; Higuchi, N. The steady-state mosaic of disturbance and succession across an old-growth Central Amazon forest landscape. *Proc. Natl. Acad. Sci. USA* **2013**, *110*, 3949–3954. [CrossRef]
45. Shen, X.; Jiang, M.; Lu, X.; Liu, X.; Liu, B.; Zhang, J.; Wang, X.; Tong, S.; Lei, G.; Wang, S. Aboveground biomass and its spatial distribution pattern of herbaceous marsh vegetation in China. *Sci. China Earth Sci.* **2021**, *64*, 1115–1125. [CrossRef]
46. Rejou-Mechain, M.; Muller-Landau, H.C.; Detto, M.; Thomas, S.C.; Le Toan, T.; Saatchi, S.S.; Barreto-Silva, J.S.; Bourg, N.A.; Bunyavejchewin, S.; Butt, N. Local spatial structure of forest biomass and its consequences for remote sensing of carbon stocks. *Biogeosciences* **2014**, *11*, 6827–6840. [CrossRef]

47. Masinda, M.M.; Li, F.; Liu, Q.; Sun, L.; Hu, T. Prediction model of moisture content of dead fine fuel in forest plantations on Maoer Mountain, Northeast China. *J. For. Res.* **2021**, *32*, 2023–2035. [CrossRef]
48. Wang, C. Biomass allometric equations for 10 co-occurring tree species in Chinese temperate forests. *For. Ecol. Manag.* **2006**, *222*, 9–16. [CrossRef]
49. Zheng, Y.; Dai, Q.; Tu, Z.; Wang, L. Guided Image Filtering-Based Pan-Sharpening Method: A Case Study of GaoFen-2 Imagery. *ISPRS Int. J. Geo-Inf.* **2017**, *6*, 404. [CrossRef]
50. Kobayashi, N.; Tani, H.; Wang, X.; Sonobe, R. Crop classification using spectral indices derived from Sentinel-2A imagery. *J. Inf. Telecommun.* **2020**, *4*, 67–90. [CrossRef]
51. López-Serrano, P.M.; Cardenas Dominguez, J.L.; Corral-Rivas, J.J.; Jiménez, E.; López-Sánchez, C.A.; Vega-Nieva, D.J. Modeling of aboveground biomass with Landsat 8 OLI and machine learning in temperate forests. *Forests* **2019**, *11*, 11. [CrossRef]
52. Dong, L.; Zhang, L.; Li, F. A compatible system of biomass equations for three conifer species in Northeast, China. *For. Ecol. Manag.* **2014**, *329*, 306–317. [CrossRef]
53. Xu, L.; Shi, Y.; Fang, H.; Zhou, G.; Xu, X.; Zhou, Y.; Tao, J.; Ji, B.; Xu, J.; Li, C. Vegetation carbon stocks driven by canopy density and forest age in subtropical forest ecosystems. *Sci. Total Environ.* **2018**, *631*, 619–626. [CrossRef] [PubMed]
54. Taddese, H.; Asrat, Z.; Burud, I.; Gobakken, T.; Ørka, H.O.; Dick, Ø.B.; Næsset, E. Use of Remotely Sensed Data to Enhance Estimation of Aboveground Biomass for the Dry Afromontane Forest in South-Central Ethiopia. *Remote Sens.* **2020**, *12*, 3335. [CrossRef]
55. Wang, F.; Yang, M.; Ma, L.; Zhang, T.; Qin, W.; Li, W.; Zhang, Y.; Sun, Z.; Wang, Z.; Li, F.; et al. Estimation of Above-Ground Biomass of Winter Wheat Based on Consumer-Grade Multi-Spectral UAV. *Remote Sens.* **2022**, *14*, 1251. [CrossRef]
56. Ding, Z.; Sun, Y.; Sun, Z. Estimation of tree biomass with GF-2. *J. Beijing Norm. Univ.* **2021**, *57*, 135–141. [CrossRef]
57. Chen, L.; Wang, Y.; Ren, C.; Zhang, B.; Wang, Z. Assessment of multi-wavelength SAR and multispectral instrument data for forest aboveground biomass mapping using random forest kriging. *For. Ecol. Manag.* **2019**, *447*, 12–25. [CrossRef]
58. Shao, Z.; Zhang, L.; Wang, L. Stacked Sparse Autoencoder Modeling Using the Synergy of Airborne LiDAR and Satellite Optical and SAR Data to Map Forest Above-Ground Biomass. *IEEE J-Stars* **2017**, *10*, 5569–5582. [CrossRef]
59. Shi, T.; Xu, H. Derivation of Tasseled Cap Transformation Coefficients for Sentinel-2 MSI At-Sensor Reflectance Data. *IEEE J-Stars* **2019**, *12*, 4038–4048. [CrossRef]
60. Xu, T.; Cao, L.; Shen, X.; She, G. Estimates of subtropical forest biomass based on airborne LiDAR and Landsat 8 OLI data. *Chin. J. Plant Ecol.* **2015**, *39*, 309–321. [CrossRef]
61. Breiman, L. Random forests. *MLear* **2001**, *45*, 5–32. [CrossRef]
62. Zeng, N.; Ren, X.; He, H.; Zhang, L.; Zhao, D.; Ge, R.; Li, P.; Niu, Z. Estimating grassland aboveground biomass on the Tibetan Plateau using a random forest algorithm. *Ecol. Indic.* **2019**, *102*, 479–487. [CrossRef]
63. Vargas-Larreta, B.; López-Martínez, J.O.; González, E.J.; Corral-Rivas, J.J.; Hernández, F.J. Assessing above-ground biomass-functional diversity relationships in temperate forests in northern Mexico. *For. Ecosyst.* **2021**, *8*, 8. [CrossRef]
64. Ferraz, A.; Saatchi, S.; Mallet, C.; Jacquemoud, S.; Gonçalves, G.; Silva, C.A.; Soares, P.; Tomé, M.; Pereira, L. Airborne Lidar Estimation of Aboveground Forest Biomass in the Absence of Field Inventory. *Remote Sens.* **2016**, *8*, 653. [CrossRef]
65. Liu, L. Simulation and correction of spatialscaling effects for leaf area index. *J. Remote Sens.* **2014**, *18*, 1158–1168. [CrossRef]
66. Hu, X.; Xu, H. A new remote sensing index for assessing the spatial heterogeneity in urban ecological quality: A case from Fuzhou City, China. *Ecol. Indic.* **2018**, *89*, 11–21. [CrossRef]
67. Gelybó, G.; Barcza, Z.; Kern, A.; Kljun, N. Effect of spatial heterogeneity on the validation of remote sensing based GPP estimations. *Agr. For. Meteorol.* **2013**, *174*, 43–53. [CrossRef]
68. Wu, L.; Liu, X.-N.; Zheng, X.; Qin, Q.-M.; Ren, H.-Z.; Sun, Y.-J. Spatial scaling transformation modeling based on fractal theory for the leaf area index retrieved from remote sensing imagery. *J. Appl. Remote Sens.* **2015**, *9*, 096015. [CrossRef]
69. Garrigues, S.; Allard, D.; Baret, F.; Weiss, M. Influence of landscape spatial heterogeneity on the non-linear estimation of leaf area index from moderate spatial resolution remote sensing data. *Remote Sens. Environ.* **2006**, *105*, 286–298. [CrossRef]
70. Whelen, T.; Siqueira, P. Coefficient of variation for use in crop area classification across multiple climates. *Int. J. Appl. Earth Observ. Geoinf.* **2018**, *67*, 114–122. [CrossRef]
71. Li, G.; Zhao, Y.; Zhang, L.; Wang, X.; Zhang, Y.; Guo, F. Entropy-based global and local weight adaptive image segmentation models. *Tsinghua Sci. Technol.* **2019**, *25*, 149–160. [CrossRef]
72. Yang, X.; He, P.; Yu, Y.; Fan, W. Stand Canopy Closure Estimation in Planted Forests Using a Geometric-Optical Model Based on Remote Sensing. *Remote Sens.* **2022**, *14*, 1983. [CrossRef]
73. Wang, D.; Wan, B.; Liu, J.; Su, Y.; Guo, Q.; Qiu, P.; Wu, X. Estimating aboveground biomass of the mangrove forests on northeast Hainan Island in China using an upscaling method from field plots, UAV-LiDAR data and Sentinel-2 imagery. *Int. J. Appl. Earth Observ. Geoinf.* **2020**, *85*, 101986. [CrossRef]
74. Zhang, R.; Zhou, X.; Ouyang, Z.; Avitabile, V.; Qi, J.; Chen, J.; Giannico, V. Estimating aboveground biomass in subtropical forests of China by integrating multisource remote sensing and ground data. *Remote Sens. Environ.* **2019**, *232*, 111341. [CrossRef]
75. Li, C.; Li, Y.; Li, M. Improving Forest Aboveground Biomass (AGB) Estimation by Incorporating Crown Density and Using Landsat 8 OLI Images of a Subtropical Forest in Western Hunan in Central China. *Forests* **2019**, *10*, 104. [CrossRef]
76. López-Serrano, P.M.; López-Sánchez, C.A.; Álvarez-González, J.G.; García-Gutiérrez, J. A Comparison of Machine Learning Techniques Applied to Landsat-5 TM Spectral Data for Biomass Estimation. *Can. J. Remote Sens.* **2016**, *42*, 690–705. [CrossRef]

77. Zhang, L.; Shao, Z.; Liu, J.; Cheng, Q. Deep learning based retrieval of forest aboveground biomass from combined LiDAR and landsat 8 data. *Remote Sens.* **2019**, *11*, 1459. [CrossRef]
78. Dong, L.; Du, H.; Han, N.; Li, X.; Zhu, D.E.; Mao, F.; Zhang, M.; Zheng, J.; Liu, H.; Huang, Z. Application of convolutional neural network on Lei Bamboo above-ground-biomass (AGB) estimation using Worldview-2. *Remote Sens.* **2020**, *12*, 958. [CrossRef]
79. Gora, E.M.; Esquivel-Muelbert, A. Implications of size-dependent tree mortality for tropical forest carbon dynamics. *Nat. Plants* **2021**, *7*, 384–391. [CrossRef]
80. Rodríguez-Veiga, P.; Wheeler, J.; Louis, V.; Tansey, K.; Balzter, H. Quantifying forest biomass carbon stocks from space. *Curr. For. Rep.* **2017**, *3*, 1–18. [CrossRef]
81. Pearson, T.R.H.; Brown, S.; Murray, L.; Sidman, G. Greenhouse gas emissions from tropical forest degradation: An underestimated source. *Carbon Balance Manag.* **2017**, *12*, 3. [CrossRef] [PubMed]
82. Chen, J.M. Spatial Scaling of a Remotely Sensed Surface Parameter by Contexture. *Remote Sens. Environ.* **1999**, *69*, 30–42. [CrossRef]
83. van Leeuwen, M.; Frye, H.A.; Wilson, A.M. Understanding limits of species identification using simulated imaging spectroscopy. *Remote Sens. Environ.* **2021**, *259*, 112405. [CrossRef]
84. Chen, Y.; Li, L.; Lu, D.; Li, D. Exploring bamboo forest aboveground biomass estimation using Sentinel-2 data. *Remote Sens.* **2018**, *11*, 7. [CrossRef]
85. Puliti, S.; Hauglin, M.; Breidenbach, J.; Montesano, P.; Neigh, C.S.R.; Rahlf, J.; Solberg, S.; Klingenberg, T.F.; Astrup, R. Modelling above-ground biomass stock over Norway using national forest inventory data with ArcticDEM and Sentinel-2 data. *Remote Sens. Environ.* **2020**, *236*, 111501. [CrossRef]
86. Li, Y.; Li, M.; Li, C.; Liu, Z. Forest aboveground biomass estimation using Landsat 8 and Sentinel-1A data with machine learning algorithms. *Sci. Rep.* **2020**, *10*, 9952. [CrossRef]
87. Jiang, J.; Ji, X.; Yao, X.; Tian, Y.; Zhu, Y.; Cao, W.; Cheng, T. Evaluation of Three Techniques for Correcting the Spatial Scaling Bias of Leaf Area Index. *Remote Sens.* **2018**, *10*, 221. [CrossRef]
88. Propastin, P. Large-scale mapping of aboveground biomass of tropical rainforest in Sulawesi, Indonesia, using Landsat ETM+ and MODIS data. *GIScі. Remote Sens.* **2013**, *50*, 633–651. [CrossRef]
89. Wang, Q.; Li, J.; Jin, T.; Chang, X.; Zhu, Y.; Li, Y.; Sun, J.; Li, D. Comparative analysis of Landsat-8, Sentinel-2, and GF-1 data for retrieving soil moisture over wheat farmlands. *Remote Sens.* **2020**, *12*, 2708. [CrossRef]
90. Wang, Z.; Wang, H.; Wang, T.; Wang, L.; Liu, X.; Zheng, K.; Huang, X. Large discrepancies of global greening: Indication of multi-source remote sensing data. *Glob. Ecol. Conserv.* **2022**, *34*, e02016. [CrossRef]
91. Shen, X.; Liu, B.; Henderson, M.; Wang, L.; Jiang, M.; Lu, X. Vegetation greening, extended growing seasons, and temperature feedbacks in warming temperate grasslands of China. *J. Clim.* **2022**, 1–51. [CrossRef]

remote sensing

Article

Spatiotemporal Variations of Forest Vegetation Phenology and Its Response to Climate Change in Northeast China

Wenrui Zheng [1], Yuqi Liu [1], Xiguang Yang [1,2,*] and Wenyi Fan [1,2]

1 School of Forestry, Northeast Forestry University, Harbin 150040, China; apollo_r@nefu.edu.cn (W.Z.); liuyq9713@nefu.edu.cn (Y.L.); fanwy@nefu.edu.cn (W.F.)
2 Key Laboratory of Sustainable Forest Ecosystem Management-Ministry of Education, Northeast Forestry University, Harbin 150040, China
* Correspondence: yangxiguang@nefu.edu.cn

Abstract: Vegetation phenology is an important indicator of vegetation dynamics. The boreal forest ecosystem is the main part of terrestrial ecosystem in the Northern Hemisphere and plays an important role in global carbon balance. In this study, the dynamic threshold method combined with the ground-based phenology observation data was applied to extract the forest phenological parameters from MODIS NDVI time-series. Then, the spatiotemporal variation of forest phenology is discussed and the relationship between phenological change and climatic factors was concluded in the northeast China from 2011 to 2020. The results indicated that the distribution of the optimal extraction threshold has spatial heterogeneity, and the changing rate was 3% and 2% with 1° increase in latitude for SOS (the start of the growing season) and EOS (the end of the growing season). This research also notes that the SOS had an advanced trend at a rate of 0.29 d/a while the EOS was delayed by 0.47 d/a. This variation of phenology varied from different forest types. We also found that the preseason temperature played a major role in effecting the forest phenology. The temperature in winter of the previous year had a significant effect on SOS in current year. Temperature in autumn of the current year had a significant effect on EOS.

Keywords: phenology; climate change; dynamic threshold method; northeast China; TIMESAT

Citation: Zheng, W.; Liu, Y.; Yang, X.; Fan, W. Spatiotemporal Variations of Forest Vegetation Phenology and Its Response to Climate Change in Northeast China. *Remote Sens.* **2022**, *14*, 2909. https://doi.org/10.3390/rs14122909

Academic Editor: Luke Wallace

Received: 29 April 2022
Accepted: 16 June 2022
Published: 17 June 2022

1. Introduction

Vegetation phenology is the subject which studies the cyclical events throughout the whole life of plants and how these events respond to environmental changes [1]. Lots of studies have clarified that global warming, with the consequence of greenhouse gases increasing, has significantly shifted the vegetation phenology in terrestrial ecosystems of the Northern Hemisphere [2,3], and the variation of vegetation phenology has greatly impacted the terrestrial ecosystem functions and structures [4,5]. Previous researches have concluded that the forest ecosystem is the main part of terrestrial ecosystem in the Northern Hemisphere, such as in China [6], America [7], Canada [8], and Europe [9], and plays an important role in the global carbon balance. Vegetation phenology may also feed back to climate changes, for example, the prolonged length of growing season (LOS) could affect the ability of forest carbon sequestration and mitigate the global temperature increase [10]. Therefore, studying the relationship of vegetation to climate is essential for enhancing the vegetation productivity, carbon storage and carbon cycle of the terrestrial ecosystem.

Phenology research dates back to ancient agricultural times. People originally obtained the timing of phenological events by observing and establishing phenology observation networks, which has been occurring since the 18th century [11]. Previous studies indicate that the spring phenological variation of most vegetation had an advanced trend proven by ground-based observations during the past decades in Northern Hemisphere. Menzel et al. concluded that the average advance of spring was 2.5 days per decade in 21 European countries between 1921 and 2000 [12]. Keenan et al. found that the temperate forest over the eastern US had a strong trend of earlier springs over combined long-term ground observations of phenology [13]. Rosbakh et al. analyzed the 67 common plant species in Siberia and found that boreal forest springs advanced 2.2 days per decade, while leaf senescence was delayed at a rate of 1.6 days per decade during 1976–2018 [14]. However, ground-based observation only recorded the timing of phenological events for species, so that it is difficult to clearly understand the seasonal changes of vegetation phenology on a regional or global scale [1]. During the past few years, remote sensing technology developed rapidly, which, as a new tool, overcomes the above limitations of ground-based observation. Data obtained from satellite remote sensing could obtain the spatially continuous information of surface, which had increasingly been used in the studying and monitoring of vegetation phenology, such as vegetation index (VI), which is a combination of two or more wavelength ranges of surface reflectance to enhance characters or details of vegetation [15]. The more commonly used remote-sensing vegetation indices include the normalized differential vegetation index (NDVI) [16], the enhanced vegetation index (EVI) [17], and the leaf area index (LAI), which is a forest structure parameter and can also be used to extract forest vegetation phenology [18].

Based on satellite data, the changes of vegetation dynamics can be studied using the vegetation indices or biophysical variables time series [19]. The quality of long-term series remote sensing data would make a big difference for the calculation of the surface vegetation phenology. Due to cloud contamination, atmospheric variability, and bi-directional effects, the long-term series remote sensing data still have a lot of noise [20]. To extract the spectral–temporal signatures accurately, many methods have been developed for reducing noise to construct high-quality VI time-series, and these can be classified into three categories: empirical methods, data transformations, and curve fitting methods [15]. The empirical methods are easy to apply, but they are determined by empirical parameters, such as the length of the sliding window. Data transformation methods use the mathematical manipulation to decompose time-series curves into seasonal, cyclical, trend, and irregular components [21], while the performance is poor in smoothing the irregular or asymmetric data [15]. Curve fitting methods fit the VI time-series to a particular function by utilizing least squares, with the advantage of effectively reducing the noise and no empirical constraints [15]. Logistic function, asymmetric Gaussian functions, and the Savitzky–Golay (S-G) filter are commonly used methods. Lara et al. compared the three smoothing methods included in TIMESAT software and concluded that the S-G filter had better performances [22]. Once the time-series curves based on remote sensing data were reconstructed, phenological parameters could be extracted.

The identify the method of the vegetation phenology from remote sensing time-series included inflection points and relative thresholds [23,24]. The inflection point method uses the inflection point of the VI time-series curve to identify the SOS (The start of the growing season) or EOS (The end of the growing season). The inflection point phenology detection algorithm usually uses a logistic function to fit the VI time-series and the results of the inflection point method relied more on the shape of the VIs time-series and the accuracy of the extracted phenology, which varied through the with and without filtering steps [25]. For the relative threshold method, the SOS or EOS was determined with a predefined percentage of VI amplitude, such as 20%, 30%, or other values [26]. Therefore, determining the relative thresholds was quite important to estimate vegetation phenological events. Wang et al. took 50% of maximum NDVI value as the threshold to extract the SOS and EOS and accessed the spatio-temporal trends of vegetation phenology, which showed dramatic spatial heterogeneity with different rates during the 1982–2012 [27]. Ding et al. found that the extraction of phenological events by using 20% of the annual NDVI amplitude was highly consistent with ground-based observation data on the Tibetan Plateau from 1982 to 2012 [28]. Xu et al. first used the fixed threshold method to extract the phenology in Tibet Plateau based on the remote sensing. The results showed that the SOS has an obvious overestimation, with about 50% error of estimation (RMSE > 50). Combined with the EC flux measurements, the SOS and EOS value of the threshold method were determined with the value of 0.17 and 0.2 in the grasslands of Inner Mongolia, while these were 0.14 and 0.29 in Tibet Plateau of China [29]. Yu studied the vegetation phenology changes of northeast of China with a threshold method and threshold of 0.2 was used in this study [30]. However, a study of phenology in the same area with a threshold of 0.3 was used in Zhao's research [31]. Fu et al. studied the effect of autumn phenology in the Greater Khingan Mountains of northeastern China, and a threshold of 0.3 was used [32]. This research area was obviously smaller, but the same value was used to extract the SOS, EOS, and LOS from the remote sensing data. Considering the spatial heterogeneity of the vegetation, the extracted phenology of the vegetation across diverse ecosystems and at different scales from satellite data might have significant differences using the fixed-threshold method. In addition, the fixed-threshold method was sensitive to non-vegetation-related variations in the VI time series, and it led to a considerable error in the phenology metrics by using remote sensing data [15]. Furthermore, it might increase the uncertain error in phenology research. Therefore, it is essential to develop a new method of threshold determination to increase the accuracy of the extracted phenological parameters.

Plant growth has been associated with temperature and precipitation to implicate climate trends in phenology shifts [33]. In turn, climate change has significantly affected vegetation phenology, which further changes the carbon, water, and energy exchange between the terrestrial ecosystem and atmosphere. Wolf et al. found that a warmer spring and earlier vegetation activity has a positive effect on the carbon cycle [34]. Xu et al. concluded that warming induced earlier greening in the Northern Hemisphere during 1982–2011 [35]. The temperature changes the activity of enzymes, and the increase in temperature can promote the activity of enzymes to accelerate vegetation phenology. Jeong et al. concluded that the warming temperature enhances vegetation photosynthesis and prolongs the LOS by advancing the SOS and delaying the EOS [36]. Liu et al. found that the warming climate prolonged the LOS of plants in the Northern Hemisphere by using the GIMMS NDVI3g [37]. Zhao et al. pointed out that over the past decades, the EOS has been delayed by 0.13 days each year in northeastern China [31]. Wang et al. discovered an advanced SOS and a delayed EOS by utilizing remote-sensing data and climate data in the northeastern China from 2011 to 2019 [38]. In addition, precipitation is also a factor which effects the phenology. Piao et al. found that precipitation played a significant role in effecting the summer NDVI in Eurasia [39]. Cong et al. found that increasing precipitation could result in the advanced SOS of broad-leaf forests in northern China by using the GIMMS NDVI3g [40]. It is not difficult to conclude from the existing research that the vegetation growth environment varies across the regions, and the response of phenology to meteorological factors is different [41]. Although the relationship between temperature and precipitation and vegetation phenology has been discussed, these are complex responses that vary according to the spatial heterogeneity of the vegetation. Therefore, it is necessary to demonstrate the relationship between the phenology and the factors of preseason, interannual, and multi-climatic factors and to conduct a comprehensive study on interactions that exist between the SOS and EOS.

In this study, we developed a dynamic thresholds method combining MODIS NDVI time-series and ground-based observation data to extract the vegetation phenological parameters in Northeast China from 2011–2020. We analyzed the changing characteristics of phenology of different forest types in northeast China during the last decade. We aimed to (1) develop a suitable dynamic threshold method to extract the SOS and EOS, combining MODIS NDVI time-series and ground-based phenology observation data; (2) summarize the spatial and temporal changing characteristics of the phenology of different forest types in northeast China; (3) study the relationship and interaction between the phenology of different forest types and climate factors on a regional scale.

We hope that this study can provide a reference to further clarify the relationship between the phenology of the different forest vegetation types and climate factors and the interaction against the backdrop of global warming.

2. Materials and Methods

2.1. Study Area

The research area of this study is the Northeast China (NEC), which includes the Heilongjiang, Jilin, and Liaoning provinces, is located from 118°50′ E to 135°09′ E and 38°42′ N to 53°35′ N (Figure 1) [42]. NEC has the considerable climatic and topographical gradients, and the main topography of NEC is mountains and plains, with mountains in the east, west and north, and plains in the middle and south [43]. Due to geographical location NEC belongs to a temperate continental monsoon climate [44], which is divided into a warm temperate zone, temperate zone, and cold temperate zone from south to north and has obvious differences in humidity from east to west [45]. As a result, NEC has a unique vegetation distribution and is one of the regions most sensitive to global change [46]. NEC has one of the largest natural forests in China, which are mainly scattered throughout the Changbai Mountains, Lesser Khingan Mountains, and Greater Khingan Mountains. The main vegetation types of NEC forests are cold-temperate deciduous coniferous forests, deciduous broad-leaved forests, and mixed coniferous broad-leaved forests. Therefore, as a main part of the boreal forest ecosystem, NEC is an ideal region for researching the forest–climate relationships of northeastern Asia.

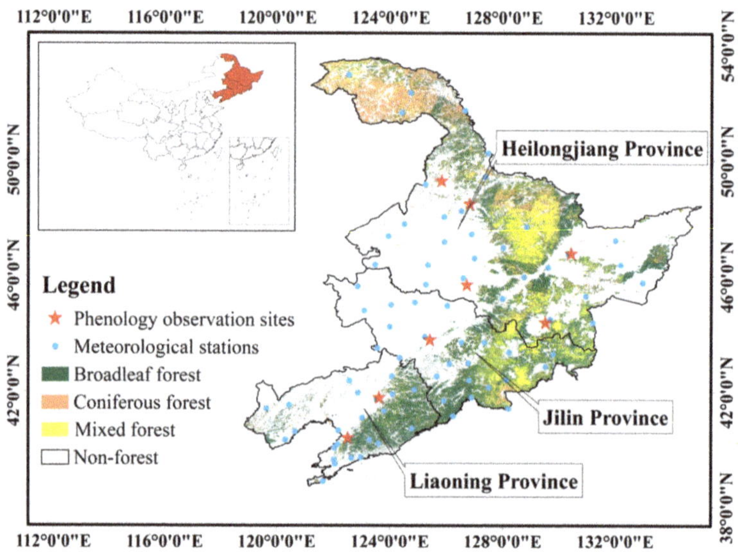

Figure 1. Forest types in northeast China and the locations of the eight phenological observation stations.

2.2. Materials

2.2.1. MODIS NDVI Dataset

The NDVI was obtained from a moderate-resolution imaging spectroradiometer (MODIS) provided by the National Aeronautics and Space Administration (NASA). Available online: https://search.earthdata.nasa.gov (accessed on 23 April 2022). MOD13Q1 (MODIS/Terra Vegetation Indices 16-Day L3 Global 250 m SIN Grid) was used in this study. The time span of the data is from 1 January 2011 to 31 December 2020. A total of 1150 images were downloaded. The spatial resolution of the NDVI data products is 250 m, and the temporal resolution is 16 days [20]. The NDVI calculation is a combined operation between the red spectral band (Red) and near-infrared spectral band (NIR) as follows [47]:

$$NDVI = \frac{NIR - Red}{NIR + Red} \tag{1}$$

where NIR is the reflectivity values in the near infrared band and *Red* is the reflectivity of the red band. The value range of NDVI is from −1 to 1.

ArcGIS software and MRT (Modis Reprojection Tool) were used to process the downloaded images. Image preprocessing included reprojection, cutting, splicing, and so on.

The software TIMESAT, which contains a S-G filter, asymmetrical Gaussian (AG) function fitting, and double logistic function fitting, was employed to reduce the noise and smooth the NDVI time-series [48]. In this study, the S-G filter, which is a weighted moving average filter proposed by Savitzky and Golay [49], was chosen for smoothing the NDVI time-series because of its better performance.

The weight of the S-G filter depends on the polynomial least squares fit in the filter window [50]. The general equation of the S-G filter for NDVI time-series smoothing can be given as follows [51]:

$$Y_j^* = \frac{\sum_{i=-m}^{i=m} C_i Y_{j+i}}{2m+1} \quad (2)$$

where Y_{j+i} represents the original value of the i-th NDVI at time j, Y_j^* represents the resultant NDVI value, C_i represents the coefficient for the i-th NDVI value of the filter, and m represents the half-width of the smoothing window.

2.2.2. Meteorological Data

The meteorological data used in this study are the Global Summary of the Day data provided by the National Oceanic and Atmospheric Administration (NOAA). Available online: https://www.ncei.noaa.gov/data/global-summary-of-the-day (accessed on 23 April 2022). First, we downloaded the daily average temperature and daily cumulative precipitation data from stations throughout the NEC and surrounding areas from 2011 to 2020. In order to maintain the continuity and effectiveness of the data, we downloaded the data from the stations of research area and around research area and deleted the stations with more than 5% of missing data and obtained 116 stations through quality control; the statistical information can be found in Table 1. Subsequently, we converted the daily data to annual data and seasonal data, that is, spring (March–May), summer (June–August), autumn (September–November), and winter (December–February (of the next year)). Finally, we obtained the interpolation grid, which had a consistent spatial resolution with the spatial resolution of NDVI, by applying the simple kriging interpolation method.

Table 1. The statistical information of meteorological stations in the research area from 2011 to 2020.

	Max	Min	Mean
annual average temperature (°C)	12.9	−4.11	5.19
annual cumulative precipitation (mm)	1499.4	192.5	586.8

2.2.3. Land Cover Dataset

It is very crucial to distinguish between vegetation and non-vegetation by using land cover data, specifically in the extraction of vegetation phenology. In this study, the land cover data FROM-GLC (Finer Resolution Observation and Monitoring of Global Land Cover, FROM-GLC), developed by group Pro. Peng Gong at Tsinghua University, was used. FROM-GLC data is a global land cover map at 30 m resolution obtained by using Landsat TM and ETM+ data with high accuracy. Available online: http://data.ess.tsinghua.edu.cn (accessed on 23 April 2022). In this study, all forest types were reclassified into coniferous forest (CF), broadleaf forest (BF), and mixed forest (MF).

2.2.4. Phenology Observation Data

In this study, we downloaded the phenology observation data from the Chinese Phenological Observation Network (CPON). Available online: http://www.geodata.cn (accessed on 23 April 2022). The phenology observation data were used to determine the threshold of the NDVI time-series and evaluate the accuracy of phenological parameters extracted by using the NDVI time-series. Eight stations in NEC were selected; these were Nenjiang, Dedu, Jiamusi, Harbin, Mudanjiang, Changchun, Shenyang, and Gaizhou stations. Combined with the geographical location of the phenological observation sites, the statistical information of the measured data after sorting can be found in Table 2.

Table 2. The mean forest phenological parameters of each station over the years.

Station Name	Latitude	Longitude	Mean SOS/DOY	Mean EOS/DOY
Gaizhou	40.4	122.5	105.5	308.9
Shenyang	41.8	123.6	115.8	305
Changchun	43.8	125.4	120.7	301.6
Mudanjiang	44.4	129.5	123.2	297
Harbin	45.7	126.7	127.3	291.3
Jiamusi	46.8	130.4	125.8	292.3
Dedu	48.5	126.8	138	282.3
Nengjiang	49.3	125.8	130.3	282.1

Descriptions of the datasets applied in our study are shown in Table 3.

Table 3. Detailed descriptions of research data.

Type	Variables	Dataset	Resolution	Source
Vegetation Index	NDVI	MODIS NDVI	250 m	NASA
Meteorological Data	Temperature, Precipitation	-	-	NOAA
Land Cover Type	Coniferous forest (CF), Broadleaf forest (BF), Mixed forest (MF).	FROM-GLC	30 m	Pro. Peng Gong at Tsinghua University
Phenology Observation Data	Nenjiang, Dedu, Jiamusi, Harbin, Mudanjiang, Changchun, Shenyang, Gaizhou	-	-	Chinese Phenological Observation Network

2.3. Method

2.3.1. Method of the Vegetation Phenology Extraction

In this study, the dynamic threshold method, also called the proportional threshold method, was used to extract SOS and EOS from NDVI time-series processed by a S-G filter. The point in time when NDVI increases to a certain percentage of the NDVI amplitude of the year is defined as the SOS, and the time when NDVI decreases to a certain percentage of the NDVI amplitude of the year is defined as the EOS (Figure 2). The threshold used in this method is not a specific vegetation index value but a dynamic ratio form, compared with the absolute threshold and difference threshold; the dynamic threshold method has better applicability in both the time and space domain [48]. The principle of this method is as follows:

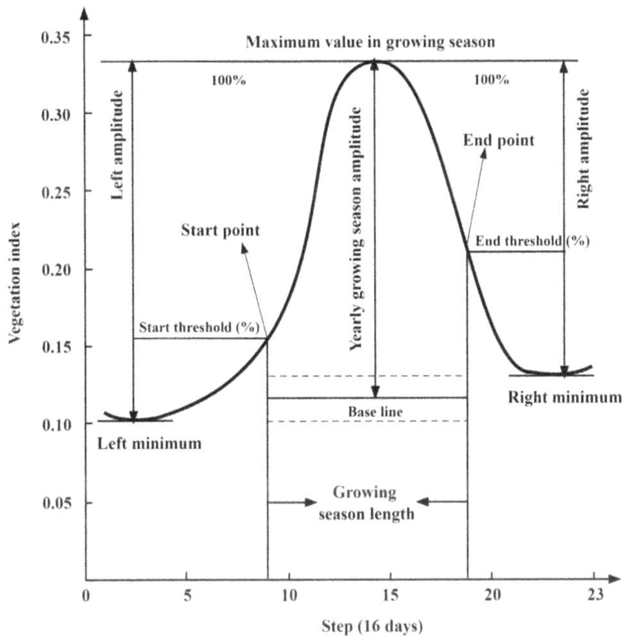

Figure 2. The principle of dynamic threshold method for extracting phenology based on vegetation index time-series curve.

The calculation formula of vegetation phenology extracted by the dynamic threshold method as follows [52]:

$$P_S = \frac{NDVI_{SOS}}{NDVI_{max} - NDVI_{\min(left)}} \tag{3}$$

$$P_E = \frac{NDVI_{EOS}}{NDVI_{max} - NDVI_{\min(right)}} \tag{4}$$

where P_S and P_E represent the extraction threshold corresponding to the SOS and EOS, respectively. $NDVI_{SOS}$ and $NDVI_{EOS}$ are the corresponding NDVI values when SOS and EOS occurred. $NDVI_{max}$ represents the maximum NDVI during the whole time-series, $NDVI_{min(left)}$ is the minimum NDVI of the first half of the time-series, and $NDVI_{min(right)}$ is the minimum NDVI of the second half of the time-series.

Firstly, we selected representative tree species in each phenological observation site and calculated the mean DOY of leaf onset and leaf senescence as SOS and EOS, respectively.

Secondly, we extracted the corresponding remote-sensing pixels of the eight phenology observation stations selected and calculated the mean NDVI of each pixel as the phenological parameters to extract the original data. Then, we brought the NDVI corresponding to the occurrence day of the observation-based phenological parameters into Formulas (3) and (4), and thus the optimal extraction threshold of each station was able to be calculated.

Thirdly, we assumed that the functional relationship between the optimal extraction threshold P of the vegetation phenology at different latitudes and latitude L as follows:

$$P = AL + B \tag{5}$$

where $P = \{P_1, P_2, \cdots, P_n\}$, which represents the optimal extraction threshold set corresponding to the eight phenology observation stations; $L = \{L_1, L_2, \cdots, L_n\}$, which represents the latitude set of the eight phenology observation stations. The values of coefficients A and B were obtained by fitting with the least square method, and the functional relationship between the optimum extraction threshold and latitude of vegetation phenology was established.

Finally, the optimal extraction threshold for each pixel can be calculated based on the central latitude value of each pixel by using the relationship established in Formula (5), and the phenological parameter can be extracted using this threshold.

2.3.2. Analysis Method

Statistical analysis is one of the most commonly used data analysis methods and is widely used in empirical modeling and the accuracy assessment of remote sensing research [53]. This parametric statistical technique requires that the data follows a continuous and normal distribution [54]. Therefore, the normal distribution test should be performed first. The data used in the study all followed the normal distribution. After that, a linear relationship between the forest phenology of different forest types and the latitude, year, and climatic factors were fitted by using the least squares method, and the changing rate of forest phenology affected by latitude, year, and climatic factors was analyzed by comparing the slope of the fitted linear function (Figure 3).

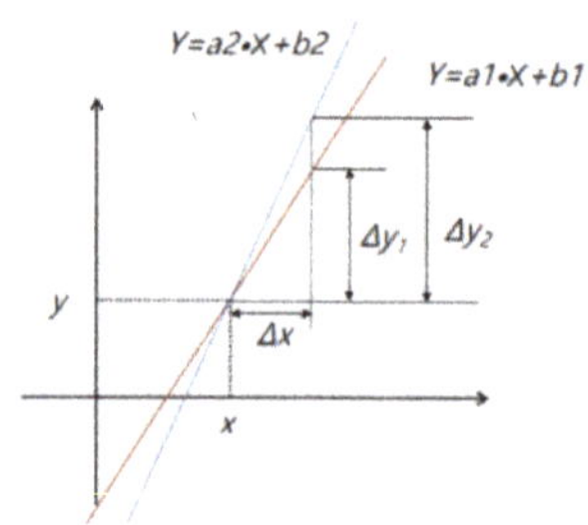

Figure 3. Schematic diagram of this method.

When x increases Δx, y increases with the increase in x, but the changes in the Δy are varied, according to the fitting function. This difference was determined by the slope of the linear function. So, when x increases Δx, the Δy_2 is larger than Δy_1 in Figure 3. Therefore, the slope of the linear function can satisfy the necessity of comparing the changing rate of forest phenology affected by latitude, year, and climatic factors.

Then, the correlation coefficient was selected to explore the relationship between forest phenology and climatic factors. The correlation coefficient was calculated as follows [55]:

$$R = \frac{\sum_{i=1}^{n}(X_i - \overline{X})(Y_i - \overline{Y})}{\sqrt{\sum_{i=1}^{n}(X_i - \overline{X})^2}\sqrt{\sum_{i=1}^{n}(Y_i - \overline{Y})^2}} \quad (6)$$

where R represents correlation coefficient between X and Y, n represents the number of samples, X and Y represent the values in the i-th year, and $\overline{X}$ and $\overline{Y}$ represent the average of values of all years, respectively. The values of $|R|$ range from 0 to 1, which is larger, meaning that the correlation relationship is stronger between two variables. In addition, we used the p-value to test the significance of the correlation coefficient.

To analyze the association between the forest phenology and climatic factors and the partial correlations of the forest phenology and monthly temperature, precipitation was calculated as follows [56]:

$$r_{ab[c]} = \frac{r_{ab} - r_{ac} \times r_{bc}}{\sqrt{(1 + r^2{}_{ac})}\sqrt{(1 + r^2{}_{bc})}} \quad (7)$$

where $r_{ab[c]}$ represents the partial correlation coefficient between phenological parameter a and climatic variable b when climatic variable c was controlled; and r_{ab}, r_{ac}, and r_{bc} represent the liner correlation coefficient between each other, respectively. n represents the number of samples and m represents the number of independent variables.

2.3.3. Validation

In this study, the determination coefficient (R^2), root mean squared error (RMSE), and mean absolute percentage error (MAPE) were selected to evaluate model accuracy. The equations are shown as follows [55]:

$$R^2 = 1 - \frac{\sum_{i=1}^{n}(y_i - \hat{y}_i)^2}{\sum_{i=1}^{n}(y_i - \overline{y})^2} \quad (8)$$

$$\text{RMSE} = \sqrt{\frac{\sum_{i=1}^{n}(y_i - \hat{y}_i)^2}{n}} \quad (9)$$

$$\text{MAPE} = \frac{1}{n}\sum_{i=1}^{n}\frac{|y_i - \hat{y}_i|}{y_i} \times 100\% \quad (10)$$

where y_i is the measured values and $\hat{y}_i$ is the predicted values for the sample i, $\overline{y}$ is the average of all the samples, and n is the number of samples.

3. Results

3.1. Determination of the Dynamic Threshold for Vegetation Phenology

In this study, the relationship between the optimal extracted threshold and the latitude of forest phenology remote-sensing based in northeast China was determined by using the S-G filter and dynamic threshold method, which was combined with NDVI time-series data and ground phenology observation data. Due to the lack of phenological observation data from stations, only eight points in 2014 was simultaneous and available during the research period. The scattering plot between the extracted threshold and latitude can be found in Figure 4. This figure showed that there was a significant relationship between the extracted threshold and latitude. Figure 4a showed the relationship between the optimal extracted threshold of SOS (P_S) and latitude. A least square method was used to fit the function. The fitted function was defined as followed.

$$P_S = 0.0286L - 0.963 \quad (11)$$

where P_S is the threshold to determine the SOS from NDVI time-series data. L is the latitude. The R^2, RMSE, and MAPE of the fitted model are 0.9589, 0.0245, and 6.617%, respectively. The models and the coefficients all passed the significance test at a 95% level of significance.

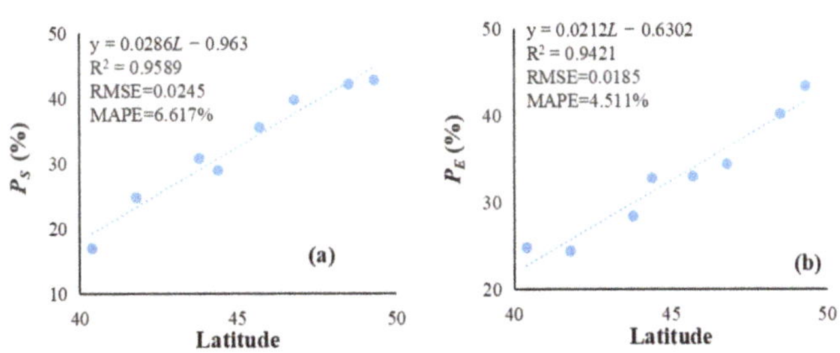

Figure 4. The variation of the extraction threshold corresponding to the observed phenology at each station with latitude (**a**) is the SOS and (**b**) is the EOS.

The linear relationship between the optimal extracted threshold of EOS (P_E) and latitude was also significant, and the fitting function is showed in Equation (12).

$$P_E = 0.0212L - 0.6302 \quad (12)$$

where P_E is the threshold to determinate the EOS from NDVI time-series data. L is the latitude. The R^2, RMSE, and MAPE of the fitted model are 0.9421, 0.0185, and 4.511%, respectively. The model and the coefficient passed the significance test at a 95% level of significance.

Then we evaluated the accuracy of the phenology extracted by using the fixed thresholds of 20% [28], 30% [31], and 50% [27] used by other scholars and dynamic threshold method developed in this study. The RMSE and MAPE between the estimated and measured SOS were calculated, and the results are shown in Table 4. The phenology extracted by using the dynamic threshold method has a better accuracy than the fixed threshold method with a RMSE and MAPE of 11.875 d and 7.623% for SOS and 9.012 d and 2.44% for EOS, respectively. However, the fixed threshold method, with the values of 20%, 30%, and 50%, had lower accuracy. The fixed threshold with the value of 20% has the larger error than other methods. The RMSE is 30.182 d for SOS and 34.846 d for EOS, and this brings the estimated error to about one month. Followed by the fixed threshold with the value of 30% with the RMSE and MAPE of 26.716 d and 16.118% for SOS and 26.528 d and 8.373% for EOS, respectively. Compared with other three methods, the fixed threshold with the value of 50% has a middling level of error, but the error is about 19 d for SOS, and for EOS and MAPE it is 14.723% and 6.118%, respectively.

Table 4. The accuracy comparation of the extracted phenology.

	SOS		EOS	
	RMSE (d)	**MAPE (%)**	**RMSE (d)**	**MAPE (%)**
Dynamic threshold	11.875	7.623	9.012	2.440
Fixed threshold = 20%	30.182	22.269	34.846	11.577
Fixed threshold = 50%	19.607	14.723	19.015	6.118

The scattering plots between measured and extracted phenology using fixed and dynamic threshold methods are shown in Figure 5. For SOS, the extracted SOS using fixed threshold has an obvious bias from the measured SOS. SOS was underestimated for the fixed threshold of 20%. By contrast, SOS was overestimated for the fixed threshold of 30% and 50%. Extracted EOS using a fixed threshold had an obvious overestimating phenomenon. It indicates that the fixed threshold method increases the estimating error and increases the uncertain error in extracted phenology analysis.

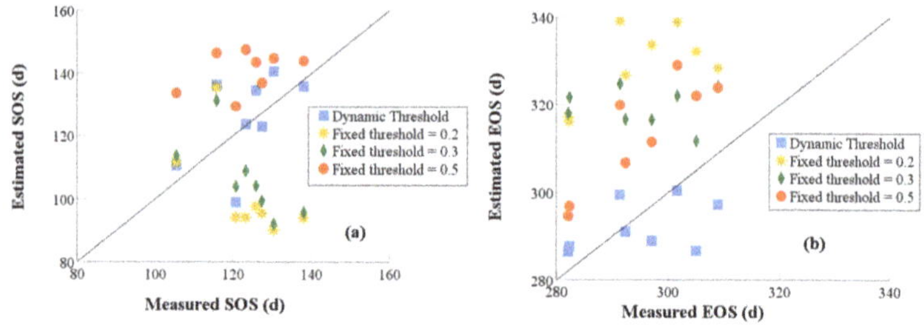

Figure 5. The scattering plot between estimated and measured value. (**a**) is the SOS and (**b**) is the EOS (black line is $y = x$).

3.2. *Characteristics of Forest Phenology in the Northeast China*

3.2.1. Spatial Distribution of the Forest Phenology

The characteristics of forest phenological variation in the northeast China from 2011 to 2020 were analyzed. The spatial pattern of the mean forest SOS in NEC from 2011 to 2020 is shown in Figure 6. The variation of the forest SOS showed significant spatial heterogeneity in the study area. The spatial distribution of the mean SOS exhibited a correlation with the latitude as the southern part was earlier than the northern part. The mean SOS in the NEC primarily occurred between 95th and 135th day, which accounted for 83.46% of the study area, and the average of the SOS in the whole research area was 116 days. The forest located 45 degrees south of the northern latitude had an earlier SOS, between the 95th and 105th day, whereas the area with an average SOS later than 135 days was mostly located in the northernmost Greater Khingan Mountains, which principally distributed in the CF and had a lower temperature. Factors leading to this spatial distribution of SOS were not only related to temperature but also to the type of tree species, because the south of NEC was dominated by broadleaf and mixed forest, while the north of NEC was dominated by coniferous and mixed forests (see Figure 1).

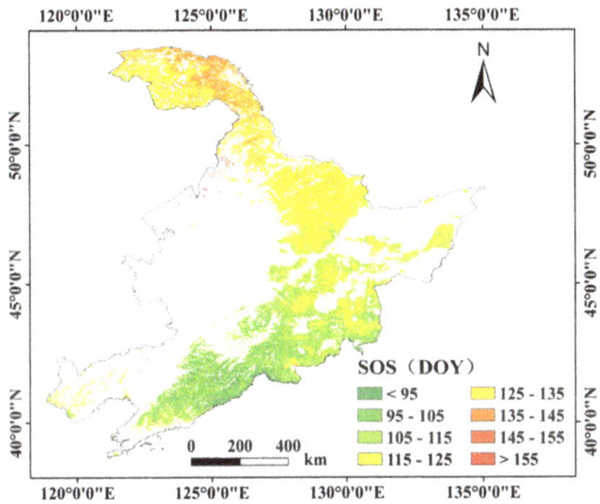

Figure 6. The spatial distribution of the average SOS (start of the growing season) from 2011 to 2020 in the northeastern China.

The mean EOS were mainly in ranges of 300 days to 330 days, which is late October and late November, and the average EOS was 315th days in the northeastern China (Figure 7). The characteristics of the forest EOS in the northeastern China from 2011 to 2020 also had obvious heterogeneity. From the northwest to the southeast of the study area, the average of the EOS was gradually delayed, which showed significantly variation according to the latitude. The forest in the southeastern Changbai Mountains, near the coast, had relatively late EOS dates, while the EOS in the northernmost Greater Khingan Mountains were earlier.

The average LOS gradually lengthened from north to south (Figure 8). The average of the LOS was mainly in ranges of 150 days to 230 days with the counting of pixels for 85.28%. The average LOS in the study area was 199 days. The LOS was longer in coastal areas at low latitudes in the east of Liaoning Province. The regions with the LOS greater than 230 days were mainly distributed in the south of 43° N and east of 122° E, accounting for 9.84% of the research area. The shortest LOS was less than 150 days in the middle of the Greater Khingan Mountains in the Heilongjiang Province.

Long-term variations of phenology could reflect the state of vegetation grown. In order to explore the relationship between forest phenological changes and latitudes in NEC from 2011 to 2020, we divided the study area into 15 parts by 1 degree latitude and calculated the average forest phenological parameters for each part. The results can be found in Figure 9. The results indicated that the SOS of the forest was sightly delayed with the increase of latitude, and the SOS was delayed by 2.33 days per latitude with the increase in latitude. The event of forest EOS would shift to an earlier time with the increase of latitude, and EOS increased by 2.22 days per latitude with the increase of latitude. The LOS of forest decreased with increasing latitude, and LOS decreased by 4.55 days per latitude with increase in latitude.

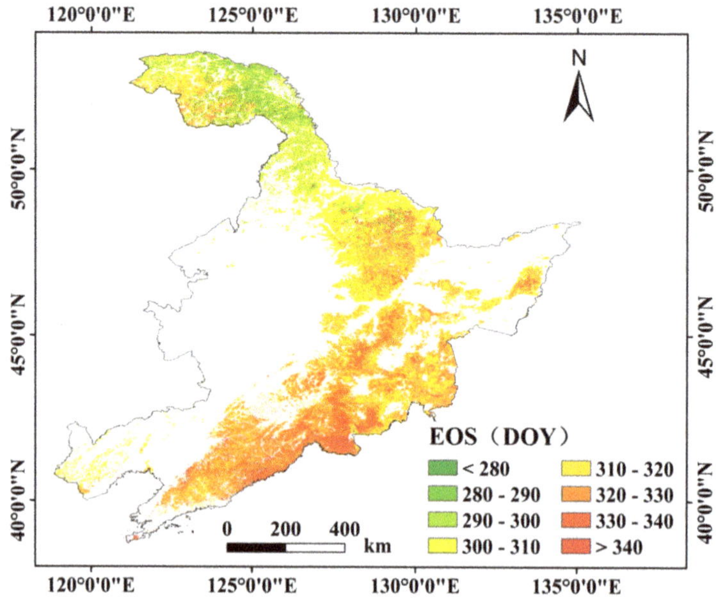

Figure 7. The spatial distribution of the average EOS (end of the growing season) from 2011 to 2020 in the northeast China.

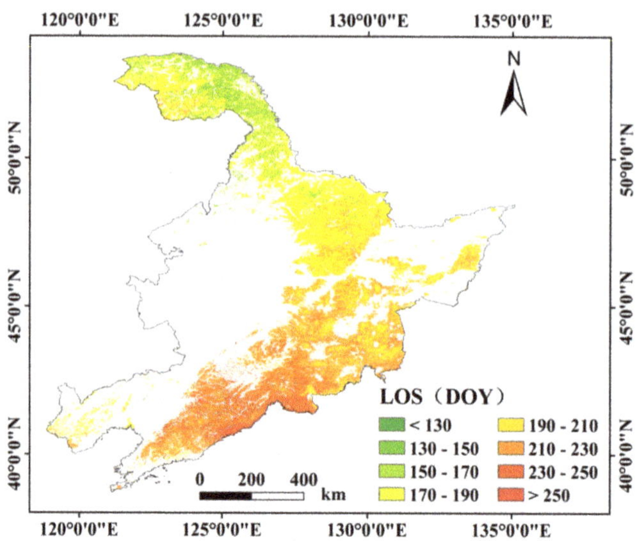

Figure 8. The spatial distribution of the average LOS (length of growing season) from 2011 to 2020 in the northeast China.

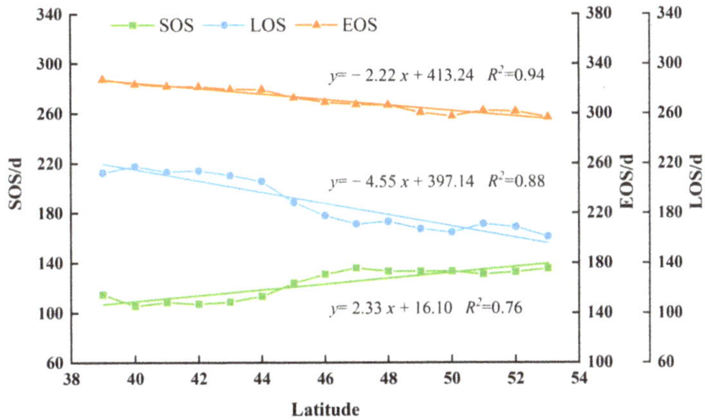

Figure 9. The trends of phenological parameters in northeast China from 2011 to 2020.

3.2.2. The Interannual Variability and Trends of Forest Phenology

Forest phenology fluctuated significantly in NEC from 2011 to 2020 and the interannual variation trend was obvious (Figure 10). The SOS of forest phenology showed a weak advancing trend of approximately 0.29 d/a. The EOS showed a weak delayed trend with a rate of 0.47 d/a. Figure 10 also showed that the variation range of LOS was larger, followed by the EOS and SOS. Overall, the LOS displayed sizeable increases of approximately 0.76 d/a. These trends may be related to global warming because the rising temperature advanced the spring and the cooling temperature trend delayed in the autumn.

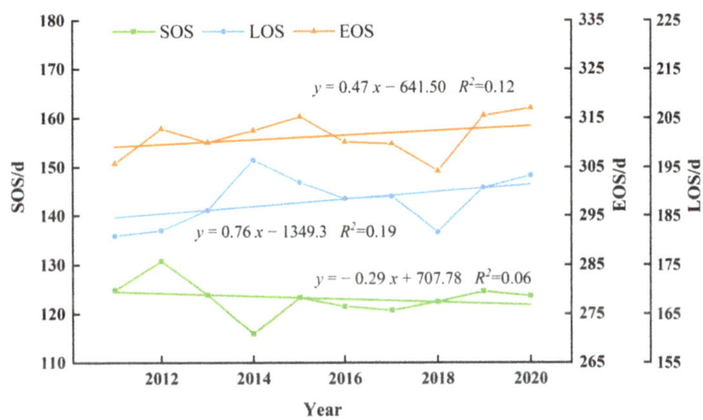

Figure 10. Interannual changes of forest phenology in northeast China from 2011 to 2020.

3.3. The Variation and Trends of Phenology in Different Forest Types

3.3.1. The Spatial Distribution of Phenology in Different Forest Types

In order to investigate whether the phenological characteristics of the different forest types changed with latitudes, we calculated the average phenological parameters of three forest types at different latitudes and analyzed and compared the results. The results showed that all three parameters of different forest types showed fluctuations with different ranges. As per the findings, the following can be discerned (Figure 11): as the latitude increased, the SOS tended to delay. It can clearly be seen that the sensitivity of BF and MF to latitude changes were significantly higher than CF. The SOS of the MF was delayed by approximately 2.51 days per latitude, while the SOS of CF delayed 1.70 days per latitude. The SOS of BF was showed a largest delaying trend with a rate of 2.68 days per latitude.

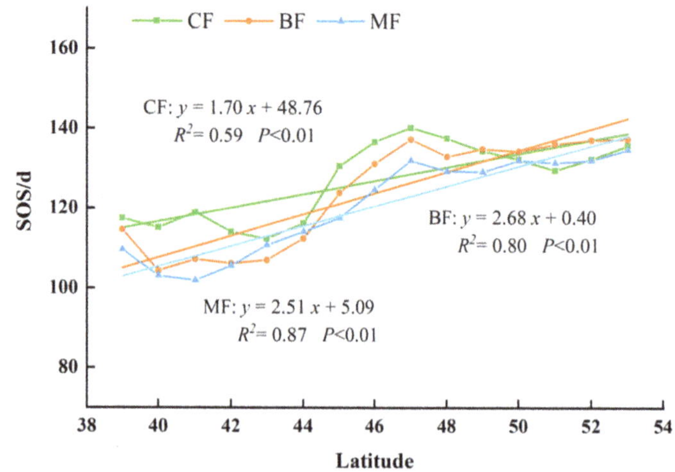

Figure 11. The variation trend of SOS in different forest types in northeast China.

The EOS of different forest types showed a significant delayed trend with the increase in the latitude (Figure 12). The EOS of BF was the greatest significant with a rate of 2.65 days per latitude. Followed by MF, the changing rate of the EOS of MF was 2.47 days per latitude. The CF had the smallest changing rate of 2.0 days per latitude, compared with other two forest types.

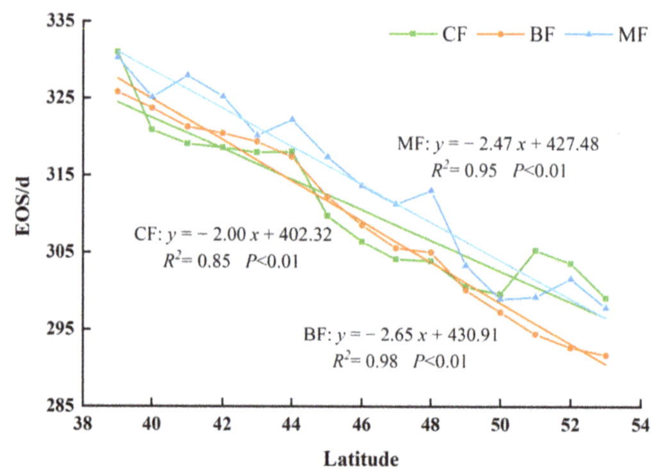

Figure 12. The variation trend of EOS in different forest types in northeast China.

The variation range of LOS was affected by the EOS and SOS. The LOS of different forest types showed a significant decreasing trend with the increase of latitude (Figure 13). The LOS of BF had the greatest changing rate of 5.33 days per latitude. The LOS of MF was 4.98 days per latitude, and the CF had the smallest changing rate of 3.69 days per latitude, compared with other two forest types.

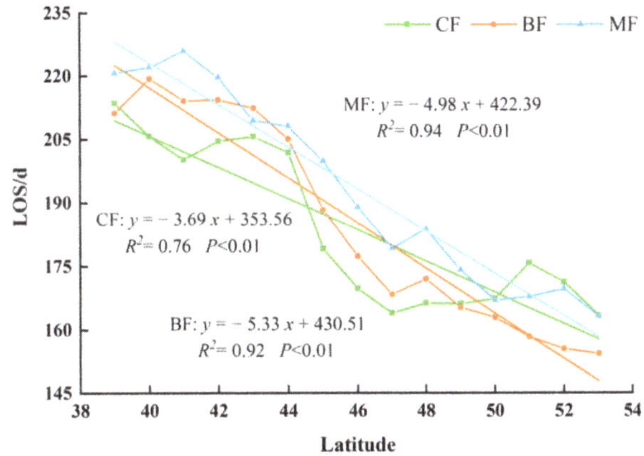

Figure 13. The variation trend of LOS in different forest types in Northeast China.

3.3.2. The Interannual Variation and Trends of Forest Phenology in Different Forest Type

The annual variation of phenology in different types of forest from 2011 to 2020 can be found in Figure 14. The SOS of all three forest types demonstrated an advancing trend year by year in the study area. The MF had the most obvious trend of advance with the rate 0.45 days per year and changing rate of BF was 0.28 days per year. While the CF changed weakly with 0.20 days per year.

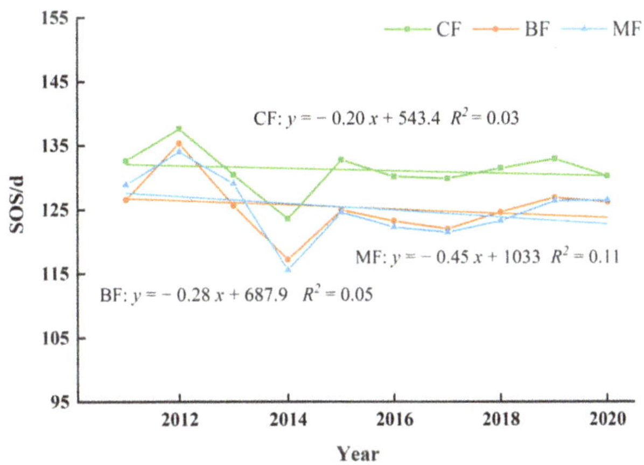

Figure 14. The interannual changes of SOS in different forest types from 2011 to 2020.

All forest types had delayed EOS, whereas MF exhibited the most considerable EOS of all with the rate of 0.58 days per year (Figure 15). The interannual changing rate of EOS of the CF was 0.46 days per year. The EOS changing rate of MS was weaker than other two forest types with a rate of 0.32 days per year.

The variation of LOS displayed an extended trend due to the combined effect of SOS and EOS, and the annual change rate of all was greater than 0.6 days per year, with the most specific change range was BF, followed by MF and CF (Figure 16). The interannual changing rate was 0.86, 0.77 and 0.66 days per year.

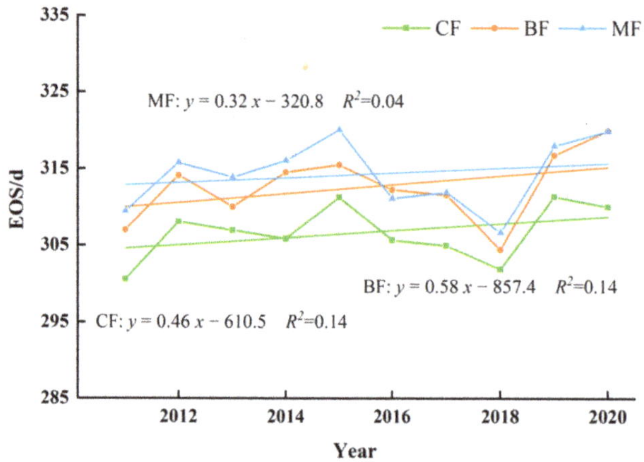

Figure 15. The annual changes of EOS in different forest types from 2011 to 2020.

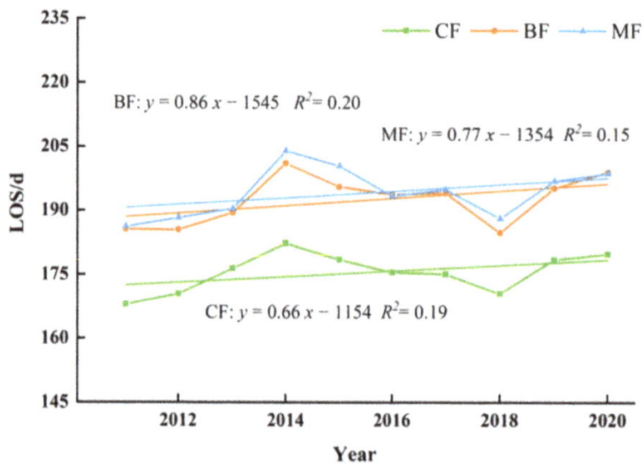

Figure 16. Interannual changes of LOS in different forest types from 2011 to 2020.

3.4. Effects of Climate Factors on Forest Phenology in the Northeast China

3.4.1. Effects of Precipitation on the Forest Phenology

Affected by geographical and climatic factors, there are significant differences in the precipitation and temperature in different regions of the NEC from 2011 to 2020. The maximum difference in the annual cumulative precipitation is 500 mm. As shown in Figure 17, the SOS had a significant correlation with the annual cumulative precipitation ($P < 0.01$). With the increase of precipitation, the phenology of forests showed a trend of advanced SOS and delayed EOS, which extended the LOS. The response of the three phenological parameters to the annual cumulative precipitation from large to small was LOS, EOS, and SOS. The SOS advanced 2.9 days per 100 mm, while the EOS and LOS delayed 4.3 days per 100 mm and 7.1 days per 100 mm, respectively.

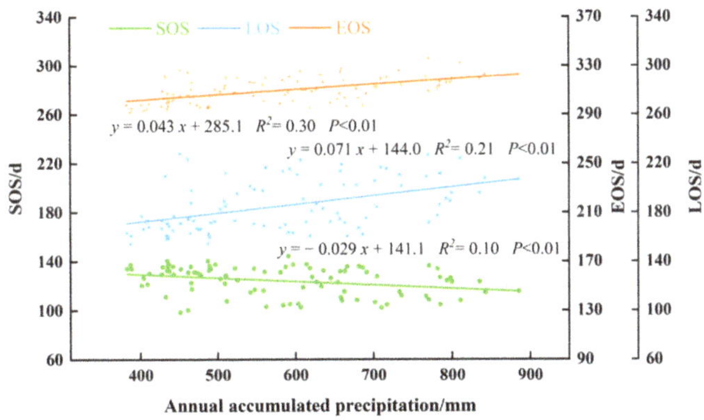

Figure 17. Response of Forest Phenology to Annual Cumulative Precipitation in northeast China from 2011 to 2020.

In this study, we analyzed the response of different forest phenological parameters to annual cumulative precipitation (Figure 18). The SOS in the BF area was obviously correlated with annual cumulative precipitation at a rate of advanced 3.7 d/100 mm ($P < 0.05$). With the increase in annual cumulative precipitation, the EOS of all forest types tended to delay, and the greatest change in MF area was approximately 4.6 d/100 mm, while the CF delayed at a rate of 3.4 d/100 mm ($P < 0.01$). The annual cumulative precipitation had significant effects on LOS of all forest types ($P < 0.01$). With the increase in precipitation, the LOS of each forest type was extended. The sensitivity of the LOS of different forest types to annual cumulative precipitation is, from high to low, BF, MF, and CF. Specifically, the LOS of BF, MF, and CF at rates of 8.5 d/100 mm, 7.5 d/100 mm, and 4.8 d/100 mm, respectively.

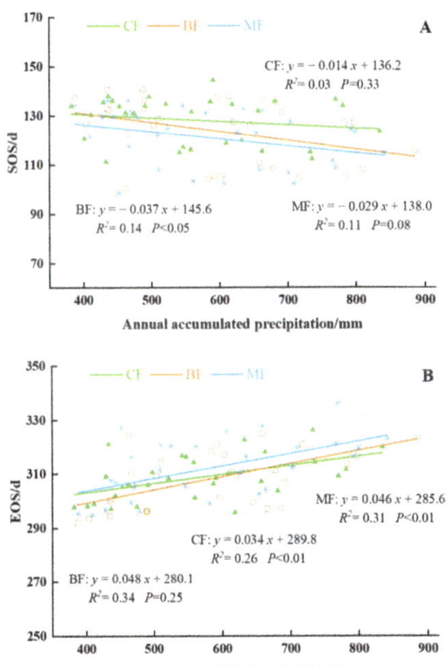

Figure 18. *Cont.*

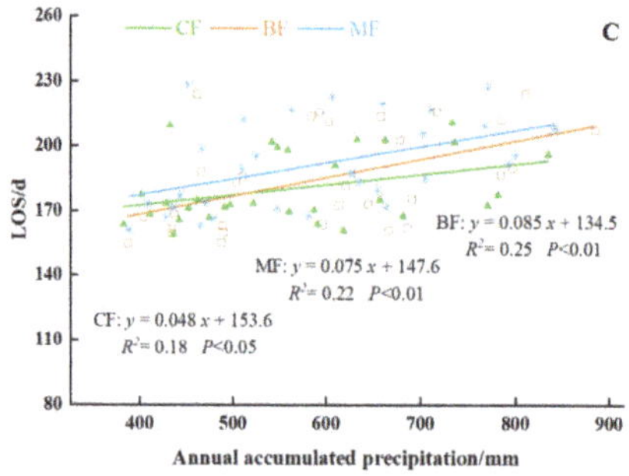

Figure 18. Responses of phenology to precipitation of different forest types in northeast China from 2011 to 2020. (**A**) SOS of different forest types; (**B**) EOS of different forest types; (**C**) LOS of different forest types.

3.4.2. Effects of Temperature on Forest Phenology

Climate change has been more evident in NEC over the past few decades [57]. Figure 19 shows the response of forest phenology to temperature in northeast China from 2011 to 2020. Compared with the annual accumulated precipitation, the impact of annual average temperature on forest phenology was more significant ($P < 0.01$). With the increase in temperature, the northeast forest showed a trend of early advanced SOS, delayed EOS, and prolonged LOS. The responses of the three phenological parameters to the annual average temperature from large to small were LOS, SOS, and EOS. With the average annual temperature increasing by 1 °C, the SOS was 2.76 days early, the EOS was delayed by 2.6 days, and the LOS was extended by 5.36 days.

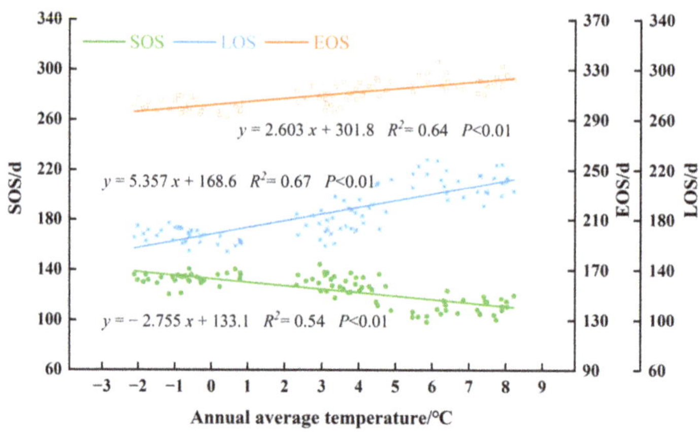

Figure 19. Responses of forest phenology to the annual mean temperature in different forest types in northeast China from 2011 to 2020.

The response of the phenological parameters to the annual average temperature changes in different forest areas are shown in Figure 20. Overall, the annual average temperature had significant effects on the three phenological parameters in all forest types ($P < 0.01$). The models and the coefficients shown in the figure passed the significance test at the 95% level of significance by using SPSS. The response of BF to the annual average temperature was the most evident, followed by MF, both of which were significantly higher than those of CF. With the increase in annual average temperature, the phenology of different forest types was characterized by early SOS, delayed EOS, and prolonged LOS. In terms of the SOS, when the temperature increases by 1 °C, BF advanced 4.03 days, MF advanced 3.59 days, CF advanced 1.69 days. The LOS, affected by the variation of SOS and EOS, had the most obvious response to the annual average temperature. When the average annual temperature increased by 1 °C, the EOS of BF was delayed by 3.51 days, the EOS of MF was delayed by 3.50 days, and the EOS of CF was delayed by 1.76 days. The lengthening of the growing season of BF is most obvious with a rate of 7.53 days when the temperature increased 1 °C. The LOS of MF extended 7.09 days with the temperature increasing by 1 °C, while the LOS of CF area was prolonged at a rate of 3.45 d per 1 °C increase.

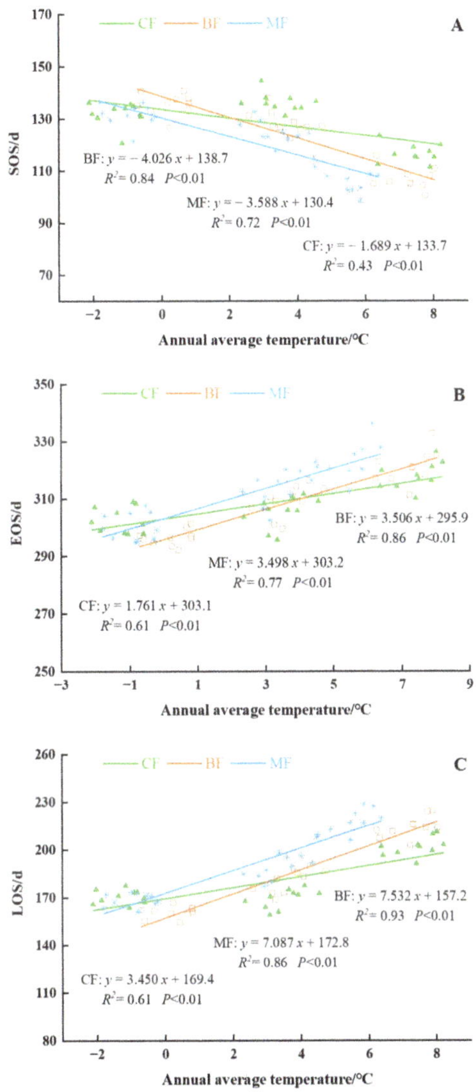

Figure 20. Responses of phenology to temperature in different forest types in northeast China from 2011 to 2020. (**A**) SOS of different forest types; (**B**) EOS of different forest types; (**C**) LOS of different forest types.

The response of the SOS and EOS to the pre-season temperature changes are shown in Figure 21. An average temperature of the past December to the current May was linearly related to the current SOS with a rate of −2.23 d/1 °C ($P < 0.01$). This means that when the pre-season temperature increases by 1 °C, the SOS was 2.23 days earlier. Similar results can be found for EOS. The EOS was obviously correlated with the average temperature of the current June to November at a rate of advance of 3.083 d/1 °C ($P < 0.01$). With the increase in the average temperature of the current June to November, the EOS of the forest tended to delay, and the EOS was delayed by 3.083 days. This result was similar to the results of the annual temperature with the SOS beginning 2.76 days earlier and EOS delayed by 2.6 days when the average annual temperature increased by 1 °C.

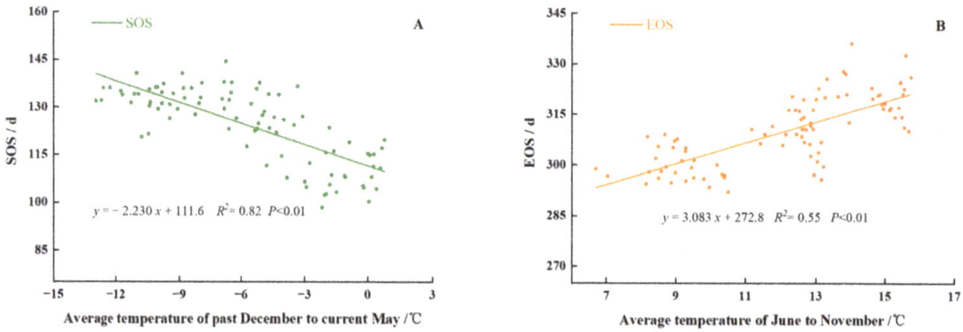

Figure 21. Responses of phenology to pre-season temperature in northeast China from 2011 to 2020. (**A**) is SOS; (**B**) is EOS.

3.5. *Time-Lag Effect of Climatic Change on the Forest Phenology*

Over the past decades, far more studies have found that the response of vegetation phenology to climatic factors have time-lag effects [58], that the phenology of vegetation could occur and change only after a period of cumulative transformation under specific climatic conditions. In addition, many scholars have demonstrated that the variation of vegetation was correlative with the preseason climatic changes. In order to study the response mechanism of forest vegetation phenology to climate change, we analyzed the correlation between the forest phenological parameters, monthly mean temperature, and the monthly accumulative precipitation of preseason. Compared with the simple linear correlation coefficient, the partial correlation coefficient can better reflect the relationship between the two variables. In this study, we investigated the correlations between the SOS and temperature and the precipitation from November of the previous year to May of the current year. The correlations between the EOS and temperature and precipitation from May to November of the current year. In order to avoid the impact of climate change in non-forest areas, monthly temperature and precipitation were extracted from areas consistent with forest distribution.

3.5.1. Time-Lag Effect of Climatic Change on Forest Phenology

The partial correlation coefficients between the SOS of forest and temperature were calculated and the results are shown in Figure 22A. The SOS of the forest had a significant negative correlation with pre-season temperature measured from the December of the previous year ($P < 0.01$); the temperature in December of the previous year and January of the current year had the greatest correlation with SOS. It could be concluded that the temperature in the winter of the previous year largely affected the SOS, which was more significant than the spring temperature. In addition, the SOS had a significantly negative correlation with precipitation in the November of the previous year and April and May of the current year. The SOS had a strongly negative correlation with precipitation in the November of the previous year (r = −0.36, $P < 0.01$) and April of the same year (r = −0.29, $P < 0.01$). However, this relationship was not significant for other months. It was notable that the SOS had a weakly positive correlation with precipitation in January of the current year, indicating that increased precipitation at the beginning of year may delay the SOS.

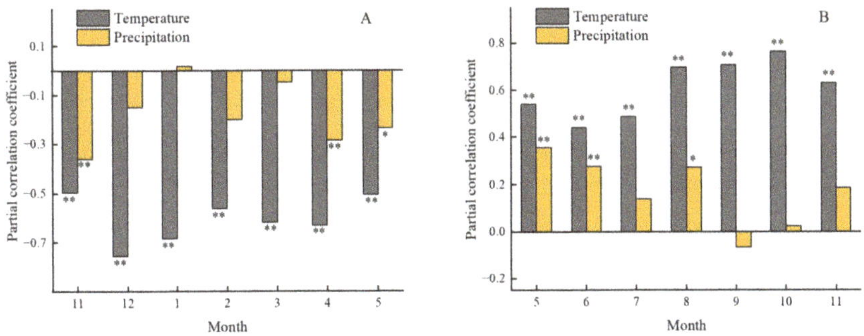

Figure 22. Partial correlation coefficient between forest phenological parameters and temperature and precipitation. (**A**) is SOS; (**B**) is EOS. Note: * $p < 0.05$, ** $p < 0.01$.

The calculated results of the partial correlation between the EOS and temperature and precipitation can be found in Figure 22B. The EOS had a significant positive correlation with temperature in seven months of the current year, which meant that the higher temperature would delay the EOS. The temperature from August to October of the current year had stronger correlation with EOS than the summer. Aside from September of the current year, other monthly precipitation had positive correlation with EOS, but only the relationship between EOS and precipitation in May, June, and August passed significance test, which indicated that more precipitation in summer would lengthen the time of the growing season of forest and lead to the delay of the event of EOS.

3.5.2. Time-Lag Effect of Climatic Change on the Phenology of Different Forest Types

We further discussed the relationship between the phenology of different forest types and climatic factors. Figure 23 demonstrated the relationship between the SOS of different forest types and temperature and precipitation. For the three forest types, there was a significant negative correlation with the pre-seasonal temperature, while the correlation coefficients between the SOS of all forest types and monthly precipitation did not pass the significance test. These results imply that the temperature increase in winter and spring, could contribute to the advanced SOS of all three forest types. BF is more sensitive to the variation of temperature than others. Compared with the temperature, precipitation also had a negative relationship with the SOS of three tree types, but this trend was not significant and did not pass the significance test. However, a very interesting fact is that the precipitation of January and March of the current year had a passive effect on the SOS. It meant that more rain in these two months would delay the beginning of the growing season of forest. A possible reason for this is that such early precipitation can slow down warm weather and thus lead to a delay in plant growth.

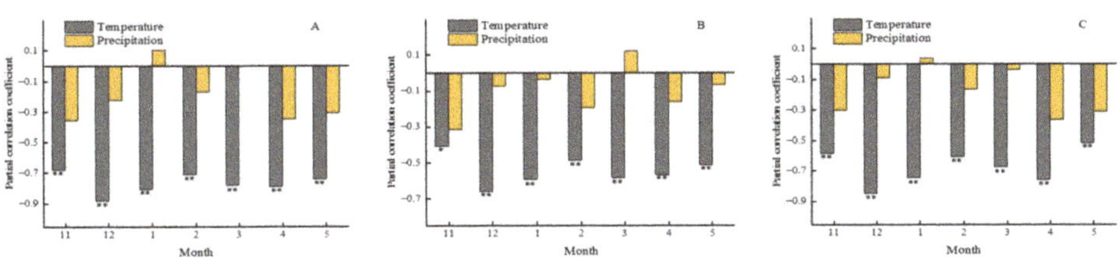

Figure 23. Partial correlation coefficient between the SOS of different forest types and temperature and precipitation. (**A**) Broadleaf Forest, (**B**) Coniferous Forest, (**C**) Mixed Forest. Note: * $p < 0.05$, ** $p < 0.01$.

Figure 24 shows the response of the EOS in different forest types to the monthly temperature and precipitation. We found that the EOS of all three types had a significant positive correlation with temperature from May to November of the current year. In particular, the higher temperature in autumn could result in the prolonged EOS of all. In terms of precipitation, the EOS of BF had a positive correlation with precipitation in May, June, and August ($P < 0.01$), while the EOS of CF had positive correlation with precipitation only in May. The results may indicate that increased precipitation in spring and summer could delay the EOS of BF. Similarly, it can also be found that the precipitation in September of current year had a negative effect on the EOS. This meant that more rain in this month would speed up the end of the growing season of the forest. This may be related to the impact of precipitation on temperature.

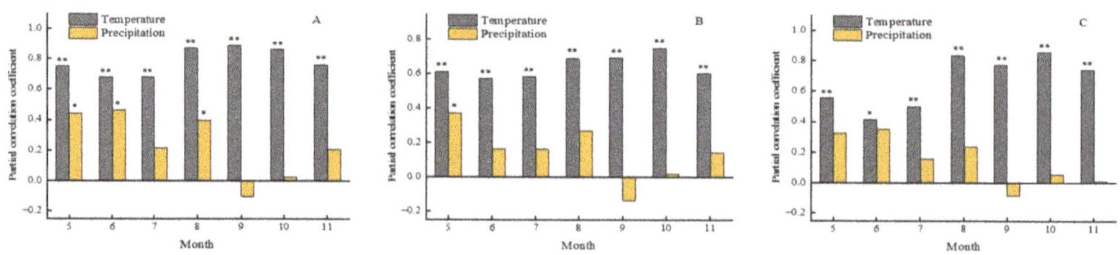

Figure 24. Partial correlation coefficient between EOS of different forest types and temperature and precipitation. (**A**) Broadleaf Forest, (**B**) Coniferous Forest, (**C**) Mixed Forest. Note: * $p < 0.05$, ** $p < 0.01$.

4. Discussion

4.1. Variation of Forest Phenology in the NEC

Vegetation phenology is an important indicator of monitoring the vegetation dynamics and changes in the climate and natural environment. More and more research on phenology by using remote sensing are emerging. The normalized difference vegetation index (NDVI) derived from remote-sensing has been widely used to detect the SOS and EOS by using NDVI time-series data. In this study, we used the MODIS NDVI products to extract the SOS and EOS of northeast China. Compared with other results, the SOS and EOS extracted results are consistent with other research. Zhao et al. extracted the SOS and EOS by using GIMMS NDVI3g dataset and concluded that the SOS ranged from 110 days to 150 days and EOS ranged from 270 days to 320 days [31]. Yu et al. concluded that the SOS in northeast China from 1982 to 2015 ranged from the 100th DOY to the 140th DOY of the year, the EOS in northeast China from 1982 to 2015 ranged from the 280th DOY to the 320th DOY [30]. These results coincide with the results of this study, which shows that the extracted phenology has a certain reference value and reliability.

Most previous studies chose fixed thresholds to extract vegetation phenology, which might result in some deviations. Li et al. defined SOS and EOS as 20% of annual LAI amplitude by using the dynamic method and found that the selection of the threshold itself has certain experience, which would affect the accuracy of phenology extraction to a certain extent in the northeast China [59]. You et al. selected the 50% as the threshold to determine the SOS and EOS of vegetation and concluded that the average of LOS was 135.2 days and significantly increased with a slope of 2.94 days per decade in the Upper Amur (Heilongjiang) River Basin in northeast Asia [60]. In this research, we mainly discussed the spatiotemporal variation of forest phenology in northeastern China from 2011 to 2020. The results of the variation of forest phenology were described in this study, which showed the varying degrees of fluctuation in the NEC from 2011 to 2020. Generally, our results demonstrated that the average SOS of the forest was primarily distributed from 90th to 135th DOY with an early trend, which is consistent with numerous former studies. Zhao et al. discovered that the mean SOS dates ranged from 115th to 140th DOY in the Changbai Mountains, Lesser Khingan Mountains, and Greater Khingan Mountains [31]. Guo et al. concluded that early SOS was distributed between 100th and 130th DOY in forest areas in the NEC from 1982 to 2014 [61]. Tang et al. found that the SOS of forests ranged between 105th to 130th DOY in Greater Khingan Mountains [62]. The EOS of forest largely displayed from 300th to 330th DOY. Qiu et al. found that EOS occurred between DOY 260 and 270 in the Greater Khingan Mountains, while the EOS of BF in southern Lesser Khingan Mountains and Changbai Mountains occurred between 280th and 300th DOY [63]. Liu et al. found that the EOS of deciduous needle-leaf forest was earlier than other vegetations in temperate China [64]. However, different studies used different datasets and methods, which resulted in differences from each other. In spite of this, all studies concluded that the EOS was advancing earlier in the Greater Khingan Mountains, while the forest in the southwestern Changbai Mountains near the coast had relatively late EOS dates. The possible reason could be that there are relatively high temperatures at lower latitude, which is beneficial for delaying leaf senescence.

From a spatial point of view, the phenology of all three forest types displayed significant spatial heterogeneity as well as differences between each other with increasing latitude in the NEC. From southeast to northwest in the study area, the multiyear average SOS advanced at a rate of 2.33 days per latitude and the multiyear mean EOS was delayed at a rate of 2.22 days per latitude, respectively, which mainly resulted in the difference of LOS. As a whole, the LOS of forests was illustrated to be longer along coastal areas at low latitude and shorter in inland areas at high latitude.

Many previous studies concentrated on the variation of mean phenology at a regional scale, while ignoring the spatial heterogeneity among different forest types. In this study, we also analyzed the changes of phenological parameters in three forest types in NEC and demonstrated that the variations of forest phenology were varied across different forest types. We fitted the relationship between phenology parameters and climate factors by using the least squares method and the slope of the linear function can be used to indicate the changing rate. Then, we compared the slope between the latitude, annual average temperature, and annual cumulative precipitation and phenology parameters of different forest types. Overall, it was pointed out that all three types of forest displayed the sightly advanced SOS and delayed EOS. Yu et al. found that the SOS of deciduous needle-leaf forests was advanced by 0.24 d/a, while the EOS was delayed at a rate of 0.36 d/a from 1982 to 2015 [30]. Zhao et al. concluded that the EOS of BF in eastern Liaoning was delayed 0.23 d/a from 1982 to 2012 [31]. Our findings were in line with previous studies. With the influence of SOS and EOS, the LOS of all forests showed a prolonged trend, with the changing rates of 0.76 d/a. To be more specific, the changing range of BF was the largest, followed by MF and CF.

4.2. The Relationship between Forest Phenology and Climatic Factors

With the increasing concern of global climate change, many studies have proposed that climate change has a substantial impact on vegetation phenology, and the variation of vegetation phenology may also feed back to climatic factors, such as temperature and precipitation. Previous research has proven that the temperature is the most important factor for the growth of vegetation. It is noteworthy that warming temperatures in spring may have an impact on the advance of SOS, especially in the Northern Hemisphere [12]. In this study, we analyzed the three phenological parameters of forest responses to the temperature and found that the SOS of forests had negative correlations with temperature, as the SOS was advanced by 2.76 days with an increase of 1 °C. It can be concluded that the warmer temperatures in spring would stimulate an early emergence from winter dormancy, resulting in an advanced phenology in the forest [13]. In addition, the average EOS of forests in NEC were delayed at a rate of 2.60 d/1 °C, which is consistent with other research. Allison et al. demonstrated that air temperature could reasonably predict the timing of leaf senescence for deciduous forests throughout the Northern Hemisphere [65]. In addition to air temperature, precipitation also contributes to the timing of forest phenological events. The SOS was advanced 2.90 d if the annual cultivate precipitation increased by 100 mm, while the EOS showed a significant positive correlation with precipitation and the LOS of forests was prolonged by 7.10 d. Tang et al. studied the relationship between the phenology and climatic factors, and concluded that the changes of both temperature and precipitation resulted in extended LOS in forest region in the Greater Khingan Mountain Area [62].

We further explored the responses of the phenology in different forest types to variations on precipitation and temperature. As a whole, BF were largely sensitive to precipitation and temperature changes, followed by MF and CF. The reason for this phenomenon may be that BF are widely distributed in the southwestern Changbai Mountains near the coast, where the temperature is warmer and humidity is higher, contributing to a higher demand for photosynthesis and water transpiration [58]. Generally, the ecosystems at high latitudes display significant correlation with temperature, while temperate areas are more correlated with precipitation [33]. Liu et al. concluded that evergreen needle-leaf forests had a later EOS due to increased temperature and precipitation based on the time-series GIMMS NDVI records from 1982 to 2011 [64].

4.3. Partial Correlation Analysis between Forest Phenology and Climatic Factors

Over the past decades, many researchers have revealed the time-lag effect while studying the responses of vegetation phenology to climatic factors [33,66]. Wu et al. proposed that the time-lag effects of different vegetation types significantly varied from the same climatic factor and that the same vegetation type also had different responses to the different climatic factors [58]. The results in this study show that the increased temperature was the main factor in delaying the SOS and EOS, and the warmer temperature in winter had a greater impact on SOS than in the spring. Fu et al. discussed the spatial correlation between the growing degree days (GDD) requirement of different vegetation types and temperature and precipitation in the winter of previous year and concluded that cold winter temperatures mainly effected the GDD, which was largely determined by the SOS [67]. Hou et al. analyzed the partial correlation between the temperature and vegetation phenology adding precipitation as a control variable and concluded that the SOS had a negative relationship with the spring temperature, and an increasing daytime temperature ensured the heat required for vegetation growth advanced the SOS [68]. In addition, compared with the summer, the warmer autumn seems to have a greater impact on the EOS. It could be concluded that the warmer temperature would result in the later autumn, which would prolong the time of both respiration and photosynthesis and delay leaf senescence [13]. Tang et al. discussed the time-lag effect of climatic factors on the forest phenology in the Greater Khingan Mountain Area and confirmed that less precipitation and warmer springs result in advanced SOS, while cumulative summer temperatures played a major role in prolonged EOS [62]. The reason that this phenomenon occurred was that the CF, largely distributed in the middle and high latitudes in the Northern Hemisphere, has a strong demand for water, while temperature and solar radiation largely affected their growth [58].

The variation of monthly precipitation weakly affected forest phenology, and while the impact of precipitation on phenology varied from month to month, the increased precipitation in summer led to delayed EOS. Huang et al. studied the effects of rain-use efficiency on vegetation phenology of the Songnen Plain and concluded that increasing precipitation would delay the EOS, particularly in the forest areas in the north, where the vegetation in arid and semiarid areas would be more sensitive to precipitation [69]. Yun et al. concluded that the increase in precipitation in winter affects the trends of vegetation growth in the spring, even in temperature-limited ecosystems [70]. The phenological variation of different forest types has a similar response to climatic factors, but BF was more sensitive to climate change. Clinton et al. studied the association of vegetation phenology with precipitation and temperature on a global scale and proposed that the boreal forest had the lowest correlations with precipitation, indicating that pre-season humidity may have stronger correlations with boreal forest than the precipitation of the same season [33].

5. Conclusions

In this study, we used the dynamic threshold method combined with ground-based data to extract the phenology of forests, using MODIS NDVI time-series data, reconstructed with the S-G filter, in the northeast China from 2011 to 2020. The results concluded that there was a relationship between threshold and latitudes, and the suitable threshold of SOS increased at a rate of 3%/1 °C, while the suitable threshold of EOS increased 2%/1 °C. The suitable threshold for detecting phenology occurred in spatial heterogeneity and varied between latitudes. Then, the spatio-temporal variations of forest phenology were discussed. The SOS of forest in northeast China was mainly concentrated between early April to mid-May and showed the spatial characteristics of occurring earlier in the south and later in the north. The EOS of forests was generally later than the end of October and showed the spatial characteristics of occurring earlier in the north and later in the south. The LOS of forests mainly ranged between 170th to 210th DOY, whereas the a longer LOS was seen in the coastal areas at low latitudes and a shorter LOS was seen in inland areas at high latitudes. In addition, the SOS of forests were advanced at a rate of 0.29 d/a, while the EOS were delayed at a rate of 0.47 d/a, so the LOS of forests had a significant extension during the past decade. Finally, the responding mechanism between the phenological change and climatic factors was considered. It was found that all forest types were significantly sensitive to the variation of temperature. Pre-seasonal temperature, especially during the previous winter had a significant effect on the SOS of the current year. The autumn temperatures of the current year were the main climatic factors affecting EOS. As a whole, the broadleaf forests and mixed forests were the most sensitive to climatic factors, followed by the conifer forest. This research can provide a reference for understanding the phenological change characteristics of the boreal forest ecosystem and reveal the phenological response mechanism of the boreal forest ecosystem against the backdrop of global warming.

Author Contributions: X.Y. conceived and designed the experiments; Y.L. and W.Z. performed the experiments and analyzed the data; X.Y. and W.Z. wrote the paper; X.Y. and W.F. reviewed and edited the paper. Y.L. and W.Z. contributed equally to this work. All authors have read and agreed to the published version of the manuscript.

Funding: This research was funded by the National Natural Science Foundation of China, grant number 31971580, 31870621; the Fundamental Research Funds for the Central Universities of China, grant number 2572019BA10, 2572021BA08, 2572019CP12; the China Postdoctoral Science Foundation, grant number 2019M661239.

Data Availability Statement: Not applicable.

Acknowledgments: We acknowledge the National Aeronautics and Space Administration (NASA) for providing the MODIS NDVI data (https://search.earthdata.nasa.gov, accessed on 22 March 2022); the meteorological data received from the National Oceanic and Atmospheric Administration (NOAA). (https://www.ncei.noaa.gov/data/global-summary-of-the-day, accessed on 22 March 2022); the land cover dataset provided by the Finer Resolution Observation and Monitoring of Global Land Cover (FROM-GLC) developed by the Pro. Peng Gong group at Tsinghua University (http://data.ess.tsinghua.edu.cn, accessed on 22 March 2022); the observation dataset provided by the Chinese Phenological Observation Network (CPON). (http://www.geodata.cn, accessed on 22 March 2022). We also acknowledge the data support from the National Earth System Science Data Center, National Science & Technology Infrastructure of China (http://www.geodata.cn, accessed on 22 March 2022).

Conflicts of Interest: The authors declare no conflict of interest.

References

1. Piao, S.; Liu, Q.; Chen, A.; Janssens, I.A.; Fu, Y.; Dai, J.; Liu, L.; Lian, X.; Shen, M.; Zhu, X. Plant phenology and global climate change: Current progresses and challenges. *Glob. Chang. Biol.* **2019**, *25*, 1922–1940. [CrossRef] [PubMed]
2. Richardson, A.D.; Keenan, T.F.; Migliavacca, M.; Ryu, Y.; Sonnentag, O.; Toomey, M. Climate change, phenology, and phenological control of vegetation feedbacks to the climate system. *Agric. For. Meteorol.* **2013**, *169*, 156–173. [CrossRef]
3. Wang, H.; Ge, Q.; Rutishauser, T.; Dai, Y.; Dai, J. Parameterization of temperature sensitivity of spring phenology and its application in explaining diverse phenological responses to temperature change. *Sci. Rep.* **2015**, *5*, 8833. [CrossRef] [PubMed]
4. Piao, S.; Wang, X.; Park, T.; Chen, C.; Lian, X.; He, Y.; Bjerke, J.W.; Chen, A.; Ciais, P.; Tommervik, H.; et al. Characteristics, drivers and feedbacks of global greening. *Nat. Rev. Earth Environ.* **2020**, *1*, 14–27. [CrossRef]
5. Pezzini, F.; Ranieri, B.; Brandão, D.; Fernandes, G.; Quesada, M.; Espírito-Santo, M.; Jacobi, C. Changes in tree phenology along natural regeneration in a seasonally dry tropical forest. *Plant Biosyst. Int. J. Dealing Aspects Plant Biosyst.* **2014**, *148*, 965–974. [CrossRef]
6. Yang, Y.; Shi, Y.; Sun, W.; Chang, J.; Zhu, J.; Chen, L.; Wang, X.; Guo, Y.; Zhang, H.; Yu, L.; et al. Terrestrial carbon sinks in China and around the world and their contribution to carbon neutrality. *Sci. China Life Sci.* **2022**, *52*, 52. [CrossRef]
7. Shiga, Y.P.; Michalak, A.M.; Fang, Y.; Schaefer, K.; Andrews, A.E.; Huntzinger, D.H.; Schwalm, C.R.; Thoning, K.; Wei, Y. Forests dominate the interannual variability of the North American carbon sink. *Environ. Res. Lett.* **2018**, *13*, 084015. [CrossRef]
8. Li, H.; Wang, S.; Bai, X.; Luo, W.; Tang, H.; Cao, Y.; Wu, L.; Chen, F.; Li, Q.; Zeng, C.; et al. Spatiotemporal distribution and national measurement of the global carbonate carbon sink. *Sci. Total Environ.* **2018**, *643*, 157–170. [CrossRef]
9. Han, Q.; Wang, T.; Jiang, Y.; Fischer, R.; Li, C. Phenological variation decreased carbon uptake in European forests during 1999–2013. *For. Ecol. Manag.* **2018**, *427*, 45–51. [CrossRef]
10. Jin, J.; Wang, Y.; Zhang, Z.; Magliulo, V.; Jiang, H.; Cheng, M. Phenology Plays an Important Role in the Regulation of Terrestrial Ecosystem Water-Use Efficiency in the Northern Hemisphere. *Remote Sens.* **2017**, *9*, 664. [CrossRef]
11. Nasahara, K.N.; Nagai, S. Development of an in situ observation network for terrestrial ecological remote sensing: The Phenological Eyes Network (PEN). *Ecol. Res.* **2015**, *30*, 211–223. [CrossRef]
12. Menzel, A.; Yuan, Y.; Matiu, M.; Sparks, T.; Scheifinger, H.; Gehrig, R.; Estrella, N. Climate change fingerprints in recent European plant phenology. *Glob. Chang. Biol.* **2020**, *26*, 2599–2612. [CrossRef]
13. Keenan, T.F.; Gray, J.; Friedl, M.A.; Toomey, M.; Bohrer, G.; Hollinger, D.Y.; Munger, J.W.; O'Keefe, J.; Schmid, H.P.; SueWing, I.; et al. Net carbon uptake has increased through warming-induced changes in temperate forest phenology. *Nat. Clim. Chang.* **2014**, *4*, 598–604. [CrossRef]
14. Rosbakh, S.; Hartig, F.; Sandanov, D.V.; Bukharova, E.V.; Miller, T.K.; Primack, R.B. Siberian plants shift their phenology in response to climate change. *Glob. Chang. Biol.* **2021**, *27*, 4435–4448. [CrossRef]
15. Zeng, L.; Wardlow, B.D.; Xiang, D.; Hu, S.; Li, D. A review of vegetation phenological metrics extraction using time-series, multispectral satellite data. *Remote Sens. Environ.* **2020**, *237*, 111511. [CrossRef]
16. Snyder, K.A.; Huntington, J.L.; Wehan, B.L.; Morton, C.G.; Stringham, T.K. Comparison of Landsat and Land-Based Phenology Camera Normalized Difference Vegetation Index (NDVI) for Dominant Plant Communities in the Great Basin. *Sensors* **2019**, *19*, 1139. [CrossRef]
17. Adole, T.; Dash, J.; Atkinson, P.M. Characterising the land surface phenology of Africa using 500 m MODIS EVI. *Appl. Geogr.* **2018**, *90*, 187–199. [CrossRef]
18. Li, X.; Du, H.; Zhou, G.; Mao, F.; Zhang, M.; Han, N.; Fan, W.; Liu, H.; Huang, Z.; He, S.; et al. Phenology estimation of subtropical bamboo forests based on assimilated MODIS LAI time series data. *ISPRS J. Photogramm. Remote Sens.* **2021**, *173*, 262–277. [CrossRef]
19. Amin, E.; Belda, S.; Pipia, L.; Szantoi, Z.; El Baroudy, A.; Moreno, J.; Verrelst, J. Multi-Season Phenology Mapping of Nile Delta Croplands Using Time Series of Sentinel-2 and Landsat 8 Green LAI. *Remote Sens.* **2022**, *14*, 1812. [CrossRef]
20. Didan, K. MOD13Q1 MODIS/Terra Vegetation Indices 16-Day L3 Global 250m SIN Grid V006. *NASA EOSDIS Land Processes DAAC* **2015**, *10*, 415. [CrossRef]
21. Zhang, G.P.; Qi, M. Neural network forecasting for seasonal and trend time series. *Eur. J. Oper. Res.* **2005**, *160*, 501–514. [CrossRef]
22. Lara, B.; Gandini, M. Assessing the performance of smoothing functions to estimate land surface phenology on temperate grassland. *Int. J. Remote Sens.* **2016**, *37*, 1801–1813. [CrossRef]
23. Workie, T.G.; Debella, H.J. Climate change and its effects on vegetation phenology across ecoregions of Ethiopia. *Glob. Ecol. Conserv.* **2018**, *13*, e00366. [CrossRef]
24. Liu, J.; Zhu, W.; Atzberger, C.; Zhao, A.; Pan, Y.; Huang, X. A phenology-based method to map cropping patterns under a wheat-maize rotation using remotely sensed time-series data. *Remote Sens.* **2018**, *10*, 1203. [CrossRef]
25. Tian, J.; Zhu, X.; Chen, J.; Wang, C.; Shen, M.; Yang, W.; Tan, X.; Xu, S.; Li, Z. Improving the accuracy of spring phenology detection by optimally smoothing satellite vegetation index time series based on local cloud frequency. *ISPRS J. Photogramm. Remote Sens.* **2021**, *180*, 29–44. [CrossRef]
26. Xin, Q.; Li, J.; Li, Z.; Li, Y.; Zhou, X. Evaluations and comparisons of rule-based and machine-learning-based methods to retrieve satellite-based vegetation phenology using MODIS and USA National Phenology Network data. *Int. J. Appl. Earth Obs. Geoinf.* **2020**, *93*, 102189. [CrossRef]

27. Kang, W.; Wang, T.; Liu, S. The response of vegetation phenology and productivity to drought in semi-arid regions of Northern China. *Remote Sens.* **2018**, *10*, 727. [CrossRef]
28. Ding, M.; Li, L.; Zhang, Y.; Sun, X.; Liu, L.; Gao, J.; Wang, Z.; Li, Y. Start of vegetation growing season on the Tibetan Plateau inferred from multiple methods based on GIMMS and SPOT NDVI data. *J. Geogr. Sci.* **2015**, *25*, 131–148. [CrossRef]
29. Xu, L.; Niu, B.; Zhang, X.; He, Y. Dynamic Threshold of Carbon Phenology in Two Cold Temperate Grasslands in China. *Remote Sens.* **2021**, *13*, 574. [CrossRef]
30. Yu, L.; Liu, T.; Bu, K.; Yan, F.; Yang, J.; Chang, L.; Zhang, S. Monitoring the long term vegetation phenology change in Northeast China from 1982 to 2015. *Sci. Rep.* **2017**, *7*, 14770. [CrossRef] [PubMed]
31. Zhao, J.; Wang, Y.; Zhang, Z.; Zhang, H.; Guo, X.; Yu, S.; Du, W.; Huang, F. The Variations of Land Surface Phenology in Northeast China and Its Responses to Climate Change from 1982 to 2013. *Remote Sens.* **2016**, *8*, 400. [CrossRef]
32. Fu, Y.; He, H.S.; Zhao, J.; Larsen, D.R.; Zhang, H.; Sunde, M.G.; Duan, S. Climate and Spring Phenology Effects on Autumn Phenology in the Greater Khingan Mountains, Northeastern China. *Remote Sens.* **2018**, *10*, 449. [CrossRef]
33. Clinton, N.; Yu, L.; Fu, H.; He, C.; Gong, P. Global-Scale Associations of Vegetation Phenology with Rainfall and Temperature at a High Spatio-Temporal Resolution. *Remote Sens.* **2014**, *6*, 7320–7338. [CrossRef]
34. Wolf, S.; Keenan, T.F.; Fisher, J.B.; Baldocchi, D.D.; Desai, A.R.; Richardson, A.D.; Scott, R.L.; Law, B.E.; Litvak, M.E.; Brunsell, N.A.; et al. Warm spring reduced carbon cycle impact of the 2012 US summer drought. *Proc. Natl. Acad. Sci. USA* **2016**, *113*, 5880–5885. [CrossRef]
35. Lian, X.; Piao, S.L.; Li, L.Z.X.; Li, Y.; Huntingford, C.; Ciais, P.; Cescatti, A.; Janssens, I.A.; Penuelas, J.; Buermann, W.; et al. Summer soil drying exacerbated by earlier spring greening of northern vegetation. *Sci. Adv.* **2020**, *6*, eaax0255. [CrossRef]
36. Jeong, S.J.; Chang-Hoi, H.O.; Gim, H.J.; Brown, M.E. Phenology shifts at start vs. end of growing season in temperate vegetation over the Northern Hemisphere for the period 1982–2008. *Glob. Chang. Biol.* **2011**, *17*, 2385–2399. [CrossRef]
37. Liu, Q.; Piao, S.; Janssens, I.A.; Fu, Y.; Peng, S.; Lian, X.; Ciais, P.; Myneni, R.B.; Peñuelas, J.; Wang, T. Extension of the growing season increases vegetation exposure to frost. *Nat. Commun.* **2018**, *9*, 426. [CrossRef]
38. Wang, C.; Jiang, Q.O.; Deng, X.; Lv, K.; Zhang, Z. Spatio-Temporal Evolution, Future Trend and Phenology Regularity of Net Primary Productivity of Forests in Northeast China. *Remote Sens.* **2020**, *12*, 3670. [CrossRef]
39. Piao, S.; Wang, X.; Ciais, P.; Zhu, B.; Wang, T.; Liu, J. Changes in satellite-derived vegetation growth trend in temperate and boreal Eurasia from 1982 to 2006. *Glob. Chang. Biol.* **2011**, *17*, 3228–3239. [CrossRef]
40. Cong, N.; Piao, S.; Chen, A.; Wang, X.; Lin, X.; Chen, S.; Han, S.; Zhou, G.; Zhang, X. Spring vegetation green-up date in China inferred from SPOT NDVI data: A multiple model analysis. *Agric. For. Meteorol.* **2012**, *165*, 104–113. [CrossRef]
41. He, Z.; Du, J.; Chen, L.; Zhu, X.; Lin, P.; Zhao, M.; Fang, S. Impacts of recent climate extremes on spring phenology in arid-mountain ecosystems in China. *Agric. For. Meteorol.* **2018**, *260*, 31–40. [CrossRef]
42. Yao, R.; Wang, L.; Huang, X.; Guo, X.; Niu, Z.; Liu, H. Investigation of Urbanization Effects on Land Surface Phenology in Northeast China during 2001–2015. *Remote Sens.* **2017**, *9*, 66. [CrossRef]
43. Jiao, Y.; Bu, K.; Yang, J.; Li, G.; Shen, L.; Liu, T.; Yu, L.; Zhang, S.; Zhang, H. Biophysical Effects of Temperate Forests in Regulating Regional Temperature and Precipitation Pattern across Northeast China. *Remote Sens.* **2021**, *13*, 4767. [CrossRef]
44. Widagdo, F.R.A.; Li, F.R.; Xie, L.F.; Dong, L.H. Intra- and inter-species variations in carbon content of 14 major tree species in Northeast China. *J. For. Res.* **2021**, *32*, 2545–2556. [CrossRef]
45. Beck, H.E.; Zimmermann, N.E.; McVicar, T.R.; Vergopolan, N.; Berg, A.; Wood, E.F. Present and future Köppen-Geiger climate classification maps at 1-km resolution. *Sci. Data* **2018**, *5*, 180214. [CrossRef]
46. Peng, C.; Zhou, X.; Zhao, S.; Wang, X.; Zhu, B.; Piao, S.; Fang, J. Quantifying the response of forest carbon balance to future climate change in Northeastern China: Model validation and prediction. *Glob. Planet.* **2009**, *66*, 179–194. [CrossRef]
47. Nse, O.U.; Okolie, C.J.; Nse, V.O. Dynamics of land cover, land surface temperature and NDVI in Uyo City, Nigeria. *Sci. Afr.* **2020**, *10*, e00599. [CrossRef]
48. Jönsson, P.; Eklundh, L. TIMESAT—A program for analyzing time-series of satellite sensor data. *Comput. Geosci.* **2004**, *30*, 833–845. [CrossRef]
49. Chen, J. A simple method for reconstructing a high-quality NDVI time-series data set based on the Savitzky-Golay filter. *Remote Sens. Environ.* **2004**, *91*, 332–344. [CrossRef]
50. Cai, Z.; Jönsson, P.; Jin, H.; Eklundh, L. Performance of smoothing methods for reconstructing NDVI time-series and estimating vegetation phenology from MODIS data. *Remote Sens.* **2017**, *9*, 1271. [CrossRef]
51. Li, S.; Xu, L.; Jing, Y.; Yin, H.; Li, X.; Guan, X. High-quality vegetation index product generation: A review of NDVI time series reconstruction techniques. *Int. J. Appl. Earth Obs. Geoinf.* **2021**, *105*, 102640. [CrossRef]
52. Ding, M.; Chen, Q.; Li, L.; Zhang, Y.; Wang, Z.; Liu, L.; Sun, X. Temperature dependence of variations in the end of the growing season from 1982 to 2012 on the Qinghai–Tibetan Plateau. *GISci. Remote Sens.* **2016**, *53*, 147–163. [CrossRef]
53. Mesa-Mingorance, J.L.; Ariza-López, F.J. Accuracy Assessment of Digital Elevation Models (DEMs): A Critical Review of Practices of the Past Three Decades. *Remote Sens.* **2020**, *12*, 2630. [CrossRef]
54. Zhu, Z.; Zhang, J.; Yang, Z.; Aljaddani, A.H.; Cohen, W.B.; Qiu, S.; Zhou, C. Continuous monitoring of land disturbance based on Landsat time series. *Remote Sens. Environ.* **2020**, *238*, 111116. [CrossRef]
55. Yang, X.; He, P.; Yu, Y.; Fan, W. Stand Canopy Closure Estimation in Planted Forests Using a Geometric-Optical Model Based on Remote Sensing. *Remote Sens.* **2022**, *14*, 1983. [CrossRef]

56. Yang, X.; Hao, Y.; Cao, W.; Yu, X.; Hua, L.; Liu, X.; Yu, T.; Chen, C. How Does Spring Phenology Respond to Climate Change in Ecologically Fragile Grassland? A Case Study from the Northeast Qinghai-Tibet Plateau. *Sustainability* **2021**, *13*, 12781. [CrossRef]
57. Ren, G.; Ding, Y.; Zhao, Z.; Zheng, J.; Wu, T.; Tang, G.; Xu, Y. Recent progress in studies of climate change in China. *Adv. Atmos. Sci.* **2012**, *29*, 958–977. [CrossRef]
58. Wu, D.; Zhao, X.; Liang, S.; Zhou, T.; Huang, K.; Tang, B.; Zhao, W. Time-lag effects of global vegetation responses to climate change. *Glob. Chang. Biol.* **2015**, *21*, 3520–3531. [CrossRef]
59. Li, Z.; Bo, Y.; He, Y. Comparison of Natural Vegetation Phenology Metrics from Remote Sensing LAI Products. *Remote Sens. Technol. Appl.* **2015**, *30*, 1103–1112. [CrossRef]
60. You, G.; Arain, M.A.; Wang, S.; McKenzie, S.; Xu, B.; He, Y.; Wu, D.; Lin, N.; Gao, J.; Jia, X. Inter-annual Climate Variability and Vegetation Dynamic in the Upper Amur (Heilongjiang) River Basin in Northeast Asia. *Environ. Res. Commun.* **2020**, *2*, 061003. [CrossRef]
61. Guo, J.; Hu, Y. Spatiotemporal Variations in Satellite-Derived Vegetation Phenological Parameters in Northeast China. *Remote Sens.* **2022**, *14*, 705. [CrossRef]
62. Tang, H.; Li, Z.; Zhu, Z.; Chen, B.; Zhang, B.; Xin, X. Variability and Climate Change Trend in Vegetation Phenology of Recent Decades in the Greater Khingan Mountain Area, Northeastern China. *Remote Sens.* **2015**, *7*, 11914–11932. [CrossRef]
63. Yue, Q.; ZHANG, L.; Deqin, F. Spatio-temporal changes of net primary productivity and its response to phenology in northeast china during 2000–2015. *Int. Arch. Photogramm. Remote Sens. Spat. Inf. Sci.* **2018**, *42*, 1453–1459. [CrossRef]
64. Liu, Q.; Fu, Y.H.; Zeng, Z.; Huang, M.; Li, X.; Piao, S. Temperature, precipitation, and insolation effects on autumn vegetation phenology in temperate China. *Glob. Chang. Biol.* **2016**, *22*, 644–655. [CrossRef]
65. Gill, A.L.; Gallinat, A.S.; Sanders-DeMott, R.; Rigden, A.J.; Gianotti, D.J.S.; Mantooth, J.A.; Templer, P.H. Changes in autumn senescence in northern hemisphere deciduous trees: A meta-analysis of autumn phenology studies. *Ann. Bot.* **2015**, *116*, 875–888. [CrossRef]
66. Rammig, A.; Wiedermann, M.; Donges, J.F.; Babst, F.; von Bloh, W.; Frank, D.; Thonicke, K.; Mahecha, M.D. Coincidences of climate extremes and anomalous vegetation responses: Comparing tree ring patterns to simulated productivity. *J. Geophys. Res.: Biogeosci.* **2015**, *12*, 373–385. [CrossRef]
67. Fu, Y.H.; Piao, S.; Zhao, H.; Jeong, S.-J.; Wang, X.; Vitasse, Y.; Ciais, P.; Janssens, I.A. Unexpected role of winter precipitation in determining heat requirement for spring vegetation green-up at northern middle and high latitudes. *Glob. Chang. Biol.* **2014**, *20*, 3743–3755. [CrossRef]
68. Hou, X.; Gao, S.; Sui, X.; Liang, S.; Wang, M. Changes in Day and Night Temperatures and Their Asymmetric Effects on Vegetation Phenology for the Period of 2001–2016 in Northeast China. *Can. J. Remote Sens.* **2018**, *44*, 629–642. [CrossRef]
69. Huang, F.; Wang, P.; Chang, S.; Li, B. Rain use efficiency changes and its effects on land surface phenology in the Songnen Plain, Northeast China. In Proceedings of the Remote Sensing for Agriculture, Ecosystems, and Hydrology, Berlin, Germany, 10 October 2018; p. 107830E.
70. Yun, J.; Jeong, S.-J.; Ho, C.-H.; Park, C.-E.; Park, H.; Kim, J. Influence of winter precipitation on spring phenology in boreal forests. *Glob. Chang. Biol.* **2018**, *24*, 5176–5187. [CrossRef]

remote sensing

MDPI

Article

Estimation and Simulation of Forest Carbon Stock in Northeast China Forestry Based on Future Climate Change and LUCC

Jianfeng Sun [1], Ying Zhang [1,*], Weishan Qin [2] and Guoqi Chai [3]

[1] College of Economics and Management, Beijing Forestry University, Beijing 100083, China; sunjf@bjfu.edu.cn
[2] College of Resource and Environment Engineering, Ludong University, Yantai 264025, China; weishan93@126.com
[3] College of Forestry, Beijing Forestry University, Beijing 100083, China; chaigq@bjfu.edu.cn
* Correspondence: zhangyin@bjfu.edu.cn

Abstract: Forest carbon sinks (FCS) play an important role in mitigating global climate change, but there is a lack of more accurate, comprehensive, and efficient forest carbon stock estimates and projections for larger regions. By combining 1980–2020 land use data from the Northeast China Forestry (NCF) and climate change data under the Shared Socioeconomic Pathway (SSP), the land use and cover change (LUCC) of NCF in 2030 and 2050 and the FCS of NCF were estimated based on the measured data of forest carbon density. In general, the forest area of NCF has not yet recovered to the level of 1980. The temporal change in the FCS experienced a U-shaped trend of sharp decline to slow increase, with the inflection point occurring in 2010. If strict ecological conservation measures are implemented, the FCS of the NCF is expected to recover to the 1980 levels by 2050. We believe that the ecological priority (EP) scenario is the most likely and suitable direction for future development of the NCF. We also advocate for more scientific and stringent management measures for NCF natural forests to unlock the huge potential for forest carbon sequestration, which is important for China to meet its carbon neutrality commitments.

Keywords: forest carbon stocks; simulation; LUCC; climate change; spatiotemporal evolution

Citation: Sun, J.; Zhang, Y.; Qin, W.; Chai, G. Estimation and Simulation of Forest Carbon Stock in Northeast China Forestry Based on Future Climate Change and LUCC. *Remote Sens.* **2022**, *14*, 3653. https://doi.org/10.3390/rs14153653

Academic Editors: Huaqiang Du, Wenyi Fan, Mingshi Li, Weiliang Fan and Fangjie Mao

Received: 23 June 2022
Accepted: 27 July 2022
Published: 29 July 2022

1. Introduction

Terrestrial ecosystems, especially forests, play an important role in the global carbon cycle and in climate change mitigation [1]. Both the IPCC and Paris Agreement concur that the substantial contribution of forests is key to achieving the Nationally Determined Contribution (NDC) goals [2]. Previous studies have shown that the increase in the forest carbon stock (FCS) in China mainly results from forest restoration and afforestation [3,4]. Carbon sinks caused by ecological projects, such as afforestation, decline as forest vegetation matures and reaches the late successional stage [5]. However, within the period of China's carbon neutrality target, forest ecosystems, especially natural forests, can still maximize their carbon sequestration effects through forest management and restoration. Therefore, it is necessary to further clarify the carbon sink capacity of forest ecosystems and accurately account for the FCS.

China has conducted extensive research in the field of FCS assessments and forecasting of future trends [1,6–10]. Current measurement methods for FCS mainly include (1) inventory-based estimation, (2) satellite-based estimation, and (3) process-based estimation. The carbon stock results calculated using different forest types, data sources, and estimation methods are significantly different [11]. The Chinese land spans a wide range of latitudes (from 18°N to 53°N). Based on natural and environmental characteristics, China's forest ecosystems can be divided into seven types [12]. The variability in the carbon sequestration capacity and carbon cycles of different types of forests makes it more difficult to accurately estimate the overall carbon stock. Large-scale FCS measurements are necessary; however, they weaken due to the spatial heterogeneity of natural environmental elements. The uncertainty in the estimation results of the FCS can be further reduced if the large-scale area is subdivided into intermediate areas with the same climatic, hydrological, and soil backgrounds for the study. Forest inventory data are considered the most reliable data source for forest carbon flux studies. Owing to its authority and comprehensiveness, most current carbon stock accounting studies, including in China, are based on national inventory data [13]. However, the national forest resources verification cycle is long, the published data have a lag, and the classification of forest types is vague, which cannot meet the requirements of real-time monitoring and rapid assessment of regional FCS [14]. In addition to natural factors, the estimation method is a key factor contributing to the uncertainty of FCS estimation [15]. Current estimation methods lack adequate response to the evolution of forest ecosystems caused by climate change. In particular, the interconversion processes between different forest stands under the stress of changing natural environmental factors need to be further clarified, which is crucial for accurate estimation of FCS. Simultaneously, the successful implementation of any CO_2 removal method requires careful consideration of other land use requirements [16]. Land use and cover change (LUCC) is a major driver of a range of ecological problems that cause carbon cycling by altering the ecosystem structure [17,18]. Therefore, it is necessary to clarify the trends of future climate change and LUCC-induced changes in forest ecosystem structure and to perform simulations and predictions of FCS to reveal its dynamic evolution pattern.

The Northeast China Forestry (NCF) is the largest natural forest area in China and is the key implementation area of China's Natural Forest Protection Project (NFPP). Compared to planted forests, natural forests can better support biodiversity conservation and achieve ecosystem services, such as surface carbon storage, soil conservation, and water conservation [19]. Over the past few decades, NCF has been an important producer of timber and forestry by-products [20]. However, if forest conservation involves timber production, policymakers must weigh environmental and production outcomes [21]. Owing to the specificity of the administrative system, the vast majority of NCF's forest resources are state-owned under the jurisdiction and development of different forestry bureaus and forest industry groups, which facilitates more efficient forest management. The main status of food production cannot be changed, and the implementation of long-term afforestation projects has resulted in very limited forest suitable land in NCF. Forest ecosystem restoration is mainly based on forest nurturing and degraded forest restoration. This indicates that the evolution of forest ecosystems in the NCF is more focused on the mutual transformation between different forest stands. Although the forest area will not expand on a large scale, the FCS may undergo significant changes.

Forest ecosystems contain four carbon pools: above ground biomass, belowground, soil, and deadfall carbon pools. Among them, the aboveground biogenic carbon pool and soil carbon pool account for the largest proportion of the total carbon stock and are the focus of research. Although deadfall only accounts for approximately 5% of the total carbon stock, it is the link between the aboveground vegetation carbon pool and the soil carbon pool [22], and is especially important for NCF, which is dominated by natural forests. Over the past few decades, researchers have made many effective attempts to estimate the FCS of the NCF [10,23–27]. However, from the results of the study, the lack of overall calculation of the four carbon pools of the forest and simulation of the process of spatial and temporal evolution of the carbon stock hinders further assessment of the ecological and economic values generated by the FCS of the NCF.

In this study, we quantified the temporal variability and spatial heterogeneity of the FCS in the NCF by specifying the interactive processes between the interior and exterior of the forest caused by LUCC in the context of future climate change. The main objectives of this study were to clarify (1) the evolutionary trends of land use in the NCF from 2030 to 2050, (2) the evolution between different forest stands within the forest, and (3) the evolutionary trends and spatial heterogeneity of the FCS.

2. Materials and Methods

2.1. Study Area and Data

2.1.1. Study Area

The National Forest Management Plan (2016–2050) prepared by China's National Forestry and Grassland Bureau divides the country into eight management zones, taking into account the status of forest resources, geographical location, forest vegetation, management status, and development direction of each region. The NCF (38°43′–53°23′N and 118°50′–135°05′E) includes the Greater-Khingan-Mountains cold temperate coniferous forest management area and the northeast middle temperate coniferous and broad-leaved mixed forest management area, involving Heilongjiang, Jilin, Liaoning, and four provinces and autonomous regions of Inner Mongolia, 244 counties (districts) (Figure 1). The NCF straddles the mid-temperate and cold temperate zones from south to north and has a temperate monsoon climate with an average annual temperature of 4.8 °C to 11.5 °C, annual precipitation of 300–1000 mm, and a large area of black soil. The total area of the existing forest land is 53.22 million hectares, the forest accumulation is 1.087 billion cubic meters, and the forest area accounts for approximately 37% of the country's total area [28]. The forests are mainly concentrated in the three major topographical areas of Greater-Khingan-Mountains, Lesser Khingan Mountains, and Changbai Mountains, and the vegetation types are mainly deciduous broad-leaved forest and coniferous forest.

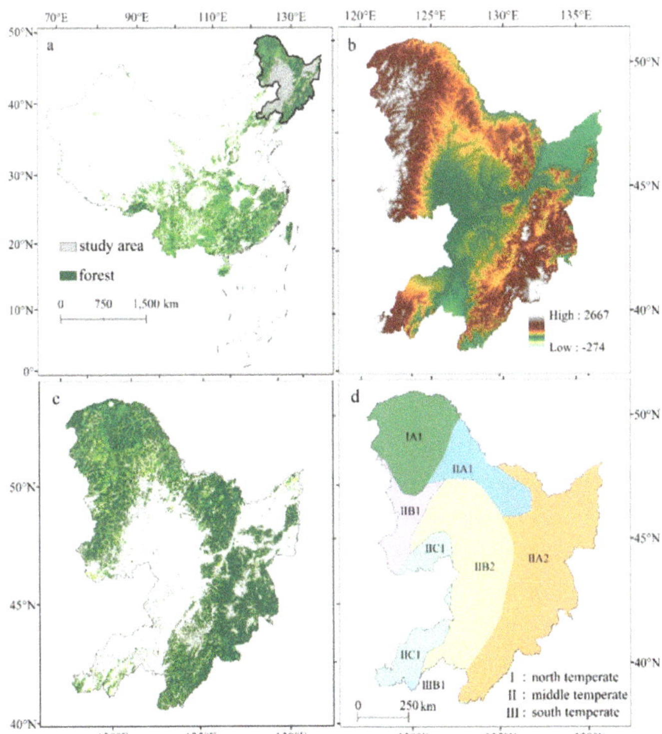

Figure 1. Main overview of the NCF (**a–d**) represents district, DEM, forest distribution, and climate zone, respectively.

2.1.2. Data Acquisition and Preprocessing

To explore the impact of LUCC on FCS in the context of future climate change, it is crucial to clarify its impact mechanism and screen the driving factors affecting LUCC

(Figure 2). The research support data mainly include land use, economic, social, climate, and soil data (Table 1). The land use data is a multi-period set from 1980 to 2020, constructed by manual visual interpretation using Landsat remote sensing images as the main information source. The dataset covers 6 major categories and 25 subcategories, and the data resolution is 30 m. Because the focus is on the interconversion between different forest stands, the land use data classifies forested land into four types according to the degree of density and tree height: closed forest land (Cl, natural and planted forests with density > 30%), shrubland (Shr, short stands and scrubland with density > 40% and height below 2 m), sparse forested land (Sp), forested land with density 10–30% and other forested land (Oth, non-forested plantations, trails, nurseries, and various types of gardens). Other land use types were reclassified as Cropland (Cult), Grassland (Gr), Water (Wat), Construction Land (Constr), and Unused Land (Un), data from the Data Center for Resources and Environmental Sciences, Chinese Academy of Sciences (DOI: 10.12078/2018070201).

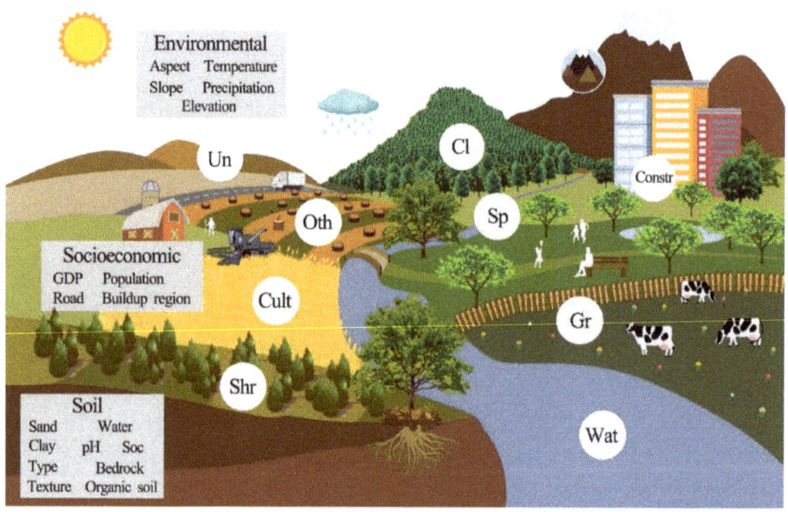

Figure 2. LUCC influence process.

Economic and social data are the main factors influencing land use change based on previous research results of 10 datasets including transportation, GDP, and population [29–31]. To eliminate the randomness of single-year climate data, we used the average values of temperature and precipitation from 1970 to 2000. Future climate change data were used under three Shared Socioeconomic Pathways (SSPs) (SSP126, SSP245, and SSP585), with 19 bioclimatic variables based on the BCC-CSM2-MR model. The DEM is derived from SRTM data measured jointly by NASA and the National Mapping Agency (NIMA) of the Department of Defense with a data resolution of 3 arc-second (~90 m). Slope and aspect data were obtained by processing DEM data using ArcGIS Pro 2.8 software. Soil is an important factor influencing changes in forest ecosystems [32], and we wanted to show the characteristics of water content, water retention, permeability, nutrients, and physicochemical properties of soil using nine indicators. A series of data preprocessing was performed in ArcGIS Pro2.8 software, including projection transformation, Euclidean distance, resampling, and clipping, and all of the above data were converted to raster data with the same projection coordinate system and 30 m spatial resolution.

Table 1. The spatial driving factors of the land use change in this study.

Category	Data	Year	Original Resolution	Data Resource
Land	Land cover	1980–2020	30 m	https://www.resdc.cn/, accessed on 28 December 2021
Soil factors	Soil water capacity	2017	250 m	https://data.isric.org/, accessed on 21 October 2021
	Soil pH Depth to bedrock Cumulative probability of organic soil Soil organic carbon stock Sand content Clay content Texture class			
	Soil type	1995	1000 m	http://www.resdc.cn/, accessed on 25 December 2021
Socioeconomic factors	Population	1990–2020	1000 m	www.worldpop.org, accessed on 28 December 2021
	GDP	1990–2020	1000 m	http://www.geodoi.ac.cn/, accessed on 28 December 2021
	Proximity to city Proximity to rural settlement	2015	30 m	
	Proximity to railway			https://www.openstreetmap.org/, accessed on 15 October 2021
	Proximity to highway Proximity to primary road Proximity to secondary road Proximity to tertiary road Proximity to quaternary road			
Climatic and environmental factors	DEM	2016	90 m	NASA SRTM1 v3.0, accessed on 25 December 2021
	Slope Aspect			
	Temperature	1970–2000	30 arc-sec	http://www.worldclim.org/, accessed on 26 October 2021
	Precipitation Bioclimatic variables	2040–2060		

2.2. Methods

2.2.1. Patch-Generation Land Use Change Simulation (PLUS) Model

Cellular automata (CA) are widely used to simulate the dynamics of complex LULC systems [33]. However, most CA models focus on the optimization of simulation techniques and the correction of transformation rules, and relatively little research has been conducted on how to deepen the analysis of potential drivers of land use, especially on the strategies of transformation rule mining and simulation of landscape dynamics, which require further clarification. The PLUS model is based on raster image data and uses a new land expansion analysis strategy (LEAS) combined with a CA model based on multiclass random patch seeding (CARS) to better simulate multiclass land use patch-level changes [34,35].

LEAS incorporates a transformation analysis strategy (TAS) and pattern analysis strategy (PAS). By extracting the parts of each type of land use expansion between the two periods of land use change and sampling, the random forest algorithm was used to mine the factors of each type of land use expansion and the corresponding driving force. Thus, the conversion probability of each type of site and the contribution of drivers to the expansion of each type of site in that period can be obtained with a better interpretation. CARS combines random seed generation and a threshold decreasing mechanism, and the

PLUS model can simulate the automatic generation of patches in a spatiotemporal dynamic manner under the constraints of transformation probability and conversion constraint (Figure 3). For the 2030 land use simulation, we used 1970–2000 data for climate factors, and for the 2050 simulation, we used SSP future climate projection data and SSP126, SSP245, and SSP585 pathways corresponding to the EP, NG, and RD scenarios, respectively.

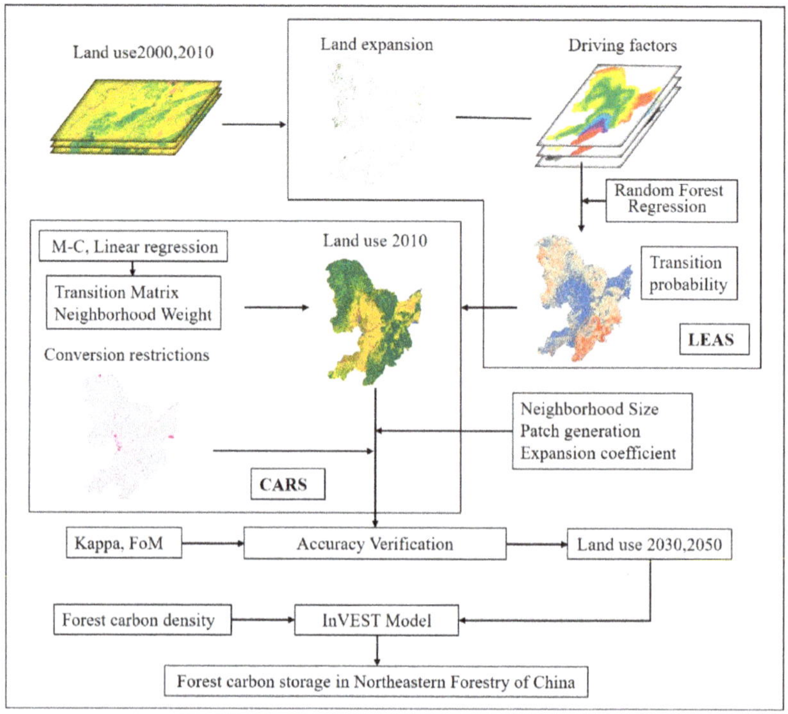

Figure 3. Calculation process of carbon storage in Northeast Forestry of China.

2.2.2. InVEST Model and Forest Carbon Density Settings

Accuracy verification is key to the land use simulation process, and we used the Kappa coefficient and Figure of merit (FoM) to estimate the accuracy of the simulation results. Usually, a Kappa coefficient greater than 0.6 indicates that the results are usable, and greater than 0.8 indicates that the simulation results are relatively accurate.

Although most of the previous studies on the PLUS model have used the Kappa coefficient to verify the accuracy of the model, the reliability of the Kappa coefficient is currently subject to many controversies [36,37]. Therefore, we introduce the FoM coefficient to further verify the accuracy of PLUS.

FoM coefficients only focus on where it has changed. FoM coefficients are superior for measuring goodness of fit in simulations of changes in landscape composition. Theoretically, FoM values range from 1% to 100%, with larger FoM values corresponding to higher simulation accuracy, but values less than 30% have been shown to be common [38]. The formula for calculating the FoM coefficient is:

$$\mathrm{FoM} = \mathrm{B}/(\mathrm{A} + \mathrm{B} + \mathrm{C} + \mathrm{D})$$

where B represents the actual area that has changed and the simulation results have also changed. A indicates that the actual area has changed, but the simulation results have not changed. C indicates that both the actual area and the simulation results have changed, but the direction of change is not consistent. D represents the actual area that has not changed, but the simulation results have changed [39].

2.2.3. InVEST Carbon Storage and Sequestration Model

The InVEST carbon storage and sequestration model uses land use raster data and stocks in four carbon pools (aboveground biomass, belowground biomass, soil, and dead organic matter) to estimate the amount of carbon currently stored in the landscape or sequestered over time. The model operates by mapping the carbon density of the carbon pools to the LUCC raster to calculate the carbon stock of each land type. Therefore, the accuracy of the InVEST model depends on the land use data and forest carbon pool data. In order to improve the accuracy of land use simulation, we refined the soil data that affect the forest evolution, involving factors such as soil physical and chemical properties, water retention, air permeability, nutrition, and root growth space, so that the simulation accuracy of closed forest land can reach more than 95%. The closed forest accounts for about 90% of the forest area in NCF, which will optimize the accuracy of the InVEST model calculation. Meanwhile, many previous studies involving the calculation of forest carbon stocks by InVEST model have classified forests as one type or included only part of the carbon pool. Obviously, the carbon density of forests with different degree of density is different. Therefore, to further improve the accuracy of the InVEST model for estimating forest carbon stocks, we divided the forest into four types of stands and included carbon density data of all carbon pools of the forest.

The InVEST model used carbon density data from four carbon pools, all of which were derived from actual measurements conducted by researchers at the NCF. Aboveground biomass carbon density measurements include the carbon density of the tree layer and carbon density of understory vegetation. Belowground biomass carbon density refers to root carbon density. Soil carbon density was replaced by a mean value of 0–100 cm in the uniform adoption.

The forested sites mainly included *Larix gmelinii*, *Pinus koraiensis*, *Pinus camphorata*, *Pinus tabulaeformis*, *Picea abies*, *Quercus mongolica*, *Betula platyphylla*, *Betula davurica*, and other dominant vegetation-building species in the northeast. The shrublands included vegetation of *Caragana korshinskii*, *Prunus sibirica*, *Ostryopsis decne*, and *Spiraea salicifolia*. The open woodlands contained *Ulmus pumila*, *Populus simonii*, and *P. davidiana*. In this study, we defined other forested lands as trails and unstocked lands to determine carbon density. All four forest stands involved the carbon density of four carbon pools, which were weighted and summed based on the area of tree species mentioned in the literature (Table 2)

Table 2. The carbon density of each stand used in the InVEST model (Mg/hm^2).

	C_above	C_below	C_soil	C_dead
Cl	68.049	1.104	129.395	5.652
Shr	6.3325	0.733	115.73	1.23
Sp	17.57	0.765	58.67	0.62
Oth	1.288	0.688	6.15	0.643

3. Results

3.1. Model Validation

To verify the reliability of the model, we combined the Markov chain (M-C) and simulated land use data for 2010 and 2020, respectively (Figure 4). The results of our random sampling (sampling rate of 0.1 and number of samples of 9,199,472), compared with the real data, show that the kappa coefficients of the simulated data in 2010 and 2020 are greater than 0.8 (Table 3). The 1990 and 2000 land use maps were selected as the initial states of the landscape pattern in 2010 and 2020, respectively. The results show that the FoM coefficients of the two simulated data are both 0.174. This study focuses on simulating the evolution of forest land in the northeastern forest region. Therefore, we reclassified the data and set other land types other than forest land to the same class. The 2010 data was selected as the initial state of the 2020 landscape pattern to validate the FoM coefficient of the 2020 simulated data. The results show that A = 0.1247442, B = 0.10471363, C = 0.469294, D = 0.4283501, FoM = 0.635722. It shows that the PLUS model has a relatively reliable accuracy for the forest land simulation in the northeast forest area.

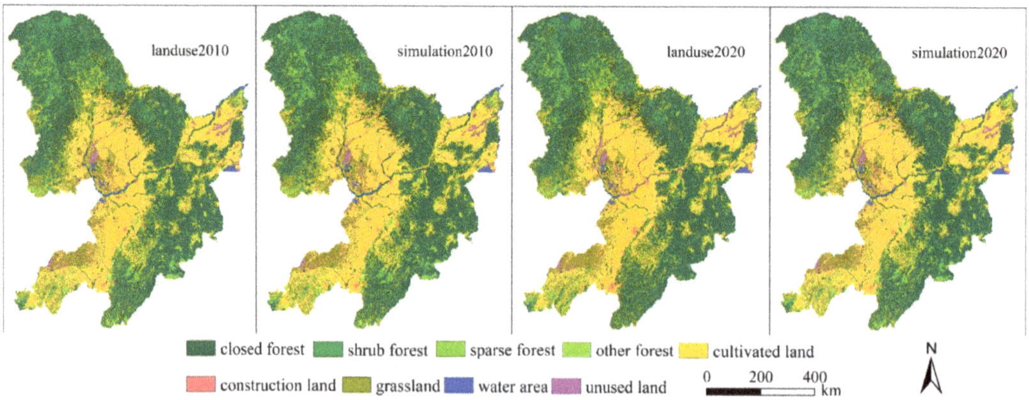

Figure 4. Land use simulation in the NCF for 2010 and 2020.

Table 3. PLUS model validation results for the NCF.

Land Use Type	User's Accuracy		Overall Accuracy		Kappa Coefficient	
	2010	2020	2010	2020	2010	2020
Closed forest land	0.976457	0.956521	0.971922	0.896424	0.960604	0.853745
Shrub forest land	0.912785	0.729021				
Sparse forest land	0.818204	0.615643				
Other forest land	0.523307	0.664626				
Cultivated land	0.990825	0.918858				
Grass land	0.989463	0.850062				
Water area	0.990503	0.452974				
Construction land	0.966698	0.787961				
Unused land	0.98088	0.884863				

3.2. Multiple Scenario Settings Based on the Amount of Land Demand

The PLUS model requires setting target values for future land use patches and assigning the changing patches to appropriate spaces according to the future land area by combining LEAS and CARS. The PLUS model provides both linear regression and Markov chain (M-C) for forecasting future land use demand. The M-C can complete the forecast using two periods of data but is more suitable for short-term forecasting. The M-C prediction results vary widely when using data from different time periods (Table 4). Our linear regression projections using NCF land used data for 10 periods from 1980 to 2020 yielded results that appear to be more in line with the NCF development expectations.

To improve the reliability of the simulation results, we set up three future development models: the ecological priority scenario (EP), natural growth scenario (NG, baseline scenario), and regional development scenario (RD). The 2030 and 2050 land use areas obtained from the linear regression projections were used as the baseline scenarios. Regarding the setting of land use areas for the two scenarios of EP and RD, two key factors need to be considered: the continuation of current RD trends and future development plans. NCF has the important task of supplying forest products and food; therefore, the area of forest and arable land should be protected first in a future development process. Construction land is the most active land type in the process of land use change and is the most direct factor affecting LUCC. It is worth noting that the northeast region has encountered a development bottleneck in the past 10 years (Figure 5). Although the Chinese government has been promoting a northeast revitalization plan, the northeast region has not met the development expectations of the central government owing to cold climate and deformed industrial structure. Owing to the early start of development and large rural population loss, the urbanization rate in the northeast is higher than the national average. It should be clear that the population loss and the decline in birth rate, as well as the late stage of urbanization development, do not imply a reduction in total urban construction land area in the future, but rather a reduction in demand [40]. Although current development trends suggest that the probability of the RD scenario is likely to be low, we set up this scenario to address possible future scenarios (Table 5).

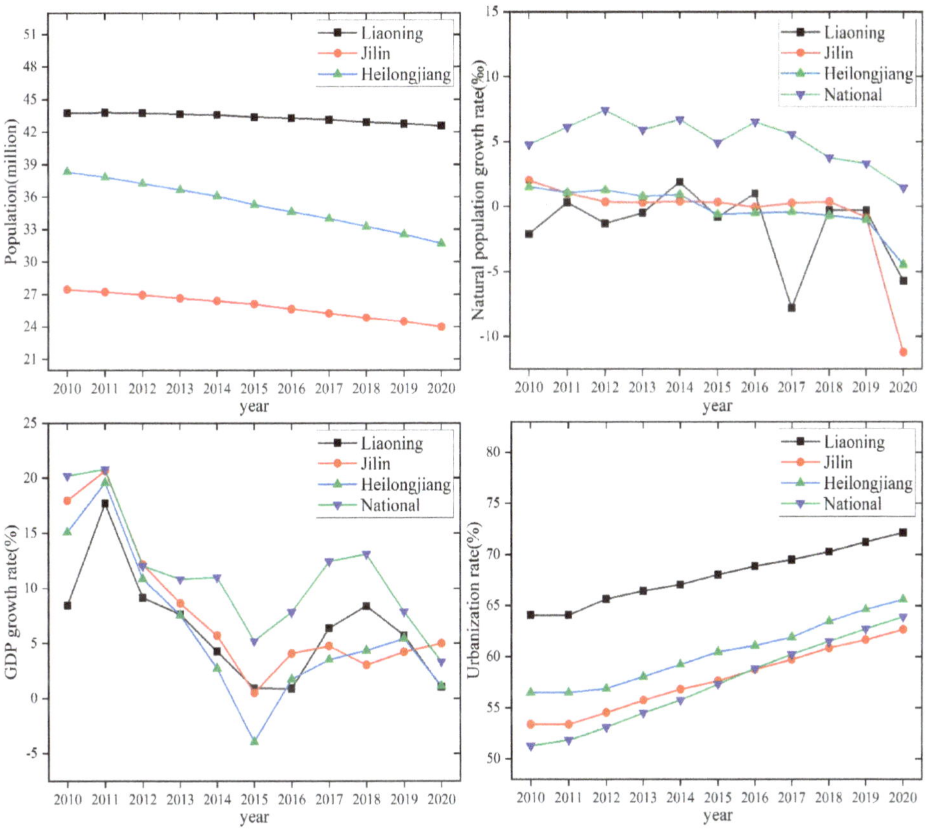

Figure 5. Changes in total population, natural population growth rate, GDP growth rate, and urbanization rate in the NCF.

Table 4. Predicted area of land calculated by Markov chain and linear regression (km^2).

	Year	Cl	Shr	Sp	Oth	Cult	Gr	Wat	Constr	Un
Current	2020	391,734	30,785	23,168	4455	298,416	99,040	16,810	24,832	50,248
M-C (2010–2020)	2030	399,589	27,308	18,205	4422	298,167	94,962	13,757	27,733	55,347
	2050	409,589	22,601	13,059	4356	297,974	88,148	10,889	31,782	61,091
M-C (2015–2020)	2030	374,833	37,365	29,642	5583	303,196	109,299	12,431	26,080	40,666
	2050	362,820	38,135	31,030	6016	312,674	111,450	10,378	27,684	38,907
Linear regression	2030	393,992	28,049	20,529	4670	304,200	95,762	17,793	25,797	48,698
	2050	402,772	22,099	12,649	4851	311,245	89,759	15,128	29,003	51,984

Table 5. Area setting of future scenarios and their changes in 2020 (km^2, %).

Scenario	Time	Cl	Shr	Sp	Oth	Cult	Gr	Wat	Constr	Un
NG	2020	391,734	30,785	23,168	4455	298,416	99,040	16,810	24,832	50,248
	2030	393,992 (0.58%)	28,049 (−8.89%)	20,529 (−11.39%)	4670 (4.83%)	304,200 (1.94%)	95762 (−3.31%)	17,793 (5.85%)	25,797 (3.89%)	48,698 (−3.08%)
	2050	402,772 (2.82%)	23,066 (−25.07%)	17,731 (−23.47%)	3826 (−14.1%)	303,334 (1.65%)	93,759 (−5.33%)	15,128 (−10.0%)	29,003 (16.80%)	50,869 (1.24%)
EP	2030	395,363 (0.93%)	27,738 (−9.90%)	20,614 (−11.02%)	4505 (1.12%)	303,010 (1.54%)	98,694 (−0.35%)	19,294 (14.78%)	25,682 (3.42%)	44,590 (−11.2%)
	2050	410,689 (4.84%)	30,329 (−1.48%)	20,343 (−12.19%)	4296 (−3.57%)	301,181 (0.93%)	89,829 (−9.30%)	19,261 (14.58%)	26,220 (5.59%)	41,342 (−17.7%)
RD	2030	392,892 (0.30%)	27,485 (−10.72%)	20,428 (−11.83%)	4345 (−2.47%)	302,019 (1.21%)	95,404 (−3.67%)	16,481 (−1.96%)	27,605 (11.17%)	51,519 (2.53%)
	2050	403,027 (2.88%)	22,601 (−26.58%)	15,922 (−31.28%)	4314 (−3.16%)	305,692 (2.44%)	88,150 (−11.0%)	15,087 (−10.3%)	31,782 (27.99%)	52,913 (5.30%)

The National Forest Management Plan has specific development requirements for NCF forest development in 2020–2050, which we followed in the setting of forest land area in the EP scenario. The RD scenario reflected more productive attributes. Rural depopulation may accelerate large-scale land-intensive production so that there are priority growth opportunities for building land and cultivated land. The grassland area would decrease to different degrees under all three scenarios. After the Third National Land Survey (2021), the central and local governments became stricter in their attitudes toward arable land protection. As a result, forestland expansion is mainly achieved through grassland and unused land conversion.

3.3. NCF Land Use Evolution Analysis

3.3.1. Historical Land Use Evolution Analysis

We used the computational change raster tool of ArcGIS Pro2.8 to comparatively analyze the quantitative relationships between land use conversions at different time periods. From 1980 to 2000 (Tables 6 and 7), there was a significant decline in closed forest land and grassland from 21.36% and 6.52% of the total area to 20.38% and 5.48%, respectively (Figure 6a). The decrease in forested land was mainly concentrated in the south-central part of Lesser Khingan Mountains and the southern part of the Sanjiang Plain, and the degradation of shrub forests in the Changbai Mountains was more obvious. The northern part of the NCF has experienced a certain expansion of forested land, which was more scattered (Figure 7a). Most of the lost forest and grassland were transformed into arable land and construction land, and the area of arable land expanded by 17.46%.

Table 6. Conversion of Land Types from 1980 to 2000 (km^2).

	Cl	Shr	Sp	Oth	Cult	Gr	Wat	Constr	Un
Cl	376,673.1	4660.43	1846.63	1422.39	13,858.13	2048.26	70.12	210.76	610.24
Shr	910.2	27,379.15	231.95	8.53	4618.52	390.24	94.45	72.51	270.6
Sp	139.94	30.82	26,721.09	15.22	1126.78	336.68	2.9	11.14	20.19
Oth	841.39	1418.53	374.15	2585.75	100.32	2157.42	0.11	1.46	2.86
Cult	1301.78	296.4	405.13	118.66	246,914.2	854.56	397.26	1913	744.18
Gr	2850.89	561.54	720.34	44.32	19,375.76	95,403.31	620.97	323.07	2639.85
Wat	11.42	42.86	12.96	0.01	608.65	228.91	20,076.5	6.93	203.48
Constr	4.03	0.89	0.73	0.06	82.26	35.81	4.25	18207.91	1.31
Un	185.1	643.1	191.78	1.1	10,392.48	1564.01	855.15	78.62	39,302.12

Table 7. Conversion of Land Types from 2000 to 2020 (km^2).

	Cl	Shr	Sp	Oth	Cult	Gr	Wat	Constr	Un
Cl	363,634.6	1956.55	2000.83	1156.92	7929.7	4131.44	889.67	412.75	805.53
Shr	3920.37	24,872.14	189.6	92.14	2209.96	2162.83	315.38	95.89	1175.43
Sp	7932.75	552.03	18,575.2	184.86	2081.31	774.25	68.56	175.24	160.57
Oth	1061.34	64.68	59.98	2745.78	154.4	68.52	7.25	19.38	14.71
Cult	8301.44	1573.53	912.24	107.58	27,2192.7	3315.28	1703.5	6855.78	2115.61
Gr	5600.76	1191.31	1310.58	126.6	4682.17	86,612.52	375.62	454.01	2665.66
Wat	356.83	356.31	21.81	2.29	1663.76	362.93	12,908.92	131.72	6315.43
Constr	232.88	68.58	36.79	7.41	3615.5	174.24	98.03	16,509.32	82.67
Un	693.08	149.48	60.64	31.26	3887.31	1438.61	443.68	178.01	36,912.84

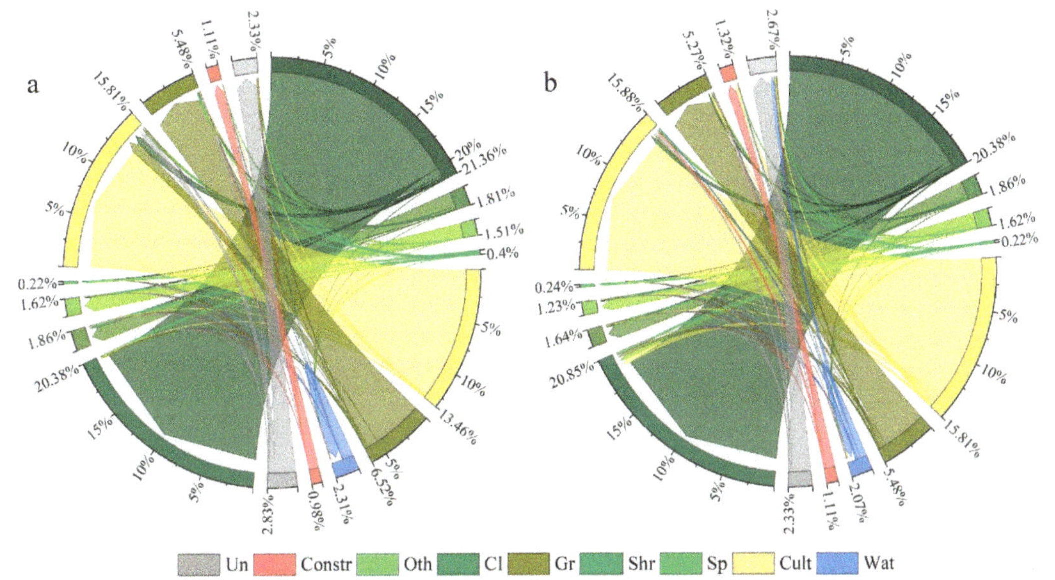

Figure 6. The relationship between land use conversion in different time periods ((**a**) represents 1980–2000, (**b**) represents 2000–2020. The beginning of the arrow indicates the land proportion in the base year, and the arrow points to the land proportion in the target year).

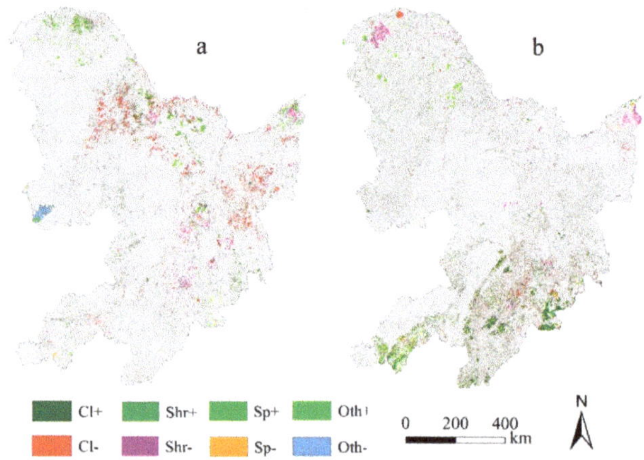

Figure 7. Expansion and decline of forest land area (**a**) represents 1980–2000, (**b**) represents 2000–2020.

The LUCC of NCF was relatively stable from 2000 to 2020. The area of closed forest land started to increase steadily, the growth rate of cropland and the decay rate of grassland both started to slow down, and the area of construction land grew at a faster rate (Figure 6b). Spatially, the Changbai Mountain region was accompanied by a relatively dramatic closed forest land evolution, but it was generally increasing. In the northern part of the Greater-Khingan-Mountains and the eastern part of the Sanjiang Plain, there was a more pronounced decrease in the area of closed forest land and shrubland (Figure 7b).

3.3.2. Simulation of Future Land Use Evolution

We simulated NCF land use under three scenarios in 2030 and 2050, calculated the expansion of each category based on 2020 (Figure 8), and selected three regions with more significant LUCC in A, B, and C for comparison (Figure 9). The relevant parameters of the PLUS model were set as follows. In the LEAS module, the number of regression trees was 50, and the sample rate was 0.01. In the CARS module, the patch-generation threshold was 0.7, the expansion coefficient was 0.3, the percentage of seeds was 0.001, and the neighborhood weights were 3.

1. Changes in the areas of the main land types

Forest conservation, food security, and urbanization are the three main developments in the NCF. Therefore, we mainly explored the changes in forest land, cultivated land, and construction land in future scenarios. All three major land types maintained growth, with closed forest land growing faster in the EP model, but cultivated land grew slightly less than in the other scenarios. Under the RD model, built-up land grew faster than in the other scenarios. Cropland and closed forest land did not grow significantly in the area in 2030 but exceeded that of the NG model by 2050 (Table 8).

Table 8. Change in area increase of major land types compared to 2020.

Time	Scenarios	Cl	Cult	Constr
2030	NG	0.58%	1.94%	3.88%
	EP	0.93%	1.54%	3.42%
	RD	0.30%	1.21%	11.17%
2050	NG	2.82%	1.65%	16.80%
	EP	4.84%	0.93%	5.59%
	RD	2.88%	2.44%	27.99%

2. Intensity of conversion between land classes.

We believe that the EP scenario was the most suitable and probable development scenario for the NCF; here, we compared the conversion relationships between the two time periods, 2020–2030EP and 2020–2050EP. Tables 9 and 10 show the converted area between land use categories at different stages, with 2020 as the base year.

Table 9. Conversion of Land Types from 2020 to 2030EP (km^2).

	Cl	Shr	Sp	Oth	Cult	Gr	Wat	Constr	Un
Cl	383,252.2	45.37	141.3	353.05	1482.82	1308.46	13.52	53.53	190.5
Shr	916.5	27,396.88	35.14	5.19	1214.73	225.88	22.36	22.47	381.8
Sp	1988.78	19.68	20,111.64	45.25	644.18	152.24	8.7	9.9	85.97
Oth	521.78	3.31	11.07	3649.02	208.11	30.76	0.5	3.21	10.66
Cult	5549.88	146.03	66.06	354.81	285,028	173.65	19.97	2090.29	1968.61
Gr	2362.87	108.99	242.17	90.86	1339.2	92,655.55	54.09	88.11	1319.92
Wat	186.3	97.98	8.3	1.21	496.13	141.46	19,879.74	55.34	236.3
Constr	168.53	11.24	3.93	1.4	864.54	30.11	4.76	23366.41	97.31
Un	602.09	6.28	3	5.24	6670.99	116.95	81.8	48.55	40,535.08

Table 10. Conversion of Land Types from 2020 to 2050EP (km^2).

	Cl	Shr	Sp	Oth	Cult	Gr	Wat	Constr	Un
Cl	376,941.9	515.61	74.16	284.3	4127.84	4854.85	50.62	27.19	248.59
Shr	321.41	28,882.61	11.11	14.78	682.76	184.08	13	8.15	117.83
Sp	1934.06	60.85	20,055.05	135.48	590.46	368.51	10.3	12.18	34.93
Oth	138.67	14.82	38.18	4295.95	3.34	70	0.21	7.26	8.06
Cult	6425.12	487.42	60.68	56.43	284,934.9	1237.56	111.78	1000.01	1139.77
Gr	22,647.28	304.62	130.7	129.49	3577.53	70,436.31	232.07	37.34	895.91
Wat	166.21	110.79	6.69	1.18	2913.05	128.31	15,474.72	112.33	2180.17
Constr	321.3	32.81	6.24	0.65	2846.75	115.81	17.39	21,095.38	112.55
Un	1950.38	44.5	4.04	0.45	4415.82	2630.77	192.48	40.11	38,791.88

NCF land use conversion will be relatively stable, with closed forest land and cultivated land being the most actively evolving land types. Grassland and construction land were the main sources of arable land expansion (Figure 10). The change in construction land in northeast China was unique. The analysis of remote sensing images and population movement data reveals that the NCF had a large rural to urban population movement, and many rural construction lands have disappeared and transformed into grassland and cropland in the past 10 years. This situation is likely to persist. Cultivated land was the main source of land for urban expansion. Although this is strictly restricted in China, the mechanism of linking land increases and decreases solves the problem. The capacity of rural construction land to be converted into cultivated land is transferred to the process of urban expansion. Although grassland has important ecological and production value, its conservation priority may be lower than that of forests and cultivated land. Therefore, in the process of future land use change, we set the grassland area to a state of continuous reduction.

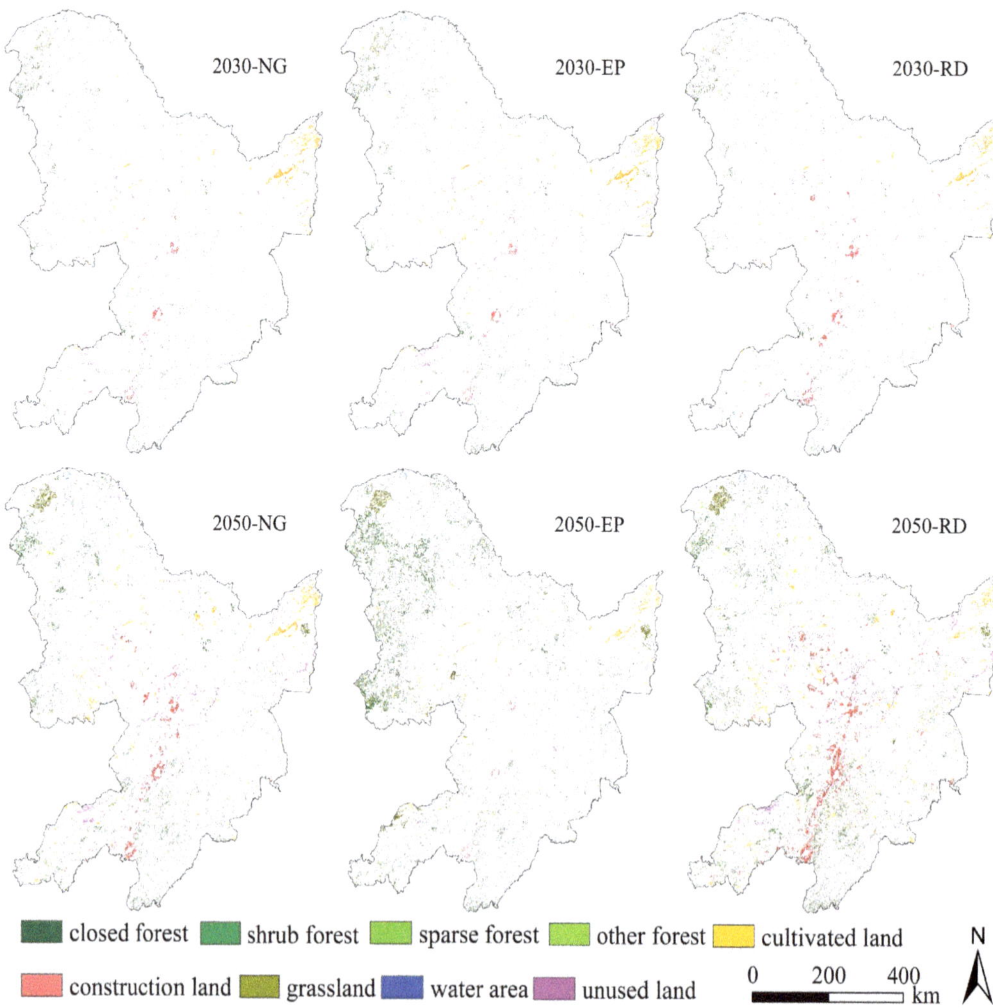

Figure 8. Future land expansion under different scenarios.

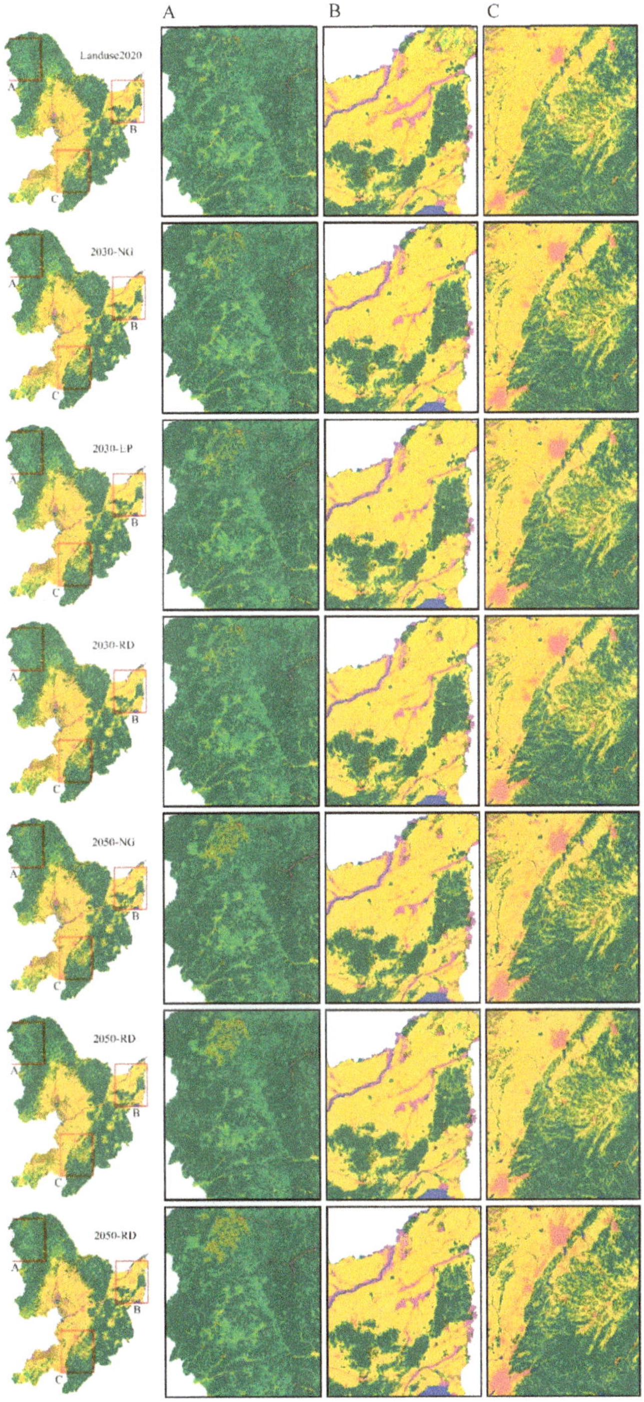

Figure 9. Future land simulations under different scenarios ((**A–C**) are three selected areas).

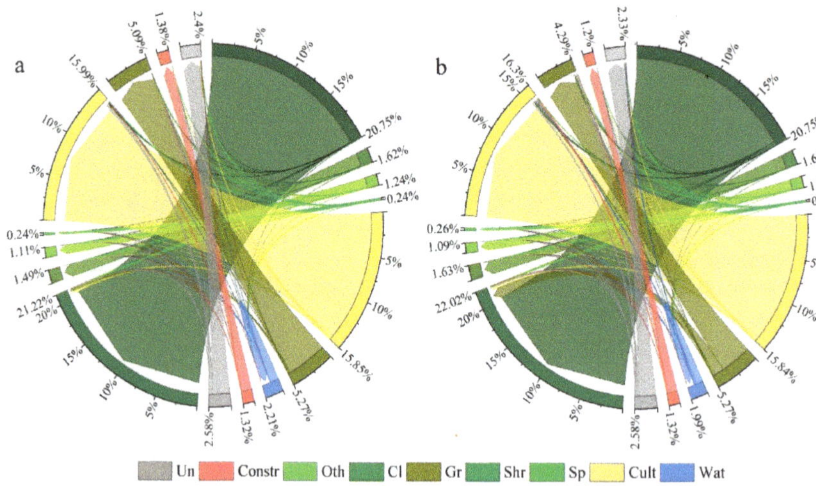

Figure 10. Future land use conversion relationship ((**a**,**b**) represents 2030, 2050, respectively. The beginning of the arrow indicates the land proportion in the base year, and the arrow points to the land proportion in the target year).

3.4. Spatial and Temporal Changes in Forest Carbon Stocks

3.4.1. FCS Evolution in a Historical Period

Based on future land use raster data simulated by the PLUS model and historical raster data, we estimated the overall carbon stock of NCF forests from 1980 to 2050 using the InVEST model (Figure 11). In 2020, the FCS of NCF was 8564.76 Mt, and the carbon stocks of the four forest stands accounted for 93.40%, 4.46%, 2.10%, and 0.05% of the total, respectively. The proportions of the four carbon pools were 31.83%, 0.56%, 64.96%, and 2.65%, respectively, and soil carbon pools were the most important components of the NCF forest ecosystem carbon stocks. In terms of spatial distribution, the FCS shows the following characteristics: Changbai Mountains > Greater Khingan Mountains > Lesser Khingan Mountains. Owing to the production attributes, the evolution of woodlands in the Lesser Khingan Mountains was more frequent, which led to the instability of FCS.

In terms of temporal trends, the change in FCS from 1980 to 2020 was roughly divided into three phases. The first phase was the rapid decline phase from to 1980–1995. During this period, the high-intensity forestry exploitation in the NCF led to a rapid decline in FCS. The second phase was the gradual slowdown phase from 1995 to 2010, when the rate of forest area decay began to slow down owing to the implementation of a series of ecological protection projects. The third stage was the rapid recovery period from 2010 to 2020. Through these efforts, the stability of the NCF ecosystem was further strengthened and the forested land area gradually recovered to the level of the 1990s.

The spatial evolution of the FCS in the NCF showed a trend from dispersion to concentration and an overall improvement. The FCS reduction had a tendency to transition from Changbai Mountains to Greater-Khingan-Mountains. From 1980 to 2000, the FCS of the three major regions decreased to varying degrees and was mainly concentrated in the Lesser Khingan Mountain area. From 2000 to 2010, it was generally stable; however, from 2010, the FCS in the Changbai Mountain area began to increase steadily, and the Lesser Khingan Mountain area became stable. The FCS in the northern part of the Greater Khingan Mountains began to decline because a large area of forest land in Mohe City was degraded to grassland (Figure 12).

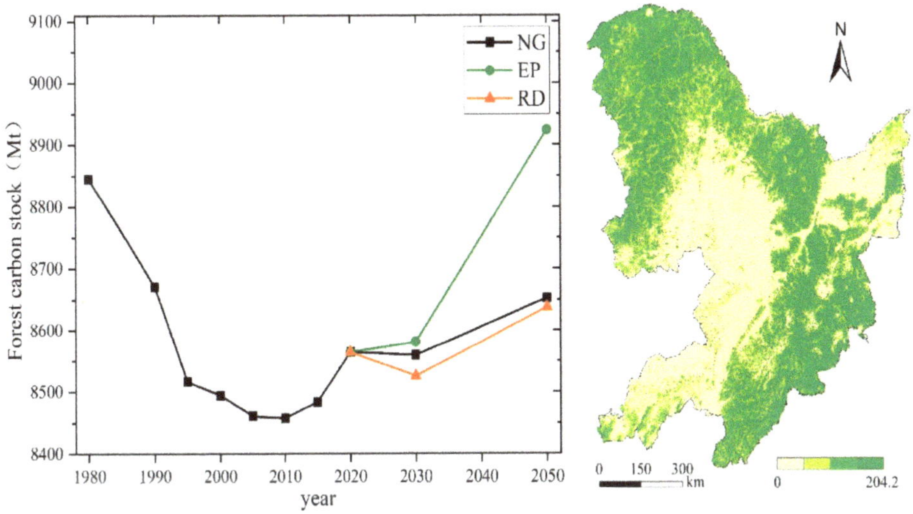

Figure 11. Change trend of forest carbon storage and spatial distribution of carbon storage in 2020 (Mg/hm^2).

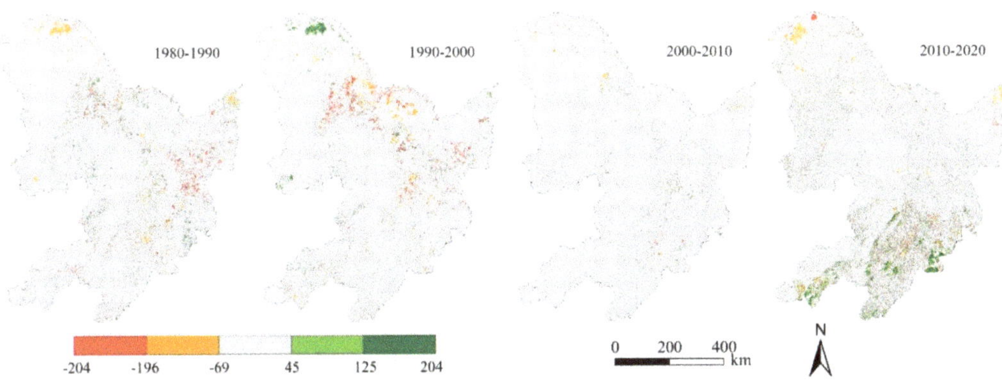

Figure 12. Stages of FCS in NCF (Mg/hm^2).

3.4.2. Characteristics of Future Changes in FCS

Compared to 2020 and 2030, the FCS evolution trend of the NCF will be basically the same. The northern part of the Greater Khingan Mountains and the eastern part of the Sanjiang Plain were the main areas where the FCS decreased. The vast area south of the line from Changchun City to Yanbian Prefecture showed a relatively clear trend of increasing carbon storage. By 2050, this trend will intensify further. Areas with relatively concentrated cities, such as the western Songnen Plain and the northern Changbai Mountains, also began to experience a decrease in FCS to varying degrees, whereas the Greater Khingan Mountains will replace Changbai Mountains as the area with the most significant increase in FCS (Figure 13). If strict natural forest protection measures are implemented, it is expected that by 2050, the FCS of the NCF will return to 1980 levels.

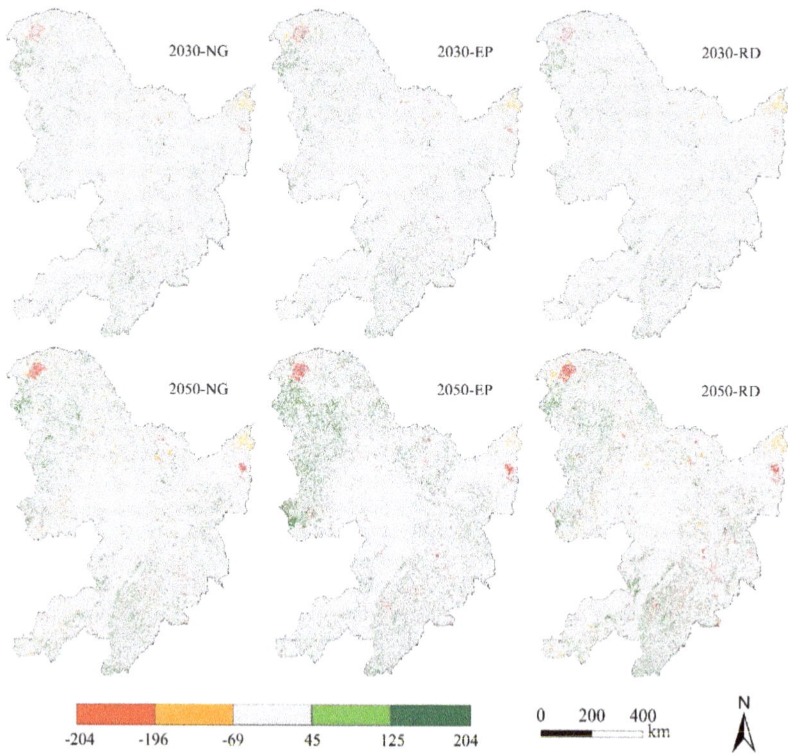

Figure 13. Spatial distribution of FCS in NCF (Mg/hm^2).

4. Discussion

4.1. Limitations, Uncertainties, and Prospects

The main limitations of this study are related to the accuracy of the data and the model. Although we divided the forest into more detailed types according to the degree of density to reflect the carbon density changes caused by different tree ages, the data accuracy directly affects the accuracy of the final carbon stock estimation, which is an unavoidable drawback of using remote sensing images for ecosystem service valuation. Another critical point is the selection of land use impact factors, which is worth discussing. The LUCC is a combination of multiple influences and complex evolutionary processes. This study focuses on the spatial evolution process of forests, but there are many factors that affect changes in forest ecosystems, such as nitrogen deposition, climate change, and CO_2 fertilization [41,42]. Although some factors are widely debated [43], for larger scale regions, scientifically available and accessible data sources remain an important support for assessing ecosystem sustainability. The PLUS model, although capable of obtaining more reliable simulation results in different future scenarios, presupposes the setting of land use target values, which gives rise to many uncertainties. Many studies have been conducted on the simulation of land use under multiple pathways of SSP, and, in general, ecological land, especially forest land, is basically reduced under the fossil-fueled development (ssp5) pathway [44–46]. Various organizations and institutions, such as the IPCC, World Bank, and IIASA, have set different development factors for different paths. However, these assessment results are for regional analyses at the national or even continental level, and the accuracy of the study results is questionable if such parameters are set at the local level without considering regional specificity. Although the InVEST model can estimate carbon stocks with less information, it assumes that none of the LULC types in the landscape are gaining or losing carbon over time. Therefore, in this study, forests with different canopy cover levels were set up instead of forests with different age classes to minimize the error, but this may be imperfect.

Simultaneously, the InVEST model is overly reliant on the carbon density values of individual land types. In this study, as much as possible, we refer to the measured values of forest ecosystem carbon stocks by many scholars, but limited to a large study area, which cannot fully take into account the variability of vegetation carbon density owing to different tree species, latitudes, and climates. For forest biomass carbon estimates, forest type and tree species have a strong influence on carbon stock estimates. Forest LULC types can be stratified by elevation, climate zone, or time interval since major disturbance. Of course, this more detailed approach requires data describing the carbon stocks in each carbon pool for each of the finer LULC categories. For soil organic carbon (SOC) and apoplastic carbon estimates, total soil C increased significantly with altitude [47]. This is because the key processes of SOC are temperature dependent. To improve SOC and apoplastic carbon estimation, surveys by biomes, climatic zones, vegetation groups, and soil groups are needed and are regularly measured with inventories such as stem volume. Thus, forest carbon stocks are closely linked to environmental conditions and the effects of seasonal and climatic variables need to be considered.

The coupled PLUS and InVEST models are process-based ecosystem models, and the approach describes the effects of forest management and human activities on the forest carbon cycle in a single way, except for the uncertainties in the model structure, parameters, and drivers. For example, we can only generalize the effects of afforestation and forest restoration on forest carbon stocks by setting different forest area. Related studies have shown that the effect of forest restoration on soil carbon varies significantly by tree species and soil properties [48], and management activities that may reduce SOC content, such as thinning or harvesting, should also be considered [49]. Considering that the recovery of forest carbon stocks in northeastern forest areas in the past decades was mainly due to ecological projects such as afforestation and forest conservation, the development of human-natural coupled ecosystem carbon cycle models is crucial to accurately assess the carbon sequestration potential of forests.

Forest carbon stock estimation methods need to be more comprehensive and accurate. With the development of technology, the integration of LiDAR and VHR satellite imaging is a good combination for better biomass mapping and spatial accuracy. With the availability of higher resolution remote sensing imagery at various scales, this integration of multisensory techniques can improve the accuracy of regional forest carbon sink estimation [50]. In particular, with further developments in the field of deep learning, some convolutional neural network algorithms (CNN) may have the ability to estimate forest carbon stocks in combination with remotely sensed images. However, optimizing and validating the accuracy of long-duration forest carbon cycle simulation models remains a great challenge and biogeochemical processes, including photosynthesis, carbon uptake, allocation and release, should be incorporated into the models.

The atmospheric inversion method has the advantage of near real-time assessment of the extreme response of large-scale terrestrial carbon sinks to climate change. However, the current limitation of atmospheric inversion of terrestrial carbon sinks in China is the lack of long-term atmospheric CO_2 concentration observation data, let alone regional-scale carbon flux estimation with high spatial resolution [51]. The main reason is the current lack of domestic scientific observation satellites to provide advanced remote sensing CO_2 column concentration data, and only TANSat satellites are currently used for this purpose. Therefore, the development of a new generation of domestic high spatial and temporal resolution greenhouse gas concentration satellites, the establishment of high-resolution radiative transfer models and molecular spectral databases, the improvement of CO_2 column concentration observation accuracy, and the enhancement of our inversion capability effectively on the calculation of our forest carbon sink.

4.2. Carbon Effects from Natural Forests

Afforestation and adaptive forest management to increase forest biomass are considered to be the most direct and effective ways to reduce atmospheric CO_2. However, with the implementation of forest ecological conservation projects in the past 30 years, the space for suitable afforestation in the NCF is extremely limited. Related studies have shown that restored primary forests can maximize biomass and capture more carbon in the long term while conserving biodiversity [52,53]. Therefore, strengthening forest tending and restoring degraded forests is an inevitable choice to significantly improve the carbon effect of NCF. Intact old-growth forests are a major long-term carbon sink because of their complex structure, over-mature forests, stable soils, and resilience to fire, drought, pests, and diseases [54]. Although governments at all levels have been strengthening NCF natural forest conservation efforts, the loss of natural forests cannot be easily compensated for by human intervention [55,56]. Most forest ecosystems require up to 100 years to recover to their original levels of ecological services after destruction [57]. Therefore, it is crucial to protect the remaining natural forests. However, NCF needs to achieve trade-offs between timber production goals and forest conservation, justifying trade-offs based on sound science and best practices to achieve the highest and best outcomes [58]. The basic principle of not harming local communities, native ecosystems, and vulnerable species should be followed to achieve synergistic production and ecological goals [59]. Natural forest conservation requires the selection of appropriate natural restoration methods for different areas, which can be broadly classified as no intervention or passive restoration, low intervention (including prevention of further damage), intermediate intervention (selective planting of missing species and auxiliary natural regeneration), and high intervention (including the framework species method and application of the nucleation method) depending on the degree of human intervention. In the northeast region, the protection and management of the original natural forests must be strictly enforced. In the key development areas of the state-owned forest area, natural over-cutting forests are protected by enclosures, and for different vegetation levels, operation methods such as strip-shaped gradual cutting, group-shaped selective felling, and single-tree selective felling are adopted to maintain continuous forest coverage and a continuous supply of wood.

4.3. Value Transformation of Forest Carbon Sequestration

Reducing emissions from deforestation and forest degradation in developing countries, coupled with sustainable forest management and the protection and enhancement of FCS (REDD+), is an important part of global efforts to mitigate climate change. The sustainability of forest restoration lies in the fact that the value of ecological services generated by forest restoration is greater than the economic and social value generated by changing forest cover. However, there are still many problems with the process of realizing ecosystem value services, but this does not change their role in achieving the UN Sustainable Development Goals (SDGs) and their bright future prospects [60]. REDD+ has made some attempts to monetize forest carbon sinks, but there have been barriers to applying REDD+ to incentivize forest restoration because of regional differences in development levels, especially the instability of carbon trading prices [61]. China has already established a national carbon emissions trading market [62], but it is still in its infancy and many trading mechanisms are still imperfect; trading is mainly focused on the energy sector and does not involve forestry. Nevertheless, it provides an opportunity to realize the economic value of forest carbon sequestration in the future. In this study, we do not hide our concern about the future economic and social development situation of NCF, and this deteriorating trend seems to show no signs of improvement. However, the practice of carbon forestry seems to offer new options for the future development of NCF [63]. At present, for NCF and even China, the main obstacle to realizing the value of forest carbon sink is the lack of a unified and perfect forest carbon trading market and a relatively controllable trading price. Many scholars have explored the relationship between forest carbon sequestration and carbon prices by drawing on international experience and related practices [64,65], but there are still some challenges that may hinder the successful implementation of these techniques. This study attempts to comprehensively estimate the FCS of the NCF, but the results obtained cannot be used as the final carbon stock of the NCF. We ignored the carbon release from wood products, harvest residues, litter, and other components, and carbon fluxes from soils are often difficult to specify. These factors contribute to the instability in forest carbon sequestration. At the same time, the FCS may have been overestimated in this study because of the uncertain effects of drought-induced tree mortality, natural disasters, insect infestation, fire, or changes in existing forest areas.

5. Conclusions

From 1980 to 2000, there was a significant decline in forested land and grasslands in the NCF. The decrease in forested land is mainly concentrated in the south-central Lesser Khingan Mountains and Changbai Mountain areas. The arable land area grew more rapidly. From 2000 to 2020, the decreasing trend in forested land was alleviated and began to show slow growth, mainly concentrated in the Changbai Mountain area. The transformation between the various land types was relatively stable. Through the simulation of future land use, it was found that the expansion preference areas of various land types in the NCF were relatively concentrated. Forest expansion was mainly concentrated in the Greater Khingan Mountains, and the probability of partial forest land conversion to cultivated land in the Lesser Khingan Mountains is relatively high. The growth of cultivated land was mainly concentrated in the Sanjiang and Songnen plains. The expansion of construction land is mainly concentrated around the three provincial capital cities, accompanied by the transformation of a large amount of rural construction land into urban construction land. Forest land and cropland in the NCF were the most active land types, and the two land types were most closely interconverted. Owing to the mandatory food production and forest conservation attributes of NCF, the grassland area was in a state of reduction in all three models. Combining the current and future development trends of NCF, we believe that the EP scenario is the most suitable and likely development model.

The FCS of NCF was mainly contributed by closed forest land, and the aboveground and soil carbon pools accounted for 96.79% of the forest carbon pool. The time change showed a U-shaped trend of decline to growth, with an inflection point occurring in 2010. The loss of FCS was mainly concentrated in the south-central Lesser Khingan Mountains and northern Greater Khingan Mountains regions, mainly resulting from forestry exploitation and forest degradation, respectively. The FCS in the Changbai Mountain region remained relatively stable and grew faster after 2010. Under the EP scenario, the FCS is expected to recover to 1980 levels in NCF by 2050. By implementing a series of natural forest conservation measures, the NCF's forest carbon sequestration capacity will be greatly enhanced, which can help the Chinese government meet its carbon neutrality commitments.

Author Contributions: Conceptualization, J.S.; Data curation, J.S.; Formal analysis, J.S., W.Q. and Y.Z.; Visualization, J.S.; Writing—original draft, J.S.; supervision, Y.Z.; funding acquisition, Y.Z.; investigation, J.S. and G.C.; resources, G.C. All authors have read and agreed to the published version of the manuscript.

Funding: This research was funded by the National Natural Science Foundation of China (grant number 72173011).

Data Availability Statement: Not applicable.

Acknowledgments: We would like to thank the editor and anonymous reviewers for their comments, which helped improve the manuscript.

Conflicts of Interest: The authors declare no conflict of interest.

References

1. Lu, F.; Hu, H.; Sun, W.; Zhu, J.; Liu, G.; Zhou, W.; Zhang, Q.; Shi, P.; Liu, X.; Wu, X.; et al. Effects of National Ecological Restoration Projects on Carbon Sequestration in China from 2001 to 2010. *Proc. Natl. Acad. Sci. USA* **2018**, *115*, 4039–4044. [CrossRef] [PubMed]
2. Harris, N.L.; Gibbs, D.A.; Baccini, A.; Birdsey, R.A.; de Bruin, S.; Farina, M.; Fatoyinbo, L.; Hansen, M.C.; Herold, M.; Houghton, R.A.; et al. Global Maps of Twenty-First Century Forest Carbon Fluxes. *Nat. Clim. Chang.* **2021**, *11*, 234–240. [CrossRef]
3. Fang, J.; Guo, Z.; Hu, H.; Kato, T.; Muraoka, H.; Son, Y. Forest Biomass Carbon Sinks in East Asia, with Special Reference to the Relative Contributions of Forest Expansion and Forest Growth. *Glob. Chang. Biol.* **2014**, *20*, 2019–2030. [CrossRef] [PubMed]
4. Hua, F.; Wang, X.; Zheng, X.; Fisher, B.; Wang, L.; Zhu, J.; Tang, Y.; Yu, D.W.; Wilcove, D.S. Opportunities for Biodiversity Gains under the World's Largest Reforestation Programme. *Nat. Commun.* **2016**, *7*, 12717. [CrossRef]
5. Piao, S.; Yue, C.; Ding, J.; Guo, Z. Perspectives on the Role of Terrestrial Ecosystems in the 'Carbon Neutrality' Strategy. *Sci. China Earth Sci.* **2022**, *65*, 1178–1186. [CrossRef]
6. Fang, J.; Wang, G.G.; Liu, G.; Xu, S. Forest Biomass of China: An Estimate Based on the Biomass–Volume Relationship. *Ecol. Appl.* **1998**, *8*, 1084–1091. [CrossRef]
7. Guo, Z.; Fang, J.; Pan, Y.; Birdsey, R. Inventory-Based Estimates of Forest Biomass Carbon Stocks in China: A Comparison of Three Methods. *For. Ecol. Manag.* **2010**, *259*, 1225–1231. [CrossRef]
8. He, N.; Wen, D.; Zhu, J.; Tang, X.; Xu, L.; Zhang, L.; Hu, H.; Huang, M.; Yu, G. Vegetation Carbon Sequestration in Chinese Forests from 2010 to 2050. *Glob. Chang. Biol.* **2017**, *23*, 1575–1584. [CrossRef]
9. Hu, H.; Wang, S.; Guo, Z.; Xu, B.; Fang, J. The Stage-Classified Matrix Models Project a Significant Increase in Biomass Carbon Stocks in China's Forests between 2005 and 2050. *Sci. Rep.* **2015**, *5*, 11203. [CrossRef]
10. Luo, W.; Kim, H.S.; Zhao, X.; Ryu, D.; Jung, I.; Cho, H.; Harris, N.; Ghosh, S.; Zhang, C.; Liang, J. New Forest Biomass Carbon Stock Estimates in Northeast Asia Based on Multisource Data. *Glob. Chang. Biol.* **2020**, *26*, 7045–7066. [CrossRef]
11. Piao, S.; Fang, J.; Ciais, P.; Peylin, P.; Huang, Y.; Sitch, S.; Wang, T. The Carbon Balance of Terrestrial Ecosystems in China. *Nature* **2009**, *458*, 1009–1013. [CrossRef]
12. Fang, J.; Shen, Z.; Tang, Z.; Wang, X.; Wang, Z.; Feng, J.; Liu, Y.; Qiao, X.; Wu, X.; Zheng, C. Forest Community Survey and the Structural Characteristics of Forests in China. *Ecography* **2012**, *35*, 1059–1071. [CrossRef]
13. Tang, X.; Zhao, X.; Bai, Y.; Tang, Z.; Wang, W.; Zhao, Y.; Wan, H.; Xie, Z.; Shi, X.; Wu, B.; et al. Carbon Pools in China's Terrestrial Ecosystems: New Estimates Based on an Intensive Field Survey. *Proc. Natl. Acad. Sci. USA* **2018**, *115*, 4021–4026. [CrossRef]
14. Piao, S.; He, Y.; Wang, X.; Chen, F. Estimation of China's Terrestrial Ecosystem Carbon Sink: Methods, Progress and Prospects. *Sci. China Earth Sci.* **2022**, *65*, 641–651. [CrossRef]
15. Sun, W.; Liu, X. Review on Carbon Storage Estimation of Forest Ecosystem and Applications in China. *For. Ecosyst.* **2019**, *7*, 4. [CrossRef]
16. Griscom, B.W.; Adams, J.; Ellis, P.W.; Houghton, R.A.; Lomax, G.; Miteva, D.A.; Schlesinger, W.H.; Shoch, D.; Siikamäki, J.V.; Smith, P.; et al. Natural Climate Solutions. *Proc. Natl. Acad. Sci. USA* **2017**, *114*, 11645–11650. [CrossRef]
17. Chang, X.; Xing, Y.; Wang, J.; Yang, H.; Gong, W. Effects of Land Use and Cover Change (LUCC) on Terrestrial Carbon Stocks in China between 2000 and 2018. *Resour. Conserv. Recycl.* **2022**, *182*, 106333. [CrossRef]
18. Homer, C.; Dewitz, J.; Jin, S.; Xian, G.; Costello, C.; Danielson, P.; Gass, L.; Funk, M.; Wickham, J.; Stehman, S.; et al. Conterminous United States Land Cover Change Patterns 2001–2016 from the 2016 National Land Cover Database. *ISPRS J. Photogramm. Remote Sens.* **2020**, *162*, 184–199. [CrossRef]
19. Hua, F.; Bruijnzeel, L.A.; Meli, P.; Martin, P.A.; Zhang, J.; Nakagawa, S.; Miao, X.; Wang, W.; McEvoy, C.; Peña-Arancibia, J.L.; et al. The Biodiversity and Ecosystem Service Contributions and Trade-Offs of Forest Restoration Approaches. *Science* **2022**, *376*, 839–844. [CrossRef]
20. Huang, W. Forest Condition Change, Tenure Reform, and Government-Funded Eco-Environmental Programs in Northeast China. *For. Policy Econ.* **2019**, *98*, 67–74. [CrossRef]
21. Naime, J.; Mora, F.; Sánchez-Martínez, M.; Arreola, F.; Balvanera, P. Economic Valuation of Ecosystem Services from Secondary Tropical Forests: Trade-Offs and Implications for Policy Making. *For. Ecol. Manag.* **2020**, *473*, 118294. [CrossRef]
22. Seibold, S.; Rammer, W.; Hothorn, T.; Seidl, R.; Ulyshen, M.D.; Lorz, J.; Cadotte, M.W.; Lindenmayer, D.B.; Adhikari, Y.P.; Aragón, R.; et al. The Contribution of Insects to Global Forest Deadwood Decomposition. *Nature* **2021**, *597*, 77–81. [CrossRef]
23. Li, Y.; Piao, S.; Li, L.Z.X.; Chen, A.; Wang, X.; Ciais, P.; Huang, L.; Lian, X.; Peng, S.; Zeng, Z.; et al. Divergent Hydrological Response to Large-Scale Afforestation and Vegetation Greening in China. *Sci. Adv.* **2018**, *4*, eaar4182. [CrossRef]
24. Wang, S.; Zhou, C.; Liu, J.; Tian, H.; Li, K.; Yang, X. Carbon Storage in Northeast China as Estimated from Vegetation and Soil Inventories. *Environ. Pollut.* **2002**, *116*, S157–S165. [CrossRef]
25. Wang, X.; Wang, S.; Dai, L. Estimating and Mapping Forest Biomass in Northeast China Using Joint Forest Resources Inventory and Remote Sensing Data. *J. For. Res.* **2018**, *29*, 797–811. [CrossRef]
26. Wei, Y.; Yu, D.; Lewis, B.J.; Zhou, L.; Zhou, W.; Fang, X.; Zhao, W.; Wu, S.; Dai, L. Forest Carbon Storage and Tree Carbon Pool Dynamics under Natural Forest Protection Program in Northeastern China. *Chin. Geogr. Sci.* **2014**, *24*, 397–405. [CrossRef]
27. Wei, Y.; Li, M.; Chen, H.; Lewis, B.J.; Yu, D.; Zhou, L.; Zhou, W.; Fang, X.; Zhao, W.; Dai, L. Variation in Carbon Storage and Its Distribution by Stand Age and Forest Type in Boreal and Temperate Forests in Northeastern China. *PLoS ONE* **2013**, *8*, e72201. [CrossRef]

28. Dong, L.; Lu, W.; Liu, Z. Developing Alternative Forest Spatial Management Plans When Carbon and Timber Values Are Considered: A Real Case from Northeastern China. *Ecol. Model.* **2018**, *385*, 45–57. [CrossRef]
29. Liang, X.; Liu, X.; Li, X.; Chen, Y.; Tian, H.; Yao, Y. Delineating Multi-Scenario Urban Growth Boundaries with a CA-Based FLUS Model and Morphological Method. *Landsc. Urban Plan.* **2018**, *177*, 47–63. [CrossRef]
30. Liang, X.; Liu, X.; Li, D.; Zhao, H.; Chen, G. Urban Growth Simulation by Incorporating Planning Policies into a CA-Based Future Land-Use Simulation Model. *Int. J. Geogr. Inf. Sci.* **2018**, *32*, 2294–2316. [CrossRef]
31. Liu, X.; Liang, X.; Li, X.; Xu, X.; Ou, J.; Chen, Y.; Li, S.; Wang, S.; Pei, F. A Future Land Use Simulation Model (FLUS) for Simulating Multiple Land Use Scenarios by Coupling Human and Natural Effects. *Landsc. Urban Plan.* **2017**, *168*, 94–116. [CrossRef]
32. Nottingham, A.T.; Meir, P.; Velasquez, E.; Turner, B.L. Soil Carbon Loss by Experimental Warming in a Tropical Forest. *Nature* **2020**, *584*, 234–237. [CrossRef]
33. Chen, G.; Li, X.; Liu, X.; Chen, Y.; Liang, X.; Leng, J.; Xu, X.; Liao, W.; Qiu, Y.; Wu, Q.; et al. Global Projections of Future Urban Land Expansion under Shared Socioeconomic Pathways. *Nat. Commun.* **2020**, *11*, 537. [CrossRef]
34. Liang, X.; Guan, Q.; Clarke, K.C.; Liu, S.; Wang, B.; Yao, Y. Understanding the Drivers of Sustainable Land Expansion Using a Patch-Generating Land Use Simulation (PLUS) Model: A Case Study in Wuhan, China. *Comput. Environ. Urban Syst.* **2021**, *85*, 101569. [CrossRef]
35. Zhang, D.; Wang, X.; Qu, L.; Li, S.; Lin, Y.; Yao, R.; Zhou, X.; Li, J. Land Use/Cover Predictions Incorporating Ecological Security for the Yangtze River Delta Region, China. *Ecol. Indic.* **2020**, *119*, 106841. [CrossRef]
36. Delgado, R.; Tibau, X.-A. Why Cohen's Kappa Should Be Avoided as Performance Measure in Classification. *PLoS ONE* **2019**, *14*, e0222916. [CrossRef]
37. Pontius, R.G.; Millones, M. Death to Kappa: Birth of Quantity Disagreement and Allocation Disagreement for Accuracy Assessment. *Null* **2011**, *32*, 4407–4429. [CrossRef]
38. Pontius, R.G.; Boersma, W.; Castella, J.-C.; Clarke, K.; de Nijs, T.; Dietzel, C.; Duan, Z.; Fotsing, E.; Goldstein, N.; Kok, K.; et al. Comparing the Input, Output, and Validation Maps for Several Models of Land Change. *Ann. Reg. Sci.* **2008**, *42*, 11–37. [CrossRef]
39. Pontius, R.G.; Peethambaram, S.; Castella, J.-C. Comparison of Three Maps at Multiple Resolutions: A Case Study of Land Change Simulation in Cho Don District, Vietnam. *Null* **2011**, *101*, 45–62. [CrossRef]
40. Wang, S.; Bai, X.; Zhang, X.; Reis, S.; Chen, D.; Xu, J.; Gu, B. Urbanization Can Benefit Agricultural Production with Large-Scale Farming in China. *Nat. Food* **2021**, *2*, 183–191. [CrossRef]
41. Wang, S.; Zhang, Y.; Ju, W.; Chen, J.M.; Ciais, P.; Cescatti, A.; Sardans, J.; Janssens, I.A.; Wu, M.; Berry, J.A.; et al. Recent Global Decline of CO_2 Fertilization Effects on Vegetation Photosynthesis. *Science* **2020**, *370*, 1295–1300. [CrossRef]
42. Zhu, Z.; Piao, S.; Myneni, R.B.; Huang, M.; Zeng, Z.; Canadell, J.G.; Ciais, P.; Sitch, S.; Friedlingstein, P.; Arneth, A.; et al. Greening of the Earth and Its Drivers. *Nat. Clim. Chang.* **2016**, *6*, 791–795. [CrossRef]
43. Schulte-Uebbing, L.F.; Ros, G.H.; de Vries, W. Experimental Evidence Shows Minor Contribution of Nitrogen Deposition to Global Forest Carbon Sequestration. *Glob. Chang. Biol.* **2022**, *28*, 899–917. [CrossRef]
44. Chen, G.; Xie, J.; Li, W.; Li, X.; Hay Chung, L.C.; Ren, C.; Liu, X. Future "Local Climate Zone" Spatial Change Simulation in Greater Bay Area under the Shared Socioeconomic Pathways and Ecological Control Line. *Build. Environ.* **2021**, *203*, 108077. [CrossRef]
45. Wang, Z.; Li, X.; Mao, Y.; Li, L.; Wang, X.; Lin, Q. Dynamic Simulation of Land Use Change and Assessment of Carbon Storage Based on Climate Change Scenarios at the City Level: A Case Study of Bortala, China. *Ecol. Indic.* **2022**, *134*, 108499. [CrossRef]
46. Zhang, S.; Zhong, Q.; Cheng, D.; Xu, C.; Chang, Y.; Lin, Y.; Li, B. Landscape Ecological Risk Projection Based on the PLUS Model under the Localized Shared Socioeconomic Pathways in the Fujian Delta Region. *Ecol. Indic.* **2022**, *136*, 108642. [CrossRef]
47. Tashi, S.; Singh, B.; Keitel, C.; Adams, M. Soil Carbon and Nitrogen Stocks in Forests along an Altitudinal Gradient in the Eastern Himalayas and a Meta-Analysis of Global Data. *Glob. Chang. Biol.* **2016**, *22*, 2255–2268. [CrossRef] [PubMed]
48. Hong, S.; Yin, G.; Piao, S.; Dybzinski, R.; Cong, N.; Li, X.; Wang, K.; Peñuelas, J.; Zeng, H.; Chen, A. Divergent Responses of Soil Organic Carbon to Afforestation. *Nat. Sustain.* **2020**, *3*, 694–700. [CrossRef]
49. Jandl, R.; Lindner, M.; Vesterdal, L.; Bauwens, B.; Baritz, R.; Hagedorn, F.; Johnson, D.W.; Minkkinen, K.; Byrne, K.A. How Strongly Can Forest Management Influence Soil Carbon Sequestration? *Geoderma* **2007**, *137*, 253–268. [CrossRef]
50. Baccini, A.; Goetz, S.J.; Walker, W.S.; Laporte, N.T.; Sun, M.; Sulla-Menashe, D.; Hackler, J.; Beck, P.S.A.; Dubayah, R.; Friedl, M.A.; et al. Estimated Carbon Dioxide Emissions from Tropical Deforestation Improved by Carbon-Density Maps. *Nat. Clim. Chang.* **2012**, *2*, 182–185. [CrossRef]
51. Wang, Y.; Wang, X.; Wang, K.; Chevallier, F.; Zhu, D.; Lian, J.; He, Y.; Tian, H.; Li, J.; Zhu, J.; et al. The Size of the Land Carbon Sink in China. *Nature* **2022**, *603*, E7–E9. [CrossRef]
52. Díaz, S.; Hector, A.; Wardle, D.A. Biodiversity in Forest Carbon Sequestration Initiatives: Not Just a Side Benefit. *Curr. Opin. Environ. Sustain.* **2009**, *1*, 55–60. [CrossRef]
53. Lewis, S.L.; Wheeler, C.E.; Mitchard, E.T.A.; Koch, A. Restoring Natural Forests Is the Best Way to Remove Atmospheric Carbon. *Nature* **2019**, *568*, 25–28. [CrossRef]
54. Maxwell, S.L.; Evans, T.; Watson, J.E.M.; Morel, A.; Grantham, H.; Duncan, A.; Harris, N.; Potapov, P.; Runting, R.K.; Venter, O.; et al. Degradation and Forgone Removals Increase the Carbon Impact of Intact Forest Loss by 626%. *Sci. Adv.* **2019**, *5*, eaax2546. [CrossRef]

55. Brancalion, P.H.S.; Chazdon, R.L. Beyond Hectares: Four Principles to Guide Reforestation in the Context of Tropical Forest and Landscape Restoration: Forest and Landscape Restoration Principles. *Restor. Ecol.* **2017**, *25*, 491–496. [CrossRef]
56. Wheeler, C.E.; Omeja, P.A.; Chapman, C.A.; Glipin, M.; Tumwesigye, C.; Lewis, S.L. Carbon Sequestration and Biodiversity Following 18 years of Active Tropical Forest Restoration. *For. Ecol. Manag.* **2016**, *373*, 44–55. [CrossRef]
57. Gibson, L.; Lee, T.M.; Koh, L.P.; Brook, B.W.; Gardner, T.A.; Barlow, J.; Peres, C.A.; Bradshaw, C.J.A.; Laurance, W.F.; Lovejoy, T.E.; et al. Primary Forests Are Irreplaceable for Sustaining Tropical Biodiversity. *Nature* **2011**, *478*, 378–381. [CrossRef]
58. Gann, G.D.; McDonald, T.; Walder, B.; Aronson, J.; Nelson, C.R.; Jonson, J.; Hallett, J.G.; Eisenberg, C.; Guariguata, M.R.; Liu, J.; et al. International Principles and Standards for the Practice of Ecological Restoration. Second Edition. *Restor. Ecol.* **2019**, *27*, S1–S46. [CrossRef]
59. Di Sacco, A.; Hardwick, K.A.; Blakesley, D.; Brancalion, P.H.S.; Breman, E.; Cecilio Rebola, L.; Chomba, S.; Dixon, K.; Elliott, S.; Ruyonga, G.; et al. Ten Golden Rules for Reforestation to Optimize Carbon Sequestration, Biodiversity Recovery and Livelihood Benefits. *Glob. Chang. Biol.* **2021**, *27*, 1328–1348. [CrossRef]
60. Li, R.; Zheng, H.; O'Connor, P.; Xu, H.; Li, Y.; Lu, F.; Robinson, B.E.; Ouyang, Z.; Hai, Y.; Daily, G.C. Time and Space Catch up with Restoration Programs That Ignore Ecosystem Service Trade-Offs. *Sci. Adv.* **2021**, *7*, eabf8650. [CrossRef]
61. Köhl, M.; Neupane, P.R.; Mundhenk, P. REDD+ Measurement, Reporting and Verification—A Cost Trap? Implications for Financing REDD+MRV Costs by Result-Based Payments. *Ecol. Econ.* **2020**, *168*, 106513. [CrossRef]
62. Jiang, W.; Chen, Y. The Time-Frequency Connectedness among Carbon, Traditional/New Energy and Material Markets of China in Pre- and Post-COVID-19 Outbreak Periods. *Energy* **2022**, *246*, 123320. [CrossRef]
63. Peng, W.; Pukkala, T.; Jin, X.; Li, F. Optimal Management of Larch (Larix Olgensis A. Henry) Plantations in Northeast China When Timber Production and Carbon Stock Are Considered. *Ann. For. Sci.* **2018**, *75*, 63. [CrossRef]
64. Dong, L.; Bettinger, P.; Liu, Z. Estimating the Optimal Internal Carbon Prices for Balancing Forest Wood Production and Carbon Sequestration: The Case of Northeast China. *J. Clean. Prod.* **2021**, *281*, 125342. [CrossRef]
65. Qin, H.; Dong, L.; Huang, Y. Evaluating the Effects of Carbon Prices on Trade-Offs between Carbon and Timber Management Objectives in Forest Spatial Harvest Scheduling Problems: A Case Study from Northeast China. *Forests* **2017**, *8*, 43. [CrossRef]

remote sensing

MDPI

Article

Forest Height Mapping Using Feature Selection and Machine Learning by Integrating Multi-Source Satellite Data in Baoding City, North China

Nan Zhang [1], Mingjie Chen [1], Fan Yang [2], Cancan Yang [1,3], Penghui Yang [1], Yushan Gao [1], Yue Shang [1] and Daoli Peng [1,*]

1 State Forestry and Grassland Administration Key Laboratory of Forest Resources & Environmental Management, College of Forestry, Beijing Forestry University, Beijing 100083, China
2 Academy of Inventory and Planning, National Forestry and Grassland Administration, Beijing 100714, China
3 Anhui Province Key Laboratory of Physical Geographic Environment, Chuzhou University, Chuzhou 239000, China
* Correspondence: dlpeng@bjfu.edu.cn

Citation: Zhang, N.; Chen, M.; Yang, F.; Yang, C.; Yang, P.; Gao, Y.; Shang, Y.; Peng, D. Forest Height Mapping Using Feature Selection and Machine Learning by Integrating Multi-Source Satellite Data in Baoding City, North China. *Remote Sens.* **2022**, *14*, 4434. https://doi.org/10.3390/rs14184434

Academic Editors: Huaqiang Du, Wenyi Fan, Weiliang Fan, Fangjie Mao and Mingshi Li

Received: 8 August 2022
Accepted: 4 September 2022
Published: 6 September 2022

Abstract: Accurate estimation of forest height is crucial for the estimation of forest aboveground biomass and monitoring of forest resources. Remote sensing technology makes it achievable to produce high-resolution forest height maps in large geographical areas. In this study, we produced a 25 m spatial resolution wall-to-wall forest height map in Baoding city, north China. We evaluated the effects of three factors on forest height estimation utilizing four types of remote sensing data (Sentinel-1, Sentinel-2, ALOS PALSAR-2, and SRTM DEM) with the National Forest Resources Continuous Inventory (NFCI) data, three feature selection methods (stepwise regression analysis (SR), recursive feature elimination (RFE), and Boruta), and six machine learning algorithms (k-nearest neighbor (k-NN), support vector machine regression (SVR), random forest (RF), gradient boosting decision tree (GBDT), extreme gradient boosting (XGBoost), and categorical boosting (CatBoost)). ANOVA was adopted to quantify the effects of three factors, including data source, feature selection method, and modeling algorithm, on forest height estimation. The results showed that all three factors had a significant influence. The combination of multiple sensor data improved the estimation accuracy. Boruta's overall performance was better than SR and RFE, and XGBoost outperformed the other five machine learning algorithms. The variables selected based on Boruta, including Sentinel-1, Sentinel-2, and topography metrics, combined with the XGBoost algorithm, provided the optimal model (R^2 = 0.67, RMSE = 2.2 m). Then, we applied the best model to create the forest height map. There were several discrepancies between the generated forest height map and the existing map product, and the values with large differences between the two maps were mostly distributed in the steep areas with high slope values. Overall, we proposed a methodological framework for quantifying the importance of data source, feature selection method, and machine learning algorithm in forest height estimation, and it was proved to be effective in estimating forest height by using freely accessible multi-source data, advanced feature selection method, and machine learning algorithm.

Keywords: forest height; multi-source data; feature selection; machine learning algorithm

1. Introduction

Forest is an important part of terrestrial ecosystems and plays a vital role in maintaining the global ecological balance, promoting global biological evolution and community succession [1–3]. As an important part of the structure parameters of the forest, forest height is not only an essential indicator for the quantitative estimation of forest biomass and terrestrial carbon circulation but also important auxiliary information for evaluating forest resources and establishing earth system models [4,5]. Traditional forest height estimation mainly depends on the means of manual field surveys. Although the ground survey method has high accuracy, it is timing and force-consuming, and it is difficult to achieve large-range and long-span forest height estimation and dynamic change monitoring [6]. The increasingly developed remote sensing technology has the advantages of multi-time phase, multi-scale, multi-sensor, and rapid macro monitoring. It has become an important way to estimate forest height by constructing empirical models combining remote sensing data and ground-measured data [7].

At present, the most recent advancement in remote sensing technology advocates producing forest height maps of large geographical areas with high resolution. Multispectral data [8–10], Light Detection and Ranging (LiDAR) [11–14], Synthetic Aperture Radar (SAR) [15,16], and other remote sensing data [17] were widely applied. LiDAR data are often regarded as the best remote sensing data source for forest structure parameters due to its direct ability to detect forest vertical structures; however, terrestrial laser scanning (TLS) and airborne laser scanning (ALS) are typically limited by high application costs [18], and it is difficult to generate wall-to-wall forest height maps in large areas due to the sparse measurements in the space of satellite LiDAR [19] Compared to lidar data, optical data are more susceptible to the influence of weather conditions and has issues such as limited sensitivity and low saturation in dense vegetation areas, SAR data are susceptible to terrain and speckle noise, and there is a problem of backscatter signal saturation in high vegetation coverage areas as well as optical data. Nevertheless, the backscattering coefficient of SAR and the rich spectral information of the optical data can also reflect the information about the structure and function of the forest [20,21]. Most importantly, optical data and SAR data can be obtained frequently, continuously, and at a low cost from various spaceborne platforms. In the past few years, numerous studies have shown that spectral reflectance, vegetation index, and spatial texture information extracted from Sentinel-2 images, backscattering coefficients, indices, and texture features calculated from Sentinel-1 C-band, ALOS-2 PALSAR-2 L-band images, and topographic metrics were effective in estimating forest canopy height and other forest parameters [22–26].

As mentioned above, there are many potential feature variables when estimating forest height using multi-source remote sensing data. High-dimensional feature variables will increase the computational load, data noise, and interference, and the problem of complex collinearity between variables will cause the redundancy of variables, which will affect the efficiency and accuracy of modeling [27,28]; therefore, the correct and efficient feature selection phase is an essential step for forest height estimation. However, because of the diverse characteristics of the sensor data and the complex biophysical environment in the forestry areas, the different feature selection methods correspond to different data structures and features, what effect of feature selection method on forest height estimation, and how to determine the best feature selection method is still poorly understood [27]. Stepwise regression analysis is the most commonly used variable selection approach in forest parameter investigations and related studies have reported positive outcomes [29–31]. In addition, the Boruta and recursive feature elimination are both well-established wrapper methods, which have been widely applied in the study of forestry research in recent years [32–35]. Several studies have been conducted to examine the impact of different feature selection strategies in predicting forest characteristics [36,37]. Nevertheless, to our knowledge, there is rarely research conducted to examine the impact of feature selection methods for different remote sensing data sources when estimating forest height.

Another key factor of forest height estimation is the regression algorithm. Currently, regression models used to estimate forest height can be divided into two categories: parametric and non-parametric algorithms. In the parametric model, there are quantitative mathematical expressions between the independent and dependent variables, which are intuitive and simple to understand. Multiple linear regression, stepwise regression, and partial least squares regression are common parametric models; however, the parameter model needs to meet the premise that the relationships between dependent and independent variables have clear model structures, while the relationship between forest height and remote sensing factors is typically quite complex, which limits the application of parametric models [27]. Compared with parametric algorithms, non-parametric algorithms based on data mining, machine learning, and other mathematical theory and methods, through the way of data-driven achieving complex nonlinear relationship prediction, are widely used in forest height estimation, including k-nearest neighbor (k-NN), support vector machine regression (SVR) and random forest (RF) [38–42]. Moreover, some decision-tree-based ensemble algorithms, such as gradient boosting decision tree (GBDT), extreme gradient boosting (XGBoost), and categorical boosting (CatBoost), have performed well in the estimation of forest aboveground biomass [43,44]; however, these algorithms are rarely employed to estimate forest height, and their efficacy has yet to be evaluated.

In summary, to address the gaps mentioned above, we proposed a methodological framework for forest height estimation and mapping using multi-source remote sensing data (Sentinel-2, Sentinel-1, ALOS PALSAR-2, SRTM DEM), three feature selection methods (SR, RFE, Boruta) and six machine learning algorithms (k-NN, SVR, RF, GBDT, XGBoost, and CatBoost) in Baoding city, north China. The purposes of this study are as follows:

(1) To examine the influence of feature selection methods of different remote sensing data sources on forest height estimation, and to explore the optimal feature selection method;

(2) To evaluate the performance of machine learning algorithms based on different feature selection methods in forest tree height estimation;

(3) To generate a forest height distribution map of 25 m spatial resolution in Baoding city, and to analyze the important factors in forest height estimation.

2. Materials and Methods

2.1. Study Area

The study area is located in Baoding city in the Midwest of Hebei province, China (38°14–39°57′N, 113°45–116°19′E), covering an area of about 2,211,200 hectares (Figure 1). It is situated near the eastern foot of the northern Taihang Mountains and on the western part of the Jizhong plain. The terrain is inclined from northwest to Southeast. The landforms in the west are mountainous, which are composed of mountains and hills; the landforms in the east region belonging to the North China Plain are flat. Baoding is in the warm temperate continental monsoon climate zone, with an annual average temperature of 12.7 °C and 2511 h of sunshine per year, accounting for 56% of total sunshine hours. The annual frost-free period is about 165–210 days. The period from June to August each year is a period of intensive precipitation, and the average annual precipitation duration is 68 days with an average precipitation of 489.9 mm. The forestry area of Baoding is nearly 590,000 hectares, accounting for approximately 28% of the administrative area of the city, and the forest stock of the whole city reaches 13.7 million cubic meters. Forest types mainly include coniferous forest, broadleaf forest, and mixed conifer-broad-leaf forest. Among them, coniferous trees are mainly Chinese pine (*Pinus tabulaeformis*) and oriental arborvitae (*Platycladus orientalis*); Broadleaf trees mainly include populus tremula (*Populus davidana*), Mongolian oak (*Quercus mongolica*), white birch (*Betula platydiana*), and acacia (*Robinia pseudoacacia*).

2.2. Methodological Framework of This Study

In this study, we proposed a methodological framework utilizing different feature selection methods and machine learning algorithms to establish forest height estimation models based on multi-source satellite data in the forest regions of Baoding city, north China. Our methodological framework consists of four primary components (Figure 2): (1) data preparation and preprocessing, (2) feature variables selection, (3) model building and assessment, and (4) forest height mapping and important factors analysis.

2.3. Data Source and Preprocessing

2.3.1. Field Data Collection

The field data utilized in this study is the ninth National Forest Resources Continuous Inventory (NFCI) data of Hebei Province. The field survey was conducted in November 2016. The sample plots were systematically arranged at an interval of 4 km × 4 km along a vertical and horizontal coordinate system. The sample plot was a square plot with a side length of 25.82 m, and each sample plot area was about 0.067 ha.

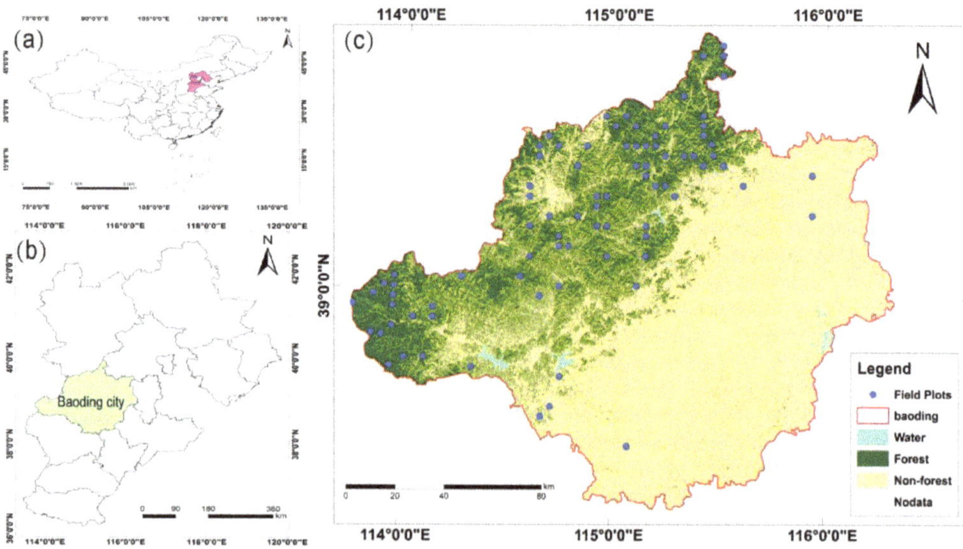

Figure 1. Overview of the study site. (**a**) Location of the Hebei province in China; (**b**) location of the Baoding city in Hebei province; (**c**) general land cover classes (forest, non-forest, and water) and distribution of field plots in Baoding city.

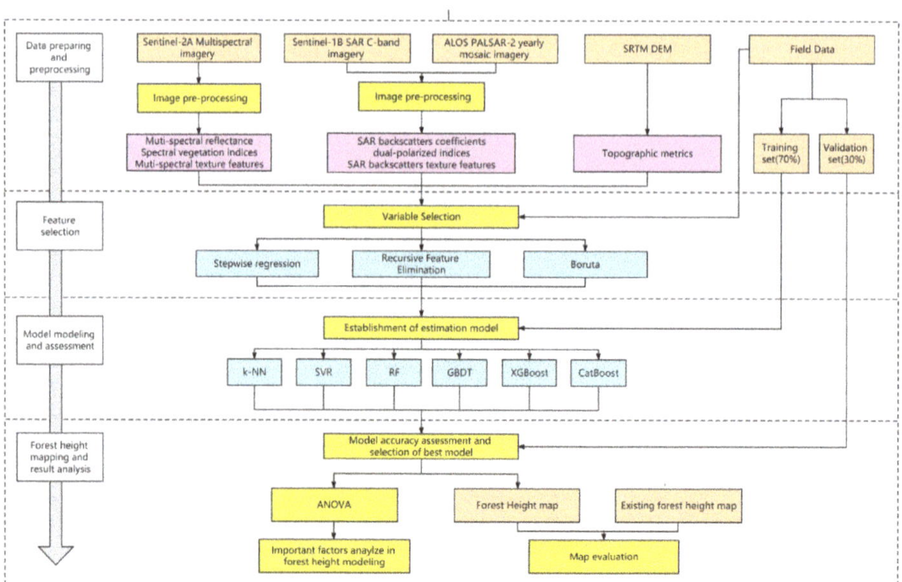

Figure 2. Flowchart of the proposed methodology for estimating forest height in Baoding city using three feature selection methods and six machine learning algorithms based on multi-source remote sensing data.

Each tree with a diameter at breast height (DBH) higher than 5 cm had its DBH, tree height, and crown height measured, as well as the land use, dominant tree species, tree species composition, average DBH, and average tree height were recorded. There were 1210 sample plots in Baoding city, and 128 sample plots were finally collected after removing the sample plots of non-forest land and inadequate information. The average tree height of the forest sample plot ranged from 3.00 m to 24.50 m, and the average, median, and standard deviation (std) were 8.57 m, 7.30 m, and 3.89 m, respectively. Among the 128 sample plots, 91 sample plots (70%) were randomly selected for training, and the remaining 37 sample plots (30%) were used as the validation data set for the machine learning model (Table 1).

Table 1. The statistics of forest height in training, testing, and total sample datasets.

Dataset	Sample Size	Min (m)	Max (m)	Mean (m)	Median (m)	Std (m)
Training	91	3.00	24.50	8.57	7.50	3.92
Validation	37	3.20	18.40	8.58	7.20	3.87
Total	128	3.00	24.50	8.57	7.30	3.89

2.3.2. Sentinel-2 Multispectral Imagery and Preprocessing

The multispectral images used in this study were Sentinel-2 satellite images from the European Space Agency (ESA). The multispectral imager instrument carried by the Sentinel-2 satellite has the advantages of high spatial resolution, excellent multispectral imaging capacity, wide wing, and short revisit cycle, which can be used to monitor the distribution and health of forests. The Sentinel-2 satellite image incorporates 13 bands, with spatial resolutions of 10 m for bands 2–4 and 8 (blue: 490 nm, green: 560 nm, red: 665 nm, and NIR: 842 nm), 20 m for bands 5–7, 8A, 11, and 12 ((red edge 1: 705 nm, red edge 2: 740 nm, red edge 3: 783 nm, narrow NIR: 865 nm, SWIR1: 1610 nm, and SWIR2: 2190 nm), and 60 m for the other three bands (coastal aerosol: 443 nm, water vapor: 940 nm, and SWIR cirrus: 1375 nm). The bands with spatial resolutions of 10 m and 20 m were employed in this study.

In order to match the time of sample plot data collection, we downloaded seven Sentinel-2 Level-1C images covering the study area with less than 10% cloud from the United States Geological Service's Earth Explorer (USGS) (https://earthexplorer.usgs.gov/ (accessed on 24 March 2022)) which were obtained in the growing season in August 2016. Since the Sentinel-2 Level-1C image is the top atmospheric reflectance image, we used the atmospheric correction processor (version 2.5.5, European Space Agency, Paris, France) of Sentinel Application Platform (SNAP) software (version 8.0, ESA, Paris, France) to acquire the Level-2A products, the bottom-of-atmosphere-corrected reflectance images. To match the field plot sizes, we resampled the preprocessed Sentinel-2 images to 25 m pixel sizes. Then, mosaicking and clipping were completed to cover the study area.

2.3.3. Synthetic Aperture Radar (SAR) Data and Preprocessing

We used synthetic aperture radar data from two different data sources, including the Sentinel-1 C-band imagery and ALOS-2 PALSAR-2 yearly mosaic imagery.

Sentinel-1 is composed of two polar-orbiting satellites, and the revisit period of a single satellite is 12 days. A total of 10 sentinel-1 ground range detected (GRD) images with good quality from October 2016 were obtained from the Google earth engine (GEE) cloud computing platform. We acquired the dual-polarization (VV and VH) images in Interferometric Wide swath (IW) mode with an ascending orbital pass. These images in GEE were already processed by the ESA Sentinel-1 toolbox, including thermal noise removal, radiometric correction, terrain correction, and conversion of the backscattering coefficient to decibels [45]. Here, we further processed them according to the framework proposed by Mullissa et al. in 2021 [46], including border noise correction, refined Lee filter for speckle filtering, and radiometric terrain normalization.

Due to the fact that PALSAR- 2 images in Baoding city were not free, the L-band SAR imagery had not been applied for this study; however, the Japan Aerospace Exploration Agency (JAXA) provides the 25 m spatial resolution ALOS/PALSAR yearly mosaic, which is produced by mosaicking SAR images measured by PALSAR-2 available each year [47]. We obtained the mosaic data in the year 2016 from GEE in this study. This SAR imagery was already ortho-rectificatied by using the 90 m SRTM Digital Elevation Model. The data were stored as 16-bit digital numbers (DN), which were converted to gamma naught values (γ_0) in decibel unit (dB) using the following equation: $\gamma_0 = 10log_{10}(DN^2) - 83.0$ dB. All of the SAR images were resampled to the same pixel sizes to ensure consistency with other data.

2.3.4. Topographic and Ancillary Data

The digital elevation model (DEM) reflects the abundant terrain information of the mountain region and provides great assistance to forest height estimation [23]. In this study, we used the Shuttle Radar Topography Mission (SRTM) V3 product, which was provided by NASA JPL at a resolution of approximately 30 m. Furthermore, we applied the FROM-GLC 2017 (Finer Resolution Observation and Monitoring of Global Land Cover at30-m resolution, 2017v1) product to define the forest regions of the study area [48].

2.4. *Feature Variable Extraction*

Based on the remote sensing data sources mentioned above, a total of 153 feature variables were extracted in this study (Table 2). For Sentinel-2 data, we extracted 10 multispectral variables from the average surface reflectance of 10 multispectral bands with spatial resolutions of 10 m and 20 m. Then, 20 vegetation indices derived from Sentinel-2 data, which were widely used in previous forest studies, were calculated [49–51]. Moreover, the texture features of 10 multispectral bands, including mean, variance, homogeneity, contrast, dissimilarity, entropy, second moment, and correlation, were calculated by using the gray level co-occurrence matrix (GLCM) with a 3×3 window. Finally, a total of 110 feature variables derived from Sentinel-2 data were obtained. As to SAR data, we extracted VH and VV backscattering coefficients from Sentinel-1 imagery and HH and HV backscattering coefficients from ALOS PALSAR-2 yearly mosaic, respectively. After that, the ratio and normalized polarized difference of VH, VV, and HV, HH were calculated as candidate variables, respectively. GLCM was also used to compute the texture features of VH, VV, HH, and HV backscattering coefficients by using a 3×3 window. Finally, 40 SAR feature variables were obtained. In addition, we extracted elevation, slope, and aspect from the DEM image as terrain factors. To analyze the impact of different data sources on forest height estimation, five combination scenarios were designed in this study (Table 3).

2.5. *Feature Variable Selection*

In this study, we employed stepwise regression analysis, recursive feature elimination, and Boruta methods to select and analyze feature variables from five combination scenarios, with all field measurements serving as a reference.

2.5.1. Stepwise Regression Analysis

In the past few decades, stepwise regression analysis (SR) has been widely used for feature selection for forest parameters estimation studies [22,52–54]. The basic principle of stepwise regression is to successively add the most contributing predictor variables in order. After adding each new variable, all variables that no longer improve the model fit were removed. The program will stop running until no variables are selected or dropped [31]. In our research, we screened the best subset of variables by iterative both-direction stepwise regression based on the Akaike information criterion (AIC) and ensured the p-values of all the selected variables were significant ($p < 0.05$) [55]. This procedure was performed in R 4.2.0 using the "MASS" package [56].

Table 2. Summary of the metrics extracted from multi-source data used in this study.

Source		Feature Variables	Description
Sentinel-2 multispectral data	Multispectral bands (10)	b2	Blue, 490 nm
		b3	Green, 560 nm
		b4	Red, 665 nm
		b5	Red edge, 705 nm
		b6	Red edge, 749 nm
		b7	Red edge, 783 nm
		b8	Near-infrared, 842 nm
		b8a	Near-infrared, 865 nm
		b11	Short-wave infrared, 1610 nm
		b12	Short-wave infrared, 2190 nm
	Vegetation indices (20)	SAVI	Soil adjusted vegetation index, 1.5 × (B8−B4)/(B8 + B4 + 0.5)
		NDVI	Normalized difference vegetation index, (B8 − B4)/(B8 + B4)
		MSAVI2	Second modified soil adjusted vegetation index, 0.5 × [2 × (B8 + 1) − sqrt[(2 × B8 + 1) × (2 × B8 + 1) − 8 × (B8 − B4)]]
		RVI	Ratio vegetation index, B8/B4
		PVI	Perpendicular vegetation index, sin(a) × B8 − cos(a) × B4(a = 45°)
		IPVI	Infrared percentage vegetation index, B8/(B8 + B4)
		WDVI	Weighted difference vegetation index, B8 − 0.5 × B4
		TNDVI	Transformed normalized difference vegetation index, sqrt[(B8 − B4)/(B8 + B4) + 0.5]
		GNDVI	Green normalized difference vegetation index, (B8 − B3)/(B8 + B3)
		CI	Color index, (B4 − B3)/(B4 + B3)
		ARVI	Atmospherically resistant vegetation index, (B8 − 2 × B4 + B2)/(B8 + 2 × B4 − B2)
		MCARI	Modified chlorophyll absorption ratio index, [(B5 − B4) − 0.2 × B5 − B3)] × (B5 − B4)
		MTCI	Meris terrestrial chlorophyll index, (B6 − B5)/(B5 − B4)
		EVI	Enhanced vegetation index, 2.5 × [(B8 − B4)/(B8 + 6 × B4 − 7.5 × B2 + 1)]
		EVI2	Enhanced vegetation index2, 2.5 × [(B8 − B4)/(B8 + 2.4 × B4 + 1)]
		NDVIre1	Normalized Difference Vegetation Index red-edge1,(B8 − B5)/(B8 + B5)
		NDVIre2	Normalized Difference Vegetation Index red-edge1, (B8cB6)/(B8 + B6)
		mNDVI	Modified normalized difference vegetation index, (B8 − B4)/(B8 + B4 − 2 × B2)
		mNDVIre	Modified red edge normalized difference vegetation index, (B8 − B5)/(B8 + B5 − 2 × B2)
		NDII	normalized difference infrared index, (B8 − B11)/(B8 + B11)
		SAVI	Soil adjusted vegetation index, 1.5 × (B8 − B4/(B8 + B4 + 0.5)
		NDVI	Normalized difference vegetation index, (B8 − B4)/(B8 + B4)
		MSAVI2	Second modified soil adjusted vegetation index, 0.5 × [2 × (B8 + 1) − sqrt[(2 × B8 + 1) × (2 × B8 + 1) − 8 × (B8 − B4)]]
		RVI	Ratio vegetation index, B8/B4
		PVI	Perpendicular vegetation index, sin(a) × B8 − cos(a) × B4, (a = 45°)
		IPVI	Infrared percentage vegetation index, B8/(B8 + B4)

Table 2. *Cont.*

Source		Feature Variables	Description
	Texture (80)	b2/b3/b4/b5/b6/b7/b8/b8a/b11/b12_con	Contrast
		b2/b3/b4/b5/b6/b7/b8/b8a/b11/b12_corr	Correlation
		b2/b3/b4/b5/b6/b7/b8/b8a/b11/b12_dis	Dissimilarity
		b2/b3/b4/b5/b6/b7/b8/b8a/b11/b12_ent	Entropy
		b2/b3/b4/b5/b6/b7/b8/b8a/b11/b12_hom	Homogeneity
		b2/b3/b4/b5/b6/b7/b8/b8a/b11/b12_mean	Mean
		b2/b3/b4/b5/b6/b7/b8/b8a/b11/b12_sm	Angular second moment
		b2/b3/b4/b5/b6/b7/b8/b8a/b11/b12_var	Variance
Sentinel-1 and PALSAR-2 mosaic	Polarization (8)	VV	Vertical transmit-vertical channel backscattering coefficients, dB
		VH	Vertical transmit-horizontal channel backscattering coefficients, dB
		HH	Horizontal transmit- horizontal channel backscattering coefficients, dB
		HV	Horizontal transmit-vertical channel backscattering coefficients, dB
		V/H	VV/VH
		s1npdi	(VV − VH)/(VV + VH)
		H/V	HH/HV
		p2npdi	(HH − HV)/(HH + HV)
	Texture (32)	VV/VH/HH/HV_con	Contrast
		VV/VH/HH/HV_corr	Correlation
		VV/VH/HH/HV_dis	Dissimilarity
		VV/VH/HH/HV_ent	Entropy
		VV/VH/HH/HV_hom	Homogeneity
		VV/VH/HH/HV_mean	Mean
		VV/VH/HH/HV_sm	Angular second moment
		VV/VH/HH/HV_var	Variance
SRTM DEM	(3)	elevation	elevation
		slope	slope
		aspect	aspect

Table 3. Different scenarios of feature variable combinations for forest height modeling.

Scenario ID	Variable Combination	Short Name
1	Sentinel-2	s2
2	Sentinel-2, SRTM DEM	s2to
3	Senitnel-1, Sentinel-2, PALSAR-2 mosaic	s1s2p2
4	Sentinel-1, PALSAR-2 mosaic, SRTM DEM	s1p2to
5	Sentinel-1, Sentinel-2, PALSAR-2 mosaic, SRTM DEM	s1s2p2to

2.5.2. Recursive Feature Elimination

Recursive feature elimination (RFE) is a wrapper-based feature-ranking algorithm for determining the best feature subset [57]. It is essentially a process that repeatedly builds a model until an optimal subset of features is selected. Based on the screening results, the features with the smallest coefficients are deleted first, and the procedure is repeated in the remaining set of features until all features are traversed by the algorithm [58]. During the process of selection, the root mean square error and standard deviation error of 10-fold cross-validation were used to determine the feature variable subset. Although many feature selection methods fusing RFE and other algorithms were proposed, previous research emphasized that RFE combined with random forest could provide unbiased and stable results and improve accuracy [59]; therefore, we used the "rfe()" function of the "caret" package in R 4.2.0 to realize the procedure with the method "Repeatedcv", repeat "10", and the function "random forests (rfFuncs)".

2.5.3. Boruta

The Boruta algorithm is a wrapper built around the random forest classification algorithm implemented in the R package "randomForest". Its core idea is to construct shadow features by shuffling the original real features and aggregate the original features and shadow features as the feature matrix for training, and then, with the feature importance score of shadow features as a reference, the feature set related to the dependent variable is selected from the original real features. The Boruta algorithm consists of the following steps: First, to create the shadow attributes by shuffling the values of the original object feature and splice the shuffled features with the original real features to form a new feature matrix. Next, use the new feature matrix as input and run the random forest classifier and compute the Z scores of the real feature and shadow feature. Thirdly, to find the maximum Z score among shadow attributes (MZSA), features that were significantly greater than MZSA were labeled as "important", significantly smaller than MSZA as "unimportant", and were permanently removed from the feature set. Lastly, to repeat the process until all the features were classified as "important" or "unimportant". This procedure was performed in R 4.2.0 using the Boruta packages [60].

2.6. Machine Learning Algorithms

In this study, we employed k-nearest neighbor (k-NN), support vector machine regression (SVR), random forest (RF), gradient boosting decision tree (GBDT), extreme gradient boosting (XGBoost), and categorical boosting (CatBoost) machine learning algorithms to model with the training data serving as the input.

2.6.1. K-Nearest Neighbor

The k-nearest neighbor (k-NN) algorithm is a simple and efficient non-parametric method, which can effectively avoid the collinearity problem of the independent variables. It applies to remote sensing data parameter estimation with non-normal distribution and unknown density function and is widely used in forestry investigations around the world [61,62]. The core idea of this algorithm is to take a point in the feature space as the reference object, record the attribute values of the k nearest sample points from the point, and calculate the average value of its inverse distance weight to get the predicted value of this object.

2.6.2. Support Vector Machine Regression

The support vector machine algorithm was proposed based on the VC dimension theory and the structural risk minimization principle [63]. It was initially applied for classification in forest applications, and recently also showed reliable advantages in forest parameter retrieving [64,65]. The basic idea of SVR is to map the features of training data to a high-dimensional feature space by defining a kernel function and finding an optimal hyperplane of linear regression in this feature space to fit the eigenvalues. In the case of limited sample information and high dimensions of feature variables, it can minimize the sampling error and has good generalization ability.

2.6.3. Random Forest

Random forest (RF) is a modified ensemble machine learning algorithm based on decision trees proposed in 2001 [66]. Numerous studies have demonstrated that RF can accurately estimate forest metrics [22,67–69]. RF constructs a series of regression trees, each of which is generated by randomly repeated sampling bootstrap training samples that can be put back, which makes some data may be used many times, while other data may not be used. Usually, 70% of the training samples are selected as the modeling samples, and the remaining 30% samples are used to evaluate the sample prediction error, which is called out-of-bag error (OOB error). At the same time, it randomly selects variables at the nodes of each tree. The procedure stops running when the trees without pruning grow to the maximum scale, and the final prediction accuracy takes the average weight of all prediction regression trees. Because of its random characteristic, this method can enhance the stability of the model, improve the prediction accuracy, and increase the robustness of the model itself to noise or overfitting phenomena to a certain extent.

2.6.4. Gradient Boosting Decision Tree

Gradient boosting decision tree (GBDT) is an integrated decision tree algorithm based on the iterative ideas of gradient boosting first proposed by Friedman [70]. It first generates a weak learner (usually a CART regression tree model), obtaining the residual of the input after training, and then trains the next learner based on the residual generated by the previous round of learners, iteratively. In the process of each iteration, each learner aims to minimize the loss function, that is, to make the loss function always reduce the residual along the descending direction of the gradient. Finally, the final prediction result is obtained by accumulating the results of all weak learners. GBDT is very robust to outliers due to the use of some robust loss functions, and in the case of relatively little tuning time, the prediction accuracy can also be relatively high. Although GBDT is very popular in the field of machine learning, it is rarely applied in the study of forest parameter estimation [43,71].

2.6.5. Extreme Gradient Boosting

Extreme gradient boosting (XGBoost) is an improved GBDT algorithm proposed by Chen et al. in the Kaggle machine learning competition [72]. Compared with GBDT, XGBoost has the following advantages: (1) Regular terms are added to the objective function to control the complexity of the model and prevent the learned model from overfitting. (2) The second-order Taylor expansion is used for the objective function, which makes the definition of the objective function more accurate and easier to find the optimal solution; (3) XGBoost builds all possible subtrees from top to bottom first and then prunes from bottom to top in reverse. In this way, it is not easy to fall into the local optimal solution. (4) XGBoost supports parallel processing. It sorts the data in advance before training and then saves it as a block structure. This structure is used repeatedly in subsequent iterations, which greatly reduces the amount of calculation. Due to the advantages of XGBoost, such as sparse data processing ability, greatly increasing algorithm speed, and reducing computational memory in large-scale data training, it has recently attracted a lot of attention. There were also some studies using XGBoost to estimate forest parameters and achieved good results [43,73–75].

2.6.6. Categorical Boosting

Categorical boosting (CatBoost), as the name suggests, consists of categorical and boosting, which is a novel gradient boosting algorithm implemented with oblivious trees as the base learner proposed by Dorogush et al. [76]. On the one hand, CatBoost builds fully symmetric trees. In each step, the leaves of the previous tree are split using the same conditions. The feature segmentation pair with the lowest loss was selected and used for nodes at all levels. This balanced tree structure facilitates an efficient CPU implementation and reduces the prediction time. On the other hand, CatBoost uses the concept of rank-lifting to train models on a subset of the data while computing the residuals on another subset, thus preventing target leakage and overfitting. Compared with other algorithms in the boosting family, CatBoost can automatically process discrete feature data, which is suitable for regression problems with multiple input features and regression data containing noisy samples. The model has stronger robustness and generalization performance and performs better in algorithm accuracy. Although CatBoost outperformed other machine learning algorithms in other fields [77,78], the effectiveness of this algorithm for forest height estimation remains to be confirmed.

2.6.7. Tuning the Hyperparameters for the Machine Learning Algorithms

When estimating the forest height, the hyperparameters of the machine learning algorithms can greatly affect the results of the model predictions; therefore, the hyperparameters must be optimized for each algorithm before doing any further examination or comparison using these algorithms. In this study, we utilized grid search technology to automatically perform hyperparameter tuning. Six machine algorithms were hyperparameter tuned based on the lowest model RMSE achieved by the 10-fold cross-validation techniques repeated 5 times on the training dataset. This procedure was performed in R 4.2.0 using the "caret" packages. Detailed information about the key tuning hyperparameters and corresponding tuning parameters configurations for each algorithm were presented in Table 4.

2.7. Model Evaluation

In our research, we randomly divided the plot data into two sets: training dataset (70%) and validation dataset (30%). The training set was used to train and develop the models, while the validation set did not participate in the model-building process and was instead used to evaluate model performance. The best model was developed based on the training set after hyperparameter tuning, and model performance metrics were produced based on the validation set. The determination coefficient (R^2, Equation (1)), root mean square error (RMSE, Equation (2)), and relative root mean square error (rRMSE, Equation (3)) were employed to evaluate the performance of different models. The higher the R^2 is, the lower the RMSE and rRMSE are, which means that the higher the prediction accuracy is, the better the estimation result is.

$$R^2 = 1 - \frac{\sum_{i=1}^{n}(y_i - \hat{y}_i)^2}{\sum_{i=1}^{n}(y_i - \bar{y})^2} \quad (1)$$

$$\text{RMSE} = \sqrt{\frac{\sum_{i=1}^{n}(y_i - \hat{y}_i)^2}{n}} \quad (2)$$

$$\text{rRMSE} = \frac{\text{RMSE}}{\bar{y}} \times 100\% \quad (3)$$

where n is the total number of sample plots, $\hat{y}_i$ is the predicted value, y_i is the field measurement value and $\bar{y}$ is the mean of the field measurement value.

Table 4. Tuning hyperparameters and corresponding configurations for each algorithm.

Algorithm	Hyperparameter	Description	Hyperparameter Configurations
k-NN	k	the number of neighbors considered.	(1–10) at intervals of 1
SVR	C	the cost of constraints violation	(1–10) at intervals of 1
	gamma	the parameter needed for all kernels except linear	(0–0.2) at intervals of 0.01
RF	mtry	the number of predictor variables randomly sampled at each split	(1–10) at intervals of 1
	ntree	the number of trees	(100–1000) at intervals of 100
GBDT	ntree	the number of trees	(100–1000) at intervals of 100
	maxdepth	the depth of the tree	(1–10) at intervals of 1
	shrinkage	the learning rate	(0.01–0.1) at intervals of 0.01
	min terminal node	the minimum samples required in a terminal node.	(1–10) at intervals of 1
XGBoost	max_depth	the depth of the tree	(1–10) at intervals of 1
	eta	the learning rate	(0.01–0.1) at intervals of 0.01
	gamma	minimum loss reduction of the tree	(0–1) at intervals of 0.1
	colsample_bytree	the number of predictor variables supplied to a tree	(0–1) at intervals of 0.1
	min_child_weight	minimum number of instances	(1–10) at intervals of 1
	subsample	the number of observations supplied to a tree	(0–1) at intervals of 0.1
CatBoost	depth	the depth of the tree	
	learning_rate	the learning rate	(0.01–0.1) at intervals of 0.01
	l2_leaf_reg	the coefficient at the L2 regularization term of the cost function	(1–10) at intervals of 1
	rsm	the percentage of features to use at each split selection	(0–1) at intervals of 0.1

2.8. ANOVA Analysis

To assess the impact of different impact factors, including data sources, feature selection methods, and modeling algorithms on forest height estimation, we applied the analysis of variance (ANOVA) to quantify the impact of each factor and to identify critical factors in forest height estimation. This procedure was performed in R 4.2.0.

2.9. Forest Height Mapping and Product Evaluation

First, the forest/non-forest mask generated from the FROM-GLC 2017 product was used to obtain the forest distribution map of the study area. Then, the optimal model was used for the wall-to-wall mapping of the forest height in Baoding city in 2016. After that, the forest height map derived from this study was compared with the existing global forest canopy height map product (RMSE = 6.6 m, R^2 = 0.62), which was generated by integrating GEDI and Landsat data by Potapov et al. [40].

3. Results

3.1. Feature Variable Selected for Forest Height Modeling

In five different scenarios, three feature variable selection methods, stepwise regression analysis, recursive feature elimination, and Boruta were compared for forest height modeling. The results of feature variable selection for different scenarios and different methods are shown in Table 5. We could see that in each different scenario, the selected feature variables based on different methods were unique. For example, in the "s1s2p2" scenario, the feature variables of stepwise regression selected were mainly the texture features of Sentinel-2 and PALSAR-2, whereas the main features chosen by RFE and Boruta included spectral band reflectance, vegetation index, and texture features of Sentinel-2 and Sentinel-1. In the "s1p2to" scenario, the SAR feature variable screened by Boruta was derived from Sentinel-1. However, this situation was just the opposite when screening variables based on SR and RFE, the selected SAR variables were from PALSAR-2, and the number of selected variables from PALSAR-2 acquired by SR and RFE was quite different.

Table 5. Five scenarios of feature variable selection result for forest height modeling.

Scenario Name	Feature Selection Method	Number of Selected Variables	Name of Selected Variables
s2	Stepwise regression analysis	9	b11, NDVIre2, b2_hom, b3_ent, b3_var, b4_ent, b4_var, b5_hom, b11_mean;
	Recursive feature elimination	10	b2, b4, b5, CI, b2_con, b2_corr, b2_hom, b2_dis, b4_ent, b4_sm;
	Boruta	16	b2, b3, b4, b5, CI, b2_con, b2_corr, b2_dis, b2_hom, b3_mean, b4_dis, b4_ent, b4_hom, b4_mean, b4_sm, b12_mean;
s2to	Stepwise regression analysis	14	b3, NDVIre2, b2_corr, b2_sm, b3_mean, b4_ent, b4_sm, b5_mean, b8_dis, b8_var, b11_var, b12_corr, b12_var, elevation;
	Recursive feature elimination	10	b2, b5, CI, b2_con, b2_corr, b2_dis, b2_hom, b4_ent, elevation, slope;
	Boruta	18	b2, b4, b5, CI, NDVI, b2_con, b2_corr, b2_dis, b2_hom, b3_mean, b4_ent, b4_hom, b4_mean, b4_sm, b4_var, b5_ent, elevation, slope;
s1s2p2	Stepwise regression analysis	8	NDVIre2, b2_hom, b4_ent, b5_sm, VV_dis, HH_con, HH_mean, HV_var;
	Recursive feature elimination	12	b2, b4, b5, b2_corr, CI, b2_con, b2_dis, b2_hom, b4_ent, VH_con, VH_dis, VH_hom;
	Boruta	21	b2, b4, b5, ARVI, CI, NDVI, b2_con, b2_corr, b2_dis, b2_hom, b2_mean, b2_sm, b3_mean, b4_ent, b4_hom, b4_sm, b4_var, b5_mean, VH_con, VH_dis, VH_hom;
s1p2to	Stepwise regression analysis	6	HH_mean, HV_con, HV_ent, HV_sm, HV_var, elevation;
	Recursive feature elimination	3	HH_con, elevation, slope;
	Boruta	3	VV_var, elevation, slope;
s1s2p2to	Stepwise regression analysis	15	NDVIre2, b2_corr, b3_ent, b3_var, b4_ent, b8_var, b11_var, b12_corr, b12_sm, VH_sm, HH_mean, HH_sm, HV_con, HV_var, slope;
	Recursive feature elimination	14	b2, b4, b5, CI, b2_con, b2_corr, b2_dis, b2_hom, b4_ent, VH_con, VH_dis, VH_hom, elevation, slope;
	Boruta	23	b2, b3, b4, b5, ARVI, CI, NDVI, NDVIre1, RVI, TNDVI, b2_con, b2_corr, b2_dis, b2_hom, b4_ent, b4_hom, b4_sm, b4_var, b5_mean, b12_mean, VH_con, elevation, slope.

Furthermore, it should be noted that in the scenarios containing terrain factors, almost the feature selection methods chose elevation and slope. In the scenarios which contained variables derived from Sentinel-2, these variables, including b2_ hom, b4_ ent, and CI were selected frequently. In the scenarios with radar-derived variables, the selected variables were different based on different methods. SR was more inclined to choose the feature variables derived from PALSAR-2, Boruta was more inclined to choose Sentinel-1, while RFE depended on specific data scenarios, and in most cases, it is preferred to choose Sentinel-1.

3.2. Forest Height Modeling Results

We applied three statistical metrics (R^2, RMSE, rRMSE) to evaluate the height models built from different variable scenarios by using the reserved 30% field plot data (Table 6).

Table 6. Performance of forest height estimation models in the validation datasets.

Data Scenario	Regression Method	Feature Selection Method								
		SR			RFE			Boruta		
		R^2	RMSE (m)	rRMSE (%)	R^2	RMSE (m)	rRMSE (%)	R^2	RMSE (m)	rRMSE (%)
s2	k-NN	0.43	2.9	33.53	0.40	3.0	34.56	0.48	2.8	32.11
s2	SVR	0.33	3.1	36.27	0.31	3.2	37.10	0.28	3.2	37.71
s2	RF	0.49	2.7	31.75	0.55	2.6	29.80	0.52	2.7	30.95
s2	GBDT	0.49	2.7	31.66	0.53	2.6	30.49	0.52	2.6	30.73
s2	XgBoost	0.55	2.6	29.91	0.56	2.5	29.66	0.57	2.5	29.10
s2	CatBoost	0.45	2.8	32.98	0.50	2.7	31.41	0.49	2.7	31.66
s1s2p2	k-NN	0.08	3.7	42.58	0.35	3.1	35.96	0.38	3.0	35.02
s1s2p2	SVR	0.33	3.2	37.72	0.27	3.3	37.94	0.39	3.0	34.82
s1s2p2	RF	0.48	2.8	32.17	0.46	2.8	32.83	0.47	2.8	32.36
s1s2p2	GBDT	0.52	2.7	30.90	0.44	2.9	33.34	0.42	2.9	33.80
s1s2p2	XgBoost	0.46	2.8	32.75	0.46	2.8	32.80	0.47	2.8	32.52
s1s2p2	CatBoost	0.48	2.8	32.14	0.44	2.9	33.42	0.46	2.8	32.65
s2to	k-NN	0.34	3.1	36.24	0.34	3.1	36.18	0.35	3.1	35.75
s2to	SVR	0.33	3.1	36.51	0.50	2.7	31.51	0.32	3.2	36.77
s2to	RF	0.51	2.7	31.02	0.57	2.5	29.18	0.56	2.5	29.44
s2to	GBDT	0.53	2.6	30.54	0.60	2.4	27.98	0.58	2.5	28.73
s2to	XgBoost	0.53	2.6	30.47	0.63	2.3	27.25	0.59	2.4	28.45
s2to	CatBoost	0.53	2.6	30.45	0.59	2.5	28.58	0.56	2.5	29.55
s1p2to	k-NN	0.31	3.2	36.98	0.21	3.4	39.61	0.27	3.3	38.10
s1p2to	SVR	0.09	3.6	42.35	0.13	3.6	41.47	0.13	3.6	41.63
s1p2to	RF	0.10	3.6	42.34	0.28	3.2	37.89	0.15	3.5	41.08
s1p2to	GBDT	0.18	3.5	40.22	0.33	3.1	36.35	0.19	3.4	40.03
s1p2to	XgBoost	0.23	3.3	39.05	0.37	3.0	35.38	0.24	3.3	38.92
s1p2to	CatBoost	0.24	3.3	38.88	0.31	3.2	36.91	0.19	3.4	40.00
s1s2p2to	k-NN	0.17	3.5	40.59	0.37	3.0	35.32	0.44	2.9	33.31
s1s2p2to	SVR	0.12	3.6	41.80	0.43	2.9	33.51	0.53	2.6	30.44
s1s2p2to	RF	0.36	3.1	35.62	0.50	2.7	31.49	0.55	2.6	29.75
s1s2p2to	GBDT	0.42	2.9	33.77	0.59	2.4	28.44	0.62	2.4	27.56
s1s2p2to	XgBoost	0.40	3.0	34.49	0.60	2.4	28.18	0.67	2.2	25.57
s1s2p2to	CatBoost	0.35	3.1	35.87	0.56	2.5	29.66	0.55	2.6	29.98

For five different data scenarios, the optimal models of five data scenarios were from different feature selection methods. In the scenario "s2" and "s1s2p2to", the models based on Boruta and XGBoost provided the best performance. In the scenario "s2to" and "s1p2to", the models based RFE and XGBoost outperformed others. In the scenario "s1s2p2", the model based on SR and GBDT was the best. Furthermore, we found that the difference in the performance between the scenario "s2", "s1s2p2", "s2to", "s1s2p2to" was not very obvious, while the scenarios combining optical and topography variables such as the "s2to" and "s1s2p2to" scenario further improved modeling accuracy overall. Compared with the other four scenarios, the scenario "s1s2p2", which contained radar and topography feature variables, provided much poorer modeling results.

Interestingly, on the basis of optical variables modeling alone, adding radar-derived variables marginally lowered the modeling accuracy of forest height, while adding topography variables improved the modeling accuracy in most situations. For instance, when combining Boruta and RF for modeling, R^2 increased by 8.95% and RMSE decreased by 4.89% after adding topography variables, while R^2 decreased by 8.69% and RMSE increased by 4.55% after adding radar variables. When topography variables and radar variables were both added to the optical variables dataset, the modeling results were connected to the technique of feature selection. While selecting feature variables based on SR, the modeling accuracy exhibited an apparent downward trend, regardless of the algorithm utilized; however, the modeling effect was improved when RFE and Boruta were used to screen feature variables, with R^2 increased from 0.31–0.56 to 0.37–0.60 based on RFE, R^2 increased from 0.28–0.57 to 0.44–0.67 based on Boruta.

Figure 3 shows the broken-line graph based on three different feature selection methods, five different data combinations, and six modeling methods (R^2 on the left and RMSE on the right). For the three different feature selection methods, the modeling performance of Boruta-based and RFE-based approaches was superior to SR. The R^2 and RMSE of SR-based ranged from 0.08 to 0.55, 2.6 to 3.7, respectively, while RFE-based R^2 varied from 0.13 to 0.63, RMSE from 2.3 to 3.6, with Boruta-based R^2 varying from 0.13 to 0.67, RMSE from 2.2 to 3.6.

For six different modeling methods, it could be seen that when the data source and the method of feature variables selection were consistent, the tree-based ensemble algorithms were always superior to k-NN (with R^2 varying from 0.08 to 0.48, RMSE varying from 2.8 to 3.7) and SVR (with R^2 varying from 0.09 to 0.53, RMSE varying from 2.6 to 3.6). Among the four ensemble machine learning algorithms, RF (with R^2 varying from 0.10 to 0.57, RMSE varying from 2.5 to 3.6), GBDT (with R^2 varying from 0.18 to 0.62, RMSE varying from 2.4 to 3.4), XGBoost (with R^2 varying from 0.23 to 0.67, RMSE varying from 2.2 to 3.3) and CatBoost (with R^2 varying from 0.19 to 0.59, RMSE varying from 2.5 to 3.4), XGBoost's overall performance was slightly better than the other three. Moreover, in all of the 90 established models, the XGBoost algorithm based on the Boruta feature selection technique in the "s1s2p2to" scenario achieved the best modeling effect (R^2 = 0.67, RMSE = 2.2 m).

3.3. Variable Importance Analysis

In order to further understand the importance of feature variables in the modeling process, we ranked the importance of "s1s2p2to" scenarios containing all types of feature variables based on the importance ranking method of XGBoost. Figure 4 displays the importance ranking of feature variables based on three distinct feature selection methods.

According to the feature selection method of Boruta and RFE, the terrain-related factors slope and elevation, vegetation index "CI" and band reflectance "b2" and "b4" had relatively high importance, accounting for approximately 40% and 60% of all the selected variables, respectively. Although there were many optical texture feature variables selected, the importance of a single feature was inferior to other features. In addition, although the radar variables selected by these two methods were very few, their significance cannot be completely ignored. Compared with Boruta and RFE, the variables selected by SR were quite different, band reflectance was not chosen, but the optical texture features and the variables derived from PALSAR-2 not considered by Boruta and RFE were taken into account. Thus, it could be seen that different feature selection methods chose different feature variables, and the importance of variables also varies according to different techniques. When using Boruta and RFE, optical variables and terrain variables were more crucial, while the importance of radar variables increased based on SR compared with Boruta and RFE.

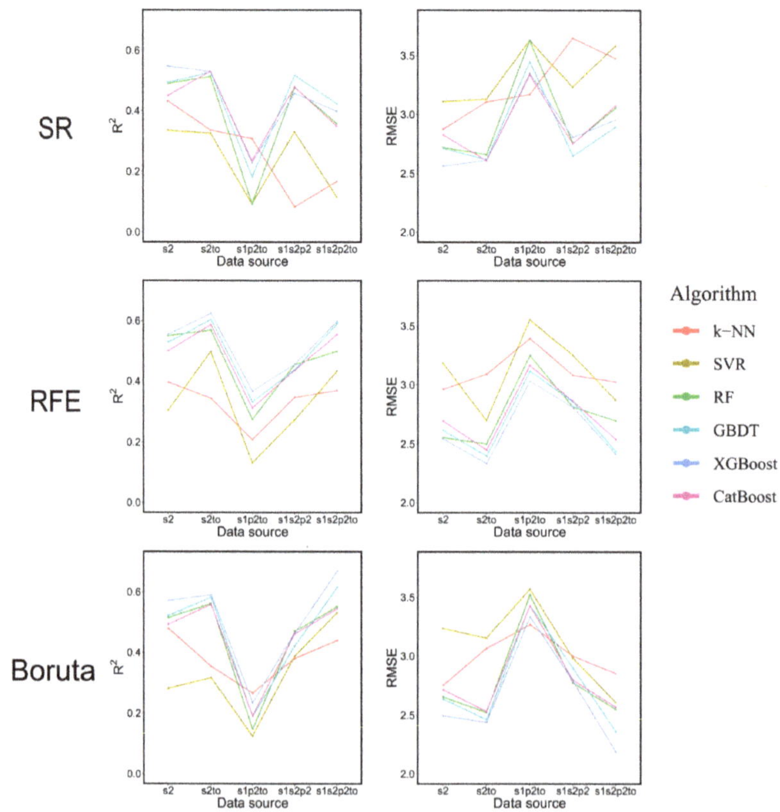

Figure 3. The broken-line graph of R^2 and RMSE based on three different feature selection methods and five different data combinations based on six modeling methods (R^2 on the **left** and RMSE on the **right**).

3.4. Forest Height Mapping and Comparison to Existing Product

Based on the modeling results, we combined the feature variables of the scenario "s1s2p2to" selected by Boruta and XGBoost algorithm to produce the forest height wall-to-wall map over Baoding city. According to our forest height map, the value of the forest height in Baoding city was 7.64 ± 1.70 m and ranged from 2.97 m to 17.91 m. We compared our results with the previously released product published by Potapov et al. [40], hereinafter called the "Pota". According to "Pota", the forest height in Baoding city was 9.15 ± 3.62 m and ranged from 3.00 m to 29.00 m (Table 7). Despite the minimum value of the two forest height products being almost identical, the average and maximum values of the "Pota" were much higher than in this study. Moreover, there were notable discrepancies in the distribution of forest height from the two maps of forest height in Baoding city (Figure 5). First, the tree height values of this study were primarily concentrated in the range of 6–8 m, with a normal distribution trend on both sides, whereas the tree height values of "Pota" were mainly distributed in the range of 7–10 m. Second, the higher values of forest height in this study were mainly distributed in the mountainous areas in the north of Baoding city, while according to "Pota", tall trees were dispersed in both north and west of Baoding. In order to explore the factors that caused the difference between the two maps, we generated a map of forest height differences between these two maps in Baoding city (Figure 6); the average value of the forest height difference was 3.25 m and ranged from 0 to 23.00 m. We found that large differences existed in the mountainous areas in the northern and midwest areas of Baoding city. From the slope distribution map (Figure 6), it could be seen that the areas with big differences were mountainous areas with large slopes and steep terrain. Further counting the difference values above the average difference value in the distribution of different slope levels, we found that the high difference values were primarily distributed in the areas with a slope above 15°, accounting for more than 80% of the total number of high difference values (Figure 7).

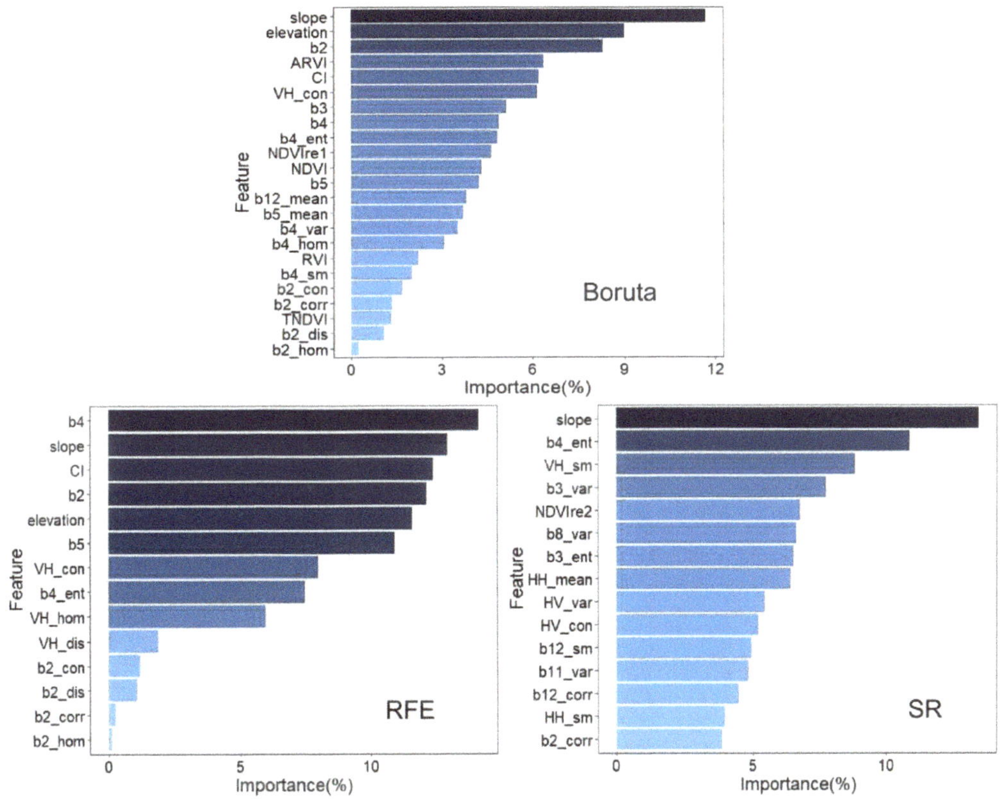

Figure 4. Variable importance ranking of XGBoost models for three feature selection methods (Boruta, RFE, and SR).

Table 7. Comparison of estimated forest heights over Baoding city.

Product	Nominal Year	Data Source	Nominal Resolution	Algorithm	Forest Height (m)			
					Min.	Max.	Mean.	Std.
Map of Potapov	2019	Landsat, GEDI, SRTM	30 m	Regression tree	3.00	29.00	9.15	3.62
Map of this study	2016	Sentinel-1, Sentinel-2, SRTM	25 m	XGBoost	2.97	17.91	7.64	1.70

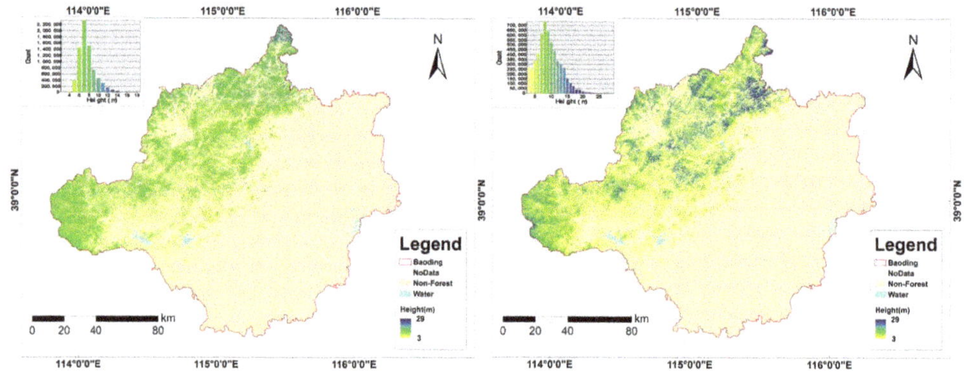

Figure 5. Map of forest height in Baoding city. Map of this study on the **left**; Potapov's map on the **right**. The inserted panels show the histogram of forest height value.

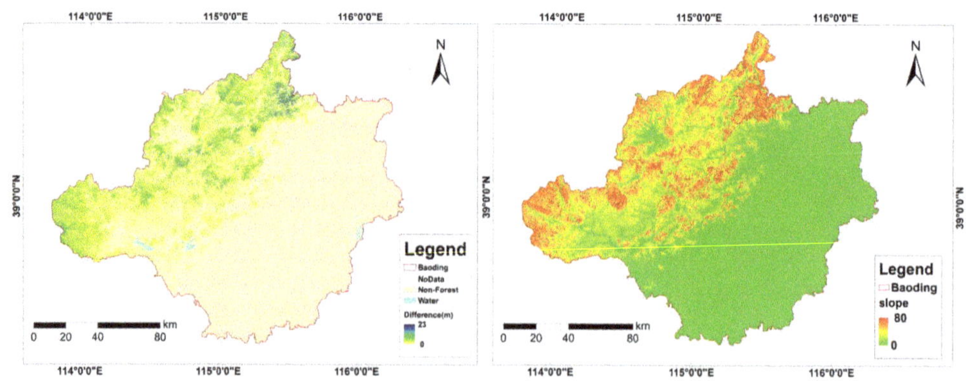

Figure 6. Map of difference between Potapov's map and map of this study in Baoding city, on the **left**. Map of slope in Baoding city, on the **right**.

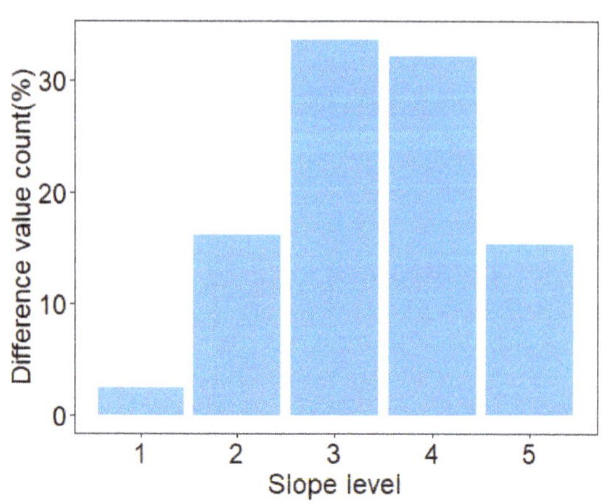

Figure 7. The percentage of the number of difference values higher than the average difference value at five slope levels (level 1: 0° < slope ≤ 5°, level 2: 5° < slope ≤ 15°, level 3: 15° < slope ≤ 25°, level 4: 0° < slope ≤ 35°, level 5: slope > 35°).

4. Discussion

4.1. Performance of Multi-Source Satellite Metrics for Forest Height Estimation

Our study used multi-source satellite data to estimate the forest height of Baoding City. First of all, from the different scenarios of various variable combinations, the variable combination of optical sensor and radar sensor was not always superior to a single optical sensor, which was consistent with the previous research findings when Li et al. applied Landsat 8 and Sentinel-1A data to estimate forest aboveground biomass [75]; however, at the same time, our study results also demonstrated that the performance of the combination of optical, radar, and terrain variables was slightly better than that of a single sensor. Secondly, according to the variables selected by three different feature selection methods and the importance ranking results, optical variables had higher potential than radar variables in estimating forest height, which was supported by Huang et al. [22]. Previous studies have shown that the variables derived from Sentinel-1 and PALSAR-2 were valuable and common predictors for forest height estimation [79,80]; however, in this study, their role was auxiliary, and the accuracy improvement of forest height estimation was not obvious. There were two potential causes to explain this phenomenon. The first was because the C-band SAR has limited penetration of the forest, and is vulnerable to topographic factors in mountainous areas. The second was that the used PALSAR-2 data did not contain the real image at the time of field data collection, but the mosaic image in 2016. The inconsistency between the ground data and the image may result in being not very inaccurate. Furthermore, terrain factors such as elevation were discovered to present good performance in estimating forest height, which was compatible with the earlier research conducted by Xi et al. [81]. Because SRTM employed an InSAR instrument, the vegetation contribution is not totally separated from the ground elevation, so the elevation may include part of the vegetation height information.

4.2. Performance of Different Feature Variable Selection Methods

We explored three different techniques to select feature variables. Table 5 showed that there were certain disparities in quantity and selected variables for different methods. In particular, the variables screened by SR were quite different from those selected by the other two methods. This might be related to the fundamentals of the three approaches. SR is based on AIC information statistics to delete or add variables accomplished by selecting the smallest AIC information statistics. It is worth noting that since the AIC tended to select more parameters than required when using small or medium samples, we mitigated the limitations of the method by removing certain non-essential variables by making the p-value of all selected variables significant ($p < 0.05$) [55]. RFE and Boruta are methods around the core idea of random forest, so the selected variables had a certain degree of similarity. Table 8 summarizes statistical data for different variable selection methods. From the mean values shown in the table, the effect of RFE and Boruta was significantly better than SR and the average value of RFE was slightly better than Boruta; however, the calculation time of executing RFE algorithm in "caret" package was much longer than that of Boruta, while its average accuracy improvement was very limited, and the optimal modeling result was also based on Boruta; therefore, from the perspective of modeling accuracy and time efficiency, we considered that Boruta was the best feature selection method in this study. Agjee et al. [82] came to the same conclusion when they compared RFE and Boruta to identify multitemporal hyperspectral data to detect the efficacy of the biocontrol agent.

4.3. Performance of Different Machine Learning Algorithms

Among six machine learning algorithms, four tree-based ensemble algorithms provided better forest height estimation accuracy than the other algorithms, and XGBoost was superior to the other three ensemble algorithms. This result was similar to the research conducted by Arjasakusuma et al. [83] when comparing MARS, SVR extra trees (ET), and extreme gradient boosting (XGB) with trees (XGbtree and XGBdart) and linear (XGBlin) classifiers for modeling forest height from the combination of LiDAR and hyperspectral data. Comparable conclusions were drawn in the studies of forest aboveground biomass estimation. Pham et al. [43] combined genetic algorithm (GA) and XGBoost to achieve optimal mangrove AGB estimation than the other four ML algorithms (RF, SRM, GBRT, and CatBoost); Li et al. [74] combined China's national forest inventory, Landsat-8 data, and LR, RF, and XGBoost algorithms to establish AGB models and found that the XGBoost model significantly improved the estimation accuracy and reduced the problem of overestimation and underestimation to a certain extent.

Table 8. Average running time and statistical of R^2, RMSE, and rRMSE for different variable selection methods.

Method	R^2				RMSE				rRMSE				Average Running Time (s)
	Min.	Max.	Mean.	Std.	Min.	Max.	Mean.	Std.	Min.	Max.	Mean.	Std.	
SR	0.08	0.55	0.36	0.15	2.6	3.7	3.0	0.4	29.91	42.58	35.38	4.09	3.68
RFE	0.13	0.63	0.44	0.13	2.2	3.6	2.8	0.3	25.57	41.46	33.13	3.81	3343.77
Boruta	0.13	0.67	0.43	0.15	2.3	3.6	2.9	0.4	27.25	41.63	33.28	4.36	17.75

The reasons why the XGBoost model performed well included two aspects. First, XGBoost is a flexible algorithm that can correct residual errors to generate a new tree based on the previous trees. Second, the XGBoost model is an advanced gradient boosting system, which improves the processing of regularization learning objectives and avoids overfitting; however, it is worth noting that all the machine learning algorithms cannot entirely address the problem of overestimation and underestimation of forest height. In the present study, XGBoost achieved the optimal solution, but its potential in the face of various geographical situations requires further investigation.

4.4. *Important Factors Analyze in Forest Height Estimation*

Numerous factors can influence the accuracy of forest height estimation. In the present study, we employed ANOVA analysis to evaluate the impact of data source, feature selection method, regression algorithm, and their interaction on forest height estimation. To better illustrate how each factor explained the total variance, we calculated the ratio of the sum of squares of each factor to the total sum of squares (η2). According to the ANOVA results (Table 9), the data source was the most influential factor, accounting for 47% of the total variance of R^2, 46% of RMSE and 46% of rRMSE. Then regression algorithm explained 24% of the total variance of R^2, 25% of RMSE and 25% of rRMSE. The influence of the feature selection method and the interaction between the three factors was relatively low, altogether accounting for approximate 20% of the total variance in R^2, RMSE, and rRMSE. However, it is worth mentioning that the feature selection method, the interaction between data source and feature selection method, and the interaction between data source and regression algorithm also had a significant effect on the results of R^2, RMSE, and rRMSE, so these three factors, including data source, feature selection, and regression algorithm could not be disregarded. In a word, it is necessary to take these three factors into account in the estimation of forest height.

4.5. *Map Product Comparison*

Previous studies had shown that complex terrain increased uncertainty in forest height estimation and the accuracy of forest height estimates decreased with increasing slope values [84,85]. In rugged mountainous areas, the radar's backscatter coefficients and optical spectral reflectance information were susceptible to terrain, and the GEDI used in Potapov's study, whose signals were also skewed by the intricate topographical conditions within its footprint. The combination of these effects led to the large difference in values between Potapov's map and our map, mainly in the areas with high slope values. Furthermore, the result of our research showed an obvious underestimation of the high forest height value. We explained this phenomenon by concentrating on two reasons. The first reason was that optical data mainly captured forest spectral information, with the SAR data of C/L-Band limited ability to penetrate forest canopy, causing their signals to appear saturated. Secondly, due to the small quantity values at the high altitude of our field plots, the high values will be underestimated in the process of machine learning modeling. Potapov reported oversampling of tall trees in their overall reference data set resulted in high values that could be overestimated to some extent. This conclusion was also verified in our study that the average and maximum tree height values in "Pota" were greater than field data.

Table 9. ANOVA results of the R^2, RMSE, and rRMSE for three different factors.

Factor	Df	R^2			RMSE			rRMSE		
		SumSq	η^2	Pr (>F)	SumSq	η^2	Pr (>F)	SumSq	η^2	Pr (>F)
Data source	4	0.90	0.47	$<2.2 \times 10^{-16}$ ***	5.30	0.46	2.571×10^{-07} ***	720.87	0.46	2.571×10^{-07} ***
Feature selection method	2	0.11	0.06	2.147×10^{-06} ***	0.70	0.06	$<2.2 \times 10^{-16}$ ***	95.02	0.06	$<2.2 \times 10^{-16}$ ***
Regression algorithm	5	0.45	0.24	1.345×10^{-12} ***	2.86	0.25	4.992×10^{-14} ***	389.54	0.25	4.992×10^{-14} ***
Data source Feature selection method	8	0.16	0.08	1.412×10^{-05} ***	1.00	0.09	1.860×10^{-06} ***	136.25	0.09	1.860×10^{-06} ***
Data source Regression algorithm	20	0.14	0.07	0.01107 *	0.85	0.07	0.003017 **	115.79	0.07	0.003017 **
Feature selection method Regression algorithm	10	0.02	0.01	0.84356	0.09	0.01	0.826854	11.96	0.01	0.826854
Residuals	40	0.12			0.62			83.68		

Signif. Codes: '***': 0; '**': 0.001; '*': 0.01.

4.6. Recent Related Works Comparison

Compared with two recent studies which used both optical and radar variables for forest tree height estimation, the similarity was that all three studies estimated forest height by constructing an empirical model between forest height and multi-source remote sensing information [22,23]. The difference was that Liu et al. [23] constructed a simple logarithmic regression to estimate forest height based on the relationship between forest height and the backscattering coefficients derived from Sentinel-1 data and the fraction of vegetation cover derived from Sentinel-2 data with the results $R^2 = 0.53414$ and RMSE = 2.9156 m, while Huang et al. [22] and our study both extracted considerable feature variables and employed different feature selection methods and regression algorithms to estimate forest height. Huang et al. systematically evaluated the performance of different remote sensing metrics, feature selection methods, and regression algorithms by dividing the extracted feature variables into ten scenarios and using two types of variable selection methods and three types of regression models; the best estimation was achieved by RF models with R^2 ranged from 0.47 to 0.52, RMSE ranged from 3.8 to 5.3 m, whereas in our study, we utilized four types of remote sensing data, three feature selection methods, and six machine learning algorithms and applied the ANOVA to quantify the importance of these factors on forest height estimation; the variables selected based on Boruta including Sentinel-1, Sentinel-2, and topography metrics, combined with the XGBoost algorithm provided the optimal model ($R^2 = 0.67$, RMSE = 2.2 m).

4.7. Limitations and Prospects

In this study, we found that all the models had the problem of high-value underestimation. From the scatter plot (Figures A1–A3), we could see intuitively the predicted value was below the center line when the tree height exceeded 15 m which meant that despite using multi-sensor datasets to decrease estimation error, the model still underestimated at higher tree heights. In light of this issue, we proposed the following potential improvement directions. (1) Optical sensor such as Sentinel-2 used in this study has some issues, such as poor sensitivity and easy saturation to dense vegetation information, and SAR data, such as Sentinel-1 and PALSAR-2, are susceptible to topography and other factors, and the backscattering information has the problem of signal saturation. As a result, lidar data with direct detection capabilities of forest vertical structures can be combined with optical and SAR data in future studies to increase the accuracy of regional forest height estimation. (2) Previous studies showed modeling based on different forest types and tree height levels can lessen the model's dependence on training samples and improve the modeling effect [81,86]. Due to a lack of sample plot data, we were unable to address forest types or tree height levels to undertake to model respectively. In the future, with sufficient plot data gathered, these strategies can be applied to minimize the uncertainty in the modeling process. (3) Since most machine learning models are black-box models, they are difficult to reflect the mechanism and process between forest parameters and remote sensing information, and the interpretability for reality is weak. The improvement of the generalizability and accuracy of forest parameter estimation by simply constructing empirical models is limited. Physical geography, bioclimatic and cultural conditions are proved to be crucial for the estimation of forest parameters [67,84]; therefore, in subsequent studies, zoning and stratification strategies or coupling remote sensing data and forest physiological process models should be emphasized to estimate forest height and other parameters.

5. Conclusions

In this study, we produced a 25 m spatial resolution wall-to-wall map of the forest height in Baoding, north China and assessed the impacts of three aspects on forest height estimation utilizing Sentinel-1, Sentinel-2, PALSAR 2 mosaic, SRTM DEM, and the NFCI data, three feature selection methods (SR, RFE, and Boruta), and six machine learning algorithms (k-NN, SVM, RF, GBDT, XGBoost, and CatBoost). The results of ANOVA analysis demonstrated that data source, feature selection method, and machine learning algorithm significantly influenced the results of forest height estimation. The accuracy with optical data alone was slightly lower than the combined data of multiple sensors, and multi-source data could improve the estimation accuracy to a certain extent. Optical and topographic indicators were proved to be more effective than that radar indicators. The subset of features screened by RFE and Boruta varied greatly from SR, and the models exhibited from the variables screened based on RFE and Boruta had better performance compared with SR. Moreover, XGBoost outperformed the other five machine learning algorithms. Ultimately, we obtained the optimal model (R^2 = 0.67, RMSE = 2.2 m) based on the combination of Sentinel-1, Sentinel-2, and topography data using Boruta and XGBoost algorithms. The generated forest height map differed from the existing map product, and the regions with large differences between the two maps were mostly distributed in the steep areas with high slope values. Overall, our findings provided a solution for the subsequent forest height mapping at larger scales (national or global) with high precision.

Author Contributions: Methodology, data curation, formal analysis, writing—original draft preparation and review and editing, N.Z.; formal analysis, software, and writing—review and editing, M.C.; investigation and data curation, F.Y.; data curation and software, C.Y., P.Y., Y.G. and Y.S.; conceptualization, project administration, and writing—review and editing, D.P. All authors have read and agreed to the published version of the manuscript.

Funding: This research was funded by the National Key R&D Program of China (2016YFD0600205), the China National Land Survey and Planning Institute Bidding Project (GXTC-A-19070081), the Key Project of Natural Science Research Project of the Education Department of Anhui Province (KJ2020A0721), and the Major Project of Natural Science Research Project of Education Department of Anhui Province (KJ2021ZD0131).

Acknowledgments: The authors are grateful to the Chinese Academy of Inventory and Planning, National Forestry, and Grassland Administration for providing the in situ data used in this study. We would also like to thank the editors and the anonymous reviewers for their valuable comments.

Conflicts of Interest: The authors declare no conflict of interest.

Appendix A

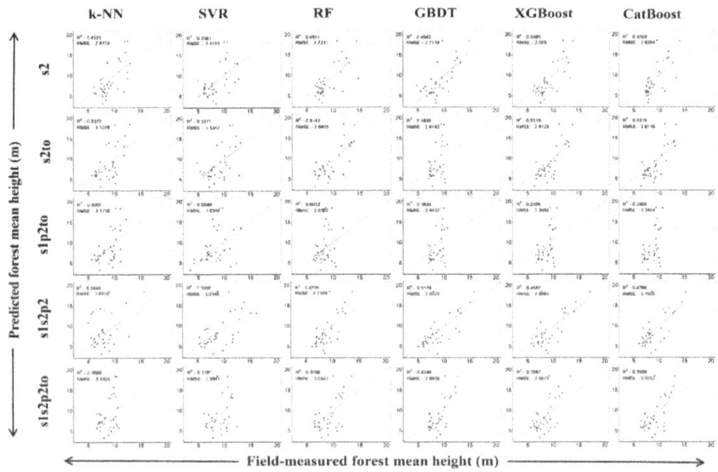

Figure A1. Scatter plot of the predicted and observed forest height for five different scenarios of the k-NN, SVR, RF, GBDT, XGBoost, and CatBoost algorithms based on the SR feature variable selection method.

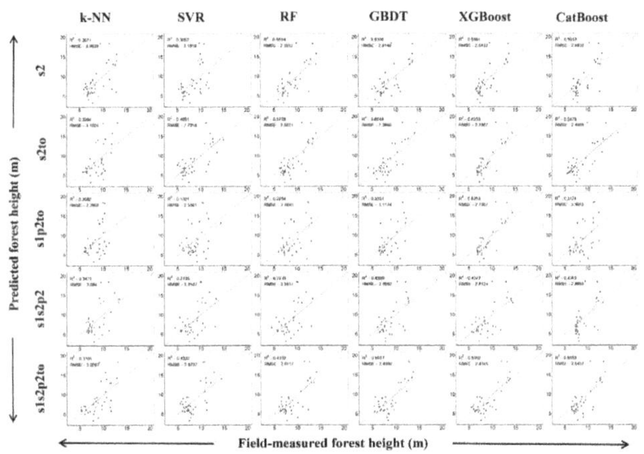

Figure A2. Scatter plot of the predicted and observed forest height for five different scenarios of the k-NN, SVR, RF, GBDT, XGBoost, and CatBoost algorithms based on the RFE feature variable selection method.

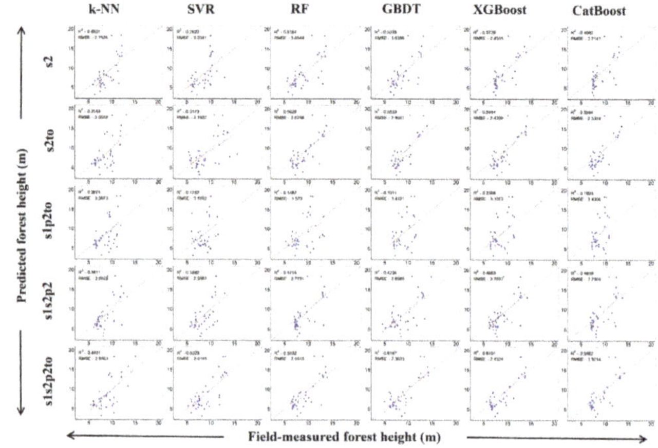

Figure A3. Scatter plot of the predicted and observed forest height for five different scenarios of the k-NN, SVR, RF, GBDT, XGBoost, and CatBoost algorithms based on the Boruta feature variable selection method.

References

1. Achard, F.; Eva, H.; Stibig, H.; Mayaux, P.; Gallego, J.; Richards, T.; Malingreau, J. Determination of Deforestation Rates of the World's Humid Tropical Forests. *Science* **2002**, *297*, 999–1002. [CrossRef] [PubMed]
2. Dong, J.; Kaufmann, R.; Myneni, R.; Tucker, C.; Kauppi, P.; Liski, J.; Buermann, W.; Alexeyev, V.; Hughes, M. Remote sensing estimates of boreal and temperate forest woody biomass: Carbon pools, sources, and sinks. *Remote Sens. Environ.* **2003**, *84*, 393–410. [CrossRef]
3. Huang, H.; Liu, C.; Wang, X.; Zhou, X.; Gong, P. Integration of multi-resource remotely sensed data and allometric models for forest aboveground biomass estimation in China. *Remote Sens. Environ.* **2019**, *221*, 225–234. [CrossRef]
4. Hurtt, G.; Zhao, M.; Sahajpal, R.; Armstrong, A.; Birdsey, R.; Campbell, E.; Dolan, K.; Dubayah, R.; Fisk, J.; Flanagan, S.; et al. Beyond MRV: High-resolution forest carbon modeling for climate mitigation planning over Maryland, USA. *Environ. Res. Lett.* **2019**, *14*, 045013. [CrossRef]
5. Herold, M.; Carter, S.; Avitabile, V.; Espejo, A.; Jonckheere, I.; Lucas, R.; McRoberts, R.; Næsset, E.; Nightingale, J.; Petersen, R.; et al. The Role and Need for Space-Based Forest Biomass-Related Measurements in Environmental Management and Policy. *Surv. Geophys.* **2019**, *40*, 757–778. [CrossRef]
6. Duncanson, L.; Armston, J.; Disney, M.; Avitabile, V.; Barbier, N.; Calders, K.; Carter, S.; Chave, J.; Herold, M.; Crowther, T.; et al. The Importance of Consistent Global Forest Aboveground Biomass Product Validation. *Surv. Geophys.* **2019**, *40*, 979–999. [CrossRef]
7. Wulder, M.; White, J.; Nelson, R.; Næsset, E.; Ørka, H.; Coops, N.; Hilker, T.; Bater, C.; Gobakken, T. Lidar sampling for large-area forest characterization: A review. *Remote Sens. Environ.* **2012**, *121*, 196–209. [CrossRef]
8. Hansen, M.; Potapov, P.; Goetz, S.; Turubanova, S.; Tyukavina, A.; Krylov, A.; Kommareddy, A.; Egorov, A. Mapping tree height distributions in Sub-Saharan Africa using Landsat 7 and 8 data. *Remote Sens. Environ.* **2016**, *185*, 221–232. [CrossRef]
9. Wolter, P.; Townsend, P.; Sturtevant, B. Estimation of forest structural parameters using 5 and 10 meter SPOT-5 satellite data. *Remote Sens. Environ.* **2009**, *113*, 2019–2036. [CrossRef]
10. Potapov, P.; Tyukavina, A.; Turubanova, S.; Talero, Y.; Hernandez-Serna, A.; Hansen, M.; Saah, D.; Tenneson, K.; Poortinga, A.; Aekakkararungroj, A.; et al. Annual continuous fields of woody vegetation structure in the Lower Mekong region from 2000–2017 Landsat time-series. *Remote Sens. Environ.* **2019**, *232*, 111278. [CrossRef]
11. Simard, M.; Pinto, N.; Fisher, J.; Baccini, A. Mapping forest canopy height globally with spaceborne lidar. *J. Geophys. Res.* **2011**, *116*, 4021. [CrossRef]
12. Liang, X.; Kankare, V.; Hyyppä, J.; Wang, Y.; Kukko, A.; Haggrén, H.; Yu, X.; Kaartinen, H.; Jaakkola, A.; Guan, F.; et al. Terrestrial laser scanning in forest inventories. *ISPRS J. Photogramm. Remote Sens.* **2016**, *115*, 63–77. [CrossRef]
13. Alexander, C.; Korstjens, A.; Hill, R. Influence of micro-topography and crown characteristics on tree height estimations in tropical forests based on LiDAR canopy height models. *Int. J. Appl. Earth Obs. Geoinf.* **2018**, *65*, 105–113. [CrossRef]
14. Almeida, D.; Broadbent, E.; Zambrano, A.; Wilkinson, B.; Ferreira, M.; Chazdon, R.; Meli, P.; Gorgens, E.; Silva, C.; Stark, S.; et al. Monitoring the structure of forest restoration plantations with a drone-lidar system. *Int. J. Appl. Earth Obs. Geoinf.* **2019**, *79*, 192–198. [CrossRef]
15. Zhang, Z.; Ni, W.; Sun, G.; Huang, W.; Ranson, K.J.; Cook, B.D.; Guo, Z. Biomass retrieval from L-band Polarimetric UAVSAR Backscatter and prism stereo imagery. *Remote Sens. Environ.* **2017**, *194*, 331–346. [CrossRef]

16. Qi, W.; Lee, S.-K.; Hancock, S.; Luthcke, S.; Tang, H.; Armston, J.; Dubayah, R. Improved Forest height estimation by fusion of simulated GEDI LIDAR data and TanDEM-X Insar Data. *Remote Sens. Environ.* **2019**, *221*, 621–634. [CrossRef]
17. Li, C.; Song, J.; Wang, J. New approach to calculating tree height at the regional scale. *For. Ecosyst.* **2021**, *8*, 24. [CrossRef]
18. Popescu, S.C. Estimating biomass of individual pine trees using airborne lidar. *Biomass Bioenergy* **2007**, *31*, 646–655. [CrossRef]
19. Lang, N.; Schindler, K.; Wegner, J.D. Country-wide high-resolution vegetation height mapping with sentinel-2. *Remote Sens. Environ.* **2019**, *233*, 111347. [CrossRef]
20. Neumann, M.; Ferro-Famil, L.; Reigber, A. Estimation of Forest Structure, Ground, and Canopy Layer Characteristics from Multibaseline Polarimetric Interferometric SAR Data. *IEEE Trans. Geosci. Remote Sens.* **2010**, *48*, 1086–1104. [CrossRef]
21. López-Serrano, P.M.; López-Sánchez, C.A.; Álvarez-González, J.G.; García-Gutiérrez, J. A Comparison of Machine Learning Techniques Applied to Landsat-5 TM Spectral Data for Biomass Estimation. *Can. J. Remote Sens.* **2016**, *42*, 690–705. [CrossRef]
22. Huang, W.; Min, W.; Ding, J.; Liu, Y.; Hu, Y.; Ni, W.; Shen, H. Forest height mapping using inventory and multi-source satellite data over Hunan Province in southern China. *For. Ecosyst.* **2022**, *9*, 100006. [CrossRef]
23. Liu, Y.; Gong, W.; Xing, Y.; Hu, X.; Gong, J. Estimation of the forest stand mean height and aboveground biomass in northeast China using SAR Sentinel-1B, multispectral sentinel-2a, and DEM imagery. *ISPRS J. Photogramm. Remote Sens.* **2019**, *151*, 277–289. [CrossRef]
24. Amini, J.; Sumantyo, J.T.S. Employing a Method on SAR and Optical Images for Forest Biomass Estimation. *IEEE Trans. Geosci. Remote Sens.* **2009**, *47*, 4020–4026. [CrossRef]
25. Forkuor, G.; Benewinde Zoungrana, J.-B.; Dimobe, K.; Ouattara, B.; Vadrevu, K.P.; Tondoh, J.E. Above-ground biomass mapping in West African dryland forest using Sentinel-1 and 2 datasets—A case study. *Remote Sens. Environ.* **2020**, *236*, e111496. [CrossRef]
26. Li, H.; Kato, T.; Hayashi, M.; Wu, L. Estimation of forest aboveground biomass of two major conifers in Ibaraki Prefecture, Japan, from palsar-2 and sentinel-2 data. *Remote Sens.* **2022**, *14*, 468. [CrossRef]
27. Lu, D.; Chen, Q.; Wang, G.; Li, G.; Moran, E. A survey of remote sensing-basedd aboveground biomass estimation methods in forest ecosystems. *Int. J. Digit. Earth.* **2016**, *9*, 63–105. [CrossRef]
28. Li, X.; Lin, H.; Long, J.; Xu, X. Mapping the growing stem volume of the coniferous plantations in north China using multispectral data from integrated GF-2 and sentinel-2 images and an optimized feature variable selection method. *Remote Sens.* **2021**, *13*, 2740. [CrossRef]
29. Li, G.; Xie, Z.; Jiang, X.; Lu, D.; Chen, E. Integration of ZiYuan-3 Multispectral and Stereo Data for Modeling Aboveground Biomass of Larch Plantations in North China. *Remote Sens.* **2019**, *11*, 2328. [CrossRef]
30. Zhao, P.; Lu, D.; Wang, G.; Liu, L.; Li, D.; Zhu, J.; Yu, S. Forest aboveground biomass estimation in Zhejiang Province using the integration of Landsat TM and ALOS PALSAR data. *Int. J. Appl. Earth Obs.* **2016**, *53*, 1–15. [CrossRef]
31. Wang, X.; Liu, C.; Lv, G.; Xu, J.; Cui, G. Integrating multi-source remote sensing to assess forest aboveground biomass in the Khingan mountains of north-eastern China using machine-learning algorithms. *Remote Sens.* **2022**, *14*, 1039. [CrossRef]
32. Purohit, S.; Aggarwal, S.P.; Patel, N.R. Estimation of forest aboveground biomass using combination of Landsat 8 and sentinel-1a data with random forest regression algorithm in Himalayan foothills. *Trop. Ecol.* **2021**, *62*, 288–300. [CrossRef]
33. Peng, X.; Zhao, A.; Chen, Y.; Chen, Q.; Liu, H.; Wang, J.; Li, H. Comparison of modeling algorithms for Forest Canopy Structures based on UAV-LIDAR: A case study in tropical China. *Forests* **2020**, *11*, 1324. [CrossRef]
34. Zhao, Q.; Yu, S.; Zhao, F.; Tian, L.; Zhao, Z. Comparison of machine learning algorithms for Forest parameter estimations and application for Forest Quality Assessments. *For. Ecol. Manag.* **2019**, *434*, 224–234. [CrossRef]
35. Chen, M.; Qiu, X.; Zeng, W.; Peng, D. Combining sample plot stratification and machine learning algorithms to improve forest aboveground carbon density estimation in northeast China using Airborne Lidar Data. *Remote Sens.* **2022**, *14*, 1477. [CrossRef]
36. Yu, G.; Lu, Z.; Lai, Y. Comparative Study on Variable Selection Approaches in Establishment of Remote Sens. Model for Forest Biomass Estimation. *Remote Sens.* **2019**, *11*, 1437. [CrossRef]
37. Luo, M.; Wang, Y.; Xie, Y.; Zhou, L.; Qiao, J.; Qiu, S.; Sun, Y. Combination of feature selection and CatBoost for prediction: The first application to the estimation of aboveground biomass. *Forests* **2021**, *12*, 216. [CrossRef]
38. Ahmed, O.S.; Franklin, S.E.; Wulder, M.A.; White, J.C. Extending airborne lidar-derived estimates of forest canopy cover and height over large areas using KNN with Landsat Time Series Data. *IEEE J. Sel. Top. Appl. Earth Obs. Remote Sens.* **2016**, *9*, 3489–3496. [CrossRef]
39. Özçelik, R.; Diamantopoulou, M.J.; Crecente-Campo, F.; Eler, U. Estimating crimean juniper tree height using nonlinear regression and artificial neural network models. *For. Ecol. Manag.* **2013**, *306*, 52–60. [CrossRef]
40. Potapov, P.; Li, X.; Hernandez-Serna, A.; Tyukavina, A.; Hansen, M.C.; Kommareddy, A.; Pickens, A.; Turubanova, S.; Tang, H.; Silva, C.E.; et al. Mapping global forest canopy height through integration of Gedi and Landsat Data. *Remote Sens. Environ.* **2021**, *253*, 112165. [CrossRef]
41. Wang, M.; Sun, R.; Xiao, Z. Estimation of forest canopy height and aboveground biomass from Spaceborne Lidar and landsat imageries in Maryland. *Remote Sens.* **2018**, *10*, 344. [CrossRef]
42. Wang, Y.; Li, G.; Ding, J.; Guo, Z.; Tang, S.; Wang, C.; Huang, Q.; Liu, R.; Chen, J.M. A combined glas and Modis estimation of the global distribution of mean forest canopy height. *Remote Sens. Environ.* **2016**, *174*, 24–43. [CrossRef]
43. Pham, T.D.; Yokoya, N.; Xia, J.; Ha, N.T.; Le, N.N.; Nguyen, T.T.T.; Dao, T.H.; Vu, T.T.P.; Pham, T.D.; Takeuchi, W. Comparison of Machine Learning Methods for Estimating Mangrove Above-Ground Biomass Using Multiple Source Remote Sens. Data in the Red River Delta Biosphere Reserve, Vietnam. *Remote Sens.* **2020**, *12*, 1334. [CrossRef]

44. Zhang, Y.; Ma, J.; Liang, S.; Li, X.; Liu, J. A stacking ensemble algorithm for improving the biases of forest aboveground biomass estimations from multiple remotely sensed datasets. *GIsci. Remote Sens.* **2022**, *59*, 234–249. [CrossRef]
45. Gorelick, N.; Hancher, M.; Dixon, M.; Ilyushchenko, S.; Thau, D.; Moore, R. Google Earth Engine: Planetary-scale geospatial analysis for everyone. *Remote Sens. Environ.* **2017**, *202*, 18–27. [CrossRef]
46. Mullissa, A.; Vollrath, A.; Odongo-Braun, C.; Slagter, B.; Balling, J.; Gou, Y.; Gorelick, N.; Reiche, J. Sentinel-1 SAR Backscatter Analysis Ready Data Preparation in Google Earth Engine. *Remote Sens.* **2021**, *13*, 1954. [CrossRef]
47. The Japan Aerospace Exploration Agency(JAXA). *Global 25m Resolution PALSAR-2/PALSAR Mosaic and Forest/Non-Forest Map (FNF) Dataset Description*; JAXA: Tsukuba, Japan, 2019.
48. Gong, P.; Liu, H.; Zhang, M.; Li, C.; Wang, J.; Huang, H.; Clinton, N.; Ji, L.; Li, W.; Bai, Y.; et al. Stable classification with limited sample: Transferring a 30-M resolution sample set collected in 2015 to mapping 10-M resolution global land cover in 2017. *Sci. Bull.* **2019**, *64*, 370–373. [CrossRef]
49. Hu, Y.; Xu, X.; Wu, F.; Sun, Z.; Xia, H.; Meng, Q.; Huang, W.; Zhou, H.; Gao, J.; Li, W.; et al. Estimating Forest Stock Volume in Hunan Province, China, by integrating in situ plot data, sentinel-2 images, and linear and machine learning regression models. *Remote Sens.* **2020**, *12*, 186. [CrossRef]
50. Frampton, W.J.; Dash, J.; Watmough, G.; Milton, E.J. Evaluating the capabilities of Sentinel-2 for quantitative estimation of biophysical variables in vegetation. *ISPRS J. Photogramm. Remote Sens.* **2013**, *82*, 83–92. [CrossRef]
51. Vaglio Laurin, G.; Pirotti, F.; Callegari, M.; Chen, Q.; Cuozzo, G.; Lingua, E.; Notarnicola, C.; Papale, D. Potential of ALOS2 and NDVI to Estimate Forest Above-Ground Biomass, and Comparison with Lidar-Derived Estimates. *Remote Sens.* **2017**, *9*, 18. [CrossRef]
52. Zhang, Y.; Liang, S.; Sun, G. Forest biomass mapping of northeastern China using GLAS and MODIS data. *IEEE J. Sel. Top. Appl. Earth Obs. Remote Sens.* **2014**, *7*, 140–152. [CrossRef]
53. Chi, H.; Sun, G.; Huang, J.; Guo, Z.; Ni, W.; Fu, A. National forest aboveground biomass mapping from ICESat/GLAS data and MODIS imagery in China. *Remote Sens.* **2015**, *7*, 5534–5564. [CrossRef]
54. Whittingham, M.J.; Stephens, P.A.; Bradbury, R.B.; Freckleton, R.P. Why do we still use stepwise modelling in ecology and behaviour? *J. Anim. Ecol.* **2006**, *75*, 1182–1189. [CrossRef]
55. Adame-Campos, R.L.; Ghilardi, A.; Gao, Y.; Paneque-Gálvez, J.; Mas, J. Variables Selection for Aboveground Biomass Estimations Using Satellite Data: A Comparison between Relative Importance Approach and Stepwise Akaike's Information Criterion. *ISPRS Int. J. Geo-Inf.* **2019**, *8*, 245. [CrossRef]
56. Venables, W.N.; Ripley, B.D.; Venables, W.N. *Modern Applied Statistics with S*; Springer: New York, NY, USA, 2002.
57. Pullanagari, R.; Kereszturi, G.; Yule, I. Integrating airborne hyperspectral, topographic, and soil data for estimating pasture quality using recursive feature elimination with random forest regression. *Remote Sens.* **2018**, *10*, 1117. [CrossRef]
58. Zhou, Q.; Zhou, H.; Zhou, Q.; Yang, F.; Luo, L. Structure damage detection based on random forest recursive feature elimination. *Mech. Syst. Signal Process.* **2014**, *46*, 82–90. [CrossRef]
59. Granitto, P.M.; Furlanello, C.; Biasioli, F.; Gasperi, F. Recursive feature elimination with random forest for ptr-ms analysis of agroindustrial products. *Chemom. Intell. Lab. Syst.* **2006**, *83*, 83–90. [CrossRef]
60. Kursa, M.B.; Rudnicki, W.R. Feature Selection with the Boruta Package. *J. Stat. Softw.* **2010**, *36*, 1–13. [CrossRef]
61. Chirici, G.; Barbati, A.; Corona, P.; Marchetti, M.; Travaglini, D.; Maselli, F.; Bertini, R. Non-parametric and parametric methods using satellite images for estimating growing stock volume in Alpine and mediterranean forest ecosystems. *Remote Sens. Environ.* **2008**, *112*, 2686–2700. [CrossRef]
62. Chirici, G.; Mura, M.; McInerney, D.; Py, N.; Tomppo, E.O.; Waser, L.T.; Travaglini, D.; McRoberts, R.E. A meta-analysis and review of the literature on the K-nearest neighbors technique for forestry applications that use remotely sensed data. *Remote Sens. Environ.* **2016**, *176*, 282–294. [CrossRef]
63. Mountrakis, G.; Im, J.; Ogole, C. Support Vector Machines in remote sensing: A Review. *ISPRS J. Photogramm. Remote Sens.* **2011**, *66*, 247–259. [CrossRef]
64. Vafaei, S.; Soosani, J.; Adeli, K.; Fadaei, H.; Naghavi, H.; Pham, T.; Tien Bui, D. Improving accuracy estimation of forest aboveground biomass based on incorporation of Alos-2 palsar-2 and sentinel-2a imagery and Machine Learning: A case study of the hyrcanian forest area (Iran). *Remote Sens.* **2018**, *10*, 172. [CrossRef]
65. Deb, D.; Deb, S.; Chakraborty, D.; Singh, J.P.; Singh, A.K.; Dutta, P.; Choudhury, A. Aboveground biomass estimation of an agro-pastoral ecology in semi-arid Bundelkhand region of India from Landsat Data: A comparison of support vector machine and traditional regression models. *Geocarto. Int.* **2020**, *37*, 1043–1058. [CrossRef]
66. Breiman, L. Random Forests. *Mach. Learn.* **2001**, *45*, 5–32. [CrossRef]
67. Su, Y.; Guo, Q.; Xue, B.; Hu, T.; Alvarez, O.; Tao, S.; Fang, J. Spatial distribution of forest aboveground biomass in China: Estimation through combination of spaceborne lidar, optical imagery, and forest inventory data. *Remote Sens. Environ.* **2016**, *173*, 187–199. [CrossRef]
68. Zhang, Y.; Ma, J.; Liang, S.; Li, X.; Li, M. An evaluation of eight machine learning regression algorithms for forest aboveground biomass estimation from multiple satellite data products. *Remote Sens.* **2020**, *12*, 4015. [CrossRef]
69. Chen, L.; Wang, Y.; Ren, C.; Zhang, B.; Wang, Z. Optimal Combination of Predictors and Algorithms for Forest Above-Ground Biomass Mapping from Sentinel and SRTM Data. *Remote Sens.* **2019**, *11*, 414. [CrossRef]
70. Friedman, J. Greedy function approximation: A gradient boosting machine. *Ann. Stat.* **2001**, *29*, 1189–1232. [CrossRef]

71. Yang, L.; Liang, S.; Zhang, Y. A New Method for Generating a Global Forest Aboveground Biomass Map From Multiple High-Level Satellite Products and Ancillary Information. *IEEE J. Sel. Top. Appl. Earth Obs. Remote Sens.* **2020**, *13*, 2587–2597. [CrossRef]
72. Chen, T.; Guestrin, C. Xgboost: A scalable tree boosting system. In Proceedings of the 22nd ACM SIGKDD International Conference on Knowledge Discovery and Data Mining, San Francisco, CA, USA, 13–17 August 2016; pp. 785–794.
73. Yu, J.-W.; Yoon, Y.-W.; Baek, W.-K.; Jung, H.-S. Forest vertical structure mapping using two-seasonal optic images and LIDAR DSM acquired from UAV platform through Random Forest, XGBoost, and support vector machine approaches. *Remote Sens.* **2021**, *13*, 4282. [CrossRef]
74. Li, Y.; Li, C.; Li, M.; Liu, Z. Influence of Variable Selection and Forest Type on Forest Aboveground Biomass Estimation Using Machine Learning Algorithms. *Forests* **2019**, *10*, 1073. [CrossRef]
75. Li, Y.; Li, M.; Li, C.; Liu, Z. Forest aboveground biomass estimation using Landsat 8 and sentinel-1a data with machine learning algorithms. *Sci. Rep.* **2020**, *10*, 9952. [CrossRef] [PubMed]
76. Dorogush, A.V.; Ershov, V.; Gulin, A. CatBoost: Gradient boosting with categorical features support. *arXiv* **2018**, arXiv:1810.11363.
77. Sun, H.; He, J.; Chen, Y.; Zhao, B. Space-Time Sea Surface PCO2 Estimation in the North Atlantic Based on CatBoost. *Remote Sens.* **2021**, *13*, 2805. [CrossRef]
78. Ahirwal, J.; Nath, A.; Brahma, B.; Deb, S.; Sahoo, U.K.; Nath, A.J. Patterns and Driving Factors of Biomass Carbon and Soil Organic Carbon Stock in the Indian Himalayan Region. *Sci. Total Environ.* **2021**, *770*, 145292. [CrossRef]
79. Li, W.; Niu, Z.; Shang, R.; Qin, Y.; Wang, L.; Chen, H. High-resolution mapping of forest canopy height using machine learning by coupling icesat-2 lidar with sentinel-1, sentinel-2 and landsat-8 data. *J. Appl. Earth Obs. Geoinf.* **2020**, *92*, 102163. [CrossRef]
80. Huang, H.; Liu, C.; Wang, X. Constructing a finer-resolution forest height in China using icesat/glas, landsat and Alos Palsar data and height patterns of natural forests and plantations. *Remote Sens.* **2019**, *11*, 1740. [CrossRef]
81. Xi, Z.; Xu, H.; Xing, Y.; Gong, W.; Chen, G.; Yang, S. Forest canopy height mapping by synergizing icesat-2, sentinel-1, sentinel-2 and topographic information based on machine learning methods. *Remote Sens.* **2022**, *14*, 364. [CrossRef]
82. Agjee, N.H.; Ismail, R.; Mutanga, O. Identifying relevant hyperspectral bands using Boruta: A temporal analysis of water hyacinth biocontrol. *J. Appl. Remote Sens.* **2016**, *10*, 042002. [CrossRef]
83. Arjasakusuma, S.; Swahyu Kusuma, S.; Phinn, S. Evaluating variable selection and machine learning algorithms for Estimating Forest Heights by combining Lidar and Hyperspectral Data. *ISPRS Int. J. Geo-Inf.* **2020**, *9*, 507. [CrossRef]
84. Fayad, I.; Baghdadi, N.; Alcarde Alvares, C.; Stape, J.L.; Bailly, J.S.; Scolforo, H.F.; Cegatta, I.R.; Zribi, M.; Le Maire, G. Terrain Slope effect on forest height and wood volume estimation from Gedi Data. *Remote Sens.* **2021**, *13*, 2136. [CrossRef]
85. Xing, Y.; de Gier, A.; Zhang, J.; Wang, L. An Improved Method for Estimating Forest Canopy Height Using ICESat-GLAS Full Waveform Data over Sloping Terrain: A Case Study in Changbai Mountains, China. *Int. J. Appl. Earth Obs. Geoinf.* **2010**, *12*, 385–392. [CrossRef]
86. Pourshamsi, M.; Xia, J.; Yokoya, N.; Garcia, M.; Lavalle, M.; Pottier, E.; Balzter, H. Tropical Forest Canopy Height Estimation from combined polarimetric SAR and Lidar using machine-learning. *ISPRS J. Photogramm. Remote Sens.* **2021**, *172*, 79–94. [CrossRef]

remote sensing

MDPI

Article

Enhancing Aboveground Biomass Estimation for Three Pinus Forests in Yunnan, SW China, Using Landsat 8

Jing Tang, Ying Liu, Lu Li, Yanfeng Liu, Yong Wu, Hui Xu and Guanglong Ou *

Key Laboratory of State Forestry Administration on Biodiversity Conservation in Southwest China, Southwest Forestry University, Kunming 650224, China
* Correspondence: olg2007621@swfu.edu.cn

Abstract: The estimation of forest aboveground biomass (AGB) using Landsat 8 operational land imagery (OLI) images has been extensively studied, but forest aboveground biomass (AGB) is often difficult to estimate accurately, in part due to the multi-level structure of forests, the heterogeneity of stands, and the diversity of tree species. In this study, a habitat dataset describing the distribution environment of forests, Landsat 8 OLI image data of spectral reflectance information, as well as a combination of the two datasets were employed to estimate the AGB of the three common pine forests (*Pinus yunnanensis* forests, *Pinus densata* forests, and *Pinus kesiya* forests) in Yunnan Province using a parametric model, stepwise linear regression model (SLR), and a non-parametric model, such as random forest (RF) and support vector machine (SVM). Based on the results, the following conclusions can be drawn. (1) As compared with the parametric model (SLR), the non-parametric models (RF and SVM) have a better fitting performance for estimating the AGB of the three pine forests, especially in the AGB segment of 40 to 200 Mg/ha. The non-parametric model is more sensitive to the number of data samples. In the case of the *Pinus densata* forest with a sample size greater than 100, RF fitting provides better fitting performance than SVM fitting, and the SVM fitting model is better suited to the AGB estimation of the *Pinus yunnanensis* forest with a sample size of less than 100. (2) Landsat 8 OLI images exhibit superior accuracy in estimating the AGB of the three pine forests using a single dataset. Variables, such as texture and vegetation index variables, which can reflect the comprehensive reflection information of ground objects, play a significant role in estimating AGBs, especially the texture variables. (3) By incorporating the combined dataset with characteristics of tree species distribution and ground object reflectance spectrum, the accuracy and stability of AGB estimation of the three pine forests can be improved. Moreover, the employment of a combined dataset is also effective in reducing the number of estimation errors in cases with AGB less than 100 Mg/ha or exceeding 150 Mg/ha.

Keywords: aboveground biomass; habitat dataset; Landsat 8-OLI images; pine forest; model comparison

Citation: Tang, J.; Liu, Y.; Li, L.; Liu, Y.; Wu, Y.; Xu, H.; Ou, G. Enhancing Aboveground Biomass Estimation for Three Pinus Forests in Yunnan, SW China, Using Landsat 8. *Remote Sens.* **2022**, *14*, 4589. https://doi.org/10.3390/rs14184589

Academic Editors: Huaqiang Du, Wenyi Fan, Mingshi Li, Weiliang Fan and Fangjie Mao

Received: 7 August 2022
Accepted: 11 September 2022
Published: 14 September 2022

1. Introduction

In addition to regulating the water supplies and climate, forests accumulate biomass by absorbing light, water, carbon dioxide, and other compounds [1]. Forests are the largest carbon source in terrestrial ecosystems [2], accounting for more than two-thirds of the total carbon sequestration annually [3,4]. Therefore, it is crucial to accurately estimate forest biomass to thoroughly understand the carbon cycle and carbon balance in terrestrial ecosystems [5].

There are two types of forest biomass: aboveground biomass (AGB) and underground biomass (UB) [6]. Because AGB accounts for 70% to 90% of the total forest biomass [7], most previous studies have focused on AGB. The estimation methods of AGB include destructive and non-destructive methods. The destructive harvest method is considered to be the most accurate AGB estimation method. However, as this method poses a threat to the flora and fauna, it is inappropriate for large-scale AGB estimation [8]. Due to the multi-scale and multi-band characteristics of remote-sensing (RS) technology, it can effectively ensure the spatial integrity and temporal continuity of data. RS has become a common method for large-scale non-destructive AGB estimation in recent decades [9]. Optical remote sensing was developed first, and a large number of data sources have been used to estimate thousands of forest AGBs [6,10–13]. The natural forest consisting of mature stands, however, is characterized by a multi-layer canopy and high density, resulting in the saturation of spectral reflectance in optical remote sensing and, in turn, leading to an underestimation of large AGB values [1]. The AGB of young forests, on the other hand, will be overestimated due to sparse canopy radiation combined with understory vegetation, soil, and other radiation information. These are the main difficulties that have arisen in estimating forest AGB using an optical RS dataset. As a result of their capability of obtaining detailed information on forest structure, radar and LiDAR have become two of the main techniques for forest AGB estimation in recent decades [13–21]. However, the number of data sources of radar and LiDAR is relatively smaller than the number of optical RS datasets; the working mechanism of these data sources also limits their application [6]. Thus, optical RS still remains an effective data source for AGB estimation in a wide area. As a result, Landsat images have been widely used in AGB estimation, in part due to their accessibility, time continuity, and moderate spatial resolution [22–25]. It should be noted that in forest AGB estimation using Landsat 8 OLI images, the problem of underestimating high values and overestimating low values when using this data source is still challenging [26].

In order to overcome the uncertainty that can be caused by using only one remote-sensing source in the estimation of forest AGBs, researchers have become increasingly aware of the necessity of multi-source data to improve the accuracy of biomass estimation in forests [6]. For example, by combining radar or UAV data with optical remote-sensing data, it is possible to overcome the challenge of AGB estimation caused by changes in the forest structure and the high heterogeneity of the forest landscapes [22,23,26–28]. The complementarity between remote-sensing data can improve the estimation accuracy of AGB to a certain extent. However, since radar and UAV data sources are limited, the combination of non-remote-sensing datasets, such as field measurement and forest surveys, with optical remote sensing can also contribute to the improvement of AGB estimation accuracy [29–33]. In recent years, some scholars have emphasized the fact that trees have a long growing season. Forest AGB estimations can be enhanced by incorporating long-term data, such as climate [34,35] and phenology [36–38], into an optical remote-sensing dataset to compensate for the lack of timeliness in remote-sensing data for forest AGB estimations.

Habitat [39] is a concept that was first proposed by Grinnel in 1917. In general, it refers to the space in which an individual, a population, or a community can complete their life processes [40]. In many studies, habitat has been shown to affect biomass accumulation and plant allocation, as well as changes in vegetation genotypes [41–44]. While adapting to the environment, vegetation also has varying carbon sequestration abilities due to the accumulation and differentiation of biomass in different parts of the foliage, resulting in the change of habitat [45]. There is a close relationship between species and habitat [46]. This relationship exists due to the statistical association of habitat information with species abundance or the probability of occurrence [47]. The abundance of species in a region is influenced by the community structure at different scales [48], and forests are no exception. Studies have confirmed that the complexity of the vertical structure of forests is related to forest biodiversity [49,50]. Thus, the habitat information that can indirectly measure the biodiversity information in the region [51] can represent the vertical structure information of the forest to a certain extent, especially on a large spatial scale [52].

Typical habitat data characteristics include climate, slope, aspect, elevation, soil type, etc. The habitat suitability value (HSV) is determined by the comprehensive effects of different habitat factors on species in a region. In ecology and conservation biology, species distribution maps are often used to indicate the HSV of species in specific areas [53,54]. However, it is difficult to produce a map of the species distribution for all species. It is especially difficult to obtain the HSV of most species in areas where there are abundant species [55]. Therefore, it is an effective method to import habitat data into species distribution models (SDMs) to obtain the HSV of species. A number of common SDMs include BIOCLIM, ENFA, HABITAT, Maxent, etc. Maxent is one of the most widely used and highly accurate SDMs [56] and has been extensively used in species conservation and management [57–59]. This model calculates the binding force of species distribution according to the environmental factors of known distribution points; then, it estimates the probability of species distribution in unknown distribution points [60] and ultimately obtains the HSV of the species. However, the number of studies that use a habitat dataset as an independent variable to estimate forest AGB is still too small [38], especially for pine forest AGBs.

Furthermore, in addition to the data used for estimation, AGB estimation is also strongly influenced by the choice of modeling algorithms [61]. There are two types of AGB modeling algorithms: parametric and non-parametric. The parametric model determines the relationship between AGB and independent variables through linear functions, power functions, exponential functions, etc. Since this model requires fewer sample data and can quantify the relationship between AGB and the variables [1], it has become the most commonly utilized biomass modeling algorithm [6]. Due to the complex correlation between the variables and AGB, the parametric model cannot provide an adequate estimation [61,62]. However, the stepwise linear regression algorithm can differentiate between the important variables in the modeling through the significance test, which, in turn, improves the estimation performance of the parametric model to a certain extent [63]. In addition, non-parametric models are also capable of handling nonlinear relationships and can be used to determine the most suitable model structure for the dependent variable from the estimated data, which has been widely used in forest AGB estimation in the past ten years [8,25,33,64]. Random forest and support vector machines are two of the most commonly used non-parametric models. However, the non-parametric model has two main shortcomings; the first is that it is sensitive to data [1,65], and the second is that its interpretation for the estimation process is less clear than that of the parametric model [6].

In this study, the AGB of the three common pine forests (*Pinus yunnanensis* forests, *Pinus densata* forests, and *Pinus kesiya* forests) was estimated in Yunnan Province, southwest China, by employing varying datasets (e.g., a Landsat 8 OLI dataset, a habitat dataset, and a combined dataset composed of them both) and modeling algorithms (e.g., stepwise linear regression (SLR), random forest (RF), and support vector machine (SVM)). Further, the performance of AGB estimations of different forests under different models was compared.

The purpose of this study was to answer the following questions:

(1) Do estimation models have an impact on the AGB estimation for pine forests?
(2) Is it possible to estimate the AGB for the three pine forests using the habitat dataset?
(3) Does the employment of a habitat dataset reduce the probability of overestimation and underestimation of the AGB estimation?

2. Materials and Methods

The AGB estimations of pine forests included in this study were carried out using different datasets from the following steps: (1) the selection of sample plots and calculation of AGBs; (2) habitat simulation of pine forests in order to obtain habitat datasets; (3) preprocessing of Landsat 8 OLI images in order to obtain RS datasets; (4) analysis of the correlation between habitat datasets and RS datasets and plot AGBs in order to select independent variables; (5) combining data from selected habitat data and RS data; (6) the development of AGB estimation models (e.g., SLR, RF, and SVM) using different datasets; (7) comparing the modeling results of AGB estimations. The specific process is shown in Figure 1.

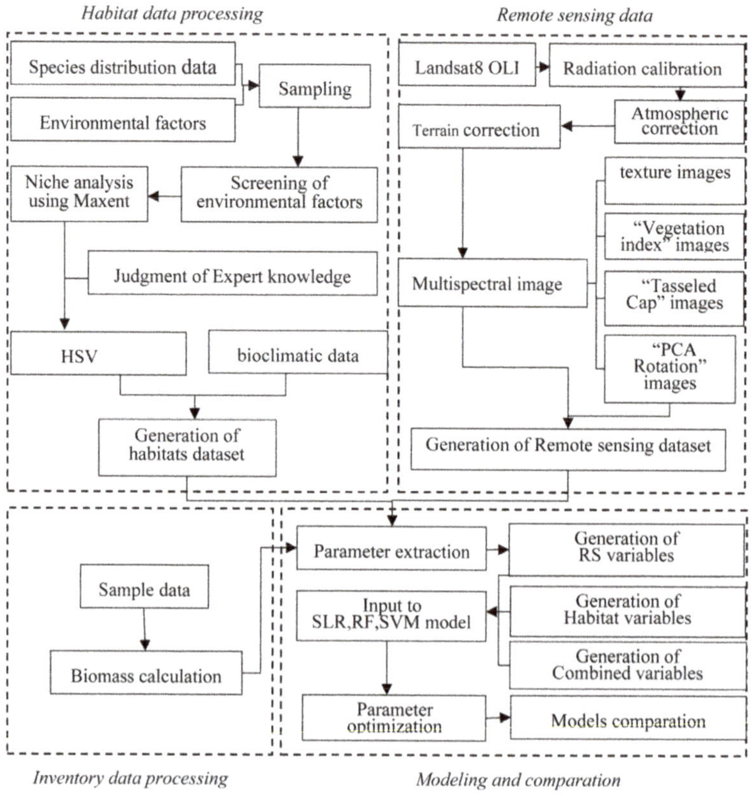

Figure 1. Flow chart of estimating aboveground biomass (AGB) of pine forest using different datasets through parametric model and non-parametric model (Note: HSV, habitat suitability value; SLR, stepwise linear regression; RF, random forest; SVM, support vector machine).

2.1. Study Area

Yunnan is located in southwest China. It is one of the three forest regions in China. The pine forests in the province mainly comprise *Pinus yunnanensis*, *Pinus densata*, and *Pinus kesiya* trees [30,65,66]. The forests play an important role in ecological services and forest carbon sinks in the region [33,67], and forest biomass is the basis for forest carbon sink estimation [68]. The study areas of this study were selected in Yongren County, Shangri-La City, and Pu'er City, where the main distribution areas of the three pine forests are located (Figure 2).

Yongren County is located in the central and northern parts of Yunnan Province. The altitude of this area ranges from 1530 to 1700 m above sea level, and the terrain is relatively flat. The climate of this region is classified as a north subtropical southwest monsoon, and the precipitation season lasts from June to October, when the dry and wet seasons are clearly distinguishable [69]. There are mainly two types of vegetation in this region: sub-humid subtropical evergreen broad-leaved forests and *Pinus yunnanensis* secondary forests. *Pinus yunnanensis* grows at an altitude of 1000–3200 m and can withstand drought as well as barren soil. This tree species is native to this area, and it is one of the pioneer tree species in southwest China. The Baima River Forest Farm in Yongren County is the largest mother forest base for *Pinus yunnanensis* in China.

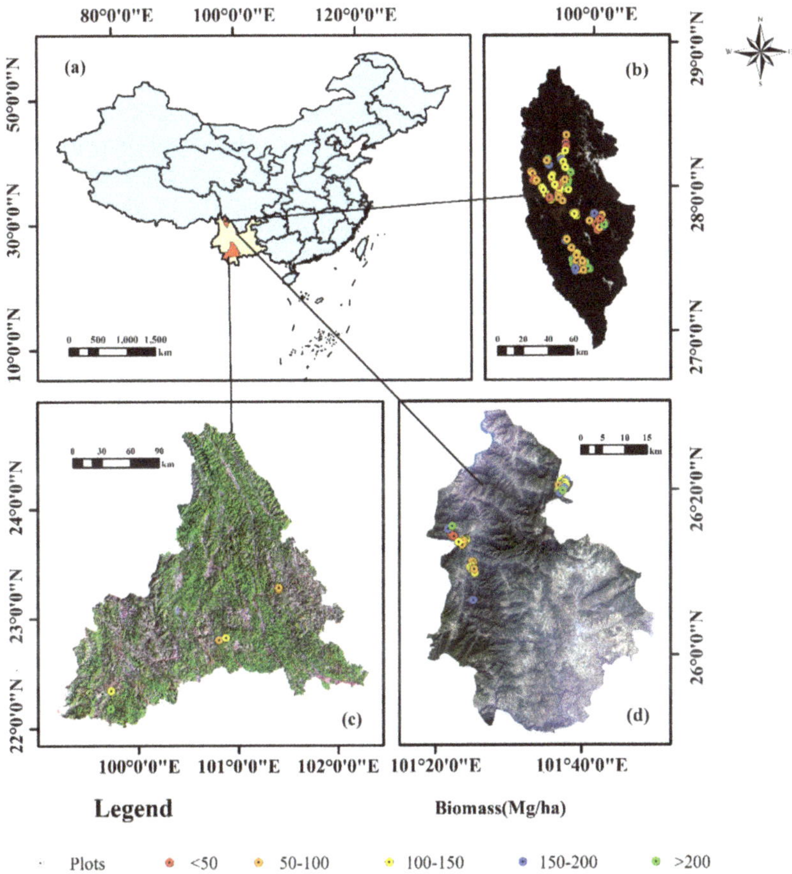

Figure 2. (**a**) Location of the study area; (**b**) RGB true color composite image of Shangri-la City and field plots of *Pinus densata* forest; (**c**) RGB true color composite image of Pu'er City and field plots of *Pinus kesiya* forest; (**d**) RGB true color composite image of Yongren County and field plots of *Pinus yunnanensis* forest.

Shangri-la, with an altitude of between 1503 and 5545 m above sea level, is located in the northwest of Yunnan Province. The annual average temperature in this area is 5.5 °C, and summer and autumn are the seasons with the most precipitation. This area is part of the World Natural Heritage site, "Parallel Flow of the Three Rivers", with rich forest resources. *Pinus yunnanensis*, *Pinus densata*, *Picea* spp, *Abies* spp, and *Quercus* spp are the dominant tree species in the Shangri-la region [70]. Among the tree species, *Pinus densata* is considered to be a unique tree species in the alpine region of western China, and its vertical distribution is slightly higher than that of *Pinus yunnanensis*.

Pu'er is located in the southwest of Yunnan Province, where the altitude ranges from 376 to 3306 m above sea level. In addition, more than 90% of the area is mountainous. In most parts of this area, there is no frost all throughout the year, and the annual precipitation is between 1100 and 2780 mm. In addition, its climate is classified as the plateau monsoon climate of the south Asian tropics. The abundant precipitation keeps the relative humidity in the region at 82% all throughout the year, resulting in favorable conditions for vegetation growth [71]. Pu'er is the second largest forest region in the Yunnan Province. Over half of the area is covered by *Pinus kesiya*, the dominant tree species. *Pinus kesiya* is also a main afforestation tree species that can be found at altitudes below 1800 m in southern, central, and western Yunnan province.

2.2. Sample Plot Data and Forest AGB

In order to obtain the AGB of the pine, destructive samples were taken from 87 *P. yunnanensis* plots, 147 *P. densata* plots, and 45 *P. kesiya* plots in the study area from 2011 to 2017. The sample plots of 30 m × 30 m were selected and established according to the stock state map based on tree age, altitude, slope, aspect, and sampling distance of 1 km. The coordinates, altitude, slope, aspect, DBH, and tree height of the sample plots were recorded in the process of sampling.

Since there is no calculation model for these three AGB of pine trees in the existing biomass model, in this study, 3 to 5 stands with the average level of stands in the sample plot were selected for cutting, and subsequently, the biomass of their trunks, bark, branches, and leaves in the aboveground part was measured. The biomass of the wood and the bark of the trees was measured by taking a 3 cm disk along the trunk of each tree at 2 m intervals, and then, the density of the samples was measured using the drainage method. The disk was dried in an oven at a constant temperature of 105 °C, and then, the biomass of the disk was determined by comparing the weight before and after drying. In addition, the volume of wood and bark from each sample tree was converted into biomass in 2 m units according to the density of the samples. By using similar methods, the branches and leaves were collected and weighed by grade, and the resulting biomass was measured accordingly. The total AGB of each sample was obtained by adding the AGB from different parts of the sampled trees. Finally, the power function was used to fit individual tree AGB data. The AGB fitting formulae of *Pinus yunnanensis* [65], *Pinus densata* [1], and *Pinus kesiya* [66] were obtained, as shown in Formula (1)–(3).

$$AGB_i = 0.048 * DBH^{1.9276} * H^{0.9638} \tag{1}$$

$$AGB_i = 0.073 * DBH^{1.739} * H^{0.880} \tag{2}$$

$$AGB_i = 0.058 * DBH^{2.12} * H^{0.4668} \tag{3}$$

In these formulae, DBH (cm) is the average diameter at breast height (1.3 m), H (m) is the average tree height, and AGB_i is the aboveground biomass of a single standing tree (kg).

In order to obtain the AGB of the sample plot, the unit was converted into the value per hectare using equation (4). The final AGB statistical data of the three pine forests are shown in Table 1.

$$AGB_p = \frac{AGB_i \times n}{30 \times 30} \times \frac{10,000}{1000} \tag{4}$$

Table 1. The statistical parameters of AGB in the sample plot.

Species	Number of Plots	Statistical Indicators	AGB (Mg/ha)
Pinus yunnanensis	87	Min.	17.901
		Max.	287.679
		Mean	114.868
Pinus densata	147	Min.	2.114
		Max.	344.382
		Mean	121.474
Pinus kesiya	45	Min.	49.063
		Max.	204.448
		Mean	116.432

In this formula, AGB_i is the biomass of individual trees, n is the number of trees in the sample plot, and AGB_p is the AGB of the sample plot (Mg/ha).

2.3. Acquisition of Remote-Sensing Datasets

The Landsat 8 OLI images and DEMs employed in this study were downloaded from http://www.gscloud.cn/ (accessed on 6 August 2022). The spatial resolution of these data was 30 m, and the coordinate system was a Universal Transverse Mercator with zone 47 north as the spatial reference frame. The specific parameters of the Landsat 8 OLI images are shown in Table 2. Radiometric calibration, FLASSH atmospheric correction, and C-correction topographic correction were performed in ENVI in order to correct the radiometric errors in the images. Subsequently, the corrected images were mosaicked and clipped to obtain the Landsat 8 OLI image of the study area.

Table 2. The specific parameters of Landsat 8 OLI images.

Study Area	Image ID	Average Cloud Cover (%)	Start Time
Yongren	LC81300422016030LGN00	0.00	30 January 2016
Shangri-la	LC81310412016325LGN00	0.40	20 November 2016
	LC81320402016348LGN00	0.73	13 December 2016
	LC81320412016348LGN00	0.76	13 December 2016
Pu'er	LC81290442015052LGN00	0.08	21 February 2015
	LC81290452015052LGN00	1.87	21 February 2015
	LC81310432015066LGN00	0.00	7 March 2015
	LC81310442015066LGN00	0.00	7 March 2015
	LC81300432015075LGN00	0.18	16 March 2015
	LC81300442016046LGN00	0.00	15 February 2016
	LC81300452016046LGN00	0.01	15 February 2016
	LC81310452016069LGN00	0.41	9 March 2016

The corrected Landsat 8 OLI image includes seven multi-spectral bands, namely Band1-Coastal, Band2-Blue, Band3-Green, Band4-Red, Band5-NIR, Band6-SWIR1, and Band7-SWIR2. In this study, for the purpose of obtaining more data for the forest AGB estimations, sixty-four conversion variables from remote-sensing images were also employed. A total of five commonly used vegetation indices were calculated, namely the normalized difference vegetation index (NDVI), simple vegetation index (SVI), enhanced vegetation index (EVI), atmospherically resistant vegetation index (ARVI), and structurally insensitive pigment index (SIPI), as well as three tasseled cap images, namely brightness, greenness, and humidity. In addition, the first, second, and third principal component data, which can reflect more than 75% of the image information, were also calculated by ENVI 5.3 software. Additionally, a total of 56 Landsat 8 OLI texture variables were also calculated under $3 * 3$ Windows based on the gray co-occurrence matrix, including homogeneity, anisotropy, mean, angle second moment, entropy, correlation, variance, and contrast. Subsequently, a Pearson correlation analysis was conducted to screen the AGB-related candidates that passed the significance test ($p \leq 0.01$) as the remote-sensing data. Finally, the RS dataset was derived from the resulting remote-sensing data. The selected remote-sensing variables and their correlation with AGB are shown in Figure 3. In the figure, the variables superscripted with "-" are factors that show a negative correlation with AGB.

2.4. Acquisition of Habitat Datasets

When the species adapt to light and precipitation after a prolonged period of time, they will form development rules that will increase their survival advantages in the habitat conditions. The 19 bioclimatic variables downloaded from WorldClim (https://www.worldclim.org/, accessed on 6 August 2022) are considered to be the most significant habitat factors because they can reflect temperature, precipitation, and some other aspects. Over 85% of habitat studies use these 19 bioclimatic factors [72], which are included in Table 3. In this study, the HSV of pine forests and 19 bioclimatic environmental variables were employed as habitat candidates to estimate the AGB.

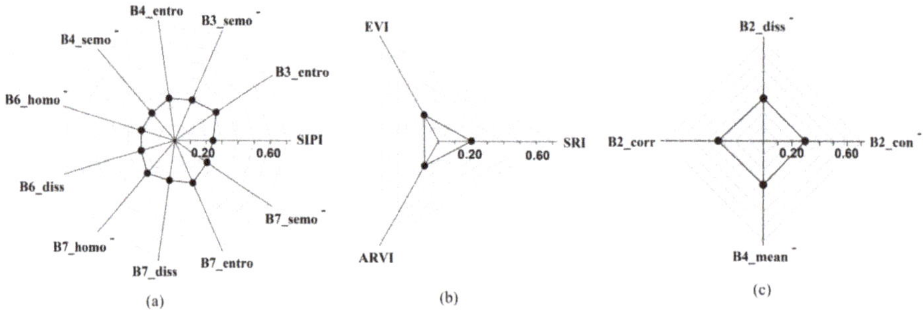

Figure 3. Radar plot of RS data associated with forest AGB (**a**) *Pinus yunnanensis* forests; (**b**) *Pinus densata* forests; (**c**) *Pinus kesiya* forests.

Table 3. The 19 bioclimatic environmental variables from WorldClim.

Variable Code	Variable Description	Variable Code	Variable Description
Bio1	Annual mean temperature	Bio2	Mean diurnal range
Bio3	Isothermality	Bio4	Temperature seasonality
Bio5	Max temperature of the warmest month	Bio6	Min temperature of the coldest month
Bio7	Range of annual temperature	Bio8	Mean temperature of the wettest quarter
Bio9	Mean temperature of the driest quarter	Bio10	Mean temperature of the warmest quarter
Bio11	Mean temperature of the coldest quarter	Bio12	Annual average precipitation
Bio13	Precipitation of the wettest month	Bio14	Precipitation of the driest month
Bio15	Precipitation seasonality	Bio16	Precipitation of the wettest quarter
Bio17	Precipitation of the driest quarter	Bio18	Precipitation of the warmest quarter
Bio19	Precipitation of the coldest quarter		

In order to avoid data disaster in subsequent analyses, a Pearson correlation analysis was conducted among the 19 bioclimatic environmental variables in SPSS 22.0 for Windows to select variables with a correlation coefficient absolute value lower than 0.9 and variables with significance to the pine habitat. The habitat constraint factors of *Pinus yunnanensis* forests included 10 variables, namely Bio1, Bio3, Bio4, Bio5, Bio6, Bio8, Bio10, Bio13, Bio14, Bio15, and Bio16, while the variables of *Pinus densata* forests were Bio2, Bio3, Bio4, Bio5, Bio6, Bio7, Bio14, Bio15, and Bio16. Furthermore, the significant variables for the growth of *Pinus kesiya* forests included Bio1, Bio2, Bio4, Bio6, Bio7, Bio10, Bio13, Bio17, and Bio18.

Species distribution point data were obtained from the Chinese Virtual Herbarium (https://www.cvh.ac.cn/, accessed on 6 August 2022) and the Global Biodiversity Information Facility (https://www.gbif.org, accessed on 6 August 2022). However, these downloaded points were difficult for unifying the collection time and the collector, and the data description was not standard. Therefore, it was necessary to screen these points and delete duplicate points and questionable points to ensure the accuracy of the data. Finally, 803 sample points were obtained, including 460 *Pinus yunnanensis*, 319 *Pinus densata*, and 24 *Pinus kesiya*.

The selected bioclimatic variables and species distribution points were imported into Maxent, while 25% of them were set as random test points, and the model was repeatedly applied ten times in order to obtain the HSV of the three pine forests located in southwest China. The anticipated results were evaluated using the methods of the receiver operating characteristic curve (ROC) and Jackknife. The results showed that the habitat fitting AUC values of the *Pinus yunnanensis* forest, *Pinus densata* forest, and *Pinus kesiya* forest were 0.9889, 0.9906, and 0.9983, respectively. It is generally accepted that the predicted result is accurate when the AUC value is greater than 0.9 [73,74]. The habitat simulation AUC values for all three pine forests were higher than 0.98. Additionally, the fitted AUC standard deviation of the *Yunnan pine* forest and *Pinus densata* forest was 0.0018, and *Pinus kesiya* forest was 0.0004, all of which were less than 0.002. The accuracy was guaranteed. By consulting with experts, the HSV variable values that were closest to the actual species distribution were selected, as expert knowledge helped improve Maxent's prediction accuracy [55].

The candidate of habitat data was derived from the HSV of the pine forests and the 19 bioclimatic variables. Subsequently, a Pearson correlation analysis was conducted to determine the AGB-related candidates that passed the significance test ($p \leq 0.01$) as the habitat data. Finally, the habitat dataset was derived from the resulting habitat data. The selected habitat variables and their correlation with AGB are shown in Figure 4. In the figure, the variables superscripted with "-" are factors that show a negative correlation with AGB.

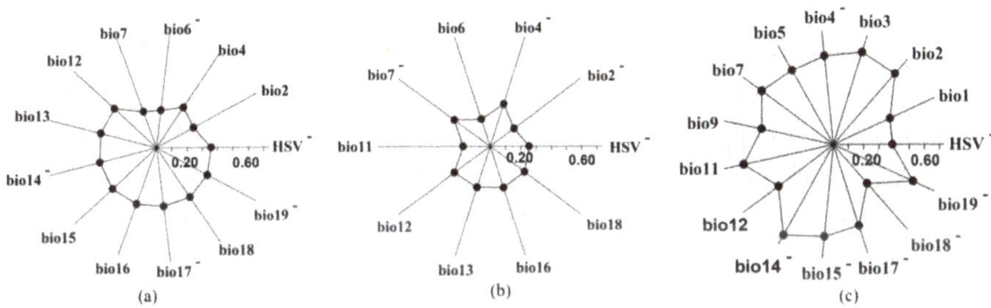

Figure 4. Radar plot of habitat data associated with forest AGB (**a**) *Pinus yunnanensis* forests; (**b**) *Pinus densata* forests; (**c**) *Pinus kesiya* forests.

2.5. Acquisition of Combined Datasets

The combined dataset employed for the AGB estimation consisted of a habitat dataset and an RS dataset, which correlated with the AGB (correlation coefficient > 0.1) and passed the significance test ($p \leq 0.01$).

The combined dataset of *Pinus yunnanensis* forests consisted of 13 habitat data and 11 remote-sensing data; the combined dataset of *Pinus densata* forests consisted of 10 habitat data and 3 remote-sensing data; and finally, the combined dataset of *Pinus kesiya* forests consisted of 15 habitat data and 4 remote-sensing data.

2.6. AGB Modeling Algorithms

In this study, a parametric model, SLR, and two additional non-parametric models, namely RF and SVM, were used for AGB modeling based on a different dataset.

2.6.1. Stepwise Linear Regression (SLR)

The stepwise linear regression model builds a prediction model by first calculating the significance of variables and then deleting the variables with low significance backward or adding the variables with significance forward to the prediction model. Thereby, it can effectively solve the collinearity problem between explanatory variables [75]. The constructed model can be expressed using Formula (5).

$$Y = b_0 + b_1 X_1 + b_2 X_2 + \cdots + b_n X_n \tag{5}$$

In this formula, $b_1, b_2, \ldots, b_n$ are the regression coefficients of the prediction variables, and b_0 is the constant of the prediction model. The stepwise backward linear regression model was utilized in this study.

2.6.2. Random Forest (RF)

A random forest is a machine-learning algorithm used for classification and regression. Its basic idea is to generate a new training sample set by repeatedly performing random extractions on two-thirds of the data from the original training set, and the remaining data that are not extracted become the out-of-bag detection data. N regression decision trees were constructed for the newly generated sample set to fully grow into a random forest. Finally, the best regression result was selected by voting on the prediction results of the decision tree.

By employing the multi-branch combinative learning method of random forests, it is possible to avoid the shortcomings of using only one classification [76], and the method has a good tolerance for outliers and noise in large-scale datasets [53]. Random forest has also been widely used in many other fields, such as agriculture and forestry, in recent years [77–79].

2.6.3. Support Vector Machine (SVM)

The SVM is a machine-learning method based on the theory of small sample statistics [64]. Its two main functions are, first, to find a hyperplane that fits the test data to the best degree, and second, to perform a two-dimensional segmentation to maximize the isolation edge of the data on both sides of the hyperplane in order to ensure the classification accuracy of the data [80]. This method can be utilized to effectively solve the problems of nonlinear data and high-dimensional pattern recognition [81]. The SVM model has a wide range of applications in image recognition [82], time series prediction [83], and so on.

In order to simulate the relationship between AGB and the estimation variables of pine forests, the "MASS", "randomForest", and "e1071" packages in R software were employed to construct the models of SLR, RF, and SVM.

2.7. Model Evaluation

An RS dataset, a habitat dataset, and a combined dataset were applied to all models of AGB estimation. In this study, the coefficient of determination (R^2), the root mean square error (RMSE), and the normalized root mean square error (NRMSE) were utilized to evaluate the accuracy of model fitting. Furthermore, the mean error (ME), mean relative error (MRE), and mean absolute relative error (MARE) were utilized to measure the overall prediction accuracy and the segment prediction accuracy with a 50 Mg/ha interval (<50, 50–100, 100–150, 150–200, >200 Mg/ha).

$$R^2 = 1 - \frac{\sum_{i=1}^{n}\left(\hat{y}_i - y_i\right)^2}{\sum_{i=1}^{n}\left(y_i - \overline{y}\right)^2} \tag{6}$$

$$RMSE = \sqrt{\frac{\sum_{i=1}^{n}\left(y_i - \hat{y}_i\right)^2}{n}} \tag{7}$$

$$NRMSE = \frac{RMSE}{\overline{y}} \tag{8}$$

$$ME = \frac{\sum_{i=1}^{n}\left(y_i - \hat{y}_i\right)}{n} \tag{9}$$

$$NRMSE = \frac{RMSE}{\overline{y}} \tag{10}$$

$$MARE = \frac{\sum_{i=1}^{n}\left|\frac{y_i - \hat{y}_i}{\hat{y}_i}\right|}{n} \times 100\% \tag{11}$$

In this table, $\hat{y}_i$ and y_i are the predicted AGB and the corresponding AGB in the sample plot; $\overline{y}$ is the mean AGB of the sample plots; and n is the number of samples.

3. Results

To compare the role of the model and the dataset in AGB estimation, the sample data were randomly divided into two groups: 70% of the sample data were fitting data for model construction, and the remaining 30% were testing data for model validation. Due to the differences in the number of sample plots in pine forests, some appropriate adjustments were made to the datasets of different pine forests. The aforementioned processing was performed in order to ensure the validity of model fitting and validation. The final statistical values of the modeling and testing samples are shown in Table 4.

Table 4. Statistics of sample plot data used in this research.

	Fitting				Testing			
Species	**Number**	**AGB Range (Mg/ha)**	**AGB Mean (Mg/ha)**	**AGB Std. Dev. (Mg/ha)**	**Number**	**AGB Range (Mg/ha)**	**AGB Mean (Mg/ha)**	**AGB Std. Dev. (Mg/ha)**
Pinus yunnanensis	57	17.9–287.7	115.1	56.9	30	40.6–270.2	114.4	53.1
Pinus densata	117	2.1–344.4	119.3	70.6	30	11.1–344.4	107.6	76.3
Pinus kesiya	30	49.1–204.4	116.2	40	15	70.1–192.2	116.8	33.6

3.1. Model Performance

AGB fitting was performed on the SLR, RF, and SVM of the three pine forests using the habitat dataset, RS dataset, and combined dataset, respectively. The R^2 by the SLR model for the AGB estimation of *Pinus yunnanensis* forests ranged from 0.1039 to 0.2514; the R^2 value of *Pinus densata* forests ranged from 0.0742 to 0.1650, and that of *Pinus kesiya* forests ranged from 0.0872 to 0.5331. The R^2 of the SLR was primarily below 0.6. When an RF model was implemented, the resulting R^2 of AGB fitting of *Pinus yunnanensis* forests ranged from 0.2028 to 0.7268, while *Pinus densata* forests ranged from 0.1903 to 0.7511, and *Pinus kesiya* forests ranged from 0.4617 to 0.8316. The R^2 distribution was mostly higher than 0.7. The R^2 of the SVM for the AGB fitting of the pine forest was mostly higher than 0.5. The R^2 of the *Pinus yunnanensis* forests ranged from 0.1791 to 0.8100; the *Pinus densata* forests ranged from 0.1559 to 0.7285; and the *Pinus kesiya* forest was concentrated between 0.5034 and 0.7956.

To further analyze the impact of different algorithms on AGB estimation, a boxplot of the three algorithms (SLR, RF, SVM) for AGB estimation was constructed, as shown in Figure 5. It can be seen that the median fitting coefficients of SLR, RF, and SVM were 0.1650, 0.7286, and 0.5361, respectively. The RF was significantly higher than the other two models, and the interquartile range (IQR) of RF was smaller than that of SVM. The median NRMSE of the three algorithms was 0.4350, 0.2634, and 0.2409, respectively. SLR had the largest error value, and RF was slightly higher than SVM. The IQR of RF was the smallest among the three models, and the RF model had the smallest estimation error dispersion. RF had the best performance of AGB estimation in the fitting data, followed by SVM and SLR, indicating that the non-parametric model had better AGB estimation characteristics for pine forests.

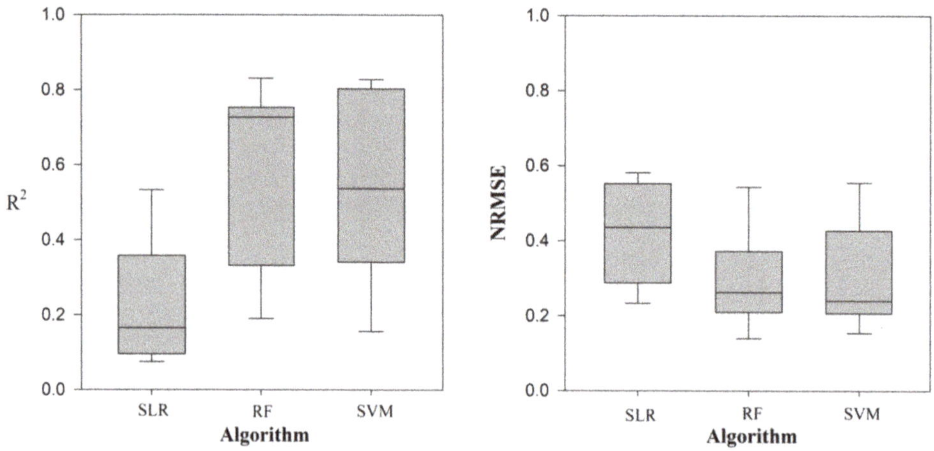

Figure 5. Boxplot of three algorithms for AGB estimation of the *Pinus* forest.

According to the results of a certain model, the dataset that provided the highest R^2 in the fitting data was applied to the AGB estimation of the test data; that is, the combined dataset was used for all, except for the RF, estimations of the *Pinus densata* forests and the SVM estimation of the *Pinus yunnanensis*, for which the RS dataset was used. Thereby, the errors of the varying forests were determined. Additionally, the independence test index of the model (Table 5) was represented by the mean value of each individual model error measurement index. It is evident that the fitting performance of AGB estimation in the test data was similar to that of the fitting data. Among the three models in question, RF had the lowest score of ME, MRE, and MARE. It was also found that the ME and MRE of the SLR were smaller than the predicted values of the SVM. However, the MARE, which is measured by the absolute value of the error, was higher than that of the SVM.

Table 5. Fitting and testing statistics of the three models.

Model	Fitting			Testing		
	R^2	RMSE (Mg/ha)	NRMSE	ME (Mg/ha)	MRE (%)	MARE (%)
SLR	0.3165	47.1631	0.4035	−3.0224	3.2818	39.8129
RF	0.7698	27.1188	0.2320	−1.8477	−1.8103	30.5218
SVM	0.7840	26.1543	0.2238	−4.5969	−3.9566	35.5108

Furthermore, the prediction performance of the models was explained by the scatter diagram (Figure 6), which was formed by utilizing the estimated value of the model and reference data. It can be seen that the fitting performance of the non-parametric models (RF and SVM) was higher than that of the parametric models (SLR). SLR had estimation errors in all AGB segments, while RF and SVM had a higher estimation accuracy in cases with 40~200 Mg/ha, but in cases with AGB < 40 Mg/ha, the estimated value was significantly higher than the actual value; and in cases with AGB > 200 Mg/ha, the estimated value was significantly lower than the actual value. Although the SVM had a good fitting performance on some data, the estimation error in other cases was considerable. Compared with the SVM, the scatter points of the RF of *Pinus yunnanensis* forests and *Pinus densata* forests were distributed within a certain range around the complete fitting line. The distribution range of the SVM was larger than that of the RF, as was expressed in Figure 6. This shows that the overall estimation performance of the RF model was better than that of the SVM. However, in the case of *Pinus kesiya* forests, the scatter distribution of the RF and SVM models was similar.

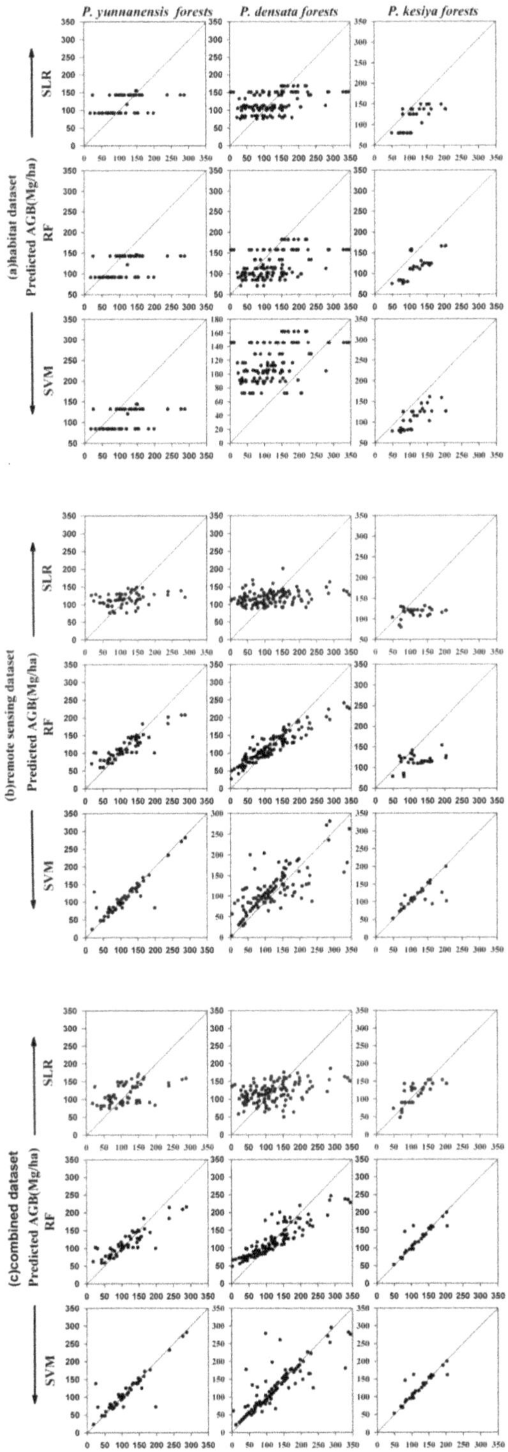

Figure 6. The predicted results of the SLR, RF, and SVM models of the different forests.

3.2. AGB Estimation Based on Different Datasets

In order to explain how the AGB estimation of pine forests is affected by a dataset, an RF with a higher accuracy performance for an AGB estimation was selected to perform AGB estimation of a habitat dataset, an RS dataset, and a combined dataset of them both.

It is evident from Table 6 that the AGB estimation performance of *Pinus yunnanensis* forests that utilized a different dataset under an RF, which was organized from high to low, was the combined dataset, the RS dataset, and the habitat dataset, respectively. In the case of *Pinus densata* forests, the habitat dataset had the lowest fitting performance. Although the fitting performance of the RS dataset was higher than the combined dataset, the error rate of the test data was also higher than the combined dataset. In the case of *Pinus kesiya* forests, the AGB estimation performance using the combined dataset was the highest, followed by the habitat dataset. However, the test error of the habitat dataset was higher than the RS dataset. On the whole, the combined dataset consisting of a habitat dataset and an RS dataset under RF significantly improved the accuracy of the AGB estimation performance.

Table 6. Fitting and testing statistics of the three datasets using RF model. Habitat: habitat dataset; RS: remote-sensing dataset; Combined: combined dataset; R^2: coefficient of determination; RMSE: root mean square error; ME: mean error; MRE: mean relative error; and MARE: mean absolute relative error.

Species	Dataset	Fitting			Testing		
		R^2	RMSE	NRMSE	ME (Mg/ha)	MRE (%)	MARE (%)
Pinus yunnanensis	Habitat	0.2028	50.8261	0.4416	4.65	4.0639	38.2588
	RS	0.7074	30.7922	0.2675	−0.3226	−0.2819	34.8031
	Combined	0.7268	29.7535	0.2585	0.0963	0.0842	35.682
Pinus densata	Habitat	0.1903	63.5056	0.5322	−13.776	−12.797	45.9524
	RS	0.7511	35.2124	0.2951	−9.4358	−8.7654	30.5785
	Combined	0.7343	36.3738	0.3048	−5.7433	−5.3352	28.1384
Pinus kesiya	Habitat	0.7553	19.7669	0.1701	4.6127	3.9489	25.6779
	RS	0.4617	29.3169	0.2522	3.9373	3.3708	23.7645
	Combined	0.8316	16.3906	0.1410	3.7964	3.2502	25.3048

The maps of the predicted AGB for the three forests were generated using three datasets (habitat dataset, RS dataset, and combined dataset) under RF, as shown in Figure 7. For the *Pinus yunnanensis* forests and *Pinus densata* forests, the estimated AGB maps using the RS and the combined dataset were more heterogeneous than the estimated AGB maps using habitat datasets. However, for the *Pinus kesiya* forests, the heterogeneity of the estimated AGB map using the habitat dataset was higher than that of the other two datasets.

In order to further analyze the impact of a dataset on AGB estimation, the means and standard deviations of the residuals of test data under RF were calculated for the overall and different AGB segments, and the results are presented in Table 7. The AGB residuals of all predicted values showed similar trends. For instance, the means of the model residuals were highest in a case where only a habitat dataset was involved, followed by an RS dataset. Additionally, the means of the model residuals were lowest in a case where a combined dataset was involved, which was also closest to the measured value of the AGB. Furthermore, in cases where a combined dataset was used for AGB estimation of pine forests, the AGB standard deviation of the *Pinus densata* forests and *Pinus kesiya* forests was in the lowest grade, while that of the *Pinus yunnanensis* forests was in the middle grade. This result concludes that the prediction results of the model involving the combined dataset were more stable.

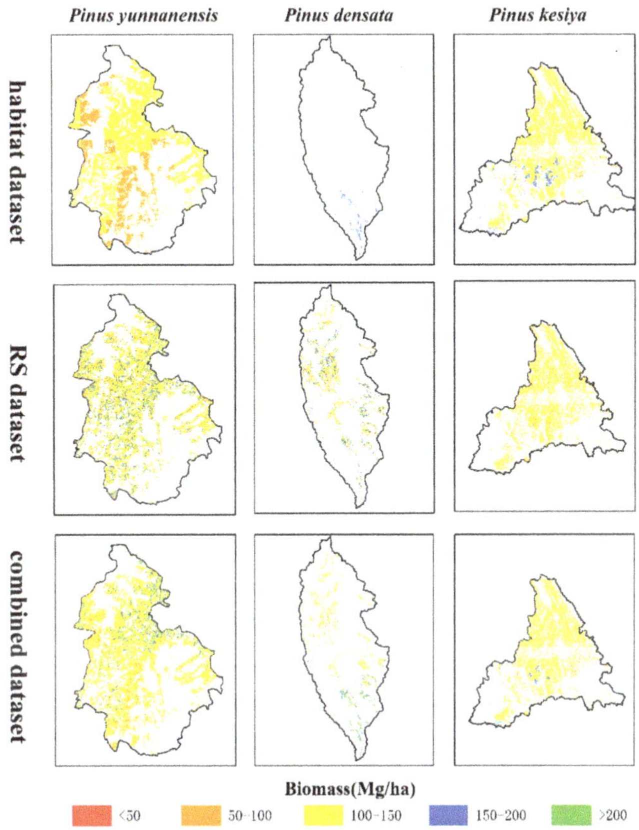

Figure 7. The spatial distributions of the predicted forest AGB values using the three datasets.

Table 7. Summary of the mean (μ) and standard deviation (σ) values of the residuals at different AGB classes for the three datasets based on the test dataset.

Species	Dataset	<50 (Mg/ha)		50–100 (Mg/ha)		100–150 (Mg/ha)		150–200 (Mg/ha)		>200 (Mg/ha)		Overall	
		μ	σ	μ	σ	μ	σ	μ	σ	μ	σ	μ	σ
Pinus yunnanensis	habitat	51.08	- - -	34.29	19.46	−13.41	31.10	−49.54	12.97	−128.17	0.67	−4.65	50.52
	RS	43.89	- - -	41.74	30.38	−10.79	25.18	−39.83	29.24	−128.59	27.03	0.32	53.68
	combined	40.67	- - -	40.00	28.85	−10.07	30.36	−41.96	19.46	−122.51	2.59	−0.11	51.86
Pinus densata	habitat	70.84	38.02	46.59	33.09	−33.36	15.85	1.08	27.10	−158.37	36.04	13.78	73.88
	RS	50.37	32.01	26.42	28.24	−13.86	19.58	−17.81	36.80	−86.73	37.73	9.44	48.52
	combined	34.57	13.82	26.42	34.54	−21.11	13.18	−7.61	31.36	−92.04	25.92	5.74	47.23
Pinus kesiya	habitat	- - -	- - -	11.71	11.99	−2.33	32.63	−61.54	6.84	- - -	- - -	−4.61	33.19
	RS	- - -	- - -	26.34	15.86	−14.74	16.23	−56.95	20.45	- - -	- - -	−3.94	32.79
	combined	- - -	- - -	12.65	12.25	−3.68	34.32	−53.54	9.46	- - -	- - -	−3.80	32.56

On the basis of the statistical value of the predicted residuals for different AGB segments, in cases where the AGB was less than 100 Mg/ha, the predicted AGB values were all higher than the actual values. With the reduction in AGB values, the overestimation errors of the model tended to increase. However, when the AGB of pine forests exceeded 100 Mg/ha, the underestimation of the AGB prediction became increasingly clear. The estimated value of the AGB was more accurate in the range of 100 to 150 Mg/ha for the

Pinus yunnanensis forests and *Pinus kesiya* forests, while that of the *Pinus densata* forests was most accurate in the range of 150 to 200 Mg/ha.

In order to further analyze the impact of a dataset on AGB estimation for different AGB segments, the segmentation of *P. kesiya* forests with the smallest AGB span among the three forests was used as the standard to redefine the AGB segment, which was divided into < 100 Mg/ha, 100—150 Mg/ha, and > 150 Mg/ha. The means of residuals under RF at different AGB segments were calculated, and the result is shown in Figure 8. It is evident that the means of residuals using a combined dataset were lower than those using an optical remote-sensing dataset only in the two segments with larger AGB estimation errors (AGB < 100 Mg/ha or AGB > 150 Mg/ha). In the range of 150 to 200 Mg/ha, the same trend was revealed for the *Pinus yunnanensis* forests and *Pinus kesiya* forests, except for the *Pinus densata* forests. Based on this result, it is concluded that incorporating the habitat dataset into the optical RS dataset will reduce the number of estimation errors in cases with AGB < 100 Mg/ha or AGB > 150 Mg/ha compared to relying only on the Landsat 8 optical remote-sensing dataset.

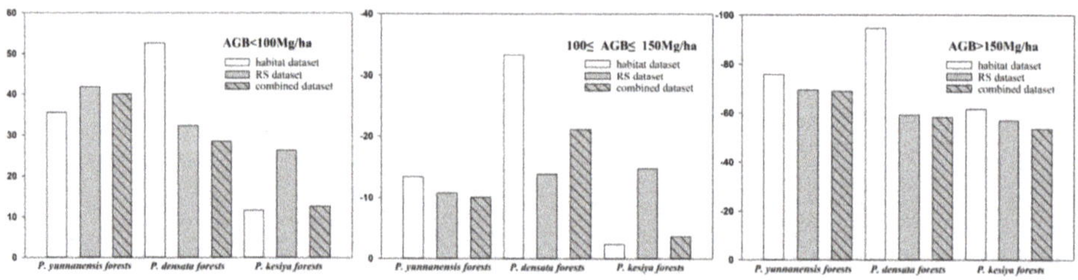

Figure 8. The means of residuals under RF at different AGB segments.

Table 8 indicates the top six most significant variables for AGB estimation by RF using a combined dataset. Thus, it also indicates the differences in significant variables for AGB estimation among tree species. The significant variables of AGB estimation of *Pinus yunnanensis* forests only came from the RS dataset; it mostly consisted of texture variables, and only one vegetation index was selected. For the *Pinus densata* forests, three important variables were obtained from the RS dataset, and the other three were obtained from the habitat dataset, which consisted of two temperature variables and one annual precipitation index. The habitat dataset was more important in the case of AGB estimation of the *Pinus kesiya* forests, since it required three precipitation indices, two seasonal variation coefficients of temperature, and one texture variable. Texture variables appeared in AGB estimation of both the *Pinus yunnanensis* forests and *Pinus kesiya* forests.

Table 8. Important variables for AGB estimation from different datasets.

Species	Variables	R^2
Pinus yunnanensis	B6_homo, B4_entro, B7_homo, SIPI, B4_semo, B7_diss	0.7268
Pinus densata	ARVI, SRI, EVI, bio4, bio7, bio12	0.7343
Pinus kesiya	B4_mean, bio4, bio14, bio17, bio19, HSV	0.8316

4. Discussion

4.1. The Selection of Modeling Algorithms

In this study, SLR was used to estimate the AGB of three pine forests with varying fitting datasets. The result showed that the estimation performance of AGB by SLR was not high, and the R^2 of more than 75% of them was below 0.3, and the highest R^2 was only 0.53. The mean NRMSE, ME, MRE, and MARE of the SLR model were larger than those of the other two non-parametric models, except for the MRE of SVM. Since a linear regression model can only analyze the relationship between a variable and independent variables from the linear point of view, it is difficult to obtain a favorable fitting performance by using a linear model [76]. Although SLR can determine the important factors for regression through a significance test, it reduces the number of modeling variables and improves the estimation accuracy of the model only to a certain extent. However, due to the complex relationship between the variables and the AGB, it is difficult to use a linear regression model to explain such a complex nonlinear relationship, and for that purpose, non-parametric machine-learning algorithms, such as RF and SVM, are necessary [84]. In this study, the mean of R^2 using the non-parametric model for AGB estimation was all above 0.5, indicating a better fitting performance than the SLR. The fitting results for the two non-parametric models (RF and SVM), including the mean, maximum, and the standard deviation of R^2 and NRMSE, showed the estimation performance using RF was slightly better than that of SVM. Additionally, this trend was also reflected in MAE and MARE, which used absolute values as the measurement standard of test data errors.

The overall estimation performance of the AGB differed by about 5% between the two non-parametric models, although there was a certain gap in the estimation of AGB for different forests. In this study, the fitting performance of AGB estimation was measured by the mean value of R^2 of different fitting datasets of a certain forest under a specific estimation model. Under SLR, the result indicates that the cases with the highest R^2 of the AGB fitting estimation were the *Pinus kesiya* forests, *Pinus yunnanensis* forests, and *Pinus densata* forests, in descending order, which was inversely proportional to the sample sizes of AGB fitting. Additionally, the R^2 difference of this model among different forests was the highest among the three models. The R^2 of AGB estimation using RF were the *Pinus kesiya* forests, *Pinus densata* forests, and *Pinus yunnanensis* forests, in descending order. The AGB estimation of the *Pinus densata* forests with a larger data volume was significantly better, and the difference of R^2 among the varying forests was the lowest among the three models. This indicates that RF has a good interpretation and a strong degree of robustness for AGB estimations. This is consistent with the study of Zhang [85]. The descending order of R^2 using SVM was *Pinus kesiya* forests, *Pinus yunnanensis* forests, and *Pinus densata* forests, respectively, which was consistent with the SLR model. However, the variation coefficient of AGB estimation using SVM was significantly smaller than that of SLR. In other words, AGB estimation using SVM was less affected by the forest type than that of SLR. This indicates that SVM is better suited for small sample data as compared to SLR and RF, which is consistent with previous research conclusions [86,87].

4.2. Selection of Suitable Variables for AGB Modeling

In cases where a single dataset was used to estimate the AGB, estimation using an RS dataset performed better, except for the case of *Pinus yunnanensis* forests using a habitat dataset under SLR and *Pinus kesiya* forests using a habitat dataset under RF. This result indicates that in comparison with a habitat dataset, an RS dataset had excellent AGB estimation ability for a large area, which is consistent with previous studies [75,87,88].

In cases where an RS dataset was used for AGB estimation, texture variables were most frequently used in model construction, followed by the vegetation indices, which were obtained through ground object reflection data, as presented in Table 9. Therefore, texture information was of great significance for AGB estimations because texture information has the ability to describe the complex canopy structure of subtropical forest [61].

Table 9. Important variables in AGB estimation using remote-sensing dataset.

Species	Model	Important Variables
P. yunnanensis	SLR	B7_homo
	RF	B6_homo, B4_entro, B7_homo
P. densata	SLR	SRI
	RF	ARVI, SRI, EVI
P. kesiya	SLR	B4_mean
	RF	B4_mean, B2_corr, B2_con

4.3. AGB Estimation by Incorporating the Habitat Dataset into the Models

The AGB fitting performance conducted through combining a habitat dataset with a remote-sensing dataset was, to a certain extent, higher than that of the cases using a single dataset. Particularly in the SLR, the R^2 of AGB estimation using a combined dataset was more than twice as high as that of an RS dataset. In the non-parametric AGB estimation using a combined dataset, only the *Pinus densata* forests based on RF and *Pinus yunnanensis* forests based on SVM were slightly lower than the R^2 of an RS dataset, and the difference in R^2 was within 0.02. Recent studies have shown that the spatial pattern of AGB distribution of vegetation is consistent with the response of its habitat [89]. Integrating a habitat dataset representing environmental characteristics into an AGB estimation can compensate for an RS dataset's problem of low time span and improve the performance of AGB estimation. As early as 1996, Phinn et al. emphasized the importance of habitat to the ecosystems [90]. Habitat is the result of long-term development of vegetation, and accurate estimation of forest biomass requires knowledge of the characteristics of long-term forest development [91].

The employment of a habitat dataset not only improved the performance of AGB estimation on the whole but also reduced the number of overestimation errors in AGB estimation. The decreasing degree of overestimation errors was greatest in the case of *Pinus densata* forests, followed by *Pinus kesiya* forests, and finally, *Pinus yunnanensis* forests, respectively. As for the underestimation errors in AGB estimations, in comparison to the RF model using only the Landsat 8 OLI optical RS dataset, the combined dataset could reduce the underestimation errors across all segments of *Pinus yunnanensis* forests and *Pinus kesiya* forests with AGB greater than 100 Mg/ha, in addition to *Pinus densata* forests with AGB higher than 150 Mg/ha. A combined dataset can significantly improve the AGB estimations of pine forests [23], and a combined dataset is not necessarily the result of combining different remote-sensing datasets but also of combining a habitat dataset and an optical remote-sensing dataset. The combination of a habitat dataset and an optical remote-sensing dataset is more suitable for low AGB estimation (AGB < 100 Mg/ha) in cases where non-parametric modeling methods are used. Based on the fact that habitat has a profound impact on the richness and spatial distribution of species in a region [92], at least to a certain extent, the habitat dataset can represent the structural information of a forest in a region. Therefore, combining a habitat dataset with an RS dataset can compensate for the unreliability of an optical remote-sensing dataset for AGB estimation. In the case of AGB estimations of pine forests using a combined dataset, the habitat variables of *Pinus yunnanensis* forests did not rank among the top six most significant variables in the RF model, and the habitat variables of *Pinus densata* forests accounted for 50%, while those of *Pinus kesiya* forests accounted for 83%. Therefore, whether the estimated differences among the forests were related to the number of habitat variables that were included still needs further discussion.

4.4. Comparison and Implication of Similar Studies

In order to further analyze the research conclusions of this paper, we compared two papers that also applied Landsat images to estimate the AGB of *Pinus densata* forests in Shangri-la. Zhang et al. [33] used Landsat time series images and national forest survey data from 1987 to 2007 to produce parametric and non-parametric AGB estimations. In Zhang's study, i.e., AGB estimation without continuous image participation, the R^2 of SLR and RF were 0.46 and 0.87, respectively, and the MAE of the validation data were 20.48 and 22.47. The R^2 of the AGB estimation with the participation of 5-year sequence images were all increased to above 0.9, and the MAEs were reduced to below 10. In this study, the R^2s of AGB estimations that utilized a combined dataset were 0.17 and 0.73, and the MAEs were 58.82 and 34.68 under SLR and RF, respectively. The R^2 was lower than that in Zhang's study, but the result of the model fitting comparison was consistent. In other words, in the case of the *Pinus densata* forests, the AGB estimation of an RF model was more accurate than that of an SLR model. The estimation performance of this study was lower than that of Zhang for the following reasons. First of all, Zhang not only used spectral variables and spectral conversion variables of RS data but also used terrain variables to reduce the impact of the slope on AGB estimation. The type of the forest ecosystem and the change in terrain both affect the estimation performance of AGB estimations [23]. However, this study only integrated the habitat dataset derived from bioclimatic data into the remote-sensing dataset for AGB estimation while ignoring the fact that the lack of topographic variables may reduce the estimation performance, especially in the case of the Shangri-La region with large terrain fluctuations. Secondly, the climate data from WorldClim used in this study were the mean values of various bioclimatic indicators obtained from 1970 to 2000. This time interval was significantly larger than the 5-year time interval estimated by the optimal AGB in Zhang's study. In addition to temporal resolution, the spatial resolution (1 km^2) of climate data was also significantly lower than that of the image data. Finally, the sample size of the *Pinus densata* forests used in this study (147 samples) was significantly larger than the 53 samples used by Zhang, which may have led to a lower estimation accuracy.

Ou et al. [30] incorporated age data as a dummy variable into a Landsat 8 optical image to estimate the AGB of *Pinus densata* forests. As a result, the RMSE of the linear regression dropped from 50.163 to 33.020, and the RMSE of RF dropped from 40.108 to 23.311. This proves that the age-fused optical RS combined dataset significantly improved the AGB estimation performance. It also indicates that the combined dataset that could improve the fitting performance of AGB estimations included the combination of different remote-sensing data sources [23,93], as well as the combination of remote-sensing data and non-remote-sensing data. For example, the combination of a habitat dataset and Landsat optical images in this study also significantly improved the overall estimation performance of the AGB and was able to reduce the number of overestimation and underestimation errors in cases where only an optical remote-sensing dataset was used for AGB estimation. Ou's study concluded that the estimation models that use age as a dummy variable perform best when AGB < 70 Mg/ha and AGB > 180 Mg/ha, while this study was the most accurate when the AGB was 150–200 Mg/ha, and the prediction accuracy was also slightly lower than Ou's study. However, in forests where the age of stands is difficult to obtain, such as pure forests with uneven ages and mixed forests with multiple dominant tree species, the habitat dataset used in this study can be regarded as an effective way to improve the performance of AGB estimation.

4.5. Limitation and Future Research

This study confirms that a nonlinear algorithm (RF and SVM) is more suitable for AGB estimation of the pine forests than SLR, and the integration of habitat information can improve the estimation accuracy of AGB estimation using Landsat optical images. However, to some extent, the following limitations still exist. Firstly, the three pine forests, which were selected for AGB estimation in this study, are located in different regions of the study area, and the environmental information, such as topography, climate, and physicochemical properties of the soil, varies. This difference will inevitably affect the habitat suitability for tree growth in the forest. Although none of the other factors play a significant role compared to the effects of climate [94], the inclusion of information on the elevation, land cover, and landscape spatial alignment, which have an impact on biodiversity [48,52,95], can reduce the uncertainty in habitat suitability calculations. Therefore, in later studies, we will try to combine more variables to improve the expressiveness of the habitat information.

Secondly, the data size of the estimated sample is not only related to the selection of the estimation model but also affects the estimation performance of the model. Due to the different distribution ranges of the three pine forests, there are differences in the number of trees obtained by the same sampling method. However, a sample size of 30 can meet the requirements for AGB estimation. Bao et al. emphasized that a minimum of 30 samples should be ensured in AGB estimation [96]. Rafaela et al. estimated the AGB of mangroves with a sample size of 30 and obtained a good estimation [97]. Moreover, in order to clarify the effect of sample size on AGB estimation using the combined dataset, we will select different sample size data to study the AGB estimation of a specific pine forest in a certain area in the next step. Technology and means such as growth cones and LiDAR can also be used to obtain more samples to facilitate the smooth progress of research.

5. Conclusions

In this study, three common pine forests (*Pinus yunnanensis* forests, *Pinus densata* forests, and *Pinus kesiya* forests) in Yunnan Province, southwest China, were taken as examples for AGB estimation. The estimation was performed under a parametric model (SLR) and a non-parametric model (RF and SVM) based on a habitat dataset, Landsat8 OLI optical remote-sensing dataset, and a combined dataset produced by combining the two. The results indicate that (1) the non-parametric models of RF and SVM are capable of predicting the AGB of the three pine forests more accurately than the parametric model of SLR. RF is suitable for AGB estimation with a large sample size, while SVM is better suited for AGB estimation with a small sample size. (2) When a single dataset is employed for AGB estimation of the three pine forests, the resulting estimation performance of the habitat dataset is lower than that of the RS dataset, and the texture variables in the RS dataset are more significant in AGB estimation. (3) As compared to the overall fitting performance of the AGB estimations, which only use a single dataset, the combined dataset resulting from the combination of the habitat dataset and the RS dataset can improve the estimation performance to a certain extent. In particular, combining datasets can reduce the number of estimation errors in cases with AGB lower than 100 Mg/ha or exceeding 150 Mg/ha using RF.

Author Contributions: J.T. participated in the collection of the field data, conducted the data analysis, and wrote the draft of the paper; Y.L. (Ying Liu), L.L. and H.X. helped with the data analysis and writing of the paper; Y.L. (Yanfeng Liu) and Y.W. participated in the collection of the field data and data analysis; G.O. supervised and coordinated the research project, designed the experiment, and revised the paper. All authors have read and agreed to the published version of the manuscript.

Funding: This research was funded by the National Natural Science Foundation of China (grant numbers 31770677 and 31660202) and the Ten-Thousand Talents Program of Yunnan Province, China (YNWR-QNBJ-2018-184).

Conflicts of Interest: The authors declare no conflict of interest.

References

1. Ou, G.; Lv, Y.; Xu, H.; Wang, G. Improving Forest Aboveground Biomass Estimation of Pinus densata Forest in Yunnan of Southwest China by Spatial Regression using Landsat 8 Images. *Remote Sens.* **2019**, *11*, 2750. [CrossRef]
2. Kramer, P.J. Carbon Dioxide Concentration, Photosynthesis, and Dry Matter Production. *BioScience* **1981**, *31*, 29–33. [CrossRef]
3. Olson, J.S.; Watts, J.A.; Allison, L.J. *Carbon in Live Vegetation of Major World Ecosystems;* Oak Ridge National Laboratory: Oak Ridge, TN, USA, 1983.
4. Woodwell, G.M.; Whittaker, R.; Reiners, W.; Likens, G.E.; Delwiche, C.; Botkin, D. The Biota and the World Carbon Budget: The terrestrial biomass appears to be a net source of carbon dioxide for the atmosphere. *Science* **1978**, *199*, 141–146. [CrossRef]
5. Xu, W.; Jin, X.; Yang, X.; Wang, Z.; Liu, J.; Wang, D.; Shan, W.; Zhou, Y. The estimation of forest vegetation biomass in China in spatial grid. *J. Nat. Resour.* **2018**, *33*, 1725–1741.
6. Lu, D.; Chen, Q.; Wang, G.; Liu, L.; Li, G.; Moran, E. A survey of remote sensing-based aboveground biomass estimation methods in forest ecosystems. *Int. J. Digit. Earth* **2016**, *9*, 63–105. [CrossRef]
7. Cairns, M.A.; Brown, S.; Helmer, E.H.; Baumgardner, G.A. Root biomass allocation in the world's upland forests. *Oecologia* **1997**, *111*, 1–11. [CrossRef]
8. Yang, S.; Feng, Q.; Liang, T.; Liu, B.; Zhang, W.; Xie, H. Modeling grassland above-ground biomass based on artificial neural network and remote sensing in the Three-River Headwaters Region. *Remote Sens. Environ.* **2018**, *204*, 448–455. [CrossRef]
9. Kumar, L.; Mutanga, O. Remote sensing of above-ground biomass. *Remote Sens.* **2017**, *9*, 935. [CrossRef]
10. Calvao, T.; Palmeirim, J. Mapping Mediterranean scrub with satellite imagery: Biomass estimation and spectral behaviour. *Int. J. Remote Sens.* **2004**, *25*, 3113–3126. [CrossRef]
11. Li, M.; Tan, Y.; Pan, J.; Peng, S. Modeling forest aboveground biomass by combining spectrum, textures and topographic features. *Front. For. China* **2008**, *3*, 10–15. [CrossRef]
12. Rahman, M.; Csaplovics, E.; Koch, B. An efficient regression strategy for extracting forest biomass information from satellite sensor data. *Int. J. Remote Sens.* **2005**, *26*, 1511–1519. [CrossRef]

13. Zheng, D.; Rademacher, J.; Chen, J.; Crow, T.; Bresee, M.; Le Moine, J.; Ryu, S.-R. Estimating aboveground biomass using Landsat 7 ETM+ data across a managed landscape in northern Wisconsin, USA. *Remote Sens. Environ.* **2004**, *93*, 402–411. [CrossRef]
14. Buckley, J.R.; Smith, A.M. Monitoring grasslands with RADARSAT 2 quad-pol imagery. In Proceedings of the 2010 IEEE International Geoscience and Remote Sensing Symposium, Honolulu, Hawaii, USA, 25–30 July 2010; pp. 3090–3093.
15. Le Toan, T.; Quegan, S.; Davidson, M.; Balzter, H.; Paillou, P.; Papathanassiou, K.; Plummer, S.; Rocca, F.; Saatchi, S.; Shugart, H. The BIOMASS mission: Mapping global forest biomass to better understand the terrestrial carbon cycle. *Remote Sens. Environ.* **2011**, *115*, 2850–2860. [CrossRef]
16. Pandey, U.; Kushwaha, S.; Kachhwaha, T.; Kunwar, P.; Dadhwal, V. Potential of Envisat ASAR data for woody biomass assessment. *Trop. Ecol.* **2010**, *51*, 117.
17. Popescu, S.C. Estimating biomass of individual pine trees using airborne lidar. *Biomass Bioenergy* **2007**, *31*, 646–655. [CrossRef]
18. Saremi, H.; Kumar, L.; Stone, C.; Melville, G.; Turner, R. Sub-compartment variation in tree height, stem diameter and stocking in a Pinus radiata D. Don plantation examined using airborne LiDAR data. *Remote Sens.* **2014**, *6*, 7592–7609. [CrossRef]
19. Saremi, H.; Kumar, L.; Turner, R.; Stone, C. Airborne LiDAR derived canopy height model reveals a significant difference in radiata pine (Pinus radiata D. Don) heights based on slope and aspect of sites. *Trees* **2014**, *28*, 733–744. [CrossRef]
20. Sarker, M.L.R.; Nichol, J.; Iz, H.B.; Ahmad, B.B.; Rahman, A.A. Forest biomass estimation using texture measurements of high-resolution dual-polarization C-band SAR data. *IEEE Trans. Geosci. Remote Sens.* **2012**, *51*, 3371–3384. [CrossRef]
21. Wempen, J.M.; McCarter, M.K. Comparison of L-band and X-band differential interferometric synthetic aperture radar for mine subsidence monitoring in central Utah. *Int. J. Min. Sci. Technol.* **2017**, *27*, 159–163. [CrossRef]
22. Dong, T.; Liu, J.; Qian, B.; He, L.; Liu, J.; Wang, R.; Jing, Q.; Champagne, C.; McNairn, H.; Powers, J. Estimating crop biomass using leaf area index derived from Landsat 8 and Sentinel-2 data. *ISPRS J. Photogramm. Remote Sens.* **2020**, *168*, 236–250. [CrossRef]
23. Li, Y.; Li, M.; Li, C.; Liu, Z. Forest aboveground biomass estimation using Landsat 8 and Sentinel-1A data with machine learning algorithms. *Sci. Rep.* **2020**, *10*, 9952. [CrossRef] [PubMed]
24. López-Serrano, P.M.; Cardenas Dominguez, J.L.; Corral-Rivas, J.J.; Jiménez, E.; López-Sánchez, C.A.; Vega-Nieva, D.J. Modeling of aboveground biomass with Landsat 8 OLI and machine learning in temperate forests. *Forests* **2019**, *11*, 11. [CrossRef]
25. Otgonbayar, M.; Atzberger, C.; Chambers, J.; Damdinsuren, A. Mapping pasture biomass in Mongolia using partial least squares, random forest regression and Landsat 8 imagery. *Int. J. Remote Sens.* **2019**, *40*, 3204–3226. [CrossRef]
26. Duncanson, L.; Niemann, K.; Wulder, M. Integration of GLAS and Landsat TM data for aboveground biomass estimation. *Can. J. Remote Sens.* **2010**, *36*, 129–141. [CrossRef]
27. Shi, Y.; Wang, Z.; Liu, L.; Li, C.; Peng, D.; Xiao, P. Improving Estimation of Woody Aboveground Biomass of Sparse Mixed Forest over Dryland Ecosystem by Combining Landsat-8, GaoFen-2, and UAV Imagery. *Remote Sens.* **2021**, *13*, 4859. [CrossRef]
28. Meng, S.; Pang, Y.; Zhang, Z.; Jia, W.; Li, Z. Mapping aboveground biomass using texture indices from aerial photos in a temperate forest of Northeastern China. *Remote Sens.* **2016**, *8*, 230. [CrossRef]
29. Næsset, E.; Bollandsås, O.M.; Gobakken, T.; Gregoire, T.G.; Ståhl, G. Model-assisted estimation of change in forest biomass over an 11 year period in a sample survey supported by airborne LiDAR: A case study with post-stratification to provide "activity data". *Remote Sens. Environ.* **2013**, *128*, 299–314. [CrossRef]
30. Ou, G.; Li, C.; Lv, Y.; Wei, A.; Xiong, H.; Xu, H.; Wang, G. Improving aboveground biomass estimation of Pinus densata forests in Yunnan using Landsat 8 imagery by incorporating age dummy variable and method comparison. *Remote Sens.* **2019**, *11*, 738. [CrossRef]
31. Puliti, S.; Hauglin, M.; Breidenbach, J.; Montesano, P.; Neigh, C.; Rahlf, J.; Solberg, S.; Klingenberg, T.; Astrup, R. Modelling above-ground biomass stock over Norway using national forest inventory data with ArcticDEM and Sentinel-2 data. *Remote Sens. Environ.* **2020**, *236*, 111501. [CrossRef]
32. Ene, L.T.; Næsset, E.; Gobakken, T.; Bollandsås, O.M.; Mauya, E.W.; Zahabu, E. Large-scale estimation of change in aboveground biomass in miombo woodlands using airborne laser scanning and national forest inventory data. *Remote Sens. Environ.* **2017**, *188*, 106–117. [CrossRef]
33. Zhang, J.; Lu, C.; Xu, H.; Wang, G. Estimating aboveground biomass of Pinus densata-dominated forests using Landsat time series and permanent sample plot data. *J. For. Res.* **2018**, *30*, 1689–1706. [CrossRef]
34. Hernández-Stefanoni, J.L.; Castillo-Santiago, M.Á.; Mas, J.F.; Wheeler, C.E.; Andres-Mauricio, J.; Tun-Dzul, F.; George-Chacón, S.P.; Reyes-Palomeque, G.; Castellanos-Basto, B.; Vaca, R. Improving aboveground biomass maps of tropical dry forests by integrating LiDAR, ALOS PALSAR, climate and field data. *Carbon Balance Manag.* **2020**, *15*, 15. [CrossRef] [PubMed]
35. Wang, L.-Q.; Ali, A. Climate regulates the functional traits–aboveground biomass relationships at a community-level in forests: A global meta-analysis. *Sci. Total Environ.* **2021**, *761*, 143238. [CrossRef] [PubMed]
36. Schucknecht, A.; Meroni, M.; Kayitakire, F.; Boureima, A. Phenology-based biomass estimation to support rangeland management in semi-arid environments. *Remote Sens.* **2017**, *9*, 463. [CrossRef]
37. Liu, N.; Harper, R.J.; Handcock, R.N.; Evans, B.; Sochacki, S.J.; Dell, B.; Walden, L.L.; Liu, S. Seasonal timing for estimating carbon mitigation in revegetation of abandoned agricultural land with high spatial resolution remote sensing. *Remote Sens.* **2017**, *9*, 545. [CrossRef]
38. Lumbierres, M.; Méndez, P.F.; Bustamante, J.; Soriguer, R.; Santamaría, L. Modeling biomass production in seasonal wetlands using MODIS NDVI land surface phenology. *Remote Sens.* **2017**, *9*, 392. [CrossRef]
39. Yan, Z.; Chen, Y. Habitat Selection in Animals. *Chin. J. Ecol.* **1998**, *17*, 43–49.

40. Sillero, N.; Arenas-Castro, S.; Enriquez-Urzelai, U.; Vale, C.G.; Sousa-Guedes, D.; Martínez-Freiría, F.; Real, R.; Barbosa, A.M. Want to model a species niche? A step-by-step guideline on correlative ecological niche modelling. *Ecol. Model.* **2021**, *456*, 109671. [CrossRef]
41. Zhang, Y.; Meng, H.; Zhou, X.; Yin, B.; Zhou, D.; Tao, Y. Biomass allocation patterns of an ephemeral species (*Erodium oxyrhinchum*) in different habitats and germination types in the Gurbantunggut Desert, China. *Arid. Zone Res.* **2022**, *39*, 541–550. [CrossRef]
42. Qiu, X.; Xu, Z.; Tu, Y.; Luo, J. Module biomass and allocation characteristics of invasive plant Tagetes minuta population modules in different habitats. *Guihaia* **2021**, *41*, 447–455. [CrossRef]
43. Hao, J.; Yao, X.; Huang, Y.; Yao, J.; Chen, Y.; Xie, H.; Chen, R. Effect of Different Habitats on the Species Diversity of Communities and Modular Biomass of Riparian Vegetation in the Wenjiang Section of the Jinma Rive. *Acta Bot. Boreal.-Occident. Sin.* **2016**, *36*, 1864–1871. [CrossRef]
44. Zhou, X.; Zuo, X.; Zhao, X.; Wang, S.; Luo, Y.; Yue, X.; Zhang, L. Effect of change in semiarid sand dune habitat on aboveground plant biomass, carbon and nitrogen. *Acta Pratacult. Sin.* **2014**, *23*, 36–44. [CrossRef]
45. He, Z. Soil microbial biomass and Its significance in nutrient cycling and environmental quality assessment. *Soils* **1997**, *29*, 61–69.
46. Chen, I.-C.; Hill, J.K.; Ohlemüller, R.; Roy, D.B.; Thomas, C.D. Rapid range shifts of species associated with high levels of climate warming. *Science* **2011**, *333*, 1024–1026. [CrossRef]
47. Gontier, M.; Balfors, B.; Mörtberg, U. Biodiversity in environmental assessment—Current practice and tools for prediction. *Environ. Impact Assess. Rev.* **2006**, *26*, 268–286. [CrossRef]
48. Steiner, N.C.; Köhler, W. Effects of landscape patterns on species richness—A modelling approach. *Agric. Ecosyst. Environ.* **2003**, *98*, 353–361. [CrossRef]
49. Hansen, A.J.; McComb, W.C.; Vega, R.; Raphael, M.G.; Hunter, M. Bird habitat relationships in natural and managed forests in the west Cascades of Oregon. *Ecol. Appl.* **1995**, *5*, 555–569. [CrossRef]
50. Imhoff, M.L.; Sisk, T.D.; Milne, A.; Morgan, G.; Orr, T. Remotely sensed indicators of habitat heterogeneity: Use of synthetic aperture radar in mapping vegetation structure and bird habitat. *Remote Sens. Environ.* **1997**, *60*, 217–227. [CrossRef]
51. Duro, D.C.; Coops, N.C.; Wulder, M.A.; Han, T. Development of a large area biodiversity monitoring system driven by remote sensing. *Prog. Phys. Geogr.* **2007**, *31*, 235–260. [CrossRef]
52. He, K.S.; Bradley, B.A.; Cord, A.F.; Rocchini, D.; Tuanmu, M.N.; Schmidtlein, S.; Turner, W.; Wegmann, M.; Pettorelli, N. Will remote sensing shape the next generation of species distribution models? *Remote Sens. Ecol. Conserv.* **2015**, *1*, 4–18. [CrossRef]
53. Chapman, D.S. Weak climatic associations among British plant distributions. *Glob. Ecol. Biogeogr.* **2010**, *19*, 831–841. [CrossRef]
54. Beale, C.M.; Lennon, J.J.; Gimona, A. Opening the climate envelope reveals no macroscale associations with climate in European birds. *Proc. Natl. Acad. Sci. USA* **2008**, *105*, 14908–14912. [CrossRef] [PubMed]
55. Loiselle, B.A.; Jørgensen, P.M.; Consiglio, T.; Jiménez, I.; Blake, J.G.; Lohmann, L.G.; Montiel, O.M. Predicting species distributions from herbarium collections: Does climate bias in collection sampling influence model outcomes? *J. Biogeogr.* **2008**, *35*, 105–116. [CrossRef]
56. Li, X.; Mao, F.; Du, H.; Zhou, G.; Xing, L.; Liu, T.; Han, N.; Liu, Y.; Zheng, J.; Dong, L. Spatiotemporal evolution and impacts of climate change on bamboo distribution in China. *J. Environ. Manag.* **2019**, *248*, 109265. [CrossRef]
57. Zhang, H.; Zhao, H.; Wang, H. Potential geographical distribution of populus euphratica in China under future climate change scenarios based on Maxent model. *Acta Ecol. Sin.* **2020**, *40*, 6552–6563. [CrossRef]
58. Hu, W.; Zhao, B.; Wang, Y.; Dong, P.; Zhang, D.; Yu, W.; Chen, G.; Chen, B. Assessing the potential distributions of mangrove forests in Fujian Province using MaxEnt model. *China Environ. Sci.* **2020**, *40*, 4029–4038.
59. Bahn, V.; McGill, B.J. Can niche-based distribution models outperform spatial interpolation? *Glob. Ecol. Biogeogr.* **2007**, *16*, 733–742. [CrossRef]
60. Yang, X.-Q.; Kushwaha, S.; Saran, S.; Xu, J.; Roy, P. Maxent modeling for predicting the potential distribution of medicinal plant, Justicia adhatoda L. in Lesser Himalayan foothills. *Ecol. Eng.* **2013**, *51*, 83–87. [CrossRef]
61. Gao, Y.; Lu, D.; Li, G.; Wang, G.; Chen, Q.; Liu, L.; Li, D. Comparative analysis of modeling algorithms for forest aboveground biomass estimation in a subtropical region. *Remote Sens.* **2018**, *10*, 627. [CrossRef]
62. Fleming, A.L.; Wang, G.; McRoberts, R.E. Comparison of methods toward multi-scale forest carbon mapping and spatial uncertainty analysis: Combining national forest inventory plot data and Landsat TM images. *Eur. J. For. Res.* **2015**, *134*, 125–137. [CrossRef]
63. Yan, W.; Zong, S.; Luo, Y.; Cao, C.; Li, Z.; Guo, Q. Application of stepwise regression model in predicting the movement of Artemisia ordosica boring insects. *J. Beijing For. Univ.* **2009**, *31*, 140–144.
64. Wang, J.; Shen, W.; Li, W.; Li, M.; Zhen, G. Performances Comparison of Multiple Non-linear Models for Estimating Plantations' Biomass Based on RapidEye Imagery. *J. Northwest For. Univ.* **2015**, *30*, 196–202. [CrossRef]
65. Liu, L. Model regression analysis of Pinus yunnanensis biomass in northwest Yunnan. *J. Shandong For. Sci. Technol.* **2015**, *5–9*, 34.
66. Chen, Q.; Zheng, Z.; Feng, Z.; Ma, Y.; Sha, L.; Xu, H. Biomass and carbon storage of Pinus kesiya var. langbianensis in Puer'Yunnan. *J. Yunnan Univ Nat. Sci.* **2014**, *36*, 439–445.
67. Ou, G.; Wang, J.; Xu, H.; Chen, K.; Zheng, H.; Zhang, B.; Sun, X.; Xu, T.; Xiao, Y. Incorporating topographic factors in nonlinear mixed-effects models for aboveground biomass of natural Simao pine in Yunnan, China. *J. For. Res.* **2015**, *27*, 119–131. [CrossRef]

68. Galidaki, G.; Zianis, D.; Gitas, I.; Radoglou, K.; Karathanassi, V.; Tsakiri–Strati, M.; Woodhouse, I.; Mallinis, G. Vegetation biomass estimation with remote sensing: Focus on forest and other wooded land over the Mediterranean ecosystem. *Int. J. Remote Sens.* **2017**, *38*, 1940–1966. [CrossRef]
69. Li, P.; Meng, G.; Wang, Y.; Li, G.; Li, G.; Cai, Y.; Wang, Q. Analysis of Growth Process Pinus yunnanensis Natural Secondary Forests in Yongren County of Yunnan Province. *J. West China For. Sci.* **2012**, *41*, 47–52.
70. Sun, X.; Xiong, X.; Xu, H.; Wei, A.; Li, C.; Lv, Y.; Zhang, B.; Ou, G. Modelling of Individual Tree Biomass Factors for Natural Pinus densata Forest. *For. Resour. Manag.* **2016**, *49–53*, 60. [CrossRef]
71. Shen, C.; Yue, C.; Mei, H. Dynamic Monitoring of Puer Land Use Change Based on Landsat Data. *For. Inventory Plan.* **2016**, *41*, 72. [CrossRef]
72. Fourcade, Y.; Besnard, A.G.; Secondi, J. Paintings predict the distribution of species, or the challenge of selecting environmental predictors and evaluation statistics. *Glob. Ecol. Biogeogr.* **2018**, *27*, 245–256. [CrossRef]
73. Swets, J.A. Measuring the accuracy of diagnostic systems. *Science* **1988**, *240*, 1285–1293. [CrossRef] [PubMed]
74. Chang, Y.; Bourque, C.P.-A. Relating modelled habitat suitability for Abies balsamea to on-the-ground species structural characteristics in naturally growing forests. *Ecol. Indic.* **2020**, *111*, 105981. [CrossRef]
75. Zhu, J.; Huang, Z.; Sun, H.; Wang, G. Mapping forest ecosystem biomass density for Xiangjiang River Basin by combining plot and remote sensing data and comparing spatial extrapolation methods. *Remote Sens.* **2017**, *9*, 241. [CrossRef]
76. Huang, J.; Hou, Y.; Su, W.; Liu, J.; Zhu, D. Mapping corn and soybean cropped area with GF-1 WFV data. *Trans. Chin. Soc. Agric. Eng.* **2017**, *33*, 164–170. [CrossRef]
77. Zhang, X.; Li, F.; Zhen, Z.; Zhao, Y. Forest Vegetation Classification of Landsat8 Remote Sensing Image Based on Random Forests Model. *J. Northeast. For. Univ.* **2016**, *44*, 53–57. [CrossRef]
78. Wang, L.; Ma, C.; Zhou, X.; Zi, Y.; Zhu, X.; Guo, W. Estimation of Wheat Leaf SPAD Value Using RF Algorithmic Model and Remote Sensing Data. *Trans. Chin. Soc. Agric. Mach.* **2015**, *46*, 259–265. [CrossRef]
79. Guo, P.; Li, M.; Luo, W.; Lin, Q.; Tang, Q.; Liu, Z. Prediction of soil total nitrogen for rubber plantation at regional scale based on environmental variables and random forest approach. *Trans. Chin. Soc. Agric. Eng.* **2015**, *31*, 194–202. [CrossRef]
80. Lin, Z.; Wu, C.; Hong, W.; Hong, T. Yield model of Cunninghamia lanceolata plantation based on back propagation neural network and support vector machine. *J. Beijing For. Univ.* **2015**, *37*, 42–47. [CrossRef]
81. Ding, S.; Qi, B.; Tan, H. An Overview on Theory and Algorithm of Support Vector Machines. *J. Univ. Electron. Sci. Technol. China* **2011**, *40*, 1–10. [CrossRef]
82. Xie, S.; Shen, F.; Qiu, X. Face Recognition Method Based on Support Vector Machine. *Comput. Eng.* **2009**, *35*, 186–188. [CrossRef]
83. Gao, W.; Wang, N. Prediction of Shallow-water Reverberation Time Series Using Support Vector Machine. *Comput. Eng.* **2008**, *34*, 25–27. [CrossRef]
84. Feng, Y.; Lu, D.; Chen, Q.; Keller, M.; Moran, E.; dos-Santos, M.N.; Bolfe, E.L.; Batistella, M. Examining effective use of data sources and modeling algorithms for improving biomass estimation in a moist tropical forest of the Brazilian Amazon. *Int. J. Digit. Earth* **2017**, *10*, 996–1016. [CrossRef]
85. Zhang, Y.; Ma, J.; Liang, S.; Li, X.; Li, M. An evaluation of eight machine learning regression algorithms for forest aboveground biomass estimation from multiple satellite data products. *Remote Sens.* **2020**, *12*, 4015. [CrossRef]
86. Liu, K.; Wang, J.; Zeng, W.; Song, J. Comparison and evaluation of three methods for estimating forest above ground biomass using TM and GLAS data. *Remote Sens.* **2017**, *9*, 341. [CrossRef]
87. Zhang, Y.; Liang, S.; Sun, G. Forest biomass mapping of northeastern China using GLAS and MODIS data. *IEEE J. Sel. Top. Appl. Earth Obs. Remote Sens.* **2013**, *7*, 140–152. [CrossRef]
88. Ali, I.; Cawkwell, F.; Dwyer, E.; Barrett, B.; Green, S. Satellite remote sensing of grasslands: From observation to management. *J. Plant Ecol.* **2016**, *9*, 649–671. [CrossRef]
89. Zhou, Z.; Yang, Y.; Chen, B. Estimating Spartina alterniflora fractional vegetation cover and aboveground biomass in a coastal wetland using SPOT6 satellite and UAV data. *Aquat. Bot.* **2018**, *144*, 38–45. [CrossRef]
90. Phinn, S.; Franklin, J.; Hope, A.; Stow, D.; Huenneke, L. Biomass distribution mapping using airborne digital video imagery and spatial statistics in a semi-arid environment. *J. Environ. Manag.* **1996**, *47*, 139–164. [CrossRef]
91. Yuan, Z.; Gazol, A.; Wang, X.; Lin, F.; Ye, J.; Zhang, Z.; Suo, Y.; Kuang, X.; Wang, Y.; Jia, S. Pattern and dynamics of biomass stock in old growth forests: The role of habitat and tree size. *Acta Oecologica* **2016**, *75*, 15–23. [CrossRef]
92. Lanham, B.S.; Gribben, P.E.; Poore, A.G. Beyond the border: Effects of an expanding algal habitat on the fauna of neighbouring habitats. *Mar. Environ. Res.* **2015**, *106*, 10–18. [CrossRef]
93. Cutler, M.; Boyd, D.; Foody, G.; Vetrivel, A. Estimating tropical forest biomass with a combination of SAR image texture and Landsat TM data: An assessment of predictions between regions. *ISPRS J. Photogramm. Remote Sens.* **2012**, *70*, 66–77. [CrossRef]
94. Sarr, D.A.; Hibbs, D.E.; Huston, M.A. A hierarchical perspective of plant diversity. *Q. Rev. Biol.* **2005**, *80*, 187–212. [CrossRef] [PubMed]
95. Rosenzweig, M.L. *Species Diversity in Space and Time*; Cambridge University Press: Cambridge, UK, 1995.
96. Huy, B.; Kralicek, K.; Poudel, K.P.; Phuong, V.T.; Van Khoa, P.; Hung, N.D.; Temesgen, H. Allometric equations for estimating tree aboveground biomass in evergreen broadleaf forests of Viet Nam. *For. Ecol. Manag.* **2016**, *382*, 193–205. [CrossRef]
97. Salum, R.B.; Souza-Filho, P.W.M.; Simard, M.; Silva, C.A.; Fernandes, M.E.; Cougo, M.F.; do Nascimento Junior, W.; Rogers, K. Improving mangrove above-ground biomass estimates using LiDAR. *Estuar. Coast. Shelf Sci.* **2020**, *236*, 106585. [CrossRef]

remote sensing

MDPI

Article

Estimating 3D Green Volume and Aboveground Biomass of Urban Forest Trees by UAV-Lidar

Lv Zhou [1,2,3,4], Xuejian Li [2,3,4], Bo Zhang [2,3,4], Jie Xuan [2,3,4], Yulin Gong [2,3,4], Cheng Tan [2,3,4], Huaguo Huang [1,*] and Huaqiang Du [2,3,4]

1 Research Center of Forest Management Engineering of State Forestry and Grassland Administration, Beijing Forestry University, Beijing 100083, China
2 State Key Laboratory of Subtropical Silviculture, Zhejiang A&F University, Hangzhou 311300, China
3 Key Laboratory of Carbon Cycling in Forest Ecosystems and Carbon Sequestration of Zhejiang Province, Zhejiang A&F University, Hangzhou 311300, China
4 School of Environmental and Resources Science, Zhejiang A&F University, Hangzhou 311300, China
* Correspondence: huaguo_huang@bjfu.edu.cn; Tel.: +86-010-62338133

Abstract: Three dimensional (3D) green volume is an important tree factor used in forest surveys as a prerequisite for estimating aboveground biomass (AGB). In this study, we developed a method for accurately calculating the 3D green volume of single trees from unmanned aerial vehicle laser scanner (ULS) data, using a voxel coupling convex hull by slices algorithm, and compared the results using voxel coupling convex hull by slices algorithm with traditional 3D green volume algorithms (3D convex hull, 3D concave hull (alpha shape), convex hull by slices, voxel and voxel coupling convex hull by slices algorithms) to estimate AGB. Our results showed the following: (1) The voxel coupling convex hull by slices algorithm can accurately estimate the 3D green volume of a single ginkgo tree (RMSE = 11.17 m^3); (2) Point cloud density can significantly affect the extraction of 3D green volume; (3) The addition of the 3D green volume parameter can significantly improve the accuracy of the model to estimate AGB, where the highest accuracy was obtained by the voxel coupling convex hull by slices algorithm (CV-R^2 = 0.85, RMSE = 11.29 kg, and nRMSE = 15.12%). These results indicate that the voxel coupling convex hull by slices algorithms can more effectively calculate the 3D green volume of a single tree from ULS data. Moreover, our study provides a comprehensive evaluation of the use of ULS 3D green volume for AGB estimation and could significantly improve the estimation accuracy of AGB.

Keywords: 3D green volume; aboveground biomass; UAV-Lidar; urban forest; random forest model

Citation: Zhou, L.; Li, X.; Zhang, B.; Xuan, J.; Gong, Y.; Tan, C.; Huang, H.; Du, H. Estimating 3D Green Volume and Aboveground Biomass of Urban Forest Trees by UAV-Lidar. *Remote Sens.* **2022**, *14*, 5211. https://doi.org/10.3390/rs14205211

Academic Editor: Francesco Pirotti

Received: 22 August 2022
Accepted: 14 October 2022
Published: 18 October 2022

1. Introduction

The three-dimensional (3D) green volume of an urban forest could be defined as the volume of space occupied by all green stems and leaves of plants in the city [1–5], which reflects the ecological functions and environmental benefits of urban forests in terms of spatial patterns. Thus, it has been well established that 3D green volume plays important roles in the estimation of aboveground biomass (AGB), the estimation of the environmental benefits of greening, and the construction of forest fire risk models, thus, effectively improving the efficiency of urban green space evaluation and green space planning [6–9]. However, urban forests are composed of scattered trees, tree belts, forests with structural diversity, and forests with broken distribution [10]. They are very different from the large, continuously distributed forests in the general sense, thus, making the monitoring and evaluation of urban forest resources complicated. How to accurately estimate the 3D green volume and AGB of urban forests has become an urgent problem to be solved.

Methods for measuring 3D green volume currently include field measurements, optical remote sensing estimation, and LiDAR estimation. The most widely used manual method for estimating canopy volume is the ellipsoidal method, in which crown diameters and canopy heights are measured, assuming an ellipsoidal shape for the canopy [9]. The disadvantage of this method is the large workload, and low efficiency, and it is difficult to carry out a wide range of extensions. Optical remote sensing estimation is mainly based on single-tree crown diameter extraction from high-resolution images combined with ground measured data and on building crown diameter–crown height models to perform 3D green volume estimation. This method has achieved good results in related research in China [11], but it is difficult to comprehensively obtain the vertical distribution of forest structure. The accuracy of the model needs to be further improved [12]. Light detection and ranging (LiDAR) are an active remote sensing technique using pulsed or continuous-wave lasers to measure the range of an object that can rapidly obtain dense 3D point clouds with high precision. Moreover, LiDAR has become a trending topic of domestic and international research because it can describe the forest canopy structure in more detail and provide advanced technical means for accurate estimation of 3D green volume by enabling the leap from two-dimensional (2D) to 3D forest ecosystem research [13–17].

A review of domestic and foreign literature found that the main algorithms for 3D green volume estimation, based on LiDAR point clouds, include 3D convex hulls, 3D concave hulls (alpha shape), convex hulls by slices, and voxels. For example, Ebadat et al. [18] used unmanned aerial vehicle laser scanner (ULS) and photogrammetric point cloud to extract 3D green volume based on a 3D convex hull algorithm. The results showed that UAV photogrammetry and LiDAR point clouds were highly correlated (R^2 = 0.99). Vauhkonen et al. [19] extracted canopy volume based on ALS data. using the alpha shape algorithm, and used it to estimate wood volume. He et al. [2] calculated the 3D green volume based on terrestrial laser scanner (TLS) data. using the convex hull by slices algorithm. and better obtained the 3D green volume of the Beijing urban forest (R^2 > 0.85). Fernández-Sarría et al. [20] extracted the 3D green volume of overhanging trees, using a voxel algorithm, based on TLS data (R^2 = 0.78), and the results showed that TLS has some potential in predicting the 3D green volume of urban forests. How to choose the appropriate algorithm and input parameters is the current problem faced by researchers.

LiDAR can accurately extract structural parameters, such as crown projection area, crown diameter, and crown height, which makes it advantageous for single-tree AGB estimation [21–23]. Three dimensional green volume characterizes the volume of space occupied by plants. Therefore, the participation of 3D green volume in the estimation of AGB has attracted extensive attention from scholars. For example, Tao et al. [24] showed that a 3D green biomass incorporation model could more accurately estimate the AGB (R^2 = 0.77, RMSE = 179.0 Mg/ha). Hauglin et al. [6] estimated the AGB of a single tree plant based on TLS extraction of voxelization parameters with higher accuracy than conventional anisotropic growth models (R^2 = 0.88, RMSE% = 32%).

LiDAR data acquisition includes TLS, airborne laser scanner (ALS), and ULS. The laser radar scanner installed on the TLS ground support obtains a high-density point cloud, but it takes a great deal of time to collect TLS data. due to its static properties, so its use cannot be widely promoted [25]. ALS can obtain a wide range of 3D point cloud data, but low point cloud density makes it impossible to accurately express stand structure [9]. Conversely, ULS can efficiently acquire large-area point cloud data, and compared with ALS, ULS flies at lower altitudes and can acquire a higher point cloud density [26]. However, there are few studies on the extraction of single tree 3D green volume based on ULS [18].

Ginkgo (*Ginkgo biloba* L.) is widely distributed in China and East Asia, with the advantages of an upright trunk, beautiful tree shape and strong resistance to diseases, etc. It is an important tree species for urban greening and has high economic and ecological value. It is of great significance to obtain accurate structural parameters of single Ginkgo trees. This study developed a new algorithm to calculate tree 3D green volume from ULS data. Coupled voxel and convex hull by slices algorithms provide a more accurate calculation of 3D green volume, compared with those achieved using conventional algorithms. The objectives of this study were the following: (1) to select the optimal algorithm for single tree 3D green volume extraction based on ULS data, (2) to evaluate the effect of point cloud density variation on 3D green volume, and (3) to compare different 3D green volume algorithms for AGB estimation.

2. Materials and Methods

2.1. Study Area and Data Acquisition

The study area, Zhejiang A&F University (Figure 1), is situated in Lin'an, Hangzhou, Zhejiang Province. The geographical coordinates of Lin'an city are 30°15′10″~30°15′30″N, 119°43′10″~119°43′40″E. The area is dominated by hilly and mountainous terrain, and the terrain slopes from west to southeast. The area has a subtropical monsoon climate, with abundant light and abundant rainfall. The average annual precipitation is 1613.9 mm, with 158 days of precipitation, and the average annual frost-free period is 237 days. The average elevation of the study area is approximately 50 m, and the area is covered with ginkgo trees on both sides of Ginkgo avenue.

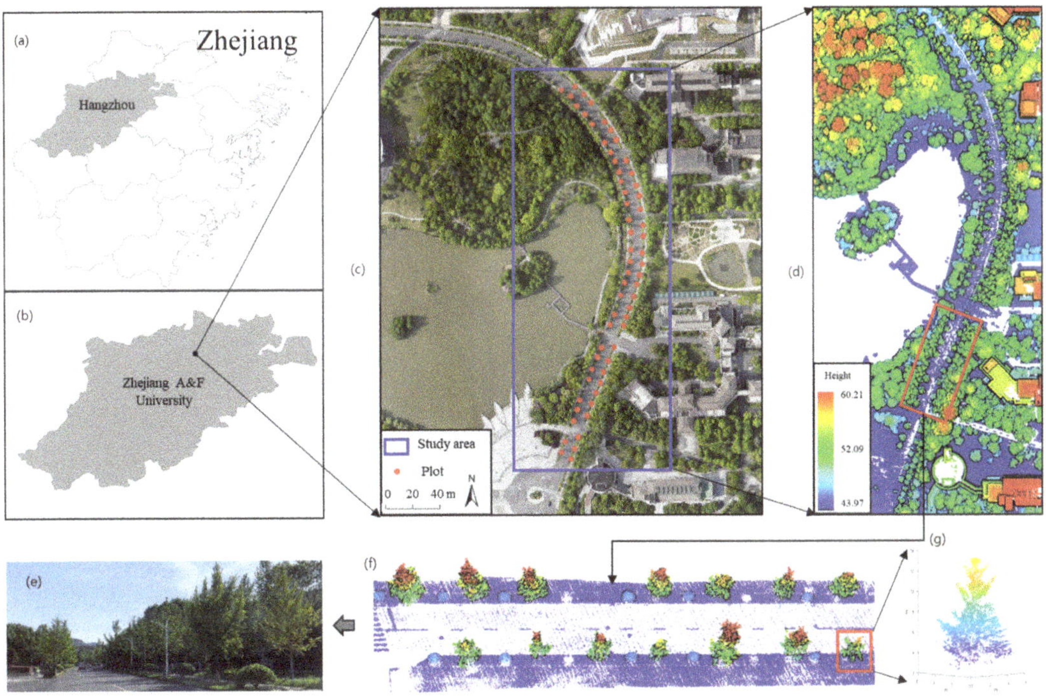

Figure 1. Overview of the study area. (**a**) location of Hangzhou, (**b**) location of the study area, (**c**) aerial photograph of the study area, (**d**) ULS point cloud of the study area, (**e**) ground image of ginkgo, (**f**) ULS point cloud of ginkgo, (**g**) point cloud of a single tree of ginkgo.

Field work was conducted in July 2021. Data from 64 single ginkgo trees were measured. The coordinates of each single tree were located using a real-time kinematic (RTK) device. The diameter at breast height (DBH) was measured using a diameter tape. The single-tree height and branch height were measured using a hypsometer, and the north–south and east–west crown diameters were measured using a tape rule [27]. Crown height is tree height minus branch height [28]. The crown diameter is the average of the east–west and north–south crown diameters. The single-tree 3D green volume values were calculated according to the crown diameter and crown height based on the volume of the geometry [9]. The single-tree AGB values were calculated according to the tree height and DBH, based on the biomass allometric model developed by [29]. The measured forest structural attributes for the single trees are summarized in Table 1.

Table 1. Descriptive statistics of field inventory data for 64 trees.

Statistics	DBH (cm)	Tree Height (m)	Crown Diameter (m)	Branch Height (m)
Minimum	14.80	8.00	6.00	1.40
Maximum	23.90	14.30	2.85	4.20
Range	9.10	6.30	3.15	2.80
SD	2.29	1.78	0.65	0.51
Average	18.76	11.01	4.45	2.37

2.2. *ULS Data*

The ULS data for this study were acquired in May 2021. The DJI Matrice 600 Pro six-rotor unmanned aerial vehicle (UAV) was used in clear-weather and low-wind conditions. Using the Velodyne Puck LITETM sensor to obtain the original ULS point cloud, the sensor records the first echo of the pulse, with a flight altitude of 60 m, flight speed of 5 m/s, swath width of 25 m, and route overlap rate of 50%, with an average point cloud density of approximately 230 pts.

2.3. *ULS Metrics*

The ULS point cloud was preprocessed using LiDAR360 software. First, the single-tree point cloud was denoised and filtered. Then, classification of ground points was carried out using the improved progressive TIN densification (IPTD) algorithm [30] and a digital elevation model (DEM), with a resolution of 0.5 m generated by irregular triangulation interpolation [31]. Finally, the point cloud data were normalized to remove topographic fluctuations from the data. Point clouds above 2 m were extracted as canopy point clouds, and three sets of metrics were computed (Table 2) [32,33].

Table 2. Description of metrics derived from ULS data.

Metrics		Description
Height-related metrics	Percentile height (H_5, H_10, H_20, H_25, H_30, H_40, H_50, H_60, H_70, H_75, H_80, H_90, H_95, H_99) Mean height (H_mean) Maximum height (H_max) Median height (H_median) Interquartile spacing (H_iq) Root mean square (H_sq) Kurtosis of height (H_kurtosis) The coefficient of variation of height (H_cv) Variance of height (H_variance)	The percentiles of the canopy height distribution (5th, 10th, 20th, 25th, 30th, 40th, 50th, 60th, 70th, 75th, 80th, 90th, 95th, 99th) of first returns Mean height above ground of all first returns Maximum height above ground of all first returns Median height above ground of all first returns The interquartile spacing of heights of all first returns The root mean square of heights of all first returns The kurtosis of heights of all first returns The coefficient of variation of heights of all first returns The variation of heights of all first returns
Density-related metrics	Canopy return density (D1,D3,D5,D7,D9)	The proportion of points above the quantiles (10th, 30th, 50th, 70th and 90th) to total number of points
Canopy-related metrics	Canopy projection area (CS) Crown diameter (CD) Crown height (CH)	The canopy projection area of all first returns $\frac{(X_{max}-X_{min})+(Y_{max}-Y_{min})}{2}$ $Z_{max} - Z_{min}$

2.4. *Green Volume Calculation Algorithm*

2.4.1. Convex Hull Algorithm

A convex hull is a concept in computational geometry defined as finding a minimal set of points such that the shape formed by the set of points can contain all points in the 2D plane or 3D space [34]. Figure 2a shows a schematic diagram of the 2D convex hull. In this study, the point cloud of a single tree canopy is projected to the 2D plane, the 2D convex hull is calculated, and the projected area of the canopy is extracted based on the convhull function in MATLAB. Figure 2b shows the results of the convex hull algorithm in

3D space, based on the quickhull algorithm to reconstruct the canopy surface and calculate the volume under the convex hull, i.e., the 3D green volume [20].

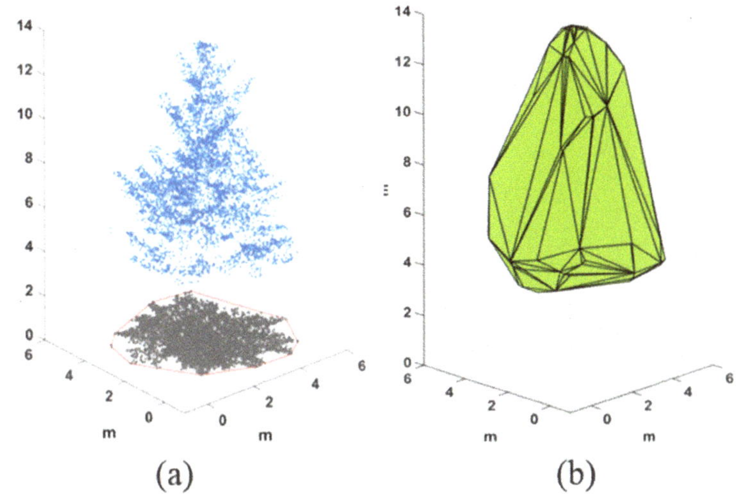

Figure 2. (**a**) A schematic diagram of a 2D convex hull algorithm, and (**b**) An example of a 3D convex hull algorithm.

2.4.2. Concave Hull Algorithm

The concave hull algorithm is another common geometric calculation method, which can be understood as an additional parameter alpha that can be set on top of the convex hull, with alpha as the diameter of the circle rolling along the boundary of the convex polyhedron. The trajectory of the rolling circle is the boundary of the concave polyhedron [19], so the method is also called alpha shape. The algorithm is shown in Figure 3a. The process of reconstructing the shape of the tree crown in the 3D concave hull does not connect vertices that are too far apart, as in the 3D convex hull. If alpha tends to infinity, the concave hull result is infinitely close to the convex hull, while a smaller alpha tends to be concave at a certain position to fit the shape of the point set more closely [35,36]. The results of the concave packet algorithm with different parameters on a tree are shown in Figure 3b. In this study, we set the alpha range as 0.1–5 m, took 0.1 m as a step and calculated the RMSE with measured 3D green volume to select the optimal scale.

2.4.3. Convex Hull by Slices Algorithm

The convex hull by slices algorithm is based on the idea of integration. First, the canopy is sliced according to the uniform thickness, and each layer is considered a table body. For each layer, the point clouds within 0.2 m of each plane are counted, these point clouds are projected to the same plane, and the projected area of the plane is calculated using the 2D convex hull algorithm. Then, the table product formula is used to calculate the volume of each layer. Finally, all the slice volumes are summed to obtain the 3D green volume (Figure 4) [27,37]. The convex hull by slices algorithm sets the height difference in the range of 0.1–5 m with a step of 0.1 m, and calculates the RMSE with measured 3D green volume to select the optimal scale. The volume of each layer of the table is calculated as follows:

$$\mathbf{V} = \sum \frac{\left(\mathbf{S_n} + \mathbf{S_{n+1}} + \sqrt{\mathbf{S_n} + \mathbf{S_{n+1}}}\right)}{3} * \mathbf{\Delta h} \quad (1)$$

where $\mathbf{S_n}$ is the projected area of the nth layer of the point cloud calculated based on the 2D convex packet algorithm and $\mathbf{\Delta h}$ is the height of the table.

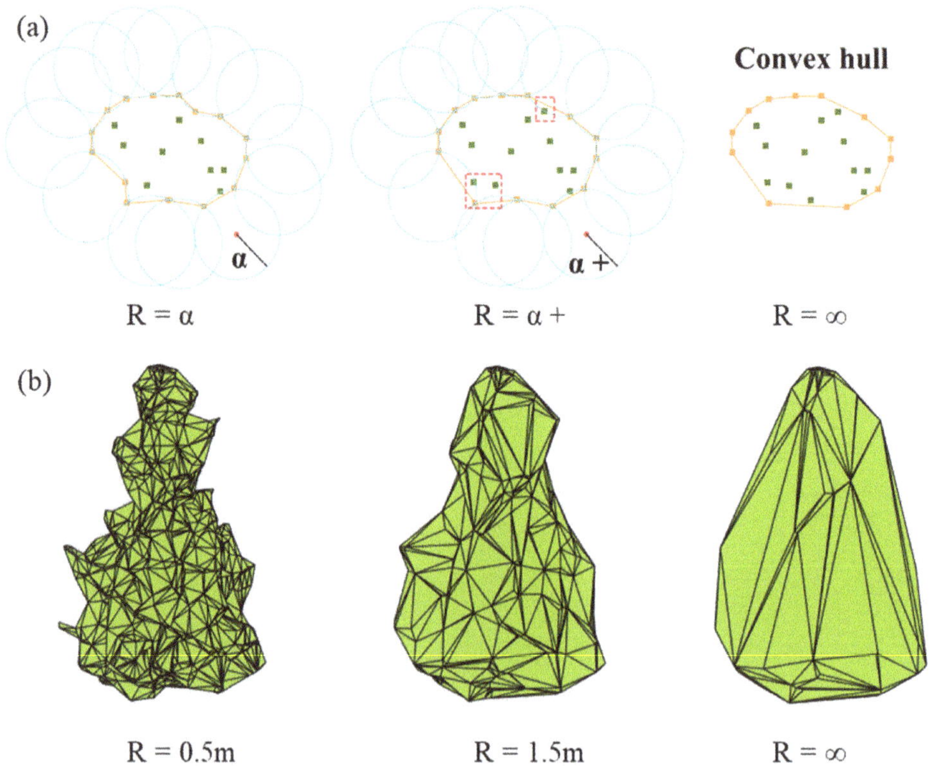

Figure 3. (**a**) The schematic diagram of the concave hull algorithm (**b**) The single tree contour constructed by the concave hull with different parameters.

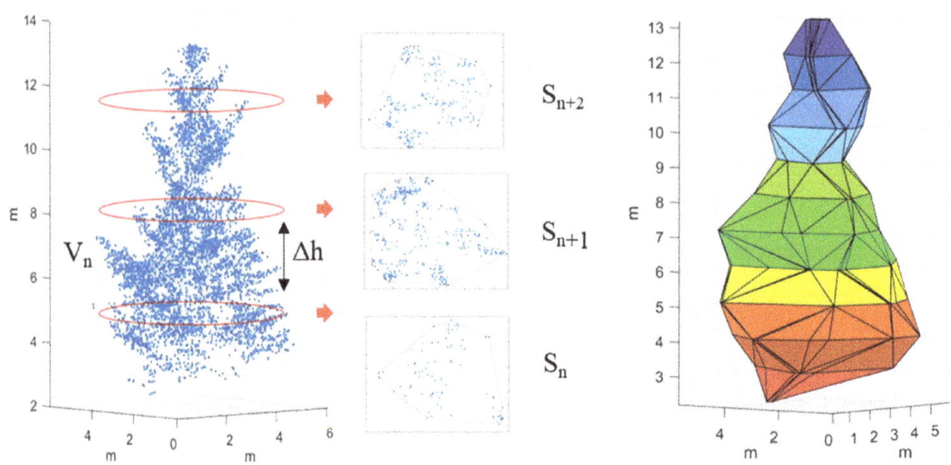

Figure 4. The schematic diagram of the convex hull by slices algorithm.

2.4.4. Voxel Algorithm

The voxel algorithm uses a regular 3D grid to partition the discrete canopy point cloud. The initialized grid is divided into N smaller voxels according to the input parameters, and the number of voxels in which at least one point exists in the statistical space is determined, based on the range of the canopy point cloud to determine the polar values of the starting grid in the XYZ coordinate directions. The sum of the space volumes occupied by the voxels is the 3D green volume (Figure 5) [28,38]. In this study, the voxel edge length was set in the range of 0.1–1 m, the step length was 0.1 m, and the RMSE was calculated with measured 3D green volume to select the optimal scale.

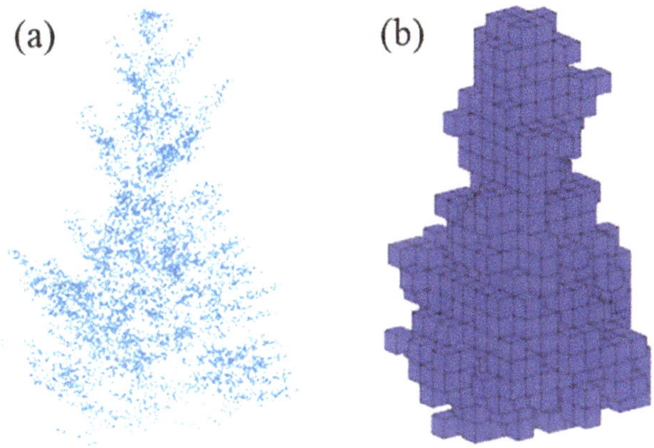

Figure 5. (**a**) The single tree point clouds (**b**) The schematic diagram of the voxel algorithm.

2.4.5. Voxel Coupling Convex Hull by Slices Algorithm

The 3D convex hull algorithm treats the canopy as a whole and cannot calculate the gaps within the canopy, and its boundary does not represent the real canopy outline, and, thus, it overestimates the measured 3D green volume [1,8]. The concave hull algorithm excessively removes voids and gaps when calculating the volume, leading to low 3D green volume estimation results [39]. Since the crown shape and size of different single trees of the same species vary greatly, uniform thickness slices in the vertical direction bring some errors to the calculation of 3D green volume based on the convex hull by slices algorithm [40]. The voxel algorithm can generate realistic canopy shapes to obtain high-accuracy 3D green volume [28,41], but the missing ULS point cloud leads to underestimation of 3D green volume [9]. Figure 6 shows a single tree ginkgo canopy for the voxel algorithm, and it can be seen that the point cloud absence increases with decreasing height.

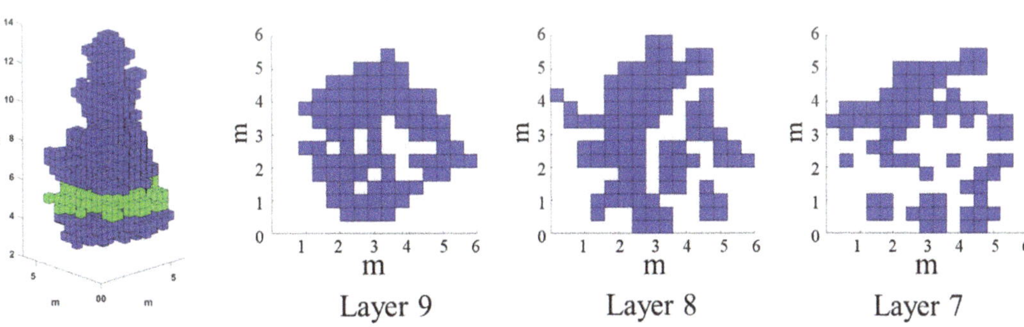

Figure 6. Single tree voxel profile analysis (Layer 7, Layer 8, Layer 9 shows the horizontal section of the tree at a crown height of 2.8 m, 3.2 m, and 3.6 m).

Based on the above analysis, a new algorithm for 3D green volume estimation, named voxel coupling convex hull by slices, is proposed in this study. The method calculates the 3D green volume by dividing the single tree canopy point cloud into two parts according to height, and the volume of the upper canopy point cloud is calculated using the voxel algorithm, to prevent the calculation error of the upper layer caused by uniform thickness slicing, while the volume of the lower canopy point cloud is calculated using the convex hull by slices algorithm, to prevent the calculation error of the green volume caused by the missing point cloud of the lower layer (Figure 7). To explore the optimal stratification ratio, this study set the stratification range, from 10% to 90%, with a step size of 10%, extracted the 3D green volume of each stratification, and calculated the RMSE with measured 3D green volume to select the optimal scale. The computational equations of this new method are shown below:

$$\mathbf{V} = \mathbf{n} * \mathbf{Vn} + \sum \frac{(\mathbf{S_n} + \mathbf{S_{n+1}} + \sqrt{\mathbf{S_n} + \mathbf{S_{n+1}}})}{3} * \mathbf{\Delta h} \tag{2}$$

where $\mathbf{n}$ is the number of voxels, $\mathbf{V_n}$ is the volume of a single voxel, $\mathbf{S_n}$ is the projected area of the nth layer of the point cloud, and $\mathbf{\Delta h}$ is the height of the table.

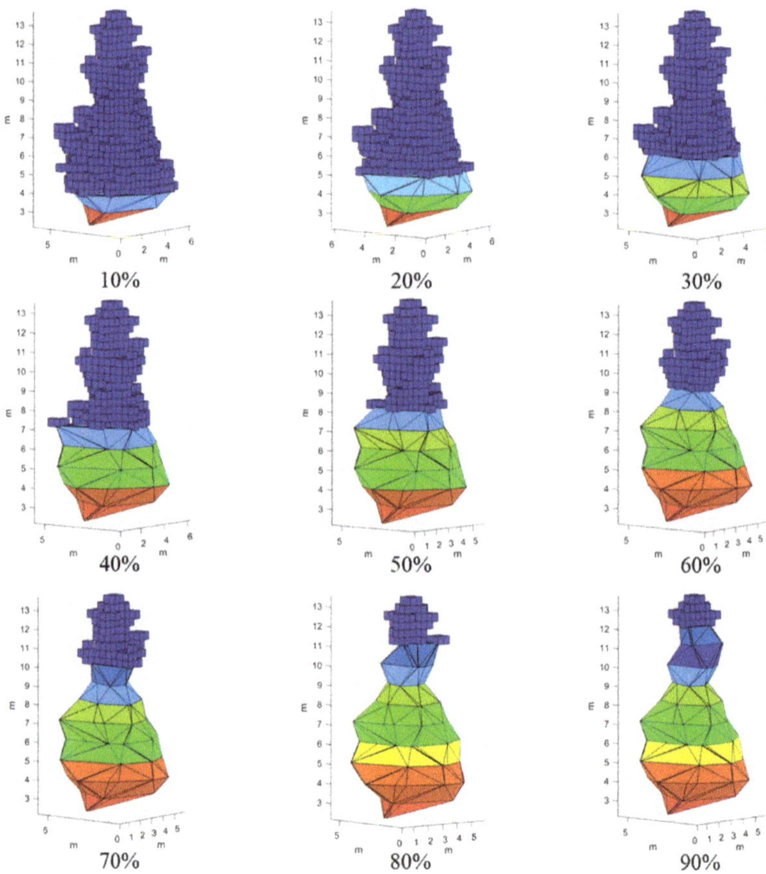

Figure 7. Schematic diagram of the voxel coupling convex hull by slices algorithm.

2.5. Sensitivity Analysis of ULS Data Density

To explore the effect of different LiDAR point cloud densities on extracting 3D green volume, the original point density (230 pts/m^2) was used to lower densities to 75% (172.5 pts/m^2), 50% (115 pts/m^2), 25% (57.5 pts/m^2), 10% (23 pts/m^2) and 5% (11.5 pts/m^2). We used a point cloud height-based algorithm that groups all point clouds by elevation and extracts reduced point clouds by percentage in each layer [15,42]. This algorithm ensured the consistency of sampling, and the extracted point cloud data could maintain a similar spatial distribution as the original point cloud data.

2.6. Random Forest Model

The RF algorithm, created by Breiman and Cutler, was developed as an integrated learning model and a basic decision tree classifier. The decision tree algorithm is an extension of the conventional framework. It improves prediction accuracy by combining multiple decision trees [43,44]. The basic idea is that by using bootstrapping with repeated sampling replacement from several random samples, and establishing a corresponding decision tree for each sample, a RF could be constituted by combining the forecasting of multiple decision trees [45].

The randomForest function under the randomForest data package in R software was used to construct the RF model. First, in the method using the random forest R language, the program determines the influence of each independent variable on the regression process and then evaluates the influence using two indexes. One is the model mean square error (%InMSE) increment when out-of-bag arguments appear, and the second is the impact of purity on the tree node model when out-of-bag arguments arise. Second, the RF algorithm has three important parameters: ntree is the number of random regression trees; nodesize is the minimum size of the terminal node, whose default value is 5; and mtry is a variable division number (the default value is one-third of the number of arguments). In this study, ntree was set to 2000, and the rest of the parameters were set as default parameters. The effect of each independent variable was determined based on the out-of-bag error (%InMSE) [43] (Figure 8).

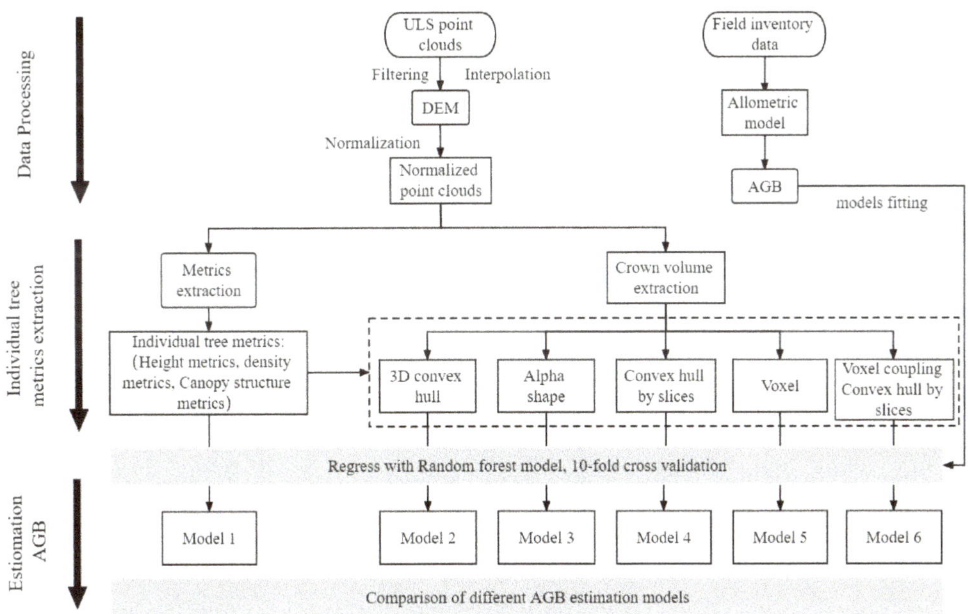

Figure 8. The flow chart of this study.

2.7. Model Verification

In this study, the R^2, RMSE, and nRMSE of 10-fold cross-validation were used to evaluate the model fit. Generally, a higher accuracy is indicated by higher values of R^2 and lower values of RMSE and nRMSE. R^2, RMSE and nRMSE were calculated as follows:

$$\mathbf{R}^2 = 1 - \frac{\sum_{i=1}^{n} (\mathbf{p_i} - \mathbf{o_i})^2}{\sum_{i=1}^{n} (\mathbf{o_i} - \overline{\mathbf{o_i}})^2} \tag{3}$$

$$\mathbf{RMSE} = \sqrt{\frac{1}{\mathbf{n}} \sum_{i=1}^{n} (\mathbf{p_i} - \mathbf{o_i})^2} \tag{4}$$

$$\mathbf{nRMSE} = \frac{\sqrt{\frac{1}{n} \sum_{i=1}^{n} (\mathbf{p_i} - \mathbf{o_i})^2}}{\overline{\mathbf{o_i}}} \tag{5}$$

where $\mathbf{o_i}$ represents the observed AGB for the ith tree, $\overline{\mathbf{o_i}}$ is the observed mean value, $\mathbf{p_i}$ is the estimated AGB for the ith tree, and n is the number of trees.

3. Results

3.1. Determination of Different 3D Green Volume Algorithm Parameters

Figure 9 shows that in the alpha shape algorithm, as the alpha parameter increased, the 3D green volume also became larger, and the alpha value tended to stabilize after 2 m. The RMSE showed first a decreasing and then an increasing trend, with a peak at the 0.6 m scale, and tended to stabilize after the alpha value was greater than 1.3 m. The RMSE under this scale was 12.20 m^3. The overall trend of the 3D green volume obtained by the convex hull by slices algorithm decreased with increasing height interval, and the RMSE did not change regularly, but, overall, the RMSE was higher when the height interval was higher, with a peak at the 0.9 m scale. The RMSE under this scale was 13.01 m^3. The 3D green volume obtained by the voxel algorithm increased linearly with increasing voxel size, and the RMSE showed a trend of first decreasing and then increasing, peaking at a scale of 0.4 m. The optimal voxel size was selected as 0.4 × 0.4 × 0.4 m, and the RMSE at this scale was 12.03 m^3. Figure 10 shows the results of the voxel coupling convex hull by slices algorithm. As the segmentation scale increased, the RMSE decreased and then increased. The optimal canopy segmentation scale was selected to be 20%, and the RMSE at this scale was 11.17 m^3, which was lower than those of the other algorithms.

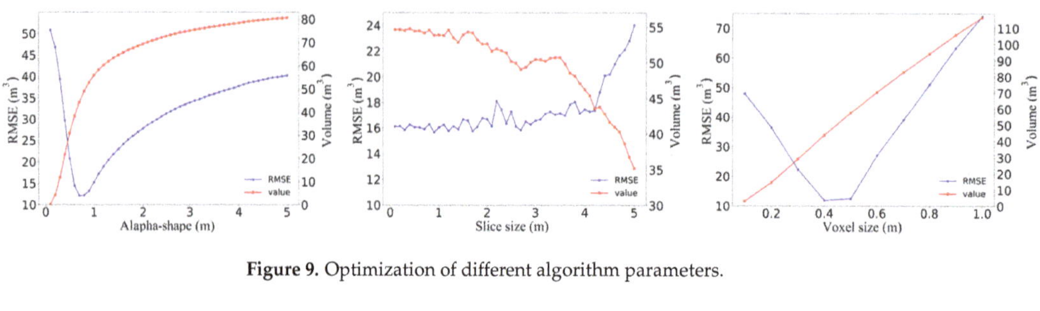

Figure 9. Optimization of different algorithm parameters.

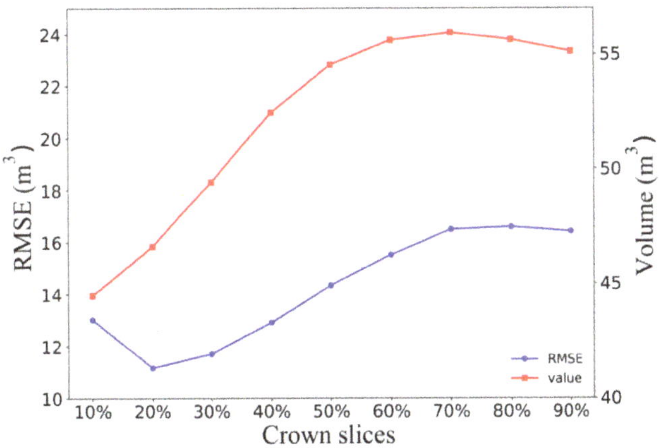

Figure 10. Stratified scale screening.

3.2. Calculation Results of 3D Green Volume

The 3D green volume of single ginkgo trees differed significantly, due to differences in growth. The results of the 3D green volume calculated using the five algorithms, 3D convex hull, 3D concave hull, convex hull by slices, voxel and voxel coupling convex hull by slices, are shown in Table 3. The average 3D green volume and RMSE calculated by the 3D convex hull algorithm was much higher than those calculated by the other models, ranging from 21.22–196.10 m^3 (mean = 84.86 m^3, RMSE = 45.31 m^3). The average 3D green volume calculated by the 3D concave hull was the lowest, ranging from 12.12–74.84 m^3 (mean = 37.72 m^3, RMSE = 12.20 m^3). The convex hull by slices algorithm overestimated 3D green volume, and ranged from 15.16–133.53 m^3 (mean = 53.85 m^3, RMSE = 13.01 m^3), and the 3D green volume calculated by the voxel algorithm ranged from 16.00–81.79 m^3 (mean = 43.13 m^3, RMSE = 12.03 m^3). The average 3D green volume calculated by the voxel coupling convex hull by slices algorithm ranged from 15.58–97.20 m^3 (mean = 46.61 m^3), and the 3D green volume calculated by this method was the minimum RMSE (11.17 m^3).

Table 3. Three dimensional green volume of single trees of ginkgo by different algorithms.

Algorithms	Min (m^3)	Max (m^3)	Mean (m^3)	RMSE (m^3)
Observed data	12.55	96.39	46.85	-
3D convex hull	21.22	196.10	84.86	45.31
3D concave hull	12.12	74.84	37.72	12.20
convex hull by slices	15.16	133.53	53.85	13.01
voxel	16.00	81.79	43.13	12.03
voxel coupling convex hull by slices	15.58	97.20	46.61	11.17

3.3. ULS Point Density Effects on the Performance of the 3D Green Volume

To investigate the effects of ULS point density on 3D green volume, we calculated using 5 algorithms with different sampling densities of 75% (172.5 pts/m^2), 50% (115 pts/m^2), 25% (57.5 pts/m^2), 10% (23 pts/m^2) and 5% (11.5 pts/m^2) and the Pearson's correlation between AGB and 3D green volume (Figure 11). The box plot in Figure 11 shows the changes in the 3D green volume values due to the decrease in point cloud density. The 3D green volume values and r values of all five algorithms decreased as the point cloud density decreased; among them, the values of the 3D convex hull and convex hull by slices algorithm decreased slowly with the point cloud density from 100% (230 pts/m^2) to 10% (23 pts/m^2), and the correlation with the AGB was stable. However, when the point density decreased to 5% (11.5 pts/m^2), there was a marked decrease in the r values and 3D green volume values (3D convex hull, r = 0.88–0.86; convex hull by slices, r = 0.84–0.80). For the alpha shape and voxel algorithms, there was a slight downward trend in the 3D green volume value and r values as the point cloud density decreased from 100% (230 pts/m^2) to 50% (115 pts/m^2), and as the point cloud density decreased from 50% (115 pts/m^2) to 5% (11.5 pts/m^2), the metric and r values decreased significantly (alpha shape, r = 0.84–0.70; voxel r = 0.87–0.80). The 3D green volume values of voxel coupling convex hull by slices algorithm decreased with point cloud density in the same way as the voxel algorithm, but the decrease in r values was lower (0.87–0.84). Therefore, this study chose to extract single-tree 3D green volume at 100% (230 pts/m^2) of point cloud density as a metric to estimate AGB.

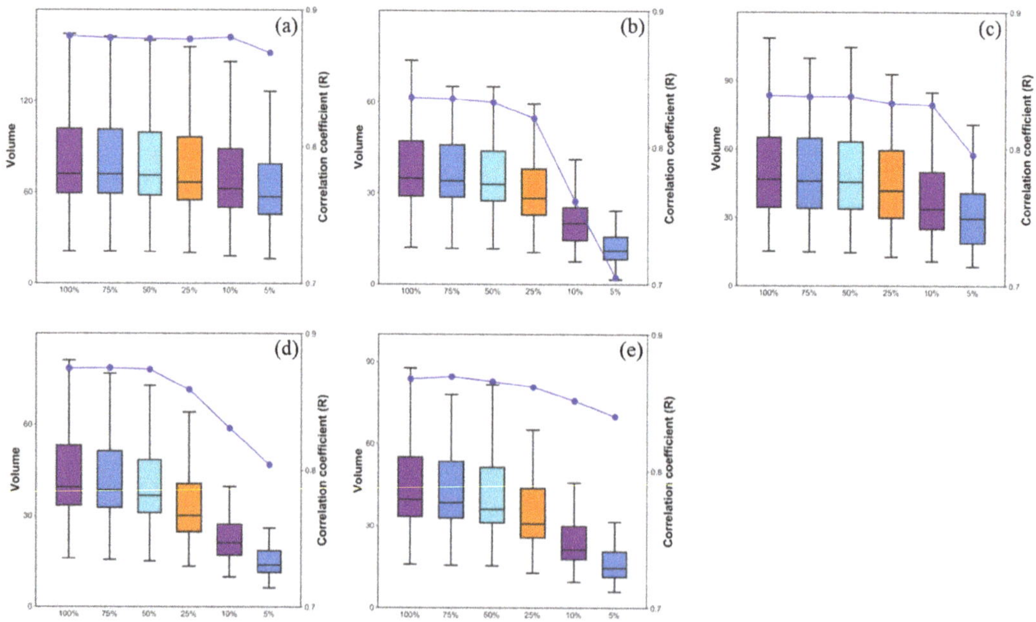

Figure 11. Distribution and correlation with AGB of 3D convex hull (**a**), 3D concave hull (**b**), convex hull by slices (**c**), voxel (**d**) and voxel coupling convex hull by slices (**e**) algorithms at different point densities (100%, 75%, 50%, 25%, 10%, 5%).

3.4. RF Variable Importance Analysis

In this study, the 3D green volume extracted by each of the five algorithms was combined with the ULS base parameters to obtain the importance scores of the input variables by adding the RF model for 100 runs. Figure 12 shows a statistical plot of the importance scores of the top 20 variables with the greatest impact on the estimated AGB. In all models, 3D green volume, H_99, and H_max were the three parameters with the highest importance; the 3D green volume in model 2 had the second highest importance after H_99; the 3D green volume in model 3 had lower importance than H_99 and H_max; and the 3D green volume in models 4, 5 and 6 were the parameters with the highest importance. The mean importance of 3D green volume in model 6 was 35.92, which was significantly higher than that of the other parameters. It indicated that 3D green volume is an important parameter for estimating single tree AGB.

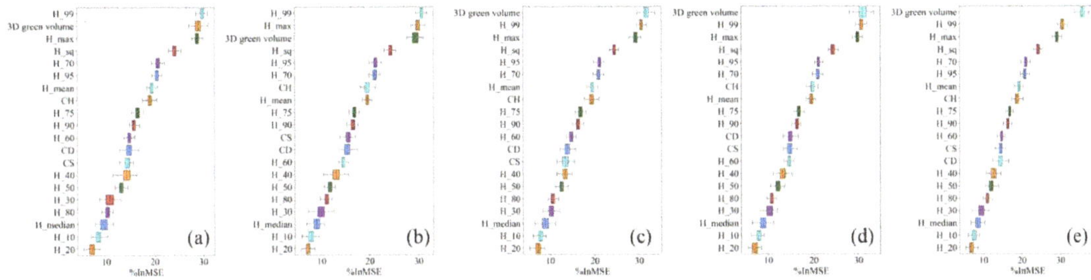

Figure 12. RF model AGB parameter importance analysis of 3D convex hull (**a**) 3D concave hull (**b**) Convex hull by slices (**c**) Voxel (**d**) Voxel coupling convex hull by slices (**e**) % in MSE for the first 20 variables running the RF model 100 times.

3.5. Single-Tree AGB Estimation

The results of different models predicting single tree AGB are shown in Figure 13 and Table 4. The accuracy of model 1, based on ULS base parameters, was CV-R^2 = 0.81, RMSE = 12.66 kg, nRMSE = 16.94%, and the AGBs estimated by models 2–5, with the addition of 3D green volume parameters, were better than that of model 1 (CV-R^2 = 0.82–0.85, RMSE = 11.29–12.54 kg, nRMSE = 15.12–16.79%). This indicated that the addition of 3D green volume could significantly improve the estimation accuracy of AGB. Model 6 had the highest accuracy (CV-R^2 = 0.85, RMSE = 11.29 kg, nRMSE = 15.12%), with an improvement in CV-R^2 of 0.04, a decrease in RMSE of 1.37 kg, and a decrease in nRMSE of 1.82%.

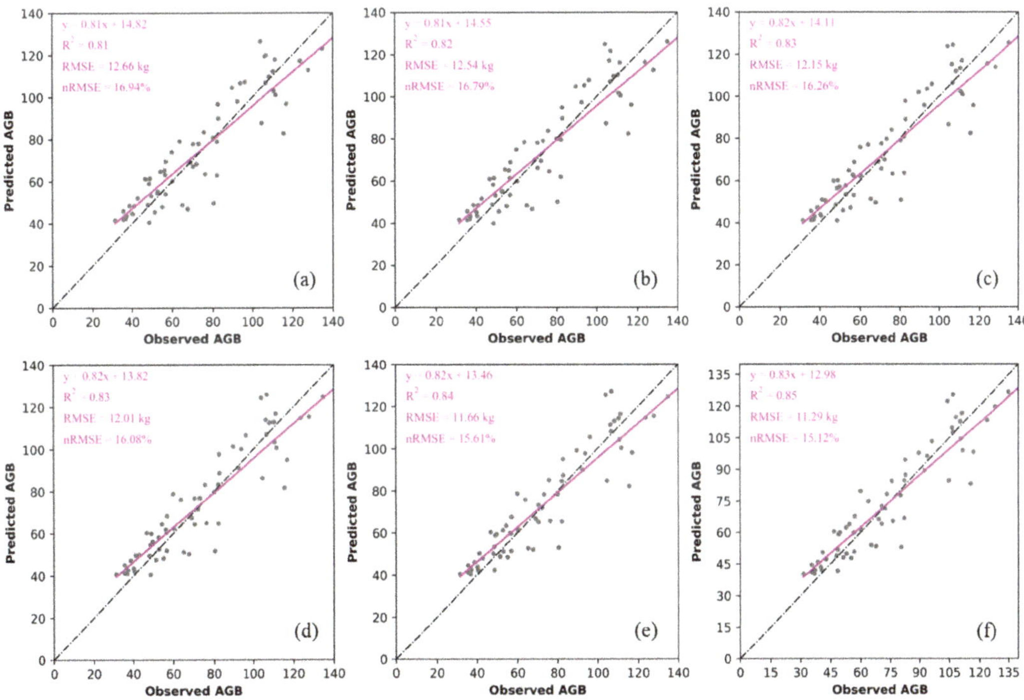

Figure 13. Different algorithms for estimating AGB (**a**) ULS basis parameters, (**b**) 3D convex hull, (**c**) 3D concave hull, (**d**) Convex hull by slices, (**e**) Voxel, (**f**) Voxel coupling convex hull by slices.

Table 4. Different algorithms for estimating AGB accuracy.

Algorithms	R^2	RMSE (kg)	nRMSE (%)
base parameters	0.81	12.66	16.94
3D convex hull	0.82	12.54	16.79
3D concave hull	0.83	12.15	16.26
convex hull by slices	0.83	12.01	16.08
voxel	0.84	11.66	15.61
voxel coupling convex hull by slices	0.85	11.29	15.12

4. Discussion

Previous studies have shown that lidar pulses are nearly vertical, resulting in the number of point clouds from the lower parts of the crown being lower than those nearer the top, and, thus, ALS tends to underestimate the 3D green volume at the bottom of the canopy [9]. Although the point cloud density obtained by ULS is higher than that of ALS, there is still a problem of missing point clouds inside the bottom layer of the canopy (Figure 6). Alpha shape and voxel algorithms are strongly affected by the point cloud integrity, resulting in an underestimation of 3D green volume [19,46]. Fernández-Sarría et al. [20] found that the 3D convex hull algorithm overestimated the canopy gap, leading to an overestimation of the 3D green volume. Yan et al. [1] argued that the 2D convex hull algorithm would overestimate the projected area of the canopy slices, leading to an overestimation of the 3D green volume by the convex hull by slices algorithm. Our results showed that the 3D concave hull and voxel algorithms underestimated the 3D green volume, and the 3D convex hull and convex hull by slices overestimated the 3D green volume, which was consistent with previous studies. In this study, we described and evaluated a method to estimate the 3D green volume of a single tree and obtained the optimal 3D green volume extraction result (RMSE = 11.17 m^3). The voxel coupling convex hull by slices algorithm calculates the 3D green volume at the bottom of the canopy using the convex hull by slices algorithm, which is more stable than the voxel algorithm [1,46] and can solve the underestimation of 3D green volume caused by the missing point cloud at the bottom of the canopy to obtain a more accurate 3D green volume.

Foreign and domestic studies have shown that changes in point cloud density have no significant effect on most ULS metrics [15,39,47]. This study also analyzed the effect of the change in point cloud density on the 3D green volume and found that as the point cloud density decreased from 100% (230 pts/m^2) to 75% (172.5 pts/m^2), 50% (115 pts/m^2), 25% (57.5 pts/m^2), 10% (23 pts/m^2) and 5% (11.5 pts/m^2), the 3D green volume values extracted by all five algorithms and the correlation with AGB decreased as the point cloud density decreased (Figure 11). Our results illustrated that the 3D concave hull algorithm and voxel algorithm were sensitive to the point cloud density; when the point cloud density decreased, the extracted 3D green volume value and correlation decreased considerably. Vauhkonen et al. [44] found that the lower the point cloud density, the lower the accuracy of the concave hull algorithm in predicting single tree features. Liu et al. [29] pointed out that the reduction in point cloud density had a significant effect on the canopy volume metrics extracted based on voxels. The 3D convex hull and convex hull by slices algorithms are more stable because these two algorithms are built based on the convex hull algorithm, which calculates the area (volume) of the entire point set based on the outermost points (planes) and is less affected by changes in point cloud density [1]. The voxel coupling convex hull by slices algorithm is more stable than the voxel algorithm in terms of r value variation, which further indicates that the algorithm solves the effect of missing point clouds in the results of the voxel algorithm.

In this study, 3D green volume was involved in AGB estimation as a parameter, and the analysis of the importance of RF variables showed that 3D green volume was one of the most important parameters for AGB estimation (Figure 11). The accuracy of AGB estimation was significantly improved with the inclusion of the 3D green volume parameter (Figure 12), indicating that 3D green volume had an important influence on AGB estimation, which was consistent with the results of previous studies [24]. The model accuracy showed a significant improvement compared to the AGB accuracy (R^2 = 0.77) estimated by [18] based on ALS. This was due to the lower flight altitude of the ULS (60 m) compared with that of the ALS, which enabled a higher density of point cloud data to be acquired, and higher density point cloud data helped to reconstruct the forest 3D structure at a more refined scale [33,48]. Through a comparative analysis, it was found that model 6 predicted AGB with the highest accuracy, because the voxel coupling convex hull by slices algorithm obtained a more accurate 3D green volume, compared to the other algorithms, which made the model fitting ability more accurate.

5. Conclusions

To better extract the 3D green volume of a single tree based on the ULS point cloud, this study proposes the voxel coupling convex hull by slices algorithm, which improves the existing algorithms. To validate the algorithm, the 3D green volume of 64 single ginkgo trees was calculated using this method, and the AGB of single trees was estimated based on the 3D green volume and compared with existing methods. The results showed that the voxel coupling convex hull by slices algorithm was most suitable for calculating the 3D green volume and estimating the AGB using ULS data. The results were as follows: (1) The choice of input parameters of different algorithms significantly affects the results for 3D green volume. Under the premise of using optimal parameters, the voxel coupling convex hull by slices algorithm provided the most accurate estimate of the 3D green volume of single ginkgo trees with RMSE = 11.17 m^3; (2). The error in calculating 3D green measures increased for all algorithms as the point cloud density decreased. The concave hull and voxel algorithms had a higher dependence on point cloud density than the other algorithms, and the correlation between AGB and the 3D convex hull, convex hull by slices and voxel coupling convex hull by slices algorithms was more stable when the point cloud density was higher than 10%; (3) The 3D green volume was the most important parameter for estimating the AGB of a single tree. The addition of the 3D green volume parameter could significantly improve the accuracy of the model to estimate AGB, where the highest accuracy was obtained with the voxel coupling convex hull by slices algorithm, which estimated AGB with CV-R^2 = 0.85, RMSE = 11.29 kg, and nRMSE = 15.12%. Our study demonstrates that ULS point cloud data can be used to accurately extract the 3D green volume of single trees in urban forests and that the 3D green volume is promising for estimating the AGB of a single tree.

Author Contributions: Conceptualization, H.D. and H.H.; methodology, L.Z.; validation, L.Z.; formal analysis, L.Z.; investigation, L.Z., B.Z., J.X., Y.G. and C.T.; data curation, L.Z.; writing—original draft preparation, L.Z.; writing—review and editing, X.L. and H.D.; visualization, L.Z.; supervision, H.D. All authors have read and agreed to the published version of the manuscript.

Funding: The authors gratefully acknowledge the support of the National Natural Science Foundation of China (U1809208, 32171785, 32201553), the State Key Laboratory of Subtropical Silviculture (No. ZY20180201), and the Key Research and Development Program of Zhejiang Province (2021C02005).

Data Availability Statement: Not applicable.

Acknowledgments: The authors gratefully acknowledge the support of various foundations. The authors are grateful to the editor and anonymous reviewers whose comments have contributed to improving the quality of this manuscript.

Conflicts of Interest: The authors declare no conflict of interest.

References

1. Yan, Z.; Liu, R.; Cheng, L.; Zhou, X.; Ruan, X.; Xiao, Y. A Concave Hull Methodology for Calculating the Crown Volume of Individual Trees Based on Vehicle-Borne LiDAR Data. *Remote Sens.* **2019**, *11*, 623. [CrossRef]
2. He, C.; Convertino, M.; Feng, Z.; Zhang, S. Using LiDAR data to measure the 3D green biomass of Beijing urban forest in China. *PLoS ONE* **2013**, *8*, e75920. [CrossRef] [PubMed]

3. Du, P. Study on the 3D Green Guantiy and Ecological Effect of the Five Kinds of Mainly Landscape Plant in Chengdu. Master's Thesis, Sichuan Agricultural University, Chengdu, China, 2009.
4. Cheng, Y. Study on Tridimensional Green Biomass Estimation and Analysis of Forest in Beijing. Master's Thesis, Beijing Forestry University, Beijing, China, 2011.
5. Xuehai, T. Estimation and Analysis of Tridimensional Green Biomass of Central Six Districts of Beijing. Ph.D. Thesis, Beijing Forestry University, Beijing, China, 2011.
6. Hauglin, M.; Astrup, R.; Gobakken, T.; Næsset, E. Estimating single-tree branch biomass of Norway spruce with terrestrial laser scanning using voxel-based and crown dimension features. *Scand. J. For. Res.* **2013**, *28*, 456–469. [CrossRef]
7. Tiit Nilson, U.P. A Forest Canopy Reflectance Model and a Test Case. *Remote Sens. Environ.* **1991**, *37*, 131–142. [CrossRef]
8. Kato, A.; Moskal, L.M.; Schiess, P.; Swanson, M.E.; Calhoun, D.; Stuetzle, W. Capturing tree crown formation through implicit surface reconstruction using airborne lidar data. *Remote Sens. Environ.* **2009**, *113*, 1148–1162. [CrossRef]
9. Korhonen, L.; Vauhkonen, J.; Virolainen, A.; Hovi, A.; Korpela, I. Estimation of tree crown volume from airborne lidar data using computational geometry. *Int. J. Remote Sens.* **2013**, *34*, 7236–7248. [CrossRef]
10. Pregitzer, K.S.; Euskirchen, E.S. Carbon cycling and storage in world forests: Biome patterns related to forest age. *Glob. Chang. Biol.* **2004**, *10*, 2052–2077. [CrossRef]
11. Chen, D.; Li, W.; Kong, W.; Shen, S. On the method of Three-Dimensional Green Volume Calculation Based on Low-altitude High-Definition Images-Case Study of the Nanjing Fotestry University Campus. *Chin. Landsc. Archit.* **2015**, *31*, 22–26.
12. Zhao, P.; Lu, D.; Wang, G.; Wu, C.; Huang, Y.; Yu, S. Examining Spectral Reflectance Saturation in Landsat Imagery and Corresponding Solutions to Improve Forest Aboveground Biomass Estimation. *Remote Sens.* **2016**, *8*, 469. [CrossRef]
13. Hyyppä, J.; Hyyppä, H.; Leckie, D.; Gougeon, F.; Yu, X.; Maltamo, M. Review of methods of small-footprint airborne laser scanning for extracting forest inventory data in boreal forests. *Int. J. Remote Sens.* **2008**, *29*, 1339–1366. [CrossRef]
14. Cao, L.; Coops, N.C.; Sun, Y.; Ruan, H.; Wang, G.; Dai, J.; She, G. Estimating canopy structure and biomass in bamboo forests using airborne LiDAR data. *ISPRS J. Photogramm. Remote Sens.* **2019**, *148*, 114–129. [CrossRef]
15. Liu, K.; Shen, X.; Cao, L.; Wang, G.; Cao, F. Estimating forest structural attributes using UAV-LiDAR data in Ginkgo plantations. *ISPRS J. Photogramm. Remote Sens.* **2018**, *146*, 465–482. [CrossRef]
16. Xu, Q.; Li, B.; Maltamo, M.; Tokola, T.; Hou, Z. Predicting tree diameter using allometry described by non-parametric locally-estimated copulas from tree dimensions derived from airborne laser scanning. *For. Ecol. Manag.* **2019**, *434*, 205–212. [CrossRef]
17. Xu, D.; Wang, H.; Xu, W.; Luan, Z.; Xu, X. LiDAR Applications to Estimate Forest Biomass at Individual Tree Scale: Opportunities, Challenges and Future Perspectives. *Forests* **2021**, *12*, 550. [CrossRef]
18. Ghanbari Parmehr, E.; Amati, M. Individual Tree Canopy Parameters Estimation Using UAV-Based Photogrammetric and LiDAR Point Clouds in an Urban Park. *Remote Sens.* **2021**, *13*, 2062. [CrossRef]
19. Vauhkonen, J.; Seppänen, A.; Packalén, P.; Tokola, T. Improving species-specific plot volume estimates based on airborne laser scanning and image data using alpha shape metrics and balanced field data. *Remote Sens. Environ.* **2012**, *124*, 534–541. [CrossRef]
20. Fernández-Sarría, A.; Martínez, L.; Velázquez-Martí, B.; Sajdak, M.; Estornell, J.; Recio, J.A. Different methodologies for calculating crown volumes of Platanus hispanica trees using terrestrial laser scanner and a comparison with classical dendrometric measurements. *Comput. Electron. Agric.* **2013**, *90*, 176–185. [CrossRef]
21. Lin, J.; Chen, D.; Wu, W.; Liao, X. Estimating aboveground biomass of urban forest trees with dual-source UAV acquired point clouds. *Urban For. Urban Green.* **2022**, *69*, 127521. [CrossRef]
22. Dalla Corte, A.P.; Rex, F.E.; Almeida, D.R.A.d.; Sanquetta, C.R.; Silva, C.A.; Moura, M.M.; Wilkinson, B.; Zambrano, A.M.A.; Cunha Neto, E.M.D.; Veras, H.F.P.; et al. Measuring Individual Tree Diameter and Height Using GatorEye High-Density UAV-Lidar in an Integrated Crop-Livestock-Forest System. *Remote Sens.* **2020**, *12*, 863. [CrossRef]
23. Suwardhi, D.; Fauzan, K.N.; Harto, A.B.; Soeksmantono, B.; Virtriana, R.; Murtiyoso, A. 3D Modeling of Individual Trees from LiDAR and Photogrammetric Point Clouds by Explicit Parametric Representations for Green Open Space (GOS) Management. *ISPRS Int. J. Geo-Inf.* **2022**, *11*, 174. [CrossRef]
24. Tao, S.; Guo, Q.; Li, L.; Xue, B.; Kelly, M.; Li, W.; Xu, G.; Su, Y. Airborne Lidar-derived volume metrics for aboveground biomass estimation: A comparative assessment for conifer stands. *Agric. For. Meteorol.* **2014**, *198–199*, 24–32. [CrossRef]
25. Gong, Y.X.; Yan, F.; Feng, Z.K.; Liu, Y.F.; Xue, W.X.; Xie, F. Extraction of crown volume using triangulated irregular network algorithm based on LiDAR. *J. Infrared Millim. Waves* **2016**, *35*, 177–183+189. [CrossRef]
26. Guo, Z.-C.; Wang, T.; Liu, S.-L.; Kang, W.-P.; Chen, X.; Feng, K.; Zhang, X.-Q.; Zhi, Y. Biomass and vegetation coverage survey in the Mu Us sandy land—Based on unmanned aerial vehicle RGB images. *Int. J. Appl. Earth Obs. Geoinf.* **2021**, *94*, 102239. [CrossRef]
27. Xu, W.; Feng, Z.; Su, Z.-F.; Xu, H.; Jiao, Y.-Q.; Deng, O. An Automatic Extraction Algorithm for Individual Tree Crown Projection Area and Volume Based on 3D Point Cloud Data. *Spectrosc. Spectr. Anal.* **2014**, *34*, 465–471. [CrossRef]
28. Wu, B.; Yu, B.; Yue, W.; Shu, S.; Tan, W.; Hu, C.; Huang, Y.; Wu, J.; Liu, H. A Voxel-Based Method for Automated Identification and Morphological Parameters Estimation of Individual Street Trees from Mobile Laser Scanning Data. *Remote Sens.* **2013**, *5*, 584–611. [CrossRef]
29. Liu, K.; Cao, L.; Wang, G.; Cao, F. Biomass allocation patterns and allometric models of Ginkgo biloba. *J. Beijing For. Univ.* **2017**, *39*, 12–20. [CrossRef]

30. Zhao, X.; Guo, Q.; Su, Y.; Xue, B. Improved progressive TIN densification filtering algorithm for airborne LiDAR data in forested areas. *ISPRS J. Photogramm. Remote Sens.* **2016**, *117*, 79–91. [CrossRef]
31. Khosravipour, A.; Skidmore, A.K.; Isenburg, M. Generating spike-free digital surface models using LiDAR raw point clouds: A new approach for forestry applications. *Int. J. Appl. Earth Obs. Geoinf.* **2016**, *52*, 104–114. [CrossRef]
32. Næsset, E.; Gobakken, T. Estimation of above- and below-ground biomass across regions of the boreal forest zone using airborne laser. *Remote Sens. Environ.* **2008**, *112*, 3079–3090. [CrossRef]
33. Thomas, V.; Treitz, P.; McCaughey, J.H.; Morrison, I. Mapping stand-level forest biophysical variables for a mixedwood boreal forest using lidar: An examination of scanning density. *Can. J. For. Res.* **2006**, *36*, 34–47. [CrossRef]
34. An, P.T.; Huyen, P.T.T.; Le, N.T. A modified Graham's convex hull algorithm for finding the connected orthogonal convex hull of a finite planar point set. *Appl. Math. Comput.* **2021**, *397*, 125889. [CrossRef]
35. Vauhkonen, J.; Tokola, T.; Packalén, P.; Maltamo, M. Identification of Scandinavian Commercial Species of Individual Trees from Airborne Laser Scanning Data Using Alpha Shape Metrics. *For. Sci.* **2009**, *55*, 37–47. [CrossRef]
36. Zhen, Z.; Quackenbush, L.; Zhang, L. Trends in Automatic Individual Tree Crown Detection and Delineation—Evolution of LiDAR Data. *Remote Sens.* **2016**, *8*, 333. [CrossRef]
37. Li, L.; Li, D.; Zhu, H.; Li, Y. A dual growing method for the automatic extraction of individual trees from mobile laser scanning data. *ISPRS J. Photogramm. Remote Sens.* **2016**, *120*, 37–52. [CrossRef]
38. Popescu, S.C.; Zhao, K. A voxel-based lidar method for estimating crown base height for deciduous and pine trees. *Remote Sens. Environ.* **2008**, *112*, 767–781. [CrossRef]
39. Vauhkonen, J.; Tokola, T.; Maltamo, M.; Packalén, P. Effects of pulse density on predicting characteristics of individual trees of Scandinavian commercial species using alpha shape metrics based on airborne laser scanning data. *Can. J. Remote Sens.* **2008**, *34*, S441–S459. [CrossRef]
40. Cheng, L.; Wu, Y.; Chen, S.; Zong, W.; Yuan, Y.; Sun, Y.; Zhuang, Q.; Li, M. A Symmetry-Based Method for LiDAR Point Registration. *IEEE J. Sel. Top. Appl. Earth Obs. Remote Sens.* **2018**, *11*, 285–299. [CrossRef]
41. Hosoi, F.; Omasa, K. Voxel-Based 3-D Modeling of Individual Trees for Estimating Leaf Area Density Using High-Resolution Portable Scanning Lidar. *IEEE Trans. Geosci. Remote Sens.* **2006**, *44*, 3610–3618. [CrossRef]
42. Magnusson, M.; Fransson, J.E.S.; Holmgren, J. Effects on Estimation Accuracy of Forest Variables Using Different Pulse Density of Laser Data. *For. Sci.* **2007**, *53*, 619–626. [CrossRef]
43. Breiman, L. Random Forests. *Mach. Learn.* **2001**, *45*, 5–32. [CrossRef]
44. Li, X.; Du, H.; Mao, F.; Zhou, G.; Chen, L.; Xing, L.; Fan, W.; Xu, X.; Liu, Y.; Cui, L.; et al. Estimating bamboo forest aboveground biomass using EnKF-assimilated MODIS LAI spatiotemporal data and machine learning algorithms. *Agric. For. Meteorol.* **2018**, *256–257*, 445–457. [CrossRef]
45. Dong, L.; Du, H.; Han, N.; Li, X.; Zhu, D.e.; Mao, F.; Zhang, M.; Zheng, J.; Liu, H.; Huang, Z.; et al. Application of Convolutional Neural Network on Lei Bamboo Above-Ground-Biomass (AGB) Estimation Using Worldview-2. *Remote Sens.* **2020**, *12*, 958. [CrossRef]
46. Lecigne, B.; Delagrange, S.; Messier, C. Exploring trees in three dimensions: VoxR, a novel voxel-based R package dedicated to analysing the complex arrangement of tree crowns. *Ann. Bot.* **2017**, *121*, 589–601. [CrossRef]
47. Silva, C.; Hudak, A.; Vierling, L.; Klauberg, C.; Garcia, M.; Ferraz, A.; Keller, M.; Eitel, J.; Saatchi, S. Impacts of Airborne Lidar Pulse Density on Estimating Biomass Stocks and Changes in a Selectively Logged Tropical Forest. *Remote Sens.* **2017**, *9*, 1068. [CrossRef]
48. Jakubowski, M.K.; Guo, Q.; Kelly, M. Tradeoffs between lidar pulse density and forest measurement accuracy. *Remote Sens. Environ.* **2013**, *130*, 245–253. [CrossRef]

remote sensing

MDPI

Article

Mapping Forest Aboveground Biomass with MODIS and Fengyun-3C VIRR Imageries in Yunnan Province, Southwest China Using Linear Regression, K-Nearest Neighbor and Random Forest

Huafang Chen [1,2,3], Zhihao Qin [4], De-Li Zhai [5], Guanglong Ou [6], Xiong Li [2,3], Gaojuan Zhao [2,3], Jinlong Fan [7], Chunliang Zhao [4] and Hui Xu [6,*]

1 Faculty of Forestry, Southwest Forestry University, Kunming 650224, China
2 Center for Mountain Futures, Kunming Institute of Botany, Chinese Academy of Sciences, Kunming 650201, China
3 Department of Economic Plants and Biotechnology, Yunnan Key Laboratory for Wild Plant Resources, Kunming Institute of Botany, Chinese Academy of Sciences, Kunming 650201, China
4 MOA Key Laboratory of Agricultural Remote Sensing, Institute of Agricultural Resources and Regional Planning, Chinese Academy of Agricultural Sciences, Beijing 100081, China
5 CAS Key Laboratory of Tropical Forest Ecology, Xishuangbanna Tropical Botanical Garden, Chinese Academy of Sciences, Mengla 666303, China
6 Key Laboratory of State Forestry Administration on Biodiversity Conservation in Southwest China, Southwest Forestry University, Kunming 650224, China
7 National Satellite Meteorological Center, China Meteorological Administration, Beijing 100081, China
* Correspondence: huixuswfu1960@swfu.edu.cn

Citation: Chen, H.; Qin, Z.; Zhai, D.-L.; Ou, G.; Li, X.; Zhao, G.; Fan, J.; Zhao, C.; Xu, H. Mapping Forest Aboveground Biomass with MODIS and Fengyun-3C VIRR Imageries in Yunnan Province, Southwest China Using Linear Regression, K-Nearest Neighbor and Random Forest. *Remote Sens.* **2022**, *14*, 5456. https://doi.org/10.3390/rs14215456

Academic Editors: Huaqiang Du, Wenyi Fan, Mingshi Li, Weiliang Fan and Fangjie Mao

Received: 3 October 2022
Accepted: 27 October 2022
Published: 30 October 2022

Abstract: The aboveground biomass (AGB) of a forest is an important indicator of the forest's terrestrial carbon storage and its relation to climate change. Due to the advantage of extensive spatial coverage and low cost, coarse-resolution remote sensing data is the main data source for wall-to-wall mapping of forest AGB at the regional scale. Despite this, improving the accuracy and efficiency of forest AGB estimation is a major challenge. In this study, two optical imageries, Moderate Resolution Imaging Spectroradiometer (MODIS) 500 m imagery and Fengyun-3C Visible and Infrared Radiometer (FY-3C VIRR) 1000 m imagery, were used and compared for forest AGB estimation in Yunnan Province, southwest China. One parametric approach, multiple linear regression (MLR), and two nonparametric approaches, k-nearest neighbor (KNN) and random forest (RF), were applied for the two imagery datasets, respectively. We evaluated the performance of the combination of remote sensing data and modeling approaches by comparing the accuracies and also explored the potential of FY-3C imagery data in forest AGB estimation at the regional scale as it was used for this purpose for the first time. We found that the machine learning models KNN and RF provided better results than MLR. From the three approaches for both MODIS and FY-3C imagery, RF performed best with R^2 values of 0.84 and 0.81 and RMSE of 23.18 and 23.43, respectively. Estimation of forest AGB based on MODIS was marginally better than the estimation based on FY-3C. FY-3C imagery could therefore be an additional optical remote sensing data source of coarse spatial resolution, comparable to MODIS data which has been widely used for regional forest AGB estimation. Indices related to forest canopy moisture levels from both types of imagery were sensitive to forest AGB. The RF model and MODIS imagery were then applied to map the spatial variation of forest AGB of Yunnan Province. As a result of our study, we determined that Yunnan Province has a total forest AGB of 2123.22 Mt, with a mean value of 58.05 t/ha for forestland in 2016.

Keywords: forest aboveground biomass (AGB); remote sensing; MODIS; FY-3C VIRR; Yunnan Province

1. Introduction

As one of the five most important ecosystems, forests contribute approximately 80% of the global terrestrial aboveground biomass (AGB) and play a key role in global carbon cycling and mitigation of climate change [1–3]. Accurate estimation of regional forest AGB and knowing the spatial distribution are essential for understanding forest carbon dynamics and carbon cycling [3,4]; thus, extensive efforts have been devoted to estimating forest AGB with greater accuracy at different scales [5,6].

With the development of space technology, remote sensing provides a more efficient way to estimate AGB at larger scales because of the repeatability of data acquisition and extensive geographical coverage compared to field surveys conducted at the plot level. The theoretical basis for remote sensing of forest AGB is that the optical reflectance of the forest canopy is highly correlated with the density of biomass within the canopy. Hence, vegetation indices (such as the normalized difference of vegetation index, NDVI) derived from spectral bands of remote sensing observations can be used as parameters for forest AGB estimation [7–9]. In recent decades, remote sensing has become the prevalent tool for forest AGB estimation [10], and various data sources have been employed for establishing the relationship between field-surveyed AGB data and spectral responses. Passive optical sensor data are widely used for monitoring forest AGB [11–18] because of low cost, easy accessibility and ease of data processing. Optical data of different spatial resolution have been extensively applied in forest AGB modeling [7], such as coarse-spatial-resolution data (>100 m), e.g., Moderate Resolution Imaging Spectroradiometer (MODIS), Advanced Very High Resolution Radiometer (AVHRR) from NOAA satellite and SPOT VEGETATION; medium-spatial-resolution data (10–100 m), e.g., Landsat Thematic Mapper (TM) and Enhanced Thematic Mapper Plus (ETM+); fine-spatial-resolution data (<5 m), e.g., IKONOS, QuickBird, WorldView, Gaofen(GF) series and ALOS/PRISM. Fine-spatial-resolution data is frequently used for modeling tree parameters, texture, or forest canopy structures [7,19–23]. Considering its high cost and limited scene coverage, it is only suitable for small areas. Medium-spatial-resolution images such as Landsat are widely used as a data source for biomass estimation at a regional scale [24], whereas coarse-spatial-resolution data are widely used in estimating forest AGB at national, continental and global scales because they have a better trade-off in the combination of image coverage, temporal and spatial resolution.

MODIS data onboard Terra and Aqua have been extensively used in environmental and ecological applications at the regional scale. The Visible and Infrared Radiometer (VIRR) onboard the Chinese second-generation earth observation satellite Fengyun 3C (FY-3C VIRR) also has the capability of global observation for many applications. FY-3C VIRR has a spectral architecture similar to that of MODIS. It can thus be rationally expected that the FY-3C VIRR may provide an alternative source of earth observation data, especially when considering that the two satellites with MODIS onboard have been in service for more than 20 years, which is far beyond their designed lifespan. The continuous mission of Fengyun 3 series satellites has provided stable earth observation data at regional and global scales for several years. In contrast to the extensive applications of MODIS data, FY-3C VIRR data still needs further exploration for multiple potential applications. Therefore, in this study, we aim to investigate the application of FY-3C VIRR and MODIS in estimating forest AGB. Our findings will enhance further understanding of the potential of FY-3C imagery in forest monitoring at regional scales.

In addition to selecting appropriate data sources, various algorithms have been developed for remote sensing-based forest AGB modeling. Parametric and nonparametric approaches have been serving as important tools for forest AGB modeling based on remote sensing data. Parametric modeling directly correlates the available known forest AGB samples with variables derived from remote sensing data to develop regression equation models for forest AGB estimation over the entire images. Linear regression with one or more remote sensing variables is one of the most frequently used parametric modeling methods [25]. However, the relationships between biomass and remote sensing variables in many cases cannot be captured directly by the parametric algorithm, especially in forest regions with eco-climatic diversity. With the development of machine learning algorithms, artificial intelligence (AI) based nonparametric approaches have been used in land surface modeling and also forest AGB mapping using remote sensing imageries at regional scales. Several machine learning algorithms have been developed to capture complicated nonlinear relationships between the input and the output variables. Commonly used nonparametric algorithms include k-nearest neighbor (KNN), artificial neural network (ANN), random forest (RF), support vector machine (SVM) and maximum entropy (MaxEnt). Due to the flexibility and data-driven manner, nonparametric algorithms have shown better performances in AGB estimation [24–26]. Moreover, the availability of forest AGB samples from the field is a prerequisite for nonparametric modeling. Forest AGB data sampled from plots or calculated from forest inventory data needs to be combined in AGB modeling as reference data. Accurate referenced AGB data has also brought challenges for AGB estimation at large scales.

A number of studies have been conducted for forest AGB mapping at a national scale in China. Yin et al. (2015) used seven single bands of MODIS and two vegetation indices NDVI and enhanced vegetation index (EVI) to map forest AGB in China for the period from 2001 to 2013 with a machine learning algorithm [1]. Chi et al. (2015) integrated ICESat/GLAS data and MODIS imagery with the national forest inventory dataset and with field measurements for mapping forest AGB for the whole of China [6]. Zhang et al. (2018) employed KNN models to estimate the species-level biomass of Chinese boreal forests through the integration of forest inventory data with MODIS spectral variables and environmental variables [27]. Lu et al. (2019) estimated the forest AGB and aboveground carbon storage (AGC) of China by volume modeling based on stand density and forest basal area of major forest types [28]. Since China has a vast territory and great eco-climatic diversity, mapping forest AGB in a specific region with unique features was highly necessary for a better understanding of the spatial variation of the forest AGB at the regional scale. Yunnan is an important forested area with high forest biomass and carbon storage. Although forest AGB has been mapped for a few specific tree species at the county level based on Landsat TM/ETM+ imagery in Yunnan Province [5,25], it has not yet been mapped at the provincial level, probably due to difficulties in processing remote sensing data caused by rugged terrains, complex forest composition and inadequate field measurements for certain tree species. It is therefore important to analyze spatially explicit forest ABG in this area to help understand the spatial distribution of forest biomass and to provide baseline data for improving forest management.

The objective of this study was to compare the quality of forest AGB prediction in Yunnan Province using MODIS and FY images and to explore the possibility of FY imagery as a substitute data source for forest AGB mapping in regional forest biomass monitoring. We also used the best combination of imagery and algorithm to map forest AGB and conducted further analysis to help understand the distribution pattern of forest AGB. One parametric approach, multiple linear regression (MLR), and two nonparametric methods, KNN and RF, were selected for forest AGB mapping with the aid of two remote sensing imageries: MODIS and FY-3C VIRR. Our study compared forest AGB mapping with each one of these three algorithms. It intends to answer the following questions: (1) What are the differences in key variables in forest AGB modeling between the two imageries? (2) Can FY-3C imagery be comparable to MODIS imagery in forest AGB estimation in a region with complex terrains, many tree species and complex forest stand composition? (3) Which modeling approach combining the required remote sensing data type achieves better accuracy in AGB mapping?

2. Materials and Methods

2.1. Study Area

Located in the southwest of China, Yunnan Province contains one of the three major forest areas in China. Situated at the juncture of the Asian Plate and Indian Plate and also at the southeast margin of the Qinghai–Tibet Plateau, the topography of Yunnan Province is characterized by mountains and plateaus. The province covers a total area of approximately 394,000 km^2 and is divided into 16 prefectures. Elevation descends from north to south and ranges from 6596 meters to 83 meters above sea level (Figure 1). Yunnan Province borders Myanmar in the west and southwest and Laos and Vietnam in the south, respectively. Mountains and plateaus account for 94% of the total land area of the province. The forest area accounts for 59.3% of the provincial total land area [29]. Yunnan Province ranks second in terms of forest area and forest stock volume, respectively, among all the provinces in China [30] and contributes a huge carbon sink. Most of the province is influenced by the tropical monsoon climate. Due to its enormous north–south span and the high elevational gradient, vegetation distribution patterns in the province are determined by both latitudinal zonation and vertical zonation, resulting in a scattered distribution of forest mosaics in the province. The range of vegetation types from the north to the south includes alpine meadows, montane and subalpine temperate forests, subtropical forests, and tropical rainforests [31]. The forests in Yunnan are categorized into five types and zones (Figure 1): ① cold-temperate coniferous forest in the northwest, ② warm evergreen broadleaved forest in the northeast, ③ warm evergreen broadleaved and coniferous forest in the central part, ④ warm–hot broadleaved and coniferous forest in the southern and central part and ⑤ tropical broadleaved forest in the south [30,32]. Broadleaved dominated forests are mainly located in the south and southwest of Yunnan Province, while most other areas of the province are dominated by coniferous forests, which cover an area of 4.53 million hectares and account for 48.6% of the total forest area in Yunnan Province [31]. The dominant coniferous tree species include *Pinus yunnanensis* (accounting for 24.1% of the total forest area), *Pinus kesiya* var. *langbianensis* (6.5%), *Pinus armandii* (3.3%), *Abies fabri* (3.16%), *Cunninghamia lanceolata* (2.2%) and *Pinus densata* (1.76%). *Pinus yunnanensis* is distributed most widely in Yunnan Province within an elevation range of 700 m to 3300 m and is dominant in the largest proportion of the forest area among the tree species. The dominant broadleaved tree species are *Quercus* (19.7%), *Alnus cremastogyne* (3.04%) and *Betula* spp. (1.0%).

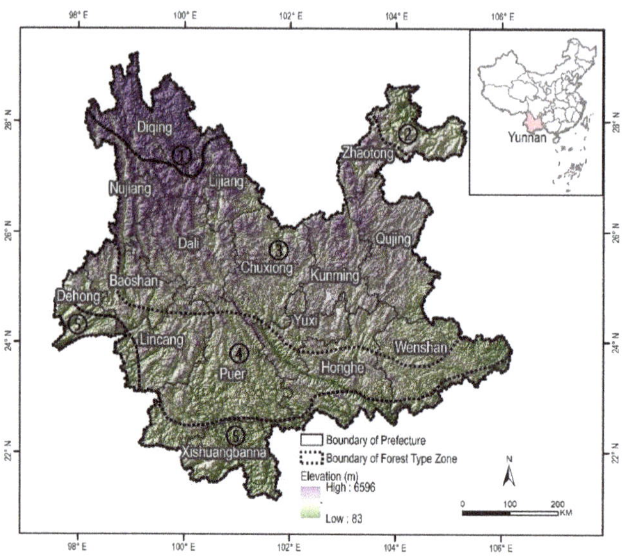

Figure 1. The location of Yunnan Province in China, topography and forest type zones: ① cold-temperate coniferous forest, ② warm evergreen broadleaved forest, ③ warm evergreen broadleaved and coniferous forest, ④ warm–hot broadleaved and coniferous forest and ⑤ tropical broadleaved forest.

Because the diverse climates and habitats harbor abundant flora and fauna, Yunnan is well known for its high biodiversity. The south, west and northwest of the province are actually located in a global biodiversity hotspot region. Yunnan Province, and in particular, the Hengduan Mountains region in the west and northwest, is one of the three zones which are most vulnerable to climate change in China [33].

2.2. Data

2.2.1. MODIS Data and Spectral Variables

The first MODIS instrument, the TERRA satellite, was launched in December 1999, and the second instrument, the AQUA satellite, was launched in May 2002. MODIS has provided multi-purpose images for monitoring large-scale changes in the biosphere in past decades. While the designed lifespan of the two satellites is six years, the MODIS sensors have been operated for over 10 years.

MODIS has a viewing swath width of 2330 km and measures 36 spectral bands between 0.405 and 14.385 μm. It acquires data at three spatial resolutions—250 m, 500 m and 1000 m with a temporal resolution of 1–2 days. Until now, MODIS and FY-3 series data are the most widely used satellite data sources for meteorological, agricultural and environmental monitoring at regional, national or continental scales [34]. Previous studies have shown that variables from MODIS land products have the spectral sensitivity to provide consistent spatial and temporal comparisons of global vegetation conditions [35]. The eight-day MOD09A1 image composite at 500 m resolution from MODIS land product for 2016 was used in this study. MOD09A1 images of tile h27v06 and h26v06 were downloaded from the EARTHDATA platform (https://search.earthdata.nasa.gov/search accessed on 2 October 2022) of the National Aeronautics and Space Administration (NASA) of the U.S. The MODIS Reprojection Tool (MRT) was used to mosaic and reproject the images.

MOD09A1 contains seven spectral bands (b1–b7, Table 1) recording surface spectral reflectance at ground level. Vegetation indices proven to correlate with vegetation characteristics and other variables were calculated using these single spectral bands (Table 2) and were then used to develop models for forest AGB mapping. Seven vegetation greenness indices and three vegetation water indices were selected and used in this study. Principal component analysis (PCA) was also performed using all the single bands to transform the multi-spectral correlated bands into a smaller set of uncorrelated image bands. While retaining as much original spectral information as possible, the first three transformed images (PC1, PC2 and PC3) were selected for the screening of predictor variables and further forest AGB modeling as they contained more than 95% of the information from the original bands. These spectral bands, vegetation indices and transformed images were used as explanatory variables for forest AGB estimation.

Table 1. The spectral characteristics of MOD09A1 bands.

Band#	Name	Spectral Range (nm)	Center Wavelength (nm)	Bandwidth (nm)
1	Red	620–670	645	50
2	Near Infrared (NIR)	841–876	859	35
3	Blue	459–479	469	20
4	Green	545–565	555	20
5	Shortwave infrared ($SWIR_{1240}$)	1230–1250	1240	20
6	Shortwave infrared ($SWIR_{1640}$)	1628–1652	1640	24
7	Shortwave infrared ($SWIR_{2130}$)	2105–2155	2130	50

Table 2. The vegetation indices derived from MOD09A1 and FY-3C VIRR imagery.

Index		Formula	MOD09A1	FY-3C VIRR	Reference
Vegetation greenness indices	NDVI	(NIR − RED)/(NIR + RED)	√	√	Rouse et al. [36]
	EVI	2.5(NIR − RED)/[(NIR + 6RED − 7.5BLUE) + 1]	√	√	Huete et al. [37]
	RVI	NIR/RED	√	√	Jordan [38]
	ARVI	[NIR − (2 RED − BLUE)]/[(NIR) + (2RED − BLUE)]	√	√	Kaufman and Tanre [39]
	SAVI	(1 + 0.5)(NIR − RED)/(NIR + RED + 0.5)	√	√	Huete [40]
	MSAVI	$[2NIR + 1 - \sqrt{(2NIR + 1)^2} - 8(NIR - RED)]/2$	√	√	Qi et al. [41]
	VARI	(GREEN − RED)/(GREEN + RED − BLUE)	√	√	Gitelson et al. [42]
Vegetation water indices	NDIIb6	$(NIR - SWIR_{1640})/(NIR + SWIR_{1640})$	√	√ *	Hunt and Rock [43]
	NDIIb7	$(NIR - SWIR_{2130})/(NIR + SWIR_{2130})$	√	NA	Hunt and Rock [43]
	NDMI	$(NIR - SWIR_{1240})/(NIR + SWIR_{1240})$	√	NA	Gao [44], Wilson [45]
	NDWI	(GREEN − NIR)/(GREEN + NIR)	√	√	Mcfeeters [46]

NDVI = normalized difference of vegetation index; EVI = enhanced vegetation index; RVI = ratio vegetation index; ARVI = atmospherically resistant vegetation index; SAVI = soil adjusted vegetation index; MSAVI = modified soil adjusted vegetation index; VARI = visible atmospherically resistant index; NDIIb6 = normalized difference of infrared index—band6; NDIIb7 = normalized difference of infrared index—band7; NDMI = normalized difference moisture index; NDWI = normalized difference of water index. √ represents available, and NA represents unavailable. * $SWIR_{1640}$ was adapted to $SWIR_{1595}$.

2.2.2. FY-3C VIRR Data and Spectral Variables

The Fengyun-3 (FY-3) series of satellites is the second generation of polar-orbit, sun-synchronous meteorological satellites of China, which have been designed for all weather, multi-spectral and three-dimensional observation of global atmospheric and geophysical elements [47]. FY-3 satellite data has been used in numerical weather prediction [48,49], climate monitoring [50] and monitoring of natural disasters. The Fengyun-3C (FY-3C) satellite was launched in September 2013. FY-3C VIRR data has 10 channels with a wavelength range of 0.43–12.50 μm, providing visible and infrared spectra. The swath width is 2,800 km, and the temporal resolution is 1 day. The 10-day FY-3C VIRR image composite with 1 km spatial resolution from 2016 was provided by the National Satellite Meteorological Center (NSMC) of the China Meteorological Administration (CMA) after radiometric calibration, atmospheric correction and geometric correction.

Similar to the development of predictor variables derived from MOD09A1 imagery, the calculation of vegetation indices and PCA were conducted using FY-3C VIRR images. The first three transformed images (PC1, PC2 and PC3) from PCA, single spectral bands and vegetation indices from FY-3C VIRR images were used to screen predictor variables. The set of spectral variables of the FY-3C VIRR imagery was slightly different from that of the MOD09A1 imagery due to a different wavelength range and available individual bands (Tables 1–3).

Table 3. The spectral characteristics of FY-3C VIRR bands used in this study.

Band#	Name	Spectral Range (nm)	Center Wavelength (nm)	Bandwidth (nm)
1	Red	580–680	630	100
2	Near Infrared (NIR)	840–890	865	50
3	Shortwave infrared ($SWIR_{1595}$)	1550–1640	1595	90
4	Blue	430–480	455	50
5	Cyan	480–530	505	50
6	Green	530–580	555	50
7	Shortwave infrared ($SWIR_{1360}$)	1325–1395	1360	70

2.2.3. Forest Inventory Data and Forest AGB Data

The forest inventory data used in this study are from the fourth Chinese National Forest Resource Inventory (NFRI) for forest management and planning, which was completed in Yunnan in 2016. This spatially explicit dataset is composed of forest stand polygons delineated on the basis of aerial images or satellite images. A forest stand is a contiguous area that contains a community of trees that are relatively homogeneous or have a common set of characteristics [6]. The spatial dataset of NFRI contains forest stand information on dominant tree species, age classes, average height, average breast–height diameters, site condition and stand volume, which was collected through field sampling following the technical protocols of NFRI. The AGB of forest stands was calculated using the biomass–volume conversion relationship [51] and used as observation values for remote sensing-based modeling for forest AGB. Forest AGB values from this dataset range from 1.09 ton/hectares (t/ha) to 631.96 t/ha with 80% falling between 12.51 t/ha and 123.30 t/ha; the mean forest AGB is 58.91 t/ha.

2.3. Sampling for Reference Data of Forest AGB

Forest stand polygons of the NFRI dataset were stratified into five classes based on AGB value using the natural breaks system. This minimizes the variation within each class and optimizes the arrangement of the sets of AGB values. AGB samples for the reference dataset were selected in proportion to the area of each class so that all the AGB value ranges were covered. In each class, sample point locations were generated randomly with an assigned minimum distance of 5 km to reduce spatial autocorrelation between samples and to make sure that these samples cover a variety of forest types in each forest zone (Figure A1).

After the sample points were generated, grids spatially aligning with pixels of the MOD09A1 or FY-3C VIRR imagery with the corresponding sample points were used for further screening of appropriate samples. The grids with more than 50% of the pixel area covered by forest stands were selected for AGB reference data to create the AGB reference dataset. Selecting AGB samples was performed with ArcGIS 10.3. The same AGB reference dataset was used for forest AGB modeling with the MOD09A1 and FY-3C VIRR imagery, respectively, so as to compare the performances of combinations of different imagery and methods. A total of 475 grids were selected as AGB references for remote sensing-based modeling, 75% of which was used for training and the remaining 25% for validation.

2.4. AGB Approaches for Estimating Forest AGB Using MODIS and FY-3C VIRR Data

One parametric approach, MLR, and two nonparametric (i.e., machine learning) methods, KNN and RF, were selected and employed to predict AGB density in this study. Parametric methods include linear regression models which have been frequently used in remote sensing-based forest AGB estimation [26,52–55]. The regression models were constructed based on the assumption that the biomass variable is linearly correlated with spectral responses and that limited correlations exist between independent variables [54]. As it is well known that variables of remote sensing are highly correlated with each other, the variance inflation factor (VIF) was used to test the multicollinearity of predictor variables of the linear regression model. Using all the predictor variables could lead to a decrease in the accuracy of the linear model coefficients; therefore, we selected the top 10 influential variables determined by the variable importance plot of RF to establish the MLR model. The KNN approach estimates dependent variables as a weighted mean of K spectrally nearest (most similar) neighbors by inverse distance weighting. No functional relationships between variables need to be formulated for this approach. One advantage of the KNN approach is that it avoids the problem of unbalanced samples [24]. RF is a tree-based assembling learning algorithm. It selects a random number of samples from the training dataset chosen by the analyst and develops decision trees based on the most important variables [56,57]. The RF algorithm has been widely used in forest AGB estimation with remote sensing data and proved to have good performance [58–60]. The parameters ntree and mtry are the two key factors affecting accuracies in RF models. They define the number of decision trees and the number of variables tried at each split of decision trees in RF models. Errors decrease and become stable with the increasing number of regression trees.

Both KNN and RF algorithms can handle nonlinear relationships between independent and dependent variables and have become increasingly popular in forest biomass studies because of the accuracy of their biomass prediction [56,61–64]. In this study, the ntree was set to 300 trees after testing for the RF models from both MOD09A1 and FY-3C VIRR imageries, and 5 and 6 were selected for the mtry numbers for the optimized models based on the variables from the two imageries, respectively. These predictive models were constructed from a training dataset (n = 351) consisting of grid-based AGB density from the NFRI dataset paired with explanatory spectral variables derived from the MOD09A1 and FY-3C VIRR imagery, respectively. We extracted raster pixel values of spectral variables from MOD09A1 and FY-3C VIRR imagery, respectively, corresponding to the selected grids for forest AGB reference.

The three models were performed using R Studio (R Version 4.2.0). The MLR model was established using the lm() and step() functions. The "Caret" package and "randomForest" package were used to generate KNN and RF models, respectively. We used the same set of spectral variables from the same remote sensing data source for MLR, KNN and RF modeling, respectively. With the varImpPlot() function of RF, the importance of each candidate variable (single spectral bands, vegetation indices and transformed images PC1, PC2 and PC3) in predicting forest AGB was assessed by computing the increase in node purity (IncNodePurity), where higher values of IncNodePurity indicate greater importance. The 10 most important predictor variables were selected and then applied to the models. The flowchart of the methodology is shown in Figure 2.

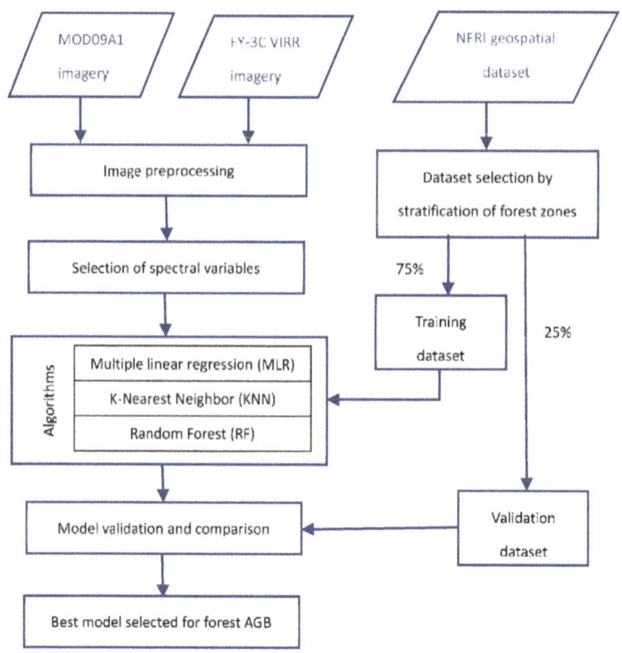

Figure 2. The flowchart of methodology.

2.5. Accuracy Assessment

The accuracies of forest AGB values predicted from the models combined with the two remote sensing data sources were evaluated, respectively, by the coefficient of determination (R^2), root mean square error (RMSE) and mean absolute error (MAE) using the validation dataset (n = 124). R^2 is a statistical calculation that indicates the degree of interrelation and dependence between two variables. RMSE measures the average distance between the predicted values from the model and the actual values. The lower the RMSE, the better a given model fits a dataset. These metrics were calculated using the following equations:

$$R^2 = 1 - \frac{\sum_{i=1}^{n}(y_i - \hat{y}_i)^2}{\sum_{i=1}^{n}(y_i - \overline{y}_i)^2}, \quad (1)$$

$$RMSE = \sqrt{\frac{1}{n}\sum_{i=1}^{n}(y_i - \hat{y}_i)^2}, \quad (2)$$

$$MAE = \frac{1}{n}\sum_{i=1}^{n}|y_i - \hat{y}_i|. \quad (3)$$

In these formulae, $\hat{y}_i$ ($i \in [1,n]$) is the predicted biomass on the ith grid, y_i is the observed biomass and $\overline{y}$ is the mean value of the observed biomass.

The model providing the best accuracy was applied to the whole study area to map the spatial distribution of forest AGB.

3. Results

3.1. Selection of Spectral Variables

The importance of spectral variables derived from MOD09A1 and FY-3C VIRR images was evaluated by the decision tree modeling of RF (Figure 3). The top 10 important variables for MOD09A1 include four indices related to the water content of vegetation canopy (NDIIb6, NDIIb7, NDWI and NDMI), three vegetation greenness indices (VARI, RVI and ARVI) and three transformed imageries (PC1, PC2 and PC3). The top 10 important variables for FY-3C VIRR include one vegetation water index (NDIIb6), two transformed imageries (PC1 and PC2), five individual spectral bands (blue, shortwave infrared (SWIR) – $SWIR_{1595}$, $SWIR_{1360}$, cyan, green) and two vegetation greenness indices (ARVI and VARI). Interestingly, NDIIb6 reflecting the moisture level of the forest canopy based on two kinds of imagery is most sensitive to forest AGB values from the NFRI dataset. Some vegetation greenness indices have a greater contribution in AGB estimation based on MOD09A1 than on FY-3C VIRR. In contrast, individual spectral bands of FY-3C VIRR play a more important role than vegetation greenness indices. Only the top 10 important variables from MOD09A1 and FY-3C VIRR were used for further AGB estimation based on the three approaches.

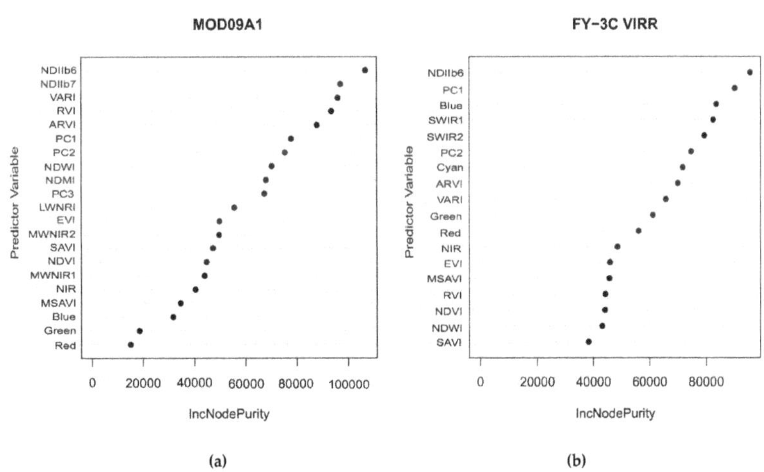

Figure 3. The variable importance plots for MOD09A1 images (**a**) and FY-3C VIRR Images (**b**).

3.2. RF Approach Outperforms KNN and MLR Approach

RF performed better than KNN and MLR in estimating forest AGB using both MOD09A1 and FY-3C VIRR imagery (Table 4). All the models were significant at a p-value < 0.001. The same set of referenced data of forest AGB from the NFRI dataset was input into the models, making it possible to compare the performances of the two satellite image products in estimating forest AGB and to help understand factors that influence accuracies. The performance ranking of the three models was the same for MOD09A1 and FY-3C VIRR imagery.

Table 4. Statistics for estimation of forest AGB using three approaches with the validation dataset (n = 124).

Model	R2		RMSE (t/ha)		MAE (t/ha)	
	MODIS	FY	MODIS	FY	MODIS	FY
MLR	0.32	0.29	49.76	51.32	43.28	46.87
KNN	0.65	0.58	36.82	40.52	33.61	37.13
RF	0.84	0.81	23.18	23.43	21.94	17.69

As a result, five variables from MOD09A1 were selected to create an MLR model after a stepwise regression, with R^2 of 0.32:

$$AGB_{MOD09A1} = 52.394 + 97.979 * NDIIb6 - 242.903 * VARI + 8.613 * RVI + 191.119 * ARVI + 230.540 * NDWI. \quad (4)$$

In contrast, the MLR model based on variables from FY-3C VIRR has a slightly lower R^2 of 0.29. Three variables were involved after a stepwise regression:

$$AGB_{FY\text{-}3C} = -28.01 + 80.95 * NDIIb6 + 312.29 * Green + 58.81 * ARVI. \quad (5)$$

After testing, the optimal KNN model based on MOD09A1 variables was established with the parameter K = 9 and an R^2 value of 0.65, and the best K value of the KNN model for FY-3C variables was 7, with an R^2 value of 0.58. The RF models for MOD09A1 and FY-3C VIRR imageries achieved R^2 values of 0.84 and 0.81, respectively, and RMSE of 23.18 and 23.43, respectively (Table 4).

The MLR models performed worst on both satellite imageries when compared to the KNN and RF models, as indicated by the comparatively lower R^2 values. This suggests that the models have low degrees of fitting to the observed values, probably due to a weak linear or nonlinear relationship between forest AGB and the remote sensing variables. All of the predictor variables from MOD09A1 and FY-3C VIRR have VIF values larger than 10. This indicates that the variables used for MLR models have serious multicollinearity, which also leads to poor performance of MLR modeling. RF modeling performed better for both imageries than the other two approaches. It showed a strong ability to avoid overfitting.

It is noticeable that variables from MOD09A1 performed marginally better than using those from FY-3C VIRR in RF models on both the training dataset and validation dataset (Figures 4 and 5). This indicates that forest AGB could achieve comparable accuracy using either MOD09A1 or FY-3C VIRR images. For both kinds of imagery, RF models predicted lower forest AGB with higher accuracy compared to predicting higher forest AGB with lower accuracy (Figures 4 and 5), which may be due to the saturation of spectral signals at higher biomass values. Further comparison was conducted to reveal the difference between the two estimations of forest AGB from MOD09A1 and FY-3C VIRR imageries, respectively, by RF models. Moreover, in this study, the RF model with the input of MOD09A1 imagery was selected for forest AGB mapping and to help understand the spatial distribution pattern of forest AGB in Yunnan Province.

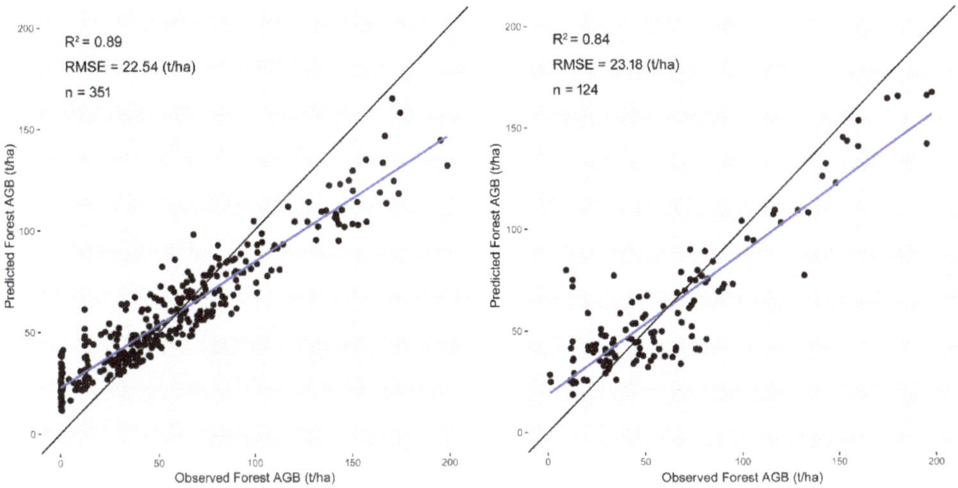

(**a**) Training (**b**) Validation

Figure 4. Observed versus predicted forest AGB applying random forest (RF) model and MOD09A1 imagery for (**a**) the training dataset and (**b**) the validation dataset.

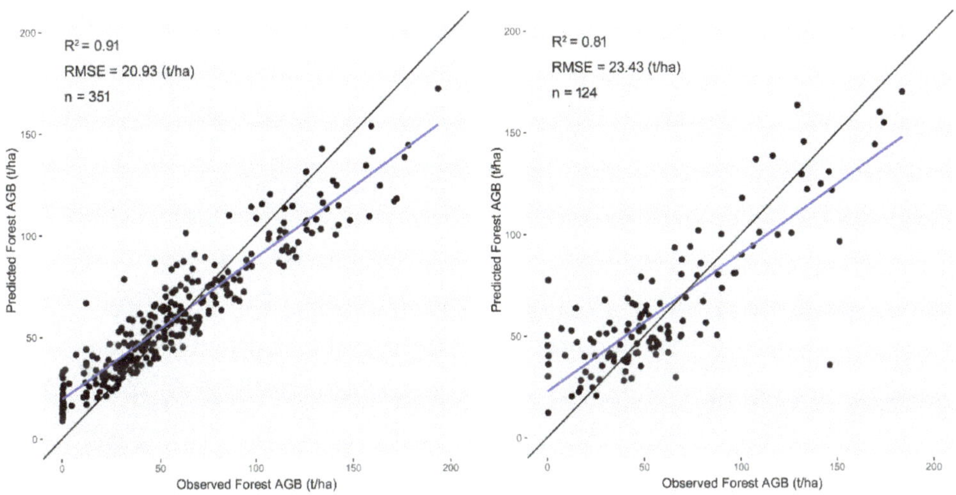

(**a**) Training (**b**) Validation

Figure 5. Observed versus predicted forest AGB applying random forest (RF) model and FY-3C VIRR imagery for (**a**) the training dataset and (**b**) the validation dataset.

3.3. Comparison of Forest AGB Estimation by RF Models Based on the Two Imageries

The total amounts of forest AGB estimated using MOD09A1 and FY-3C VIRR imageries by RF methods were greater than that of the NFRI dataset (Figure 6), by 4.13% and 6.25%, respectively. The total mean of forest AGB from FY-3C imagery (62.10 t/ha) was higher than the means from the NFRI dataset and MOD09A1 imagery (58.91 t/ha and 58.05 t/ha, respectively). This is probably attributable to the larger estimation in both the mean value of forest AGB (47.77 t/ha) and the proportion of the largest share of forest AGB range (30–60 t/ha) by FY-3C VIRR imagery. The mean AGB, at the 0–30 t/ha range, obtained from MOD09A1 and FY-3C VIRR was similar (24.52 t/ha and 23.77 t/ha, respectively) and was the same for the ranges of 30–60 t/ha and 60–90 t/ha (47.80 t/ha and 47.77 t/ha, 84.25 t/ha and 84.58 t/ha, respectively). The total amounts of forest AGB at the two ranges below 60 t/ha from the two imageries were almost identical. The mean values of forest AGB estimated using the two imageries were higher than that of the NRFI dataset at the three ranges lower than 90 t/ha. The underestimation in forest AGB modeled by remote sensing data at the AGB ranges higher than 90 t/ha was due to signal saturation in optical remote sensing data at high forest AGB. Higher forest AGB was associated with a greater discrepancy between estimations derived from remote sensing data and observation values.

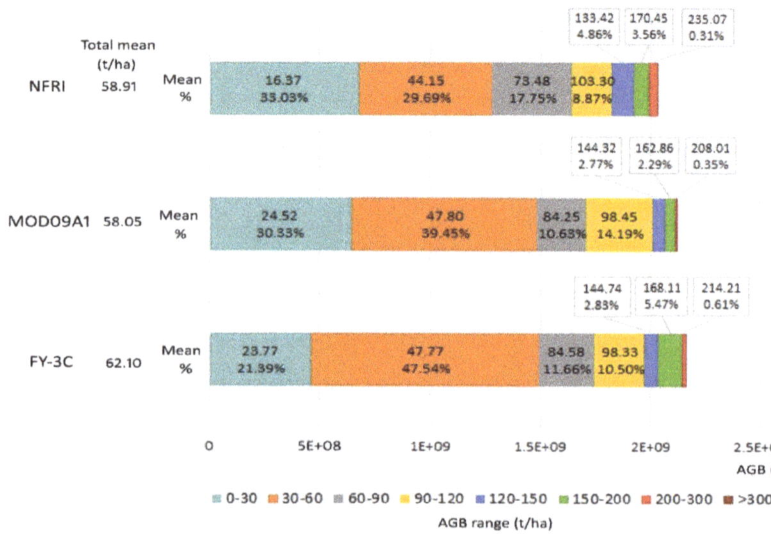

Figure 6. Distribution of forest AGB in the Chinese National Forest Resource Inventory (NFRI) dataset and estimated forest AGB based on MOD09A1 and FY-3C VIRR imageries by RF models.

3.4. Mapping of Forest AGB Distribution by Forest Zones and Dominant Tree Species

Figure 7 shows that the forest AGB distribution in Yunnan mapped by MOD09A1 and FY-3C VIRR imageries is similar. AGB was found to be high mainly at the northwestern, western, southwestern and southern periphery of the province, in contrast to low AGB, which was mainly located in the central and eastern parts of the province. The color scheme of Figure 7 shows that the prediction of forest AGB by FY-3C VIRR imagery had larger variation. Because the RF algorithm based on MOD09A1 images has achieved the highest accuracy in forest AGB mapping, we employed this combination to estimate forest AGB. Then, we produced a map of its spatial distribution in Yunnan Province and used it for further statistical analysis to show the spatial distribution pattern of forest AGB.

The total forest AGB in Yunnan Province is 2123.22 Mt, with a mean value of 58.05 t/ha. The cold-temperate coniferous forest zone in the northwest has the highest mean forest AGB density, 76.08 t/ha, followed by the tropical broadleaved forest zone at 75.40 t/ha. These two forest zones contribute 8.39% and 18.08% of the total forest AGB of Yunnan Province, respectively. The lowest mean forest AGB density (43.52 t/ha) occurs in the warm evergreen broadleaved forest zone. The warm evergreen broadleaved and coniferous forest zone has a relatively low AGB density (49.45 t/ha), which accounts for the largest proportion (39.93%) of the total forest AGB in the study area due to the large area that it covers. The distribution pattern of high forest AGB is consistent with the spatial distribution of the key areas of forest conservation and biodiversity conservation in recent decades, e.g., national and provincial protected areas and ecological forest protection.

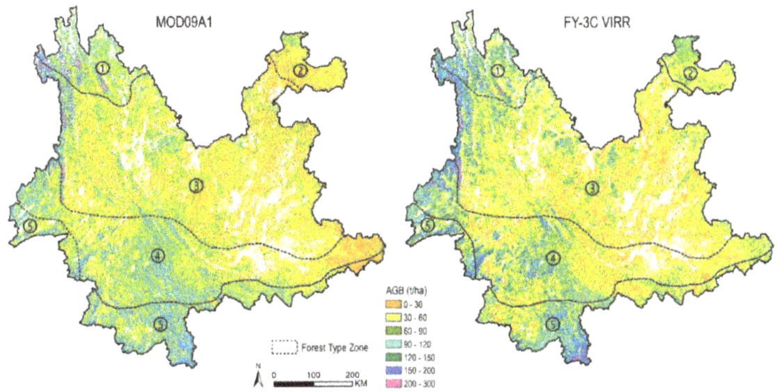

(a) (b)

Figure 7. Forest AGB maps derived from (**a**) MOD09A1 and (**b**) FY-3C imageries based on RF models. Forest type zones: ① cold-temperate coniferous forest, ② warm evergreen broadleaved forest, ③ warm evergreen broadleaved and coniferous forest, ④ warm–hot broadleaved and coniferous forest and ⑤ tropical broadleaved forest.

Abies fabri, mainly located in northwest Yunnan, had the highest AGB density among all the dominant tree species at 115.93 t/ha, followed by *Quercus* spp. at 88.52 t/ha, other broadleaved species (a group of broadleaved tree species except for *Quercus* spp., *Alnus cremastogyne* and *Betula* spp.) forest at 85.61 t/ha and *Pinus kesiya* var. *langbianensis* at 81.23 t/ha. *Pinus densata*, *Pinus yunnanensis* and *Pinus armandii* had relatively low AGB density, with values of 61.35 t/ha, 55.42 t/ha and 51.63 t/ha, respectively. *Quercus* spp., *Pinus yunnanensis*, other broadleaved forests and *Pinus kesiya* var. *langbianensis* were the top four important forests, accounting for 14.50%, 13.20%, 12.35% and 5.83% of the total forest AGB in Yunnan Province.

4. Discussion

In this study, we mapped the forest AGB in Yunnan Province, an important region for carbon storage, using three approaches with MODIS and FY-3C imageries. The results indicate that RF performed best out of all the approaches for forest AGB modeling. The thematic map by MOD09A1 images achieved marginally better accuracy than the map by FY-3C VIRR images. Although MODIS imagery has been used for national forest AGB mapping in China and other regions, this study was the first to adopt FY series imagery for this purpose and to conduct a comparison of these two imageries. Our results indicate that FY-3C VIRR imagery achieved acceptable accuracy, relative to MOD09A1 imagery, in mapping forest AGB at the regional scale.

4.1. Contribution of Spectral Index Variables

We found that although MOD09A1 and FY-3C VIRR imageries have different spatial resolutions, some common features can be detected in the predictor variables. Among the predictor variables derived, NDIIb6 outperformed all the other variables for both imageries, respectively, and both ARVI and VARI also ranked in the top 10 of important variables. This could be one of the reasons for the consistency of predictions of forest AGB between MOD09A1 and FY-3C VIRR imageries ($R^2 > 0.80$).

Indices of vegetation water content often correlate with vegetation health and vigor and indicate vegetation biomass [65]. Our study demonstrated a close relationship between these indices and forest AGB. NDIIb6 and NDIIb7 derived from near-infrared (NIR) and SWIR centering at 1640 nm and 2130 nm bands, respectively, showed great potential for vegetation water content estimation in previous studies [65,66], as the water content in vegetation was very important for photosynthesis leading to biomass formation. These

NDII indices were named the normalized difference water index (NDWI), NDIIb6 as $NDWI_{1640}$ and NDIIb7 as $NDWI_{2130}$. The fact that NDII was less sensitive to the bandwidth of different sensors, as proven by Chen et al. [67], was also confirmed by this study. The vegetation–water content-related index NDMI based on NIR (860 nm) and SWIR (1240 nm) is also one of the top 10 important predictor variables from MOD09A1 imagery. Gao (1996) demonstrated the sensitivity of this index to changes in the water content of vegetation canopies and showed that it is less sensitive to atmospheric effects than NDVI [44]. A positive correlation between vegetation water content and vegetation biomass was also revealed by Xing et al. [68] and Momen et al. [69]. ARVI and VARI are the two common and important indices related to vegetation greenness from MOD09A1 and FY-3C VIRR imageries, both enhanced by the presence of the blue channel to minimize the atmospheric effects.

4.2. The Ability of MOD09A1 and FY-3C VIRR to Map Forest AGB

The ability of remote sensing data to map forest AGB depends on the sensitivity of the predictor variables selected for mapping forest AGB and on the complexity of the vegetation structure. Our results showed acceptable accuracies of forest AGB estimation by both imageries, although the performance of MOD09A1 was marginally better than that of FY-3C VIRR by RF modeling. This indicates that the suitability of the two imageries for forest AGB prediction is comparable. The comparable applicability of MOD09A1 and FY-3C VIRR imageries for forest AGB at the regional scale can be attributable to their similarity in spectral architectures, temporal resolution and geographical coverage. MOD09A1 and FY-3C VIRR have spectral compositions for visible light and near-infrared, as well as for the overlapped range of infrared, which makes it possible to calculate vegetation indices for modeling. Both imageries have short revisit cycles of 1–2 days and large geographical coverage extents of over 2000 km^2. These features enable efficient image processing for monitoring the dynamics of forest AGB at regional and global scales.

The difference in the performance of AGB estimation can be attributed to the difference in the spectrum ranges and spatial resolution of the two imageries. The spectrum range of MOD09A1 used for calculating variables to model forest AGB is slightly wider than that of FY-3C VIRR, particularly in SWIR. MOD09A1 has two indices related to vegetation water content (NDIIb7 and NDMI), more than FY-3C VIRR. These two indices are absent for FY-3C due to the lack of two specific SWIR bands for index calculation. Thus, to some extent, MOD09A1 provides more spectral information in forest AGB estimation.

The higher spatial resolution of MOD09A1 (500 m) could account for the higher accuracy of forest AGB mapping compared to FY-3C VIRR (1000 m). This finding is consistent with previous findings of forest AGB mapping by finer resolution satellite imagery with higher accuracy [70,71]. Because the average area of forest stand polygons in the NFRI dataset of Yunnan Province is approximately 9 hectares, which is smaller than single pixels of MOD09A1 and FY-3C VIRR imageries, the spectral information from individual bands and vegetation indices used for modeling was "aggregated" and "averaged". Higher spatial resolutions are expected to achieve higher accuracies [71].

4.3. Performance of Parametric and Nonparametric Approaches

In our study, the adopted parametric approach, the MLR model, was found to have performed poorly in forest AGB prediction, which was also confirmed by other studies [5]. Linear regression is used for modeling when remote sensing variables have a strong linear relationship with biomass and a weak relationship with selected remote sensing variables themselves. However, biomass is often nonlinearly related to remote sensing variables [24,54]; thus, MLR models may lead to low accuracy in predictions.

Our results found that nonparametric approaches achieved higher accuracy compared to the parametric approach in mapping forest AGB, which is consistent with previous studies [7,24,56,72]. Lu et al. (2016) suggested that nonparametric approaches should be explored if large representative field datasets exist for calibration [24]. In our study, we selected forest AGB reference data by stratified forest-type zones as an effective way to ensure the representativeness of reference datasets.

5. Conclusions

In this study, we compared forest AGB estimation for Yunnan Province, southwest of China, with MOD09A1 and FY-3C VIRR imagery by applying one parametric approach, MLR, and two nonparametric approaches, KNN and RF. Reference data of forest AGB from the NFRI dataset, individual spectral bands and the derived vegetation indices were used to establish the models. The results indicated that (1) RF models outperformed the MLR and KNN models for both imageries using the same sampled forest AGB reference dataset from NFRI data. (2) Among all the remote sensing variables, NDIIb6 related to the moisture of vegetation canopy was the most sensitive to forest AGB for both imageries. Vegetation greenness indices contributed more to AGB prediction based on MOD09A1 than those based on FY-3C VIRR, while individual spectral bands of FY-3C played a more important role than vegetation greenness indices. (3) FY-3C VIRR imagery had high potential to be an alternative data source substituting the MODIS data for forest AGB mapping at regional scales.

This study examined overall forest AGB estimation using MODIS and FY imageries. However, spectral saturation at high forest AGB is a challenge for AGB mapping with optical satellite imagery; further studies involving the sensitivity of spectral variables of these imageries to forest AGB need to be conducted, and as FY imagery is a new option as a data source for AGB mapping, solutions need to be explored for improving mapping accuracy. Moreover, the real-time forest AGB monitoring or quantifying AGB change at regional scales could be an applicative aspect for FY series satellite imagery.

Author Contributions: H.C. and Z.Q. designed the study; H.C. conducted the data analysis, modeling forest AGB and wrote the draft of the manuscript; H.C., Z.Q. and D.-L.Z. improved the manuscript; G.O. obtained NFRI dataset and provided support to the study; X.L. and G.Z. conducted statistical analysis; J.F. and C.Z. collected and preprocessed the FY-3C VIRR data. H.X. supervised and coordinated the research project. All authors have read and agreed to the published version of the manuscript.

Funding: This study was financially supported by the National Natural Science Foundation of China (grant number 31770677).

Data Availability Statement: Data from this study will be available upon request to the authors.

Acknowledgments: We sincerely thank Dietrich Schmidt-Vogt for checking the English grammar of the manuscript.

Conflicts of Interest: The authors declare no conflict of interest.

Appendix A

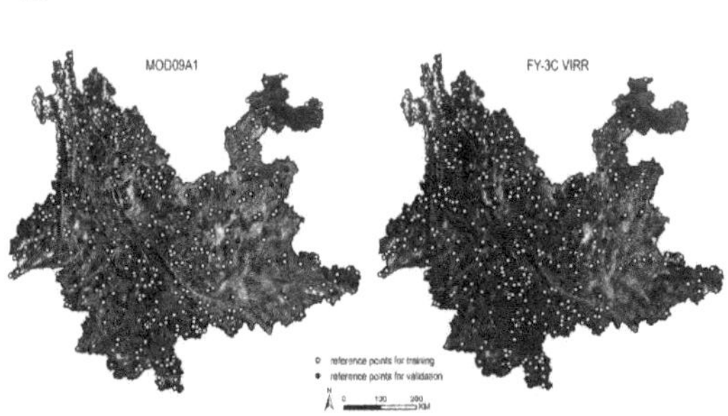

(a) (b)

Figure A1. The true-color MOD09A1 (**a**) and FY-3C VIRR (**b**) imageries of Yunnan Province, with the dots showing the sampling points to estimate forest AGB.

References

1. Yin, G.; Zhang, Y.; Sun, Y.; Wang, T.; Zeng, Z.; Piao, S. MODIS Based Estimation of Forest Aboveground Biomass in China. *PLoS ONE* **2015**, *10*, e0130143. [CrossRef] [PubMed]
2. Goodale, C.L.; Apps, M.J.; Birdsey, R.A.; Field, C.B.; Heath, L.S.; Houghton, R.A.; Jenkins, J.C.; Kohlmaier, G.H.; Kurz, W.; Liu, S.R.; et al. Forest carbon sinks in the Northern Hemisphere. *Ecol. Appl.* **2002**, *12*, 891–899. [CrossRef]
3. Houghton, R.A. Aboveground forest biomass and the global carbon balance. *Glob. Chang. Biol.* **2005**, *11*, 945–958. [CrossRef]
4. Pan, Y.D.; Birdsey, R.A.; Fang, J.Y.; Houghton, R.; Kauppi, P.E.; Kurz, W.A.; Phillips, O.L.; Shvidenko, A.; Lewis, S.L.; Canadell, J.G.; et al. A Large and Persistent Carbon Sink in the World's Forests. *Science* **2011**, *333*, 988–993. [CrossRef] [PubMed]
5. Ou, G.L.; Lv, Y.Y.; Xu, H.; Wang, G.X. Improving Forest Aboveground Biomass Estimation of Pinus densata Forest in Yunnan of Southwest China by Spatial Regression using Landsat 8 Images. *Remote Sens.* **2019**, *11*, 2750. [CrossRef]
6. Chi, H.; Sun, G.Q.; Huang, J.L.; Guo, Z.F.; Ni, W.J.; Fu, A.M. National Forest Aboveground Biomass Mapping from ICESat/GLAS Data and MODIS Imagery in China. *Remote Sens.* **2015**, *7*, 5534–5564. [CrossRef]
7. Lu, D.S. The potential and challenge of remote sensing-based biomass estimation. *Int. J. Remote Sens.* **2006**, *27*, 1297–1328. [CrossRef]
8. Dong, J.R.; Kaufmann, R.K.; Myneni, R.B.; Tucker, C.J.; Kauppi, P.E.; Liski, J.; Buermann, W.; Alexeyev, V.; Hughes, M.K. Remote sensing estimates of boreal and temperate forest woody biomass: Carbon pools, sources, and sinks. *Remote Sens. Environ.* **2003**, *84*, 393–410. [CrossRef]
9. Calvao, T.; Palmeirim, J.M. Mapping Mediterranean scrub with satellite imagery: Biomass estimation and spectral behaviour. *Int. J. Remote Sens.* **2004**, *25*, 3113–3126. [CrossRef]
10. Zhang, Y.; Liang, S.; Yang, L. A Review of Regional and Global Gridded Forest Biomass Datasets. *Remote Sens.* **2019**, *11*, 2744. [CrossRef]
11. Li, X.; Zhang, M.; Long, J.; Lin, H. A Novel Method for Estimating Spatial Distribution of Forest Above-Ground Biomass Based on Multispectral Fusion Data and Ensemble Learning Algorithm. *Remote Sens.* **2021**, *13*, 3910. [CrossRef]
12. Cooper, S.; Okujeni, A.; Pflugmacher, D.; van der Linden, S.; Hostert, P. Combining simulated hyperspectral EnMAP and Landsat time series for forest aboveground biomass mapping. *Int. J. Appl. Earth Obs. Geoinf.* **2021**, *98*, 102307. [CrossRef]
13. Kumar, L.; Mutanga, O. Remote Sensing of Above-Ground Biomass. *Remote Sens.* **2017**, *9*, 935. [CrossRef]
14. Foody, G.M.; Boyd, D.S.; Cutler, M.E.J. Predictive relations of tropical forest biomass from Landsat TM data and their transferability between regions. *Remote Sens. Environ.* **2003**, *85*, 463–474. [CrossRef]
15. Steininger, M.K. Satellite estimation of tropical secondary forest above-ground biomass: Data from Brazil and Bolivia. *Int. J. Remote Sens.* **2000**, *21*, 1139–1157. [CrossRef]
16. Baccini, A.; Friedl, M.A.; Woodcock, C.E.; Warbington, R. Forest biomass estimation over regional scales using multisource data. *Geophys. Res. Lett.* **2004**, *31*, L10501. [CrossRef]
17. Lu, D. Aboveground biomass estimation using Landsat TM data in the Brazilian Amazon. *Int. J. Remote Sens.* **2005**, *26*, 2509–2525. [CrossRef]
18. Rahman, M.M.; Csaplovics, E.; Koch, B. An efficient regression strategy for extracting forest biomass information from satellite sensor data. *Int. J. Remote Sens.* **2005**, *26*, 1511–1519. [CrossRef]
19. Zhou, J.J.; Zhao, Z.; Zhao, Q.X.; Zhao, J.; Wang, H.Z. Quantification of aboveground forest biomass using Quickbird imagery, topographic variables, and field data. *J. Appl. Remote Sens.* **2013**, *7*, 073484. [CrossRef]
20. Gomez, J.A.; Zarco-Tejada, P.J.; Garcia-Morillo, J.; Gama, J.; Soriano, M.A. Determining Biophysical Parameters for Olive Trees Using CASI-Airborne and Quickbird-Satellite Imagery. *Agron. J.* **2011**, *103*, 644–654. [CrossRef]
21. Gomez, C.; Wulder, M.A.; Montes, F.; Delgado, J.A. Modeling Forest Structural Parameters in the Mediterranean Pines of Central Spain using QuickBird-2 Imagery and Classification and Regression Tree Analysis (CART). *Remote Sens.* **2012**, *4*, 135–159. [CrossRef]
22. Zhu, Y.H.; Liu, K.; Liu, L.; Wang, S.G.; Liu, H.X. Retrieval of Mangrove Aboveground Biomass at the Individual Species Level with WorldView-2 Images. *Remote Sens.* **2015**, *7*, 12192–12214. [CrossRef]
23. Qiu, P.H.; Wang, D.Z.; Zou, X.Q.; Yang, X.; Xie, G.Z.; Xu, S.J.; Zhong, Z.Q. Finer Resolution Estimation and Mapping of Mangrove Biomass Using UAV LiDAR and WorldView-2 Data. *Forests* **2019**, *10*, 871. [CrossRef]
24. Lu, D.S.; Chen, Q.; Wang, G.X.; Liu, L.J.; Li, G.Y.; Moran, E. A survey of remote sensing-based aboveground biomass estimation methods in forest ecosystems. *Int. J. Digit. Earth* **2016**, *9*, 63–105. [CrossRef]
25. Zhang, J.L.; Lu, C.; Xu, H.; Wang, G.X. Estimating aboveground biomass of Pinus densata-dominated forests using Landsat time series and permanent sample plot data. *J. For. Res.* **2019**, *30*, 1689–1706. [CrossRef]
26. Fassnacht, F.E.; Hartig, F.; Latifi, H.; Berger, C.; Hernandez, J.; Corvalan, P.; Koch, B. Importance of sample size, data type and prediction method for remote sensing-based estimations of aboveground forest biomass. *Remote Sens. Environ.* **2014**, *154*, 102–114. [CrossRef]
27. Zhang, Q.L.; He, H.S.; Liang, Y.; Hawbaker, T.J.; Henne, P.D.; Liu, J.X.; Huang, S.L.; Wu, Z.W.; Huang, C. Integrating forest inventory data and MODIS data to map species-level biomass in Chinese boreal forests. *Can. J. For. Res.* **2018**, *48*, 461–479. [CrossRef]
28. Lu, J.; Feng, Z.; Zhu, Y. Estimation of Forest Biomass and Carbon Storage in China Based on Forest Resources Inventory Data. *Forests* **2019**, *10*, 650. [CrossRef]

29. Forestry Department of Yunnan Province. *Report of Forest Resource Survey in Yunnan Province*; Yunnan Science and Technology Press: Kunming, China, 2017.
30. Forestry Department of Yunnan Province. *Forest Resources in Yunnan*; Yunnan Science and Technology Press: Kunming, China, 2018.
31. Chen, F.; Niu, S.; Tong, X.; Zhao, J.; Sun, Y.; He, T. The Impact of Precipitation Regimes on Forest Fires in Yunnan Province, Southwest China. *Sci. World J.* **2014**, 326782. [CrossRef]
32. Editting Committee of Yunnan Forest. *Yunnan Forest*; China Forestry Press: Beijing, China; Yunnan Science and Technology Press: Kunming, China, 1986.
33. Weng, E.S.; Zhou, G.S. Modeling distribution changes of vegetation in China under future climate change. *Env. Model Assess* **2006**, *11*, 45–58. [CrossRef]
34. Yongqian, W.; Dejun, Z.; Liang, S.; Shiqi, Y.; Tang, S.; Yanghua, G.; Qinyu, Y.; Hao, Z. Evaluating FY3C-VIRR reconstructed land surface temperature in cloudy regions. *Eur. J. Remote Sens.* **2021**, *54*, 266–280. [CrossRef]
35. Guo, N.; Wang, X.; Cai, D.; Yang, J. Comparison and evaluation between MODIS vegetation indices in Northwest China. In Proceedings of the IEEE International Geoscience and Remote Sensing Symposium (IGARSS), Barcelona, Spain, 23–27 July 2007; p. 3366.
36. Rouse, J.W.; Haas, R.H.; Schell, J.A.; Deering, D.W. Monitoring vegetation systems in the Great Plains with ERTS. *NASA Spe.* **1974**, *351*, 309–317.
37. Huete, A.; Didan, K.; Miura, T.; Rodriguez, E.P.; Gao, X.; Ferreira, L.G. Overview of the radiometric and biophysical performance of the MODIS vegetation indices. *Remote Sens. Environ.* **2002**, *83*, 195–213. [CrossRef]
38. Jordan, C.F. Derivation of leaf-area index from quality of light on forest floor. *Ecology* **1969**, *50*, 663. [CrossRef]
39. Kaufman, Y.J.; Tanre, D. Atmospherically resistant vegetation index (ARVI) for EOS-MODIS. *IEEE Trans. Geosci. Remote Sens.* **1992**, *30*, 261–270. [CrossRef]
40. Huete, A.R. A soil-adjusted vegetation index (SAVI). *Remote Sens. Environ.* **1988**, *25*, 295–309. [CrossRef]
41. Qi, J.; Chehbouni, A.; Huete, A.R.; Kerr, Y.H.; Sorooshian, S. A modified soil adjusted vegetatiob index. *Remote Sens. Environ.* **1994**, *48*, 119–126. [CrossRef]
42. Gitelson, A.A.; Kaufman, Y.J.; Stark, R.; Rundquist, D. Novel algorithms for remote estimation of vegetation fraction. *Remote Sens. Environ.* **2002**, *80*, 76–87. [CrossRef]
43. Hunt, E.R.; Rock, B.N. Detection of Changes in Leaf Water-Content Using near-Infrared and Middle-Infrared Reflectances. *Remote Sens. Environ.* **1989**, *30*, 43–54.
44. Gao, B.C. NDWI—A normalized difference water index for remote sensing of vegetation liquid water from space. *Remote Sens. Environ.* **1996**, *58*, 257–266. [CrossRef]
45. Wilson, E.H.; Sader, S.A. Detection of forest harvest type using multiple dates of Landsat TM imagery. *Remote Sens. Environ.* **2002**, *80*, 385–396. [CrossRef]
46. McFeeters, S.K. The use of the normalized difference water index (NDWI) in the delineation of open water features. *Int. J. Remote Sens.* **1996**, *17*, 1425–1432. [CrossRef]
47. Bi, Y.; Yang, Z.; Zhang, P.; Sun, Y.; Bai, W.; Du, Q.; Yang, G.; Chen, J.; Liao, M. An introduction to China FY3 radio occultation mission and its measurement simulation. *Adv. Sp. Res.* **2012**, *49*, 1191–1197. [CrossRef]
48. Dong, P.M.; Huang, J.P.; Liu, G.Q.; Zhang, T. Assimilation of FY-3A microwave observations and simulation of brightness temperature under cloudy and rainy condition. *J. Trop. Meteorol.* **2014**, *30*, 302–310.
49. Yang, Y.M.; Du, M.B.; Zhang, J. Experiments of assimilating FY-3A microwave data in forecast of typhoon Morakot. *J. Trop. Meteorol.* **2012**, *28*, 23–30.
50. Wang, W.; Zhang, X.; An, X.; Zhang, Y.; Huang, F.; Wang, Y.; Wang, Y.; Zhang, Z.; Lue, J.; Fu, L.; et al. Analysis for retrieval and validation results of FY-3 Total Ozone Unit (TOU). *China Sci. Bull.* **2010**, *55*, 3037–3043. [CrossRef]
51. Fang, J.Y.; Wang, G.G.; Liu, G.H.; Xu, S.L. Forest biomass of China: An estimate based on the biomass-volume relationship. *Ecol. Appl.* **1998**, *8*, 1084–1091.
52. Mitchard, E.T.A.; Saatchi, S.S.; Lewis, S.L.; Feldpausch, T.R.; Woodhouse, I.H.; Sonke, B.; Rowland, C.; Meir, P. Measuring biomass changes due to woody encroachment and deforestation/degradation in a forest-savanna boundary region of central Africa using multi-temporal L-band radar backscatter. *Remote Sens. Environ.* **2011**, *115*, 2861–2873. [CrossRef]
53. Sun, G.; Ranson, K.J.; Guo, Z.; Zhang, Z.; Montesano, P.; Kimes, D. Forest biomass mapping from lidar and radar synergies. *Remote Sens. Environ.* **2011**, *115*, 2906–2916. [CrossRef]
54. Lu, D.; Chen, Q.; Wang, G.; Moran, E.; Batistella, M.; Zhang, M.; Laurin, G.V.; Saah, D. Aboveground Forest Biomass Estimation with Landsat and LiDAR Data and Uncertainty Analysis of the Estimates. *Int. J. For. Res.* **2012**, *2012*, 436537. [CrossRef]
55. Zhao, P.; Lu, D.; Wang, G.; Liu, L.; Li, D.; Zhu, J.; Yu, S. Forest aboveground biomass estimation in Zhejiang Province using the integration of Landsat TM and ALOS PALSAR data. *Int. J. Appl. Earth Obs. Geoinf.* **2016**, *53*, 1–15. [CrossRef]
56. Lee, H.; Wang, J.; Leblon, B. Using Linear Regression, Random Forests, and Support Vector Machine with Unmanned Aerial Vehicle Multispectral Images to Predict Canopy Nitrogen Weight in Corn. *Remote Sens.* **2020**, *12*, 2071. [CrossRef]
57. Breiman, L. Random Forests. *Mach. Learn.* **2001**, *45*, 5–32. [CrossRef]
58. Ahmad, A.; Gilani, H.; Ahmad, S.R. Forest Aboveground Biomass Estimation and Mapping through High-Resolution Optical Satellite Imagery-A Literature Review. *Forests* **2021**, *12*, 914. [CrossRef]

59. Li, Z.; Bi, S.; Hao, S.; Cui, Y. Aboveground biomass estimation in forests with random forest and Monte Carlo-based uncertainty analysis. *Ecol. Indic.* **2022**, *142*, 109246. [CrossRef]
60. Li, Y.; Li, M.; Li, C.; Liu, Z. Forest aboveground biomass estimation using Landsat 8 and Sentinel-1A data with machine learning algorithms. *Sci. Rep.* **2020**, *10*, 9952. [CrossRef]
61. Esteban, J.; McRoberts, R.E.; Fernandez-Landa, A.; Luis Tome, J.; Naesset, E. Estimating Forest Volume and Biomass and Their Changes Using Random Forests and Remotely Sensed Data. *Remote Sens.* **2019**, *11*, 1944. [CrossRef]
62. Zeng, N.; Ren, X.; He, H.; Zhang, L.; Zhao, D.; Ge, R.; Li, P.; Niu, Z. Estimating grassland aboveground biomass on the Tibetan Plateau using a random forest algorithm. *Ecol. Indic.* **2019**, *102*, 479–487. [CrossRef]
63. Yang, H.; Li, F.; Wang, W.; Yu, K. Estimating Above-Ground Biomass of Potato Using Random Forest and Optimized Hyperspectral Indices. *Remote Sens.* **2021**, *13*, 2339. [CrossRef]
64. McRoberts, R.E.; Naesset, E.; Gobakken, T. Optimizing the k-Nearest Neighbors technique for estimating forest aboveground biomass using airborne laser scanning data. *Remote Sens. Environ.* **2015**, *163*, 13–22. [CrossRef]
65. Zhang, F.; Zhou, G. Estimation of vegetation water content using hyperspectral vegetation indices: A comparison of crop water indicators in response to water stress treatments for summer maize. *BMC Ecol.* **2019**, *19*, 18. [CrossRef] [PubMed]
66. Yi, Y.; Yang, D.; Chen, D.; Huang, J. Retrieving crop physiological parameters and assessing water deficiency using MODIS data during the winter wheat growing period. *Can. J. Remote Sens.* **2007**, *33*, 189–202. [CrossRef]
67. Chen, X.; Wang, S.; Zhang, L.; Jiang, H. Accuracy and Sensitivity of Retrieving Vegetation Leaf Water Content. *Remote Sens. Inf.* **2016**, *31*, 48–57.
68. Xing, M.; He, B.; Li, X. Integration method to estimate above-ground biomass in arid prairie regions using active and passive remote sensing data. *J. Appl. Remote Sens.* **2014**, *8*, 083677. [CrossRef]
69. Momen, M.; Wood, J.D.; Novick, K.A.; Pangle, R.; Pockman, W.T.; McDowell, N.G.; Konings, A.G. Interacting Effects of Leaf Water Potential and Biomass on Vegetation Optical Depth. *J. Geophys. Res. Biogeo.* **2017**, *122*, 3031–3046. [CrossRef]
70. Salajanu, D.; Jacobs, D.M. Accuracy assessment of biomass and forested area classification from modis, landstat-tm satellite imagery and forest inventory plot data. In Proceedings of the ASPRS 2007 Annual Conference, Tampa, FL, USA, 7–11 May 2007.
71. Jha, N.; Tripathi, N.K.; Barbier, N.; Virdis, S.G.P.; Chanthorn, W.; Viennois, G.; Brockelman, W.Y.; Nathalang, A.; Tongsima, S.; Sasaki, N.; et al. The real potential of current passive satellite data to map aboveground biomass in tropical forests. *Remote Sens. Ecol. Conserv.* **2021**, *7*, 504–520. [CrossRef]
72. Tian, X.; Su, Z.; Chen, E.; Li, Z.; van der Tol, C.; Guo, J.; He, Q. Estimation of forest above-ground biomass using multi-parameter remote sensing data over a cold and arid area. *Int. J. Appl. Earth Obs. Geoinf.* **2012**, *14*, 160–168.

remote sensing

MDPI

Article

Research on the Spatiotemporal Evolution of Mangrove Forests in the Hainan Island from 1991 to 2021 Based on SVM and Res-UNet Algorithms

Chang Fu [1,2,3,†], Xiqiang Song [2,3,†], Yu Xie [3], Cai Wang [3], Jianbiao Luo [3], Ying Fang [3], Bing Cao [1] and Zixuan Qiu [1,2,3,*]

1 Hainan Yazhou Bay Seed Laboratory, Sanya Nanfan Research Institute, Hainan University, Sanya 572025, China
2 Key Laboratory of Genetics and Germplasm Innovation of Tropical Special Forest Trees and Ornamental Plants, Ministry of Education, College of Forestry, Hainan University, Haikou 570228, China
3 Intelligent Forestry Key Laboratory of Haikou City, College of Forestry, Hainan University, Haikou 570228, China
* Correspondence: zixuanqiu@hainanu.edu.cn; Tel.: +86-156-0080-4604
† These authors contributed equally to this work.

Citation: Fu, C.; Song, X.; Xie, Y.; Wang, C.; Luo, J.; Fang, Y.; Cao, B.; Qiu, Z. Research on the Spatiotemporal Evolution of Mangrove Forests in the Hainan Island from 1991 to 2021 Based on SVM and Res-UNet Algorithms. *Remote Sens.* **2022**, *14*, 5554. https://doi.org/10.3390/rs14215554

Academic Editors: Mingming Jia, Huaqiang Du, Wenyi Fan, Weiliang Fan, Fangjie Mao and Mingshi Li

Received: 9 September 2022
Accepted: 1 November 2022
Published: 3 November 2022

Abstract: Mangrove ecosystems play a dominant role in global, tropical, and subtropical coastal wetlands. Remote sensing plays a central role in mangrove conservation, as it is the preferred tool for monitoring changes in spatiotemporal distribution. To improve correlated estimation accuracies and explore the influencing mechanisms based on the mangrove ground survey, this study used a support vector machine (SVM) machine learning and Res-UNet deep learning algorithms to identify the land area of mangrove forests and the crown surface cover area of mangrove forests in the Hainan Island from 1991 to 2021. Both classification techniques were verified by a confusion matrix, which from 1991 to 2021, revealed overall accuracies of 93.11 ± 1.54% and 96.43 ± 1.15% for SVM and Res-UNet, respectively. Res-UNet was more accurate in identifying the crown surface cover area, whereas SVM was more suitable for obtaining the land area. Furthermore, based on the crown surface cover area of the mangrove forests on the Hainan Island, influencing mechanisms were analyzed through dynamic changes and landscape patterns. Since 1991, the Hainan Island mangrove forest area has increased, with the center of mass moving from coastal areas to the ocean and increasing the overall landscape fragmentation. Moreover, the change in the mangrove forests area was correlated with economic development and the increasingly urban population of the entire island. Altogether, the reliable assessment of the tropical mangrove forest land area and crown surface cover provides an important research foundation for the protection and restoration plans of tropical mangrove forests.

Keywords: mangrove forests; Hainan Island; deep learning; spatiotemporal evolution; influential mechanism

1. Introduction

Mangrove forests are an important type of coastal wetland that contain woody plant communities mainly distributed in the intertidal zones of tropical and subtropical regions [1]. These biomes constitute one of the most productive ecosystem types worldwide and maintain substantial social, ecological, and economic values for the natural environment and human society [2]. Specifically, mangrove forests play an important role in maintaining the ecological balance of coastlines and protecting the land from erosion [3]. Recently, these forests have also been recognized as the main contributor to "blue carbon sinks" in the global coastal zone, playing an important role in the suppression of ever-increasing atmospheric carbon dioxide concentrations [4]. Before the 21st century, mangrove forest areas were continuously reduced and degraded due to increasing socioeconomic threats, making them one of the most threatened ecosystems on the planet [5]. From 2000 to 2016, as government departments turned their attention to greater protection, the damaging human activities against global mangrove forests gradually decreased; however, the number of mangrove forests lost because of natural factors increased during the same period [6]. To further protect and manage mangrove forests, it remains necessary to understand their spatiotemporal evolution, as well as their response and adaptation mechanisms to population growth, economic development, climate, and other factors [7]. Mangrove forests are usually located within a large area of inaccessible mudflats, which complicates any corresponding groundwork efforts [8]. Alternatively, the history of mapping the range of mangrove forests with remote sensing data dates to the 1970s [9]; advances in sensor technology have offered increasingly improved effective mapping and monitoring techniques.

Remote sensing has been widely used for the multi-scale and long-term monitoring of environments and natural resources [10]. Over the past three decades, optical and radar satellites commonly used in mangrove research have included Landsat, SPOT, IRS 1C, IRS 1D, ASTER, IKONOS, QuickBird, RADARSAT-1 SAR, ENVISAT ASAR, ERS-1 SAR, JERS-1, AIRSAR, and ALOS PALSAR. The first civilian Earth satellite, Landsat, was launched in 1972, and the first commercial satellite, SPOT, was launched in 1986 [11]. Optical satellites are used more frequently than radar satellites in mangrove studies. For example, Hauser et al. [12] studied the spatiotemporal dynamics of mangrove forests on the Ga Mau Peninsula, Vietnam, from 2004 to 2013 using SPOT satellite imagery; moreover, Proisy et al. [13] used IKONOS, QuickBird, and WorldView satellite images to map the evolution of mangrove forests within an abandoned aquaculture estuary area in India from 2001 to 2015. Landsat time series are often the most common satellite data used to monitor ecosystem change at larger scales [14]. For example, Gaw et al. [15] used Landsat remote sensing imagery to study the dynamics of mangrove forests in Tanintharyi, Myanmar, from 1989 to 2014. Hu et al. [16] showed that Landsat remote sensing image data are the most commonly used data for mangrove forest feature classification because: (1) Landsat imagery data of a 30 m medium resolution can effectively extract spatial information from mangrove forests; (2) it contains rich waveform information; (3) it requires relatively short time intervals for image acquisition; (4) it maintains a long history (>30 years); and (5) it is characterized by relatively low imaging costs. Therefore, Landsat imagery data were used in this study to ensure the spatiotemporal integrity of the mangrove forest data to the maximum extent possible.

Although remote sensing technology can provide continuous spatiotemporal data for monitoring ecosystem changes, the accuracy of information extraction is influenced by image classification techniques and sensor resolution [17]. In land cover classification studies, nascent shrubs and herbs remain difficult to classify due to their similar spectral properties [18]. Similarly, a separate study in China showed that agricultural lands, inland dwarf tree forests, shrub forests, and aquatic plants with highly similar spectral characteristics to mangrove forests were easily misclassified [19]. Abdi [20] found that the support vector machine (SVM) machine-learning algorithm produced the highest accuracy for distinguishing regenerating shrubs and herbaceous plants (overall accuracy, OA = 76%). Guo et al. [21] found that the U-Net deep learning algorithm obtained good classification results for mangrove forest identification by multilayer convolutional operations (OA = 81%); however, the image elements in a small area near the mangrove forests' boundary were also misclassified. In response to the degradation problem exposed by deep learning algorithms during network training, Li et al. [22] proposed a residual learning framework ResNet, which, in a classification study of tree species, achieved a classification accuracy of 90.9% for ResNet-18. Moreover, deep residual U-Net is also widely used in remote sensing image classification [23]. Cao and Zhang [24] proposed the Res-UNet network, which combines U-Net and Resnet, to extract multi-scale spatial features that can effectively improve the accuracy of tree species classification. Therefore, in this study, to address issues with mangrove forest misclassification in large-scale feature analyses, the SVM and Res-UNet algorithms were chosen to remotely monitor the mangrove forest distribution on Hainan Island and conduct a comparative analysis.

Remote sensing technology can also accurately explore the dynamic changes in small-scale mangrove reserves. For example, Ibharim et al. [25] used remote sensing techniques to monitor changes in the Matang Mangrove Reserve, Malaysia, between 1993 and 2011, proposing conservation recommendations in terms of species distributions and patch types. Similarly, Son et al. [26] studied changes in the Can Gio Biosphere mangrove reserve in Vietnam from 1989 to 2014 using Landsat imagery data, finding that ~24% of the mangrove forests in the reserve were converted to aquaculture farms during this period and providing suggestions to assist local managers with reserve development. Zhen et al. [27] used radar and optical satellites combined with an improved mangrove classification method to assess the spatial distribution and dynamics of mangrove forests in the Dongzhai Port Mangrove Reserve, China, to improve the development of conservation and management policies. Such small-scale mangrove reserve studies can provide more accurate data on species distribution and land type but are limited in their ability to capture the effects of climate, social benefits, and economic benefit changes. Therefore, exploring the large-scale spatiotemporal evolution of mangroves can provide ideas for upstream planning, an important component of their sustainable development. For example, Gilani et al. [28] used Landsat imagery to monitor changes in the mangrove cover and fragmentation in Pakistan to assess the conservation and sustainability of mangrove forests. Giri et al. [29] used similar techniques at two spatial resolutions to study the proportion, patterns, causes, and consequences of changes in mangrove cover in South Asia, which can regularly monitor and manage mangroves in this region. Considering the advantages and limitations of different research scopes, this study systematically explored the response and adaptation mechanisms between the spatiotemporal evolution of tropical mangrove forests, climate, and socioeconomic changes at the provincial/city/county levels hierarchically.

This study aimed to improve the accuracy of remote sensing estimates of tropical mangrove forest spatial distributions and to explore the influential mechanisms of the spatiotemporal evolution of tropical mangrove forests. Furthermore, this study aimed to achieve the following three research objectives: (1) compare the advantages and limitations of SVM machine-learning and Res-UNet deep learning algorithms for extracting spatial information from the mangrove forest; (2) explore the spatiotemporal evolution of tropical mangrove forests on the Hainan Island from 1991 to 2021; and (3) analyze the response and adaptation mechanisms between the spatiotemporal evolution of tropical mangrove forests and changes in climate, environment, and socioeconomic benefits.

2. Materials and Methods

2.1. Materials

2.1.1. Study Area

The Hainan Island is located at the southernmost tip of China, on the northern edge of the tropics, between 108°37′ to 111°03′ E and 18°10′ to 20°10′ N. The island covers an area of ~3.54 × 10^4 km^2, with a coastline of 1944.4 km, and maintains a tropical maritime monsoon climate. It has the richest mangrove species and the most extensive mangrove forest area in China, including 26 species of true mangrove plants and 12 species of semi-mangrove plants [29]. The mangrove forests of the Hainan Island are mainly distributed along the coastal areas of 12 cities and counties in the northeast, south, east, and west, including the cities of Haikou, Wenchang, Danzhou, and Sanya.

2.1.2. Ground Survey Data Sources

From January 2020 to September 2020, we organized over 200 people to identify the range of the mangrove forests and record the distribution of dominant tree species on the Hainan Island during the ground survey (Figure 1 and Table A1). In addition, during past ground surveys, other members of our team recorded site data for mangrove forests on the Hainan Island as follows: in 1991 (429 sites), in 1996 (423 sites), in 2000 (441 sites), in 2007 (510 sites), in 2010 (485 sites), in 2015 (508 sites); site data for other land types were: in 1991 (1747 sites), in 1996 (1777 sites), in 2000 (1814 sites), in 2007 (1816 sites), in 2010 (1809 sites), in 2015 (1744 sites). Based on the range of the mangrove forests in the Hainan Island in 2020 and the site data previously surveyed by other members of the team, in October 2021, we conducted another ground survey and recorded data from 504 mangrove sites and 1805 other land-type sites. In addition, in each ground survey, all members of our team used a handheld GPS and Google Earth (Google Inc., Santa Clara County, CA, USA) to collect site data. The size of each site was 30 m × 30 m.

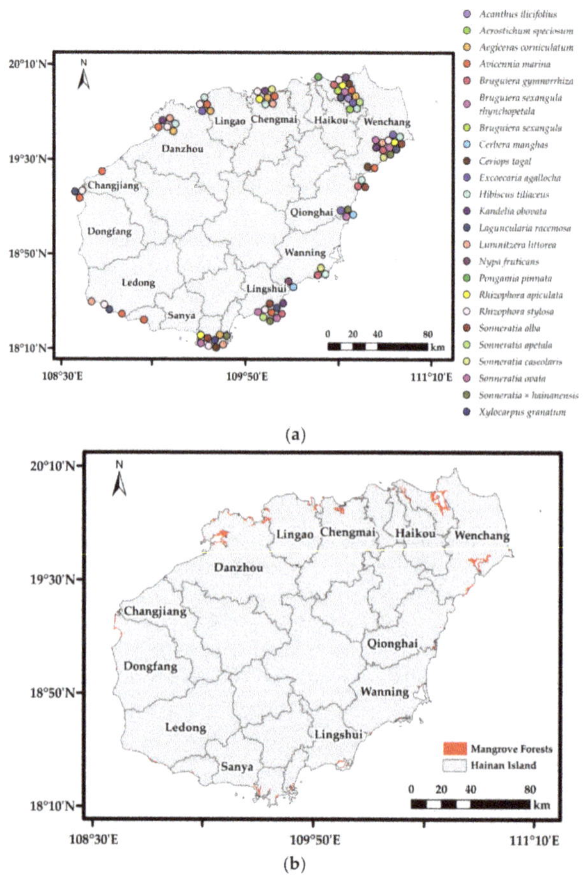

Figure 1. Ground survey data in 2020. (**a**) Distribution of dominant mangrove forest tree species in the Hainan Island. (**b**) Distribution range of mangrove forests in the Hainan Island.

2.1.3. Landsat Data Sources and Preprocessing

Landsat satellite image data were downloaded from the United States Geological Survey (USGS) for Earth Resources Observation and Science (https://www.usgs.gov/, accessed on 9 September 2022), from which the spatial resolution was 30 m. This study required images with a cloud coverage of less than 20%, and thus, compared and selected the Landsat satellite data obtained in 1991 (Landsat-5 TM), 1996 (Landsat-5 TM), 2000 (Landsat-5 TM), 2007 (Landsat-5 TM), 2010 (Landsat-5 TM), 2015 (Landsat-8 OLI), and 2021 (Landsat-8 OLI) (Table 1). Considering the large study area and complexity of the landscape, images of the same area were collected from adjacent years to reduce data loss related to cloudiness. The selected Landsat remote sensing image data were pre-processed with atmospheric correction, band combination, and image cropping. Because mangrove forests have more distinct spectral features in remote sensing image data, especially a strong reflectance in the near-infrared (NIR) band, they are more easily classified than other land cover types [30]. To distinguish the mangrove forests, Landsat-5 TM usually uses B4 (NIR, 0.76–0.90 μm), B3 (Red, 0.63–0.69 μm), and B2 (Green, 0.52–0.60 μm) bands to synthesize standard false color feature images. In such standard false color feature images, mangrove forests typically appear as deep red. However, B5 (NIR, 0.85–0.89 μm), B4 (Red, 0.63–0.68 μm), and B3 (Green, 0.53–0.60 μm) bands were used from Landsat 8 OLI.

Table 1. Information about the Landsat data images used in the study.

Year	Landsat Data Acquisition Times					Satellite Sensor	Standard False Color
1991	15 June	20 August	30 October	30 October	16 April 1992	Landsat-5 TM	B4 (NIR, 0.76–0.90 μm), B3 (Red, 0.63–0.69 μm), B2 (Green, 0.52–0.60 μm)
1996	14 July	14 December	23 December	23 December	23 September 1995		
2000	28 March	20 April	20 April	7 November	24 March 2001		
2007	6 July	13 July	15 July	15 July	22 July		
2010	7 February	24 March	7 July	16 September	21 August 2009		
2015	16 April	16 April	5 September	17 November	8 March 2016	Landsat-8 OLI	B5 (NIR, 0.85–0.89 μm), B4 (Red, 0.63–0.68 μm), B3 (Green, 0.53–0.60 μm)
2021	1 January	1 January	11 March	13 June	19 June		

2.1.4. Population, Economy, and Climate Data Sources

This study was taken from the WorldClim data website (https://www.worldclim.org/data/index.html, accessed on 9 September 2022) where a spatial resolution of 2.5 m of monthly weather data over 1990–2018 years of history was downloaded. Then, the CNRM-CM6-1 model and the sustainable development scenario (SSP226) were selected in CMIP6 to download the monthly climate data with a spatial resolution of 2.5 m. The average minimum temperature (°C), average maximum temperature (°C), and total precipitation (mm) in 1991, 1996, 2000, 2007, 2010, 2015, and 2021 were sorted out in the TIFF format climate data set. Furthermore, the total population, urban population, rural population, GDP, and gross output fishery value were obtained from the Annual Statistical Report of Hainan Province in 1991, 1996, 2000, 2007, 2010, 2015, and 2021, respectively.

2.2. Methods

2.2.1. Support Vector Machine

The SVM machine-learning algorithm used here for supervised classification is based on the statistical learning theory and was originally developed to solve dichotomous classification problems [21]. SVM tries to identify the optimal thresholds that maximize the separation or bounds between the support vectors [31]. In another way, this requires finding the best hyperplane in a multidimensional space that splits two sets of vectors so that the vectors closest to the hyperplane (i.e., the support vectors) are as far away as possible from the hyperplane (Figure 2). Assuming that the Euclidean distance of the vector to the hyperplane is d_i, the minimum value of d_i is required to represent the shortest distance of this vector to the hyperplane. Accordingly, the mathematical expressions for the hyperplane $g(x)$ and d_i are defined by Equations (1) and (2):

$$g(x) = w^T \cdot x + b;\ w,\ x \in R^n \tag{1}$$

$$d_i = \frac{|g(x)|}{\|w\|} \tag{2}$$

where w and x are vectors in the n-dimensional space. x is a function variable and w is a normal vector. $\|w\|$ is the parametrization of the hyperplane.

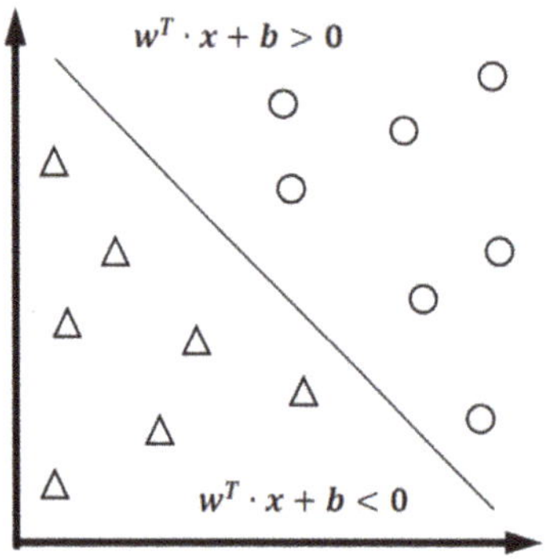

Figure 2. Visualization of the hyperplane separating the two types of vectors (assuming $g(x) = 0$).

In this study, the radial basis function (RBF) was selected as the kernel function, Gamma = 1, and the system default values were used for other parameters when SVM was used to establish the mangrove forests distribution model in the Hainan Island. The input band parameters were as follows: Landsat-5 TM: B4-Near IR (0.76–0.90 μm), B3-Red (0.63–0.69 μm), B2-Green (0.52–0.60 μm); Landsat-8 OLI: B5-Near IR (0.85–0.89 μm), B4-Red (0.63–0.68 μm), B3-Green (0.53–0.60 μm). The number of training samples selected for SVM machine-learning each year is as follows: in 1991 (312 mangrove sites and 1245 non-mangrove sites), in 1996 (304 mangrove sites and 1260 non-mangrove sites), in 2000 (318 mangrove sites and 1280 non-mangrove sites), in 2007 (380 mangrove sites and 1299 non-mangrove sites), in 2010 (351 mangrove sites and 1303 non-mangrove sites), in 2015 (374 mangrove sites and 1244 non-mangrove sites), and in 2021 (353 mangrove sites and 1296 non-mangrove sites). The size of a single sample is 30 × 30 m.

2.2.2. Res-UNet

U-Net was first applied to medical image segmentation [32]. Later, it was also widely used in remote sensing image classification [33]. The deep residual network ResNet can avoid the problem of gradient degradation in the process of network deepening [34]. This study used U-Net to equip the ResNet-18 backbone to train deep learning models (Figure 3) in order to increase the feature expression ability of the model [35]. Among them, ResNet-18 is a two-level ResNet residual unit, and the network structure of residual learning can be seen in Figure 4.

During model training, the average cross-entropy loss was used to calculate the model loss via the function presented in Equation (3):

$$loss = -\frac{1}{n}\sum_{i=1}^{n} y_i log a_i + (1 - y_i) log(1 - a_i) \quad (3)$$

where n represents the batch size; y_i and a_i are the predicted and true values of the ith sample in each batch, respectively. For the loss of the model, the network parameters were optimized using the Adam optimizer proposed by Kingma and Ba [36], according to Equation (4):

$$\theta_t = \theta_{t-1} - \alpha * \hat{m}_t / \left(\sqrt{\hat{v}_t} + \varepsilon\right) \quad (4)$$

where t is the number of training iterations, α is the learning rate, m is the exponential moving average of the gradient, and v is the exponential moving average of the gradient squared. The "ε" is usually a constant with a value of 1×10^{-8}.

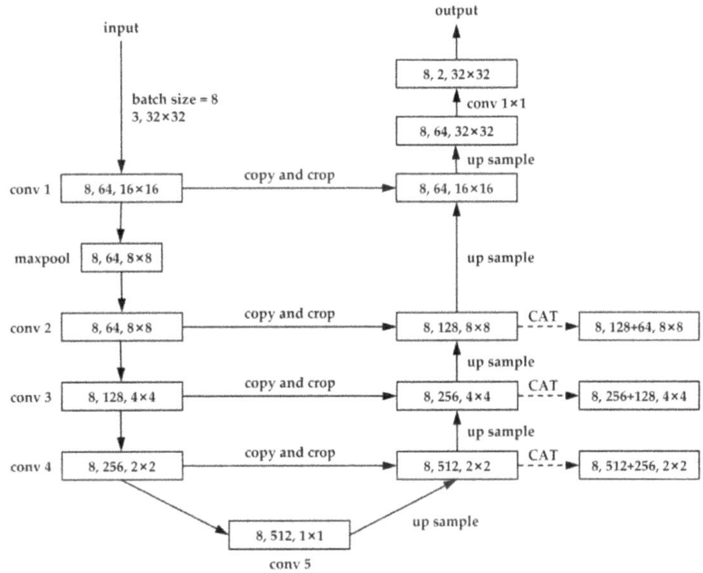

Figure 3. Network structure of Res-UNet. "conv" referring to the feature map. "CAT" is a method of concatenating feature maps with different numbers of channels.

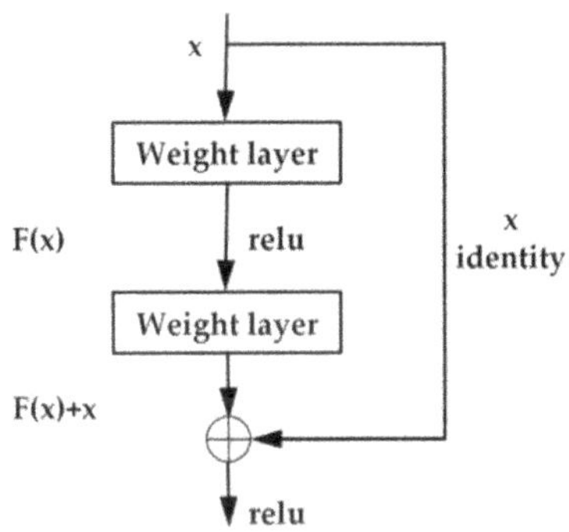

Figure 4. Network structure of residual learning. "F(x)" refers to the residual and "x" is the feature mapping of the output of the previous layer ResNet.

This study used Python based on the TensorFlow deep learning framework. The hardware configuration of this operating platform included a Lenovo ThinkStation P620 AMD3955WX 64G and an NVIDIA Quadro RTX4000 8G GPU. Based on the site data from the ground survey and the distribution range of mangrove forests from the ground survey in 2020, the distribution range of mangrove forests in the Haikou, Wenchang, and Danzhou cities were mapped by visual interpretation in Landsat remote sensing images. When Res-UNet trained the model of mangrove forest distribution on the Hainan Island, the Landsat remote sensing images were cut, referring to the visually interpreted mangrove distribution range. The slice size was set to 32 × 32 pixels, the batch size = 8, the backbone model was set to ResNet–18, and the default values for other parameters were used. The input band parameters were as follows: Landsat-5 TM: B4-Near IR (0.76–0.90 μm), B3-Red (0.63–0.69 μm), and B2-Green (0.52–0.60 μm); Landsat-8 OLI: B5-Near IR (0.85–0.89 μm), B4-Red (0.63–0.68 μm), and B3-Green (0.53–0.60 μm). Finally, the training samples for Res-UNet deep learning were obtained as follows: in 1991 (780 sites), in

1996 (873 sites), in 2000 (767 sites), in 2007 (875 sites), in 2010 (928 sites), in 2015 (1050 sites), and in 2021 (1201 sites). The size of a single sample is 32 × 32 pixels.

2.2.3. Accuracy Assessment

Here, two metrics, the Kappa coefficient, and the OA were used to evaluate the classification accuracy of SVM and Res-UNet, respectively. Both metrics were calculated based on a confusion matrix, which provided a clear picture of the number of features correctly and incorrectly classified [37]. Specifically, the Kappa coefficient is typically used to test the consistency of results and measure the effectiveness of classifications, whereas OA is the ratio of correctly classified categories to the total category number [21]. Of the sites obtained each year, approximately 30% were selected as a validation sample. The sample types were divided into mangrove and non-mangrove (other land types). The number of validation samples selected each year is as follows: in 1991 (117 mangrove sites and 502 non-mangrove sites), in 1996 (119 mangrove sites and 517 non-mangrove sites), in 2000 (123 mangrove sites and 534 non-mangrove sites), in 2007 (130 mangrove sites and 517 non-mangrove sites), in 2010 (134 mangrove sites and 506 non-mangrove sites), in 2015 (134 mangrove sites and 500 non-mangrove sites), and in 2021 (151 mangrove sites and 509 non-mangrove sites). the size of a single sample is 30 × 30 m. The confusion matrix was used to evaluate the classification results of the model, and the precise equations for *OA* and *Kappa*, are presented in Equations (5) and (6):

$$OA = \frac{\sum_{i=1}^{2} a_{ii}}{N} \tag{5}$$

$$Kappa = \frac{OA - \frac{\sum_{i=1}^{2} a_{i+} * a_{+i}}{N^2}}{1 - OA}, a_{i+} = \sum_i a_{ij}, a_{+i} = \sum_j a_{ij} \tag{6}$$

where a_{ii} denotes the accurate values of i predicted to be i, a_{ij} denotes the values of i predicted to be j, and N is the total number of samples.

2.2.4. Dynamic Change and Landscape Pattern Analysis

Here, the area of mangrove forest cover change was evaluated and compared based on the mangrove distribution of 1991, 1996, 2000, 2007, 2010, 2015, and 2021. The annual rate of change in the crown surface area was used to analyze the mangrove forest changes over the last 30 years for six stages: 1991–1996, 1996–2000, 2000–2007, 2007–2010, 2010–2015, and 2015–2021. Specifically, the annual rate of change in the crown surface area was calculated using the formula proposed by Puyravaud [38]:

$$r = \frac{1}{t_2 - t_1} ln \frac{A_2}{A_1} \tag{7}$$

where r is the annual percentage change rate; t_1 and t_2 are the starting and ending years at the time of calculation, respectively; and A_1 and A_2 are the corresponding areas in t_1 and t_2, respectively.

In evaluating the spatiotemporal changes in the landscape patterns of mangrove forests, landscape indices, such as shape complexity and patch fragmentation, can further reveal the impacts of human activities [39]. Five landscape indices were selected based on the actual situation of the study area: the number of patches (NP), patch density (PD), maximum patch index (LPI), landscape shape index (LSI) and aggregation index (AI), where NP reflects the spatial pattern of the landscape; PD describes the degree of landscape fragmentation; LPI indicates the expansion or fragmentation of the largest mangrove forest patches, reflecting the health of the mangrove forests in the core area; LSI determines the shape changes of the patch, corresponding to the resistance abilities of the mangrove forests to external disturbances; and AI reflects the connectivity and degree of aggregation and dispersion within mangrove forest patches [40]. PD, LSI, and AI were calculated according to Equations (8)–(10) [39]:

$$PD = \frac{NP}{A}, \ PD > 0 \tag{8}$$

$$LSI = \frac{0.25E}{\sqrt{A}}, \ LSI \geq 1 \tag{9}$$

$$AI = \frac{p_{ij}}{max \ p_{ij}} \times 100 \tag{10}$$

where A is the total landscape area (ha), E is the total length of the edge in the landscape, and p_{ij} represents the number of adjacent patches in patches of the same type as the landscape, i represents a landscape type, and j represents patches of the same type as i.

The mangrove mass center offset trajectory can reflect the spatial distribution of mangrove forests over different years, an important factor when studying the dynamic changes over certain periods of time. Here, the principle was to adopt the change in the mass center coordinates of the landscape patches to reflect the change laws of the mangrove area mass center distributions. The center of mass formula was derived from Li et al. [41] (Equation (11)):

$$X_t = \frac{\sum_{i=1}^{N}(C_{ti}X_i)}{\sum_{i=1}^{N}C_{ti}}, \ Y_t = \frac{\sum_{i=1}^{N}(C_{ti}Y_i)}{\sum_{i=1}^{N}C_{ti}} \tag{11}$$

where X_t and Y_t denote the latitude and longitude coordinates of the landscape mass center in year t, respectively; X_i and Y_i are the latitude and longitude coordinates of the mass center of the ith patch of a landscape, respectively; C_{ti} is the area of the ith patch, and N is the total number of landscape patches.

2.2.5. Statistical Analysis of Driving Forces

Zheng and Takeuchi [42] showed that mangroves vary over space and time, with changes related to the climate, environment, and socioeconomic benefits. To quantify the main drivers affecting the evolution of mangrove landscapes, this study conducted a Pearson bivariate correlation analysis of the mangrove area with socioeconomic and natural environmental indicators. Eight indicators were selected for the study area: total population, urban population, rural population, GDP, gross production fishery value, average annual rainfall, minimum temperature, and maximum temperature.

3. Results

3.1. Analysis of the Classification Results

3.1.1. Classification Results of SVM Machine Learning

The SVM classification results are shown in Figure A1 in Appendix A. Confusion matrix calculations were used to summarize the producer accuracy (PA), user accuracy (UA), OA metrics, and Kappa coefficients. The SVM classification accuracy was the highest in 1996 and 2021 (Table 2), with the OA and Kappa coefficients at >94% and >0.80, respectively. The lowest classification accuracy was recorded in 2010 (OA and Kappa coefficients of 91.6% and 0.71, respectively). The primary classification task was to identify the mangrove forest presence; however, the spectral information of other land types can influence the classification results. The highest PA of the mangrove forests was recorded in 1996 (77.3%), and although the overall classification results of SVM were high, the identification results of the mangrove forests remained relatively inaccurate as the probability of the mangrove forests being misclassified persisted.

Table 2. Accuracy assessment of SVM classification results from mangrove forests in the Hainan Island during 1991–2021, where μ depicts the average values.

Period	Classified	Ground-Truth			Summary	
		Mangrove	Non-Mangrove	Total	PA	UA
1991	Mangrove	79	7	86	67.5%	91.9%
	Non-Mangrove	38	495	533	98.6%	92.9%
	Total	117	502	619	83.1μ	92.4μ
					OA = 92.7%	Kappa = 0.74
1996	Mangrove	92	7	99	77.3%	92.9%
	Non-Mangrove	27	510	537	98.7%	95.0%
	Total	119	517	636	88.0μ	94.0μ
					OA = 94.65%	Kappa = 0.81
2000	Mangrove	79	2	81	64.2%	97.5%
	Non-Mangrove	44	532	576	99.6%	92.4%
	Total	123	534	657	81.9μ	95.0μ
					OA = 93.00%	Kappa = 0.74
2007	Mangrove	80	2	82	61.5%	97.6%
	Non-Mangrove	50	515	565	99.6%	91.2%
	Total	130	517	647	80.6μ	94.4μ
					OA = 92.0%	Kappa = 0.71
2010	Mangrove	83	3	86	61.9%	96.5%
	Non-Mangrove	51	503	554	99.4%	90.8%
	Total	134	506	640	80.7μ	93.7μ
					OA = 91.6%	Kappa = 0.71
2015	Mangrove	96	2	98	71.6%	98.0%
	Non-Mangrove	38	498	536	99.6%	92.9%
	Total	134	500	634	85.6μ	95.4μ
					OA = 93.7%	Kappa = 0.79
2021	Mangrove	114	1	115	75.5%	99.1%
	Non-Mangrove	37	508	545	99.8%	93.2%
	Total	151	509	660	87.7μ	96.2μ
					OA = 94.2%	Kappa = 0.82

3.1.2. Classification Results of Res-UNet Deep Learning

The classification results of the Res-UNet deep learning algorithm are shown in Figure A2 in Appendix A. The extracted sample labels were divided into two categories: mangrove and non-mangrove forests. The confusion matrix was selected for accuracy verification, besides the PA, UA, OA, and Kappa coefficient calculations, for the classification results of the mangrove forests. When comparing the validation results across different years (Table 3), it was found that Res-UNet produced superior classification results (OA, 95%; Kappa coefficients, >0.80). Among them, the best classification accuracy was achieved in 2021 (OA and Kappa coefficient of 97.6% and 0.93, respectively), and the worst classification accuracy appeared in 1996 (OA and Kappa coefficient values of 95.3% and 0.83, respectively). Few mangrove forests were misclassified (low errors) using this deep learning algorithm (Table 3), resulting in high PA values in all the mangrove forest classes (reaching a maximum of 93.4% in 2021).

Table 3. Accuracy assessment of Res-UNet classification results of mangrove forests in the Hainan Island during 1991–2021, where μ depicts the averaged values.

Period	Classified	Ground-Truth			Summary	
		Mangrove	Non-Mangrove	Total	PA	UA
1991	Mangrove	92	3	95	78.6%	96.8%
	Non-Mangrove	25	499	524	99.4%	95.2%
	Total	117	502	619	89.0μ	96.0μ
					OA = 95.5%	Kappa = 0.84
1996	Mangrove	93	4	97	78.2%	95.9%
	Non-Mangrove	26	513	539	99.2%	95.2%
	Total	119	517	636	88.7μ	95.5μ
					OA = 95.3%	Kappa = 0.83
2000	Mangrove	106	4	110	86.2%	96.4%
	Non-Mangrove	17	530	547	99.3%	96.9%
	Total	123	534	657	92.7μ	96.6μ
					OA = 96.8%	Kappa = 0.89
2007	Mangrove	108	1	109	83.1%	99.1%
	Non-Mangrove	22	516	538	99.8%	95.9%
	Total	130	517	647	91.4μ	97.5μ
					OA = 96.5%	Kappa = 0.88
2010	Mangrove	115	3	118	85.8%	97.5%
	Non-Mangrove	19	503	522	99.4%	96.4%
	Total	134	506	640	92.6μ	96.9μ
					OA = 96.6%	Kappa = 0.89
2015	Mangrove	112	1	113	83.6%	99.1%
	Non-Mangrove	22	499	521	99.8%	95.8%
	Total	134	500	634	91.7μ	97.5μ
					OA = 96.4%	Kappa = 0.88
2021	Mangrove	141	6	147	93.4%	95.9%
	Non-Mangrove	10	503	513	98.8%	98.1%
	Total	151	509	660	96.1μ	97.0μ
					OA = 97.6%	Kappa = 0.93

The cross-entropy loss curve of the Res-UNet model is shown in Figure 5. Under the optimal model, the batch size was eight. After ~5000 training iterations, the loss of Res-UNet stabilized at 0.1, where the model weights gained certainty.

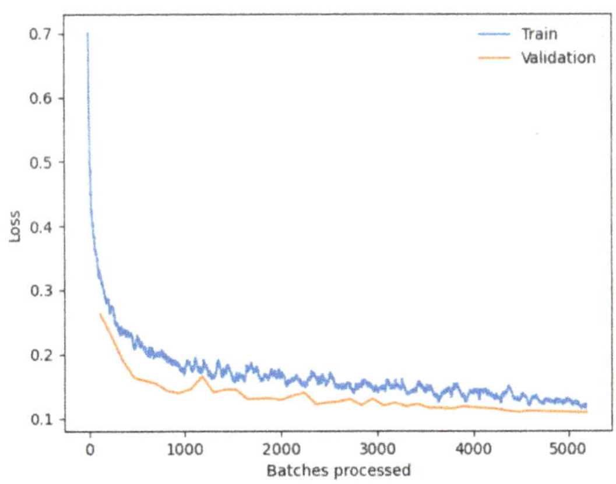

Figure 5. Res-UNet loss curve, where the x-axis indicates the number of training iterations.

3.1.3. Comparison of Mapping Results between SVM Machine Learning and Res-UNet Deep Learning

The SVM and Res-UNet classification results were compared with ground truth remote sensing imagery data to analyze the ability of the two algorithms to identify the distribution range of mangrove forests on the Hainan Island (Figure 6). Here, it was found that Res-UNet was more accurate in identifying the crown surface cover area, whereas SVM better reproduced the land area covered by entire wetlands or protected areas, including mangrove forests. The inability of SVM to distinguish mangrove forests from water and aquatic plants was evident (Figure 6a,b); therefore, the extracted mangrove distributions were more contiguous. Furthermore, numerous pixels with mixed low trees and shrubs were misclassified as "mangrove forests" (Figure 6e,f). In Figure 6c,d, an under-classification is observed due to the non-recognition of mangrove forests. In contrast, Res-UNet more accurately distinguished mangrove forests from other feature types with similar spectral information within mixed vegetation areas, greatly reducing the probability of mangrove forest misclassification on a large scale.

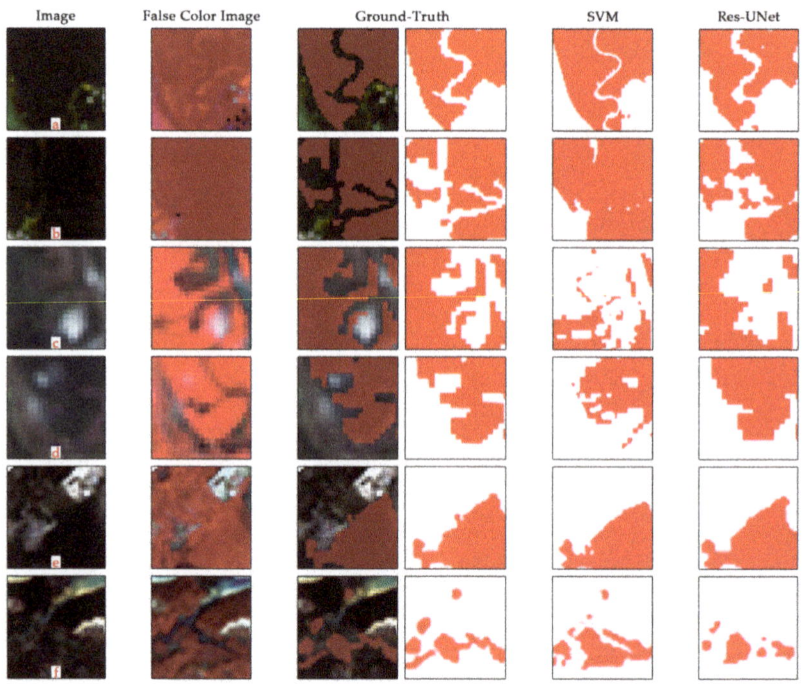

Figure 6. Illustrative examples of the classification method limitations for SVM and Res-UNet: (**a**,**b**) Haikou City, (**c**,**d**) Wenchang City, (**e**,**f**) Danzhou City; red represents mangrove forests, and white represents all other land types, each square is captured from a 30 m Landsat remote sensing image, and the side length is about 975 × 975 m.

3.2. Analysis of Spatiotemporal Changes of Mangrove Forests in the Hainan Island

3.2.1. Change in Mangrove Forest Crown Surface Cover Area during 1991–2021

Based on the validation of the Res-UNet algorithm, this trained model was applied to a large-scale mangrove forest crown surface mapping to compare the extent of the mangrove forest crown coverage changes in 1991, 1996, 2000, 2007, 2010, 2015, and 2021 on the Hainan Island. The crown surface cover area of the total mangrove forests on the Hainan Island in each of these seven periods was 1740.15, 2076.66, 1984.68, 2371.59, 2694.78, 2233.80, and 3438.63 ha, respectively (Figure 7 and Table 4). The mangrove forests were mainly distributed over 12 cities and counties around the coast of the Hainan Island. By 2021, the forest crown surface cover areas in the cities of Haikou, Wenchang, and Danzhou increased, whereas those of the Wanning City, Ledong Li Autonomous County, and Changjiang Li Autonomous County receded during the analysis period. In addition, Wanning City, Ledong Li Autonomous County, Lingshui Li Autonomous County, and Changjiang Li Autonomous County were characterized by the disappearance of mangrove forests in individual years, with mangrove forests in Changjiang Li Autonomous County only present in 2000 and 2010. Overall, the mangrove forest crown surface cover area in the Hainan Island showed an increasing trend over the last three decades, with a net increase of 1698.48 ha from 1991 to 2021, representing an annual change rate of 2.27% (Table 4). The highest growth rate of the surface cover area of the mangrove forest crown was recorded throughout the analysis period in Dongfang City (16.24%), whereas the annual change rate peaked in the autonomous Ledong Li County from 2007 to 2021 ($\leq$35.50%).

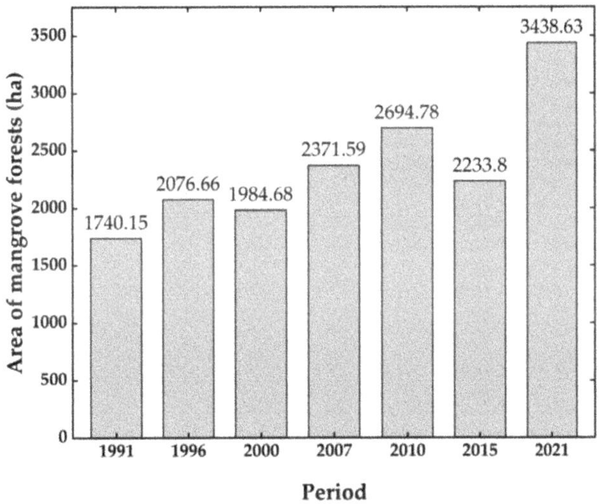

Figure 7. Trends of mangrove forest crown cover area in the Hainan Island during 1991–2021.

Table 4. Mangrove forest crown surface cover area (ha) and annual rate of area change (%) in the Hainan Island and in each city/county for every year of analysis.

City/County	Mangrove Forests Crown Cover (ha)							Annual Rate of Change (%)
	1991	1996	2000	2007	2010	2015	2021	
Haikou	898.20	1259.73	1221.12	1343.07	1294.74	1233.09	1183.59	0.92
Sanya	49.50	14.76	3.06	4.95	35.91	13.14	57.96	0.53
Wenchang	286.83	552.24	356.94	598.95	755.19	449.82	1083.42	4.43
Qionghai	25.02	1.26	11.61	6.12	41.31	1.80	32.13	0.83
Wanning	4.95	0.00	2.43	0.90	0.00	0.00	6.3	0.80
Chengmai	48.06	63.45	53.01	36.54	66.78	98.82	191.97	4.62
Lingao	46.71	9.9	24.39	42.3	80.01	40.23	129.96	3.41
Danzhou	377.73	155.07	277.29	319.86	369.18	364.86	610.56	1.60
Dongfang	0.63	18.54	33.84	16.29	9.36	30.51	82.17	16.24
Ledong	0.00	0.00	0.00	0.09	3.51	0.36	12.96	35.50 *
Lingshui	2.61	0.00	0.27	2.52	38.07	1.17	47.61	9.68
Changjiang	0.00	0.00	0.72	0.00	0.72	0.00	0.00	0.00 *
Total Area	1740.15	2076.66	1984.68	2371.59	2694.78	2233.80	3438.63	2.27

* Monitoring time starts from the year that mangrove forests appeared.

The changes in the surface cover area of the mangrove forest crown were compared and analyzed for each city and county in the Hainan Island across six periods: 1991–1996, 1996–2000, 2000–2007, 2007–2010, 2010–2015, and 2015–2021. Although the surface cover area increased for all cities and counties in the Hainan Island over the analysis period (Table 4), the observed growth was unstable in terms of phase changes. The surface cover area decreased in the Hainan Island from 1996 to 2000 and 2010 to 2015 (annual rates of change: −1.13% and −3.75%, respectively; Table 5), while the highest growth rate in the island was observed from 2015 to 2021 (7.19%·yr^{-1}). However, the area of surface coverage of the mangrove forest crown in Haikou City decreased in the three phases from 2007 to 2021. Conversely, the crown surface cover area increased in Chengmai County from 2007 to 2021. In addition, the mangrove forest crown surface cover area in Lingao County and Danzhou City increased between 1996 and 2010, and the crown surface coverage of Lingshui Li Autonomous County also showed an increasing trend from 2000 to 2010.

Table 5. Annual rate of change (%) in the crown surface cover area of mangrove forests in the Hainan Island during 1991–1996, 1996–2000, 2000–2007, 2007–2010, 2010–2015, and 2015–2021.

City/County	Annual Rate of Change					
	1991–1996	1996–2000	2000–2007	2007–2010	2010–2015	2015–2021
Haikou	6.77	−0.78	1.36	−1.22	−0.98	−0.68
Sanya	−24.20	−39.34	6.87	66.05	−20.11	24.73
Wenchang	13.10	−10.91	7.39	7.73	−10.36	14.65
Qionghai	−59.77	55.52	−9.15	63.65	−62.67	48.03
Wanning	0.00	0.00	−14.19	0.00	0.00	0.00
Chengmai	5.56	−4.49	−5.32	20.10	7.84	11.07
Lingao	−31.03	22.54	7.87	21.25	−13.75	19.54
Danzhou	−17.81	14.53	2.04	4.78	−0.24	8.58
Dongfang	67.64	15.04	−10.44	−18.47	23.63	16.51
Ledong	0.00	0.00	0.00	122.12	−45.55	59.73
Lingshui	0.00	0.00	31.91	90.51	−69.65	61.77
Changjiang	0.00	0.00	0.00	0.00	0.00	0.00
Hainan Island	3.54	−1.13	2.54	4.26	−3.75	7.19

3.2.2. Spatial Distribution and Changes in Mangrove Forests during 1991–2021

The landscape-level pattern index can reflect the corresponding change characteristics of the entire study area (Figure 8). From 1991 to 2021, the NP, PD, and LSI of mangrove forests in the Hainan Island showed repeated trends of decreasing, followed by an increase. NP and LSI both reached a maximum in 2021, with 732 and 30.03%, respectively. This indicates that the patch shape of mangrove forests was complex as the NP increased. LPI and AI also fluctuated from an increase to a decrease several times, with LPI reaching at least 6.42 in 2021. In conclusion, the edge shape of the mangrove patch in the Hainan Island in 2021 is complex, with low connectivity and substantial fragmentation. At the city and county levels, only Haikou City and Dongfang City displayed relatively reduced landscape fragmentation and strong landscape connectivity by 2021.

The spatial distribution of the surface cover area of the mangrove forest crown in the 12 cities and counties along the coast of the Hainan Island was used to investigate the path of mass center offsets across the six periods. From 1991 to 2021, most centers of the mangrove forest mass in the Hainan Island showed a trend of coastal movement toward the ocean or inlets, the distance of movement in the first stage being the largest (Figure 9). Specifically, the mass center of Changjiang Li Autonomous County moved in a unidirectional line, as mangrove forests were only positively identified in two of the analysis years; moreover, the mass centers of the mangrove forests in Sanya, Danzhou, and Wanning cities also moved unidirectionally until 2021, when they showed a folded-back trend. The movement trajectories in all the remaining locations appeared circular or crossed and overlapped, indicating the factors influencing mangrove forest survival.

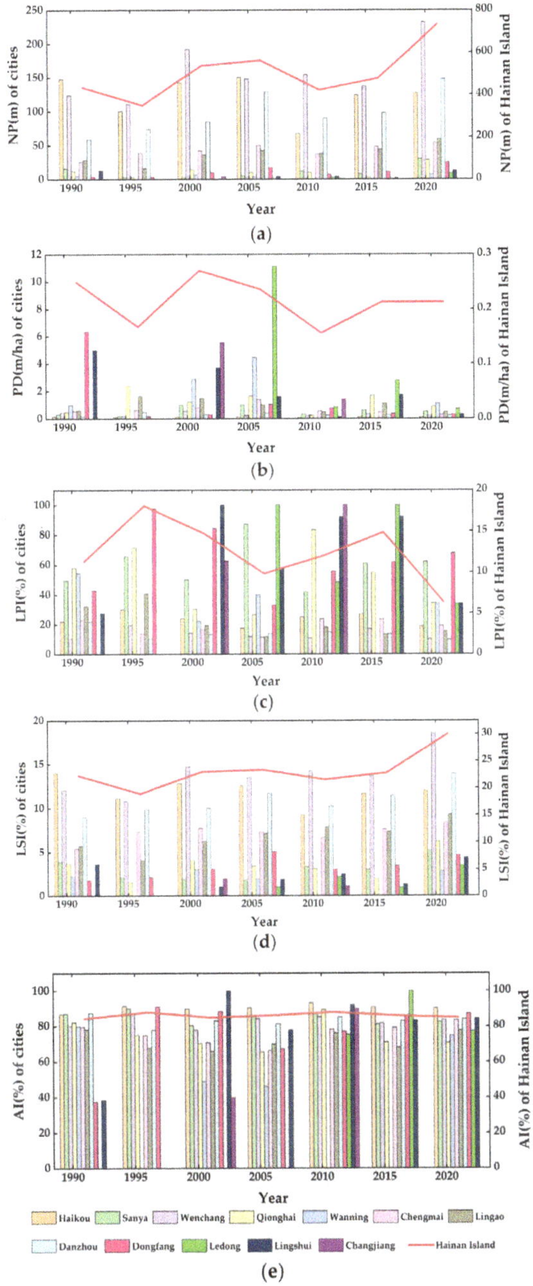

Figure 8. Landscape pattern index of mangrove forests in the Hainan Island during 1991–2021. (**a**) NP (m) index; (**b**) PD (m/ha) index; (**c**) LPI (%) index; (**d**) LSI (%) index; (**e**) AI (%) index.

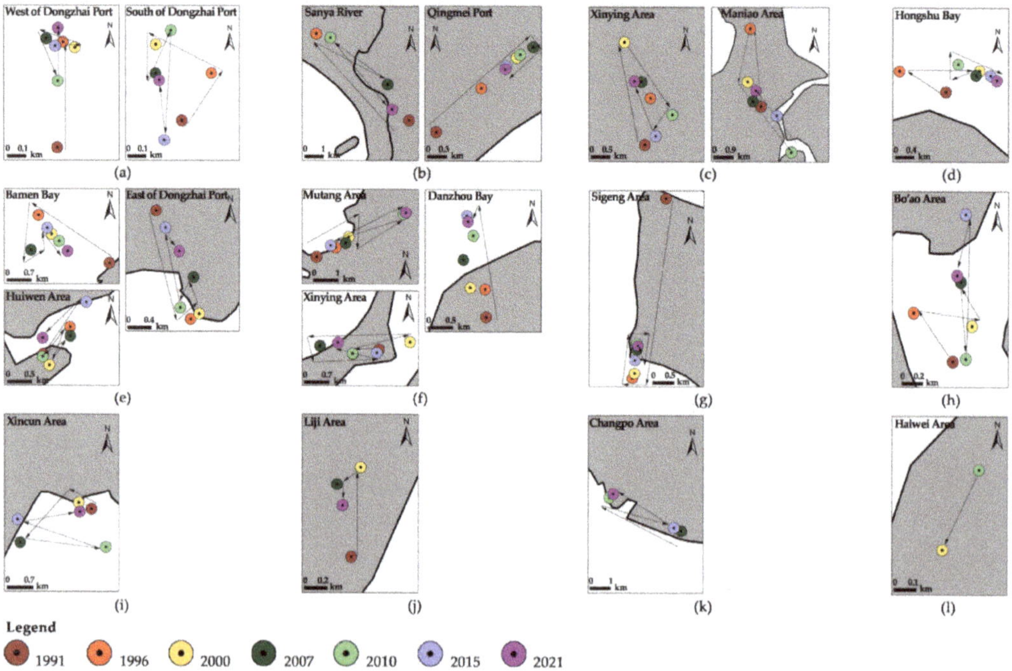

Figure 9. Mass center offset maps of mangrove forests across the Hainan Island for (**a**) Haikou, (**b**) Sanya, (**c**) Lingao, (**d**) Chengmai, (**e**) Wenchang, (**f**) Danzhou, (**g**) Dongfang, (**h**) Qionghai, (**i**) Lingshui, (**j**) Wanning, (**k**) Ledong, and (**l**) Changjiang.

3.2.3. Influential Mechanisms of Mangrove Forest Landscape Evolution

From 1991 to 2021, the total and urban populations of the Hainan Island grew continuously, whereas the rural population slowly decreased. Furthermore, the GDP of the island increased from 10.793 billion yuan in 1991 to 553.229 billion yuan in 2021, from which the value of fishery rose from 836 million yuan to 39.080 billion yuan (Figure 10). However, the overall patterns of average annual rainfall and minimum and maximum temperatures throughout the study period were complex, although all increased (Table 6).

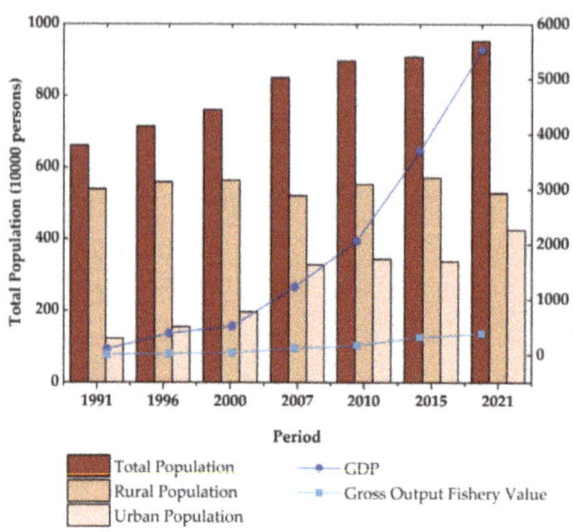

Figure 10. Population and socioeconomic development dynamics of the Hainan Island during 1991–2021.

Table 6. Climate and environmental indicator dynamics in the Hainan Island during 1991–2021.

Period	Average Annual Rainfall (mm)	Average Annual Minimum Temperature (°C)	Average Annual Maximum Temperature (°C)
1991	1289.12	21.65	28.5
1996	1531.08	21.12	27.68
2000	1804.22	21.6	27.76
2007	1334.88	21.72	28.01
2010	1507.96	21.45	27.88
2015	1554.69	22.27	28.66
2021	1548.06	22.44	29.11
Linear Fit	$y = 18.849x + 1434.6$	$y = 0.1614x + 21.104$	$y = 0.1396x + 27.67$
R^2	$R^2 = 0.0585$	$R^2 = 0.5760$	$R^2 = 0.3163$

According to the correlation analyses with mangrove forest crown surface cover areas in the Hainan Island (Table 7), positive correlations were observed with the socioeconomic factors of the total population, GDP, and the gross output value of fisheries ($p < 0.05$). Moreover, the change in the surface cover of the mangrove forest crown showed a significant positive correlation with the urban population ($p < 0.01$). Specifically, in the correlation analysis of the mangrove forest crown surface cover area change in each city and county, Wenchang City and Lingshui Li Autonomous County showed a significant positive correlation between mangrove forest crown surface cover area and urban population; the growth of the mangrove forest crown surface cover area in Wenchang City, Chengmai County, Lingao County, Danzhou City, Dongfang City, and Ledong Li Autonomous County displayed significant positive correlations with the local GDP; whereas that of Chengmai County showed a highly significant positive correlation with both GDP and the gross output value of fisheries. Regarding climatic factors, all correlations with the mangrove forest crown surface cover area across the Hainan Island were positive but weak; however, analyses at city and county levels found that the crown surface cover area changes in Chengmai County and Danzhou City were significantly positively correlated with both the average annual minimum and maximum temperatures.

Table 7. Pearson correlation analysis results of mangrove forest crown surface cover area with socioeconomic and climatic factors over the Hainan Island during 1991–2021.

City/County	Total Pop [1]	Rural Pop.	Urban Pop.	GDP	Gross Output Fishery Value	Average Annual Rainfall	Average Annual Minimum Temperature	Average Annual Maximum Temperature
Hainan Island	0.836 *	−0.42	0.875 **	0.853 *	0.801 *	0.09	0.56	0.52
Haikou	0.51	0.37	0.49	0.19	0.29	0.28	−0.16	−0.37
Sanya	0.21	−0.70	0.39	0.47	0.30	−0.48	0.33	0.63
Wenchang	0.72	−0.764 *	0.901 **	0.797 *	0.788 *	0.11	0.35	0.41
Qionghai	0.29	0.14	0.18	0.28	0.21	−0.04	0.07	0.24
Wanning	−0.15	−0.52	0.15	0.27	0.23	−0.20	0.44	0.63
Chengmai	0.61	−0.775 *	0.73	0.922 **	0.885 **	0.21	0.767 *	0.797 *
Lingao	0.59	0.57	0.58	0.772 *	0.71	−0.14	0.66	0.73
Danzhou	0.53	0.51	0.47	0.800 *	0.75	−0.15	0.842 *	0.867 *
Dongfang	0.57	−0.30	0.73	0.770 *	0.64	0.49	0.71	0.62
Ledong	0.62	0.58	0.40	0.810 *	0.67	0.00	0.63	0.64
Lingshui	0.60	−0.40	0.825 *	0.69	0.62	−0.03	0.39	0.38
Changjiang	0.19	0.49	−0.01	−0.17	−0.21	0.62	−0.16	−0.47

[1] Pop. refers to the population; * $p < 0.05$; ** $p < 0.01$.

4. Discussion

4.1. Comparative Analysis of Mangrove Classification Methods

In this study, the accuracy of the SVM and Res-UNet algorithms used to identify the distribution range of mangrove forests in the Hainan Island from 1991 to 2021 produced OA values of 93.11 ± 1.54% and 96.43 ± 1.15%, respectively; the PA of Res-UNet was resultantly much greater than SVM. It was observed that the Res-UNet algorithm based on a convolutional neural network produced a higher correct classification rate for the crown surface cover area of the mangrove forest.

In their examination of the Pichavaram mangrove wetland that spans 2335.5 ha, Singh et al. [43] achieved the highest overall classification accuracy by using an SVM to identify mangrove images (94.53%). Zhen et al. [27] used an SVM to classify the land use of the Dongzhai Port National Nature Reserve in Hainan, finding that OA could reach 83.5%; thus, it has been shown that SVMs can delineate the distributions of mangrove forests in small-scale wetland parks or natural reserves. Similarly, this study found that SVM can effectively extract the land area of mangrove forests in the Hainan Island, which has the advantage of identifying mangrove forest land areas at a large scale.

However, the SVM algorithm often failed to accurately distinguish spectrally similar mangrove forests from aquatic herbs and water surfaces. Hu et al. [19] found that spectral-temporal variability metrics could distinguish mangrove forests from agricultural fields or other natural terrestrial vegetation with high spectral similarity, but some aquatic plants were still misclassified. Alternatively, Jia et al. [44] used K-nearest neighbor (KNN) for object-based classification; however, mangrove forests were still incorrectly distinguished from water surfaces. Thus, the results show that machine-learning algorithms have yet to clearly resolve the misclassification problem of mangrove forest land cover classifications.

In the ground survey of mangrove forests on the Hainan Island, in areas with high mangrove mortality, the local government would usually plant mangrove seedlings frequently, which caused the mangrove forests in most areas to be at the seedling stage. However, Landsat satellite data with a 30 m spatial resolution were not effective in identifying mangrove forests at the seedling stage, which caused the mangrove forest land area identified by SVM and the mangrove crown surface cover identified by Res-UNet to be smaller than the studies of Hu et al. [19] and Jia et al. [44] (Table 8). Furthermore, the area of the mangrove crown surface cover identified by Res-UNet was more detailed and could better reflect the characteristics of the distribution of the patches of mangrove forests while offering more advantages for analyzing the fragmentation of mangrove forests. The Res-UNet deep learning not only produced a high OA but also significantly reduced misclassifications. Specifically, most of the mixed pixels containing spectrally similar aquatic plants and water surfaces to the mangrove forests were correctly separated by this algorithm. Therefore, in mangrove areas difficult to access in the surface cover of the ground survey, the mangrove forest crown could be obtained with the help of Res-UNet deep learning. The Res-UNet is more effective in identifying a large-scale area of mangrove crown surface cover area. In addition, the mangrove forest crown surface cover area is helpful for us to explore changes in mangrove biomass and carbon storage.

Table 8. Comparison of mangrove forest areas in the Hainan Island among different studies.

Name	Classification Algorithm	Mangrove Forests Area (ha)						
		1991	1996	2000	2007	2010	2015	2021
Mangrove forest land area in this study	SVM	3081	2917	2851	3030	3072	3493	3827
Mangrove forests crown surface cover area in this study	Res-UNet	1740	2077	1985	2372	2695	2234	3439
Mangrove forest land area Hu et al. [19]	RF	1990	1995	2000	2005	2010	2015	
		3701	3141	3235	3305	3623	3702	
Mangrove forest land area Jia et al. [44]	KNN	1990		2000		2010	2015	
		4809		3978		3576	4017	

4.2. Spatiotemporal Evolution of Mangrove Forests in the Hainan Island

During 1991–2021, the total area of the mangrove forest crown surface coverage on the Hainan Island showed a net increase of 1698.48 ha, corresponding to an annual change rate of 2.27%$\cdot$yr^{-1}. Related studies have shown that since the early 1990s, China has paid increasing attention to wetland conservation, with the government enacting a series of corresponding protective laws and regulations, including the China Biodiversity Conservation Action Plan (State Environmental Protection Administration, 1994), Agenda 21 Forestry Action Plan (State Forestry Administration, 1995; 1996), Ecological Environmental Protection Plan (State Council, 1998), and the Wetland Conservation Action Plan (State Forestry Administration, 2000) [42]. Combining the change in center mass, population trends, and socioeconomic developments in the Hainan Island, it was further revealed that mangrove forests near landed areas were rapidly decreasing and expanding to the mudflats by the sea due to population growth and urbanization.

This study revealed that in both 2000 and 2015, the area of the mangrove forest crown surface cover on the Hainan Island decreased, NP increased, and both LPI and AI decreased, indicative of the continued deterioration and fragmentation of the mangrove forest connectivity during these two phases. Changes in LSI indicated that landscape shape complexity was also increasing. With the gradual progress of urbanization, the interference of human activities on the landscape pattern also proved to be increasing; therefore, in landscape pattern evolution, fragmentation levels are growing, leading to the increased complexity of landscape patches. Although the area of the mangrove forest crown surface cover in the Hainan Island has increased over the past three decades largely due to the intensification of mangrove forest restoration efforts, negative growth occurred approximately every 10 to 15 years throughout the study period. Especially in 2015, the crown surface cover area of the mangrove forest on the Hainan Island decreased significantly. However, in six years, it added more than a thousand hectares. The reason for this can be found in the Annual Statistical Report of the Hainan Province. From 2015 to 2021, the total area of shelter forests planted on the Hainan Island reached 14,661 hectares. This shows that the increase in planted mangrove forests based on conservation strategies and the decrease in naturally occurring mangrove forests may cause increased landscape fragmentation, and the landscape shape is single. Furthermore, the survival rate of artificially planted mangrove forests is low [44], indicating a relatively low overall conservation efficiency. Therefore, future mangrove protection and management should be based on protection and supplemented by restoration, as maintaining the current health of existing mangrove ecosystems to improve their resilience is usually more time efficient and economical than planting large amounts of new mangrove forests [45].

Spatially, the arial changes in the mangrove forest crown surface cover observed in each city or county over the 30-year analysis period followed the overall growth trends. In addition, the landscape patterns in Haikou and Danzhou cities showed a significant improvement. According to the preliminary analysis, this results from the excellent landscape patterns in these cities due to the presence of mangrove nature reserves [46]. The expansion of the mangrove forest crown surface coverage in the Hainan Island was positively correlated with the development of the whole island economy, fishery production, and expanding urban population. This suggests that the mangrove forest crown surface cover area in the Hainan Island will increase as the rural population shifts toward urban areas with greater socioeconomic development. The rapid development of this tertiary industry and the shift of the rural population to cities have reduced the damage to mangrove forests caused by agricultural practices, such as constructing coastal lands. Furthermore, because mangrove forests maintain their natural purification ability and can provide a constant source of organic debris and other food sources for benthic organisms, organized fish farming activities may play a certain role in promoting the growth of mangrove forest areas. Therefore, the local government and residents' awareness of mangrove forest protection should be increased while focusing on maintaining the ecological environment of mangrove forests; furthermore, the benefits of resources should be optimized for sustainable fish farming, so a synergistic effect between ecological protection and economic development can be achieved.

5. Conclusions

Using Landsat imagery data in this study alongside employed machine-learning (SVM) and deep learning algorithms (Res-UNet) to extract information from tropical mangrove forests meant that the accuracy of these two methods could be analyzed and compared. The OA for the extraction of the mangrove forest spatial distribution extraction produced values of 93.11 $\pm$ 1.54 and 96.43 $\pm$ 15% for SVM and Res-UNet, respectively. The superior classification results were produced by the deep learning algorithm compared to machine learning, as the proposed model of Res-UNet combined a semantic segmentation network (U-Net) and the feature extraction network ResNet-18. This method effectively resolved previous issues regarding the misclassification of spectrally similar pixels in large-scale study areas. Moreover, the Res-UNet algorithm was more efficient and accurate at extracting the crown surface cover area of mangrove forests, providing an important foundation for the refined calculation of the carbon sequestration potential for these forests.

The present study analyzed the spatiotemporal changes in the tropical mangrove landscape patterns on the Hainan Island over the past 30 years from multiple perspectives, including the corresponding changes in crown surface cover, landscape fragmentation, mass centering offsets, as well as anthropogenic and climatic factors. The results revealed that mangrove forests in most areas underwent an overall trend of growth. Although there were various spatial differences among cities and counties, the recorded changes to the mangrove forests were mainly influenced by an increase in landscape fragmentation due to human disturbance. Additionally, this study assessed the relationships between changes to the tropical mangrove forested land area or crown surface coverage as responses to mechanisms of shifting climate and socioeconomic factors across the Hainan Island. Although this study focused on the socioeconomic factors affecting mangrove forest dynamics, and climatic and environmental factors, it also investigated how these factors contributed to these corresponding changes. For example, it was found that the average annual rainfall, as well as average annual minimum and maximum temperatures, were positively correlated with mangrove forest crown surface cover area changes in the Hainan Island; however, these correlations were not significant. Only the growth of mangrove forests in Chengmai County and Danzhou City was significantly correlated with climatic factors. Because, compared with human activity disturbances, the process of climate factors affecting mangrove wetlands has an inherent lag component, the impacts of more gradual environmental changes on mangrove ecosystems appear relatively insignificant [47]. Furthermore, the strong interference of human activities makes the evolutionary mechanisms that affect mangrove landscapes highly complex; therefore, it is necessary to obtain additional data related to the influencing factors for in-depth analyses. Therefore, more field surveys and remote sensing monitoring data are required to further study the integrated driving forces of mangrove forest dynamics. More detailed and perfect suggestions must be presented for mangrove forest nature reserve-related landscape planning to provide more appropriate ideas for tropical mangrove forest protection.

Author Contributions: Conceived the research route, Z.Q. and B.C.; designed, and performed the experiments, Z.Q., C.F., X.S., Y.X., C.W., J.L. and Y.F.; analyzed the data and wrote the main manuscript, Z.Q., C.F. and X.S. All authors have read and agreed to the published version of the manuscript.

Funding: This research was funded by the "National Natural Science Foundation of China (Grant number 32160364)", the "Hainan Provincial Key Research and Development Plan of China (Grant number ZDYF2021SHFZ110)", the "Hainan Provincial Natural Science Foundation of China (Grant number 320QN185)", the "Scientific Research Staring Foundation of Hainan University (Grant number KYQD(ZR)20056)" and the "Science and Technology Project of Haikou City, China (Grant number 2020-057)".

Data Availability Statement: These data can be found here: http://doi.org/10.6084/m9.figshare.21405531 (accessed on 8 September 2022).

Acknowledgments: The authors thank those students who assisted with fieldwork and data collection, and the instructor for their constructive comments on the improvement of this study.

Conflicts of Interest: The authors declare no conflict of interest.

Appendix A

Appendix A.1

In this study, a handheld GPS and Google Earth were used to survey the main distribution areas of the mangrove forests in 12 cities and counties along the coast of the Hainan Island in October 2021. The dominant tree species in each city are shown in Table A1.

Table A1. Information on dominant tree species of mangrove forests in the Hainan Island obtained from a ground survey.

Distribution of Dominant Mangrove Forests Tree Species in Hainan Island, 2021		
City/County	**Tree Species**	
Haikou	*Acanthus ilicifolius* L. *Acrostichum speciosum* Will. *Aegiceras corniculatum* (Linn.) Blanco *Avicennia marina* (Forsk) Vierh. *Bruguiera gymnorrhiza* (Linn.) Sav. *Bruguiera sexangula* (Lour.) Poir. *Bruguiera sexangula* (Lour.) Poir. var. *rhynchopetala* Ko *Ceriops tagal* (Perr.) C. B. Rob.	*Excoecaria agallocha* Linn. *Hibiscus tiliaceus* Linn. *Kandelia obovata* Sheue, Liu et Yong *Laguncularia racemosa* Gaertn. f. *Pongamia pinnata* (Linn.) Pierre *Rhizophora apiculata* Blume *Rhizophora stylosa* Griff *Sonneratia apetala* Buch. -Ham.
Sanya	*Aegiceras corniculatum* (Linn.) Blanco *Avicennia marina* (Forsk) Vierh. *Ceriops tagal* (Perr.) C. B. Rob. *Lumnitzera racemosa* Willd *Rhizophora apiculata* Blume	*Rhizophora stylosa* Griff. *Sonneratia* × *hainanensis* Ko, E. Y. Chen et W. Y. Chen *Sonneratia alba* J. Smith *Sonneratia ovata* Backer *Xylocarpus granatum* J. Koenig
Wenchang	*Avicennia marina* (Forsk) Vierh. *Bruguiera gymnorrhiza* (Linn.) Sav. *Bruguiera sexangula* (Lour.) Poir. var. *rhynchopetala* Ko *Ceriops tagal* (Perr.) C. B. Rob. *Excoecaria agallocha* Linn. *Hibiscus tiliaceus* Linn. *Kandelia obovata* Sheue, Liu et Yong *Laguncularia racemosa* Gaertn. f.	*Lumnitzera littorea* (Jack) Voigt *Rhizophora apiculata* Blume *Rhizophora stylosa* Griff. *Sonneratia* × *hainanensis* Ko, E. Y. Chen et W. Y. Chen *Sonneratia alba* J. Smith *Sonneratia caseolaris* (Linn.) Engl. *Sonneratia ovata* Backer
Qionghai	*Bruguiera gymnorrhiza* (Linn.) Sav. *Cerbera manghas* L. *Hibiscus tiliaceus* Linn.	*Sonneratia* × *hainanensis* Ko, E. Y. Chen et W. Y. Chen *Sonneratia alba* J. Smith *Sonneratia ovata* Backer
Wanning	*Bruguiera gymnorrhiza* (Linn.) Sav. *Cerbera manghas* L. *Excoecaria agallocha* Linn.	*Hibiscus tiliaceus* Linn. *Nypa fruticans* Wurmb. *Sonneratia caseolaris* (Linn.) Engl.
Chengmai	*Aegiceras corniculatum* (Linn.) Blanco *Avicennia marina* (Forsk) Vierh. *Hibiscus tiliaceus* Linn. *Kandelia obovata* Sheue, Liu et Yong	*Lumnitzera littorea* (Jack) Voigt *Rhizophora apiculata* Blume *Rhizophora stylosa* Griff. *Sonneratia caseolaris* (Linn.) Engl.
Lingao	*Aegiceras corniculatum* (Linn.) Blanco *Avicennia marina* (Forsk) Vierh. *Excoecaria agallocha* Linn.	*Hibiscus tiliaceus* Linn. *Rhizophora stylosa* Griff.
Danzhou	*Aegiceras corniculatum* (Linn.) Blanco *Avicennia marina* (Forsk) Vierh. *Hibiscus tiliaceus* Linn.	*Kandelia obovata* Sheue, Liu et Yong *Lumnitzera littorea* (Jack) Voigt *Rhizophora stylosa* Griff.
Dongfang	*Avicennia marina* (Forsk) Vierh.	*Laguncularia racemosa* Gaertn. f.
Ledong	*Rhizophora stylosa* Griff. *Lumnitzera littorea* (Jack) Voigt	*Avicennia marina* (Forsk) Vierh. *Laguncularia racemosa* Gaertn. f.

Table A1. *Cont.*

Distribution of Dominant Mangrove Forests Tree Species in Hainan Island, 2021		
City/County	**Tree Species**	
Lingshui	*Avicennia marina* (Forsk) Vierh. *Bruguiera gymnorrhiza* (Linn.) Sav. *Bruguiera sexangula* (Lour.) Poir. var. *rhynchopetala* Ko *Kandelia obovata* Sheue, Liu et Yong *Laguncularia racemosa* Gaertn. f.	*Rhizophora stylosa* Griff. *Sonneratia* × *hainanensis* Ko, E. Y. Chen et W. Y. Chen *Sonneratia alba* J. Smith *Sonneratia apetala* Buch. -Ham. *Sonneratia ovata* Backer
Changjiang	*Avicennia marina* (Forsk) Vierh.	*Rhizophora stylosa* Griff

Appendix A.2

Figure A1 indicates the classification results of the SVM machine-learning algorithm, and Figure A2 indicates the classification results of the Res-UNet deep learning algorithm.

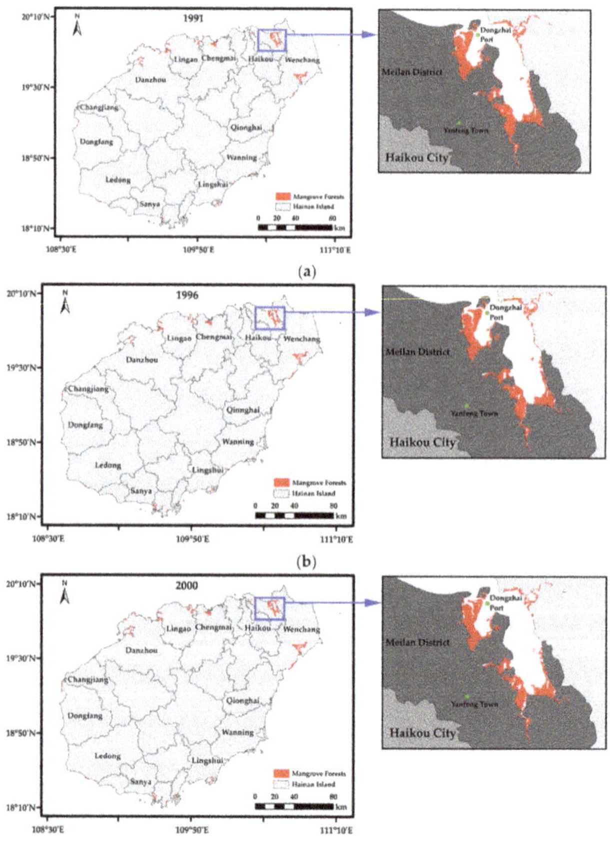

Figure A1. *Cont.*

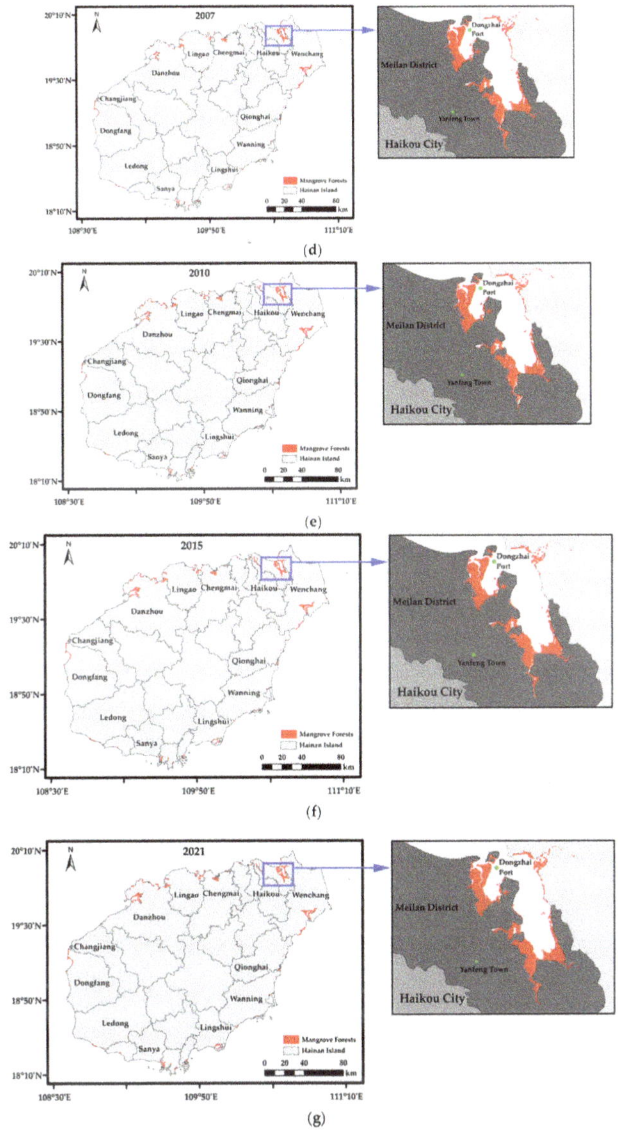

Figure A1. SVM machine-learning classification results. (**a**) 1991; (**b**) 1996; (**c**) 2000; (**d**) 2007; (**e**) 2010; (**f**) 2015; (**g**) 2021.

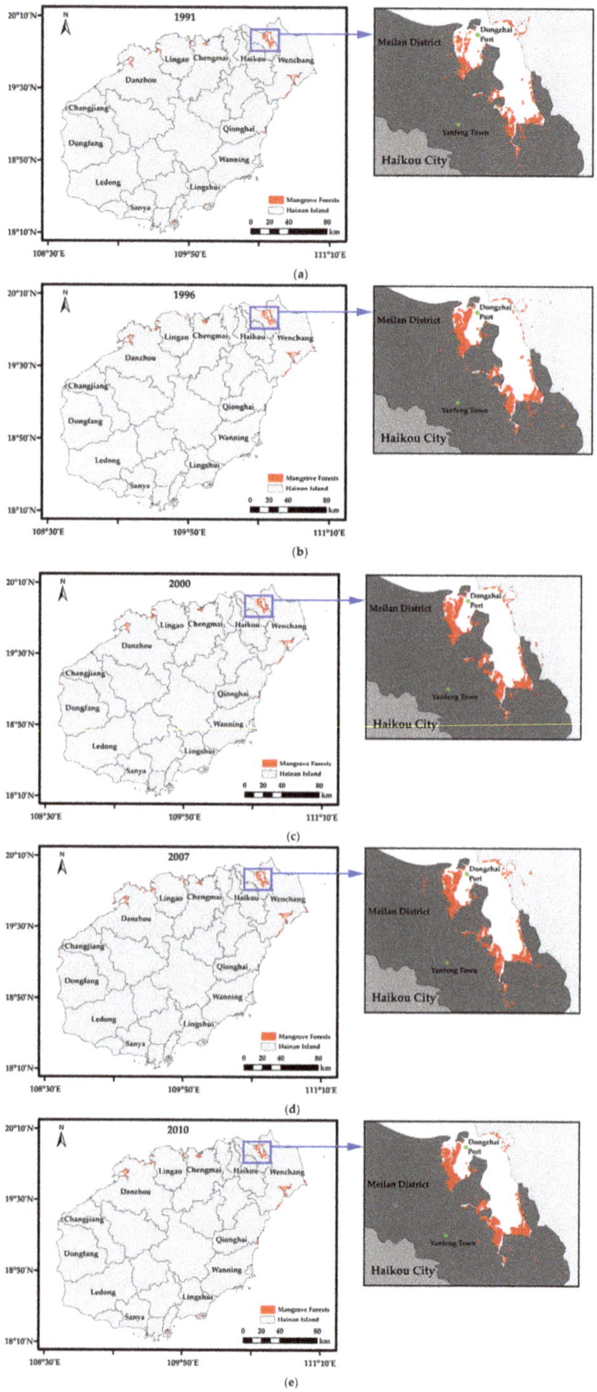

Figure A2. *Cont.*

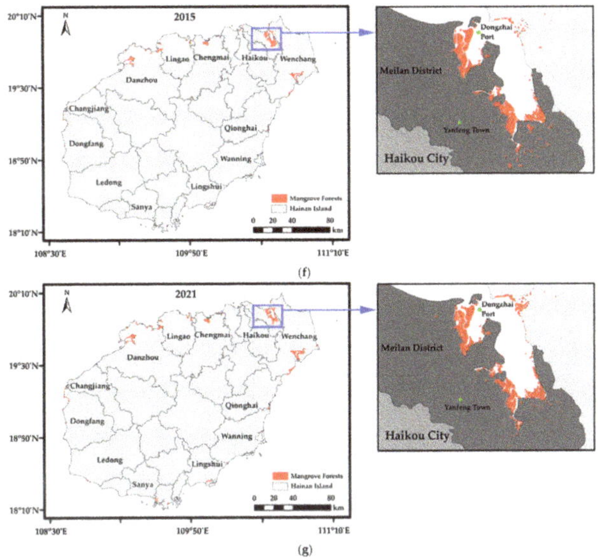

Figure A2. Classification results of Res-UNet machine-learning. (**a**) 1991; (**b**) 1996; (**c**) 2000; (**d**) 2007; (**e**) 2010; (**f**) 2015; (**g**) 2021.

References

1. Tomlinson, P.B. *The Botany of Mangroves*; Cambridge University Press: Cambridge, MA, USA, 2016.
2. Simard, M.; Fatoyinbo, L.; Smetanka, C.; Rivera-Monroy, V.H.; Castañeda-Moya, E.; Thomas, N.; Van der Stocken, T. Mangrove canopy height globally related to precipitation, temperature and cyclone frequency. *Nat. Geosci.* **2019**, *12*, 40–45. [CrossRef]
3. Getzner, M.; Islam, M.S. Ecosystem services of mangrove forests: Results of a meta-analysis of economic values. *Int. J. Environ. Res. Public Health* **2020**, *17*, 5830. [CrossRef] [PubMed]
4. Donato, D.C.; Kauffman, J.B.; Murdiyarso, D.; Kurnianto, S.; Stidham, M.; Kanninen, M. Mangroves among the most carbon-rich forests in the tropics. *Nat. Geosci.* **2011**, *4*, 293–297. [CrossRef]
5. Lee, S.Y.; Primavera, J.H.; Dahdouh-Guebas, F.; McKee, K.; Bosire, J.O.; Cannicci, S.; Diele, K.; Fromard, F.; Koedam, N.; Marchand, C. Ecological role and services of tropical mangrove ecosystems: A reassessment. *Glob. Ecol. Biogeogr.* **2014**, *23*, 726–743. [CrossRef]
6. Goldberg, L.; Lagomasino, D.; Thomas, N.; Fatoyinbo, T. Global declines in human-driven mangrove loss. *Glob. Chang. Biol.* **2020**, *26*, 5844–5855. [CrossRef]
7. Friess, D.A.; Rogers, K.; Lovelock, C.E.; Krauss, K.W.; Hamilton, S.E.; Lee, S.Y.; Lucas, R.; Primavera, J.; Rajkaran, A.; Shi, S. The state of the world's mangrove forests: Past, present, and future. *Annu. Rev. Environ. Resour.* **2019**, *44*, 89–115. [CrossRef]
8. Murray, N.J.; Phinn, S.R.; DeWitt, M.; Ferrari, R.; Johnston, R.; Lyons, M.B.; Clinton, N.; Thau, D.; Fuller, R.A. The global distribution and trajectory of tidal flats. *Nature* **2019**, *565*, 222–225. [CrossRef]
9. Wang, L.; Jia, M.; Yin, D.; Tian, J. A review of remote sensing for mangrove forests: 1956–2018. *Remote Sens. Environ.* **2019**, *231*, 111223. [CrossRef]
10. Liu, X.; Yang, X.; Zhang, T.; Wang, Z.; Zhang, J.; Liu, Y.; Liu, B. Remote sensing based conservation effectiveness evaluation of mangrove reserves in china. *Remote Sens.* **2022**, *14*, 1386. [CrossRef]
11. Purnamasayangsukasih, P.R.; Norizah, K.; Ismail, A.A.; Shamsudin, I. A review of uses of satellite imagery in monitoring mangrove forests. Proceeding of 8th IGRSM International Conference and Exhibition on Geospatial & Remote Sensing (IGRSM 2016), Kuala Lumpur, Malaysia, 13–14 April 2016; 37, p. 012034. [CrossRef]
12. Hauser, L.T.; Vu, G.N.; Nguyen, B.A.; Dade, E.; Nguyen, H.M.; Nguyen, T.T.Q.; Le, T.Q.; Vu, L.H.; Tong, A.T.H.; Pham, H.V. Uncovering the spatio-temporal dynamics of land cover change and fragmentation of mangroves in the Ca Mau peninsula, Vietnam using multi-temporal SPOT satellite imagery (2004–2013). *Appl. Geogr.* **2017**, *86*, 197–207. [CrossRef]
13. Proisy, C.; Viennois, G.; Sidik, F.; Andayani, A.; Enright, J.A.; Guitet, S.; Gusmawati, N.; Lemonnier, H.; Muthusankar, G.; Olagoke, A. Monitoring mangrove forests after aquaculture abandonment using time series of very high spatial resolution satellite images: A case study from the perancak estuary, bali, indonesia. *Mar. Pollut. Bull.* **2018**, *131*, 61–71. [CrossRef] [PubMed]
14. Pham, T.D.; Xia, J.; Ha, N.T.; Bui, D.T.; Le, N.N.; Takeuchi, W. A review of remote sensing approaches for monitoring blue carbon ecosystems: Mangroves, seagrasses and salt marshes during 2010–2018. *Sensors* **2019**, *19*, 1933. [CrossRef] [PubMed]
15. Gaw, L.Y.; Linkie, M.; Friess, D.A. Mangrove forest dynamics in tanintharyi, myanmar from 1989–2014, and the role of future economic and political developments. *Singap. J. Trop. Geogr.* **2018**, *39*, 224–243. [CrossRef]

16. Hu, L.; Li, W.; Xu, B. The role of remote sensing on studying mangrove forest extent change. *Int. J. Remote Sens.* **2018**, *39*, 6440–6462. [CrossRef]
17. Thakur, S.; Mondal, I.; Ghosh, P.; Das, P.; De, T. A review of the application of multispectral remote sensing in the study of mangrove ecosystems with special emphasis on image processing techniques. *Spat. Inf. Res.* **2020**, *28*, 39–51. [CrossRef]
18. Buck, O.; Millán, V.E.G.; Klink, A.; Pakzad, K. Using information layers for mapping grassland habitat distribution at local to regional scales. *Int. J. Appl. Earth Obs. Geoinf.* **2015**, *37*, 83–89. [CrossRef]
19. Hu, L.; Li, W.; Xu, B. Monitoring mangrove forest change in china from 1990 to 2015 using landsat-derived spectral-temporal variability metrics. *Int. J. Appl. Earth Obs. Geoinf.* **2018**, *73*, 88–98. [CrossRef]
20. Abdi, A.M. Land cover and land use classification performance of machine learning algorithms in a boreal landscape using sentinel-2 data. *GIScience Remote Sens.* **2020**, *57*, 1–20. [CrossRef]
21. Guo, Y.; Liao, J.; Shen, G. Mapping large-scale mangroves along the maritime silk road from 1990 to 2015 using a novel deep learning model and landsat data. *Remote Sens.* **2021**, *13*, 245. [CrossRef]
22. Li, H.; Hu, B.; Li, Q.; Jing, L. Cnn-based individual tree species classification using high-resolution satellite imagery and airborne lidar data. *Forests* **2021**, *12*, 1697. [CrossRef]
23. Zhang, Z.; Liu, Q.; Wang, Y. Road extraction by deep residual u-net. *IEEE Geosci. Remote Sens. Lett.* **2018**, *15*, 749–753. [CrossRef]
24. Cao, K.; Zhang, X. An improved res-unet model for tree species classification using airborne high-resolution images. *Remote Sens.* **2020**, *12*, 1128. [CrossRef]
25. Ibharim, N.; Mustapha, M.; Lihan, T.; Mazlan, A. Mapping mangrove changes in the matang mangrove forest using multi temporal satellite imageries. *Ocean Coast. Manag.* **2015**, *114*, 64–76. [CrossRef]
26. Son, N.T.; Thanh, B.X.; Da, C.T. Monitoring mangrove forest changes from multi-temporal landsat data in can gio biosphere reserve, vietnam. *Wetlands* **2016**, *36*, 565–576. [CrossRef]
27. Zhen, J.; Liao, J.; Shen, G. Mapping mangrove forests of dongzhaigang nature reserve in china using landsat 8 and radarsat-2 polarimetric sar data. *Sensors* **2018**, *18*, 4012. [CrossRef]
28. Gilani, H.; Naz, H.I.; Arshad, M.; Nazim, K.; Akram, U.; Abrar, A.; Asif, M. Evaluating mangrove conservation and sustainability through spatiotemporal (1990–2020) mangrove cover change analysis in pakistan. *Estuar. Coast. Shelf Sci.* **2021**, *249*, 107128. [CrossRef]
29. Giri, C.; Long, J.; Abbas, S.; Murali, R.M.; Qamer, F.M.; Pengra, B.; Thau, D. Distribution and dynamics of mangrove forests of south asia. *J. Environ. Manag.* **2015**, *148*, 101–111. [CrossRef]
30. Liao, J.; Zhen, J.; Zhang, L.; Metternicht, G. Understanding dynamics of mangrove forest on protected areas of hainan island, china: 30 years of evidence from remote sensing. *Sustainability* **2019**, *11*, 5356. [CrossRef]
31. Vapnik, V. *The Nature of Statistical Learning Theory*; Springer Science & Business Media: Berlin/Heidelberg, Germany, 1999.
32. Maxwell, A.E.; Warner, T.A.; Fang, F. Implementation of machine-learning classification in remote sensing: An applied review. *Int. J. Remote Sens.* **2018**, *39*, 2784–2817. [CrossRef]
33. Ronneberger, O.; Fischer, P.; Brox, T. In U-net: Convolutional networks for biomedical image segmentation. In Proceedings of the International Conference on Medical Image Computing and Computer-Assisted Intervention, Munich, Germany, 5–9 October 2015; Springer: Berlin/Heidelberg, Germany, 2015; pp. 234–241. Available online: https://lmb.informatik.uni-freiburg.de/people/ronneber/u-net/ (accessed on 1 September 2022).
34. He, K.; Zhang, X.; Ren, S.; Sun, J. Deep residual learning for image recognition. In Proceedings of the IEEE Conference on Computer Vision and Pattern Recognition, Las Vegas, NV, USA, 27–30 June 2016; pp. 770–778. Available online: https://openaccess.thecvf.com/content_cvpr_2016/html/He_Deep_Residual_Learning_CVPR_2016_paper.html (accessed on 1 September 2022).
35. Chaurasia, A.; Culurciello, E. Linknet: Exploiting encoder representations for efficient semantic segmentation. In Proceedings of the 2017 IEEE Visual Communications and Image Processing (VCIP), St. Petersburg, FL, USA, 10–13 December 2017; pp. 1–4. [CrossRef]
36. Kingma, D.P.; Ba, J. Adam: A method for stochastic optimization. *arXiv* **2014**, arXiv:1412.6980. [CrossRef]
37. Anand, A. *Unit-14 Accuracy Assessment. Processing and Classification of Remotely Sensed Images. Remote Sensing and Image Interpretaion*; Indiara Gandhi National Open University: Delhi, India, 2017; pp. 59–78.
38. Puyravaud, J.-P. Standardizing the calculation of the annual rate of deforestation. *For. Ecol. Manag.* **2003**, *177*, 593–596. [CrossRef]
39. Zhang, J.; Yang, X.; Wang, Z.; Zhang, T.; Liu, X. Remote sensing based spatial-temporal monitoring of the changes in coastline mangrove forests in china over the last 40 years. *Remote Sens.* **2021**, *13*, 1986. [CrossRef]
40. Fu, F.; Deng, S.; Wu, D.; Liu, W.; Bai, Z. Research on the spatiotemporal evolution of land use landscape pattern in a county area based on ca-markov model. *Sustain. Cities Soc.* **2022**, *80*, 103760. [CrossRef]
41. Li, H.; Man, W.; Li, X.; Ren, C.; Wang, Z.; Li, L.; Jia, M.; Mao, D. Remote sensing investigation of anthropogenic land cover expansion in the low-elevation coastal zone of liaoning province, china. *Ocean Coast. Manag.* **2017**, *148*, 245–259. [CrossRef]
42. Zheng, Y.; Takeuchi, W. Quantitative assessment and driving force analysis of mangrove forest changes in china from 1985 to 2018 by integrating optical and radar imagery. *ISPRS Int. J. Geo-Inf.* **2020**, *9*, 513. [CrossRef]
43. Singh, S.K.; Srivastava, P.K.; Gupta, M.; Thakur, J.K.; Mukherjee, S. Appraisal of land use/land cover of mangrove forest ecosystem using support vector machine. *Environ. Earth Sci.* **2014**, *71*, 2245–2255. [CrossRef]

44. Jia, M.; Wang, Z.; Zhang, Y.; Mao, D.; Wang, C. Monitoring loss and recovery of mangrove forests during 42 years: The achievements of mangrove conservation in china. *Int. J. Appl. Earth Obs. Geoinf.* **2018**, *73*, 535–545. [CrossRef]
45. Schmitt, K.; Duke, N.C. Mangrove management, assessment and monitoring. In *Tropical Forestry Handbook*; Springer: Berlin/Heidelberg, Germany, 2015; pp. 1–29. [CrossRef]
46. Chen, L.; Wang, W.; Zhang, Y.; Lin, G. Recent progresses in mangrove conservation, restoration and research in china. *J. Plant Ecol.* **2009**, *2*, 45–54. [CrossRef]
47. Wang, Q.; Wang, H.; Zhang, W.; Wang, Z.; Xiao, D. The correlations between wetland landscape and social-natural factors on Northwestern Yunnan Plateau. *Acta Ecol. Sin.* **2019**, *39*, 726–738. Available online: https://kns.cnki.net/kcms/detail/11.2031.Q.20181018.1458.050.html (accessed on 1 September 2022). (In Chinese).

remote sensing

Article

Landsat-8 and Sentinel-2 Based Prediction of Forest Plantation C Stock Using Spatially Varying Coefficient Bayesian Hierarchical Models

Tsikai Solomon Chinembiri [1], Onisimo Mutanga [1,*] and Timothy Dube [2]

[1] College of Agricultural, Earth and Environmental Sciences, University of KwaZulu-Natal, Private Bag X01, Scottsville 3209, South Africa
[2] Institute of Water Studies, Department of Earth Sciences, University of the Western Cape, Private Bag X17, Bellville 7535, South Africa
* Correspondence: mutangao@ukzn.ac.za

Abstract: This study sought to establish the performance of Spatially Varying Coefficient (SVC) Bayesian Hierarchical models using Landsat-8, and Sentinel-2 derived auxiliary information in predicting plantation forest carbon (C) stock in the eastern highlands of Zimbabwe. The development and implementation of Zimbabwe's land reform program undertaken in the year 2000 and the subsequent redistribution and resizing of large-scale land holdings are hypothesized to have created heterogeneity in aboveground forest biomass in plantation ecosystems. The Bayesian hierarchical framework, accommodating residual spatial dependence and non-stationarity of model predictors, was evaluated. Firstly, SVC models utilizing Normalized Difference Vegetation Index (NDVI), Soil Adjusted Vegetation Index (SAVI), and Enhanced Vegetation Index (EVI), derived from Landsat-8 and Sentinel-2 data and 191 sampled C stock observations were constructed. The SVC models built for each of the two multispectral remote sensing data sets were assessed based on the goodness of fit criterion as well as the predictive performance using a 10-fold cross-validation technique. The introduction of spatial random effects in the form of Landsat-8 and Sentinel-2 derived covariates to the model intercept improved the model fit and predictive performance where residual spatial dependence was dominant. For the Landsat-8 C stock predictive model, the RMSPE for the non-spatial, Spatially Varying Intercept (SVI) and SVC models were 8 $MgCha^{-1}$, 7.77 $MgCha^{-1}$, and 6.42 $MgCha^{-1}$ whilst it was 7.85 $MgCha^{-1}$, 7.69 $MgCha^{-1}$ and 6.23 $MgCha^{-1}$ for the Sentinel-2 C stock predictive models, respectively. Overall, the Sentinel-2-based SVC model was preferred for predicting C stock in plantation forest ecosystems as its model provided marginally tighter credible intervals, [1.17–1.60] $MgCha^{-1}$ when compared to the Landsat-8 based SVC model with 95% credible intervals of [1.13–1.62] $Mg\ Cha^{-1}$. The built SVC models provided an understanding of the performance of the multispectral remote sensing derived predictors for modeling C stock and thus provided an essential foundation required for updating the current carbon forest plantation databases.

Keywords: Bayesian hierarchical modelling; geostatistics; *Eucalyptus grandis*; *Eucalyptus camaldulensis*; *Pinus patula*; spatial random effects; spatially varying coefficient

Citation: Chinembiri, T.S.; Mutanga, O.; Dube, T. Landsat-8 and Sentinel-2 Based Prediction of Forest Plantation C Stock Using Spatially Varying Coefficient Bayesian Hierarchical Models. *Remote Sens.* **2022**, *14*, 5676. https://doi.org/10.3390/rs14225676

Academic Editors: Huaqiang Du, Wenyi Fan, Mingshi Li, Weiliang Fan and Fangjie Mao

Received: 19 October 2022
Accepted: 3 November 2022
Published: 10 November 2022

1. Introduction

Since the onset of the Fast Track Land Reform (FTLRP) program in Zimbabwe and the subsequent redistribution and resizing of large-scale land holdings in the year 2000, plantation forests within and around the neighborhood of resettlement areas continuously faced physical distress [1]. Zimbabwe lost close to 224,000 ha of tree cover, which is equivalent to a forest degradation of 17% and 88 Mt of CO_2 emissions from 2000 to 2021 [2]. The steady decline of land under forest over the years has therefore been mainly driven by the activities of the year 2000 taking place on resettled land, which included wildfires, illegal logging, mining, and agriculture expansion. Yet the amount of additional biomass that can be accumulated in these areas depends much on the forest condition and land management practices.

A number of studies modeling the relationship between C stock and remote sensing-derived information adopt a standard approach where the effects of independent variables, including vegetation indices on C stock, remain constant over space in the model [3,4]. Given the history of Zimbabwe's land reform program, it is well known that management and conservation practices of plantation forests have been changing over the years, particularly since the adoption of the program in the year 2000. Greater portions of formally designated commercial plantation forests were occupied and replaced by subsistence farmers across the country. It is upon this premise that it cannot be assumed that the impact of covariates on C stock is constant in these ecosystems. The carbon sequestration potential in various regions of the occupied forest plantations, therefore, remains unknown as the management and conservation practices of the managed ecosystems have been largely modified and altered [5]. Yet government and other stakeholders in the timber industry are obliged under the 2015 Paris climate agreement to provide accurate accounts of the country's carbon sequestration potential for managed and natural forest ecosystems.

The literature proffers methods used in the estimation of AGB, either as direct or indirect approaches from forest inventory data. These methods either use allometric equations, conversion factors such as wood density, or biomass expansion factors (BEF) [1]. In spite of the advantages of using conventional methods given in [6,7] as generally providing accurate AGB estimates, these techniques are also regarded as time-consuming and environmentally unfriendly. The inaccessibility of some areas due to complicated topography and forest conditions also makes conventional methods less attractive to AGB estimation, especially in extensive areas [8]. However, remote sensing is emerging as a promising technique free of the aforementioned limitations as it offers cost-effective methods of AGB monitoring through stratification of canopy density and forest types. Repeated application of remote sensing leads to the generation of historical data critical for change detection analysis and incorporation into a Geographic Information System (GIS) for integration with other datasets [9]. However, remote sensing methods are also not immune from limitations as they usually face limitations in areas of bad weather conditions, especially in areas with cloud cover, and require a certain level of training for effective use and application [7].

Remote sensing C stock data derived from satellite imagery has significantly grown over the years. This data has been the basis for informing international policy agreements associated with CO_2 emissions into the atmosphere, mainly from deforestation and other land use land cover changes (LULCC) [10]. Remote sensing is regarded as a powerful tool for deriving AGB and forest structure as it offers practical means of acquiring spatially-distributed forest carbon from local, continental, and global scales [5,11]. Forest biomass can generally be measured from three broad categories of remote sensing (RS) data, namely, passive optical remote sensing, radio detection and ranging, microwave (radar), and light detection and ranging (LiDAR) [8,10]. Passive optical spectral reflectance is responsive to vegetation structural attributes (tree density, leaf area index, and crown size), shadow, and texture [8,12]. Crown size, tree density, and leaf area index (LAI) are strongly correlated with AGB. Radar remote sensing measure geometrical and dielectric attributes of forests. LiDAR RS methods of biomass measurement characterise forest vertical structure and height. Remote sensing (RS) and ancillary technologies such as Geographical Information System (GIS) are practical and cost-time effective, allowing for imaging and research on extensive and inaccessible areas.

The estimation of plantation forest C stock, together with other structural parameters using new regeneration multispectral remote sensing data, is a relatively new ground in the climate change and carbon monitoring arena. Research in this field has demonstrated the potential for remote sensing data as tools for developing estimates of forest attributes, amongst them, being Above Ground Biomass (AGB), either as a standalone or coupled with other earth observation techniques [11,13]. Spatial regression models applied in mapping C or biomass using new-generation multispectral remote sensing data may fail to clearly make room for residual spatial dependence [14,15]. Modeling of natural resource variables without explicitly accommodating spatial variation can be justified if covariates can account for all the spatially structured dependence. Such assumptions do not hold in many practical applications involving georeferenced data.

Failure to account for spatial dependence in the modeling of regionalized variables can lead to inaccurate model parameters and incorrect predictions [9]. Subsequently, disregarding spatial correlation in electing model choices can result in higher prediction uncertainty for inference of the outcome variable. In addition to the drawbacks highlighted in the foregoing, non-Bayesian spatial modeling can further lead to underestimation of uncertainty [9,16] as traditional spatial regression estimation methods assume stationarity of the covariance matrix, $\mathbf{\Sigma}$. The ultimate effect is the derivation of standard error estimates that are unable to take all the uncertainties in the parameters into account. Checking for spatially correlated residuals when spatial data are employed in the modeling of aboveground biomass is therefore critical.

Some attention has been given to spatially varying coefficient (SVC) models in the literature [17–19]. The Bayesian framework of statistical inference is the bedrock of SVC models in which analyses make use of samples derived from Markov Chain Monte Carlo (MCMC) techniques from the posterior distribution of model parameters [20]. What makes Spatially Varying Intercept (SVI) and SVC models unique in applied geostatistical and the remote sensing literature is their ability to consider the residual spatial dependence and the non-homogeneity in model parameters differently than ordinary geostatistical approaches. SVI models assume the model intercept is spatially varying, whilst the SVC models assume all the model regression coefficients to be spatially varying [21]. A number of applications and methodologies utilizing spatially varying coefficient models are documented in the literature. Amongst them are the geographically-weighted regression (GWR) by [21], who employed the technique of canopy height prediction using remote sensing data. Application of these models resulted in significant improvement in model fit. On the other hand, [22] made provisions for spatial dependence and parameter non-homogeneity by exploring geostatistical kriging variants for forest canopy height prediction. Co-kriging and regression kriging models resulted in significant improvements in model fit.

However, it has been shown in recent times that GWR might not be robust to correlation among predictors and can potentially lead to inaccurate results when complex correlation structures are involved [18,23]. Again, from an inferential viewpoint, GWR can present problems when drawing inferences regarding prediction uncertainty and model parameters. The lack of valid underlying probability models in GWR makes prediction uncertainty, and standard parameter error estimates difficult to justify. For instance, uncertainty maps produced from kriging and co-kriging techniques make no consideration of the uncertainty in the variogram-based spatial covariance parameters. This is an established and common weakness encountered when using these geostatistical approaches [24]. It is possible to estimate spatial covariance parameters within the SVC and SVI models using a Bayesian hierarchical construction. Such an approach allows the propagation of uncertainty to the prediction of the response variable [25]. In such scenarios, a better and statistically defendable map of uncertainty can be produced than would else be produced when GWR or traditional kriging methods are utilized.

Landsat-8 and Sentinel-2 are multispectral platforms categorized as new generation remote sensing sensors with enhanced spectral and spatial properties than previous missions of the Landsat series. It is this perceived improvement in earth observation properties that we expect to give a dividend to C stock predictive models for applications in carbon accounting and inventorying under the United Nations Framework Convention on Climate Change (UNFCC) [8]. We employed a Bayesian hierarchical framework with spatially varying coefficients (SVC) using predictor information derived from the aforementioned sensors for predicting C stock. The modeling framework considered the non-stationarity and residual spatial dependence of model covariates through the inclusion of spatial random effects into the SVC models. We, therefore, developed and compared the performance of C stock predictive models under spatially varying regression coefficients derived from Landsat-8 and Sentinel-2 predictors. Prediction accuracy and uncertainty quantification for the REDD collaborative program in the developing world is a critical aspect of the Carbon Measurement, Reporting, and Verification Systems (MRVs) of the United Nations (UN-REDD, 2009; CMS, 2014). Thus, in this paper, we explored how spatially varying coefficient models constructed using a Bayesian hierarchical set-up with aiding information from Landsat-8 and Sentinel-2 multispectral sensors and implemented through Markov Chain Monte Carlo (MCMC), perform in C stock prediction in plantation forest ecosystems.

2. Materials and Methods

2.1. Study Area

The study was carried out at lot 75 A of Nyanga Downs in Nyanga district in the Eastern Highlands of Zimbabwe (Figure 1). The study area is dominated by *Eucalyptus grandis, Eucalyptus camaldulensis,* and *Pinus patula* plantation forest species which have some of its patches being used for agriculture, grazing, and gold panning and is located between latitude 32°40′E and 32°54′E and 18°10′12″S and 18°25′4″S longitude as illustrated in Figure 1 [5,26]. Grazing, agriculture, and gold panning activities came after part of the commercially owned plantation forests were redistributed to small and medium-sized indigenous farmers in 2000. This development has increased the interface between settlements and timber plantations in all forests originally designated under forest plantations in Zimbabwe. The study area covers an area of approximately 2766 ha. Rainfall amounts are varied, with a mean annual precipitation range of 741 mm to 2997 mm. Annual mean temperatures range from a minimum of 9 °C to 12 °C to a maximum of 25 °C to 28 °C. The weather is very hot, and extensive wildfires occur in the high-elevation grasslands from August to November when the grasses are dry [26].

As illustrated in Figure 1c, *Eucalyptus camaldulensis* and *Pinus patula* are the most dominant species in the study area. Pockets of cultivated land within the plantations are evident and are partly responsible for the present biomass density in the sampled region. Greater portions of former plantation vegetation have been cleared by pockets of resettled small-scale and medium-scale farmers venturing into coffee and tea plantations and, in some cases, for dairy farming. Patches of unprotected forest plantations are still present but remain vulnerable to attack for agriculture by settled farmers in the area.

2.2. Sampling Design

We employed the spatial coverage sampling scheme that exploits the Mean Squared Shortest Distance (MSSD) for the optimization of data locations in the study domain. The k-means clustering scheme for equal area coverage was therefore used [27] for obtaining a representative sample from the studied region. The work of [28] demonstrated how the mapping and estimation of mean spatial problems could be resolved through the employment of a uniform coverage sampling scheme. MSSD is particularly suited for areas where sampling campaigns cannot be extended beyond a single phase. As illustrated in Figure 2, the smallest separation distance between samples was approximately 8 m whilst the largest separation distance between samples was 2500 m.

2.3. Above-Ground Biomass Measurements and C Stock Derivation

We sampled and collected measurements of Diameter at Breast Height (DBH) for Above Ground Biomass (AGB) estimation for all trees with a DBH of more than 10 cm (1.3 m) using 500 m^2 circular plots from the 19 September 2021 to 24 October 2021 in Manicaland Province of Zimbabwe as illustrated in Figure 1. The sampling program meant to measure DBH for subsequent AGB and C stock estimation utilized the MSSD optimization function, resulting in 191 sampling plots being measured from the study area. The 191 sample plot measurements of DBH were then transformed into per plot C stock data using allometric equations of [4] for the Pinus species, whilst the allometric equations of [10] were used for deriving C stock for the Eucalyptus species. The aforementioned allometric equations were also applied by [26].

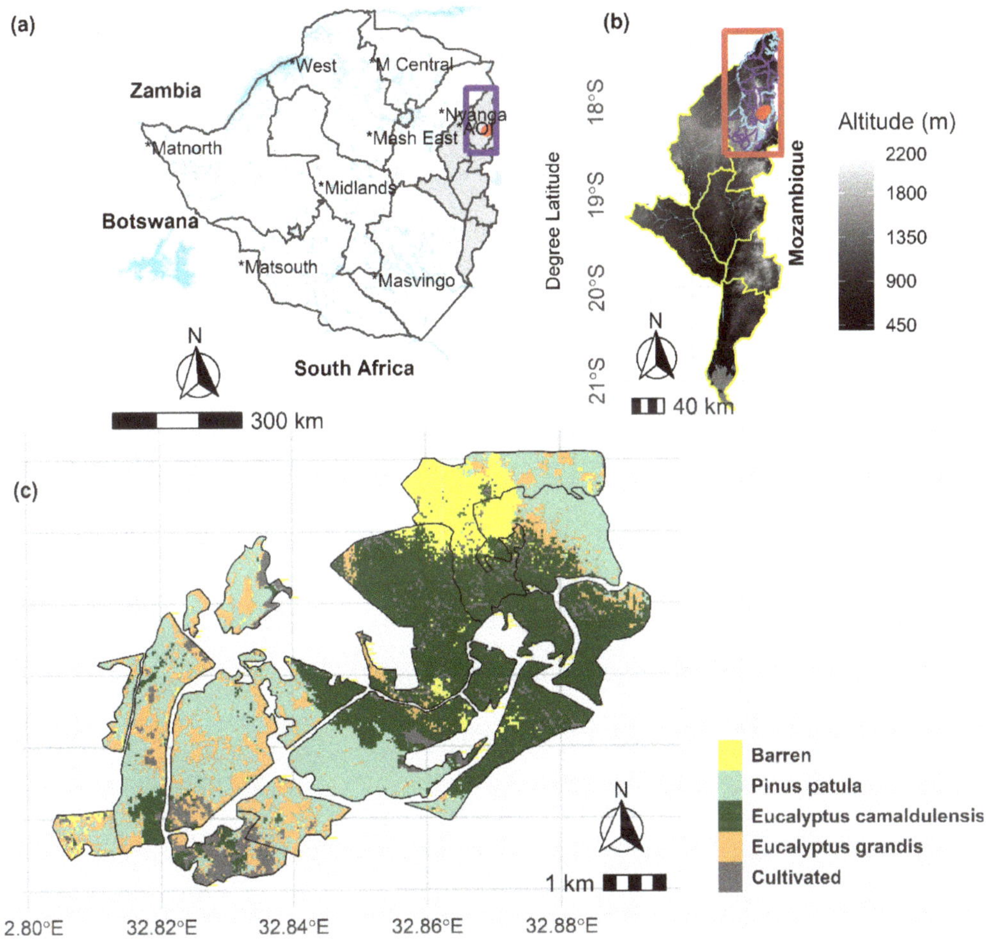

Figure 1. Map of the study area indicating (**a**) the province where samples were derived, (**b**) the study area location within the particular province, and (**c**) the spatial distribution of major plantation tree species in the studied region. * refers to Provinces in Zimbabwe whilst the box encompasses the studied area in this research.

Allometric equations shown in Equations (1) and (2) were used for the calculation of Above Ground Biomass (AGB) for the *Pinus patula and the Eucalyptus grandis* and *Eucalyptus camaldulensis* species, respectively [1]. A default conversion factor of 0.47 used by the IPCC was applied to derive AGB to C stock.

$$tDw = e^{(-1.170+2.119\times ln(dbh))} \tag{1}$$

$$tDw = 0.39 \times (dbh)^{2.142} \tag{2}$$

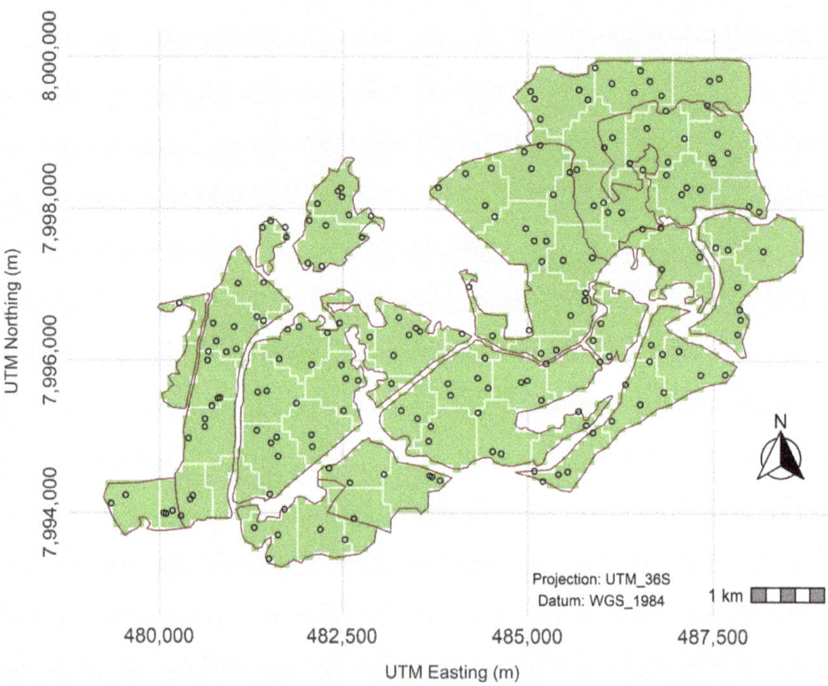

Figure 2. Study area sampling design.

2.4. Modelling Framework

It is a common geostatistical practice to assume at location $S \in D \subseteq \mathbb{R}^2$ where s is a vector of observed x, y coordinates within the domain D. A Gaussian response variable $y(s)$ is therefore modeled through the regression model as in Equation (3):

$$y(\boldsymbol{s}) = x(\boldsymbol{s})'\beta + \widetilde{x}(\boldsymbol{s})'\boldsymbol{w}(\boldsymbol{s}) + \varepsilon(\boldsymbol{s}) \quad (3)$$

$x(\boldsymbol{s})'$ denotes a set of p covariates in the model. In this case, the linear mean structure accounting for wide-scale variation in the response is comprised of $px1$ vectors of $\boldsymbol{x}(\boldsymbol{s})$ which include an intercept and spatially varying georeferenced predictors together with an associated column vector of model coefficients $\boldsymbol{\beta} = (\beta_0, \beta_1, \ldots, \beta_{p-1})'$. The $\widetilde{x}(\boldsymbol{s})$ in the model represents a q x 1 vector accommodating the intercept and those predictors from $x(s)$ whose impact on the response is assumed to vary spatially. The space-varying impact is obtained from the vector of spatial random effects $\boldsymbol{w}(\boldsymbol{s}) = (w_1(\boldsymbol{s}), w_2(\boldsymbol{s}), \ldots, w_q(\boldsymbol{s}))'$. The specification of different combinations of $\widetilde{x}(\boldsymbol{s})$ and the associated $\boldsymbol{w}(\boldsymbol{s})$ leads to the formation of different sub-models. We model $\varepsilon(\boldsymbol{s})$ as an independent white noise process that takes care of the micro-scale (measurement error) variation. As such, with the collection of n C stock locations, $S = \boldsymbol{s_1}, \boldsymbol{s_2}, \ldots, \boldsymbol{s_n}$, we assume the $\varepsilon(\boldsymbol{s_i})$'s are independent and identically distributed (*iid*) as provided by $N(0, \tau^2)$ where τ^2 is the nugget.

The spatial structure of this model is generally introduced by way of a multivariate Gaussian process (GP) [24,25,29] in which a cross-covariance function clearly models the covariance of $w(\boldsymbol{s})$ within and among data points. The added flexibility in this model is documented in the literature [17,30]. We assume in this study that elements of $\boldsymbol{w}(\boldsymbol{s})$ emanate from q independent univariate GPs. Precisely, the process associated with the $k-$th prediction is $w_k(s) \sim GP(0, (.,.;\boldsymbol{\theta}_k))$ where $C(\boldsymbol{s}, \boldsymbol{s}^*; \boldsymbol{\theta}_k) = Cov(w_k(\boldsymbol{s}), w_k(\boldsymbol{s}^*))$ is a valid spatial covariance function modelling the covariance related to a pair of observations s and s^*. The process outcomes are gathered into an $n \times 1$ vector, say $w_k = (w_k(s_1), \ldots, w_k(s_n))'$ which permits a multivariate normal distribution $MVN(0, \Sigma_k)$, in which Σ_k is an $n \times n$ covariance matrix of w_k with the (i,j) – th element provided by $C(s_i, s_j; \theta_k)$. Evidently, $C(s, s^*; \theta_k)$ cannot just be a function, but guarantees that the resulting Σ_k matrix is positive definite and symmetric. Functions of this type are regarded as positive definite and are referred to as the characteristic function of a symmetric stochastic variable [24,25,31].

We denote $C(s, s^*; \theta_k) = \sigma_k^2 \rho(s, s^*; \theta_k)$ with $\theta_k = \{\sigma_k^2, \phi_k\}, \rho(.; \phi_k)$ to be a valid spatial correlation function in which ϕ_k computes the correlation decay rate and $var(w_k) = \sigma_k^2$. We assumed for all the accompanying analyses an exponential correlation function $\rho(||s - s^*||; \phi_k) = exp(-\phi_k ||s - s^*||)$, where $||s - s^*||$ represents the Euclidean distance between location s and location s^*. Completion of the Bayesian modeling framework requires specification and assignment of prior distributions to the parameters of the model, where inference proceeds by sampling from the posterior distribution of the modeled parameters. As standard practice, we assume $\beta \sim MVN(\mu_\beta, \Sigma_\beta)$ prior where $\mu_\beta = 0$ and $\Sigma_\beta = 10,000 I_p$, whilst the spatial variance components σ_k^2's and the measurement error variance τ^2 are designated inverse-Gamma, $IG(a,b)$ priors. The spatial decay parameters $\phi_k \sim Unif(a,b)$ with the lower and upper bounds of the distribution covering the geographic domain of the sampled study region.

Applying notation similar to the ones used by [32], we can specify the model parameter posterior distribution as $p(\Omega|y)$ where:

$$\Omega = \left\{ \beta, w_1, w_2, \ldots, w_q, \sigma_1^2, \sigma_2^2, \ldots, \sigma_q^2, \phi_1, \phi_2, \ldots, \phi_q, \tau^2 \right\}$$

$y = (y(s_1), \ldots y(s_n))'$ and is proportional to:

$$\prod_{k=1}^{q} Unif\left(\phi_k \middle| a_{\phi k}, b_{\phi k}\right) x \prod_{k=1}^{q} IG\left(\sigma_k^2 \middle| a_{\sigma k}, b_{\sigma k}\right) \; x \; N(\beta|\mu_\beta) x IG\left(\tau^2 \middle| a_\tau, b_\tau\right) \; x$$

$$\prod_{k=1}^{q} N(w_k|0, \Sigma_k) x \prod_{i=1}^{n} N(y(s_i)|x(s_i)'\beta + \tilde{x}\left(s\right)' w(s), \tau^2\Big) \quad (4)$$

An effective Markov Chain Monte Carlo (MCMC) function for the estimation of Equation (4) is derived through updating of β from its full conditional and utilizing Metropolis procedures for the remainder of the parameters. Reparameterization of the model is an alternative way of ensuring the spatial random effects w do not require direct sampling [33]. The spatial random effects could represent other independent variables that are spatially structured and not taken into consideration in the current modeling approach. Nonetheless, the MCMC process produces posterior samples of the parameter space, Ω.

From a prediction point of view, if $S_0 = \{s_{0,1}, s_{0,2}, \ldots, s_{0,m}\}$ is a set of m new sites, the spatial random effects posterior predictive distribution corresponding to the $k - th$ regression coefficient is provided by:

$$p(w_{k,0}|y) \propto \int p(w_{k,0}|w_k, \Omega, y) p(w_k|\Omega, y) p(\Omega|y) d\Omega w_k \quad (5)$$

where $w_{k,0} = (w_k(s_{0,1}), w_k(s_{0,2}), \ldots, w_k(s_{0,m}))'$.

Since we are making use of MCMC sample-based inference, the integral in Equation (5) does not need to be evaluated precisely. Instead, given L posterior samples for the parameter space (Ω), that is, $\left\{\Omega^{(l)}\right\}_{l=1}^{L}$, composition sampling can be used to derive this distribution [33] by first drawing w_k^l followed by $w_{k,0}^{(l)}$ for each l from $p(w_{k,0}|w_k^{(l)}, \Omega^{(l)}, y)$. The last distribution is multivariate normal as it is a derivative of a conditional distribution from a multivariate normal distribution. More specifically, the process realizations over the new sites are conditionally independent of the measured outcomes given the values over the sampled locations and the process parameters. Expressed differently,

$p(w_{k,0}|w_k, \Omega, y) = p(w_{k,0}|w_k, \Omega)$ is a multivariate distribution with mean and variance given by

$$E[\boldsymbol{w}_{k,0}|\boldsymbol{w}_k, \Omega] = Cov(\boldsymbol{w}_{k,0}, \boldsymbol{w}_k)var^{-1}(\boldsymbol{w}_k)\boldsymbol{w}_k = R_0(\phi_k)'R(\phi_k)^{-1}\boldsymbol{w}_k \quad (6)$$

and

$$var[\boldsymbol{w}_{k,0}|\boldsymbol{w}_k, \Omega] = \sigma_k^2\left\{R(\phi_k) - R_0(\phi_k)'R(\phi_k)^{-1}R_0(\phi_k)\right\} \quad (7)$$

where $R_0(\phi_k$ is an $n \ x \ n$ matrix with (i,j) − th element specified as $\rho(\boldsymbol{s}_{0,i}, \boldsymbol{s}_j; \theta_k)$ and $R(\phi_k$ is an $\boldsymbol{n} \ \boldsymbol{x} \ \boldsymbol{n}$ matrix with (i,j) − th element provided by $\rho(\boldsymbol{s}_i, \boldsymbol{s}_j; \phi_k)$. Repetition of the same procedure results in the generation of samples for all the $\boldsymbol{w}_{k,0}$'s. Lastly, for a set of predictors at unsampled locations $\boldsymbol{s}_0$, posterior predictive distribution samples regarding the outcome variable $y(\boldsymbol{s}_0)^l$, are derived from $N(\boldsymbol{x}(\boldsymbol{s})_0'\beta^{(l)} + \tilde{\boldsymbol{x}}(\boldsymbol{s}_0)\prime\boldsymbol{w}_0^{(l)}, \tau^{2(l)})$ for $l = 1, 2, \ldots, L$.

We evaluated 95% credible interval widths ($CIWs$) using the posterior predictive distribution of Landsat-8 and Sentinel-2 C stock models by calculating the difference between the 2.5% and the 97.5% quantile bounds. The 95% $CIWs$ were therefore used as summaries of the C stock prediction uncertainty for the Landsat-8 and Sentinel-2 C stock-based spatially varying coefficient models.

2.5. Competing Models

We derived five candidate models for each of the two multispectral remote sensing-based C stock models using Equation (1) using $NDVI$ and $SAVI$ as predictors of C stock [34]. The models included the non-spatial where w_k's is fixed to zero; the spatially varying intercept (SVI) in which we only included the spatial random effects related to the model intercept; the complete SVC model in which all predictors have associated spatial random effects; the $SVC - \beta_1$ which included the spatial random effects for the intercept and $NDVI$ predictor variables and the $SVC - \beta_2$ which included the spatial random effects for the intercept.

We utilized empirical semivariograms modeled for the residuals of the independent error model as guidelines for candidate model IG and $Unif$ hyperprior specifications. Precisely, for the variance parameters, the Inverse Gamma hyperprior a was set equal to 1.76, which would result in a mean prior distribution equal to b and infinite variance [35]. To add on, the models' b hyperpriors for the τ^2 and σ^2's were calibrated in accordance with the sample variograms of the simple linear regression model residuals derived from Landsat-8 and Sentinel-2 sensor's nugget and partial sill. We programmed the prior for the spatial decay parameter ϕ's to $Unif$ (0.38, 0.0012) which, adopting the exponential covariance function, equates to support an effective range spanning between $\sim$ 8 and 2500 m. We define the effective spatial range as the distance where the correlation equals 0.05 [18].

Three Markov Chain Monte Carlo (MCMC) chains were run for 20,000 iterations, each with the computationally demanding model requiring approximately 30 min to complete a single MCMC chain. We diagnosed convergence using the CODA library in the R Statistical and Computing environment by monitoring the mixing of chains and the Gelman–Rubin statistic [36]. Acceptable convergence was established within 10,000 iterations for all the models. The posterior inference was premised on a post-burn-in subsample of 15,000 iterations, that is, every third sample from the last 15,000 iterations of each chain. SVC and SVI models were fit using the spBayes R Statistical and Computing Library version 0.4.3. We, therefore, utilized the spBayes R statistical package for fitting all the predictive models for both Landsat-8 and Sentinel-2 SVC models.

2.6. Landsat OLI and Sentinel-2 MSI Imagery Derived Covariates

Landsat-8 has a revisit period of 16 days and offers nine spectral bands with a spatial resolution of 30 m for Bands 1 to 7 and 9 [6,37]. The panchromatic band, Band 8, has a spatial resolution of 15 m. On the other hand, Sentinel-2 has thirteen spectral bands where four bands are configured at 10 m spatial resolution, six bands at 20 m, and three bands at 60 m spatial resolution [37]. Common vegetation indices utilized in the estimation of biophysical variables of Absorbed photosynthetically active radiation (APAR), Leaf Area Index (LAI), and biomass are Normalised Difference Vegetation Index (NDVI), Enhanced Vegetation Index (EVI) and Soil Adjusted Vegetation Index (SAVI) [5]. We utilized the same Vegetation Indices (VIs) in this study. Readily available and cost-effective new-generation sensors (Landsat-8 and Sentinel-2) with improved spectral and spatial resolution were therefore utilized in the modeling of C stock under the spatially varying coefficients assumption.

We obtained Landsat 8 imagery from the United States Geological Survey Earth Explorer (http://earthexplorer.usgs.gov) as data ready for analysis (ARDs) on the 20 of September 2020. All the datasets were riddled with cloud cover and shadow cloud cover limits set to smaller than 10%. We acquired Sentinel-2 cloud-free images on 20 September 2020 at the same time as the Landsat 8 OLI data collection covering the entire area coinciding with the dates Landsat-8 OLI was collected covering the domain of interest. The multispectral instrument is the main imaging instrument used for Sentinel-2, a push broom scanner that measures the terrestrial Top of the Atmosphere (TOA) reflectance in thirteen spectral bands, that is, 443 nm to 2190 nm. We derived Sentinel-2 spectral data as level-1C 12-bit automated TOA reflectance values. Orthorectification and pre-processing of the TOA-derived products were performed using the sen2r package of the R Statistical and Computing environment [38].

Soil Adjusted Vegetation Index ($SAVI$), Enhanced Vegetation Index (EVI), and Normalized Difference Vegetation Index ($NDVI$ were used as Landsat-8 and Sentinel-2 derived covariates for C stock prediction in the plantation forest of the eastern highlands of Zimbabwe. The literature is replete with studies that have utilized the aforementioned vegetation indices [3,39] as ABG predictors in biomass estimation. In this study, predictor variables supplied by both Landsat-8 and Sentinel-2 are used specifically for comparing SVC Bayesian hierarchical geostatistical models predicting C stock in a disturbed environment in Zimbabwe. Given information regarding C stock distribution at each location, we fit the model in Equation (1) SVC with $p = 2$. Hence, we have two processes, an intercept and one slope process relating to $NDVI$.

2.7. Model Fit and Prediction Accuracy Evaluation

We assessed the performance of the models using the commonly used Deviance Information Criterion (DIC) to categorize models in terms of how well they fit the data [40]. The sum of the Bayesian Deviance and the effective number of model parameters make up the DIC criterion. The Bayesian deviance measurement, which assesses model goodness of fit, and the effective number of model parameters which penalizes model complexity, are measured by D and pD, respectively. Attractive models have lower DIC values.

Predictive performance was evaluated through a $k - fold$ cross-validation technique. C stock was predicted from observations within each subset, given the estimated parameters from the remaining subsets. We employed the Root Mean Squared Prediction Error (RMSPE) as a metric from the R Statistical and Computing environment to calculate the sampled C stock data values and the accompanying median of the posterior predictive distribution (PPD). Models with lower RMSPE signify more accurate C stock predictions.

3. Results

3.1. Multispectral Remote Sensing C Stock Derived Predictors

Employed predictors in C stock prediction using a Bayesian hierarchical framework with spatially varying coefficients showed NDVI as a significant predictor of C stock as illustrated in Table 1. 95% credible intervals of SAVI and EVI contained zero and hence, rendering them as insignificant predictors of C stock. These predictors were therefore excluded in the final prediction and mapping of C stock distribution.

Table 1. Landsat-8 and Sentinel-2 derived predictors of C stock.

Parameter	Lansat-8 OLI C Stock Model				Sentinel-2 MSI C Stock Model			
	Mean	s.d	2.5%	97.5%	Mean	s.d	2.5%	97.5%
Intercept	−2.65	1.02	−4.76	−0.83	−2.40	0.31	−3.01	−1.80
NDVI	2.47	0.98	0.78	4.62	4.90	0.23	4.52	5.38
SAVI	−0.57	0.67	−2.04	0.63	−0.55	0.34	−1.25	0.17
EVI	−0.62	0.55	−1.57	0.50	−0.002	0.096	−0.17	0.24
σ_w^2	1.25	0.26	0.72	1.72	0.075	0.016	0.046	0.092
σ_ε^2	0.35	0.17	0.074	0.54	0.0043	0.0030	0.0005	0.011
ϕ	0.0016	0.0003	0.0012	0.0018	0.0015	0.0002	0.0014	0.0014

3.2. Candidate Models and Parameter Estimates

Model posterior estimates of the regression coefficients for the Landasat-8 and Sentinel-2-based C stock models are illustrated in Tables 2 and 3, respectively, for the non-spatial, SVI, and SVC models. Spatial autocorrelation is modeled in the residuals in the SVI, SVC, and all the $SVC - \beta_k$ variant models. This may entail differences in the posterior regression coefficient parameter estimates, depending on the C stock model parameter structure. Credible intervals (95% CI) for β_0 and β_{NDVI} including zero for both non-spatial and SVI models would hint at a non-significant relationship between the C stock observations and predictor variables. However, we cannot apply the same reasoning and interpretation for the SVC models as the predictor-specific spatially varying coefficient maps of $\beta_{NDVI} + w_{NDVI}(s)$ should be considered and ascertain whether location-specific CIs include zero.

Table 2. Landsat-8-based SVC model median parameter estimates alongside their 95% credible intervals.

		Non-Spatial	SVI	SVC−NDVI	SVC
Parameter C.I 50% (2.5%, 97.5%)	β_0	−2.5 (−4.1, −0.8)	−2.7 (−4.6, −0.8)	−2.9 (−5.0, −0.9)	−3.5 (−5.3, −1.7)
	$\widetilde{\beta}_{NDVI}(s)$	5.4 (3.3, 7.5)	2.8 (0.8, 4.7)	3.3 (1.2, 5.1)	3.6 (1.4, 5.8)
	τ^2	1.5 (1.2, 2.0)	0.5 (0.3, 0.8)	0.12 (0.03, 0.40)	0.1 (0.02, 0.2)
	$\frac{3}{\phi_0}$	-	0.0014 (0.0012, 0.0021)	0.003 (0.002, 0.003)	0.1 (0.07, 0.16)
	$\frac{3}{\phi_{NDVI}}$	-	-	0.13 (0.06, 0.22)	0.03 (0.02, 0.05)
	σ_0^2	-	1.0 (0.5, 1.5)	1.1 (0.8, 1.6)	0.24 (0.12, 0.48)
	σ_{NDVI}^2	-	-	1.0 (0.2, 3.0)	0.45 (0.14, 1.43)
Fit Statistics	D	436	140	41.8	65.9
	pD	4.1	68.9	87.0	108.9
	DIC	207	105.6	−77.9	29.0
	$RMSPE$ (MgCha^{-1})	8	7.77	7.54	6.42

C.I means 95% Credible Interval.

Table 3. Sentinel-2-based SVC model median parameter estimates alongside 95% credible intervals.

		Non-Spatial	SVI	SVC−NDVI	SVC
Parameter C.I 50% (2.5%, 97.5%)	β_0	−2.5 (−4.1, −0.8)	−2.8 (−4.6, −0.9)	−2.9 (−4.9, −0.7)	−3.5 (−5.4, −1.6)
	$\widetilde{\beta}_{NDVI}(s)$	5.4 (3.3, 7.5)	3.0 (1.2, 4.9)	2.9 (0.7, 4.9)	3.7 (1.5, 5.7)
	τ^2	1.5 (1.2, 2.0)	0.5 (0.3, 0.9)	0.32 (0.15, 0.55)	0.2 (0.11, 0.42)
	$\frac{3}{\phi_0}$	-	0.0015 (0.0014, 0.0016)	0.0016 (0.0015, 0.0018)	0.06 (0.018, 0.076)
	$\frac{3}{\phi_{NDVI}}$	-	-	0.09 (0.036, 0.13)	0.0015 (0.0014, 0.0017)
	σ_0^2	-	0.7 (0.4, 1.1)	1.1 (0.8, 1.9)	0.23 (0.16, 0.47)
	σ_{NDVI}^2	-	-	0.7 (0.43, 1.1)	0.65 (0.41, 1.05)
Fit Statistics	D	436.6	159	93.3	17.9
	pD	4.1	58.0	89.8	106.1
	DIC	207.6	112.5	65.6	−177.3
	$RMSPE$ (MgCha^{-1})	7.85	7.69	7.46	6.23

C.I means 95% Credible Interval.

Posterior estimates for the Landsat-8 and Sentinel-2 based $\beta_{NDVI} + w_{NDVI}(s)$ coefficients are shown in Figures 3 and 4, with a significant relationship between the outcome variable and $NDVI$ being evident over the entire study domain. In both cases, there is significant variability in the values of $\beta_{NDVI} + w_{NDVI}(s)$ in the studied region. This could imply more accessibility to plantation resources by communities settling in the plantation areas. The same trend is noticeable for the Sentinel-2-based C stock-based SVC model (Figures 5 and 6) for both predictor coefficient values of $\beta_{NDVI} + w_{NDVI}(s)$ and for the same region as observed in the Landsat-8 based SVC model in Figure 3. Communities settled within certain areas of the plantation forest have more access to forest resources than in other areas, rendering low C stock density in these areas as demonstrated by the corresponding low $NDVI$ values in the same region for the Sentinel-2-based SVC model. Enhanced detail in the spatial resolution of Sentinel-2 based $\beta_{NDVI} + w_{NDVI}(s)$ vindicates variability in $\beta_{NDVI} + w_{NDVI}(s)$ coefficient values in the same region over those derived from generalized Landsat-8 multispectral data [41].

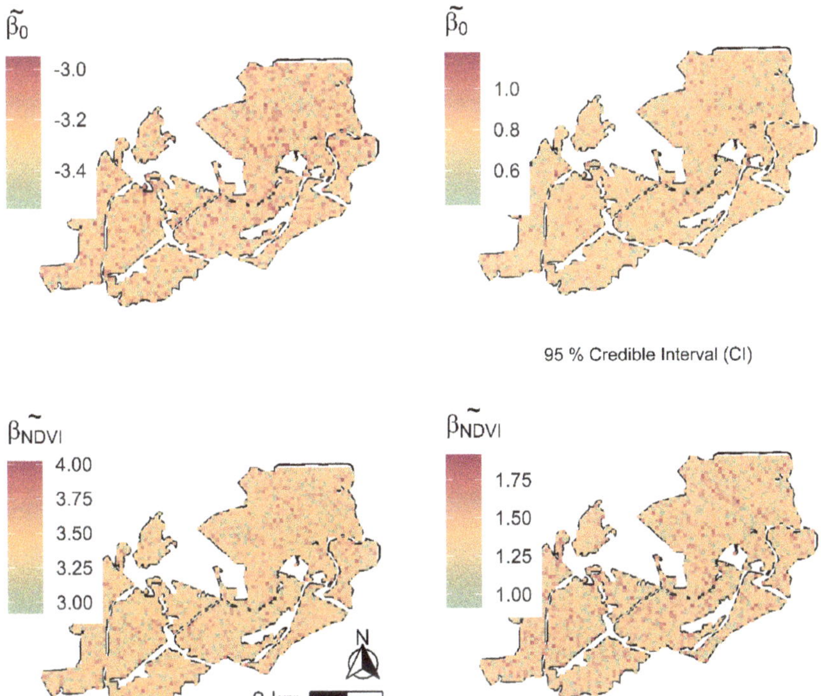

Figure 3. Landsat-8-based C stock spatially varying coefficient maps alongside their 95% credible intervals for the SVC model.

Estimates of $\beta_0 + w_0(s)$ for the Landsat-8 and Sentinel-2-based SVI and SVC models are shown in Figures 3 and 4, respectively. The $\beta_0 + w_0(s)$ pattern for the SVI is the same for both Landsat-8 and Sentinel-2 spatially varying coefficient models whilst the $\beta_0 + w_0(s)$ SVC pattern in Landsat-8 is the same for Landsat-8-based SVC ($\beta_{NDVI} + w_{NDVI}(s)$) process. On the other hand, the $\beta_0 + w_0(s)$ SVI pattern for Sentinel-2 based model is different for Sentinel-2 $\beta_{NDVI} + w_{NDVI}(s)$ SVC model. However, the same is not true for Sentinel-2-based SVI as the partitioning of w_0 into w_0 and w_{NDVI} in Sentinel-2 SVC is detailed with enhanced spatial resolution compared to Landsat-8 based w_{NDVI} SVC model.

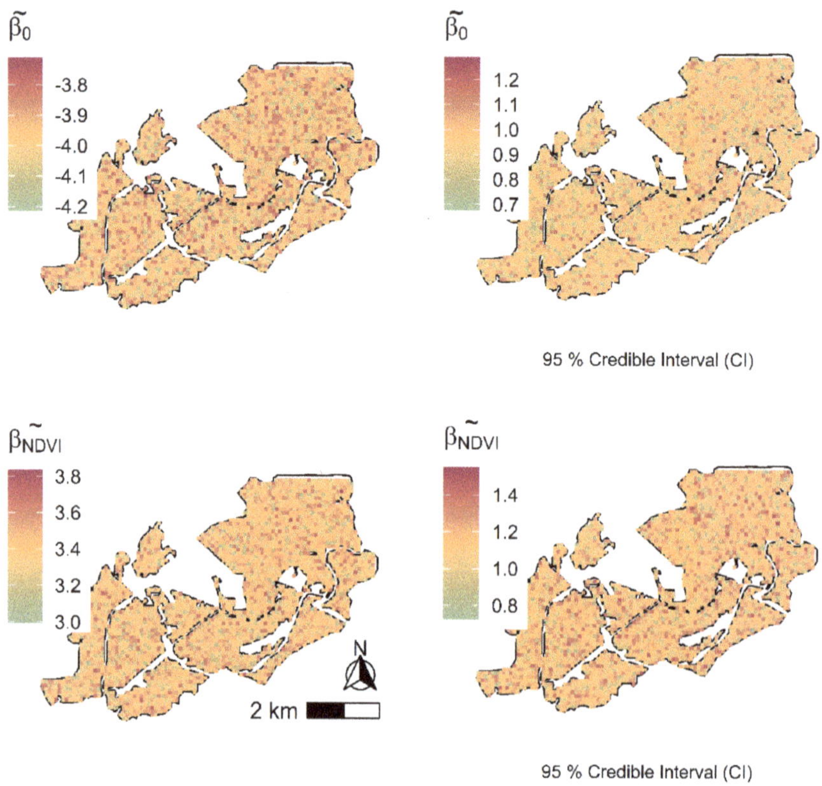

Figure 4. Sentinel-2-based C stock spatially varying coefficient maps alongside their 95% credible intervals for the SVC model.

Tables 2 and 3 illustrate the posterior estimates of the uncorrelated residual variance, τ^2. The uncorrelated residual variance is largest in the non-spatial model, whilst it is small for the SVC-variant models for both Landsat-8 and Sentinel-2-based SVC. The SVI and the SVC variant models incorporate a spatially varying correlated random effect, w_0 with variance σ_0^2. The SVC model variants further incorporate more spatially varying correlated random effects w_{NDVI} and variance σ^2_{NDVI}. For both Landsat-8 and Sentinel-2 based models, the w_0 and w_k explained much of the residual variability and hence, a reduced τ^2. The implication is higher predictive accuracy for the SVC-variant models in both sensors, making the SVI and SVC more attractive over the error-independent models. This is supported by the goodness of fit diagnostics for the SVI and the SVC-variant models illustrated in Tables 2 and 3 for the Landsat-8 and Sentinel-2 derived regression coefficients, respectively.

In comparison to the SVI model, the SVC model based on both remote sensing-derived covariates reduced the non-spatial residual spatial dependence by incorporating the space-varying impact of β_{NDVI}. Estimates of the spatial process parameters have a big difference. In particular, the spatial process parameters for the Landsat-8-based (Figure 3) SVC point estimates of $\sigma_0^2 = 0.24$ and the effective range of 30 m (i.e., $\approx -log(0.05/\phi) = -log(0.05/0.1)$ denote a less variable and significantly shorter effective spatial range than the spatial process of the SVI model. The same pattern recurs in the Senitnel-2-based SVI and SVC models, where the effective spatial range reduces from $\approx$ 1800 m in the SVI to $\approx$ 100 m in the SVC model (Table 3). The non-negligible spatial process parameter estimates of σ^2_{NDVI} and ϕ_{NDVI} in the SVC model denote a potentially space-varying relationship between C stock and multispectral remote sensing derived covariates.

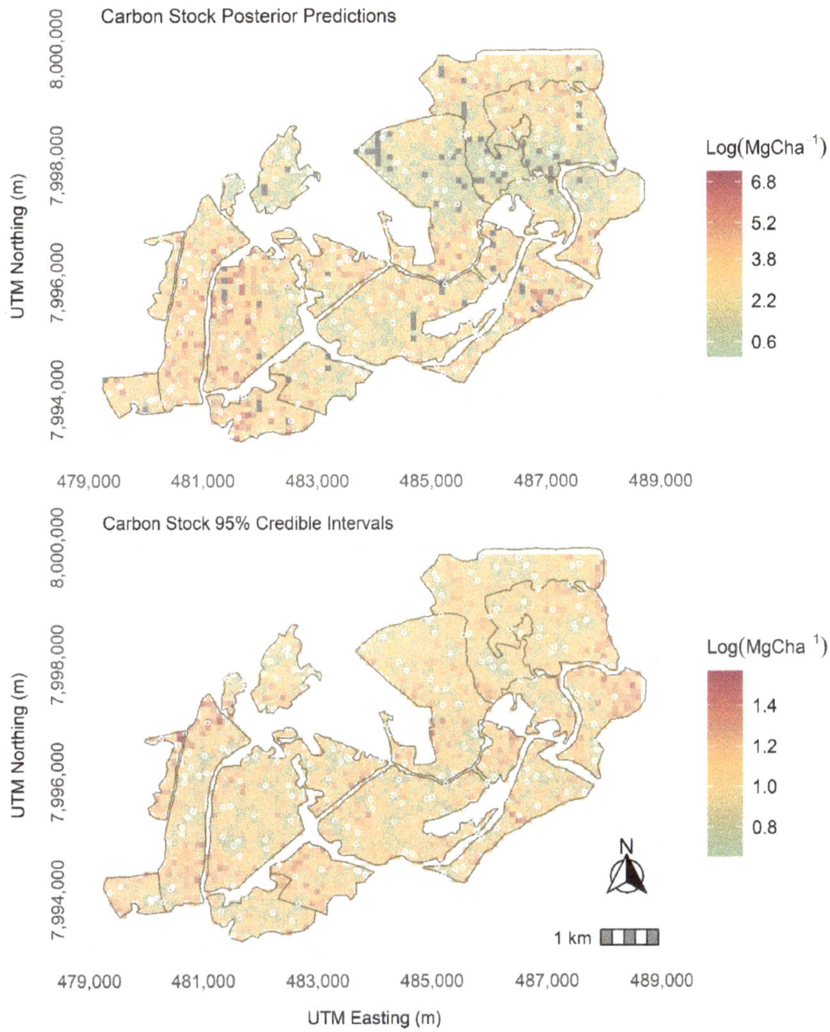

Figure 5. Landsat-8-based C stock posterior predictions.

3.3. *Landsat-8 and Sentinel-2 C Stock-Based Predictions*

The entire SVC model based on Landsat-8 and Sentinel-2-based C stock models generated the lowest DIC, D, and RMSPE, as illustrated in Tables 2 and 3, respectively. Landsat-8-based SVC 95% CIW is much shorter when compared to the non-spatial fitted model values. A 10% improvement in the RMSPE is seen in the Landsat-8-based model when moving from the error-independent to the SVC model. In the same vein, the Sentinel-2-based model had a 12% improvement from the SVI to the SVC model (Table 3). Predictions produced for both Landsat-8 and Sentinel-2-based C stock models mirrored the observed C stock data in the studied region. Observed C stock data values ranged from log (0.2–6) $MgCha^{-1}$ (Figure 1). Variability in C stock uncertainty is fairly constant across the studied domain for both Landsat-8 and Sentinel-2 SVC-based models.

The almost similar variability in the density of C stock predicted by both sensors could be attributed to the inadequacy of covariates in the modeling framework, which when the range of modeled covariates is broadened, could accurately depict the variability of C stock in these disturbed plantation forest ecosystems [5].

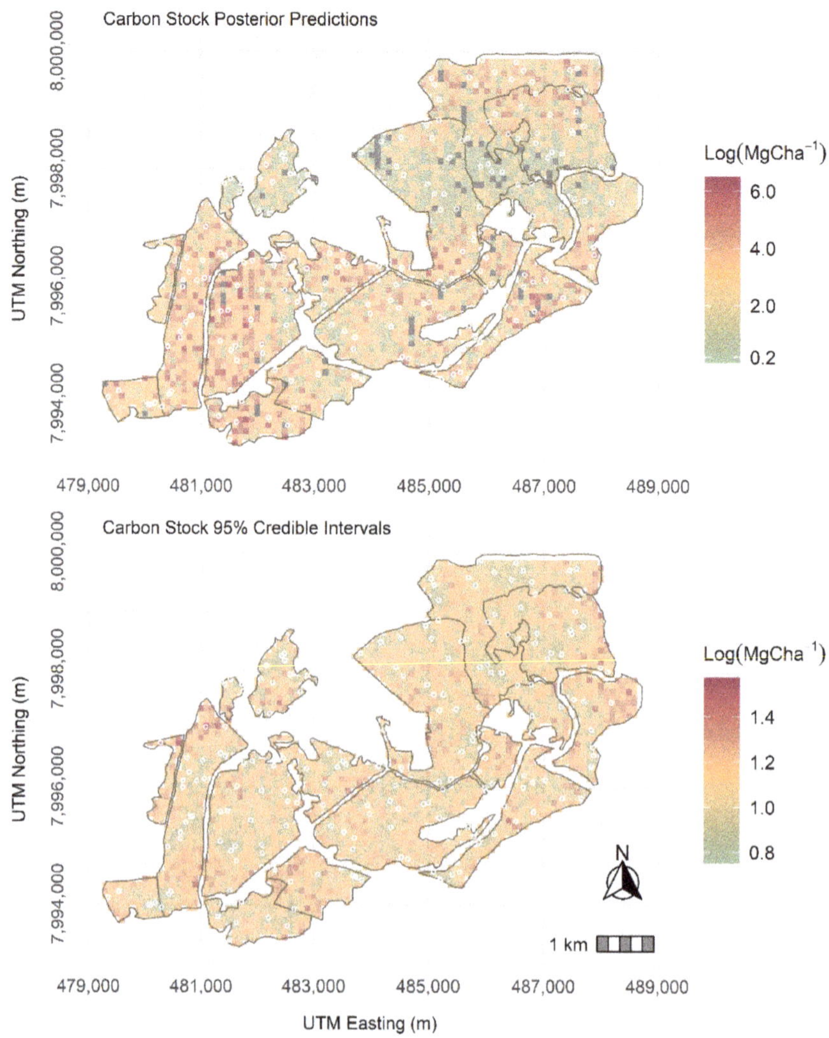

Figure 6. Sentinel-2-based C stock posterior predictions.

3.4. C Stock Model Prediction Assessment

Scatterplots observed against predicted C stock alongside the 95% intervals are illustrated in Figure 7. Slight improvement in the performance of Landsat-8 and Sentinel-2 C stock-based predictions can be validated through a visual inspection of the results. Estimated Root Mean Square Prediction Error (RMSPE) for Landsat-8 (6.42 Mg C ha^{-1}) and Sentinel-2 (6.23 Mg C ha^{-1}) based C stock prediction further reinforces the model prediction performance for the two sensors illustrated in Figure 7.

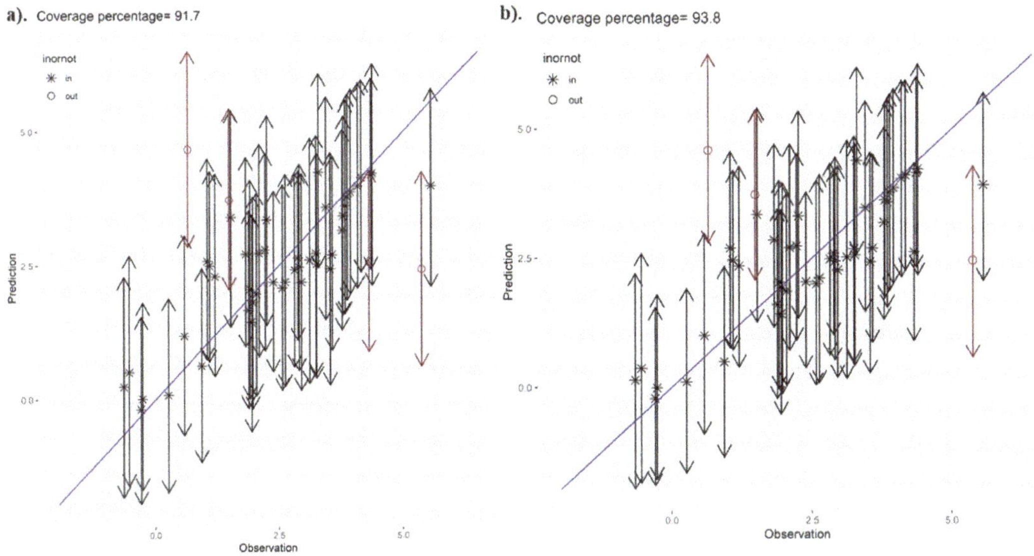

Figure 7. Landsat-8 and Sentinel-2 C stock-based predictions vs. C stock observations alongside 95% intervals.

4. Discussion

Space varying coefficient models constructed from different but related new-generation multispectral remote sensing platforms were used to predict C stock in a managed plantation forest ecosystem in Zimbabwe. The RMSPE is marginally higher in Sentinel-2-based C stock SVC model than in the Landsat-8-based SVC counterpart. Our findings with regard to the performance of Landsat-8 and Sentinel-2 sensors are also congruent with the work of [12,37]. SVC models for both new-generation remote sensing-derived predictors showed preference over the error-independent models. However, estimates from the two data sources were marginally different from each other on the prediction of C stock, as illustrated in validation diagnostics [42]. The adaptable structure of the SVC permitted the residuals spatial variability to be apportioned between the random slope and the random intercept. This provided additional benefits not available in the SVI models from the two predictor sources. The SVC permitted reconnaissance of the C stock observations and the $NDVI$ predictor. Furthermore, the SVC models in both multispectral data sources had better representation of the processes thereby yielding C stock predictions with reduced variance.

Evaluation of the models utilizing predictors from both remote sensing data sources showed SVC to fit the data better than the SVI models. Kriged maps for C stock using Landsat-8 and Sentinel-2 data were not significantly different from each other, with the Landsat-8 SVC displaying a slightly wider 95% CI compared to the Sentinel-2-based SVC model. Again, this is partly because the study employed conventional bands (indices) that are calculated in a similar fashion in both Landsat-8 and Sentinel-2, and hence, the differences in prediction only emanated from the spatial resolutions. Explicit accommodation of residual spatial dependence through spatially correlated random effects gathered better predictions as the SVC models using the two data sources as regression coefficients borrowed additional information from neighboring C stock observations [43,44]. Precise estimates of covariance parameters are not easy to derive with small inventory sample sizes [45]. Consequently, we might anticipate some impact on the accuracy of the predictions when uncertainty in the covariance ensues to the posterior predictive distributions. Predictions had limited information to borrow from because of the sparsity of C stock inventory observations. This was further worsened by the reduction in the overall sample size through cross-validation.

Similar to previous studies predicting ABG and C stock, our findings establish Sentinel-2 as a better source of RS data for predicting C stock in disturbed environments [37] compared Sentinel-2 and Landsat-8 imagery for forest biomass prediction and showed Sentinel-2 outperforms Landsat-8 because of the enhanced spatial resolution in the former in comparison to Landsat-8. Most studies comparing Sentinel-2 and Landsat-8 for predicting AGB prefer Sentinel-2 over Landsat-8. This is further justified by the work of [13], who compared Worldview-3, Sentinel-2, and Landsat-8 for representing AGB in a forest environment in Thailand and demonstrated Worldview-3 and Sentinel-2 as better data sources and, therefore, predictors than Landsat-8 due to the red-edge and the improved spatial and spectral properties of Worldview-3 and Sentinel-2. Furthermore, [16] utilized LiDAR derived covariates from establishing the prediction performance of SVC models using forest inventory data in North America. The researchers established significant improvement in biomass prediction accuracy in the presence of residual spatial dependence deriving from the finer resolution LiDAR covariates. In most cases, the effectiveness of SVC models in these studies is usually strengthened by the solid non-stationary relationships between the response variable and the predictor variables influenced by unobserved ecological factors operating at broad geographical scales. Such ecological factors were also seen in the present study as NDVI was established to be a statistically significant predictors of C stock in the studied region. The biggest drivers to these factors are the presumed activities of forest disturbance due to human encroachment into plantation forests that subsequently impact the density and distribution of ABG biomass.

4.1. Limitations of the Study

Apart from the vegetation indices influencing the spatial distribution and density of AGB employed in this study, it is also known and acknowledged in the literature that climate and topographic variables play a part in the distribution of C stock. For example, [37,38] have shown elevation and aspect accounting for the bigger portion of the spatial variability of C stock in a mountainous region of Nepal. As such, the limitation of our study is the application of vegetation indices as predictors of C stock, and this may, therefore, not be representative of the general C stock dynamics in the studied region. We, therefore, recommend the integration of topo-climatic factors with new generation remote sensing-derived vegetation indices for future research in order to obtain a more accurate global overview of the C stock density and distribution in the studied region.

4.2. Conclusions

The study presented a hierarchical Bayesian geostatistical spatially varying coefficient model for determining the relationship between sampled C stock data and multispectral remote sensing derived predictors. There was a marginal improvement in model fit, and prediction accuracy in both Landsat-8 and Sentinel-2-based SVC models in comparison to the error-independent models. The SVC model permitted exploration of the observed C stock locations where the models performed well or poorly, which was missing in the SVI models. This provided an understanding of the performance of the multispectral remote sensing derived predictors for modeling C stock and hence, sets the foundation for the updating of the carbon forest plantation database for forest practitioners in the country and utilized as a monitoring tool in the long term. The Sentinel-2-based SVC model was preferred for prediction in the plantation forest ecosystems as its model provided tighter credible intervals compared to the Landsat-8-based C stock SVC model.

The small sample size of the data utilized in the present research enabled the modeling approach to be computationally feasible. When inventory plots are comprised of bigger sample sizes, the matrix operations of immense dimensions are needed for computing model parameter estimates of higher magnitude and may not be achievable through ordinary PCs. Our future work is therefore aimed at exploring algorithms for resolving the dimensionality curse when fitting spatially varying coefficient models. The problem of dimensionality can also get complicated if many predictors are involved in the SVC modeling framework. Resolving dimensionality issues is needed as forest C stock is typically modeled with predictors from many data sources, chief amongst them being topographic, bioclimatic, and anthropogenic variables.

Author Contributions: T.S.C.: research design, writing, and data processing. O.M.: supervision and funding of research activity. T.D.: supervision and technical assistance. All authors have read and agreed to the published version of the manuscript.

Funding: The study was partially supported by the South African Research Chair Initiative (SARChI) in Land Use Planning and Management (Grant no. 84157).

Institutional Review Board Statement: Not applicable.

Data Availability Statement: All relevant data are within the paper.

Acknowledgments: The authors thank Joseph. Z.Z. Matowanyika and Nixon Kutsaranga for allowing the researchers access to forest plantations at Lot 75 A of Nyanga Downs in Nyanga District of Manicaland province. The researchers also thank Trylee Matongera for assistance with the formatting of the paper.

Conflicts of Interest: The authors declare no conflict of interest.

References

1. Matose, F. Trends in forest ownership, institutional arrangements and the impact on forest management and poverty reduction. *Cbneih* **2008**, *13*, 373.
2. Newsday. Forestry Commision decentralise issuance of timber movement. *Newsday*, 24 July 2017.
3. Bordoloi, R.; Das, B.; Tripathi, O.; Sahoo, U.; Nath, A.; Deb, S.; Das, D.; Gupta, A.; Devi, N.; Charturvedi, S.; et al. Satellite based integrated approaches to modelling spatial carbon stock and carbon sequestration potential of different land uses of Northeast India. *Environ. Sustain. Indic.* **2022**, *13*, 100166. [CrossRef]
4. Brown, S. *Estimating Biomass and Biomass Change of Tropical Forests: A Primer*; Food & Agriculture Organization: Rome, Italy, 1997; Volume 134.
5. Zvobgo, L.; Tsoka, J. Deforestation rate and causes in Upper Manyame Sub-Catchment, Zimbabwe: Implications on achieving national climate change mitigation targets. *Trees For. People* **2021**, *5*, 100090. [CrossRef]
6. Wang, Q.; Li, J.; Jin, T.; Chang, X.; Zhu, Y.; Li, Y.; Sun, J.; Li, D. Comparative analysis of Landsat-8, Sentinel-2, and GF-1 data for retrieving soil moisture over wheat farmlands. *Remote Sens.* **2020**, *12*, 2708. [CrossRef]
7. Fuller, A.; Dawson, T.; Helmuth, B.; Hetem, R.S.; Mitchell, D.; Maloney, S.K. Physiological mechanisms in coping with climate change. *Physiol. Biochem. Zool.* **2010**, *83*, 713–720. [CrossRef]
8. Mutanga, O.; Dube, T.; Ahmed, F. Progress in remote sensing: Vegetation monitoring in South Africa. *S. Afr. Geogr. J.* **2016**, *98*, 461–471. [CrossRef]
9. Hoeting, J.A. The importance of accounting for spatial and temporal correlation in analyses of ecological data. *Ecol. Appl.* **2009**, *19*, 574–577. [CrossRef]
10. Zunguze, A.X. *Quantificação de Carbono Sequestrado em Povoamentos de Eucalyptus spp na Floresta de Inhamacari-Manica*; Universidade Eduardo Mondlane: Maputo, Mozambique, 2012. (In Portuguese)
11. Gelfand, A.E.; Kim, H.-J.; Sirmans, C.; Banerjee, S. Spatial modeling with spatially varying coefficient processes. *J. Am. Stat. Assoc.* **2003**, *98*, 387–396. [CrossRef]
12. Chrysafis, I.; Mallinis, G.; Siachalou, S.; Patias, P. Assessing the relationships between growing stock volume and Sentinel-2 imagery in a Mediterranean forest ecosystem. *Remote Sens. Lett.* **2017**, *8*, 508–517. [CrossRef]
13. Green, E.J.; Finley, A.O.; Strawderman, W.E. *Introduction to Bayesian Methods in Ecology and Natural Resources*; Springer Nature: Berlin/Heidelberg, Germany, 2020.
14. Popescu, S.C.; Zhao, K.; Neuenschwander, A.; Lin, C. Satellite lidar vs. small footprint airborne lidar: Comparing the accuracy of aboveground biomass estimates and forest structure metrics at footprint level. *Remote Sens. Environ.* **2011**, *115*, 2786–2797. [CrossRef]
15. Tonolli, S.; Dalponte, M.; Neteler, M.; Rodeghiero, M.; Vescovo, L.; Gianelle, D. Fusion of airborne LiDAR and satellite multispectral data for the estimation of timber volume in the Southern Alps. *Remote Sens. Environ.* **2011**, *115*, 2486–2498. [CrossRef]
16. Babcock, C.; Finley, A.O.; Bradford, J.B.; Kolka, R.; Birdsey, R.; Ryan, M.G. LiDAR based prediction of forest biomass using hierarchical models with spatially varying coefficients. *Sens. Environ.* **2015**, *169*, 113–127. [CrossRef]
17. Gelfand, A.E.; Schmidt, A.M.; Banerjee, S.; Sirmans, C. Nonstationary multivariate process modeling through spatially varying coregionalization. *Test* **2004**, *13*, 263–312. [CrossRef]

18. Finley, A.O.; Banerjee, S.; MacFarlane, D.W. A hierarchical model for quantifying forest variables over large heterogeneous landscapes with uncertain forest areas. *J. Am. Stat. Assoc.* **2011**, *106*, 31–48. [CrossRef]
19. Finley, A.O.; Banerjee, S.; McRoberts, R.E. Hierarchical spatial models for predicting tree species assemblages across large domains. *Ann. Appl. Stat.* **2009**, *3*, 1052. [CrossRef] [PubMed]
20. Schabenberger, O.; Gotway, C.A. *Statistical Methods for Spatial Data Analysis: Texts in Statistical Science*; Chapman and Hall/CRC: London, UK, 2017.
21. Fotheringham, A.S.; Brunsdon, C.; Charlton, M. *Geographically Weighted Regression: The Analysis of Spatially Varying Relationships*; John Wiley & Sons: Hoboken, NJ, USA, 2003.
22. Hudak, A.T.; Lefsky, M.A.; Cohen, W.B.; Berterretche, M. Integration of lidar and Landsat ETM+ data for estimating and mapping forest canopy height. *Remote Sens. Environ.* **2002**, *82*, 397–416. [CrossRef]
23. Wheeler, D.C.; Waller, L.A. Comparing spatially varying coefficient models: A case study examining violent crime rates and their relationships to alcohol outlets and illegal drug arrests. *J. Geogr. Syst.* **2009**, *11*, 1–22. [CrossRef]
24. Cressie, N. *Statistics for Spatial Data*; Wiley Series in Probability and Statistics; John Wiley & Sons, Inc.: Hoboken, NJ, USA, 1993.
25. Banerjee, S.; Carlin, B.P.; Gelfand, A.E. *Hierarchical Modeling and Analysis for Spatial Data*; Chapman and Hall/CRC: London, UK, 2003.
26. Lisboa, S.N.; Guedes, B.S.; Ribeiro, N.; Sitoe, A. Biomass allometric equation and expansion factor for a mountain moist evergreen forest in Mozambique. *Carbon Balance Manag.* **2018**, *13*, 1–16. [CrossRef]
27. Walvoort, D.J.; Brus, D.; De Gruijter, J. An R package for spatial coverage sampling and random sampling from compact geographical strata by k-means. *Comput. Geosci.* **2010**, *36*, 1261–1267. [CrossRef]
28. Brus, D.; De Gruijter, J.; Van Groenigen, J. Designing spatial coverage samples using the k-means clustering algorithm. *Dev. Soil Sci.* **2006**, *31*, 183–192.
29. Wackernagel, H. *Multivariate Geostatistics: An Introduction with Applications*; Springer Science & Business Media: Berlin/Heidelberg, Germany, 2003.
30. Banerjee, S.; Finley, A.O.; Waldmann, P.; Ericsson, T. Hierarchical spatial process models for multiple traits in large genetic trials. *J. Am. Stat. Assoc.* **2010**, *105*, 506–521. [CrossRef] [PubMed]
31. Chiles, J.-P.; Delfiner, P. *Geostatistics: Modeling Spatial Uncertainty*; John Wiley & Sons: Hoboken, NJ, USA, 2009.
32. Kaplan, D.; Depaoli, S. *Bayesian Structural Equation Modeling*; Guilford Press: New York, NY, USA, 2012.
33. Banerjee, S.; Gelfand, A.E.; Finley, A.O.; Sang, H. Gaussian predictive process models for large spatial data sets. *J. R. Stat. Soc. Ser. B* **2008**, *70*, 825–848. [CrossRef] [PubMed]
34. Mansfield, E.R.; Helms, B.P. Detecting multicollinearity. *Am. Stat.* **1982**, *36*, 158–160.
35. Gelman, A.; Carlin, J.B.; Stern, H.S.; Rubin, D.B. *Bayesian Data Analysis*; Chapman and Hall/CRC: London, UK, 1995.
36. Gelman, A.; Rubin, D.B. Inference from iterative simulation using multiple sequences. *Stat. Sci.* **1992**, *7*, 457–472. [CrossRef]
37. Korhonen, L.; Packalen, P.; Rautiainen, M. Comparison of Sentinel-2 and Landsat 8 in the estimation of boreal forest canopy cover and leaf area index. *Remote Sens. Environ.* **2017**, *195*, 259–274. [CrossRef]
38. Ranghetti, L.; Boschetti, M.; Nutini, F.; Busetto, L. "sen2r": An R toolbox for automatically downloading and preprocessing Sentinel-2 satellite data. *Comput. Geosci.* **2020**, *139*, 104473. [CrossRef]
39. Li, C.; Li, X. Hazard rate and reversed hazard rate orders on extremes of heterogeneous and dependent random variables. *Stat. Probab. Lett.* **2019**, *146*, 104–111. [CrossRef]
40. Spiegelhalter, D.J.; Best, N.G.; Carlin, B.P.; Van Der Linde, A. Bayesian measures of model complexity and fit. *J. R. Stat. Soc. Ser. B* **2002**, *64*, 583–639. [CrossRef]
41. Forkuor, G.; Dimobe, K.; Serme, I.; Tondoh, J.E. Landsat-8 vs. Sentinel-2: Examining the added value of sentinel-2's red-edge bands to land-use and land-cover mapping in Burkina Faso. *GIScience Remote Sens.* **2018**, *55*, 331–354. [CrossRef]
42. Chen, G.; Zhao, K.; McDermid, G.J.; Hay, G.J. The influence of sampling density on geographically weighted regression: A case study using forest canopy height and optical data. *Int. J. Remote Sens.* **2012**, *33*, 2909–2924. [CrossRef]
43. Gelfand, A.E. Hierarchical modeling for spatial data problems. *Spat. Stat.* **2012**, *1*, 30–39. [CrossRef] [PubMed]
44. Babcock, C.; Finley, A.O.; Cook, B.D.; Weiskittel, A.; Woodall, C.W. Modeling forest biomass and growth: Coupling long-term inventory and LiDAR data. *Remote Sens. Environ.* **2016**, *182*, 1–12. [CrossRef]
45. Baumer, B.S.; Kaplan, D.T.; Horton, N.J. *Texts in Statistical Science: Modern Data Science with R*; Chapman and Hall/CRC: London, UK, 2017.

remote sensing

MDPI

Article

Assessment of the Impact of Rubber Plantation Expansion on Regional Carbon Storage Based on Time Series Remote Sensing and the InVEST Model

Chong Huang [1,2,†], Chenchen Zhang [1,†] and He Li [1,*]

1 State Key Laboratory of Resources and Environmental Information System, Institute of Geographic Sciences and Natural Resources Research, Chinese Academy of Sciences, Beijing 100101, China
2 CAS Engineering Laboratory for Yellow River Delta Modern Agriculture, Institute of Geographic Sciences and Natural Resources Research, Chinese Academy of Sciences, Beijing 100101, China
* Correspondence: lih@lreis.ac.cn; Tel.: +86-10-6488-9042
† These authors contributed equally to this work.

Abstract: Rubber plantations in southeast Asia have grown at an unprecedented rate in recent decades, leading to drastic changes in regional carbon storage. To this end, this study proposes a systematic approach for quantitatively estimating and assessing the impact of rubber expansions on regional carbon storage. First, using Sentinel-1 and Sentinel-2 satellite data, the distributions of forest and rubber, respectively, were extracted. Then, based on the Landsat time series (1999–2019) remote sensing data, the stand age estimation of rubber plantations was studied with the improved shapelet algorithm. On this basis, the Ecosystem Services and Tradeoffs model (InVEST) was applied to assess the regional carbon density and storage. Finally, by setting up two scenarios of actual planting and hypothetical non-planting of rubber forests, the impact of the carbon storage under these two scenarios was explored. The results of the study showed the following: (1) The area of rubber was 1.28×10^5 ha in 2019, mainly distributed at an elevation of 200–400 m (accounting for 78.47% of the total of rubber). (2) The average age of rubber stands was 13.85 years, and the total newly established rubber plantations were converted from cropland and natural forests, accounting for 54.81% and 45.19%, respectively. (3) With the expansion of rubber plantations, the carbon density increased from only 2.25 Mg·C/ha in 1999 to more than 15 Mg·C/ha in 2018. Among them, the carbon sequestration increased dramatically when the cropland was replaced by rubber, while deforestation and replacement of natural forests will cause a significant decrease. (4) The difference between the actual and the hypothetical carbon storage reached −0.15 million tons in 2018, which means that the expansion of rubber led to a decline in carbon storage in our study area. These research findings can provide a theoretical basis and practical application for sustainable regional rubber forest plantation and management, carbon balance maintenance, and climate change stabilization.

Keywords: rubber plantation; time series; shapelet; carbon storage; InVEST model

Citation: Huang, C.; Zhang, C.; Li, H. Assessment of the Impact of Rubber Plantation Expansion on Regional Carbon Storage Based on Time Series Remote Sensing and the InVEST Model. *Remote Sens.* **2022**, *14*, 6234. https://doi.org/10.3390/rs14246234

Academic Editors: Huaqiang Du and Inge Jonckheere

Received: 15 October 2022
Accepted: 5 December 2022
Published: 9 December 2022

1. Introduction

Carbon sequestration in terrestrial ecosystems is critical to the effects of carbon dioxide (CO_2)-driven global climate change [1–3]. As an important part of the terrestrial ecosystem and the largest carbon pool, the annual carbon sequestration of forests accounts for about two-thirds of the entire terrestrial ecosystem, playing an important role in reducing the rise in atmospheric CO_2 concentration and stabilizing global climate change [4,5]. Therefore, the study of carbon storage in the forest ecosystem is a hotspot of carbon neutralization research and focus of current global climate change research [6]. However, today's research is mainly focused on the carbon stock of primary natural forests, and relatively less research has been conducted on the carbon sink balance of the artificial forests, which accounts for seven percent of the world's overall forest area [5,7]. Therefore, it is necessary to study the change in carbon storage in key plantation forest types.

Rubber forests, as the second largest tropical plantation ecosystem after oil palm [8], are planted for their large economic value, and they have a strong carbon sequestration capacity that plays an important role in the economic development and ecosystem service value [9,10]. Southeast Asia is the world's major natural rubber growing region because of its suitable climate and growing conditions, accounting for more than 80% of the global natural rubber forest plantation area [11,12]. With the rapid development of economic globalization, the importance of rubber products in the national economy is increasing, the rigid demand for natural rubber is growing, and the planting area of rubber is expanding [13,14]. According to the statistics, the area planted with rubber forests in Thailand increased by nearly 800% from about 400,000 ha in 1961 to more than 3 million ha in 2017 [15]. However, most rubber forests are planted at the expense of the primary tropical rainforests or secondary forests, which inevitably leads to a certain loss of carbon sinks [7,16,17]. In addition, the deforestation of the tropical rainforest caused by rubber forest plantations will inevitably lead to the continuous reduction in the tropical rainforest-based biological habitat, the gradual reduction of soil and water conservation capacity, regional environmental degradation, and serious damage to biodiversity and the ecological environment [18–21]. Therefore, it is of great significance to understand the quantitative impact of rubber forests on carbon storage for the rational development of rubber plantations, the protection of forest ecosystems, the maintenance of the carbon sink balance, and the stability of climate change.

The age information of rubber is the key parameter of carbon storage assessment and can improve the accuracy of rubber carbon storage estimation [7]. Traditional rubber forest age monitoring is mainly based on the ground survey of sampling theory, which is time-consuming, labor-intensive, not comprehensive, and may be less accurate for regional estimations [10,22–25]. Access to information about the age of rubber on each plantation is even more difficult in areas where rubber plantations are small, fragmented, flexible cropping systems, with high variability in planting status, and which are mainly established and managed by smallholders, e.g., in Thailand, where an estimated 90% of rubber is produced by smallholders [16]. With the development of earth observation technology, remote sensing, which is macro, rapid, dynamic, and rich in information acquisition capacity, has been used to map rubber plantations and become an effective means of extracting age information from rubber in recent years [10,22,26,27].

The methods for age identification of rubber forests using remote sensing can commonly be divided into four main categories: post-classification comparison (PCC) [28], threshold method [29,30], regression method [10,23,25], and trajectory analysis method [14,31]. The PCC method first extracts rubber results at different times and then analyzes the classifications by superposition and statistics to obtain the age of rubber forests [32]. However, images of key phenological periods are often not available owing to the frequent cloud cover in tropical regions. In most cases, because of the extreme spectral similarity between rubber and natural forests, classification errors may result in high uncertainty of the rubber forest change detection and age extraction [15,33,34]. The threshold method sets certain thresholds to extract change information according to the pattern of vegetation indices (VIs) over time [29,30]. However, the threshold value of the VIs may fluctuate owing to the different rubber species, phenological period, or geographical environment, and setting a specific threshold value may cause some uncertainty in the extraction of the age information [18]. The regression method estimates stand age by establishing a regression model between spectral bands, vegetation indices or backscatter coefficients of synthetic aperture radar (SAR) images, and stand age of rubber [23,25]. However, because the coefficients saturate after a certain stand age, the reflectance values of young and open canopy stands are likely to be influenced by ground cover crops, and the regression method may seriously overestimate young stands and underestimate old stands. Compared with the previous three methods, the trajectory analysis method is considered more robust to inherent noise in the data (e.g., interannual variation) and has become an important research hotspot for the extraction of land-related type change information [14,22]. The trajectory analysis method also contains many algorithms, such as Landtrendr [35] and shapelet [14] to extract the age of rubber forests. However, these algorithms require a large amount of storage and incur high computational costs. Fortunately, with the free sharing of the Google Earth Engine (GEE) remote sensing cloud computing platform [36,37], which provides fast processing and analysis of massive remote sensing data, there is strong technical support for the real-time processing of large-scale and long-term remote sensing data. Therefore, combined with long time series satellite remote sensing data and GEE cloud platform, the relevant algorithms of the trajectory analysis method can be developed to achieve the identification of rubber tree age and pre-conversion land cover in large areas.

For the carbon storage of rubber forests, traditional estimation methods, such as the stockpile method, biomass method, and box method for the field monitoring of carbon storage, are clear and explicit, easy to apply, and more widely used [38,39]. However, because of the inconsistency of measurement methods, sampling locations, and study scales, the results vary and cannot accurately reflect the changes in carbon storage over long periods of time and large scales. With the development of information technology, remote sensing biomass estimation transformation methods and remote sensing-driven model simulation methods have emerged [40–43]. However, these methods either need extensive ground survey data to support them or have problems with complicated model-driven data [40,44]. With the carbon storage model of the Integrated Valuation of Ecosystem Services and Tradeoffs model (InVEST) proposal [45,46], more and more scholars at home and abroad have started to use the carbon storage module [47] of the InVEST model to estimate regional carbon storage in terrestrial ecosystems [48–50]. Compared with the traditional carbon storage estimation methods, the carbon module of InVEST has the advantages of simple and easy access to driving data (e.g., types and carbon densities of the land use/land cover (LULC)), simple operation, fast running speed, and strong visibility of output results. It can realize mapping of spatial distribution and dynamic changes of carbon storage [41] and reflect the relationship between land-use change and carbon storage under different scenarios [9]. However, existing studies [50–52] generally focus on the effects of all types of LULC on total carbon stock, and relatively little research has been conducted on the effects of single land-use types (e.g., rubber forests) on total carbon storage, which may be important for forest carbon neutralization [19]. In addition, the existing studies [50–52] on carbon storage estimation in rubber forests are basically converted from biomass by assuming a constant value or approximating the multi-year average rate, which will inevitably lead to large uncertainties in the actual carbon storage simulations. Therefore, using the age data of rubber, the InVEST model can improve the temporal dynamic evolution of carbon storage in the rubber forest, improving the accuracy of carbon storage simulation to a certain extent.

Thus, this study takes northeast Thailand, where rubber plantations are expanding rapidly, as an example to assess the impact of spatial and temporal changes in rubber planting on carbon storage over the past two decades. The objectives of this study are (1) to develop an algorithm for mapping the stand ages of rubber plantations and identify the land-cover types prior to rubber plantation conversion; (2) to analyze the spatial and temporal patterns of carbon sequestration under the expansion of rubber plantations using the InVEST model; and (3) to explore the differences in carbon sequestration processes between planted and non-rubber-planted conditions.

2. Materials and Methods

2.1. Study Area

The province of Loei is located in northeast Thailand with elevation ranging from 100 to 1798 m (Figure 1). This area features a humid subtropical monsoon climate with two main seasons: a rainy season from May to October and a dry season from November to

April. The southwest monsoon brings abundant precipitation to the study area, and the heavy rainfall is concentrated in August or September [36].

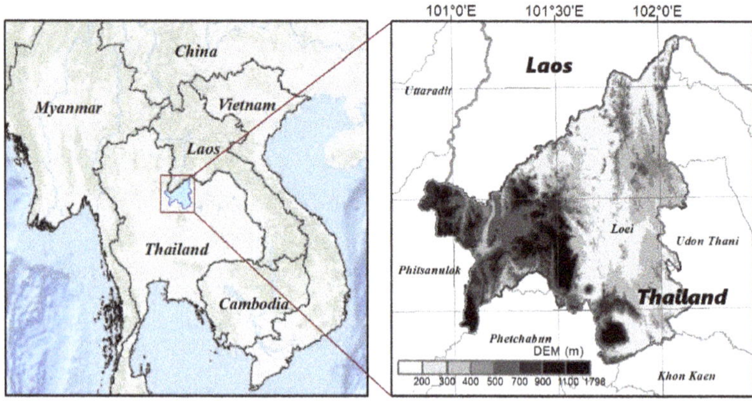

Figure 1. The location and spatial distribution of digital elevation model (DEM) of Loei Province, Thailand.

Traditionally, northeastern Thailand has been an important cultivation area with fewer rubber plantations. Encouraged by an active government policy for rubber plantations since 2003, the rubber plantation area has expanded rapidly in northeastern Thailand [33]. As a result, rapid land use and land cover change has taken place in most of its territory. Many patches of natural forest and cropland have been encroached on by rubber plantations, which can nowadays be found all over Loei from the highland areas down to the low-lying plains [16]. Therefore, it is of great practical value to assess the rubber plantations impact on carbon storage in the study area for the development of sustainable rubber plantations and forest conservation.

2.2. Data

2.2.1. Remote Sensing Data and Preprocessing

- Sentinel-1 data and preprocessing

A total of 170 scenes of Sentinel-1A and Sentinel-1B interferometric wide swath ground range detected (GRD) images of 2019 from the GEE platform [53] were used to generate a forest map (including rubber plantations). The Sentinel-1 data in GEE were pre-processed with the Sentinel-1 Toolbox using orbit metadata update, GRD border noise removal, thermal noise removal, radiometric calibration, and terrain correction [36]. The final terrain-corrected digital number (DN) values were converted to decibels (dB) in each pixel via log scaling $10\log_{10}(DN)$. A Refined Lee filter was applied to de-speckle the images. Two additional indices, including (1) the ratio of the dual polarization of vertical transmit and vertical receive (VV) to vertical transmit and horizontal receive (VH) dB data (Ratio = dB_{VV}/dB_{VH}), and (2) the difference between VV and VH dB data (Difference = $dB_{VV} - dB_{VH}$), were also calculated for each image. The annual mean value indicators (i.e., VV_mean, VH_mean, Ratio_mean, and Difference_mean) were generated for forest mapping, since the mean images can reduce the geometric and radiometric distortion of Sentinel-1 SAR images [54].

- Sentinel-2 data and preprocessing

Four Sentinel-2 Level 1C data (tile: 47QQU, 47QQV, 47QRU, 47QRV) were downloaded from the European Space Agency's (ESA) Copernicus Scihub [55]. The acquired Sentinel-2 data were obtained on 23 March 2019. Radiometric and geometric corrections were conducted to acquire top-of-atmosphere (TOA) reflectance. We conducted atmospheric correction and obtained surface reflectance using ESA's Sen2Cor in Sentinel Application Platform (SNAP) 7.0 software [56]. The spatial resolution of Sentinel-2 data varies from 10 to 60 m. Bands of 1, 9, and 10 were excluded from the dataset owing to their sensitivity to aerosol and clouds and their spatial resolution (60 m). Then, the images were resampled at 10 m using a bilinear method [56]. In all, 33 spectral indices were calculated based on the surface reflectance [57], including the Normalized Difference Vegetation Index (NDVI) [58], the Enhanced Vegetation Index (EVI) [59], and the Red-edge Normalized Difference Vegetation Index 1 (NDVIre1) [60]. A full list detailing all spectral indices can be found in the Supplementary Material (Table S1). Eight textural features were derived using a gray-level co-occurrence matrix (GLCM) [61], including mean (MEAN), variance (VAR), homogeneity (HOM), contrast (CON), dissimilarity (DIS), entropy (ENT), angular second moment (ASM), and correlation (COR).

- Landsat data and preprocessing

We obtained cloud-free Landsat thematic mapper (TM), enhanced thematic mapper plus (ETM+), operational land imager (OLI) images spanning 1999–2019 with one image per year for the study region (worldwide reference system 2 (WRS-2) path/row = 129/48) (Table 1). To reduce the errors and uncertainties caused by different months and clouds/rain, cloud-free satellite images were only collected for the dry season (mid-October to mid-May of the following year). The Landsat data were downloaded from the USGS [62]. Radiometric calibration, atmospheric correction, and geometric correction were conducted for each image. All acquired data were georeferenced in the WGS_84_UTM_ZONE_47N, and additional relative geometric corrections were also conducted to improve the geometric consistency of image time series. For the year 2012, we gap-filled the Landsat 7 scan lines corrector off (SLC-off) data using the neighborhood similar pixel interpolator method [63]. The NDVI was calculated for each image to build the interannual time-series image stack. Several studies [11,32,64,65] have utilized the NDVI as a monitoring indicator of tropical forest disturbance.

Table 1. List of Landsat images used to build the time-series stack.

Year	Date	Sensor	Year	Date	Sensor
1999	1999/11/16	ETM+	2010	2010/2/23	TM
2000	2000/11/10	TM	2011	2011/1/25	TM
2001	2001/1/05	ETM+	2012	2012/4/25	ETM+
2002	2002/11/08	ETM+	2013	2013/11/30	OLI
2003	2003/3/16	ETM+	2014	2014/1/17	OLI
2004	2004/11/5	TM	2015	2015/1/4	OLI
2005	2005/11/24	TM	2016	2016/4/12	OLI
2006	2006/11/27	TM	2017	2017/2/10	OLI
2007	2007/1/14	TM	2018	2018/2/13	OLI
2008	2008/3/5	TM	2019	2019/4/21	OLI
2009	2009/3/8	TM			

2.2.2. Ground Reference Data

Ground reference data were acquired from random sample points generated in ArcGIS 10.5 [9] and were checked by visual interpretation based on Google Earth high resolution images. A total of 1700 sample points were generated, and those at the boundary between the two categories were eliminated. Finally, a total of 1628 sample points were selected for training and validation, including 352 natural forest points, 504 rubber plantation points, 458 cropland points, 74 water body points, and 240 built-up points.

In the mapping of the forest, the sample points of the rubber forest and natural forest were merged into "forest" sample points. Of the sample points, 70% were used to train the forest extraction algorithm for mapping the forest/non-forest base map, and the remaining 30% were used for accuracy verification [34,36]. After generating the forest base map, 70% of the rubber forest and natural forest samples were used to train the rubber forest extraction algorithm, and the remaining 30% were used for accuracy verification of the rubber forest extraction results.

The validation of the rubber forest age was difficult because of the few year-by-year Google Earth high-resolution images from 2000 to 2019. Therefore, the age of rubber was divided into five groups for accuracy verification, with the age composition of 1–5 years (2014–2018), 6–10 years (2009–2013), 10–15 years (2004–2008), 16–19 years (2000–2003), and ≥20 years (before 2000). In combination with Google Earth historical high-resolution images, 3678 validation points were randomly selected for the accuracy validation of rubber forest age results.

2.2.3. Carbon Density Data of Different Land Cover Types in Different Years

Because it is difficult to measure carbon storage in rubber forests and primary tropical rainforests, this study referred to the Intergovernmental Panel on Climate Change's (IPCC) 2006 methodology [66] for determining greenhouse gas inventories in the agriculture, forestry, and other land use (AFOLU) sector, the calculation methods and result criteria for carbon storage of agricultural and forestry land (updated and refined in 2019) [67], and the data from related studies on rubber forest carbon storage [7,68].

Since this study focused on the annual change in carbon storage in the process of rubber planting, the aboveground biomass and belowground biomass ratios of rubber forests were variable dynamic values with reference to the relevant calculation methods and result criteria in the IPCC report in the forest sector [67]. The annual increases in the aboveground and belowground biomass of natural forests were set as 3 Mg·C/ha and 1 Mg·C/ha, respectively, and the type of cropland was set as a fixed value after referring to the relevant literature [69]. Finally, the reference carbon density values for different planting years of rubber forests, the natural forest, and cropland in this study were formed and are displayed in Table S2.

2.2.4. Auxiliary Data

The digital elevation model (DEM) data were available from the NASA SRTM V3 digital elevation products [30]. The spatial resolution of these data was 30 m. These DEM products were directly used to analyze the distribution characteristics of rubber forests in terms of altitude.

2.3. Methods

A systematic approach for quantitatively assessing the impact of rubber expansions on regional carbon storage was proposed in this study. Our method consisted of five main stages (Figure 2): (1) mapping the forest distribution of 2019 using Sentinel-1 time-series satellite data; (2) extracting the rubber forest by integrating Sentinel-1 and Sentinel-2 data; (3) identifying the rubber planting year and pre-conversion land cover; (4) estimating the carbon storage using the InVEST model; and (5) assessing the impact of rubber expansions on regional carbon storage under actual planted and hypothetical non-rubber-planted scenarios.

2.3.1. Rubber Plantation Delineation by Integrating Sentinel-1 and Sentinel-2 Data

The random forest (RF) algorithm [70] and Sentinel-1 10 m data (VV_mean, VH_mean, Ratio_mean, and Difference_mean images) were used to classify forest from other land-cover types (cropland, water, and built-up). The number of decision trees was set at 100, and the number of variables per split was set at the square root of the number of variables. Finally, the forest map with 10 m resolution was obtained by merging non-forest categories and then used to derive rubber plantations from natural forests in 2019.

The annual mean value indicators of Sentinel-1 data, spectral indices, and the textural features of Sentinel-2 data were combined to derive rubber plantations. The mean decrease in Gini (MDG) was used to measure a feature importance, and the out-of-bag (OOB) score determined which Sentinel-2 features were involved in the classification process. Finally, 13 spectral indices (NDWI2, BAI, B3, NDVIre1, B8A, GI, NDVIre2, LAnthoC, B4, B5, Chlorgreen, DVI, and SR-BlueRededge1—the full list detailing the feature importance of spectral indices can be found in Figure S1), and 4 textural features (HOM, ENT, ASM and

COR) were selected to identify rubber plantations combined with Sentinel-1 mean features in 2019.

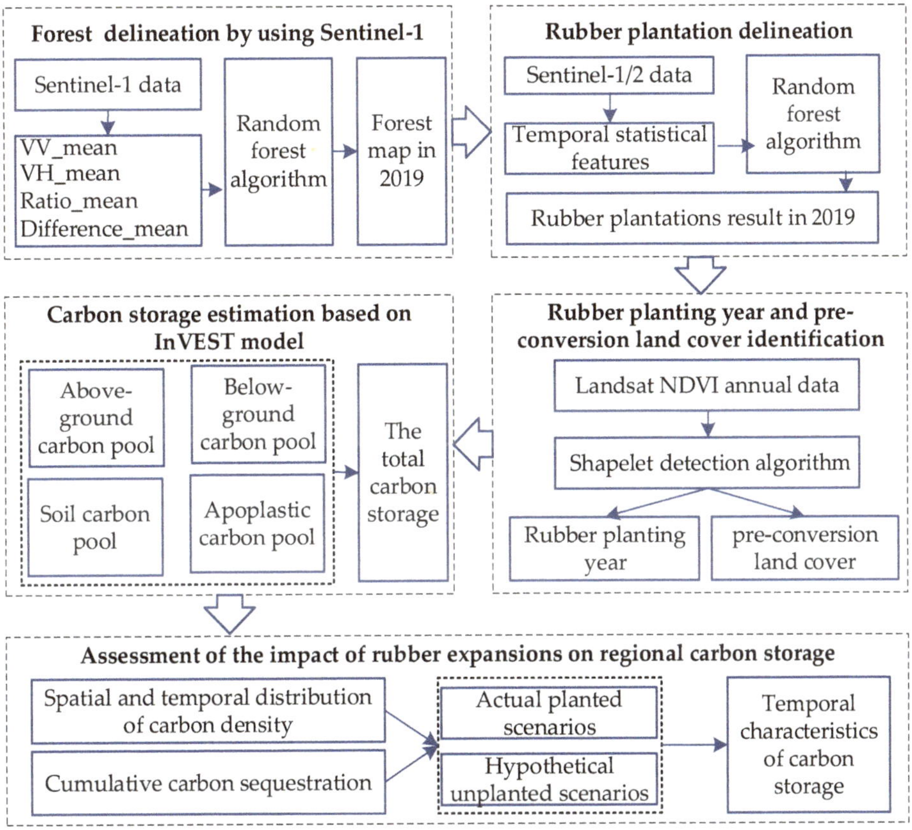

Figure 2. Workflow of the method adopted for this study to assess the impact of rubber expansions on regional carbon storage.

Ground reference data described in Section 2.2.2 were used in confusion matrices [71] to assess the accuracy of forest/non-forest and rubber plantation maps, including overall accuracy, kappa coefficient, producer accuracy, and user accuracy.

2.3.2. Shapelet-Based Planting Monitoring of Rubber Plantations

Once the rubber plantation mask was conducted, each rubber plantation pixel possessed an NDVI interannual time series with 21 time points $T = \{T_1, T_2, \ldots, T_{20}\}$ from 1999 to 2019. $T_1, T_2, \ldots, T_{21}$ were arranged in chronological order. Figure 3 shows the temporal changes in the annual NDVI for three scenes. A rubber pixel NDVI series may be characterized as stable high (i.e., rubber planted before 2000) (Figure 3a) or could suddenly decline and then increase owing to the land clearing and planting preparation and the typical open canopy period of the juvenile rubber tree cover (Figure 3b,c). Therefore, the unique subset of the NDVI time series representing the time period of planting event characterization was used to distinguish the rubber plantations that lasted for 20 years from the rubber plantations converted from other land-cover types.

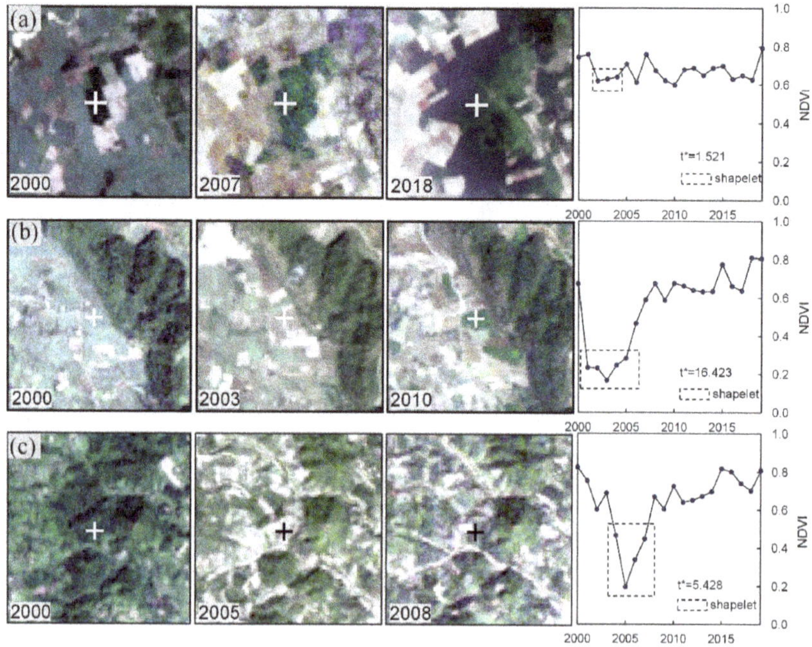

Figure 3. Inter-annual NDVI time series with related Landsat imagery (RGB = 3-2-1 for Landsat TM/ETM+, RGB = 4-3-2 for Landsat OLI) for (**a**) rubber plantations that last for 20 years, (**b**) rubber plantations converted from cropland, and (**c**) rubber plantations planted after natural forests are cut down.

A shapelet algorithm was applied to detect the unique characteristics of clear-cut fields and newly cultivated rubber plantations in the rubber NDVI time series [11]. The shapelet consisted of two main steps: (1) shapelet detection and (2) time-series classification. The detection step found the most representative "shapelet" of a time-series category by searching all possible shapelet locations in one image time series, whereas the time-series classification distinguished between rubber plantations that have remained intact for 20 years and those where planting activities have occurred. A candidate shapelet is characterized by two time-position parameters: s is the starting point of the shapelet, and w is the width of the shapelet. A shapelet is a continuous subsequence of a time series, and the remaining time points belong to a non-shapelet. Both the minimum shapelet width and the minimum non-shapelet width were set to 3 time points (3 years). A separation metric called GAP [72] (i.e., the difference between the mean and standard deviation of the non-shapelet group and the candidate shapelet group) was used to find the final "shapelet" among the shapelet candidates. A paired-sample t test was used to detect the discrepancy between the shapelet and non-shapelet and to construct a decision tree (for details, see [11]). For this study area, we used $\alpha = 0.01$, and the threshold was $t_{(18,0.99)} = 2.552$. The parameter α is the significant level for the t-test. A lower α value means that the time series for a rubber pixel has a higher discrepancy between its shapelet and non-shapelet segments, i.e., rubber plantation activity has occurred.

As shown in Figure 3, the dashed rectangles represent the detected shapelet of the rubber time series. Because of the absence of disturbance in the intact rubber plantation example, the NDVI value of the shapelet was very similar to that of the non-shapelet (Figure 3a). Accordingly, the discrepancy between its shapelet and non-shapelet segments was low with $t^* = 1.521$, which was lower than the threshold ($t = 2.552$). Figure 3b shows an example of rubber plantations converted from cropland and Figure 3c shows rubber plantations planted after natural forests were cut down. These two rubber plantation conversion scenarios had a relatively longer period of consistently and significantly low NDVI values of non-vegetated timespan, which is related to the land clearing, land preparation, and open canopy period typical of juvenile rubber tree cover. As a result, the *t* test showed that the rubber plantation pixels with planting events had a t statistic (t * = 16.423 and t * = 5.428) higher than the threshold.

Using the shapelet as the smallest unit of analysis captures the continuous process of true change or the constant state and reduces the effects of noise owing to seasonal and radiometric changes. For further details and applications of shapelets, see [11].

In the same way as the verification of the forest mapping, the accuracy of rubber age was validated based on the confusion matrix [71] and reference data.

2.3.3. Rubber Planting Year and Pre-Conversion Land Cover Identification

After the shapelet detection and time-series classification algorithms were performed, each time-series rubber plantation pixel was assigned a shapelet. Year of deforestation (YD) was defined as the starting time point of a shapelet. The last vertex in the shapelet was recorded as the year of rubber planting (YRP) because only the latest vertex was associated with a rubber planting event, where "vertex" was defined as a point that met the condition of that the value of the point is smaller than that of both the previous and the next time points ($\{T_x | T_x < T_{x-1}$ and $T_x < T_{x+1}\}$).

In order to identify the pre-conversion land cover (PCLC), the interval between YD and YRP was calculated and called the planting temporal interval (PTI). The PTI varied owing to the different land-cover status (i.e., natural forest or cropland) prior to planting rubber [11]. If the PTI was short enough (Figure 3c showed only 5.2 years), it implied that the rubber plantation was established shortly after the deforestation events, and therefore, the PCLC was "natural forest". Otherwise, the non-forest status existed before the rubber was planted, so the PCLC was "non-forest", i.e., cropland. The threshold for PTI (3 years) was determined based on the manual statistical analysis of extensive NDVI time-series (1387 samples). However, PTI alone cannot fully describe the complex rubber planting process (flexible cropping system, high variability in planting status, and fragmentation) [13]. Therefore, we introduced $NDVI_{initial}$, the NDVI value of the first time point before the shapelet, to the PCLC identification process. The PCLC was identified with the decision rule that if PTI is less than or equal to 3 years and $NDVI_{initial}$ is greater than or equal to 0.5869 then the PCLC is natural forest, otherwise PCLC is cropland.

The threshold for $NDVI_{initial}$ (0.5869) was selected based on the statistical analysis of 50 regions of interest (ROIs) for rubber plantations (25 ROIs converted from natural forest and the remaining 25 from cropland). The $NDVI_{initial}$ range of cropland fluctuates widely in southeast Asia because of the flexible cropping system, while the $NDVI_{initial}$ value of the natural forest is more stable owing to consistently high cover, so we set the lowest value of $NDVI_{initial}$ of the natural forest (0.5869) as the $NDVI_{initial}$ threshold.

2.3.4. Carbon Storage Estimation Based on InVEST Model

The simulation and assessment of carbon storage in rubber forest in this study were implemented using the InVEST model [9,45,73], which has been widely used to estimate various ecosystem services [45,47]. The carbon storage calculation in the model took the land-use/cover type as the assessment unit and calculated the carbon storage of the ecosystem according to different land-use/cover types in the study area. It divided the carbon storage of each land use/land cover (LULC) type into four basic carbon pools: aboveground carbon pool, belowground carbon pool, soil carbon pool, and apoplastic carbon pool (i.e., dead organic carbon pool). The outputs of the model were carbon density and carbon storage, which were calculated by the formula [11,47]:

$$C_i = C_{i\_above} + C_{i\_below} + C_{i\_soil} + C_{i\_dead} \quad (1)$$

$$C_{total} = \sum_{i=1}^{n}(C_i \times S_i) \quad (2)$$

where i represents a certain land-use/cover type; C_i is the carbon density of the i-th type; C_{i_above}, C_{i_below}, C_{i_soil} and C_{i_dead} are the aboveground, belowground, soil, and apoplastic carbon densities of the i-th LULC type, respectively. The unit is megagrams·carbon/hectare (Mg·C/ha). C_{total} represents the total carbon storage in the study area (tons/year, t/a), n represents the number of land-use/cover types in the study area, S_i is the area of the ith type of area (hectare, ha).

For the acquisition and definition of carbon density data for rubber forests, primary forests and croplands in different years were detailed in Section 2.2.3 and Table S2.

2.3.5. Defining Different Scenarios

In order to further analyze the differences in the influences of the regional carbon storage under planted and non-planted rubber forest conditions, two different scenarios were set: (1) the actual planted scenario of rubber forest, i.e., the rubber forest expansion occupied cropland and deforestation and land reclamation. The carbon density and carbon storge evolution characteristics under actual planted rubber forest conditions over 20 years were simulated using the age data of the rubber forest from remote sensing time-series data and carbon density obtained in different years; (2) the hypothetical non-rubber-planted scenario, i.e., the rubber forest, was unexpanded over 20 years. That is, there is no occupation of cropland or deforestation and land reclamation. Under such conditions, the regional carbon density and carbon storge were directly simulated using the maps of rubber and related land types in 1999 and the carbon density of different land-cover/use types in different years (Table S2).

3. Results

3.1. Forest and Rubber Plantation Mapping for 2019

The forest map produced with Sentinel-1 data is shown in Figure 4a. Forests were mainly concentrated in higher elevations of the hilly area. The overall accuracy of the forest map was 0.92 with a kappa coefficient of 0.84 (Table S3). The forest had reasonably good accuracy, with both high user accuracy (92.58%) and producer accuracy (92.22%), implying that the resultant Sentinel-1 forest map could be used as a reliable base map for rubber plantation identification.

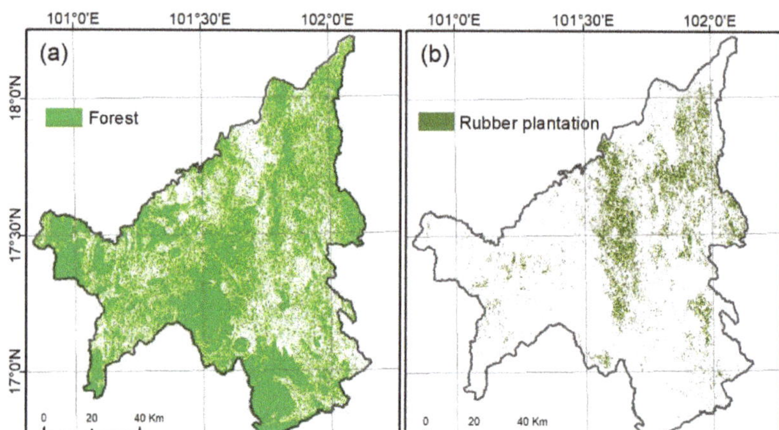

Figure 4. Distribution of the forest (**a**) and rubber plantations (**b**) of Loei Province in 2019.

The spatial distribution of rubber plantations is shown in Figure 4b. Most rubber plantations were distributed in the central and northeastern parts in the study area. The overall accuracy was 0.91 and the kappa coefficient was 0.82. The interpretation accuracy for the rubber plantations was high with both user and producer accuracies greater than 90%. The forest area was estimated at 6.67×10^5 ha in 2019, while the rubber plantation area was 1.28×10^5 ha in Loei Province.

According to the results of rubber plantation areas at different elevations in 2019 (Figure 5), most rubber forests were distributed in hilly areas at 200–400 m, accounting for 78.47% of the total area of rubber forests, while there were few rubber plantations at the other altitudes.

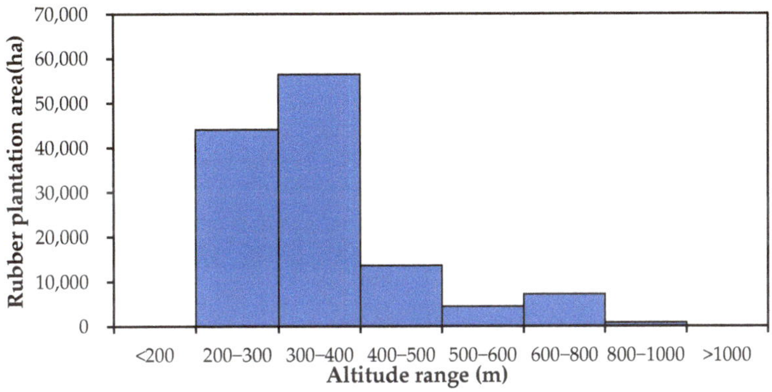

Figure 5. Statistics of rubber plantation areas at different elevations in 2019.

3.2. *Age Estimation and Pre-Conversion Land-Cover Identification of Rubber Plantations*

Based on the 2019 rubber plantation mask (Figure 4b), the rubber plantations where planting activities occurred were identified from the inter-annual NDVI time series using the shapelet approach. The results (Table S4) showed that the overall accuracy was 0.83 and the kappa coefficient was 0.78. This indicated that the automatic identification of rubber planting years had good estimation accuracy.

Then, YRP and PCLC maps (Figure 6) were produced from the shapelet segment. Figure 7 shows the statistics for annual planting area and pre-conversion land-cover types for each year. The area of rubber plantations increased nearly 8-fold from 0.14×10^5 ha or 1.3% of Loei Province before 2000 to 1.28×10^5 ha or 12.2% in 2019, showing clear expansion trends from centralization to scattering. The average plantation age in Loei Province was 13.85 years (assuming an age of 20 years for all plantations older than 19 years). Rubber plantations in Loei Province were mainly planted before 2008, where the areas planted between 2004 and 2006 accounted for 61.8% of all new planted areas, reflecting a close relationship with the increase in rubber price before 2009 and the vigorous support of the first phase (2004–2006) of the Thai government's promotion of rubber forest plantations in the northeast. Plantations began to increase slightly again after 2012, at an average rate of about 2767 ha/year. Spatially, the newly established rubber plantations were distributed on the periphery of the existing rubber plantations.

Most of the total newly established rubber plantations in Loei Province were converted from cropland, accounting for 54.81% (61,708.75 ha), while 45.19% were converted from natural forests (50,871.24 ha). Before 2004, rubber was planted mainly by encroaching on cropland. The conversion from natural forests began to increase after 2004, and the conversion area of natural forests was larger than that of cropland in 2005–2007 and 2013–2017.

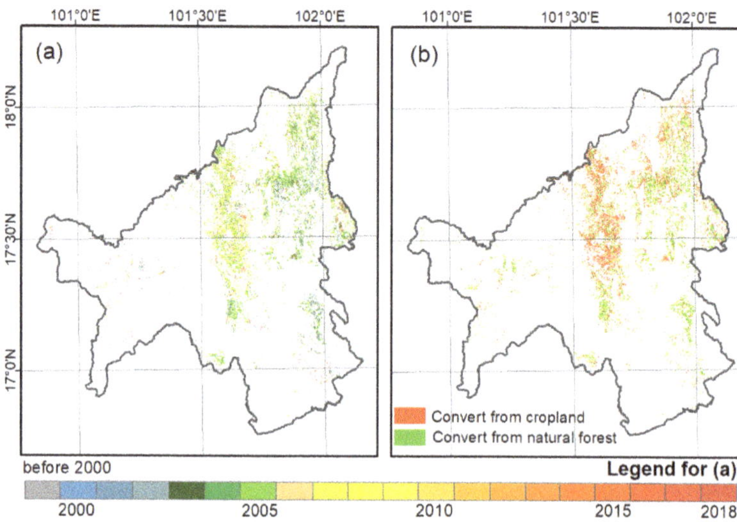

Figure 6. (**a**) Planting year estimation, and (**b**) pre-conversion land-cover types of cropland and natural forest in Loei Province.

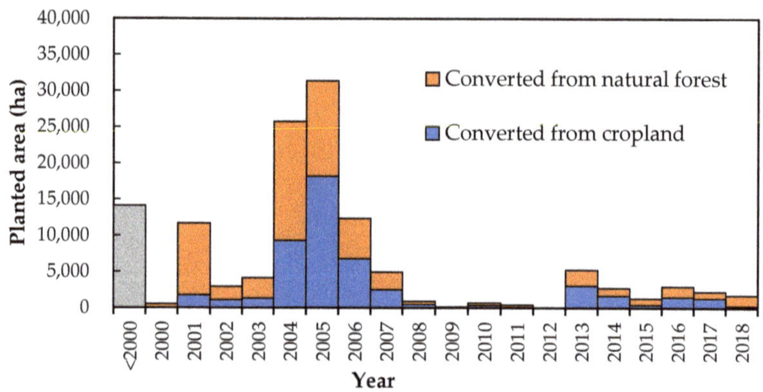

Figure 7. The statistics for annual planting area and pre-conversion land-cover types of cropland and natural forest in Loei Province for each year. Year < "2000" indicates rubber planted before 2000.

We assumed natural forests in 2019 to be stable natural forests that had not been disturbed in the last 20 years. The natural forests in 2000 were obtained by combining the 2019 natural forests with the natural forests encroached upon by rubber plantations in 2000–2018 (rubber plantations established in 2019 were ignored). The proportion of natural forest disturbance in Loei Province related to rubber plantations was 6.01% in 2000–2012, and 7.21% in 2000–2018.

3.3. *Spatial and Temporal Distribution of Carbon Density and Cumulative Carbon Sequestration*

With the continuous expansion of rubber plantations over the past 20 years, the carbon storage capacity has varied significantly among these years (Figure 8a–e shows only the spatial distribution of carbon density for five years (1999, 2004, 2009, 2014, and 2018)).

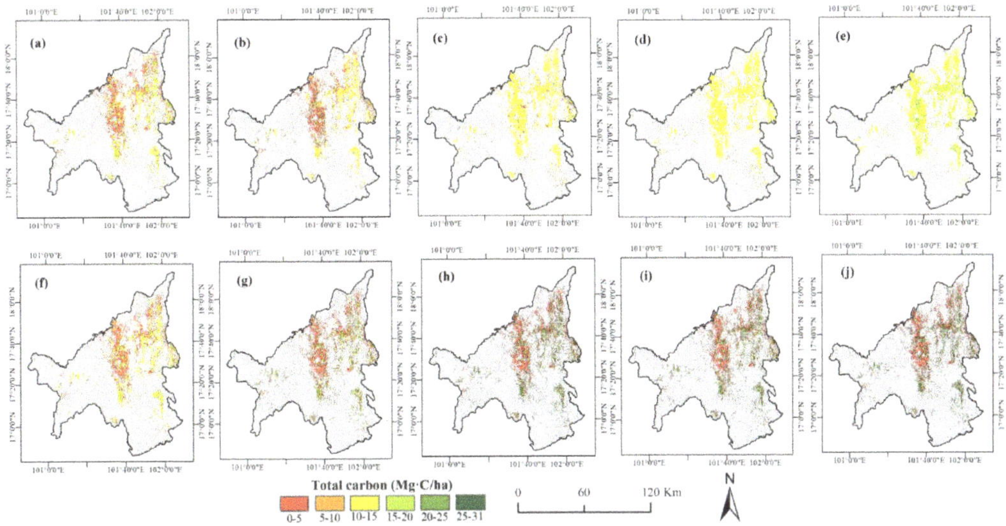

Figure 8. Spatial and temporal pattern distribution of carbon density in Loei Province (where (**a–e**)/(**f–j**) correspond to 1999, 2004, 2009, 2014, and 2018, respectively; (**a–e**) and (**f–j**) represent the scenarios under actual rubber plantations and hypothetical non-rubber plantations, respectively).

The results indicated that the central and northeastern parts of Loei Province were mainly cultivated before 2000, with a low carbon storage capacity and a carbon density value of only 2.25 Mg·C/ha, while the western, southern, and eastern areas were mainly natural forests with a high carbon storage capacity and a carbon density value of approximately 20.00 Mg·C/ha. The carbon sequestration capacity of some cropland in the northeast area was greatly increased after it was converted to rubber forest in 2004. On the contrary, the carbon sequestration capacity of natural forests in central and eastern areas was decreased after they were converted to rubber plantations, resulting in carbon density within the threshold range of 10–15 Mg·C/ha for rubber plantations. By 2009, most of the cropland in the central part of the area had been converted to rubber forests, which further increased the carbon sequestration capacity, but the overall carbon density was still mostly distributed in the 10–15 Mg·C/ha range. Up to 2014, the expansion of rubber gradually slowed down, mainly by encroachment on natural forests, and the carbon density values of scattered distributed rubber forests increased to 15–20 Mg·C/ha; subsequently, up to 2018, the carbon density of rubber forests exceeded 15 Mg·C/ha, except for some newly planted rubber plantations.

Comparing the changes in the carbon density of rubber forests in the study area over 20 years from 1999–2018 (Figure 8a–e), we found that the carbon storage capacity of each type was significantly different. The carbon storage capacity of cropland was the smallest at only 2.25 Mg·C/ha and did not change over time. By 2009, most of the cropland was replaced by rubber forests, and the carbon sequestration increased dramatically. On the contrary, the deforestation and replacement of natural forests—which have high carbon storage—with rubber reduced the carbon sequestration, causing the carbon density to drop from about 20.00 Mg·C/ha to 10–15 Mg·C/ha. However, with the growth of rubber forests, their carbon sequestration capacity gradually increased again and basically returned to the carbon sequestration level before deforestation in 2018.

For the scenario assuming no rubber planting over 20 years, we recalculated the results of carbon density from 1999–2018 (Figure 8f–j) and found that under this scenario, the study area was mainly dominated by cropland and natural forest types, and the carbon sequestration of cropland remained unchanged over time, while that of the natural forest increased from 15–20 Mg·C/ha to 25–31 Mg·C/ha over the past 20 years.

Comparing the changes in carbon storage under the actual rubber plantation scenario (Figure 8a–e), we found that the carbon storage gradually increased over time in the area where the cropland was replaced by rubber forest. The carbon storage in the area where the natural forest land was replaced by rubber showed a process of first decreasing and then gradually increasing and returning to the carbon sequestration level before the forests were cut down. However, it was still much smaller than that under the no-rubber-planting scenario (Figure 8f–j), and because of that the carbon sequestration of the original forest is also gradually increasing.

Based on the results of carbon density, the spatiotemporal evolution pattern of cumulative carbon sequestration was further obtained over a 20-year period by using the difference calculations (Figure 9a–d). Before 2004, the cumulative carbon sequestration was mainly distributed between −10 and 15 Mg·C/ha. In the eastern and south-central regions, the forest was replaced by rubber forests resulting in carbon emissions, and carbon sequestration ranged from −10 to −5 Mg·C/ha. The carbon sink in the cultivated area replaced by rubber increased, and the distribution ranged from 10 to 15 Mg·C/ha. By 2009, the carbon sequestration of rubber plantations was distributed in the range of 10–15 Mg·C/ha in most of the occupied croplands, and the carbon sequestration in deforested rubber plantations was the same as in the previous five years. By 2014, with the expansion of rubber forests and increasing carbon sequestration, the distribution of carbon sequestration was more complex, with that in the central region distributed at 10–15 Mg·C/ha, in the southern region at −10 to −5 Mg·C/ha, and in the eastern region at −5 to 0 Mg·C/ha. By 2018, carbon sequestration was mainly distributed at 10–15 Mg·C/ha in the central area and −5 to 0 Mg·C/ha in the southern and eastern areas. There were some areas of −15 to −10 Mg·C/ha where short-term deforestation and land reclamation occurred and 15–20 Mg·C/ha obtained from sporadic early planted rubber forests accumulated over time.

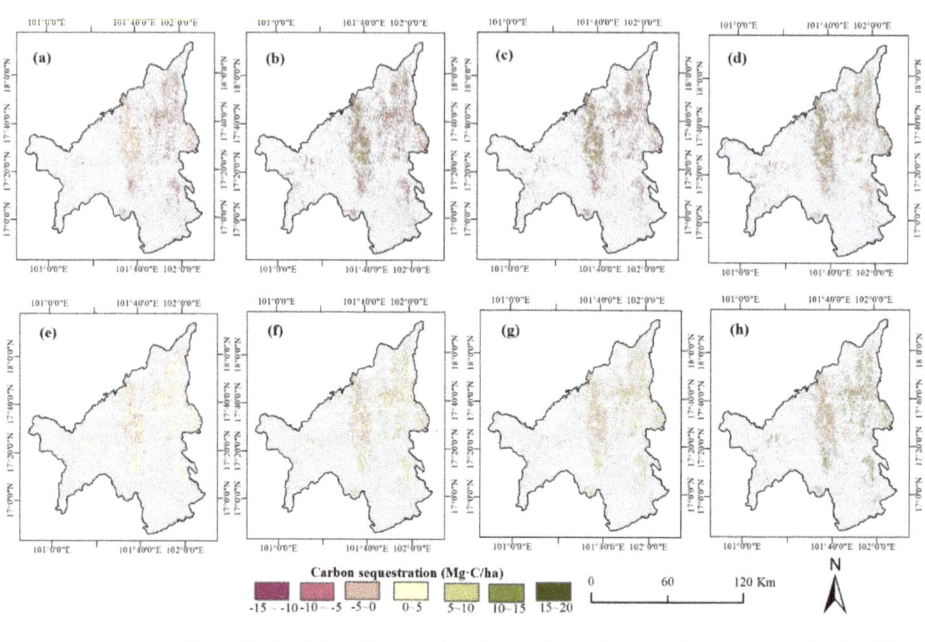

Figure 9. Spatial and temporal patterns of cumulative carbon sequestration in different periods (corresponding to 1999–2004, 2005–2009, 2010–2014, and 2015–2018; (**a**–**d**) and (**e**–**h**) represent the cumulative carbon sequestration under actual rubber planting and hypothetical no-rubber-planting scenarios, respectively).

For the hypothetical no-rubber-planting scenario, the evolution pattern of carbon sequestration in the past 20 years is shown in Figure 9e–h. For cropland, because the default carbon sequestration capacity remained unchanged and at zero over 20 years, we focused here on the natural forest. From 1999 to 2004, the cumulative carbon sequestration was mainly distributed in the range of 0 to 5 Mg·C/ha. In 2009, the cumulative carbon sequestration increased to 5–10 Mg·C/ha. From then to 2014, although the carbon sequestration gradually increased, it did not exceed the 5–10 Mg·C/ha range. Until 2018, the cumulative carbon sequestration increased to 10–15 Mg·C/ha.

Compared with the actual rubber forest plantation scenario, we found that rubber forest expansion occupied cropland, which can improve the level of carbon sequestration to a certain extent, but deforestation and land reclamation remained the main reasons for the decrease in overall cumulative carbon sequestration in our study.

3.4. Temporal Characteristics of Carbon Storage

In the past 20 years, the actual storage decreased only slightly in 2001 and 2004 and showed a fluctuating growth in the other years (Figure 10). The overall carbon storage gradually increased from 1.58 million tons to 2.12 million tons. The carbon storage increased by 0.54 million tons in the past 20 years, with an average annual growth rate of 2.69×10^4 t/a, while under the hypothetical no-rubber-planting scenario, the overall carbon storage increased from 1.57 million tons to 2.27 million tons, and the carbon storage increased by 0.69 million tons over the past 20 years, with an annual average growth rate of 3.46×10^4 t/a. Comparison between the actual planting and the hypothetical no-rubber-planting scenarios indicated that rubber plantations caused a decrease in carbon storage in all years except 1999 and 2000, and the difference between the actual and the hypothetical carbon storage reached −0.15 million tons in 2018.

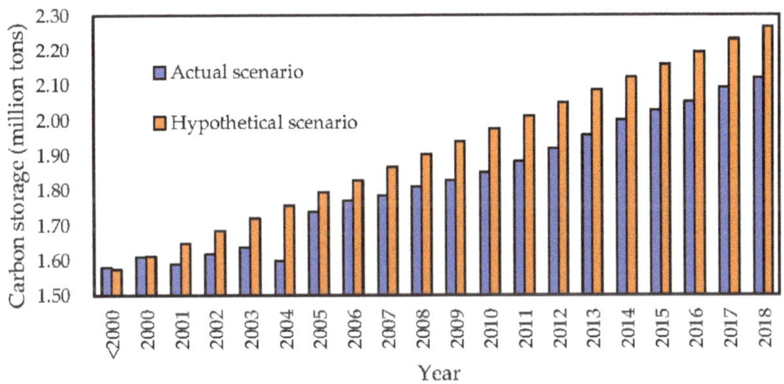

Figure 10. Annual changes in carbon storage for Loei Province from 1999 to 2018 under actual planted and hypothetical non-rubber-planted scenarios.

In order to further analyze the inter-annual variation pattern of carbon storage caused by rubber plantations, we considered the difference in the inter-annual cumulative value of carbon storage under the actual and hypothetical scenarios (Figure 11). The results showed that before 2000, the planting of rubber did not cause a reduction in carbon storage. However, since 2001, the annual cumulative difference in carbon storage was less than 0 t, and the overall trend was decreasing. The annual cumulative difference in carbon storage from 2004 to 2006 increased significantly from −0.24 million tons to −0.12 million tons, mainly because of the large occupation of cropland, which increased the level of carbon sequestration. Subsequently, the annual cumulative difference of carbon storage after 2006 gradually decreased to −0.29 million tons in 2018.

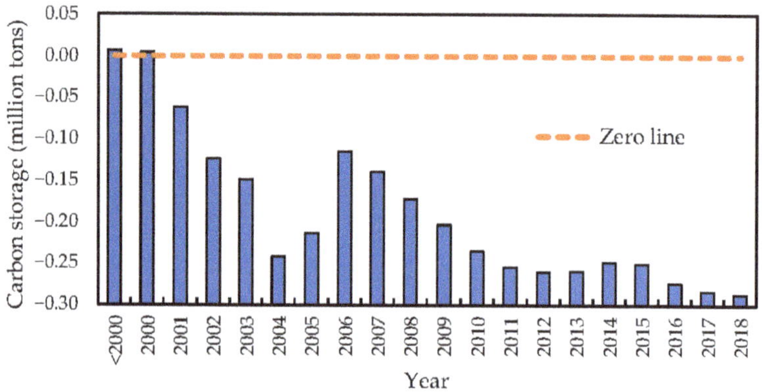

Figure 11. The cumulative differences in carbon storage between the actual and hypothetical scenarios.

4. Discussion

4.1. Potential of the Optical and SAR Imagery-Based Approach for Identifying and Mapping Rubber Plantations

An accurate rubber forest base map is a prerequisite for accurate age estimation and pre-conversion land-cover identification of rubber plantations. Previous studies combining optical (e.g., MODIS and Landsat) and SAR data (e.g., PALSAR) for rubber mapping mostly classified forested versus non-forested land types based on differences in the backscatter coefficients of SAR data, while the distinction between rubber and natural forest relied more on spectral features [22,74]. However, because rubber plantations at their peak of growth or after the stand reaches a certain age have similar spectral characteristics to natural forests [10,18], greater uncertainty exists in the extraction of rubber. In addition, these two different types of satellite sensors differ greatly in terms of spatial resolution and image acquisition time, which limits the synergistic application effect to a certain extent.

In order to reduce the spectral confusion between rubber and natural forests, this study extended the identification features of rubber forests from the spectral dimension to the spectral, spatial (texture), and structural (backscattering structural features) dimensions. The spatial information can effectively reduce the spectral confusion between rubber and natural forests, effectively reduce the "pretzel phenomenon", and improve the integrity and classification accuracy of patches [34,75]. In addition, SAR data can provide additional information including vegetation surface information and surface roughness, which are highly sensitive to differences in forest structure, such as biomass, density, and vertical stratification in different stands [76], and can improve the discrimination between rubber forest and natural forest. By adding distinguishing features from spatial and structural dimensions and integrating the advantages of different sensors, the limitations of any sensor can be overcome or supplemented to obtain a more accurate and spatially finer forest and rubber forest base map in our study, which provides the basis for accurate identification of rubber age and pre-conversion land-cover types.

4.2. Advantages of Time-Series Remote Sensing Methods for Age Estimation and Pre-Conversion Land-Cover Identification of Rubber Plantations

The identification of forest age and initial land state is the basis for the fine estimation of carbon storage in rubber plantation areas. Our results indicated some advantages to utilizing the time-series Landsat-based method using shapelets to identify the establishment year of rubber plantations and pre-conversion land-cover types. First, the shapelet makes full use of temporal information and eliminates the cumulative error caused by the underestimation of young rubber forests by single-period classification [22]. Second, the consistency requirement of the shapelet algorithm is not very strict for the time-series data, and it ignores the influence of cloud noise and other factors over a short period of time [14].

Finally, the shapelet algorithm does not consider the saturation of spectral coefficients and SAR backscattering coefficients in the regression methods [10], which improves the stand age differentiation of mature rubber forests. In addition, the shapelet algorithm focuses on detecting the changes in only one specific type, i.e., the conversion of natural forest (or cropland) to rubber forest. With fewer parameters, it can determine when the change occurred and the specific type of change. Compared with the Landtrendr algorithm [31,35], the Breaks for Additive Season and Trend Monitor (BFAST) algorithm [77], the Vegetation Change Tracker (VCT) [65], and other timing change detection algorithms, the identification process is simpler and more efficient.

Here, we refined the shapelet algorithm in two ways: (a) by using a rubber plantation mask instead of a forest mask in 2019; and (b) by adding statistical boundary constraints for the NDVI (i.e., $NDVI_{initial}$) when identifying the pre-conversion land-cover (PCLC) types. By selecting the rubber plantations in 2019 as the mask, we avoided the possibility that the intact rubber plantation pixels were mistakenly detected as natural forests. In addition, the introduction of $NDVI_{initial}$ can provide the greenness difference information between different PCLC types (i.e., natural forest and cropland), and it has been widely used in the forest and cropland classification process [75,78]. The combination of planting event and greenness features provides an acceptable method for a flexible cropping system and high variability in planting status and fragmentation areas, and is expected to be effective for identifying the PCLC in other similar rubber-planting regions [25].

4.3. Changes in Carbon Storage Because of Rubber Forest Expansion

Over the past two decades, rubber forests in northeast Thailand have expanded rapidly, gradually making it one of the largest natural rubber production bases in Thailand. However, smallholders dominate rubber production in this region, producing 90% of the rubber [16]. Attractive economic returns and agricultural extension interventions are the most important drivers of land-use conversion to rubber plantations [79]. Therefore, the expansion of rubber plantations was more complex and fragmented in the study area, owing to the combined influences of natural rubber market price fluctuations, government interventions, and the flexible cropping system and high variability in planting status [80], which posed a greater challenge to the calculation of regional rubber carbon storage.

The InVEST carbon storage model was applied to calculate the carbon stocks of rubber plantations under actual planted and hypothetical non-rubber-planted conditions in this study. However, this model is based on a simplified carbon cycle, where the carbon storage is a static inventory, assuming that each hectare of land is identical and constant [52]. This may bias the carbon storage estimates for rubber and natural forests, whose carbon storage gradually accumulates and increases with time [80]. To this end, based on obtaining the rubber forest stand age and initial land-cover types before rubber planting and referring to the IPCC calculation methods and result criteria for carbon storage of agricultural and forestry land [67], each carbon storage component of rubber and natural forests in different years over time was given a new definition and assignment, and then was added to the InVEST carbon storage model to improve the accuracy of carbon storage simulation to some extent. This theoretical approach has not been reported in previous related InVEST carbon storage simulation studies [9,11,51].

According to the carbon storage results under the actual rubber forest plantation scenario, we found that rubber forest plantation can increase regional carbon density when occupying cropland, but through deforestation and clearing it would cause a rapid decrease in carbon density, and then gradually increase and recover to the storage stock level before deforestation. This result was consistent with the findings of some existing studies [9,69,81]. By comparing the hypothetical scenario of no rubber planting, we found that rubber planting reduced the regional carbon storage and caused regional carbon emissions. Moreover, with the expansion of rubber forests, the difference between the actual carbon storage and the hypothetical scenario became larger and larger, causing a large carbon storage gap of −0.15 million tons in 2018. This result was also consistent with the findings of some studies on carbon emissions caused by rubber plantations in tropical regions [7,80,82,83]. In order to achieve sustainable rubber forest plantation and management, the carbon storage gap caused by deforestation and clearing must be bridged by partial replacement of cropland with rubber in the future to achieve the carbon balance of rubber forest plantation and environmental sustainability.

4.4. Limitations

In this study, the impact of rubber forest expansion on regional carbon storage was investigated through rubber forest extraction, forest age estimation, pre-planting land cover type identification, carbon storage estimation, and different scenario simulations. However, it should be noted that eight spatial texture features and four structural features of the optical and SAR images were combined in the rubber forest mapping. How to further combine the diverse textural and structural features of both to improve the accuracy of fine rubber forest mapping will be the focus of later research.

Second, the tree age estimation of a rubber forest relies only on the remote sensing NDVI to construct time-series curves. During the planting and growing process of rubber forests, the land-cover state undergoes a transformation from natural forest/cropland to bare land, rubber with a low canopy cover, and rubber with a high canopy cover. Different land-cover states also have obvious differences in texture characteristics, and the application of texture characteristic time series to the tree age estimation of rubber forests will be explored in the future.

Finally, the application of the InVEST model for carbon storage estimation is more sensitive to the input data. Researchers generally collect carbon density data from the literature or field experimental observations in their study area [9,11,52]; because of the lack of observation data, only the relevant research algorithms and results of the IPCC [66,67] were used as input for carbon storage simulation in this study. This may cause uncertainty in the simulation results, and it poses difficulties in the validation of the model results owing to the lack of measured data [11,19,50]. Although the results simulated by the carbon storage module of the InVEST model have been shown to be reasonably accurate [51] and representative [9,11], validation of the measured data can increase the effect of the carbon density simulation to some extent [52]. In the future, when conditions permit, partial field observations will be carried out to compare the field data with IPCC data for testing whether the data consistent, to then further correct the results of the carbon density-related components and improve the simulation accuracy. On the other hand, the carbon accumulation process of cropland was not considered in the carbon storage simulation. Later, the law and process of cropland soil carbon sequestration in tropical regions will be further explored to improve the simulation accuracy and eventually improve the simulation accuracy of regional carbon storage caused by rubber expansion.

5. Conclusions

Based on the multi-source satellite time series data of Sentinel-1, Sentinel-2, and Landsat, coupled with random forest, shapelet, and InVEST carbon storage model, a systematic approach was proposed for estimating the carbon storage of regional rubber forests, and then explored the impact of rubber forest expansion on regional carbon storage. The conclusions of the study are as follows.

1. High accuracy extractions of forest and rubber forest were achieved, by using the Sentinel-1/2 time-series satellite images, extended spectral, spatial, and structural features, and random forest algorithm. The overall accuracies are 0.92 and 0.91, respectively, which provide accurate background data for tree age and carbon storage estimation.
2. Using Landsat time-series satellite imagery, combined with the improved shapelet algorithm, the high accuracy extraction of rubber tree age can be achieved. The overall accuracy was 0.83 and the kappa coefficient was 0.78. The average age of rubber stands was 13.85 years (assuming that all plantations older than 19 years are 20 years

old). Before 2004, rubber was mainly grown through encroachment on cropland. After that, rubber conversion from natural forests started to increase.

3. Regional carbon storage estimation of rubber forest was achieved using the InVEST model. The carbon density increased from only 2.25 Mg·C/ha in 1999 to more than 15 Mg·C/ha in 2018, except for some newly planted rubber plantations. The use of cropland for rubber plantations will increase carbon storage, while for deforestation the carbon storage will decrease, then gradually increase, and recover to the storage stock level before deforestation.
4. The expansion of rubber caused a decline in regional carbon storage. The difference and annual cumulative difference between the actual and the hypothetical carbon storage reached −0.15 million tons and −0.29 million tons in 2018, respectively.

Supplementary Materials: The following supporting information can be downloaded at: https://www.mdpi.com/article/10.3390/rs14246234/s1. Table S1. List of spectral indices generated for Sentinel-2 imagery; Table S2: Carbon density of each component in different land-cover types in different years (units: t/ha); Table S3: Confusion matrix of the forest and rubber plantation mapping; Table S4: Confusion matrix of identification results of rubber planting years.; Figure S1. Importance statistics of spectral features. References [84–94] are citied in the Supplementat Materials.

Author Contributions: Conceptualization, C.H. and H.L.; methodology, C.Z. and H.L.; data curation, C.Z. and H.L.; writing—original draft preparation, H.L.; writing—review and editing, C.H.; supervision, C.H.; project administration, C.H.; funding acquisition, C.H. All authors have read and agreed to the published version of the manuscript.

Funding: This study was supported by the National Natural Science Foundation of China (42130508) and the CAS Earth Big Data Science Project (XDA19060302).

Acknowledgments: We acknowledge all who contributed to the data collection and processing, as well as the constructive and insightful comments by the editor and anonymous reviewers.

Conflicts of Interest: The authors declare no conflict of interest.

References

1. Sasmito, S.D.; Taillardat, P.; Clendenning, J.N.; Cameron, C.; Friess, D.A.; Murdiyarso, D.; Hutley, L.B. Effect of land-use and land-cover change on mangrove blue carbon: A systematic review. *Glob. Chang. Biol.* **2019**, *25*, 4291–4302. [CrossRef] [PubMed]
2. Luo, Y.; Keenan, T.F.; Smith, M. Predictability of the terrestrial carbon cycle. *Glob. Chang. Biol.* **2015**, *21*, 1737–1751. [CrossRef] [PubMed]
3. Hararuk, O.; Smith, M.J.; Luo, Y. Microbial models with data-driven parameters predict stronger soil carbon responses to climate change. *Glob. Chang. Biol.* **2015**, *21*, 2439–2453. [CrossRef] [PubMed]
4. Yu, Z.; Ciais, P.; Piao, S.; Houghton, R.A.; Lu, C.; Tian, H.; Agathokleous, E.; Kattel, G.R.; Sitch, S.; Goll, D.; et al. Forest expansion dominates China's land carbon sink since 1980. *Nat. Commun.* **2022**, *13*, 5374. [CrossRef]
5. Brinck, K.; Fischer, R.; Groeneveld, J.; Lehmann, S.; Dantas De Paula, M.; Pütz, S.; Sexton, J.O.; Song, D.; Huth, A. High resolution analysis of tropical forest fragmentation and its impact on the global carbon cycle. *Nat. Commun.* **2017**, *8*, 14855. [CrossRef]
6. Cook-Patton, S.C.; Leavitt, S.M.; Gibbs, D.; Harris, N.L.; Lister, K.; Anderson-Teixeira, K.J.; Briggs, R.D.; Chazdon, R.L.; Crowther, T.W.; Ellis, P.W.; et al. Mapping carbon accumulation potential from global natural forest regrowth. *Nature* **2020**, *585*, 545–550. [CrossRef]
7. Blagodatsky, S.; Xu, J.; Cadisch, G. Carbon balance of rubber (*Hevea brasiliensis*) plantations: A review of uncertainties at plot, landscape and production level. *Agric. Ecosyst. Environ.* **2016**, *221*, 8–19. [CrossRef]
8. Warren-Thomas, E.M.; Edwards, D.P.; Bebber, D.P.; Chhang, P.; Diment, A.N.; Evans, T.D.; Lambrick, F.H.; Maxwell, J.F.; Nut, M.; Kelly, H.J.O.; et al. Protecting tropical forests from the rapid expansion of rubber using carbon payments. *Nat. Commun.* **2018**, *9*, 911. [CrossRef]
9. Li, Y.X.; Liu, Z.S.; Li, S.J.; Li, X. Multi-Scenario Simulation Analysis of Land Use and Carbon Storage Changes in Changchun City Based on FLUS and InVEST Model. *Land* **2022**, *11*, 647. [CrossRef]
10. Azizan, F.A.; Kiloes, A.M.; Astuti, I.S.; Abdul Aziz, A. Application of Optical Remote Sensing in Rubber Plantations: A Systematic Review. *Remote Sens.* **2021**, *13*, 429. [CrossRef]
11. Babbar, D.; Areendran, G.; Sahana, M.; Sarma, K.; Raj, K.; Sivadas, A. Assessment and prediction of carbon sequestration using Markov chain and InVEST model in Sariska Tiger Reserve, India. *J. Clean Prod.* **2021**, *278*, 123333. [CrossRef]
12. Kusakabe, K.; Myae, A.C. Precarity and Vulnerability: Rubber Plantations in Northern Laos and Northern Shan State, Myanmar. *J. Contemp. Asia.* **2019**, *49*, 586–601. [CrossRef]

13. Ye, S.; Rogan, J.; Sangermano, F. Monitoring rubber plantation expansion using Landsat data time series and a Shapelet-based approach. *Isprs-J. Photogramm. Remote Sens.* **2018**, *136*, 134–143. [CrossRef]
14. Dong, J.; Xiao, X.; Sheldon, S.; Biradar, C.; Xie, G. Mapping tropical forests and rubber plantations in complex landscapes by integrating PALSAR and MODIS imagery. *Isprs-J. Photogramm. Remote Sens.* **2012**, *74*, 20–33. [CrossRef]
15. *Statistical Database of the Food and Agricultural Organization of the United Nations*; Food and Agriculture Organization (FAO): Rome, Italy, 2020.
16. Fox, J.; Castella, J. Expansion of rubber (*Hevea brasiliensis*) in Mainland Southeast Asia: What are the prospects for smallholders? *J. Peasant. Stud.* **2013**, *40*, 155–170. [CrossRef]
17. Ziegler, A.D.; Fox, J.M.; Xu, J.C. The Rubber Juggernaut. *Science* **2009**, *324*, 1024–1025. [CrossRef]
18. Gao, S.; Liu, X.; Bo, Y.; Shi, Z.; Zhou, H. Rubber Identification Based on Blended High Spatio-Temporal Resolution Optical Remote Sensing Data: A Case Study in Xishuangbanna. *Remote Sens.* **2019**, *11*, 496. [CrossRef]
19. von Essen, M.; Do Rosário, I.T.; Santos-Reis, M.; Nicholas, K.A. Valuing and mapping cork and carbon across land use scenarios in a Portuguese montado landscape. *PLoS ONE* **2019**, *14*, e212174. [CrossRef]
20. Chaya Sarathchandra, Y.A.A.F.; Wijerathne, L.; Ma, H.; Yingfeng, B.; Jiayu, G.; Chen, H.; Yan, Q.; Geng, Y.; Weragoda, D.S.; Li, L.; et al. Impact of land use and land cover changes on carbon storage in rubber dominated tropical Xishuangbanna, South West China. *Ecosyst. Health Sustain.* **2021**, *7*, 1915183. [CrossRef]
21. Li, Y.; Liu, C.; Zhang, J.; Zhang, P.; Xue, Y. Monitoring Spatial and Temporal Patterns of Rubber Plantation Dynamics Using Time-Series Landsat Images and Google Earth Engine. *IEEE J. Sel. Top. Appl. Earth Observ. Remote Sens.* **2021**, *14*, 9450–9461. [CrossRef]
22. Dong, J.; Xiao, X.; Chen, B.; Torbick, N.; Jin, C.; Zhang, G.; Biradar, C. Mapping deciduous rubber plantations through integration of PALSAR and multi-temporal Landsat imagery. *Remote Sens. Environ.* **2013**, *134*, 392–402. [CrossRef]
23. Chen, B.; Cao, J.; Wang, J.; Wu, Z.; Tao, Z.; Chen, J.; Yang, C.; Xie, G. Estimation of rubber stand age in typhoon and chilling injury afflicted area with Landsat TM data: A case study in Hainan Island, China. *For. Ecol. Manag.* **2012**, *274*, 222–230. [CrossRef]
24. Liu, X.; Jiang, L.; Feng, Z.; Li, P. Rubber Plantation Expansion Related Land Use Change along the Laos-China Border Region. *Sustainability* **2016**, *8*, 1011. [CrossRef]
25. Chen, G.; Thill, J.C.; Anantsuksomsri, S.; Tontisirin, N.; Tao, R. Stand age estimation of rubber (*Hevea brasiliensis*) plantations using an integrated pixel- and object-based tree growth model and annual Landsat time series. *Isprs-J. Photogramm. Remote Sens.* **2018**, *144*, 94–104. [CrossRef]
26. Trisasongko, B.H. Mapping stand age of rubber plantation using ALOS-2 polarimetric SAR data. *Eur. J. Remote Sens.* **2017**, *50*, 64–76. [CrossRef]
27. Koedsin, W.; Huete, A. Mapping Rubber Tree Stand Age using Pléiades Satellite Imagery: A Case Study in Talang District, Phuket, Thailand. *Eng. J.* **2015**, *19*, 45–56. [CrossRef]
28. Xiao, C.; Li, P.; Feng, Z.; Liu, X. An updated delineation of stand ages of deciduous rubber plantations during 1987-2018 using Landsat-derived bi-temporal thresholds method in an anti-chronological strategy. *Int. J. Appl. Earth Obs. Geoinf.* **2019**, *76*, 40–50. [CrossRef]
29. Kou, W.; Xiao, X.; Dong, J.; Gan, S.; Zhai, D.; Zhang, G.; Qin, Y.; Li, L. Mapping Deciduous Rubber Plantation Areas and Stand Ages with PALSAR and Landsat Images. *Remote Sens.* **2015**, *7*, 1048–1073. [CrossRef]
30. Beckschäfer, P. Obtaining rubber plantation age information from very dense Landsat TM & ETM + time series data and pixel-based image compositing. *Remote Sens. Environ.* **2017**, *196*, 89–100.
31. Grogan, K.; Pflugmacher, D.; Hostert, P.; Kennedy, R.; Fensholt, R. Cross-border forest disturbance and the role of natural rubber in mainland Southeast Asia using annual Landsat time series. *Remote Sens. Environ.* **2015**, *169*, 438–453. [CrossRef]
32. Xiao, C.; Li, P.; Feng, Z. Monitoring annual dynamics of mature rubber plantations in Xishuangbanna during 1987–2018 using Landsat time series data: A multiple normalization approach. *Int. J. Appl. Earth Obs. Geoinf.* **2019**, *77*, 30–41. [CrossRef]
33. Li, Z.; Fox, J.M. Mapping rubber tree growth in mainland Southeast Asia using time-series MODIS 250 m NDVI and statistical data. *Appl. Geogr.* **2012**, *32*, 420–432. [CrossRef]
34. Zhang, C.; Huang, C.; Li, H.; Liu, Q.; Li, J.; Bridhikitti, A.; Liu, G. Effect of Textural Features in Remote Sensed Data on Rubber Plantation Extraction at Different Levels of Spatial Resolution. *Forests* **2020**, *11*, 399. [CrossRef]
35. Kennedy, R.E.; Yang, Z.; Cohen, W.B. Detecting trends in forest disturbance and recovery using yearly Landsat time series: 1. LandTrendr—Temporal segmentation algorithms. *Remote Sens. Environ.* **2010**, *114*, 2897–2910. [CrossRef]
36. Li, H.; Fu, D.; Huang, C.; Su, F.; Liu, Q.; Liu, G.; Wu, S. An Approach to High-Resolution Rice Paddy Mapping Using Time-Series Sentinel-1 SAR Data in the Mun River Basin, Thailand. *Remote Sens.* **2020**, *12*, 3959. [CrossRef]
37. Gong, P.; Liu, H.; Zhang, M.; Li, C.; Wang, J.; Huang, H.; Clinton, N.; Ji, L.; Li, W.; Bai, Y.; et al. Stable classification with limited sample: Transferring a 30-m resolution sample set collected in 2015 to mapping 10-m resolution global land cover in 2017. *Sci. Bull.* **2019**, *64*, 370–373. [CrossRef]
38. Sun, W.; Liu, X. Review on carbon storage estimation of forest ecosystem and applications in China. *For. Ecosyst.* **2020**, *7*, 4. [CrossRef]
39. Houghton, R.A. Counting terrestrial sources and sinks of carbon. *Clim. Chang.* **2001**, *48*, 525–534. [CrossRef]
40. Issa, S.; Dahy, B.; Ksiksi, T.; Saleous, N. A Review of Terrestrial Carbon Assessment Methods Using Geo-Spatial Technologies with Emphasis on Arid Lands. *Remote Sens.* **2020**, *12*, 2008. [CrossRef]

41. Gómez, C.; Alejandro, P.; Hermosilla, T.; Montes, F.; Pascual, C.; Ruiz, L.A.; Álvarez-Taboada, F.; Tanase, M.; Valbuena, R. Remote sensing for the Spanish forests in the 21st century: A review of advances, needs, and opportunities. *For. Syst.* **2019**, *28*, R1. [CrossRef]
42. Brunet Navarro, P.; Jochheim, H.; Muys, B. Modelling carbon stocks and fluxes in the wood product sector: A comparative review. *Glob. Chang. Biol.* **2016**, *22*, 2555–2569. [CrossRef] [PubMed]
43. Pan, Y.; Birdsey, R.A.; Fang, J.; Houghton, R.; Kauppi, P.E.; Kurz, W.A.; Phillips, O.L.; Shvidenko, A.; Lewis, S.L.; Canadell, J.G.; et al. A large and persistent carbon sink in the world's forests. *Science* **2011**, *333*, 988–993. [CrossRef] [PubMed]
44. Goetz, S.; Dubayah, R. Advances in remote sensing technology and implications for measuring and monitoring forest carbon stocks and change. *Carbon Manag.* **2011**, *2*, 231–244. [CrossRef]
45. Project, N.C. InVEST: A Tool for Integrating Ecosystem Services into Policy and Decision-Making. *Int. J. Biodivers. Sci. Ecosyst. Serv. Manag.* **2015**, *11*, 205–215.
46. Nelson, E.; Polasky, S.; Lewis, D.J.; Plantinga, A.J.; Lonsdorf, E.; White, D.; Bael, D.; Lawler, J.J. Efficiency of incentives to jointly increase carbon sequestration and species conservation on a landscape. *Proc. Natl. Acad. Sci. USA* **2008**, *105*, 9471–9476. [CrossRef]
47. Natural Capital Project. *Land-based carbon offsets with InVEST*; Stanford University: Stanford, CA, USA, 2015. Available online: https://naturalcapitalproject.stanford.edu/sites/default/files/publications/investinpractice_carbon.pdf (accessed on 1 May 2022).
48. Xiao, D.; Niu, H.; Guo, J.; Zhao, S.; Fan, L. Carbon Storage Change Analysis and Emission Reduction Suggestions under Land Use Transition: A Case Study of Henan Province, China. *Int. J. Environ. Res. Public Health* **2021**, *18*, 1844. [CrossRef]
49. Chaplin-Kramer, R.; Ramler, I.; Sharp, R.; Haddad, N.M.; Gerber, J.S.; West, P.C.; Mandle, L.; Engstrom, P.; Baccini, A.; Sim, S.; et al. Degradation in carbon stocks near tropical forest edges. *Nat. Commun.* **2015**, *6*, 10158. [CrossRef]
50. Ouyang, Z.; Zheng, H.; Xiao, Y.; Polasky, S.; Liu, J.; Xu, W.; Wang, Q.; Zhang, L.; Xiao, Y.; Rao, E.; et al. Improvements in ecosystem services from investments in natural capital. *Sci. Am. Assoc. Adv. Sci.* **2016**, *352*, 1455–1459. [CrossRef]
51. Nel, L.; Boeni, A.F.; Prohaszka, V.J.; Szilagyi, A.; Kovacs, E.T.; Pasztor, L.; Centeri, C. InVEST Soil Carbon Stock Modelling of Agricultural Landscapes as an Ecosystem Service Indicator. *Sustainability* **2022**, *14*, 9808. [CrossRef]
52. Trisasongko, B.H.; Paull, D. A review of remote sensing applications in tropical forestry with a particular emphasis in the plantation sector. *Geocarto Int.* **2020**, *35*, 317–339. [CrossRef]
53. Gorelick, N.; Hancher, M.; Dixon, M.; Ilyushchenko, S.; Thau, D.; Moore, R. Google Earth Engine: Planetary-scale geospatial analysis for everyone. *Remote Sens. Environ.* **2017**, *202*, 18–27. [CrossRef]
54. Quin, G.; Pinel-Puyssegur, B.; Nicolas, J.; Loreaux, P. MIMOSA: An Automatic Change Detection Method for SAR Time Series. *Ieee Trans. Geosci. Remote Sens.* **2014**, *52*, 5349–5363. [CrossRef]
55. Copernicus Open Access Hub. Available online: https://scihub.copernicus.eu/ (accessed on 1 May 2022).
56. Zuhlke, M.; Fomferra, N.; Brockmann, C.; Peters, M.; Veci, L.; Malik, J.; Regner, P. SNAP (Sentinel Application Platform) and the ESA Sentinel 3 Toolbox.:Sentinel-3 for Science Workshop. *Sentin. -3 Sci. Workshop* **2015**, *734*, 21.
57. Zeng, Y.; Hao, D.; Huete, A.; Dechant, B.; Berry, J.; Chen, J.M.; Joiner, J.; Frankenberg, C.; Bond-Lamberty, B.; Ryu, Y.; et al. Optical vegetation indices for monitoring terrestrial ecosystems globally. *Nat. Rev. Earth Environ.* **2022**, *3*, 477–493. [CrossRef]
58. Broge, N.H.; Mortensen, J.V. Deriving green crop area index and canopy chlorophyll density of winter wheat from spectral reflectance data. *Remote Sens. Environ.* **2002**, *81*, 45–57. [CrossRef]
59. Huete, A.; Didan, K.; Miura, T.; Rodriguez, E.P.; Gao, X.; Ferreira, L.G. Overview of the radiometric and biophysical performance of the MODIS vegetation indices. *Remote Sens. Environ.* **2003**, *83*, 195–213. [CrossRef]
60. Gitelson, A.; Merzlyak, M.N. Spectral Reflectance Changes Associated with Autumn Senescence of *Aesculus hippocastanum* L. and *Acer platanoides* L. Leaves. Spectral Features and Relation to Chlorophyll Estimation. *J. Plant Physiol.* **1994**, *143*, 286–292. [CrossRef]
61. Berberoğlu, S.; Akin, A.; Atkinson, P.M.; Curran, P.J. Utilizing image texture to detect land-cover change in Mediterranean coastal wetlands. *Int. J. Remote Sens.* **2010**, *31*, 2793–2815. [CrossRef]
62. United States Geological Survey. Available online: https://earthexplorer.usgs.gov/ (accessed on 30 April 2022).
63. Chen, J.; Zhu, X.; Vogelmann, J.E.; Gao, F.; Jin, S. A simple and effective method for filling gaps in Landsat ETM+ SLC-off images. *Remote Sens. Environ.* **2011**, *115*, 1053–1064. [CrossRef]
64. Tsalyuk, M.; Kelly, M.; Getz, W.M. Improving the prediction of African savanna vegetation variables using time series of MODIS products. *Isprs-J. Photogramm. Remote Sens.* **2017**, *131*, 77–91. [CrossRef]
65. Huang, C.; Goward, S.N.; Masek, J.G.; Thomas, N.; Zhu, Z.; Vogelmann, J.E. An automated approach for reconstructing recent forest disturbance history using dense Landsat time series stacks. *Remote Sens. Environ.* **2010**, *114*, 183–198. [CrossRef]
66. Intergovernmental Panel on Climate Change. *2006 IPCC Guidelines for National Greenhouse Gas Inventories*; IPCC: Geneva, Switzerland, 2006.
67. Intergovernmental Panel on Climate Change. *2019 Refinement to the 2006 IPCC Guidelines for National Greenhouse Gas Inventories*; IPCC: Geneva, Switzerland, 2019.
68. de Blecourt, M.; Brumme, R.; Xu, J.; Corre, M.D.; Veldkamp, E. Soil carbon stocks decrease following conversion of secondary forests to rubber (*Hevea brasiliensis*) plantations. *PLoS ONE.* **2013**, *8*, e69357. [CrossRef] [PubMed]

69. Yang, X.; Blagodatsky, S.; Lippe, M.; Liu, F.; Hammond, J.; Xu, J.; Cadisch, G. Land-use change impact on time-averaged carbon balances: Rubber expansion and reforestation in a biosphere reserve, South-West China. *For. Ecol. Manage.* **2016**, *372*, 149–163. [CrossRef]
70. Breiman, L. Random Forests. *Mach. Learn.* **2001**, *5*, 5–32. [CrossRef]
71. Foody, G.M. Status of land cover classification accuracy assessment. *Remote Sens. Environ.* **2002**, *80*, 185–201. [CrossRef]
72. Zakaria, J.; Mueen, A.; Keogh, E. Clustering Time Series using Unsupervised-Shapelets. In Proceedings of the 12th IEEE International Conference on Data Mining (ICDM), Brussels, Belgium, 10–13 December 2012.
73. Sharp, R.; Douglass, J.; Wolny, S. *VEST 3. 9. 0 User's Guide*; The Natural Capital Project; Stanford University: Stanford, CA, USA, 2020.
74. Gutiérrez-Vélez, V.H.; DeFries, R. Annual multi-resolution detection of land cover conversion to oil palm in the Peruvian Amazon. *Remote Sens. Environ.* **2013**, *129*, 154–167. [CrossRef]
75. Huang, C.; Zhang, C.; He, Y.; Liu, Q.; Li, H.; Su, F.; Liu, G.; Bridhikitti, A. Land Cover Mapping in Cloud-Prone Tropical Areas Using Sentinel-2 Data: Integrating Spectral Features with Ndvi Temporal Dynamics. *Remote Sens.* **2020**, *12*, 1163. [CrossRef]
76. Torbick, N.; Ledoux, L.; Salas, W.; Zhao, M. Regional Mapping of Plantation Extent Using Multisensor Imagery. *Remote Sens.* **2016**, *8*, 236. [CrossRef]
77. Verbesselt, J.; Hyndman, R.; Newnham, G.; Culvenor, D. Detecting trend and seasonal changes in satellite image time series. *Remote Sens. Environ.* **2010**, *114*, 106–115. [CrossRef]
78. Lunetta, R.S.; Shao, Y.; Ediriwickrema, J.; Lyon, J.G. Monitoring agricultural cropping patterns across the Laurentian Great Lakes Basin using MODIS-NDVI data. *Int. J. Appl. Earth Obs. Geoinf.* **2010**, *12*, 81–88. [CrossRef]
79. Kenney-Lazar, M. Plantation rubber, land grabbing and social-property transformation in southern Laos. *J. Peasant. Stud.* **2012**, *39*, 1017–1037. [CrossRef]
80. Arunyawat, S.; Shrestha, R. Assessing Land Use Change and Its Impact on Ecosystem Services in Northern Thailand. *Sustainability* **2016**, *8*, 768. [CrossRef]
81. Charoenjit, K.; Zuddas, P.; Allemand, A.P.; Sura Pattanakiat, C.A.K.P. Estimation of biomass and carbon stock in Para rubber plantations using object-based classification from Thaichote satellite data in Eastern Thailand. *J. Appl. Remote Sens.* **2015**, *9*, 096072. [CrossRef]
82. Fang, Z.; Bai, Y.; Jiang, B.; Alatalo, J.M.; Liu, G.; Wang, H. Quantifying variations in ecosystem services in altitude-associated vegetation types in a tropical region of China. *Sci. Total Environ.* **2020**, *726*, 138565. [CrossRef]
83. Liu, S.; Yin, Y.; Cheng, F.; Hou, X.; Dong, S.; Wu, X. Spatio-temporal variations of conservation hotspots based on ecosystem services in Xishuangbanna, Southwest China. *PLoS ONE.* **2017**, *12*, e189368. [CrossRef]
84. Bolyn, C.; Michez, A.; Gaucher, P.; Lejeune, P.; Bonnet, S. Forest mapping and species composition using supervised per pixel classification of Sentinel-2 imagery. *Biotechnol. Agron. Soc.* **2018**, *22*, 172–187. [CrossRef]
85. Wulf, H.; Stuhler, S. Sentinel-2: Land cover, preliminary user feedback on Sentinel-2a data. In Proceedings of the Sentinel-2a Expert Users Technical Meeting, Frascati, Italy, 29–30 September 2015.
86. Radoux, J.; Chomé, G.; Jacques, D.; Waldner, F.; Bellemans, N.; Matton, N.; Lamarche, C.; D Andrimont, R.; Defourny, P. Sentinel-2's Potential for Sub-Pixel Landscape Feature Detection. *Remote Sens.* **2016**, *8*, 488. [CrossRef]
87. Fernández-Manso, A.; Fernández-Manso, O.; Quintano, C. SENTINEL-2A red-edge spectral indices suitability for discriminating burn severity. *Int. J. Appl. Earth Obs. Geoinf.* **2016**, *50*, 170–175. [CrossRef]
88. le Maire, G.; François, C.; Dufrêne, E. Towards universal broad leaf chlorophyll indices using PROSPECT simulated database and hyperspectral reflectance measurements. *Remote Sens. Environ.* **2004**, *89*, 1–28. [CrossRef]
89. Lichtenthaler, H.K.; Lang, M.; Sowinska, M.; Heisel, F.; Miehé, J.A. Detection of Vegetation Stress Via a New High Resolution Fluorescence Imaging System. *J. Plant Physiol.* **1996**, *148*, 599–612. [CrossRef]
90. Immitzer, M.; Vuolo, F.; Atzberger, C. First Experience with Sentinel-2 Data for Crop and Tree Species Classifications in Central Europe. *Remote Sens.* **2016**, *8*, 166. [CrossRef]
91. Gitelson, A.A.; Gritz, Y.; Merzlyak, M.N. Relationships between leaf chlorophyll content and spectral reflectance and algorithms for non-destructive chlorophyll assessment in higher plant leaves. *J. Plant Physiol.* **2003**, *160*, 271–282. [CrossRef] [PubMed]
92. VanDeventer, A.; Ward, A.; Gowda, P.; Lyon, J. Using Thematic Mapper data to identify contrasting soil plains and tillage practices. *Photogramm. Eng. Remote Sens.* **1997**, *63*, 87–93.
93. Sripada, R.P.; Heiniger, R.W.; White, J.G.; Meijer, A.D. Aerial Color Infrared Photography for Determining Early In-Season Nitrogen Requirements in Corn. *Agron. J.* **2006**, *98*, 968–977. [CrossRef]
94. Gao, B. NDWI—A normalized difference water index for remote sensing of vegetation liquid water from space. *Remote Sens. Environ.* **1996**, *58*, 257–266. [CrossRef]

remote sensing

MDPI

Article

Modelling the Dynamics of Carbon Storages for *Pinus densata* Using Landsat Images in Shangri-La Considering Topographic Factors

Yi Liao [1], Jialong Zhang [1,*], Rui Bao [2], Dongfan Xu [3] and Dongyang Han [4]

1 College of Forestry, Southwest Forestry University, Kunming 650224, China
2 Institute of Southwest Survey and Planning, National Forestry and Grassland Administration, Kunming 650021, China
3 Ministry of Education Key Laboratory for Biodiversity Science and Ecological Engineering, Institute of Biodiversity Science, Fudan University, Shanghai 200438, China
4 Research Institute of Forestry Policy and Information, Chinese Academy of Forestry, Beijing 100091, China
* Correspondence: jialongzhang@swfu.edu.cn; Tel.: +86-138-8802-1540

Abstract: Accurate estimation of forest carbon storage is essential for understanding the dynamics of forest resources and optimizing decisions for forest resource management. In order to explore the changes in the carbon storage of *Pinus densata* in Shangri-La and the influence of topography on carbon storage, two dynamic models were developed based on the National Forest Inventory (NFI) and Landsat TM/OLI images with a 5-year interval change and annual average change. The three modelling methods used were partial least squares (PLSR), random forest (RF) and gradient boosting regression tree (GBRT). Various spectral and texture features of the images were calculated and filtered before modelling. The terrain niche index (TNI), which is able to reflect the combined effect of elevation and slope, was added to the dynamic model, the optimal model was selected to estimate the carbon storage, and the topographic conditions in areas of change in carbon storage were analyzed. The results showed that: (1) The dynamic model based on 5-year interval change data performs better than the dynamic model with annual average change data, and the RF model has a higher accuracy compared to the PLSR and GBRT models. (2) The addition of TNI improved the accuracy, in which R^2 is improved by up to 10.48% at most, RMSE is reduced by up to 7.32% at most, and MAE is reduced by up to 8.89% at most, and the RF model based on the 5-year interval change data has the highest accuracy after adding TNI, with an R^2 of 0.87, an RMSE of 3.82 $t\text{-}C\cdot ha^{-1}$, and a MAE of 1.78 $t\text{-}C\cdot ha^{-1}$. (3) The direct estimation results of the dynamic model showed that the carbon storage of *Pinus densata* in Shangri-La decreased in 1987–1992 and 1997–2002, and increased in 1992–1997, 2002–2007, 2007–2012, and 2012–2017. (4) The trend of increasing or decreasing carbon storage in each period is not exactly the same on the TNI gradient, according to the dominant distribution, as topographic conditions with lower elevations or gentler slopes are favorable for the accumulation of carbon storage, while the decreasing area of carbon storage is more randomly distributed topographically. This study develops a dynamic estimation model of carbon storage considering topographic factors, which provides a solution for the accurate estimation of forest carbon storage in regions with a complex topography.

Keywords: Landsat; *Pinus densata*; terrain niche index; dynamic model; carbon storage

Citation: Liao, Y.; Zhang, J.; Bao, R.; Xu, D.; Han, D. Modelling the Dynamics of Carbon Storages for *Pinus densata* Using Landsat Images in Shangri-La Considering Topographic Factors. *Remote Sens.* **2022**, *14*, 6244. https://doi.org/10.3390/rs14246244

Academic Editor: Wataru Takeuchi

Received: 12 November 2022
Accepted: 6 December 2022
Published: 9 December 2022

1. Introduction

Forests are the mainstay of terrestrial ecosystems, which store 60% of the carbon in the terrestrial ecosystem [1]. Forests play an indispensable role in balancing and regulating CO_2 in the atmosphere with their powerful carbon sink function. Global warming is a great threat to humanity today, and the excessive emissions of greenhouse gases, of which CO_2 is an important member, is one of the main reasons for global warming. Atmospheric CO_2

concentrations have risen dramatically over the past 100 years (they have risen by approximately 110 ppm) [2], a phenomenon that has accelerated the rate of global warming. The accurate estimation of forest carbon storage is related to the ability to reduce atmospheric CO_2 concentration. Against this background, the necessity of the study of forest carbon storage is revealed and its accurate estimation is an important basis for regional and even global carbon change studies.

The methods for estimating forest carbon storage includes both traditional direct measurement methods and indirect estimation methods based on remote sensing [3,4]. The traditional method is based on sample site survey and data statistics, which require a large investment of human and financial resources [5], and due to cost and time constraints, there are significant limitations regarding the large range of study objectives and long study periods. In contrast, remote sensing-based methods have the characteristics of being fast, low-cost, large-scale, and less destructive [1,6]. With the rapid development of quantitative remote sensing technology, the results of carbon storage estimation based on remote sensing have become more and more stable and reliable. Therefore, the use of remote sensing methods for the quantitative estimation of forest carbon storage on a larger scale has become a major current trend [7,8].

At present, there is a significant amount of research in remote sensing estimation, and while scholars in various fields have constructed different remote sensing models [9], most of these are static models. The static model is a model developed directly from the attribute values of the sample sites and the remotely sensed feature values of the corresponding points, which are uncalculated state data. If these values are further processed and calculated to produce different types of change values (periodic change, annual average change, rate of change, etc.), then the model developed using these change data is the dynamic model. Change data can describe the different processes of change in forest ecosystems at a given time and have the ability to monitor forest dynamics [10]. Gómez et al. [11] and Zhang et al. [12] compared static models with dynamic models and showed that dynamic models have higher accuracy and better predictive power than static models, with the highest R^2 for dynamic models in their study being 0.90 and 0.94, respectively. In the study of forest dynamics, the dynamic model is able to directly estimate the corresponding level of forest change and directly respond to forest dynamics, which improves the efficiency of the long-term forest dynamics study. The calculation of change data needs to be supported by multi-period data. The current National Forest Inventory (NFI) projects in many countries provide the possibility for long-term forest dynamics modelling studies. The ongoing NFI projects in many countries and open-access Landsat time series data offer the potential for long-term forest dynamics modelling studies.

Trees accumulate carbon mainly through photosynthesis during growth, and topographical factors influence the photosynthetic effect by affecting light and water conditions in the environment, which eventually affects the carbon storage of trees [13]. Therefore, topographic factors are an important concern in forest carbon storage studies. The digital elevation model (DEM) provides valuable topographic information including elevation, slope, and aspect [14]. These topographic factors are often used as independent factors in various studies because they do not require complex calculations and processing, are relatively easy to obtain, and reflect the effects of topography on different aspects of vegetation to some extent [15]. However, the influence of topography on forest carbon storage is the result of the combined effect of various topographic factors, and it is difficult to reflect this combined effect by a single elevation or slope. The terrain niche index (TNI) combines elevation and slope information, which is able to reflect the spatial variability of regional elevation and slope [16,17]. Therefore, in past studies, it has been frequently applied to land use change, landscape patterns and ecological effects [18,19]. The topography of forest ecosystems is complex and diverse, so the comprehensive influence of topographic factors should also be considered when modelling carbon storage. The change data can reflect the degree of change in forest carbon storage, and TNI can reflect the comprehensive influence of topographic factors on forest carbon storage, so the combination of the two is important for the accurate estimation of change in forest carbon storage in areas with complex topography.

Shangri-La, Yunnan Province, is rich in forest resources, and *Pinus densata* is one of the major tree species, which is widely distributed and has an essential impact on the carbon balance of the region. This region is located at the edge of the Tibetan Plateau, with an overall high altitude, and the terrain comprises high mountains and deep valleys with large topographic undulations, making it more difficult to estimate the forest carbon storage in this region accurately. Therefore, in this study, we used the remote sensing dynamic model combined with TNI to estimate the carbon storage of *Pinus densata* in Shangri-La. The main objectives are the following: (1) developing the carbon storage dynamics model based on two different kinds of change data; (2) exploring the impact of TNI on carbon storage models; and (3) estimating the historical carbon storage changes of *Pinus densata* in Shangri-La and analyzing the topographic effects of carbon storage changes.

2. Materials and Methods

2.1. Study Area and Study Process

The study area, Shangri-La, is located in Yunnan Province in southwestern China (Figure 1), at the border of the Yunnan Province and the Sichuan Province, and the geographical coordinates range from 99°20′ to 100°19′ eastern longitude and 26°52′ to 28°52′ northern latitude, the total area is 1.16×10^6 ha. The area is located in the southeastern part of the Tibetan Plateau (the average elevation is 3459 m), with the Hengduan Mountains running north and south through the whole territory, the regional topography is undulating [20], and the whole terrain is high in the northwest and low in the southeast. Shangri-La has a remarkable monsoon climate, with rainfall concentrated in the months of June to October each year [21], the average annual precipitation is 268~945 mm, and the annual sunshine is 1742.9~2186.6 h. This region is very rich in forest resources, with 76% of forest cover [22]. *Pinus densata*, *Pinus yunnanensis*, and *Picea asperata* are the main tree species in the region [23]. The target species for this study is the *Pinus densata*, which includes both artificial and natural forests distributed within the study area.

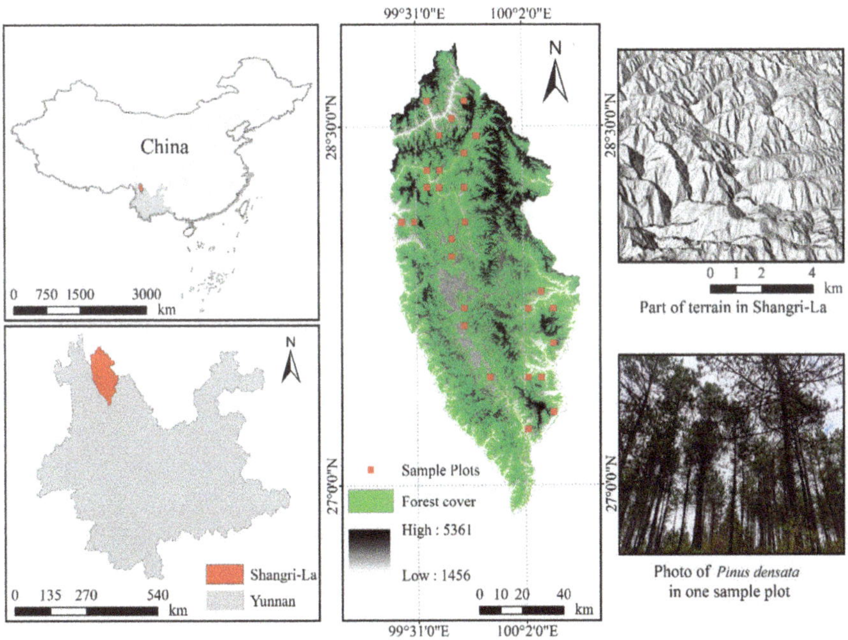

Figure 1. Study area and sample plots.

The main content of the study is shown in Figure 2. The research process began with the pre-processing of all the data; then, two types of change data (annual average change and 5-year interval change) were calculated, three methods were used to construct the model, and TNI was added to the model. Finally, in order to compare the effect of the dynamic model with different change data and to compare the effect of the model before and after the addition of TNI, different indicators were used to evaluate the accuracy.

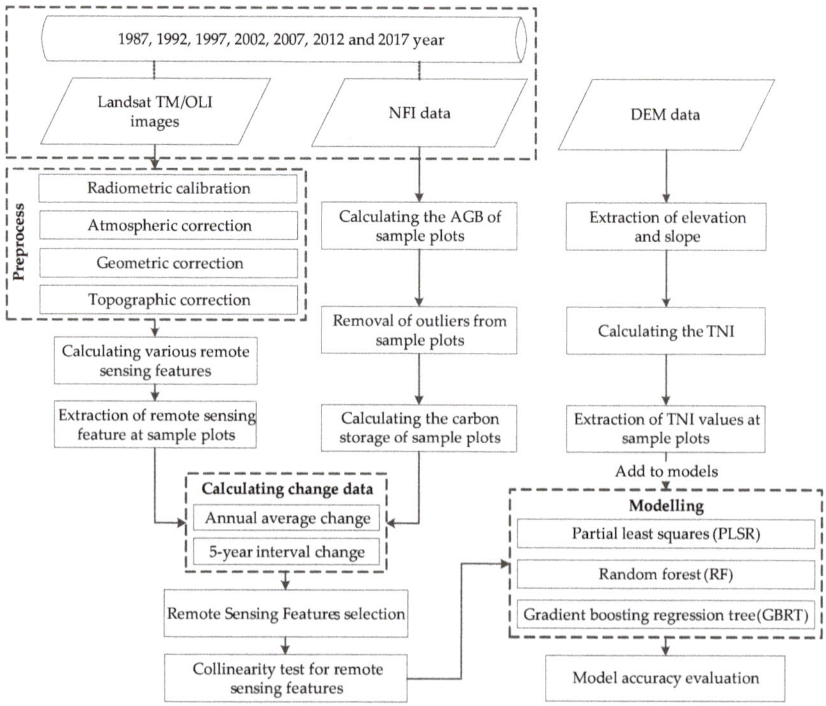

Figure 2. Flow chart of the main content in this study.

2.2. Ground Survey Data and Carbon Storage Calculation

This study focused on *Pinus densata*, and the ground survey data were obtained from the National Forestry Inventory, which contains a total of 136 *Pinus densata* sample plots (comprising pure forests of *Pinus densata* or forests with *Pinus densata* as the main species). The years of the collection were 1987, 1992, 1997, 2002, 2007, 2012 and 2017, where the number of sample plots was 19, 22, 23, 16, 16, 17 and 23 for each year, respectively (Figure 3). The dataset records information on tree height, diameter at breast height (DBH), number of trees, coordinate location, and major tree species. The distance between the sample plots consisted of regular distributions of 6 km × 8 km, and each sample plot was a rectangle of 28.28 m × 28.28 m (0.08 ha).

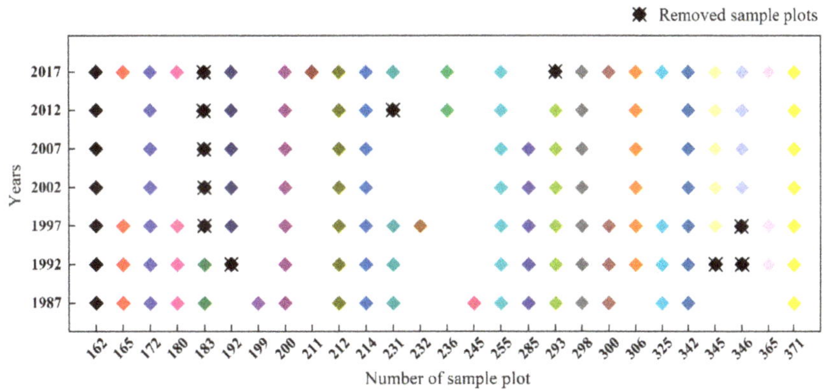

Figure 3. Sample plot data records by year: different colors represent sample plots with different locations (different numbered sample plots).

The aboveground biomass (AGB) of the sample plots was calculated based on the allometric growth equation of *Pinus densata* [24]. The average AGB was first calculated using the average tree height and average DBH of the sample plots. Then, the total AGB of the sample plots was calculated based on the average AGB and the number of *Pinus densata*. The allometric growth equation is as follows:

$$AGB_1 = 0.073 \times DBH^{1.739} \times H^{0.880} \quad (1)$$

where AGB_1 is single wood aboveground biomass (t), DBH is the diameter at breast height (cm), and H is tree height (m). We filtered the sample plot data. In this process, five sample plots with an AGB (AGB < 1 t·ha^{-1}) that was too small were removed, followed by six outliers that were screened out and removed using the Pauta criterion [25]. According to the Pauta criterion, if a value is outside the range of three times the standard deviation of the mean (outside the range of $\overline{x} \pm 3\sigma$, where $\overline{x}$ is the mean and σ is the standard deviation), it is considered an outlier. This is a common method for outlier screening [26]. In the end, 125 sample plots of *Pinus densata* remained.

The carbon storage of the sample sites was calculated using the AGB multiplied by the carbon content coefficient. According to the guidelines for measuring carbon storage in forest ecosystems issued by the State Forestry and Grassland Administration of China [27], the average carbon content in the dry matter of *Pinus densata* is 0.501. The equation is as follows:

$$CS = 0.501 \times AGB_2 \quad (2)$$

where CS is the carbon storage (t-C·ha^{-1}), and AGB_2 is the aboveground biomass of the sample plots (t·ha^{-1}). The basic sample data is shown in Figure 4. The box plots (Figure 4) depict information on the diameter at breast height, tree height, and carbon storage for each year of the sample plots. The overall range of DBH is 6.0–91.7 cm, with a maximum annual mean of 32.0 cm and a maximum annual standard deviation of 27.8 cm; the overall range of tree height is 1.5–24.8 m, with a maximum annual mean of 11.4 m and a maximum annual standard deviation of 5.5 m; and the overall range of carbon storage is 0.7–84.2 t-C·ha^{-1}, with a maximum annual mean of 38.6 t-C·ha^{-1} and a maximum annual standard deviation of 41.9 t-C·ha^{-1}.

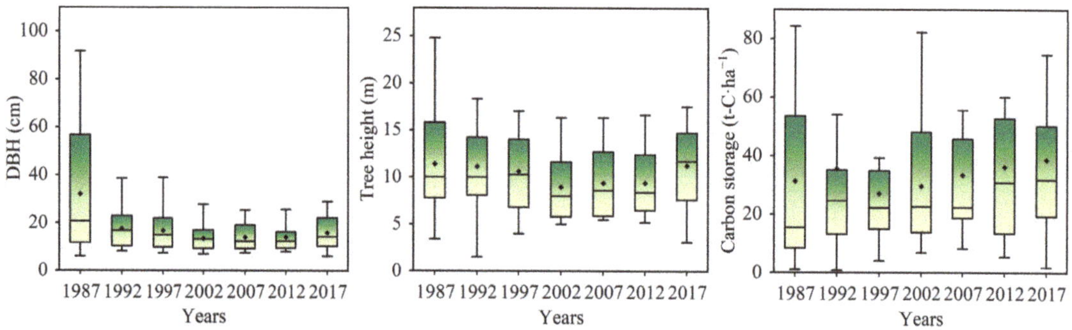

Figure 4. Boxplots of basic information on sample plots by year.

2.3. *Remote Sensing Images and Obtained Features*

2.3.1. Remote Sensing Data

The remote sensing images used in this study were obtained from the USGS website (http://glovis.usgs.gov/ (accessed on 28 October 2021)). Landsat 5 TM images from 1987, 1992, 1997, 2002, 2007, and 2012 and Landsat 8 OLI images from 2017 of the Shangri-La region were downloaded from the website (Table 1), comprising 21 views in total and a spatial resolution of 30 m. When an image contains too much cloud or is of poor quality, we chose an image from a neighboring timepoint to replace it. Most of the selected images contained less than 5% cloud, and the cloud did not cover the study area when the cloud content was greater than 5%.

Table 1. Landsat time-series of imagery used in the study.

Year	Landsat/Sensor	Path/Row	Image Acquisition Date
	5/TM	132/040	30 December 1987
1987	5/TM	132/041	30 December 1987
	5/TM	131/041	23 December 1987
	5/TM	132/041	7 November 1991
1992	5/TM	132/040	7 November 1991
	5/TM	131/041	16 November 1991
	5/TM	132/041	7 November 1997
1997	5/TM	132/040	6 October 1997
	5/TM	131/041	16 November 1997
	5/TM	132/041	5 January 2002
2002	5/TM	132/040	5 January 2002
	5/TM	131/041	29 October 2002
	5/TM	132/041	15 October 2006
2007	5/TM	132/040	3 January 2007
	5/TM	131/041	1 March 2007
	5/TM	132/041	13 October 2011
2012	5/TM	132/040	14 January 2011
	5/TM	131/041	7 January 2011
	8/OLI	132/041	16 December 2017
2017	8/OLI	132/040	16 December 2017
	8/OLI	131/041	25 December 2017

In order to improve the image quality, all the images were preprocessed: firstly, radiometric calibration was performed, which converts the original DN (digital number) value of the image to a consistent radiometric brightness in order to eliminate the influence of the sensor [28]; secondly, the Fast Line-of-sight Atmospheric Analysis of Spectral Hypercubes (FLAASH) method was used to eliminate the influence of the atmosphere [29]; thirdly, geometric correction was performed on each image with reference to the existing standard SPOT-5 images to eliminate geometric errors, where the error in geometric correction was less than one pixel; and finally, in order to eliminate the change of image grey value caused by the topographic undulation, topographic correction was performed using the slope matching model [30]. After topographic correction, the brightness of the illuminated side was suppressed and the brightness of the shadow side was enhanced, which better restored the reflectance of the hidden features in the shadow of the original image [31].

2.3.2. Remote Sensing Features

The remote sensing pixel values were evaluated by considering bands with the same name and close wavelengths in Landsat 8 and Landsat 5. In this study, two types of remote sensing features, spectral features and texture features, are calculated. Spectral features are the most basic and direct features of remote sensing images, but the phenomena of "the same objects with different spectra" and "different objects with the same spectra" commonly exist in remote sensing images, so it may be difficult to obtain good research results when only relying on spectral features [32,33]. The texture features can describe the grey scale information of an image, express the spatial distribution of the grey scale of pixels in the image, and reflect the structural information of the image. Compared with spectral features, texture features are less affected by the environment and more stable in information representation [34]. Therefore, the combination of spectral and textural features can effectively reflect the characteristics of the ground objects and their changes [35,36]. A total of 35 spectral features and 540 texture features were obtained in this study (the calculations of various types of textures from all single bands of each image in odd windows from 1 to 19) (Table 2).

Table 2. Extracted remote sensing feature information.

Categories		Information on Remote Sensing Characteristics
Spectral features	Original bands	B1; B2; B3; B4; B5; B7
	Vegetation indices [37]	$NDVI = (B4 - B3)/(B4 + B3)$; $ND32 = (B3 - B2)/(B3 + B2)$; $ND54 = (B5 - B4)/(B5 + B4)$; $ND53 = (B5 - B3)/(B5 + B3)$; $ND57 = (B5 - B7)/(B5 + B7)$; $ND452 = (B4 + B5 - B2)/(B4 + B5 + B2)$; $DVI = B4 - B3$; $RVI = B4/B3$; $ARVI = (B4 - (2B3 - B1))/(B4 + (2B3 - B1))$ $EVI = 2.5 \times \frac{B4-B3}{B4+6B3-7.5B1+1}$;
	Band combination [12]	$B4/B2$; $B5/B3$; $B5/B4$; $B5/B7$; $B7/B3$; $B4 \times (B3/B7)$
	Image enhancement [38,39]	$Albedo = B1 + B2 + B3 + B4 + B5 + B7$; $VIS123 = B1 + B2 + B3$; $MID = B5 + B7$; $FVC = \frac{EVI-EVI_{min}}{EVI_{max}-EVI_{min}}$; Principal component analysis (PCA1, PCA2, PCA3, PCA4, PCA5, PCA7); Tasseled cap transform (TCT1, TCT2, TCT3)
Texture features	Grey level co-occurrence matrix [34]	Homogeneity (HO); Dissimilarity (DI); Mean (ME); Angular second moment (SM); Entropy (EN); Correlation (CC); Variance (VA); Contrast (CO)
	Filtering of probabilistic statistics [40]	Skewness (SK)

Where B1 is the blue band, B2 is the green band, B3 is the red band, B4 is the near-infrared band, B5 is the shortwave infrared-1 band, and B7 is the shortwave infrared-2 band.

2.4. Terrain Niche Index

To explore the influence of elevation and slope on the carbon storage dynamics model, the terrain niche index (TNI), a composite factor combining elevation and slope, was established in this study [41]. Elevation and slope were extracted from the digital elevation model (DEM). The DEM used in this study was an ASTER GDEM data product with a spatial resolution of 30 m. The DEM was downloaded from the Geospatial Data Cloud website (http://www.gscloud.cn/ (accessed on 18 August 2022)). The equation for TNI is as follows:

$$\text{TNI} = \log\left[\left(\frac{E}{\overline{E}} + 1\right) \times \left(\frac{S}{\overline{S}} + 1\right)\right] \tag{3}$$

where E and S are the elevation and slope of any point, respectively, and $\overline{E}$ and $\overline{S}$ are the average elevation and average slope of the study area, respectively. A larger TNI value indicates a higher elevation and slope; a smaller TNI value indicates a lower elevation and gentler slope; and a medium TNI value indicates a high elevation gentle slope, or a low elevation steep slope, or a medium elevation and slope [42].

2.5. Distribution Index

The distribution index (DI) is often used to show the distribution of various land use types. In this study, the areas of the increasing and decreasing carbon storage of *Pinus densata* were considered as two different land classes. Then, the DI was used to explore the distribution of carbon storage changes in different TNI gradients. The DI is a standardized, dimensionless metric that eliminates the effects of area differences and allows the comparison of distribution characteristics between carbon storage changes of different area proportions [18,43]. The equation is as follows:

$$P = (S_{ie}/S_i)/(S_e/S) \tag{4}$$

where P is the DI, e is the gradient of TNI, S_{ie} is the area of i changes on the e TNI gradient, S_i is the area of the i change, S_e is the area of the TNI for gradient e, and S is the total area of the study area. A larger P indicates a higher frequency of this type of change, and $P > 1$ indicates that the type of change under this TNI gradient belongs to the dominant distribution. Therefore, the analysis of the distribution of each carbon storage change type on different TNI levels is able to reveal the influence of different topographic conditions on carbon storage change.

2.6. Modelling Process

2.6.1. Calculation of Change Data

In this study, two kinds of change data were used to develop the forest carbon storage model, one is the 5-year interval change, and the other is the annual average change. The change data were calculated based on a common sample plot for both years: we calculate the change value of a sample plot when data were recorded in both adjacent years, otherwise the change value for the sample plot during that specific time period is not calculated. The equations are as follows:

$$\Delta I = I_n - I_m \tag{5}$$

$$\Delta R = \frac{\Delta I}{n - m} \tag{6}$$

where ΔI is the interval change value, n and m are two different years in which the change is calculated ($n > m$), I_n and I_m are the data values for years n and m, respectively, and ΔR is the average annual change value. The above two equations were used to calculate the change data of carbon storage and the corresponding remote sensing features in the sample sites. The calculation of change values is supported by the NFI data records every 5 years from 1987–2017. The shortest time interval for which change data can be calculated is 5 years and the longest time interval is 30 years. In generally, the shorter the time interval, the stronger the continuity of the data and the more accurately the dynamic process of forest change can be expressed [10,44]. Zhang et al. [12] tested the change data for 5, 10, and 15 year intervals and found that the 5 year interval resulted in the smallest error.

Lunetta et al. [45] compared the effect of 3, 5, and 7 year intervals on monitoring change and showed that the 3 year interval had the highest accuracy. Therefore, only the change data for the shortest interval (5 years) were calculated in this study: the 5-year interval change and the annual average change during the 5-year period.

2.6.2. Remote Sensing Features Selection

The efficiency of the model can be affected by the number of features, because too many features reduce the speed of model fitting [46]. In addition, the original features contain some redundant information, which has a negative impact on the model accuracy [47]. Therefore, the feature screening was performed. In this study, 575 extracted remote sensing features were screened using the feature importance assessment method of the random forest [48]. In order to show the combined effect of spectral features and texture features in modelling, the spectral feature and texture feature were screened independently in this study. These two features were then combined and used for modelling.

Since most of the texture values of the extracted 1 × 1 windows were 0 or 1, they were not usable and thus removed, and the remaining 486 texture features were used for the subsequent screening. The 5-year interval change data and the annual average change data were calculated using Equations (5) and (6), respectively. Then, the feature importance assessment method of the random forest was used to filter the feature variables in the two different change types, and the top five features with contributions greater than 5% were selected from the spectral features and the texture features, respectively. Considering the problem of collinearity between features, the variance inflation factor (VIF) was used to check all the features selected in this study. When the VIF value is less than 3, it means that the collinearity between features is weak [49]. According to this principle, two features with VIF values greater than 3 in the 5-year interval change were removed, and the 5-year interval change and annual average change resulted in 8 and 10 features for the modelling study, respectively (Table 3).

Table 3. Results of remote sensing feature screening.

Change Type	Remote Sensing Features
5-year interval change	B4 × (B3/B7), B5/B4, PCA2, FVC, PCA4, R9B1EN, R17B7CO, R5B4SM
Annual average change	R3B3VA, R15B5VA, R17B1VA, R9B4HO, R15B5ME, DVI, PCA2, EVI, B4 × (B3/B7), B7/B3

The expression of the texture features is "RXBYTT", RX is the size of the texture window, BY is a certain single band of the image, and TT is the abbreviation of a certain texture feature.

2.6.3. Modelling Methods

In this study, carbon storage dynamics models were developed using change data, and the modelling methods included partial least squares regression (PLSR), gradient boosting regression tree (GBRT), and random forest (RF).

PLSR is a method used to study the correlation and quantitative relationship between the response variable Y and a set of explanatory variables, $X = x_1, x_2, \cdots, x_n$, and the set of independent variables X can also be used to predict Y [50]. The modelling principle of PLSR is the development of multivariate linear regression and principal component analysis. Compared with these two methods, PLSR has better stability and can solve the problem of collinearity between multiple explanatory variables X [51–53]. The implementation of the PLSR model in this study was based on the Minitab 20 software [54].

GBRT is an algorithm for iterative regression trees proposed by Friedman [55,56]. All the regression trees in this model are interconnected, and at its core, each tree is fitted based on the residuals and conclusions of the previous tree [57]. GBRT uses a forward-distributed algorithm to minimize the loss function by selecting the appropriate decision tree function from the current model and the fitted function. Based on the characteristics described above, the GBRT model has low computational complexity, is able to reduce errors, and has the ability to handle unevenly distributed data [58]. The implementation of the GBRT model in this study is based on the "Gradient Boosting Regressor" algorithm provided in the "Scikit-learn" package for the Python language.

RF is an integrated learning model, first proposed by Breiman, which consists of many aggregated randomly generated trees [59,60]. In the classification problem, RF outputs the type with the most votes, and in the regression problem, RF outputs the mean of all decision trees [61]. RF can reduce variance and effectively reduce overfitting by assembling different trees, and it has excellent classification and regression performance; thus, it is currently used in many fields of research [58,61,62]. The implementation of the RF model in this study is based on the "Random Forest Regressor" algorithm provided in the "Scikit-learn" package for the Python language.

In this study, the above three methods were used to develop 5-year interval change and annual average change models, respectively. In order to explore the influence of TNI on the dynamic model of forest carbon storage, the TNI was added to each model in this study, then the accuracy of different dynamic models before and after adding the TNI was compared, and the optimal model was selected to estimate the carbon storage of *Pinus densata* in Shangri-La.

2.7. Accuracy Evaluation

Ultimately, this study obtained 92 groups of change values of each type of change; 70% (64 groups) of the data were randomly selected for model fitting, and the remaining 30% (28 groups) were used for validation, and cross-validation was performed during model fitting. The indicators used to evaluate the accuracy of the model were the coefficient of determination (R^2), the root mean square error (RMSE), and the mean absolute error (MAE). In order to ensure that the model results were as objective as possible, each model was fitted 20 times in this study to allow take the mean values of evaluation indicators to be used for comparison. The equations are as follows:

$$R^2 = 1 - \frac{\sum_{i=1}^{n}(y_i - \hat{y}_i)^2}{\sum_{i=1}^{n}(y_i - \overline{y})^2} \tag{7}$$

$$RMSE = \sqrt{\frac{\sum_{i=1}^{n}(y_i - \hat{y}_i)^2}{n}} \tag{8}$$

$$MAE = \frac{1}{n}\sum_{i=1}^{n}|y_i - \hat{y}_i| \tag{9}$$

where y_i is the observed value of the sample site, $\hat{y}_i$ is the predicted value of the model, $\overline{y}$ is the observed mean value of the sample site, and n is the number of samples.

3. Results

3.1. Analysis of Modelling Results

The PLSR, GBRT, and RF methods were used to develop the 5-year interval change and annual average change models of carbon storage, respectively, TNI was added to each model subsequently, and the results of accuracy evaluation indexes (20 times fitted mean values) of each type of model are shown in Table 4. As can be seen from Table 4, among the models developed with the same type of change data, the PLSR model performed the worst and the RF model performed the best. The change of model indicators (percentage) after adding TNI is shown in Figure 5, from which it can be seen that R^2 increases and RMSE decreases for all models, and the MAE values decreased for all models except for the PLSR model with annual average change. The maximum increase in R^2 is 10.48%, the maximum decrease in RMSE is 7.32%, and the maximum decrease in MAE is 8.89%. Considering the changes of all accuracy indicators together, the addition of TNI improves the model accuracy. The mean values of carbon storage calculated for the 5-year interval change and the mean annual change of the sample plot data were 5.14 t-C·ha^{-1} and 1.03 t-C·ha^{-1}, respectively. After comparison, it can be seen that the RMSE and MAE values of both the RF and GBRT models are smaller than the mean values, except for the PLSR model, indicating that the errors of the RF and GBRT models are smaller and the model results are more reasonable.

Table 4. Comparison of modelling results.

Change Type	Model	Fitting		Validation
		R^2	RMSE/t-C·ha^{-1}	MAE/t-C·ha^{-1}
5-year interval change	PLSR	0.18	9.61	9.09
	GBRT	0.81	4.58	2.51
	RF	0.85	4.09	1.83
	$PLSR_{TNI}$	0.18	9.50	8.36
	$GBRT_{TNI}$	0.83	4.24	2.43
	RF_{TNI}	0.87	3.82	1.78
Annual average change	PLSR	0.23	1.83	1.64
	GBRT	0.80	0.94	0.45
	RF	0.83	0.87	0.36
	$PLSR_{TNI}$	0.25	1.82	1.67
	$GBRT_{TNI}$	0.82	0.89	0.41
	RF_{TNI}	0.84	0.86	0.35

$PLSR_{TNI}$, $GBRT_{TNI}$, and RF_{TNI} in the table indicate the PLSR model, GBRT model, and RF model after adding TNI, respectively.

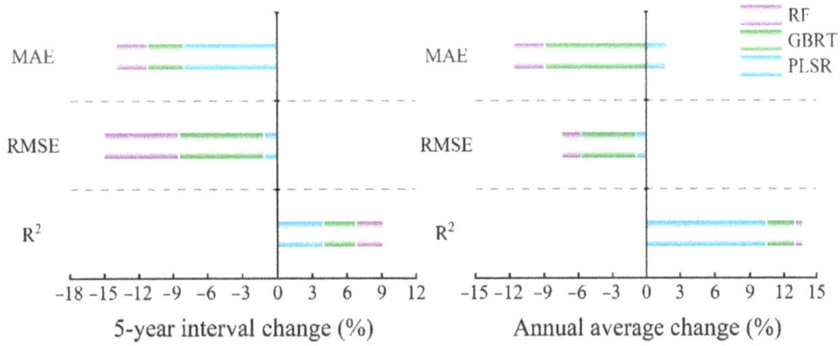

Figure 5. Changes in model indicators after adding TNI: positive values on the right side of the *y*-axis represent an increase, and negative values on the left side of the *y*-axis represent a decrease.

The 5-year interval change represents the change in carbon storage over five years, and the annual average change represents the change in carbon storage in one year. In order to compare the effects of dynamic models with 5-year interval changes and annual average changes, the RMSE and MAE values of the annual average change model were adjusted to five times their original values in this study, so that the error indicators of both models are on the same time scale. The adjusted results are shown in Table 5. Comparing the accuracy indicators of the two models in Tables 4 and 5, it can be seen that the R^2 of all 5-year interval change models is higher than that of the annual average change model, the RMSE values are lower than those of the 5-year interval change model except for the PLSR model, and the difference in MAE values is not significant. Therefore, the 5-year interval change model works better overall.

According to the confidence interval plot (Figure 6) of the 20 fittings for each model, the confidence range of the RF model among the types is smaller than that of the GBRT and PLSR models, indicating that the RF model in this study has the least uncertainty and the results obtained are more stable and reliable. In summary, the accuracy of 5-year interval change RF model with the addition of TNI is the highest.

Table 5. Error indicator for the adjusted annual average change.

Change Type	Model	Fitting	Validation
		RMSE/t-C·ha^{-1}	MAE /t-C·ha^{-1}
Annual average change	PLSR	9.16	8.2
	GBRT	4.69	2.25
	RF	4.37	1.8
	$PLSR_{TNI}$	9.075	8.35
	$GBRT_{TNI}$	4.465	2.05
	RF_{TNI}	4.295	1.75

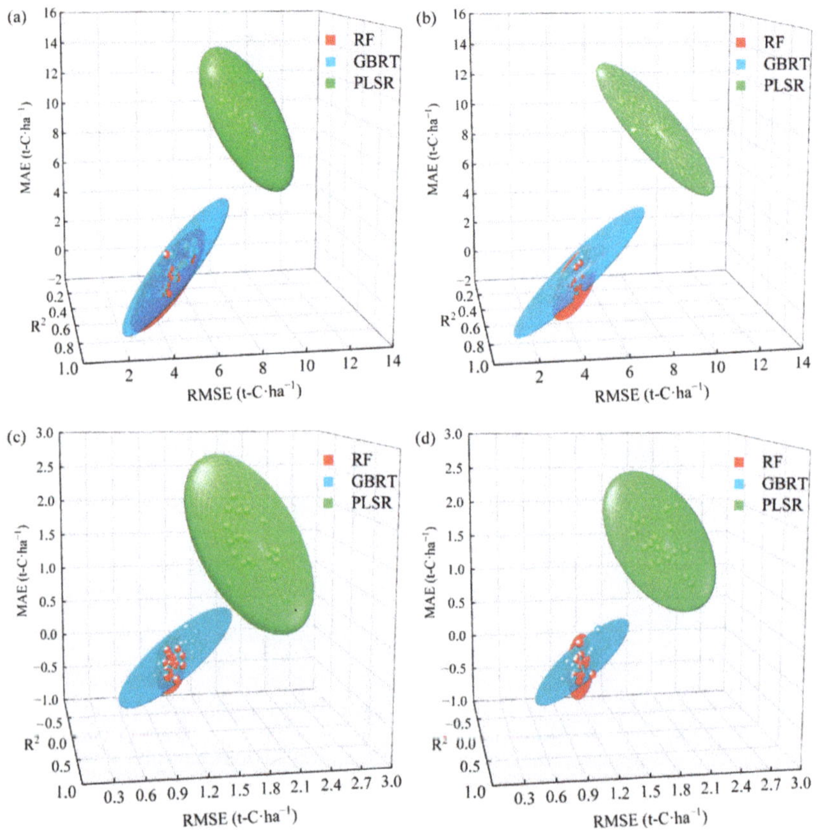

Figure 6. Confidence intervals (95%) of the accuracy evaluation indicators for different models: (**a**) the model of 5-year interval change before adding TNI; (**b**) the model of 5-year interval change after adding TNI; (**c**) the model of annual average change model before adding TNI; and (**d**) the model of annual average change after adding TNI.

3.2. Mapping Carbon Storage

According to the model evaluation results, the 5-year interval change RF model with the highest accuracy was used to estimate the carbon storage change. The scatter plot of this estimation model is shown in Figure 7. The direct estimation result of the model is the 5-year interval change of carbon storage of *Pinus densata* in Shangri-La. The values of the carbon storage change for the six change periods are $\Delta CS_1 = CS_{1992} - CS_{1987}$, $\Delta CS_2 = CS_{1997} - CS_{1992}$, $\Delta CS_3 = CS_{2002} - CS_{1997}$, $\Delta CS_4 = CS_{2007} - CS_{2002}$, $\Delta CS_5 = CS_{2012} - CS_{2007}$ and $\Delta CS_6 = CS_{2017} - CS_{2012}$.

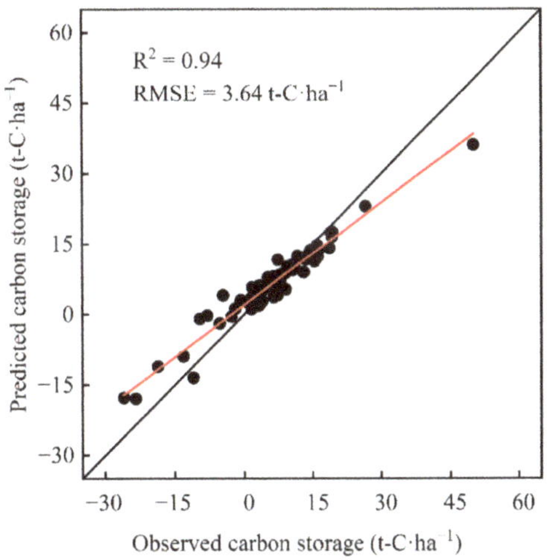

Figure 7. Scatter plot of the estimation model.

The estimated carbon storage changes in the six periods are as follows: $\Delta CS_1 = -53.327 \times 10^4$ t, $\Delta CS_2 = 13.887 \times 10^4$ t, $\Delta CS_3 = -94.041 \times 10^4$ t, $\Delta CS_4 = 14.602 \times 10^4$ t, $\Delta CS_5 = 20.869 \times 10^4$ t, and $\Delta CS_6 = 5.602 \times 10^4$ t. A positive value indicates an increase in total carbon storage, and a negative value indicates a decrease in total carbon storage. The estimation results indicate that the carbon storage of *Pinus densata* in Shangri-La decreased in the periods 1987–1992 and 1997–2002 and increased in the remaining four periods.

In order to obtain carbon storage maps for each year, the published AGB estimates for each year from our group were used [63], which contain the aboveground biomass estimates of *Pinus densata* in Shangri-La for 1987, 1992, 1997, 2002, 2007, 2012, and 2017. These data were combined with the results of carbon storage change values estimated in this paper to calculate the total carbon storage values for each year. Firstly, the AGB of each year in the published study was converted to carbon storage by multiplying the carbon content coefficient (0.501) and the converted values were expressed as T_{1987}, T_{1992}, T_{1997}, T_{2002}, T_{2007}, T_{2012}, and T_{2017}; then the carbon storage change values ΔCS_1, ΔCS_2, ΔCS_3, ΔCS_4, ΔCS_5 and ΔCS_6 were added to the conversion values of the smaller years in each change period in order to obtain the carbon storage values for the six years 1992, 1997, 2002, 2007, 2012 and 2017, respectively (e.g., $CS_{1992} = \Delta CS_1 + T_{1987}$), and the carbon storage value for 1987 was obtained by subtracting ΔCS_1 from T_{1992} ($CS_{1992} = T_{1987} - \Delta CS_1$). The statistical results for carbon storage are shown in Table 6, and Figures 8 and 9 (the spatial resolution is 30 m) were obtained by mapping the values of the carbon storage change for each period and each year according to the *Pinus densata* distribution range. The distribution range and area of *Pinus densata* are derived from Forest Manager Inventory (FMI) data.

Table 6. Statistics of *Pinus densata* carbon storage.

Year	Area of *Pinus densata* (ha)	Total CS (Million Tons)	Average CS (t-C·ha^{-1})
1987	171560.28	5.30	30.91
1992	171560.28	4.24	24.69
1997	170589.86	4.77	27.97
2002	170589.86	3.12	18.30
2007	174179.37	3.99	22.93
2012	174213.12	4.03	23.12
2017	184815.84	3.80	20.53

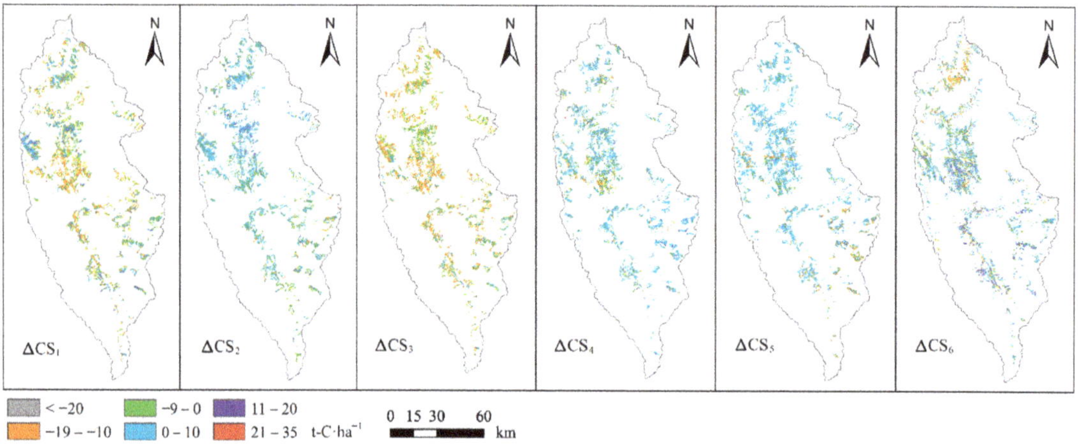

Figure 8. Change values for carbon storage for six time periods.

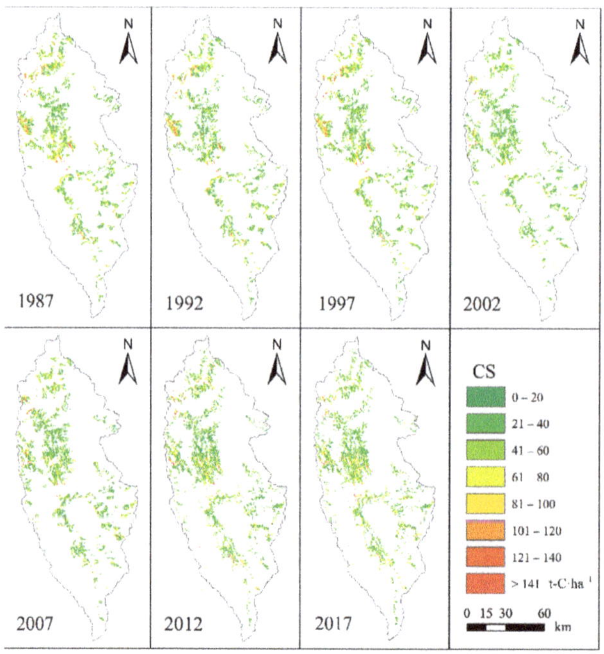

Figure 9. Carbon storage values for seven years.

3.3. Spatial Distribution Characteristics for the Changes in Carbon Storage on Different TNI Gradients

In this study, the estimated changes in carbon storage were classified as increases (positive change values) and decreases (negative change values), and the TNI (0.30 to 0.93) was classified into ten levels from low to high using the natural breaks (Jenks) method: low (1–3), medium–low (4–5), medium–high (6–7), high (8–10). The DI was then calculated according to Equation (4) and used to explore the distribution of carbon storage changes along different TNI gradients. The distribution results are shown in Figure 10.

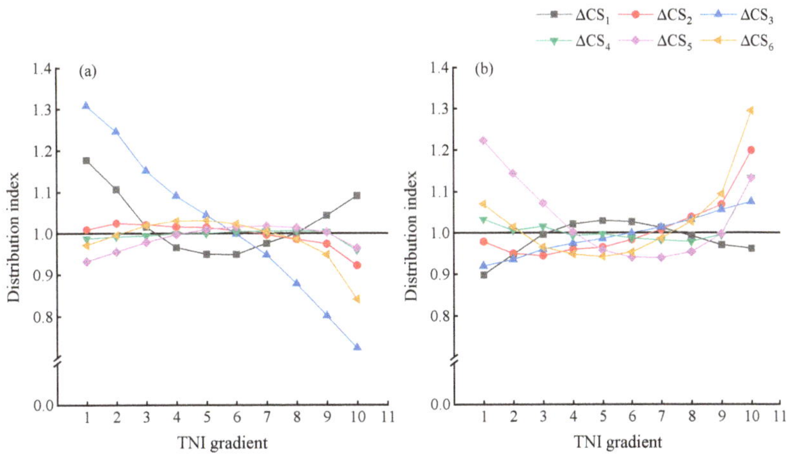

Figure 10. Distribution of increased carbon storage (**a**) and decreased carbon storage (**b**) on the TNI gradient.

As can be seen from Figure 10, the increase and decrease in the carbon storage of *Pinus densata* in Shangri-La at each period have a certain regularity on the TNI gradient. The DI curve for ΔCS_1(1987–1992) decreases and then rises when carbon storage increases (Figure 10a) with a dominant distribution in the low and high TNI gradients; the DI curve for ΔCS_2 (1992–1997), ΔCS_4 (2002–2007), ΔCS_5 (2007–2012), and ΔCS_6 (2012–2017) rises and then reduces with the dominant distribution areas of ΔCS_2 being in the low and medium TNI gradients, and the dominant distribution areas of ΔCS_4, ΔCS_5, and ΔCS_6 being concentrated in the medium–high TNI gradients; and ΔCS_3 (1997–2002) reduces with the TNI gradient, and its dominant distribution area is in the low and medium–low gradients. The corresponding curve change period displays a decreasing trend when the carbon storage decreases (Figure 10b). Overall, the dominant distribution areas for each period when carbon storage increases are mainly found in the lower or middle TNI gradients, while the dominant distribution areas for each period when carbon storage decreases are found in all TNI gradients.

4. Discussion

In long time series data, the uncertainty of the historical data, due to the events that took place at the time being taken into account, cause some of the data to display large deviations in values, which are known as anomalies or outliers [64]. The stability of the model can be affected by outliers, and at the same time, the prediction accuracy of the model may be reduced, so the detection and handling of outliers is very important in the modelling process [65]. The sample plot data were screened twice in this study, the first time removing plots with an AGB of less than 1 $t{\cdot}ha^{-1}$ because they had an average diameter at breast height of less than 5 cm and an average tree height of less than 1.5 m. They are young forests, too low in canopy density, and poorly characterized by the appearance of the forest floor (similar to bare ground) on 30 m resolution imagery. Values exceeding the range of $\overline{x} \pm 3\sigma$ were selected for the second time using the Pauta criterion. By removing values that are too small and too large from the data in these two steps, respectively, the overall uncertainty of the data is reduced and the reliability of the data is improved.

The remote sensing-based estimation model is one of the main methods used in current carbon storage studies [7]. Based on the accumulation of ground survey data, the results of remote sensing estimation models are considered to have certain advantages and reliability in forest carbon storage studies. Time series images also provide an important basis for describing change [66]. The NFI and Landsat time series data over a 30-year period were combined to develop a dynamic model to study forest carbon storage in this paper. Although dynamic models based on change data are currently less commonly used in forestry, their advantages in terms of accuracy have been demonstrated [11,12]. In this study, although the effects of dynamic models and static models are not directly compared through experiments, we can refer to some of the existing studies related to forest carbon storage modelling in recent years [8,67,68]. The R^2 of the remote sensing static models developed in these studies ranges from 0.64 to 0.73 as the highest value, while the R^2 of the optimal model in this study is 0.87. Thus, the advantages of the dynamic model in terms of accuracy can be seen.

The DI is an area evaluation indicator. When the distribution curve is flatter, it indicates that the distribution of such changes deviates less from the standard distribution, and its adaptability to the terrain is wider [16]. In this study, the areas of increasing or decreasing *Pinus densata* carbon storages were calculated separately, and the main topography of these two areas was evaluated according to the DI. The results of this study demonstrate that the increase in carbon storage in the periods ΔCS_4, ΔCS_5, and ΔCS_6 is much greater than the decrease, leading to an increase in total carbon storage from 2002 to 2017, which is related to the policy of "returning farmland to the forest" that has been implemented in Shangri-La since 2000. In the area of increasing carbon storage, the distribution curves of these three periods are relatively flat, showing a general adaptation to different levels of TNI and suggesting that the implementation of this government policy has led to the widespread planting of *Pinus densata* in Shangri-La, thus resulting in a more balanced topographic distribution of increasing carbon storage across the region.

Different TNI levels reflect different elevation and slope conditions, and the dominant distribution reflects the main topographic range of carbon storage changes. Although the dominant distribution of carbon storage changes over the TNI gradient can be clearly seen throughout a particular period of change, the six periods of change in carbon storage are more complex, and the distribution trends across the periods of change are not entirely consistent along the TNI gradient. What can be seen from the dominant distribution is that the regions where carbon storage increased are mainly located at low altitudes or on gently sloping terrain or both, as such terrain is conducive to artificial tending. In contrast, in regions with decreased carbon storage, the dominant distribution occurs across the TNI gradient, and the overall dominance of the terrain is not obvious, which is due to the randomicity of deforestation or forest destruction. In the past, TNI has not been applied to separate forest land class studies, but this study shows that its combination with the DI can be used to evaluate the increase and decrease effects of a single land type.

Some studies have suggested that human activities are the main driver of forest carbon [69], but as Shangri-La is a highland region with a low population density, the impact of human life on the surrounding forests is also relatively low, and it is environmental factors, including topography, that mainly impact forest carbon storage. Therefore, it is essential to consider topographic factors when studying forest carbon storage in this region. The results of this paper show that adding TNI to the dynamic model can reflect the topographic effect of the model and enable the improvement of the dynamic model accuracy, and by analyzing the distribution of forest carbon storage changes along the TNI gradient, it is possible to reveal the changes in forest carbon storage under different topographic conditions.

At present, the main approach to the study of historical forest dynamics is still based on the estimation of forest biomass or carbon storage over several single years, and the values of changes in different periods are calculated in order to derive forest dynamics results [70]. The level of periodic change in the forest can be estimated directly from models developed from interval change data, and due to the advantages of dynamic models in terms of accuracy, the resulting change values are estimated more precisely. Different dynamic models can be obtained for different types of change data. In this paper, we only compared two dynamic models for the 5-year interval change data and annual average change data, and more types of dynamic models are yet to be explored.

Although it was difficult to derive carbon storage results for a single year in this study without the support of carbon storage inventory data or graphs related to the study years, the results of this study demonstrate that the dynamic model is able to estimate the change in forest carbon storage and is well suited for the study of forest carbon storage change. Dynamic models are capable of quantitatively describing changes in specific properties, so they need not be limited to the field of forestry research but have the potential for application in other areas of natural or manufactured landscapes where regular variation exists. This study has so far only considered the influence of topographic factors, and there remains the possibility of other environmental factors, such as climate and soil, also influencing the model, which could be the next direction for future research.

5. Conclusions

In this study, the carbon storage dynamic model was developed based on NFI data and Landsat time series images over a 30-year period. PLSR, RF, and GBRT were used for modelling, and the accuracy of the two dynamic models was compared; then TNI, which represents topographic factors, was introduced into the model, and the distribution of carbon storage changes in Shangri-La on different TNI gradients was analyzed. The main conclusions are as follows: (1) the model effects of the non-parametric methods, RF and GBRT, are much better than those of the parametric method, PLSR, and the accuracy of the 5-year interval change model is better than that of the annual average change model; (2) TNI can improve the accuracy of the dynamic carbon storage estimation model, and the accuracy of the dynamic model with RF is the highest after adding TNI; (3) the dynamic model displays good performance regarding the estimation of carbon storage changes, and the results of the interval change model show that the total carbon storage of *Pinus densata* in Shangri-La decreased in 1987–1992 and 1997–2002, and increased in 1992–1997, 2002–2007, 2007–2012, and 2012–2017; and (4) the DI can be used to evaluate the main topography for the regions that display a change in *Pinus densata* carbon storage, the predominant topography in the regions of increasing carbon storage is that of low elevations or low slopes, or a combination of both conditions, while the topography in regions of decreasing carbon storage is more random. The results of this study can be used as a reference for forest carbon storage estimation using Landsat images in areas with complex topography.

Author Contributions: Y.L. participated in the data analysis and wrote the draft article, J.Z. helped with the idea and draft revision, R.B. helped with the data analysis, and D.X. and D.H. participated in the draft revision. All authors have read and agreed to the published version of the manuscript.

Funding: This research was funded by the National Natural Science Foundation of China (Nos. 31860207 and 32260390), and the "Young Top Talents" special project of the high-level talent training support program of Yunnan province, China, in 2020 (No. YNWR-QNBJ-2020-164).

Data Availability Statement: The data presented in this study are available on request from the corresponding author. The data are not publicly available due to the confidentiality of the NFI dataset.

Acknowledgments: We would like to thank the editor and anonymous reviewers for their comments, which helped improve the manuscript. We also would like to acknowledge all the others individuals who contributed to this paper.

Conflicts of Interest: The authors declare no conflict of interest.

References

1. Shao, W.; Cai, J.; Wu, H.; Liu, J.; Zhang, H.; Huang, H. An Assessment of Carbon Storage in China's Arboreal Forests. *Forests* **2017**, *8*, 110. [CrossRef]
2. Solomon, S.D.; Qin, D.; Manning, M.; Chen, Z.; Miller, H.L. Climate Change 2007: The Physical Science Basis. In *Working Group I Contribution to the Fourth Assessment Report of the IPCC*; Cambridge University Press: Cambridge, UK, 2007.
3. Luo, K. Spatial Pattern of Forest Carbon Storage in the Vertical and Horizontal Directions Based on HJ-CCD Remote Sensing Imagery. *Remote Sens.* **2019**, *11*, 788. [CrossRef]
4. Li, T.; Li, M.-Y.; Tian, L. Dynamics of Carbon Storage and Its Drivers in Guangdong Province from 1979 to 2012. *Forests* **2021**, *12*, 1482. [CrossRef]

5. Long, Y.; Jiang, F.G.; Sun, H.; Wang, T.H.; Zou, Q.; Chen, C.S. Estimating vegetation carbon storage based on optimal bandwidth selected from geographically weighted regression model in Shenzhen City. *Acta Ecol. Sin.* **2022**, *42*, 4933–4945.
6. Zhang, P.P.; Li, Y.H.; Yin, H.R.; Chen, Q.T.; Dong, Q.D.; Zhu, L.Q. Spatio-temporal variation and dynamic simulation of ecosystem carbon storage in the north-south transitional zone of China. *J. Nat. Resour.* **2022**, *37*, 1183–1197. [CrossRef]
7. Yan, E.; Lin, H.; Wang, G.; Sun, H. Improvement of Forest Carbon Estimation by Integration of Regression Modeling and Spectral Unmixing of Landsat Data. *IEEE Geosci. Remote Sens. Lett.* **2015**, *12*, 2003–2007.
8. Safari, A.; Sohrabi, H.; Powell, S.; Shataee, S. A comparative assessment of multi-temporal Landsat 8 and machine learning algorithms for estimating aboveground carbon stock in coppice oak forests. *Int. J. Remote Sens.* **2017**, *38*, 6407–6432. [CrossRef]
9. Sun, W.; Liu, X. Review on carbon storage estimation of forest ecosystem and applications in China. *For. Ecosyst.* **2019**, *7*, 4. [CrossRef]
10. Gómez, C.; White, J.C.; Wulder, M.A. Characterizing the state and processes of change in a dynamic forest environment using hierarchical spatio-temporal segmentation. *Remote Sens. Environ.* **2011**, *115*, 1665–1679. [CrossRef]
11. Gómez, C.; White, J.C.; Wulder, M.A.; Alejandro, P. Historical forest biomass dynamics modelled with Landsat spectral trajectories. *ISPRS J. Photogramm. Remote Sens.* **2014**, *93*, 14–28. [CrossRef]
12. Zhang, J.; Lu, C.; Xu, H.; Wang, G. Estimating aboveground biomass of Pinus densata-dominated forests using Landsat time series and permanent sample plot data. *J. For. Res.* **2019**, *30*, 1689–1706. [CrossRef]
13. Liu, C.; Zhang, L.; Li, F.; Jin, X. Spatial modeling of the carbon stock of forest trees in Heilongjiang Province, China. *J. For. Res.* **2014**, *25*, 269–280. [CrossRef]
14. Oltean, G.S.; Comeau, P.G.; White, B. Linking the Depth-to-Water Topographic Index to Soil Moisture on Boreal Forest Sites in Alberta. *For. Sci.* **2016**, *62*, 154–165. [CrossRef]
15. Deng, Y.; Chen, X.; Kheir, R.B. An Exploratory Procedure Defining a Local Topographic Index for Mountainous Vegetation Conditions. *GIScience Remote Sens.* **2007**, *44*, 383–401. [CrossRef]
16. Wu, A.B.; Qin, Y.J.; Zhao, Y.X. Terrain composite index and its application in terrain gradient effect analysis of land use change: A case study of Taihang Hilly areas. *Geogr. Geo. Inf. Sci.* **2018**, *34*, 93–99+118.
17. Wang, Q.; Yang, K.; Li, L.; Zhu, Y. Assessing the Terrain Gradient Effect of Landscape Ecological Risk in the Dianchi Lake Basin of China Using Geo-Information Tupu Method. *Int. J. Environ. Res. Public Health* **2022**, *19*, 9634. [CrossRef] [PubMed]
18. Gong, W.; Wang, H.; Wang, X.; Fan, W.; Stott, P. Effect of terrain on landscape patterns and ecological effects by a gradient-based RS and GIS analysis. *J. For. Res.* **2017**, *28*, 1061–1072. [CrossRef]
19. Zhao, Y.; Cao, J.; Zhang, X.; Zhang, M. Analyzing the characteristics of land use distribution in typical village transects at Chinese Loess Plateau based on topographical factors. *Open Geosci.* **2022**, *14*, 429–442. [CrossRef]
20. Li, C.; Chen, D.; Wu, D.; Su, X. Design of an EIoT system for nature reserves: A case study in Shangri-La County, Yunnan Province, China. *Int. J. Sustain. Dev. World Ecol.* **2015**, *22*, 184–188. [CrossRef]
21. Pan, J.; Wang, J.; Gao, F.; Liu, G. Quantitative estimation and influencing factors of ecosystem soil conservation in Shangri-La, China. *Geocarto Int.* **2022**, 1–15. [CrossRef]
22. Yu, Y.; Wang, J.; Liu, G.; Cheng, F. Forest Leaf Area Index Inversion Based on Landsat OLI Data in the Shangri-La City. *J. Indian Soc. Remote Sens.* **2019**, *47*, 967–976. [CrossRef]
23. Chen, Y.; Wang, J.; Liu, G.; Yang, Y.; Liu, Z.; Deng, H. Hyperspectral Estimation Model of Forest Soil Organic Matter in Northwest Yunnan Province, China. *Forests* **2019**, *10*, 217. [CrossRef]
24. Sun, X.L. Study on Biomass Estimation of Pinus Densata in Shangri-La Based on Landsat8-OLI. Master's Thesis, Southwest Forestry University, Kunming, China, 2016.
25. Yao, X.L.; Chen, H.L.; Zhao, X.; Guo, S.J. Weak link determination of anti-shock performance of shipboard equipments based on Pauta criterion. *Chin. J. Ship Res.* **2007**, *2*, 10–14.
26. Xie, H.; Zhao, A.; Huang, S.; Han, J.; Liu, S.; Xu, X.; Luo, X.; Pan, H.; Du, Q.; Tong, X. Unsupervised Hyperspectral Remote Sensing Image Clustering Based on Adaptive Density. *IEEE Geosci. Remote Sens. Lett.* **2018**, *15*, 632–636. [CrossRef]
27. *LY/T 2988-2018*; Guideline on Carbon Stock Accounting in Forest Ecosystem. State Forestry and Grassland Administration of China: Beijing, China, 2018; pp. 1–16.
28. Chander, G.; Markham, B.L.; Helder, D.L. Summary of current radiometric calibration coefficients for Landsat MSS, TM, ETM+, and EO-1 ALI sensors. *Remote Sens. Environ.* **2009**, *113*, 893–903. [CrossRef]
29. Yu, K.; Liu, S.; Zhao, Y. CPBAC: A quick atmospheric correction method using the topographic information. *Remote Sens. Environ.* **2016**, *186*, 262–274. [CrossRef]
30. Nichol, J.; Hang, L.K.; Sing, W.M. Empirical correction of low Sun angle images in steeply sloping terrain: A slope-matching technique. *Int. J. Remote Sens.* **2006**, *27*, 629–635. [CrossRef]
31. Zhang, J.; Pham, T.-T.-H.; Kalacska, M.; Turner, S. Using Landsat Thematic Mapper records to map land cover change and the impacts of reforestation programmes in the borderlands of southeast Yunnan, China: 1990–2010. *Int. J. Appl. Earth Obs. Geoinf.* **2014**, *31*, 25–36. [CrossRef]
32. Carleer, A.P.; Wolff, E. Urban land cover multi-level region-based classification of VHR data by selecting relevant features. *Int. J. Remote Sens.* **2008**, *27*, 1035–1051. [CrossRef]
33. Wang, W.; Jiang, Y.; Wang, G.; Guo, F.; Li, Z.; Liu, B. Multi-Scale LBP Texture Feature Learning Network for Remote Sensing Interpretation of Land Desertification. *Remote Sens.* **2022**, *14*, 3486. [CrossRef]

34. Zhang, C.; Huang, C.; Li, H.; Liu, Q.; Li, J.; Bridhikitti, A.; Liu, G. Effect of Textural Features in Remote Sensed Data on Rubber Plantation Extraction at Different Levels of Spatial Resolution. *Forests* **2020**, *11*, 399. [CrossRef]
35. Yuan, J.; Wang, D.; Li, R. Remote Sensing Image Segmentation by Combining Spectral and Texture Features. *IEEE Trans. Geosci. Remote Sens.* **2014**, *52*, 16–24. [CrossRef]
36. Duan, M.; Song, X.; Liu, X.; Cui, D.; Zhang, X. Mapping the soil types combining multi-temporal remote sensing data with texture features. *Comput. Electron. Agric.* **2022**, *200*, 107230. [CrossRef]
37. Rui, B.; Zhang, J.; Lu, C.; Chen, P. Estimating above-ground biomass of Pinus densata Mast. Using best slope temporal segmentation and Landsat time series. *J. Appl. Remote Sens.* **2021**, *15*, 024507.
38. Powell, S.L.; Cohen, W.B.; Healey, S.P.; Kennedy, R.E.; Moisen, G.G.; Pierce, K.B.; Ohmann, J.L. Quantification of live aboveground forest biomass dynamics with Landsat time-series and field inventory data: A comparison of empirical modeling approaches. *Remote Sens. Environ.* **2010**, *114*, 1053–1068. [CrossRef]
39. Wei, H.L.; Qi, Y.J. Analysis of grassland degradation of the Tibet Plateau and human driving forces based on remote sensing. *Pratacultural Sci.* **2016**, *33*, 2576–2586.
40. Zhang, G.L. Spatial distribution characteristics of carbon storage of urban forests in Shanghai based on remote sensing estimation. *Ecol. Environ. Sci.* **2021**, *30*, 1777–1786.
41. Yu, H.; Zeng, H.; Jiang, Z.Y. Study on Distribution Characteristics of Landscape Elements along the Terrain Gradient. *Scientia Geogr. Sin.* **2001**, *1*, 64–69.
42. Emamian, A.; Rashki, A.; Kaskaoutis, D.G.; Gholami, A.; Opp, C.; Middleton, N. Assessing vegetation restoration potential under different land uses and climatic classes in northeast Iran. *Ecol. Indic.* **2021**, *122*, 107325. [CrossRef]
43. Chen, L.D.; Yang, S.; Feng, X.M. Land use change characteristics along the terrain gradient and the spatial expanding analysis: A case study of Haidian District and Yanqing County, Beijing. *Geogr. Res.* **2008**, *27*, 1225–1234+1481.
44. Huang, C.; Goward, S.N.; Schleeweis, K.; Thomas, N.; Masek, J.G.; Zhu, Z. Dynamics of national forests assessed using the Landsat record: Case studies in eastern United States. *Remote Sens. Environ.* **2009**, *113*, 1430–1442. [CrossRef]
45. Lunetta, R.S.; Johnson, D.M.; Lyon, J.G.; Crotwell, J. Impacts of imagery temporal frequency on land-cover change detection monitoring. *Remote Sens. Environ.* **2004**, *89*, 444–454. [CrossRef]
46. Yin, S.-L.; Zhang, X.-L.; Liu, S. Intrusion detection for capsule networks based on dual routing mechanism. *Comput. Netw.* **2021**, *197*, 108328. [CrossRef]
47. Lian, Y.; Luo, J.; Xue, W.; Zuo, G.; Zhang, S. Cause-driven Streamflow Forecasting Framework Based on Linear Correlation Reconstruction and Long Short-term Memory. *Water Resour. Manag.* **2022**, *36*, 1661–1678. [CrossRef]
48. Gregorutti, B.; Michel, B.; Saint-Pierre, P. Correlation and variable importance in random forests. *Stat. Comput.* **2017**, *27*, 659–678. [CrossRef]
49. Zuur, A.F.; Ieno, E.N.; Elphick, C.S. A protocol for data exploration to avoid common statistical problems. *Methods Ecol. Evol.* **2010**, *1*, 3–14. [CrossRef]
50. Farifteh, J.; Van der Meer, F.; Atzberger, C.; Carranza, E.J.M. Quantitative analysis of salt-affected soil reflectance spectra: A comparison of two adaptive methods (PLSR and ANN). *Remote Sens. Environ.* **2007**, *110*, 59–78. [CrossRef]
51. Paul, G.; Kowalski, B.R. Partial least-squares regression: A tutorial. *Anal. Chim. Acta* **1986**, *185*, 1–17.
52. Wold, S.; Sjöström, M.; Eriksson, L. PLS-regression: A basic tool of chemometrics. *Chemom. Intell. Lab. Syst.* **2001**, *58*, 109–130. [CrossRef]
53. Song, G.Y. Research on Partial Least Squares Regression. Master's Thesis, Zhejiang University, Hangzhou, China, 2009.
54. Ryan, T.A., Jr.; Joiner, B.L. Minitab: A Statistical Computing System for Students and Researchers. *Am. Stat.* **1973**, *27*, 222–225. [CrossRef]
55. Friedman, J.H. Greedy Function Approximation: A Gradient Boosting Machine. *Ann. Stat.* **2001**, *29*, 1189–1232. [CrossRef]
56. Friedman, J.H. Stochastic gradient boosting. *Comput. Stat. Data Anal.* **2002**, *38*, 367–378. [CrossRef]
57. Lawrence, R.; Bunn, A.; Powell, S.; Zambon, M. Classification of remotely sensed imagery using stochastic gradient boosting as a refinement of classification tree analysis. *Remote Sens. Environ.* **2004**, *90*, 331–336. [CrossRef]
58. Wang, L.; Zhang, Y.; Yao, Y.; Xiao, Z.; Shang, K.; Guo, X.; Yang, J.; Xue, S.; Wang, J. GBRT-Based Estimation of Terrestrial Latent Heat Flux in the Haihe River Basin from Satellite and Reanalysis Datasets. *Remote Sens.* **2021**, *13*, 1054. [CrossRef]
59. Breiman, L. Random Forests. *Mach. Learn.* **2001**, *45*, 5–32. [CrossRef]
60. van der Meer, D.; Hoekstra, P.J.; van Donkelaar, M.; Bralten, J.; Oosterlaan, J.; Heslenfeld, D.; Faraone, S.V.; Franke, B.; Buitelaar, J.K.; Hartman, C.A. Predicting attention-deficit/hyperactivity disorder severity from psychosocial stress and stress-response genes: A random forest regression approach. *Transl. Psychiatry* **2017**, *7*, e1145. [CrossRef]
61. Sun, J.; Zhong, G.; Huang, K.; Dong, J. Banzhaf random forests: Cooperative game theory based random forests with consistency. *Neural Netw.* **2018**, *106*, 20–29. [CrossRef]
62. Wang, X.; Liu, T.; Zheng, X.; Peng, H.; Xin, J.; Zhang, B. Short-term prediction of groundwater level using improved random forest regression with a combination of random features. *Appl. Water Sci.* **2018**, *8*, 125. [CrossRef]
63. Xu, D.; Zhang, J.; Bao, R.; Liao, Y.; Han, D.; Liu, Q.; Cheng, T. Temporal and Spatial Variation of Aboveground Biomass of Pinus densata and Its Drivers in Shangri-La, CHINA. *Int. J. Environ. Res. Public Health* **2021**, *19*, 400. [CrossRef]
64. Li, J.; Sun, G.Z.; Xu, Y.L. Application and anomaly detection of Application and anomaly detection of predictive model based on time series predictive model based on time series. *Comput. Aided Eng.* **2006**, *15*, 49–54+84.

65. Ranjan, K.G.; Tripathy, D.S.; Prusty, B.R.; Jena, D. An improved sliding window prediction-based outlier detection and correction for volatile time-series. *Int. J. Numer. Model. Electron. Netw. Devices Fields* **2021**, *34*, e2816. [CrossRef]
66. Goodwin, N.R.; Coops, N.C.; Wulder, M.A.; Gillanders, S.; Schroeder, T.A.; Nelson, T. Estimation of insect infestation dynamics using a temporal sequence of Landsat data. *Remote Sens. Environ.* **2008**, *112*, 3680–3689. [CrossRef]
67. Zhang, M.; Du, H.; Zhou, G.; Li, X.; Mao, F.; Dong, L.; Zheng, J.; Liu, H.; Huang, Z.; He, S. Estimating Forest Aboveground Carbon Storage in Hang-Jia-Hu Using Landsat TM/OLI Data and Random Forest Model. *Forests* **2019**, *10*, 1004. [CrossRef]
68. Zhang, X.; Sun, Y.; Jia, W.; Wang, F.; Guo, H.; Ao, Z. Research on the Temporal and Spatial Distributions of Standing Wood Carbon Storage Based on Remote Sensing Images and Local Models. *Forests* **2022**, *13*, 346. [CrossRef]
69. Pyles, M.V.; Magnago, L.F.S.; Maia, V.A.; Pinho, B.X.; Pitta, G.; de Gasper, A.L.; Vibrans, A.C.; Santos, R.M.D.; van Den Berg, E.; Lima, R.A. Human impacts as the main driver of tropical forest carbon. *Sci. Adv.* **2022**, *8*, eabl7968. [CrossRef]
70. Nguyen, T.H.; Jones, S.D.; Soto-Berelov, M.; Haywood, A.; Hislop, S. Monitoring aboveground forest biomass dynamics over three decades using Landsat time-series and single-date inventory data. *Int. J. Appl. Earth Obs. Geoinf.* **2020**, *84*, 101952. [CrossRef]

remote sensing

MDPI

Article

Estimation of Urban Forest Characteristic Parameters Using UAV-Lidar Coupled with Canopy Volume

Bo Zhang [1,2,3], Xuejian Li [1,2,3], Huaqiang Du [1,2,3,*], Guomo Zhou [1,2,3], Fangjie Mao [1,2,3], Zihao Huang [1,2,3], Lv Zhou [1,2,3,4], Jie Xuan [1,2,3], Yulin Gong [1,2,3] and Chao Chen [1,2,3]

1 State Key Laboratory of Subtropical Silviculture, Zhejiang A & F University, Hangzhou 311300, China
2 Key Laboratory of Carbon Cycling in Forest Ecosystems and Carbon Sequestration of Zhejiang Province, Zhejiang A & F University, Hangzhou 311300, China
3 School of Environmental and Resources Science, Zhejiang A & F University, Hangzhou 311300, China
4 Research Center of Forest Management Engineering of State Forestry and Grassland Administration, Beijing Forestry University, Beijing 100083, China
* Correspondence: dhqrs@126.com

Abstract: The estimation of characteristic parameters such as diameter at breast height (DBH), aboveground biomass (AGB) and stem volume (V) is an important part of urban forest resource monitoring and the most direct manifestation of the ecosystem functions of forests; therefore, the accurate estimation of urban forest characteristic parameters is valuable for evaluating urban ecological functions. In this study, the height and density characteristic variables of canopy point clouds were extracted as Scheme 1 and combined with the canopy structure variables as Scheme 2 based on unmanned aerial vehicle lidar (UAV-Lidar). We analyzed the spatial distribution characteristics of the canopies of different tree species, and multiple linear regression (MLR), support vector regression (SVR), and random forest (RF) models were used to estimate the DBH, AGB, and V of urban single trees. The estimation accuracy of different models was evaluated based on the field-measured data. The results indicated that the model accuracy of coupling canopy structure variables (R^2 = 0.69–0.85, rRMSE = 9.87–24.67%) was higher than that of using only point-cloud-based height and density characteristic variables. The comparison of the results of different models shows that the RF model had the highest estimation accuracy (R^2 = 0.76–0.85, rRMSE = 9.87–22.51%), which was better than that of the SVR and MLR models. In the RF model, the estimation accuracy of AGB was the highest (R^2 = 0.85, rRMSE = 22.51%), followed by V, with an accuracy of R^2 = 0.83, rRMSE = 18.51%, and the accuracy of DBH was the lowest (R^2 = 0.76, rRMSE = 9.87%). The results of the study provide an important reference for the estimation of single-tree characteristic parameters in urban forests based on UAV-Lidar.

Keywords: urban forest; UAV-Lidar; canopy volume; diameter at breast height (DBH); aboveground biomass (AGB); stem volume (V)

Citation: Zhang, B.; Li, X.; Du, H.; Zhou, G.; Mao, F.; Huang, Z.; Zhou, L.; Xuan, J.; Gong, Y.; Chen, C. Estimation of Urban Forest Characteristic Parameters Using UAV-Lidar Coupled with Canopy Volume. *Remote Sens.* **2022**, *14*, 6375. https://doi.org/10.3390/rs14246375

Academic Editor: Klaus Scipal

Received: 6 November 2022
Accepted: 13 December 2022
Published: 16 December 2022

1. Introduction

Urban forests are an important part of urban ecosystems, and they are the foundation and guarantee of urban sustainable development [1,2]. They can effectively reduce the urban heat island effect and improve air quality and other environmental conditions as well as ecosystem services [3,4]. The single tree is the basic unit of the forests [5]; their characteristic parameters such as tree height (H), diameter at breast height (DBH), crown width, aboveground biomass (AGB), and tree volume (V) can effectively reflect the growth status, spatial distribution, and structural characteristics of forest resources, which are important elements of forest resource investigation and reliable diversity indicators of forest succession stages [6], as well as being the focus of research on urban ecosystems and their functions [7].

Single trees in cities are highly fragmented and unevenly distributed in urbanized areas with high population densities and high concentrations of artificial landscapes [8], so it is time-consuming and labor-intensive to obtain information on the characteristic parameters of urban single trees through traditional forest resource investigation methods [9]. Remote sensing technology has rapidly developed and can be used to quickly and accurately obtain multiscale and multitemporal information on forest structure characteristics, effectively making up for the shortcomings of traditional forest resource monitoring methods and greatly improving the work efficiency [10]. However, passive remote sensing technology only provides spectral and textural information of the forest canopy surface, which is susceptible to atmospheric conditions and other factors, and it is difficult to obtain the three-dimensional structure of the vegetation canopy [11,12].

Light detection and ranging (LiDAR) is an active remote sensing technology that obtains the distance between a sensor and a target by calculating the time difference between the laser pulse emitted by the sensor and the received echo pulse. Because LiDAR has strong penetration into the forest, it can accurately obtain the three-dimensional structure information of forest tree height and canopy [13–15], thus realizing the leap from two-dimensional to three-dimensional forest canopy structure information, resulting in the emergence of canopy parameters extracted from lidar data as a hot research topic globally. Previous studies have shown that LiDAR-based estimation of diameter at breast height, biomass, stock volume, and forest distribution mapping can be effective [16–19]. For example, Cao et al. [20] used full-waveform unmanned aerial vehicle (UAV) lidar data to extract point cloud metrics and waveform metrics calculated based on voxel-based methods to estimate the single tree AGB of plantation forests in the coastal region of east China. The results indicated that full-waveform lidar data can effectively estimate the AGB of single trees. Liu et al. [21] used the constant allometric ratio model to estimate the forest single tree biomass based on UAV lidar data to obtain single tree canopy characteristic parameters (tree height, crown width, canopy projection area, and canopy volume) and achieved a good fit and high prediction accuracy. Qin et al. [22] used UAV lidar to estimate the subtropical single tree carbon stock in Shenzhen, southern China, and the results indicated that the height variable can explain the variation of tree carbon stock and estimate the single tree carbon stock well. Therefore, it is important to examine the application of UAV lidar for estimating single tree characteristic parameters in urban areas.

Forest canopy structure includes the horizontal and vertical directions of branches and leaves, canopy width and height, and canopy light transmission [23,24]. The canopy width and cross-sectional area can be used to measure the horizontal extension size of the canopy. The vertical structure of the canopy is mainly the spatial distribution and hierarchical characteristics of the forest vegetation, and the performance is comparatively complex [25–27]. To quantify the three-dimensional structure of the forest canopy, Lefsky et al. proposed a voxel-based canopy volume model (CVM) to characterize the differences in the volume and vertical spatial distribution of the canopy [28]. The basic principle of the CVM model is reflecting the spatial heterogeneity of the forest structure arising from the difference in the light environment within the canopy by dividing the canopy into two parts, the photosynthetically active zone and the inactive zone, to realize the spatial arrangement of elements within the canopy structure and the distinction of volume structure [29–32]. Therefore, the CVM model is an important method for obtaining the parameters of tree canopy structure.

In summary, the estimation of DBH, AGB, and stem volume of trees is an important element of urban forest resource monitoring and the most direct manifestation of the ecosystem functions of forests, and lidar is an advanced technical tool for detecting the three-dimensional structure of forests. Therefore, in this study, three urban tree species, Ginkgo (*Ginkgo biloba* L.), Cinnamomum camphora (*Cinnamomum camphora (Linn.) Presl*), and Metasequoia glyptostroboides (*Metasequoia glyptostroboides Hu et Cheng*) were used as examples. UAV-Lidar data were used to obtain the canopy point clouds of the three kinds of single trees, coupled with point clouds and canopy structure variables, and three methods, multiple linear regression (MLR), support vector regression (SVR) and random forest (RF), were used to establish models for estimating the DBH, AGB, and V of single trees in urban forests based on UAV-Lidar data. The model and estimation results were validated using ground-measured data. This provides an important technical tool for rapid and accurate monitoring of single tree parameters in urban forests.

2. Materials and Methods

2.1. Study Area

The study area is in Lin'an (29°56′ to 30°23′N latitude and 118°51′ to 119°52′E longitude) (Figure 1), Hangzhou City, Zhejiang Province. Lin'an belongs to the subtropical monsoon climate with warm and humid conditions, abundant light and rainfall, and four distinct seasons. The average annual temperature is 16.4 °C, the frost-free period is 237 days, the sunshine time is 1847.3 h, and the annual precipitation is 1613.9 mm. The area is dominated by hills and mountains, the terrain inclines from northwest to southeast, and the three-dimensional climate is obvious. The climax vegetation is subtropical evergreen broad-leaved forest, and the main tree species planted in the urban area of Lin'an City include *Metasequoia glyptostroboides* (*M. glyptostroboides*), *Ginkgo biloba* (*G. biloba*), *Cinnamomum camphora* (*C. camphora*), etc.

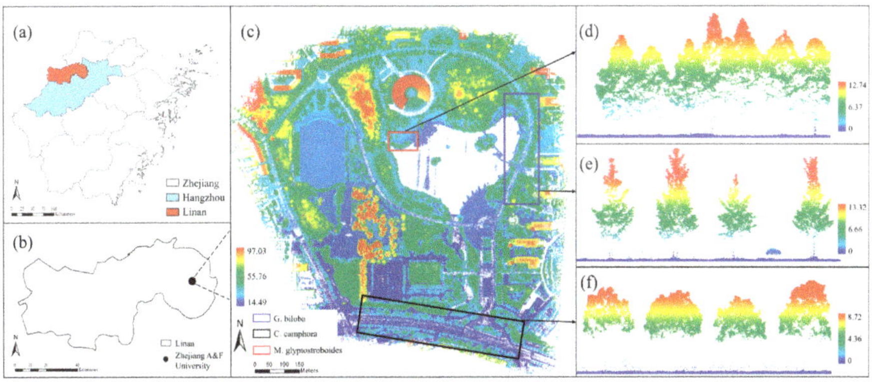

Figure 1. Overview of the study area: (**a**) location of Lin'an, (**b**) location of the study area, (**c**) UAV-Lidar point clouds of the study area, (**d**) Metasequoia glyptostroboides point cloud profile, (**e**) Ginkgo biloba point cloud profile, (**f**) Cinnamomum camphora point cloud profile.

2.2. Field Measurements

In July 2021, the DBH, tree height, crown height, and crown width of 64 stems of *G. biloba*, 74 stems of *C. camphora*, and 55 stems of *M. glyptostroboides* in the study area were measured in detail, as shown in Figure 1. The single trees were positioned with Huace Smart Real Time Kinematic. The AGB consisted of stem biomass (W_S), branch biomass (W_B), and foliage biomass (W_F). In this study, the biomass of each component was calculated according to the biomass allometric equations of different tree species and summed to obtain the AGB of a single tree [33–35], as shown in Table 1. The stem volume was calculated according to the single-entry stem volume table of Zhejiang Province and the measured single tree DBH data [36]. Table 2 shows the statistical characteristics of the parameters of the three tree species.

Table 1. Biomass allometric equations for each biomass component of the three tree species.

Tree Species	Biomass Components	Biomass Allometric Equations	Reference
G. biloba	Stem biomass (W_S) Branch biomass (W_B) Foliage biomass (W_F)	$\ln(W_S) = -3.84 + 0.95 \times \ln(DBH^2H)$ $\ln(W_B) = -9.38 + 1.46 \times \ln(DBH^2H)$ $\ln(W_F) = -6.95 + 1.03 \times \ln(DBH^2H)$	[34]
C. camphora	Stem biomass (W_S) Branch biomass (W_B) Foliage biomass (W_F)	$\ln(W_S) = -3.175 + 0.948 \times \ln(DBH^2H)$ $\ln(W_B) = -6.690 + 1.195 \times \ln(DBH^2H)$ $\ln(W_F) = -7.601 + 1.287 \times \ln(DBH^2H)$	[33]
M. glyptostroboides	Stem biomass (W_S) Branch biomass (W_B) Foliage biomass (W_F)	$W_S = 0.01749 \times (DBH^2H)^{0.9608}$ $W_B = 0.03037 \times (DBH^2H)^{0.7082}$ $W_F = 0.11079 \times (DBH^2H)^{0.4607}$	[35]

Note: H is tree height (m), DBH is diameter at breast height (cm).

Table 2. Summary of information on measured characteristics parameters of the three tree species.

Parameters	*G. biloba* (*n* = 64)			*C. camphora* (*n* = 74)			*M. glyptostroboides* (*n* = 55)		
	Range	Mean	SD	Range	Mean	SD	Range	Mean	SD
H/m	8–14.3	11.01	1.78	6.9–11	8.58	0.84	7.8–18.8	11.63	2.70
DBH/cm	14.8–23.9	18.76	2.29	16.3–29.3	22.88	2.65	11.8–33.9	19.94	5.60
AGB/kg	31.4–154.6	75.26	29.31	73.72–345.43	186.84	59.54	21.87–231.29	77.51	50.77
V/m^3	0.074–0.25	0.138	0.041	0.077–0.355	0.183	0.054	0.047–0.688	0.206	0.154

2.3. Lidar Data

The DJI Matrice 600 Pro six-rotor UAV with a lightweight Velodyne Puck LITE™ laser scanner was used to acquire the original lidar point clouds in the study area (Figure 2). The flight height of the UAV is 60 m above ground level, with a flight speed of 8 m/s, a route spacing of 25 m, and a lateral overlap rate of data sampling of 50%. The sensor records the first echo information of the laser pulse with a wavelength of 903 nm, a maximum scanning angle of ±15°, a scanning frequency of 20 Hz, and a scanning speed of 300,000 points/s. The final average point cloud density obtained is approximately 230 points/m^2.

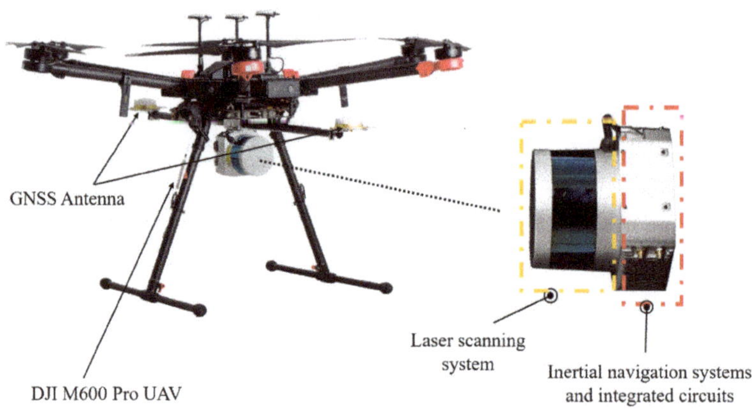

Figure 2. UAV and LiDAR system.

2.3.1. Lidar Data Preprocessing

The original lidar point cloud data were denoised using the height thresholding method, and the point cloud data after noise removal were filtered and separated into ground points and nonground points. First, the ground points were extracted by filtering with the improved progressive TIN densification algorithm [37]. Then, the height average of the laser points within a cell was calculated via the inverse distance weighting method to obtain a digital elevation model (DEM) with a spatial resolution of 0.5 m. Finally, the DEM was used to normalize the point cloud data to obtain the normalized point cloud data. In this study, the point cloud segmentation algorithm was used to segment individual trees based on normalized point cloud data [38,39]. This algorithm identified single trees via region growing combined with thresholding, and then identified the top of the tree to determine the distance between the surrounding points and the vertex, and expanded the region to segment the first tree. Successive iterations were made on this basis until all trees were segmented. The characteristic variables of a single tree were extracted based on the segmented single tree point cloud.

2.3.2. Lidar Metrics

The characteristic variables based on the lidar data can be used to estimate the forest characteristic parameters, and the point cloud characteristics extracted from the first returns have a remarkable correlation with the height, which is more suitable for estimating the forest characteristic parameters [13,40]. Of course, to reduce the influence of low ground vegetation on the data, the data after filtering the point clouds below 2 m were used as the crown point clouds, and characteristic variables were extracted from the first returns of the lidar point cloud [41,42]. In this study, the lidar data characteristic variables included: height-based metrics (HB) describing the parameters related to the lidar point cloud height; density-based metrics (DB) describing the canopy return density variable, which is the ratio of the number of height percentile point clouds to the total number of point clouds; the canopy area (S), which is the projected area of the canopy point cloud calculated based on the two-dimensional convex packet algorithm; the crown diameter (CD), which is the average of the east–west and north–south crown diameters of the point clouds. The canopy volume variables include OG, CG, EV, and OV. The metrics and descriptions are shown in Table 3.

2.3.3. Calculation of the Volume of Single Tree Canopy

Urban trees are frequently pruned and truncated, [43], making the crown of the pruned single tree change, often with a special crown shape. Therefore, canopy volume was extracted as a metric for single tree characteristic parameter estimation, as shown in Table 3. As shown in Figure 3, in this study, the voxel-based canopy volume method was used to calculate the canopy volume metrics for lidar point clouds of tree crowns:

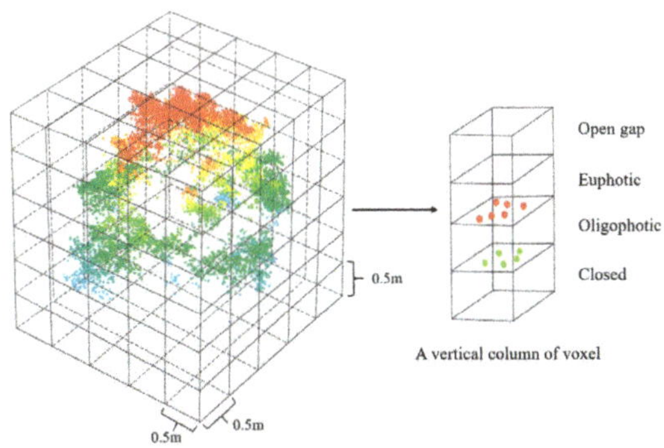

Figure 3. Illustration of the voxel-based canopy volume model.

The space where the canopy point clouds was located was divided into 0.5 × 0.5 × 0.5 m voxels [44], which were divided into vertical columns, and each column was further layered into four canopy structures. First, each voxel was classified as "filled" or "empty" according to whether there was a point cloud in the voxel, that is, the volume in unit area (m^3/m^2).

Table 3. Description of metrics derived from lidar data.

Metrics		Description	Reference
Height-based metrics(HB)	Height percentiles (H_5, H_{25}, H_{50}, H_{75}, H_{95}, and H_{99})	The percentiles of the canopy height distribution (5th, 25th, 50th, 75th, 95th, and 99th) of first returns	[25,45,46]
	The coefficient of variation of height (Hcv)	The coefficient of variation of heights of all first returns	
	Maximum height (Hmax)	Maximum height above ground of all first returns	
	Variance of height (Hva)	The variation in heights of all first returns	
	Standard deviation of height (Hstd)	The standard deviation of heights of all first returns	
	Median height (Hmed)	Median height above ground of all first returns	
	Mean height (Hmean)	Mean height above ground of all first returns	
	Interquartile distance of height (H_{IQ})	The interquartile distance of height of all first returns	
	Root mean square of height (Hsq)	The root mean square of height of all first returns	
	Cube mean of height (Hcm)	The cube mean of height of all first returns	
Density-based metrics(DB)	Canopy return density (D_3, D_5, D_7, D_9)	The proportion of points above the quantiles (30th, 50th, 70th and 90th) to total number of points	[47]
Canopy structure metrics(CS)	Canopy projection area (S)	Canopy projection area calculated using two-dimensional convex hull algorithm	[21]
	Crown diameter (CD)	Average diameter of crown point cloud $\frac{(X_{max}-X_{min})+(Y_{max}-Y_{min})}{2}$	
	Open gap volume (OG) and closed gap volume (CG) of CVM	The volume of empty voxels located above and below the filled canopy, respectively	[32]
	Euphotic volume (EV) and oligophotic volume (OV) of CVM	The volume of filled voxels located 65% above and 35% below of all filled grid cells of that column	

Then, according to the distribution of the filled position, the upper 65% of the filled zone was defined as "euphotic", in every column of voxels, and the remaining 35% was defined as "oligophotic". According to the spatial distribution and location of the empty voxel, in each voxel column, the empty voxels between the top of the canopy and the first filled voxels were defined as the "open gap", and the empty voxels between the filled voxels and the ground were called the "closed gap". The three-dimensional canopy volume distribution was converted into a two-dimensional canopy volume profile (CVP) according to the percentage of the volume of the four classified canopy volume characteristics in each height interval. The canopy volume distribution indicated the distribution of elements arranged in the vertical spatial extent of the canopy [30].

2.4. Model Construction Methods and Scheme

In this study, three modeling methods, MLR, SVR and RF, were used to construct the estimation models of urban single tree characteristic parameters based on the obtained lidar data characteristic variables. To study the influence of canopy structure on the accuracy of single tree characteristic parameters, the models were constructed in two schemes. The model excluding canopy structure variables is referred to as "Scheme 1", and the model including canopy structure variables is referred to "Scheme 2".

2.4.1. MLR Model

MLR is the most commonly used parameterization method for estimating forest characteristic parameters from remote sensing information, and can quickly establish a linear relationship between two or more independent variables and dependent variables to achieve parameter estimation. The MLR is generally expressed as follows:

$$Y = a_0 + a_1x_1 + a_2x_2 + \ldots + a_nx_n \quad (1)$$

where a_0 is the constant term; $a_1, a_2, \ldots, a_n$ are regression coefficients representing the degree of contribution of the respective variables to the dependent variable; and $x_1, x_2, \ldots, x_n$ are the independent variables, which are the characteristic variables shown in Table 3. Y is the dependent variable, which is the estimated characteristic parameter of this study. In this study, all possible combinations of variables were evaluated using "all subsets" regression, and the best combination of variables was selected to build the MLR model to estimate the three characteristic parameters [48].

2.4.2. SVR Model

SVR is a machine learning model that uses support vector machines to perform regression analysis [49,50]. It applies classification methods to solve regression problems with finite samples, mainly based on a given sample data set, by seeking a function to fit all sample points so that the total variance of sample points from the hyperplane is minimized [51]. SVR transforms the nonlinear problem into a linear problem in high-dimensional space via kernel functions for nonlinear separable samples in low-dimensional input space, replacing the inner product operation in high-dimensional space, and ensuring good generalization ability [52]. Therefore, the SVR model has high accuracy, good ability to handle high-dimensional and small sample data, good generalization ability, and robustness. In this study, four kernel function models, including linear, polynomial, radial basis function (RBF), and multilayer perceptron (Sigmoid), were used, and the best penalty coefficient (C) with gamma value (g) was selected via grid search cross-validation to construct the SVR model to estimate the three characteristic parameters.

2.4.3. RF Model

The random forest algorithm is another commonly used machine learning method. The algorithm is based on modified nonparametric modeling of decision trees [53], and constructs a decision tree by bootstrapping from the original sample set with put-back randomly selected N samples to predict the results. The RF algorithm has good noise resistance and can handle high-dimensional data with relatively high prediction accuracy. An unbiased estimate of the error can be generated during the RF calculation, and the importance of each variable involved in the model can be evaluated. There are three important parameters in the estimation of single tree characteristic parameters using the RF algorithm: Mtry is the number of variables used randomly at the nodes of each tree, and Ntree is the number of regression trees in the RF. Nodesize is the minimum number of terminal nodes in the regression analysis, and the default value is 5 [54,55]. In this study, three characteristic parameters estimation models are developed based on the optimization of RF parameters.

2.5. The Flow Chart and Accuracy Validation

The flow chart of this study is shown in Figure 4. First, field measurements, lidar data processing, and characteristic variable extraction were conducted. Second, two-thirds of all measured samples were selected into training samples, and one-third were divided into test samples. Finally, three modeling methods, MLR, SVR and RF, were used to construct single tree characteristic estimation models according to Scheme 1 and Scheme 2, and the accuracy of the models was evaluated.

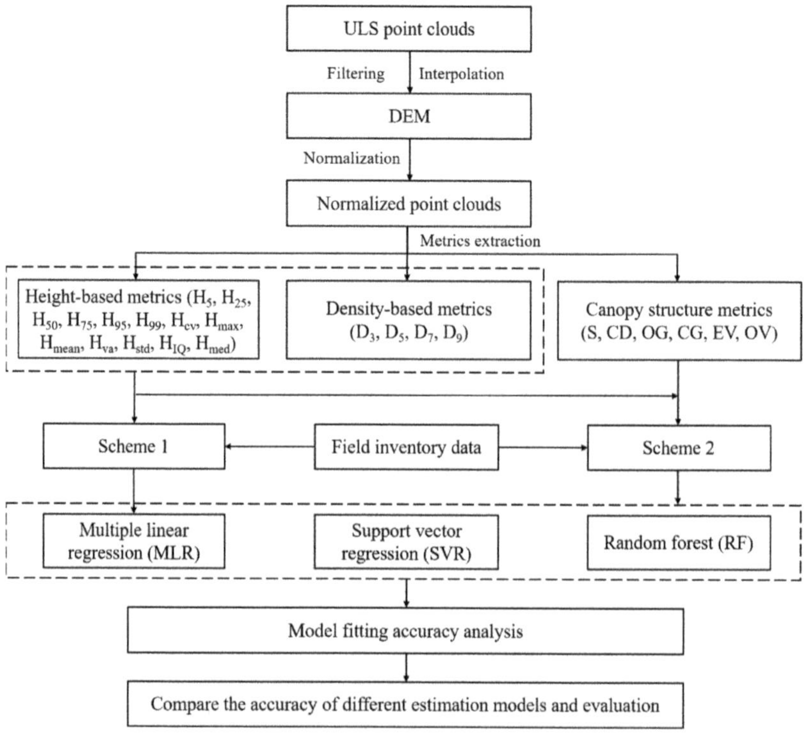

Figure 4. The flow chart of this study.

The model accuracy evaluation metrics include the determination coefficient (R^2), root mean square error (RMSE), and relative root mean square error (rRMSE). Generally, higher values of R^2 and lower values of RMSE and rRMSE indicate better performance of the model. R^2, RMSE, and rRMSE are calculated as follows:

$$R^2 = 1 - \frac{\sum_{i=1}^{n}(\hat{x}_i - x_i)^2}{\sum_{i=1}^{n}(x_i - \overline{x})^2} \quad (2)$$

$$\text{RMSE} = \sqrt{\frac{1}{n}\sum_{i=1}^{n}(x_i - \hat{x}_i)^2} \quad (3)$$

$$\text{rRMSE} = \frac{\text{RMSE}}{\overline{x}} \times 100\% \quad (4)$$

where n is the number of samples, $\overline{x}$ is the measured value of the sample canopy characteristic parameters, $\overline{x}$ is the mean value of the sample canopy characteristic parameters, and $\hat{x}_i$ is the predicted value of the sample canopy characteristic parameters.

3. Results

3.1. Canopy Volume and Profile Analysis

The canopy volume profile can directly show the spatial heterogeneity of the forest canopy structure and the distribution and change in elements in the vertical direction. Comparing the canopy volumes and profiles of the three species, the euphotic volume was significantly larger than the oligophotic volume in the filled volume of the *G. biloba* canopy (Figure 5); the closed gap volume was larger than the open gap volume in the empty volume, and the closed gap volume occupied the largest volume. The canopy distribution of *C. camphora* was consistent with that of *G. biloba* (Figure 6), and the euphotic volume was greater than the oligophotic volume; the closed gap volume was larger than the open gap volume in the empty volume, and the closed gap volume occupied the largest volume. In the canopy distribution of *M. glyptostroboides* (Figure 7), the euphotic volume was close to the oligophotic volume; the closed gap volume was slightly larger than the open gap volume in the empty volume.

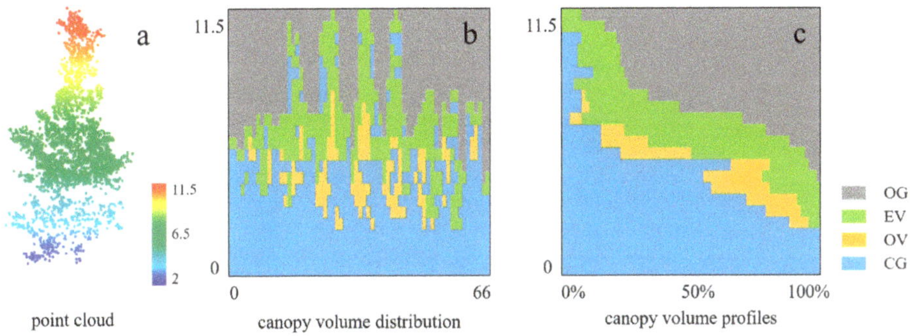

Figure 5. (**a**) *G. biloba* point cloud, (**b**) canopy volume distribution, which shows the distribution of canopy structure classes after the expansion of all columns in the panel; and (**c**) canopy volume profile, which shows the volume percentage of each class of total volume in each height interval.

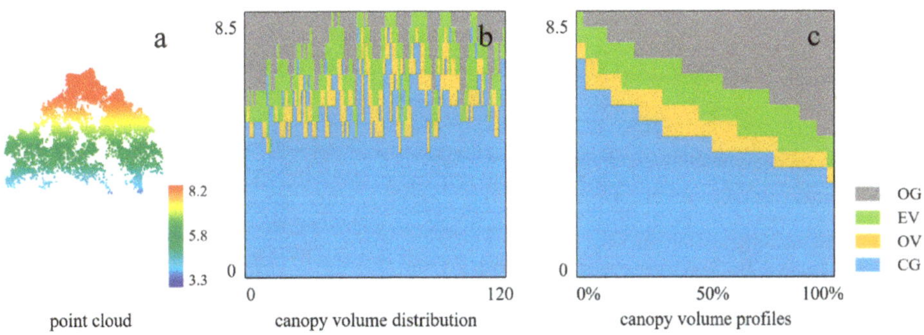

Figure 6. (**a**) *C. camphora* point cloud, (**b**) canopy volume distribution, which shows the distribution of canopy structure classes after the expansion of all columns in the panel, and (**c**) canopy volume profile, which shows the volume percentage of each class of total volume in each height interval.

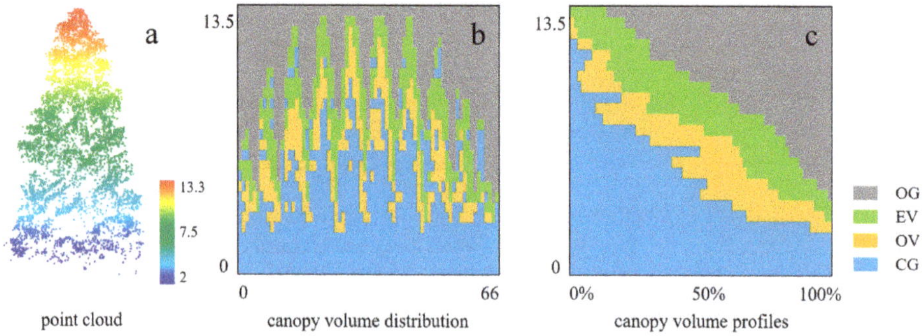

Figure 7. (**a**) *M. glyptostroboides* point cloud, (**b**) canopy volume distribution, which shows the distribution of canopy structure classes after the expansion of all columns in the panel; and (**c**) canopy volume profile, which shows the volume percentage of each class of total volume in each height interval.

3.2. Variable Importance Analysis

In this study, the point cloud characteristic variables extracted from lidar data were run 100 times using the RF model according to two modeling schemes. Figures 8 and 9 show the importance scores of Scheme 1 and Scheme 2 input variables, respectively. As shown in Figure 8, the importance score of the height characteristic variable in Scheme 1 is significantly higher than that of the density characteristic variable. Among the three single tree characteristic parameter models, Hcv was the largest in DBH and AGB, with values of 27.89 and 44.05, respectively; Hsq was the largest in V, with a value of 21.12. As shown in Figure 9, the three parameters with the highest importance in Scheme 2 were all canopy structure variables. In the DBH and V models, CG had the highest influence, with 31.96 and 28.97, respectively, and in the AGB model, CD had the highest influence, with 40.57. The results show that the canopy structure variable was significantly more important than other variables and was the key variable in the estimation of single tree structure parameters.

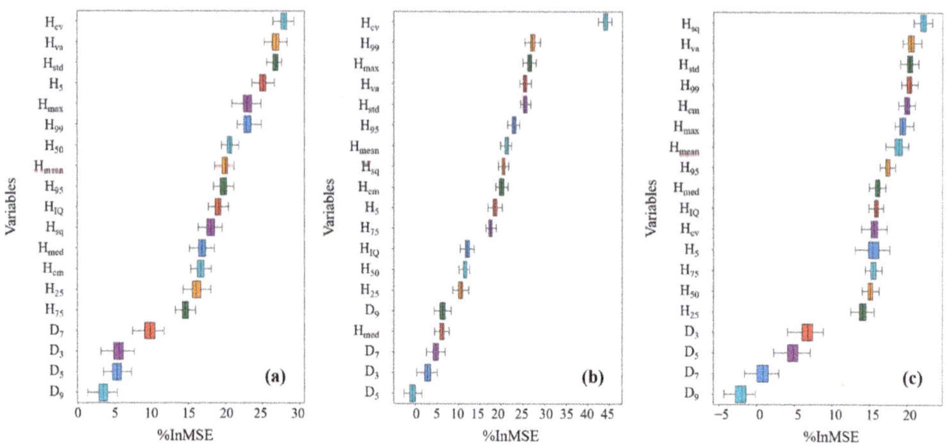

Figure 8. Importance of input variables based on Scheme 1 (%InMSE: the percentage increase in the mean square error, ((**a**): DBH, (**b**): AGB, (**c**): V)).

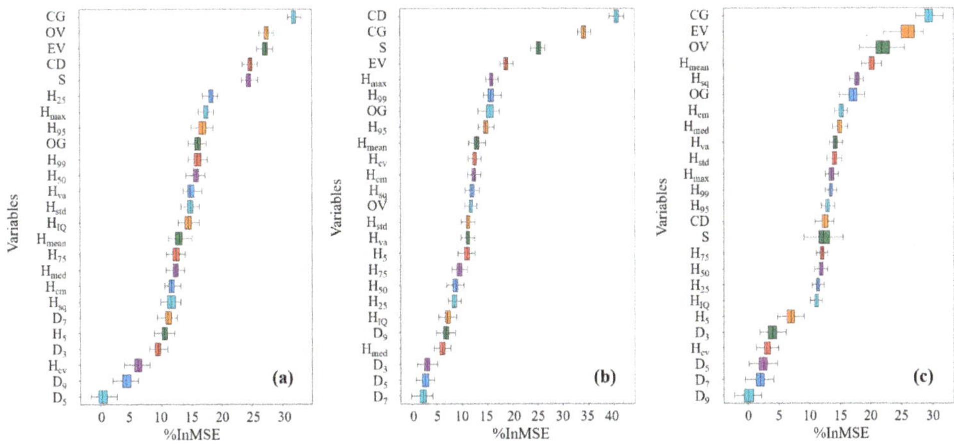

Figure 9. Importance of input variables based on Scheme 2 (%InMSE: the percentage increase in the mean square error, ((**a**): DBH, (**b**): AGB, (**c**): V)).

Figure 10 shows the path coefficients (absolute values) calculated using the structural equations for the direct effect of the three characteristic variables of height characteristics, density characteristics, and canopy structure on the characteristic parameters. As shown in Figure 10, CS had the largest direct effect on DBH and AGB, while HD had the largest direct effect on V, and DB had the smallest effect on characteristic parameters. For DBH, the path coefficient values, in order, were as follows: CS (0.647) > HD (0.318) > DB (−0.095). For AGB, the path coefficient values were as follows: CS (0.871) > HD (0.116) > DB (−0.101). For V, the path coefficient values were as follows: HD (0.593) > CS (0.389) > DB (0.038). The canopy structure variables had important influence on the estimation of the three single tree characteristics parameters.

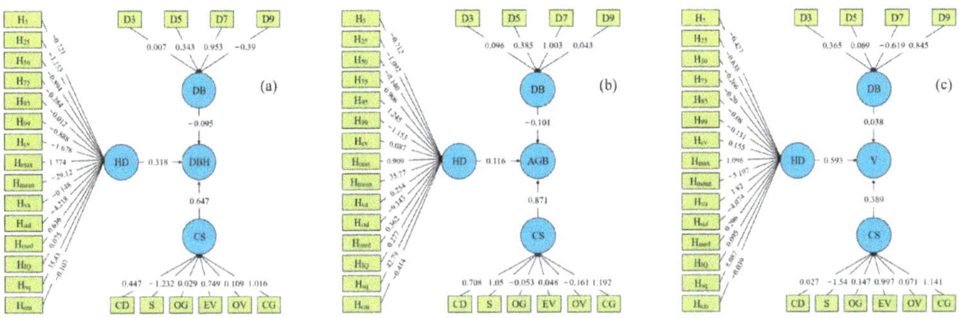

Figure 10. Relative importance of input variables to characteristic parameters based on Scheme 2 ((**a**): DBH, (**b**): AGB, (**c**): V).

3.3. Model Construction and Evaluation

3.3.1. MLR Estimation Results

To evaluate the accuracy of the established prediction models of single tree characteristic parameters, three independent variables were selected to construct the MLR models in this study (Table 4). Among all the metrics selected in the MLR model of Scheme 1, the height percentiles (H_{cv}, H_{mean}, H_{sq}, H_{std}, H_{99}, and Hvar) were frequently selected by the models. Among all the metrics selected by the MLR model in Scheme 2, height percentile (H_{cm}, H_{95}, H_{cv} and H_{va}r), canopy structure variables (CD, CG, EV) were frequently selected by the models. The accuracy value R^2 of the models was improved after coupling the canopy structure variables. The canopy structure variables played an important role in the construction of the model, indicating that these variables are sensitive in estimating forest structure.

Table 4. The MLR prediction models and their accuracy assessment under different schemes.

Scheme	Parameters	Equations	R^2
Scheme 1	DBH	$36.2 - 108.85 \times H_{cv} - 56.78 \times H_{mean} - 56.65 \times H_{sq}$	0.76
	AGB	$-63.11 - 183.5 \times H_{mean} + 193 \times H_{sq} - 56.92 \times H_{std}$	0.78
	V	$0.11 + 0.04 \times H_{99} - 0.36 \times H_{std} + 0.08 \times H_{var}$	0.83
Scheme 2	DBH	$5.18 + 1.32 \times H_{cm} + 0.09 \times EV + 0.61 \times CD$	0.78
	AGB	$-101.75 + 10.44 \times H_{95} + 0.201 \times CG + 22.71 \times CD$	0.85
	V	$0.11 - 0.51 \times H_{cv} + 0.04 \times H_{var} + 5.7 \times 10 - 5 \times CG$	0.84

Figures 11 and 12 show the correlations between the predicted and field-measured values of characteristic parameters estimated using the MLR models of Scheme 1 and Scheme 2, respectively. As shown in Figure 11, the R^2 values of the model training accuracy for single tree DBH, AGB, and V in Scheme 1 were 0.76, 0.78, and 0.83, respectively, with rRMSE = 9.67 to 27.23%; the R^2 values of the model testing accuracy were 0.64, 0.69, and 0.62, respectively, with rRMSE = 11.02 to 39.01%. As shown in Figure 12, the training and testing accuracy of Scheme 2 single tree characteristic parameters were improved. The R^2 values of the model training accuracy of DBH, AGB, and V were 0.78, 0.85, and 0.84, respectively, with rRMSE = 9.33 to 30.82%. The R^2 values of the model testing accuracy were 0.69, 0.82, and 0.80, respectively, with rRMSE = 10.41–24.67%. The comparison of the results shows that the R^2 values of the estimation accuracy of the characteristic parameters of Scheme 2 were all improved after adding the canopy structure variable. Among them, the estimation accuracy of V improved by 29.0%, which was the greatest improvement, followed by AGB with an accuracy improvement of 18.8% and DBH with an accuracy improvement of 7.8%. In addition, the rRMSE values of the AGB and V estimation results also decreased by 36.8% and 27.7%, respectively, which were large decreases.

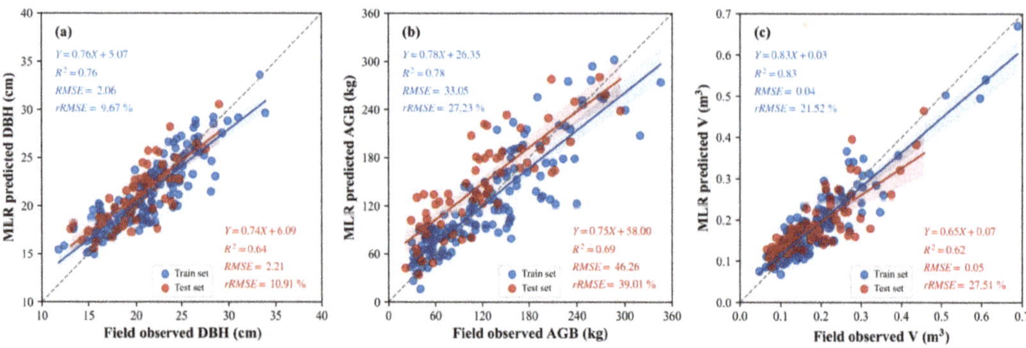

Figure 11. Correlation between measured and estimated values of single tree DBH (**a**), AGB (**b**), and V (**c**) of the MLR model based on Scheme 1.

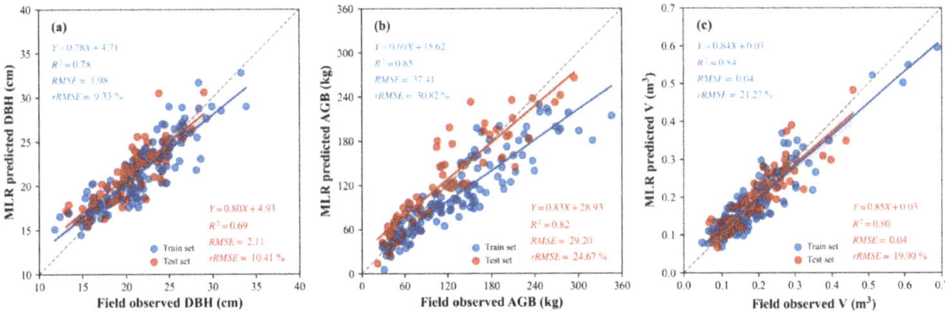

Figure 12. Correlation between measured and estimated values of single tree DBH (**a**), AGB (**b**), and V (**c**) of the MLR model based on Scheme 2.

3.3.2. SVR Estimation Results

Figure 13 shows the comparison of the R^2 values of the estimated characteristic parameters for the four kernel function SVR models under the two modeling schemes. As shown in Figure 13, the R^2 values of the single tree characteristic parameters in both Scheme 1 and Scheme 2 were the highest for the SVR model with RBF as the kernel function. Therefore, the RBF kernel function was chosen to construct the SVR model in this study.

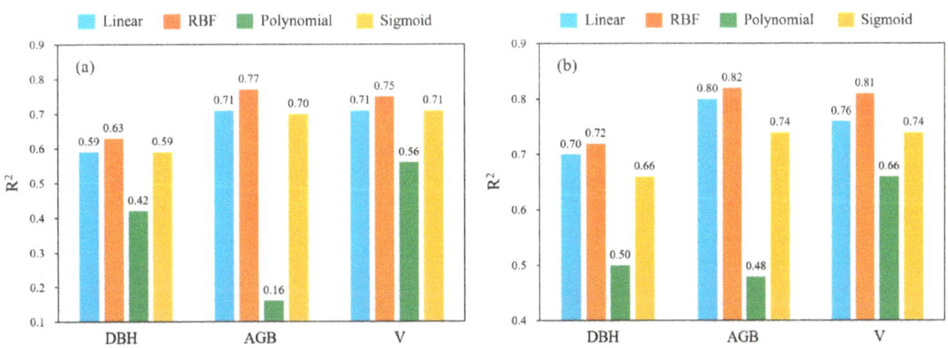

Figure 13. Results of single tree characteristic parameters using different kernel functions ((**a**): Scheme 1, (**b**): Scheme 2).

Figures 14 and 15 show the correlations between the predicted and field-measured values of characteristic parameters estimated using the SVR models of two modeling schemes, respectively. As shown in Figure 14, the R^2 values of the model training accuracy for single tree DBH, AGB, and V in Scheme 1 were 0.83, 0.84, and 0.85, respectively, with rRMSE = 8.14 to 24.10%; the R^2 values of the model testing accuracy were 0.67, 0.77, and 0.75, respectively, with rRMSE = 10.76 to 26.78%. As shown in Figure 15, the R^2 values of the model testing accuracy of DBH, AGB, and V were 0.72, 0.82, and 0.81, respectively, with rRMSE = 10.24 to 23.37%. The comparison of the results shows that the R^2 values of the training and testing accuracy of single tree characteristic parameters of the SVR model were all improved by adding canopy structure variables in Scheme 2. Among them, the estimation accuracy of V improved by 8%, which was the greatest improvement, followed by DBH with an accuracy improvement of 7.4% and AGB with an accuracy improvement of 6.5%. In addition, the rRMSE values of the AGB and V estimation results also decreased by 12.7% and 11.6%, respectively, which were large decreases, indicating that coupling canopy parameters can improve the estimation accuracy of urban single tree parameters.

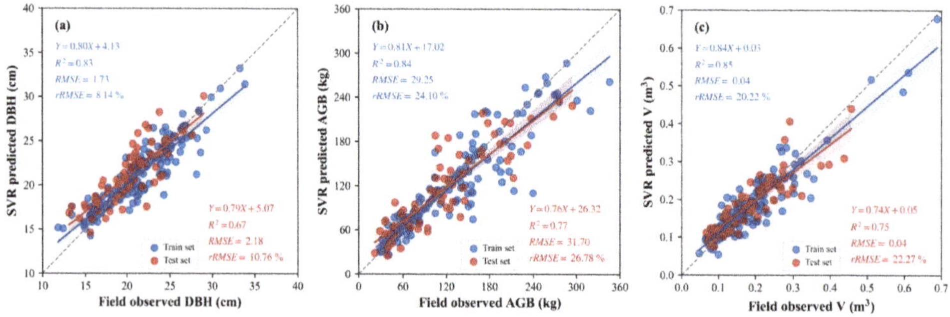

Figure 14. Correlation between measured and estimated values of single tree DBH (**a**), AGB (**b**), and V (**c**) of the SVR model based on Scheme 1.

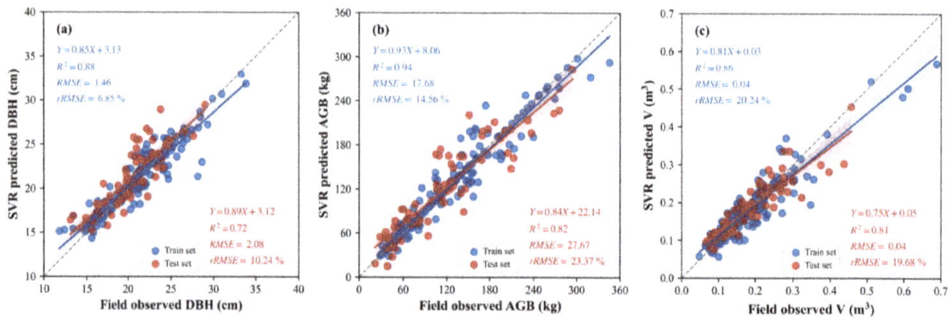

Figure 15. Correlation between measured and estimated values of single tree DBH (**a**), AGB (**b**), and V (**c**) of the SVR model based on Scheme 2.

3.3.3. RF Estimation Results

The training data were input into the random forest model to traverse all variable values and eventually obtain the optimal parameters. Figure 16 shows Mtry, which was used to determine the minimum variable for each tree in the RF model, and the minimum Mtry value was required when the model error was minimal. As shown in Figure 17, the RMSE of the model error tended to be stable after Ntree reached 1500. Therefore, the values of Ntree in the optimized random forest model were set to 1500 in this study. Table 5 lists the specific settings for different parameter values for the two schemes.

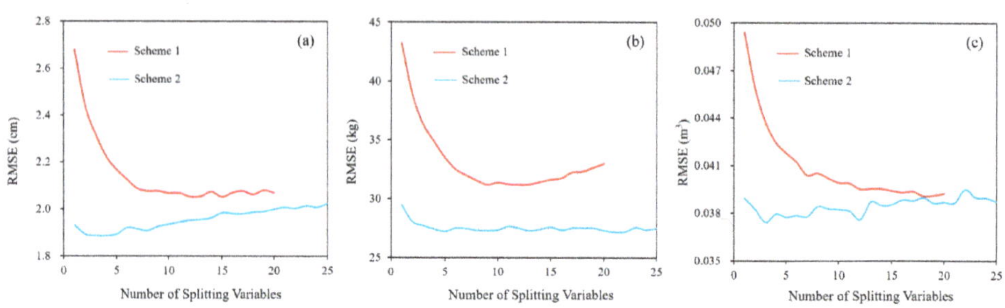

Figure 16. Influence of Mtry on model error ((**a**): DBH, (**b**): AGB, (**c**): V).

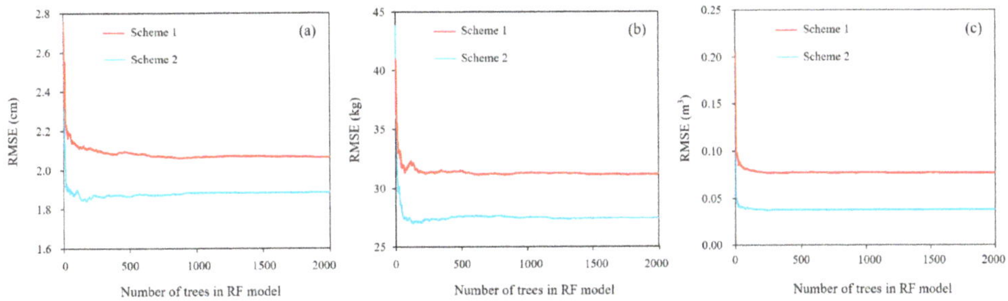

Figure 17. Influence of Ntree on model error ((**a**): DBH, (**b**): AGB, (**c**): V).

Table 5. Results of the optimization of model parameters for different schemes.

Scheme	Parameters	Nodesize	Mtry	Ntree	Number of Variables
Scheme 1	DBH	5	15	1500	19
	AGB	5	12	1500	19
	V	5	18	1500	19
Scheme 2	DBH	5	4	1500	25
	AGB	5	21	1500	25
	V	5	3	1500	25

Figures 18 and 19 show the correlations between the predicted and field-measured values of characteristic parameters estimated using the RF models of Scheme 1 and Scheme 2, respectively. As shown in Figure 18, the R^2 values of the model training accuracy for single tree DBH, AGB, and V in Scheme 1 were 0.74, 0.80, and 0.81, respectively, with rRMSE = 10.46 to 26.36%; the R^2 values of the model testing accuracy were 0.67, 0.74, and 0.76, respectively, with rRMSE = 10.95 to 29.69%. As shown in Figure 19, the training and testing accuracy of Scheme 2 single tree characteristic parameters were improved. The R^2 values of the model training accuracy of DBH, AGB, and V were 0.80, 0.89, and 0.86, respectively, with rRMSE = 8.9 to 20.77%. The R^2 values of the model testing accuracy were 0.76, 0.85, and 0.83, respectively, with rRMSE = 9.87–22.51%. Comparing the RF model results, the R^2 values of the estimation accuracy of the characteristic parameters of Scheme 2 were all improved after adding the canopy structure variable. Among them, the estimation accuracy of AGB improved by 14.9%, which was the greatest improvement, followed by DBH with an accuracy improvement of 13.4% and V with an accuracy improvement of 9.2%. In addition, the rRMSE values of the AGB and V estimation results also decreased by 24.2% and 15.2%, respectively, which were large decreases.

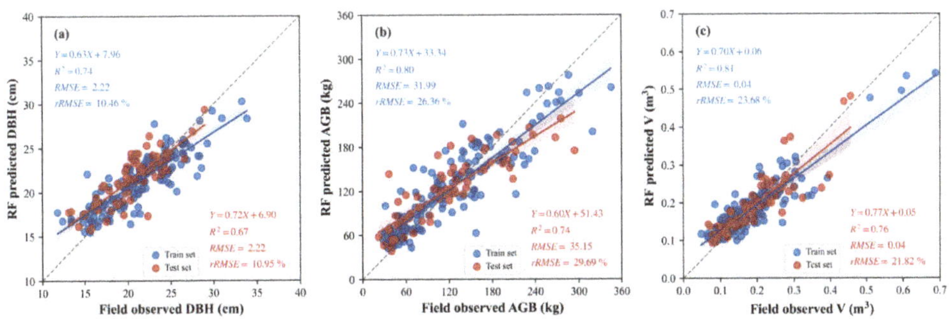

Figure 18. Correlation between measured and estimated values of single tree DBH (**a**), AGB (**b**), and V (**c**) of the RF model based on Scheme 1.

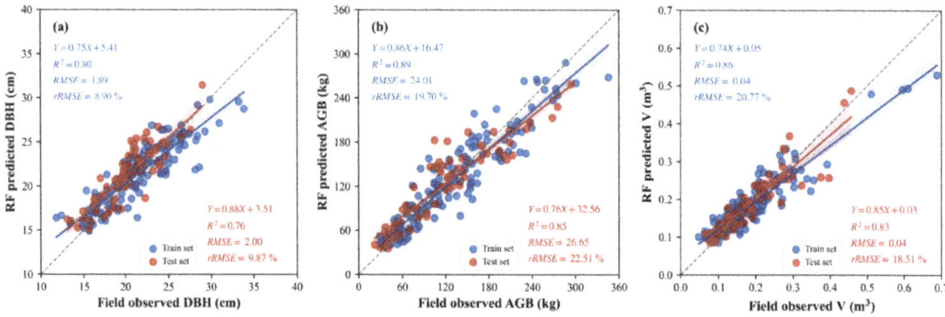

Figure 19. Correlation between measured and estimated values of single tree DBH (**a**), AGB (**b**), and V (**c**) of the RF model based on Scheme 2.

3.4. Comparison of Model Results

Table 6 shows the summary of the field measured data and the estimation results of different model characteristic parameters. Compared with the field-measured data, the estimated CV values of MLR, SVR, and RF models range from 16.6 to 18.11%, the estimated CVs of AGB range from 49.15 to 54.22%, and the estimated CVs of V range from 44.67 to 46.55%, with smaller variations and mean values closer to the measured values. Appendix A (Table A1) shows the comparison of the model training accuracy and testing accuracy of three models for estimating three single tree parameters in two schemes. An analysis of the results in Appendix A (Table A1) shows that all models of the two schemes achieved higher accuracy estimation of single tree parameters in urban forests. Figure 20 shows the distribution of the normalized residuals for the testing phase of the characteristic parameters for different models. The normalized residuals of the characteristic parameters of the testing samples of the three models were in the range of −2 to 2, indicating that all models had good stability and reliability in predicting the characteristic parameters of a single tree.

Table 6. Summary of information on measured and predicted characteristic parameters of the samples.

Method	DBH				AGB				V			
	Min	Max	Mean	CV(%)	Min	Max	Mean	CV(%)	Min	Max	Mean	CV(%)
Measured	11.80	33.90	20.97	19.00	21.87	345.43	120.46	57.49	0.047	0.688	0.189	50.79
MLR	14.46	32.79	21.31	16.86	10.08	283.90	123.09	51.85	0.067	0.590	0.190	46.22
SVR	13.82	33.54	21.13	18.11	16.20	297.22	121.77	54.22	0.078	0.613	0.188	46.55
RF	14.73	30.39	21.28	16.60	36.99	286.66	121.73	49.15	0.08	0.536	0.190	44.67

However, for the two schemes, the model accuracy and testing accuracy of the single tree parameters for all models of Scheme 2 were improved, and the error was decreased compared with Scheme 1, indicating that the coupled canopy parameters could improve the estimation accuracy of the urban single tree parameter. In addition, from the three models, the performance of the two machine learning models was better than that of the MLR model, where the training accuracy of single tree characteristic parameters of the SVR model was slightly higher than that of the RF model, but the testing accuracy of single tree characteristic parameters of the RF model was better than that of the SVR model.

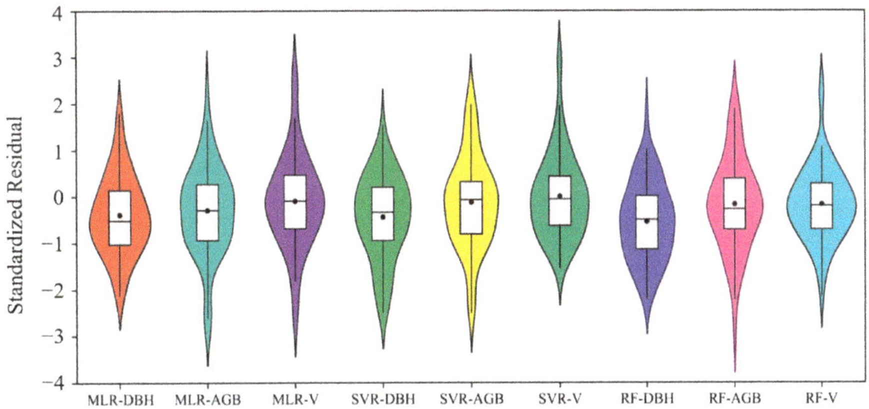

Figure 20. Distribution of normalized residuals for the testing phase of characteristic parameters for different models.

4. Discussion

The forest canopy is a key component of forests that affects ecosystem processes and functions [56], and the quantification and analysis of canopy distribution is one of the methods used to characterize the spatial structure of forests. In this study, voxel-based canopy volume was used to characterize the canopy spatial structure of different urban trees, and the canopy volume profile was derived to more intuitively reflect the spatial heterogeneity of the canopy structure and the variation in the arrangement of elements within the canopy. As shown in Figures 5–7, the open gap volume of *G. biloba* was the largest, while the closed gap volume of *C. camphora* was larger than that of both *G. biloba* and *M. glyptostroboides*. First, this had a direct influence on the structure and shape of the canopy, which was relatively regular for *C. camphora* and *M. glyptostroboides*, while the canopy shape of *G. biloba* was more complex and the distribution of branches and leaves was more dispersed, resulting in a larger volume share in the open gap of *G. biloba*. In addition, closed gap volume was also strongly related to proper pruning in cities, which causes higher crown base height [43], thus leading to a larger percentage of closed gap volume. On the other hand, coniferous trees allow more light penetration into the lower canopy compared to broadleaved trees [57], and *M. glyptostroboides* belongs to the coniferous species, thus allowing *M. glyptostroboides* to form more areas of oligophotic zone.

Lidar data can provide parameter information directly connected to forest canopy structure, and the estimation of forest characteristic parameters using different regression methods can produce satisfactory prediction results [58]. The results of this study show that the estimation accuracy of single tree characteristic parameters with coupled canopy structural variables was improved compared to using only height and density characteristic variables. The model accuracy was higher than the prediction accuracy of stand stock and aboveground biomass in urban broadleaf forest areas estimated using ALS data [59]. This is due to the fact that high-density ULS data had richer canopy structure information. Therefore, high-density lidar data were more advantageous in estimating stand volume and aboveground biomass.

This research shows that the performance of the SVR and RF machine learning models was better than that of the MLR model. As a statistical regression model, MLR is not suitable for representing the complexity of high data and is sensitive to noise, and the MLR model is often prone to underfitting, making the model performance poor. SVR seeks linear regression hyperplanes and solves nonlinear problems in low-dimensional spaces by mapping kernel functions from low to high dimensions. Although the SVR model fits the training data well, there may be overfitting of the model. In addition, the SVR model also needs to find the optimal penalty coefficient and gamma value to obtain the optimal model [60,61]. The RF was able to handle high-dimensional data and had good noise resistance. During the model operations, unbiased estimates of errors can be obtained, and the importance of each variable can be evaluated in the RF model [62]. The accuracy of training the single tree characteristic parameters of the SVR model in this study was slightly higher than that of the RF model, but the validation accuracy was not as good as that of the RF model in the study, which may be caused by this reason.

Related studies around the world also indicate that RF has good predictive ability in forest parameter estimation [58,60,63]. For example, Zhou et al. [64] used a RF model to estimate the AGB of urban single trees based on UAV lidar data and achieved high estimation accuracy. Zhang et al. [65] combined lidar and high-resolution remote sensing images by comparing different models (SLR, LNN, BPNN, SVR, RF) for the quantitative estimation and inversion of biomass, and the results indicated that the RF model had the highest fitting accuracy. Cao et al. [63] indicated that the accuracy of the RF model was higher than that with SVR, backpropagation neural networks, k-nearest neighbor, and the generalized linear mixed model in the remote sensing estimation of forest biomass based on satellite remote sensing. Peng Xi et al. [66] established different models for estimating the s characteristic parameters of tropical forests in China based on UAV lidar data, and the study indicated that the RF model had good accuracy in estimating forest characteristic parameters. Although this study provides a reference for the application of UAV lidar in urban forest characteristic parameter estimation, there are still some limitations. The urban forest is unevenly distributed, and the estimation of large-scale urban forest characteristic parameters using UAV lidar is still a challenge.

5. Conclusions

In this study, we used UAV lidar to obtain three kinds of single tree canopy point clouds coupled with point cloud and canopy structure variables. MLR, SVR, and RF models were used to estimate the characteristic parameters of DBH, AGB, and V of single trees in urban forests based on UAV-Lidar data. The results indicate that canopy volume profiles can visualize the spatial heterogeneity of forest canopy structure and the distribution and variation of elements in the vertical direction, and canopy structure variables such as CG, OV, EV, S, and CD had important effects on single tree characteristic parameters. The model training accuracy and testing accuracy of the single tree parameters of the MLR, SVR, and RF models were improved by incorporating canopy structure variables. In comparison, the two machine learning models, SVR and RF, outperformed MLR, but the testing accuracy of single tree characteristic parameters of the RF model was better than that of the SVR model. The results of the study provide an important reference for the estimation of single tree characteristic parameters in urban forests based on UAV-Lidar data, which is necessary and useful for urban managers to understand the functions and values of urban forests and to maximize the environmental benefits of urban forests.

Author Contributions: Conceptualization, H.D.; methodology, B.Z.; validation, B.Z.; formal analysis, B.Z. and Z.H.; investigation, B.Z., L.Z., J.X., Y.G. and C.C.; data curation, B.Z.; writing—original draft preparation, B.Z.; writing—review and editing, X.L., F.M., G.Z. and H.D.; visualization, B.Z.; supervision, H.D. All authors have read and agreed to the published version of the manuscript.

Funding: The authors gratefully acknowledge the support of the National Natural Science Foundation of China (U1809208, 32171785), the Key Research and Development Program of Zhejiang Province (2021C02005), and the State Key Laboratory of Subtropical Silviculture (No. ZY20180201).

Data Availability Statement: Not applicable.

Acknowledgments: The authors gratefully acknowledge the support of various foundations. The authors are grateful to the editor and anonymous reviewers whose comments have contributed to improving the quality of this manuscript.

Conflicts of Interest: The authors declare no conflict of interest.

Appendix A

Table A1. Accuracy evaluation of the parametric model of a single tree characteristic for different schemes.

	Model	Scheme	DBH			AGB			V		
			R^2	RMSE	rRMSE (%)	R^2	RMSE	rRMSE (%)	R^2	RMSE	rRMSE (%)
Train	MLR	Scheme 1	0.76	2.06	9.67	0.78	33.05	27.23	0.83	0.04	21.52
		Scheme 2	0.78	1.98	9.33	0.85	37.41	30.82	0.84	0.04	21.27
	SVR	Scheme 1	0.83	1.73	8.14	0.84	29.25	24.10	0.85	0.04	20.22
		Scheme 2	0.88	1.46	6.85	0.94	17.68	14.56	0.86	0.04	20.24
	RF	Scheme 1	0.74	2.22	10.46	0.80	31.99	26.36	0.81	0.04	23.68
		Scheme 2	0.80	1.89	8.90	0.89	24.01	19.70	0.86	0.04	20.77
Test	MLR	Scheme 1	0.64	2.21	10.91	0.69	46.26	39.01	0.62	0.05	27.51
		Scheme 2	0.69	2.11	10.41	0.82	29.20	24.67	0.80	0.04	19.90
	SVR	Scheme 1	0.67	2.18	10.76	0.77	31.70	26.78	0.75	0.04	22.27
		Scheme 2	0.72	2.08	10.24	0.82	27.67	23.37	0.81	0.04	19.68
	RF	Scheme 1	0.67	2.22	10.95	0.74	35.15	29.69	0.76	0.04	21.82
		Scheme 2	0.76	2.00	9.87	0.85	26.65	22.51	0.83	0.04	18.51

References

1. Baumeister, C.F.; Gerstenberg, T.; Plieninger, T.; Schraml, U. Exploring cultural ecosystem service hotspots: Linking multiple urban forest features with public participation mapping data. *Urban For. Urban Green.* **2020**, *48*, 126561. [CrossRef]
2. Cheng, W.; Chunju, C.; Kanghua, T. The Concept, Range and Researh Area of Urban Forest. *World For. Res.* **2004**, *17*, 23–27. [CrossRef]
3. Escobedo, F.J.; Nowak, D.J. Spatial heterogeneity and air pollution removal by an urban forest. *Landsc. Urban Plan* **2009**, *90*, 102–110. [CrossRef]
4. Liu, L.; Coops, N.C.; Aven, N.W.; Pang, Y. Mapping urban tree species using integrated airborne hyperspectral and LiDAR remote sensing data. *Remote Sens. Environ.* **2017**, *200*, 170–182. [CrossRef]
5. Zengyuan, L.; Qingwang, L.; Yong, P. Review on forest parameters inversion using LiDAR. *Natl. Remote Sens. Bull.* **2016**, *20*, 1138–1150.
6. Jones, M.S.D. Characterizing forest ecological structure using pulse types and heights of airborne laser scanning. *Remote Sens. Environ.* **2010**, *114*, 1069–1076.
7. Li, X.; Liu, J.; Jiang, S.; Jia, B. Analysis of the urban tree canopy and community structure of hospitals in urban areas of Beijing. *Acta Ecol. Sin.* **2019**, *39*, 12.
8. Wang, C.; Peng, Z.; Tao, K. Characteristics and development of urban forest in China. *Chin. J. Ecol.* **2004**, *23*, 5.
9. Gao, L.; Chai, G.; Zhang, X. Above-Ground Biomass Estimation of Plantation with Different Tree Species Using Airborne LiDAR and Hyperspectral Data. *Remote Sens.* **2022**, *14*, 2568. [CrossRef]
10. Wulder, M.A.; White, J.C.; Nelson, R.F.; Nsset, E.; Gobakken, T. LiDAR sampling for large-area forest characterization: A review. *Remote Sens. Environ.* **2012**, *121*, 196–209. [CrossRef]
11. Hirata, Y.; Furuya, N.; Saito, H.; Pak, C.; Leng, C.; Sokh, H.; Ma, V.; Kajisa, T.; Ota, T.; Mizoue, N. Object-Based Mapping of Aboveground Biomass in Tropical Forests Using LiDAR and Very-High-Spatial-Resolution Satellite Data. *Remote Sens.* **2018**, *10*, 438. [CrossRef]
12. Torres de Almeida, C.; Gerente, J.; Rodrigo dos Prazeres Campos, J.; Caruso Gomes Junior, F.; Providelo, L.A.; Marchiori, G.; Chen, X. Canopy Height Mapping by Sentinel 1 and 2 Satellite Images, Airborne LiDAR Data, and Machine Learning. *Remote Sens.* **2022**, *14*, 4112. [CrossRef]
13. Cao, L.; Coops, N.C.; Sun, Y.; Ruan, H.; Wang, G.; Dai, J.; She, G. Estimating canopy structure and biomass in bamboo forests using airborne LiDAR data. *ISPRS J. Photogramm. Remote Sens.* **2019**, *148*, 114–129. [CrossRef]
14. Yin, T.; Cook, B.D.; Morton, D.C. Three-dimensional estimation of deciduous forest canopy structure and leaf area using multi-directional, leaf-on and leaf-off airborne lidar data. *Agric. For. Meteorol.* **2022**, *314*, 108781. [CrossRef]
15. Zhao, Y.; Liu, X.; Wang, Y.; Zheng, Z.; Zheng, S.; Zhao, D.; Bai, Y. UAV-based individual shrub aboveground biomass estimation calibrated against terrestrial LiDAR in a shrub-encroached grassland. *Int. J. Appl. Earth Obs. Geoinf.* **2021**, *101*, 102358. [CrossRef]
16. Jiang, X.; Li, G.; Lu, D.; Chen, E.; Wei, X. Stratification-Based Forest Aboveground Biomass Estimation in a Subtropical Region Using Airborne Lidar Data. *Remote Sens.* **2020**, *12*, 1101. [CrossRef]
17. Lin, W.; Lu, Y.; Li, G.; Jiang, X.; Lu, D. A comparative analysis of modeling approaches and canopy height-based data sources for mapping forest growing stock volume in a northern subtropical ecosystem of China. *GIScience Remote Sens.* **2022**, *59*, 568–589. [CrossRef]

18. Nguyen, T.H.; Jones, S.D.; Soto-Berelov, M.; Haywood, A.; Hislop, S. Monitoring aboveground forest biomass dynamics over three decades using Landsat time-series and single-date inventory data. *Int. J. Appl. Earth Obs. Geoinf.* **2020**, *84*, 101952. [CrossRef]
19. Zhao, K.; Suarez, J.C.; Garcia, M.; Hu, T.; Wang, C.; Londo, A. Utility of multitemporal lidar for forest and carbon monitoring: Tree growth, biomass dynamics, and carbon flux. *Remote Sens. Environ.* **2018**, *204*, 883–897. [CrossRef]
20. Cao, L.; Gao, S.; Li, P.; Yun, T.; Shen, X.; Ruan, H. Aboveground Biomass Estimation of Individual Trees in a Coastal Planted Forest Using Full-Waveform Airborne Laser Scanning Data. *Remote Sens.* **2016**, *8*, 729. [CrossRef]
21. Liu, H.; Fan, W.; Xu, Y.; Lin, W. Single tree biomass estimation based on UAV LiDAR point cloud. *J. Cent. South Univ. For. Technol.* **2021**, *41*, 8.
22. Qin, H.; Zhou, W.; Yao, Y.; Wang, W. Estimating Aboveground Carbon Stock at the Scale of Individual Trees in Subtropical Forests Using UAV LiDAR and Hyperspectral Data. *Remote Sens.* **2021**, *13*, 4969. [CrossRef]
23. Hao, L.; Zhengnan, Z.; Lin, C. Estimating forest stand characteristics in a coastal plain forest plantation based on vertical structure profile parameters derived from ALS data. *Natl. Remote Sens. Bull.* **2018**, *22*, 17.
24. McElhinny, C.; Gibbons, P.; Brack, C.; Bauhus, J. Forest and woodland stand structural complexity: Its definition and measurement. *For. Ecol. Manag.* **2005**, *218*, 1–24. [CrossRef]
25. Camarretta, N.; Ehbrecht, M.; Seidel, D.; Wenzel, A.; Zuhdi, M.; Merk, M.S.; Schlund, M.; Erasmi, S.; Knohl, A. Using Airborne Laser Scanning to Characterize Land-Use Systems in a Tropical Landscape Based on Vegetation Structural Metrics. *Remote Sens.* **2021**, *13*, 4794. [CrossRef]
26. Zhao, J.; Li, J.; Liu, Q. Review of forest vertical structure parameter inversion based on remote sensing technology. *Natl. Remote Sens. Bull.* **2013**, *17*, 20.
27. Zheng, J.; Zhang, C.; Zhou, J.; Zhao, X.; Yu, X.; Qin, Y. Study on Vertical Structue of Forest Communities in Yunmengshan. *For. Res.* **2007**, *20*, 7.
28. Lefsky, M.A.; Cohen, W.B.; Acker, S.A.; Parker, G.G.; Spies, T.A.; Harding, D. Lidar Remote Sensing of the Canopy Structure and Biophysical Properties of Douglas-Fir Western Hemlock Forests. *Remote Sens. Environ.* **1999**, *70*, 339–361. [CrossRef]
29. Coops, N.C.; Hilker, T.; Wulder, M.A.; St-Onge, B.; Newnham, G.; Siggins, A.; Trofymow, J.A. Estimating canopy structure of Douglas-fir forest stands from discrete-return LiDAR. *Trees* **2007**, *21*, 295–310. [CrossRef]
30. Fu, X.; Zhang, Z.; Cao, L.; Coops, N.C.; Wu, X. Assessment of approaches for monitoring forest structure dynamics using bi-temporal digital aerial photogrammetry point clouds. *Remote Sens. Environ.* **2021**, *255*, 112300. [CrossRef]
31. Hilker, T.; Leeuwen, M.V.; Coops, N.C.; Wulder, M.A.; Newnham, G.J.; Jupp, D.; Culvenor, D.S. Comparing canopy metrics derived from terrestrial and airborne laser scanning in a Douglas-fir dominated forest stand. *Trees* **2010**, *24*, 819–832. [CrossRef]
32. Zhang, Z.; Cao, L.; She, G. Estimating Forest Structural Parameters Using Canopy Metrics Derived from Airborne LiDAR Data in Subtropical Forests. *Remote Sens.* **2017**, *9*, 940. [CrossRef]
33. Cao, Z. Biomass and Distribution Pattern of Cinnamomum camphora in Yangzhou. *For. Sci. Technol.* **2020**, *10*, 69–71.
34. Liu, K.; Cao, L.; Wang, G.; Cao, F. Biomass allocation patterns and allometric models of Ginkgo biloba. *J. Beijing For. Univ.* **2017**, *39*, 12–20. [CrossRef]
35. Zhuang, H.; Becuwe, X.; Xiao, C.; Wang, Y.; Wang, H.; Yin, S.; Liu, C. Alometric Equation-Based Estimation of Biomass Carbon Sequestration in Metasequoia glyptostroboides Plantations in Chongming Island, Shanghai. *J. Shanghai Jiaotong Univ. (Agric. Sci.)* **2012**, *30*, 8.
36. Xingan, L. Mathematical Model of Tree Volume Table in Zhejiang Province. *J. Zhejiang For. Sci. Technol.* **1986**, *4*, 25–30.
37. Zhao, X.q.; Guo, Q.h.; Su, Y.j.; Xue, B.l. Improved progressive TIN densification filtering algorithm for airborne LiDAR data in forested areas. *ISPRS J. Photogramm. Remote Sens.* **2016**, *117*, 79–91. [CrossRef]
38. Chen, Q.; Gao, T.; Zhu, J.; Wu, F.; Li, X.; Lu, D.; Yu, F. Individual Tree Segmentation and Tree Height Estimation Using Leaf-Off and Leaf-On UAV-LiDAR Data in Dense Deciduous Forests. *Remote Sens.* **2022**, *14*, 2787. [CrossRef]
39. Li, W.; Guo, Q.; Jakubowski, M.K.; Kelly, M. A New Method for Segmenting Individual Trees from the Lidar Point Cloud. *Photogramm. Eng. Remote Sens.* **2012**, *78*, 75–84. [CrossRef]
40. Kim, Y.; Yang, Z.; Cohen, W.B.; Pflugmacher, D.; Lauver, C.L.; Vankat, J.L. Distinguishing between live and dead standing tree biomass on the North Rim of Grand Canyon National Park, USA using small-footprint lidar data. *Remote Sens. Environ.* **2009**, *113*, 2499–2510. [CrossRef]
41. Liu, H.; Cao, F.; She, G.; Cao, L. Extrapolation Assessment for Forest Structural Parameters in Planted Forests of Southern China by UAV-LiDAR Samples and Multispectral Satellite Imagery. *Remote Sens.* **2022**, *14*, 2677. [CrossRef]
42. Pang, Y.; Li, Z. Inversion of biomass components of the temperate forest using airborne Lidar technology in Xiaoxing' an Mountains, Northeastern of China. *Chin. J. Plant Ecol.* **2012**, *36*, 1095–1105. [CrossRef]
43. Dong, Y.H.; Li, Y.Q.; Sun, D.; Li, P.P.; Fan, H.L. Street Tree Information Extraction and Dynamic Analysis Based on Vehicle-Borne LiDAR Data. *Geogr. Geo-Inf. Sci.* **2018**, *34*, 46–51+82.
44. Zięba-Kulawik, K.; Skoczylas, K.; Wężyk, P.; Teller, J.; Mustafa, A.; Omrani, H. Monitoring of urban forests using 3D spatial indices based on a LiDAR point clouds and voxel approach. *Urban For. Urban Green.* **2021**, *65*, 127324. [CrossRef]
45. Liu, K.; Shen, X.; Cao, L.; Wang, G.; Cao, F. Estimating forest structural attributes using UAV-LiDAR data in Ginkgo plantations. *ISPRS J. Photogramm. Remote Sens.* **2018**, *146*, 465–482. [CrossRef]
46. Estornell, J.; Hadas, E.; Martí, J.; López-Cortés, I. Tree extraction and estimation of walnut structure parameters using airborne LiDAR data. *Int. J. Appl. Earth Obs. Geoinf.* **2021**, *96*, 102273. [CrossRef]

47. Michałowska, M.; Rapiński, J. A Review of Tree Species Classification Based on Airborne LiDAR Data and Applied Classifiers. *Remote Sens.* **2021**, *13*, 353. [CrossRef]
48. Lumley, T.; Miller, A. Leaps: Regression Subset Selection. *EMBO J.* **2009**, *12*, 4657–4666.
49. Ding, J.; Huang, W.; Liu, Y.; Hu, Y. Estimation of Forest Aboveground Biomass in Northwest Hunan Province Based on Machine Learning and Multi-Source Data. *Sci. Silvae Sin.* **2021**, *57*, 36–48.
50. Poorazimy, M.; Shataee, S.; McRoberts, R.E.; Mohammadi, J. Integrating airborne laser scanning data, space-borne radar data and digital aerial imagery to estimate aboveground carbon stock in Hyrcanian forests, Iran. *Remote Sens. Environ.* **2020**, *240*, 111669. [CrossRef]
51. Zhao, X. *Research on Forest Aboveground Biomass Estimation Based on Airborne LiDAR Data*; Xi'an University of Science and Technology: Xi'an, China, 2020.
52. Xiao, Y. *Research on Estimation Method of Forest Volume Wangyedian Forest Farm Based on Multi-Source Remote Sensing Data*; Central South University of Forestry and Technology: Changsha, China, 2021.
53. Breiman, L. Random forests. *Mach. Learn.* **2001**, *45*, 5–32. [CrossRef]
54. Dong, L.; Du, H.; Han, N.; Li, X.; He, S. Application of Convolutional Neural Network on Lei Bamboo Above-Ground-Biomass (AGB) Estimation Using Worldview-2. *Remote Sens.* **2020**, *12*, 958. [CrossRef]
55. Li, X.; Du, H.; Mao, F.; Zhou, G.; Chen, L.; Xing, L.; Fan, W.; Xu, X.; Liu, Y.; Cui, L. Estimating bamboo forest aboveground biomass using EnKF-assimilated MODIS LAI spatiotemporal data and machine learning algorithms—ScienceDirect. *Agric. For. Meteorol.* **2018**, *256*, 445–457. [CrossRef]
56. Leiterer, R.; Torabzadeh, H.; Furrer, R.; Schaepman, M.E.; Morsdorf, F. Towards Automated Characterization of Canopy Layering in Mixed Temperate Forests Using Airborne Laser Scanning. *Forests* **2015**, *6*, 4146–4167. [CrossRef]
57. Ishii, H. The role of crown architecture in promoting complementary use of light among coexisting species in temperate forests. *Ecol. Res.* **2010**, *25*, 715–722. [CrossRef]
58. Almeida, C.; Galvo, L.S.; Sato, Y.; Lopes, A.P.; Longo, M. Combining LiDAR and hyperspectral data for aboveground biomass modeling in the Brazilian Amazon using different regression algorithms. *Remote Sens. Environ.* **2019**, *232*, 111323. [CrossRef]
59. Giannico, V.; Lafortezza, R.; John, R.; Sanesi, G.; Chen, J. Estimating Stand Volume and Above-Ground Biomass of Urban Forests Using LiDAR. *Remote Sens.* **2016**, *8*, 339. [CrossRef]
60. Gao, Y.; Dengsheng, L.; Guiying, L.; Guangxing, W.; Qi, C.; Lijuan, L.; Dengqiu, L. Comparative Analysis of Modeling Algorithms for Forest Aboveground Biomass Estimation in a Subtropical Region. *Remote Sens.* **2018**, *10*, 627. [CrossRef]
61. Shen, B.; Ding, L.; Ma, L.; Li, Z.; Pulatov, A.; Kulenbekov, Z.; Chen, J.; Mambetova, S.; Hou, L.; Xu, D.; et al. Modeling the Leaf Area Index of Inner Mongolia Grassland Based on Machine Learning Regression Algorithms Incorporating Empirical Knowledge. *Remote Sens.* **2022**, *14*, 4196. [CrossRef]
62. Chen, M.; Qiu, X.; Zeng, W.; Peng, D. Combining Sample Plot Stratification and Machine Learning Algorithms to Improve Forest Aboveground Carbon Density Estimation in Northeast China Using Airborne LiDAR Data. *Remote Sens.* **2022**, *14*, 1477. [CrossRef]
63. Cao, L.; Pan, J.; Li, R.; Li, J.; Li, Z. Integrating Airborne LiDAR and Optical Data to Estimate Forest Aboveground Biomass in Arid and Semi-Arid Regions of China. *Remote Sens.* **2018**, *10*, 532. [CrossRef]
64. Zhou, L.; Li, X.; Zhang, B.; Xuan, J.; Gong, Y.; Tan, C.; Huang, H.; Du, H. Estimating 3D Green Volume and Aboveground Biomass of Urban Forest Trees by UAV-Lidar. *Remote Sens.* **2022**, *14*, 5211. [CrossRef]
65. Zhang, Y.; Shao, Z. Assessing of Urban Vegetation Biomass in Combination with LiDAR and High-resolution Remote Sensing Images. *Int. J. Remote Sens.* **2021**, *42*, 964–985. [CrossRef]
66. Peng, X.; Zhao, A.; Chen, Y.; Chen, Q.; Liu, H.; Wang, J.; Li, H. Comparison of Modeling Algorithms for Forest Canopy Structures Based on UAV-LiDAR: A Case Study in Tropical China. *Forests* **2020**, *11*, 1324. [CrossRef]

remote sensing

MDPI

Article

High-Resolution Remote Sensing Images Can Better Estimate Changes in Carbon Assimilation of an Urban Forest

Qing Huang [1], Xuehe Lu [2,*], Fanxingyu Chen [2], Qian Zhang [3,4] and Haidong Zhang [5]

1 School of Environmental Science, Nanjing Xiaozhuang University, Nanjing 211171, China
2 School of Geography Science and Geomatics Engineering, Suzhou University of Science and Technology, Suzhou 215009, China
3 School of Geomatics Science and Technology, Nanjing Tech University, Nanjing 211816, China
4 State Key Laboratory of Soil Erosion and Dryland Farming on the Loess Plateau, Institute of Soil and Water Conservation, Northwest A&F University, Yangling 712100, China
5 Institute of Agricultural Sciences in Taihu Area of Jiangsu, Suzhou 215155, China
* Correspondence: luxh@usts.edu.cn

Abstract: Urban forests have the potential to sink atmospheric CO_2. With the improvement of coverage of vegetation in urban environments, more attention has been paid to the carbon sequestration potential of the urban forest. However, the high fragmentation of urban forests makes it difficult to evaluate their carbon budget on a regional scale. In this study, the GPP-NIRv relationship model was employed to estimate GPP in Suzhou by MODIS, Landsat-8 and Sentinel-2 remote sensing data, and to further explore what kind of remote images can figure out the spatial-temporal pattern of GPP in urban forests. We found that the total GPP of the terrestrial ecosystem in Suzhou reached 8.43, 8.48, and 9.30 Tg C yr-1 for MODIS, Landsat-8, and Sentinel-2, respectively. Monthly changes of GPP were able to be derived by MODIS and Sentinel-2, with two peaks in April and July. According to Sentinel-2, urban forests accounted for the majority of total GPP, with an average of about 44.63%, which was larger than the results from GPP products with coarser resolutions. Additionally, it is clear from the high-resolution images that the decline of GPP in May was due to human activities such as the rotation of wheat and rice crops and the pruning of urban forests. Our results improve the understanding of the contribution of the urban forest to the carbon budget and highlight the importance of high-resolution remote sensing images for estimating urban carbon assimilation.

Keywords: gross primary productivity; near-infrared reflectance of vegetation; urban forest; carbon budget

Citation: Huang, Q.; Lu, X.; Chen, F.; Zhang, Q.; Zhang, H. High-Resolution Remote Sensing Images Can Better Estimate Changes in Carbon Assimilation of an Urban Forest. *Remote Sens.* **2023**, *15*, 71. https://doi.org/10.3390/rs15010071

Academic Editor: Arturo Sanchez-Azofeifa

Received: 13 November 2022
Revised: 19 December 2022
Accepted: 20 December 2022
Published: 23 December 2022

1. Introduction

Gross Primary Productivity (GPP) quantifies the total amount of carbon assimilated by plants through photosynthesis per unit time. As the critical variable of carbon cycling of terrestrial ecosystems, GPP is the initial amount of energy and material entering the terrestrial ecosystem and plays an essential role in regulating the global carbon cycle [1–3]. GPP can be observed at the leaf level or ecosystem level [4–6], and also can be simulated by process-based terrestrial ecosystem models [7–9] or estimated through remote sensing [10–13]. Benefiting from the eddy-covariance techniques and the multisource data of remote sensing, satellite-based GPP models were developed to estimate regional and global GPP by establishing the empirical relationship between the vegetation index (VI) and surface observation data [14–18]. Previous studies have shown that the correlation coefficient between Enhanced Vegetation Index (EVI) and surface observation data in global GPP estimation can reach 0.52–0.92 [19], and the correlation coefficient between Solar-induced Chlorophyll Fluorescence (SIF) and GPP for crops and grass can achieve 0.87 [20]. However, there are several limitations in these methods; for example the adaptability of the empirical method is affected by different ecosystem structures and climate conditions [21].

Additionally, the coarse spatiotemporal resolution of satellite SIF is generally a limiting factor in the regional estimation of GPP [22,23].

In recent years, Near-infrared Reflectance of Vegetation (NIRv), which integrates the advantages of NDVI and near-infrared reflectance, has provided a new avenue for GPP estimation [24–27]. NDVI can intuitively reflect vegetation coverage, which is tightly correlated to vegetation productivity, while near-infrared reflectance can embody the information on vegetative canopy structure and leaf area [24,28]. Therefore, NIRv takes into consideration both the pigment and structure of vegetation in remote sensing images to avoid the saturation effects of NDVI [29–31]. Meanwhile, a strong and stable correlation was found between NIRv and GPP in the same ecosystem [21,23,32]. In the study of regional and global carbon budgets of terrestrial ecosystems, NIRv has been suggested as the effective substitution for satellite SIF, based on theoretical derivations and radiative transfer simulations, and has gained a great deal of attention in recent years [24,33].

Urban forest is considered as a large green infrastructure in the urban area, which consists of trees, shrubs and grasses in streets, parks, gardens, alongside rivers and so on [34–36]. Additionally, urban forest in this study is also defined as the above description. Urban forest usually plays an indispensable role for people living in urban environments, such as reducing air pollution and heat islands effect [37], and increasing biodiversity [38]. As global CO_2 concentrations continue to increase, there is also increasing interest in the carbon storage and sequestration of the urban forest [39–41]. Vegetation in urban areas is offered for environments with higher temperatures and CO_2 concentrations, highly intensive use of water and pesticides [42,43], and its photosynthetic activity would play an important role in the carbon cycle of terrestrial ecosystems.

However, in some studies, the vegetation productivity in urban areas was set as zero [44]; the role of urban vegetation might be underestimated in the global carbon budget. With the improvement of urban green coverage (e.g., the green coverage in the urban built areas in Jiangsu province increased from 19.5% in 1990 to 42.1% in 2020), more attention should be paid to the effect of the urban forest in the global and regional carbon assimilation [45–47]. Although the ability of carbon assimilation can be estimated through GPP, the contribution of the urban forest to regional carbon assimilation is still unclear. A number of difficulties in the current estimations of urban GPP need to be overcome; for example the lack of surface observation data in urban areas makes it difficult to conduct empirical estimation on the urban scale by using traditional VI. At the same time, the lack of driving and parameterized data with high spatiotemporal resolution significantly reduced the performance of process-based ecological models in the simulation of GPP in urban areas.

As NIRv has successfully estimated GPP in other ecosystems, it may be able to provide a new approach to estimating carbon assimilation in urban forests. To the best of our knowledge, the contribution of urban forests to regional carbon assimilation has not yet been studied according to this method. Meanwhile, it is not yet clear which resolution of satellite data is appropriate for determining urban forest carbon assimilation capacity. Therefore, in this study, the GPP-NIRv relationship model is employed to estimate the GPP of terrestrial ecosystems in Suzhou by MODIS, Landsat-8, and Sentinel-2 remote sensing data, and to further explore what kind of remote images can figure out the spatiotemporal pattern of GPP of urban forests. Then, we evaluate the contribution of urban forest to the regional carbon budgets and the impact of human activities on regional GPP. The expected results will improve our understanding of the ability of urban forests to affect carbon assimilation and will provide an appropriate reference for selecting the right resolution of remote sensing images for carbon budget research in urban areas.

2. Materials and Methods

2.1. Study Area

Suzhou is one of the most economically developed cities in eastern China, and is located in the southeast of Jiangsu and the middle of the Yangtze River Delta, with the region ranging from eastern longitude 119°55′ to 121°20′ and northern latitude from 30°47′ to 32°02′. Suzhou connects to Shanghai to the east, Jiaxing and Huzhou of Zhejiang province to the north, Wuxi to the west, and the Yangtze River to the north, with a total area of 8657.32 km^2. As one of the largest industrial cities in China, Suzhou has an urban population of 12.84 million, and the gross output value of all the above designated-size industrial enterprises in Suzhou exceeded CNY 4 trillion in 2021 (http://tjj.suzhou.gov.cn/sztjj/tjnj/2021/zk/indexce.htm (accessed on 1 May 2022)).

Suzhou lies in the subtropical monsoon climate zone, where abundant precipitation and warm temperatures are suitable for vegetation growth. The annual precipitation and mean temperature in the year 2021 were 1318.6 mm and 18.3 °C, respectively. Suzhou is low and even, with a general elevation range from 3.5 to 5 m above sea level. The southeast of Suzhou is lower, with the lowest elevation below 2 m, and the southwest is a hilly area where vegetation grows well. The terrain of this area is shown in Figure 1. Suzhou is a famous water country region with a dense river network and numerous lakes, and rivers, lakes, tidal flats and wetlands account for 34.6% of the total area. Cropland is the dominant land use type in Suzhou, with an area of about 2871.6 km^2, accounting for 33.17% of the total area.

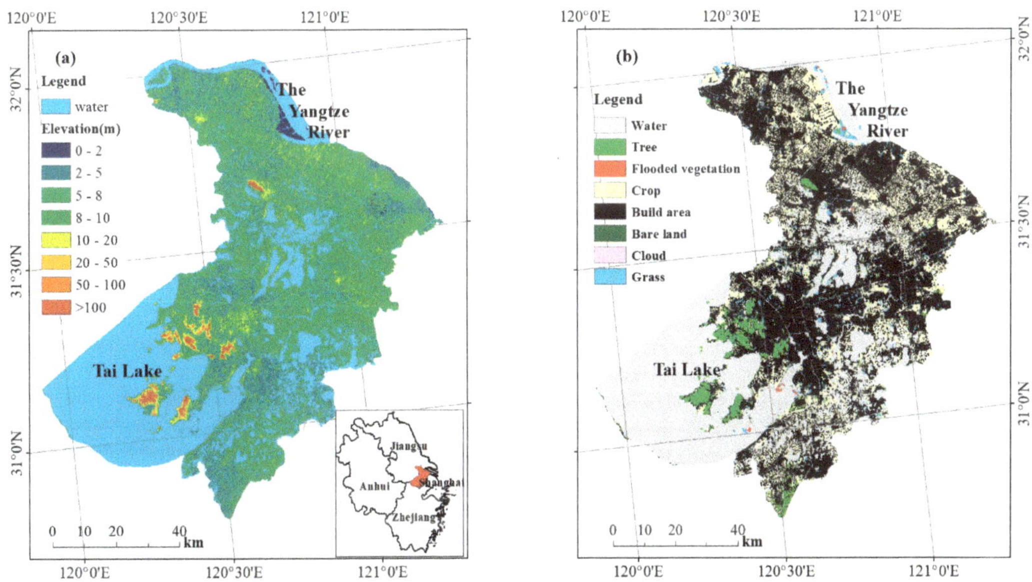

Figure 1. The geographic location (**a**) and land cover (**b**) derived from Sentinel-2 by ESRI of Suzhou in Jiangsu province in China.

2.2. Data Sources and Processing

Remote sensing data, climate data and land cover data were used in this research to estimate GPP in Suzhou and analyze the importance of urban forests for the regional carbon budget. Three kinds of resolution remote sensing images are used to estimate GPP. Of these, 30 m-Landsat 8 and 10 m-Sentinel 2 surface reflectance datasets were downloaded from the United States Geological Survey (https://earthexplorer.usgs.gov/ (accessed on 10 April 2022)) and the Copernicus Open Access Hub (https://scihub.copernicus.eu/dhus/#/home (accessed on 15 April 2022)) in 2021 to calculate NIRv. In order to avoid the impact of cloud on data quality, the maximum value composite (MVC) method was set to generate the monthly NIRv.

MODIS 500 m surface reflectance data (MOD09A1) and NDVI data (MOD13A1) downloaded from NASA's Distributed Active Archive Center (DAAC) have eliminated the impact of cloud. In order to maintain the consistency of the production of NIRv, the MVC method was also used to generate the MODIS monthly NIRv.

Climate data, including monthly temperature and precipitation, were used to investigate GPP changes in response to the local climate; climate data were obtained from the National Meteorological Science Data Center (https://data.cma.cn/ (accessed on 2 May 2022)).

MODIS, Landsat and Sentinel-2 land cover datasets were downloaded to summarize the GPP of different land cover types according to the corresponding estimation of GPP with the same spatial resolution. A MODIS land cover data (MCD12Q1) product was downloaded from NASA DAAC. The MCD12Q1 product supplies maps of land cover at annual steps, and the land cover in Suzhou was classified according to the International Geosphere Biosphere Programme (IGBP) land cover classification scheme as urban and built-up lands, croplands, grasslands, water bodies, evergreen broadleaf forests (EBF), deciduous broadleaf forest (DBF), mixed forest (MF) and so on.

Global land cover data producing 30 m resolution were downloaded from http://data.ess.tsinghua.edu.cn/ (accessed on 17 April 2022). This product used the amount of training samples across the world to optimize many kinds of classifiers, eg. maximum likelihood, decision tree, random forest, etc. A unique land-cover classification system was used in this product. In Suzhou, the typical land cover types were crop, forest, grass, shrub, water, and impervious. Additionally, impervious was considered as urban built-up areas in this study.

ESRI generated a global map of land use and land cover (https://livingatlas.arcgis.com/landcover/ (accessed on 17 April 2022)) derived from Sentinel-2 imagery at 10 m resolution by using a deep learning AI land classification model trained by billions of human-labeled image pixels [48]. This product has 9 classes, of which water, tree, crop, shrub land, built area and grass are the dominant ones in Suzhou.

In order to better calculate the changes in carbon assimilation of urban forest, the urban area of Suzhou was divided into built area (black areas in Figure 1b) and non-built area (remaining areas besides built area).

2.3. Estimation of GPP

NIRv has been found to accurately capture both the seasonal and annual variation in GPP at flux sites [25,26]. GPP correlates linearly with NIRv among different vegetation types (Table S1), and global GPP was estimated with high accuracy on a monthly basis by upscaling the relationships between NIRv and GPP [23]. In this study, we also use these correlations to determine the Suzhou monthly GPP based on NIRv, and the calculation of NIRv and GPP are shown in Equations (1) and (2):

$$NIRv = \frac{NIR - R}{NIR + R} \times NIR \tag{1}$$

$$GPP = a \times NIRv + b \tag{2}$$

R and NIR are the red and the near-infrared bands, a and b are derived from linear regression, and the values can be found in Table S1. We used the MVC method to generate the monthly NIRv. Landsat-8 has a limited frequency of revisits, so some images were lost during the rainy season, i.e., in July and August. MODIS and Sentinel-2 had sufficient data to produce a full image each month.

For different kinds of forests, such as deciduous broadleaf forests (DBF), evergreen broadleaf forests (EBF), evergreen needle forests (ENF), and mixed forests (MF), Wang et al. [23] used different coefficients for calculating GPP. However, the land cover data of Sentinel-2 used in this study do not distinguish forest type. As a result, based on the studies of [49], changes in tree NDVI between summer and winter were examined to determine the forest types, including deciduous broadleaf forest, mixed forest and evergreen broadleaf forest, according to the criteria listed in Criteria 2 of Table 1. Additionally, we did not include evergreen needle forest in calculating GPP, for the reason that its distribution in

Suzhou is limited and hard to separate from evergreen broadleaf forests by only the change in NDVI. Considering that the grassland and different kinds of forests in urban area are always integrated but still able to be identified according to dominated coverage of plants, vegetation types of urban forest in built area were also identified as grass (GRA), DBF, MF and EBF.

Table 1. NDVI for identifying vegetation type in land cover of tree and built area.

NO.	Criteria 1	Criteria 2	TR	BA	Veg
1	$NDVI_8 < 0.2$			C1 and 2	BRE
2	$0.2 \leq NDVI_8 < 0.5$			C1 and 2	GRA
3	$0.5 \leq NDVI_8 \leq 1$	$(NDVI_8\text{-}NDVI_{12})/NDVI_8 > 0.35$	C2	C1 and 2	DBF
4	$0.5 \leq NDVI_8 \leq 1$	$0.2 < (NDVI_8\text{-}DVI_{12})/NDVI_8 < 0.35$	C2	C1 and 2	MF
5	$0.5 \leq NDVI_8 \leq 1$	$(NDVI_8\text{-}NDVI_{12})/NDVI_8 < 0.2$	C2	C1 and 2	EBF

Note: The subscript is month, $NDVI_8$ and $NDVI_{12}$ are monthly NDVI in August and December. C1 and C2 are criteria 1 and criteria 2, respectively. Veg is the vegetation type. TR and BA are the land cover of tree and built area. BRE and GRA are bare land and grass land, respectively.

In the built area, the contribution of urban forests to total regional carbon assimilation cannot be ignored as they are widely distributed in Suzhou. However, they are usually buried in lower-resolution imagery. Considering the fragmentation of urban forests, three different resolutions of remote images, i.e., MODIS, Landsat and Sentinel-2, were used to evaluate the ability to detect urban forests. For calculating GPP in urban forest, NDVI changes in the built area were used to determine vegetation types as shown in Criterions 1 and 2 of Table 1.

2.4. Data Analysis

Based on our estimation, the GPPs of different land cover types were summarized according to land cover data. Then, we identified the contribution of urban forests to the regional total GPP. Vegetation in urban areas is inevitably affected by human activities. Therefore, the monthly changes of GPP were examined to find out the abnormal fluctuation of monthly GPP and identify what kind of human activities disturbed the change in GPP.

3. Results

3.1. Variance of NIRv in Different Remote Sensing Data with Distinct Spatial Resolutions

The NIRv can be calculated based on MODIS, Landsat-8 and Sentinel-2 by using Equation (1). Figure 2 shows the spatial distribution of Sentinel-2 NIRv monthly.

The area along the Yangtze River in the northern part of Suzhou was a high-value area of NIRV. Additionally, the changes in NIRv in this area were significantly greater than in other parts of the city. During the growing season, in April and from July to September, NIRv reached more than 0.45. The high NIRv value in this area was a result of the high photosynthetic ability of crops planted in this region, for example, wheat and rice. Furthermore, around Tai Lake, the large water body in the southwest of the city, there was another area with a high NIRv. Forests and crops were spread throughout this area (Figure 1b), and the peak NIRv was around 0.4 from June to August.

The NIRv in Suzhou's built area ranged from 0 to 0.1, generally, without an evidently monthly dynamic. Nevertheless, there were many points scattered in the built area with high NIRvs (ranging from 0.2 to 0.35) and also seasonal changes. The urban forest in built areas contributed to the scattered high-value points. This kind of vegetation is usually found in small urban green spaces, and was represented as discrete points by remote sensing data with high resolution.

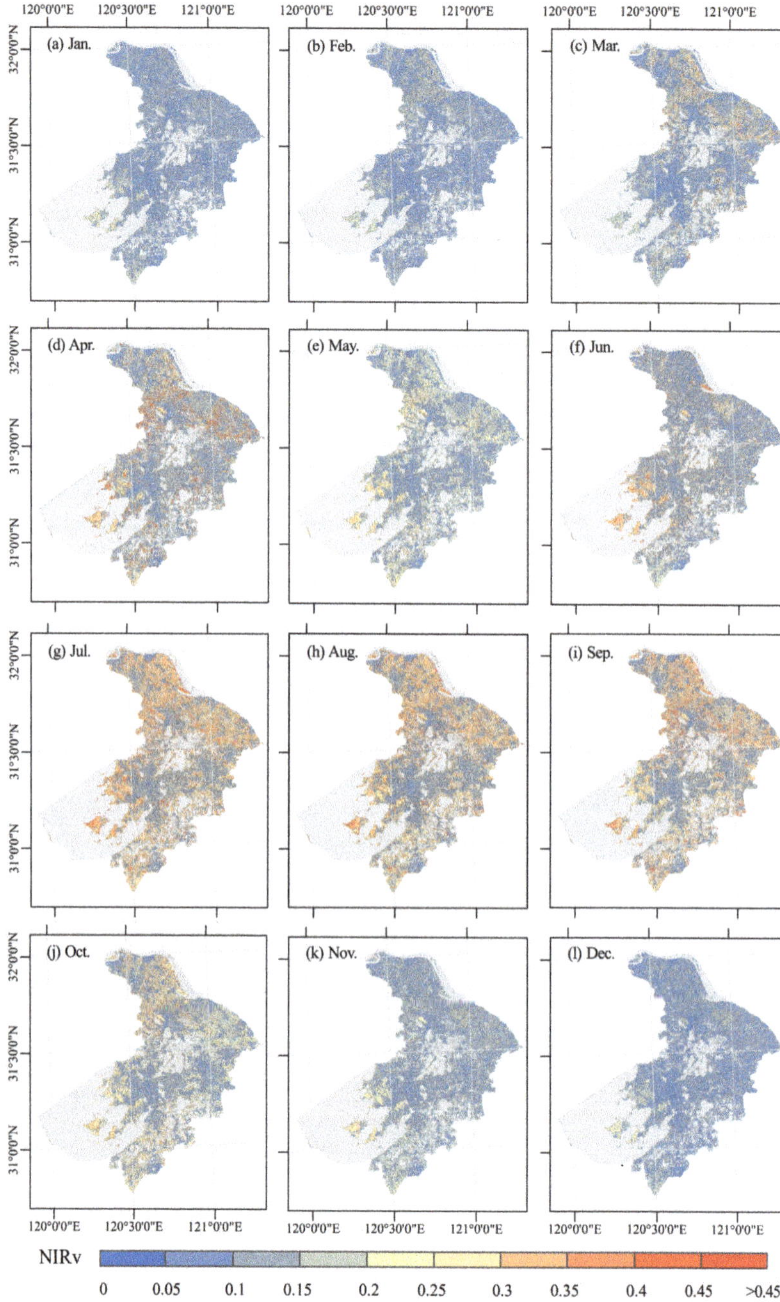

Figure 2. Spatial distribution of monthly Sentinel-2 NIRv in 2021, Suzhou.

For comparing the differences of NIRv derived from MODIS, Landsat-8 and Sentinel-2, the monthly averages and standard deviations of NIRv were listed in Figure 3. Sentinel-2 NIRv showed two peaks in April and July, with values of 0.13 and 0.18, respectively. The values of MODIS NIRv were generally lower than Sentinel-2 and had similar trends to

Sentinel-2. Due to the impact of the cloud, the continuous change of Landsat-8 NIRv was hard to obtain. Additionally, current valid data of Landsat-8 were similar to MODIS NRIv and also showed a peak in September.

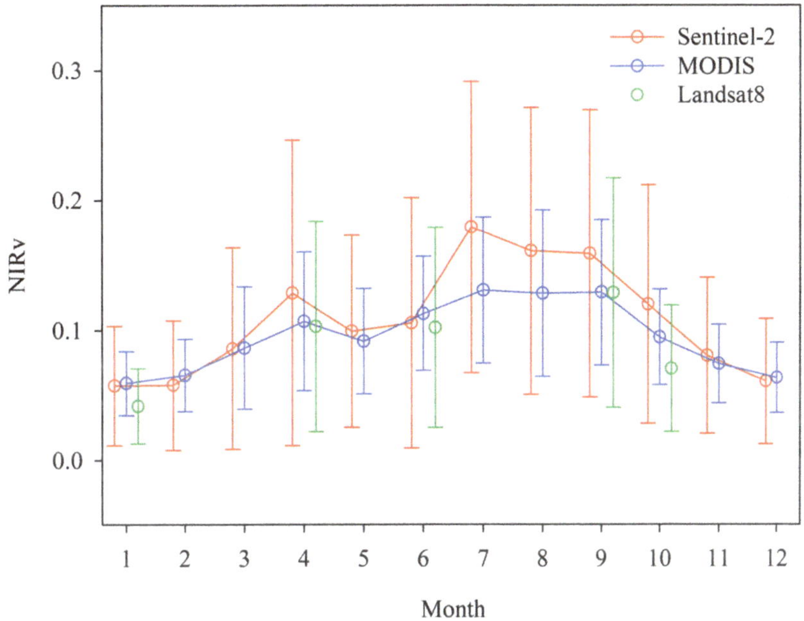

Figure 3. Monthly averages (red, blue and green cycle) and standard deviations (red, blue and green bar) of NIRv in Suzhou derived from MODIS, Landsat-8 and Sentinel-2.

3.2. Comparing the GPPs Estimated by Different Resolutions of Remote Sensing Data

The GPPs estimated by NIRv derived from MODIS, Landsat-8 and Sentinel-2 were shown in Table 2. The total GPP in Suzhou in 2021 estimated by MODIS, Landsat-8 and Sentinel-2 was 8.43, 8.48, and 9.30 Tg C yr^{-1} (Tg = 10^{12} g), respectively. Our estimations of GPP were higher than the results of MOD17A2 (4.37 Tg C yr^{-1}), which ignores the contribution of the urban forest in built areas (black areas in Figure 4a). In contrast, our results included this ignoring component and fell within the range of previous studies which considered the contribution of urban forest (Table 2).

Table 2. Comparison of urban contribution to GPP in different GPP products.

Product	Resolution	Time	Model	Total GPP (Tg C)	Remark
This Study	10 m	2021	NIRv-GPP	9.30	Sentinel-2
	30 m	2021		8.48	Landsat8
	500 m	2021		8.43	MODIS
MOD17A2	500 m	2021	LUE	4.37	
Zhang [50]	0.050°	2016	VPM	9.37	
Ju [51]	0.073°	2019	BEPS	6.72	

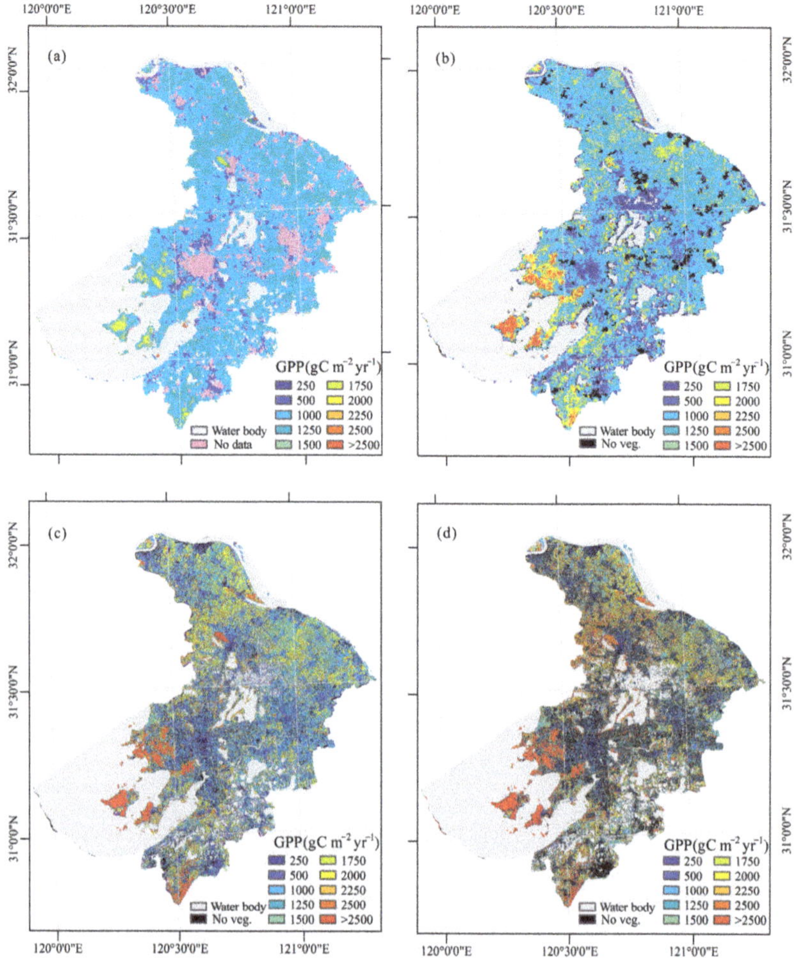

Figure 4. The spatial distribution of GPP from MOD17A2 product (**a**) and estimated by MODIS (**b**), Landsat-8 (**c**), Sentinel-2 (**d**). Areas in panel (**a**) with no data are due to the MODIS product not calculating GPP if the land cover type is built area. Additionally, the no vegetation (veg.) areas in panels (**b**–**d**) are the regions where the NDVI is less than 0.2 and, as a result, is recognized as bare land with no GPP.

The spatial patterns of GPP in Suzhou estimated by MODIS, Landsat-8 and Sentinel-2 in 2021 were shown in Figure 4b–d. All three kinds of GPP showed the same high GPP region located around Tai Lake (grey area in the southwest of Figure 4). In many areas of this region, the GPP ranged from 2500 to 3000 g C m^{-2} yr^{-1}. The estimated GPPs had some differences in the northeast, along the Yangtze River (grey area in the northeast of Figure 4). In this region, the GPP of MODIS was lower than the GPP of Landsat-8 and Sentinel-2. According to our MODIS GPP results, most built areas were capable of photosynthesis from the vegetation within them. Additionally, GPP in built area of Suzhou ranged from 250 to 500 g C m^{-2} yr^{-1}. Ignoring the contribution of GPP from the built area (black area in Figure 4a) induced the lower GPP in MOD17A2. Furthermore, high-resolution results from Landsat-8 and Sentinel-2 also indicated the contribution of urban forests, which were scattered over the built area due to their spatial distribution being highly discrete in Suzhou (Figure 4c,d).

3.3. Monthly Change of GPP in the Year 2021

The Landsat-8 GPP failed to overcome the effects of clouds every month since its long revisit. Additionally, with the help of the MVC method, monthly NIRvs derived by MODIS and Sentinel-2 were generated. As a result, Figure 5 shows only the MODIS and Sentinel-2 GPPs, as well as the changes in monthly temperature and precipitation for the year 2021.

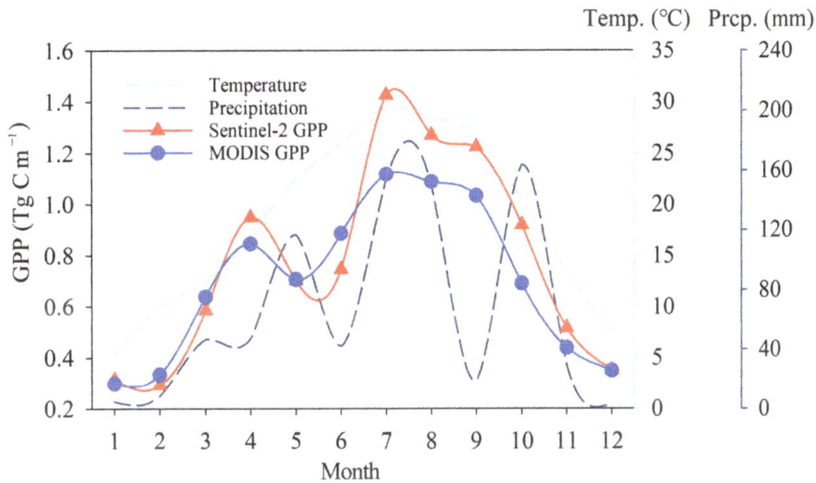

Figure 5. The monthly change of Suzhou's GPP in the year 2021.

The monthly GPP of MODIS and Sentinel-2 showed two peaks in 2021. One peak was in April, with GPPs 0.85 and 0.95 Tg C m^{-1} for MODIS and Sentinel-2, respectively. The other peak was in July, when the GPPs of MODIS and Sentinel-2 reached their summit for the whole year: 1.43 and 1.12 Tg C m^{-1}, respectively. Between the two peaks was an evident decrement in GPP in May. GPP decreased by 16.4% and 25.9% according to the estimations of MODIS and Sentinel-2, respectively.

In the growing season, the MODIS GPP was generally lower than Sentinel-2 GPP. For example, from July to September, the average of Sentinel-2 GPP was 1.31 Tg C m^{-1}, which was 21.3% higher than MODIS GPP in the same period. Additionally, a similar difference was also found in the growing peak in April. Moreover, in winter and early spring MODIS GPP and Sentinel-2 GPP were similar.

As Figure 6 showed, a significant ($p < 0.001$) linear relationship was found between the air temperature and GPP. R^2 for the temperature–GPP relationships were 0.81 and 0.89 for Sentinel-2 and MODIS GPP, respectively (Figure 6a,b), which was higher than the precipitation–GPP relationship (Figure 6c,d). Therefore, the change in GPP was mainly determined by the air temperature in the year 2021. Additionally, the positive coefficient of slopes for the linear relationships indicated that temperature and precipitation had positive effects on GPP. However, the GPP declined when temperature and precipitation increased in May. This decrement in GPP implies that, besides climatic factors, some other factors determined the change in GPP in May. In urban areas, the impact of anthropogenic activities on vegetation cannot be ignored.

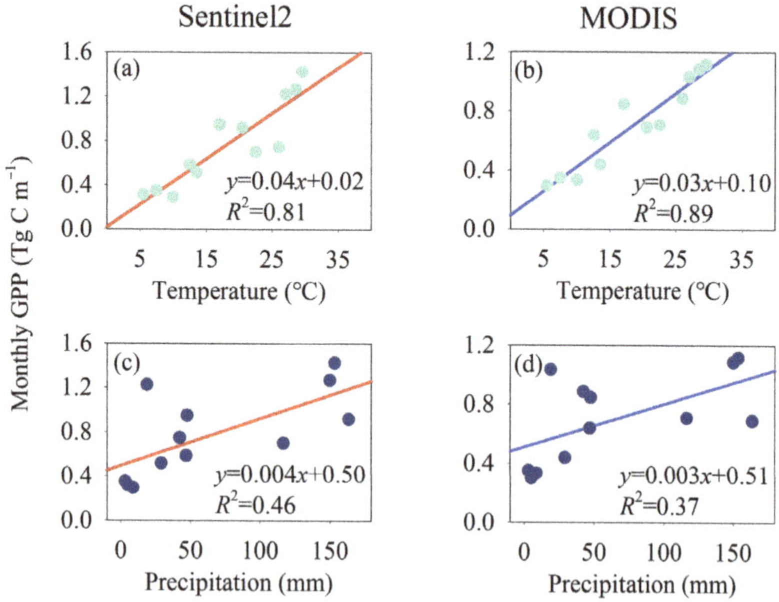

Figure 6. The relationship between the estimated GPP and climate factors. (**a**) and (**b**) are the correlations between temperature and monthly GPP for Sentinel2 and MODIS, respectively. (**c**) and (**d**) are the correlations between precipitation and monthly GPP for Sentinel2 and MODIS, respectively.

3.4. Changes in GPP of Different Land Cover Types

To better understand the importance of urban forests to the regional GPP, the Sentinel-2 and MODIS GPP changes in different land cover types were shown in Figure 7. According to the result of Sentinel-2, the average monthly GPP of vegetation in the built area (e.g., urban forest) and non-built area (e.g., crop, tree, grassland) was 0.35 Tg C and 0.43 Tg C (0.32 for crops, 0.08 for trees, and 0.03 Tg C for grassland), respectively. Additionally, for MODIS, GPP of the built area and non-built area was 0.28 Tg C and 0.42 Tg C (0.18, 0.01, and 0.23 Tg C for crop, tree, and grassland, respectively). The MODIS GPP of grassland was 0.20 Tg C higher than Sentinel-2 GPP, and the MODIS GPP of cropland was 0.14 Tg C lower than Sentinel-2 GPP. This may be caused by the use of different land cover datasets in statistics. MODIS land cover products might be misplaced between farmland and grassland due to the lower spatial resolution. In fact, as an industrialized city, Suzhou does not have a lot of grasslands.

The GPP of each land cover type varied throughout the year (Figure 7a,c). Similar to the change in total GPP, each land cover type had two peaks in spring and summer. Additionally, they also showed decrements in GPP in May. These decrements in GPP for crops and urban forests were obviously about 32.33% and 23.57%, respectively, according to Sentinel-2 GPP.

Sentinel-2 results indicated that crops and urban forests accounted for a majority of the total GPP in Suzhou in 2021, with an average of about 38.15% and 44.43%, respectively. Trees and grasses in non-built area made relatively small contributions to total GPP, accounting for approximately 11.87% and 4.17%, respectively. According to MODIS GPP, urban forests still accounted for the majority (39.66%) of total GPP, followed by grassland and crops at 33.66% and 25.60%, respectively.

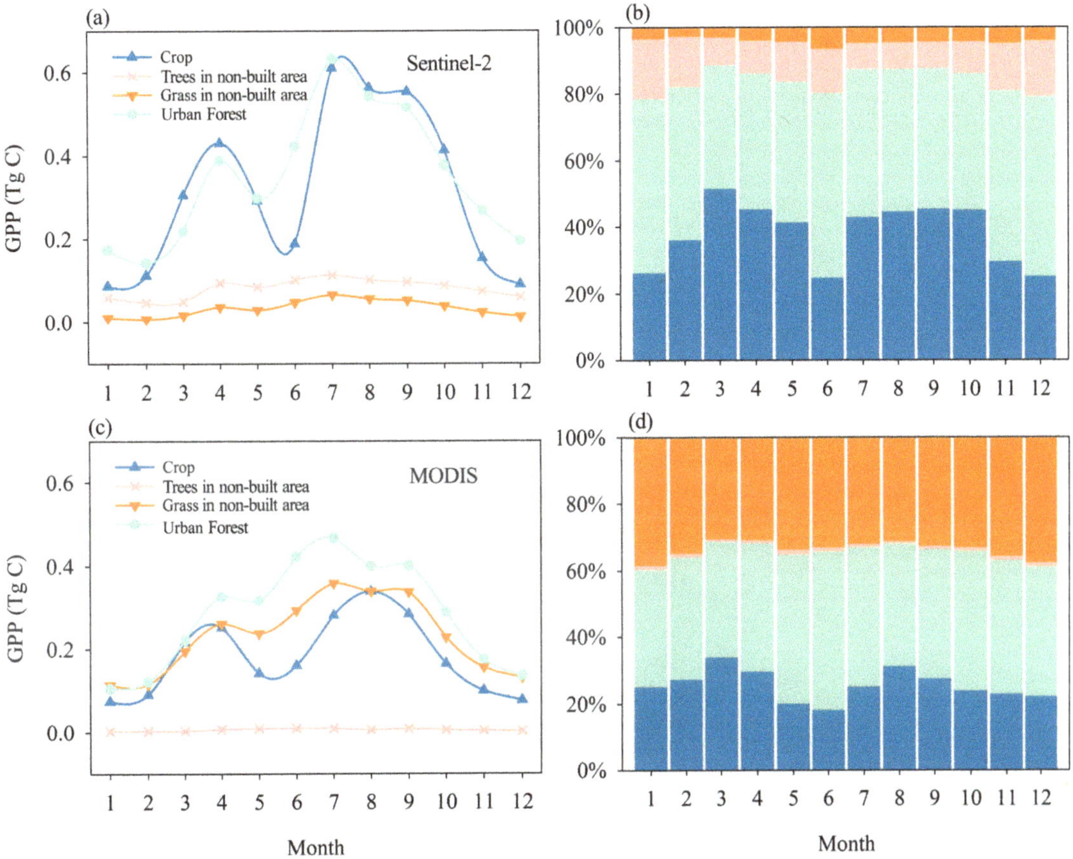

Figure 7. The Sentinel-2 and MODIS GPP changes in different land cover types in 2021 (**a**,**c**) and their contribution to total GPP (**b**,**d**).

The contribution of different land cover types also varied throughout the year (Figure 7b,d). Taking the most contributed urban forests as an example, their Sentinel-2 GPP exceeded 50% of the total GPP for five months of the year, with the highest percentage at 55.39% in June. Additionally, the contribution made by crops was also large, even more than the contribution of urban forests in growing peak seasons, such as April and August.

3.5. Changes in GPP by Anthropogenic Factors

The spatial distribution of this decrement in GPP in May is shown in Figure 8. The evident decrement of GPP for the crop was mainly in the rice and wheat rotation area along the Yangtze River. In May, as wheat matures, its photosynthetic capacity also decreases significantly. Additionally, in June, after the wheat harvest, rice has just been planted. In consequence, the photosynthetic capacity had not yet recovered.

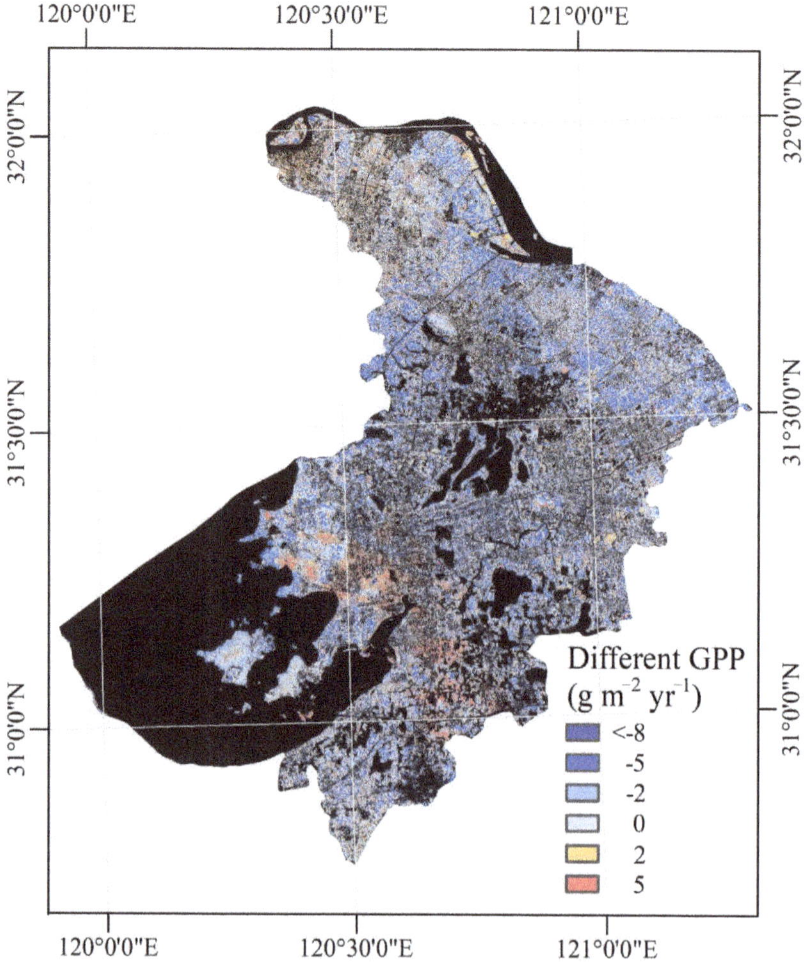

Figure 8. The difference in GPP between April and May in Suzhou.

Furthermore, urban forests' GPP also decreased significantly in May, but recovered quickly in June (Figure 7a,c). For making urban vegetation more attractive, urban forests in Suzhou undergo a pruning process in May. This process removes a considerable amount of leaves from the canopy. Usually, these leaves are current-year leaves and are located at the top of the canopy. Additionally, the photosynthetic capacity of these kinds of leaves is greater than that of old leaves and lower canopy leaves [52,53]. The pruning process is usually conducted on the grassland and shrub land, which were the main parts of urban green spaces in Suzhou, approximately 55.67% (Figure 9) according to our criteria in Table 1. Haberl [54] proposed that the averaged biomass loss during gardening (such as pruning) or park and infrastructure maintenance amounted to 50% of the aboveground Net Primary Productivity (the remainder of GPP deducts autotrophic respiration). Therefore, the pruning process removes the most photosynthetic part of the canopy, which results in the decline of GPP of urban forests.

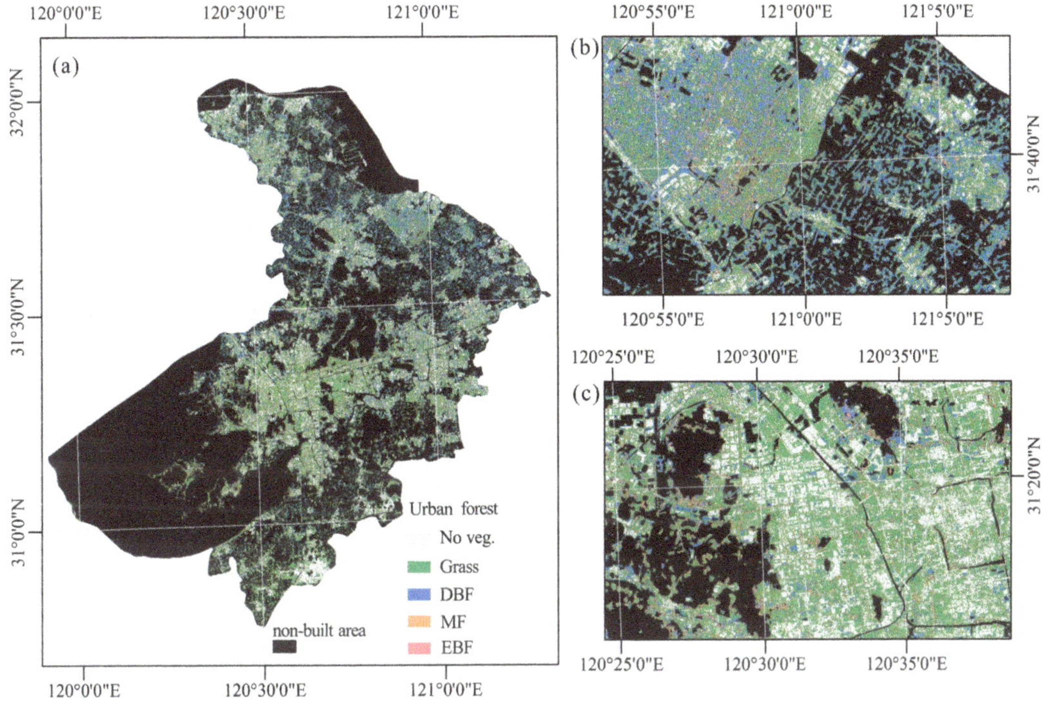

Figure 9. The distribution of urban forests in Suzhou (**a**) and its detailed views (**b**,**c**). Veg. is vegetation. DBF, MF, and EBF are deciduous broadleaf forests, mixed forests, and evergreen broadleaf forests.

4. Discussion

4.1. Uncertainty of Estimating GPP by NIRv

Although, as we know, few studies have examined vegetation productivity in Suzhou, studies in similar cities may provide useful information. From 2000 to 2014, the average rate of GPP in Shanghai, the city next to Suzhou, was between 800 and 1050 g C m^{-2} yr^{-1} based on the simulations by Vegetation Photosynthesis Model (VPM) [55]. This result is close to our 973.74, 979.51, and 1074.21 g C m^{-2} yr^{-1} for MODIS, Landsat-8, and Sentinel-2, respectively.

According to the MOD17A2 product, the previous study also indicated the GPP of crops along the Yangtze River in the northeast of Suzhou as around 1200–1400 g C m^{-2} yr^{-1} [56], which is lower than our Sentinel-2 GPP, 1250–2000 g C m^{-2} yr^{-1}, in the same region (Figure 4d). Limited by mixed pixel effects, the MOD17A2 product usually overestimates in lower value and underestimates in higher value [57,58]. The high resolution of Sentinel-2 eliminated the many effects of the mixed pixel. Therefore, the Sentinel-2 GPP was obviously higher than the GPP calculated by MODIS reflectance. Furthermore, Landsat-8 GPP would have to be greater than MODIS GPP if it did not miss images during July and August, when carbon assimilation rates are at their peak.

Compared to deciduous forests, evergreen forests are inflexible to short-term changes in environmental conditions. As a result, their NIRv is suitable for predicting GPP over a longer period, such as 90 days [25]. Consequently, the NIRv–GPP relationship is a challenge in estimating monthly GPP for evergreen forests. Additionally, for this reason, our GPP of evergreen forests in the southwest of Suzhou was more than 2500 g C m^{-2} yr^{-1}, which is higher than the previously observed and simulated results [59,60]. Additionally, the original NIRv-GPP relationships across the different land cover types are generated by 0.05 degree AVHRR reflectance [23]. The uncertainty induced by the different remote sensing data in terms of field of view angle, signal-to-noise ratio, and spectral width needs to be further evaluated.

4.2. Evaluation of High-Resolution Remote Sensing Images in Urban Carbon Research

Compared to Landsat-8 and Sentinel-2, the lower resolution of MODIS lost many details inside the city (Figure 4b–d). In the built areas, vegetation usually accounts for a small proportion at 500 m resolution. As a result, the reflection characteristics of vegetation will be affected by the background. Meanwhile, due to the variety of land cover types within a built area, low-resolution remote sensing images fail to accurately depict the spatial variability of urban surfaces (Figure 10), thereby excluding changes in vegetation characteristics within the city. As a result, the uncertainty in estimating GPP from low remote sensing images increases. As Figure 11 shows, with the increment of spatial resolution, the contribution of the urban forest to total GPP was increased. Therefore, the photosynthetic ability of urban forests can be better figured out by fine-resolution remote sensing images. Landsat having similar performance as MODIS is partially due to its coarser resolution and lower NDVI relative to Sentinel-2, and also partially owing to missing data of high values in summer with large fractions of cloudy days.

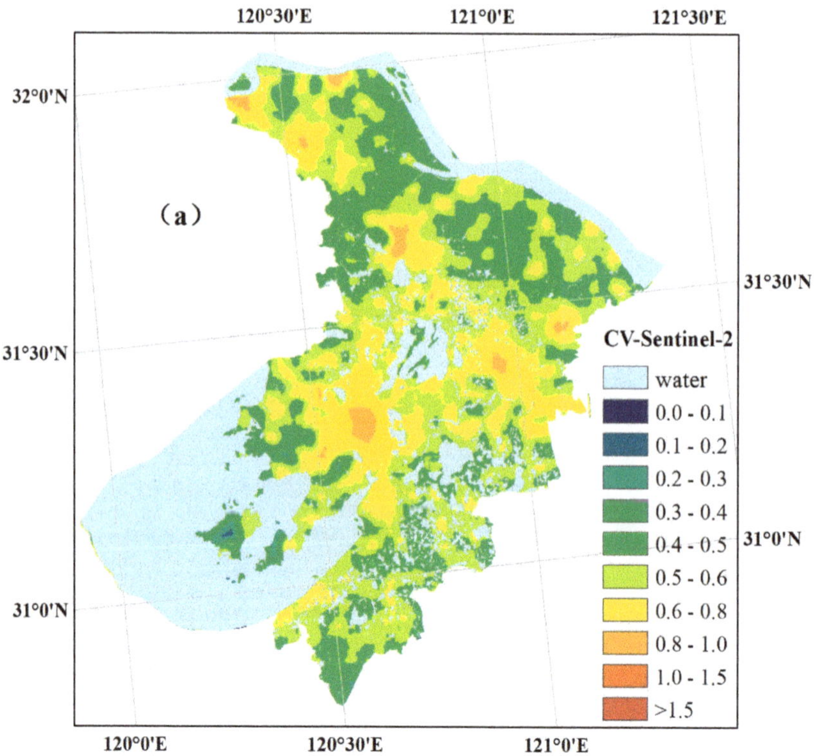

Figure 10. *Cont.*

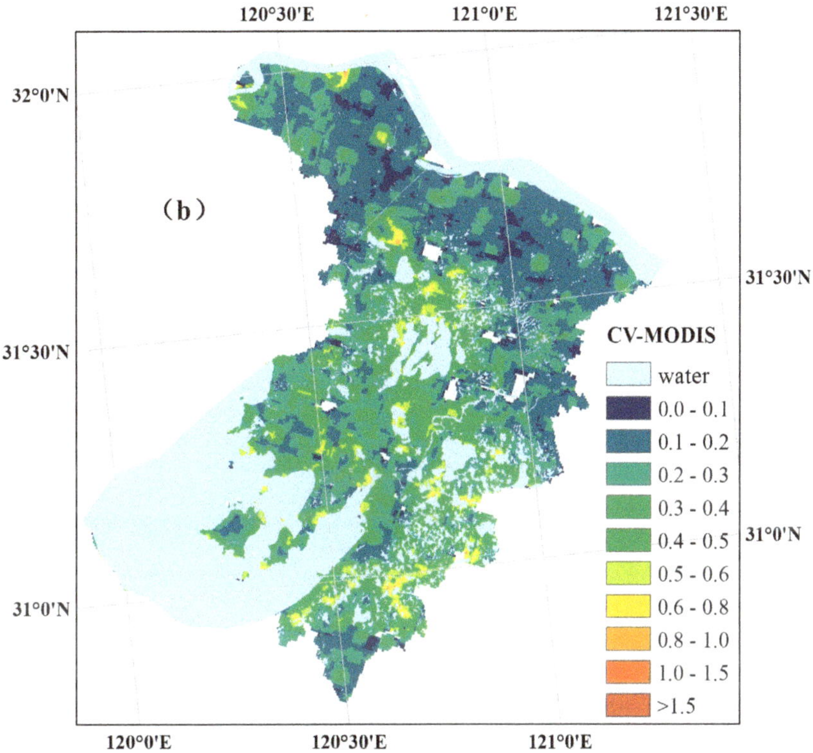

Figure 10. The GPP variation coefficient (CV) of Sentinel-2 (**a**) and MODIS (**b**).

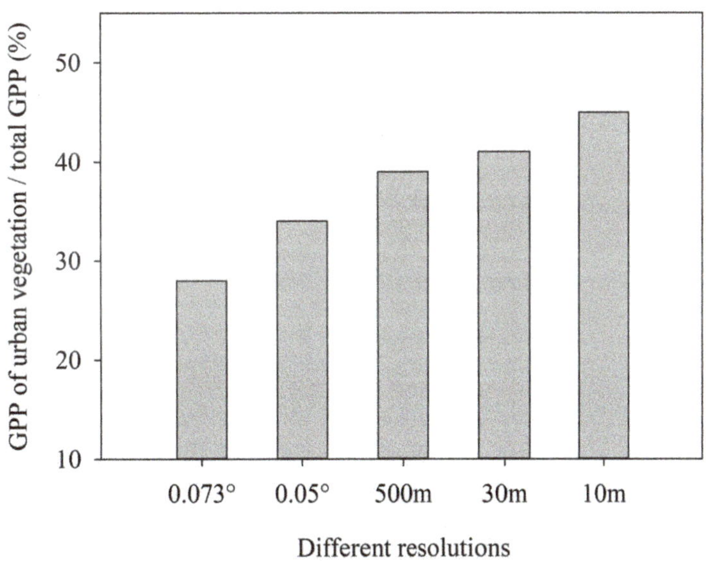

Figure 11. The contribution of urban forests to total GPP. The percentages of 0.073° and 0.05° are from Ju [51] and Zhang [50], respectively. Additionally, results of 500, 30, and 10 m resolutions were calculated by MODIS, Landsat-8 and Sentinel-2 in this study.

With its 10 m resolution, the Sentinel-2 GPP can provide more information about vegetation dynamics in built areas. Despite the fact that Landsat-8 GPP has a relatively high resolution to explore the dynamic of urban forests, the 16-day revisiting period made the images of Landsat-8 subject to cloud cover. For example, for July and August of 2021, the whole regional average NIRv of Landsat 8 cannot be generated. Vegetation is at a growing peak during these two months, which is vital for the yearly total GPP. Therefore, the 5-day revisiting period and higher spatial resolution of Sentinel-2 images make them ideal for studying GPP on an urban scale.

4.3. *Importance of Urban Forests for Regional Carbon Budget*

In the land process-based model, as well as the remote sensing model, urban built areas are usually masked according to land cover data based on the hypothesis that the photosynthesis of urban forest is weak [45,47]. However, our study illustrated that the photosynthetic capacity of urban forest was also considerable (Table 3). Meanwhile, in Suzhou, the coverage of urban forest was 2253.1 km^2, which was significantly higher than the 371.9 km^2 of vegetation in non-built area. As a result, about 44.43% of the total GPP in Suzhou was contributed by urban forests (Figure 3). Therefore, the GPP of urban forest is an important part of the regional carbon budget and should not be ignored.

Table 3. Differences in yearly GPP between vegetation in non-built areas and built area.

Veg. Type	Non-Built Area (g C m^{-2} yr^{-1})	Urban Forest (g C m^{-2} yr^{-1})
GRA	2275.9 ± 949.8	841.5 ± 599.2
DBF	2500.9 ± 749.0	1589.7 ± 722.0
MF	2306.1 ± 605.6	1478.4 ± 624.4
EBF	3206.6 ± 507.4	2640.6 ± 478.8

Note: GRA, DBF, MF, and EBF are grassland, deciduous broadleaf forest, mixed forest, and evergreen broadleaf forest, respectively. Urban forest is the GRA, DBF, MF, and EBF in the built area.

The average GPP of urban forests was significantly lower (about 36.3%) than the average GPP of vegetation in the non-built area (Table 3). This reduction of GPP can primarily be attributed to many factors related to human activities in built areas [61,62]. From an ecological perspective, urban forests are highly fragmented and, as a result, they are vulnerable to human activities [63]. Additionally, in view of the atmosphere environment, urban forests are usually exposed to a high concentration of PM2.5 and O_3, which damages leaf tissue and affects photosynthesis [64]. However, the urban environment also includes some positive factors for carbon assimilation, for example, the increment in temperature due to the heat island effect [61,65], high CO_2 concentration from greenhouse gas emission [45,66], and enhanced nitrogen deposition from fossil fuel [67]. If we can improve the management of urban forests to maximize positive factors that promote carbon assimilation, urban forests will be able to play a more significant role in regional carbon budgets, which will help China achieve the target of carbon neutrality.

5. Conclusions

In this study, MODIS, Landsat-8 and Sentinel-2 images were used to generate regional GPP in one of China's most economically developed cities, Suzhou, to identify the applicability of different kinds of remote sensing data on urban scale studies and the characteristics of urban carbon budgets. The results of GPP estimated by MODIS, Landsat-8, and Sentinel-2 images were 8.43, 8.48, and 9.30 Tg C yr^{-1}, respectively. The monthly dynamic of GPP exhibited two peaks in April and September. In May, the harvest of wheat and the pruning process conducted on urban forests made a pronounced decline in total GPP by about 25.93% according to Sentinel-2. Accordingly, climate factors as well as anthropogenic factors contribute to the change of urban GPP. As the spatial resolution rose, the contribution of the urban forest to regional total GPP increased, reaching about 44.63% according to the 10 m sentinel-2 images. Since the distribution of urban forests is highly fragmental, high-resolution remote sensing images can better figure out the dynamic changes in the GPP of the urban forest. The results of our study demonstrate the importance of using high-resolution remote sensing images for estimating the GPP of the urban forest and for improving our understanding of the urban carbon budget.

Supplementary Materials: The following supporting information can be downloaded at: https://www.mdpi.com/article/10.3390/rs15010071/s1, Table S1. The relationship between GPP and NIRv.

Author Contributions: Conceptualization, X.L., Q.H. and Q.Z., methodology, X.L. and Q.H., software, F.C.; validation, Q.H. and X.L.; formal analysis, X.L., Q.H., H.Z. and Q.Z.; resources, X.L. and Q.H.; data curation, F.C. and X.L.; writing—original draft preparation, Q.H. and X.L.; writing—review and editing, Q.H., X.L., Q.Z. and H.Z.; visualization, X.L., funding acquisition, Q.H. All authors have read and agreed to the published version of the manuscript.

Funding: This study was funded by the National Natural Science Foundation of China (Grant no. 41871334, Grant number. 42071392), the Natural Science Foundation of Nanjing Xiaozhuang University (Grant number. 2020NXY15), and the Suzhou Agricultural Science and Technology Innovation project (Grant no. SNG2020072).

Data Availability Statement: The research does not involve this issue.

Acknowledgments: We really appreciate the editors and anonymous reviewers for their meaningful comments for improving our manuscript.

Conflicts of Interest: The authors declare no conflict of interest.

References

1. Anav, A.; Friedlingstein, P.; Beer, C.; Ciais, P.; Harper, A.; Jones, C.; Murray-Tortarolo, G.; Papale, D.; Parazoo, N.C.; Peylin, P.; et al. Spatiotemporal patterns of terrestrial gross primary production: A review. *Rev. Geophys.* **2015**, *53*, 785–818. [CrossRef]
2. Zheng, Y.; Shen, R.; Wang, Y.; Li, X.; Liu, S.; Liang, S.; Chen, J.M.; Ju, W.; Zhang, L.; Yuan, W. Improved estimate of global gross primary production for reproducing its long-term variation, 1982–2017. *Earth Syst. Sci. Data* **2020**, *12*, 2725–2746. [CrossRef]
3. Jian, J.; Bailey, V.; Dorheim, K.; Konings, A.G.; Hao, D.; Shiklomanov, A.N.; Snyder, A.; Steele, M.; Teramoto, M.; Vargas, R.; et al. Historically inconsistent productivity and respiration fluxes in the global terrestrial carbon cycle. *Nat. Commun.* **2022**, *13*, 1733. [CrossRef] [PubMed]
4. Baldocchi, D.; Falge, E.; Gu, L.; Olson, R.; Hollinger, D.; Running, S.; Anthoni, P.; Bernhofer, C.; Davis, K.; Evans, R.; et al. FLUXNET: A new tool to study the temporal and spatial variability of ecosystem-scale carbon dioxide, water vapor, and energy flux densities. *Bull. Am. Meteorol. Soc.* **2001**, *82*, 2415–2434. [CrossRef]
5. Beer, C.; Reichstein, M.; Tomelleri, E.; Ciais, P.; Jung, M.; Carvalhais, N.; Rödenbeck, C.; Arain, M.A.; Baldocchi, D.; Bonan, G.B.; et al. Terrestrial gross carbon dioxide uptake: Global distribution and covariation with climate. *Science* **2010**, *329*, 834–838. [CrossRef] [PubMed]
6. Jung, M.; Reichstein, M.; Margolis, H.A.; Cescatti, A.; Richardson, A.D.; Arain, M.A.; Arneth, A.; Bernhofer, C.; Bonal, D.; Chen, J.; et al. Global patterns of land-atmosphere fluxes of carbon dioxide, latent heat, and sensible heat derived from eddy covariance, satellite, and meteorological observations. *J. Geophys. Res. Biogeosci.* **2011**, *116*, G00J07. [CrossRef]
7. Jiang, C.; Ryu, Y. Multi-scale evaluation of global gross primary productivity and evapotranspiration products derived from Breathing Earth System Simulator (BESS). *Remote Sens. Environ.* **2016**, *186*, 528–547. [CrossRef]
8. Qiu, B.; Chen, J.M.; Ju, W.; Zhang, Q.; Zhang, Y. Simulating emission and scattering of solar-induced chlorophyll fluorescence at far-red band in global vegetation with different canopy structures. *Remote Sens. Environ.* **2019**, *233*, 111373. [CrossRef]
9. Wu, Z.; Ahlström, A.; Smith, B.; Ardö, J.; Eklundh, L.; Fensholt, R.; Lehsten, V. Climate data induced uncertainty in model-based estimations of terrestrial primary productivity. *Environ. Res. Lett.* **2017**, *12*, 064013. [CrossRef]
10. Running, S.W.; Nemani, R.R.; Heinsch, F.A.; Zhao, M.; Reeves, M.; Hashimoto, H. A continuous satellite-derived measure of global terrestrial primary production. *BioScience* **2004**, *54*, 547–560. [CrossRef]
11. Zhang, Z.; Zhang, Y.; Zhang, Y.; Gobron, N.; Frankenberg, C.; Wang, S.; Li, Z. The potential of satellite FPAR product for GPP estimation: An indirect evaluation using solar-induced chlorophyll fluorescence. *Remote Sens. Environ.* **2020**, *240*, 111686. [CrossRef]
12. Damm, A.; Elbers, J.A.N.; Erler, A.; Gioli, B.; Hamdi, K.; Hutjes, R.; Kosvancova, M.; Meroni, M.; Miglietta, F.; Moersch, A.; et al. Remote sensing of sun-induced fluorescence to improve modeling of diurnal courses of gross primary production (GPP). *Glob. Chang. Biol.* **2010**, *16*, 171–186. [CrossRef]
13. Zhang, Y.; Guanter, L.; Berry, J.A.; van der Tol, C.; Yang, X.; Tang, J.; Zhang, F. Model-based analysis of the relationship between sun-induced chlorophyll fluorescence and gross primary production for remote sensing applications. *Remote Sens. Environ.* **2016**, *187*, 145–155. [CrossRef]

14. Wu, C.; Chen, J.M.; Huang, N. Predicting gross primary production from the enhanced vegetation index and photosynthetically active radiation: Evaluation and calibration. *Remote Sens. Environ.* **2011**, *115*, 3424–3435. [CrossRef]
15. Wu, C.; Niu, Z.; Gao, S. Gross primary production estimation from MODIS data with vegetation index and photosynthetically active radiation in maize. *J. Geophys. Res. Atmos.* **2010**, *115*, D12127. [CrossRef]
16. Guanter, L.; Zhang, Y.; Jung, M.; Joiner, J.; Voigt, M.; Berry, J.A.; Frankenberg, C.; Huete, A.R.; Zarco-Tejada, P.; Lee, J.E.; et al. Global and time-resolved monitoring of crop photosynthesis with chlorophyll fluorescence. *Proc. Natl. Acad. Sci. USA* **2014**, *111*, E1327–E1333. [CrossRef]
17. Zhang, J.; Xiao, J.; Tong, X.; Zhang, J.; Meng, P.; Li, J.; Liu, P.; Yu, P. NIRv and SIF better estimate phenology than NDVI and EVI: Effects of spring and autumn phenology on ecosystem production of planted forests. *Agric. For. Meteorol.* **2022**, *315*, 108819. [CrossRef]
18. Wang, R.; Gamon, J.A.; Emmerton, C.A.; Springer, K.R.; Yu, R.; Hmimina, G. Detecting intra-and inter-annual variability in gross primary productivity of a North American grassland using MODIS MAIAC data. *Agric. For. Meteorol.* **2020**, *281*, 107859. [CrossRef]
19. Shi, H.; Li, L.; Eamus, D.; Huete, A.; Cleverly, J.; Tian, X.; Yu, Q.; Wang, S.; Montagnani, L.; Magliulo, V.; et al. Assessing the ability of MODIS EVI to estimate terrestrial ecosystem gross primary production of multiple land cover types. *Ecol. Indic.* **2017**, *72*, 153–164. [CrossRef]
20. Li, X.; Xiao, J. TROPOMI observations allow for robust exploration of the relationship between solar-induced chlorophyll fluorescence and terrestrial gross primary production. *Remote Sens. Environ.* **2022**, *268*, 112748. [CrossRef]
21. Baldocchi, D.D.; Ryu, Y.; Dechant, B.; Eichelmann, E.; Hemes, K.; Ma, S.; Sanchez, C.R.; Shortt, R.; Szutu, D.; Valach, A.; et al. Outgoing near-infrared radiation from vegetation scales with canopy photosynthesis across a spectrum of function, structure, physiological capacity, and weather. *J. Geophys. Res. Biogeosci.* **2020**, *125*, e2019JG005534. [CrossRef]
22. Zhang, Y.; Joiner, J.; Gentine, P.; Zhou, S. Reduced solar-induced chlorophyll fluorescence from GOME-2 during Amazon drought caused by dataset artifacts. *Glob. Chang. Biol.* **2018**, *24*, 2229–2230. [CrossRef] [PubMed]
23. Wang, S.; Zhang, Y.; Ju, W.; Qiu, B.; Zhang, Z. Tracking the seasonal and inter-annual variations of global gross primary production during last four decades using satellite near-infrared reflectance data. *Sci. Total Environ.* **2021**, *755*, 142569. [CrossRef]
24. Badgley, G.; Field, C.B.; Berry, J.A. Canopy near-infrared reflectance and terrestrial photosynthesis. *Sci. Adv.* **2017**, *3*, e1602244. [CrossRef]
25. Badgley, G.; Anderegg, L.D.; Berry, J.A.; Field, C.B. Terrestrial gross primary production: Using NIRV to scale from site to globe. *Glob. Chang. Biol.* **2019**, *25*, 3731–3740. [CrossRef] [PubMed]
26. Wu, G.; Guan, K.; Jiang, C.; Peng, B.; Kimm, H.; Chen, M.; Yang, X.; Wang, S.; Suyker, A.E.; Bernacchi, C.J.; et al. Radiance-based NIRv as a proxy for GPP of corn and soybean. *Environ. Res. Lett.* **2020**, *15*, 034009. [CrossRef]
27. Mengistu, A.G.; Mengistu Tsidu, G.; Koren, G.; Kooreman, M.L.; Boersma, K.F.; Tagesson, T.; Ardö, J.; Nouvellon, Y.; Peters, W. Sun-induced fluorescence and near-infrared reflectance of vegetation track the seasonal dynamics of gross primary production over Africa. *Biogeosciences* **2021**, *18*, 2843–2857. [CrossRef]
28. Qiao, K.; Zhu, W.; Xie, Z.; Li, P. Estimating the seasonal dynamics of the leaf area index using piecewise LAI-VI relationships based on phenophases. *Remote Sens.* **2019**, *11*, 689. [CrossRef]
29. Camps-Valls, G.; Campos-Taberner, M.; Moreno-Martínez, Á.; Walther, S.; Duveiller, G.; Cescatti, A.; Mahecha, M.D.; Muñoz-Marí, J.; García-Haro, F.J.; Guanter, L.; et al. A unified vegetation index for quantifying the terrestrial biosphere. *Sci. Adv.* **2021**, *7*, eabc7447. [CrossRef]
30. Yang, R.; Wang, J.; Zeng, N.; Sitch, S.; Tang, W.; McGrath, M.J.; Cai, Q.; Liu, D.; Lombardozzi, D.; Tian, H.; et al. Divergent historical GPP trends among state-of-the-art multi model simulations and satellite-based products. *Earth Syst. Dyn.* **2022**, *13*, 833–849. [CrossRef]
31. Merrick, T.; Pau, S.; Detto, M.; Broadbent, E.N.; Bohlman, S.A.; Still, C.J.; Almeyda Zambrano, A.M. Unveiling spatial and temporal heterogeneity of a tropical forest canopy using high-resolution NIRv, FCVI, and NIRvrad from UAS observations. *Biogeosciences* **2021**, *18*, 6077–6091. [CrossRef]
32. Dechant, B.; Ryu, Y.; Badgley, G.; Zeng, Y.; Berry, J.A.; Zhang, Y.; Goulas, Y.; Li, Z.; Zhang, Q.; Kang, M.; et al. Canopy structure explains the relationship between photosynthesis and sun-induced chlorophyll fluorescence in crops. *Remote Sens. Environ.* **2020**, *241*, 111733. [CrossRef]
33. Zeng, Y.; Badgley, G.; Dechant, B.; Ryu, Y.; Chen, M.; Berry, J.A. A practical approach for estimating the escape ratio of near-infrared solar-induced chlorophyll fluorescence. *Remote Sens. Environ.* **2019**, *232*, 111209. [CrossRef]
34. Escobedo, F.J.; Wagner, J.E.; Nowak, D.J.; De la Maza, C.L.; Rodriguez, M.; Crane, D.E. Analyzing the cost effectiveness of Santiago, Chile's policy of using urban forests to improve air quality. *J. Environ. Manag.* **2008**, *86*, 148–157. [CrossRef]
35. Escobedo, F.J.; Adams, D.C.; Timilsina, N. Urban forest structure effects on property value. *Ecosyst. Serv.* **2015**, *12*, 209–217. [CrossRef]
36. Lee, S.J.; Longcore, T.; Rich, C.; Wilson, J.P. Increased home size and hardscape decreases urban forest cover in Los Angeles County's single-family residential neighborhoods. *Urban For. Urban Green.* **2017**, *24*, 222–235. [CrossRef]
37. Ren, Z.; He, X.; Pu, R.; Zheng, H. The impact of urban forest structure and its spatial location on urban cool island intensity. *Urban Ecosyst.* **2018**, *21*, 863–874. [CrossRef]

38. Roeland, S.; Moretti, M.; Amorim, J.H.; Branquinho, C.; Fares, S.; Morelli, F.; Niinemets, Ü.; Paoletti, E.; Pinho, P.; Sgrigna, G.; et al. Towards an integrative approach to evaluate the environmental ecosystem services provided by urban forest. *J. For. Res.* **2019**, *30*, 1981–1996. [CrossRef]
39. Davies, Z.G.; Edmondson, J.L.; Heinemeyer, A.; Leake, J.R.; Gaston, K.J. Mapping an urban ecosystem service: Quantifying above-ground carbon storage at a city-wide scale. *J. Appl. Ecol.* **2011**, *48*, 1125–1134. [CrossRef]
40. Myeong, S.; Nowak, D.J.; Duggin, M.J. A temporal analysis of urban forest carbon storage using remote sensing. *Remote Sens. Environ.* **2006**, *101*, 277–282. [CrossRef]
41. Agbelade, A.D.; Onyekwelu, J.C. Tree species diversity, volume yield, biomass and carbon sequestration in urban forests in two Nigerian cities. *Urban Ecosyst.* **2020**, *23*, 957–970. [CrossRef]
42. Wang, S.; Ju, W.; Peñuelas, J.; Cescatti, A.; Zhou, Y.; Fu, Y.; Huete, A.; Liu, M.; Zhang, Y. Urban–rural gradients reveal joint control of elevated CO_2 and temperature on extended photosynthetic seasons. *Nat. Ecol. Evol.* **2019**, *3*, 1076–1085. [CrossRef] [PubMed]
43. Zhang, Y.; Meng, W.; Yun, H.; Xu, W.; Hu, B.; He, M.; Mo, X.; Zhang, L. Is urban green space a carbon sink or source?—A case study of China based on LCA method. *Environ. Impact Assess. Rev.* **2022**, *94*, 106766. [CrossRef]
44. Zhao, M.S.; Running, S.; Heinsch, F.A.; Nemani, R. MODIS-Derived Terrestrial Primary Production. In *Land Remote Sensing and Global Environmental Change*; Ramachandran, B., Justice, C.O., Abrams, M.J., Eds.; Springer: New York, NY, USA, 2010; pp. 635–660.
45. Zhao, T.; Brown, D.G.; Bergen, K.M. Increasing gross primary production (GPP) in the urbanizing landscapes of southeastern Michigan. *Photogramm. Eng. Remote Sens.* **2007**, *73*, 1159–1167. [CrossRef]
46. Ding, Z.; Zheng, H.; Li, H.; Yu, P.; Man, W.; Liu, M.; Tang, X.; Liu, Y. Afforestation-driven increases in terrestrial gross primary productivity are partly offset by urban expansion in Southwest China. *Ecol. Indic.* **2021**, *127*, 107641. [CrossRef]
47. Cui, Y.; Xiao, X.; Dong, J.; Zhang, Y.; Qin, Y.; Doughty, R.B.; Wu, X.; Liu, X.; Joiner, J.; Moore, B., III. Continued Increases of Gross Primary Production in Urban Areas during 2000–2016. *J. Remote Sens.* **2022**, *2022*, 9868564. [CrossRef]
48. Karra, K.; Kontgis, C.; Statman-Weil, Z.; Mazzariello, J.C.; Mathis, M.; Brumby, S.P. Global land use/land cover with Sentinel 2 and deep learning. In Proceedings of the 2021 IEEE International Geoscience and Remote Sensing Symposium IGARSS, Brussels, Belgium, 11–16 July 2021; pp. 4704–4707.
49. Li, X.B.; Shi, P. Sensitivity analysis of variation in NDVI, temperature and precipitation in typical vegetation types a across China. *Chin. J. Plant Ecol.* **2000**, *24*, 379.
50. Zhang, Y.; Xiao, X.; Wu, X.; Zhou, S.; Zhang, G.; Qin, Y.; Dong, J. A global moderate resolution dataset of gross primary production of vegetation for 2000–2016. *Sci. Data* **2015**, *4*, 170165. [CrossRef]
51. Ju, W.M.; Zhou, Y.L. *Global Daily GPP Simulated Data Products from 1981 to 2019 [DB/OL]*; National Ecosystem Science Data Center: Beijing, China, 2021. [CrossRef]
52. He, M.; Ju, W.; Zhou, Y.; Chen, J.; He, H.; Wang, S.; Wang, H.; Guan, D.; Yan, J.; Li, Y.; et al. Development of a two-leaf light use efficiency model for improving the calculation of terrestrial gross primary productivity. *Agric. For. Meteorol.* **2013**, *173*, 28–39. [CrossRef]
53. Zhou, H.; Xu, M.; Pan, H.; Yu, X. Leaf-age effects on temperature responses of photosynthesis and respiration of an alpine oak, Quercus aquifolioides, in southwestern China. *Tree Physiol.* **2015**, *35*, 1236–1248. [CrossRef]
54. Haberl, H.; Erb, K.H.; Krausmann, F.; Gaube, V.; Bondeau, A.; Plutzar, C.; Gingrich, S.; Lucht, W.; Fischer-Kowalski, M. Quantifying and mapping the human appropriation of net primary production in earth's terrestrial ecosystems. *Proc. Natl. Acad. Sci. USA* **2007**, *104*, 12942–12947. [CrossRef] [PubMed]
55. Cui, Y.; Xiao, X.; Zhang, Y.; Dong, J.; Qin, Y.; Doughty, R.B.; Zhang, G.; Wang, J.; Wu, X.; Qin, Y.; et al. Temporal consistency between gross primary production and solar-induced chlorophyll fluorescence in the ten most populous megacity areas over years. *Sci. Rep.* **2017**, *7*, 14963. [CrossRef] [PubMed]
56. Wang, F.; Jiang, H.; Zhang, X. Spatial–temporal dynamics of gross primary productivity, evapotranspiration, and water-use efficiency in the terrestrial ecosystems of the Yangtze River Delta region and their relations to climatic variables. *Int. J. Remote Sens.* **2015**, *36*, 2654–2673. [CrossRef]
57. Wang, L.; Zhu, H.; Lin, A.; Zou, L.; Qin, W.; Du, Q. Evaluation of the latest MODIS GPP products across multiple biomes using global eddy covariance flux data. *Remote Sens.* **2017**, *9*, 418. [CrossRef]
58. Zhang, F.; Chen, J.M.; Chen, J.; Gough, C.M.; Martin, T.A.; Dragoni, D. Evaluating spatial and temporal patterns of MODIS GPP over the conterminous US against flux measurements and a process model. *Remote Sens. Environ.* **2012**, *124*, 717–729. [CrossRef]
59. Chen, Y.; Gu, H.; Wang, M.; Gu, Q.; Ding, Z.; Ma, M.; Liu, R.; Tang, X. Contrasting performance of the remotely-derived GPP products over different climate zones across China. *Remote Sens.* **2019**, *11*, 1855. [CrossRef]
60. Liu, Z.; Wang, L.; Wang, S. Comparison of different GPP models in China using MODIS image and ChinaFLUX data. *Remote Sens.* **2014**, *6*, 10215–10231. [CrossRef]
61. Zhang, L.; Yang, L.; Zohner, C.M.; Crowther, T.W.; Li, M.; Shen, F.; Guo, M.; Qin, J.; Yao, L.; Zhou, C. Direct and indirect impacts of urbanization on vegetation growth across the world's cities. *Sci. Adv.* **2022**, *8*, eabo0095. [CrossRef]
62. Nuarsa, I.W.; As-syakur, A.R.; Gunadi, I.G.A.; Sukewijaya, I.M. Changes in Gross Primary Production (GPP) over the past two decades due to land use conversion in a tourism city. *ISPRS Int. J. Geo-Inf.* **2018**, *7*, 57. [CrossRef]
63. Ren, Y.; Yan, J.; Wei, X.; Wang, Y.; Yang, Y.; Hua, L.; Xiong, Y.; Niu, X.; Song, X.J. Effects of rapid urban sprawl on urban forest carbon stocks: Integrating remotely sensed, GIS and forest inventory data. *J. Environ. Manag.* **2012**, *113*, 447–455. [CrossRef]

64. Li, Y.; Wang, Y.; Wang, B.; Wang, Y.; Yu, W. The response of plant photosynthesis and stomatal conductance to fine particulate matter (PM2. 5) based on leaf factors analyzing. *J. Plant Biol.* **2019**, *62*, 120–128. [CrossRef]
65. Peng, S.; Piao, S.; Ciais, P.; Friedlingstein, P.; Ottle, C.; Bréon, F.M.; Nan, H.; Zhou, L.; Myneni, R.B. Surface urban heat island across 419 global big cities. *Environ. Sci. Technol.* **2012**, *46*, 696–703. [CrossRef] [PubMed]
66. Marcotullio, P.J.; Sarzynski, A.; Albrecht, J.; Schulz, N.; Garcia, J. The geography of global urban greenhouse gas emissions: An exploratory analysis. *Clim. Chang.* **2013**, *121*, 621–634. [CrossRef]
67. Townsend, A.R.; Braswell, B.H.; Holland, E.A.; Penner, J.E. Spatial and temporal patterns in terrestrial carbon storage due to deposition of fossil fuel nitrogen. *Ecol. Appl.* **1996**, *6*, 806–814. [CrossRef]

remote sensing

MDPI

Article

An Improved Forest Height Model Using L-Band Single-Baseline Polarimetric InSAR Data for Various Forest Densities

Ao Sui [1], Opelele Omeno Michel [2], Yu Mao [3] and Wenyi Fan [1,*]

1 Key Laboratory of Sustainable Forest Ecosystem Management—Ministry of Education, School of Forestry, Northeast Forestry University, Harbin 150040, China
2 Department of Natural Resources Management, Faculty of Agricultural Sciences, University of Kinshasa, Kinshasa 117, Democratic Republic of the Congo
3 International Institute for Earth System Sciences, School of Geography and Ocean Science, Nanjing University, Nanjing 210000, China
* Correspondence: fanwy@nefu.edu.cn; Tel.: +86-139-4605-5384

Abstract: Forest density affects the inversion of forest height by influencing the penetration and attenuation of synthetic aperture radar (SAR) signals. Traditional forest height inversion methods often fail in low-density forest areas. Based on L-band single-baseline polarimetric SAR interferometry (PolInSAR) simulation data and the BioSAR 2008 data, we proposed a forest height optimization model at the stand scale suitable for various forest densities. This optimization model took into account shortcomings of the three-stage inversion method by employing height errors to represent the mean penetration depth and SINC inversion method. The relationships between forest density and extinction coefficient, penetration depth, phase, and magnitude were also discussed. In the simulated data, the inversion height established by the optimization method was 17.35 m, while the RMSE value was 3.01 m when the forest density was 100 stems/ha. This addressed the drawbacks of the conventional techniques including failing at low forest density. In the real data, the maximum RMSE of the optimization method was 2.17 m as the stand density increased from 628.66 stems/ha to 1330.54 stems/ha, showing the effectiveness and robustness of the optimization model in overcoming the influence of stand density on the inversion process in realistic scenarios. This study overcame the stand density restriction on L-band single baseline PolInSAR data for forest height estimation and offered a reference for algorithm selection and optimization. The technique is expected to be extended from the stand scale to a larger area for forest ecosystem monitoring and management.

Keywords: L-band PolInSAR; RVoG model; forest height; three-stage inversion method; forest density; terrain slope; coherence; extinction coefficient; signal penetration

Citation: Sui, A.; Michel, O.O.; Mao, Y.; Fan, W. An Improved Forest Height Model Using L-Band Single-Baseline Polarimetric InSAR Data for Various Forest Densities. *Remote Sens.* **2023**, *15*, 81. https://doi.org/10.3390/rs15010081

Academic Editor: Arturo Sanchez-Azofeifa

Received: 28 October 2022
Revised: 21 December 2022
Accepted: 22 December 2022
Published: 23 December 2022

1. Introduction

Forests are significant contributors to the global carbon cycle. Forest height is a crucial input variable for building biomass models and assessing the condition of forest resources. Traditional forest surveys obtain forest height through field measurements. This approach can only obtain small-scale data on a point-by-point basis, while at the same time consuming both human and material resources. It is challenging to obtain forest data on a large scale, but remote sensing is best suited to address this challenge. The commonly used remote sensing techniques include optical remote sensing, Lidar, synthetic aperture radar (SAR) methods, and so on. Optical remote sensing can obtain information on forest biomass, species, and biochemical properties. However, optical remote sensing is not penetrating and is susceptible to cloud cover and weather, thus limiting its use [1]. LiDAR can provide high-precision information about the vertical structure of the forest, but it is also affected by weather, such as clouds, fog, and rain [2]. Additionally, the observation area is small, and the acquisition cost is high [3]. SAR signals have strong penetration and are not affected by the weather. In recent years, SAR has developed the capability to monitor the vertical and horizontal structure of features continuously and has excellent potential to estimate forest height and above-ground biomass (AGB) on a global scale [4,5].

Polarimetric SAR interferometry (PolInSAR) is one of the SAR techniques that has received much attention in recent years. PolInSAR uses the interferometric phase to observe the vertical structure of the forests. The technique uses the sensitivity of polarization to the shape, dielectric properties, texture, and orientation of the scatterers to identify different scattering mechanisms of the target [6]. Based on PolInSAR inversion models, several researchers have effectively achieved the inversion of forest height and AGB [7,8]. The DEM difference method [9], RVoG ground phase method [10], SINC inversion method [11], phase and coherence inversion method [12], and three-stage inversion method [13] are the popular PolInSAR inversion models. Among these, the three-stage inversion method based on the random volume over ground (RVoG) model minimizes the complexity of the inversion model, expanding the application of the RVoG method.

The RVoG model uses a mathematical model to connect the forest parameters to the SAR system parameters. Most current RVoG models and their improvement methods are mainly concerned with the optimization of model parameters. The key points and difficulties of the vegetation scattering model are how to estimate the ground phase accurately and effectively separate the ground scattering contribution from the canopy scattering contribution. Many researchers enhance the separation of two-phase centers using the singular value decomposition (SVD) [14] and phase diversity (PD) coherence optimization method [15] to precisely determine the ground phase.

On the other hand, some studies have focused on the effect of scenes on the vegetation scattering model. Since the RVoG model initially focused on airborne data, it did not consider the impact of terrain and temporal decorrelation [10,16]. Temporal decorrelation and terrain distortion in real scenarios lead to the limitations of traditional models in practical applications [17–19]. Lu et al. suggested the sloped random volume over ground (S-RVoG) model, which accounts for the influence of terrain slope on PolInSAR data inversion. S-RVoG rectifies the forest parameters by including the range slope in the RVoG model and demonstrating the validity using L-band simulation data [20]. XIE et al. showed that the S-RVoG model corrected terrain slope and improved forest height inversion accuracy using P-band dual-baseline PolInSAR data [21]. Cloude et al. proposed the random volume over ground with volume temporal decorrelation (RVoG+VTD) model. The RVoG+VTD model only considers the effect of temporal decorrelation on the coherence amplitude [22]. Later, various researchers presented the temporal decorrelation random volume over ground with volume (TD-RVoG) model, the random motion over ground (RMOG) model, and the semi-empirical forest height inversion approach to transform temporal decorrelation into a complex form [23–26].

However, the forest density also degrades the inversion results, making the ability of the vegetation scattering model less effective. Forest density affects forest characteristics (extinction coefficient, ground phase, forest height) and SAR system parameters (amplitude, phase) by influencing the penetration and attenuation of the SAR signal. Thick forests diminish the degree of complex coherence separation, resulting in erroneous estimation of the ground phase [13]. Xie et al. discovered that high forest density had a substantial canopy scattering contribution, resulting in an underestimating of the inversion findings after adjusting terrain slope [27]. Meanwhile, Garestier et al. employed X-band data to invert sparse coniferous forests. Low forest density causes extensive canopy gaps, allowing short wavelength PolInSAR data to penetrate the canopy and result in forest height inversion [28]. Wang et al. used simulated and P-band BioSAR 2008 data to show that the traditional three-stage inversion method failed to invert in sparsely vegetated areas. Noteworthy, forest density is positively correlated with ground phase estimation accuracy and negatively correlated with ground-to-volume scattering ratio [29]. Most current vegetation scattering model studies focuses on the qualitative relationship between forest density and model parameters. In this paper, we present an improved method applicable to various forest densities based on a study of the influence of forest density on SAR system parameters and forest characteristics.

In this study, we first analyze the forest height inversion results of five commonly utilized PolInSAR inversion models using simulated data of different forest densities. By comprehensively analyzing the relationships and regulations of forest density and forest parameters, we propose the coherence magnitude and three-stage hybrid theoretical method applicable to various forest densities. The model uses the forest height error to represent the average penetration depth of various forest densities. The adjustment coefficients are selected iteratively according to different forest stands' characteristics and penetration depth. Finally, the SINC inversion method, three-stage inversion method, and adjustment coefficients constitute the model. Comparing the inversion results of the improved model with those of the traditional model shows that the hybrid iterative theoretical process achieves single-baseline PolInSAR forest height inversion for different forest densities. In particular, it overcomes the shortcomings of conventional approaches to inversion failure at low forest density. The improved model was validated using actual SAR data. Furthermore, a slope correction algorithm and PD coherence optimization algorithm were introduced to the hybrid iterative theoretical method to form the coherence magnitude and three-stage hybrid iterative application model (hybrid iterative application model). The optimization model achieves high accuracy inversion in real scenarios with different forest densities and enables a reference for future research.

2. Datasets and Pre-Processing

2.1. PolSARproSim Simulated Datasets

This study aims to investigate the effects of different forest densities on the widely employed PolInSAR inversion methods and the effectiveness of the improved model under various forest densities, which requires controlling the same forest height and reducing the impact of temporal decoherence. The experimental conditions are more stringent, and obtaining ideal airborne SAR data is more complicated. The ESA's PolSARpro software allows the user to flexibly set the platform configuration, forest/ground surface configuration, and other parameters for data simulation to build the ideal experimental conditions [30,31].

To study the effect of forest density on the tree height inversion technique, PolSARproSim built a simulated airborne L-band (central frequency 1.3 GHz) PolInSAR dataset, which contains nine groups of forest densities ranging from 100 stems/ha to 900 stems/ha. The other parameters in the simulated dataset are similar and kept fixed, as shown in Table 1. The forest density indicator in the data set is the tree number per hectare (stems/ha), and the forest density of 100 stems/ha, 200 stems/ha, 300 stems/ha, 400 stems/ha, 500 stems/ha, 600 stems/ha, 700 stems/ha, 800 stems/ha, and 900 stems/ha coniferous forest, the slope in both azimuth and range directions is 0. The ground roughness and soil moisture are also 0. Terrain, shrub layer, and temporal decoherence do not influence the echo signal or inversion technique. Figure 1 illustrates scenario images for nine groups of simulated datasets, and Figure 2 displays Pauli-basis images for nine groups of simulated datasets.

Table 1. ESA's PolSARproSim module simulated nine groups of stand-specific parameters. Coniferous forest densities ranged from 100 stems/ha to 900 stems/ha. The reference heights are all 18 m, the azimuth and range slope are all 0, and the ground roughness and soil wetness are 0. Terrain, shrub layer, and temporal decoherence do not influence the echo signal or inversion technique.

Platform Configuration	Parameter	Forest/Ground Surface Configuration	Parameter
Platform Altitude	3000 m	Tree Species	Pine
Horizontal/Vertical Baseline	10 m,1 m	Surface Properties/Ground Moisture Content	0,0
Incidence Angle	45°	Azimuth/Range Ground Slope	0
Centre Frequency	1.3 GHZ	Tree Height	18 m

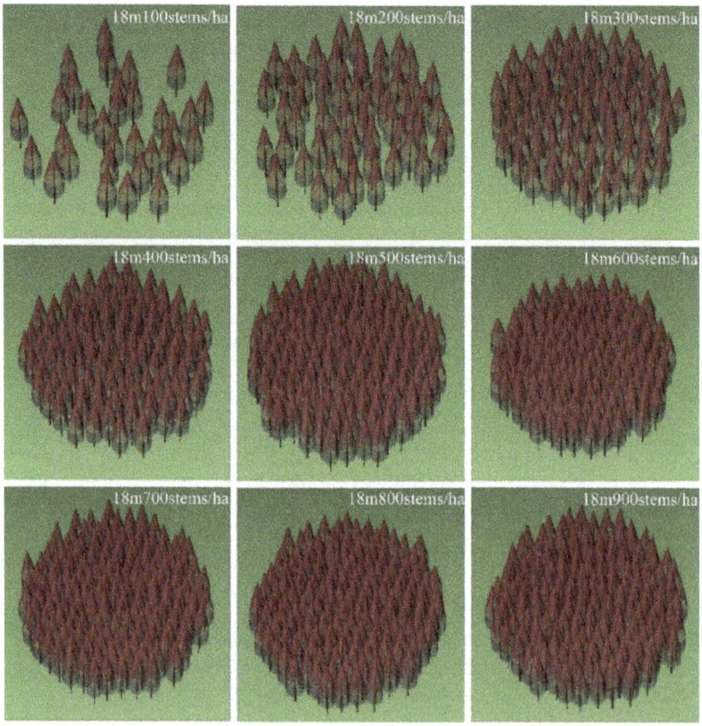

Figure 1. Scenarios of simulation for nine groups of simulated data. The reference height for all nine datasets is 18 m. Forest density ranges from 100 to 900 stems/ha.

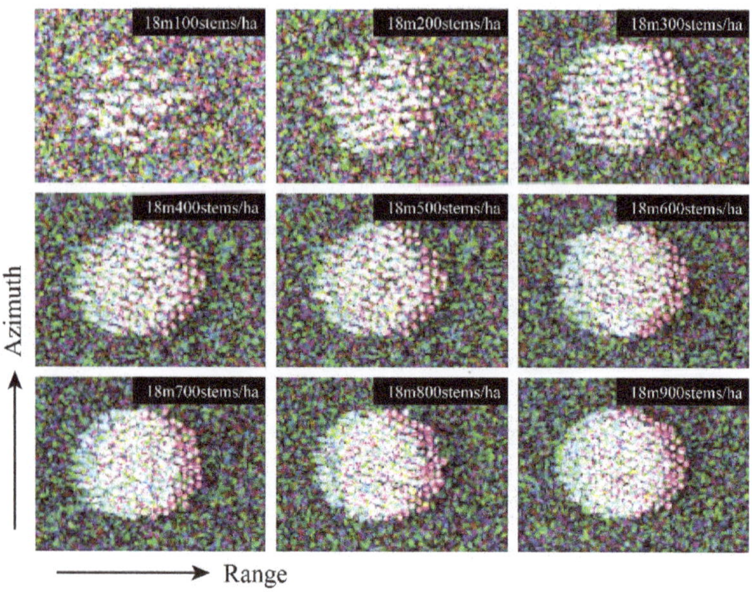

Figure 2. Pauli-basis (HV, HH + VV, and HH − VV) composite images of nine simulated data sets.

2.2. The BioSAR 2008 Datasets

Airborne L-band PolInSAR data from the Krycklan catchment in northern Sweden, taken during ESA's BioSAR 2008 experiment, were used for model validation in this study. The BioSAR 2008 campaign acquired high-latitude boreal forest data with terrain impacts to investigate the BIOMASS mission's potential for estimating biomass in boreal forests. The test site (64°14′N, 19°46′E) is in Vindeln municipality, situated in Sweden (Figure 3a), with elevation changes from 100 m to 300 m and mixed coniferous forest as the dominating forest type. The dataset is available on the ESA by application and includes airborne SAR data, field inventory data such as Lidar data, DEM data, and 31 forest stands.

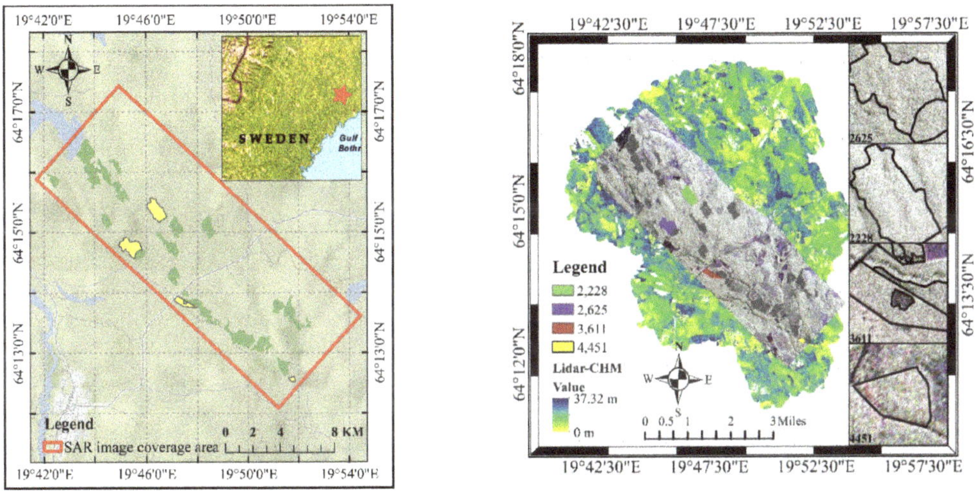

(**a**) (**b**)

Figure 3. The extent of the study area and the several products included in the BioSAR 2008 dataset. (**a**) The red pentagram in the upper right corner of the thumbnail depicts the location of the study area in Sweden. The red rectangle region shows the SAR image. The green and yellow polygons represent the 31 forest stands in the red rectangle region; the four yellow polygons represent the forest stands employed in this study. (**b**) The green background image is a Lidar image with a grid size resampling of 1 m × 1 m. Pauli-basis image is the extent of SAR images. The Pauli-basis image (on the left) shows the locations of the four forest stands. The shape of four forest stands on the Pauli-basis image is shown on the right. All products on the map are geocoded to WGS84 UTM Zone 34N.

In 2008, 31 forest stands were field-surveyed, and ArcGIS software was used to generate vector contour polygons (yellow polygons in Figure 3a). The stand-level tree height, volume, and biomass were estimated and corrected using measured data and static functions. The guidebook explains the specific prediction technique and field survey process [32]. Because this research investigates the influence of forest density on the inversion of PolInSAR data, we consider tree species composition, stand mean height, and forest density as indicators to select among 31 forest stands. This was also to ensure that the experimental conditions of the realistic and simulated datasets were comparable. In this study, the four pure coniferous stands (stand numbers 4451, 2625, 3611, and 2228) with forest density of 628.660 stems/ha, 840.340 stems/ha, 1149.100 stems/ha, and 1330.540 stems/ha, respectively, and the average tree height measured in the field (nearly 18 m) were nearly similar to a real-world scenario for algorithm validation. The gap in forest density between neighboring forest stands is 200 stems/ha. The range of forest density was vast, with a difference of 700 stems/ha between the lowest and the highest forest density, which was consistent with the experimental goal of this work.

The two images of L-band PolInSAR data collected on 15 October 2008, were taken as real-world data for testing the inversion model in this work. The SceneIDs were 08BioSAR0201 × 1 and 08BioSAR0205 × 1, respectively, as shown in Table 2. The pre-processing of the master and slave images mainly includes coregistration, removing the flat-earth phase, multilooking, 7′7 LEE refined speckle filtering [33], and using 50 m × 50 m DEM for geocoding [34]. Figure 3b shows the SAR image and LiDAR image of WGS84 UTM Zone 34N on the left side and the shapes of the four forest stands on the right side.

Table 2. BioSAR 2008 L-band PolInSAR data specific parameter information. The sceneIDs of master and slave images are 08BioSAR0201 × 1 and 08BioSAR0205 × 1, respectively.

Scene ID	Baseline (m)	Kz	Band	Polarization
08BioSAR0201	Master 0 m	Master	L	Quad
08BioSAR0205	Slave 12 m	0.046–0.370	L	Quad

3. Methodology

This study uses five common inversion methods in the simulated dataset to investigate the effect of forest density on the PolInSAR inversion process. According to the features of the three-phase inversion technique and SINC inversion method, a coherence magnitude and three-stage hybrid iterative inversion method are suited for various forest densities. The terrain-corrected incidence angle and vertical wavenumber were introduced into the coherence amplitude and three-phase hybrid iterative inversion model to reduce terrain's impact on the inversion model. The model without terrain correction is called the coherence magnitude and three-stage hybrid iterative inversion theory method. The coherence magnitude and three-stage hybrid iterative inversion application model is the model with terrain correction. Figure 4 depicts the flowchart of the study.

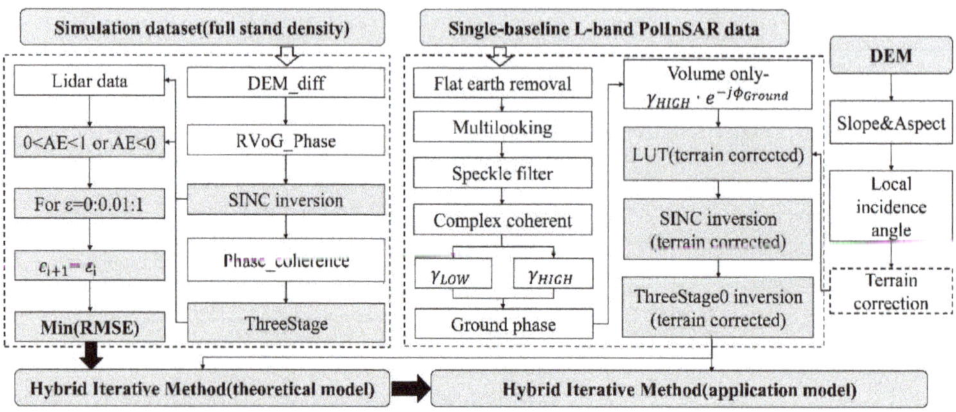

Figure 4. Flow chart. DEM_diff is the DEM difference method, SINC inversion is the SINC inversion method, RVoG_Phase represents the RVoG ground phase method, ThreeStage represents the traditional three-stage inversion method, Phase_Coherence is the phase and coherence inversion method, Hybrid Iterative Method (theoretical model) represents coherence amplitude and three-stage hybrid iterative theoretical model. The hybrid Iterative Method (application model) describes the coherence amplitude and three-stage hybrid iterative application model.

3.1. *Typical Models for the PolInSAR Technique of Forest Height Inversion*

The DEM Difference Model determines the forest height h_v by comparing the phase differences and vertical wavenumbers k_z of the two polarization modes representing the canopy and ground scattering centers. The key to this technique is determining the two polarization states representing forest canopy scattering and ground scattering, as shown in Equation (1) [6].

$$h_v = \frac{\arg(\gamma_{w_v}) - \arg(\gamma_{w_g})}{k_z} \tag{1}$$

$$k_z = \frac{4\pi\Delta\theta}{\lambda\sin\theta} \approx \frac{4\pi B_\perp}{\lambda R\sin\theta} \tag{2}$$

where w_g and γ_{w_g} mean the ground scattering mechanism and the ground scattering complex coherence, respectively. γ_{w_v} is the volume scattering complex coherence and w_v is the forest canopy scattering mechanism, respectively. θ is the radar angle of incidence; λ represents the radar wavelength; R represents the radar slant range, and $B_\perp$ represents the perpendicular baseline. Although this approach is theoretically simple, the results often underestimate 1/3 of the actual tree height [35].

The **Three-Stage Inversion Method** represents complex coherence coefficients of different polarizations that are linearly distributed in the complex unit circle. The RVoG model is decomposed into a linear structure. The quantitative inversion process of vegetation parameters is divided into three stages (Equation (3)) [10,28].

Using the least squares algorithm to fit a straight line in the complex plane;

Estimating the actual ground phase point based on the intersection of the fitted line with the complex unit circle;

Inverting the tree height and extinction coefficient using a height-extinction coefficient two-dimensional look-up table (LUT).

$$\begin{aligned} \gamma(w) &= e^{j\phi_0}\frac{\gamma_v + m(w)}{1+m(w)} \\ &= e^{j\phi_0}\left[\gamma_v + \frac{m(w)}{1+m(w)}(1-\gamma_v)\right], L_w = \frac{m(w)}{1+m(w)} \\ &= e^{j\phi_0}[\gamma_v + L_w(1-\gamma_v)] \end{aligned} \tag{3}$$

$$\begin{aligned} \gamma_v &= \frac{\int_0^{h_v} f(z)e^{jk_z z}dz}{\int_0^{h_v} f(z)dz} = \frac{2\sigma}{\cos\left(e^{2\sigma h_v/\cos\theta}-1\right)}\int_0^{h_v} e^{jk_z z}e^{\frac{2\sigma z}{\cos\theta}}dz \\ &= \frac{2\sigma}{2\sigma + jk_z\cos\theta}\cdot\frac{\exp(2\sigma h_v/\cos\theta + jk_z h_v)-1}{\exp(2\sigma h_v/\cos\theta)-1} \end{aligned} \tag{4}$$

where γ_v is the pure complex coherence of the forest canopy, and L_w is the ground scattering ratio for a given polarization state w. $m(w)$ is the ground-to-volume scattering ratio. σ is the extinction coefficient, representing the energy loss of electromagnetic waves through the medium. $f(z)$ is the vegetation vertical structure function, which is usually simplified to an exponential function in the RVoG model.

The **RVoG Ground Phase Method** (Equation (5)) is an improved model of the DEM difference method. The RVoG model is used to compute L_{w_g}, which is then introduced in the DEM difference method to estimate forest height (Equations (6) and (7)). The approach inverses forest height is better than the DEM difference method, but it still underestimates forest height [7].

$$h_v = \frac{\arg(\gamma_{w_v}) - \phi_0}{k_z} \tag{5}$$

$$\begin{aligned} \phi_0 &= \arg\left[\gamma_{w_g} - \gamma_{w_v}\left(1 - L_{w_g}\right)\right] \\ &\Rightarrow \left\{\begin{array}{c} \gamma_{w_v} = e^{j\phi_0}\gamma_v \\ \gamma_{w_g} = e^{j\phi_0}\left[\gamma_v + \frac{m(w_g)}{1+m(w_g)}(1-\gamma_v)\right] \end{array}\right. , L_{w_g} = \frac{m(w_g)}{1+m(w_g)}, 0 \le L_{w_g} \le 1 \\ &\Rightarrow \gamma_{w_g} = \gamma_{w_v} + L_{w_g}\cdot e^{j\phi_0} - L_{w_g}\cdot\gamma_{w_v} \\ &\Rightarrow e^{j\phi_0} = \frac{\gamma_{w_g} - \gamma_{w_v}\left(1-L_{w_g}\right)}{L_{w_g}} \end{aligned} \tag{6}$$

$$\begin{array}{l} AL^2{}_{w_g} + BL_{w_g} + C = 0 \Rightarrow L_{w_g} = \frac{-B-\sqrt{B^2-4AC}}{2A} \\ A = |\gamma w_v|^2 - 1, B = 2Re\left(\left(\gamma w_g - \gamma w_v\right)\gamma^*{}_{w_v}\right), C = \left|\gamma w_g - \gamma w_v\right|^2 \end{array} \tag{7}$$

The **SINC Inversion Method** assumes a zero extinction coefficient, ignores ground scattering, and inverts forest height directly using coherence magnitude representing canopy scattering coherence. So, the SINC inversion method is also known as the coherence magnitude inversion method [7,8]. The RVoG model (3) is simplified to a random volume (RV) scattering model (Equation (4)). The forest height h_v and volume scattering complex coherence γ_{w_v} show a SINC function connection under this assumption, as seen in Equation (8). This approach generally overestimates tree height.

$$\begin{array}{rl} \gamma w_v & = \lim\limits_{\sigma \to 0}\left[\frac{\int_0^{h_v} f(z)e^{jk_z z}dz}{\int_0^{h_v} f(z)dz}\right] = e^{\frac{1}{2}jk_z h_v}\frac{\sin\left(\frac{1}{2}k_z h_v\right)}{\frac{1}{2}k_z h_v} \\ \Rightarrow h_v & = \frac{2\cdot\text{sinc}^{-1}(|\gamma w_v|)}{k_z} \end{array} \tag{8}$$

The **Phase and Coherence Inversion Method** combines the RVoG ground phase method (first term of Equation (5)), which underestimates tree height, and the SINC inversion method (second term of Equation (8)), which overestimates tree height. The model has a more robust structure and low computational complexity. Some researchers compared the phase and coherence inversion method to other methods and discovered that the phase and coherence inversion method has the best inversion property and is easier to use. Additionally, it is vulnerable to the accuracy of the RVoG ground phase method [9]. The correction factor ε in this investigation sets to 0.4 due to the absence of a priori knowledge.

$$h_v = \frac{\arg(\gamma w_v) - \phi_0}{k_z} + \varepsilon \cdot \frac{2\text{sinc}^{-1}(|\gamma w_v|)}{k_z} \tag{9}$$

3.2. *Coherence Amplitude and Three-Stage Hybrid Iterative Model*

3.2.1. Coherence Amplitude and Three-Stage Hybrid Iterative Theoretical Model

The classic SINC Inversion Method often overestimates the height of the forest because it assumes that the ground scattering contribution is zero and that the volume scattering contribution dominates the interferometric coherence. The three-stage inversion method has greater accuracy than the SINC inversion method, DEM difference method, and RVoG ground phase method. It simplifies the complexity of the height inversion model. However, several studies have shown that the results of the three-stage inversion method remain to underestimate the height of the forest. The three-stage inversion method's results typically underestimate [36,37] because the algorithm assumes that the ground-to-volume scattering ratio of HV polarization is zero. Additionally, all polarizations in real scenarios contain some ground scattering contribution.

Additionally, this inversion approach takes more time. The LUT steps can be used to produce more precise results for forest height and extinction coefficients, but this will result in a massive rise in the model's inversion time because the LUT must invert each pixel. Another disadvantage is that the images' uniformity impacts the traditional three-stage inversion method. Forest density influences the attenuation of SAR signals in the canopy layer, which causes errors in the inversion. The RVoG two-layer scattering model assumes that the volume layer is a random medium with uniform density (see, for example, Figure 5). Still, low forest density is no longer considered a homogeneous medium, and applying the RVoG model under such inhomogeneous vegetation conditions affects the accuracy of the inversion results [26].

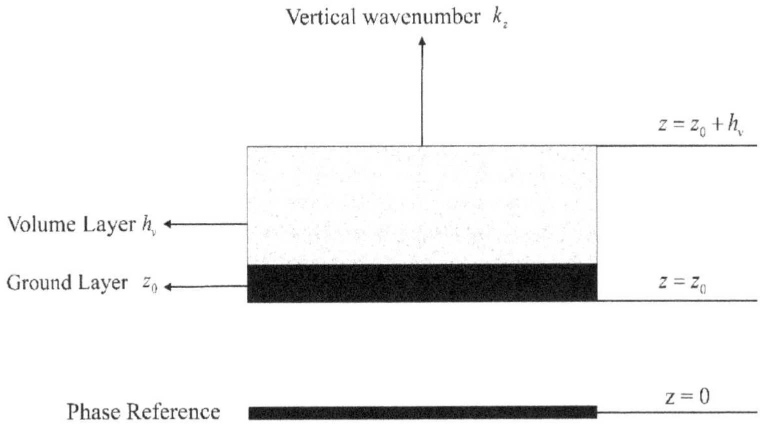

Figure 5. Schematic representation of the RVoG two-layer scattering model. The two-layer model consists of a volume layer and a ground layer, where the volume layer is supposed to be a medium with uniform random density.

This study improves the forest height inversion model by combining the three-stage inversion method and the SINC inversion method based on the model structure of the phase and coherence inversion approach. The traditional three-stage inversion method is not suitable for the inversion of sparse vegetation conditions and underestimates forest height. According to previous research, both the RVoG ground phase method and the three-stage inversion method underestimate forest height. The three-stage inversion method is used in place of the RVoG ground phase method in the traditional phase and coherence inversion method, where $ThreeStage$ represents the three-stage inversion method. The three-stage inversion method's relative error AE and the magnitude term $(\frac{2\text{sinc}^{-1}(|\gamma_{wv}|)}{k_z})$ compensate for $ThreeStage$ to solve the problem of forest height underestimation under sparse vegetation conditions, which is associated with the three-stage inversion method (Equation (10)). When the pixel's result is overestimated, $AE < 0$; and when the pixel's effect is underestimated, $AE > 0$. The average tree height ($realvalue$) represents the average level in a forest stand and it is used in inversion to decrease the complexity of collecting real-world data. In this paper, we used average tree height to obtain penetration depth. Section 5.3 discusses the relationship between penetration depth and forest density.

$$\begin{gathered} AE < 0 \\ h_v = ThreeStage + AE \cdot \varepsilon_i \cdot \frac{2\text{sinc}^{-1}(|\gamma_{wv}|)}{k_z} \end{gathered} \tag{10}$$

$$AE = (realvalue - ThreeStage)/realvalue \tag{11}$$

The traditional phase and coherence inversion approach needs the model adjustment coefficients to be generated based on a priori information. The same weighting factor ε is employed for each pixel in the inversion process to correct the magnitude term for the phase term. Although using a single ε for the entire image simplifies the inversion process, it has obvious disadvantages when the SAR image contains forest conditions with varying forest density or forest types. This study uses the adjustment coefficients ε_i to control the minimization of inversion error RMSE by instructing each pixel to select the adjustment coefficient with the minimum forest height error $\varepsilon = \arg(RMSE_{\min})$ in i iterations. The new weighting term $AE \cdot \varepsilon_i$ strengthens the entire model's structure. It overcomes the drawbacks of the three-stage inversion method, which is better suited for inversion under whole-forest density situations, but fails under sparse forest conditions. The following summarizes the inversion procedure of the coherence amplitude and three-stage hybrid iterative approaches.

Linear least-squares fitting of all complex coherence coefficients in the complex unit circle. This stage follows the traditional three-stage inversion method.

Obtaining of the actual ground phase. The fitted line and the complex unit circle intersect at (ϕ_1, ϕ_2). The HV channel usually represents the volume scattering complex coherence and HH-VV channel represents the ground scattering complex. The intersection farthest from HV represents the actual ground phase ϕ_{ground};

$$\phi_{ground} = \begin{cases} \phi_1 \Leftarrow & \left|e^{i\phi_1} - \gamma_{HV}\right| > \left|e^{i\phi_1} - \gamma_{HH-VV}\right| \\ \phi_2 \Leftarrow & else \end{cases} \tag{12}$$

Calculation of forest height. The iteration range of forest height and extinction coefficient are first set, then the pure volume scattering complex coherence $LUT(h_v, \sigma)$ is constructed using Equation (13). The distance between the volume scattering complex coherence and the LUT whose ground phase has been removed is calculated. The smallest distance calculated on the LUT is the most appropriate forest height and extinction coefficient;

$$h_{ThreeStage} = \underset{(height,\sigma)}{\text{argmin}}\left\{\left|\gamma_{HV}e^{-i\phi_{ground}} - LUT(h_v, \sigma)\right|\right\} \tag{13}$$

Equation (8) generates the SINC inversion method's findings with an extinction coefficient of 0;

Determining the relative error of the height AE of each image pixel from the tree height and the three-stage Inversion method's forest height, and applying a weighting coefficient $\varepsilon_i (0 \le \varepsilon_i \le 1)$ to each image pixel. Equation (10) includes the weighting factor $AE \cdot \varepsilon_i$. The image pixel is iterated by image element based on the minimized RMSE, and the appropriate $\varepsilon = \arg(RMSE_{\min})$ is determined.

3.2.2. Coherence Magnitude and Three-Stage Hybrid Iterative Application Model

The coherence magnitude and three-stage hybrid iterative theoretical model (Equation (10)) apply to ideal conditions, i.e., flat areas with no terrain effects or temporal decoherence, which is compatible with the traditional RVoG model. A large number of studies, however, have demonstrated that the conventional RVoG model is sensitive to terrain effects: the higher the slope of the terrain, the higher the inversion error; the higher the forest height, the higher the inversion error [17,18,24]. As a result, this work employs a widely used slope correction approach to improve the theoretical model [15]. In non-flat locations, vegetation is distributed along the surface's slope, and the traditional RVoG model assumes that the vegetation is in a flat area. Because of the presence of the terrain slope α, the local coordinate system of the ground surface patch (Γ) is corrected from yoz to $y'o'z'$, as illustrated in Figure 6. When the terrain slope facing the radar sensor is positive (Figure 6a), the corrected local incidence angle θ_0 is smaller than the original radar incidence angle θ, and the vertical wavenumber increases. When the terrain slope facing the radar sensor is negative (Figure 6b), the local incidence angle θ_0 is larger than the original radar incidence angle θ, and the vertical wavenumber decreases. The terrain slope affects the volume complex coherence by influencing the pixel's radiometric brightness and vertical wavenumber [15]. The radar incidence angle may be corrected using the terrain slope to produce the local incidence angle θ_0, while the original k_z uses the local incidence angle θ_0 to correct it.

$$\theta_0 = \theta - \alpha \tag{14}$$

$$k_{z0} = \frac{4\pi B_\perp}{\lambda R \sin\theta_0} \tag{15}$$

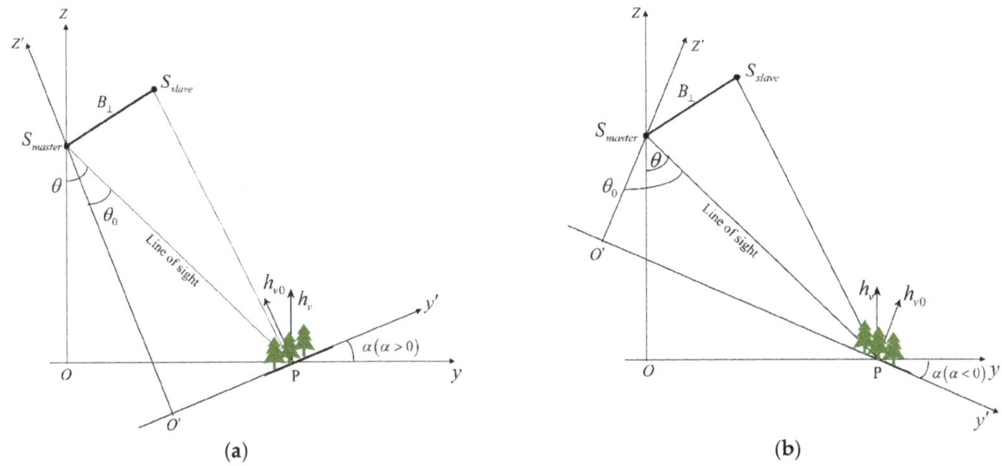

Figure 6. The geometric model of forest scattering is affected by terrain slope. (**a**) The landscape distribution faces the radar sensor when the terrain slope is positive. (**b**) When the terrain slope is negative, the landscape distribution is at the back of the radar sensor. yoz and $y'o'z'$ are the coordinate system under flat terrain and the slope-corrected coordinate system, respectively. $B_\perp$ means the vertical baseline. S_{master} and S_{slave} are the master image and the slave image, respectively. P is the ground object (e.g., forest), θ is the radar incidence angle, θ_0 is the local incidence angle, α is the range slope, h_{v0} is the slope-corrected forest height under the $y'o'z'$ coordinate system, and h_v is the forest height under the yoz coordinate system.

The RVoG model's volume coherence is then corrected using k_{z0}:

$$\gamma_v = \frac{2\sigma}{2\sigma + jk_{z0}\cos\theta_0} \cdot \frac{\exp((2\sigma h_v/\cos\theta_0 + jk_{z0}h_v)\cdot\cos\alpha) - 1}{\exp(2\sigma h_v/\cos\theta_0 \cdot \cos\alpha) - 1} \tag{16}$$

The forest heights acquired by inversion were all along the $y'o'z'$ coordinate system, and the terrain-corrected forest heights were h_{v0} as illustrated in in Figure 6. Thus, it is essential to project h_{v0} to the yoz coordinate system to obtain the forest heights $h_{ThreeStage}$ according to Equation (17) [17,38]. Meanwhile, some studies have shown that the range component of terrain slope is the dominant effect, and the azimuth slope is a minor one. Therefore, this study only considers the range slope correction. Figure 7 shows the slope and DEM of the four forest stands.

$$h_{ThreeStage0} = \frac{h_{v0}}{\cos|\alpha|} \tag{17}$$

The hybrid iterative algorithm is modified at this point to use the terrain-corrected three-stage inversion method $ThreeStage0$ to replace the first term of the theoretical model. The vertical wavenumber k_{z0} was corrected to reduce the impact of terrain on the SINC inversion method (Equation (18)).

$$h_v = ThreeStage0 + AE \cdot \varepsilon_i \cdot \frac{2\cdot \mathrm{sinc}^{-1}(|\gamma_{w_v}|)}{k_{z0}} \tag{18}$$

The traditional SINC inversion method uses the coherence amplitude of a particular polarization channel, such as HV or PD_{HIGH}. It assumes that its ground scattering contribution is zero to invert the forest height. In reality, all polarization channels have ground-scattering components. The extra ground scattering coherence amplitude causes overestimation of the SINC Inversion Method. The second stage of the three-stage inversion method obtained the actual ground phase, which was removed from the assumed pure volume complex coherence. At this stage, the pure-volume complex coherence is the most suitable input variable for the SINC inversion method. As a result, we employ the pure volume complex coherence to replace the second term of Equation (18), obtaining the application model of the coherence magnitude and the three-phase hybrid iterative approach (Equation (19)).

$$h_v = ThreeStage0 + AE \cdot \varepsilon_i \cdot \frac{2 \cdot \mathrm{sinc}^{-1}\left(\left|\gamma_{w-assu} \cdot e^{-j\phi_{ground}}\right|\right)}{k_{z0}} \tag{19}$$

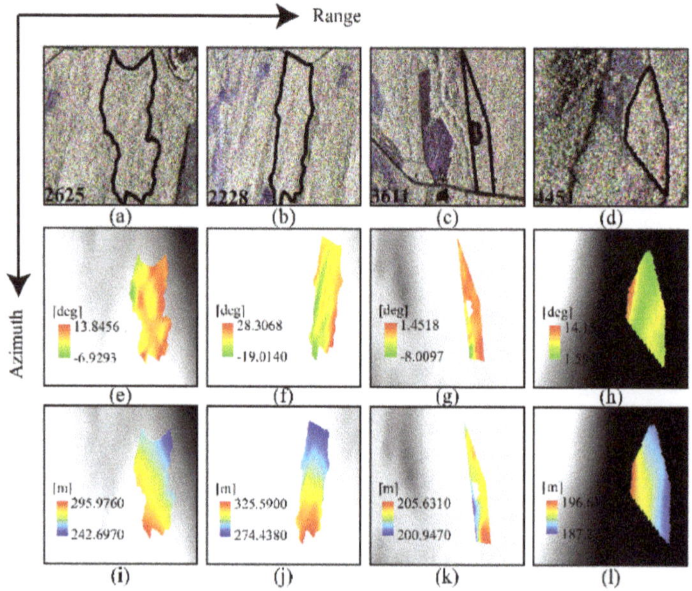

Figure 7. Images of the four forest stands' terrain slope and DEM. (**a**–**d**) Pauli-basis of four stands in the SAR coordinate system. (**e**–**h**) Terrain slope (**i**–**l**) DEM.

4. Results

4.1. Results of the Forest Height Inversion for the Simulated Dataset

Simulation of the forest height inversion was performed under ideal conditions where nine groups of stands with various forest densities and a mean tree height of 18 m were utilized. HV polarization represents the volume scattering complex coherence w_v, and HH-VV polarization represents the ground scattering complex w_g. Comparison experiments used six forest inversion methods. Table 3 and Figure 8 show each algorithm's performance and the specific statistical results.

Out of the six inversion algorithms used in the inversion of forest height, the DEM difference method (red line) and RVoG ground phase method (yellow line) performed the worst. Both phase difference algorithms significantly underestimated forest height in sparsely vegetated areas with forest density below 400 stems/ha. The inversion error decreased once the forest density exceeded 400 stems/ha, but the RMSE was consistently greater than 1/3 of the actual tree height. Figure 9 displayed the relative height of the DEM difference method and RVoG ground phase method, respectively. The canopy phase center gradually moved toward the top of the canopy, and the ground phase center also slowly increased. The HV phase center improvement rate was higher than HH-VV polarization in 100–300 stems/ha, and the inversion accuracy was gradually improved. However, when the stand density exceeds 600 stems/ha, the forest gap area decreases, and the canopy phase center no longer increases at this point. The HV and HH-VV polarization phase center heights are also slowly saturated at this point. As a result, the DEM differential method and RVoG ground phase method are subsequently saturated. It also suggested that in sparsely vegetated regions, the canopy phase and ground phase separations represented by HV polarization and HH-VV polarization practically fail, significantly impacting the algorithm's performance.

Table 3. Using simulated datasets of 100–900 stems/ha, six forest height inversion techniques produced forest height and error results.

Forest Density (stems/ha)	100	200	300	400	500	600	700	800	900
				DEM Difference Method					
MEAN	6.82	8.18	8.10	8.36	9.38	9.64	9.10	9.167	8.84
RMSE	11.41	10.15	10.19	9.93	8.99	8.78	9.27	9.19	9.56
				SINC Inversion Method					
MEAN	20.04	18.94	18.09	17.16	16.56	16.72	17.25	17.31	16.52
RMSE	5.19	4.94	4.20	4.40	4.43	4.59	4.63	4.46	4.72
				RVoG Ground Phase Method					
MEAN	8.46	9.70	9.97	9.73	10.66	11.01	10.37	10.32	10.47
RMSE	9.87	8.59	8.28	8.48	7.67	7.31	7.92	7.94	7.84
				Three-Stage Inversion Method					
MEAN	13.73	15.65	15.66	15.69	15.93	16.16	15.63	16.15	15.92
RMSE	6.30	4.90	4.19	4.28	4.17	3.900	4.30	4.07	4.21
				Phase and Coherence Inversion Method					
MEAN	15.66	16.86	16.87	16.24	16.87	17.22	16.79	16.80	15.66
RMSE	4.85	4.13	3.32	3.76	3.57	3.51	3.53	3.71	4.85
				Coherence amplitude and three-stage hybrid iteration method					
MEAN	17.35	16.79	17.23	16.92	16.85	17.36	16.99	16.96	16.88
RMSE	3.01	3.52	2.81	3.50	3.27	3.01	3.19	3.34	3.27

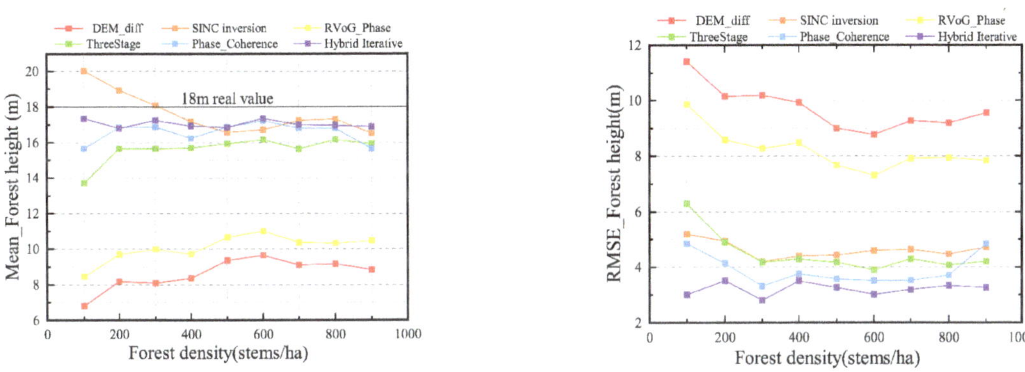

(**a**) (**b**)

Figure 8. Results graphs for the six forest height inversion techniques for simulated data sets with 100–900 stems/ha. (**a**) Line graph of forest height (**b**) Line graph of forest height error. The red line means the DEM difference method (DEM_diff), and the orange line represents the SINC inversion method (SINC inversion). The yellow line describes the RVoG ground phase method (RVoG_Phase), and the green line represents the traditional three-stage inversion method (ThreeStage). The blue line is the phase and coherence inversion method (Phase_Coherence), and the purple line represents the coherence magnitude and three-stage hybrid iterative method (Hybrid Iterative).

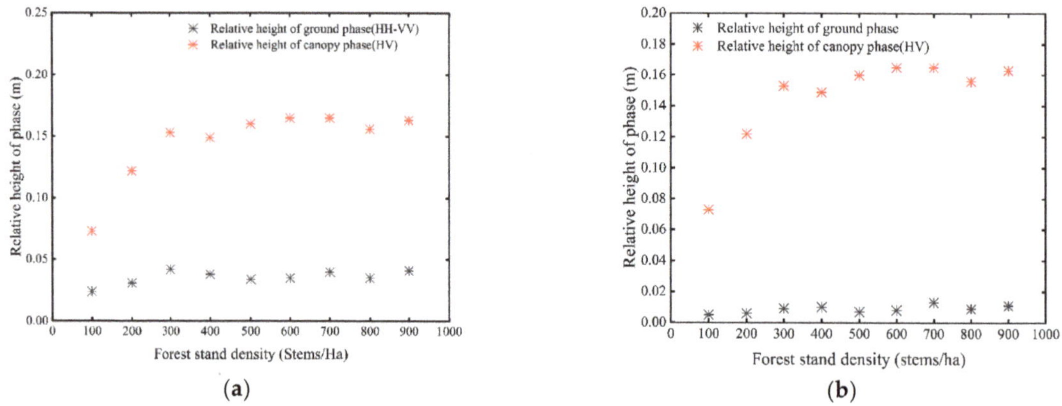

Figure 9. Relative heights of canopy phase center and ground phase center. (**a**) DEM difference method (**b**) RVoG ground phase method.

The orange line represented the SINC inversion method. The overestimation of this algorithm was apparent when the forest density was below 400 stems/ha. However, the RMSE gradually decreased as the forest density increased and always maintained the RMSE at less than 30% of the actual height. The SINC inversion method ignored the effect of ground phase information on the tree height inversion. As a result, the magnitude impacted the effectiveness of the SINC inversion method. One of the significant characteristics of the SINC inversion method is that the average magnitude is inversely related to the tree height. Figure 10 displayed two-dimensional images of tree height and amplitude representing the canopy polarization channel. When the stand density is low, the amplitude of the HV polarization is small, and the forest height inversion is high. The amplitude of HV polarization increases as the forest density approaches 400 stems/ha, and the forest height inversion error gradually decreases. However, as the forest density increases further, the amplitude of HV polarization reaches the saturation state, which limits the SINC inversion method's ability to invert and causes the forest height inversion error to fluctuate steadily.

At stand densities lower than 400 stems/ha, the traditional three-stage inversion method (green line) demonstrated a significantly high forest height RMSE. In particular, the forest height was greatly overestimated in the sparse forest stand scenarios below 200 stems/ha, and the error was greater than 30% of the actual forest height. With an increase in forest density, the three-stage inversion method's advantage was gradually more apparent, and the RMSE was below 25% of the actual measurement. The sparse vegetation broke the assumption of a random homogeneous medium, preventing the RVoG model from exploiting it. On the other hand, the pure canopy phase incorporates a portion of the ground component, causing mistakes in the calculated ground phase.

The inversion error of the phase and coherence inversion method (blue line) was smaller than in the four approaches for low stand density situations. However, the RMSE of this approach was much greater than that of the other four methods at sufficiently high stand density. The main reason was that the results of the SINC inversion method and the RVoG ground phase method were stable enough but increased data redundancy.

The coherence magnitude and three-stage hybrid iterative method (purple line) had significantly better inversion effects under various forest densities when compared to the other five commonly used algorithms. The model's inversion effect was stable without apparent overestimation or underestimation. The inversion error was about 15% of the actual forest height under both sparse and dense vegetation. In a forest stand with 100 stems/ha, the inversion height was 17.3507 m, and the RMSE was 3.014 m, overcoming the drawbacks of the SINC inversion method and the three-stage inversion method, which fail at low stand densities. The forest height RMSE was 0.5–1 m lower than the phase and coherence inversion method and 1–3 m lower than the three-stage inversion method.

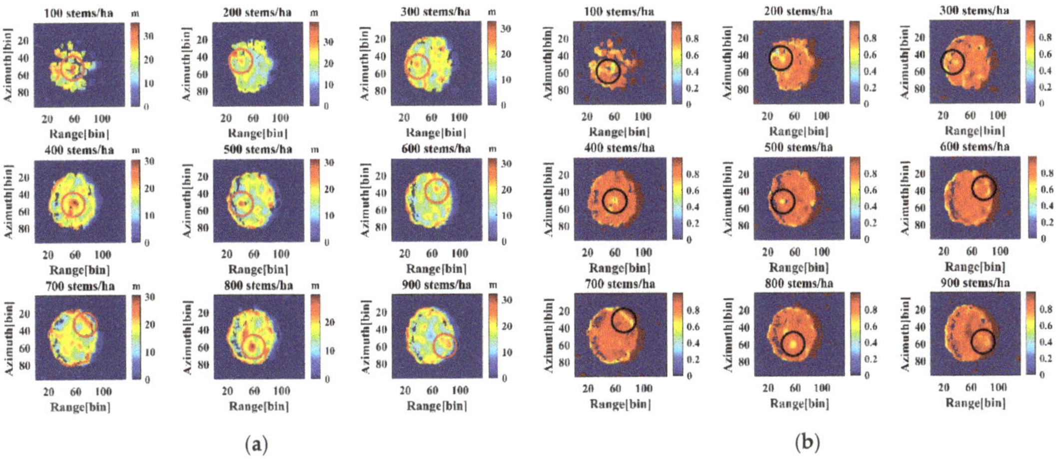

Figure 10. In the SINC inversion method, two-dimensional images of tree height and HV polarization intensity are generated for nine groups of forest stands. (**a**) A two-dimensional map of forest height for nine groups of forest stands. (**b**) A two-dimensional map of the HV magnitude for nine groups of forest stands. For each of the nine groups of forest stands, the black circles indicate the regions with low amplitude values. The red and black circles are positioned in the same location in both images.

4.2. Results of Forest Height Inversion for a Real Dataset

Considering that the forest conditions in the simulated dataset were ideal, the algorithm was applied to four forest stands in the BioSAR 2008 L-band dataset in this study to test the applicability of the improved method in actual data. The real dataset had four forest stands with a forest density of 628.66 stems/ha, 840.34 stems/ha, 1149.10 stems/ha, and 1330.54 stems/ha. Since the actual forest height was often lower than the field measurement of stand tree height, the Laser-100th was used as the actual forest height for model inversion in this work. Table 4 displays the forest density, average stand height measured in the field, and average laser radar stand height for the four stands.

Table 4. The forest density and height of four realistic scenario forest stands. The mean height was measured manually in the field, and the mean height from Lidar is Laser-100th.

Forest Stand Number	Forest Density (stems/ha)	Mean Height (m)	Mean Height from Lidar (m)
4451	628.66	18.72	20.99
2625	840.34	18.06	22.45
3611	1149.10	17.36	21.44
2228	1330.54	17.69	20.50

The algorithm's performance was affected by the terrain fluctuations in the area covered by the airborne data. So, the hybrid iterative theoretical model used the terrain correction to determine the coherence magnitude and three-stage hybrid iterative application model (Equation (22)), which was then applied to the BioSAR 2008 L-band data. This study introduced the PD coherence optimization approach to enhance the conventional fitting method. γPD_{HIGH} and γPD_{LOW} represented the two ends of the long axis of the coherence region, which were used in linear fitting to improve ground phase inversion accuracy. For the inversion of forest height, the pure volume complex coherence after elimination of the ground phase, the terrain-corrected vertical wavenumber, and the local incidence angle were concurrently input into the SINC Inversion Method and the terrain-corrected LUT. Table 5 displays the results of their forest inversion.

Table 5. Coherence magnitudes and three-stage hybrid iterative application model inversion outcomes for four realistic forest stands. The hybrid iterative algorithm height results from the coherence magnitude and three-stage hybrid iterative application model.

Forest Stand Number	Forest Density (stems/ha)	Hybrid Iterative Algorithm Height (m)	RMSE (m)	MAPE (%)	STD (m)	VAR
4451	628.66	21.21	1.14	3.99	1.11	1.22
2625	840.34	22.19	1.60	6.20	1.05	1.11
3611	1149.10	21.54	1.83	5.86	1.83	3.34
2228	1330.54	20.89	2.17	7.70	1.51	2.27

The method partially addressed the effects of terrain slope and forest density on the single-baseline PolInSAR inversion and produced good inversion results for all four stands. The standard deviations were all controlled at about 1.5 m, ensuring slight fluctuation when the method was applied to stands with different forest densities. Figure 11 displays the inversion results for the four forest stands. As the forest density increased, the present method was found to progressively raise the RMSE. However, the RMSE < 3 m was not noticeably overestimated or underestimated. In particular, the RMSE was 1.14 m, and the error was lower when the stand density was 628.66 stems/ha. The present technique will be extended from the stand scale to a broader region to estimate forest height inversion, biomass, and carbon stock information across a greater area.

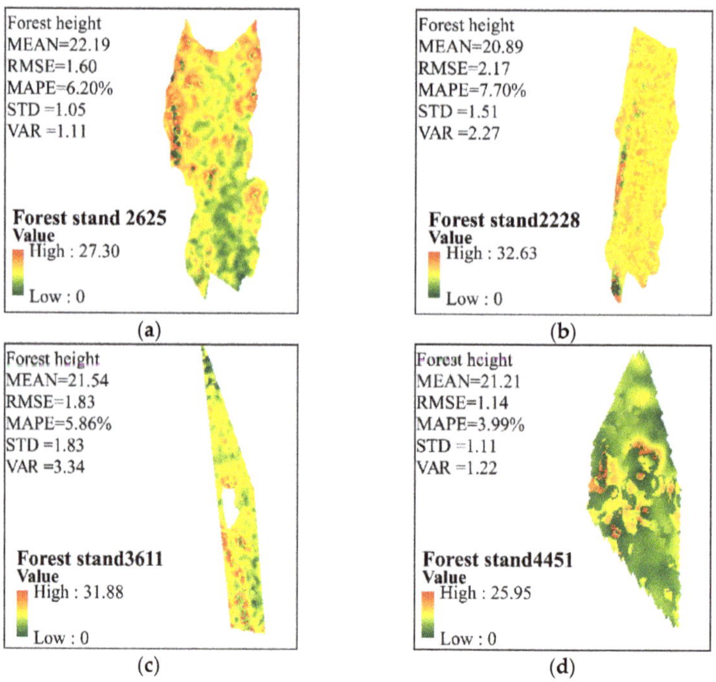

Figure 11. Forest height mapping in four forest stands was accomplished using coherence magnitude and three-stage hybrid iterative application algorithms. (**a**) Stand 2625, (**b**) stand 2228, (**c**) stand 3611, and (**d**) stand 4451.

5. Discussion

5.1. Effect of Forest Density on Phase

The relative height of the ground phase centers for the RVoG ground phase method was significantly lower than the DEM difference method as illustrated in Figure 9a,b. The phase separation was more significant in the RVoG ground phase method than in the DEM difference method. Figure 12 shows the ground phase error bar of the DEM difference method and RVoG ground phase method, respectively. However, when the stand density exceeds 600 stems/ha, the standard deviation of the ground phase center gradually rises. The HV and HH-VV polarization phase center heights also slowly saturated (Figure 9). As a result, the ground phase is no longer accurately represented by HH-VV.

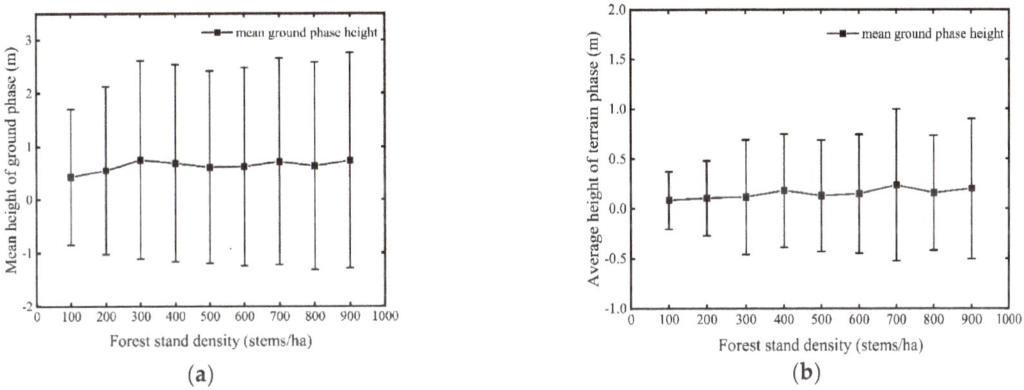

Figure 12. Error bar plot of ground phase center relative heights. (**a**) DEM difference method (**b**) RVoG ground phase method.

In general, the height of the volume scattering phase center steadily rises with increasing forest density at the top of the canopy until the saturation of the polarization channel that represents volume scattering. The center height of the ground phase of the DEM difference method or the RVoG ground phase method gradually rises to saturation as the density of the forest stand increases. Improving the precision of forest height inversion is however difficult to achieve since accurate estimation of ground phase in high-density stands is often a challenge. This is because the increase in stand density causes an increase in ground phase inaccuracy. Although the ground phase error in low-density stands is low, the distance between the canopy phase center and the ground phase center is too close, making it impossible to efficiently discern between the two. As such, increasing the accuracy of forest height inversion is quite a challenge.

5.2. Effect of Forest Density on the Magnitude

The SINC inversion method is typically a case where the extinction coefficient is 0, as shown by the black line in Figure 13a. The average extinction coefficient depends on the wavelength and the characteristics of the medium (such as height and density). When the signal frequency is fixed, the average extinction coefficient rises, indicating the existence of an effective scattering layer at the top of the bulk layer that attenuates the signal and causes a decrease in penetration depth. As a result, the average extinction coefficient and penetration depth are inversely proportional [39]. The relationship between the volume coherence and the average extinction coefficient is shown in Figure 13b. The volume coherence increases gradually for a fixed tree height as the average extinction coefficient increases. The volume layer is confined into a small space at the top of the volume, indicating a relationship between the volume coherence and the penetration depth. Volume coherence increases with increasing extinction coefficient, while penetration depth in the volume layer decreases. Figure 13c illustrates the relationship between phase, tree height, and extinction coefficient. As the extinction coefficient increases, the relative height of the phase center gradually rises, and the center of the volume scattering phase slowly moves toward the top of the canopy [8]. Extinction coefficient, volume coherence, and volume scattering phase are therefore related. As the extinction coefficient increases, the volume coherence increases, the volume scattering phase increases, while the penetration depth decreases.

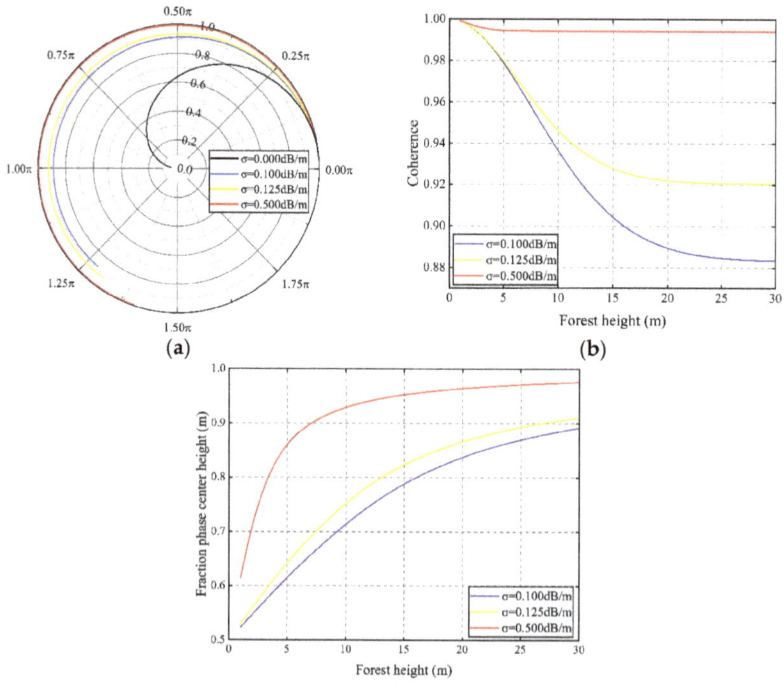

Figure 13. A presentation of the relationship between extinction coefficient, coherence amplitude, and phase. (**a**) The relationship between complex coherence and extinction coefficient in the complex unit circle. (**b**) The relationship between tree height, amplitude, and extinction coefficient. (**c**) The relationship between extinction coefficient and phase center height, tree height ($\sigma = 0\ dB/m$ is the black line, $\sigma = 0.1\ dB/m$ is the blue line, $\sigma = 0.125\ dB/m$ is the yellow line, and $\sigma = 0.5\ dB/m$ is the red line).

Forest density is one of the forest features. The penetration and attenuation of the signal impacted by the forest density can change the forest stand's average extinction coefficient. Low forest density causes a wide gap in forest vegetation [40], which allows many electromagnetic wave signals to pass through and reach the ground. It reduces signal attenuation in the volume layer, increases penetration depth, and reduces the extinction coefficient. Figure 14 shows two-dimensional images of nine sets of forest stands' extinction coefficients, with the red circles at the same location on each set of images. For forest stands with densities of 100 stems/ha and 200 stems/ha, the extinction coefficient of the volume layer is lower as shown in Figure 14. The attenuation of the signal into the stand rises as the stand density grows, the extinction coefficient increases, and the effective volume layer develops at the top of the canopy, impacting the computation of complex coherence. When the density grows from 100 stems/ha to 600 stems/ha, the average extinction coefficient in the red circle increases progressively (Figure 14). However, when the stand density reaches 700–900 stems/ha, the extinction coefficient begins to fluctuate, causing fluctuations in the mean phase and amplitude at higher stand densities.

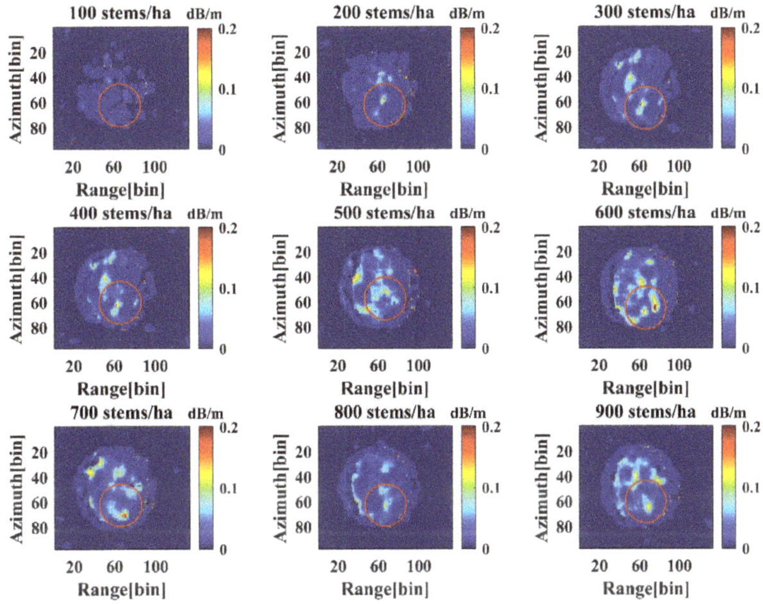

Figure 14. The two-dimensional maps depict the extinction coefficients of nine groups of stands inverted using the traditional three-stage inversion method, with red circles representing the same locations of the nine groups of forest stands.

In summary, for L-band, the forest density impacts the inversion of forest height by influencing the extinction coefficient. The effect of the extinction coefficient on the inversion process is reflected in both amplitude and phase. When the stand density is low, the extinction coefficient within the stand is also low. The volume scattering contains a significant ground scattering contribution, and the incomplete separation of the volume phase and ground phase invalidates the forest height method. The extinction coefficient inside the stand increases as stand density rises, the amplitude and phase saturation threshold increases, the distance between ground and canopy phases gradually widens, and the forest height inversion error gradually reduces. When the stand density is high enough, the amplitude and phase tend to saturate. The accuracy of the ground phase estimation influences the accuracy of forest height inversion, and the extinction coefficient stops increasing continuously. Single-baseline PolInSAR inversion approaches can employ coherence optimization algorithms (e.g., SVD, PD, MCD) to increase the distance between ground and canopy phases, or they can use an external DTM to choose the correct ground phase [34,41]. These techniques still result in inversion errors for some ground phases, which raise the inversion errors for forest height. The accuracy of the three-stage inversion method has been increased by restricting the extinction coefficient following the signal penetration depth [42]. Additionally, the multi-baseline PolInSAR approach has been used to overcome the RVoG model's shortcomings by reducing the extinction coefficient's impact [42] or applying a tomographic technique to reconstruct the forest height model [43]. These are strategies to overcome the constraints of single-baseline PolInSAR forest height inversion due to factors such as forest density.

5.3. Discussion of Coherence Magnitude and Three-Stage Hybrid Iterative Model

(1) The physical significance of the improved model.

The coherence amplitude and three-stage hybrid iterative model also accounts for the variations in forest density and changes in the extinction coefficient and forest structure. The model is based on the stand density variation, which accounts for the variation in forest structure in relation to stand density. This variation is represented mathematically as the relative error of forest height and physically as the relative height of signal penetration depth. The inversion results are overestimation if AE is negative and underestimated if AE is positive. Higher $|AE|$ indicates deeper penetration, a lower canopy phase position, and less forest density. The smaller the $|AE|$, the lower the penetration depth, the higher the canopy phase position, and the denser the forest stand.

$$\begin{aligned} AE &= (realvalue - ThreeStage)/realvalue \\ &= 1 - \frac{ThreeStage}{realvalue} \\ &= \frac{penetration\ depth}{realvalue} \end{aligned} \tag{20}$$

When the canopy phase position Y is lower at lower stand densities, the extinction coefficient is lower, and the penetration canopy depth d is larger (Figure 15a). The canopy phase position Y is close to the top of the canopy when the forest density increases. The extinction coefficient is higher, and the penetration canopy depth d is smaller (Figure 15b). Figure 15c depicts the relative errors (penetration depths) for the nine groups of forest stands. At lower forest density, substantial forest gaps emerge, and signal penetration depth under the canopy is high, with relative errors ranging from 0.4 to 0.8. As forest density increases, the forest gap reduces, forest structure becomes homogeneous, canopy phase reaches the top of the canopy, and the signal penetration depth within the forest gradually declines to a value falling between −0.2 and 0.2.

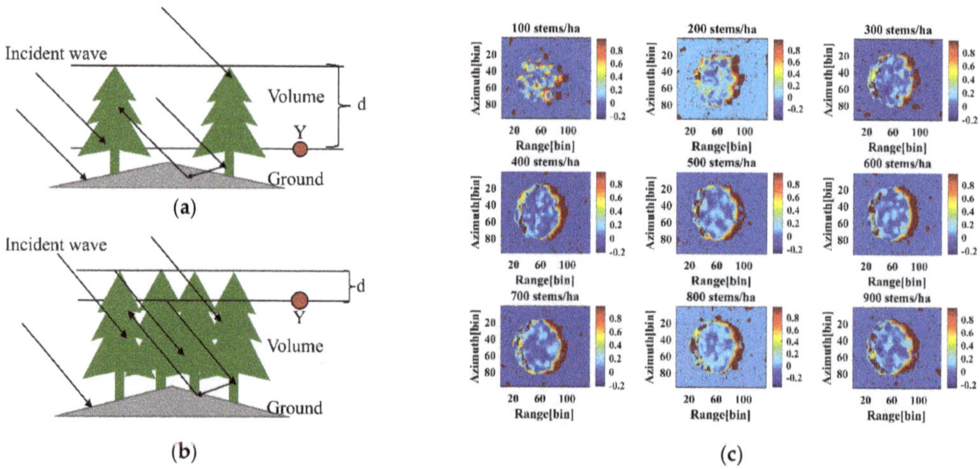

Figure 15. Diagram showing the relative tree height error, penetration depth, and forest density. (**a**) Schematic representation of signal penetration at low forest density, where penetration depth d is higher, and canopy phase center Y is lower. (**b**) Schematic illustration of signal penetration at high forest density showing that the penetration depth d is less and the canopy phase center Y is more elevated and closer to the canopy's top. (**c**) The relative error (penetration depth) of the nine groups of forest stand inversions using the traditional three-stage inversion method.

The coherence magnitude and three-stage hybrid iterative model provides the top of vegetation with the traditional three-stage algorithm replacing the phase term, and compensate for the top height of the canopy compressed by the phase information with the SINC inversion method and canopy penetration depth. Because penetration depth is involved, the coherence magnitude and three-stage hybrid iterative algorithm no longer selects the adjustment factor using fixed empirical parameters. Iterative correction coefficients ε_i are selected pixel-by-pixel based on the stand's features to reduce the tree height inversion error. As a result, the coherence magnitude and three-stage hybrid iterative method is more reliable and compatible with the structural traits of the target forest stands, increasing the accuracy of forest height inversion. The two critical assumptions that the volume coherence is independent of polarization and that the ground-to-volume scattering ratio is zero for the volume scattering complex coherence still place restrictions on the model because it is based on the conventional three-stage inversion method. The model's applicability to X-band and P-band data still requires particular studies because the signal frequency is also a crucial factor impacting the extinction coefficient.

(2) The structure of the improved model

According to the traditional three-stage inversion method, the ideal coherence region comprises all the complex coherence points (the gray elliptical area in Figure 16). The extinction coefficient is often empirically adjusted as a constant to increase the effectiveness of the whole image inversion, which reduces the number of spirals in the LUT to one (the red circle spiral in Figure 16). The hypothetical pure-volume scattering complex coherence point γ_{V-pure}, indicated by the red dot in Figure 16a, is located on the ellipse's long axis furthest from the ground phase ϕ_{ground}. In an ideal situation, finding the closest point (green dot in Figure 16) to γ_{V-pure} in the LUT should yield suitable tree height and extinction coefficient values. The distance between γ_{V-pure} and $\gamma_{V-ideal}$ becomes too large when the coherence region is too distant from the LUT or if the coherence region's long axis is not long enough. It means $\gamma_{V-wrong}$ (the blue dots in Figure 16b) is closer to γ_{V-pure} than $\gamma_{V-ideal}$, which causes a significant under- or overestimation of forest height.

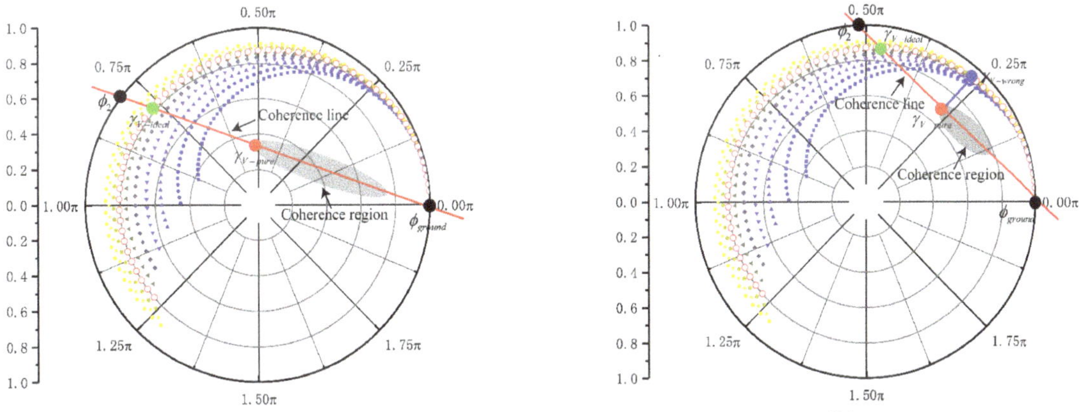

Figure 16. An illustration of ideal geometric structure of the conventional three-stage inversion method and the situation of a failed geometric structure. (**a**) The ideal geometry of the traditional three-stage inversion method. (**b**) Either the coherence region's long axis is too short or too far away from the LUT for the geometric structure to fail.

The above ambiguous solutions happen when the ground phase has made an incorrect selection, or the separation between the ground phase and canopy phase is insufficient. For example, in realistic situations with sparse vegetation, overly low extinction coefficients lead to poor coherence and affect the inversion of the LUT. The single-baseline PolInSAR technique, with its simple algorithm, is ineffective, but the multi-baseline PolInSAR approach is complex and requires numerous sets of data input. And the coherent magnitude and three-stage hybrid iterative model uses the penetration depth to alter the weighting factors of pixels to make the overall model more robust. A balance is struck between the simplicity and the validity of the model.

Although the coherence magnitude and three-stage hybrid iterative application model is adjusted for terrain effects, there is still a need to evaluate whether temporal decoherence can be used in the present study's technique. Temporal decoherence affects forest stands in various ways. The level of landform disturbance varies in PolInSAR data with different temporal baselines [9,20,44]. For various frequencies or various temporal baselines, many temporal decoherence techniques have been developed [19,21–23]. The impact of temporal decoherence on various forest densities or stand structures varies [45]. There is currently no standardized algorithm to achieve tree height inversion for all forest densities and all temporal baselines. As a result, data characteristics and forest features need to be considered to execute temporal decoherence correction for the coherence magnitude and three-stage hybrid iterative model.

6. Conclusions

Forest density affects signal penetration and attenuation, leading to the change of extinction coefficient and penetration depth, which affects the inversion performance of the traditional forest height models. When the forest density is sparse, the forest is no longer a uniform medium, and the inversion error is more significant. With the increase in forest density, the inversion error decreases. When the forest density is enormous, the extinction coefficient no longer increases continuously. At this time, the amplitude and phase of the electromagnetic wave are saturated, and the inversion error will increase. Meanwhile, many forest gaps within the forest affect signal penetration, extinction coefficient, and penetration depth.

The coherence magnitude and three-stage hybrid iterative model solves the problem of forest height inversion failure in low forest density regions. It quantifies the penetration depth of different forest stands and compensates for the compression at the top of the canopy. The inversion results do not have significant overestimation or underestimation. The coherence magnitude and three-stage hybrid iterative model can achieve high precision inversion of various forest densities at the stand scale and overcome the underestimation effect of low forest density on the traditional model.

In future work, we intend to will extend the model from the stand scale to a larger scale and realize the high-precision inversion of various forest densities at large scales. The coherence magnitude and three-stage hybrid iterative model introduce the terrain correction algorithm. However, the correction of temporal decoherence must consider the radar frequency, temporal resolution, and other system parameters and stand characteristics. The fusion of the temporal decoherence correction algorithm and the improved algorithm in this paper will also be considered in future work. Nevertheless, the proposed method is simple, robust, and easy to extend, which addresses the failure problems associated with traditional inversion methods in sparse forest areas.

Author Contributions: Conceptualization, methodology, project administration and resources, W.F. and A.S.; Data curation, software, validation, and visualization, A.S.; Formal analysis, A.S. and Y.M.; Funding acquisition, W.F.; Writing—original draft preparation, A.S. and O.O.M.; writing—review and editing, A.S., O.O.M. and W.F. All authors have read and agreed to the published version of the manuscript.

Funding: This research was funded by the National Natural Science Foundation of China (contract no. 31971654) and the Civil Aerospace Technology Advance Research Project (contract no. D040114).

Data Availability Statement: The E-SAR data was provided by the European Space Agency (ESA) under the BIOSAR 2008 campaign (ESA EO Project Campaign ID 69713) at https://earth.esa.int/eogateway/campaigns/biosar-2, (accessed on 26 September 2021).

Conflicts of Interest: The authors declare no conflict of interest.

References

1. Ghasemi, N.; Sahebi, M.R.; Mohammadzadeh, A. A Review on Biomass Estimation Methods Using Synthetic Aperture Radar Data. *Int. J. Geomat. Geosci.* **2011**, *1*, 776–788.
2. Mora, B.; Wulder, M.A.; White, J.C.; Hobart, G. Modeling Stand Height, Volume, and Biomass from Very High Spatial Resolution Satellite Imagery and Samples of Airborne LIDAR. *Remote Sens.* **2013**, *5*, 2308–2326. [CrossRef]

3. Cao, C.; Bao, Y.; Xu, M.; Chen, W.; Zhang, H.; He, Q.; Li, Z.; Guo, H.; Li, J.; Li, X.; et al. Retrieval of Forest Canopy Attributes Based on a Geometric-Optical Model Using Airborne LiDAR and Optical Remote-Sensing Data. *Int. J. Remote Sens.* **2012**, *33*, 692–709. [CrossRef]
4. Wenxue, F.; Huadong, G.; Xinwu, L.; Bangsen, T.; Zhongchang, S. Extended Three-Stage Polarimetric SAR Interferometry Algorithm by Dual-Polarization Data. *IEEE Trans. Geosci. Remote Sens.* **2016**, *54*, 2792–2802. [CrossRef]
5. Askne, J.I.H.; Ulander, L.M.H.; Soja, M.J. Biomass Estimation in a Boreal Forest from TanDEM-X Data, Lidar DTM, and the Interferometric Water Cloud Model. *Remote Sens. Environ.* **2017**, *196*, 265–278. [CrossRef]
6. Treuhaft, R.N.; Madsen, S.N.; Moghaddam, M.; Zyl, J.J. Van Vegetation Characteristics and Underlying Topography from Interferolnetric Radar. *Radio Sci.* **1996**, *31*, 1449–1485. [CrossRef]
7. Denbina, M.; Simard, M.; Hawkins, B. Forest Height Estimation Using Multibaseline PolInSAR and Sparse Lidar Data Fusion. *IEEE J. Sel. Top. Appl. Earth Obs. Remote Sens.* **2018**, *11*, 3415–3433. [CrossRef]
8. Cloude, S.R. Polarimetric Sar Interferometry. *IEEE Trans. Geosci. Remote Sens.* **1998**, *36*, 1551–1565. [CrossRef]
9. Askne, J.I.H.; Dammert, P.B.G.; Ulander, L.M.H.; Smith, G. C-Band Repeat-Pass Interferometric SAR Observations of the Forest. *IEEE Trans. Geosci. Remote Sens.* **1997**, *35*, 25–35. [CrossRef]
10. Zhou, Z.B.; Ma, H.Z.; Zhu, X.B.; Sun, L. Comparative Analysis of Forest Height Retrieval Methods by Polarimetric SAR Interferometry. *Adv. Mater. Res.* **2013**, *726–731*, 4686–4689. [CrossRef]
11. Cloude, S. *Polarisation: Applications in Remote Sensing*; OUP: Oxford, UK, 2009.
12. Cloude, S.R. Polarization Coherence Tomography. *Radio Sci.* **2006**, *41*, 1–27. [CrossRef]
13. Cloude, S.R.; Papathanassiou, K.P. Three-Stage Inversion Process for Polarimetric SAR Interferometry. *IEE Proc.-Radar Sonar Navig.* **2003**, *150*, 125–134. [CrossRef]
14. Cloude, S.R.; Papathanassiou, K.P. Polarimetric Radar Interferometry. *Opt. Sci. Eng. Instrum.* **1997**, *3120*, 224–235.
15. Tabb, M.; Orrey, J.; Flynn, T.; Carande, R. Phase Diversity: A Decomposition for Vegetation Parameter Estimation Using Polarimetric SAR Interferometry. In Proceedings of the European Conference on Synthetic Aperture Radar Conference, Cologne, Germany, 4–6 June 2002; pp. 721–724.
16. Zhang, Q.; Mercer, J.B.; Cloude, S.R. Forest Height Estimation from Indrex-II L-Band Polarimetric InSAR Data. In Proceedings of the International Archives of the Photogrammetry, Remote Sensing and Spatial Information Sciences, Beijing, China, 3–11 July 2008.
17. Zebker, H.A.; Villasenor, J. Decorrelation in Interferometric Radar Echoes. *IEEE Trans. Geosci. Remote Sens.* **1992**, *30*, 950–959. [CrossRef]
18. Ulander, L.M.H. Radiometric Slope Correction of Synthetic-Aperture Radar Images. *IEEE Trans. Geosci. Remote Sens.* **1996**, *34*, 1115–1122. [CrossRef]
19. Sun, X.; Wang, B.; Xiang, M.; Fu, X.; Zhou, L.; Li, Y. S-RVoG Model Inversion Based on Time-Frequency Optimization for P-Band Polarimetric SAR Interferometry. *Remote Sens.* **2019**, *11*, 1033. [CrossRef]
20. Lu, H.; Suo, Z.; Guo, R.; Bao, Z. S-RVoG Model for Forest Parameters Inversion over Underlying Topography. *Electron. Lett.* **2013**, *49*, 618–620. [CrossRef]
21. Xie, Q.; Zhu, J.; Wang, C.; Fu, H.; Lopez-Sanchez, J.M.; Ballester-Berman, J.D. A Modified Dual-Baseline PolInSAR Method for Forest Height Estimation. *Remote Sens.* **2017**, *9*, 819. [CrossRef]
22. Papathanassiou, K.P.; Cloude, S.R. The Effect of Temporal Decorrelation on the Inversion of Forest Parameters from Polinsar Data. *Int. Geosci. Remote Sens. Symp.* **2003**, *3*, 1429–1431. [CrossRef]
23. Lee, S.K.; Kugler, F.; Papathanassiou, K.P.; Hajnsek, I. Quantifying Temporal Decorrelation over Boreal Forest at L- And P-Band. In Proceedings of the 7th European Conference on Synthetic Aperture Radar, Friedrichshafen, Germany, 2–5 June 2008.
24. Lavalle, M.; Simard, M.; Pottier, E.; Solimini, D. PolInSAR Forestry Applications Improved by Modeling Height-Dependent Temporal Decorrelation. In Proceedings of the 2010 IEEE International Geoscience and Remote Sensing Symposium, Honolulu, HI, USA, 25–30 July 2010; pp. 4772–4775.
25. Lei, Y.; Siqueira, P. Estimation of Forest Height Using Spaceborne Repeat-Pass L-Band InSAR Correlation Magnitude over the US State of Maine. *Remote Sens.* **2014**, *6*, 10252–10285. [CrossRef]
26. Lavalle, M.; Simard, M.; Solimini, D.; Pottier, E. Height-Dependent Temporal Decorrelation for POLINSAR and TOMOSAR Forestry Applications. In Proceedings of the 8th European Conference on Synthetic Aperture Radar, Aachen, Germany, 7–10 June 2010; pp. 1–4.
27. Qinghua, X.I.E.; Changcheng, W.; Jianjun, Z.H.U.; Haiqiang, F.U. Forest Height Inversion by Combining S-RVOG Model with Terrain Factor and PD Coherence Optimization. *Acta Geod. Cartogr. Sin.* **2015**, *44*, 686.
28. Garestier, F.; Dubois-Fernandez, P.C.; Papathanassiou, K.P. Pine Forest Height Inversion Using Single-Pass X-Band PolInSAR Data. *IEEE Trans. Geosci. Remote Sens.* **2008**, *46*, 59–68. [CrossRef]
29. Wang, C.; Wang, L.; Fu, H.; Xie, Q.; Zhu, J. The Impact of Forest Density on Forest Height Inversion Modeling from Polarimetric InSAR Data. *Remote Sens.* **2016**, *8*, 291. [CrossRef]
30. Pottier, E.; Ferro-Famil, L.; Allain, S.; Cloude, S.R.; Hajnsek, I.; Papathanassiou, K.; Moreira, A.; Williams, M.; Minchella, A.; Lavalle, M. Overview of the PolSARpro v4. 0 Software New Updates of the Educational Toolbox for Polarimetric and Interferometric Polarimetric SAR Data Processing. In Proceedings of the POLinSAR 2009, Frascati, Italy, 26 January 2009; p. CD-ROM.
31. Papathanassiou, K.P.; Cloude, S.R. Single-Baseline Polarimetric SAR Interferometry. *IEEE Trans. Geosci. Remote Sens.* **2001**, *39*, 2352–2363. [CrossRef]

32. Hajnsek, I.; Scheiber, R.; Keller, M.; Horn, R.; Lee, S.; Ulander, L.; Gustavsson, A.; Sandberg, G.; Le Toan, T.; Tebaldini, S. *BIOSAR 2008: Final Report*; ESA-ESTEC: Noordwijk, Netherlands, 2009; Volume 22052.
33. Neumann, M.; Neumann, M.; De, T.; Neumann, M. Remote Sensing of Vegetation Using Multi-Baseline Polarimetric SAR Interferometry: Theoretical Modeling and Physical Parameter Retrieval. Ph.D. Thesis, Université Rennes, Rennes, France, 2009.
34. Soja, M.J.; Sandberg, G.; Ulander, L.M.H. Regression-Based Retrieval of Boreal Forest Biomass in Sloping Terrain Using P-Band SAR Backscatter Intensity Data. *IEEE Trans. Geosci. Remote Sens.* **2012**, *51*, 2646–2665. [CrossRef]
35. Yamada, H.; Yamaguchi, Y.; Rodriguez, E.; Kim, Y.; Boerner, W.M. Polarimetric SAR Interferometry for Forest Canopy Analysis by Using the Super-Resolution Method. In Proceedings of the 2001 International Geoscience and Remote Sensing Symposium (Cat. No. 01CH37217), Sydney, NSW, Australia, 9–13 July 2001; Volume 3, pp. 1101–1103.
36. Mette, T.; Kugler, F.; Papathanassiou, K.; Hajnsek, I. Forest and the Random Volume over Ground-Nature and Effect of 3 Possible Error Types. In Proceedings of the European Conference on Synthetic Aperture Radar (EUSAR), Dresden, Germany, 16–18 May 2006; pp. 1–4.
37. Mao, Y.; Michel, O.O.; Yu, Y.; Fan, W.; Sui, A.; Liu, Z.; Wu, G. Retrieval of Boreal Forest Heights Using an Improved Random Volume over Ground (RVoG) Model Based on Repeat-Pass Spaceborne Polarimetric SAR Interferometry: The Case Study of Saihanba, China. *Remote Sens.* **2021**, *13*, 4306. [CrossRef]
38. Liao, Z.; He, B.; Quan, X.; van Dijk, A.I.J.M.; Qiu, S.; Yin, C. Biomass Estimation in Dense Tropical Forest Using Multiple Information from Single-Baseline P-Band PolInSAR Data. *Remote Sens. Environ.* **2019**, *221*, 489–507. [CrossRef]
39. Managhebi, T.; Maghsoudi, Y.; Zoej, M.J.V. An Improved Three-Stage Inversion Algorithm in Forest Height Estimation Using Single-Baseline Polarimetric Sar Interferometry Data. *IEEE Geosci. Remote Sens. Lett.* **2018**, *15*, 887–891. [CrossRef]
40. Liu, F.; Yang, Z.; Zhang, G. Canopy Gap Characteristics and Spatial Patterns in a Subtropical Forest of South China after Ice Storm Damage. *J. Mt. Sci.* **2020**, *17*, 1942–1958. [CrossRef]
41. Zhang, J.; Zhang, Y.; Fan, W.; He, L.; Yu, Y.; Mao, X. A Modified Two-Steps Three-Stage Inversion Algorithm for Forest Height Inversion Using Single-Baseline L-Band PolInSAR Data. *Remote Sens.* **2022**, *14*, 1986. [CrossRef]
42. Cloude, S.R.; Williams, M.L. A Coherent EM Scattering Model for Dual Baseline POLInSAR. In Proceedings of the IGARSS 2003. 2003 IEEE International Geoscience and Remote Sensing Symposium. Proceedings (IEEE Cat. No. 03CH37477), Toulouse, France, 21–25 July 2003; Volume 3, pp. 1423–1425.
43. Tebaldini, S.; Rocca, F. Multibaseline Polarimetric SAR Tomography of a Boreal Forest at P-and L-Bands. *IEEE Trans. Geosci. Remote Sens.* **2011**, *50*, 232–246. [CrossRef]
44. Rocca, F. Modeling Interferogram Stacks. *IEEE Trans. Geosci. Remote Sens.* **2007**, *45*, 3289–3299. [CrossRef]
45. Simard, M.; Hensley, S.; Lavalle, M.; Dubayah, R.; Pinto, N.; Hofton, M. An Empirical Assessment of Temporal Decorrelation Using the Uninhabited Aerial Vehicle Synthetic Aperture Radar over Forested Landscapes. *Remote Sens.* **2012**, *4*, 975–986. [CrossRef]

remote sensing

MDPI

Article

Forest Structure Simulation of Eucalyptus Plantation Using Remote-Sensing-Based Forest Age Data and 3-PG Model

Yi Zhang [1,2], Dengsheng Lu [1,2], Xiandie Jiang [1,2], Yunhe Li [1,2] and Dengqiu Li [1,2,*]

1 Institute of Geography, Fujian Normal University, Fuzhou 350007, China
2 Fujian Provincial Key Laboratory for Subtropical Resources and Environment, Fujian Normal University, Fuzhou 350007, China
* Correspondence: lidengqiu@fjnu.edu.cn

Abstract: Eucalyptus plantations play an important role in the timber supply and global warming mitigation around the world. Forest age is a critical factor for evaluating and modeling forest structure (e.g., diameter at breast height (DBH), height (H), aboveground carbon stocks (ACS)) and their dynamics. Recently, the spatial distribution of forest age at different scales based on time series remote sensing data has been widely investigated. However, it is unclear whether such data can effectively support the simulation and assessment of forest structure, especially in fast-growing plantation forests. In this study, the physiological principles in predicting growth (3-PG) model was firstly optimized and calibrated using survey and UAV lidar data at the sample plot (SP) scale, and was then applied at the forest sub-compartment (FSC) scale by designing different simulation scenarios driven by different forest age data sources and adjustments. The sensitivity of the simulated forest structure parameters to forest age was assessed at the SP and FSC levels. The results show that both the survey forest age data and the remote-sensing-derived forest age data could accurately estimate the DBH, H, and ACS of eucalyptus plantations with the coefficients of determination (R^2) ranging from 0.87 to 0.94, and the relative root mean square error (RRMSE) below 20% at SP level. At the FSC level, the simulation results based on remotely sensed forest age data are significantly better than FSC forest age data from surveys by forestry bureaus, with R^2 of ACS 0.7, RMSE 9.12 Mg/ha, and RRMSE 28.24%. The results of the sensitivity analysis show that the DBH, H, and ACS show different degrees of variation under different adjusted forest ages at SP and FSC level. The maximum difference in ACS is 82.91% at the SP scale if the forest age decreases 12 months and 41.23% at the FSC scale if the forest age increases 12 months. This study provides an important reference for future studies using forest age data obtained by remote sensing to drive the forest carbon model in a large spatial scale.

Keywords: 3-PG model; eucalyptus; forest age; forest structure; remote sensing; sensitivity

Citation: Zhang, Y.; Lu, D.; Jiang, X.; Li, Y.; Li, D. Forest Structure Simulation of Eucalyptus Plantation Using Remote-Sensing-Based Forest Age Data and 3-PG Model. *Remote Sens.* **2023**, *15*, 183. https://doi.org/10.3390/rs15010183

Academic Editor: Arturo Sanchez-Azofeifa

Received: 24 October 2022
Revised: 19 December 2022
Accepted: 27 December 2022
Published: 29 December 2022

1. Introduction

Forests as an important component of the terrestrial carbon pool play a vital role in regulating regional and global carbon balances and slowing down the increase in atmospheric CO_2 concentration [1]. A lot of research work was performed to quantify the carbon stocks, carbon density, and potential carbon sink of forest ecosystems [2]. Accurate estimation of these carbon variables of forest ecosystems is an important goal pursued by ecologists and geographers, and also an important basis for achieving carbon neutralization.

Forest age is an important stand parameter of the forest ecosystem, which not only represents the planting time and succession stage of trees or stands, but also has important impacts on the physiological and ecological parameters in the carbon and water cycle models [3]. It is a critical factor that determines the temporal and spatial distribution of carbon pool and flux of the forest ecosystem, and corresponding management measures in forest plantations [4]. Previous studies show that net primary productivity (NPP) increases with the increase in stand age in the early stage of the forest ecosystems, reaches the maximum in the middle stage after canopy closure, and gradually declines in the later stage [5]. This relationship makes most carbon cycle variables, such as biomass, carbon stocks, gross primary productivity (GPP), and net ecosystem productivity (NEP), closely related to forest age [6,7]. Therefore, forest age is the key data to accurately estimate and simulate the carbon uptake dynamics of the forest ecosystem [8], and many carbon cycle models take forest age as known information [9,10]. However, there is often a lack of accurate, timely, and high-spatial-resolution information on the spatial distribution of forest age in regional forest carbon cycle research, which makes it difficult for the models to conduct forest carbon stock simulation and estimation [11].

Traditionally, the way to obtain forest age was mainly through forest inventory at sample plot (SP) by inquiring, professional experience, or tree cones [12], which were very costly, long cycle, and easily subject to geographical restrictions. It is difficult to obtain large-scale and long-term spatial forest age data. Taking China as an example, a three-level forest resources inventory system has been established: national forest continuous inventory (NFCI), forest management planning inventory (FMPI), and forest operation design inventory (FODI) [13]. FODI is a very detailed survey conducted at the smallest forest management compartment (FSC), and is the only spatial data of forest age from a manual survey. However, the survey is conducted every five years, meaning the forest age information is relatively lagged and full of uncertainties. Satellite remote sensing has the advantage of continuous monitoring of land surface change information over long distances and large areas. It has become an important and effective means to obtain the spatial distribution of forest age [14]. There are two main strategies to retrieve forest age from remote sensing data. One is to establish a forest age estimation model based on single or multi-period remote sensing data (e.g., spectral, vegetation index, tree height product), combining with ground survey, and meteorological and other data. This method has been used to extract the spatial distribution of forest age at global, national, and regional scales [8,11,14–16]. The second is to extract forest disturbance year based on time-series remote sensing data change detection [17,18]. Recently, Li et al. [19] proposed a random localization segmentation-based method to map the spatial distribution of successive plantation generation and forest age for these short-rotation eucalyptus plantations based on time series Landsat data. These remote-sensing-based forest age products provided the valuable input data for forest ecosystem carbon models. However, due to the limitations of remote sensing data and algorithms, such as cloud snow, noise, spatial and temporal resolution, and saturation of remote sensing signals [20], there are often some errors and uncertainties in the obtained forest age data, with R^2 ranging from 0.7 to 0.92 and RMSE ranging from 1.2 to 2.91 years, especially in tropical and subtropical regions [3,19,21].

The carbon cycle model based on tree growth and ecological process is an effective method to simulate forest growth, biomass, and carbon stocks [22,23]. It can be grouped into two categories: patch-scale carbon cycle model and regional-scale carbon cycle model, according to the simulation spatial scale [24]. The patch-scale carbon cycle model can be further divided into individual tree-based and stand-based carbon cycle models. The prior can simulate the growth and mortality of each tree, and predict the diameter distribution of the stand. These models usually require lots of input data, computationally intensive, and most are conducted locally [23]. The stand-based patch carbon model can simulate the forest carbon cycle at different time scales (day, month, or year) by assuming that the trees are spatially uniformly distributed in a stand [23]. The stand-based carbon models, such as spatial production allocation model (SPAM), the individual-based forest landscape and disturbance (iLand) model, and 3-PG model [25–27] are widely used to simulate forest growth and carbon cycle, and make management measures plans [28–30]. These models can be easily extended to the regional scale [24].

The roles of forest age in the forest carbon cycle research mainly focus on the use of forest age to analyze the impact of forest management on carbon sinks and to improve carbon estimates in the terrestrial carbon models [1]. The utilization of forest age data can effectively improve the accuracy of the simulations, but the uncertainty in the forest age might also bring much ambiguity to the carbon cycle model, and few studies have assessed the impact of such uncertainties on the carbon simulation results, especially for the remote-sensing-based forest age data. For example, many researchers conducted sensitivity analyses on different parameters in 3-PG models for variety research purposes, including soil fertility (FR), age at canopy closure (fullCanAge), maximum canopy quantum efficiency, maximum canopy conductance, aWS (constant in stem mass v diam. relationship), and nWS (power in stem mass v diam. relationship) [22,31]. Few studies selected stand age parameters for sensitivity analysis, because most of the studies performed their research at plot level with accurate and known forest age.

Eucalyptus is a globally important plantation tree species with fast growth rate, short harvest rotation, and strong carbon sequestration capacity [32]. Eucalyptus was introduced into China in 1890 and has been planted for more than 130 years, making China the second largest plantation country in the world [33]. Eucalyptus plantations have greatly alleviated the shortage of timber supply from plantation forests in China, but the very short rotation cycle (about 6 years) and intensive management have led to many ecological problems [34]. Some studies show that the large-scale plantation of eucalyptus plantations has resulted in soil fertility degradation and soil erosion, limited growth of understory vegetation, and decline in biodiversity, while some studies show that eucalyptus plantations have an important role in promoting the ecological environment [33,35,36]. In the context of achieving the goal of carbon neutrality, people pay more attention to the carbon stock and carbon sequestration potential of eucalyptus, and accurate estimation of the carbon dynamics of eucalyptus has become an important issue.

This study selected the eucalyptus plantations in Yuanling Forestry Farm, Zhangzhou City, Fujian Province, China as the research object. We comprehensively used SP survey data, forest inventory data, meteorological data, UAV lidar data, forest age obtained based on time-series remote sensing data, and a 3-PG model to simulate the forest structure of the study area, and assess the simulation accuracy. Specifically, the following two questions remain to be answered: (1) Can the forest age data obtained from remote sensing data support the 3-PG model to accurately simulate forest structural parameters at the SP scale and FSC scale? (2) How sensitive are the simulation results of the 3-PG model to the forest age data at the two scales?

2. Materials and Methods

2.1. Study Area

The study area is located in Yuanling State Forestry Farm in Yunxiao County, Zhangzhou City, Fujian Province, China (Figure 1). It has a typical southern subtropical maritime monsoon climate with an average annual temperature of 21.2 °C and annual precipitation of 1730.6 mm. The planting history of the study area in recent decades can be summarized as: rubber trees was planted in the beginning of the 1980s, and were gradually replaced by fruit trees (such as longan) from 1993 due to the declined economic value of rubber trees, and eucalyptus was introduced around 2005, and then widely planted in the study area. Some Chinese fir and Pinus elliottii forests were also gradually replaced by eucalyptus during the period 2007–2010. The main species of eucalyptus were eucalyptus grandis x urophylla and eucalyptus urophylla S.T. Blake.

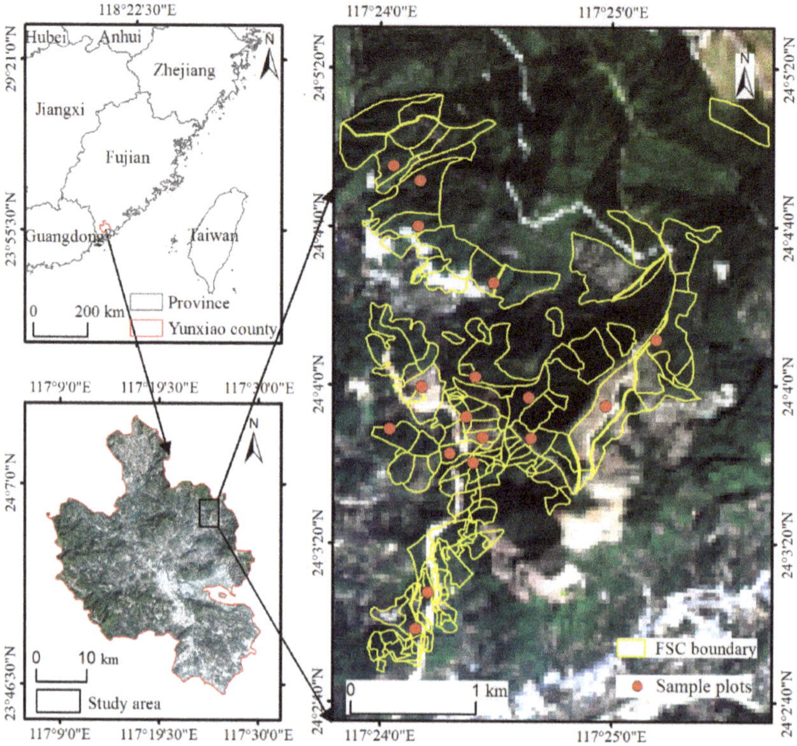

Figure 1. Location of the study area, spatial distribution of sample plots, and forest sub-compartment of eucalyptus. The base map is a true color composite of Sentinel-2 image.

2.2. Data collection and Processing

2.2.1. Field Survey Data

The survey data include the FSC data carried out by government departments in 2017, and SP data surveyed in 2021. The main information of the FSC data includes average diameter at breast height (DBH), average tree height (H), stand age, survey date, stand volume per hectare, number of trees per hectare, elevation, depth of soil, etc. The FSC data were surveyed in 2017 and are the latest available forestry survey data. We chose 140 eucalyptus FSC, a total area of 379.2 ha, to carry out our simulation with the 3-PG model (Figure 1). The forest age of these FSC in 2017 was mostly concentrated in 0–4 and 9–12 years (Table 1).

We investigated 17 eucalyptus plots with an area of 20 m × 20 m in the study area. The forest age of the survey plots ranges from 1 to 13 years, the average DBH is 3.62–16.26 cm, and the average H is 4.02–19.69 m. We measured and recorded DBH and H for each tree with DBH greater than 5 cm in the plots. The planting time, management history, and environment information were also recorded through asking the owner. The model developed by [37] was used to calculate the biomass of each organ of each tree (Table 2). The biomass of each tree was summed to obtain the aboveground biomass of SP, and then converted to ACS by multiply carbon coefficient (0.4764) [37]. Considering eucalyptus has a very rapid growth rate and 3-PG model can simulate the forest structure monthly, the SP were surveyed about every six months (January 2021, July 2021, and December 2021). The data from the three surveys were used to verify the simulation accuracy of the model at the SP level. Some plots were harvested when conducting the second and third survey, and some plots were only measured for DBH. Finally, we collected a total of 44 DBH observations and 41 H observations for these plots after the three surveys.

Table 1. Basic information of the 140 eucalyptus FSC.

Age (Year)	Number (n)	Mean DBH (cm)	Mean H (m)	Total Area (ha)
≤4	55	<9.5	2.5–10.8	105.38
5–8	26	10.9–17.6	12.3–22	115.75
9–12	53	11.5–24.4	14.3–29.2	143.25
13–17	6	20.8–24.6	21.7–28.3	14.82
Total	140	0–27.6	2.5–29.2	379.2

Table 2. Model for estimating aboveground biomass (stem, branch, bark, and foliage) of eucalyptus.

Organ	Fitting Equation	R^2
Stem	$W = 0.0259 \times DBH^{2.8762}$	0.978
Branch	$W = 0.0263 \times DBH^{2.2471}$	0.887
Bark	$W = 0.0539 \times DBH^{1.7802}$	0.949
Foliage	$W = 0.1785 \times DBH^{1.1753}$	0.871

2.2.2. Meteorology Data

We calculated monthly minimum temperature (°C), maximum temperature (°C), average temperatures (°C), and precipitation (mm) based on the hourly recorded data from 2008–2021 that were acquired from the meteorological station nearby the study area. Considering some FSC have an older forest age, the temperature and precipitation data were extended to 1997–2007 using the data provided by National Aeronautics and Space Administration (NASA). Solar radiation data from 1997 to 2021 were also acquired from the website (https://power.larc.nasa.gov/data-access-viewer/, accessed on 10 January 2022) due to a lack of local observations. These data have been proven to be accurate enough to provide reliable meteorological and solar radiation data in areas where site observations are sparse [38,39]. Compared with the data of the same year from the meteorological station, the two source data products have high consistency and can be used together for the model simulation.

2.2.3. UAV Lidar Data

The UAV lidar data were acquired in July 2021 with an average point cloud density of 60 points/m^2. The process of lidar data mainly includes filtering, denoising, normalization, and generating CHM data [40]. The Lidar360 software was used to remove noise in the point cloud data, such as bird points, low points, and utility poles. The discrete point cloud echo points were divided into ground and non-ground points. The ground points were used to generate a digital elevation model (DEM) by inverse distance weighted interpolation method. All non-ground points were interpolated to a digital surface model (DSM) with a spatial resolution of 1 m. Then, the canopy height model (CHM) was obtained by subtracting DEM from DSM. The lidar data obtained in July 2021 and sample plot data surveyed at the same time were used to establish ACS prediction model, that is, 17 observations were used in the model. Stepwise regression method was used to establish a carbon stock estimation model with the variables from the CHM acquired from lidar with a resolution of 20m × 20m (the same as plot size) [41]. Two variables (mean CHM and skewness) for ACS prediction were identified using stepwise regression method. Then, leave-one-out cross-validation was used in the evaluation processes, with 16 samples used to train the model, and the established model was used to predict the ACS value of the one observation left out of the model. The validation shows that the model works quite well, with R^2 (coefficient of determination), RMSE (root mean square error), and RRMSE (relative root mean square error) values of 0.87, 8.73 Mg/ha, and 18.72%, respectively. The average H and average ACS of each FSC were calculated based on the modelled data, and used to assess the 3-PG model simulation results (Figure 2a,b).

2.2.4. Forest Age Data from Landsat Time Series Data

The forest age data of each FSC based on Landsat-based forest age data were provided by [19] (Figure 2c). The forest age of short rotation eucalyptus plantations was developed using a random localization segmentation algorithm and all available Landsat time series data. The Chow test and random forest continuous classification were used to obtain the spatial distribution of eucalyptus forest age at 30 m × 30 m spatial resolution with RMSE of 13 months in 2021. In our study area, the forest age error was about 12 months compared with the survey data. The simulation unit of this study was at FSC scale, and the average age of each FSC was obtained through zonal statistics.

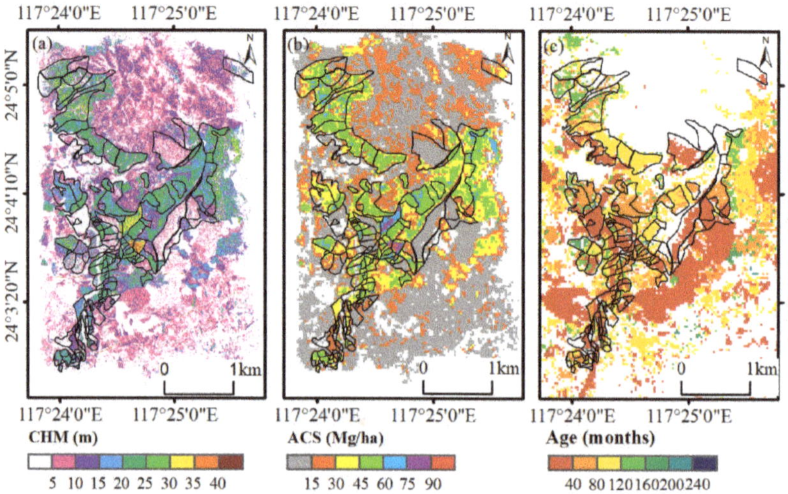

Figure 2. CHM (**a**), aboveground carbon stocks (**b**) based on UAV lidar, and forest age data (**c**) from Landsat time-series data.

2.3. 3-PG Model and Parameter Setting

The 3-PG model is a physiological–ecological process model based on allometric equations and a monthly time scale [27]. The model has a relatively simple structure and few input parameters [42]. It can simulate many tree species including eucalyptus, and is widely used in Australia, Brazil, Canada, and China [30,43–45]. The model was initially developed to simulate even-aged evergreen forest species, and now is able to simulate deciduous, uneven-aged, and mixed forest, and assess the forest growth under different management measures [46]. Many studies utilized the model to simulate forest growth of eucalyptus, Masson pine, and larch at the plot level [31,47,48]. The model has four submodules: the light sub-model, the biomass production and allocation sub-model, the water balance sub-model, and the mortality sub-model. More details about 3-PG model are provided in [27,45]. The tree growth was simulated at monthly intervals by inputting monthly meteorological data (maximum and minimum temperatures, average temperature, precipitation, and solar radiation), site conditions (latitude, soil class, and soil fertility), planting time, and initial organ biomass, management measures, and parameters for the tree species. The model can output many variables such as GPP, NPP, DBH, H, organ biomass (monthly), etc. The DBH, H, and ACS (calculated from biomass) were selected for output and evaluation in this study. All the simulations were performed with the r3PG package in the R platform [49].

2.3.1. Model Parameters

The 3-PG model provided a complete set of parameter values for eucalyptus, which was a very useful reference for the parameter setting of this study. For the stem biomass parameters, the key parameters aWS (0.0259) and nWS (2.8762) for eucalyptus in our study area were obtained by fitting the allometric equation $W_S = aWS \times DBH^{nWS}$ (W_S is the stem biomass) ($R^2 = 0.9998$) based on the DBH and stem biomass obtained from the survey data. The model simulation for each FSC started from its planting time, and the initialized biomass values of stem, root, and leaf were set to 1 Mg/ha, 2 Mg/ha, and 0.5 Mg/ha, respectively [50]. Soil class and soil moisture data were acquired by the Second National Soil Survey data, and other parameters were set following the reference [51]. See Appendix A Table A1 for details.

2.3.2. Simulation Scheme Design

Figure 3 shows the overall flowchart of the study. We used SP data, FSC data, Landsat age data, meteorological data, and site conditions to calibrate and drive the 3-PG model. Then, the surveyed forest age and forest age from Landsat were used to drive the 3-PG model and simulate the DBH, H, and ACS of eucalyptus plantations at the plot scale, and 17 sample plots with three sets of investigation data were used for validation. The impacts of historical management information on the accuracy of simulation results were also evaluated. Similarly, the FSC age and forest age from Landsat were used to simulate the DBH, tree height, and carbon storage of the eucalyptus plantations at the FSC scale, and validated by UAV lidar data. Finally, we explored the sensitivity of the simulated forest structure to the forest age on two scales.

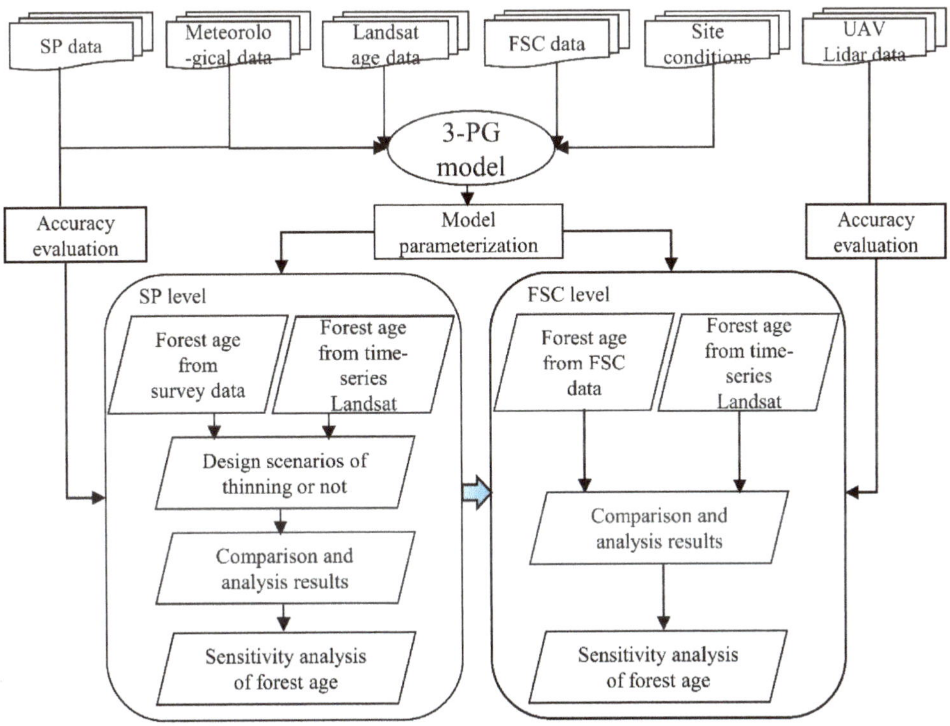

Figure 3. The overall work flow chart of the article.

Simulation Scheme Based on the SP Level

The carbon stocks for the 17 sample plots were simulated using the parameterized 3-PG model based on the surveyed forest data and forest age data from Landsat (Figure 4). The sample plot was chosen to represent a certain area that had similar planation history and management. The forest age information of the plot was obtained by extracting the age of the pixel where the plot was located in the Landsat pixels. The impacts of historical management information on the simulation results were also evaluated by considering the selective cutting or not (acquired during the survey). It was difficult for Landsat time-series data to accurately detect the thinning activities, so the management was not considered in the forest age from Landsat-based simulation. All the biomass variables from the model were conversed to carbon stock by multiplying the carbon coefficient and obtaining the ACS. The simulation accuracy of three forest structure variables (DBH, H, ACS) were assessed by the survey data.

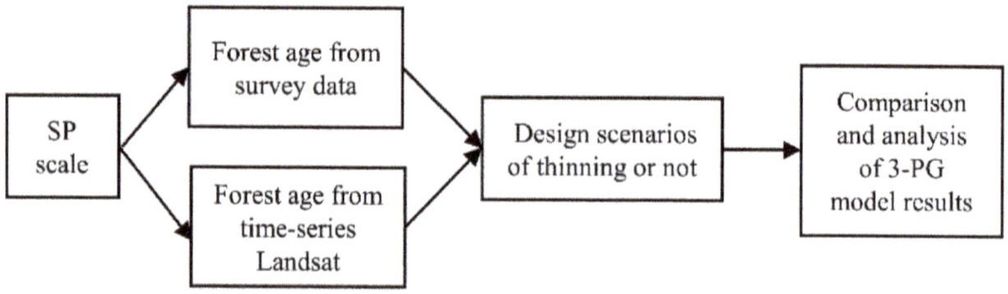

Figure 4. Simulation scheme design at the sample plot level.

Simulation Scheme Based on the FSC Level

The simulations were then carried out for the 140 FSC of eucalyptus plantation. The following three scenarios were designed to evaluate the impact of forest age data on the carbon stocks simulation.

(a) Simulation based on FSC age information. The forest age from 2017 FSC survey data was used as the input data to drive the 3-PG model. Considering that FSC age was obtained from 2017, and the validation data from lidar were obtained in July 2021, some FSC may have been harvested during the period, but the FSC data may have lagged. Therefore, the FSC were divided into two groups for evaluation: FSC planted before 2015 and FSC planted after 2015 (eucalyptus plantation harvested age mainly $\geq$ 6 years in the study area);

(b) Simulation based on forest age data from Landsat. The forest age data (introduced in Section 2.2.3) extracted from Landsat time-series data in January 2021 were used as the input to drive the 3-PG model. It should be noted that the simulation was performed for the 140 FSC, but not for each pixel due to lack of high spatial resolution data of soil, meteorological, tree density, etc.;

(c) Simulation based on the adjusted forest age data. As the forest age based on remote sensing data has many uncertainties, we adjusted the forest age by $\pm$3 months, $\pm$6 months, and $\pm$12 months to test the sensitivity of the model simulation results for both the SP and FSC.2.3.3. accuracy evaluation.

The simulated ACS, DBH, and H of SP were evaluated by the surveyed data. The simulated ACS and H of FSC were evaluated by the data calculated from UAV lidar (introduced in Section 2.2.4). The coefficient of determination (R^2), root mean square error (RMSE), and relative root mean square error (RRMSE) were used to evaluate the simulation accuracy of the model. They were calculated as follows:

$$R^2 = \frac{\sum_{i=1}^{n}(\hat{y}_i - \overline{y})^2}{\sum_{i=1}^{n}(y_i - \overline{y})^2} \tag{1}$$

$$\text{RMSE} = \sqrt{\sum_{i=1}^{n}\frac{(y_i - \hat{y}_i)^2}{n}} \tag{2}$$

$$\text{RRMSE} = \frac{\sqrt{\sum_{i=1}^{n} \frac{(y_i - \hat{y}_i)^2}{n}}}{\sum_{n=1}^{n} \frac{y_i}{n}} \tag{3}$$

where n is the number of observations, y_i is the observed value of plot i, $\hat{y}_i$ is the simulated value of plot i, and $\overline{y}_i$ is the mean value of all sample plots.

3. Results

3.1. Simulation Results at SP

The results based on the surveyed forest age show that the 3-PG model can accurately simulate DBH, H, and ACS of eucalyptus plantations (Figure 5), with R^2 values ranging from 0.80 to 0.93. Taking thinning information into account can further improve the simulation accuracy. The R^2 of ACS, DBH, and H increase by 0.09, 0.06, and 0.07, respectively, and the RRMSE decreases by 8.54%, 6.75%, and 4.2%, respectively. The simulation results based on forest age data from Landsat also achieve high accuracy, with R^2 of DBH, H, and carbon stock all higher than 0.85, and RRMSE less than 20% (Figure 6). They are generally better than the simulated results not considering thinning, and are closer to the simulated results with thinning information considered.

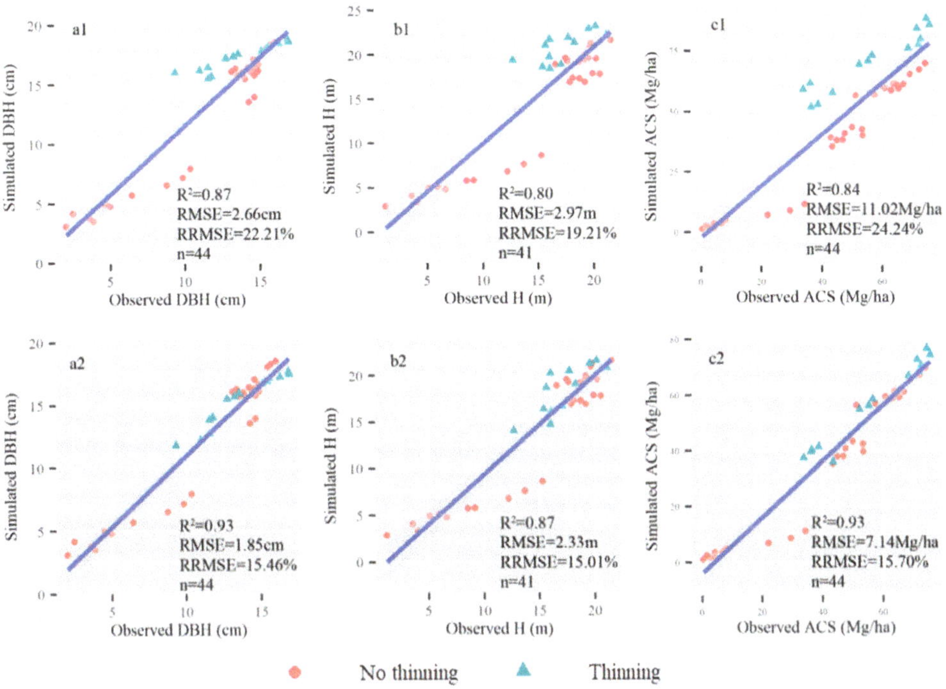

Figure 5. Validation of simulated forest structure at the SP. Plots (**a1–c1**) are simulated diameter at breast height, height, aboveground carbon stock of not considering thinning in the model; (**a2–c2**) are simulated diameter at breast height, height, aboveground carbon stock of considering thinning. No thinning and thinning denote whether the SP thinned or not during the growth.

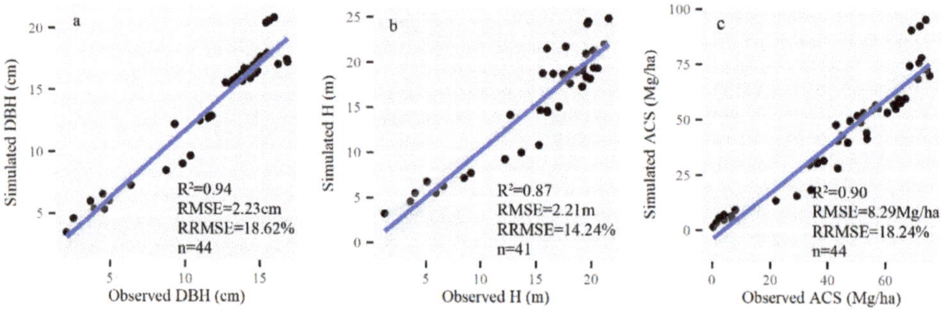

Figure 6. Validation of simulated diameter at breast height (**a**), height (**b**), and aboveground carbon stock (**c**) based on remote sensing stand age data at SP scale.

3.2. Simulation Results at FSC

The simulation results based on FSC age data for the 140 FSC show very low R^2 and high RMSE compared with the ACS and H data estimated from UAV lidar. However, for the FSC planted after 2015, the simulation results are quite well-matched (Figure 7), with RMSE of H and ACS of 2.91 m and 14.22 Mg/ha, respectively. Obviously, for the FSC planted before 2015, there is no significant relationship between the simulated results and validation data, due to the unknown harvest information and inaccurate forest age data, and the RMSE of H and ACS are 14.06 m and 80.78 Mg/ha, respectively. This suggests that accurate and timely updating of forest age is critical for the model simulation of eucalyptus plantations. The accuracies of simulated H and ACS using forest age data from Landsat significantly increase for the 140 FSC compared to the results based on FSC age data (Figure 8). The forest age data from Landsat are very effective for driving the 3-PG model. Both the simulated H and ACS show high R^2 and low RRMSE, but the accuracy is not so good as the SP level.

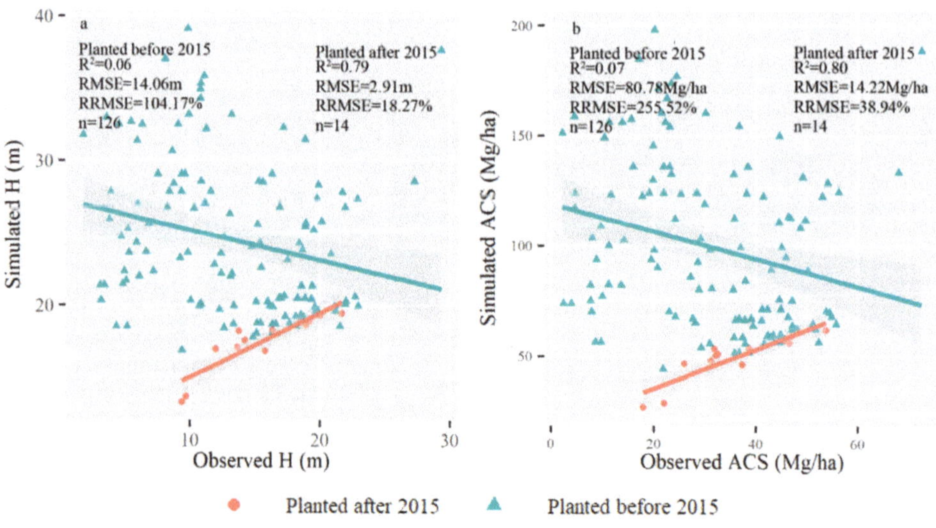

Figure 7. Validation of simulated height (**a**) and aboveground carbon stocks (**b**) based on FSC forest age data.

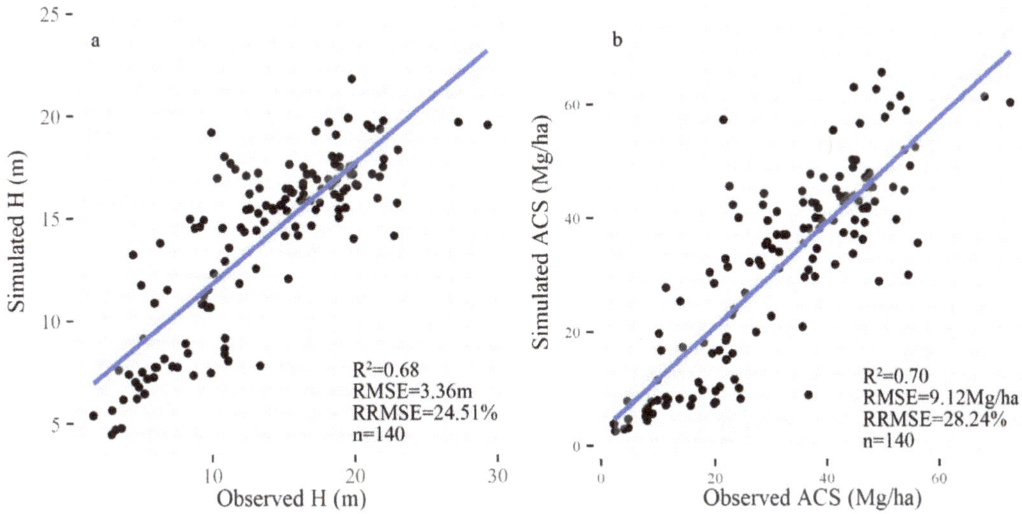

Figure 8. Validation of simulated height (**a**) and aboveground carbon stocks (**b**) based on forest age data from Landsat.

3.3. Sensitivity of the Simulation Results to Forest Age

3.3.1. Sensitivity Analysis of the 3-PG Model at the SP Level

The simulation results from different adjusted forest ages at sample plot level show that DBH, H, and ACS exhibit different degrees of variation (Table 3). The largest variation occurs in DBH with the RMSE increasing by 33.45% when the forest age increases by 12 months. The H and ACS have consistent change trends. The changes in RMSE are small when the forest age increases, but become larger when the forest age decreases from 3 months to 12 months. The RMSE of H and ACS increase by 42.92% and 82.91% as the forest age decreases 12 months. It should be noted that the sensitivity analysis shows the highest R^2 and the lowest RMSE are not consistent for DBH, H, and ACS in these adjusted forest age designs. For example, the lowest RMSE of simulated DBH occurs when the forest age decreases 6 months, while for H and ACS this occurs in increased by 6 months design. In addition, the highest R^2 and lowest RMSE occur in forest age adjusted designs that are inconsistent.

Table 3. Comparison of model predictions of the diameter at breast height (DBH), height (H), and aboveground carbon stocks (ACS) with observations of DBH, H, and ACS under different adjusted forest ages based at SP level.

Variables	DBH			H			ACS		
	R^2	RMSE (cm)	Change Degree of RMSE	R^2	RMSE (m)	Change Degree of RMSE	R^2	RMSE (Mg/ha)	Change Degree of RMSE
−3 months	0.94	1.74	5.95%	0.86	2.50	7.30%	0.94	8.07	13.03%
−6 months	0.93	1.68	9.19%	0.85	2.77	18.88%	0.93	9.56	33.89%
−12 months	0.90	1.72	7.03%	0.79	3.33	42.92%	0.90	13.06	82.91%
No change	0.93	1.85	0	0.87	2.33	0	0.93	7.14	0
+3 months	0.94	2.09	12.97%	0.87	2.26	3%	0.94	6.15	13.86%
+6 months	0.94	2.29	23.78%	0.87	2.25	3.43%	0.95	5.92	17.09%
+12 months	0.94	2.78	33.45%	0.88	2.42	3.86%	0.95	7.41	3.78%

3.3.2. Sensitivity Analysis of the 3-PG Model at the FSC Level

The impacts of adjusted forest age on the simulated H and ACS for the 140 FSC show that the largest deviation for both occurs in the scenario of increased age of 12 months (Table 4), with RMSE increasing by 12.2% and 41.23%, respectively. The lowest RMSE of H and ACS are observed in the scenario of decreased age of 3 months and no adjustment scenario, respectively. The decreased forest age does not lead to much variation in ACS simulation at the SP level.

Table 4. Comparison of model predicted height (H) and aboveground carbon stocks (ACS) with lidar inverse H and ACS under different stand age conditions based on FSC scale.

Variables	H			ACS		
	R^2	RMSE (m)	Change Degree of RMSE	R^2	RMSE (Mg/ha)	Change Degree of RMSE
−3 months	0.74	3.04	−10.53%	0.75	9.22	1.1%
−6 months	0.73	3.04	−10.53%	0.74	9.27	1.64%
−12 months	0.72	3.19	5.06%	0.71	10.33	13.27%
No change	0.68	3.36	0	0.70	9.12	0
+3 months	0.74	3.25	−3.27%	0.77	10.25	12.39%
+6 months	0.74	3.4	1.19%	0.77	11	20.61%
+12 months	0.74	3.77	12.2%	0.77	12.88	41.23%

4. Discussion

4.1. High Accuracy Can Be Realized Based on the Forest Age Data from Landsat

The 3-PG model has been widely used to estimate forest growth parameters such as DBH, H, biomass, and NPP. In addition, the model can also output other parameters, such as forest volume, stand basal area, and stand density, which are required by forest managers. In this study, we estimated and evaluated the simulated DBH, H, and ACS of eucalyptus at the SP level and FSC level, and analyzed their sensitivity to the forest age data. During the simulation, we adopted most of the default parameters that have been established for eucalyptus (except the allometric growth equations) [51]. Both the measured data and estimated data from UAV lidar show that the 3-PG model has high simulation accuracy as long as high-quality forest age is provided. The management information can further improve the simulation accuracy. Our study shows that the forest age data from Landsat data have similar simulation accuracy with the scenario of using surveyed forest age and thinning data together. The reasons might be the uncertainties of surveyed forest age data and the minor impact of thinning measures on final ACS. In fact, it is difficult to acquire the exact planting time of eucalyptus (e.g., month), especially under the condition of the coexistence of coppice and planting. Eucalyptus has a very rapid growth in the early stage and reaches canopy closure within 2–3 years [19]. It is very difficult to obtain such high precision planting time. The forest planting time in the model was needed to be set at month, which might be difficult to simulate the early growth process of eucalyptus. The assimilation of more dense time series remote sensing data or products (e.g., LAI from Landsat or Sentinel) might improve these processes.

As an important parameter of forest carbon cycle model, forest age represents the planting time of trees/stands and reflects the current growth stage. For physiological–ecological process models, changes in stand age inevitably affect factors such as stomatal conductivity and hydraulic conductivity, which, in turn, affect physiological processes in trees, such as photosynthesis and root turnover rates [27,52]. In addition, trees at different ages have different sensitivities to parameters [53], e.g., trees have a high sensitivity to parameters such as soil fertility in the young stage and a low sensitivity to stand density in the mature stage. Therefore, it is necessary to obtain accurate and reliable information on the age of the forest during the carbon cycle, and will be the fundamental to optimize and parameterize the regional carbon models.

The constant (aWS) and power (nWS) in the allometric equation of stem biomass play an important role in the biomass allocation sub-model. Previous studies show that the main reason for the poor simulation of the 3-PG model is not using the local biomass allocation and allometric growth parameters [54]. This parameter was also observed to have the greatest influence on the simulated volume and DBH of Chinese fir in Nanping, Fujian [55]. The remaining parameters of the model can also affect the simulation accuracy of the model. For example, Hua et al. [56] found that the simulation accuracy can be further improved by fitting the maximum canopy conductance and canopy quantum efficiency based on the corrected aWS and nWS. Deciduous species have distinct growing seasons and non-growing seasons, which can be set through several parameters such as temperature, gammaF1 (maximum litterfall rate), gammaF0 (litterfall rate at $t = 0$), leafgrow, and leaffall. For example, for deciduous species, gammaF0 and gammaF1 can be set to 0, because all of the foliage will disappear at the end of the growing season. Eucalyptus is an evergreen tree species and previous research with 3-PG models seldom considered the difference between growing season and non-growing season [48,51,57,58]. However, further studies should pay more attention to the growth characteristics and responses to extreme climate events in different seasons.

4.2. Impact of Spatial Heterogeneity on Modelling Results

The simulated carbon stock for FSC using remote-sensing-based forest age data is significantly improved compared to the results based on FSC forest age data. However, some FSC still deviate greatly from the observed data, which is probably caused by the spatial heterogeneity of the FSC. At the beginning, the boundary of FSC was determined by the homogeneity within the forest stand, and similar management was performed. As time goes on, the same FSC might experience different management measures (thinning, fertilization, tree species, etc.) and disturbances (fires, diseases, typhoons), which causes the FSC to be more heterogeneous (for example in Figure 9). Both the CHM and aerial maps (Figure 9a1,a2) show that H in the northeast of the FSC is high, up to 30 m, but is low in the northwest of the FSC, and the maximum difference reaches 20 m. Obviously, the ACS also shows high spatial heterogeneity in this FSC. In Figure 9b, the H of the FSC is generally high, but the heterogeneity within the FSC is more obvious, and the difference between high and low trees is close to 25 m. This spatial heterogeneity could easily lead to overestimation or underestimation of the simulation results in the simulation process. Therefore, it is necessary to redraw the FSC and determine new boundaries to reduce the heterogeneity in future study, which will, potentially, significantly improve the accuracy of simulation results.

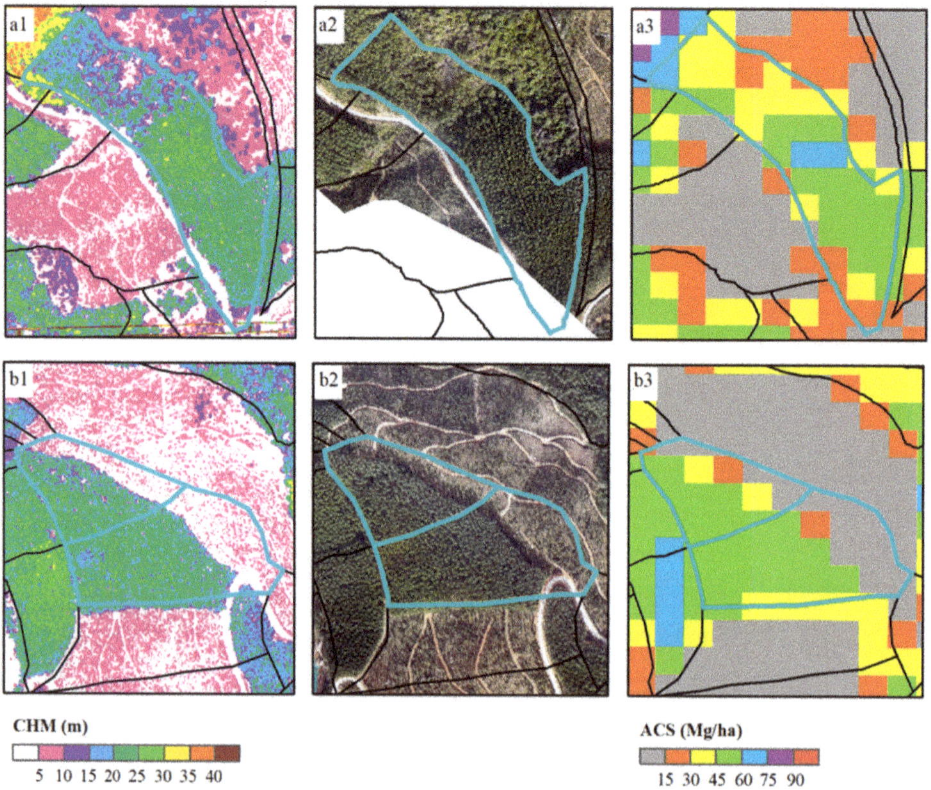

Figure 9. Spatial heterogeneity in FSC. Plots (**a1**,**b1**) are CHM; (**a2**,**b2**) are aerial photo; (**a3**,**b3**) are aboveground carbon stocks (ACS) estimated by UAV lidar data with the spatial resolution of 20 m.

4.3. Limitations and Potential Improvement

The 3-PG model was used to estimate DBH, H, and ACS of eucalyptus based on forest age data, meteorological data, and site conditions in the study area, and obtained a high simulation accuracy. The model can not only simulate the normal growing forest, but also estimate the growth state of the forest under different management measures such as thinning. Through thinning management, forests can achieve the goal of adjusting stand density, changing stand structure, and reducing competition among individual tree species, thus, changing the normal growth of trees. The simulation results at the SP scale show that the model captured well the thinning effects on forest growth. Considering thinning information can improve the simulation accuracy, which is consistent with the research results of Xie et al. [10]. However, the response of NPP to thinning measures has not been well-explored, and positive, negative, and neutral impacts coexist in different studies [59–61]. This should be better considered in future simulations. Landsat time-series data-based forest age data fails to monitor management such as thinning in eucalyptus plantations due to its coarse resolution in spatial and temporal data. This may reduce the accuracy of model simulation. Therefore, these subtle changes in forest dynamics should be better characterized through spectral mixture analysis or the use of higher spatial–temporal resolution data (such as Sentinel, Gaofen).

5. Conclusions

In this study, a process-based physiological–ecological 3-PG model was used to predict the forest structure of eucalyptus plantations at the local scale by combining remotely sensed stand age data. The results show that the 3-PG model can achieve satisfactory simulation results at the SP and FSC scales. The results of sensitivity analysis show that forest age has a significant effect on forest carbon stocks, with a maximum difference of 82.91% and 41.23% in ACS between different stand age conditions at the SP scale and FSC scale, respectively. The fact that thinning information can improve the simulation accuracy, but that the information is difficult to obtain, especially for the remote sensing data, must be considered. More subtle changes can be further acquired by integrating more efficient change detection algorithms and high spatial–temporal resolution data. This study was carried out in a local forestry farm, but our method can be easily extended to large regions with the time-series remote-sensing-acquired forest age data. The impact of uncertainty in the remotely sensed forest age data provides a useful reference for regional forest carbon cycle simulations based on forest age products.

Author Contributions: Conceptualization, D.L. (Dengqiu Li) and D.L. (Dengsheng Lu); methodology, Y.Z., D.L. (Dengqiu Li) and D.L. (Dengsheng Lu); software, Y.Z.; validation, Y.Z., D.L. (Dengqiu Li), X.J. and Y.L.; formal analysis, Y.Z. and D.L. (Dengqiu Li); investigation, Y.Z., X.J., and Y.L.; resources, D.L. (Dengqiu Li) and D.L. (Dengsheng Lu); data curation, Y.Z. and D.L. (Dengqiu Li); writing—original draft preparation, Y.Z. and D.L. (Dengqiu Li); writing—review and editing, D.L. (Dengqiu Li) and D.L. (Dengsheng Lu); visualization, Y.Z. and D.L. (Dengqiu Li); supervision, D.L. (Dengqiu Li) and D.L. (Dengsheng Lu); project administration, D.L. (Dengqiu Li) and D.L. (Dengsheng Lu); funding acquisition, D.L. (Dengqiu Li) and D.L. (Dengsheng Lu). All authors have read and agreed to the published version of the manuscript.

Funding: This research was financially supported by the Natural Science Foundation of Fujian Province, grant number 2022J01640, Public welfare projects of Fujian Provincial Science and Technology Department, grant number 2021R1002008, and the National Natural Science Foundation of China, grant number 41701490.

Data Availability Statement: Due to confidentiality agreements, supporting data can only be made available to bona fide researchers subject to a non-disclosure agreement. Details of the data and how to request access are available from Professor Dengqiu Li at Fujian Normal University.

Acknowledgments: The authors thank reviewers for their help in improving our manuscript.

Conflicts of Interest: The authors declare no conflict of interest.

Appendix A

Table A1. Description of parameters, unit, source, and their values.

Parameter Name	Description	Unit	Source	Value
pFS2	Foliage:stem partitioning ratio at DBH = 2 cm	-	D	1
pFS20	Foliage:stem partitioning ratio at DBH = 20 cm	-	D	0.15
aWS	Constant in stem mass vs. DBH relationship	-	F	0.0259
nWS	Power in stem mass vs. DBH relationship	-	F	2.8762
pRx	Maximum fraction of NPP to roots	-	D	0.8
pRn	Minimum fraction of NPP to roots	-	D	0.25
gammaF0	Litterfall rate at t = 0 month	$month^{-1}$	D	0.001

Table A1. *Cont.*

Parameter Name	Description	Unit	Source	Value
gammaF1	Litterfall rate for mature stands	$month^{-1}$	D	0.027
tgammaF	Age at which litterfall rate has median value	$month^{-1}$	D	12
Rttover	Average monthly root turnover rate	$month^{-1}$	D	0.015
Tmin	Minimum temperature for growth	°C	F	10
Topt	Optimum temperature for growth	°C	F	20
Tmax	Maximum temperature for growth	°C	F	36
MaxAge	Maximum stand age used in age modifier	yr	D	50
nAge	Power of relative age in fage	-	D	4
rAge	Relative age to give fage = 0.5	-	D	0.95
MinCond	Minimum canopy conductance	$m\ s^{-1}$	D	0
MaxCond	Maximum canopy conductance	$m\ s^{-1}$	D	0.02
LAIgcx	LAI for maximum canopy conductance	$m^2\ m^{-2}$	D	3.33
thinPower	Power in self-thinning rule	-	D	1.5
SLA0	Specific leaf area at age 0	$m^2\ kg^{-1}$	D	11
SLA1	Specific leaf area for mature stands	$m^2\ kg^{-1}$	D	4
tSLA	Age at which specific leaf area = (SLA0+SLA1)/2	yr	D	2.5
K	Extinction coefficient for absorption of PAR by canopy	-	D	0.5
fullCanAge	Age at full canopy cover	yr	D	3
alphaCx	Maximum canopy quantum efficiency	-	D	0.06
Y	Ratio NPP/GPP	-	D	0.47
fracBB0	Branch and bark fraction at age 0	-	D	0.75
fracBB1	Branch and bark fraction for mature stands	-	D	0.15
tBB	Age at which pBB = 1/2(PBB0 + PBB1)	yr	D	2
aH	Constant in the stem H relationship	-	F	1.4022
nHB	Power of DBH in stem H relationship	-	F	0.7079
nHN	Power of competition in stem H relationship	-	F	0.2492

References

1. Pan, Y.; Chen, J.M.; Birdsey, R.; McCullough, K.; He, L.; Deng, F. Age Structure and Disturbance Legacy of North American Forests. *Biogeosciences* **2011**, *8*, 715–732. [CrossRef]
2. Piao, S.; He, Y.; Wang, X.; Chen, F. Estimation of China's Terrestrial Ecosystem Carbon Sink: Methods, Progress and Prospects. *Sci. China Earth Sci.* **2022**, *65*, 641–651. [CrossRef]
3. Tang, S.; Tian, Q.; Xu, K.; Xu, N.; Yue, J. Age Information Retrieval of Larix Gmelinii Forest Using Sentinel-2 Data. *Natl. Remote Sens. Bull.* **2020**, *24*, 1511–1524.
4. Koedsin, W.; Huete, A. Mapping Rubber Tree Stand Age Using Pléiades Satellite Imagery: A Case Study in Thalang District, Phuket, Thailand. *Eng. J.* **2015**, *19*, 45–56. [CrossRef]

5. He, L.; Chen, J.M.; Pan, Y.; Birdsey, R.; Kattge, J. Relationships between Net Primary Productivity and Forest Stand Age in U.S. Forests. *Glob. Biogeochem. Cycles* **2012**, *26*, GB3009. [CrossRef]
6. Haywood, A.; Stone, C. Estimating Large Area Forest Carbon Stocks-a Pragmatic Design Based Strategy. *Forests* **2017**, *8*, 99. [CrossRef]
7. Ju, W.; Wang, X.; Sun, Y. Age Structure Effects on Stand Biomass and Carbon Storage Distribution of Larix Olgensis Plantation. *Acta Ecol. Sin.* **2011**, *31*, 1139–1148.
8. Yu, Z.; Zhao, H.; Liu, S.; Zhou, G.; Fang, J.; Yu, G.; Tang, X.; Wang, W.; Yan, J.; Wang, G.; et al. Mapping Forest Type and Age in China's Plantations. *Sci. Total Environ.* **2020**, *744*, 140790. [CrossRef]
9. Pugh, T.A.M.; Lindeskog, M.; Smith, B.; Poulter, B.; Arneth, A.; Haverd, V.; Calle, L. Role of Forest Regrowth in Global Carbon Sink Dynamics. *Proc. Natl. Acad. Sci. USA* **2019**, *116*, 4382–4387. [CrossRef]
10. Xie, Y.; Wang, H.; Lei, X. Simulation of Climate Change and Thinning Effects on Productivity of Larix Olgensis Plantations in Northeast China Using 3-PGmix Model. *J. Environ. Manag.* **2020**, *261*, 110249. [CrossRef]
11. Zhang, Y.; Yao, Y.; Wang, X.; Liu, Y.; Piao, S. Mapping Spatial Distribution of Forest Age in China. *Earth Space Sci.* **2017**, *4*, 108–116. [CrossRef]
12. Li, F.; Li, M.; Shi, Z.; Jiang, H.; An, J. Estimate Stand Age Distribution Based on Forest Survey and Remote Sensing Data. *For. Eng.* **2018**, *34*, 30–34. [CrossRef]
13. Xie, X.; Wang, Q.; Dai, L.; Su, D.; Wang, X.; Qi, G.; Ye, Y. Application of China's National Forest Continuous Inventory Database. *Environ. Manag.* **2011**, *48*, 1095–1106. [CrossRef] [PubMed]
14. Dai, M.; Tao, Z.; Lingling, Y.; Jia, G. Spatial Pattern of Forest Ages in China Retrieved from National-Level Inventory and Remote Sensing Imageries. *Geogr. Res.* **2011**, *30*, 172–184.
15. Zhang, C.; Ju, W.; Chen, J.; Li, D.; Wang, X.; Fan, W.; Li, M.; Zan, M. Mapping Forest Stand Age in China Using Remotely Sensed Forest Height and Observation Data. *J. Geophys. Res. Biogeosci.* **2014**, *119*, 1163–1179. [CrossRef]
16. Besnard, S.; Koirala, S.; Santoro, M.; Weber, U.; Nelson, J.; Gütter, J.; Herault, B.; Kassi, J.; N'Guessan, A.; Neigh, C.; et al. Mapping Global Forest Age from Forest Inventories, Biomass and Climate Data. *Earth Syst. Sci. Data* **2021**, *13*, 4881–4896. [CrossRef]
17. Ma, S.; Zhou, Z.; Zhang, Y.; An, Y.; Yang, G. Bin Identification of Forest Disturbance and Estimation of Forest Age in Subtropical Mountainous Areas Based on Landsat Time Series Data. *Earth Sci. Inform.* **2022**, *15*, 321–334. [CrossRef]
18. Zhao, F.; Sun, R.; Zhong, L.; Meng, R.; Huang, C.; Zeng, X.; Wang, M.; Li, Y.; Wang, Z. Monthly Mapping of Forest Harvesting Using Dense Time Series Sentinel-1 SAR Imagery and Deep Learning. *Remote Sens. Environ.* **2022**, *269*, 112822. [CrossRef]
19. Li, D.; Lu, D.; Wu, Y.; Luo, K. Retrieval of Eucalyptus Planting History and Stand Age Using Random Localization Segmentation and Continuous Land-Cover Classification Based on Landsat Time-Series Data. *GISci. Remote Sens.* **2022**, *59*, 1426–1445. [CrossRef]
20. Zhang, Q.; Pavlic, G.; Chen, W.; Latifovic, R.; Fraser, R.; Cihlar, J. Deriving Stand Age Distribution in Boreal Forests Using SPOT VEGETATION and NOAA AVHRR Imagery. *Remote Sens. Environ.* **2004**, *91*, 405–418. [CrossRef]
21. Spracklen, B.; Spracklen, D.V. Synergistic Use of Sentinel-1 and Sentinel-2 to Map Natural Forest and Acacia Plantation and Stand Ages in North-Central Vietnam. *Remote Sens.* **2021**, *13*, 185. [CrossRef]
22. Pérez-Cruzado, C.; Muñoz-Sáez, F.; Basurco, F.; Riesco, G.; Rodríguez-Soalleiro, R. Combining Empirical Models and the Process-Based Model 3-PG to Predict Eucalyptus Nitens Plantations Growth in Spain. *For. Ecol. Manag.* **2011**, *262*, 1067–1077. [CrossRef]
23. Zhao, J.; Liu, D.; Zhu, Y.; Peng, H.; Xie, H. A Review of Forest Carbon Cycle Models on Spatiotemporal Scales. *J. Clean. Prod.* **2022**, *339*, 130692. [CrossRef]
24. Wang, P. Forest Carbon Cycle Model: A Review. *Chin. J. Appl. Ecol.* **2009**, *20*, 1505–1510.
25. Frolking, S.; Goulden, M.L.; Wofsy, S.C.; Fan, S.M.; Sutton, D.J.; Munger, J.W.; Bazzaz, A.M.; Daube, B.C.; Crill, P.M.; Aber, J.D.; et al. Modelling Temporal Variability in the Carbon Balance of a Spruce/moss Boreal Forest. *Glob. Chang. Biol.* **1996**, *2*, 343–366. [CrossRef]
26. Seidl, R.; Rammer, W.; Scheller, R.M.; Spies, T.A. An Individual-Based Process Model to Simulate Landscape-Scale Forest Ecosystem Dynamics. *Ecol. Model.* **2012**, *231*, 87–100. [CrossRef]
27. Landsberg, J.J.; Waring, R.H. A Generalised Model of Forest Productivity Using Simplified Concepts of Radiation-Use Efficiency, Carbon Balance and Partitioning. *For. Ecol. Manag.* **1997**, *95*, 209–228. [CrossRef]
28. Cai, Y.; Guan, K.; Lobell, D.; Potgieter, A.B.; Wang, S.; Peng, J.; Xu, T.; Asseng, S.; Zhang, Y.; You, L.; et al. Integrating Satellite and Climate Data to Predict Wheat Yield in Australia Using Machine Learning Approaches. *Agric. For. Meteorol.* **2019**, *274*, 144–159. [CrossRef]
29. Seidl, R.; Rammer, W. Climate Change Amplifies the Interactions between Wind and Bark Beetle Disturbances in Forest Landscapes. *Landsc. Ecol.* **2017**, *32*, 1485–1498. [CrossRef]
30. Chang, X.; Xing, Y.; Wang, X.; You, H.; Xu, K. Application of 3PG Carbon Production Model in the Gross Primary Productivity Estimation of Broadleaved Korean Pine Forest in Changbai Mountain, China. *Chin. J. Appl. Ecol.* **2019**, *30*, 1599–1607.
31. Xie, Y.; Wang, H.; Lei, X. Application of the 3-PG Model to Predict Growth of Larix Olgensis Plantations in Northeastern China. *For. Ecol. Manag.* **2017**, *406*, 208–218. [CrossRef]
32. Zhang, Y.X.; Wang, X.J. Geographical Spatial Distribution and Productivity Dynamic Change of Eucalyptus Plantations in China. *Sci. Rep.* **2021**, *11*, 19764. [CrossRef] [PubMed]

33. Huang, G.; Zhao, Q. The History, Status Quo, Ecological Problems and Countermeasures of Eucalyptus Plantations in Guangxi. *Acta Ecol. Sin.* **2014**, *34*, 5142–5152. [CrossRef]
34. Wen, Y.; Zhou, X.; Yu, S.; Zhu, H. The Predicament and Countermeasures of Development of Global Eucalyptus Plantations. *Guangxi Sci.* **2018**, *25*, 107–116, 229.
35. Zaiton, S.; Sheriza, M.R.; Ainishifaa, R.; Alfred, K.; Norfaryanti, K. Eucalyptus in Malaysia: Review on Environmental Impacts. *J. Landsc. Ecol. Repub.* **2020**, *13*, 79–94. [CrossRef]
36. Bayle, G.K. Ecological and Social Impacts of Eucalyptus Tree Plantation on the Environment. *J. Biodivers. Conserv. Bioresour. Manag.* **2019**, *5*, 93–104. [CrossRef]
37. Shi, Y.; Wei, G.; Zhang, L.; Du, A. Patterns of Vegetation Carbon Storage in Eucalyptus Urophylla X E.grandis Plantations of Different Ages. *Eucalypt Sci. Technol.* **2017**, *34*, 24–27. [CrossRef]
38. White, J.W.; Hoogenboom, G.; Wilkens, P.W.; Stackhouse, P.W.; Hoel, J.M. Evaluation of Satellite-Based, Modeled-Derived Daily Solar Radiation Data for the Continental United States. *Agron. J.* **2011**, *103*, 1242–1251. [CrossRef]
39. Zhang, T.; Stackhouse, P.W.; Macpherson, B.; Mikovitz, J.C. A Solar Azimuth Formula That Renders Circumstantial Treatment Unnecessary without Compromising Mathematical Rigor: Mathematical Setup, Application and Extension of a Formula Based on the Subsolar Point and atan2 Function. *Renew. Energy* **2021**, *172*, 1333–1340. [CrossRef]
40. Chen, Q.; Wang, X.; Hang, M.; Li, J. Research on the Improvement of Single Tree Segmentation Algorithm Based on Airborne LiDAR Point Cloud. *Open Geosci.* **2021**, *13*, 705–716. [CrossRef]
41. Jiang, X.; Li, G.; Lu, D.; Chen, E.; Wei, X. Stratification-Based Forest Aboveground Biomass Estimation in a Subtropical Region Using Airborne Lidar Data. *Remote Sens.* **2020**, *12*, 1101. [CrossRef]
42. Gupta, R.; Sharma, L.K. The Process-Based Forest Growth Model 3-PG for Use in Forest Management: A Review. *Ecol. Model.* **2019**, *397*, 55–73. [CrossRef]
43. Grace, P.R.; Basso, B. Offsetting Greenhouse Gas Emissions through Biological Carbon Sequestration in North Eastern Australia. *Agric. Syst.* **2012**, *105*, 1–6. [CrossRef]
44. Almeida, A.C.; Landsberg, J.J.; Sands, P.J. Parameterisation of 3-PG Model for Fast-Growing Eucalyptus Grandis Plantations. *For. Ecol. Manag.* **2004**, *193*, 179–195. [CrossRef]
45. Jégo, G.; Thibodeau, F.; Morissette, R.; Crépeau, M.; Claessens, A.; Savoie, P. Estimating the Yield Potential of Short-Rotation Willow in Canada Using the 3PG Model. *Can. J. For. Res.* **2017**, *47*, 636–647. [CrossRef]
46. Forrester, D.I.; Tang, X. Analysing the Spatial and Temporal Dynamics of Species Interactions in Mixed-Species Forests and the Effects of Stand Density Using the 3-PG Model. *Ecol. Model.* **2015**, *319*, 233–254. [CrossRef]
47. Qu, L.H.; Zhao, X.H.; Zhang, C.Y. Application of 3-PG Model in the Prediction of Growth Factors in Natural Larix Gmelinii Forest. *For. Res.* **2022**, *35*, 158–165. [CrossRef]
48. Elli, E.F.; Sentelhas, P.C.; de Freitas, C.H.; Carneiro, R.L.; Alvares, C.A. Assessing the Growth Gaps of Eucalyptus Plantations in Brazil–Magnitudes, Causes and Possible Mitigation Strategies. *For. Ecol. Manag.* **2019**, *451*, 117464. [CrossRef]
49. Trotsiuk, V.; Hartig, F.; Forrester, D.I. r3PG—An R Package for Simulating Forest Growth Using the 3-PG Process-Based Model. *Methods Ecol. Evol.* **2020**, *11*, 1470–1475. [CrossRef]
50. Wang, B.; Waters, C.; Anwar, M.R.; Cowie, A.; Liu, D.L.; Summers, D.; Paul, K.; Feng, P. Future Climate Impacts on Forest Growth and Implications for Carbon Sequestration through Reforestation in Southeast Australia. *J. Environ. Manag.* **2022**, *302*, 113964. [CrossRef]
51. Sands, P.J.; Landsberg, J.J. Parameterisation of 3-PG for Plantation Grown Eucalyptus Globulus. *For. Ecol. Manag.* **2002**, *163*, 273–292. [CrossRef]
52. Stape, J.L.; Ryan, M.G.; Binkley, D. Testing the Utility of the 3-PG Model for Growth of Eucalyptus Grandis X Urophylla with Natural and Manipulated Supplies of Water and Nutrients. *For. Ecol. Manag.* **2004**, *193*, 219–234. [CrossRef]
53. Song, X.; Bryan, B.A.; Almeida, A.C.; Paul, K.I.; Zhao, G.; Ren, Y. Time-Dependent Sensitivity of a Process-Based Ecological Model. *Ecol. Model.* **2013**, *265*, 114–123. [CrossRef]
54. Fontes, L.; Landsberg, J.; Tomé, J.; Tomé, M.; Pacheco, C.A.; Soares, P.; Araujo, C. Calibration and Testing of a Generalized Process-Based Model for Use in Portuguese Eucalyptus Plantations. *Can. J. For. Res.* **2006**, *36*, 3209–3221. [CrossRef]
55. Liu, C.; Zheng, X.; Ren, Y. Parameter Optimization of the 3PG Model Based on Sensitivity Analysis and a Bayesian Method. *Forests* **2020**, *11*, 1369. [CrossRef]
56. Hua, L.; Jiang, X.; He, X. Application of 3-PG Model in Eucalyptus Urophylla Plantations of Southern China. *J. Beijing For. Univ.* **2007**, *29*, 100–104. [CrossRef]
57. Rodríguez-Suárez, J.A.; Soto, B.; Iglesias, M.L.; Diaz-Fierros, F. Application of the 3PG Forest Growth Model to a Eucalyptus Globulus Plantation in Northwest Spain. *Eur. J. For. Res.* **2010**, *129*, 573–583. [CrossRef]
58. Caldeira, D.R.M.; Alvares, C.A.; Campoe, O.C.; Hakamada, R.E.; Guerrini, I.A.; Cegatta, Í.R.; Stape, J.L. Multisite Evaluation of the 3-PG Model for the Highest Phenotypic Plasticity Eucalyptus Clone in Brazil. *For. Ecol. Manag.* **2020**, *462*, 117989. [CrossRef]
59. Li, R.S.; Yang, Q.P.; Zhang, W.D.; Zheng, W.H.; Chi, Y.G.; Xu, M.; Fang, Y.T.; Gessler, A.; Li, M.H.; Wang, S.L. Thinning Effect on Photosynthesis Depends on Needle Ages in a Chinese Fir (Cunninghamia Lanceolata) Plantation. *Sci. Total Environ.* **2017**, *580*, 900–906. [CrossRef]

60. Borys, A.; Suckow, F.; Reyer, C.; Gutsch, M.; Lasch-Born, P. The Impact of Climate Change under Different Thinning Regimes on Carbon Sequestration in a German Forest District. *Mitig. Adapt. Strat. Glob. Chang.* **2016**, *21*, 861–881. [CrossRef]
61. Sabatia, C.O.; Will, R.E.; Lynch, T.B. Effect of Thinning on Aboveground Biomass Accumulation and Distribution in Naturally Regenerated Shortleaf Pine. *South. J. Appl. For.* **2009**, *33*, 188–192. [CrossRef]

remote sensing MDPI

Article

Comparison of Canopy Clumping Index Measuring Methods and Analysis of Their Impact

Zhiguo Liang [1,2], Ying Yu [1,2], Xiguang Yang [1,2,*] and Wenyi Fan [1,2]

1 School of Forestry, Northeast Forestry University, Harbin 150040, China
2 Key Laboratory of Sustainable Forest Ecosystem Management-Ministry of Education, Northeast Forestry University, Harbin 150040, China
* Correspondence: yangxiguang@nefu.edu.cn

Abstract: The clumping index (CI) is a commonly used vegetation dispersion parameter used to characterize the spatial distribution of the clumping or random distribution of leaves in canopy environments, as well as to determine the radiation transfer of the canopy, the photosynthesis of the foliage, and hydrological processes. However, the method of CI estimation using the measurement instrument produces uncertain values in various forest types. Therefore, it is necessary to clarify the differences in CI estimation methods using field measurements with various segment lengths in different forest types. In this study, three 100 m $\times$ 100 m plots were set, and the CI and leaf area index (LAI) values were measured. The CI estimation results were compared. The results show that the accuracy of CI estimation was affected by different forest types, different stand densities, and various segment lengths. The segment length had a significant effect on CI estimation with various methods. The CI estimation accuracy of the LX and CLX methods increased alongside a decrease in the segment length. The CI evidently offered spatial heterogeneity among the different plots. Compared with the true CI, there were significant differences in the CI estimation values with the use of various methods. Moreover, the spatial distribution of the CI estimation values using the Ω_{CMN} method could more effectively describe the spatial heterogeneity of the CI. These results can provide a reference for CI estimation in field measurements with various segment lengths in different forest types.

Keywords: clumping index; estimation; impact analysis; field measurement

Citation: Liang, Z.; Yu, Y.; Yang, X.; Fan, W. Comparison of Canopy Clumping Index Measuring Methods and Analysis of Their Impact. *Remote Sens.* **2023**, *15*, 471. https://doi.org/10.3390/rs15020471

Academic Editors: Dengsheng Lu and Gherardo Chirici

Received: 9 December 2022
Revised: 4 January 2023
Accepted: 11 January 2023
Published: 13 January 2023

1. Introduction

As a common phenomenon in natural forests, canopy clumping can affect both gap fraction and canopy radiation transfer [1–3]. Meanwhile, it can cause the leaf area index (LAI) to be underestimated without considering canopy clumping [4–6]. Therefore, it is essential to quantify the non-random distribution characteristics of the forest canopy [7].

The canopy clumping index (CI) is a commonly used vegetation dispersion parameter used to characterize the spatial distribution of leaves or needles within the forest canopy [8]. The CI is often defined as the ratio of the effective leaf area index (LAI_e) to the real leaf area index (LAI_r) [9]. The LAIr is defined as the total area of plant leaves per unit land area, accounting for half of the land area [10,11]. The non-randomness level of foliage distribution in the forest canopy can be quantified by the CI in real scenarios. The CI is equal to 1.0; there is a random distribution of foliage in canopy environments, i.e., larger than 1.0 when the canopy offers a regular distribution and less than 1.0 when the canopy offers an aggregated distribution [12]. An exploration into the canopy clumping effect can not only help improve understanding around canopy efficiency in order to intercept light, but can also quantitatively calculate the carbon capture of vegetation in the ecosystem and the proportion of chlorophyll fluorescence photons escaping from the canopy [13]. Therefore, accurately acquisitioning the CI is of great importance in order to understand the distribution characteristics of leaves in the canopy and gas exchange in the ecosystem [14]. Meanwhile, a lack of consideration of the CI can lead to an LAI error of up to 70% [15]. Therefore, it is essential to correctly estimate the LAI when using an indirect optical approach that accounts for the clumping effect [16].

The current CI is a ratio of the canopy gap fraction under real and random conditions, and it is a quantified ratio of the effective leaf area index (LAI_e) to the real leaf area index (LAI_r). In many studies, the CI is used to efficiently quantify the transmittance and interception of light and precipitation in canopy environments. Furthermore, the primary ways of obtaining the CI include field and remote sensing methods. In field measurements, the CI can be estimated both directly or indirectly using commercial optical instruments [16,17]. There are usually two steps available to retrieve the forest canopy clumping index, including the gap size estimation phase and the CI estimation phase [7]. The gap size distributions can be obtained using commercial optical instruments, such as tracing radiation and architecture of canopy (TRAC) and digital hemispherical photography (DHP) in-field measurements. For the CI estimation phase, there are several well-developed and accepted methods used to quantitatively calculate the clumping degree. The finite-length averaging method (LX) proposed by Lang and Xiang was the earliest method used for CI estimation [18]. In this method, the whole scene was divided into different segments according to the clustering effect of the whole scene, and the canopy was assumed as random distributions in each segment. The LAI_r in each segment was calculated by the gap rate model and the LAI_e of the whole scene was calculated by the gap rate model. Then, the CI was calculated by the ratio of the LAI_e to the LAI_r. Evidently, the size of the segments using this method will significantly affect the accuracy of CI estimation [19,20]. Chen et al. proposed a gap size distribution model that can be used to calculate the CI. In this method, large gaps from the measured gap size accumulation curve were sequentially removed until the pattern of gap size accumulation resembled a random spatial distribution, and the CI was calculated using the logarithmic gap size averaging method (the CC method) [21–23]. Unlike the LX method, this method is not limited by light conditions and has strong applicability [17]. Additionally, this method is more commonly used for CI estimation [24]. Pisek et al. improved the CC method for CI estimation based on the original Miller's law, known as the CMN method. The difference between the CMN and CC methods is that the CMN method does not consider the normalization after removing large gaps [19]. Leblanc combined the LX and CC methods to calculate the CI from hemispheric photography (HP) images, in what is called the CLX method [25]. As a combination of the two methods, the CLX method was disadvantageous. It was sensitive to the segment length (similar to the LX method) and failed to identify and eliminate small gaps, which caused increases in the CI estimation error (similar to the CC method) [26]. To conclude, the advantages and disadvantages of various methods in different scenes are still important to consider when taking actual measurements.

In field measurements, the CI can also be estimated by obtaining the LAI_e and LAI_r and then calculating the ratio, which can retrieve the CI. The LAI_r generally adopts destructive sampling [27], allometric growth [28], and litter collection [29]. The destructive sampling method is used to pick leaves of all vegetation in the study area and individually measure the leaf areas, before then calculating the LAI_r. This method has certain impacts on the ecological environment, making it unsuitable for forests with complex structures [30]. The allometric technique depends on the relationship between the leaf area and any dimension of the element of the woody plant, such as the green leaf biomass, the stem diameter, the diameter at breast height, the tree height, or the volume [8]. Moreover, this relationship is determined via destructive sampling. The allometric equation can be used to estimate the LAI_r within the study area. However, this method is disadvantageous given that it can destroy the samples. The allometric equation is also restricted because of its site specificity, and the relationship is stand-specific and dependent on the season, site fertility, local climate, and canopy structure [31,32]. The measured result may be less than the LAI, as measured by the optical instrument method [33]. The litter collection method retrieves fallen leaves during the leaf-falling period using litter traps in the study area and the LAI_r can be determined from the litter using the weight method. This method is very useful for LAI_r measurements, especially in deciduous forests. The accuracy of LAI_r measurements was 95% within a bias of 10% with respect to the mean with an appropriate spatial and temporal sampling scheme [34,35]. The LAI_e is generally measured with an optical instrument, such as Demon, LAI-2200, TRAC, SunScan, and AccuPAR, or with digital hemispheric photography (DHP). These optical instruments have usually been used to monitor the LAI status and any small-scale dynamic changes in forest ecosystems [36–38]. Denmon is an instrument used for log-averaging the transmittance of solar beams, and its sensor has a filter that can have a filtering effect on other scattered light measurements and can allow the light to be measured at a wavelength of only 430 nm. However, it required several repeated time observations and there were more complex operating procedures compared with LAI-2200 [8]. The LAI-2200 instrument, equipped with a fisheye lens and five concentric conical rings (7°, 23°, 38°, 53°, and 68°), was used to record incident light intensities. Similar to Denmon, the LAI-2200 calculated the LAI_e by comparing the different measurements among the above and below canopies. The result measured with the LAI-2200 was usually sensitive to different light conditions, so it was used often to take measurements before dawn or after dusk [39]. The LAI-2200 has been successfully used to estimate LAI_e in continuous and homogeneous canopies; however, the potential of the LAI-2200 instrument is restricted by a general tendency of LAI_e underestimation [6,16]. The TRAC technique can also be used to calculate the LAI by combining the average leaf width, the needle cluster ratio, and the woody leaf area ratio of the study area [21,23]. SunScan and AccuPAR were used to calculate the LAI by measuring the solar transmittance of the upper and lower parts of the canopy, but these two methods were not suitable for measurements within coniferous forests [8]. The DHP method generally uses a fisheye lens and a digital camera to measure the canopy gap ratio and the LAI. However, the accuracy of this method depended on recognizing the algorithm of the gap ratio and woody parts, meaning that the accuracy could be further improved by the optimization algorithm [17]. To summarize, the accuracy of LAI measurements varied among different instruments, as did CI estimation when using the ratio of the LAI_e and the LAI_r.

Global- and regional-scale CI estimation methods have been generated by remote sensing technology. Optical remote sensing, such as with POLDER, MODIS, and MISR satellite data, has been successfully used for CI estimation purposes [40–42]. An empirical relationship between the CI and vegetation index, such as the normalized difference between the hotspot and dark-spot (NDHD) models, was established to obtain a clumping index [43]. Chen et al. (2005) generated a monthly CI global mapping model from POLDER with a 6 km resolution [44]. He et al. proposed a global CI mapping model based on the NDHD model at a 500 m resolution by utilizing the MODIS BRDF product [40]. Fang et al. obtained the global CI distribution data from 2000 to 2020 by calculating the NDHD based on MODIS data and implemented the retrieval service in Google Earth Engine. The results indicate that the global clumping index range is about 0.3–1.0 [45]. With the development of laser ranging technology, light detection and ranging (LiDAR) data have been used to estimate the CI through calculating the gap rate or gap fraction by CC, CLX, or other methods [7,46,47]. Unfortunately, there is no effective and robustness algorithm used to retrieve the clumping index in a wide range with spaceborne lidar. Therefore, optical remote sensing is still the primary data source for CI estimation on the large spatial and multi-temporal scale [16].

Regardless of field measurements or remote sensing estimations, accurately measuring the CI is essential. Some studies have focused on researching the availability and accuracy of CI calculation methods in field measurements [17,19,48]. The first factor that will have a potential impact on the measurement results is the segment length. Segment length is an artificially assumed variable used to estimate the CI. Before CI estimation, a cell that is small enough for the assumption of leaf distribution randomness within a cell should be assumed. Meanwhile, the size of this cell should be large enough so that the statistics of the gap fraction are meaningful. Moreover, this segment size is usually called the segment length. A theoretical analysis of this problem suggests that the segment length should be at least 10 times the width of a leaf [16]. However, varied segment lengths can lead to different and random canopy situations within each segment. In the case of relatively long segments, the canopy may have a clustered distribution. This results in a large error caused by the calculation method [17]. Leblanc et al. compared the differences in CI estimation methods and showed that the CLX method was less sensitive to the segment length compared to other methods, and the LX method was more sensitive compared with other methods [26]. Pisek et al. compared the CI estimation results using the LX method and the CLX method under various segment lengths, and showed that the segment length can have an uncertain impact on CI estimation, though the optimal segment length for CI estimation was not determined [19]. Woodgate found that the CLX method was better than the CC and LX methods for CI estimation in eucalyptus forests [49]. However, Pisek et al. found that the CLX method was more suitable for CI estimation than the CC, LX, and CMN methods in Scots pine forests [19]. This indicates that the various types of forest will increase the uncertainty of CI measurements [50]. Therefore, it is necessary to evaluate the differences in CI estimation methods when taking in-field measurements with various segment lengths in different forest types.

To sum up, the studies on the differences among CI field measuring methods were insufficient and the influence of the segment lengths, different forest types, and tree density in the plot on CI field measurement was also not clear. To this aim, three 100 × 100 m plots of different forest types were set in the research region, and the CI was estimated using the measurements from TRAC, LAI-2200, DHP, and litter collection methods. Then, the CI estimation results were compared and the uncertainty of CI estimation was analyzed at various segment lengths with the LX, CC, CLX, P, and CMN methods. These results could provide a reference for CI estimation in field measurements with various segment lengths in different forest types.

2. Materials and Methods

2.1. Study Area

The research area is located in Maoer Mountain Experimental Forest Farm of Northeast Forestry University in Shangzhi City, Heilongjiang Province, northeast of China, with longitude of 127°29′–127°44′E and a latitude of 45°14′–45°29′N (Figure 1). The area of the Maoer Mountain forest farm is about 26.496 km^2. The study area is a low-mountain and hilly area with an average altitude of 300 m. The research region has a mid-temperate continental monsoon climate with an average annual temperature of 2.7 °C [51]. The hottest month is in July with an average temperature of 21.8 °C. The coldest month is in January with an average temperature of −19.9 °C [52]. The average annual precipitation is 700–800 mm.

The vegetation type of the Maoer Mountain Experimental Forest Farm is mainly natural secondary forest and artificial forest. The average forest coverage rate is 95%, and the total forest volume is 3.5 million m^3. The main tree species are *Quercus mongolica Fisch., Betula platyphylla Suk., Pinus koraiensis Sieb., Fraxinus mandschurica Rupr., Phellodendron amurense Rupr., Populus davidiana Dode, Tilia amurensis Rupr., Larix gmelini Rupr.,* and *Betula costata Trautv.* [52,53].

2.2. Field Data

2.2.1. Field Measurement

In this study, three 100 m × 100 m plots with different forest types were set. In order to minimize the uncertain error caused by terrain, the three selected plots were set at the region with a flat slope. These were broad-leaved forest, coniferous forest, and mixed forest plots. The measured forest parameters included the diameter at breast height (DBH), the tree height, the tree density, the tree species, the relative dominance, and the specific leaf area (SLA). The DBH above 5 cm was measured and recorded. Then, the geographical coordinate of each tree in one plot was recorded using the SOUTH RTK (SOUTH Inc., Guangzhou, China) instrument.

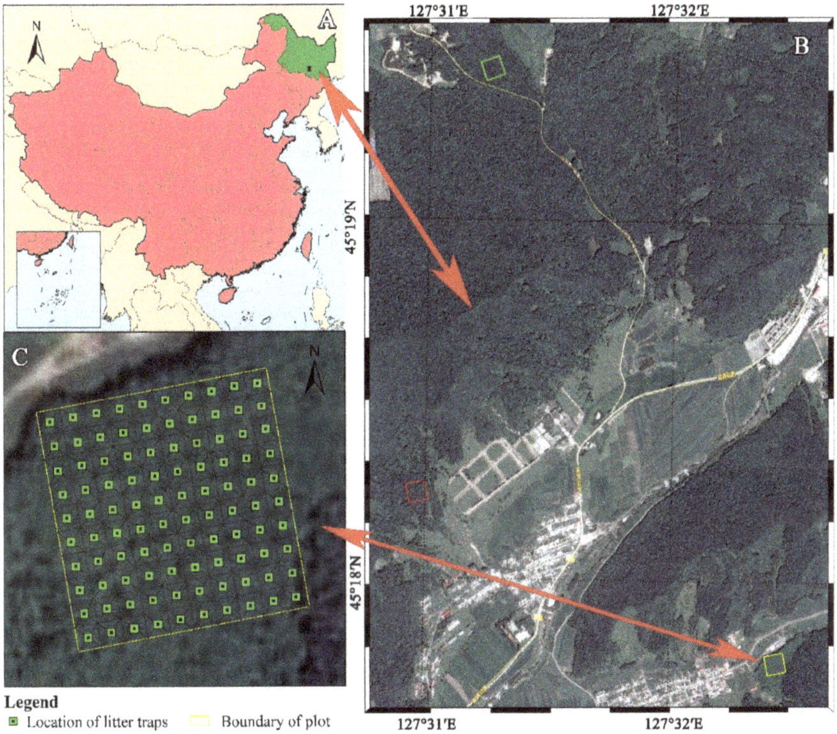

Figure 1. Study area map and LAI-2200 measurement points, TRAC transects, and leaf litter collection measurement points. ((**A**) was the overview of the research regeion; (**B**) was the overview of the plots; (**C**) was the positions of the measurements. Note: in (**B**), the yellow rectangle is the broadleaf forest, the red rectangle is the coniferous forest, and the green rectangle is the mixed forest).

The plot was divided into 100 squares with an area of 10 m × 10 m. Moreover, a 1 m × 1 m litter trap was set at the center of each square (Figure 2). The leaves were then collected twice in each month from the beginning of the season to the end of the season. Afterwards, the area and the weight of the fresh leaves were measured. The Li-3000 was used to measure the leaf area of the deciduous tree species. Then, the needle surface area of the coniferous trees was measured using the volume displacement method, as reported in Chen's publication in 1996 [54]. All fresh leaf samples were separated by species type and were subsequently oven-dried at 70 °C for 24 h. The mass of the dried leaf samples was recorded. The specific leaf area (SLA) was then calculated as follows [55]:

$$SLA = \frac{S_a}{W} \quad (1)$$

where S_a is the average fresh leaf area (cm^2) and W is the dried weight of leaves (g). The statistical information of the measurements can be found in Table 1.

Figure 2. Photos of the field measurement process.

Table 1. Basic tree information table of the three plots.

Sample Type	Major Species	Number of Trees (Number)	Mean DBH (cm)	Relative Dominance (%)	SLA ($cm^2 \times g^{-1}$)
Broad-leaved forest	*Juglans mandshurica Maxim.*	1508	11.87	100	213.62
Coniferous forest	*Pinus sylvestris Linn.*	496	26.84	68.19	67
	Pinus koraiensis Sieb.	192	15.91	14.37	78.91
	Fraxinus mandschurica Rupr.	38	21.26	5.08	305.96
	Ulmus pumila L.	200	41.99	8.94	245.48
	Others	213	11.85	3.41	——
	Total	1139	26.6	100	——
Mixed forest	*Betula platyphylla Suk.*	357	27.84	39.85	195.46
	Fraxinus mandschurica Rupr.	244	33.41	16.61	305.96
	Pinus koraiensis Sieb.	256	31.75	5.74	78.91
	Quercus mongolica Fisch.	98	51.2	10.84	253.94
	Larix gmelini Rupr.	112	44.07	11.66	159.22
	Ulmus pumila L.	233	34.55	8.19	245.48
	Others	87	14.5	7.11	——
	Total	1387	237.31	100	——

Leaves collected from the litter trap were separated by species type and the wet weight was measured and recorded. Then, the samples were oven-dried at 70 °C for 24 h and the weights of the samples were measured. This drying process was repeated until the measured weight of the samples was less than 0.01 g. Subsequently, the ratio of leaf dry mass to fresh mass was calculated based on the measured dry and wet weight of samples, following Equation (2):

$$a = \frac{M_0}{M_1} \tag{2}$$

where M_0 is the dried weight of leaf samples with a unit of g, M_1 is the wet weight of leaves with a unit of g, and a is the ratio of leaf dry mass to fresh mass. This calculation was made based on the types of tree species and varied with different types of tree species.

After that, the real leaf area index of each type of tree species in the plot could be calculated based on the specific leaf area and the ratio of dry weight to fresh weight using litter collection method. The equation used to calculate the LAI was as follows:

$$LAI_{litter} = a * M * SLA \tag{3}$$

where LAI_{litter} is the real leaf area index (cm^2/cm^2), a is the ratio of dry to fresh mass (g), M is the wet weight of the samples (g), and SLA is the specific leaf area (cm^2/g). With this method, we could obtain the LAI_r of each litter trap, each tree species in one plot, and the total LAI_r in the plot for further analysis.

2.2.2. Measurements of the LAI_e and the CI

The LAI_e and the CI were measured using the LAI-2200, TRAC, and DHP methods. Two pieces of LAI-2200 (LI-COR Inc., Lincoln, Nebraska, USA) equipment were used to record the light penetration into the canopy and the above canopy; as a result, the LAI_e could be calculated. Each measurement was repeated twice and a 90° view cap was used to shield the sensor from the operator during the measurement stage. The measurements were conducted near sunset or under overcast conditions. It was used to minimize the measuring error under direct illumination [56]. The position of the measured LAI can be found in Figure 3.

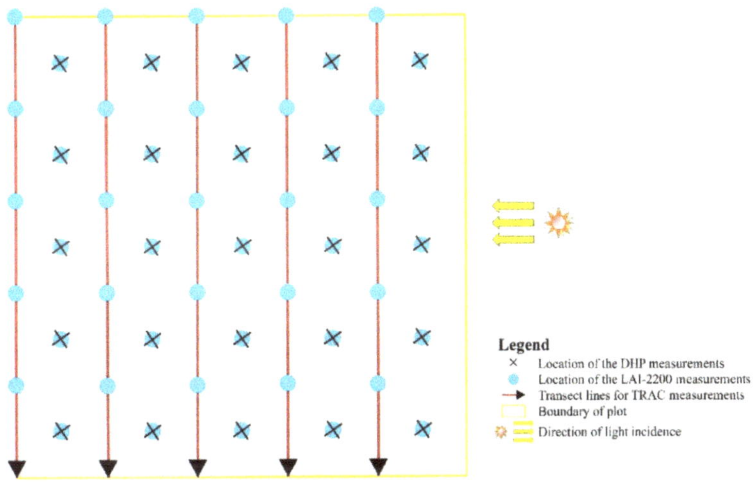

Figure 3. Spatial distribution of measuring points of LAI-2200 and DHP and the TRAC transect line.

The DHP images were taken with a fixed azimuth angle using Nikon D800 (Nikon, Tokyo, Japan) with a 4.5 mm F2.8 EX DC circular fisheye converter. The position of the DHP measurement can be found in Figure 3. The equipment was mounted and levelled using a bubble level before the measurements were taken. Then, the hemispherical photographs were taken. The effective LAI could be derived using digital hemispherical photography (DHP) software (Natural Resources Canada, Ottawa, Canada). The measurements were conducted under overcast conditions or the conditions of diffuse skylight were used to minimize the error of the direct illumination. More details can be found in the Digital Hemispherical Photography Manual [57].

The TRAC equipment was used to record the photosynthetic photon flux density (PPFD) and to retrieve the CI of the forest canopy. In this study, five transect lines of 100 m, with an interval distance of 20 m, were set. Then, we collected the TRAC-based PPFD gradient values along the line transects, which are perpendicular to the incident directions of the solar beams in the plot. To compare the differences in the results with various segment lengths, the segment lengths for TRAC measurements were set to 1 m, 2 m, 4 m, 5 m, 10 m, 20 m, 50 m, and 100 m, respectively. It helped us to create different datasets based on the original TRAC data when transects were segmented by 1 m, 2 m, 4 m, 5 m, 10 m, 20 m, 50 m, 100 m, respectively. At last, the forest canopy CI was computed using the TRACWin software (Natural Resources Canada, Ottawa, ON, Canada). To minimize the influence of the different zenith angles on the CI measurements, a zenith angle of 57.5° was used during the measurement stage [56–58].

2.3. Methodology of CI Estimation

In this study, we compared the CI estimation performances using the LX, P, CC, CMN, and CLX methods. Moreover, the theory of these methods is described as follows.

2.3.1. Theory of the LX Method

The LX method was the first method used for LAI estimation, which took the logarithmic averages with a solid statistical background [25]. Thus, LAI estimation can have greater accuracy when the canopy transmittance is logarithmically averaged in a discontinuous or clumped canopy, referred to as method of LX. The LX method assumed that the foliage within the finite length was random and that the segment contained gaps.

$$\Omega_{\mathrm{LX}}(\theta) = \frac{\ln\left[\overline{P(\theta)}\right]}{\overline{\ln[P(\theta)]}} \tag{4}$$

where $\overline{P(\theta)}$ is the average of the canopy gap fraction and $\overline{\ln[P(\theta)]}$ is the logarithmic mean gap fractions for all segments. This method may give erroneous results due to the short length of the segments in a clumped canopy [59,60]. The short length of the segments is also called segment length, and is defined as the various lengths of all the sunlit segments occurring along this line that can be expressed as a statistical distribution function under given canopy [61]. In addition, the segment length is considered an essential factor for CI estimation [62].

2.3.2. Theory of the P method

The p method is based on the gap size distribution [63]. The gap size distribution of the canopy with a random canopy spatial distribution can be described as follows:

$$P(\iota) = e^{-L_P(1+\iota/w_{ep})} \tag{5}$$

where $P(\iota)$ is the probability that a probe with length ι will completely fall into a light spot. w_{ep} is the mean width of the element shadows cast on the transect. L_P is the projected foliage element area.

The $P(l)$ for the clumped canopy can be determined with the following formula:

$$P(l) = P_c(l) + P_{E1}(l)P_{c1} + P_{E2}(l)P_{c2} + \ldots + P_{En}(l)P_{cn} = P_c(l) + \sum_{i=1}^{n} P_{Ei}(l)P_{ci} \tag{6}$$

The term $P_c(l)$ refers to the sunfleck size distribution beneath a canopy with opaque clumps. P_{ci} is the probability of i number of clumps overlapping in the sun's direction, and the P_{Ei} is a sunfleck size distribution within the intersection of i clumps. These terms are defined as follows:

$$P_c(l) = \exp\left[-L_{c\theta}(1 + l/w_{ep})\right] \tag{7}$$

$$P_{Ei}(l) = \exp\left[-iL_{EP}(1 + l/w_{ep})\right] \tag{8}$$

$$P_{ci} = \frac{\exp(-L_{c\theta}) \times L_{c\theta}^{n}}{i!} \tag{9}$$

where $L_{e\theta}$ is the intercept of the plot of gap size distribution for a clumped canopy and $L_{c\theta}$ is the intercept found from extrapolating the straight portion of the curve at large l value. In addition, L_{E_p} was calculated using $L_{e\theta}$ and $L_{c\theta}$ from Equation (10).

$$L_{E_p} = \ln\left(\frac{(1+\alpha)L_{c\theta}\exp(-L_C\theta)}{\sqrt{2(1+\alpha)\exp[-(L_{e\theta}+L_{c\theta})] - (1+2\alpha)\exp(-2L_{c\theta}) - \exp(-L_{c\theta})}}\right) \tag{10}$$

In this equation, $\alpha = L_{c\theta}P_{E1}(0)/3$ can be derived from Equation (6) after being truncated at $i = 2$.

The gaps can be measured along the transects and $P(l)$ can be calculated following methods proposed by Chen and Cihlar (1995) [22,23]. Meanwhile, the clumping index of the leaves can be determined from:

$$\Omega_{\mathrm{P}} = \frac{L_{e\theta}}{L_{EP}L_{c\theta}} \tag{11}$$

2.3.3. Theory of the CC Method and CMN Method

The Ω_{CC} method is improved on the basis of the F-Approach proposed by Chen et al. [22,23]. It measures the width of the sunlit patches on the transect when the light enters the canopy. The width of the canopy gaps on the transect can be calculated with consideration of the penumbra effect. Then, the accumulated gap size distribution $F(\lambda)$ can be formed using the calculated gaps and by sorting them in ascending order based on their size. The random distribution of canopy gaps can be described as follows:

$$F(\lambda) = \left(1 + L_p \cdot \frac{\lambda}{w_{ep}}\right)\exp\left[-L_p\left(1 + \frac{\lambda}{w_{ep}}\right)\right] \tag{12}$$

where $F(\lambda)$ is the faction of the transect occupied by the gap larger than or equal to λ. λ is the size of the gaps. w_{ep} is the mean width of the element shadows cast on the transect. L_P is the projected foliage element area.

In a clumped canopy, a measured gap size distribution $F_m(\lambda)$ will deviate from $F(\lambda)$. The non-randomness of the gaps can also be removed by comparing the difference between $F_m(\lambda)$ and $F(\lambda)$. A new distribution of gaps denoted by $F_{mr}(\lambda)$ can be formed after the gap removal, i.e., $F_m(\lambda)$ closest to $F(\lambda)$. In this case, $F_{mr}(0)$ is the total gap fraction in the canopy as if the canopy is random, and the clumping index for the clumped canopy can be calculated as:

$$\Omega_{\mathrm{CC}} = \frac{\ln[F_m(0,\theta)]}{\ln[F_{mr}(0,\theta)]} \cdot \frac{1 - F_{mr}(0,\theta)}{1 - F_m(0,\theta)} \tag{13}$$

If the normalization factor after the removal of large gaps in Equation (14) is neglected, the element clumping might be calculated simply as [19]:

$$\Omega_{\mathrm{CMN}} = \frac{\ln[F_m(0,\theta)]}{\ln[F_{mr}(0,\theta)]} \tag{14}$$

This simplified equation of CI estimation was named the CMN method.

2.3.4. Theory of the CLX Method

Leblanc et al. developed a CI estimation method that combined the gap size distribution and the finite-length methods to address the problems of segment length associated with the finite-length method. In this method, the gap size distribution method is used to assess the foliage heterogeneity within a larger segment due to the non-homogeneous canopy in a larger segment. This method is called the CLX method. The clumping index can then be calculated as follows:

$$\Omega_{\mathrm{CLX}} = \frac{n\,\ln\left[\overline{P(\theta)}\right]}{\sum_{k=1}^{n}\ln[P_k(\theta)]/\Omega_{\mathrm{CCk}}(\theta)} \tag{15}$$

where n is the number of segments, $P_k(\theta)$ is the gap fraction of the k-th segment, and $\Omega_{\mathrm{CCk}}(\theta)$ is the elements' clumping index of the segment k using the CC method.

2.3.5. Comparation of Different CI Estimating Methods

When comparing various methods for CI estimation, it is found that using different methods leads to them making an error in CI estimation. To clarify this error and increase the accuracy of CI measurement, we performed CI measurement experiments at three plots with different forest types. We measured CI at various segment lengths by using TRAC. Then CI value was extracted using the Ω_{LX}, Ω_{CC}, Ω_{CLX}, Ω_{P}, and Ω_{CMN} methods. Then the results were compared with litter collection method and DHP. The error was calculated and the effect of the different CI estimating methods, measuring strategy, and forest type on CI measurement was analyzed.

2.4. Verification and Analysis

The real leaf area index was measured by the litter collection method, and the LAI_e was obtained by the LAI-2200. The real CI within the canopy was estimated by the ratio between the LAI_e and the LAI_r. Then, the estimated CI value was extracted using the Ω_{LX}, Ω_{CC}, Ω_{CLX}, Ω_{P}, and Ω_{CMN} methods with the help of the TRAC, DHP, and LAI-2200, respectively. Then, the CI estimating results from different forest types, different section lengths, and different tree densities were evaluated using a normal distribution hypothesis test. After that, the difference and accuracy of CI estimation values using various methods with different segment lengths and tree densities in various forest types were compared using the one-way analysis of variance assay (ANOVA). In addition, the relative error was used to compare the results using various methods. The relative error is calculated using the following equation:

$$e_{relative\ error} = \frac{(\Omega_{\text{method}} - \Omega_{\text{r}})}{\Omega_{\text{r}}} \times 100\% \quad (16)$$

where Ω_{method} is the estimated CI using the Ω_{LX}, Ω_{CC}, Ω_{CLX}, Ω_{P}, and Ω_{CMN} methods, and Ω_r is the real CI.

Boxplots were used to compare the differences in the results using the LX, CC, CLX, P, and CMN methods; kernel density analysis was used to represent the plant number densities of broad-leaved, coniferous, and mixed forests; and contour plots were used to represent the relationship between the stand density and the CI.

3. Results

3.1. Comparison of CI Estimation Results Using Various Methods

We set five transects, 100 m in length, to obtain the gap size distributions of foliage elements using the TRAC and DHP methods (Figure 3). The segment length was set to a width of 20 m with the suggestion of TRAC manuals and publications [64]. Then, the clumping index of the plot was estimated using the methods mentioned above. Moreover, the estimated CI was compared with the real clumping index (Ω_r) using various methods in different forest types. These results show that the CI estimation results varied depending on the method selected (Figure 4). For DHP measurements, the relative errors of CI estimation using the Ω_{LX}, Ω_{CC}, Ω_{CLX}, Ω_{P}, and Ω_{CMN}, and Ω_r methods were 53.6%, 44%, 28.1%, 45.3%, and 36.4%, respectively. For TRAC measurements, the relative errors of CI estimation using the Ω_{LX}, Ω_{CC}, Ω_{CLX}, Ω_{P}, and Ω_{CMN}, and Ω_r methods were 50.2%, 36.2%, 29.8%, 40.4%, and 31.2%, respectively. CI estimation using the Ω_{CLX} method exhibited the best accuracy among the five methods. The relative errors of CI estimation using the Ω_{CLX} method were 28.1% and 29.8% for the DHP and TRAC measurements, respectively. Moreover, the accuracy values of CI estimation using the TRAC measurements were better than those of the DHP measurements, indicating that the TRAC measurements have better robustness for CI estimation.

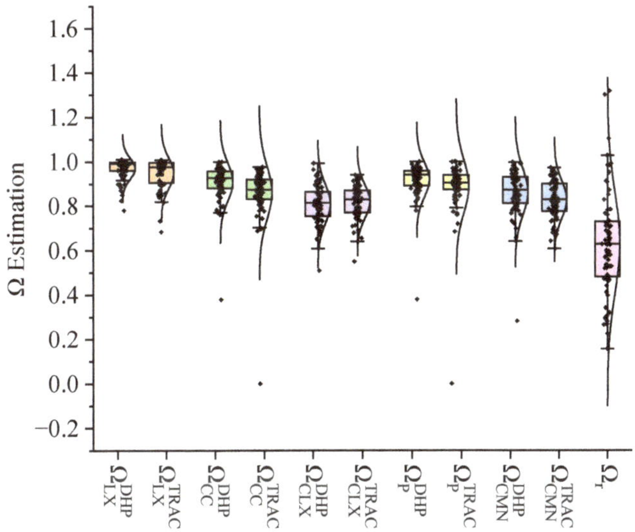

Figure 4. Comparison of CI estimation values using different methods. (Note: the black curve on the right indicates that the data are normally distributed.)

3.2. Comparison of CI Estimation Results Using Various Methods

We compared the CI estimation results using various methods in different forest types. These results can be found in Figure 5. For broadleaf forests, CI estimation values using the Ω_{LX}, Ω_{CC}, Ω_{CLX}, Ω_{P}, and Ω_{CMN} methods from TRAC measurements were 0.97 ± 0.04, 0.88 ± 0.05, 0.83 ± 0.06, 0.91 ± 0.05, and 0.82 ± 0.08, respectively. The CI estimation of the Ω_{LX}, Ω_{CC}, Ω_{CLX}, Ω_{P}, and Ω_{CMN} methods exhibited an overestimating trend with relative errors of 59.6%, 45.7%, 37%, 51.1%, and 36.4%, respectively. CI estimation values using the Ω_{LX}, Ω_{CC}, Ω_{CLX}, Ω_{P}, and Ω_{CMN} methods from DHP measurements were 0.98 ± 0.03, 0.93 ± 0.04, 0.89 ± 0.05, 0.94 ± 0.04, and 0.87 ± 0.06, respectively. The relative errors of CI estimation using the Ω_{LX}, Ω_{CC}, Ω_{CLX}, Ω_{P}, and Ω_{CMN} methods from DHP measurements were 62.1%, 53.9%, 47.5%, 55.1%, and 44%, respectively.

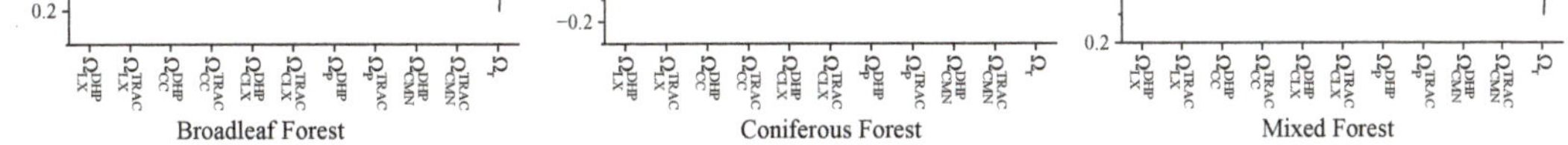

Figure 5. A comparison of CI estimation values using various methods in different forest types. (Note: the black curve on the right indicates that the data are normally distributed.).

For coniferous forests, CI estimation values using the Ω_{LX}, Ω_{CC}, Ω_{CLX}, Ω_{P}, and Ω_{CMN} methods from TRAC measurements were 0.92 ± 0.09, 0.83 ± 0.19, 0.8 ± 0.1, 0.84 ± 0.19, and 0.83 ± 0.1, respectively. The relative errors of CI estimation using the Ω_{LX}, Ω_{CC}, Ω_{CLX}, Ω_{P}, and Ω_{CMN} methods were 74.6%, 56.2%, 51.1%, 57.9%, and 56.8%, respectively. CI estimation values using the Ω_{LX}, Ω_{CC}, Ω_{CLX}, Ω_{P}, and Ω_{CMN} methods from DHP measurements were 0.95 ± 0.06, 0.88 ± 0.12, 0.75 ± 0.08, 0.89 ± 0.12, and 0.85 ± 0.14, respectively. The relative errors of CI estimation using the Ω_{LX}, Ω_{CC}, Ω_{CLX}, Ω_{P}, and Ω_{CMN} methods from DHP measurements were 80%, 66.9%, 41%, 67.3%, and 60.1%. At the same time, CI measurement values from TRAC and DHP were overestimated compared with the true values.

For mixed forests, CI estimation values using the Ω_{LX}, Ω_{CC}, Ω_{CLX}, Ω_{P}, and Ω_{CMN} methods from TRAC measurements were 0.96 ± 0.06, 0.87 ± 0.06, 0.83 ± 0.06, 0.91 ± 0.05, and 0.83 ± 0.08, respectively. The relative errors of CI estimation using the Ω_{LX}, Ω_{CC}, Ω_{CLX}, Ω_{P}, and Ω_{CMN} methods were 25.9%, 14.8%, 9.3%, 19.7%, and 9%, respectively. CI estimation values using the Ω_{LX}, Ω_{CC}, Ω_{CLX}, Ω_{P}, and Ω_{CMN} methods from DHP measurements were 0.98 ± 0.04, 0.91 ± 0.07, 0.79 ± 0.05, 0.93 ± 0.05, and 0.86 ± 0.09, respectively. The relative errors of CI estimation using the Ω_{LX}, Ω_{CC}, Ω_{CLX}, Ω_{P}, and Ω_{CMN} methods from DHP measurements were 28.4%, 20%, 3.6%, 22.1%, and 13.4%, respectively.

A comparison between the CI estimation results using both TRAC and DHP methods shows that there were no significant differences, even for the same CI estimation methods. For broadleaf forests, the relative errors of CI estimation using the Ω_{LX}, Ω_{CC}, Ω_{CLX}, Ω_{P}, and Ω_{CMN} methods between different pieces of equipment were 1.5%, 5.4%, 7.2%, 2.6%, and 5.6%, respectively. Moreover, the differences in CI estimation for coniferous forests were 6.9%, 3%, 6.7%, 5.9%, and 5.5%, respectively. For mixed forests, these values changed to 4.5%, 2%, 5.2%, 2%, and 3.9%, respectively.

3.3. *The Effect of the Segment Length on CI Estimation Using Different Methods*

Optical methods were non-destructive and cheaper, but the CI estimation results were affected by the segment length on the transect [21]. Meanwhile, the segment size in the field measurements was usually arbitrarily decided, whereas the difference in the choices were derived from the difference in the CI estimation values [16]. Thus, we compared the CI estimation result with various segment lengths and discussed the influence of the segment length on CI estimation.

We estimated CI values with segment lengths of 1 m, 2 m, 4 m, 5 m, 10 m, 20 m, 50 m, and 100 m. CI estimation values using the Ω_{P} method at various segment lengths were 0.97 ± 0.03, 0.97 ± 0.03, 0.96 ± 0.04, 0.93 ± 0.04, 0.99 ± 0.04, 0.92 ± 0.03, 0.93 ± 0.03, and 0.85 ± 0.03, respectively. CI estimation values using the Ω_{CC} method at various segment lengths were 0.98 ± 0.04, 0.98 ± 0.04, 0.98 ± 0.03, 0.95 ± 0.04, 0.99 ± 0.02, 0.92 ± 0.06, 0.91 ± 0.03, and 0.84 ± 0.02, respectively. CI estimation values using the Ω_{LX} method at various segment lengths were 0.73 ± 0.1, 0.69 ± 0.06, 0.73 ± 0.09, 0.79 ± 0.09, 0.90 ± 0.01, 0.88 ± 0.03, 0.93 ± 0.04, and 1.0 ± 0.01 respectively. CI estimation values using the Ω_{CLX} method at various segment lengths were 0.72 ± 0.01, 0.62 ± 0.03, 0.79 ± 0.02, 0.72 ± 0.05, 0.92 ± 0.04, 0.81 ± 0.04, 0.85 ± 0.03, and 0.84 ± 0.03, respectively. CI estimation values using the Ω_{CMN} method at various segment lengths were 0.97 ± 0.04, 0.98 ± 0.04, 0.97 ± 0.05, 0.84 ± 0.04, 1.0 ± 0.03, 0.91 ± 0.05, 0.89 ± 0.02, and 0.8 ± 0.04, respectively. The figure shows the estimated CI values with segment lengths of 1 m, 2 m, 4 m, 5 m, 10 m, 20 m, 50 m, and 100 m (Figure 6).

The results show that the CI estimations using the Ω_{LX} and Ω_{CLX} methods were more sensitive to the changes in the segment length compared with other methods. Furthermore, the ability to estimate the average CI was most stable when the segment length was between 10 and 50 m (the real CI Ω_r = 0.532). CI estimation using the other three methods was less affected by the segment length compared to the results derived from the Ω_{LX} and Ω_{CLX} methods. This is because other methods do not rely on the random situation of the canopy within each segment. However, the CLX method can achieve the assumption of a random canopy in the segment by eliminating large light spots in the segment, indicating that the CLX method is less sensitive to the segment length compared to the LX method [17,19]. Compared with DHP measurements, there were similar results taken from the TRAC method. However, compared with DHP measurements, CI estimation values using the Ω_{CC}, Ω_{P}, and Ω_{CMN} methods from the TRAC method were more stable. This was because the three methods had similar principles for CI estimation. Moreover, these methods were less affected by the segment length. Meanwhile, the estimated CI values using the Ω_{LX} and Ω_{CLX} methods were close to the real clumping index when the segment length was 2 m. However, the CI estimation values became closer and closer to the true value with an increase in the segment length. This is because the smaller the segment length, the more random the canopy distribution within the segment. However, if the segment length is too small, the possibility of a zero gap or a full gap in the segment increases, resulting in errors in the algorithm processing [48]. In addition, we found the smallest deviation with a segment length in the range of 10 m–50 m among different methods. The CI estimation values deviated the most when the segment length was 2 m. Furthermore, similar results can be found in different forest-type plots. Therefore, the segment length can affect CI estimation and the segment length should be determined by CI estimation methods.

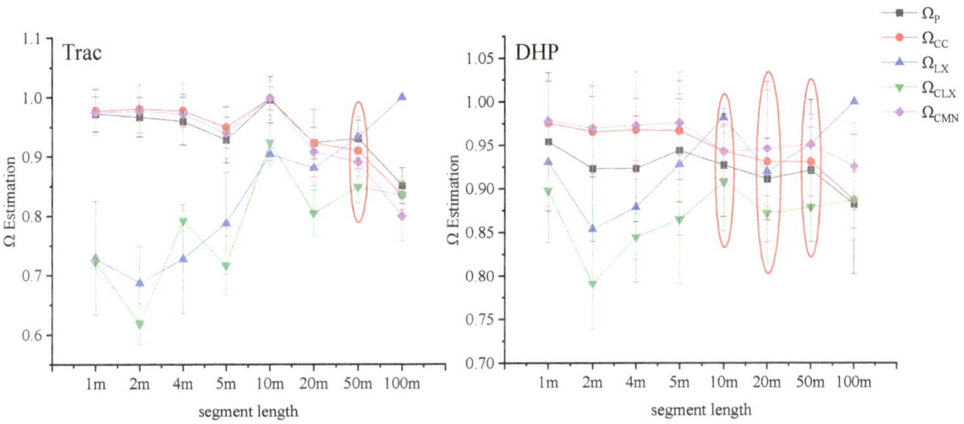

Figure 6. A comparison of CI estimation values with various segment lengths using the Ω_{LX}, Ω_{CC}, Ω_{CLX}, Ω_{P}, and Ω_{CMN} methods.

3.4. The Effect of the Tree Density on CI Estimation

Various tree density values affect the transmission of direct sunlight in canopy environments and canopy gap distributions [19]. This may also have an impact on CI estimation using different methods. We calculated the tree density of plots in the broadleaf forest, the coniferous forest, and the mixed forest. Additionally, the tree density ranged from 0 to 0.08 (tree/m^2), 0.08 to 0.16 (tree/m^2), 0.16 to 0.24 (tree/m^2), and 0.24 to 0.32 (tree/m^2). The figure shows that the distribution of the trees had an obviously spatial heterogeneity effect in the different plots (Figure 7). This affects the sunlight transmitted to the canopy and leads to uncertainties in CI estimation when using optical equipment. Therefore, we compared the results of CI estimation among the scale plots in the different forest types.

We calculated the true clumping index using the litter trap, and then CI estimation was performed using the five methods mentioned above with TRAC measurements. Next, the contour lines of CI estimation using various methods were extracted with ARCGIS software. Finally, the results were overlapped in different tree density mapping studies, as shown in Figure 7. The results show that the real CI had an obviously spatial heterogeneity effect among the different plots too, and this feature was related to the spatial distribution of trees in the plot. The real CI increased when the tree density increased in the plots. This was because the distribution of the trees becomes random with an increase in the number of trees in one plot. In contrast, when the number of trees is smaller, the aggregated distribution becomes more obvious. According to the results shown in Figure 8, the CI estimation values varied with different methods. The relative error of CI estimation using the Ω_{LX} method ranged from 41% to 67%. The relative error of CI estimation using the Ω_{CC} method ranged from 31% to 62%. This value changed from 18% to 56% when the

Ω_{CLX} method was used. The relative error of CI estimation using the Ω_P method ranged from 41% to 56%. This value changed from 18% to 60% when the Ω_{CMN} method was used. In contrast, the accuracy of CI estimation using the Ω_{CLX} method was better than that of other methods. The CI value estimated by the Ω_{CLX} method was more similar to the real CI, followed by that of the Ω_{LX}, Ω_{CC}, Ω_P, and Ω_{CMN} methods. At the same time, the difference in the spatial distribution of the CI in plot scale was also obvious. Compared with the results used by other methods, the spatial distribution of CI estimation using the Ω_{CMN} method was richer for describing the spatial heterogeneity of the CI. However, the accuracy of this method was not as good as the Ω_{CLX} method.

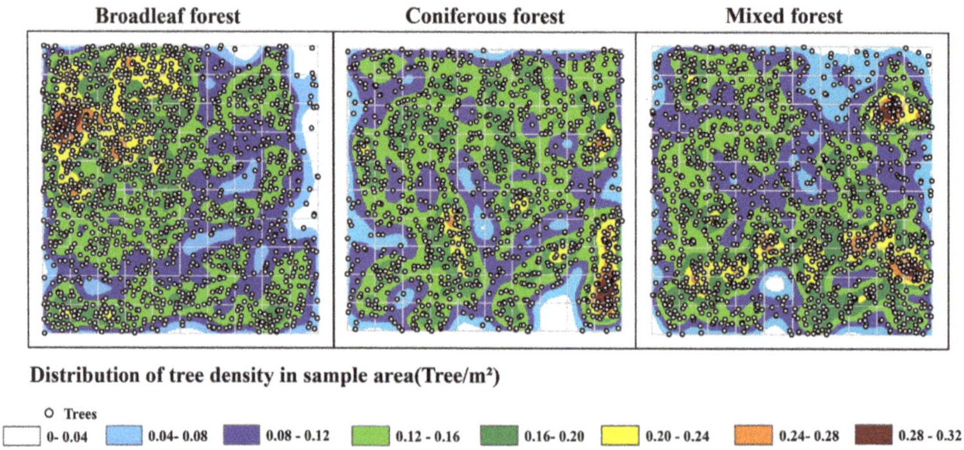

Figure 7. Distribution of the trees and their density in the plots.

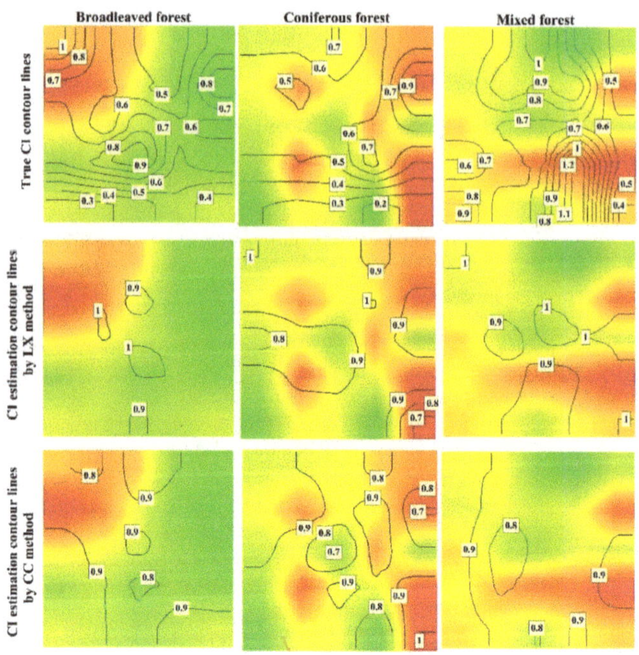

Figure 8. *Cont*.

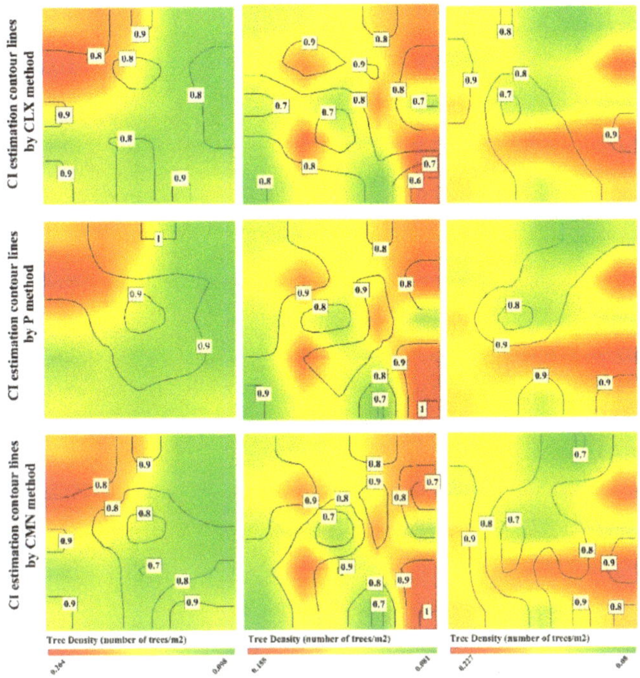

Figure 8. The spatial distribution of CI estimation using different methods in the plot (note: the contour lines show the CI estimation using various methods).

4. Discussion

As a parameter used to describe the distribution of canopy foliage elements, the clumping index (CI) is a measurement of the clumping or random distribution of canopy environments in space, and it is very important to determine the radiation transfer of the canopy, the photosynthesis of the foliage, and the hydrological processes. At the moment, there are many methods used to obtain the CI, such as commercial optical instruments or satellite data [18]. The DHP, LAI-2200, and TRAC methods have frequently been used to estimate the CI during field measurements [16,17,21]. However, CI estimation varies depending on the estimation method and the accuracy of the different methods is still unverified. Meanwhile, the choice of a specific method varies depending on the vegetation type and field conditions [12]. Therefore, we compared the CI estimation values with the CI value (calculated by the litter collection method) using the LX, CC, CLX, P, and CMN methods. The advantages and disadvantages of different methods were compared and the influence factors were analyzed.

4.1. Differences in CI Estimation Methods

In this study, the accuracy values of CI estimation using the LX, CC, CLX, P, and CMN methods were compared. The results show different CI estimation methods lead to huge differences in the estimation results. Zou et al. conducted a comparative study of CI estimation on three algorithms, namely, the gap size distribution method, the finite-length average method, and the segregation coefficient method [17]. The results showed that there were great differences among the three algorithms, and that the gap size distribution method and the segregation coefficient method were the most stable. However, the results of the segregation coefficient method were seriously low compared with the other two methods. Similar results can also be found in this study. We compared the accuracy of CI estimation using the LX, CC, CLX, P, and CMN methods. The results indicate that CI estimation using the Ω_{CLX} method exhibited better results compared to the other methods. Similar results can also be found in previous research [21–23]. Chen mentioned that the CI calculated by the CC method was different from that of the P method [22,23]. Chu et al. compared the CI estimation values using the finite-length average method, the gap size distribution method, and the path length distribution method. The results indicate an overestimating trend for the LX method in the low clumping area or during the low clumping growth period. This was because this method can fail when there is no gap in the finite length. Instead, in the high clumping area or during the high clumping growth period, the gap size distribution method will underestimate the clumping effect [58].

In addition, we found that the accuracy values of CI estimation using TRAC measurements were different from those for DHP measurements. This was because the DHP method offered a directional sampling of canopy gaps and because the accuracy of CI estimation from the DHP method was dependent on the subjective classification procedure of plant pixels and gaps [8]. However, the TRAC method calculated the canopy gap size distribution using transmitted sunlight recordings along a transect. These results are influenced by a solar zenith angle, a limited field-of-view light beam, a gap threshold, and a gap removal procedure [16,22,23].

4.2. *Effecting Factors of the CI Estimation Method*

Indirect optical methods for CI estimation were non-destructive and cost-effective, but these common instruments such as DHP, LAI-2200, and TRAC were subject to the influence of segment selection. In this study, we estimated the CI with segment lengths of 1 m, 2 m, 4 m, 5 m, 10 m, 20 m, 50 m, and 100 m, and compared the CI estimation results with various segment lengths. The results indicate that the Ω_{LX} and Ω_{CLX} methods were more sensitive to the changes in the segment length compared with other methods. This result was similar to previous research findings. Pisek et al. concluded that both the LX and CLX methods were highly sensitive to segment length compared to the actual measurements of DHP and TRAC, and the CI estimation error when using the DHP and TRAC methods decreased with a decrease in the segment length [19]. Woodgate found that the LX method was more sensitive to segment length than the CC or CLX methods, among others [49]. In addition, we found that the accuracy of CI estimation decreased with a decrease in the segment length. Gonsamo et al. measured the canopy gap with DHP and CI estimation using the LX method. The results indicate that the CI estimation values decreased with a decrease in the segment size from 15° to 2.5° [65]. This was because the random assumption for low plants was not true and because it was not feasible to measure the small segment size with the high probability of obtaining a zero-gap fraction. To solve the problem, some scholars proposed to add sky pixels into the fragment [25,66]. Gonsamo et al. improved the LX method by merging the gap-free part with the adjacent gap-filled part, but this improvement affected the choice of optimal segment length and thus impacted CI estimation [65].

The various tree densities affect the transmission of direct sunlight and gap distributions in canopy environments [19]. Therefore, this may have an impact on CI estimation when different methods are used. In this study, we calculated CI estimation values using different methods and compared the results among various tree densities in the broadleaf forest, the coniferous forest, and the mixed forest. The results indicate that the clumping index had significant spatial heterogeneity. The estimation results in different forest types also varied. Some research results have indicated that different stand types have different CI estimation values when different methods are used [17,19,67]. Craig Macfarlane et al. found that the CLX method had significant advantages over the CC method in a jarrah forest in Australia [68]. Woodgate found that the CLX method provided better CI estimation values than the CC and LX methods in eucalyptus forests [49]. Pisek et al. found that the CLX method performed better compared to the CC, LX, and CMN methods in Scots pine forests [19]. Similar results can also be found in this study. In general, the CLX method is better than the other methods used in this study. This is because the CLX method can not only eliminate the problem of the non-randomness in the canopy inside the segment, but can also normalize the whole line, resulting in more obvious random effects at the transect level [19]. In addition, the Ω_{CMN} method can effectively describe the spatial distribution of the CI. Moreover, it can reflect the change characteristics of the clumping index with various tree densities. Therefore, in field measurements, the CI estimation method should be decided after considering the light conditions, the solar zenith, the segment length or size, and the stand types [69,70].

5. Conclusions

In this study, we set three 100 × 100 m plots of different forest types and estimated the clumping index using the measurements from TRAC, LAI-2200, DHP, and litter collection methods. Then, the results of CI estimation at various segment lengths using the LX, CC, CLX, P, and CMN methods were compared. The results indicate the following:

(1) The segment length has a significant effect on CI estimation with various methods. The CI estimation accuracy values of the LX and CLX methods increase with a decrease in segment lengths. The CI estimation results using the CC, P, LX, and CLX methods are the most similar under the segment lengths in the range of 10 m to 50 m. Moreover, CI estimation using the CLX method is most effective at a segment length of 2 m.

(2) The CI has an obviously spatial heterogeneity effect in the different plots. Compared with the true CI, there is significant difference in CI estimation when using various methods. Moreover, the spatial distribution of the CI, estimated using the Ω_{CMN} method, is more useful when describing the spatial heterogeneity patterns of the CI.

Author Contributions: Y.Y. and X.Y. conceived and designed the experiments; Z.L. performed the experiments and analyzed the data; X.Y. and Z.L. wrote the paper; X.Y., Y.Y. and W.F. reviewed and edited the paper. All authors have read and agreed to the published version of the manuscript.

Funding: This research was funded by the National Natural Science Foundation of China (grant numbers 31870621 and 31971580); the Fundamental Research Funds for the Central Universities of China (grant numbers 2572021BA08, 2572019BA10, and 2572019CP12); and the China Postdoctoral Science Foundation (grant number 2019M661239).

Data Availability Statement: Not application.

Conflicts of Interest: The authors declare no conflict of interest.

References

1. Peng, J.; Fan, W.; Wang, L.; Xu, X.; Li, J.; Zhang, B.; Tian, D. Modeling the Directional Clumping Index of Crop and Forest. *Remote Sens.* **2018**, *10*, 1576. [CrossRef]
2. Duthoit, S.; Demarez, V.; Gastellu-Etchegorry, J.-P.; Martin, E.; Roujean, J.-L. Assessing the effects of the clumping phenomenon on BRDF of a maize crop based on 3D numerical scenes using DART model. *Agric. For. Meteorol.* **2008**, *148*, 1341–1352. [CrossRef]
3. Qi, J.; Xie, D.; Jiang, J.; Huang, H. 3D radiative transfer modeling of structurally complex forest canopies through a lightweight boundary-based description of leaf clusters. *Remote Sens. Environ.* **2022**, *283*, 113301. [CrossRef]
4. Nilson, T. Inversion of gap frequency data in forest stands. *Agric. For. Meteorol.* **1999**, *98*, 437–448. [CrossRef]
5. Fassnacht, K.S.; Gower, S.T.; Norman, J.M.; McMurtric, R.E. A comparison of optical and direct methods for estimating foliage surface area index in forests. *Agric. For. Meteorol.* **1994**, *71*, 183–207. [CrossRef]
6. Jonckheere, I.; Fleck, S.; Nackaerts, K.; Muys, B.; Coppin, P.; Weiss, M.; Baret, F. Review of methods for in situ leaf area index determination: Part I. Theories, sensors and hemispherical photography. *Agric. For. Meteorol.* **2004**, *121*, 19–35. [CrossRef]
7. Ma, L.; Zheng, G.; Wang, X.; Li, S.; Lin, Y.; Ju, W. Retrieving forest canopy clumping index using terrestrial laser scanning data. *Remote Sens. Environ.* **2018**, *210*, 452–472. [CrossRef]
8. Chen, J.M.; Black, T. Defining leaf area index for non-flat leaves. *Plant Cell Environ.* **1992**, *15*, 421–429. [CrossRef]
9. Nilson, T. A theoretical analysis of the frequency of gaps in plant stands. *Agric. Meteorol.* **1971**, *8*, 25–38. [CrossRef]
10. Hilty, J.; Muller, B.; Pantin, F.; Leuzinger, S. Plant growth: The What, the How, and the Why. *New Phytol.* **2021**, *232*, 25–41. [CrossRef]
11. Fang, H.; Baret, F.; Plummer, S.; Schaepman-Strub, G. An Overview of Global Leaf Area Index (LAI): Methods, Products, Validation, and Applications. *Rev. Geophys.* **2019**, *57*, 739–799. [CrossRef]
12. Fang, H.; Liu, W.; Li, W.; Wei, S. Estimation of the directional and whole apparent clumping index (ACI) from indirect optical measurements. *ISPRS J. Photogramm. Remote Sens.* **2018**, *144*, 1–13. [CrossRef]

13. Yang, K.; Ryu, Y.; Dechant, B.; Berry, J.A.; Hwang, Y.; Jiang, C.; Kang, M.; Kim, J.; Kimm, H.; Kornfeld, A. Sun-induced chlorophyll fluorescence is more strongly related to absorbed light than to photosynthesis at half-hourly resolution in a rice paddy. *Remote Sens. Environ.* **2018**, *216*, 658–673. [CrossRef]
14. Béland, M.; Baldocchi, D. Is foliage clumping an outcome of resource limitations within forests? *Agric. For. Meteorol.* **2020**, *295*, 108185. [CrossRef]
15. Woodgate, W.; Disney, M.; Armston, J.D.; Jones, S.D.; Suarez, L.; Hill, M.J.; Wilkes, P.; Soto-Berelov, M.; Haywood, A.; Mellor, A. An improved theoretical model of canopy gap probability for Leaf Area Index estimation in woody ecosystems. *For. Ecol. Manag.* **2015**, *358*, 303–320. [CrossRef]
16. Fang, H. Canopy clumping index (CI): A review of methods, characteristics, and applications. *Agric. For. Meteorol.* **2021**, *303*, 108374. [CrossRef]
17. Zou, J.; Zhuang, Y.; Chianucci, F.; Mai, C.; Lin, W.; Leng, P.; Luo, S.; Yan, B. Comparison of Seven Inversion Models for Estimating Plant and Woody Area Indices of Leaf-on and Leaf-off Forest Canopy Using Explicit 3D Forest Scenes. *Remote Sens.* **2018**, *10*, 1297. [CrossRef]
18. Lang, A.; Yueqin, X. Estimation of leaf area index from transmission of direct sunlight in discontinuous canopies. *Agric. For. Meteorol.* **1986**, *37*, 229–243. [CrossRef]
19. Pisek, J.; Lang, M.; Nilson, T.; Korhonen, L.; Karu, H. Comparison of methods for measuring gap size distribution and canopy nonrandomness at Järvselja RAMI (RAdiation transfer Model Intercomparison) test sites. *Agric. For. Meteorol.* **2011**, *151*, 365–377. [CrossRef]
20. Demarez, V.; Duthoit, S.; Baret, F.; Weiss, M.; Dedieu, G. Estimation of leaf area and clumping indexes of crops with hemispherical photographs. *Agric. For. Meteorol.* **2008**, *148*, 644–655. [CrossRef]
21. Leblanc, S.G. Correction to the plant canopy gap-size analysis theory used by the Tracing Radiation and Architecture of Canopies instrument. *Appl. Opt.* **2002**, *41*, 7667–7670. [CrossRef]
22. Chen, J.M.; Cihlar, J. Quantifying the effect of canopy architecture on optical measurements of leaf area index using two gap size analysis methods. *IEEE Trans. Geosci. Remote Sens.* **1995**, *33*, 777–787. [CrossRef]
23. Chen, J.M.; Cihlar, J. Plant canopy gap-size analysis theory for improving optical measurements of leaf-area index. *Appl. Opt.* **1995**, *34*, 6211–6222. [CrossRef]
24. Ryu, Y.; Sonnentag, O.; Nilson, T.; Vargas, R.; Kobayashi, H.; Wenk, R.; Baldocchi, D.D. How to quantify tree leaf area index in an open savanna ecosystem: A multi-instrument and multi-model approach. *Agric. For. Meteorol.* **2010**, *150*, 63–76. [CrossRef]
25. Leblanc, S.G.; Chen, J.M.; Fernandes, R.; Deering, D.W.; Conley, A. Methodology comparison for canopy structure parameters extraction from digital hemispherical photography in boreal forests. *Agric. For. Meteorol.* **2005**, *129*, 187–207. [CrossRef]
26. Leblanc, S.; Fournier, R. Hemispherical photography simulations with an architectural model to assess retrieval of leaf area index. *Agric. For. Meteorol.* **2014**, *194*, 64–76. [CrossRef]
27. Chen, J.M. Optically-based methods for measuring seasonal variation of leaf area index in boreal conifer stands. *Agric. For. Meteorol.* **1996**, *80*, 135–163. [CrossRef]
28. Bréda, N.J. Ground-based measurements of leaf area index: A review of methods, instruments and current controversies. *J. Exp. Bot.* **2003**, *54*, 2403–2417. [CrossRef]
29. Jurik, T.W.; Briggs, G.M.; Gates, D.M. A comparison of four methods for determining leaf area index in successional hardwood forests. *Can. J. For. Res.* **1985**, *15*, 1154–1158. [CrossRef]
30. Liu, Z.; Chen, J.M.; Jin, G.; Qi, Y. Estimating seasonal variations of leaf area index using litterfall collection and optical methods in four mixed evergreen–deciduous forests. *Agric. For. Meteorol.* **2015**, *209*, 36–48. [CrossRef]
31. Le Dantec, V.; Dufrêne, E.; Saugier, B. Interannual and spatial variation in maximum leaf area index of temperate deciduous stands. *For. Ecol. Manag.* **2000**, *134*, 71–81. [CrossRef]
32. Mencuccini, M.; Grace, J. Climate influences the leaf area/sapwood area ratio in Scots pine. *Tree Physiol.* **1995**, *15*, 1–10. [CrossRef] [PubMed]
33. Smith, N. Predicting radiation attenuation in stands of Douglas-fir. *For. Sci.* **1991**, *37*, 1213–1223.
34. Marshall, J.; Waring, R. Comparison of methods of estimating leaf-area index in old-growth Douglas-fir. *Ecology* **1986**, *67*, 975–979. [CrossRef]
35. Liu, Z.; Jin, G.; Chen, J.M.; Qi, Y. Evaluating optical measurements of leaf area index against litter collection in a mixed broadleaved-Korean pine forest in China. *Trees* **2015**, *29*, 59–73. [CrossRef]
36. Guindin-Garcia, N.; Gitelson, A.A.; Arkebauer, T.J.; Shanahan, J.; Weiss, A. An evaluation of MODIS 8-and 16-day composite products for monitoring maize green leaf area index. *Agric. For. Meteorol.* **2012**, *161*, 15–25. [CrossRef]
37. Myneni, R.B.; Ramakrishna, R.; Nemani, R.; Running, S.W. Estimation of global leaf area index and absorbed PAR using radiative transfer models. *IEEE Trans. Geosci. Remote Sens.* **1997**, *35*, 1380–1393. [CrossRef]
38. Curran, P. Multispectral remote sensing for the estimation of green leaf area index. *Philos. Trans. R. Soc. Lond. Ser. A Math. Phys. Sci.* **1983**, *309*, 257–270.
39. Chen, J.M.; Rich, P.M.; Gower, S.T.; Norman, J.M.; Plummer, S. Leaf area index of boreal forests: Theory, techniques, and measurements. *J. Geophys. Res. Atmos.* **1997**, *102*, 29429–29443. [CrossRef]
40. He, L.; Chen, J.M.; Pisek, J.; Schaaf, C.B.; Strahler, A.H. Global clumping index map derived from the MODIS BRDF product. *Remote Sens. Environ.* **2012**, *119*, 118–130. [CrossRef]

41. Gonsamo, A.; Chen, J.M. Improved LAI algorithm implementation to MODIS data by incorporating background, topography, and foliage clumping information. *IEEE Trans. Geosci. Remote Sens.* **2013**, *52*, 1076–1088. [CrossRef]
42. Chen, J.M.; Mo, G.; Pisek, J.; Liu, J.; Deng, F.; Ishizawa, M.; Chan, D. Effects of foliage clumping on the estimation of global terrestrial gross primary productivity. *Glob. Biogeochem. Cycles* **2012**, *26*, GB1019. [CrossRef]
43. Dong, Y.; Jiao, Z.; Yin, S.; Zhang, H.; Zhang, X.; Cui, L.; He, D.; Ding, A.; Chang, Y.; Yang, S. Influence of snow on the magnitude and seasonal variation of the clumping index retrieved from MODIS BRDF products. *Remote Sens.* **2018**, *10*, 1194. [CrossRef]
44. Chen, J.; Menges, C.; Leblanc, S. Global mapping of foliage clumping index using multi-angular satellite data. *Remote Sens. Environ.* **2005**, *97*, 447–457. [CrossRef]
45. Li, Y.; Fang, H. Real-Time Software for the Efficient Generation of the Clumping Index and Its Application Based on the Google Earth Engine. *Remote Sens.* **2022**, *14*, 3837. [CrossRef]
46. Bao, Y.; Ni, W.; Wang, D.; Yue, C.; He, H.; Verbeeck, H. Effects of tree trunks on estimation of clumping index and LAI from HemiView and Terrestrial LiDAR. *Forests* **2018**, *9*, 144. [CrossRef]
47. Li, Y.; Guo, Q.; Su, Y.; Tao, S.; Zhao, K.; Xu, G. Retrieving the gap fraction, element clumping index, and leaf area index of individual trees using single-scan data from a terrestrial laser scanner. *ISPRS J. Photogramm. Remote Sens.* **2017**, *130*, 308–316. [CrossRef]
48. Woodgate, W. *In-Situ Leaf Area Index Estimate Uncertainty in Forests: Supporting Earth Observation Product Calibration and Validation*; RMIT University Melbourne: Melbourne, VIC, Australia, 2015.
49. Woodgate, W.; Armston, J.D.; Disney, M.; Jones, S.D.; Suarez, L.; Hill, M.J.; Wilkes, P.; Soto-Berelov, M. Quantifying the impact of woody material on leaf area index estimation from hemispherical photography using 3D canopy simulations. *Agric. For. Meteorol.* **2016**, *226*, 1–12. [CrossRef]
50. Yin, S.; Jiao, Z.; Dong, Y.; Zhang, X.; Cui, L.; Xie, R.; Guo, J.; Li, S.; Zhu, Z.; Tong, Y. Evaluation of the Consistency of the Vegetation Clumping Index Retrieved from Updated MODIS BRDF Data. *Remote Sens.* **2022**, *14*, 3997. [CrossRef]
51. Zhao, Y.; Ma, Y.; Quackenbush, L.J.; Zhen, Z. Estimation of Individual Tree Biomass in Natural Secondary Forests Based on ALS Data and WorldView-3 Imagery. *Remote Sens.* **2022**, *14*, 271. [CrossRef]
52. Masinda, M.M.; Li, F.; Liu, Q.; Sun, L.; Hu, T. Prediction model of moisture content of dead fine fuel in forest plantations on Maoer Mountain, Northeast China. *J. For. Res.* **2021**, *32*, 2023–2035. [CrossRef]
53. Wang, C. Biomass allometric equations for 10 co-occurring tree species in Chinese temperate forests. *For. Ecol. Manag.* **2006**, *222*, 9–16. [CrossRef]
54. Chen, J.M.; Cihlar, J. Retrieving leaf area index of boreal conifer forests using Landsat TM images. *Remote Sens. Environ.* **1996**, *55*, 153–162. [CrossRef]
55. Eriksson, H.; Eklundh, L.; Hall, K.; Lindroth, A. Estimating LAI in deciduous forest stands. *Agric. For. Meteorol.* **2005**, *129*, 27–37. [CrossRef]
56. Baret, F.; de Solan, B.; Lopez-Lozano, R.; Ma, K.; Weiss, M. GAI estimates of row crops from downward looking digital photos taken perpendicular to rows at 57.5° zenith angle: Theoretical considerations based on 3D architecture models and application to wheat crops. *Agric. For. Meteorol.* **2010**, *150*, 1393–1401. [CrossRef]
57. Nomura, K.; Saito, M.; Kitayama, M.; Goto, Y.; Nagao, K.; Yamasaki, H.; Iwao, T.; Yamazaki, T.; Tada, I.; Kitano, M. Leaf area index estimation of a row-planted eggplant canopy using wide-angle time-lapse photography divided according to view-zenith-angle contours. *Agric. For. Meteorol.* **2022**, *319*, 108930. [CrossRef]
58. Wei, S.; Fang, H. Estimation of canopy clumping index from MISR and MODIS sensors using the normalized difference hotspot and darkspot (NDHD) method: The influence of BRDF models and solar zenith angle. *Remote Sens. Environ.* **2016**, *187*, 476–491. [CrossRef]
59. Whitford, K.; Colquhoun, I.; Lang, A.; Harper, B. Measuring leaf area index in a sparse eucalypt forest: A comparison of estimates from direct measurement, hemispherical photography, sunlight transmittance and allometric regression. *Agric. For. Meteorol.* **1995**, *74*, 237–249. [CrossRef]
60. Chen, J.; Black, T. Foliage area and architecture of plant canopies from sunfleck size distributions. *Agric. For. Meteorol.* **1992**, *60*, 249–266. [CrossRef]
61. Miller, E.E.; Norman, J.M. A Sunfleck Theory for Plant Canopies I. Lengths of Sunlit Segments along a Transect 1. *Agron. J.* **1971**, *63*, 735–738. [CrossRef]
62. Hu, R.; Yan, G.; Mu, X.; Luo, J. Indirect measurement of leaf area index on the basis of path length distribution. *Remote Sens. Environ.* **2014**, *155*, 239–247. [CrossRef]
63. Yan, G.; Hu, R.; Luo, J.; Weiss, M.; Jiang, H.; Mu, X.; Xie, D.; Zhang, W. Review of indirect optical measurements of leaf area index: Recent advances, challenges, and perspectives. *Agric. For. Meteorol.* **2019**, *265*, 390–411. [CrossRef]
64. Kuusk, A.; Pisek, J.; Lang, M.; Märdla, S. Estimation of Gap Fraction and Foliage Clumping in Forest Canopies. *Remote Sens.* **2018**, *10*, 1153. [CrossRef]
65. Gonsamo, A.; Walter, J.-M.N.; Pellikka, P. Sampling gap fraction and size for estimating leaf area and clumping indices from hemispherical photographs. *Can. J. For. Res.* **2010**, *40*, 1588–1603. [CrossRef]
66. Van Gardingen, P.; Jackson, G.; Hernandez-Daumas, S.; Russell, G.; Sharp, L. Leaf area index estimates obtained for clumped canopies using hemispherical photography. *Agric. For. Meteorol.* **1999**, *94*, 243–257. [CrossRef]

67. Kucharik, C.J.; Norman, J.M.; Gower, S.T. Characterization of radiation regimes in nonrandom forest canopies: Theory, measurements, and a simplified modeling approach. *Tree Physiol.* **1999**, *19*, 695–706. [CrossRef]
68. Macfarlane, C.; Hoffman, M.; Eamus, D.; Kerp, N.; Higginson, S.; McMurtrie, R.; Adams, M. Estimation of leaf area index in eucalypt forest using digital photography. *Agric. For. Meteorol.* **2007**, *143*, 176–188. [CrossRef]
69. Leblanc, S.G.; Chen, J.M. A practical scheme for correcting multiple scattering effects on optical LAI measurements. *Agric. For. Meteorol.* **2001**, *110*, 125–139. [CrossRef]
70. Zou, J.; Yan, G.; Chen, L. Estimation of canopy and woody components clumping indices at three mature picea crassifolia forest stands. *IEEE J. Sel. Top. Appl. Earth Obs. Remote Sens.* **2015**, *8*, 1413–1422. [CrossRef]

remote sensing

MDPI

Article

Reduction in Uncertainty in Forest Aboveground Biomass Estimation Using Sentinel-2 Images: A Case Study of *Pinus densata* Forests in Shangri-La City, China

Lu Li [1], Boqi Zhou [1], Yanfeng Liu [1], Yong Wu [1], Jing Tang [1], Weiheng Xu [1,2], Leiguang Wang [1,2] and Guanglong Ou [1,*]

1 Key Laboratory of State Forestry Administration on Biodiversity Conservation in Southwest China, Southwest Forestry University, Kunming 650224, China
2 Institute of Big Data and Artificial Intelligence, Southwest Forestry University, Kunming 650233, China
* Correspondence: olg2007621@swfu.edu.cn

Abstract: The uncertainty from the under-estimation and over-estimation of forest aboveground biomass (AGB) is an urgent problem in optical remote sensing estimation. In order to more accurately estimate the AGB of *Pinus densata* forests in Shangri-La City, we mainly discuss three non-parametric models—the artificial neural network (ANN), random forests (RFs), and the quantile regression neural network (QRNN) based on 146 sample plots and Sentinel-2 images in Shangri-La City, China. Moreover, we selected the corresponding optical quartile models with the lowest mean error at each AGB segment to combine as the best QRNN (QRNNb). The results showed that: (1) for the whole biomass segment, the QRNNb has the best fitting performance compared with the ANN and RFs, the ANN has the lowest R^2 (0.602) and the highest RMSE (48.180 Mg/ha), and the difference between the QRNNb and RFs is not apparent. (2) For the different biomass segments, the QRNNb has a better performance. Especially when AGB is lower than 40 Mg/ha, the QRNNb has the highest R^2 of 0.961 and the lowest RMSE of 1.733 (Mg/ha). Meanwhile, when AGB is larger than 160 Mg/ha, the QRNNb has the highest R^2 of 0.867 and the lowest RMSE of 18.203 Mg/ha. This indicates that the QRNNb is more robust and can improve the over-estimation and under-estimation in AGB estimation. This means that the QRNNb combined with the optimal quantile model of each biomass segment provides a method with more potential for reducing the uncertainties in AGB estimation using optical remote sensing images.

Keywords: Sentinel-2 images; artificial neural network; random forests; quantile regression neural network; *Pinus densata* forests

Citation: Li, L.; Zhou, B.; Liu, Y.; Wu, Y.; Tang, J.; Xu, W.; Wang, L.; Ou, G. Reduction in Uncertainty in Forest Aboveground Biomass Estimation Using Sentinel-2 Images: A Case Study of *Pinus densata* Forests in Shangri-La City, China. *Remote Sens.* **2023**, *15*, 559. https://doi.org/10.3390/rs15030559

Academic Editors: Huaqiang Du, Wenyi Fan, Weiliang Fan, Fangjie Mao and Mingshi Li

Received: 27 November 2022
Revised: 10 January 2023
Accepted: 10 January 2023
Published: 17 January 2023

1. Introduction

Forest biomass is a crucial factor in carbon storage in terrestrial ecosystems and plays an essential role in protecting the ecological environment and biodiversity [1]. The biomass harvesting method is time-consuming and labor-intensive; thus, it is not available for large-scale data acquisition [2]. Along with the development of remote sensing technology, more and more researchers are using remote sensing data combined with ground survey data to estimate large-scale forest biomass [3,4].

Three types of remote sensing data are available for biomass estimation: optical images, active sensor radar data, and light detection and ranging (LiDAR) data [5,6]. The main LiDAR technology used in forest biomass estimation is backpack LiDAR and airborne LiDAR. Backpack LiDAR is hard to use for large-area assessment because the terrain and forestland accessibility easily influence it. Although airborne LiDAR is not limited by the terrain and can capture three-dimensional structure information; thus, it has a better performance for forest biomass estimation by improving the saturation problem in biomass estimation using optical remote sensing data [7,8]. However, it still needs to be more suitable for large-area forest biomass estimation due to the limitation of the battery capacity and the increased imaging cost, etc. Moreover, LiDAR has no infrared signals, a limiting factor for vegetation analysis [9]. Radar has an intense penetration in vegetation, but the data processing is quite complicated, and the forest AGB has a different sensitivity to its wavelength [10,11]. The most accurate radar systems operate with short wavelengths (i.e., X- and C-bands). However, the radar signal does not reach the ground because it is mainly backscattered by the canopy of the upper layer [12–14]. Using long radar wavelengths (mainly L- and P-bands), the radar signal can penetrate the different layers from the top of the canopy to the ground. However, P-band radar imagery is expected to be available with the ESA BIOMASS mission to be launched in 2024 [15,16]. High- and medium-resolution optical remote sensing are used for AGB estimation commonly. Generally, high-resolution optical images are too expensive, and the images are quite hard to obtain even though they have more accurate results of AGB estimation than medium-resolution optical images [17]. Therefore, the medium-resolution satellite images (e.g., Landsat and Sentinel-2) are a better choice for forest biomass evaluation by different spatial scales due to their free accessibility and high suitability to landscape scale analysis [18]. However, reducing the uncertainties is still a significant difficulty for AGB estimation using optical remote sensing data, especially when the study area has a high canopy [19,20]. The European Space Agency launched a high-resolution and multi-spectral imaging satellite, Sentinel-2A, in 2015 and Sentinel-2B in 2017. In addition, the spatial resolutions are 10 m, 20 m, and 60 m, respectively. Sentinel-2 can revisit an area in 5 days by two satellites and it has a wide swath at 290 km with 13 multi-spectral bands, including four additional spectral bands strategically positioned in the red-edge region, which is a more sensitive band to vegetation [21,22]. It is expected to improve the uncertainties of AGB estimation [22–24].

To reduce the saturation impact on forest biomass estimation, vegetation indices (VI) have been employed in lots of research [25–27]. The VI has been shown to be related to photosynthesis to some extent and directly proportional to biomass or yield [28]. The normalized difference vegetation index (NDVI), atmospherically resistant vegetation index (ARVI), difference vegetation index (DVI), simple ratio index (RVI), etc., were extracted from images, which were used in AGB assessment [1,28,29]. With the development of the research, the researchers found that the vegetation index changed little after the biomass reached a specific value [17,18]; in particular, tropical and subtropical woodlands with high coverage and structural complexity are more likely to lead to insensitivity [30]. Moreover, the researchers found that the texture features are more sensitive to the horizontal structure of the canopy and the shadow, which may be suitable for improving the prediction precision of forest AGB biomass estimation. Some studies have been found to apply textures in forest biomass assessment [31,32], and the image texture has excellent potential to enhance the accuracy of AGB estimation [33–35]. Therefore, variables screening is vital for reducing the impact on the multi-collinearity and increasing the accuracy in the AGB remote sensing estimation [36–38].

The accuracy of forest AGB estimation is not only affected by the survey data but also impacts the methodology of the assessment model [39,40]. Two kinds of algorithms were applied for forest AGB estimation, including parametric and non-parametric algorithms [41]. The parametric method can quantitatively describe the relationship between AGB and the variables, in which h contains linear, logarithm, exponential, and other functions [42]. In contrast, the artificial neural network (ANN), K-nearest neighbor (KNN), support vector machine (SVM), random forests (RF), etc., were counted into a not-parametric model [43–46]. The relationship between AGB and variables cannot easily be analyzed by fixed quantity due to the complex relationship between AGB and forest construction. A lot of research has been conducted to compare the accuracy of parametric and non-parametric algorithms, and the result have shown that non-parametric algorithms exhibited excellent performance [46]. The artificial neural network (ANN) is a supervised learning algorithm in machine learning which has adaptability and improves the precision of updated data. It has been used widely to demonstrate the complex relationships between independent and dependent variables [47]. The ANN frequently uses AGB estimation due to the parallelism, fault tolerance, generalization capability, and multiple input multiple output architecture. Then, it can reveal a solid ability to fit data [44]. Moreover, ANN was applied to predict the AGB in natural forest ecosystems, with it showing that it offered a higher accuracy result than traditional protocols [48–50].

The quantile regression neural network (QRNN) is a non-parametric, nonlinear model that is combined with a neural network (NN) and quantile regression (QR) approach, which was introduced by Taylor [51]. It centralizes the advantages of both the ANN and the QR. The QR was created by Koenker [52], and it can more accurately describe the influence of independent variables on the range of dependent variables and the shape of the conditional distribution. It is not impacted by abnormal data such as sharp peaks, discrete values, and heavy-tailed allocations [53,54]. When independent variables have different effects on the distribution of dependent variables in different parts, such as skewness on the left or right, it can describe the characteristics of the distribution more comprehensively [53,54]. The QRNN is a suitable methodology for predicting mixed discrete–continuous variables. It has already been applied in ecological environments [55–57]. Rarely have studies been found using the QRNN in forest AGB estimation.

In general, the forest resources of Shangri-La City are characterized by extensive forestry land and are identified as one of the species genetic pools [58,59]. Meanwhile, Yunnan is known as the kingdom of plants and animals; thus, the forest resources status of Yunnan in China or around the world is self-evident [60,61]. Given this, it is significant to emphasize the precision of forest AGB assessment in Shangri-La City to protect forest resources and improve the ecological environment. In this study, we will estimate forest AGB by combining the measured sample data, Sentinel-2 images, vegetation index, and texture value extracted from the images. We screened the correlation variables with AGB using RF, then RF, the ANN, and the QRNN were selected to compare the fitting performance. The significant contributions of this work are:

(1) To compare different biomass estimation models—the ANN, RF, and the QRNN for estimating the biomass of *Pinus densata* forests using Sentinel-2 images in Shangri-La City.
(2) To explore the optimal quantile model on each biomass segment to improve the AGB estimation accuracy, and then provide a method to reduce the uncertainties from over-estimation and under-estimation of forest AGB estimation.

2. Materials and Methods

2.1. Study Site

The study area is located in Shangri-La City, northwestern Yunnan, China. The coordinates of Shangri-La City are: latitude 26°52′~28°52′N and longitude 99°20′~100°29′E (Figure 1). The elevation range is from 3350 to 3696 m above sea level, the annual mean temperature is 4.7–16.5 °C, and the extreme maximum and minimum temperature are 25.1 and −20.1 °C, respectively. The dry and wet seasons are distinct, and the four seasons are not apparent in Shangri-La City. For the rainfall time concentrates from June to September, the mean annual precipitation is 607 mm and the average annual evaporation is 1643.6 mm [62]. The particular geographical environment and complex ecological conditions have created a unique natural landscape and rich natural resources. The original forest area with the sub-alpine coniferous forest is the main forest area that is well preserved in China.

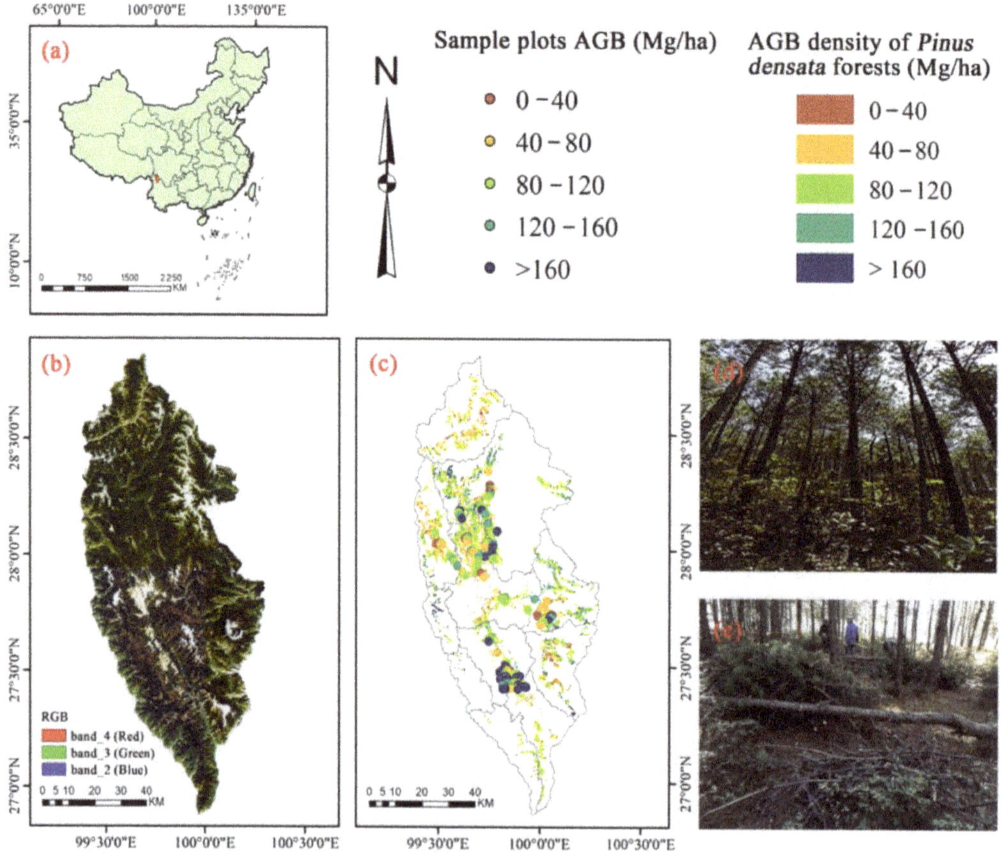

Figure 1. (**a**) Location of Shangri-La City in China; (**b**) The Sentinel-2 images of the study area; (**c**) the spatial distribution of *Pinus densata* forests according to the forest management inventory (FMI) data in 2016 and the sample plots investigated in 2016; (**d**) the typical stand structure of *Pinus densata* forests in the study area; and (**e**) the field investigation of AGB.

Pinus densata, one of the barren tolerance pioneer tree species of sub-alpine coniferous, is light-loving and cold-resistant in Shangri-La City. *Pinus densata* forests are single-storied stands with even age in common, and most of the study areas were conducted in pure *Pinus densata* forests [60,61,63] (Figure 1).

2.2. Flow Chart

In Figure 2, the methodological framework of this study was described in the following steps: (1) collecting the sample plots and tree biomass data and the Sentinel images data; (2) calculating the plot AGB; (3) pre-processing of the Sentinel images; (4) correlation between spectral variables and AGB; (5) developing the model: the artificial neural network (ANN), random forests (RF), and the quantile regression neural network(QRNN); and (6) assessing the models.

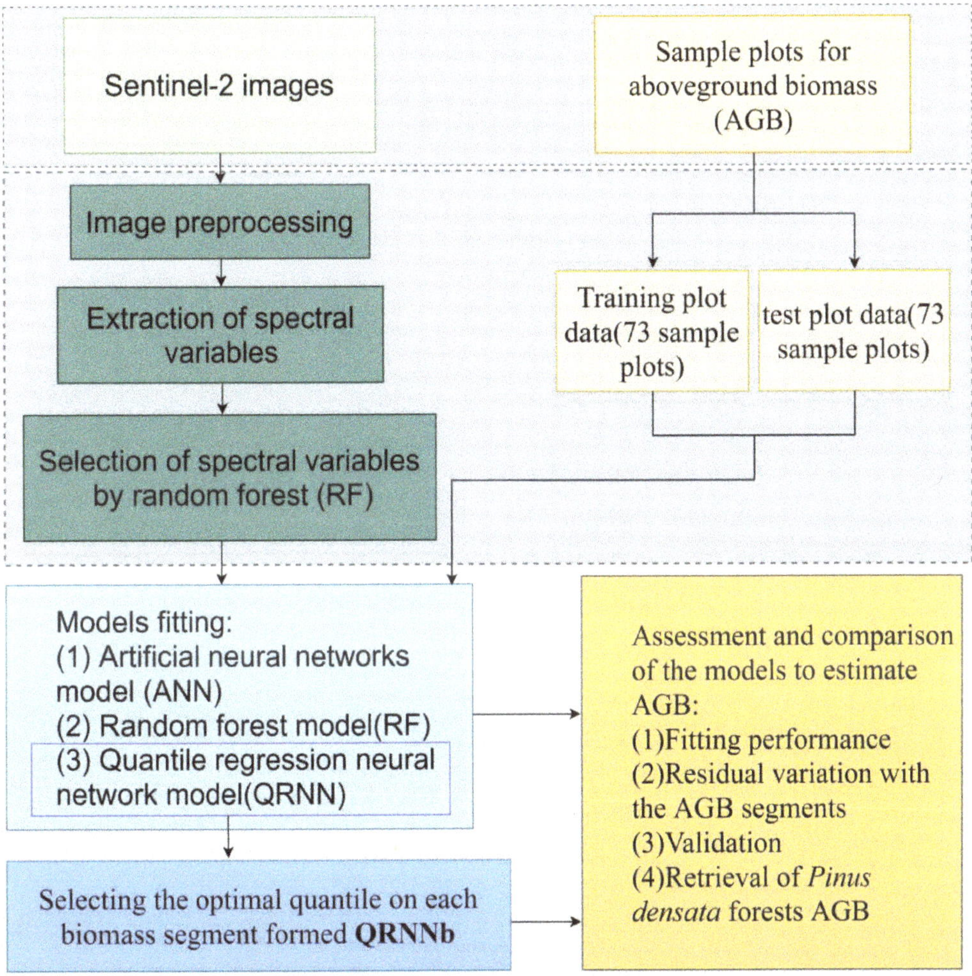

Figure 2. The methodological framework of estimating the forest above-ground biomass (AGB). RF is the random forests, ANN is the artificial neural network, QRNN is the quantiles regression neural network, and QRNNb is the quantile regression neural network with the best fitting performance in each biomass segment.

2.3. Field Data Collection and Aboveground Biomass Calculation

Field data collection work was conducted in August 2016, and in situ data from over 146 sample plots were collected. The plot size was 30 m × 30 m. A GPS was used to measure and record the coordinates and elevation. All of the trees of each plot with a diameter at breast height 1.3 m above ground (DBH) >5 cm were measured. The trees on the south and west boundary of the sample plot were recorded. Three to five trees with a similar average stand DBH were chosen, and the height of the selected trees was measured to calculate the average height of the stand in each plot. The other information in the plot needed to be recorded, such as forest site conditions, origin, age, soil, and the trees' health situation, etc.

The process of investigation, sampling, determination, and individual tree biomass construction has been detailed in the literature [63]. The equation for the tree AGB was as below:

$$AGB_i = 0.073 \cdot DBH^{1.739} \cdot H^{0.880} \quad (1)$$

where *DBH* is the tree diameter at the breast height >5 cm, *H* is the tree height, and AGB_i is the AGB of the individual tree in the plot (kg).

Equation (2), as below, was the sample plot AGB (Mg/ha). To ensure enough comparable sample plot datasets at each biomass segment for the fitting test and validation test, the sample numbers of the two datasets were the same; then, 146 sample plots were randomly divided into a fitting dataset of 73 plots and a test dataset of 73 plots, and the statistical information is listed in Table 1. In addition, there were no significant statistical differences in the mean and standard deviation values between the fitted and the test datasets.

$$AGB_s = \frac{\sum_{i=1}^{n} AGB_i}{900} \cdot 10,000/1000 \quad (2)$$

where AGB_s is the AGB of a plot, AGB_i is the AGB of individual trees, and *n* is the number of trees within each plot.

Table 1. The statistical parameters of sample plot datasets. *H* is the average tree height, *Dg* is the average diameter at breast height (1.3 m), and AGB the is above-ground biomass.

Variables		Fitting Data (*n* = 73)	Test Data (*n* = 73)	All Data (*n* = 146)
Minimum	*H* (m)	2.2	2.9	2.2
	Dg (cm)	2.9	4.9	2.9
	AGB (Mg/ha)	2.1	11.1	2.1
Maximum	*H* (m)	24.3	19.5	24.3
	Dg (cm)	41.3	24.7	41.3
	AGB (Mg/ha)	335.9	344.4	344.4
Mean	*H* (m)	10.0	10.3	10.1
	Dg (cm)	14.6	15.0	14.8
	AGB (Mg/ha)	120.7	122.2	121.5
Standard deviation	*H* (m)	3.8	3.7	3.7
	Dg (cm)	6.3	4.5	5.5
	AGB (Mg/ha)	67.5	79.9	73.7

2.4. Remote Sensing Data and Variables

2.4.1. Pre-Processing of Sentinel-2 Images

Five Sentinel-2 images obtained from the European Space Agency (ESA) were used in this study (Table 2). Since there were no level-2A products before May 2017, level-1C products with UTM/WGS 84 ortho-images were downloaded, and they were orthorectified top-of-atmosphere reflectance products. Bottom-of-atmosphere reflectance product L2A needed to be obtained by atmospheric correction. Thereby, the Sen2Cor (version 02.05) plugin under the toolbox in SNAP was installed to create L2A products, and the open-access software of SNAP was downloaded from http://step.esa.int/main/download/snap-download/ and accessed on 10 October 2022. Then, we resampled all of the bands with a 10 m resolution under cubic convolution interpolation by using the resample tool in SNAP. Finally, we resampled all of the bands with a 30 m resolution to meet the plot size of the field AGB survey, and the images were cropped and spliced in ENVI.

Table 2. The parameters of five Sentinel-2 images.

Image ID	Acquisition Date	Central Longitude (Degree)	Central Latitude (Degree)	Solar Elevation	Solar Azimuth	Mean Cloud Amount (%)
S2A_MSIL1C_ 20161124T040102_ N0204_R004_ T47RNK_ 20161124T040118	24 November 2016	99.5513	26.6257	1.0249	162.1176	12.6
S2A_MSIL1C_ 20161124T040102_ N0204_R004_ T47RNL_ 20161124T040118	24 November 2016	99.5557	27.5287	1.0249	162.2853	25.6
S2A_MSIL1C_ 20161124T040102_ N0204_R004_ T47RNM_ 20161124T040118	24 November 2016	99.5604	28.4315	1.0249	162.4446	41.7
S2A_MSIL1C_ 20161124T040102_ N0204_R004_ T47RPL_ 20161124T040118	24 November 2016	100.5684	27.5209	1.0249	163.4582	15.1
S2A_MSIL1C_ 20161124T040102_ N0204_R004_ T47RPL_ 20161124T040118	24 November 2016	100.5815	28.4235	1.0249	163.6144	38.5

2.4.2. Extraction Feature Variables from Remote Sensing

The vegetation index and conversion factor have been widely used to estimate forest AGB [27,28]. The texture feature is an essential feature of remote sensing images, and it reflects the properties of the object itself and helps to distinguish two different objects [28]. First and foremost, texture features have been proven to have essential contributions to increasing the accuracy of AGB estimation because they can describe complex forest structures with high accuracy [17,28,31]. Therefore, this study extracted 134 remote sensing variables, including 11 spectral bands, 21 vegetation indices, 6 image conversion algorithms, and 96 texture measurements (Table 3).

Table 3. Spectral variables derived from Sentinel-2 images.

Data Sources	SV	Definitions of SV	Number of SV
Sentinel-2	Original band	b2—blue, b3—green, b4—red, b5—vegetation red edge, b6—vegetation red edge, b7—vegetation red edge, b8—NIR, b9—water vapor, b10—SWIR-cirrus, b11—SWIR, b12—SWIR	11
	Vegetation indices	Normalized difference vegetation index (NDVI), atmospherically resistant vegetation index (ARVI), difference vegetation index (DVI), ratio vegetation index (RVI), vegetation index of soil adjustment ratio (SARV), oil adjusted vegetation index (SAVI), modified soil vegetation index (MSAVI), short infrared temperature vegetation index (MVI5), mid-infrared temperature vegetation index (MVI7), transformation vegetation index (TVI), nonlinear vegetation index (NLI), perpendicular vegetation Index (PVI), infrared vegetation index (II), optimization simple ratio index (MSR), simple vegetation index (SR), brightness vegetation index (B), temperature vegetation index (W), greenness vegetation index (G), normalized difference vegetation index using R and G bands (ND43), normalized difference vegetation index using band 6 and band 7 (ND67), normalized difference vegetation index using band 5, band 6, and band 3 (ND563)	21
	Image transformations	The first three components from the tasseled cap transform (K T transform) and the first three principal components of principal component analysis (PCA)	6
	Texture measures	Grey-level co-occurrence matrix-based texture measures including the mean, angular second moment, contrast, correlation, dissimilarity, entropy, homogeneity, and variance using moving window sizes of 3 × 3, 5 × 5, and 7 × 7 pixels	96

2.4.3. Variables Screening

A total of 134 variables were extracted, but not all of them were sensitive to AGB. Random forests (RF) technology was chosen to analyze the correlation between derived variables and the field-based AGB data to gain a set of parsimonious and valid variables for the AGB model. Then, the final vital performed variables were selected for building the regression model. RFs is an ensemble machine-learning algorithm that was first proposed by Breiman [36]. The keys to an RF construction include the random selection of the decision tree and features (subset). Two thirds of variables are randomly selected from the original dataset by the bootstrap sampling method to avoid over-fitting so that the training dataset of the decision tree and the data amount of all training datasets is consistent with the amount of original data [64]. The features (the other 1/3 of the original data) are selected as the nodes of each decision tree and they were also chosen randomly. The features are split based on the Gini criterion. The remaining features are the out-of-bag (OOB) data used as validation samples. OOB data can be used to calculate the unbiased estimate of prediction error by comparing the dataset with the out-of-bag data. Meanwhile, they can also be applied to determine the importance of the variables. The optimal solution is obtained by voting according to the principle that the minority is subordinate to the majority. Moreover, the quality of the RFs model is related to the mean square errors (MSE) between the decision tree and the features, and the smaller the MSE is, the better [36]. In this study, 80% of the original data was used as the training dataset, and the left data were the test dataset. Random forest recursive feature elimination (RF-RFE) was used to remove variables that did not contribute significantly to model accuracy. This experiment was conducted in the sklearn.assembly module of Python 3.7, used the RandomizedSearchCV and GridSearchCV functions to optimize the model parameters and variables screening.

2.5. Modeling Methods

2.5.1. Random Forests Modeling (RF)

Random forests (RF) are an accurate methodology for classification and a validation way to predict the AGB [65]. The two parameters that must be set are the number of trees for growing (ntree) and the split variables for selecting randomly. Balancing the two parameters is the most critical work for avoiding the lowest generalization error [36]. Different numbers for the ntree and minimum sample split (mtry) and the other factors, such as the max-depth (the sample depth that contains the minimum sample) and min-sample-leaf (the minimum number of samples at the leaf node), were chosen to compare the R^2 in Python. The highest R^2 was finally obtained. The parameters were set as follows: the maximum number of iterations was 200, the max-depth was 10, the min-samples-leaf was 1, and the min-samples-split was 2. In this study, 80% of the original data was used as the training dataset, and the other 20% was the test dataset. Ten-fold cross-validation was applied to prevent over-fitting and to prevent it affecting the accuracy and stability of the model.

2.5.2. Artificial Neural Networks Model (ANN)

The artificial neural network (ANN) is a mathematical model for information processing using similar structures to the synaptic connections in the brain [66]. It consists of a large number of nodes (or neurons) that are connected. Each node represents a specific output function called the activation function. Each connection between two nodes represents the weighted value of the signal passing through the link. The training of the neural network model is the process of modifying the connection weight between the neuron and the neuron deviation according to the training data. ANN comprises three essential elements: the processing unit, network topology, and training rules. The processing unit is the basic unit of artificial neural network operation. A processing unit has multiple input and output paths. The network topology determines the information transmission between each processing unit and each layer, generally composed of an input layer, a hidden layer, and an output layer [44]. The number of hidden layer nodes has been paid a lot attention to because the network cannot have the necessary learning ability and information prediction ability if the number of hidden layer nodes is too small. On the contrary, it will increase the complexity of the network structure and make the network fall into a partial minimum or lead to overfitting [67]. Training rules are trained and adjusted repeatedly to achieve the required accuracy. It mainly uses the transformation function to weigh and sum the processed data and trains the network system to carry on the pattern recognition.

The back propagation neural network (BPNN), one of the most widely used neural network algorithms, was applied in this study. The BP neural network model was constructed by using the R language package. Firstly, the initially hidden node was set to 4. It was found that the average error decreased at the beginning and then increased with the number of hidden nodes increasing. When the number of hidden nodes was 7, the average error was minimum. Ten-fold cross-validation was used to test the accuracy; 80% of the modeling samples was used as a training set, and 20% was used as a test set. Each test data will yield a corresponding rate of accuracy (or error rate). The average of the accuracy (or the error rate) of the ten modeling results was used as an estimate of the accuracy of the algorithm [68,69].

2.5.3. Quantile Regression Neural Network (QRNN)

The nonlinear relationship between a dependent and independent variable is very complex and challenging to describe. Taylor used neural network structure to establish neural network quantile regression (QRNN) [51]. The model combines the nonparametric, nonlinear quantile regression method and achieves a nonlinear mapping of conditional quantiles from dependent variables to independent variables. As the artificial neural networks, the number of hidden layer nodes has an essential effect on the complexity of the model. The number of hidden layer nodes should be manageable because it would cause the fitting time to be too long, which may add non-regular content and this leads to over-fitting [55].

QRNN was constructed with the QRNN function package in R software. Three hidden layers and seven hidden nodes were used that were the same as the ANN. At the same time, 10-fold cross-validation was also carried out to prevent over-fitting or reduce errors from affecting the accuracy and stability of the model. The scale of the training and test dataset was the same as in the ANN.

Moreover, the corresponding optical quartile models with the lowest mean error at each AGB segment were combined as the best QRNN (QRNNb). In addition, the AGB segments were 0–40 Mg/ha, 40–80 Mg/ha, 80–120 Mg/ha, 120–160 Mg/ha, and greater than 160 Mg/ha. Therefore, the QRNNb represents a complete biomass estimation model formed by selecting the highest precision of the five quantitative models corresponding to the QRNN on each of the five biomass segments.

2.6. Assessment and Validation of the Models

It is critical to obtain the AGB model and the assessment values during the process of AGB model building. Coefficient determination (R^2) and mean square root error (RMSE) were used to estimate the AGB prediction model and assessment value. R^2 and RMSE were applied to compare the accuracy of prediction values from different estimate models based on fitting plots data in Table 1.

Linear regression between AGB predicted values of different biomass segments and the observed data was used to assess models' performance using 73 test plots. In addition, except for R^2 and RMSE, the mean absolute error (MAE) and mean error (ME) were added to test the validation of each model by using the test dataset. The AGB segments were 0–40 Mg/ha, 40–80 Mg/ha, 80–120 Mg/ha, 120–160 Mg/ha, and greater than 160 Mg/ha [41].

$$R^2 = 1 - \frac{\sum_{i=1}^{n}\left(\hat{y}_i - y_i\right)^2}{\sum_{i=1}^{n}(y_i - \overline{y})^2} \tag{3}$$

$$RMSE = \sqrt{\frac{\sum_{i=1}^{n}\left(y_i - \hat{y}_i\right)^2}{n}} \tag{4}$$

$$MAE = \frac{\sum_{i=1}^{n}|y_i - \hat{y}_i|}{n} \tag{5}$$

$$ME = \frac{\sum_{i=1}^{n}(y_i - \hat{y}_i)}{n} \tag{6}$$

where $\hat{y}_i$ and y_i are the predicted AGB and the corresponding AGB in the sample plot, $\overline{y}$ is the mean AGB of the sample plots, and n is the number of sample plots.

3. Results

3.1. Results of Spectral Variables Screening

The prediction accuracy would decrease if all of the biomass prediction variables extracted from the images were applied to build the AGB estimation model, and information redundancy would also occur. The function of AGB assessment would be reduced as some variables may have a weak association with biomass. Thus, screening suitable and strong correlation variables was the critical step. In this study, RFs was used to screen the characteristic variables according to the sort of variable importance. In addition, the first ten important variables were VA3_2, VA5_12, CO7_8, DI5_8, VA5_2, HO5_3, VA3_12, VA7_12, ME5_12, and SE5_3. To prevent the selected variables from displaying multi-collinearity and thus reducing the accuracy of biomass estimation, we performed collinearity tests on selected variables using the Kappa functions in R software. The results indicated less collinearity between the variables as the Kappa coefficient value was 11.72709, which was less than 100. The correlation between forest AGB and the characteristic variables of *Pinus densata* forests is shown in Figure 3.

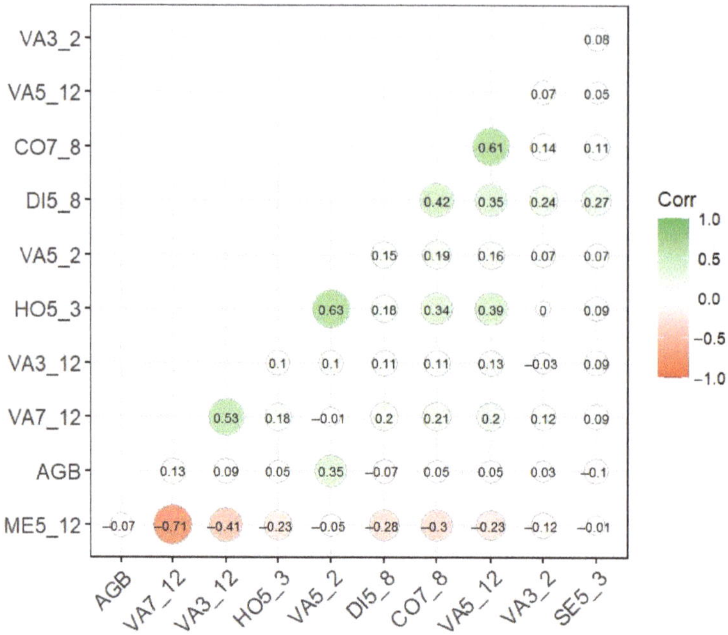

Figure 3. The correlation between AGB and the characteristic variables of *Pinus densata* forests. Corr is the correlation coefficient between the characteristic variables and AGB. VA3_2 is the variance on band 2 with the window size 3 × 3, VA5_12 is the variance on band 12 with the window size 5 × 5,

CO7_8 is the correlation on band 8 with the window size 7 × 7, DI5_8 is the dissimilarity on band 8 with the window size 5 × 5, VA5_2 is the variance on band 2 with the window size 5 × 5, HO5_3 is the third homogeneity on band 3 with the window size 5 × 5, VA3_12 is the variance on band 12 with the window size 3 × 3, VA7_12 is the variance on band 12 with the window size 7 × 7, ME5_12 is the mean on band 12 with the window size 5 × 5, and SE5_3 is the second moment on band 5 with the window size 3 × 3.

3.2. Model Comparison of the Model

3.2.1. Model Fitting

Scatter plots of AGB and predicted biomass based on the ANN and QRNN models based on ten variables are shown in Figure 4. It was shown that the ANN's fitting performance was not significantly different from that of QRNN at 0.1, 0.25, and 0.5 percentiles. However, when the optimal quantile model for each biomass segment was integrated into a complete QRNNb (Figure 4c), the fitting accuracy of the QRNNb was significantly improved. The R^2 and RMSE of the ANN were 0.722 and 31.0689, respectively. In addition, the R^2 and RMSE of the QRNNb were 0.962 and 13.9326, respectively. The results also demonstrated that the fitting performance of RFs (R^2 = 0.934, RMSE = 11.3305) was quite similar to that of the QRNNb. RFs and the QRNNb had a better fitting performance than the ANN and the QRNN, which means both RFs and the QRNNb had a higher accuracy than the ANN and the QRNN.

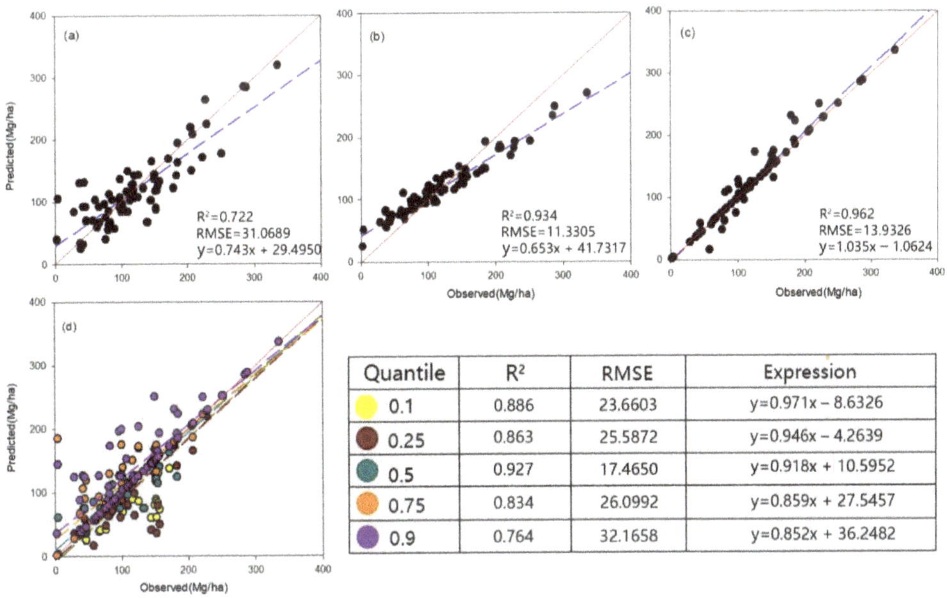

Quantile	R^2	RMSE	Expression
0.1	0.886	23.6603	y=0.971x − 8.6326
0.25	0.863	25.5872	y=0.946x − 4.2639
0.5	0.927	17.4650	y=0.918x + 10.5952
0.75	0.834	26.0992	y=0.859x + 27.5457
0.9	0.764	32.1658	y=0.852x + 36.2482

Figure 4. Scatter plots of the ground-observed and estimated biomass values for the (**a**) artificial neural network model (ANN); (**b**) the random forests model (RF); (**c**) the quantile regression neural network model (QRNN), and the quartiles groups are 0.1, 0.25, 0.5, 0.75, and 0.9; and (**d**) the quantile regression neural network with the best fitting performance in each biomass segment (QRNNb).

Compared to the scatter plots of the ANN, RFs, and the QRNN at each quantile, the scatter plot of the QRNNb was narrower and looked more similar to the line of y = x. The ranked absolute intercept value of each model was: RFs (41.7317) > QRNN0.9 (36.2482) > ANN (29.4950) > QRNN0.75 (27.5457) > QRNN0.5 (10.5952) > QRNN0.1 (8.6326) > QRNN0.25 (4.2639) > QRNNb (1.0624). The larger the intercept, the greater the angle with y = x, indicating the greater the degree of deviation. Figure 4a shows that the ANN had an excellent fitting performance at the middle biomass level. Still, it would overestimate the lower biomass value and underestimate the higher one as it had a greater intercept value. Similarly, RFs (Figure 4b) and the QRNN (Figure 4d) at each quantile showed the same phenomenon. Figure 4c indicates that the QRNNb had an excellent accuracy because it had a smaller intercept.

3.2.2. Method Validation

The biomass prediction accuracy of each model for each biomass segment was further validated by comparing the R^2 and RMSE (Table 4). These results indicated that QRNNb has a higher R^2 (0.943) and a higher RMSE (18.203) in the three models, especially as the AGB segment was 0–40 Mg/ha and >160 Mg/ha.

Table 4. Summary of R^2, RMSE, ME, and MAE at the different AGB segments based on the test dataset. ANN is the artificial neural network, RFs is the random forests, and QRNNb is the best quantile regression neural network in each biomass segment.

Indices		**Models**		
		ANN	**RFs**	**QRNNb**
R^2	0–40	0.105	0.402	0.961
	40–80	0.043	0.094	0.757
	80–120	0.167	0.598	0.430
	120–160	0.277	0.385	0.671
	>160	0.480	0.857	0.867
	Total	0.602	0.936	0.943
RMSE (Mg/ha)	0–40	8.341	6.818	1.733
	40–80	11.948	11.624	6.019
	80–120	10.421	7.242	9.851
	120–160	11.915	10.987	8.034
	>160	43.555	23.215	22.052
	Total	48.180	19.396	18.203
ME (Mg/ha)	0–40	−44.364	−30.845	1.035
	40–80	−33.623	−19.38	7.029
	80–120	−0.338	2.093	2.683
	120–160	13.741	8.230	−6.861
	>160	44.386	34.321	−11.617
	Total	−1.507	1.927	−1.419
MAE (Mg/ha)	0–40	48.400	30.846	1.035
	40–80	36.041	19.438	7.090
	80–120	11.213	5.720	5.926
	120–160	18.874	18.482	9.202
	>160	47.465	34.321	11.618
	Total	32.066	21.271	8.357

For the ME values, there were significant differences among the three models in different biomass segments, and QRNNb had no significant difference from zero at each biomass segment. The ANN and RFs showed negative mean errors in the 0–40 Mg/ha biomass segment. They were significantly different from zero at the significance level of 0.01, which means significant overestimation in the AGB segment. The ANN and RFs had a positive mean error as the segment was at 80–120 Mg/ha and >160 Mg/ha, demonstrating a significantly different value from zero, which would give a lower estimate at a higher biomass value, especially as the AGB was greater than 160 Mg/ha.

The MAE values showed that QRNNb was not significantly different from zero, and the MAE value was 8.359. QRNNb had a small MAE at the lower and higher biomass segments, which means the prediction values at these two segments were close to the observed value. The MAE showed that the prediction value from RFs and the ANN models at 80–120 Mg/ha had a minor error compared with the other biomass segments, while the MAE for the RFs and ANN models at 0–40 Mg/ha and >160 Mg/ha showed that the prediction value had a substantial deviation from zero. The bias of the ANN and RFs models for all of the biomass segments except for the 80–120 Mg/ha segment was relatively high. The highest MAE of ANN and RFs was 48.400 Mg/ha and 30.846 Mg/ha at the biomass segment of 0–40 Mg/ha, and 47.465 Mg/ha and 34.321 Mg/ha at the AGB >160 Mg/ha, respectively. In addition, RFs and the ANN showed significant deviations from zero.

In sum, QRNNb was more accurate than the ANN and RFs in biomass estimation, especially in the low-biomass segment and the high-biomass segments. QRNNb can improve the problem of low-value overestimation and high-value underestimation and it has a very stable prediction effect.

The AGB maps of the *Pinus densata* forests are shown in Figure 5, which was inverted by using three models. The high heterogeneity of the AGB distribution can be seen using the model of QRNNb, which means the model of QRNNb has an excellent prediction of AGB biomass value at each of the AGB segments. On the contrary, the ANN had a higher count at the segment with 120–160 Mg/ha and 40–80 Mg/ha, which means that the ANN cannot capture the AGB at the lower biomass segment, which would lead to an over-estimation of the low AGB biomass. Meanwhile, some of the higher AGB biomass values (>160 Mg/ha) may be counted into 120–160 Mg/ha, leading to an under-estimation of the high AGB biomass. The prediction AGB biomass values of RFs were more concentrated at 40–80 Mg/ha and 80–120 Mg/ha than the ANN. This proved that the high precision of RFs was at the cost of discarding high accuracy.

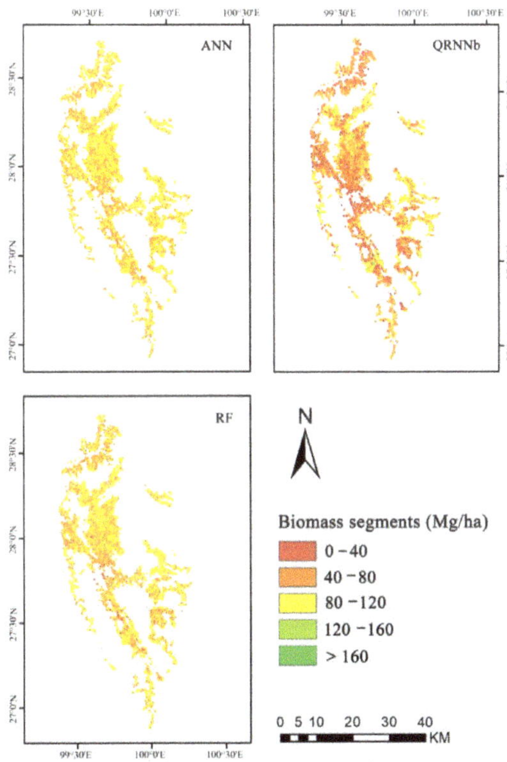

Figure 5. The spatial distributions of the predicted aboveground biomass (AGB) values of the *Pinus densata* forests using four models. ANN is the artificial neural network, RF is the random forests, and QRNNb is the best quantile regression neural network in each biomass segment.

4. Discussion

4.1. Accuracy Comparison

Shettles et al. [69] found that model uncertainty is the main element affecting the accuracy of AGB estimation, and model uncertainty accounts for 55% of total uncertainty. Thus, improving model accuracy is still the main challenge for AGB estimation using optical remote sensing data. This research attempted to promote the accuracy of biomass assessment by comparing three non-parametric model regression models. The results have shown that the ranked fitting performance for the three models was the QRNNb > RFs > the ANN. From the values of R^2 and RMSE in the fitting model using the observed and predicted values, the accuracy for the RFs was slightly better than the QRNNb. Still, the intercept for the QRNNb was 1.0624 Mg/ha, which means the prediction value was much closer to the observed value. In contrast, RFs had apparent phenomena of under-estimation at higher biomass segments and over-estimation at lower biomass segments with a high intercept. The value reached 41.7317 Mg/ha, affecting the entire forest AGB assessment value. Thus, the QRNNb has the best performance among the three models. Moreover, RFs has a higher R^2 value and a lower RMSE in this study. Many studies have shown that RFs exhibited excellent performance [70,71]. Then, RFs was the most optimal model with the highest accuracy under the premise of considering only the overall situation.

Furthermore, for the different biomass segments, the results showed RFs at the lower and higher biomass segment was significantly worse than the QRNNb, the R^2 values for RFs at AGB < 40 Mg/ha and >160 Mg/ha were lower than for the QRNNb, and the RMSE values at both biomass segments for RFs were extremely larger than the QRNNb. This reveals that the QRNNb could promote biomass estimation accuracy, especially at the lower and higher biomass segments. The QRNNb could describe the complete conditional distribution of biomass with more stability and it is not easily affected by the extreme value. Then, the QRNNb would be an excellent method to reduce the uncertainties from over-estimation and under-estimation in the AGB estimation using optical remote sensing data.

In addition, the Sentinel-2 images were resampled with 30 m × 30 m corresponding with the plot size of the field survey in this study. The mismatch between the former image spatial resolution and field size would affect the AGB estimation accuracy. We performed AGB estimation using the resampled Sentinel-2 image product with a spatial resolution of 10 m. Similar fitting and validation results for the three models were obtained, and the QRNNb was more accurate than the ANN and RFs in biomass estimation, especially in the low-biomass segments and the high-biomass segments (see Appendices A and B). This further illustrates the availability of the proposed method for reducing the uncertainties of AGB estimation using optical remote sensing.

4.2. Data Resource and Variables

The information extracted from optical remote sensing is the radiation information of the canopy surface, which is easily affected by the complexity of forest crown layers. Therefore, the precision problem is the biggest challenge of optical remote sensing in current remote sensing biomass estimation [19,20]. Using high-resolution and hyperspectral remote sensing images will enhance biomass estimation accuracy, but the high price limits such data being widely utilized [17,72]. Researchers prefer to choose free, open-source data, such as Landsat or Sentinel-2. Even though those two are both optical remote sensing, Sentinel-2 has a double-satellite orbit and has four more bands than Landsat. It is the unique one with three bands of data in the red edge range, which can efficiently obtain more rich geographical information [21]. Studies have shown that Sentinel-2 is more suitable than Landsat for improving estimation accuracy [73]. Although the vegetation index will bring a saturation problem, the vegetation index extracted from near-infrared and red edge can strengthen the estimation accuracy [74]. This study found that band 2 (blue), band 3 (green), band 5 (vegetation red edge), band 8 (NIR), and band 12 (SWIR) of Sentinel-2 had a strong correlation with biomass. Because the vegetation index is affected by the saturation value in biomass estimation, the texture feature has been introduced as a variable. Then, the biomass value is more sensitive to the texture feature [75,76]. This study also extracted the textural features of different window sizes (3 × 3, 5 × 5, and 7 × 7) to model. After screening and analysis, it was found that the texture information of entropy and the correlation with various window sizes and bands (VA5_2 and VA7_12) strongly correlated with biomass.

Moreover, Shangri-La City has a cold-temperate monsoon climate with altitudes ranging from 3350 to 3696 m above sea level. The cloud and snow significantly affect the spectral bands of optical remote sensing [77]. Lacking high-quality images with a lower cover of cloud and snow corresponding to the field investigation date, we only obtained five Sentinel-2 images from the ESA. The image acquisition date is 24 November 2016. The time difference between the survey data (August) and the remote sensing data (November) is about three months. To avoid or reduce the impact of the time mismatch between image acquisition and the field survey, we obtained the bottom-of-atmosphere reflectance product by atmospheric correction to normalize as a common reference [78]. Furthermore, *Pinus densata* is an evergreen coniferous tree distributed in the alpine and sub-alpine areas in China, and it grows slowly within one to two years [79]. Therefore, the tree growth and forest structure are almost unchanged; then, the change of image reflectance caused by forest growth in the three months will have a negligible impact on the AGB estimation in this study.

4.3. Limitation and Future Research

Although QRNNb obtained a high-accuracy estimation in the different biomass segments, this study still has some limitations. Firstly, Sentinel-2 can yield an accurate biomass estimation [23]. Still, some studies have shown that mixed remote sensing data are more precise than single-source data, especially in tropical and subtropical regions where the stand structures and tree species are complex [80–82]. Secondly, the accuracy of AGB estimation is highly dependent on prediction methods [83]. Therefore, other models for biomass estimation in subsequent studies should be considered to improve the precision of biomass estimation, for instance, combining quantile regression and random forests to form quantile random forests (QRF), the convolutional neural networks (CNN), the gradient boost regression tree (GBRT) [84–86], etc. Thirdly, the best combination of different vegetation indices is expected to predict the AGB of vegetation at different stages [87].

Moreover, we only selected *Pinus densata* forests as the research area. They are mainly distributed over the subalpine and alpine areas in southern Qinghai, western Sichuan, northwestern Yunnan, and southeastern Tibet in China. In addition, *Pinus densata* forests are single-storied stands with even age in common [60,61]. Therefore, the proposed method can be applied to improve the forest AGB estimation for even-aged or single-storied forests. The applicability in the multi-storied stands or the uneven-aged forests with complex stand structures would be further explored.

5. Conclusions

To reduce uncertainties from under-estimation and over-estimation, optical remote sensing was applied to assess forest AGB. In this study, Sentinel-2 was used to explore the potential and capability of three non-parametric models of the ANN, RF, and the QRNN for *Pinus densata* in Shangri-La City. In addition, the biomass was segmented, and the quantile regression neural network with the best fitting performance in each biomass segment was selected to combine an integrity model named QRNNb. The results showed: (1) from the whole biomass data, the performance of QRNNb and RFs was a priority over the ANN. The corresponding R^2 and RMSE were QRNNb: 0.943, 18.203 Mg/ha; RF: 0.936, 19.396 Mg/ha; ANN: 0.602, 48.180 Mg/ha. (2) The prediction accuracy of QRNNb at different biomass segments was higher than the ANN and RF. It had the highest R^2 and the smallest RMSE when AGB < 40 Mg/ha and AGB > 160 Mg/ha. The R^2 at values those two biomass segments were 0.961 and 0.867, and the RMSE values for those two were 1.733 Mg/ha and 22.052 Mg/ha. This demonstrated that QRNNb could efficiently improve the under-estimation at higher biomass values and the over-estimation at lower biomass values compared with the ANN and RF. QRNNb was sensitive to extreme values and could express low biomass values and high biomass values wholly and effectively. This means that QRNNb combined with the optimal quantile model of each biomass segment provides a more suitable method for estimating AGB for even-aged or single-storied forests.

Author Contributions: L.L. and B.Z. participated in the collection of the field data, conducted the data analysis, and wrote the draft of the paper; Y.L. and J.T. helped with the data analysis and the writing of the paper; Y.L., Y.W., W.X. and L.W. participated in collecting and analyzing the field data; G.O. supervised and coordinated the research project, designed the experiment, and revised the paper. All authors have read and agreed to the published version of the manuscript.

Funding: This research was funded by the National Natural Science Foundation of China (grant numbers 31770677 and 31760206) and the Ten-Thousand Talents Program of Yunnan Province, China (YNWR-QNBJ-2018-184).

Data Availability Statement: The data used in this study can be acquired from the corresponding author upon reasonable request.

Conflicts of Interest: The authors declare no conflict of interest.

Appendix A

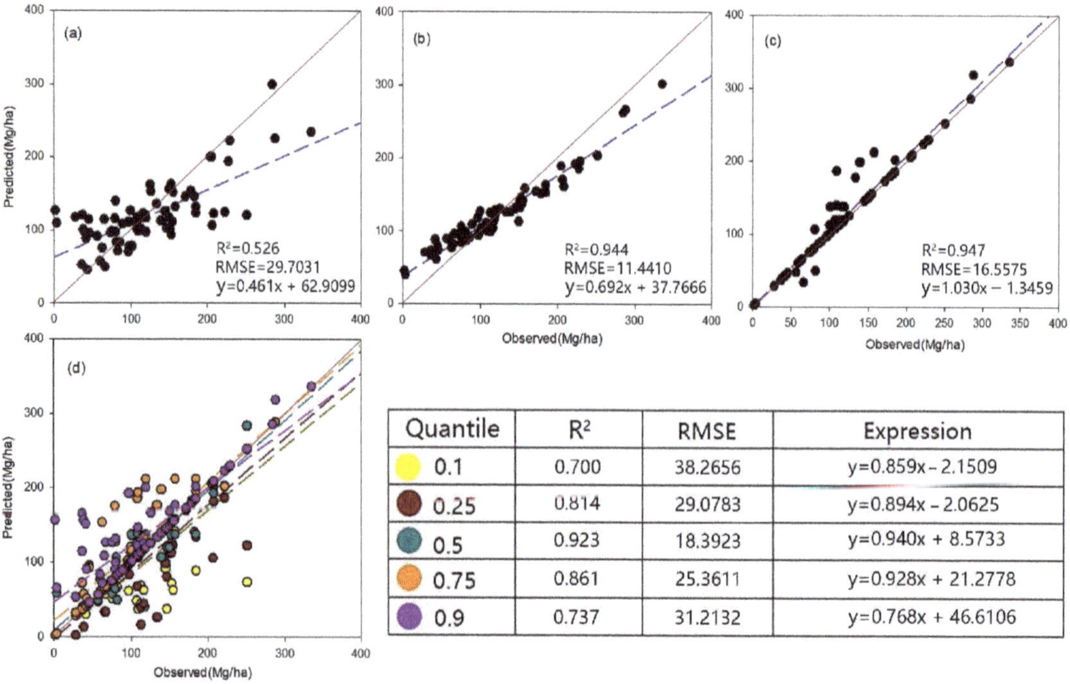

Quantile	R^2	RMSE	Expression
0.1	0.700	38.2656	y=0.859x−2.1509
0.25	0.814	29.0783	y=0.894x − 2.0625
0.5	0.923	18.3923	y=0.940x + 8.5733
0.75	0.861	25.3611	y=0.928x + 21.2778
0.9	0.737	31.2132	y=0.768x + 46.6106

Figure A1. Scatter plots of the ground-observed and estimated biomass values using the resampled Sentinel-2 image product with a spatial resolution of 10 m. (**a**) The artificial neural network model (ANN); (**b**) the random forests model (RF); (**c**) the quantile regression neural network model (QRNN), and the quartiles groups are 0.1, 0.25, 0.5, 0.75, and 0.9, respectively; (**d**) and the quantile regression neural network with the best fitting performance in each biomass segment (QRNNb).

Appendix B

Table A1. Summary of R^2, RMSE, ME, and MAE at the different AGB segments based on the test dataset using the resampled Sentinel-2 image product with a spatial resolution of 10 m. ANN is the artificial neural network, RFs is the random forests, and QRNNb is the best quantile regression neural network in each biomass segment.

Indices		Models		
		ANN	RFs	QRNNb
R^2	0–40	0.384	0.076	0.958
	40–80	0.022	0.200	0.889
	80–120	0.050	0.241	0.430
	120–160	0.031	0.302	0.234
	>160	0.257	0.861	0.968
	Total	0.549	0.932	0.956
RMSE (Mg/ha)	0–40	6.919	8.474	1.799
	40–80	12.075	10.926	4.068
	80–120	11.050	9.952	8.621
	120–160	13.586	11,708	12.258
	>160	52.062	22.510	2.774
	Total	51.310	19.960	16.063
ME (Mg/ha)	0–40	−58.615	−31.676	−0.597
	40–80	−37.525	−18.002	1.489
	80–120	−7.131	−1.365	−7.131
	120–160	10.183	7.239	−4.231
	>160	60.937	42.327	0.200
	Total	−0.454	2.275	−1.211
MAE (Mg/ha)	0–40	61.077	31.676	0.601
	40–80	39.856	18.243	0.941
	80–120	20.327	6.905	1.455
	120–160	25.946	10.825	5.084
	>160	65.955	42.327	0.308
	Total	42.060	22.555	2.396

References

1. Houghton, R.A.; Hackler, J.L.; Lawrence, K.T. The U.S. Carbon budget: Contributions from land-use change. *Science* **1999**, *285*, 574–578. [CrossRef] [PubMed]
2. Feng, H.; Chen, Q.; Hu, Y.; Du, Z.; Lin, G.; Wang, C.; Huang, Y. Estimation of forest aboveground biomass by using a mixed-effects model. *Int. J. Remote Sens.* **2021**, *42*, 8675–8690. [CrossRef]
3. Sun, S.; Wang, Y.; Song, Z.; Chen, C.; Zhang, Y.; Chen, X.; Chen, W.; Yuan, W.; Wu, X.; Ran, X.; et al. Modelling aboveground biomass carbon stock of the Bohai rim coastal wetlands by integrating remote sensing, terrain, and climate data. *Remote Sens.* **2021**, *13*, 4321. [CrossRef]
4. Wulder, M.A.; Roy, D.P.; Radeloff, V.C.; Loveland, T.R.; Anderson, M.C.; Johnson, D.M.; Healey, S.; Zhu, Z.; Scambos, T.A.; Pahlevan, N.; et al. Fifty years of Landsat science and impacts. *Remote Sens. Environ.* **2022**, *280*, 113195. [CrossRef]
5. Foody, G.M.; Boyd, D.S.; Cutler, M.E.J. Predictive relations of tropical forest biomass from Landsat TM data and their transferability between regions. *Remote Sens. Environ.* **2003**, *85*, 463–474. [CrossRef]
6. Puliti, S.; Solberg, S.; Næsset, E.; Gobakken, T.; Zahabu, E.; Mauya, E.; Malimbwi, R. Modelling aboveground biomass in *Tanzanian miombo* woodlands using TanDEM-X world DEM and field data. *Remote Sens.* **2017**, *9*, 984. [CrossRef]
7. Xue, B.W. Lidar and Machine Learning Estimation of Hardwood Forest Biomass in Mountainous and Bottomland Environments. Master's Thesis, University of Arkansas, Fayetteville, NC, USA, 2015; p. 1274.
8. Tian, Y.; Huang, H.; Zhou, G.; Zhang, Q.; Tao, J.; Zhang, Y.; Lin, J. Aboveground mangrove biomass estimation in Beibu Gulf using machine learning and UAV remote sensing. *Sci. Total Environ.* **2021**, *781*, 146816. [CrossRef]
9. Vaglio, L.G.; Chen, Q.; Lindsell, J.A.; Coomes, D.A.; Frate, F.D.; Guerriero, L.; Pirotti, F.; Valentini, R. Above ground biomass estimation in an African tropical forest with lidar and hyperspectral data. *ISPRS J. Photogramm. Remote Sens.* **2014**, *89*, 49–58. [CrossRef]
10. Listopad, C.M.C.S.; Drake, J.B.; Masters, R.E.; Weishampel, J.F. Portable and airborne small footprint LiDAR: Forest canopy structure estimation of fire managed plots. *Remote Sens.* **2011**, *3*, 1284–1307. [CrossRef]

11. Spriggs, R.; Coomes, D.; Jones, T.; Caspersen, J.; Vanderwel, M. An alternative approach to using LiDAR remote sensing data to predict stem diameter distributions across a temperate forest landscape. *Remote Sens.* **2017**, *9*, 944. [CrossRef]
12. Cutler, M.E.J.; Boyd, D.S.; Foody, G.M.; Vetrivel, A. Estimating tropical forest biomass with a combination of SAR image texture and Landsat TM data: An assessment of predictions between regions. *ISPRS J. Photogramm. Remote Sens.* **2012**, *70*, 66–77. [CrossRef]
13. El Hage, M.; Villard, L.; Huang, Y.; Ferro-Famil, L.; Koleck, T.; Le Toan, T.; Polidori, L. Multicriteria accuracy assessment of digital elevation models (DEMs) produced by airborne P-band polarimetric SAR tomography in tropical rainforests. *Remote Sens.* **2022**, *14*, 4173. [CrossRef]
14. Quegan, S.; Le Toan, T.; Chave, J.; Dall, J.; Exbrayat, J.-F.; Minh, D.H.T.; Lomas, M.; D' Alessandro, M.M.; Paillou, P.; Papathanassiou, K.; et al. The European space agency BIOMASS Mission: Measuring forest above-ground biomass from space. *Remote Sens. Environ.* **2019**, *227*, 44–60. [CrossRef]
15. El Idrissi Essebtey, S.; Villard, L.; Borderies, P.; Koleck, T.; Burban, B.; Le Toan, T. Long-term trends of P-Band temporal decorrelation over a tropical dense forest-experimental results for the BIOMASS mission. *IEEE Trans. Geosci. Remote Sens.* **2022**, *60*, 1–15. [CrossRef]
16. Le Toan, T.; Quegan, S.; Davidson, M.W.J.; Balzter, H.; Paillou, P.; Papathanassiou, K.; Plummer, S.; Rocca, F.; Saatchi, S.; Shugart, H.; et al. The BIOMASS mission: Mapping global forest biomass to better understand the terrestrial carbon cycle. *Remote Sens. Environ.* **2011**, *115*, 2850–2860. [CrossRef]
17. Papathanassiou, K.P.; Cloude, S.R.; Pardini, M.; Quiñones, M.J.; Hoekman, D.; Ferro-Famil, L.; Goodenough, D.; Chen, H.; Tebaldini, S.; Neumann, M.; et al. Forest Applications. In *Polarimetric Synthetic Aperture Radar: Principles and Application;* Remote Sensing and Digital Image, Processing; Hajnsek, I., Desnos, Y.-L., Eds.; Springer International Publishing: Cham, Germany, 2021; pp. 59–117. ISBN 978-3-030-56504-6.
18. López-Serrano, P.M.; Cárdenas, D.; José, L.; Corral-Rivas, J.J.; Jiménez, E.; López-Sánchez, C.A.; Vega-Nieva, D.J. Modeling of aboveground biomass with Landsat 8 OLI and machine learning in temperate forests. *Forests* **2019**, *11*, 11. [CrossRef]
19. Zeng, N.; Ren, X.L.; He, H.L.; Zhang, L.; Zhao, D.; Ge, R.; Li, P.; Niu, Z.G. Estimating grassland aboveground biomass on the Tibetan Plateau using a random forest algorithm. *Ecol. Indic.* **2019**, *102*, 479–487. [CrossRef]
20. Sagang, L.B.T.; Ploton, P.; Sonké, B.; Poilvé, H.; Couteron, P.; Barbier, N. Airborne Lidar sampling pivotal for accurate regional AGB predictions from multispectral images in forest-Savanna landscapes. *Remote Sens.* **2020**, *12*, 1637. [CrossRef]
21. SUHET. *Sentinel-2 User Handbook;* European Space Agency: Paris, France, 2013.
22. Chen, Y.; Guerschman, J.; Shendryk, Y.; Henry, D.; Harrison, M.T. Estimating pasture biomass using Sentinel-2 imagery and machine learning. *Remote Sens.* **2021**, *13*, 603. [CrossRef]
23. Castillo, J.A.A.; Apan, A.A.; Maraseni, T.N.; Salmo, S.G. Estimation and mapping of above-ground biomass of mangrove forests and their replacement land uses in the Philippines using Sentinel imagery. *ISPRS J. Photogramm. Remote Sens.* **2017**, *134*, 70–85. [CrossRef]
24. Abdullah, H.; Skidmore, A.K.; Darvishzadeh, R.; Heurich, M.; Pettorelli, N.; Disney, M. Sentinel-2 accurately maps green-attack stage of European spruce bark beetle (*Ips typographus* L.) compared with Landsat-8. *Remote Sens. Ecol. Conserv.* **2018**, *5*, 87–106. [CrossRef]
25. Huete, A.R.; Liu, H.Q.L.; Batchily, K.B.; Leeuwen, W.V. A comparison of vegetation indices over a global set of TM images for EOS-MODIS. *Remote Sens. Environ.* **1997**, *59*, 440–451. [CrossRef]
26. Kalaitzidis, C.; Zianis, D.; Heinzel, V. A Review of Vegetation Indices for the Estimation of Biomass. In *Proceedings of the 29th Symposium of the European Association of Remote Sensing Laboratories, Chania, Greece;* IOS Press Ebook: Amsterdam, The Netherlands, 2009; pp. 201–208. [CrossRef]
27. Safari, A.; Sohrabi, H. Integration of synthetic aperture radar and multispectral data for aboveground biomass retrieval in Zagros oak forests, Iran: An attempt on Sentinel imagery. *Int. J. Remote Sens.* **2020**, *41*, 8069–8095. [CrossRef]
28. Luan, P.V.; Everardo, C.M.; do Amaral, C.H.; Christopher, M.U.; NealeZution, G.I.; Filgueiras, R.; Fernando, C.E. Potential of using spectral vegetation indices for corn green biomass estimation based on their relationship with the photosynthetic vegetation sub-pixel fraction. *Agric. Water Manag.* **2020**, *236*, 106155. [CrossRef]
29. David, R.M.; Rosser, N.J.; Donoghue, D.N.M. Improving above ground biomass estimates of Southern Africa dryland forests by combining Sentinel-1 SAR and Sentinel-2 multispectral imagery. *Remote Sens. Environ.* **2022**, *282*, 113232. [CrossRef]
30. Lu, D.S. The potential and challenge of remote sensing-based biomass estimation. *Int. J. Remote Sens.* **2007**, *27*, 1297–1328. [CrossRef]
31. Kuplich, T.M.; Curran, P.J.; Atkinson, P.M. Relating SAR image texture to the biomass of regenerating tropical forests. *Int. J. Remote Sens.* **2011**, *26*, 4829–4854. [CrossRef]
32. Wang, X.Q.; Pang, Y.; Zhang, Z.J.; Yuan, Y. Forest Aboveground Biomass Estimation Using SPOT-5 Texture Indices and Spectral Derivatives. In Proceedings of the 2014 IEEE Geoscience and Remote Sensing Symposium, Quebec, QC, Canada, 13–18 July 2014; pp. 2830–2833. [CrossRef]
33. Ghosh, S.M.; Behera, M.D. Aboveground biomass estimation using multi-sensor data synergy and machine learning algorithms in a dense tropical forest. *Appl. Geogr.* **2018**, *96*, 29–40. [CrossRef]
34. Jolliffe, I.T.; Cadima, J. Principal component analysis: A review and recent developments. *Philos. Trans. Ser. A Math. Phys. Eng. Sci.* **2016**, *374*, 20150202. [CrossRef]

35. Su, H.Y.; Shen, W.J.; Wang, J.R.; Ali, A.; Li, M. Machine learning and geostatistical approaches for estimating aboveground biomass in Chinese subtropical forests. *For. Ecosyst.* **2020**, *7*, 64. [CrossRef]
36. Breiman, L. Random forests. *Mach. Learn.* **2001**, *34*, 5–32. [CrossRef]
37. Belgiu, M.; Drăguţ, L. Random forest in remote sensing: A review of applications and future directions. *ISPRS J. Photogramm. Remote Sens.* **2016**, *114*, 24–31. [CrossRef]
38. Arsalan, G.; Soheil, Z.; Reza, M.A.; Meisam, A.; Mohammadzadeh, A.; Sadegh, J. Ghrobanian-Mangrove ecosystem mapping using Sentinel-1 and Sentinel-2 Satellite images and random forest algorithm in Google Earth Engine. *Remote Sens.* **2021**, *13*, 2565. [CrossRef]
39. Jiang, F.; Kutia, M.; Ma, K.; Chen, S.; Long, J.; Sun, H. Estimating the aboveground biomass of coniferous forest in Northeast China using spectral variables, land surface temperature and soil moisture. *Sci. Total Environ.* **2021**, *785*, 147335. [CrossRef]
40. Zeng, P.; Zhang, W.; Li, Y.; Shi, J.; Wang, Z. Forest total and component above-ground biomass (AGB) estimation through C- and L-band polarimetric SAR data. *Forests* **2022**, *13*, 442. [CrossRef]
41. Lourenço, P.; Godinho, S.; Sousa, A.; Gonçalves, A.C. Estimating tree aboveground biomass using multispectral satellite-based data in Mediterranean agroforestry system using random forest algorithm. *Remote Sens. Appl. Soc. Environ.* **2021**, *23*, 100560. [CrossRef]
42. Ou, G.L.; Lv, Y.Y.; Xu, H.; Wang, G.X. Improving forest aboveground biomass estimation of *Pinus densata* forest in Yunnan of southwest China by spatial regression using Landsat 8 images. *Remote Sens.* **2019**, *11*, 2750. [CrossRef]
43. Axelsson, C.; Skidmore, A.K.; Schlerf, M.; Fauzi, A.; Verhoef, W. Hyperspectral analysis of mangrove foliar chemistry using PLSR and support vector regression. *Int. J. Remote Sens.* **2012**, *34*, 1724–1743. [CrossRef]
44. Wang, S.; Wang, D.; Sun, J. Artificial neural network-based ionospheric delay correction method for satellite-based augmentation systems. *Remote Sens.* **2022**, *14*, 676. [CrossRef]
45. Beaudoin, A.; Hall, R.J.; Castilla, G.; Filiatrault, M.; Villemaire, P.; Skakun, R.; Guindon, L. Improved k-NN mapping of forest attributes in northern Canada using spaceborne L-Band SAR, multispectral and LiDAR data. *Remote Sens.* **2022**, *14*, 1181. [CrossRef]
46. Yadav, S.; Padalia, H.; Sinha, S.K.; Srinet, R.; Chauhan, P. Above-ground biomass estimation of Indian tropical forests using X band Pol-InSAR and Random Forest. *Remote Sens. Appl. Soc. Environ.* **2021**, *21*, 100462. [CrossRef]
47. Joshua, O.L.; Chinenye, A.L.; Adewale, G.A. Multi-layer perceptron artificial neural network (MLP-ANN) prediction of biomass higher heating value (HHV) using combined biomass proximate and ultimate analysis data. *Model. Earth Syst. Environ.* **2022**, *8*, 3177–3191. [CrossRef]
48. Vahedi, A.A. Artificial neural network application in comparison with modeling allometric equations for predicting above-ground biomass in the Hyrcanian mixed-beech forests of Iran. *Biomass Bioenergy* **2016**, *88*, 66–76. [CrossRef]
49. Xie, Y.; Sha, Z.; Yua, M.; Bai, Y.; Zhang, L. A comparison of two models with Landsat data for estimating above-ground grassland biomass in Inner Mongolia, China. *Ecol. Model.* **2009**, *220*, 1810–1818. [CrossRef]
50. Yang, S.; Feng, Q.; Liang, T.; Liu, B.; Zhang, W.; Xie, H. Modeling grassland above-ground biomass based on artificial neural network and remote sensing in the Three-River Headwaters Region. *Remote Sens. Environ.* **2018**, *204*, 448–455. [CrossRef]
51. Taylor, J.W. A quantile regression neural network approach to estimating the conditional density of multiperiod returns. *J. Forecast.* **2000**, *19*, 299–311. [CrossRef]
52. Koenker, R.; Bassett, G. Regression Quantiles. *Econometrica* **1978**, *46*, 33–50. [CrossRef]
53. Cade, B.S.; Noon, B.R. A gentle introduction to quantile regression for ecologists. *Front. Ecol. Environ.* **2003**, *1*, 412–420. [CrossRef]
54. Julien, L. A quantile regression study of climate change in Chicago, 1960–2010. *SIAM Undergrad. Res. Online* **2012**, *5*, 148–165. [CrossRef]
55. Cannon, A.J. Quantile regression neural networks: Implementation in R and application to precipitation downscaling. *Comput. Geosci.* **2011**, *37*, 1277–1284. [CrossRef]
56. Cannon, A.J. Non-crossing nonlinear regression quantiles by monotone composite quantile regression neural network, with application to rainfall extremes. *Stoch. Environ. Res. Risk Assess.* **2018**, *32*, 3207–3225. [CrossRef]
57. He, Y.; Li, H. Probability density forecasting of wind power using quantile regression neural network and kernel density estimation. *Energy Convers. Manag.* **2018**, *164*, 374–384. [CrossRef]
58. Dong, M.; Wu, D.; Fu, X.; Deng, H.; Wu, G. Regional-scale analysis on the strengths, weaknesses, opportunities, and threats in sustainable development of Shangri-La County. *Int. J. Sustain. Dev. World Ecol.* **2014**, *22*, 171–177. [CrossRef]
59. Guo, Y.; Li, Y.; Huang, Y.; Jarvis, D.; Sato, K.; Kato, K.; Tsuyuzaki, H.; Chen, L.; Long, C. Genetic diversity analysis of hulless barley from Shangri-la region revealed by SSR and AFLP markers. *Genet. Resour. Crop Evol.* **2011**, *59*, 1543–1552. [CrossRef]
60. Wang, B.; Mao, J.F.; Jie, G.; Wei, Z.; Wang, X.R. Colonization of the Tibetan plateau by the homoploid hybrid pine *Pinus densata*. *Mol. Ecol.* **2011**, *20*, 3796–3811. [CrossRef] [PubMed]
61. Compilation Committee of Yunnan Forest. *Yunnan Forest*; Yunnan Science and Technology Press: Kunming, China; China Forestry Publishing House: Beijing, China, 1986.
62. Zhang, J.; Lu, C.; Xu, H.; Wang, G. Estimating aboveground biomass of *Pinus densata*-dominated forests using Landsat time series and permanent sample plot data. *J. For. Res.* **2019**, *30*, 1689–1706. [CrossRef]

63. Ou, G.L.; Li, C.; Lv, Y.Y.; Wei, A.C.; Xiong, H.X.; Xu, H.; Wang, G.X. Improving aboveground biomass estimation of *Pinus densata* forests in Yunnan using Landsat 8 imagery by incorporating age dummy variable and method comparison. *Remote Sens.* **2019**, *11*, 738. [CrossRef]
64. Ghosh, A.; Fassnacht, F.E.; Joshi, P.K.; Koch, B. A framework for mapping tree species combining hyperspectral and LiDAR data: Role of selected classifiers and sensor across three spatial scales. *Int. J. Appl. Earth Obs. Geoinf.* **2014**, *26*, 49–63. [CrossRef]
65. Wang, Z.; Zhang, X.Q.; Chhin, S.; Zhang, J.; Duan, A. Disentangling the effects of stand and climatic variables on forest productivity of Chinese fir plantations in subtropical China using a random forest algorithm. *Agric. For. Meteorol.* **2021**, *304–305*, 108412. [CrossRef]
66. Roy, N.; Pal, A.; Dey, A.; Das, S. Applications of artificial intelligence in machine learning: Review and Prospect. *Int. J. Comput. Appl.* **2015**, *115*, 31–41. [CrossRef]
67. Stathakis, D. How many hidden layers and nodes? *Int. J. Remote Sens.* **2009**, *30*, 2133–2147. [CrossRef]
68. Tiryaki, S.; Aydın, A. An artificial neural network model for predicting compression strength of heat treated woods and comparison with a multiple linear regression model. *Constr. Build. Mater.* **2014**, *62*, 102–108. [CrossRef]
69. Babcock, C.; Finley, A.O.; Bradford, J.B.; Kolka, R.; Birdsey, R.; Ryan, M.G. LiDAR-based prediction of forest biomass using hierarchical models with spatially varying coefficients. *Remote Sens. Environ.* **2015**, *169*, 113–127. [CrossRef]
70. Wang, Y.; Wu, G.; Deng, L.; Tang, Z.; Wang, K.; Sun, W.; Shangguan, Z. Prediction of aboveground grassland biomass on the Loess Plateau, China, using a random forest algorithm. *Sci. Rep.* **2017**, *7*, 6940. [CrossRef] [PubMed]
71. Sun, X.; Li, G.; Wang, M.; Fan, Z. Analyzing the uncertainty of estimating forest aboveground biomass using optical imagery and spaceborne LiDAR. *Remote Sens.* **2019**, *11*, 722. [CrossRef]
72. Zhu, X.L.; Liu, D.S. Improving forest aboveground biomass estimation using seasonal Landsat NDVI time-series. *J. Photogramm. Remote Sens.* **2015**, *102*, 222–231. [CrossRef]
73. Sibanda, M.; Mutanga, O.; Rouget, M. Examining the potential of Sentinel-2 MSI spectral resolution in quantifying above-ground biomass across different fertilizer treatments. *ISPRS J. Photogramm. Remote Sens.* **2015**, *110*, 55–65. [CrossRef]
74. Mutanga, O.; Adam, E.; Cho, M.A. High-density biomass estimation for wetland vegetation using WorldView-2 imagery and random forest regression algorithm. *Int. J. Appl. Earth Obs. Geoinf.* **2012**, *18*, 399–406. [CrossRef]
75. Pandit, S.; Tsuyuki, S.; Dube, T. Exploring the inclusion of Sentinel-2 MSI texture metrics in above-ground biomass estimation in the community forest of Nepal. *Geocarto Int.* **2019**, *35*, 1832–1849. [CrossRef]
76. Kelsey, K.; Neff, J. Estimates of aboveground biomass from texture analysis of Landsat imagery. *Remote Sens.* **2014**, *6*, 6407–6422. [CrossRef]
77. Xu, H.; Yue, C. *Study on Forest Landscape Change and Forest Biomass Estimation in Shangri-La Based on Remote Sensing Technology*; Yunnan Science and Technology Press: Kunming, China, 2014.
78. Powell, S.L.; Cohen, W.B.; Healey, S.P.; Kennedy, R.E.; Moisen, G.G.; Pierce, K.B.; Ohmann, J.L. Quantification of live aboveground forest biomass dynamics with Landsat time-series and field inventory data: A comparison of empirical modeling approaches. *Remote Sens. Environ.* **2010**, *114*, 1053–1068. [CrossRef]
79. Xie, F.; Shu, Q.; Zi, L.; Wu, R.; Wu, Q.; Wang, H.; Liu, Y.; Ji, Y. Remote sensing estimation of *Pinus densata* aboveground biomass based on k-NN nonparametric model. *Acta Agric. Univ. Jiangxiensis* **2018**, *40*, 743–750.
80. Li, L.; Zhou, X.S.; Chen, L.; Chen, L.; Zhang, Y.; Liu, Y. Estimating urban vegetation biomass from Sentinel-2A image data. *Forests* **2020**, *11*, 125. [CrossRef]
81. Lu, D.S.; Batistella, M.; Moran, E. Satellite estimation of aboveground biomass and impacts of forest stand structure. *Photogramm. Eng. Remote Sens.* **2005**, *71*, 967–974. [CrossRef]
82. Chang, J.S.; Shoshany, M. Mediterranean Shrublands Biomass Estimation Using Sentinel-1 and Sentinel-2. In Proceedings of the International Geoscience and Remote Sensing Symposium (IGARSS), Beijing, China, 10–15 July 2016. [CrossRef]
83. Masjedi, A.; Zhao, J.Q.; Thompson, A.M.; Yang, K.W.; Flatt, J.E.; Crawford, M.M.; Ebert, D.S.; Tuinstra, M.R.; Hammer, G.; Chapman, S. Sorghum Biomass Prediction Using UAV-Based Remote Sensing Data and Crop Model Simulation. In Proceedings of the IGARSS 2018—2018 IEEE International Geoscience and Remote Sensing Symposium, Valencia, Spain, 22–27 July 2018.
84. Freeman, E.A.; Moisen, G.G. An Application of Quantile Random Forests for Predictive Mapping of Forest Attributes. In Proceedings of the New Directions in Inventory Techniques & Applications Forest Inventory & Analysis (FIA) Symposium, Portland, OR, USA, 8–10 December 2015; p. 362.
85. Pham, T.D.; Le, N.N.; Ha, N.T.; Nguyen, L.V.; Xia, J.; Yokoya, N.; To, T.T.; Trinh, H.X.; Kieu, L.Q.; Takeuchi, W. Estimating Mangrove Above-Ground Biomass Using Extreme Gradient Boosting Decision Trees Algorithm with Fused Sentinel-2 and ALOS-2 PALSAR-2 Data in Can Gio Biosphere Reserve, Vietnam. *Remote Sens.* **2020**, *12*, 777. [CrossRef]
86. Kattenborn, T.; Leitloff, J.; Schiefer, F.; Hinz, S. Review on Convolutional Neural Networks (CNN) in vegetation remote sensing. *ISPRS J. Photogramm. Remote Sens.* **2021**, *173*, 24–49. [CrossRef]
87. Xu, M.; Liu, R.; Chen, J.M.; Liu, Y.; Shang, R.; Ju, W.; Wu, C.; Huang, W. Retrieving leaf chlorophyll content using a matrix-based vegetation index combination approach. *Remote Sens. Environ.* **2019**, *224*, 60–73. [CrossRef]

MDPI
St. Alban-Anlage 66
4052 Basel
Switzerland
Tel. +41 61 683 77 34
Fax +41 61 302 89 18
www.mdpi.com

Remote Sensing Editorial Office
E-mail: remotesensing@mdpi.com
www.mdpi.com/journal/remotesensing

www.ingramcontent.com/pod-product-compliance
Lightning Source LLC
Chambersburg PA
CBHW041658270926
51850CB00026B/47
* 9 7 8 3 0 3 6 5 7 2 0 8 6 *